Handbook of Systems Biology
Concepts and Insights

Handbook of Systems Biology
Concepts and Insights

A.J. Marian Walhout
University of Massachusetts Medical School,
Worcester, MA

Marc Vidal
Dana-Farber Cancer Institute,
Boston, MA

Job Dekker
University of Massachusetts Medical School,
Worcester, MA

AMSTERDAM • BOSTON • HEIDELBERG • LONDON • NEW YORK • OXFORD
PARIS • SAN DIEGO • SAN FRANCISCO • SINGAPORE • SYDNEY • TOKYO

Academic Press is an Imprint of Elsevier

Academic Press is an imprint of Elsevier
32 Jamestown Road, London NW1 7BY, UK
225 Wyman Street, Waltham, MA 02451, USA
525 B Street, Suite 1800, San Diego, CA 92101-4495, USA

First edition 2013

Copyright © 2013 Elsevier Inc. All rights reserved.

No part of this publication may be reproduced, stored in a retrieval system or transmitted in any form or by any means electronic, mechanical, photocopying, recording or otherwise without the prior written permission of the publisher. Permissions may be sought directly from Elsevier's Science & Technology Rights Department in Oxford, UK: phone (+44) (0) 1865 843830; fax (+44) (0) 1865 853333; email: permissions@elsevier.com. Alternatively, visit the Science and Technology Books website at www.elsevierdirect.com/rights for further information

Notice
No responsibility is assumed by the publisher for any injury and/or damage to persons or property as a matter of products liability, negligence or otherwise, or from any use or operation of any methods, products, instructions or ideas contained in the material herein.

Because of rapid advances in the medical sciences, in particular, independent verification of diagnoses and drug dosages should be made.

British Library Cataloguing-in-Publication Data
A catalogue record for this book is available from the British Library

Library of Congress Cataloging-in-Publication Data
A catalog record for this book is available from the Library of Congress

ISBN: 978-0-12-385944-0

For information on all Academic Press publications
visit our website at elsevierdirect.com

Typeset by TNQ Books and Journals Pvt Ltd.
www.tnq.co.in

Printed and bound in United States of America

12 13 14 15 16   10 9 8 7 6 5 4 3 2 1

Working together to grow
libraries in developing countries

www.elsevier.com | www.bookaid.org | www.sabre.org

ELSEVIER   BOOK AID International   Sabre Foundation

Contents

List of Contributors   vii
Preface   xi
Reviewers   xiii

Section I
Components of Biological Systems

1. Proteomic Analysis of Cellular Systems   3

   Marco Y. Hein, Kirti Sharma, Jürgen Cox and Matthias Mann

2. The Coding and the Non-coding Transcriptome   27

   Roderic Guigó

Section II
Network Properties of Biological Systems

3. Interactome Networks   45

   Anne-Ruxandra Carvunis, Frederick P. Roth, Michael A. Calderwood, Michael E. Cusick, Giulio Superti-Furga and Marc Vidal

4. Gene Regulatory Networks   65

   Martha L. Bulyk and A.J. Marian Walhout

5. Analyzing the Structure, Function and Information Flow in Signaling Networks using Quantitative Cellular Signatures   89

   Meghana M. Kulkarni and Norbert Perrimon

6. Genetic Networks   115

   Michael Costanzo, Anastasia Baryshnikova, Benjamin VanderSluis, Brenda Andrews, Chad L. Myers and Charles Boone

7. The Spatial Architecture of Chromosomes   137

   Job Dekker and Bas van Steensel

8. Chemogenomic Profiling: Understanding the Cellular Response to Drug   153

   Anna Y. Lee, Gary D. Bader, Corey Nislow and Guri Giaever

9. Graph Theory Properties of Cellular Networks   177

   Baruch Barzel, Amitabh Sharma and Albert-László Barabási

Section III
Dynamic and Logical Properties of Biological Systems

10. Boolean Models of Cellular Signaling Networks   197

    Zhongyao Sun and Réka Albert

11. Transcriptional Network Logic: The Systems Biology of Development   211

    Isabelle S. Peter and Eric H. Davidson

12. Reconstruction of Genome-Scale Metabolic Networks   229
Hooman Hefzi, Bernhard O. Palsson and Nathan E. Lewis

13. Genotype Networks and Evolutionary Innovations in Biological Systems   251
Andreas Wagner

14. Irreversible Transitions, Bistability and Checkpoint Controls in the Eukaryotic Cell Cycle: A Systems-Level Understanding   265
John J. Tyson and Béla Novák

15. Phenotypes and Design Principles in System Design Space   287
Michael A. Savageau

16. System Biology of Cell Signaling   311
Chiara Mariottini and Ravi Iyengar

17. Spatial Organization of Subcellular Systems   329
Malte Schmick, Hernán E. Grecco and Philippe I.H. Bastiaens

Section IV
Systems and Biology

18. Yeast Systems Biology: Towards a Systems Understanding of Regulation of Eukaryotic Networks in Complex Diseases and Biotechnology   343
Juan I. Castrillo, Pinar Pir and Stephen G. Oliver

19. Systems Biology of *Caenorhabditis elegans*   367
Andrew Fraser and Ben Lehner

20. *Arabidopsis* as a Model for Systems Biology   391
Philip N. Benfey and Ben Scheres

21. The Role of the Circadian System in Homeostasis   407
Anand Venkataraman, Heather Ballance and John B. Hogenesch

22. Biological and Quantitative Models for Stem Cell Self-Renewal and Differentiation   427
Huilei Xu, Dmitri Papatsenko, Avi Ma'ayan and Ihor Lemischka

Section V
Multi-Scale Biological Systems, Health and Ecology

23. Systems Medicine and the Emergence of Proactive P4 Medicine: Predictive, Preventive, Personalized and Participatory   445
Leroy Hood, Mauricio A. Flores, Kristin R. Brogaard and Nathan D. Price

24. Cancer Systems Biology: A Robustness-Based Approach   469
Hiroaki Kitano

25. Systems Immunology: From Cells and Molecules to a Dynamic Multi-Scale System   481
Mark M. Davis and Shai S. Shen-Orr

26. Causal Inference and the Construction of Predictive Network Models in Biology   499
Eric E. Schadt

27. Social Networks, Contagion Processes and the Spreading of Infectious Diseases   515
Bruno Gonçalves, Nicola Perra and Alessandro Vespignani

Index   529

List of Contributors

Réka Albert, Department of Physics, The Pennsylvania State University, University Park, PA 16802, USA; The Huck Institutes of the Life Sciences, The Pennsylvania State University, University Park, PA 16802, USA

Brenda Andrews, Banting and Best Department of Medical Research, The Donnelly Center for Cellular and Biomolecular Research, University of Toronto, Toronto, Ontario M5S 3E1, Canada; Department of Molecular Genetics, University of Toronto, Toronto, Ontario M5S 3E1, Canada

Gary D Bader, The Donnelly Centre for Cellular and Biomolecular Research, University of Toronto, Toronto, Ontario M5S 3E1, Canada; Department of Molecular Genetics, University of Toronto, Toronto, Ontario M5S 3E1, Canada

Heather Ballance, Department of Pharmacology, Penn Center for Bioinformatics, Penn Genome Frontiers Institute, Institute for Translational Medicine and Therapeutics, University of Pennsylvania Perelman School of Medicine, Philadelphia, PA 19104, USA

Albert-László Barabási, Center for Complex Network Research, Department of Physics, Northeastern University, 360 Huntington avenue, Boston, MA 02115, USA; Center for Cancer System Biology (CCSB) and Department of Cancer Biology, The Dana-Farber Cancer Institute and Department of Genetics, Harvard Medical School, 44 Binney Street, Boston, MA 02215, USA

Baruch Barzel, Center for Complex Network Research, Department of Physics, Northeastern University, 360 Huntington Avenue, Boston, MA 02115, USA; Center for Cancer System Biology (CCSB) and Department of Cancer Biology, The Dana-Farber Cancer Institute and Department of Genetics, Harvard Medical School, 44 Binney Street, Boston, MA 02215, USA

Anastasia Baryshnikova, Banting and Best Department of Medical Research, The Donnelly Center for Cellular and Biomolecular Research, University of Toronto, Toronto, Ontario M5S 3E1, Canada; Department of Molecular Genetics, University of Toronto, Toronto, Ontario M5S 3E1, Canada

Philippe I.H. Bastiaens, Department of Systemic Cell Biology, Max Planck Institute of Molecular Physiology, Otto-Hahn-Str. 11, 44227 Dortmund, Germany

Philip N. Benfey, Department of Biology and Duke Center for Systems Biology, Duke University, Durham, NC 27708, USA

Charles Boone, Banting and Best Department of Medical Research, The Donnelly Center for Cellular and Biomolecular Research, University of Toronto, Toronto, Ontario M5S 3E1, Canada; Department of Molecular Genetics, University of Toronto, Toronto, Ontario M5S 3E1, Canada

Kristin R. Brogaard, Institute for Systems Biology, 401 N. Terry Ave, Seattle, WA 98121, USA

Martha L. Bulyk, Division of Genetics, Department of Medicine; Department of Pathology; Brigham & Women's Hospital and Harvard Medical School, Boston, MA 02115, USA & Harvard-MIT Division of Health Sciences and Technology (HST), Harvard Medical School, Boston, MA 02115, USA

Michael A. Calderwood, Center for Cancer Systems Biology (CCSB) and Department of Cancer Biology, Dana-Farber Cancer Institute, Boston, MA 02215, USA; Department of Genetics, Harvard Medical School, Boston, MA 02115, USA

Anne-Ruxandra Carvunis, Center for Cancer Systems Biology (CCSB) and Department of Cancer Biology, Dana-Farber Cancer Institute, Boston, MA 02215, USA; Department of Genetics, Harvard Medical School, Boston, MA 02115, USA

Juan I. Castrillo, Cambridge Systems Biology Centre and Department of Biochemistry, University of Cambridge, Sanger Building, 80 Tennis Court Road, Cambridge CB2 1GA, UK

Michael Costanzo, Banting and Best Department of Medical Research, The Donnelly Center for Cellular and Biomolecular Research, University of Toronto, Toronto, Ontario M5S 3E1, Canada

Jürgen Cox, Department of Proteomics and Signal Transduction, Max-Planck-Institute of Biochemistry, Am Klopferspitz 18, 82152 Martinsried, Germany

Michael E. Cusick, Center for Cancer Systems Biology (CCSB) and Department of Cancer Biology, Dana-Farber Cancer Institute, Boston, MA 02215, USA; Department of Genetics, Harvard Medical School, Boston, MA 02115, USA

Eric H. Davidson, Division of Biology, MC 156–29, California Institute of Technology, Pasadena, CA 91125, USA

Mark M. Davis, Department of Microbiology & Immunology, Howard Hughes Medical Institute, Beckman Center, Room B221, Stanford University School of Medicine, Stanford, CA 94305, USA

Job Dekker, Program in Systems Biology, Department of Biochemistry and Molecular Pharmacology, University of Massachusetts Medical School, Worcester, MA 01605, USA

Mauricio A. Flores, P4 Medicine Institute, 401 N. Terry Ave, Seattle, WA 98121, USA

Andrew Fraser, The Donnelly Centre, University of Toronto, 160 College Street, Ontario M5S 3E1, Canada

Guri Giaever, The Donnelly Centre for Cellular and Biomolecular Research, University of Toronto, Toronto, Ontario M5S 3E1, Canada; Department of Molecular Genetics, University of Toronto, Toronto, Ontario M5S 3E1, Canada; Department of Pharmaceutical Sciences, University of Toronto, Toronto, Ontario M5S 3M2, Canada

Bruno Gonçalves, College of Computer and Information Sciences and Bouvé College of Health Sciences, Northeastern University, Boston, MA 02115, USA

Hernán E. Grecco, Department of Systemic Cell Biology, Max Planck Institute of Molecular Physiology, Otto-Hahn-Str. 11, 44227 Dortmund, Germany

Roderic Guigó, Centre de Regulació Genòmica Universitat Pompeu Fabra, Dr Aiguader, 88, E-08003 Barcelona, Catalonia, Spain; Departament de Ciències Experimentals i de la Salut, Universitat Pompeu Fabra, Dr Aiguader, 88, E-08003 Barcelona, Catalonia, Spain

Hooman Hefzi, Bioengineering Department, University of California San Diego, La Jolla, CA 92093-0412, USA

Marco Y. Hein, Department of Proteomics and Signal Transduction, Max-Planck-Institute of Biochemistry, Am Klopferspitz 18, 82152 Martinsried, Germany

John B. Hogenesch, Department of Pharmacology, Penn Center for Bioinformatics, Penn Genome Frontiers Institute, Institute for Translational Medicine and Therapeutics, University of Pennsylvania Perelman School of Medicine, Philadelphia, PA 19104, USA

Leroy Hood, Institute for Systems Biology, 401 N. Terry Ave, Seattle, WA 98121, USA

Ravi Iyengar, Department of Pharmacology and Systems Therapeutics and Systems Biology Center, New York Mount Sinai School of Medicine, New York, NY 10029, USA

Hiroaki Kitano, The Systems Biology Institute, Tokyo, Japan; Okinawa Institute of Science and Technology; Sony Computer Science Laboratories, Inc., Tokyo, Japan; Department of Cancer Systems Biology, The Cancer Institute, Tokyo, Japan

Meghana M. Kulkarni, Department of Genetics, Harvard Medical School, 77 Avenue Louis Pasteur, Boston, MA 02115, USA

Anna Y. Lee, The Donnelly Centre for Cellular and Biomolecular Research, University of Toronto, Toronto, Ontario M5S 3E1, Canada

Ben Lehner, European Molecular Biology Laboratory (EMBL)-Centre for Genomic Regulation (CRG) Systems Biology Unit and Institució Catalana de Recerca i Estudis Avançats (ICREA), CRG, Universitat Pompeu Fabra (UPF), c / Dr Aiguader, 88, Barcelona 08003, Spain

Ihor Lemischka, Department of Regenerative and Developmental Biology, Mount Sinai School of Medicine, One Gustave L. Levy Place, New York, NY 10029, USA; Black Family Stem Cell Institute, Mount Sinai School of Medicine, One Gustave L. Levy Place, New York, NY 10029, USA

Nathan E. Lewis, Bioengineering Department, University of California San Diego, La Jolla, CA 92093-0412, USA; Department of Genetics, Harvard Medical School, Boston, MA 02115, USA

Avi Ma'ayan, Department of Pharmacology and System Therapeutics, Mount Sinai School of Medicine, Systems Biology Center New York, One Gustave L. Levy Place, New York, NY 10029, USA; Black Family Stem Cell Institute, Mount Sinai School of Medicine, One Gustave L. Levy Place, New York, NY 10029, USA

Matthias Mann, Department of Proteomics and Signal Transduction, Max-Planck-Institute of Biochemistry, Am Klopferspitz 18, 82152 Martinsried, Germany

Chiara Mariottini, Department of Pharmacology and Systems Therapeutics and Systems Biology Center, New York Mount Sinai School of Medicine, New York, NY 10029, USA

Chad L. Myers, Department of Computer Science and Engineering, University of Minnesota, Minneapolis, MN 55455, USA

Corey Nislow, The Donnelly Centre for Cellular and Biomolecular Research, University of Toronto, Toronto, Ontario M5S 3E1, Canada; Department of Molecular Genetics, University of Toronto, Toronto, Ontario M5S 3E1, Canada; Banting and Best Department of Medical Research, University of Toronto, Toronto, Ontario M5S 3E1, Canada

Béla Novák, Centre for Integrative Systems Biology, Department of Biochemistry, Oxford University, Oxford OX1 3QU, UK

Stephen G. Oliver, Cambridge Systems Biology Centre and Department of Biochemistry, University of Cambridge, Sanger Building, 80 Tennis Court Road, Cambridge CB2 1GA, UK

Bernhard O. Palsson, Bioengineering Department, University of California San Diego, La Jolla, CA 92093-0412, USA

Dmitri Papatsenko, Department of Regenerative and Developmental Biology, Mount Sinai School of Medicine, One Gustave L. Levy Place, New York, NY 10029, USA; Black Family Stem Cell Institute, Mount Sinai School of Medicine, One Gustave L. Levy Place, New York, NY 10029, USA

Isabelle S. Peter, Division of Biology, MC 156-29, California Institute of Technology, Pasadena, CA 91125, USA

Nicola Perra, College of Computer and Information Sciences and Bouvé College of Health Sciences, Northeastern University, Boston, MA 02115, USA

Norbert Perrimon, Department of Genetics and Howard Hughes Medical Institute, Harvard Medical School, 77 Avenue Louis Pasteur, Boston, MA 02115, USA

Pinar Pir, Cambridge Systems Biology Centre and Department of Biochemistry, University of Cambridge, Sanger Building, 80 Tennis Court Road, Cambridge CB2 1GA, UK

Nathan D. Price, Institute for Systems Biology, 401 N. Terry Ave, Seattle, WA 98121, USA

Frederick P. Roth, Center for Cancer Systems Biology (CCSB) and Department of Cancer Biology, Dana-Farber Cancer Institute, Boston, MA 02215, USA; Donnelly Centre for Cellular & Biomolecular Research, University of Toronto, Toronto, Ontario M5S-3E1, Canada, & Samuel Lunenfeld Research Institute, Mt. Sinai Hospital, Toronto, Ontario M5G-1X5, Canada

Michael A. Savageau, Biomedical Engineering Department and Microbiology Graduate Group, University of California, One Shields Avenue, Davis, CA 95616, USA

Eric E. Schadt, Department of Genetics and Genomic Sciences, Mount Sinai School of Medicine, New York City, NY 10029, USA

Ben Scheres, Department of Biology, Utrecht University, Padualaan 8, 3584 CH Utrecht, The Netherlands

Malte Schmick, Department of Systemic Cell Biology, Max Planck Institute of Molecular Physiology, Otto-Hahn-Str. 11, 44227 Dortmund, Germany

Amitabh Sharma, Center for Complex Network Research, Department of Physics, Northeastern University, 360 Huntington Avenue, Boston, MA 02115, USA; Center for Cancer System Biology (CCSB) and Department of Cancer Biology, The Dana-Farber Cancer Institute and Department of Genetics, Harvard Medical School, 44 Binney Street, Boston, MA 02215, USA

Kirti Sharma, Department of Proteomics and Signal Transduction, Max-Planck-Institute of Biochemistry, Am Klopferspitz 18, 82152 Martinsried, Germany

Shai S. Shen-Orr, Department of Immunology, Faculty of Medicine, Technion, 1 Efron St. Haifa, 31096, Israel

Zhongyao Sun, Department of Physics, The Pennsylvania State University, University Park, PA 16802, USA

Giulio Superti-Furga, Research Center for Molecular Medicine of the Austrian Academy of Sciences, 1090 Vienna, Austria

John J. Tyson, Department of Biological Sciences, Virginia Polytechnic Institute & State University, Blacksburg, VA 24061, USA

Benjamin VanderSluis, Department of Computer Science and Engineering, University of Minnesota, Minneapolis, MN 55455, USA

Bas van Steensel, Division of Gene Regulation, Netherlands Cancer Institute, Amsterdam, The Netherlands

Anand Venkataraman, Department of Pharmacology, Penn Center for Bioinformatics, Penn Genome Frontiers Institute, Institute for Translational Medicine and Therapeutics, University of Pennsylvania Perelman School of Medicine, Philadelphia, PA 19104, USA

Alessandro Vespignani, College of Computer and Information Sciences and Bouvé College of Health Sciences, Northeastern University, Boston, MA 02115, USA; Institute for Scientific Interchange (ISI) Foundation, Via Alassio 11/c, 10126 Torino, Turin, Italy

Marc Vidal, Center for Cancer Systems Biology, (CCSB) and Department of Cancer Biology, Dana-Farber Cancer Institute, Boston, MA 02215, USA; Department of Genetics, Harvard Medical School, Boston, MA 02115, USA

Andreas Wagner, University of Zurich, Institute of Evolutionary Biology and Environmental Studies, Y27-J-54, Winterthurerstrasse 190, CH-8057 Zurich, Switzerland

A.J. Marian Walhout, Programs in Systems Biology, and Molecular Medicine, University of Massachusetts Medical School, Worcester, MA 01605, USA

Huilei Xu, Department of Pharmacology and System Therapeutics, Mount Sinai School of Medicine, Systems Biology Center New York, One Gustave L. Levy Place, New York, NY 10029, USA; Black Family Stem Cell Institute, Mount Sinai School of Medicine, One Gustave L. Levy Place, New York, NY 10029, USA

Preface

Life is complicated. Even seemingly simple organisms such as bacteria display highly sophisticated behaviors and processes that allow them to reproduce, develop and grow, to respond to their milieu and to fend off pathogens and insults. These behaviors have fascinated generations of scientists, and understanding how the complexity of living systems arises continues to be one of the principal goals of biology today.

Whereas the second half of the 20th century was dominated by molecular biology, genetics, cell biology and biochemistry to unravel the detailed mechanisms of biological processes, the years since the turn of the 21st century and onwards have been marked by development of new technologies to probe the full complexity of living systems. These include powerful imaging, proteomic, genomic and high-throughput screening approaches to systematically identify and locate biological components, as well as large-scale methods for detecting how these components, such as proteins, genes and metabolites, are dynamically organized in space and how they functionally and physically interact. Combined with new computational, statistical and mathematical tools, these studies are leading to insights into the processes by which living organisms attain their organization and respond to environmental signals and conditions. Further, human disease states are increasingly considered to be caused not by singular biochemical alterations, but instead result from wider disruptions of the complex interplay between the many molecular components and processes that make up the human body.

The current interest in systems biology reflects a renewed appreciation of the amazing complexity of living organisms. Perhaps not surprisingly, the field of systems biology itself is highly diverse and wide-ranging, encompassing interlinked experimental, computational and mathematical approaches, datasets and models. In this book we have attempted to capture that diversity, while also emphasizing how these studies converge to comprehensive and consistent views of biological systems.

We have brought together leading experts in the field who have contributed individual chapters that cover a range of topics and technologies. Several chapters provide overviews of complex biological systems such as the cell cycle, development, cell signaling, tumorigenesis, nuclear organization, and the immune system. Several contributions cover the mapping and analysis of different types of network, including protein—protein interaction networks, protein—DNA interactions and transcriptional networks, and genetic interaction networks. Other chapters cover methods to analyze networks, such as graph theory and network reconstruction, as well as methods to gain insights into network dynamics, evolution and spatial organization.

We hope this book will serve both as a valuable introduction to the field of systems biology and as a general reference for those interested to enter the field as well as for experts.

We are grateful to all the authors for their enthusiasm for this project and their outstanding contributions. We thank the editorial staff at Elsevier for their excellent help in preparing this book, and members of our laboratories for their advice and input.

A.J. Marian Walhout, Marc Vidal and Job Dekker

Reviewers

The editors would like to extend their thanks to the following reviewers, who reviewed select chapters over and above the editors' review:

Marc Brehme,
Center for Cancer Systems Biology (CCSB), Dana-Farber Cancer Institute (DFCI), Boston MA, USA

Michael A. Calderwood,
CCSB, DFCI

Anne-Ruxandra Carvunis,
CCSB, DFCI

Michael E. Cusick,
CCSB, DFCI

David E. Hill,
CCSB, DFCI

Sam Pevzner,
CCSB, DFCI

Nidhi Sahni,
CCSB, DFCI

Jean Vandenhaute,
Unité de Recherche en Biologie Moléculaire, Facultés Universitaires Notre-Dame de la Paix, Belgium

Roseann Vidal,
On parenting leave

Song Yi,
CCSB, DFCI

Section I

Components of Biological Systems

Chapter 1

Proteomic Analysis of Cellular Systems

Marco Y. Hein, Kirti Sharma, Jürgen Cox and Matthias Mann
Department of Proteomics and Signal Transduction, Max-Planck-Institute of Biochemistry, Am Klopferspitz 18, 82152 Martinsried, Germany

Chapter Outline

Introduction	3
MS-Based Proteomics Workflow	4
Computational Proteomics	7
Deep Expression Proteomics	11
Interaction Proteomics	13
Large-Scale Determination of Post-Translational Modifications	15
Outlook and Future Challenges	18
References	19

INTRODUCTION

A prerequisite for a system-wide understanding of cellular processes is a precise knowledge of the principal actors involved, which are biomolecules such as oligonucleotides, proteins, carbohydrates and small molecules. Ever more sophisticated methods to measure the identity and amount of such biomolecules were an integral component of most of the biological breakthroughs of the last century. At the level of the genome, DNA sequencing technology can now give us a complete inventory of the basic set of genetic instructions of any organism of interest. Furthermore, recent breakthroughs in next-generation sequencing are promising to allow large-scale comparison of the genomes of individuals. However, genomic sequences and their variations between individuals are completely uninterpretable without knowledge of the encoded genes as well as the biological processes in which they are involved. Therefore, the growing ability to obtain genetic data provides an increasing need and impetus to study the functions of gene products individually (classic molecular biology) and at a large scale (systems biology). The first such system-wide studies were performed at the level of mRNA ('transcriptomics'). They enable an unbiased and increasingly comprehensive view of which parts of the genome are actually expressed in a given situation. Transcriptomics also revealed that the relationship between the genomic coding sequences and their corresponding RNA molecules can be exceedingly complex. However, in terms of cellular function, the transcriptome still represents only a middle layer of information transmission, with no or little function of its own. The actual 'executives' of the cell are the proteins, which perform myriad roles, from orchestrating gene expression to catalyzing chemical reactions, directing the information flow of the cell and performing structural roles in cells and organisms. This crucial role of proteins is also underlined by the fact that diseases always involve malfunctioning proteins, and that drugs are almost invariably directed against proteins or modify their expression levels.

Unfortunately, given the central importance of proteins, until recently there were no methods of protein measurement that were comparable to the powerful sequencing, hybridization or amplification-based methods to characterize oligonucleotides. This is finally beginning to change owing to the introduction of mass spectrometry, first in protein science and later for the large-scale study of proteins, a field called mass spectrometry (MS)-based proteomics [1].

The proteome of a cell designates the totality of all expressed proteins in a given biological situation, and is therefore a dynamic entity. It encompasses not only the identity and amount of all proteins but also their state of modification, their turnover, location in the cell, interaction partners and — by some definitions — their structures and functions. Clearly, the proteome of the cell is the most complex and functionally most relevant level of cellular regulation and function.

Accordingly, in systems biology it is usually the proteome that is the object of modeling. Typically only very small subsets of all proteins — those participating in a defined function of interest — are included in these models. Even then, reliable and relevant information on these few proteins has been hard to come by. This has meant that systems biological models suffered from a paucity of hard parameters, and instead usually had to make do with very rough estimates of the identities, abundances, localization and modification states of the involved proteins. Modern MS-based proteomics is now ready to change this situation completely.

Its success in protein analysis comes as the last chapter in the very long history of mass spectrometry, which began with the observation of *Kanalstrahlen* (anode rays) by Eugen Goldstein in 1886 and the construction of the first mass spectrometer by Francis William Aston in 1919. The first application to amino acids by Carl-Ove Andersson dates back to 1958. Later, both the quadrupole and the three-dimensional ion traps were developed by Wolfgang Paul, for which in 1989 he received the Nobel Prize in Physics, together with Hans Georg Dehmelt. However, the breakthrough for MS in biology came with the development of soft ionization technologies that enabled gentle transfer of peptides or proteins into the mass spectrometer, for which the Nobel Prize in Chemistry in 2002 was awarded. The emergence of MS as a powerful 'omics' discipline was also enabled by continuous developments in sample preparation, separation technologies and breakthroughs in the capabilities of the mass spectrometers themselves, some of which are detailed below. In parallel with these improvements on the 'wet side', data analysis and computational strategies on the 'in silico side' over the last 20 years have been just as important, as they allow the identification of peptides in sequence databases from a minimum of mass and fragmentation information. Originally applied to one peptide at a time in a manual fashion, these algorithms now deal with hundreds of thousands of peptides in multifaceted projects and require large-scale data management issues to be addressed which are just as demanding as they are in the other 'omics' technologies.

The development of relative and absolute quantification methods over the last decade has been particularly crucial to proteomics. Using the latest proteomics technologies, it is now possible to quantify essentially complete proteomes of model organisms such as yeast [2]. More complex organisms are also coming within reach [3–5]. However, quantitative proteomics not only permits precise proteome quantification in one state compared to another (termed 'expression proteomics' and providing data conceptually similar to transcriptomics) but also enables 'functional proteomics', when combined with appropriate biochemical workflows. This can, for example, identify specific protein interactions or reveal the composition of subcellular structures [6–8]. Together, these methods allow the proteome to be studied in space and time, something that cannot easily be done on a large scale and in an unbiased manner by other technologies [9]. The resulting proteomic data perfectly complement large-scale studies following individual proteins in single cells, for instance by means of immunostaining [10] or protein tagging [11].

One of the most important areas for MS-based proteomics is the analysis of post-translational modifications (PTMs) [12,13]. During recent years, MS-based proteomics has revealed an unexpected diversity and extent of protein modifications. For example, phosphorylation turns out to occur not only on a few key proteins but on thousands of them, which possibly also applies to less studied PTMs. How to model their regulatory roles will long be a key challenge for systems biology.

MS-based proteomics now for the first time opens up the entire universe of cellular proteins to detailed study. Protein amount, localization, modification state, turnover and interactions can all be measured with increasing precision and increasingly sophisticated approaches, as detailed below. There is a unique opportunity to employ these data as a crucial underpinning for building accurate and comprehensive models of the cell [14].

MS-BASED PROTEOMICS WORKFLOW

The analysis of complex protein mixtures is very difficult. Accordingly, the field of MS-based proteomics has been made possible by seminal advances in technology that have helped to overcome a number of critical challenges. Together, they have resulted in a generic and general 'shotgun' workflow that can be applied to any source of proteins and almost any problem that can be addressed by MS-based proteomics (Figure 1.1). Here we explain the principles of this workflow, but also point out variations to the general theme.

Until the 1980s proteins or peptides were largely incompatible with MS, as they could not be transferred into the vacuum of the mass spectrometer without being destroyed. Two alternative approaches solved this fundamental problem: electrospray ionization (ESI), for which a share of the 2002 Nobel Prize in Chemistry was awarded to John B. Fenn, and matrix-assisted laser desorption/ionization (MALDI). MALDI involves embedding the analyte in a solid matrix of an organic compound, followed by transfer into the vacuum system. A laser pulse then excites the matrix molecules, leading to their desorption along with the ionized analyte molecules, whose mass is measured in a time-of-flight (TOF) analyzer [15]. In contrast, in electrospray a stream of liquid is dispersed into a charged aerosol when high voltage is applied to the emitter. Solvent molecules in aerosol droplets rapidly evaporate, and charged analyte molecules are then transferred into the vacuum of the mass spectrometer, where they finally arrive as 'naked' ions [16].

Even with appropriate ionization techniques at hand, large intact proteins are usually difficult to handle, therefore the standard MS-based proteomic workflow follows

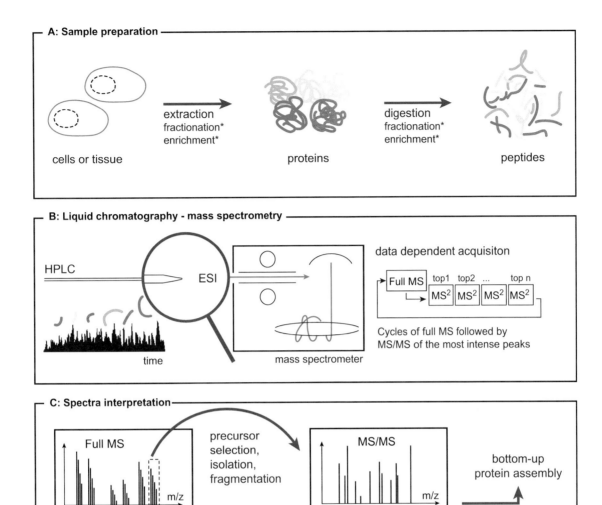

FIGURE 1.1 **Outline of a typical shotgun proteomics workflow.** A: Sample preparation: Proteins extracted from tissues or cells are digested into peptides using proteases such as trypsin. A fractionation step may be applied at either the protein or peptide level to improve the coverage and dynamic range. Peptides bearing specific post-translational modifications can be enriched using specialized approaches (see Figure 1.5A). B: Liquid chromatography-mass spectrometry: Peptides are separated by high-performance liquid chromatography (HPLC) and electrosprayed directly into the mass spectrometer. Peptide ions are measured at high resolution in a data-dependent mode: after each full MS scan, the most intense peptide ions are fragmented to generate MS/MS spectra. C: Spectra interpretation: The full MS spectra provide information about the peptide mass, intensity, presence of a PTM and stable isotope pairs. The mass of each fragmented peptide together with its fragment ion pattern is searched against databases for peptide identification and bottom-up protein assembly.

the bottom-up principle: proteins are first digested to peptides using a sequence-specific endoprotease. This is typically trypsin, which cleaves C-terminal to arginine or lysine. These peptides are analyzed by MS and afterwards proteins are reconstructed in silico. For the general purpose of identifying and quantifying proteins with high sensitivity and in complex mixtures, this 'bottom-up' approach is extremely powerful. This is due to the convenience of handling peptides and the much superior characteristics of the MS analysis of peptides compared to proteins. The complementary 'top-down' approach omits the enzymatic digestion step and analyzes intact protein species instead [17]. Its principal merit is that it retains information about the entire protein (such as co-occurring modifications), but this advantage comes at the cost of vastly increased experimental effort [18].

For an unbiased and comprehensive analysis of the proteome, the cell or tissue lysis method must ensure complete solubilization of all proteins contained in the sample. This is particularly challenging with membrane proteins, which demand a detergent-based solubilization method even though detergents are known to interfere with subsequent MS analysis. Furthermore, endoproteinases work optimally in a detergent-free environment. The first MS sample preparation methods successfully employed on biological samples used detergent-mediated solubilization followed by SDS polyacrylamide gel electrophoresis and in-gel enzymatic digestion of proteins [19]. 'In-solution' digestion methods employed detergent-free protein extraction using strong chaotropic agents such as urea, and digestion of proteins under denaturing conditions. In the early days of applying MS to protein identification, stained protein bands were excised from one-dimensional gel electrophoresis runs, in-gel digested and analyzed directly in the mass spectrometer by MALDI or electrospray. For samples containing peptides from only one or a few proteins, the combination of several peptide masses may be sufficient for identification. This technique is called 'mass fingerprinting' and it is still often used in conjunction with two-dimensional gel electrophoresis (2D-GE). However, both mass fingerprinting and 2D-GE have serious analytical limitations in the dynamic range of protein abundances that they can handle, as well as many other issues, and they are no longer generally used in proteomics. Today the inherent complexity of proteomic samples is being addressed by a combination of fractionation techniques as well as fast and sensitive mass spectrometers, but it remains a major challenge when the goal is to define complete proteomes [20]. For these very complex mixtures, electrospray, and not MALDI, is the ionization method of choice. This is because electrospray handles analytes in solution, which allows it to be coupled directly or 'on-line' to liquid chromatography (LC) by applying the spray voltage to the end of the chromatographic column. LC is arguably the most powerful separation technique available for peptides, which can then be analyzed sequentially as they elute from the column. Current developments in peptide LC aim at further improvements in separation as well as decreasing flow rates and column diameters, which increases sensitivity [21].

In addition, a multitude of gel-based and gel-free fractionation techniques have been developed that are applied on either the protein or the peptide level prior to the liquid chromatography step [22–26]. While increasing the number of separation steps generally increases the depth of coverage of the proteome, it also increases the sample processing and MS-measurement time, as well as requirements for sample amount. Therefore, proteomics experiments should be planned with the minimum number of fractionation steps possible. This is especially important when several conditions are to be measured and compared.

Although online coupling of LC with MS via electrospray is clearly the method of choice for complex protein mixtures, the MALDI method still offers advantages in specific situations. For instance, in principle the spatial resolution of the MALDI laser spot makes it possible to 'image' analytes in situ, e.g. on tissue slices treated with appropriate MALDI matrices [27, 28].

Once peptides have been transferred into the vacuum of the mass spectrometer, their mass-to-charge ratio (m/z) and intensity have to be measured. For unambiguous identification, it is additionally necessary to fragment each peptide in turn and to record the resulting mass spectrum, a technique called MS/MS, MS^2 or tandem mass spectrometry. In the data-dependent 'shotgun' approach, the most abundant peptide species eluting from the LC column at a given time are isolated one at a time and activated in the mass spectrometer, usually by collisions at low pressure of an inert gas. Peptides mainly dissociate at the amide bonds, generating overlapping series of N-terminal and C-terminal fragments (called b-ions and y-ions, respectively) [29]. In principle, the peptide sequence can be reconstructed 'de novo' from a complete fragment ion series. In practice, it is much easier to match uninterpreted fragment information to a comprehensive protein sequence database of the organism under investigation. There are many different algorithms and search engines for this (see section Computational Proteomics), but virtually all are based on the comparison of measured masses with the theoretical masses of expected peptides and their fragments.

Determining accurate masses is a key step in this procedure, and advances in mass spectrometric technology in recent years have made significant contributions to the achievable depth of analysis. Key characteristics of a high-performance mass spectrometer are resolution, mass accuracy, speed, sensitivity and dynamic range [30]. High resolution is the ability to distinguish two peaks of only slightly different m/z ratio, while mass accuracy describes the difference between measured and theoretical mass [31]. Sensitivity is the capacity to detect low abundant analytes whereas dynamic range of an instrument denotes the difference between the lowest and highest abundant species that are detected. Together, the aforementioned properties should allow a high-performance instrument to perform peptide sequencing at sufficiently high speed to obtain adequate coverage of the complexity of the sample within the timeframe of analysis. The Orbitrap mass analyzer is a particularly powerful instrument in proteomics [32–34], but modern TOF-based analyzers are also popular [35,36].

However, even today's best mass spectrometers are technically unable to isolate and sequence all peptide species present in an LC-MS run, resulting in extensive undersampling of the observable peptides [37]. This leads to a certain degree of stochastic behavior between shotgun runs, which can complicate analysis, especially in systems biology applications. In such cases, it is often attractive to measure only a subset of peptides — such as those of a few key proteins — but to ensure that they are measured in each of multiple conditions. This requirement has led to so-called targeted approaches, where the mass spectrometer is fed with a list of predefined peptide species and their corresponding fragments. It then simply records series of transitions from precursor to fragment ions; this is referred to as multiple or selected reaction monitoring (SRM or MRM) [38]. Both shotgun and targeted approaches have their advantages and drawbacks: the shotgun approach does not require prior development of peptide-specific assays and in principle can measure the entire proteome. Therefore, it is the method of choice for the discovery phase of proteomic studies. However, it may require extensive measurement time and proteins of interest may be missed. In contrast, the targeted approach can be performed rapidly and in principle without pre-fractionation, but is necessarily biased in the sense that only predefined peptides are measured.

The most promising approach is probably a hybrid one, which is facilitated by the latest generation of mass spectrometric hardware: a combination of general shotgun sequencing with targeted sequencing of a list of preselected candidates. Another interesting hybrid approach has been called SWATH-MS and involves the acquisition of fragment data for all precursor masses in consecutive mass windows of 25 m/z units (termed 'swaths') across the entire mass scale in rapid succession. When combined with targeted data extraction, this enables repeated scanning of the same fragment ion maps for quantification of proteins or peptides of interest [39].

Relative or absolute quantification has increasingly become the focus of proteomics experiments and has largely replaced the initial goal of only generating accurate and complete lists of identified proteins [40]. This is a challenging task because mass spectrometry is not inherently quantitative. A number of elegant approaches have been developed that now make MS the most quantitatively accurate protein technology by far; these are summarized in Box 1.1.

The correct identification and quantification of peptides by MS/MS sequencing, the assembly of a series of peptide sequences into protein identifications and the integration of peptide quantification into protein quantification becomes increasingly challenging as the complexity of the sample increases. It can only be dealt with correctly using rigorous statistical methods. To this end, a plethora of software tools and mass spectra search engines have been developed, which are discussed in the next section.

COMPUTATIONAL PROTEOMICS

An important aspect of high-throughput technologies is the availability of suitable computational workflows supporting the analysis and interpretation of the large-scale datasets that are routinely generated in current systems biology. Modern MS-based proteomics measurements produce data at similar rates as deep sequencing experiments of cellular DNA and RNA. For all of these technologies it is a challenging task to produce condensed representations of the data in a form and amount suitable for biologic interpretation in a reasonable timeframe within the constraints of the available computer hardware. In the early days of MS-based proteomics, the interpretation of spectra for the purpose of identification and quantification of peptides and proteins was done in a manual or semi-automatic fashion [41]. Nowadays, however, a single mass spectrometer can generate a million mass spectra per day [42]. Obviously, it is impractical to interpret the entire raw data in a 'one spectrum at a time' fashion by a human expert. Therefore, it is a necessity to employ reliable and efficient computational workflows for the identification and quantification of these enormous amounts of spectral data. Of particular importance is the control of false positives, e.g. by calculating and enforcing false discovery rates (FDRs) by statistical methods that take into account the multiple hypotheses testing nature inherent in large MS datasets.

Historically, computational proteomics started from the development of peptide search engines, and for this reason software tools have evolved around them. Furthermore, vendors strive to provide software enabling the computational analysis of the output of their instruments. These often interface with popular peptide search engines. There is much activity in software development for MS-based proteomics and dedicated reviews have been published [43–46].

All-encompassing end-to-end computational workflow solutions have also been developed, for instance the freely available trans-proteomic pipeline [47] and MaxQuant software packages [48]. MaxQuant contains a comprehensive set of data analysis functionalities and will be the basis of the subsequent discussion. Furthermore, there is a plethora of individual solutions for more specialized tasks. As examples, ProSight assists in the analysis of top-down protein fragmentation spectra [49,50], special search engines have been developed to identify cross-linked peptides [51,52], and commercial software for the 'de novo' interpretation of fragmentation spectra is available [53,54].

Here we focus on the computational steps that are needed to generate quantitative protein expression values from the raw data. Later chapters in this book focus on subsequent analysis of this kind of expression data in terms of multivariate data analysis, in the context of biomolecular interaction networks or in the modeling of biochemical reaction pathways. This initial part of the shotgun proteomics data analysis pipeline can roughly be subdivided into four main components (Figure 1.2): (a) feature detection and processing, (b) peptide identification, (c) protein identification and (d) quantification. Each of these consists of several sub-tasks, some of which are obligatory constituents of the generic data analysis workflow whereas others address specific questions in particular datasets.

Box 1.1 Quantification in MS-Based Proteomics

Mass spectrometric approaches providing relative and absolute quantification have been a focus of many recent developments in the field. MS-based quantification is non-trivial because for different peptide species there is no proportionality between their respective amounts and the signal intensities they generate in the mass spectrometer. This is due to the very diverse chemicophysical properties of peptides with different sequences, resulting in widely varying ionization efficiencies. For chemically identical peptides, however, signal intensity is proportional to the amount — within the linear range of the instrument — and this is the basis of all isotope labeling methods as well as of many label-free quantification approaches. In addition, it is often assumed that the most readily detected peptides of each protein have roughly similar ionization efficiencies across all proteins, and that their signal is therefore a proxy for the protein amount.

Label-free approaches are appealing because they can be used on any sample and do not require any additional experimental steps. A basic version of label-free quantification is called spectral counting, and simply compares the number of times a peptide has been fragmented. Since there is a stochastic tendency of shotgun proteomics to fragment more abundant peptides more often, this provides a rough measure of peptide abundance [180,181]. In a more accurate version of label-free quantification, the MS signals of each peptide identifying a protein are added and this protein intensity value is compared between the different experiments [75]. Ideally, the intensities of the same peptide species are directly compared across experiment. Challenges in label-free workflows are day-to-day variations in instrument performance or slight variations in sample preparation, which can reduce accuracy. Nevertheless, they are gaining ground owing to the increasing availability of high-resolution mass spectrometers and the development of sophisticated algorithms. They are best suited to cases where at least several-fold changes in protein or peptide intensities are expected.

The most accurate methods for quantification by MS make use of the fact that the MS response of the same compound in different isotopic states is the same. This principle has been employed for decades in the small molecule field, where it is sometimes called isotope dilution MS, and it has also been used for many years with peptides. In proteomics, the peptide populations from two different samples are labeled by the introduction of light or heavy stable isotopes such as ^{12}C vs. ^{13}C and ^{14}N vs. ^{15}N, mixed and analyzed together. The mass spectrometer easily distinguishes heavy peptides from light peptides by their mass shift, but since they are chemically equivalent they behave the same during chromatographic separation and ionization. The ratio of the heavy and light peak intensities therefore represents the relative amounts of the corresponding proteins in the samples to be compared. There are many different methods of introducing labels, for example metabolic labeling methods such as SILAC [182], or chemical ones such as TMT [183], iTRAQ [184] and di-methyl labeling [185,186]. The metabolic methods have the principal advantage that the two populations to be compared can be mixed at a very early stage of sample preparation. All variations in sample preparation are then experienced by both samples equally, leading to very high quantitative accuracy. Chemical methods are usually applied at a later stage, by which time quantitative differences due to separate sample preparation may already be established. Furthermore, care has to be taken that the chemical labeling procedure proceed to the same degrees of completeness in the different samples and that chemical side reactions are minimized.

Metabolic methods almost always quantify the peptide in the intact form in the MS spectrum, whereas some of the chemical methods use differentially isotope labeled fragment ions ('reporter ions') to determine the relative ratios from the MS/MS spectra. A disadvantage of the latter methods is that, in complex mixtures, peptides apart from the intended one are co-fragmented. These also contribute their identical reporter ions, distorting measured ratios [77].

Targeted approaches (SRM or MRM) are also fragmentation-based quantification methods but they aim to monitor only transitions from precursor to sequence-specific fragments. Several such transitions are monitored in rapid succession for a single peptide and several peptides can be targeted at any given elution time. This ensures that the recorded signal is due to the intended peptide. For quantification, an isotope-labeled, synthetic peptide standard for each peptide of interest needs to be introduced. However, since synthesis, purification and storage of many labeled peptides are resource-intensive, the label-free transition data is often used for approximate quantification. In general, MRM-based quantification methods require extensive method development because the most sensitive and specific transitions need to be determined for each peptide separately. There are therefore a number of large-scale projects to construct such data on a global, organism-wide scale [187–189].

Apart from relative quantification of two or more proteomes, it is in many instances necessary to estimate the absolute amounts of proteins. If a known amount of a synthetic, labeled peptide is added to the sample, the ratio of the heavy to the light version of the peptide immediately yields the absolute amount of the endogenous peptide present (absolute quantification or 'AQUA' method [190]). If the extraction and digestion efficiency of the protein in the sample is also known, this furthermore yields the absolute amount of protein in the sample. The same principle also applies to spiking in known amounts of proteins, except that this automatically controls for digestion efficiency, including the tendency of the enzyme to produce peptides with missed cleavages [191].

Absolute protein amounts can be converted into copy numbers per cell, an important parameter for modeling. Evidently, it is impractical to spike in reference peptides or proteins for an entire proteome. Therefore, in the simplest case, the MS signals of peptides identifying a given protein are summed up and divided by the total MS signal of all proteins. This procedure can be calibrated by the estimated total protein amount or with the help of reference peptides or proteins for a select subset of proteins across the dynamic range [2,86,91].

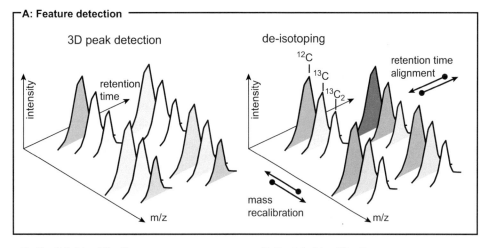
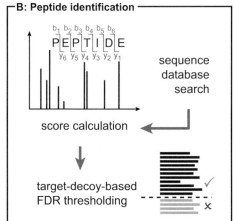
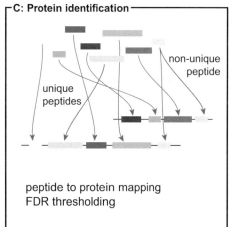
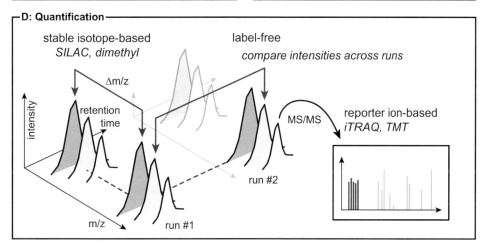

FIGURE 1.2 Overview of the main components of the computational workflow of shotgun proteomics. A: Detection and processing of peptide features in LC-MS runs. B: Identification of peptides based on their characteristic fragmentation patterns. C: Assembly of peptides into proteins. D: Quantification of peptides and proteins based on stable isotope labeling or by label-free quantification.

The first group of tasks is concerned with extracting features from the raw data that correspond to peptides in the MS spectra and to peptide fragments in the MS/MS spectra (Figure 1.2A). Depending on the specific details of the MS technology employed, pre-processing steps may be required, for instance subtraction of a background level, or smoothing and filtering of the raw data [43]. Then, peaks are detected, which in LC-MS constitute three-dimensional 'hills' over the mass-retention time plane. These 3D peaks usually occur in co-eluting isotope patterns that correspond to peptides with a given charge. For a peptide this pattern is mainly due to the natural content of ^{13}C atoms. In the case of stable-isotope-based quantification methods the peptide exists in different labeling states, such as heavy and light SILAC partners. These have to be assigned to each other based on characteristic mass differences and similarity of elution profiles.

Often there are systematic errors in the measured masses of these peptide features that vary continuously with mass and retention time. Algorithms can be applied to recalibrate the mass measurements and thereby remove these systematic errors, resulting in very accurate mass measurements with solely non-systematic and small random errors remaining [55]. From the standard deviations in the mass measurements one can calculate individual mass tolerances for each peptide, which aid peptide identification by restricting the possible molecules to elemental compositions that are consistent with the individual peptide mass tolerance. Similar to this recalibration of mass measurements, the retention times of peptides can also be recalibrated. Here the goal is to make the retention times in different LC-MS runs as comparable to each other as possible using computational means.

The next important computational block is concerned with the identification of peptides from fragmentation spectra [29,56] (Figure 1.2B). Here one can follow one of two approaches: The de novo approach starts with interpreting mass differences between fragment peaks as amino acids and tries to build up amino acid sequences, often by representing MS/MS spectra as spectrum graphs [57]. This either results in a de novo sequence of the whole peptide or in a sequence tag within the peptide[58]. In the database search approach one first digests the protein sequences of an organism in-silico to obtain a list of peptides that a certain protease, typically trypsin, can potentially generate. Peptides are then identified by scoring MS/MS spectra against the sequences from the database, either with a cross-correlation approach as used in SEQUEST[59], or with a probability-based strategy as used by the Mascot [60] and Andromeda [61] search engines, for instance. In the latter case, for each peptide—spectrum comparison the probability is calculated to obtain the observed number of matching peaks between the spectrum and the theoretical fragment series derived from the peptide sequence by chance. The peptide identification rate can be further improved by taking into account peptide properties such as sequence length, number of missed cleavages, and others, either with the help of bayesian methods [48,62] or with machine learning techniques [63]. A false discovery rate (FDR) can be imposed on the peptide identification process by modeling of the score distribution [62] or by the target-decoy approach [64]. Statistical techniques controlling the FDR [65] are superior to simple ad hoc methods, such as using a fixed score cutoff, since they properly take into account multiple hypothesis testing and incorporate properties of the search space. Peptide identifications can be transferred between LC-MS runs based on highly accurate mass measurements and optimally aligned retention-time values. We recently developed a method for determining an FDR for this procedure [66]. Post-translational modifications of proteins can be identified by incorporating them into the database search in the form of variable modifications. In principle, scoring is similar to the scoring of unmodified peptides. However, the search space may increase dramatically, especially when considering several modifications at once. Additionally, the specific amino acid that has been modified needs to be pinpointed. This positioning of the identified modifications can be carried out in several ways, which usually provide scores that reflect the certainty of the localization [67–71]. A larger class of modifications can be detected with peptide sequence tags [58], the error-tolerant search in Mascot [72], or with the completely unbiased dependent peptide search [73], which does, however, require the unmodified peptide to have been fragmented and identified as well. Finally, in order to validate the identification of peptides or proteins of particular interest it is useful to visualize and export their fragmentation spectra. At this stage extensive peak annotation, including peaks resulting from neutral losses and originating from other peptide chemistry reactions, can also be provided.

Once peptides and their modifications are identified they need to be assigned to proteins, a non-trivial task that has been termed the 'protein inference problem' (Figure 1.2C) [74]. The basic challenge is that a peptide may occur in several proteins. The reason might be that these proteins result from alternative splicing and therefore share common exons, or that the proteins originate from distinct genomic locations that encode homologous genes with very high sequence identity. A pragmatic approach to the protein redundancy problem is to assemble proteins into groups of non-distinguishable members based on the sets of identified peptides either being identical between these members or containing each other. Additionally, one can map peptides to the protein coding regions of known transcripts and investigate whether unknown splice variants can be detected by identifying peptides that span new splice junctions.

Most proteomics datasets that are generated with current equipment are sufficiently large that one needs to take care of the FDR on the protein level as well as the peptide level. Approaches that have only a peptide-based control of false identifications, even if it is quite stringent, will accumulate false positive protein identifications if sufficiently large amounts of data are measured, and should therefore be used with caution.

After peptides and proteins are identified, the absolute or relative amounts of proteins in different samples usually need to be calculated, which requires quantification of the identified peptides (Box 1.1 and Figure 1.2D). In stable isotope labeling approaches that produce pairs or higher multiples of peptide isotope patterns in the MS spectra, one can use algorithms that provide very precise estimates of peptide abundance ratios. In MaxQuant this is done by comparing the full elution profiles and isotope patterns of the labeled partners. Once peptide ratios are calculated they need to be combined in appropriate ways to obtain protein ratios. In isobaric labeling techniques the relative peptide abundances are read out at specific mass values in the MS/MS spectra [75]. Here special attention needs to be devoted to the distortion of the signal by co-fragmented peptides and to filtering the peptides accordingly for quantification [76–78]. Finally, samples can be measured without isotopic labeling, which is referred to as 'label-free quantification'. In this case optimal alignment of the runs should be performed, and further normalization steps should be included to make peptide signals from different LC-MS runs comparable to each other. This is computationally challenging, in particular if the samples are each pre-fractionated into several LC-MS runs.

In addition to the basic workflow described so far, which provides quantitative protein expression data, several additional downstream computational tasks need to be performed. Fortunately, once the proteomic expression data matrix has been obtained, many statistical and computational methods that were developed for microarray data analysis can be re-used for proteomics. For instance, clustering, principal component analysis, tests for differential regulation, time series, pathway and ontology enrichment analysis and many other methods can be applied just as well to proteomics data. The Perseus module of MaxQuant assembles these capabilities into a single, user-friendly platform for high-resolution proteomic data. Modeling in systems biology has so far relied on either mRNA levels as proxies for protein expression or on small-scale protein data that monitored only a few different molecular species. In the future, modeling will surely benefit from the increasing availability of large-scale and precise proteomics data.

DEEP EXPRESSION PROTEOMICS

One of the limitations of proteomics so far has been its inability to probe the proteome in great depth. Over the last few decades, 2D gel electrophoresis, for instance, has produced gels that visualized hundreds or thousands of spots. Upon identification, however, they generally proved to derive from a very small number of highly expressed genes. The difficulties in exploring the proteome in depth are mostly related to the 'dynamic range problem', that is, the difficulty of measuring extremely low abundance proteins in the presence of very high abundant ones. Until a few years ago many thought that this problem would be unsolvable even in principle [79]. Fortunately, it has now become clear that the dramatic improvements in the proteomic workflow do indeed allow complete characterization of proteomes.

Like its genome, the proteome of the yeast model system was the first to be completely analyzed [2]. Haploid and diploid yeasts were SILAC-labeled, mixed and measured together. With a combination of different approaches, 4400 yeast proteins were identified with 99% confidence, a larger number than detected either by genome-wide TAP (tandem affinity purification) tagging or GFP (green fluorescent protein) tagging of all yeast open reading frames [11,80]. The most regulated genes belonged to the yeast mating pathway, most of which are expressed at very low levels and are only functionally relevant in haploid yeast. However, not all members of this pathway were differentially regulated, immediately highlighting that they must have additional roles in other cellular processes. The total dynamic range of the yeast proteome under these basal conditions turned out to be between 10^4 and 10^5.

A targeted analysis of the yeast proteome likewise identified proteins across its entire dynamic range [81]. SRM assays were developed on triple quadrupole instruments for members of the glycolysis pathway, and expression changes upon metabolic shifts were measured across multiple time points in relatively short LC-MS runs.

Recently, our group has proposed 'single-shot proteomics' as a complement to the shotgun and targeted approaches: single-shot proteomics simply means the analysis of as much of the proteome as possible by a single LC MS/MS run [82]. Its attractions are that sample consumption and measurement times are very low, while still preserving the large-scale, unbiased and systems biology character of the measurement. Employing recent advances in chromatography, mass spectrometry and bioinformatics, the yeast proteome can now be covered almost completely in this mode. This was illustrated by investigating the heat-shock response of the yeast proteome in quadruplicate measurements with nearly complete coverage and with about a day of total measurement time [83].

The human proteome is more complex than the yeast proteome (Figure 1.3A), but until very recently it was unknown how many different proteins a single cell line actually expresses. Using deep shotgun proteomics approaches two different human cancer cell lines have recently been investigated in depth by MS-based proteomics [3,5]. Both studies found that such cell lines contain at least 10 000 different proteins. Saturation analysis [5] or comparison to deep RNA-seq data [3] suggested that this number is not very far from the total number of expressed proteins with functional roles in these cells. A subsequent study of 11 commonly used cell lines also identified

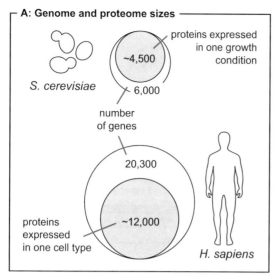 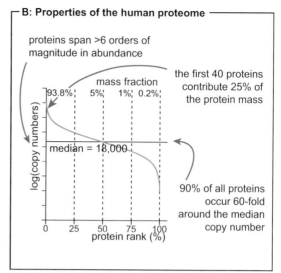

FIGURE 1.3 **Properties of complete proteomes.** A: Comparison of genome and proteome sizes in yeast and human. With increasing complexity of the organism, a smaller fraction of the genome appears to be expressed in individual cells. B: The human cell line proteome spans more than six orders of magnitude in the abundance of individual proteins. However, 90% of all proteins occur within 60-fold above or below the median copy number.

>10 000 proteins each [84]. Although none of the above studies employed accurate quantification strategies, the summed and normalized peptide intensities nevertheless allow important insights into the proteome of cancer cell lines. One such conclusion from the 11 cell lines study, and an earlier study that also used deep transcriptome sequencing and large-scale imaging with an antibody collection [85], was that cellular proteomes are remarkably similar in terms of the identity of their expressed proteins. The expression levels even of household proteins, however, often vary quite significantly across different cell lines [84]. The dynamic range of protein expression was larger than that of the yeast proteome and was estimated to be more than 10^6 (Figure 1.3B), but at the same time, about 90% of the proteome lies within a 60-fold expression range compared to the median level in the HeLa proteome [84]. Rather than being estimated indirectly from total proteome measurements, copy numbers have also been measured by more direct methods in microorganisms [86] or in human cell lines [87]. In the latter study, copy numbers for 40 proteins were determined in HeLa cells and ranged from 20×10^6 for the cytoskeletal protein vimentin to 6000 copies for the transcription factor FOS. Such data can now be generated quite accurately and readily, and should greatly assist in estimating parameters for systems biologic models.

Although proteomics is still in the process of approaching comprehensiveness, by its nature it can answer many questions that are outside of the scope of transcript-based gene expression studies. The reason for this is that the proteome integrates the effects of post-transcriptional regulation as well as regulation by targeted protein degradation. As an example, two studies have used proteomics to delineate the effects of micro-RNAs on expression levels of their targets [88,89]. These studies concluded that these effects were relatively small and dispersed to many substrates for each different micro-RNA.

The availability of deep and accurate proteome data also sheds new light on the longstanding question of the extent of correlation of transcript levels with the corresponding protein levels. Many early studies had found very poor correlation between levels of mRNA and protein. However, this seems to have been caused in large part by the relatively primitive state of the art of transcriptomics, and especially proteomics, at the time. The technical imperfections of the two technologies frequently led to incorrectly measured protein or transcript levels; however, because they are independent of each other, this suggested artificially low correlation of message and protein levels. Recent studies have revealed higher correlation coefficients for steady-state levels, generally in the range of 0.6. The correlation of mRNA changes with protein changes is even higher [85,90]. This level of correlation is biologically plausible, given the flow of genetic information from mRNA to protein. Nevertheless, even when there is good correlation, the level of protein change cannot easily be predicted from the level of transcript change. Interestingly, a recent cell line-based study has shown that the discrepancies between message and protein levels can mostly be explained by differences in mRNA translation rates [91]. However, these translation rates are themselves subject to regulation, which cannot easily be measured without determining protein levels and protein turnover.

More fundamentally, a major potential of proteomics is that it can measure the protein expression levels as a function of subcellular compartment, as well as the redistribution of the proteome between compartments as a function of stimulus [92–94]. Given increasing coverage of the proteome, even the isoform specific regulation of the proteome can be investigated [95].

INTERACTION PROTEOMICS

Specific interactions of proteins with other proteins, with nucleic acids, lipids, carbohydrates, and metabolites or other small molecules, orchestrate all aspects of life at the molecular level. The dissection of molecular assemblies has been a longstanding goal of modern biology, which requires identification of the constituent partners as the first step. This is a field at which MS-based proteomics has excelled from its early days. The ultimate goal is the delineation of the 'interactome', which is defined as the sum of all molecular interactions of a biological system. The size of the interactome of a given organism is a matter of debate and of how the definition is interpreted, but it is undoubtedly far more complex than the genome or proteome; current interactome datasets likely merely scratch its surface [96].

Mass spectrometry has the unique ability to identify very small amounts of any protein without prior knowledge, and in principle it can therefore directly unravel the protein composition of any molecular assembly. Alternative methods of unbiased interaction detection, such as phage display [97] or the yeast two hybrid assay (Y2H) [98], use genetic readouts that test for direct binding but do not involve the formation of actual multi-protein complexes. All approaches in MS-based interaction proteomics are based on the assumption that a molecular interaction is the result of an affinity that can be exploited to purify or enrich the assembly from a crude mixture. Typically, one molecule serves as the 'bait' which is coupled to an affinity matrix. This can be done via an antibody or a genetically encoded tag in the case of proteins or via chemical synthesis in the case of peptides, nucleic acids or small molecules. Mass spectrometry is then used to identify the 'prey' proteins that interact with the bait. This workflow is known as affinity purification followed by mass spectrometry (AP-MS) (Figure 1.4A).

The first application of this methodology was the identification of the members of protein complexes [99], classically defined as entities that can be purified biochemically. This was fuelled by the development of the tandem affinity purification (TAP) tag, which resulted in clean preparations of protein complexes from endogenous sources by two consecutive purification steps [99,100]. This technology was mostly used for the generation of the first large-scale AP-MS interaction datasets of model organisms such as the budding yeast [101–104].

These datasets allowed the first comparisons of AP-MS data with each other and with previously available large-scale Y2H datasets [105,106]. The overlap turned out to be surprisingly low, pointing to technical limitations of the individual approaches and emphasizing that, despite being large-scale, all datasets were non-saturating, sampling different parts of a vast interactome [107]. In addition, it also emphasizes the fundamentally different, but highly complementary nature of AP-MS and Y2H [108]: Y2H data consist of binary combinations of proteins with mutual affinity, including weak interactions, which can be recorded as long as they lead to activation in the genetic readout. In contrast, AP-MS data provide lists of proteins that co-occur in protein complexes — including indirect binders — but provide no direct topological information. Comparison of data from both sources therefore requires the conversion of co-complex members into binary contacts, which can be done using different models [107].

Weak or transient interactors tend to be under-represented in AP-MS datasets because they easily get lost during washing steps in the sample preparation workflow. These washing steps are necessary to reduce the number of proteins that bind non-specifically to the affinity matrix. Unspecific interactors have been the bane of interaction proteomics and were originally dealt with by extensive blacklisting of proteins that were identified across many different affinity purifications. However, this is a less-than-ideal solution as it inevitably leads to lower true positive rates while also failing to remove many false positives.

Virtually all of these drawbacks have been overcome by the advent of quantitative proteomics: specific interactors can easily be distinguished from unspecific background binders by directly comparing their quantities in affinity purifications vs. controls [109,110]. This paved the way for second-generation quantitative interaction proteomic studies (Figure 1.4A). Isotopic labeling techniques allowed the detection of interactions in the presence of high amounts of background binders, and of molecular assemblies which could not be purified extensively. Importantly, this principle is applicable to any conceivable bait molecule that can be immobilized on the affinity matrix. For instance, early highlights include the identification of proteins that interact with specific post-translational modifications represented as modified, synthetic peptides [111,112]. Such assays can be streamlined and used to probe the biological relevance of large-scale PTM datasets. For instance, synthetic peptides corresponding to phosphotyrosine sites with potential key functions as molecular switches have been synthesized and their cellular interaction partners have been determined [113,114].

In a similar fashion, oligonucleotides can be immobilized to identify proteins binding to specific DNA [115] or RNA sequences [116]. In this way, quantitative interaction proteomics identified the transcriptional repressor responsible for the difference between a fat and a lean pig genotype, which is caused by a single nucleotide mutation [117,118].

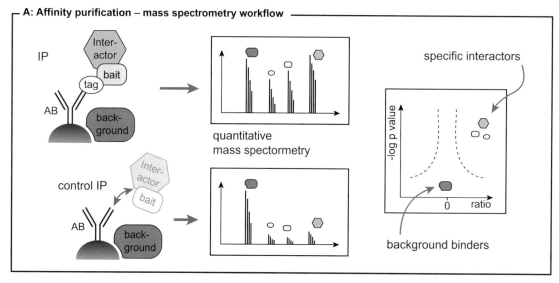
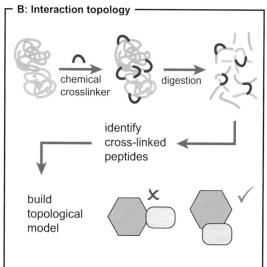
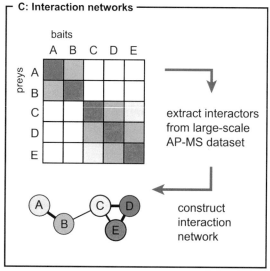

FIGURE 1.4 **Interaction proteomics.** A: Generic scheme of the affinity purification–mass spectrometry workflow. Quantitative comparison of the amounts of proteins in affinity purifications vs. control purifications distinguishes specific interactors from background binders. B: Protein complexes treated with chemical cross-linkers. The identification of cross-linked peptides yields spatial restraints that can be used to infer the topology of interactions and to map binding sites. C: Construction of interaction networks from large-scale AP-MS datasets.

'Chemical proteomics' approaches make use of immobilized small molecule inhibitors to capture and identify their cellular binding proteins [119]. Although this constitutes a powerful and generic approach, synthesizing a suitable, immobilizable derivative of an individual small molecule of interest can be challenging and in some cases impossible. Alternatively, broadly selective inhibitors can be used for affinity-capture of a target protein class. This has been successfully applied for profiling inhibitors targeting kinases [120,121] and more recently histone deacetylases [122]. Inhibitor affinity towards its binding partners can be measured by quantitative, dose-dependent assays by monitoring the binding response to different concentrations of the free molecule. Quantitative drug affinity purification experiments thereby provide a conceptual framework for identifying the protein targets that mediate drug responsiveness and those that potentially cause side effects. Proteome-wide determination of drug targets may also reveal alternate therapeutic uses.

Quantitative MS-based approaches furthermore enable researchers to determine the proteomes of subcellular structures or organelles, which can only be enriched from a whole-cell preparation, but not be purified biochemically [123,124]. One principle is to profile proteins along gradients or across different enrichment steps and to classify them by correlation to known marker proteins. This approach — termed protein correlation profiling — was used to assign proteins to their respective compartments [125], and even substructures of organelles, such as the contact sites between outer and inner mitochondrial membranes, can be distinguished [126]. Another study integrated such data with other MS-based datasets to comprehensively identify chromosome-associated proteins across different phases of the cell cycle [127].

For systems biological applications it is attractive to generate datasets of sufficient size to capture a reasonably large part of the interactome. Because isotopic labeling, which helps to ensure accurate quantitative data, is more challenging to perform at a very large scale, recent high-throughput protein—protein interaction datasets mostly employed simple label-free quantification methods, such as counting the number of times peptides belonging to a certain protein have been sequenced as a proxy for its abundance (see Box 1.1). Based on this technology, large scale protein—protein interaction datasets of human [128,129] and *Drosophila* [130] have been published. Today, high-resolution MS data are routinely available and can be analyzed with very sophisticated label-free quantification algorithms. As a result, datasets with much higher true positive and lower false negative rates should become available for systems biological modeling.

Beyond the goal of accurate and comprehensive mapping of the interactome into lists of proteins associated in complexes, the next challenge is to provide additional functionally relevant information such as topology and stoichiometry. One future direction involves the use of chemical cross-linkers in combination with bioinformatic algorithms to help deduce the three-dimensional architecture of protein complexes [131,132] (Figure 1.4B). This technology is still under development and currently limited to purified complexes, but it has the potential to be extended towards more complex samples, ultimately offering the vision of 'the interactome in a single experiment'. MS-based interaction proteomics is also uniquely suited to measuring the dynamics of protein interactions in response to stimuli. Such effects range from subtle modulations of the interaction network, e.g. in the case of autophagy-related proteins when that process was triggered [133], or the changing composition of Wnt signaling complexes [134], to extensive disruption of complexes, e.g. of the Bcr-Abl kinase complex after drug treatment [135].

Accurate quantification is paramount when it comes to dynamic interactions, and various groups have addressed this with isotope-labeled reference peptides and absolute quantification, which also allows estimation of the stoichiometries of interacting proteins [136,137]. The ultimate challenge in interaction proteomics is to achieve high throughput and coverage while maintaining very high quality standards. With the proteomics methods evolving and quantification being increasingly accurate, the biological samples from which the interactions are determined should also represent the in vivo situation in the best possible way. For protein—protein interaction data, one critical parameter is the expression level of the bait protein. Ideally, this should be adjusted to near-physiologic levels to avoid aberrant localization and to ensure that bona fide interaction partners are present in appropriate amounts compared to the bait [138]. This can be achieved by tagging the endogenous locus encoding the bait, which is straightforward in lower organisms such as yeasts, but much more complicated in human cell lines. A recent method based on bacterial artificial chromosome (BAC) transgenes alleviates these limitations and allows the expression of GFP-tagged proteins under fully endogenous gene regulatory control [139]. In addition to providing a subcellular localization tool via the fluorescent tag, this method can easily be combined with quantitative interaction analysis, for example to allow the splice isoform-specific interaction partners of a bait protein to be identified [140]. BAC—GFP interaction data have also been combined with phenotypic data from RNA interference screens to place genes involved in mitosis into the context of protein complexes [141]. This study showed how physical interactions derived from proteomics integrate beautifully with other omics data, providing functional relationships of genes or proteins. Consolidating physical with functional interaction data will ultimately allow the placing of proteins into complexes and arranging complexes into dynamic pathways and networks.

In this way ever-growing large-scale datasets will become increasingly useful for biologists and systems biologists alike (Figure 1.4C). Systems biologists will better understand the intricate interplay of molecules inside the cell, while biologists will find new interaction partners of their protein of interest, and they will be able to place specific genes into pathways, helping to explain observed phenotypes.

However, the complete characterization of a mammalian protein—protein interactome and its integration with other omics data of the same scale is a vision of the future and is just coming into reach for the most primitive organisms [142—144].

LARGE-SCALE DETERMINATION OF POST-TRANSLATIONAL MODIFICATIONS

Post-translational modifications (PTMs) of proteins are a key regulatory mechanism in signal transmission that controls nearly all aspects of cellular function. Traditionally, signaling processes are perceived as discrete linear pathways that transduce external signals via the post-translational modification of a few key sites. For example, a specific phosphorylation event might regulate the function of a crucial pathway. These pathways have typically been studied in the conventional, reductionist manner, with researchers focusing on the characterization of individual components and causal interactions in an individual cascade. However, this biochemical simplicity as usually visualized fails to account for the systems properties that are an inherent part of any pathway. It has become increasingly clear that the specificity of signal–response events, for example for individual receptor pathways, does not rely on a single protein or gene that is responsible for signaling specificity. Instead, it has been shown that pathways are extensively connected and embedded in signaling 'networks' rather than 'linear pathways' [145]. Therefore, the analysis of complex networks as large functional ensembles may be necessary to infer their behavior. Engineering techniques such as control theory, which were developed to analyze self-regulating technological systems, have become popular for describing complex and dynamic cellular control mechanisms. Among these, a key mechanism is the regulation of the expression levels of proteins through the gene expression program. However, cells also extensively use PTMs, which constitute an important class of molecular switches, for signal propagation to control the activity, structure, localization and interactions of proteins. Often a signal to the cell will initially lead to a cascade of PTM changes, which can happen very rapidly, and later to a change in the expression of a specific set of proteins. The specificity as well as the robustness of biological control mechanisms is largely determined by a combinatorial system of regulated post-translational modifications, the resulting protein–protein interactions, and protein expression of downstream signaling components along the temporal and spatial axes. An example illustrating the specificity of PTM-induced cell decisions is the classic case of stimulation of ERK activity in PC12 cells: when these neuronal cells are stimulated for a short time they proliferate, whereas a longer-term activation of the same pathway leads to their differentiation [146].

As a first step towards understanding the overwhelmingly complex circuitry of signaling networks, the PTMs should be identified and quantified in an unbiased and global manner (Figure 1.5). For this purpose, modern quantitative mass spectrometry has proved an ideal platform because it is a highly precise yet generic method for detecting PTMs: MS directly measures the presence of a PTM by a defined corresponding shift in the mass of the modified peptide. MS-based mapping and quantification of PTMs is set to revolutionize signaling research and is already providing large-scale information on the extent and diversity of different PTMs in the expressed proteome and their regulation in response to perturbations [147]. To date, about 300 different types of protein modifications have been reported to occur physiologically, and yet more are being discovered [148]. However, just a few PTMs have accounted for the majority of classic and MS-based investigations. Representative examples include phosphorylation [67,149,150], lysine acetylation [151,152], glycosylation [153,154], ubiquitylation [155–158] and methylation [159]. Remarkably, these reports often expanded the known universe of the PTM under investigation 10–100-fold compared to the previous non-MS-based state [147].

Despite this impressive progress in MS-based PTM proteomics, exhaustive mapping of protein modifications is challenging for a number of reasons: (i) modified peptides are present in sub-stoichiometric amounts in complex mixtures; (ii) the peptides carrying certain PTMs display more complicated MS/MS fragmentation patterns that can be difficult to interpret; (iii) the effective database search space explodes when the search program is allowed to consider potential PTMs at each modifiable amino acid residue; and (iv) in addition to identifying the modified peptide, the PTM needs to be placed with single amino acid accuracy in the sequence.

To address the sub-stoichiometric amounts of PTMs, much effort has been put into improving pre-fractionation and specific enrichment of PTM-carrying peptides. In this way, more input material is used, leading to higher amounts of modified peptides and improving their mass-spectrometric analysis. At the same time the sample complexity is reduced, facilitating proteome-wide mapping of modifications [160] (Figure 1.5A). These methods can be evaluated by the enrichment factor with respect to the starting peptide mixture, or the enrichment efficiency, which refers to the fraction of modified peptides in the enriched population. For phosphorylation, strong cation exchange chromatography or metal affinity complexation allow up to 100-fold enrichment and often close to 100% efficiency of phosphopeptide enrichment [161]. At the other extreme, methylated peptides are only enriched a few fold, and enrichment efficiency is about 5% using antibodies directed towards Lys-acetylated peptides. Analysis of methylation, acetylation and many others has lagged behind those PTMs for which very specific tools have been available. Illustrating the importance of such reagents, the recent development of a monoclonal antibody to profile lysine ubiquitylation has dramatically boosted our knowledge about the extent of this PTM [156–158]. This antibody recognizes peptides containing lysine residues modified by diglycine, a ubiquitin remnant at the modification site after trypsin digestion of the sample.

Confident localization of the PTM on modified peptides requires the presence of the relevant fragment ions in the MS/MS spectra. Usually, as mentioned above, algorithms for the analysis of modified peptides provide a PTM localization score, which indicates how much confidence should be placed in the site assignment (Figure 1.5B). Because of the technical challenges in mapping PTMs they are particularly prone to being undersampled, i.e., to be missing in certain runs, thereby leading to incomplete

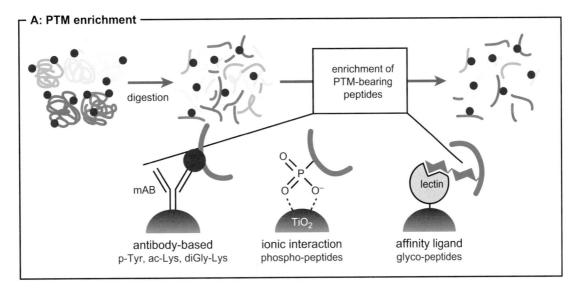
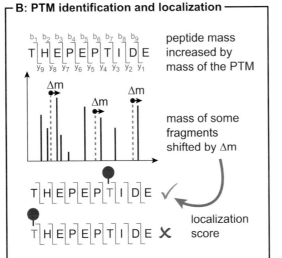
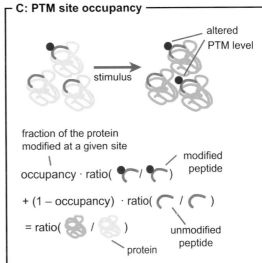

FIGURE 1.5 **Analysis of post-translational modifications by MS.** A: PTM enrichment: Substoichiometric PTM-bearing proteins or peptides are enriched using various strategies, including PTM-directed antibodies, metal ion complexation and affinity ligands. B: PTM identification and localization: MS directly measures the presence of a PTM by a defined corresponding shift in mass of the peptide and PTM location within the peptide is obtained by the MS/MS pattern with single amino acid resolution. C: PTM site occupancy represents the fraction of a protein that is modified at a given PTM site. Site occupancies can be calculated if one can quantify changing amounts of a modified peptide, the corresponding unmodified peptide and the entire protein in a perturbed system.

datasets. In principle, this can be addressed by targeted methods in which the mass spectrometer is directed to acquire data for a particular set of modified peptides [162].

Quantification of PTM sites is achieved in the same ways as for non-modified peptides. However, this becomes more complicated when a single peptide has multiple modification sites. By employing quantification at different time points, kinetic maps of PTM-site dynamics in response to various perturbations can be obtained [67,163,164]. PTM-level information can be combined with information on protein levels, as we have recently shown in a combined phosphoproteomic and proteomic analysis of the cell cycle [67]. However, when measuring early signaling changes, for example downstream of receptor tyrosine kinase activation, one usually assumes that proteomic changes will be minimal and that observed quantitative changes in phosphopeptides can directly be attributed to changes at the modification site level. Furthermore, it may be desirable in a systems biology context to quantify not only the relative change of a modification site but also the fraction of the protein that is modified at this site (Figure 1.5C). First reports of the large-scale determination of phosphorylation stoichiometry have recently appeared [165,166]. Thus, with ever improving technology, proteomics can now deliver key parameters on PTMs that are important for cellular modeling, such as PTM site occupancy together with kinetics upon perturbation.

Even after correctly and comprehensively measuring the phosphorylation changes upon cellular perturbation, the question remains which kinase or kinases are responsible for a given phosphorylation site. A variety of combinations of quantitative phosphoproteomics and chemical genetics approaches can answer this question by identifying direct kinase substrates. For instance, this can involve controlled inhibition of a genetically engineered cellular kinase by a small molecule [167]. In an alternative approach, phosphorylation patterns in 124 kinase and phosphatase yeast deletion strains have been measured to globally extract kinase—substrate relationships [168]. To understand the circuitry that underpins cellular information flow, the changes in PTM dynamics can additionally be overlaid with direct protein—protein interaction datasets such as those of kinases and phosphatases from yeast [169]. Clearly, it would also be important to understand how the dynamic kinase—substrate interactions vary under different growth and stress conditions.

From a cellular control perspective, a significant increase in information content can be achieved if many proteins are multiply modified, especially if these PTMs acted combinatorially. In fact, it has become increasingly clear that a number of cellular processes are regulated by PTM cross-talk, as exemplified by phosphorylation and ubiquitylation [170]. Another example is the intimate interplay of different histone modification marks in the histone code, which represents one of the most important epigenetic regulatory mechanisms governing the structure and function of the genome [171]. To understand such PTM cross-talk codes that mediate cellular control, MS-based proteomics is an excellent large-scale method. However, owing to the fact that correlating PTMs may occur on different peptides, specialized MS strategies may have to be used, such as top-down proteomics [172,173].

As described above, MS-based PTM analysis is uncovering an unexpectedly large extent and diversity of PTMs that occur on multiple but specific residues on most proteins. These large-scale PTM studies now serve as an information-rich resource to the community. For example, biological researchers can focus on regulatory PTM sites in high-quality MS data for their proteins or processes of interest. The data can also be used to investigate basic characteristics of particular PTMs, such as their evolutionary conservation [154,174,175] and preferential localization across secondary structures of proteins [154,176].

In addition, in vivo maps of many PTMs are beginning to emerge [150,157,177,178] and a first example of large-scale PTM quantification in a mouse organ after perturbation has been described [179]. This is now unlocking the opportunity to study PTM dynamics in tissues to characterize the physiological or pathological responses of different organs in mammals. With their key roles in cellular control, MS-enabled PTM signatures also hold great potential as prognostic and therapeutic biomarkers.

OUTLOOK AND FUTURE CHALLENGES

As detailed in this chapter, MS-based proteomics is a technology-driven discipline that has made tremendous progress during recent years. These advances affect the entire proteomics workflow, starting with sample preparation and ending with computational proteomics. The advent of high-resolution high-accuracy MS data, combined with sophisticated quantification strategies, has been especially important in obtaining biologically relevant information from MS-based proteomics. This technology has now clearly become the method of choice for studying

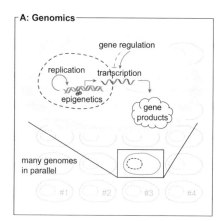
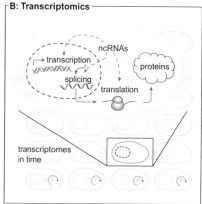
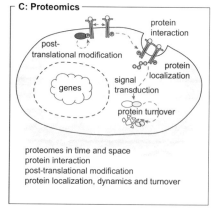

FIGURE 1.6 Unique contribution of different 'omics' technologies to systems biology. A: Genomics investigates the sequences of genomes and their epigenetic marks for many genomes in parallel, but does not provide direct information about the fate of the gene products. B: Transcriptomics measures the gene expression program and allows the comparison of changes of gene expression in different cellular states or over time. C: Proteomics strives to provide a complete picture of all proteins, the primary agents of cellular processes. Proteins can be monitored over time and with sub-cellular resolution along with their post-translational modifications, interactions and turnover.

endogenous proteins, either at a small scale involving one or a few different proteins or involving entire cellular proteomes. It provides a crucial layer of information on the proteins that previously had to be inferred indirectly from other measurements, or was absent altogether.

Given the increasingly mature proteomics toolbox, an ever larger set of cell biological and biomedical problems can now be tackled. For instance, we expect many more reports of essentially complete proteome measurements, as well as highly accurate comparative transcriptome and proteome studies. It will be interesting to see whether MS-based proteomics can make inroads into the clinical area, for instance in classifying cancer patients by their protein expression patterns.

Despite these promises, major challenges with MS-based proteomics remain. Foremost among these is the limited community access to high-accuracy in-depth proteomics. Compared to transcriptomics and the current massive investments into deep-sequencing based technologies, the area of MS-based proteomics remains tiny. There are also entire areas, such as body fluid-based biomarker discovery, where MS-based proteomics could in principle make a revolutionary impact but where our current technology fails woefully to live up to expectations. On the other hand, this means that MS-based proteomics will offer exciting opportunities for young researchers for years to come.

From a systems biology perspective, the ability of proteomics to detect not only the presence of but to also to estimate copy numbers of virtually all proteins in a proteome will be crucial in modeling the cellular proteome. Equally important, proteomics is now poised to deliver increasingly comprehensive lists of the major PTMs, including phosphorylation, ubiquitylation, acetylation, glycosylation and many more. This is a precondition for determining their function, which will be a monumental task for the years ahead, and for accurate models of information processing in the cell. Identification and quantification of protein isoforms is still a challenge for MS-based proteomics, but it is becoming increasingly accessible due to more extensive sequence coverage of the identified proteins. The direct analysis of undigested proteins by MS ('top-down' proteomics) will also contribute to this question.

Figure 1.6 summarizes the indispensable role of proteomics in the context of other large-scale methods of genomics and gene expression analysis. Both genomics and transcriptomics benefit from the current revolution in next-generation sequencing methods. We expect deep-sequencing data to be readily accessible for essentially every situation of interest in systems biology in the near future. This includes the genomes of different individuals as well as differences between normal and cancer genomes in the same individuals. Likewise, deep sequencing will contribute tremendously to accurate and comprehensive mapping of the abundance of mRNA molecules, an early step in the gene expression program. However, this is still only half of the story. Proteomics can give us a detailed picture of the end product of the gene expression cascade, the mature, active and fully modified protein form. It also measures regulation directly at the expression level of all proteins, which cannot be predicted from transcript levels. In contrast to genomics and transcriptomics, it can characterize gene expression at subcellular resolution, i.e., by analyzing the proteomes of different cellular compartments. Furthermore, the interactions and dynamics of the proteome can likewise be studied either at a whole cell level or in individual subcellular compartments. In conclusion, despite the technological challenges it faces, MS-based proteomics is crucial to a systems-level understanding of cellular function, and is ready to make even more extensive contributions to the field in the future.

REFERENCES

[1] Aebersold R, Mann M. Mass spectrometry-based proteomics. Nature 2003;422:198–207.

[2] de Godoy LM, Olsen JV, Cox J, Nielsen ML, Hubner NC, Frohlich F, et al. Comprehensive mass-spectrometry-based proteome quantification of haploid versus diploid yeast. Nature 2008;455:1251–4.

[3] Nagaraj N, Wisniewski JR, Geiger T, Cox J, Kircher M, Kelso J, et al. Deep proteome and transcriptome mapping of a human cancer cell line. Mol Syst Biol 2011;7:548.

[4] Munoz J, Low TY, Kok YJ, Chin A, Frese CK, Ding V, et al. The quantitative proteomes of human-induced pluripotent stem cells and embryonic stem cells. Mol Syst Biol 2011;7:550.

[5] Beck M, Schmidt A, Malmstroem J, Claassen M, Ori A, Szymborska A, et al. The quantitative proteome of a human cell line. Mol Syst Biol 2011;7:549.

[6] Yates 3rd JR, Gilchrist A, Howell KE, Bergeron JJ. Proteomics of organelles and large cellular structures. Nat Rev Mol Cell Biol 2005;6:702–14.

[7] Au CE, Bell AW, Gilchrist A, Hiding J, Nilsson T, Bergeron JJ. Organellar proteomics to create the cell map. Curr Opin Cell Biol 2007;19:376–85.

[8] Walther TC, Mann M. Mass spectrometry-based proteomics in cell biology. J Cell Biol 2010;190:491–500.

[9] Lamond AI, Uhlen M, Horning S, Makarov A, Robinson CV, Serrano L, et al. Advancing cell biology through Proteomics in Space and Time (PROSPECTS). Mol Cell Proteomics 2012;11:O112.017731.

[10] Berglund L, Bjorling E, Oksvold P, Fagerberg L, Asplund A, Szigyarto CA, et al. A genecentric Human Protein Atlas for expression profiles based on antibodies. Mol Cell Proteomics 2008;7:2019–27.

[11] Huh WK, Falvo JV, Gerke LC, Carroll AS, Howson RW, Weissman JS, et al. Global analysis of protein localization in budding yeast. Nature 2003;425:686–91.

[12] Jensen ON. Interpreting the protein language using proteomics. Nat Rev Mol Cell Biol 2006;7:391–403.

[13] Wolf-Yadlin A, Hautaniemi S, Lauffenburger DA, White FM. Multiple reaction monitoring for robust quantitative proteomic analysis of cellular signaling networks. Proc Natl Acad Sci U S A 2007;104:5860–5.

[14] Cox J, Mann M. Quantitative, high-resolution proteomics for data-driven systems biology. Annu Rev Biochem 2011;80:273–99.

[15] Karas M, Hillenkamp F. Laser desorption ionization of proteins with molecular masses exceeding 10,000 daltons. Anal Chem 1988;60:2299–301.

[16] Fenn JB, Mann M, Meng CK, Wong SF, Whitehouse CM. Electrospray ionization for mass spectrometry of large biomolecules. Science 1989;246:64–71.

[17] Breuker K, Jin M, Han X, Jiang H, McLafferty FW. Top-down identification and characterization of biomolecules by mass spectrometry. J Am Soc Mass Spectrom 2008;19:1045–53.

[18] Tran JC, Zamdborg L, Ahlf DR, Lee JE, Catherman AD, Durbin KR, et al. Mapping intact protein isoforms in discovery mode using top-down proteomics. Nature 2011;480:254–8.

[19] Shevchenko A, Wilm M, Vorm O, Mann M. Mass spectrometric sequencing of proteins silver-stained polyacrylamide gels. Analytical chemistry 1996;68:850–8.

[20] Beck M, Claassen M, Aebersold R. Comprehensive proteomics. Curr Opin Biotechnol 2011;22:3–8.

[21] Altelaar AM, Heck AJ. Trends in ultrasensitive proteomics. Curr Opin Chem Biol 2012.

[22] Washburn MP, Wolters D, Yates 3rd JR. Large-scale analysis of the yeast proteome by multidimensional protein identification technology. Nat Biotechnol 2001;19:242–7.

[23] Herbert B, Righetti PG. A turning point in proteome analysis: sample prefractionation via multicompartment electrolyzers with isoelectric membranes. Electrophoresis 2000;21:3639–48.

[24] Shen Y, Berger SJ, Anderson GA, Smith RD. High-efficiency capillary isoelectric focusing of peptides. Anal Chem 2000;72:2154–9.

[25] Peng J, Elias JE, Thoreen CC, Licklider LJ, Gygi SP. Evaluation of multidimensional chromatography coupled with tandem mass spectrometry (LC/LC-MS/MS) for large-scale protein analysis: the yeast proteome. J Proteome Res 2003;2:43–50.

[26] Horth P, Miller CA, Preckel T, Wenz C. Efficient fractionation and improved protein identification by peptide OFFGEL electrophoresis. Mol Cell Proteomics 2006;5:1968–74.

[27] Caldwell RL, Caprioli RM. Tissue profiling by mass spectrometry: a review of methodology and applications. Mol Cell Proteomics 2005;4:394–401.

[28] Schwamborn K, Caprioli RM. Molecular imaging by mass spectrometry – looking beyond classical histology. Nat Rev Cancer 2010;10:639–46.

[29] Steen H, Mann M. The ABC's (and XYZ's) of peptide sequencing. Nat Rev Mol Cell Biol 2004;5:699–711.

[30] Domon B, Aebersold R. Mass spectrometry and protein analysis. Science 2006;312:212–7.

[31] Zubarev R, Mann M. On the proper use of mass accuracy in proteomics. Mol Cell Proteomics 2007;6:377–81.

[32] Hu Q, Noll RJ, Li H, Makarov A, Hardman M, Graham Cooks R. The Orbitrap: a new mass spectrometer. J Mass Spectrom 2005;40:430–43.

[33] Olsen JV, Schwartz JC, Griep-Raming J, Nielsen ML, Damoc E, Denisov E, et al. A dual pressure linear ion trap orbitrap instrument with very high sequencing speed. Mol Cell Proteomics 2009;8:2759–69.

[34] Michalski A, Damoc E, Lange O, Denisov E, Nolting D, Mueller M, et al. Ultra high resolution linear ion trap Orbitrap mass spectrometer (Orbitrap Elite) facilitates top down LC MS/MS and versatile peptide fragmentation modes. Mol Cell Proteomics 2012;11:O111.013698.

[35] Silva JC, Gorenstein MV, Li GZ, Vissers JP, Geromanos SJ. Absolute quantification of proteins by LCMSE: a virtue of parallel MS acquisition. Mol Cell Proteomics 2006;5:144–56.

[36] Geromanos SJ, Vissers JP, Silva JC, Dorschel CA, Li GZ, Gorenstein MV, et al. The detection, correlation, and comparison of peptide precursor and product ions from data independent LC-MS with data dependant LC-MS/MS. Proteomics 2009;9:1683–95.

[37] Michalski A, Cox J, Mann M. More than 100,000 detectable peptide species elute in single shotgun proteomics runs but the majority is inaccessible to data-dependent LC-MS/MS. J Proteome Res 2011;10:1785–93.

[38] Lange V, Picotti P, Domon B, Aebersold R. Selected reaction monitoring for quantitative proteomics: a tutorial. Mol Syst Biol 2008;4:222.

[39] Gillet LC, Navarro P, Tate S, Roest H, Selevsek N, Reiter L, et al. Targeted data extraction of the MS/MS spectra generated by data independent acquisition: a new concept for consistent and accurate proteome analysis. Mol Cell Proteomics 2012;11:O111.016717.

[40] Ong SE, Mann M. Mass spectrometry-based proteomics turns quantitative. Nat Chem Biol 2005;1:252–62.

[41] Biemann K. Four decades of structure determination by mass spectrometry: from alkaloids to heparin. J Am Soc Mass Spectrom 2002;13:1254–72.

[42] Michalski A, Damoc E, Hauschild JP, Lange O, Wieghaus A, Makarov A, et al. Mass spectrometry-based proteomics using Q Exactive, a high-performance benchtop quadrupole Orbitrap mass spectrometer. Mol Cell Proteomics 2011;10. M111.011015.

[43] Listgarten J, Emili A. Statistical and computational methods for comparative proteomic profiling using liquid chromatography-tandem mass spectrometry. Mol Cell Proteomics 2005;4:419–34.

[44] Deutsch EW, Lam H, Aebersold R. Data analysis and bioinformatics tools for tandem mass spectrometry in proteomics. Physiol Genomics 2008;33:18–25.

[45] Mueller LN, Brusniak MY, Mani DR, Aebersold R. An assessment of software solutions for the analysis of mass spectrometry based quantitative proteomics data. J Proteome Res 2008;7:51–61.

[46] Kumar C, Mann M. Bioinformatics analysis of mass spectrometry-based proteomics data sets. FEBS Lett 2009;583:1703–12.

[47] Keller A, Shteynberg D. Software pipeline and data analysis for MS/MS proteomics: the trans-proteomic pipeline. Methods Mol Biol 2011;694:169–89.

[48] Cox J, Mann M. MaxQuant enables high peptide identification rates, individualized p.p.b.-range mass accuracies and proteome-wide protein quantification. Nat Biotechnol 2008;26:1367–72.

[49] Taylor GK, Kim YB, Forbes AJ, Meng F, McCarthy R, Kelleher NL. Web and database software for identification of

[49] ...intact proteins using 'top down' mass spectrometry. Anal Chem 2003;75:4081–6.
[50] Zamdborg L, LeDuc RD, Glowacz KJ, Kim YB, Viswanathan V, Spaulding IT, et al. ProSight PTM 2.0: improved protein identification and characterization for top down mass spectrometry. Nucleic Acids Res 2007;35:W701–6.
[51] Rinner O, Seebacher J, Walzthoeni T, Mueller LN, Beck M, Schmidt A, et al. Identification of cross-linked peptides from large sequence databases. Nat Methods 2008;5:315–8.
[52] Rappsilber J. The beginning of a beautiful friendship: cross-linking/mass spectrometry and modelling of proteins and multi-protein complexes. J Struct Biol 2011;173:530–40.
[53] Ma B, Zhang K, Hendrie C, Liang C, Li M, Doherty-Kirby A, et al. PEAKS: powerful software for peptide de novo sequencing by tandem mass spectrometry. Rapid Commun Mass Spectrom 2003;17:2337–42.
[54] Ma B, Johnson R. De novo sequencing and homology searching. Mol Cell Proteomics 2012;11. O111 014902.
[55] Cox J, Michalski A, Mann M. Software lock mass by two-dimensional minimization of peptide mass errors. Journal of the American Society for Mass Spectrometry 2011;22:1373–80.
[56] Sadygov RG, Cociorva D, Yates 3rd JR. Large-scale database searching using tandem mass spectra: looking up the answer in the back of the book. Nat Methods 2004;1:195–202.
[57] Dancik V, Addona TA, Clauser KR, Vath JE, Pevzner PA. De novo peptide sequencing via tandem mass spectrometry. Journal of computational biology: a journal of computational molecular cell biology 1999;6:327–42.
[58] Mann M, Wilm M. Error-tolerant identification of peptides in sequence databases by peptide sequence tags. Anal Chem 1994;66:4390–9.
[59] Eng JK, McCormack AL, Yates JR. An approach to correlate tandem mass spectral data of peptides with amino acid sequences in a protein database. J Am Soc Mass Spectrom 1994;5:976–89.
[60] Perkins DN, Pappin DJ, Creasy DM, Cottrell JS. Probability-based protein identification by searching sequence databases using mass spectrometry data. Electrophoresis 1999;20:3551–67.
[61] Cox J, Neuhauser N, Michalski A, Scheltema RA, Olsen JV, Mann M. Andromeda: a peptide search engine integrated into the MaxQuant environment. Journal of proteome research 2011;10:1794–805.
[62] Keller A, Nesvizhskii AI, Kolker E, Aebersold R. Empirical statistical model to estimate the accuracy of peptide identifications made by MS/MS and database search. Analytical chemistry 2002;74:5383–92.
[63] Brosch M, Yu L, Hubbard T, Choudhary J. Accurate and sensitive peptide identification with Mascot Percolator. J Proteome Res 2009;8:3176–81.
[64] Elias JE, Gygi SP. Target-decoy search strategy for increased confidence in large-scale protein identifications by mass spectrometry. Nat Methods 2007;4:207–14.
[65] Nesvizhskii AI, Vitek O, Aebersold R. Analysis and validation of proteomic data generated by tandem mass spectrometry. Nat Methods 2007;4:787–97.
[66] Geiger T, Wehner A, Schaab C, Cox J, Mann M. Comparative proteomic analysis of eleven common cell lines reveals ubiquitous but varying expression of most proteins. Mol Cell Proteomics 2012;11(3):M111.014050.
[67] Olsen JV, Blagoev B, Gnad F, Macek B, Kumar C, Mortensen P, et al. Global, in vivo, and site-specific phosphorylation dynamics in signaling networks. Cell 2006;127:635–48.
[68] Beausoleil SA, Villen J, Gerber SA, Rush J, Gygi SP. A probability-based approach for high-throughput protein phosphorylation analysis and site localization. Nat Biotechnol 2006;24:1285–92.
[69] Bailey CM, Sweet SMM, Cunningham DL, Zeller M, Heath JK, Cooper HJ. SLoMo: Automated Site Localization of Modifications from ETD/ECD Mass Spectra. J Proteome Res 2009;8:1965–71.
[70] Savitski MM, Lemeer S, Boesche M, Lang M, Mathieson T, Bantscheff M, et al. Confident phosphorylation site localization using the Mascot Delta Score. Mol Cell Proteomics 2012;11(3):M110.003830.
[71] Taus T, Kocher T, Pichler P, Paschke C, Schmidt A, Henrich C, et al. Universal and confident phosphorylation site localization using phosphoRS. J Proteome Res 2011;10:5354–62.
[72] Creasy DM, Cottrell JS. Error tolerant searching of uninterpreted tandem mass spectrometry data. Proteomics 2002;2:1426–34.
[73] Savitski MM, Nielsen ML, Zubarev RA. ModifiComb, a new proteomic tool for mapping substoichiometric post-translational modifications, finding novel types of modifications, and fingerprinting complex protein mixtures. Mol Cell Proteomics 2006;5:935–48.
[74] Nesvizhskii AI, Aebersold R. Interpretation of shotgun proteomic data: the protein inference problem. Mol Cell Proteomics 2005;4:1419–40.
[75] Bantscheff M, Schirle M, Sweetman G, Rick J, Kuster B. Quantitative mass spectrometry in proteomics: a critical review. Anal Bioanal Chem 2007;389:1017–31.
[76] Ow SY, Salim M, Noirel J, Evans C, Rehman I, Wright PC. iTRAQ underestimation in simple and complex mixtures: 'the good, the bad and the ugly'. Journal of proteome research 2009;8:5347–55.
[77] Mertins P, Udeshi ND, Clauser KR, Mani DR, Patel J, Ong SE, et al. iTRAQ labeling is superior to mTRAQ for quantitative global proteomics and phosphoproteomics. Mol Cell Proteomics 2012;11:M111.014423.
[78] Christoforou A, Lilley KS. Taming the isobaric tagging elephant in the room in quantitative proteomics. Nat Methods 2011;8:911–3.
[79] Malmstrom J, Lee H, Aebersold R. Advances in proteomic workflows for systems biology. Curr Opin Biotechnol 2007;18:378–84.
[80] Ghaemmaghami S, Huh WK, Bower K, Howson RW, Belle A, Dephoure N, et al. Global analysis of protein expression in yeast. Nature 2003;425:737–41.
[81] Picotti P, Bodenmiller B, Mueller LN, Domon B, Aebersold R. Full dynamic range proteome analysis of *S. cerevisiae* by targeted proteomics. Cell 2009;138:795–806.
[82] Thakur SS, Geiger T, Chatterjee B, Bandilla P, Frohlich F, Cox J, et al. Deep and highly sensitive proteome coverage by LC-MS/MS without prefractionation. Mol Cell Proteomics 2011;10. M110.003699.
[83] Nagaraj N, Kulak NA, Cox J, Neuhaus N, Mayr K, Hoerning O, et al. Systems-wide perturbation analysis with near complete coverage of the yeast proteome by single-shot UHPLC runs on

[84] Geiger T, Wehner A, Schaab C, Cox J, Mann M. Comparative proteomic analysis of eleven common cell lines reveals ubiquitous but varying expression of most proteins. Mol Cell Proteomics 2012;11:M111.014050.

[85] Lundberg E, Fagerberg L, Klevebring D, Matic I, Geiger T, Cox J, et al. Defining the transcriptome and proteome in three functionally different human cell lines. Mol Syst Biol 2010;6:450.

[86] Malmstrom J, Beck M, Schmidt A, Lange V, Deutsch EW, Aebersold R. Proteome-wide cellular protein concentrations of the human pathogen *Leptospira interrogans*. Nature 2009;460:762–5.

[87] Zeiler M, Straube WL, Lundberg E, Uhlen M, Mann M. A protein epitope signature Tag (PrEST) library allows SILAC-based absolute quantification and multiplexed determination of protein copy numbers in cell lines. Mol Cell Proteomics 2012;11: O111.009613.

[88] Selbach M, Schwanhausser B, Thierfelder N, Fang Z, Khanin R, Rajewsky N. Widespread changes in protein synthesis induced by microRNAs. Nature 2008;455:58–63.

[89] Baek D, Villen J, Shin C, Camargo FD, Gygi SP, Bartel DP. The impact of microRNAs on protein output. Nature 2008;455:64–71.

[90] Bonaldi T, Straub T, Cox J, Kumar C, Becker PB, Mann M. Combined use of RNAi and quantitative proteomics to study gene function in Drosophila. Mol Cell 2008;31:762–72.

[91] Schwanhausser B, Busse D, Li N, Dittmar G, Schuchhardt J, Wolf J, et al. Global quantification of mammalian gene expression control. Nature 2011;473:337–42.

[92] Andersen JS, Lam YW, Leung AK, Ong SE, Lyon CE, Lamond AI, et al. Nucleolar proteome dynamics. Nature 2005;433:77–83.

[93] Boisvert FM, Lam YW, Lamont D, Lamond AI. A quantitative proteomics analysis of subcellular proteome localization and changes induced by DNA damage. Mol Cell Proteomics 2010;9:457–70.

[94] Boisvert FM, Ahmad Y, Gierlinski M, Charriere F, Lamont D, Scott M, et al. A quantitative spatial proteomics analysis of proteome turnover in human cells. Mol Cell Proteomics 2010; 9:457–70.

[95] Ahmad Y, Boisvert FM, Lundberg E, Uhlen M, Lamond AI. Systematic analysis of protein pools, isoforms and modifications affecting turnover and subcellular localisation. Mol Cell Proteomics 2012;11:M111.013680.

[96] Stumpf MP, Thorne T, de Silva E, Stewart R, An HJ, Lappe M, et al. Estimating the size of the human interactome. Proc Natl Acad Sci U S A 2008;105:6959–64.

[97] Smith GP. Filamentous fusion phage: novel expression vectors that display cloned antigens on the virion surface. Science 1985;228:1315–7.

[98] Fields S, Song O. A novel genetic system to detect protein–protein interactions. Nature 1989;340:245–6.

[99] Rigaut G, Shevchenko A, Rutz B, Wilm M, Mann M, Seraphin B. A generic protein purification method for protein complex characterization and proteome exploration. Nat Biotechnol 1999;17:1030–2.

[100] Burckstummer T, Bennett KL, Preradovic A, Schutze G, Hantschel O, Superti-Furga G, et al. An efficient tandem affinity purification procedure for interaction proteomics in mammalian cells. Nat Methods 2006;3:1013–9.

[101] Gavin AC, Bosche M, Krause R, Grandi P, Marzioch M, Bauer A, et al. Functional organization of the yeast proteome by systematic analysis of protein complexes. Nature 2002;415:141–7.

[102] Ho Y, Gruhler A, Heilbut A, Bader GD, Moore L, Adams SL, et al. Systematic identification of protein complexes in *Saccharomyces cerevisiae* by mass spectrometry. Nature 2002;415: 180–3.

[103] Gavin AC, Aloy P, Grandi P, Krause R, Boesche M, Marzioch M, et al. Proteome survey reveals modularity of the yeast cell machinery. Nature 2006;440:631–6.

[104] Krogan NJ, Cagney G, Yu H, Zhong G, Guo X, Ignatchenko A, et al. Global landscape of protein complexes in the yeast *Saccharomyces cerevisiae*. Nature 2006;440:637–43.

[105] Uetz P, Giot L, Cagney G, Mansfield TA, Judson RS, Knight JR, et al. A comprehensive analysis of protein–protein interactions in *Saccharomyces cerevisiae*. Nature 2000;403:623–7.

[106] Ito T, Chiba T, Ozawa R, Yoshida M, Hattori M, Sakaki Y. A comprehensive two-hybrid analysis to explore the yeast protein interactome. Proc Natl Acad Sci U S A 2001;98:4569–74.

[107] Bader GD, Hogue CW. Analyzing yeast protein–protein interaction data obtained from different sources. Nat Biotechnol 2002;20:991–7.

[108] Yu H, Braun P, Yildirim MA, Lemmens I, Venkatesan K, Sahalie J, et al. High-quality binary protein interaction map of the yeast interactome network. Science 2008;322:104–10.

[109] Blagoev B, Kratchmarova I, Ong SE, Nielsen M, Foster LJ, Mann M. A proteomics strategy to elucidate functional protein–protein interactions applied to EGF signaling. Nat Biotechnol 2003;21:315–8.

[110] Ranish JA, Yi EC, Leslie DM, Purvine SO, Goodlett DR, Eng J, et al. The study of macromolecular complexes by quantitative proteomics. Nat Genet 2003;33:349–55.

[111] Schulze WX, Mann M. A novel proteomic screen for peptide–protein interactions. J Biol Chem 2004;279:10756–64.

[112] Vermeulen M, Mulder KW, Denissov S, Pijnappel WW, van Schaik FM, Varier RA, et al. Selective anchoring of TFIID to nucleosomes by trimethylation of histone H3 lysine 4. Cell 2007;131:58–69.

[113] Schulze WX, Deng L, Mann M. Phosphotyrosine interactome of the ErbB-receptor kinase family. Mol Syst Biol 2005;1: 2005.0008.

[114] Hanke S, Mann M. The phosphotyrosine interactome of the insulin receptor family and its substrates IRS-1 and IRS-2. Mol Cell Proteomics 2009;8:519–34.

[115] Mittler G, Butter F, Mann M. A SILAC-based DNA protein interaction screen that identifies candidate binding proteins to functional DNA elements. Genome Res 2009;19:284–93.

[116] Butter F, Scheibe M, Morl M, Mann M. Unbiased RNA-protein interaction screen by quantitative proteomics. Proc Natl Acad Sci U S A 2009;106:10626–31.

[117] Markljung E, Jiang L, Jaffe JD, Mikkelsen TS, Wallerman O, Larhammar M, et al. ZBED6, a novel transcription factor derived from a domesticated DNA transposon regulates IGF2 expression and muscle growth. PLoS Biol 2009;7:e1000256.

[118] Butter F, Kappei D, Buchholz F, Vermeulen M, Mann M. A domesticated transposon mediates the effects of a single-nucleotide

polymorphism responsible for enhanced muscle growth. EMBO Rep 2010;11:305–11.
[119] Bantscheff M, Eberhard D, Abraham Y, Bastuck S, Boesche M, Hobson S, et al. Quantitative chemical proteomics reveals mechanisms of action of clinical ABL kinase inhibitors. Nat Biotechnol 2007;25:1035–44.
[120] Rix U, Superti-Furga G. Target profiling of small molecules by chemical proteomics. Nat Chem Biol 2009;5:616–24.
[121] Sharma K, Weber C, Bairlein M, Greff Z, Keri G, Cox J, et al. Proteomics strategy for quantitative protein interaction profiling in cell extracts. Nat Methods 2009;6:741–4.
[122] Bantscheff M, Hopf C, Savitski MM, Dittmann A, Grandi P, Michon AM, et al. Chemoproteomics profiling of HDAC inhibitors reveals selective targeting of HDAC complexes. Nat Biotechnol 2011;29:255–65.
[123] Andersen JS, Wilkinson CJ, Mayor T, Mortensen P, Nigg EA, Mann M. Proteomic characterization of the human centrosome by protein correlation profiling. Nature 2003;426: 570–4.
[124] Dunkley TP, Watson R, Griffin JL, Dupree P, Lilley KS. Localization of organelle proteins by isotope tagging (LOPIT). Mol Cell Proteomics 2004;3:1128–34.
[125] Foster LJ, de Hoog CL, Zhang Y, Xie X, Mootha VK, Mann M. A mammalian organelle map by protein correlation profiling. Cell 2006;125:187–99.
[126] Harner M, Korner C, Walther D, Mokranjac D, Kaesmacher J, Welsch U, et al. The mitochondrial contact site complex, a determinant of mitochondrial architecture. EMBO J 2011;30:4356–70.
[127] Ohta S, Bukowski-Wills JC, Sanchez-Pulido L, Alves Fde L, Wood L, Chen ZA, et al. The protein composition of mitotic chromosomes determined using multiclassifier combinatorial proteomics. Cell 2010;142:810–21.
[128] Ewing RM, Chu P, Elisma F, Li H, Taylor P, Climie S, et al. Large-scale mapping of human protein-protein interactions by mass spectrometry. Mol Syst Biol 2007;3:89.
[129] Malovannaya A, Lanz RB, Jung SY, Bulynko Y, Le NT, Chan DW, et al. Analysis of the human endogenous coregulator complexome. Cell 2011;145:787–99.
[130] Guruharsha KG, Rual JF, Zhai B, Mintseris J, Vaidya P, Vaidya N, et al. A protein complex network of Drosophila melanogaster. Cell 2011;147:690–703.
[131] Leitner A, Walzthoeni T, Kahraman A, Herzog F, Rinner O, Beck M, et al. Probing native protein structures by chemical cross-linking, mass spectrometry, and bioinformatics. Mol Cell Proteomics 2010;9:1634–49.
[132] Chen ZA, Jawhari A, Fischer L, Buchen C, Tahir S, Kamenski T, et al. Architecture of the RNA polymerase II-TFIIF complex revealed by cross-linking and mass spectrometry. EMBO J 2010;29:717–26.
[133] Behrends C, Sowa ME, Gygi SP, Harper JW. Network organization of the human autophagy system. Nature 2010;466: 68–76.
[134] Hilger M, Mann M. Triple SILAC to determine stimulus specific interactions in the wnt pathway. J Proteome Res 2012;11:982–94.
[135] Brehme M, Hantschel O, Colinge J, Kaupe I, Planyavsky M, Kocher T, et al. Charting the molecular network of the drug target Bcr-Abl. Proc Natl Acad Sci U S A 2009;106:7414–9.

[136] Wepf A, Glatter T, Schmidt A, Aebersold R, Gstaiger M. Quantitative interaction proteomics using mass spectrometry. Nat Methods 2009;6:203–5.
[137] Bennett EJ, Rush J, Gygi SP, Harper JW. Dynamics of cullin-RING ubiquitin ligase network revealed by systematic quantitative proteomics. Cell 2010;143:951–65.
[138] Gingras AC, Gstaiger M, Raught B, Aebersold R. Analysis of protein complexes using mass spectrometry. Nat Rev Mol Cell Biol 2007;8:645–54.
[139] Poser I, Sarov M, Hutchins JR, Heriche JK, Toyoda Y, Pozniakovsky A, et al. BAC TransgeneOmics: a high-throughput method for exploration of protein function in mammals. Nat Methods 2008;5:409–15.
[140] Hubner NC, Bird AW, Cox J, Splettstoesser B, Bandilla P, Poser I, et al. Quantitative proteomics combined with BAC TransgeneOmics reveals in vivo protein interactions. J Cell Biol 2010;189:739–54.
[141] Hutchins JR, Toyoda Y, Hegemann B, Poser I, Heriche JK, Sykora MM, et al. Systematic analysis of human protein complexes identifies chromosome segregation proteins. Science 2010;328:593–9.
[142] Kuhner S, van Noort V, Betts MJ, Leo-Macias A, Batisse C, Rode M, et al. Proteome organization in a genome-reduced bacterium. Science 2009;326:1235–40.
[143] Guell M, van Noort V, Yus E, Chen WH, Leigh-Bell J, Michalodimitrakis K, et al. Transcriptome complexity in a genome-reduced bacterium. Science 2009;326:1268–71.
[144] Yus E, Maier T, Michalodimitrakis K, van Noort V, Yamada T, Chen WH, et al. Impact of genome reduction on bacterial metabolism and its regulation. Science 2009;326:1263–8.
[145] Jordan JD, Landau EM, Iyengar R. Signaling networks: the origins of cellular multitasking. Cell 2000;103:193–200.
[146] Marshall CJ. Specificity of receptor tyrosine kinase signaling: transient versus sustained extracellular signal-regulated kinase activation. Cell 1995;80:179–85.
[147] Choudhary C, Mann M. Decoding signalling networks by mass spectrometry-based proteomics. Nature reviews. Molecular cell biology 2010;11:427–39.
[148] Witze ES, Old WM, Resing KA, Ahn NG. Mapping protein post-translational modifications with mass spectrometry. Nature methods 2007;4:798–806.
[149] Ficarro S, Chertihin O, Westbrook VA, White F, Jayes F, Kalab P, et al. Phosphoproteome analysis of capacitated human sperm. Evidence of tyrosine phosphorylation of a kinase-anchoring protein 3 and valosin-containing protein/p97 during capacitation. J Biol Chem 2003;278:11579–89.
[150] Huttlin EL, Jedrychowski MP, Elias JE, Goswami T, Rad R, Beausoleil SA, et al. A tissue-specific atlas of mouse protein phosphorylation and expression. Cell 2010;143:1174–89.
[151] Kim SC, Sprung R, Chen Y, Xu Y, Ball H, Pei J, et al. Substrate and functional diversity of lysine acetylation revealed by a proteomics survey. Mol Cell 2006;23:607–18.
[152] Choudhary C, Kumar C, Gnad F, Nielsen ML, Rehman M, Walther TC, et al. Lysine acetylation targets protein complexes and co-regulates major cellular functions. Science 2009;325: 834–40.
[153] Kaji H, Kamiie J, Kawakami H, Kido K, Yamauchi Y, Shinkawa T, et al. Proteomics reveals N-linked glycoprotein

diversity in *Caenorhabditis elegans* and suggests an atypical translocation mechanism for integral membrane proteins. Mol Cell Proteomics 2007;6:2100–9.

[154] Zielinska DF, Gnad F, Wisniewski JR, Mann M. Precision mapping of an in vivo N-glycoproteome reveals rigid topological and sequence constraints. Cell 2010;141:897–907.

[155] Peng J, Schwartz D, Elias JE, Thoreen CC, Cheng D, Marsischky G, et al. A proteomics approach to understanding protein ubiquitination. Nat Biotechnol 2003;21:921–6.

[156] Xu G, Paige JS, Jaffrey SR. Global analysis of lysine ubiquitination by ubiquitin remnant immunoaffinity profiling. Nat Biotechnol 2010;28:868–73.

[157] Wagner SA, Beli P, Weinert BT, Nielsen ML, Cox J, Mann M, et al. A proteome-wide, quantitative survey of in vivo ubiquitylation sites reveals widespread regulatory roles. Mol Cell Proteomics 2011;10:M111.013284.

[158] Kim W, Bennett EJ, Huttlin EL, Guo A, Li J, Possemato A, et al. Systematic and quantitative assessment of the ubiquitin-modified proteome. Mol Cell 2011;44:325–40.

[159] Ong SE, Mittler G, Mann M. Identifying and quantifying in vivo methylation sites by heavy methyl SILAC. Nat Methods 2004;1:119–26.

[160] Zhao Y, Jensen ON. Modification-specific proteomics: strategies for characterization of post-translational modifications using enrichment techniques. Proteomics 2009;9:4632–41.

[161] Macek B, Mann M, Olsen JV. Global and site-specific quantitative phosphoproteomics: principles and applications. Annual review of pharmacology and toxicology 2009;49:199–221.

[162] Wolf-Yadlin A, Hautaniemi S, Lauffenburger DA, White FM. Multiple reaction monitoring for robust quantitative proteomic analysis of cellular signaling networks. Proc Natl Acad Sci U S A 2007;104:5860–5.

[163] Salomon AR, Ficarro SB, Brill LM, Brinker A, Phung QT, Ericson C, et al. Profiling of tyrosine phosphorylation pathways in human cells using mass spectrometry. Proc Natl Acad Sci U S A 2003;100:443–8.

[164] Blagoev B, Ong SE, Kratchmarova I, Mann M. Temporal analysis of phosphotyrosine-dependent signaling networks by quantitative proteomics. Nat Biotechnol 2004;22:1139–45.

[165] Olsen JV, Vermeulen M, Santamaria A, Kumar C, Miller ML, Jensen LJ, et al. Quantitative phosphoproteomics reveals widespread full phosphorylation site occupancy during mitosis. Sci Signal 2010;3:ra3.

[166] Wu RH, Haas W, Dephoure N, Huttlin EL, Zhai B, Sowa ME, et al. A large-scale method to measure absolute protein phosphorylation stoichiometries. Nat Methods 2011;8:677–83.

[167] Bishop AC, Buzko O, Shokat KM. Magic bullets for protein kinases. Trends in Cell Biology 2001;11:167–72.

[168] Bodenmiller B, Wanka S, Kraft C, Urban J, Campbell D, Pedrioli PG, et al. Phosphoproteomic analysis reveals interconnected system-wide responses to perturbations of kinases and phosphatases in yeast. Sci Signal 2010;3(153). rs4.

[169] Breitkreutz A, Choi H, Sharom JR, Boucher L, Neduva V, Larsen B, et al. A global protein kinase and phosphatase interaction network in yeast. Science 2010;328:1043–6.

[170] Hunter T. The age of crosstalk: phosphorylation, ubiquitination, and beyond. Molecular Cell 2007;28:730–8.

[171] Lee JS, Smith E, Shilatifard A. The language of histone crosstalk. Cell 2010;142:682–5.

[172] Wu SL, Kim J, Bandle RW, Liotta L, Petricoin E, Karger BL. Dynamic profiling of the post-translational modifications and interaction partners of epidermal growth factor receptor signaling after stimulation by epidermal growth factor using Extended Range Proteomic Analysis (ERPA). Mol Cell Proteomics 2006;5:1610–27.

[173] Siuti N, Kelleher NL. Decoding protein modifications using top-down mass spectrometry. Nature methods 2007;4:817–21.

[174] Gnad F, Ren S, Cox J, Olsen JV, Macek B, Oroshi M, et al. PHOSIDA (phosphorylation site database): management, structural and evolutionary investigation, and prediction of phosphosites. Genome Biol 2007;8:R250.

[175] Zhang J, Sprung R, Pei J, Tan X, Kim S, Zhu H, et al. Lysine acetylation is a highly abundant and evolutionarily conserved modification in *Escherichia coli*. Mol Cell Proteomics 2009;8:215–25.

[176] Collins MO, Yu L, Campuzano I, Grant SGN, Choudhary JS. Phosphoproteomic analysis of the mouse brain cytosol reveals a predominance of protein phosphorylation in regions of intrinsic sequence disorder. Mol Cell Proteomics 2008;7:1331–48.

[177] Villen J, Beausoleil SA, Gerber SA, Gygi SP. Large-scale phosphorylation analysis of mouse liver. Proc Natl Acad Sci U S A 2007;104:1488–93.

[178] Wisniewski JR, Nagaraj N, Zougman A, Gnad F, Mann M. Brain phosphoproteome obtained by a FASP-based method reveals plasma membrane protein topology. J Proteome Res 2010;9:3280–9.

[179] Monetti M, Nagaraj N, Sharma K, Mann M. Large-scale phosphosite quantification in tissues by a spike-in SILAC method. Nat Methods 2011;8(8):655–8.

[180] Liu H, Sadygov RG, Yates 3rd JR. A model for random sampling and estimation of relative protein abundance in shotgun proteomics. Anal Chem 2004;76:4193–201.

[181] Gao J, Opiteck GJ, Friedrichs MS, Dongre AR, Hefta SA. Changes in the protein expression of yeast as a function of carbon source. J Proteome Res 2003;2:643–9.

[182] Mann M. Functional and quantitative proteomics using SILAC. Nat Rev Mol Cell Biol 2006;7:952–8.

[183] Thompson A, Schafer J, Kuhn K, Kienle S, Schwarz J, Schmidt G, et al. Tandem mass tags: a novel quantification strategy for comparative analysis of complex protein mixtures by MS/MS. Anal Chem 2003;75:1895–904.

[184] Ross PL, Huang YN, Marchese JN, Williamson B, Parker K, Hattan S, et al. Multiplexed protein quantitation in *Saccharomyces cerevisiae* using amine-reactive isobaric tagging reagents. Mol Cell Proteomics 2004;3:1154–69.

[185] Hsu JL, Huang SY, Chow NH, Chen SH. Stable-isotope dimethyl labeling for quantitative proteomics. Anal Chem 2003;75:6843–52.

[186] Boersema PJ, Raijmakers R, Lemeer S, Mohammed S, Heck AJ. Multiplex peptide stable isotope dimethyl labeling for quantitative proteomics. Nat Protoc 2009;4:484–94.

[187] Picotti P, Lam H, Campbell D, Deutsch EW, Mirzaei H, Ranish J, et al. A database of mass spectrometric assays for the yeast proteome. Nat Methods 2008;5:913–4.

[188] Picotti P, Rinner O, Stallmach R, Dautel F, Farrah T, Domon B, et al. High-throughput generation of selected reaction-monitoring assays for proteins and proteomes. Nat Methods 2010;7:43—6.

[189] Farrah T, Deutsch EW, Kreisberg R, Sun Z, Campbell DS, Mendoza L, et al. PASSEL: The PeptideAtlas SRM Experiment Library. Proteomics 2012;12(8):1170—5.

[190] Gerber SA, Rush J, Stemman O, Kirschner MW, Gygi SP. Absolute quantification of proteins and phosphoproteins from cell lysates by tandem MS. Proc Natl Acad Sci U S A 2003;100: 6940—5.

[191] Hanke S, Besir H, Oesterhelt D, Mann M. Absolute SILAC for accurate quantitation of proteins in complex mixtures down to the attomole level. J Proteome Res 2008;7:1118—30.

Chapter 2

The Coding and the Non-coding Transcriptome

Roderic Guigó[1,2]

[1]*Centre de Regulació Genòmica Universitat Pompeu Fabra, Dr Aiguader, 88, E-08003 Barcelona, Catalonia, Spain*
[2]*Departament de Ciències Experimentals i de la Salut, Universitat Pompeu Fabra, Dr Aiguader, 88, E-08003 Barcelona, Catalonia, Spain*

Chapter Outline

Introduction	27
The Pathway from DNA to Protein Sequences	28
Methods to Determine the Reference Transcriptome	29
Experimental Methods	29
Random cDNA Cloning	29
DNA Microarrays	30
RNASeq	30
Computational Methods	31
Integrated Computational Gene Prediction	32
The Use of Chromatin Marks	33
Assessing the Reference Transcriptome	33
The Human Transcriptome	**33**
The Number of Human Genes	33
Human Genome Reference Gene Sets	34
The Protein-coding Transcriptome	35
The Long Non-coding RNA Transcriptome	36
The Expressed Transcriptome	37
The Small RNA Transcriptome	38
Conclusions and Future Challenges	**38**
References	**38**

INTRODUCTION

The unfolding of the instructions encoded in the genome is triggered by the transcription of DNA into RNA, and the subsequent processing of the resulting primary RNA transcripts into functional mature RNAs. The population of all RNAs in the cell, the so-called transcriptome, is therefore the first phenotypic manifestation of the genome, and at the same time, the determinant of the higher-order phenotypes of the cell, necessarily mediating all phenotypic changes at the organism level caused by changes in the DNA sequence. Until very recently it was assumed that the main role of RNA was merely that of a messenger, mediating the transfer of information from DNA to proteins. These were assumed to be the main effectors of biological function. Indeed, proteins are essential parts of organisms and participate in virtually every process within cells. They catalyze the biochemical reactions in metabolic pathways, and also have structural and mechanical functions. In recent years, however, a plethora of novel RNA species have been discovered in all organisms. Some of these species correspond to novel splice forms — often of extraordinary complexity — of known protein-coding genes [1,2], but many do not appear to code for proteins and belong to novel families of small [3–7] or longer multi-exonic non-coding RNAs [8–12]. Although the precise role of the vast majority of these species is yet unknown, many seem involved in the regulation of gene expression. It is thus becoming clear that RNA plays a complex, multifaceted role in cellular homeostasis, rather than merely serving as a carrier of information from DNA to proteins.

Traditional methods for global transcriptome characterization, such as EST sequencing and DNA microarrays, have recently been complemented by deep RNA sequencing (RNASeq) using massively parallel sequencing instruments. Profiling of RNA by RNASeq, across multiple conditions (cell types, species, individuals, cellular compartments, RNA classes, perturbations, etc.) is revealing a eukaryotic transcriptome of unanticipated complexity. Combined with the profiling of other players involved in RNA synthesis and processing (DNA methylation status, chromatin structure and modifications, transcription factors, RNA polymerase, splicing regulators, etc., — also greatly facilitated by massively parallel sequencing) we can now obtain a global, holistic view of the transcriptional activity of the genome within the cell, allowing for the first time a systems biology approach to the understanding and modeling of the pathways involved in RNA synthesis and processing.

In this chapter we will first briefly review these cellular pathways. We will then review the experimental and computational methods that are commonly used to characterize and monitor cellular transcriptomes, and then summarize the efforts leading to the definition of the human reference transcriptome. We use the term reference transcriptome to denote the set of all genes and transcripts potentially encoded in a given genome. We will next describe the characteristic features of the human transcriptome. We will deal separately with the protein-coding transcriptome — that is, the set of genes and transcripts that code for proteins — and the non-coding transcriptome — the set of genes and transcripts that are not translated into proteins. Many of the characteristic features of the human transcriptome can be extrapolated to other mammalian, vertebrate, and even metazoan genomes. In the final section we will review recent work, mostly based on RNASeq profiling, contributing to the characterization of the expressed transcriptome. In contrast to the reference transcriptome, the expressed transcriptome refers to the set of genes and transcripts that are expressed in a given condition, and which are therefore responsible for cellular specificity.

THE PATHWAY FROM DNA TO PROTEIN SEQUENCES

The pathway leading from DNA to protein and functional RNA sequences includes a number of steps, which are relatively well characterized. The first step is the transcription of DNA into RNA. During transcription, RNA polymerase copies the DNA template into a complementary RNA molecule. Specific DNA sequences in the 5′ upstream region of genes — the so-called promoter region — act as binding sites for proteins called transcription factors that recruit the RNA polymerase. Transcription factors interact in these regions with sequence-specific elements or motifs, the transcription factor-binding sites. These are typically 5—8 nucleotides long, and one promoter region usually contains many of them to harbor different transcription factors. The interplay between these factors is not well understood, but in eukaryotes the motifs appear to be arranged in specific configurations that confer on each gene an individualized spatial and temporal transcription program. In many eukaryotic promoters between 10% and 20% of all genes contain a TATA box (sequence TATAAA), which in turn binds a TATA-binding protein, which assists in the formation of the RNA polymerase transcriptional complex. Specific modifications in the histones (the proteins that form the nucleosomes, the basic DNA packaging unit) regulate the binding of transcription factors to the promoter region, in this way contributing to control gene expression.

Whereas transcription of prokaryotic protein-coding genes creates a messenger RNA molecule (mRNA) which is ready for translation, transcription of eukaryotic genes produces a primary transcript of RNA (pre-mRNA) which undergoes a series of modifications before becoming a mature mRNA. These include 5′ capping, which involves a set of enzymatic reactions that modify the 5′ end of the pre-mRNA and thus protects the RNA from degradation by exonucleases. Another modification is 3′ cleavage and polyadenylation. They occur if the polyadenylation signal sequence (5′- AAUAAA-3′) is present near the 3′ end of the pre-mRNA sequence. The pre-mRNA is first cleaved and then a series of about 200 adenines are added to form a 3′ poly(A) tail which protects the RNA from degradation. The poly(A) tail is bound by multiple poly(A)-binding proteins necessary for mRNA export to the cytosol.

The most notable modification in eukaryotic pre-mRNA is RNA splicing. During the process of splicing, an RNA—protein catalytic complex known as the spliceosome catalyzes two *trans*-esterification reactions in which intervening sequences in the pre-mRNA (the introns) are excised and released in the form of lariat structures, and neighboring sequences (the exons) are concatenated together to form the mature mRNA. Often, introns or exons can be either removed or retained in mature mRNA. This so-called alternative splicing creates series of different transcripts originating from a single gene, increasing the transcriptional complexity beyond that simply reflected in gene number. Until very recently, splicing was assumed to occur mostly in pre-mRNA sequences destined to be translated into proteins (see below). However, an emerging class of long RNA molecules that lack protein-coding capacity — the long non-coding RNAs, lncRNAs — seem to be subjected to the same splicing process as mRNAs

In eukaryotes, mRNAs are usually exported to the cytoplasm, where they are translated into proteins. LncRNAs, although mostly of nuclear function, may also be transported to the cytosol and localize to specific subcellular compartments. During translation, mRNAs are decoded by the ribosome to produce specific amino acid chains, or polypeptides. Initiation of translation involves the ribosome binding to the 5′ end of the mRNAs with the help of a number of proteins known as initiation factors. The nucleotides in the RNA are then 'read' by the ribosome in consecutive non-overlapping triplets, known as codons. Each codon is translated to a specific amino acid. The equivalence of codons and amino acids, known as the genetic code, is implemented through the collection of tRNAs in the cell. Each tRNA carries at one end the so-called anticodon sequence, a triplet that will base-pair with the codons in the mRNA sequence, and is charged with the corresponding specific amino acid at the other end. The ribosome induces the binding of tRNA with complementary anticodon sequences to the triplet sequences in the mRNA.

The amino acid carried out by the tRNA is then incorporated into the nascent polypeptide chain. Termination of translation happens when the ribosome faces a stop codon (usually UAA, UAG, or UGA) that induces the binding of a release factor protein that prompts the disassembly of the entire ribosome–mRNA complex. After synthesis, proteins fold and are transported to subcellular compartments where they localize and exert their function.

In contrast to prokaryotes, in eukaryotes the presence of the nuclear membrane leads to cellular compartmentalization, where RNA processing in the nucleus is physically separated from translation in the cytosol. Within the nucleus, the different steps of RNA processing are also often thought as separate processes occurring in a stepwise manner: transcription, capping, polyadenylation and splicing. However, there is increasing evidence of strong interconnections among all of these processes. Indeed, on the one hand there appears to be coordinated regulation of alternative splicing and alternative polyadenylation and cleavage [13]; and on the other hand splicing appears to occur essentially during transcription [14]. These observations provide a natural explanation for a number of recent observations that link chromatin structure to splicing regulation [15].

Some families of small non-coding RNA genes (sncRNA) follow specific processing pathways, distinct from those of long RNAs. They are usually transcribed as precursors by variants of the RNA polymerase different from those involved in the transcription of long RNAs; they then undergo further processing. Ribosomal RNAs (rRNA) are often transcribed as pre-rRNAs that contain one or more rRNAs. The pre-rRNA is cleaved and modified at specific sites by small nucleolus-restricted RNA species called snoRNAs (small nucleolar RNAs). SnoRNAs themselves, also involved in the processing of transfer RNAs (tRNAs) and small nuclear RNAs (snRNAs), are usually encoded in the introns of proteins involved in ribosome synthesis or translation. In tRNAs, the 5′ and 3′ ends of the precursor are removed to generate the mature RNA. Micro-RNAs (miRNAs) are first transcribed as primary transcripts or pri-miRNAs with a cap and poly(A) tail and processed to short, 70-nucleotide stem–loop structures known as pre-miRNAs in the nucleus by the enzymes Drosha and Pasha. After being exported to the cytosol, they are further processed to mature miRNAs by the endonuclease Dicer, which also initiates the formation of the RNA-induced silencing complex (RISC), where the miRNA and its mRNA target interact.

METHODS TO DETERMINE THE REFERENCE TRANSCRIPTOME

We will use the term reference transcriptome to refer to the set of all mature transcript RNA sequences encoded in a given genome. Delineating this set of genes and transcripts is invariably the first step after the completion of the sequence of a genome, since the vast majority of the biology of a genome is initially inferred from the set of genes predicted to be encoded in its sequence. To delineate the reference transcriptome of a given genome, usually a combination of high throughput RNA sequencing and computational gene and transcript modeling is carried out. In some selected genomes, such as that of human, full-length sequencing of targeted RNA species may also be attempted.

Experimental Methods

Random cDNA Cloning

A complementary DNA (cDNA) is a DNA sequence synthesized by reverse transcription using an RNA molecule as a template. cDNAs obtained from RNAs of interest can be cloned in an appropriate vector, and the resulting clones can be sequenced. Using oligo-dT probes that are complementary to the poly(A) tail of long eukaryotic RNAs, a cDNA library containing copies of many different transcripts expressed in a given cell type or condition can be created. Systematic in-depth sequencing of cDNA libraries is likely the most powerful approach for the identification of genes and transcripts. Traditionally, one popular strategy is 'single-pass' sequencing of random cDNA clones that results in the production of short partial sequences identifying a specific transcript, known as expressed sequence tags [16] (ESTs). In turn, ESTs can be used to identify clones suitable for targeted full-length sequencing, as in the Mammalian Gene Collection initiative [17]. However, unless very large sequencing capacity is available, the wide dynamic range of mRNA abundances in the eukaryotic cell makes random sequencing of cDNA clones inefficient for discovering relatively rare transcripts, because highly abundant cDNAs will predominantly be sequenced. In the absence of such sequencing capacity, procedures that increase the likelihood of sampling rare or tissue-specific transcripts, such as normalization and subtraction, were developed [18,19]. These rely on denaturation of cDNA, rapid re-hybridization in the presence of a 'driver' sample (the same sample in normalization; a different one in subtraction), extraction of the double-stranded portion of the sample, and finally sequencing of a sample of the remaining single-stranded sequences. The rapid re-hybridization step favors duplex formation of abundant species, therefore the remaining single-stranded sample is enriched for rare transcripts. This has been the preferred approach during the last two decades for generating large EST collections used to delineate the reference transcriptomes of many species. Normalization, however, destroys all quantitative information. Therefore, while random sequencing of cDNA clones can be successfully used to delineate the set of genes and transcripts present in a given genome or cellular condition, they can not be used to monitor transcript abundances.

DNA Microarrays

DNA microarrays (or DNA chips) have been the most commonly used technique during the last two decades to globally monitor cellular abundances of transcript species. A DNA microarray is a collection of microscopic DNA spots attached to a solid surface. Each DNA spot contains many thousands of copies of a specific DNA sequence, known as probes. These usually correspond to a short section of a gene — generally at the 3' end. Each microarray includes one or a few probe sets for each interrogated gene. These are used to hybridize a cDNA sample (the target) under high-stringency conditions. Probe−target hybridization is usually detected and quantified by detection of fluorophore-, silver-, or chemiluminescence-labeled targets to determine the relative abundance of transcripts in the target sample. Data on about 700 000 sample hybridizations performed on DNA microarrays are accessible through the databases Gene Expression Omnibus (GEO) at NCBI, and ArrayExpress at EBI.

Because DNA microarrays require spotting of nucleotide probes corresponding to known transcripts, only the abundances of these transcripts can be monitored. Quantification of previously unknown transcripts — often specific to the particular cell type being interrogated, and therefore particularly relevant to the phenotype of this cell type — is impossible. Moreover, probes are usually shared between multiple splice forms of the same gene, and unless specific array designs are employed, it is impossible to deconvolute the abundances of individual alternative transcript isoforms from overall gene expression.

RNASeq

Thanks to impressive technological advances during the last decade, massively parallel sequencing appears to be providing the sequencing throughput required for direct sequencing of cDNA libraries (RNASeq). In the most cost-effective and popular approaches for transcriptome characterization (i.e., those using the Illumina platform), many very short sequence tags or reads (usually 50−100 bp long) are obtained along the transcripts in the interrogated RNA population. Sequenced reads are used to reconstruct transcript sequences. They are also used to quantify transcript abundances, under the assumption that the number of sequence reads originating from a given transcript is roughly proportional to transcript abundance (see Wang et al. [20] and Mortazavi et al. [21] for introductions). RNASeq has a dynamic range similar to or larger than that of DNA microarrays [22], allowing in addition for the discovery of novel genes and the quantification of alternative splice isoforms. It combines both transcript discovery, as in EST projects, and transcript quantification, as in DNA microarrays.

Since the first RNASeq experiments, sample preparations and protocols have been developed to target specific RNA populations within the cell (i.e., short vs. long RNAs, polyadenylated vs. non-polyadenylated, etc.) or to enrich for specific domains within the transcripts (5' or 3' ends, etc.). These different protocols share a common set of elementary components. With a few exceptions, massively parallel sequencing instruments are unable to directly sequence RNA, and therefore preparation of a cDNA library from the RNA sample of interest is, as with EST sequencing and DNA microarrays, the first step in the experimental protocol. In contrast to ESTs and microarrays, random priming is a commonly used alternative to polydT priming. This usually leads to a more uniform representation along the entire length of the transcript sequence in the library, minimizing the 3' bias common to polydT priming. Next, fragmentation of the cDNA sequences is necessary owing to the technological limitations of current sequencing technologies and their inability to obtain long (>100 bp) sequence reads. The most popular approaches for cDNA fragmentation are enzymatic digestion, nebulization, and hydrolysis. To minimize biases arising from reverse transcription, some protocols postpone this step until after fragmentation. Then, during final library preparation, adapter sequences are ligated to both ends of double-stranded cDNA molecules. These mediate the binding of fragments to beads in the sequencing medium, and harbor primer binding sites for amplification. Before sequencing, the primary library is amplified using polymerase chain reaction (PCR), as most instruments cannot sequence single nucleic acid molecules. To keep amplification biases under control, usually a size selection step that homogenizes fragment length is performed prior to amplification. Size selection is generally implemented by gel electrophoresis. In the sequencing step one arbitrary end (single reads) or both ends (paired end reads) of the cDNA fragments in the library are sequenced.

Sequence reads obtained after sequencing are used to infer transcript sequences and transcript abundances. Broadly, reads are first mapped to the genome and to the transcriptome of the organism investigated. If the genome is not available, reads may be assembled de novo into transcript contigs. From the mapped reads, it is possible to infer novel transcriptional elements (splice junctions, exons, transcripts and genes), and to quantify transcript abundances. Gene and transcript abundances are usually measured in reads per kilobase per million mapped reads (RPKM) [21]. This allows for normalization across experiments sequenced at very different depths. Although it is difficult to extrapolate the number of RNA copies per cell from the RPKM values in the absence of knowledge of the quantity of RNA in the cellular fraction being analyzed, Mortazavi et al. [21] estimate that 1 RPKM corresponds to 1−3 transcripts per cell. Given the nature of the data produced by RNASeq experiments (millions of very short sequence fragments), both transcript reconstruction and quantification are challenging and have triggered active research in computational biology. A number of tools are currently available to perform such tasks (see next section), but their accuracy and reliability have not yet been properly benchmarked.

Computational Methods

The contribution of computational methods is essential to infer the gene and transcript set characteristic of a given genome. First, data produced by technologies to globally monitor transcriptomes, such as those reviewed in the previous section, cannot be processed without sophisticated computational tools. Second, experimental approaches provide only information on the location and exonic structure of the genes expressed in the conditions that have been surveyed. Unless a very large panel of heterogeneous conditions has been monitored, transcribed sequences detected in this way capture only a fraction of the reference transcriptome.

Current computational methods use a variety of heterogeneous sources of information, which are processed and integrated through complex computational and statistical models (Figure 2.1). There are three broad sources of information that can be used to pinpoint the location and exonic structure of genes in genomic sequences.

- **Sequence comparisons across genomes** (Panel 1 in Figure 2.1). Functional regions in genomes are subjected to natural selection, and therefore are more conserved through evolution than non-functional ones. Regions that code for proteins are among the most conserved through evolution. Moreover, they have a specific mode of conservation, which can be computationally detected — for instance, because of the degeneracy of the genetic code, in coding regions the third codon position is usually less conserved than the first and second positions. Conservation is also a function of evolutionary distance. Conservation in genomes of closely related species extends beyond functional regions; on the other hand, it may have already vanished in these regions when distantly related species are compared.
- **Intrinsic sequence features in the genome.** There are sequence features in the genomic sequence that are revealing of the existence of genes. These features are of two types: intrinsic sequence signals involved in gene specification (panel 2), and statistical bias in the genome sequence specific to protein-coding regions (panel 3).

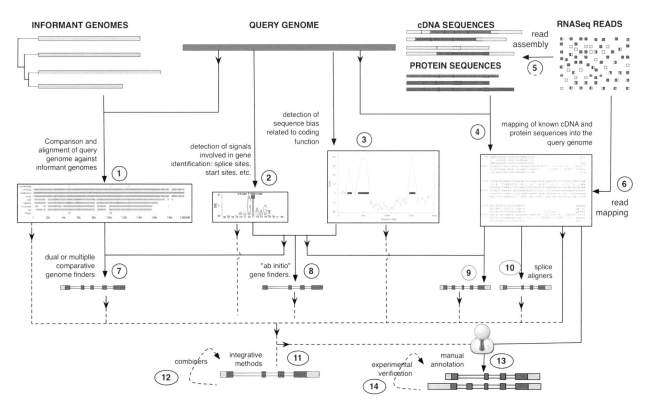

FIGURE 2.1 **Methods to determine reference transcriptomes.** See text for details. The yellow portion in the input cDNA sequences represents the UTRs. Vertical lines in cDNA sequences as well as in input protein sequences represent the location of the exons. Arrows indicate methods. Dashed lines indicate methods that depend on previously constructed gene models. *Adapted from Harrow et al. [83].*

Sequence signals, (patterns or motifs) here are defined as short functional DNA elements involved in the definition of exons: the translational start site, the 5′ or donor splice site, the 3′ or acceptor splice site and the translational stop codon. Typically, these sequence signals are represented by so-called position weight matrices (PWMs). In these matrices, the probability of each nucleotide at each position is computed from a set of known functional signals. The matrices can then be used to compute at each position along a problem sequence the likelihood of the corresponding site to be a functional signal. In the figure (panel 2) a PWM for splicing donor sites is represented as a sequence logo. In these pictorial representations, the relative frequency of each nucleotide along each position of the motif is represented by the height of the letter representing the nucleotide. The total height at each position measures the information content of the position — that is, how relevant the position is in defining the pattern.

Protein-coding regions, on the other hand, exhibit characteristic DNA sequence composition bias, which is absent from non-coding regions. The bias is a consequence of the uneven usage of amino acids in real proteins, and of the uneven usage of synonymous codons (Figure 2.2). To discriminate protein-coding from non-coding regions, a number of content measures can be computed to detect this bias. Such content measures — also known as coding statistics — can be defined as functions that compute a real number related to the likelihood that a given DNA sequence codes for a protein (or a fragment of a protein). Hexamer frequencies, usually in the form of codon position-dependent fifth-order Markov models [23]; appear to offer the maximum discriminative power, and are at the core of most popular gene finders today. In the figure (panel 3), we have computed a simple measure of codon bias along a 2000 bp-long stretch of the human genome encoding the β-globin 3-exons gene. We have used the known human codon usage table (Figure 2.2) to compute the likelihood that an observed sequence occurs in a protein-coding region. We have then used a sliding window to record the likelihood at each position along the investigated genome sequence. Peaks in the resulting distribution correspond to protein-coding exons and valleys correspond to introns.

- **Transcribed sequences.** Transcribed sequences corresponding to the genome being investigated are the most powerful and reliable source of information to locate genes in genomes (panel 4). In addition to cDNA sequencing — through either ESTs or RNASeq (see previous section) — genomic similarity to known protein-coding sequences may also provide strong evidence of protein-coding function.

Integrated Computational Gene Prediction

During the last two decades a plethora of programs and strategies have been developed to combine these sources of information in order to obtain reliable gene predictions (see Brent and Guigo [24] for a review). Programs exist that can combine, using a variety of frameworks often related to hidden Markov models (see Borodovsky and McIninch [25] and references therein for an introduction), the 'intrinsic' evidence from sequence signals and statistical bias to produce gene predictions (panel 8). These

UUU	Phe	0.46	17.6	UCU	Ser	0.19	15.2	UAU	Tyr	0.44	12.2	UGU	Cys	0.46	10.6
UUC	Phe	0.54	20.3	UCC	Ser	0.22	17.7	UAC	Tyr	0.56	15.3	UGC	Cys	0.54	12.6
UUA	Leu	0.08	7.7	UCA	Ser	0.15	12.2	UAA	STOP	0.30	1.0	UGA	STOP	0.47	1.6
UUG	Leu	0.13	12.9	UCG	Ser	0.05	4.4	UAG	STOP	0.24	0.8	UGG	Trp	1.00	13.2
CUU	Leu	0.13	13.2	CCU	Pro	0.29	17.5	CAU	His	0.42	10.9	CGU	Arg	0.08	4.5
CUC	Leu	0.20	19.6	CCC	Pro	0.32	19.8	CAC	His	0.58	15.1	CGC	Arg	0.18	10.4
CUA	Leu	0.07	7.2	CCA	Pro	0.28	16.9	CAA	Gln	0.27	12.3	CGA	Arg	0.11	6.2
CUG	Leu	0.40	39.6	CCG	Pro	0.11	6.9	CAG	Gln	0.73	34.2	CGG	Arg	0.20	11.4
AUU	Ile	0.36	16.0	ACU	Thr	0.25	13.1	AAU	Asn	0.47	17.0	AGU	Ser	0.15	12.1
AUC	Ile	0.47	20.8	ACC	Thr	0.36	18.9	AAC	Asn	0.53	19.1	AGC	Ser	0.24	19.5
AUA	Ile	0.17	7.5	ACA	Thr	0.28	15.1	AAA	Lys	0.43	24.4	AGA	Arg	0.21	12.2
AUG	Met	1.00	22.0	ACG	Thr	0.11	6.1	AAG	Lys	0.57	31.9	AGG	Arg	0.21	12.0
GUU	Val	0.18	11.0	GCU	Ala	0.27	18.4	GAU	Asp	0.46	21.8	GGU	Gly	0.16	10.8
GUC	Val	0.24	14.5	GCC	Ala	0.40	27.7	GAC	Asp	0.54	25.1	GGC	Gly	0.34	22.2
GUA	Val	0.12	7.1	GCA	Ala	0.23	15.8	GAA	Glu	0.42	29.0	GGA	Gly	0.25	16.5
GUG	Val	0.46	28.1	GCG	Ala	0.11	7.4	GAG	Glu	0.58	39.6	GGG	Gly	0.25	16.5

FIGURE 2.2 The human codon usage table. For each codon (first column in each sub-table), the table lists the encoded amino acid (second column), the relative frequency with which the amino acid is encoded by the codon (for instance, 46% of the codons encoding Phe are UUU, while 54% are UUC), and the per-thousand usage of the codon in human coding regions (for instances, 17.6 out of every 1000 codons in human coding regions is UUU). Sub-tables in the table are sorted by first codon position (y-axis) and second codon position (x-axis). Within each sub-table, rows are sorted by third-codon position. Updated values taken from http://www.kazusa.or.jp/codon/.

programs are often referred to as 'ab initio' or 'de novo' gene finders. They are the programs of choice in the absence of known transcript and protein sequences for the interrogated genome, and of other phylogenetically related genomes. If the sequence of related genomes is available, programs exist that combine this intrinsic information with the patterns of genomic sequence conservation. These programs are often referred to as comparative (or dual- or multi-genome) gene finders (panel 7). The most sophisticated of such programs use an underlying phylogenetic tree to appropriately weight sequence conservation depending on evolutionary distance [26–29].

If cDNA, ESTs or RNASeq data are available, these often take priority over other existing sources of information. In these cases, the initial map of the transcript or protein sequences of a genome can be obtained using a variety of popular tools, such as BLAT [30] or BLAST [31]; or other tools devoted specifically to the mapping of short sequence reads. This map can be refined using more sophisticated 'splice alignment' algorithms, whose explicit splice site models allow more precise alignment across gaps corresponding to introns (panel 10). In the figure, we show a protein to genome alignment produced by the GENEWISE [32] program. Each amino acid in the protein sequence aligns with a codon, and the large intron gaps are delimited by the canonical GT-AG splice dinucleotides. Alternatively, cDNA and protein information can be fed into an ab initio gene finder algorithm to inform the exons included in the prediction (panel 10). RNASeq reads can also be used. A number of methods have been developed to delineate gene structures from RNASeq reads and to quantify transcript abundances [22,33,34]. Two general approaches are possible. Reads can be assembled before mapping them to the genome (panel 5). In this way, contigs similar to ESTs can be created, and pipelines to deal with ESTs can be then used. Alternatively, reads can be directly mapped to the genome (if available for the species under investigation), and used to inform gene prediction programs (panel 6).

Often cDNA and protein evidence for a given genome is only partial; in such cases the initial reliable gene and transcript set may be extended with more hypothetical models derived from ab initio or comparative gene finders, or from the genome mapping of cDNA and protein sequences from other species. Pipelines have been derived that automate this multistep process (panel 9). More recently, programs have been developed that combine the output of many individual gene finders (panel 11). The underlying assumption in these 'combiners' is that consensus across programs increases the likelihood of the predictions. Thus, the predictions are weighted according to the particular features of the program producing them. The most general of such frameworks allow for the integration of a great variety of types of predictions, which include not only gene predictions, but also predictions of individual sites and exons. In spite of all development efforts in computational gene finding, the most reliable and complete gene annotations are still obtained after the initial alignments of cDNA and proteins with the genome sequence have been inspected manually to establish the exon boundaries of genes and transcripts (panel 13). This is the task carried out by the HAVANA team at the Sanger Institute. The initial manual annotation can be even further refined by subsequent experimental verification of those transcript models lacking sufficiently strong evidence, such as in the GENCODE project (panel 14, and see below).

The Use of Chromatin Marks

Recently an entirely different approach has been employed to identify genes in genome sequences on the basis of exploiting information on chromatin structure. Efficient methods have been developed to create genome-wide chromatin-state maps using chromatin immunoprecipitation followed by massively parallel sequencing (ChIP-Seq). These maps have revealed that genes actively transcribed by RNA polymerase II (Pol II) are marked by trimethylation of lysine 4 of histone H3 (H3K4me3) at their promoter, and trimethylation of lysine 36 of histone H3 (H3K36me3) along the length of the transcribed region. Computational methods have been developed to identify these so-called K4–K36 domains in genome-wide chromatin-state maps. Using these methods, Guttman et al. discovered thousands of novel long non-coding RNAs (lncRNAs) [10].

Assessing the Reference Transcriptome

Standard metrics and data sets have been developed to benchmark the accuracy of computational gene-finding methods [35]. Community-wide assessment projects, in which gene predictions obtained by different groups in a benchmark set of genomic sequences are evaluated in an unbiased way, are also very popular in the field. In 1999 GASP was organized to evaluate gene prediction programs in the *Drosophila melanogaster* genome [36]. It was continued in 2005 with EGASP, to evaluate gene prediction programs in the human genome [37]. Other gene-finding community assessment projects are NGASP [38] to evaluate gene prediction in the *Caenorhabditis elegans* genome, and RGASP to evaluate methods to produce reference transcriptomes using RNASeq data (http://www.gencodegenes.org/rgasp/).

THE HUMAN TRANSCRIPTOME

The Number of Human Genes

The human transcriptome has been intensely investigated. During recent decades hundreds of EST libraries have been generated and thousands of microarray experiments have been performed, monitoring gene expression in many different cell types, individuals, conditions, etc., providing an extremely rich characterization of the transcriptional diversity and complexity of the human genome. It may therefore come as a surprise that the exact number of genes in the human genome is not yet precisely known, much less the number of transcripts encoded by these genes. Prior to the publication of the human genome sequence, the number of human genes had been estimated to be in the range of 70 000–100 000 [39]. These numbers were mostly based on the experimental estimation of the number of CpG islands that exist in the human genome. CpG islands are genomic regions that contain a high frequency of CG dinucleotides. They are in and near approximately 60% of human promoters [40] and through methylation of the cytosine they may play a role in the regulation of gene expression. Their distinctive properties allow their physical separation from bulk DNA. In this way, Antequera and Bird estimated approximately 45 000 CpG islands in the human genome, which produced a rough estimate of about 80 000 human genes.

This number was widely accepted by the scientific community. Therefore, one of the surprises of the publication of the draft human genome sequences was that the number of human genes was estimated to be considerably lower. Indeed, Lander et al. [41] estimated the number of protein-coding genes to be in the range of 30 000–40 000, and Venter et al. [42] estimated it to be in the range of 27 000–38 000. The publication of the human genome draft sequence did not end the dispute, and just before and after the publication of this sequence estimates with very discrepant numbers ranging from 30 000 to 120 000 genes were published [43–45]. Since then, some consensus appears to have been reached, and although the exact number of human genes is not yet known, there is agreement that the number of protein-coding genes in the human genome is likely to be between 20 000 and 21 000 [46]. However, evidence has since emerged for an unanticipated large number of lncRNAs. Whereas a few years ago only a few dozen examples were known, currently about 10 000 long non-coding RNA genes are annotated on the human genome. This number, however, is likely to be an underestimate, and the overall number of human genes (coding and non-coding, long and small) may in the end be not too distant from that in the earlier estimates.

Human Genome Reference Gene Sets

Since the publication of the draft human genome sequence in 2001 [47,48] a number of human gene reference sets have been created using either computational prediction, manual annotation, or a hybrid mixture of the two methods. The Ensembl project was initially set up to warehouse and annotate the large amount of unfinished genomic data being produced as part of the public Human Genome Project, as well as to provide browser capacity for both sequences and annotations. Ensembl has expanded and now generates automatic predictions for over 35 species. The Ensembl gene build process is based on alignments of protein and cDNA sequences to produce a highly accurate gene set with low false positives [49]. For human, Ensembl has recently merged with the GENCODE project (see below).

Another site supplying sequence and annotation data for a large number of genomes is the University of California, Santa Cruz (UCSC) genome browser database [50]. The UCSC Genes track includes both protein-coding genes and non-coding RNA genes. Both types of gene can produce non-coding transcripts, but non-coding RNA genes do not produce protein-coding transcripts. This is a moderately conservative set of predictions. Transcripts of protein-coding genes require the support of one RefSeq RNA, or one GenBank RNA sequence plus at least one additional line of evidence. The latest release (February 2012) of the UCSC genes includes 80 922 transcripts corresponding to 31 227 genes.

Manual annotation still plays a significant role in annotating high-quality finished genomes. Currently the National Center for Biotechnology Information (NCBI) reference sequences (RefSeq) collection provide a highly manually curated resource of multispecies transcripts and includes plant, viral, vertebrate and invertebrate sequences [51,52]. These are, as their name indicates, transcript orientated and usually rely on full-length cDNAs for reliable curation, although the dataset also contains predictions using ESTs and partial cDNAs aligned against genomic sequence using the Gnomon prediction program. RefSeq is a very reliable but also conservative gene reference set. Other reference sets usually include RefSeq, but extend it substantially. Compared to RefSeq, the UCSC gene set, for instance, has about 10% more protein-coding genes, approximately four times as many putative non-coding genes, and about twice as many splice variants.

The Havana group at the Wellcome Trust Sanger Institute produces its annotation on vertebrate genomes by mapping transcript evidence (mostly from known cDNA sequences) onto the genome, and manually curating the resulting alignments. Currently only three vertebrate genomes — human, mouse and zebrafish — are sequenced to a quality that merits manual annotation [53]. In the case of the human genome, the work by the Havana team is at the core of the GENCODE annotation produced [9] within the framework of the ENCODE project [55,56]. The GENCODE annotation extends the Havana manually curated transcript models with computational predictions that are experimentally validated. GENCODE is the most comprehensive human genome gene set to date, and is becoming the standard reference in many large-scale genome projects, such as the 1000 genomes project [57]; the International Cancer Genome Consortium [58]; etc. Version 7 (2011) includes approximately 138 000 transcript models at 20 687 protein-coding and 9640 long non-coding RNA loci.

In 2006 the groups mentioned above (NCBI (RefSeq), UCSC, WTSI (Havana) and Ensembl) identified a need to collaborate and produce a consensus gene set for the human reference genome, since there was still no official agreement between the different databases on the human protein-coding genes. Referred to as the Consensus Coding Sequence Set (CCDS), it only contains coding transcripts that are equivalent in each database's gene build from start codon to stop codon. Version 37.3 of CCDS (September 2011) contains 26 473 transcripts that correspond to 18 471 genes. The CCDS set constitute the most solid set of protein-coding gene sequences available for the human genome.

The Protein-coding Transcriptome

Current gene numbers in human (and other species) should be taken only as indications. Although the number of human protein-coding genes is unlikely to change substantially — albeit not the number of transcripts generated from these loci — the number of long non-coding RNA loci is essentially unknown, and as RNASeq analysis is performed in an increasingly large number of tissues and cell types it is likely to increase substantially.

Protein-coding and long non-coding RNAs, as well as other classes of small RNAs, are organized along the genome in a complex network of interleaving transcripts, challenging our long-prevailing notion of genes as separate and well defined entities (Figure 2.3). Indeed, about 8500 genes annotated in GENCODE encode transcripts that

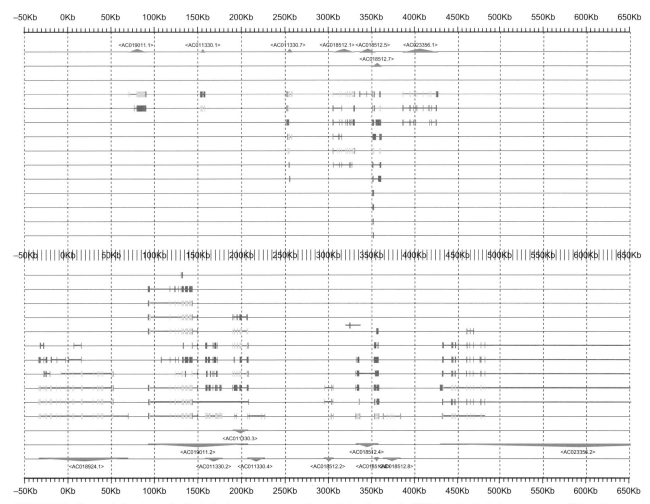

FIGURE 2.3 **Transcriptional complexity in the human genome.** Transcriptional map of a 650 Mb region in the human genome. This region starts approximately at position 41 520 000 on human chromosome 15, and corresponds roughly to the region corresponding to the region referred to as ENr233 in the pilot phase of the ENCODE project [54]. Blue triangles represent gene loci, and connected boxes represent transcripts. Each box corresponds to an exon. Green boxes correspond to protein-coding exons. Transcripts corresponding to loci encoded in the forward strand of the DNA sequence are displayed above the *x*-axis at the center of the display. Transcripts corresponding to loci in the reverse strand are displayed below. The map illustrates the transcriptional complexity of the human genome, with loci encoding a mixture of coding and non-coding transcripts, and transcripts themselves often overlapping multiple loci.

overlap transcripts from other genes, in both the sense and the antisense direction. Often 5′ ends of genes are found very distant from the body of the gene, overlapping exons from other distal genes [1,59].

The 20 687 protein-coding loci in GENCODE encode 122 099 transcripts. Therefore, there are on average six alternative transcripts per protein-coding locus. Most transcripts at protein-coding genes (>98%) are multi-exonic. On average there are 8.1 exons per transcript. Surprisingly, only 76 052 (62%) of the transcripts at protein-coding genes appear to code for proteins (on average 3.7 coding transcripts per locus). The role of the non-coding transcripts associated with protein-coding loci is mostly unknown, although evidence of a regulatory role is known for some of them. For instance, Sox2 is a transcription factor and plays a key role in the maintenance of the undifferentiated state of embryonic and adult neural stem cells. The Sox2 gene encodes a long non-coding RNA that shares the same transcriptional orientation as Sox2 [60]. The Sox2 lncRNA is expressed in the neurogenic region of the adult mouse brain, and is dynamically regulated during vertebrate development of the central nervous system, implying a role in regulating self-renewal and neurogenesis of neural stem cells.

Furthermore, within a given locus different protein-coding transcripts often differ only in their untranslated sequences. Therefore, the effective number of different protein sequences encoded at protein-coding loci is less than two on average [9]. Moreover, many of these protein variants appear to contain truncated functional domains having markedly different structures and functions from their more constitutively spliced counterparts [61]. For the vast majority of these alternative isoforms, little evidence exists to suggest they have a role as functional proteins. Therefore, only a small fraction of the enormous complexity observed at the transcriptional level appears to be translated into protein complexity.

The Long Non-coding RNA Transcriptome

Although they are poorly understood, long non-coding RNAs (lncRNAs) are emerging as central players in cell biology. LncRNAs are long, multi-exonic transcripts, often polyadenylated, and the loci that encode them exhibit the epigenetic marks typical of transcribed regions [10]. Over the last two decades the role of a small number of lncRNAs has been extensively researched in key epigenetic circuits. The 19 kb *XIST* RNA is necessary and sufficient for the silencing in *cis* of an entire X-chromosome in placental mammals [62]. In genomic imprinting at least three lncRNAs are known to stably silence specific parental alleles through both *cis* and *trans* mechanisms (Air [63]; H19 [63]; Kcnq1ot1 [64]. All of these have been shown to interact with known epigenetic regulatory proteins to regulate gene expression.

The discovery and study of lncRNAs is of significant relevance to human biology and disease, as they represent a huge, largely unexplored, and functional component of the genome [65,66]. It has been proposed that these may explain human-specific traits [67,68] and more practically, may underlie the deficiencies in rodent models of human diseases. There is diverse evidence that lncRNAs are intimately involved in gene networks underlying cancer: an antisense lncRNA that is overexpressed in leukemia represses expression of the p15 tumor suppressor [69]. Also, in genome-wide association studies the long non-coding RNA *CDKN2B-AS1* (also called *ANRIL*) has been associated with diverse diseases such as diabetes, glioma and basal cell carcinoma [70]. Elsewhere, an intergenic lncRNA, linc-p21, functions as a downstream effector of the p53 tumor suppressor [71]. And MEG3 activates p53 through an unknown mechanism [72]. Evidence is also mounting that numerous neurological diseases involve components of toxic RNA gain-of-function mutations (particularly trinucleotide repeat disorders) [73]; or involve misregulation of coding genes by antisense transcripts [74]. Given the lack of lncRNA annotation in the human genome until very recently, it is likely that many 'intergenic' disease-associated loci discovered in genome-wide association studies in fact modify the regulation or function of lncRNAs.

In the early 2000s the FANTOM consortium pioneered the genome-wide discovery of lncRNAs in mouse, publishing a set of 34 030 lncRNAs based on cDNA sequencing [75]. Recently, a catalogue of 5446 human lncRNAs has been created by Jia et al. [76] based on a computational pipeline of sequenced cDNAs. Meanwhile, the large intervening non-coding RNAs ('lincRNAs') [10], discovered through epigenetic annotation of human and mouse genomes, represent a useful set of RNAs but omit the many lncRNAs that reside within or overlap protein-coding loci. The GENCODE consortium within the ENCODE project has for several years been manually annotating a comprehensive set of human lncRNAs. Early releases of the GENCODE annotation have already been used to investigate the potential function of these transcripts (see, for instance, Ørom et al. [77]). GENCODE constitutes the most exhaustive collection of human lncRNAs available to date. Version 7 includes 9640 long non-coding RNA loci producing 14 880 transcripts.

A number of large-scale analyses of the GENCODE and other lncRNA collections [78,79] have revealed a number of features characteristic of this class of RNAs. LncRNAs are generated through pathways similar to those of protein-coding genes. They appear to be under similar transcriptional regulatory control, as their promoters are marked by the same set of histone modifications (e.g. histone H3 lysine 4 methylation), and the majority are spliced by canonical splicing signals, exhibiting exon and intron lengths typical of protein-coding genes. In contrast to protein-coding genes, however, lncRNAs display a striking bias towards two-exon transcripts, they are predominantly localized in the nucleus, and a fraction of them appear to be preferentially post-processed into small RNAs. They are under significant negative evolutionary selection — which is weaker in exons, but not in promoters, compared to what is observed in protein-coding genes. A relatively large subset seems to have arisen within the primate lineage. A subset of lncRNAs with high evolutionary conservation but ambiguous coding potential may function as non-coding RNAs, but, alternatively, encode small peptides. LncRNAs are consistently expressed at lower levels than protein-coding genes and display more tissue-specific expression patterns, with a large fraction of tissue-specific lncRNAs expressed in the brain. Correlation of expression analysis indicates that lncRNAs show particularly striking positive correlation with the expression of overlapping antisense coding genes. Finally, a few hundred lncRNAs reside within intergenic regions previously associated with specific diseases/traits by genome-wide association studies, and they could be candidates for future disease-focused studies.

The Expressed Transcriptome

Cellular specificity is achieved by the regulated expression of selected sets of genes and transcripts. Studies using expressed sequence tags have yielded relatively low estimates of tissue specificity, but, as we have pointed out, have limited statistical power to detect differences in isoform levels because of normalization. Microarray analyses have achieved more consistent coverage of tissues, but are constrained in their limited ability to distinguish closely related mRNA isoforms, and their inability to identify novel transcribed elements. RNASeq has the potential to circumvent these limitations. Recent RNASeq analysis performed in 15 different cell lines and subcellular compartments in the context of the ENCODE project provides a rich unbiased survey of the transcriptional landscape of the human genome [80]. These analyses reveal that, within a given cell line, on average expression is detected for about 40% of the genes. The range of gene expression values within a cell covers about six orders of magnitude for protein-coding and non-coding transcripts (from 10^{-2} to 10^4 RPKM) measured in the polyadenylated RNA fraction of the cell, but it is consistently lower for lncRNAs. Cumulatively, about 65% of all annotated genes (91% of all protein-coding genes) are detected in the panel of 15 investigated cell lines. Overall, about 14% of the genes are specific to one cell line, and 37% are detected in all cell lines and could be considered constitutively expressed. However, the behavior is strikingly different for protein-coding genes and lncRNAs. A considerable fraction (29%) of all expressed lncRNAs are restricted to only one of the cell lines, while only about 2% are expressed in all cell lines. Conversely, although a large fraction (55%) of expressed protein-coding genes are expressed in all cell lines, only about 7% are cell-line specificity. Variability of gene expression across cell lines is also greater for annotated non-coding RNAs than protein-coding RNAs. Expression variability is larger in the nucleus than in the cytosol, possibly indicating that the contents of the cytosolic subcompartment are buffered against stochastic transcription through selective export in the nucleus. All these results are consistent with the possibility that lncRNAs may contribute more to cell line specificity than protein-coding genes.

In principle, by attributing the sequence reads to the underlying RNA isoforms, RNASeq allows for the quantification of the abundance of individual transcript species. As we have pointed out, however, read deconvolution from RNASeq reads is a difficult problem, given the fact that different isoforms from the same gene share a substantial fraction of their sequence. The accuracy of methods to quantify the abundance of individual transcript isoforms is not currently established, and therefore the results of the analyses based on transcript quantification should be considered only indicative. Still, a number of trends appear to emerge [80]. When monitored in a homogeneous population of cells, genes tend to express many isoforms simultaneously. Isoforms, however, are not expressed at similar levels, and usually one dominant isoform captures a large fraction of the total expression of a given gene. The dominant isoform often changes between conditions and cell types. By comparing changes in gene expression and changes in individual transcript abundances across multiple conditions, it is possible to investigate whether changes in the abundances of individual transcripts across cell lines are mostly due to changes in overall gene expression or rather to changes in the splicing ratios within a gene. RNASeq analysis of individuals and the ENCODE cell lines reveals that between 60% and 70% of the variability in the abundance of transcript isoforms can be explained by variability in gene expression [80,81]. Thus, regulation at the level of gene expression appears to have a higher impact on the abundance of spliced isoforms than does regulation at the level of splicing.

Beyond contributing to the quantification of known genes and transcripts, RNASeq experiments usually uncover additional evidence for sites of transcription, both within and outside the boundaries of annotated genes [13,80,82]. For instance, although only about 85% of the GENCODE genes are annotated with multiple transcripts, Wang et al. [13] used RNASeq data to estimate that as many as 92–94% of human genes undergo alternative splicing. Overall, RNASeq studies underline that the transcriptional complexity of the human genome is likely to be much larger than that captured even in the richest of all human transcriptome reference sets. Interestingly, only a small fraction of this previously undetected transcriptional activity appears to exhibit protein-coding capacity. This suggests that the complexity of the functional non-coding transcriptome maybe much larger than the protein-coding one.

Transcriptome variation across individuals within human populations has also recently been monitored through RNASeq [22,82], leading to the discovery of mutations that alter both gene expression and alternative splicing. Transcriptome variation across individuals appears not to be as large as variation across cell lines and tissues [13], and alternative splicing ratios appear to be more constant across individuals than gene expression. LncRNAs show higher expression variability than protein-coding genes. Gonzalez-Porta et al. estimated that about 10% of the human genes exhibit population specific splicing ratios [81].

The Small RNA Transcriptome

Version 7 of GENCODE reports a total of 8713 annotated short RNAs. Four main classes represent about 85% of all of them. Small nuclear RNAs (snRNAs, 1944 genes) are found within the nucleus of eukaryotic cells. They are involved in RNA splicing, in regulation of transcription factors or RNA polymerase II, and in the maintenance of telomeres. They are always associated with specific proteins, and the complexes are referred to as small nuclear ribonucleoproteins. Small nucleolar RNAs (snoRNAs, 1521 genes) are involved in the processing of rRNAs, tRNAs and snRNAs, and are located in the nucleolus and Cajal bodies of eukaryotic cells. MicroRNAs (miRNAs, 1756 genes) are post-transcriptional regulators that bind to complementary sequences on target mRNAs, usually resulting in translational repression or target degradation and gene silencing. There are also 624 tRNA genes. Whereas most of these annotated short RNAs (65%) are located in intergenic regions, approximately one-third map within the boundaries of annotated genes.

CONCLUSIONS AND FUTURE CHALLENGES

Thanks to recent technological advances we are able to survey the transcriptional activity of genomes with unprecedented resolution. The picture that emerges is of a transcriptional landscape of unanticipated complexity. The genome appears almost as a continuum of transcription, where boundaries between genes — and the very concept of a gene as a defined unit — become increasingly fuzzy. Many functional RNA species — maybe most — appear to play a role not directly related to protein production. Characterizing the function of these transcripts, and the processing pathways through which they are synthesized, will be the focus of molecular biology for the next few decades.

Despite technological progress, however, our instruments to monitor the output of transcription — the transcriptome — are extraordinarily primitive. There are no working technologies allowing for the full-length sequencing of all transcripts present in a given condition, and methods for transcript reconstruction and quantification from short sequence reads are only approximate. High-throughput techniques to detect and measure protein abundances are lagging even further behind those for transcript characterization, as current methods based on mass spectrometry suffer from both poor sensitivity and poor specificity. Moreover, although all these methods generally provide a global overview of the cellular abundances of RNAs and proteins, they have little resolution to survey subcellular compartments, as experimental protocols to enrich for subcellular populations of RNAs and proteins tend to be challenging. To fully understand the function of these RNA species, however, knowledge of their subcellular localization is essential.

Furthermore, current technologies are in general unable to interrogate individual cells. Therefore, transcriptome characterizations emerging from microarray or RNASeq surveys are very coarse, providing information of the average behavior of ensembles of tens, hundreds of thousands or even millions of cells. When these ensembles are relatively uniform — as in cultures of immortalized cells — average behaviors may be confidently extrapolated to the behaviors of individual cells. However, if interrogating complex tissues, the behavior of the ensemble may be very different of that of individual cells — which may even be regulated at the individual cell level, such in neuronal tissues. Single-cell protocols exist, but monitoring RNA abundances transcriptome-wide still requires RNA amplification. Because amplification does not proceed uniformly across different RNA species, distortions are expected when the abundances of different transcripts are compared. Although single-cell single-molecule genome-wide protocols may be in sight, they will still require the destruction of the biological material. Following the in vivo fate of the entire transcriptome within an individual cell may remain forever in the realm of science fiction, but technological advances in the near future are likely to significantly increase the resolution at which we can monitor cellular transcriptomes.

REFERENCES

[1] Denoeud F, Kapranov P, Ucla C, Frankish A, Castelo R, Drenkow J, et al. Prominent use of distal 5′ transcription start sites

and discovery of a large number of additional exons in ENCODE regions. Genome Res 2007;17:746—59.
[2] Djebali S, Kapranov P, Foissac S, Lagarde J, Reymond A, Ucla C, et al. Efficient targeted transcript discovery via array-based normalization of RACE libraries. Nat Methods 2008;5:629—35.
[3] Seto AG, Kingston RE, Lau NC. The coming of age for Piwi proteins. Mol Cell 2007;26:603—9.
[4] Kapranov P, Cheng J, Dike S, Nix DA, Duttagupta R, Willingham AT, et al. RNA maps reveal new RNA classes and a possible function for pervasive transcription. Science 2007; 316:1484—8.
[5] Borel C, Gagnebin M, Gehrig C, Kriventseva EV, Zdobnov EM, Antonarakis SE. Mapping of small RNAs in the human ENCODE regions. Am J Hum Genet 2008;82:971—81.
[6] Lu C, Jeong DH, Kulkarni K, Pillay M, Nobuta K, German R, et al. Genome-wide analysis for discovery of rice microRNAs reveals natural antisense microRNAs (nat-miRNAs). Proc Natl Acad Sci U S A 2008;105:4951—6.
[7] Borsani O, Zhu J, Verslues PE, Sunkar R, Zhu JK. Endogenous siRNAs derived from a pair of natural cis-antisense transcripts regulate salt tolerance in Arabidopsis. Cell 2005; 123:1279—91.
[8] Carninci P, Kasukawa T, Katayama S, Gough J, Frith MC, Maeda N, et al. The transcriptional landscape of the mammalian genome. Science 2005;309:1559—63.
[9] Harrow J, Denoeud F, Frankish A, Reymond A, Chen CK, Chrast J, et al. GENCODE: producing a reference annotation for ENCODE. Genome Biol 2006;7(Suppl. 1). S4 1—9.
[10] Guttman M, Amit I, Garber M, French C, Lin MF, Feldser D, et al. Chromatin signature reveals over a thousand highly conserved large non-coding RNAs in mammals. Nature 2009;458:223—7.
[11] Marques AC, Ponting CP. Catalogues of mammalian long non-coding RNAs: modest conservation and incompleteness. Genome Biol 2009;10:R124.
[12] Khalil AM, Guttman M, Huarte M, Garber M, Raj A, Rivea Morales D, et al. Many human large intergenic noncoding RNAs associate with chromatin-modifying complexes and affect gene expression. Proc Natl Acad Sci U S A 2009;106:11667—72.
[13] Wang ET, Sandberg R, Luo S, Khrebtukova I, Zhang L, Mayr C, et al. Alternative isoform regulation in human tissue transcriptomes. Nature 2008;456:470—6.
[14] Khodor YL, Rodriguez J, Abruzzi KC, Tang CH, Marr 2nd MT, Rosbash M. Nascent-seq indicates widespread cotranscriptional pre-mRNA splicing in Drosophila. Genes Dev 2011;25:2502—12.
[15] Luco RF, Pan Q, Tominaga K, Blencowe BJ, Pereira-Smith OM, Misteli T. Regulation of alternative splicing by histone modifications. Science 2010,327(5968).996—1000.
[16] Adams MD, Soares MB, Kerlavage AR, Fields C, Venter JC. Rapid cDNA sequencing (expressed sequence tags) from a directionally cloned human infant brain cDNA library. Nat Genet 1993;4:373—80.
[17] Gerhard DS, Wagner L, Feingold EA, Shenmen CM, Grouse LH, Schuler G, et al. The status, quality, and expansion of the NIH full-length cDNA project: the Mammalian Gene Collection (MGC). Genome Res 2004;14:2121—7.
[18] Bonaldo MF, Lennon G, Soares MB. Normalization and subtraction: two approaches to facilitate gene discovery. Genome Res 1996;6:791—806.

[19] Soares MB, Bonaldo MF, Jelene P, Su L, Lawton L, Efstratiadis A. Construction and characterization of a normalized cDNA library. Proc Natl Acad Sci U S A 1994;91:9228—32.
[20] Wang Z, Gerstein M, Snyder M. RNA-Seq: a revolutionary tool for transcriptomics. Nat Rev Genet 2009;10:57—63.
[21] Mortazavi A, Williams BA, McCue K, Schaeffer L, Wold B. Mapping and quantifying mammalian transcriptomes by RNA-Seq. Nat Methods 2008;5:621—8.
[22] Montgomery SB, Sammeth M, Gutierrez-Arcelus M, Lach RP, Ingle C, Nisbett J, et al. Transcriptome genetics using second generation sequencing in a Caucasian population. Nature 2010; 464:773—7.
[23] Borodovsky M, McIninch J. Recognition of genes in DNA sequence with ambiguities. Biosystems 1993;30:161—71.
[24] Brent MR, Guigo R. Recent advances in gene structure prediction. Curr Opin Struct Biol 2004;14:264—72.
[25] Eddy S. What is a hidden Markov model? Nat Biotechnol. 2004;22:1315—6.
[26] Pedersen JS, Hein J. Gene finding with a hidden Markov model of genome structure and evolution. Bioinformatics 2003;19:219—27.
[27] Holmes I, Bruno WJ. Evolutionary HMMs: a Bayesian approach to multiple alignment. Bioinformatics 2001;17:803—20.
[28] Siepel AC, Haussler D. Combining Phylogenetic and Hidden Markov Models in Biosequence Analysis. In: Miller W, Vingron M, Istrail S, Pevzner P, Waterman MS, editors. RECOMB 2003. ACM Press; 2003. p. 277—87.
[29] Siepel AC, Haussler D. San Diego, CA: RECOMB; 2004.
[30] Kent WJ. BLAT — the BLAST-like alignment tool. Genome Res 2002;12:656—64.
[31] Altschul SF, Gish W, Miller W, Myers EW, Lipman DJ. Basic local alignment search tool. J Mol Biol 1990;215:403—10.
[32] Birney E, Clamp M, Durbin R. GeneWise and Genomewise. Genome Res 2004;14:988—95.
[33] Trapnell C, Williams BA, Pertea G, Mortazavi A, Kwan G, van Baren MJ, et al. Transcript assembly and quantification by RNA-Seq reveals unannotated transcripts and isoform switching during cell differentiation. Nat Biotechnol 2010;28:511—5.
[34] Guttman M, Garber M, Levin JZ, Donaghey J, Robinson J, Adiconis X, et al. Ab initio reconstruction of cell type-specific transcriptomes in mouse reveals the conserved multi-exonic structure of lincRNAs. Nat Biotechnol 2010;28:503—10.
[35] Burset M, Guigo R. Evaluation of gene structure prediction programs. Genomics 1996;34:353—67.
[36] Reese MG, Hartzell G, Harris NL, Ohler U, Abril JF, Lewis SE. Genome annotation assessment in *Drosophila melanogaster*. Genome Res 2000;10:483—501.
[37] Guigo R, Flicek P, Abril J, Reymond A, Lagarde J, Denoeud F, et al. EGASP: the human ENCODE Genome Annotation Assessment Project. Genome biology 2006;7(Suppl. 1). 2—1.
[38] Coghlan A, Fiedler TJ, McKay SJ, Flicek P, Harris TW, Blasiar D, et al. nGASP — the nematode genome annotation assessment project. BMC Bioinformatics 2008;9:549.
[39] Antequera F, Bird A. Number of CpG islands and genes in human and mouse. Proc Natl Acad Sci U S A 1993;90:11995—9.
[40] Fatemi M, Pao MM, Jeong S, Gal-Yam EN, Egger G, Weisenberger DJ, et al. Footprinting of mammalian promoters: use of a CpG DNA methyltransferase revealing nucleosome positions at a single molecule level. Nucleic Acids Res 2005;33:e176.

[41] Lander ES, Linton LM, Birren B, Nusbaum C, Zody MC, Baldwin J, et al. Initial sequencing and analysis of the human genome. Nature 2001;409:860–921.

[42] Venter JC, Adams MD, Myers EW, Li PW, Mural RJ, Sutton GG, et al. The sequence of the human genome. Science 2001;291: 1304–51.

[43] Ewing B, Green P. Analysis of expressed sequence tags indicates 35,000 human genes. Nat Genet 2000;25:232–4.

[44] Liang F, Holt I, Pertea G, Karamycheva S, Salzberg SL, Quackenbush J. Gene index analysis of the human genome estimates approximately 120,000 genes. Nat Genet 2000;25: 239–40.

[45] Hogenesch JB, Ching KA, Batalov S, Su AI, Walker JR, Zhou Y, et al. A comparison of the Celera and Ensembl predicted gene sets reveals little overlap in novel genes. Cell 2001;106:413–5.

[46] Clamp M, Fry B, Kamal M, Xie X, Cuff J, Lin MF, et al. Distinguishing protein-coding and noncoding genes in the human genome. Proc Natl Acad Sci U S A 2007;104:19428–33.

[47] Venter J, Adams M, Myers E, Li P, Mural R, Sutton G, et al. The sequence of the human genome. Science 2001;291:1304–51.

[48] Lander ES, Linton LM, Birren B, Nusbaum C, Zody MC, Baldwin J, et al. Initial sequencing and analysis of the human genome. Nature 2001;409:860–921.

[49] Flicek P, Aken BL, Beal K, Ballester B, Caccamo M, Chen Y, et al. Ensembl 2008. Nucleic acids research 2008;36:D707–14.

[50] Karolchik D, Hinrichs AS, Kent WJ, et al. The UCSC Genome Browser. Current protocols in bioinformatics/editoral board, Andreas D. Baxevanis. [et al.]; 2007. Chapter 1, Unit 1.4.

[51] Pruitt KD, Tatusova T, Maglott DR. NCBI reference sequences (RefSeq): a curated non-redundant sequence database of genomes, transcripts and proteins. Nucleic Acids Res 2007;35:D61–5.

[52] Maglott DR, Katz KS, Sicotte H, Pruitt KD. NCBI's LocusLink and RefSeq. Nucleic Acids Res 2000;28:126–8.

[53] Wilming LG, Gilbert JGR, Howe K, Trevanion S, Hubbard T, Harrow JL. The vertebrate genome annotation (Vega) database. Nucleic Acids Res 2008;36:D753–60.

[54] Birney E, Stamatoyannopoulos JA, Dutta A, Guigo R, Gingeras TR, Margulies EH, et al. Identification and analysis of functional elements in 1% of the human genome by the ENCODE pilot project. Nature 2007;447:799–816.

[55] Myers RM, Stamatoyannopoulos J, Snyder M, Dunham I, Hardison RC, Bernstein BE, et al. A user's guide to the encyclopedia of DNA elements (ENCODE). PLoS Biol 2011;9:e1001046.

[56] TheENCODEProjectConsortium. An integrated encyclopedia of DNA elements in the human genome. Nature (in review) 2012.

[57] The1000GenomesProjectConsortium. A map of human genome variation from population-scale sequencing. Nature 2010;467: 1061–73.

[58] Hudson TJ, Anderson W, Artez A, Barker AD, Bell C, Bernabe RR, et al. International network of cancer genome projects. Nature 2010;464:993–8.

[59] Djebali S, Lagarde J, Kapranov P, Lacroix V, Borel C, Mudge JM, et al. Evidence for transcript networks composed of chimeric RNAs in human cells. PLoS One 2012;7:e28213.

[60] Bian S, T., S.. Functions of noncoding RNAs in neural development and neurological diseases. Mol Neurobiol 2011;44:359–73.

[61] Tress ML, Martelli PL, Frankish A, Reeves GA, Wesselink JJ, Yeats C, et al. The implications of alternative splicing in the ENCODE protein complement. Proc Natl Acad Sci U S A 2007;104:5495–500.

[62] Brockdorff N, Ashworth A, Kay GF, McCabe VM, Norris DP, Cooper PJ, et al. The product of the mouse Xist gene is a 15 kb inactive X-specific transcript containing no conserved ORF and located in the nucleus. Cell 1992;71:515–26.

[63] Gabory A, Ripoche MA, Le Digarcher A, Watrin F, Ziyyat A, Forne T, et al. H19 acts as a trans regulator of the imprinted gene network controlling growth in mice. Development 2009;136:3413–21.

[64] Mancini-Dinardo D, Steele SJ, Levorse JM, Ingram RS, Tilghman SM. Elongation of the Kcnq1ot1 transcript is required for genomic imprinting of neighboring genes. Genes Dev 2006;20: 1268–82.

[65] Ponting CP, Oliver PL, Reik W. Evolution and functions of long noncoding RNAs. Cell 2009;136:629–41.

[66] Mattick JS. The genetic signatures of noncoding RNAs. PLoS Genet 2009;5:e1000459.

[67] Lipovich L, Johnson R, Lin CY. MacroRNA underdogs in a microRNA world: evolutionary, regulatory, and biomedical significance of mammalian long non-protein-coding RNA. Biochim Biophys Acta 2010;1799:597–615.

[68] Mattick JS, Mehler MF. RNA editing, DNA recoding and the evolution of human cognition. Trends Neurosci 2008;31:227–33.

[69] Yu W, Gius D, Onyango P, Muldoon-Jacobs K, Karp J, Feinberg AP, et al. Epigenetic silencing of tumour suppressor gene p15 by its antisense RNA. Nature 2008;451:202–6.

[70] Pasmant E, Sabbagh A, Vidaud M, Bieche I. ANRIL, a long, noncoding RNA, is an unexpected major hotspot in GWAS. FASEB J 2011;25:444–8.

[71] Huarte M, Guttman M, Feldser D, Garber M, Koziol MJ, Kenzelmann-Broz D, et al. A large intergenic noncoding RNA induced by p53 mediates global gene repression in the p53 response. Cell 2010;142:409–19.

[72] Zhou Y, Zhong Y, Wang Y, Zhang X, Batista DL, Gejman R, et al. Activation of p53 by MEG3 non-coding RNA. J Biol Chem 2007;282:24731–42.

[73] Daughters RS, Tuttle DL, Gao W, Ikeda Y, Moseley ML, Ebner TJ, et al. RNA gain-of-function in spinocerebellar ataxia type 8. PLoS Genet 2009;5:e1000600.

[74] Faghihi MA, Modarresi F, Khalil AM, Wood DE, Sahagan BG, Morgan TE, et al. Expression of a noncoding RNA is elevated in Alzheimer's disease and drives rapid feed-forward regulation of beta-secretase. Nat Med 2008;14:723–30.

[75] Maeda N, Kasukawa T, Oyama R, Gough J, Frith M, Engstrom PG, et al. Transcript annotation in FANTOM3: mouse gene catalog based on physical cDNAs. PLoS Genet 2006;2:e62.

[76] Jia H, Osak M, Bogu GK, Stanton LW, Johnson R, Lipovich L. Genome-wide computational identification and manual annotation of human long noncoding RNA genes. RNA 2010;16:1478–87.

[77] Orom UA, Derrien T, Beringer M, Gumireddy K, Gardini A, Bussotti G, et al. Long noncoding RNAs with enhancer-like function in human cells. Cell 2010;143:46–58.

[78] Cabili MN, Trapnell C, Goff L, Koziol M, Tazon-Vega B, Regev A, et al. Integrative annotation of human large intergenic noncoding RNAs reveals global properties and specific subclasses. Genes Dev 2011;25:1915–27.

[79] Derrien T, Johnson R, Bussoti G, Tanzer A, Djebali S, Tilgner H, et al. The GENCODE v7 catalogue of human long non-coding

RNAs: analysis of their gene structure, evolution and expression. Genome Res (in revision) 2012.
[80] TheENCODERNAConsortium. Landscape of transcription in human cells. Nature (in revision) 2012.
[81] Gonzalez-Porta M, Calvo M, Sammeth M, Guigo R. Estimation of alternative splicing variability in human populations. Genome Res 2012.
[82] Pickrell JK, Marioni JC, Pai AA, Degner JF, Engelhardt BE, Nkadori E, et al. Understanding mechanisms underlying human gene expression variation with RNA sequencing. Nature 2010;464: 768−72.
[83] Harrow J, Nagy A, Reymond A, Alioto T, Patthy L, Antonarakis SE, et al. Identifying protein-coding genes in genomic sequences. Genome Biol 2009;10:201.

Section II

Network Properties of Biological Systems

Chapter 3

Interactome Networks

Anne-Ruxandra Carvunis[1,2], Frederick P. Roth[1,3], Michael A. Calderwood[1,2], Michael E. Cusick[1,2], Giulio Superti-Furga[4] and Marc Vidal[1,2]

[1]*Center for Cancer Systems Biology (CCSB) and Department of Cancer Biology, Dana-Farber Cancer Institute, Boston, MA 02215, USA,* [2]*Department of Genetics, Harvard Medical School, Boston, MA 02115, USA,* [3]*Donnelly Centre for Cellular & Biomolecular Research, University of Toronto, Toronto, Ontario M5S-3E1, Canada, & Samuel Lunenfeld Research Institute, Mt. Sinai Hospital, Toronto, Ontario M5G-1X5, Canada,* [4]*Research Center for Molecular Medicine of the Austrian Academy of Sciences, 1090 Vienna, Austria*

Chapter Outline

Introduction	45
Life Requires Systems	45
Cells as Interactome Networks	46
Interactome Networks and Genotype–Phenotype Relationships	47
Mapping and Modeling Interactome Networks	47
Towards a Reference Protein–Protein Interactome Map	48
Strategies for Large-Scale Protein–Protein Interactome Mapping	48
Large-Scale Binary Interactome Mapping	49
Large-Scale Co-Complex Interactome Mapping	50
Drawing Inferences from Interactome Networks	51
Refining and Extending Interactome Network Models	51
Predicting Gene Functions, Phenotypes and Disease Associations	53
Assigning Functions to Individual Interactions, Protein Complexes and Network Motifs	54
Towards Dynamic Interactomes	55
Towards Cell-Type and Condition-Specific Interactomes	55
Evolutionary Dynamics of Protein–Protein Interactome Networks	56
Concluding Remarks	57
Acknowledgements	58
References	58

INTRODUCTION

Life Requires Systems

What is Life? The answer to the question posed by Schrödinger in a short but incisive book published in 1944 remains elusive more than seven decades later. Perhaps a less ambitious, but more pragmatic question could be: what does Life require? Biologists agree on at least four fundamental requirements, among which three are palpable and easily demonstrable, and a fourth is more intangible (Figure 3.1). First, Life requires *chemistry*. Biomolecules, including metabolites, proteins and nucleic acids, mediate the most elementary functions of biology. Life also requires *genes* to encode and 'reproduce' biomolecules. For most organisms, *cells* provide the fundamental medium in which biological processes take place. The fourth requirement is *evolution* by natural selection. Classically, these 'four great ideas of biology' [1] have constituted the main intellectual framework around which biologists formulate biological questions, design experiments, interpret data, train younger generations of scientists and attempt to design new therapeutic strategies.

The next question then should be: even if we fully understood each of these four basic requirements of biology, would we be anywhere near a complete understanding of how Life works? Would we be able to fully explain genotype–phenotype relationships? Would we be able to fully predict biological behaviors? How close would we be to curing or alleviating suffering from human diseases? It is becoming clear that even if we knew everything there is to know about the four currently accepted requirements of biology, the answer to 'What Life is' would remain elusive.

The main reason is that biomolecules do not function in isolation, nor do cells, organs or organisms, or even ecosystems and sociological groups. Rather, biological entities are involved in intricate and dynamic interactions, thereby forming 'complex systems'. In the last decade, novel biological questions and answers have surfaced, or resurfaced, pointing to *systems* as a fifth fundamental

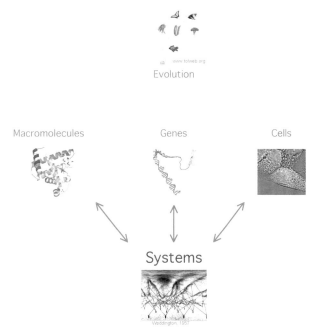

FIGURE 3.1 Systems as a fifth requirement for Life.

requirement for Life [2]. Although conceptual, systems may turn out to be as crucial to biology as chemistry, genes, cells or evolution (Figure 3.1).

Cells as Interactome Networks

The study of biological systems, or 'systems biology', originated more than half a century ago, when a few pioneers initially formulated a theoretical framework according to which multiscale dynamic complex systems formed by interacting biomolecules could underlie cellular behavior. To explain cellular differentiation, Delbrück hypothesized the existence of positive feedback circuits required for 'bistability', a model in which systems would remain stably activated after having been turned on, and conversely, remain steadily inactive once turned off [3]. Empirical evidence for feedback regulation in biology first emerged in the 1950s. The Umbarger and Pardee groups uncovered enzymatic feedback inhibition [4,5], and Novick and Weiner described the positive feedback circuit regulating the *lac* operon [6]. Monod and Jacob subsequently proposed how negative feedback circuits could account for homeostasis and other oscillatory phenomena observed in many biological processes [7]. These teleonomic arguments were later formalized by René Thomas and others in terms of requirements for cellular and whole organism differentiation based on positive and negative feedback circuits of regulation, using Boolean modeling as powerful simplifications of cellular systems [8] (see Chapter 10).

Equally enlightening theoretical systems properties were imagined beyond small-scale regulatory mechanisms composed of just a few molecules. Waddington introduced the metaphor of 'epigenetic landscape', whereby cells respond to genetic, developmental and environmental cues by following paths across a landscape containing peaks and valleys dictated by interacting genes and gene products [9]. This powerful idea, together with theoretical models of 'randomly constructed genetic nets' by Kauffman [10], suggested that a cellular system could be described in terms of 'states' resulting from particular combinations of genes, gene products, or metabolites, all considered either active or inactive at any given time. Complex wiring diagrams of functional and logical interconnectivity between biomolecules and genes acting upon each other could be imagined to depict how systems 'travel' from state to state over time throughout a 'state space' determined by intricate, sophisticated combinations of genotype, systems properties and environmental conditions. These concepts, elaborated at a time when the molecular components of biology were poorly described, remained largely ignored by molecular biologists until recently (see Chapter 15).

Over the past two decades, scientific knowledge of the biomolecular components of biology has dramatically increased. In particular, sequencing and bioinformatics have allowed prediction of coding and non-coding gene products at genome scale. Transcriptome sequencing approaches have revealed the existence of transcripts that had escaped prediction and which often remain of unknown function (see Chapter 2). Additionally, the list of known molecular components of cellular systems, including nucleic acids, gene products and metabolites, is lengthening and becoming increasingly detailed. With these advances came a humbling realization, best summed up as 'too much data, too few drugs' (see Chapter 8). It has become clearer than ever that knowing everything there is to know about each biomolecule in the cell is not sufficient to predict how the cell will react as a whole to particular external or internal perturbations.

Functional interactions, perhaps more so than individual components, mediate the fundamental requirements of the cell. Consequently, one needs to consider biological phenomena as the product of ensembles of interacting components with emergent properties that go beyond those of their individual components considered in isolation. One needs to step back and measure, model, and eventually perturb nearly all functional interactions between cellular components to fully understand how cellular systems work. In analogy to the word genome, the union of all interactions between all cellular components is termed the 'interactome'. Our working hypothesis is that interactomes exhibit local and global properties that relate to biology in general, and to genotype—phenotype relationships in particular.

Interactome Networks and Genotype−Phenotype Relationships

Since drafts of a composite reference human genome sequence were released 10 years ago, powerful technological developments, such as next-generation sequencing, have started a true revolution in genomics [11−14]. With time, most human genotypic variations will be described, together with large numbers of phenotypic associations. Unfortunately, such knowledge cannot translate directly into new mechanistic understanding or therapeutic strategies, in part because the 'one-gene/one-enzyme/one-function' concept developed by Beadle and Tatum oversimplifies genotype−phenotype relationships [15]. In fact, Beadle and Tatum themselves state so in the introduction of their groundbreaking paper that initiated reductionism in molecular biology: 'An organism consists essentially of an integrated system of chemical reactions controlled in some manner by genes. Since the components of such systems are likely to be interrelated in complex ways, it would appear that there must exist orders of directedness of gene control ranging from simple one-to-one relations to relations of great complexity.'

So-called complex traits provide the most compelling evidence of complexity between genotypes and phenotypes in human disease. Genome-wide association studies have revealed many more contributing loci than originally anticipated, with some loci contributing as little as a few percent to the heritability of the phenotype(s) of interest. Simple Mendelian traits are not immune to complex genotype−phenotype relationships either. Incomplete penetrance, variable expressivity, differences in age of onset, and modifier mutations are more frequent than generally appreciated. These discrepancies in the one-gene/one-function model appear across all phyla. In worms, for example, where self-fertilization is possible and growth conditions are easily controllable, it is not uncommon to observe that a significant proportion of mutant animals exhibit a near wild-type phenotype [16,17] (see Chapter 19).

Genome-wide functional genomic and proteomic experiments also point to greater complexity than anticipated, leading one to ask fundamental questions about gene function and evolution. How could a linear view of gene function account for the seemingly small proportions of essential genes [18]? How do genes become essential during evolution in the first place? How to explain that genes that are genetically required for particular biologic processes are not necessarily transcriptionally regulated during those same processes [19]? How to account for increasing reports of 'protein moonlighting', where specific gene products appear to be necessary for multiple biological processes [20−22]? We argue that viewing and modeling cells as interactome networks will help unravel

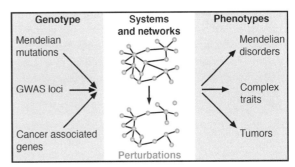

FIGURE 3.2 Interactome networks and genotype−phenotype relationships.

the complexity of genotype−phenotype relationships, including susceptibility to human disease (Figure 3.2).

Mapping and Modeling Interactome Networks

In the path towards deciphering the mechanisms underlying genotype−phenotype relationships, it helps to abstractly simplify the complexity of interactomes by modeling. Interactomes can be effectively modeled as network representations of biological relationships among biomolecules. This abstraction converts a complex web of biological relationships into a graph, allowing the application of intuition and mathematical concepts of graph theory.

The power of graph theory in revealing emergent properties of complex systems is exemplified by a social science experiment of the 1960s. Stanley Milgram attempted to follow the path of letters sent through the mail in order to measure the average number of 'degrees of separation' between people, thereby providing a network representation of human society. He famously found that humans are connected on average at a distance of six degrees from each other, and our vision of the human population on earth immediately became that of a 'small-world' [23]. From politics to social media to modern journalism, Milgram's discovery still resonates today, and probably will for years to come.

In a network representation of interactomes, nodes represent biomolecules and an edge between two nodes indicates a biological relationship between the corresponding biomolecules. In the cell, multiple types of biomolecule, e.g., genes, proteins, RNAs, regulatory elements, or small-molecule metabolites, can be connected by multiple types of physical or functional biological relationship. These can be combined into 'multicolor' interactome network representations, where node colors represent biomolecule types and edge colors represent biological relationship types, refining complex biological processes [24]. Binding of transcription factors to DNA regulatory elements (physical relationship), regulation of target genes by transcription factors or micro-RNAs (direct functional relationship), and similarity of expression profiles of genes across multiple conditions (profile-based functional relationship) form interrelated interactome networks. Study of these interactome networks, individually or together, is contributing key insights into the cellular control of gene expression (see Chapters 2 and 4).

Material flow, such as in metabolic reactions, and information flow, such as in transduction pathways, can be represented mathematically by edge direction (see Chapters 4, 5 and 11), while edge thickness, or weight, can symbolize the relative strength of biological relationships. Additionally, interactome network models can incorporate logic (see Chapters 10 and 11) or dynamics (see Chapters 12, 13, 14 and 16). Eventually, the aim is to understand how different interactomes are integrated together to form the cellular systems that we believe underlie genotype−phenotype relationships.

For this aim to be reached, complete single-color interactome network maps first need to be assembled. Physical interactions between proteins, or protein interactions, constitute the fundamental backbone of the cell and are instrumental for most biological processes, including signaling, differentiation and cell fate determination. This chapter describes the mapping and modeling of protein−protein interactome networks, where edges connect pairs of proteins that physically associate with one another directly or indirectly.

TOWARDS A REFERENCE PROTEIN−PROTEIN INTERACTOME MAP

Most individual proteins execute their biological functions by interacting with one or several other proteins. Protein interactions can form large protein complexes such as the proteasome, in which ~50 protein subunits act together to degrade other proteins and play a key role in cell protein homeostasis. The existence of such molecular machines, performing functions that no single protein can assume, demonstrates that protein−protein interactomes exhibit emergent properties beyond the sum of all individual protein interactions.

There is no such thing as a typical protein interaction. Protein interactions occur *in vivo* with a wide range of dissociation constants and dynamics. Proteins associated by strong and permanent interactions with low dissociation constants tend to form protein complexes. Protein interactions may also be simultaneously strong and transient, when controlled by the expression level of either interacting partner using 'just-in-time' assembly, by a change in subcellular location of one protein or the other, or by conformation changes induced by post-translational modification. For example, GTP hydrolysis controls the interaction between the α and β subunits of G-proteins, which in turn rapidly switches entire signaling pathways on or off [25]. Many protein interactions are weak and transient with high dissociation constants, as are associations between membrane receptors and extracellular matrix proteins that assist cellular motility [25].

Forthcoming models of protein−protein interactomes will undoubtedly involve sophisticated network representations integrating weighted and dynamic protein interactions [26]. For protein interactions such as those involved in signaling, interactome network models can also incorporate edge direction. Many protein interactions, such as subunit−subunit interactions within protein complexes, are best described with undirected edges [27−29]. It is not possible yet to assemble a proteome-scale interactome network model that integrates strength, dynamics and direction of edges because available technology is only beginning to allow experimental measurements of such interaction properties. Even a catalog of all possible protein interactions has not yet been compiled for any single species. Today's challenge lies in obtaining nearly complete but static, undirected and unweighted reference protein−protein interactome maps.

Strategies for Large-Scale Protein−Protein Interactome Mapping

Three fundamentally different but complementary strategies have been deployed towards this goal: i) curation of protein interaction data already available in the scientific literature [30]; ii) computational predictions of protein interactions based on available orthogonal information, such as sequence similarity or the co-presence of genes in sequenced genomes [31]; and iii) systematic, unbiased high-throughput experimental mapping strategies applied at the scale of whole genomes or proteomes [32].

Literature-curated interactome maps present the advantage of using already available, experimentally derived information, but are limited by the inherently variable quality of the curation process [33−35]. A randomly chosen set of literature-curated protein interactions supported by a single publication was shown to be approximately three times less reproducible than a reference set of manually curated protein interactions supported by multiple publications [36]. Another caveat of literature-curated interactome maps is that they mostly derive from hypothesis-driven research, which often focuses on a few proteins deemed to be scientifically interesting [37]. Some 'star proteins', such as the cancer-associated product of the *TP53* gene [38], are interrogated for protein interactions much more often than other proteins, resulting in an artificial increase of their apparent connectivity relative to other proteins. For these reasons, literature-curated maps cannot be viewed as representative samples of the underlying interactome, and inferring systems-level properties from literature-curated protein–protein interactome maps can be misleading [39]. Nevertheless, literature-curated protein–protein interactome maps are instrumental in deriving hypotheses about focused biological mechanisms.

Computational predictions have the advantage of being applicable at genome or proteome scale for only a moderate cost. We discuss the numerous computational strategies that have been designed to predict protein interactions in the section of this chapter entitled 'Drawing inferences from interactome networks'. In brief, computational predictions apply 'rules' learned from current knowledge to infer new protein interactions. Albeit potent, this approach is also intrinsically limiting since the rules governing biological systems in general, and protein interactions in particular, remain largely undiscovered. Therefore, predicted protein–protein interactome maps, like literature-curated interactome maps, should be handled with caution when modeling biological systems.

High-throughput experimental interactome mapping approaches attempt to describe unbiased, systematic and well-controlled biophysical interactions. Two complementary approaches are currently in widespread use for high-throughput experimental interactome mapping (Figure 3.3): i) testing all combinations of protein pairs encoded by a given genome to find all binary protein interactions that can take place among them and uncover the 'binary interactome'; and ii) interrogating *in vivo* protein complexes in one or several cell line(s) or tissue(s) to expose the 'co-complex interactome'. Binary interactome maps contain mostly direct physical interactions, an unknown proportion of which may never take place *in vivo* despite being biophysically true. On the other hand, co-complex interactome datasets are composed of many indirect, and some direct, associations that mostly do take place *in vivo*.

Large-Scale Binary Interactome Mapping

The technologies that enabled large-scale binary interactome mapping were first developed in the 1990s by a few groups [40–45]. The following years saw significant progress towards assembling binary protein–protein interactome maps for model organisms such as the yeast *Saccharomyces cerevisiae* [39,46–49], the worm *Caenorhabditis elegans* [50–53], the fly *Drosophila melanogaster* [54], and most recently the plant *Arabidopsis thaliana* [55–57]. Similar efforts have been deployed to map the human binary interactome [36,58–62].

Large-scale binary interactome mapping is amenable to only a few existing experimental assays [47,48,63] and is carried out primarily by ever-improving variants on the yeast two-hybrid (Y2H) system [64,65]. The Y2H system is based on the reconstitution of a yeast transcription factor through the expression of two hybrid proteins, one fusing the DNA-binding (DB) domain to a protein X (DB-X) and the other fusing the activation domain (AD) to a protein Y (AD-Y) [65]. In the last 20 years the technique has been streamlined to increase throughput and quality controlled to avoid foreseeable artifacts [32,43,45,53,66–69]. Today, Y2H can interrogate hundreds of millions of protein pairs for binary interactions, in a manner that is both highly efficient and highly reliable.

The contemporary Y2H-based high-throughput binary interactome mapping pipeline consists of two essential stages: primary screening and secondary verification [64,70,71]. Large collections of cloned genes are transferred into DB-X and AD-Y expression vectors, then efficiently screened using either a pooling or a pairwise strategy [49,51,64,72–74]. All protein pairs identified in

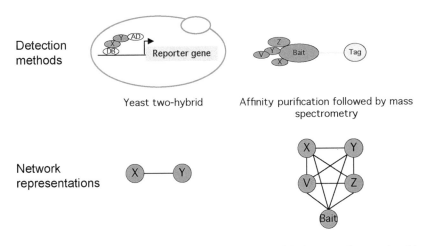

FIGURE 3.3 Binary and co-complex protein-interaction detection methods and network representations.

the primary screen are re-tested independently by Y2H using fresh yeast cells (verification). In our most recent Y2H-based binary interactome mapping efforts, DB-X/AD-Y pairs that score positive in at least three out of four replicates during the verification step were considered high-quality verified Y2H interactions [55,57]. Further sequencing can be undertaken to confirm the identity of the reported protein pairs [55]. Ultimately, this pipeline can systematically generate large numbers of highly reproducible binary protein interactions.

Reproducibility, however, does not necessarily guarantee high dataset quality. Technical artifacts could result in protein pairs appearing reproducibly positive despite actually being false positives. To obtain a quantitative estimate of the quality of a protein–protein interactome dataset, an integrated empirical framework for quality control of interactome mapping was proposed in 2009 [36]. At the heart of this endeavor was the recognition that high-throughput interactome mapping needed to be rigorously calibrated, like any well-controlled reliable small-scale experiment. Using this framework, protein–protein interactome maps generated with the mapping pipeline described here were shown to have high precision (80–100% of reported protein pairs are true positives) but low sensitivity (~10–15% of all interrogated true positive interactions are captured in the experiment) [36,39,52,55]. Because of this low sensitivity, binary interactome maps generated so far represent small fractions of the underlying true interactomes. This explains why only a marginal number of protein interactions are found in multiple binary interactome maps assembled independently for the same organism [36,39,55].

How far along is the journey towards a complete binary protein–protein interactome reference map? The answer requires an estimation of the size of such a reference map for any given species. Many statistical methods have been designed to this end, often based on dataset overlap and hypergeometric distributions [36,75–82]. Mapping of the binary interactome of the model organism *S. cerevisiae* is estimated to be the closest to completion, with ~6–30% coverage already obtained (~2900 binary protein interactions of demonstrated high technical quality detected, out of an estimated total of 10 000–45 000, assuming one splicing isoform per gene) [39,46,49]. However, most of the task remains to be accomplished.

How can this daunting challenge be overcome? Inspiration is drawn from the history of genome sequencing, which underwent a disruptive shift in the late 1990s. Like sequencing at that time, Y2H-based mapping is currently seeing more efficient automation, stricter quality control and innovative technology development which together are increasing productivity while reducing cost [36,62,83]. It is unlikely that Y2H-based mapping alone will be sufficient to complete a comprehensive reference binary protein–protein interactome map. No single interaction assay may ever be capable of doing so [84,85]. Individual protein interaction assays seem optimized for the detection of a certain subtype of binary interactions, although the biochemical and structural biases of these assays remain poorly understood. Intuitively, interactions involving membrane proteins would be expected to perform better in the cell membrane environment of the split-ubiquitin system than in the nuclear environment of the Y2H system [65,47]. It will therefore be necessary to join forces and use multiple assays to fully map the reference binary interactome of any organism. We are confident that this challenge can be overcome within the next decade.

Large-Scale Co-Complex Interactome Mapping

To fully map the reference interactome, it is operationally helpful to go beyond binary protein interactions and identify protein complexes within cells [86]. Protein complexes typically contain five to six different proteins, within a wide range from two to hundreds in a variety of stoichiometries [87]. The concentration and binding affinity of the protein subunits determine complex assembly according to the law of mass action [88]. Two proteins in isolation may have only weak or no propensity to form a binary interaction. Owing to cooperative and allosteric effects, a third protein may have a high affinity to both simultaneously, so that the resulting protein complex is considerably more stable than the sum of its component affinities [89]. Hence, even if a reference binary protein–protein interactome had been fully mapped, co-complex interactome network maps would still provide novel protein interactions and bring a fundamentally different view of interactome organization. The characterization of entire protein complexes, as they assemble in cells, is a necessary route to gather information on gene function and biological systems [90–94].

The most common methodologies currently used for the mapping of co-complex interactomes rely on protein complex purification followed by identification of constituent proteins by mass spectrometry. These experiments necessitate a trade-off between throughput, reproducibility and physiological setting. Cellular proteins range in their abundance up to seven orders of magnitude in humans [95] and five in yeast [96]. Protein complexes therefore need to be purified from the soup of cellular extracts without losing too many components, while at the same time avoiding those proteins that are extremely abundant and co-purify artificially [97–100]. A fraction of protein complexes may consist of dedicated elements, but most complexes also include abundant proteins that participate in several other complexes, such as chaperone proteins. Purification strategies tailored for individual complexes typically make use of high-affinity antibodies directed against a specific complex member, or use other affinity matrices such as DNA, RNA, metabolites, or drugs, inspired by the specific biochemical properties of the complex. These approaches have the important advantage of targeting endogenous, natural forms of the complexes, but they are not readily amenable to proteome scale.

Proteome-scale co-complex interactome mapping employs a variety of strategies that attach epitope tags to bait proteins. These include DNA engineering as well as post-translational protein engineering [101–106]. The purified protein complexes are then systematically treated with proteases to release peptides, which then are fractionated by one- or two-dimensional liquid chromatography. Amino acid sequences are then imputed based on the mass and charge of the resulting peptides using mass spectrometry readouts. If applied rigorously and with attention to statistical significance, proteins containing these peptide sequences can be derived with a low false discovery rate. Proteins may also go undetected for a number of reasons, leading to false negatives in co-complex interactome maps. Quantitative estimation of dataset quality using a framework analogous to the one implemented for binary interactome maps [36] has now been implemented once for a D. melanogaster co-complex interactome map, which demonstrated high sensitivity [107].

How should we interpret the long lists of proteins that typically are the readouts of the mass spectrometry analyses of co-purified proteins? All successful bioinformatics approaches to assign co-complex memberships to co-purified proteins rely on network analyses, considering each protein as a node and each co-purification relationship as an edge. Algorithms can isolate subnetworks that are highly interconnected or completely interconnected, and then compute affiliation to one subnetwork compared to overall frequency to determine the most likely co-complex associations [91,108–110]. High-quality datasets are obtained from multiple redundant purifications over a single search space, which may encompass whole genomes, as was done for S. cerevisiae, Escherichia coli, Mycoplasma pneumonia, and D. melanogaster [87,91,104,107,111,112], or selected pathways and subnetworks as was done for human [113–118]. The more redundant the dataset, the more reliable complex prediction will be, leading to ever finer granularity of the resolution of the map.

When two proteins belong to the same protein complex they may not necessarily be in direct physical contact (Figure 3.3). Hence, edges in a network representation of co-complex interactomes have a very different meaning than edges in a network representation of binary interactomes. Most literature-curation databases are struggling to design the appropriate infrastructure that will allow users to distinguish intuitively between these two types of edge [119]. Curation, storage, representation and analysis of co-complex interactome data are key challenges in computational systems biology.

Regardless, the emerging picture of mass spectrometry-derived co-complex interactome maps is that of a modular organization. Protein complexes appear to assemble from a limited number of core modules, with small sub-complexes as well as individual proteins binding to the core modules and to each other. Modular organization creates the possibility of achieving functional diversity through combinatorial effects while maintaining highly interdependent central parts of molecular machines invariable [120–123].

DRAWING INFERENCES FROM INTERACTOME NETWORKS

Combining analyses of network topology with exogenous data integration can help make sense of complex systems. This ability is well illustrated in a famous study of high-school friendships across the USA [124]. In the topological structure of networks where nodes are students and edges are friendships, communities of tightly linked high-school friends emerge (see Chapter 9). When nodes in these topological communities are colored with ethnicity information, the extent of ethnic segregation in each high school is revealed (Figure 3.4). Similarly, binary and co-complex protein–protein interactome network maps provide 'scaffold' information about cellular systems. When interactome maps are analyzed in terms of topology and integrated with orthogonal functional information, the resulting knowledge allows investigators to imagine novel hypotheses and answer basic questions of biology (Figure 3.4) [125,126].

Refining and Extending Interactome Network Models

A major aim of the analysis of interactome network maps is to obtain better representations of the underlying interactome itself, since available maps are imperfect and incomplete. Given topological and exogenous biological data, which proteins are most likely to interact with a given protein of interest for which few or no protein interactions have been described? Which binary interactions and co-complex associations are the most reliable, and therefore worthy of mechanistic follow-up?

When there are multiple sources of experimental evidence supporting a particular protein interaction, the evidence can be combined to generate a confidence score for this interaction. This integration can be restricted to

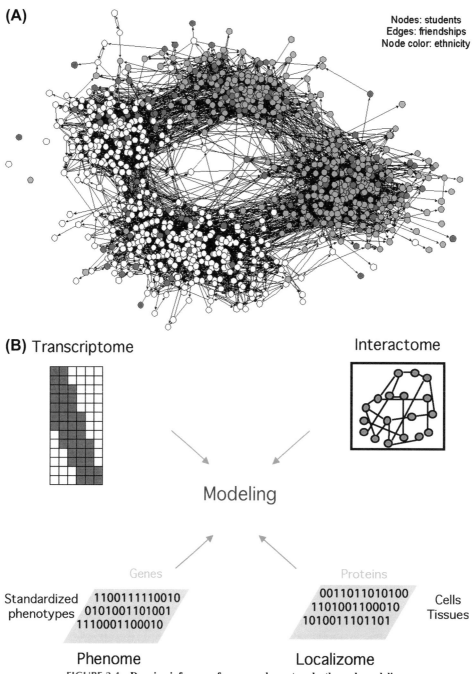

FIGURE 3.4 Drawing inferences from complex networks through modeling.

experimental evidence that directly relates to the protein interaction [84], or can include a broader range of biological relationships. Methods are emerging to score the likelihood that a given protein pair will physically interact, given all the other known single-color relationships about that protein pair. Prediction strategies that model single-color relationships as being independent of one another have proved worthwhile [54,127,128]. More sophisticated approaches that explicitly model the interdependency between multiple interaction types further improve the quality of interaction predictions [128−133].

With such confidence scores, interactome networks can be represented as probabilistic networks in which each edge is assigned a weight representing the posterior probability that the edge is real. Sampling many 'deterministic' subgraphs from the probabilistic network measures the fraction of sampled networks containing a path between a known protein complex and a candidate protein member of that complex, thus predicting new members of partially defined protein complexes [133,134].

The topology of interactome networks can be exploited to predict novel protein interactions. Network motifs are patterns of interconnection involving more than two nodes [135,136]. Triangles and larger densely connected subgraphs are more frequent in most protein–protein interactome networks than would be expected by chance, in turn rendering candidate protein interactions that complete many new triangles in the network more likely to be true interactions [137,138]. Accordingly, identification of densely connected subgraphs in an interactome network can help identify protein complexes [108,139–142]. Other methods exploiting network topology parameters have proved useful in predicting physical interaction networks of other organisms [54].

A promising strategy for interaction prediction is to produce multicolor network motifs derived by integrating protein–protein interactome networks with genetic interaction networks or phenotypic profile similarity networks (see Chapters 5 and 6). Protein interactions tend to connect genes with related phenotypes, as was first discerned in small-scale studies of protein interactions, and later demonstrated for large-scale interactome maps and systematic phenotyping data [50,143–147]. It follows that genetic interaction profiles can be used to predict protein complexes [18,143,146,148–150]. On the basis of the strength of the physical and genetic interactions of a particular protein pair it is possible to assess the likelihood of that protein pair operating either within a protein complex or connecting two functionally related protein complexes [151]. The predictive power of these integrative approaches lies in the systems organization through which interactomes underlie genotype–phenotype relationships.

Predicting Gene Functions, Phenotypes and Disease Associations

Early in the implementation of Y2H, Raf kinase was imputed as an oncogene based on its specific interaction with H-Ras [152]. This type of reasoning is behind the principles of 'guilt-by-association' and 'guilt-by-profiling', whereby a functional annotation can be transferred from one gene/protein to another 'across' biological relationships, or 'across' profiles of biological relationships [153–157]. Function prediction methods based on these principles either make assumptions about the independence of evidence types, or model the interdependencies between edge types [158]. Protein A, of unknown function, can be said with some likelihood to be involved in the same biological process as protein B, of known function, if A and B belong to the same protein complex, or if their profile of membership in modules identified from the interactome network are similar [141]. Protein interactions, mostly transferred by orthology from human, allow the most precise predictions of gene function in mouse genes among several data types [159].

Topological modules in interactome networks most likely correspond to specific biological processes or functions [160]. It follows that identifying modules containing genes/proteins of both known and unknown function can help assign function to uncharacterized genes/proteins. Combining topological modules with other genomic or proteomic data in multicolor networks brings about more biologically coherent units [128,161–165].

That protein interactions mediate protein functions, and that protein interactions tend to connect genes/proteins with related phenotypes just as they tend to connect genes/proteins with related functions, suggests that protein interactions can be used to predict new disease genes. Mutations causative for ataxia, a neurodegenerative disorder, affect proteins that share interacting partners. A subset of these shared partners have been found to be associated with neurodegeneration in animal models [59,166]. The ability of interactome maps to highlight new candidate disease genes and disease modifier genes had been anticipated in the early large-scale binary interactome maps [60,61]. With large-scale interactome maps available for human, various computational efforts systematically prioritize potential human disease genes based on the patterns of protein interactions [93,167–171]. Current efforts to predict human gene function and disease phenotype are now striving to combine several orthogonal large-scale genomic and proteomic data types [172,173].

Progress in this area will depend increasingly on efforts to establish benchmark data to allow rigorous comparisons among the evolving methods. Benchmarking has happened preliminarily for gene function prediction [159]. For prediction of phenotype or disease most current methods rely upon a handful of known 'training examples' — small sets of genes known to be associated with the phenotype or disease. This strategy has its worth, but eventually methods that can predict disease genes in the absence of known examples will also be needed. Genome-wide association (GWA) studies provide an emerging example where predictions can be attempted without training examples. Although GWA studies serve to identify a genomic locus associated with a disease, they often cannot pinpoint which single gene of several or many resident within the locus is the actual disease gene. Where multiple GWA loci are linked to a single disorder or trait, the subset of genes that exhibit between-locus protein interactions, or other types of biological relationships, may be the most likely to be the causal disease genes [168,169,174].

It may be possible to predict phenotypes and prioritize disease genes based on local network topology alone. One particularly informative and well-studied network topology parameter is node connectivity, or 'degree'. Protein−protein interactome networks follow a power-law distribution, where most nodes have a low connectivity and a few nodes, the network 'hubs', have a high connectivity (see Chapter 9). This 'scale-free' degree distribution is also observed in many other networks, the internet being a particularly noteworthy example, and has important consequences for network robustness. The overall structure of a scale-free network is hardly affected by the removal of random nodes, but is highly sensitive to removal of hubs [175]. This scale-free behavior has clear applications in the design of electrical power grids, but what does it mean for protein−protein interactome networks?

About a decade ago, it was reported that hubs in the protein−protein interactome of *S. cerevisiae* are preferentially essential, meaning that knockouts of the corresponding genes tend to be unviable [176]. This initial observation was biased by the inclusion of literature-curated interactome maps, where 'star' proteins that have an artificially high connectivity also tend to be essential [39,176]. There is instead unbiased evidence of a correlation between degree and essentiality in co-complex interactome networks, but not in binary interactome networks [39,177]. Deeper examination clearly demonstrated a correlation between connectivity in an interactome network map and functional pleiotropy in *S. cerevisiae* [39]. Thus, the connectivity of a protein does relay information about the phenotype of its corresponding gene. Together with other topological information, network connectivity may one day be used to predict new disease genes, as suggested by the observation that proteins associated with cancer are preferentially hubs in the human interactome [178].

Assigning Functions to Individual Interactions, Protein Complexes and Network Motifs

Functional genomics experiments and function prediction algorithms are typically designed to uncover the biological roles of genes and gene products. We argue that if interactome networks underlie genotype−phenotype relationships, then edges (protein interactions) should be associated with functions and phenotypes just as nodes (proteins) are.

Fanconi anemia (FA) is a rare chromosome instability disorder associated with congenital defects and susceptibility to cancer. Of the 13 genes genetically associated with FA, seven encode members of a core FA protein complex [179,180]. This example and others show high interconnectivity between proteins associated to a particular disease, which suggests that the disease phenotype may result from genetically induced perturbation of protein interactions [59,166,178,181−183]. Similarly, when a single gene is linked to multiple disorders, it often seems to be because distinct mutations of this gene affect specific individual interactions with different partners [94,184]. It follows that looking for 'disease interactions' rather than for 'disease genes' should assist the delineation of disease mechanisms and aid efforts to rationally interrupt disease progression.

From an interactome network perspective the effects of genetic variations are traditionally modeled as complete loss of gene products ('node removal'). While such interpretations are generally suitable for nonsense or frameshift mutations occurring early in the protein, large insertions or deletions, or complete gene knockouts, the node removal model may not readily apply to truncations that preserve specific autonomous protein domains, or to single amino acid substitutions. Such genetic variations could instead lead to perturbations of specific interactions ('edge removal'), or 'edgetic perturbations' [184,185].

The systematic isolation of genetic variants associated with edgetic perturbations, or edgetic alleles, and their characterization *in vivo*, represent a promising strategy for investigating the function of specific interactions, particularly with regard to human disease [185]. Two complementary strategies, 'forward and reverse edgetics', reminiscent of the time-tested dichotomy of forward and reverse genetics [186], allow systematic investigation of the phenotypic outcomes of perturbations of specific binary protein interactions [187]. Taking a set of mutations in a gene associated with particular phenotypes, such as disease-associated mutations, the forward edgetics approach uses Y2H to determine the interaction defects of proteins where the mutations have been introduced by site-directed mutagenesis [184]. Reciprocally, reverse edgetics starts from a set of interactions for a protein of interest, and aims to systematically isolate alleles encoding proteins with desired specific interaction defects by reverse Y2H selections [44,68,188]. The edgetic alleles that are thus selected can be reintroduced *in vivo* into a model organism to investigate the phenotypic consequences of specifically altering the corresponding molecular interaction(s) [185].

Besides individual edges, higher-level topological structures such as network motifs can also be associated to specific biological functions [135,136]. Different types of networks exhibit distinctive profiles of the relative abundance of network motifs, so network motif profiles can be used to characterize and compare networks. Neuron networks or regulatory networks are enriched in feedforward loops, whereas food webs are enriched in bi-parallel motifs [135,136,189]. These distinctive enrichments suggest that interactions that are part of feedforward loops are likely to be involved in information processing, while interactions that are part of bi-parallel motifs are likely to be involved in energy transfer. We anticipate that analytical tools utilizing network motifs, particularly multicolor network motifs, will help in deciphering the function of many interactome network edges and local structures.

Multicolor triangle motifs containing two nodes linked by protein interactions with both of these nodes connected to a third node by a genetic interaction appear enriched in the S. cerevisiae interactome [24]. Consequently, if A, B and C are three genes such that the translation products of A and B physically interact, and the A and C genes are linked by a genetic interaction, then a genetic interaction between B and C can be predicted [131]. This particular motif suggests a 'compensatory complex' theme wherein two proteins/complexes/processes function in parallel. An excellent example of compensatory complexes is the pairing of endoplasmic reticulum (ER) protein-translocation sub-complex [190] and the Gim complex [191], with each complex connected to the other by multiple genetic interactions [24,150]. The Gim complex facilitates the folding of actin and tubulin components of the cytoskeleton. The genetic interactions between the Gim complex and ER protein translocation suggests that defects in moving proteins into the ER are ameliorated by a fully functioning cytoskeleton, whereas the trafficking of protein via lipid vesicles requires the cytoskeleton to act as a 'molecular train track'.

TOWARDS DYNAMIC INTERACTOMES

Towards Cell-Type and Condition-Specific Interactomes

The cell interior is a constantly changing environment. Biomolecules and cellular processes respond dynamically to intra- and extracellular cues. Available protein—protein interactome maps are, regrettably, mostly static, representing the union of protein interactions that may occur in all locations, times and environments. Analysis of protein—protein interactome networks will continue to contribute profound insights to systems biology only by reaching the temporal and spatial resolution necessary to dynamically model coordination of biological processes across the cell and the organism.

For a protein complex to be active at the right time and place in the cell, and at a controlled concentration, the cell has to undertake a large number of parallel and successive decisions. For each complex subunit, the cognate gene needs to be transcribed (chromatin opening, initiation and elongation of transcription ...) and the mRNA processed, exported, and translated. Complex subunits often also need to be post-translationally modified, controlled in quality and quantity by chaperones, and actively targeted to the required site of action. There, protein complex assembly can require a specific order of addition to reach stability. Because each step leading to a protein complex is potentially subject to regulation, co-complex interactome networks are dense with accumulated information on the cell dynamics.

Interrogating the dynamics of complex assembly at proteome scale is not yet feasible experimentally. It is, however, becoming possible to compare proteome expression across cell types, thanks to technological innovations developed throughout the last decade, such as stable isotope labeling by amino acids in cell culture [192] (see Chapter 1). We can now interrogate interactome networks for node dynamics (at least partially), but not yet for edge dynamics. The first dynamic measures of protein complex membership successfully followed a single protein, GRB2, as it dynamically associated with multiple complexes [193]. Pending increases in throughput and further advances, the modeling of co-complex interactome network dynamics will need to rely on computational analyses. Empirical measurements of binary interaction dynamics are also lacking. The LUMIER technology has paved the way [194], but most binary interactome maps remain static, and attempts at dynamical modeling also rely on computational analyses.

Computational integration of interactome maps and expression profiles can identify biological conditions whereby two proteins that can interact, according to an interactome map, are also co-present, according to their expression profiles. This additional knowledge allows the inference of spatiotemporal 'interaction territories' marking where or when the interaction can take place, e.g., during cell cycle or organism development [52,122]. To what extent is the expression of interacting proteins transcriptionally coordinated in cellular systems? Physically interacting proteins are more likely to exhibit similar expression patterns than would be expected by chance [39,129,195]. Most interacting proteins are not co-expressed, however, and some pairs are even anti-correlated in expression. Interactome dynamics therefore appear to be under tight transcriptional control, with most protein interactions being transient.

Transient protein interactions involved in signaling and intercomplex connections are enriched in binary interactome maps compared to co-complex interactome maps [39]. Members of a given protein complex can be co-regulated by a common transcription factor, when a transcription factor is connected by transcription regulatory edges directed towards several interacting and co-expressed proteins, forming 'regulonic complexes' [24,196,197]. In response to extracellular perturbations, protein complexes generally remain stable, but the functional connections between these complexes are substantially reorganized, as reflected by genetic interaction changes [151,198,199].

By overlaying transcriptome patterns with a binary interaction network, Han and colleagues discovered two types of highly connected hub protein. On the one hand, 'party' hubs are strongly co-expressed with most of their interacting partners; on the other hand, 'date' hubs connect to different partners at different times or contexts [200]. These interactome dynamics differences are reflected in the structures of hub proteins from the two groups. Relative to date hubs, where different interacting partners may utilize the same surface of interaction at different times, party hubs tend to contain less disordered regions and to display more interaction interfaces at their surface, as would be expected for proteins with many simultaneous interacting partners [201–203]. The observation that date hubs are more strongly associated with breast cancer phenotypes was used to develop a co-expression signature that strongly differentiated breast cancer patients on the basis of disease outcome [204].

Despite these initial successes, interactome dynamics modeling will need to move beyond computational analyses. Cell-type-specific transcriptome data may provide intuitive approximation of protein expression levels, but such estimations are bound to be imprecise. Detection of a transcript does not necessarily imply that the corresponding protein is present and stable, and the absence of a transcript does not necessarily imply absence of the corresponding protein, as proteins can remain stable long past transcript degradation and can transit from cell to cell. Relative protein concentrations must also be considered when modeling interactome dynamics. Protein concentrations influence the affinity of proteins for one another due to mass action, and the effect of cell crowding on protein–protein interactomes remains unexplored. The intracellular environment also affects protein interactions. Proteins can be restricted to particular organelles bound by membranes, as are the nuclei or mitochondria, or localized in less sharply delimited regions such as nucleoli. Cellular localization data are available at genomic scale for several organisms [205,206], but information about the dynamic movements of proteins across cellular compartments is lacking.

These caveats limit the scope of computational approaches in modeling interactome dynamics. Experimentally measuring protein–protein interactomes at high resolution both in space, across subcellular locations and across cell and tissue types, and in time, for example through the course of development, may still appear a distant goal, but this goal deserves to be actively pursued. Conversely, evidence that the expression of interacting proteins is tightly regulated shows that co-expression should not be used as a benchmark for protein interaction reliability.

Evolutionary Dynamics of Protein–Protein Interactome Networks

A central hypothesis of systems biology is that genotype–phenotype relationships are mediated through physical and functional interactions between genes and gene products that form intricate molecular networks within cells. Genotype–phenotype relationships are also governed by natural selection. Hence, understanding the principles driving the evolution of molecular networks would contribute to a deeper understanding of Life. Is there a 'core interactome' shared by every form of Life? Do the constraints of natural selection enforce constraints on interactome network structure? Do interactomes grow over evolutionary time? Does interactome complexity scale with organism complexity? Are interactomes more stable or more variable than genomes? Given these fundamental questions [207], it is not surprising that the evolutionary dynamics of protein–protein interactome networks have been a focus of investigation ever since the first large-scale protein–protein interactome maps appeared.

If protein–protein interactomes were evolutionarily stable systems, interactions between orthologous protein pairs from distinct species should be largely conserved. However, the observed fraction of interactions corresponding to such 'interolog' pairs is consistently low across several species [53,208–211]. The incompleteness of available interactome maps, and/or the difficulties of orthology mapping, may explain these apparently low proportions of conserved interactions. Still, as even these low proportions would not be expected at random, they are consistent with natural selection acting on the conservation of at least a subset of interactions throughout evolutionary time. A complementary interpretation would be that protein–protein interactomes are evolutionarily dynamic systems, constantly changing under the action of natural selection.

Cross-species comparisons indicate that $\sim 10^{-5}$ interactions are lost or gained per protein pair per million years, leaving aside the interactome remodeling that necessarily follows gene death and gene birth events [25,212,213]. This corresponds to approximately 10^3 interaction changes per 10^6 years in the evolution of the human lineage. Different types of protein interaction are rewired at different rates. Transient interactions appear more evolutionarily volatile than the more lasting interactions forming protein complexes [213–215], and protein–peptide interactions appear to change more rapidly than interactions between long proteins [212]. Evolutionary variation is observed even for protein complexes participating in the cell cycle. These complexes are globally conserved across several yeast species, but differ in their regulatory subunit composition and timing of assembly [216]. Incidentally, this dynamic rewiring of interactome networks during evolution is bound to limit the reliability of predictions of protein interactions based on sequence similarity measures across species [217].

Single amino acid changes can result in edgetic perturbations of protein—protein interactome networks [184]. Fixation of such sequence changes under selective constraints is expected to shape protein interaction interfaces. In agreement, the sequences of hub proteins appear under tighter constraints than the sequences of non-hub proteins, with intra-module hubs significantly more constrained than inter-module hubs [202,218,219]. The yeast protein Pbs2, which endogenously interacts specifically with a single yeast SH3 domain, is able to promiscuously interact with many non-yeast SH3 domains [220]. At an equivalent level of sequence similarity, protein interactions are more conserved within species, when considering paralogous protein pairs originating from duplication events, than across species when considering orthologous protein pairs [217]. It seems that tinkering with interaction interfaces and specificity causes protein interactions to co-evolve dynamically within biological systems.

Immediately following gene duplication events, paralogous proteins are expected to have identical protein sequences and to share all of their interactors. Empirical observations have revealed that the fraction of interactors shared by paralogous proteins decreases over evolutionary time, likely reflecting the well-described functional divergence of retained paralogous proteins [221]. Such evolutionary dynamics may explain the origin of the scale-free degree distribution that protein—protein interactome networks invariably follow, via an evolutionary version of the 'rich-get-richer' principle [222,223] (see Chapter 9). These evolutionary dynamics may also lead to an elevated clustering in protein—protein interactome networks if self-interactions are taken into account, as their duplication enables the formation of novel complexes of paralogous proteins [224,225] (see Chapter 9). The proteasome complex likely evolved from an ancestral homodimeric interaction through multiple successive duplication events [25,226].

Attempts to estimate the interactome rewiring rate following duplication events have yielded conflicting results [212,213,221—223,227—230]. These contradictions may be reconciled by a model according to which rewiring does not occur at a constant rate, but rather in a rapid-then-slow fashion [55]. Similar rapid-then-slow dynamics characterize protein sequence divergence following duplication events, likely reflecting relaxed-then-tight selective constraints on the function of the duplicated proteins. Signatures of neo-functionalization, sub-functionalization and asymmetric edge-specific divergence have been observed in protein—protein interactome networks [219,231—233]. Edgetic rewiring of protein— protein interactome networks following duplication events thus appears associated with the Darwinian selection of the functions of the corresponding proteins.

How to reconcile the dynamic rewiring of protein—protein interactome networks with the existence of universal processes found within all forms of Life? Beyond the union of individual interactions, interactome networks exhibit higher-level organizational properties, such as signaling pathways, or other types of functional module. Several pathways and modules appear evolutionarily conserved, as measured by orthology-based network alignment algorithms [234,235]. Similarly, topology-based network alignment algorithms have revealed considerable similarities in the local wiring of cellular networks across evolutionarily distant organisms [236]. The global network topology of binary interactomes of organisms as diverse as humans, plants, worms and yeasts appear qualitatively similar, characterized by a scale-free distribution of degrees and small-world structures [55] (see Chapter 9). Likewise, the estimated ratio of interacting pairs among all possible protein pairs in these organisms, with genomes encoding anywhere from 6000 to 30 000 proteins, appears surprisingly stable, with 5—10 interactions per 10 000 protein pairs [55]. It is possible that these high-level systems properties are ultimately the object of evolutionary conservation and so unify all forms of Life.

In summary, natural selection seems to shape the dynamic evolution of protein—protein interactome networks. Regulatory interactome networks seem to evolve faster than protein—protein interactome networks [213,228]. More refined models of the evolution of biological systems, including population size effects and the concept of 'genotype networks', are being investigated [237] (see chapter by A. Wagner). Life could be perceived as a system containing genotypes and phenotypes, with genotypes shaping phenotypes through the prism of interactomes, and phenotypes shaping genotypes through the feedback of evolution by natural selection.

CONCLUDING REMARKS

It is becoming increasingly clear that protein—protein interactome mapping and modeling will be key to understanding cellular systems and genotype—phenotype relationships. In this chapter we have described the state-of-the-art experimental and computational strategies currently used to detect and predict binary and co-complex protein interactions at proteome scale, and outlined the major achievements of the field so far. We covered the new concepts that will need to emerge, and the new technologies that will need to be developed, so that complete reference protein—protein interactome maps can materialize for several organisms in the near future. With such maps in hand, the principles governing interactome dynamics will be deciphered and causal paths between genotype and phenotype will be drawn.

An additional layer of complexity lies in the number of protein isoforms resulting from alternative splicing of transcripts for each individual gene. Protein–protein interactome maps available so far have mostly disregarded isoforms, opting for a gene-centered approach for simplicity and because differentiating between protein isoforms is technically challenging. Isoforms of the same protein may exhibit distinct combinations of protein interaction interfaces, leading to distinct local interactome networks. It will therefore be crucial to differentiate isoforms in future interactome mapping and modeling. This will undoubtedly shed light on protein interactome dynamics, as isoform expression is expected to be highly regulated across different cell types and conditions.

Mechanistic understanding of biological systems will also require quantitative estimation of interaction strength. To this end, systematic measures of the affinity of proteins for each other, in binary as well as in higher-order interactions, would generate a tremendous impetus to mathematically model biological processes. It will also be essential to achieve the systematic integration of three-dimensional structural data, whether derived experimentally or by computational modeling [238,239]. Eventually, three-dimensional mapping of the sequence variations found in populations and their association with traits may allow the almost seamless reconstruction of genotype–phenotype relationships through edgetic modeling of protein–protein interactomes.

ACKNOWLEDGEMENTS

We thank past and current members of the Vidal Lab and the Center for Cancer Systems Biology (CCSB) for their help and constructive discussions over the course of developing our binary interaction mapping strategies, framework, and protocols. We thank Dr David E. Hill for insightful editing of this book chapter and Dr Robin Lee for providing an image of cells. This work was supported by National Human Genome Research Institute grants R01-HG001715 awarded to M.V. D.E.H, and F.P.R, P50-HG004233 awarded to M.V., R01-HG006061 to M.V., D.E.H. and M.E.C., RC4-HG006066 to M.V and D.E.H.; National Heart, Lung and Blood Institute grant U01-HL098166 (M.V. subcontract); National Institute of Environmental Health Sciences R01-ES015728 to M.V.; National Cancer Institute grant R33-CA132073 to M.V.; National Science Foundation PGRP grant DBI-0703905 to M.V. and D.E.H.; and by Institute Sponsored Research funds from the Dana-Farber Cancer Institute Strategic Initiative awarded to M.V. and CCSB. M.V. is a 'Chercheur Qualifié Honoraire' from the Fonds de la Recherche Scientifique (FRS-FNRS, French Community of Belgium).

REFERENCES

[1] Nurse P. The great ideas of biology. Clin Med 2003;3(6):560–8.
[2] Vidal M. A unifying view of 21st century systems biology. FEBS Lett 2009;583(24):3891–4.
[3] Delbrück M. Unités biologiques douées de continuité génétique. Paris: Editions du Centre National de la Recherche Scientifique; 1949. 33–35.
[4] Pardee AB, Yates RA. Control of pyrimidine biosynthesis in *Escherichia coli* by a feed-back mechanism. J Biol Chem 1956;221(2):757–70.
[5] Umbarger HE. Evidence for a negative-feedback mechanism in the biosynthesis of isoleucine. Science 1956;123(3202):848.
[6] Novick A, Weiner M. Enzyme induction as an all-or-none phenomenon. Proc Natl Acad Sci U S A 1957;43(7):553–66.
[7] Monod J, Jacob F. Teleonomic mechanisms in cellular metabolism, growth, and differentiation. Cold Spring Harb Symp Quant Biol 1961;26:389–401.
[8] Thomas R. Boolean formalization of genetic control circuits. J Theor Biol 1973;42(3):563–85.
[9] Waddington CH. The Strategy of the Genes. London: George Allen & Unwin; 1957.
[10] Kauffman SA. Metabolic stability and epigenesis in randomly constructed genetic nets. J Theor Biol 1969;22(3):437–67.
[11] Lander ES, Linton LM, et al. Initial sequencing and analysis of the human genome. Nature 2001;409(6822):860–921.
[12] Pettersson E, Lundeberg J, et al. Generations of sequencing technologies. Genomics 2009;93(2):105–11.
[13] Schuster SC. Next-generation sequencing transforms today's biology. Nat Methods 2008;5(1):16–8.
[14] Venter JC, Adams MD, et al. The sequence of the human genome. Science 2001;291(5507):1304–51.
[15] Beadle GW, Tatum EL. 'Genetic control of biochemical reactions in *Neurospora*. Proc Natl Acad Sci U S A 1941;27(11):499–506.
[16] Burga A, Casanueva MO, et al. Predicting mutation outcome from early stochastic variation in genetic interaction partners. Nature 2011;480(7376):250–3.
[17] Horvitz HR, Sulston JE. Isolation and genetic characterization of cell-lineage mutants of the nematode *Caenorhabditis elegans*. Genetics 1980;96(2):435–54.
[18] Giaever G, Chu AM, et al. Functional profiling of the *Saccharomyces cerevisiae* genome. Nature 2002;418(6896):387–91.
[19] Begley TJ, Rosenbach AS, et al. Damage recovery pathways in *Saccharomyces cerevisiae* revealed by genomic phenotyping and interactome mapping. Mol Cancer Res 2002;1(2):103–12.
[20] Cusick ME, Klitgord N, et al. Interactome: gateway into systems biology. Hum Mol Genet 2005;14(Spec No. 2):R171–81.
[21] Huberts DH, van der Klei IJ. Moonlighting proteins: an intriguing mode of multitasking. Biochim Biophys Acta 2010;1803(4):520–5.
[22] Jeffery CJ. Moonlighting proteins. Trends Biochem. Sci 1999;24(1):8–11.
[23] Milgram S. The small world problem. Psychol Today 1967;1(1):60–7.
[24] Zhang LV, King OD, et al. Motifs, themes and thematic maps of an integrated *Saccharomyces cerevisiae* interaction network. J Biol 2005;4(2):6.
[25] Levy ED, Pereira-Leal JB. Evolution and dynamics of protein interactions and networks. Curr Opin Struct Biol 2008;18(3):349–57.

[26] Kaushansky A, Allen JE, et al. Quantifying protein–protein interactions in high throughput using protein domain microarrays. Nat Protoc 2010;5(4):773–90.
[27] Linding R, Jensen LJ, et al. Systematic discovery of in vivo phosphorylation networks. Cell 2007;129(7):1415–26.
[28] Ma'ayan A, Jenkins SL, et al. Formation of regulatory patterns during signal propagation in a mammalian cellular network. Science 2005;309(5737):1078–83.
[29] Ptacek J, Devgan G, et al. Global analysis of protein phosphorylation in yeast. Nature 2005;438(7068):679–84.
[30] Roberts PM. Mining literature for systems biology. Brief Bioinform 2006;7(4):399–406.
[31] Marcotte E, Date S. Exploiting big biology: integrating large-scale biological data for function inference. Brief Bioinform 2001;2(4):363–74.
[32] Walhout AJ, Vidal M. Protein interaction maps for model organisms. Nat Rev Mol Cell Biol 2001;2(1):55–62.
[33] Cusick ME, Yu H, et al. Literature-curated protein interaction datasets. Nat Methods 2009;6(1):39–46.
[34] Turinsky AL, Razick S, et al. Literature curation of protein interactions: measuring agreement across major public databases. Database 2010 (Oxford) 2010: baq026.
[35] Turinsky AL, Razick S, et al. Interaction databases on the same page. Nat Biotechnol 2011;29(5):391–3.
[36] Venkatesan K, Rual JF, et al. An empirical framework for binary interactome mapping. Nat Methods 2009;6(1):83–90.
[37] Edwards AM, Isserlin R, et al. Too many roads not taken. Nature 2011;470(7333):163–5.
[38] Vousden KH, Lane DP. p53 in health and disease. Nat Rev Mol Cell Biol 2007;8(4):275–83.
[39] Yu H, Braun P, et al. High-quality binary protein interaction map of the yeast interactome network. Science 2008;322(5898):104–10.
[40] Bartel PL, Roecklein JA, et al. A protein linkage map of Escherichia coli bacteriophage T7. Nat Genet 1996;12(1):72–7.
[41] Finley Jr RL, Brent R. Interaction mating reveals binary and ternary connections between Drosophila cell cycle regulators. Proc Natl Acad Sci U S A 1994;91(26):12980–4.
[42] Fromont-Racine M, Rain JC, et al. Toward a functional analysis of the yeast genome through exhaustive two-hybrid screens. Nat Genet 1997;16(3):277–82.
[43] Vidal M. The reverse two-hybrid system. In: Fields S, Bartel P, editors. The yeast two-hybrid system. New York, NY: Oxford University Press; 1997. p. 109–47.
[44] Vidal M, Brachmann RK, et al. Reverse two-hybrid and one-hybrid systems to detect dissociation of protein–protein and DNA–protein interactions. Proc Natl Acad Sci U S A 1996;93(19):10315–20.
[45] Walhout AJ, Vidal M. A genetic strategy to eliminate self-activator baits prior to high-throughput yeast two-hybrid screens. Genome Res 1999;9(11):1128–34.
[46] Ito T, Chiba T, et al. A comprehensive two-hybrid analysis to explore the yeast protein interactome. Proc Natl Acad Sci U S A 2001;98(8):4569–74.
[47] Miller JP, Lo RS, et al. Large-scale identification of yeast integral membrane protein interactions. Proc Natl Acad Sci U S A 2005;102(34):12123–8.
[48] Tarassov K, Messier V, et al. An in vivo map of the yeast protein interactome. Science 2008;320(5882):1465–70.
[49] Uetz P, Giot L, et al. A comprehensive analysis of protein–protein interactions in Saccharomyces cerevisiae. Nature 2000;403(6770):623–7.
[50] Li S, Armstrong CM, et al. A map of the interactome network of the metazoan C. elegans. Science 2004;303(5657):540–3.
[51] Reboul J, Vaglio P, et al. C. elegans ORFeome version 1.1: experimental verification of the genome annotation and resource for proteome-scale protein expression. Nat Genet 2003;34(1):35–41.
[52] Simonis N, Rual JF, et al. Empirically controlled mapping of the Caenorhabditis elegans protein–protein interactome network. Nat Methods 2009;6(1):47–54.
[53] Walhout AJ, Sordella R, et al. Protein interaction mapping in C. elegans using proteins involved in vulval development. Science 2000;287(5450):116–22.
[54] Giot L, Bader JS, et al. A protein interaction map of Drosophila melanogaster. Science 2003;302(5651):1727–36.
[55] Arabidopsis interactome mapping consortium. Evidence for network evolution in an Arabidopsis interactome map. Science 2011;333(6042):601–7.
[56] Lalonde S, Sero A, et al. A membrane protein/signaling protein interaction network for Arabidopsis version AMPv2. Front Physiol 2010;1:24.
[57] Mukhtar MS, Carvunis AR, et al. Independently evolved virulence effectors converge onto hubs in a plant immune system network. Science 2011;333(6042):596–601.
[58] Colland F, Jacq X, et al. Functional proteomics mapping of a human signaling pathway. Genome Res 2004;14(7):1324–32.
[59] Lim J, Hao T, et al. A protein–protein interaction network for human inherited ataxias and disorders of Purkinje cell degeneration. Cell 2006;125(4):801–14.
[60] Rual JF, Venkatesan K, et al. Towards a proteome-scale map of the human protein–protein interaction network. Nature 2005;437(7062):1173–8.
[61] Stelzl U, Worm U, et al. A human protein–protein interaction network: a resource for annotating the proteome. Cell 2005;122(6):957–68.
[62] Yu H, Tardivo L, et al. Next-generation sequencing to generate interactome datasets. Nat Methods 2011;8(6):478–80.
[63] Zhu H, Bilgin M, et al. Global analysis of protein activities using proteome chips. Science 2001;293(5537):2101–5.
[64] Dreze M, Monachello D, et al. High-quality binary interactome mapping. Methods Enzymol 2010;470:281–315.
[65] Fields S, Song O. A novel genetic system to detect protein-protein interactions. Nature 1989;340(6230):245–6.
[66] Durfee T, Becherer K, et al. The retinoblastoma protein associates with the protein phosphatase type 1 catalytic subunit. Genes Dev 1993;7(4):555–69.
[67] Gyuris J, Golemis E, et al. Cdi1, a human G1 and S phase protein phosphatase that associates with Cdk2. Cell 1993;75(4):791–803.
[68] Vidal M, Braun P, et al. Genetic characterization of a mammalian protein–protein interaction domain by using a yeast reverse two-hybrid system. Proc Natl Acad Sci U S A 1996;93(19):10321–6.
[69] Walhout AJ, Vidal M. High-throughput yeast two-hybrid assays for large-scale protein interaction mapping. Methods 2001;24(3):297–306.
[70] Armstrong CM, Li S, Vidal M. Modular scale yeast two-hybrid screening. In: Celis JE, editor. Cell biology: A laboratory handbook, 3rd ed. San Diego, CA: Academic Press; 2005. p. IV. 39. 1–9.

[71] Walhout AJ, Boulton SJ, et al. Yeast two-hybrid systems and protein interaction mapping projects for yeast and worm. Yeast 2000;17(2):88–94.

[72] Grove CA, De Masi F, et al. A multiparameter network reveals extensive divergence between *C. elegans* bHLH transcription factors. Cell 2009;138(2):314–27.

[73] Lamesch P, Li N, et al. hORFeome v3.1: a resource of human open reading frames representing over 10,000 human genes. Genomics 2007;89(3):307–15.

[74] Rual JF, Hirozane-Kishikawa T, et al. Human ORFeome version 1.1: a platform for reverse proteomics. Genome Res 2004; 14(10B):2128–35.

[75] D'Haeseleer P, Church GM. Estimating and improving protein interaction error rates. Proc IEEE Comput Syst Bioinform Conf 2004:216–23.

[76] Grigoriev A. On the number of protein–protein interactions in the yeast proteome. Nucleic Acids Res 2003;31(14):4157–61.

[77] Hart GT, Ramani AK, et al. How complete are current yeast and human protein-interaction networks? Genome Biol 2006; 7(11):120.

[78] Huang H, Jedynak BM, et al. Where have all the interactions gone? Estimating the coverage of two-hybrid protein interaction maps. PLoS Comput Biol 2007;3(11):e214.

[79] Reguly T, Breitkreutz A, et al. Comprehensive curation and analysis of global interaction networks in *Saccharomyces cerevisiae*. J Biol 2006;5(4):11.

[80] Sambourg L, Thierry-Mieg N. New insights into protein–protein interaction data lead to increased estimates of the *S. cerevisiae* interactome size. BMC Bioinformatics 2010;11:605.

[81] Sprinzak E, Sattath S, et al. How reliable are experimental protein–protein interaction data? J Mol Biol 2003;327(5):919–23.

[82] Stumpf MP, Thorne T, et al. Estimating the size of the human interactome. Proc Natl Acad Sci U S A 2008;105(19):6959–64.

[83] Worseck JM, Grossmann A, et al. A stringent yeast two-hybrid matrix screening approach for protein–protein interaction discovery. Methods Mol Biol 2012;812:63–87.

[84] Braun P, Tasan M, et al. An experimentally derived confidence score for binary protein–protein interactions. Nat Methods 2009; 6(1):91–7.

[85] Chen YC, Rajagopala SV, et al. Exhaustive benchmarking of the yeast two-hybrid system. Nat Methods 2010;7(9):667–8.

[86] Alberts B. The cell as a collection of protein machines: preparing the next generation of molecular biologists. Cell 1998;92(3): 291–4.

[87] Gavin AC, Bosche M, et al. Functional organization of the yeast proteome by systematic analysis of protein complexes. Nature 2002;415(6868):141–7.

[88] Kuriyan J, Eisenberg D. The origin of protein interactions and allostery in colocalization. Nature 2007;450(7172):983–90.

[89] Williamson JR. Cooperativity in macromolecular assembly. Nat Chem Biol 2008;4(8):458–65.

[90] Fraser HB, Plotkin JB. Using protein complexes to predict phenotypic effects of gene mutation. Genome Biol 2007;8(11): R252.

[91] Gavin AC, Aloy P, et al. Proteome survey reveals modularity of the yeast cell machinery. Nature 2006;440(7084):631–6.

[92] Ideker T, Sharan R. Protein networks in disease. Genome Res 2008;18(4):644–52.

[93] Lage K, Karlberg EO, et al. A human phenome-interactome network of protein complexes implicated in genetic disorders. Nat Biotechnol 2007;25(3):309–16.

[94] Wang PI, Marcotte EM. It's the machine that matters: predicting gene function and phenotype from protein networks. J Proteomics 2010;73(11):2277–89.

[95] Beck M, Schmidt A, et al. The quantitative proteome of a human cell line. Mol Syst Biol 2011;7:549.

[96] Ghaemmaghami S, Huh WK, et al. Global analysis of protein expression in yeast. Nature 2003;425(6959):737–41.

[97] Aebersold R, Mann M. Mass spectrometry-based proteomics. Nature 2003;422(6928):198–207.

[98] Gingras AC, Gstaiger M, et al. Analysis of protein complexes using mass spectrometry. Nat Rev Mol Cell Biol 2007;8(8): 645–54.

[99] Kocher T, Superti-Furga G. Mass spectrometry-based functional proteomics: from molecular machines to protein networks. Nat Methods 2007;4(10):807–15.

[100] Puig O, Caspary F, et al. The tandem affinity purification (TAP) method: a general procedure of protein complex purification. Methods 2001;24(3):218–29.

[101] Burckstummer T, Bennett KL, et al. An efficient tandem affinity purification procedure for interaction proteomics in mammalian cells. Nat Methods 2006;3(12):1013–9.

[102] de Boer E, Rodriguez P, et al. Efficient biotinylation and single-step purification of tagged transcription factors in mammalian cells and transgenic mice. Proc Natl Acad Sci U S A 2003; 100(13):7480–5.

[103] Glatter T, Wepf A, et al. An integrated workflow for charting the human interaction proteome: insights into the PP2A system. Mol Syst Biol 2009;5:237.

[104] Ho Y, Gruhler A, et al. Systematic identification of protein complexes in *Saccharomyces cerevisiae* by mass spectrometry. Nature 2002;415(6868):180–3.

[105] Poser I, Sarov M, et al. BAC TransgeneOmics: a high-throughput method for exploration of protein function in mammals. Nat Methods 2008;5(5):409–15.

[106] Rigaut G, Shevchenko A, et al. A generic protein purification method for protein complex characterization and proteome exploration. Nat Biotechnol 1999;17(10):1030–2.

[107] Guruharsha KG, Rual JF, et al. A protein complex network of *Drosophila melanogaster*. Cell 2011;147(3):690–703.

[108] Bader GD, Hogue CW. An automated method for finding molecular complexes in large protein interaction networks. BMC Bioinformatics 2003;4:2.

[109] Choi H, Larsen B, et al. SAINT: probabilistic scoring of affinity purification-mass spectrometry data. Nat Methods 2011;8(1): 70–3.

[110] Hart GT, Lee I, et al. A high-accuracy consensus map of yeast protein complexes reveals modular nature of gene essentiality. BMC Bioinformatics 2007;8:236.

[111] Hu P, Janga SC, et al. Global functional atlas of *Escherichia coli* encompassing previously uncharacterized proteins. PLoS Biol 2009;7(4):e1000096.

[112] Kuhner S, van Noort V, et al. Proteome organization in a genome-reduced bacterium. Science 2009;326(5957):1235–40.

[113] Behrends C, Sowa ME, et al. Network organization of the human autophagy system. Nature 2010;466(7302):68–76.

[114] Bouwmeester T, Bauch A, et al. A physical and functional map of the human TNF-alpha/NF-kappa B signal transduction pathway. Nat Cell Biol 2004;6(2):97−105.

[115] Brehme M, Hantschel O, et al. Charting the molecular network of the drug target Bcr-Abl. Proc Natl Acad Sci U S A 2009;106(18): 7414−9.

[116] Ewing RM, Chu P, et al. Large-scale mapping of human protein−protein interactions by mass spectrometry. Mol Syst Biol 2007;3:89.

[117] Sowa ME, Bennett EJ, et al. Defining the human deubiquitinating enzyme interaction landscape. Cell 2009;138(2): 389−403.

[118] Vermeulen M, Eberl HC, et al. Quantitative interaction proteomics and genome-wide profiling of epigenetic histone marks and their readers. Cell 2010;142(6):967−80.

[119] Orchard S, Kerrien S, et al. Protein interaction data curation: the International Molecular Exchange (IMEx) consortium. Nat Methods 2012;9(4):345−50.

[120] Aloy P, Bottcher B, et al. 'Structure-based assembly of protein complexes in yeast. Science 2004;303(5666):2026−9.

[121] Bork P, Jensen LJ, et al. Protein interaction networks from yeast to human. Curr Opin Struct Biol 2004;14(3):292−9.

[122] de Lichtenberg U, Jensen LJ, et al. Dynamic complex formation during the yeast cell cycle. Science 2005;307(5710): 724−7.

[123] Schadt EE, Sachs A, et al. Embracing complexity, inching closer to reality. Sci STKE 2005;295:e40.

[124] Moody J. Race, school integration, and friendship segregation in America. Am J Sociol 2001;107(3):679−716.

[125] Vidal M. A biological atlas of functional maps. Cell 2001;104(3): 333−9.

[126] Vidal M. Interactome modeling. FEBS Lett 2005;579(8):1834−8.

[127] Bader JS, Chaudhuri A, et al. Gaining confidence in high-throughput protein interaction networks. Nat Biotechnol 2004;22(1):78−85.

[128] Jansen R, Yu H, et al. A Bayesian networks approach for predicting protein−protein interactions from genomic data. Science 2003;302(5644):449−53.

[129] Jansen R, Greenbaum D, et al. Relating whole-genome expression data with protein−protein interactions. Genome Res 2002;12(1): 37−46.

[130] Qi Y, Klein-Seetharaman J, et al. Random forest similarity for protein−protein interaction prediction from multiple sources. Pac Symp Biocomput 2005:531−42.

[131] Wong SL, Zhang LV, et al. Combining biological networks to predict genetic interactions. Proc Natl Acad Sci U S A 2004; 101(44):15682−7.

[132] Yan H, Venkatesan K, et al. A genome-wide gene function prediction resource for *Drosophila melanogaster*. PLoS ONE 2010;5(8):e12139.

[133] Zhang LV, Wong SL, et al. Predicting co-complexed protein pairs using genomic and proteomic data integration. BMC Bioinformatics 2004;5:38.

[134] Asthana S, King OD, et al. Predicting protein complex membership using probabilistic network reliability. Genome Res 2004; 14(6):1170−5.

[135] Milo R, Shen-Orr S, et al. Network motifs: simple building blocks of complex networks. Science 2002;298(5594):824−7.

[136] Shen-Orr SS, Milo R, et al. Network motifs in the transcriptional regulation network of *Escherichia coli*. Nat Genet 2002;31(1): 64−8.

[137] Albert I, Albert R. Conserved network motifs allow protein−protein interaction prediction. Bioinformatics 2004;20(18): 3346−52.

[138] Goldberg DS, Roth FP. Assessing experimentally derived interactions in a small world. Proc Natl Acad Sci U S A 2003;100(8): 4372−6.

[139] King OD. Comment on 'Subgraphs in random networks'. Phys Rev E Stat Nonlin Soft Matter Phys 2004;70(5 Pt 2):058101.

[140] Li W, Liu Y, et al. Dynamical systems for discovering protein complexes and functional modules from biological networks. IEEE/ACM Trans Comput Biol Bioinform 2007;4(2):233−50.

[141] Rives AW, Galitski T. Modular organization of cellular networks. Proc Natl Acad Sci U S A 2003;100(3):1128−33.

[142] Spirin V, Mirny LA. Protein complexes and functional modules in molecular networks. Proc Natl Acad Sci U S A 2003;100(21): 12123−8.

[143] Boulton SJ, Gartner A, et al. Combined functional genomic maps of the *C. elegans* DNA damage response. Science 2002;295(5552): 127−31.

[144] Dezso Z, Oltvai ZN, et al. Bioinformatics analysis of experimentally determined protein complexes in the yeast *Saccharomyces cerevisiae*. Genome Res 2003;13(11):2450−4.

[145] Ge H, Walhout AJ, et al. Integrating 'omic' information: a bridge between genomics and systems biology. Trends Genet 2003; 19(10):551−60.

[146] Gunsalus KC, Ge H, et al. Predictive models of molecular machines involved in *Caenorhabditis elegans* early embryogenesis. Nature 2005;436(7052):861−5.

[147] Walhout AJ, Reboul J, et al. Integrating interactome, phenome, and transcriptome mapping data for the *C. elegans* germline. Curr Biol 2002;12(22):1952−8.

[148] Costanzo M, Baryshnikova A, et al. The genetic landscape of a cell. Science 2010;327(5964):425−31.

[149] Piano F, Schetter AJ, et al. Gene clustering based on RNAi phenotypes of ovary-enriched genes in *C. elegans*. Curr Biol 2002;12(22):1959−64.

[150] Tong AH, Lesage G, et al. Global mapping of the yeast genetic interaction network. Science 2004;303(5659):808−13.

[151] Bandyopadhyay S, Kelley R, et al. 'Functional maps of protein complexes from quantitative genetic interaction data. PLoS Comput Biol 2008;4(4):e1000065.

[152] Vojtek AB, Hollenberg SM, et al. Mammalian Ras interacts directly with the serine/threonine kinase Raf. Cell 1993;74(1):205−14.

[153] Karaoz U, Murali TM, et al. Whole-genome annotation by using evidence integration in functional-linkage networks. Proc Natl Acad Sci U S A 2004;101(9):2888−93.

[154] Lee I, Date SV, et al. A probabilistic functional network of yeast genes. Science 2004;306(5701):1555−8.

[155] Letovsky S, Kasif S. Predicting protein function from protein/protein interaction data: a probabilistic approach. Bioinformatics 2003;19(Suppl. 1):i197−204.

[156] Oliver S. Guilt-by-association goes global. Nature 2000; 403(6770):601−3.

[157] Schwikowski B, Uetz P, et al. A network of protein−protein interactions in yeast. Nat Biotechnol 2000;18(12):1257−61.

[158] Tian W, Zhang LV, et al. Combining guilt-by-association and guilt-by-profiling to predict *Saccharomyces cerevisiae* gene function. Genome Biol 2008;9(Suppl. 1):S7.

[159] Pena-Castillo L, Tasan M, et al. A critical assessment of *Mus musculus* gene function prediction using integrated genomic evidence. Genome Biol 2008;9(Suppl. 1):S2.

[160] Barabasi AL, Oltvai ZN. Network biology: understanding the cell's functional organization. Nat Rev Genet 2004;5(2):101–13.

[161] Bader GD, Hogue CW. Analyzing yeast protein–protein interaction data obtained from different sources. Nat Biotechnol 2002;20(10):991–7.

[162] Bar-Joseph Z, Gerber GK, et al. Computational discovery of gene modules and regulatory networks. Nat Biotechnol 2003;21(11):1337–42.

[163] Ihmels J, Friedlander G, et al. Revealing modular organization in the yeast transcriptional network. Nat Genet 2002;31(4):370–7.

[164] Stuart JM, Segal E, et al. A gene-coexpression network for global discovery of conserved genetic modules. Science 2003;302(5643):249–55.

[165] Tornow S, Mewes HW. Functional modules by relating protein interaction networks and gene expression. Nucleic Acids Res 2003;31(21):6283–9.

[166] Lim J, Crespo-Barreto J, et al. Opposing effects of polyglutamine expansion on native protein complexes contribute to SCA1. Nature 2008;452(7188):713–8.

[167] Aerts S, Lambrechts D, et al. Gene prioritization through genomic data fusion. Nat Biotechnol 2006;24(5):537–44.

[168] Franke L, van Bakel H, et al. Reconstruction of a functional human gene network, with an application for prioritizing positional candidate genes. Am J Hum Genet 2006;78(6):1011–25.

[169] George RA, Liu JY, et al. Analysis of protein sequence and interaction data for candidate disease gene prediction. Nucleic Acids Res 2006;34(19):e130.

[170] Oti M, Snel B, et al. Predicting disease genes using protein–protein interactions. J Med Genet 2006;43(8):691–8.

[171] Tiffin N, Adie E, et al. Computational disease gene identification: a concert of methods prioritizes type 2 diabetes and obesity candidate genes. Nucleic Acids Res 2006;34(10):3067–81.

[172] Lee I, Blom UM, et al. Prioritizing candidate disease genes by network-based boosting of genome-wide association data. Genome Res 2011;21(7):1109–21.

[173] Tasan M, Drabkin HJ, et al. A resource of quantitative functional annotation for *Homo sapiens* genes. G3 (Bethesda) 2012;2(2):223–33.

[174] Raychaudhuri S, Plenge RM, et al. Identifying relationships among genomic disease regions: predicting genes at pathogenic SNP associations and rare deletions. PLoS Genet 2009;5(6):e1000534.

[175] Albert R, Jeong H, et al. Error and attack tolerance of complex networks. Nature 2000;406(6794):378–82.

[176] Jeong H, Mason SP, et al. Lethality and centrality in protein networks. Nature 2001;411(6833):41–2.

[177] Zotenko E, Mestre J, et al. Why do hubs in the yeast protein interaction network tend to be essential: reexamining the connection between the network topology and essentiality. PLoS Comput Biol 2008;4(8):e1000140.

[178] Goh KI, Cusick ME, et al. The human disease network. Proc Natl Acad Sci U S A 2007;104(21):8685–90.

[179] D'Andrea AD. Susceptibility pathways in Fanconi's anemia and breast cancer. N Engl J Med 2010;362(20):1909–19.

[180] Wang W. Emergence of a DNA-damage response network consisting of Fanconi anaemia and BRCA proteins. Nat Rev Genet 2007;8(10):735–48.

[181] Camargo LM, Collura V, et al. Disrupted in Schizophrenia 1 interactome: evidence for the close connectivity of risk genes and a potential synaptic basis for schizophrenia. Mol Psychiatry 2007;12(1):74–86.

[182] Kaltenbach LS, Romero E, et al. Huntingtin interacting proteins are genetic modifiers of neurodegeneration. PLoS Genet 2007;3(5):e82.

[183] Sakai Y, Shaw CA, et al. Protein interactome reveals converging molecular pathways among autism disorders. Sci Transl Med 2011;3(86). 86ra49.

[184] Zhong Q, Simonis N, et al. Edgetic perturbation models of human inherited disorders. Mol Syst Biol 2009;5:321.

[185] Dreze M, Charloteaux B, et al. 'Edgetic' perturbation of a *C.elegans* BCL2 ortholog. Nat Methods 2009;6(11):843–9.

[186] Griffiths A, Miller J, et al. An Introduction to Genetic Analysis. New York, W. H. Freeman; 2000.

[187] Charloteaux B, Zhong Q, et al. Protein-protein interactions and networks: forward and reverse edgetics. Methods Mol Biol 2011;759:197–213.

[188] Endoh H, Walhout AJ, et al. A green fluorescent protein-based reverse two-hybrid system: application to the characterization of large numbers of potential protein-protein interactions. Methods Enzymol 2000;328:74–88.

[189] Milo R, Itzkovitz S, et al. Superfamilies of evolved and designed networks. Science 2004;303(5663):1538–42.

[190] Hanein D, Matlack KE, et al. Oligomeric rings of the Sec61p complex induced by ligands required for protein translocation. Cell 1996;87(4):721–32.

[191] Geissler S, Siegers K, et al. A novel protein complex promoting formation of functional alpha- and gamma-tubulin. EMBO J 1998;17(4):952–66.

[192] Choudhary C, Mann M. Decoding signalling networks by mass spectrometry-based proteomics. Nat Rev Mol Cell Biol 2010;11(6):427–39.

[193] Bisson N, James DA, et al. Selected reaction monitoring mass spectrometry reveals the dynamics of signaling through the GRB2 adaptor. Nat Biotechnol 2011;29(7):653–8.

[194] Barrios-Rodiles M, Brown KR, et al. High-throughput mapping of a dynamic signaling network in mammalian cells. Science 2005;307(5715):1621–5.

[195] Ge H, Liu Z, et al. Correlation between transcriptome and interactome mapping data from *Saccharomyces cerevisiae*. Nat Genet 2001;29(4):482–6.

[196] Simonis N, Gonze D, et al. Modularity of the transcriptional response of protein complexes in yeast. J Mol Biol 2006;363(2):589–610.

[197] Simonis N, van Helden J, et al. Transcriptional regulation of protein complexes in yeast. Genome Biol 2004;5(5):R33.

[198] Hannum G, Srivas R, et al. Genome-wide association data reveal a global map of genetic interactions among protein complexes. PLoS Genet 2009;5(12):e1000782.

[199] Ideker T, Krogan NJ. Differential network biology. Mol Syst Biol 2012;8:565.
[200] Han JD, Bertin N, et al. Evidence for dynamically organized modularity in the yeast protein-protein interaction network. Nature 2004;430(6995):88–93.
[201] Ekman D, Light S, et al. What properties characterize the hub proteins of the protein–protein interaction network of *Saccharomyces cerevisiae*? Genome Biol 2006;7(6):R45.
[202] Kim PM, Sboner A, et al. The role of disorder in interaction networks: a structural analysis. Mol Syst Biol 2008;4:179.
[203] Kim SH, Yi SV. Correlated asymmetry of sequence and functional divergence between duplicate proteins of *Saccharomyces cerevisiae*. Mol Biol Evol 2006;23(5):1068–75.
[204] Taylor IW, Linding R, et al. Dynamic modularity in protein interaction networks predicts breast cancer outcome. Nat Biotechnol 2009;27(2):199–204.
[205] Huh WK, Falvo JV, et al. Global analysis of protein localization in budding yeast. Nature 2003;425(6959):686–91.
[206] Kumar A, Agarwal S, et al. Subcellular localization of the yeast proteome. Genes Dev 2002;16(6):707–19.
[207] Kiemer L, Cesareni G. Comparative interactomics: comparing apples and pears? Trends Biotechnol 2007;25(10):448–54.
[208] Cesareni G, Ceol A, et al. Comparative interactomics. FEBS Lett 2005;579(8):1828–33.
[209] Gandhi TK, Zhong J, et al. Analysis of the human protein interactome and comparison with yeast, worm and fly interaction datasets. Nat Genet 2006;38(3):285–93.
[210] Matthews LR, Vaglio P, et al. Identification of potential interaction networks using sequence-based searches for conserved protein–protein interactions or 'interologs'. Genome Res 2001;11(12):2120–6.
[211] Suthram S, Sittler T, et al. The *Plasmodium* protein network diverges from those of other eukaryotes. Nature 2005;438(7064):108–12.
[212] Beltrao P, Serrano L. Specificity and evolvability in eukaryotic protein interaction networks. PLoS Comput Biol 2007;3(2):e25.
[213] Shou C, Bhardwaj N, et al. Measuring the evolutionary rewiring of biological networks. PLoS Comput Biol 2011;7(1):e1001050.
[214] Roguev A, Bandyopadhyay S, et al. Conservation and rewiring of functional modules revealed by an epistasis map in fission yeast. Science 2008;322(5900):405–10.
[215] Teichmann SA. The constraints protein–protein interactions place on sequence divergence. J Mol Biol 2002;324(3):399–407.
[216] Jensen LJ, Jensen TS, et al. Co-evolution of transcriptional and post-translational cell-cycle regulation. Nature 2006;443(7111):594–7.
[217] Mika S, Rost B. Protein-protein interactions more conserved within species than across species. PLoS Comput Biol 2006;2(7):e79.
[218] Fraser HB. Modularity and evolutionary constraint on proteins. Nat Genet 2005;37(4):351–2.
[219] Kim PM, Lu LJ, et al. Relating three-dimensional structures to protein networks provides evolutionary insights. Science 2006;314(5807):1938–41.
[220] Zarrinpar A, Park SH, et al. Optimization of specificity in a cellular protein interaction network by negative selection. Nature 2003;426(6967):676–80.
[221] Wagner A. The yeast protein interaction network evolves rapidly and contains few redundant duplicate genes. Mol Biol Evol 2001;18(7):1283–92.
[222] Pastor-Satorras R, Smith E, et al. Evolving protein interaction networks through gene duplication. J Theor Biol 2003;222(2):199–210.
[223] Vazquez A. Growing network with local rules: preferential attachment, clustering hierarchy, and degree correlations. Phys Rev E Stat Nonlin Soft Matter Phys 2003;67(5 Pt 2):056104.
[224] Musso G, Zhang Z, et al. Retention of protein complex membership by ancient duplicated gene products in budding yeast. Trends Genet 2007;23(6):266–9.
[225] Pereira-Leal JB, Levy ED, et al. Evolution of protein complexes by duplication of homomeric interactions. Genome Biol 2007;8(4):R51.
[226] Bochtler M, Ditzel L, et al. The proteasome. Annu Rev Biophys Biomol Struct 1999;28:295–317.
[227] Ispolatov I, Krapivsky PL, et al. Duplication-divergence model of protein interaction network. Phys Rev E Stat Nonlin Soft Matter Phys 2005;71(6 Pt 1):061911.
[228] Maslov S, Sneppen K, et al. Upstream plasticity and downstream robustness in evolution of molecular networks. BMC Evol Biol 2004;4:9.
[229] Presser A, Elowitz MB, et al. The evolutionary dynamics of the *Saccharomyces cerevisiae* protein interaction network after duplication. Proc Natl Acad Sci U S A 2008;105(3):950–4.
[230] Wagner A. How the global structure of protein interaction networks evolves. Proc Biol Sci 2003;270(1514):457–66.
[231] Conant GC, Wagner A. Asymmetric sequence divergence of duplicate genes. Genome Res 2003;13(9):2052–8.
[232] He X, Zhang J. Rapid subfunctionalization accompanied by prolonged and substantial neofunctionalization in duplicate gene evolution. Genetics 2005;169(2):1157–64.
[233] Wagner A. Asymmetric functional divergence of duplicate genes in yeast. Mol Biol Evol 2002;19(10):1760–8.
[234] Sharan R, Ideker T. Modeling cellular machinery through biological network comparison. Nat Biotechnol 2006;24(4):427–33.
[235] Wuchty S, Oltvai ZN, et al. Evolutionary conservation of motif constituents in the yeast protein interaction network. Nat Genet 2003;35(2):176–9.
[236] Kuchaiev O, Milenkovic T, et al. Topological network alignment uncovers biological function and phylogeny. J R Soc Interface 2010;7(50):1341–54.
[237] Fernandez A, Lynch M. Non-adaptive origins of interactome complexity. Nature 2011;474(7352):502–5.
[238] O'Donoghue SI, Goodsell DS, et al. Visualization of macromolecular structures. Nat Methods 2010;7(Suppl. 3):S42–55.
[239] Stein A, Mosca R, et al. Three-dimensional modeling of protein interactions and complexes is going 'omics. Curr Opin Struct Biol 2011;21(2):200–8.

Chapter 4

Gene Regulatory Networks

Martha L. Bulyk[1] and A.J. Marian Walhout[2]

[1]*Division of Genetics, Department of Medicine; Department of Pathology; Brigham & Women's Hospital and Harvard Medical School, Boston, MA 02115, USA & Harvard-MIT Division of Health Sciences and Technology (HST), Harvard Medical School, Boston, MA 02115, USA,* [2]*Programs in Systems Biology, and Molecular Medicine, University of Massachusetts Medical School, Worcester, MA 01605, USA*

Chapter Outline

Cells are Computers	65
Gene Regulation	65
Transcription	65
Gene Regulatory Networks (GRNs)	66
GRN Nodes: Transcription Factors	67
GRN Nodes: *Cis*-Regulatory Elements	67
TF-Binding Sites	69
Cis-Regulatory Modules (CRMs)	69
GRN Edges: Physical Interactions Between TFs and DNA	73
GRNs: Visualization	74
GRNs: Data Quality	74
GRN Structure and Function	75
GRNs: Model Organisms	79
Future Challenges	**81**
Acknowledgements	**82**
References	**82**

CELLS ARE COMPUTERS

Cells constantly interpret their environment during development, in response to different physiological and environmental conditions and in disease. In response to various inputs, they exert appropriate outputs such as proliferation and differentiation during development, the generation of defense molecules to combat infection by harmful parasites or microbes, or the synthesis of hormones such as leptin or insulin to attain the appropriate physiological state upon feeding. Cells have evolved a repertoire of sophisticated regulatory mechanisms that measure and process environmental input in order to exert the appropriate biological output. One of the most important mechanisms is to modulate the expression of genes that encode proteins involved in any of the cellular responses. Understanding how such differential gene regulation is orchestrated to deliver a biological output based on a particular input is a primary focus in the larger field of systems biology.

GENE REGULATION

The central dogma in molecular biology (Figure 4.1A) depicts the flow of information in the generation of a protein though transcription (the synthesis of an RNA molecule from a DNA template by an RNA polymerase) and translation (the subsequent synthesis of a polypeptide from an RNA template by the ribosome) [1]. The balance between synthesis and breakdown of both the RNA and the polypeptide dictates protein expression levels unique to each cell in response to different developmental, physiological or pathological conditions. The human genome contains ~25 000 protein-coding genes, as well as numerous small and long non-coding RNA genes (see Chapter 2). The total complement of RNAs encoded by the genome is referred to as the transcriptome, and the full repertoire of proteins is called the proteome (Figure 4.1B). The two synthesis processes, transcription and translation, are most pivotal in gene regulation and are tightly controlled by different types of regulatory networks. Transcription is regulated by proteins called transcription factors (TFs) (see below), as well as by non-coding RNAs, whereas translation is regulated by non-coding RNAs and RNA-binding proteins. In this chapter, we focus mainly on transcriptional regulation in the context of gene regulatory networks (GRNs).

TRANSCRIPTION

Transcriptional regulation differs between prokaryotes and eukaryotes. For instance, in contrast to prokaryotes,

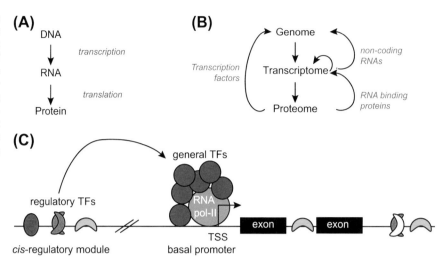

FIGURE 4.1 Transcription plays a pivotal role in the regulation of gene expression. (A) The central dogma in molecular biology. (B) Different types of networks regulate gene expression: TFs are proteins that control gene expression by interacting with the genome; non-coding RNAs affect both the genome and the transcriptome and RNA binding proteins affect gene expression post-transcriptionally. (C) Cartoon depicting transcriptional regulation for an individual eukaryotic gene.

eukaryotes have three different RNA polymerases: protein-coding genes are transcribed by RNA polymerase II, while RNA polymerase I and III transcribe rRNAs and tRNAs. The most important mechanism of gene regulation is by changing the levels of gene transcription. In addition to RNA polymerase II, a set of general TFs is required for all eukaryotic genes to be transcribed at a basal level (Figure 4.1C). For each individual gene, this basal level of transcription can be enhanced or reduced by regulatory TFs (hereafter referred to simply as TFs) that are characterized by the presence of a DNA-binding domain that directly interacts with short DNA sequences located within gene promoters, enhancers or other types of complex regulatory sequences (see below). Through additional protein domains, TFs interact with chromatin remodelers and/or members of the basal transcriptional machinery to achieve transcriptional activation or repression. Individual genes are regulated by distinct sets of TFs, which leads to the exquisitely specific spatiotemporal gene expression patterns and levels that are required in biological processes ranging from development to pathologies.

It is well known that TFs are of critical importance during the development of complex multicellular organisms. For instance, several TFs have been identified as master regulators of the formation of particular organs and tissues during development. Master regulators are classically defined as being necessary and sufficient to induce organogenesis or other developmental processes. Examples include Pax6, a key regulator of eye development in many different species [2], and ELT-2, which is central to intestinal development in the nematode *Caenorhabditis elegans* [3]. TFs are important not only during development, but also in adult organisms. Interestingly, most TFs are expressed throughout the lifetime of an organism, suggesting that many may play broad and important roles in the responses to stress and other physiological, environmental or pathological cues. While some TFs, such as master regulators, have been studied in detail using biochemical and/or genetic approaches, the functions of the vast majority of eukaryotic TFs remain unknown. This is in part because most TFs likely have rather subtle regulatory and biological effects compared to master regulators, and may act redundantly with other TFs. As a result, they are less likely to be identified in genetic screens to function in a particular biological process. Currently, many efforts aim to gain insight into how many TFs interact together in GRNs to give rise to a desired biological output.

GENE REGULATORY NETWORKS (GRNs)

GRNs are circuits composed of physical and/or regulatory interactions between the two types of components that orchestrate transcriptional outputs: TFs and *cis*-regulatory elements (CREs) located in the genome. GRNs can be represented as graphs that depict interactions between genes (to which CREs are ascribed) and their regulators (Figure 4.2). Network graphs are composed of nodes and edges (see Chapter 9). GRNs are bipartite, which means that there are two types of nodes: TFs and their target genes. GRNs are also directional, as the edges indicate a relationship *from* a TF *to* a target gene or CRE (Figure 4.2). Different types of edges can be included in GRNs. First, direct physical interactions between TFs and DNA elements attributed to a gene (e.g., promoter, intron, or more distally located *cis*-regulatory module or CRM) can be included. The second type of edge that can be included is a regulatory interaction (logical relationship), which can be inferred from experiments that identify which genes are positively or negatively affected upon loss (or overexpression) of a TF. Before discussing GRNs in more detail, we will first focus on the two types of node: TF proteins and DNA targets.

GRN Nodes: Transcription Factors

TFs can be grouped into families based on their DNA-binding domain(s). There are different types of DNA-binding domains, some of which are lineage specific and others that are more ubiquitous. Surprisingly, only three types of DNA-binding domains are found in all kingdoms of life: cold shock, helix-turn-helix (HTH) type 3, and HTH psq [4]. Some DNA-binding domains are clearly involved in direct DNA interactions, including basic helix-loop-helix (bHLH), basic leucine zipper (bZip) and nuclear hormone receptor (NHR)-type zinc fingers, whereas other potential DNA-binding domains, most notably Cys_2His_2 (C_2H_2)-type zinc fingers, can also be involved in RNA binding or mediate protein−protein interactions [5,6]. This complicates the annotation of TFs and emphasizes the need for the application of systematic assays to determine, for individual domains in individual proteins, whether they mediate nucleic acid or protein binding.

Identifying all the genes in a genome that encode TFs is not a trivial task because it can be challenging to recognize protein domains based on sequence alone, and because of ambiguity in proposed protein functionality, as described above. Databases such as InterPro, Pfam and SMART [7−9] can be used to predict which genes in a genome encode TFs, based on how well their amino acid sequence matches canonical DNA-binding domains. However, the use of such tools is limited by how predictive the different domains are of DNA binding, how accurately proteins are annotated to possess a particular domain, as well as the quality of the computational domain model. The accuracy and coverage of TF predictions can be greatly improved by manually curating protein sequences to ensure that they indeed possess a likely DNA-binding domain [10,11]. The final number of TFs encoded in the genome is likely to change for most organisms based on either improved curation or experimental identification of sequence-specific protein−DNA interactions (see below). Overall, between 5% and 10% of all protein-coding genes of most organisms are predicted to encode a TF, illustrating the overall importance of this class of proteins [10−13].

The number of TF-encoding genes in a genome is not the sole determinant of the total number of functional TFs that occur in an organism. Alternative splicing can result in multiple TFs encoded by a single gene, and these TFs can have different functional or biochemical properties. For instance, some TF genes encode multiple DNA-binding domains that can be included or excluded in the protein product through the use of different gene promoters or by alternative splicing. Alternative splicing increases with organismal complexity, and therefore, even though many eukaryotic genomes have similar numbers of TF-encoding genes, more complex metazoans such as mammals have effectively more TFs than relatively simple multicellular organisms such as C. elegans [10]. A second mechanism of obtaining functionally distinct TFs that are encoded by the same gene is by post-translational modifications (e.g., phosphorylation) that can alter the activity of the TF. Indeed, many TFs are the functional endpoint of signaling pathways that, for instance, instruct differentiation programs during embryonic development, or in response to environmental cues. An example is the DAF-16/Foxo TF, which is the downstream target of the insulin/IgF signaling pathway in worms, flies and mammals. When the pathway is active, DAF-16 is phosphorylated and prevented from translocating into the nucleus to activate relevant target genes [14]. Finally, differential dimerization can also greatly affect the number of active TF entities in an organism because many TFs bind DNA as obligatory or facultative dimers. For instance, bHLH TFs bind their cognate target sequences only as dimers [15], whereas NHRs can often bind either as monomers or as dimers [16]. Differential dimerization can greatly affect the number of active TF entities. For instance, if TF-X and TF-Y function as both monomers, homodimers and heterodimers, this is equivalent to five distinct TF entities [17]. High-throughput system-level protein−protein interaction studies have been applied to delineate TF dimerization. For instance, yeast two-hybrid assays (see Chapter 3) have been used to identify dimers within the C. elegans bHLH family [18], and mammalian two-hybrid assays were applied to large sets of human and mouse TFs [19]. However, for a complete picture of TF dimerization, multiple protein−protein interaction assays need to be applied to all TFs and in a variety of model organisms.

The comprehensive study of TFs is facilitated not only by the prediction of which genes in a genome encode putative TFs, but also by the generation of clone resources that enable their characterization in a variety of experimental assays (Box 4.1). The Gateway cloning system [20,21] has provided a versatile method for open reading frame (ORF) cloning, and large collections of clones (referred to as 'ORFeomes') are available for a variety of organisms, including C. elegans and human [22,23]. These resources form the basis for comprehensive TF clone collections [24−27] that can be used in functional assays to test for protein−protein interactions using yeast two-hybrid assays; to test for DNA binding using yeast one-hybrid (Y1H) assays and protein-binding microarrays (PBMs); or to test for in vivo function using RNA interference (RNAi) (Box 4.1).

GRN Nodes: *Cis*-Regulatory Elements

CREs serve as the DNA-binding sites for sequence-specific TFs that either activate or repress gene expression. The term

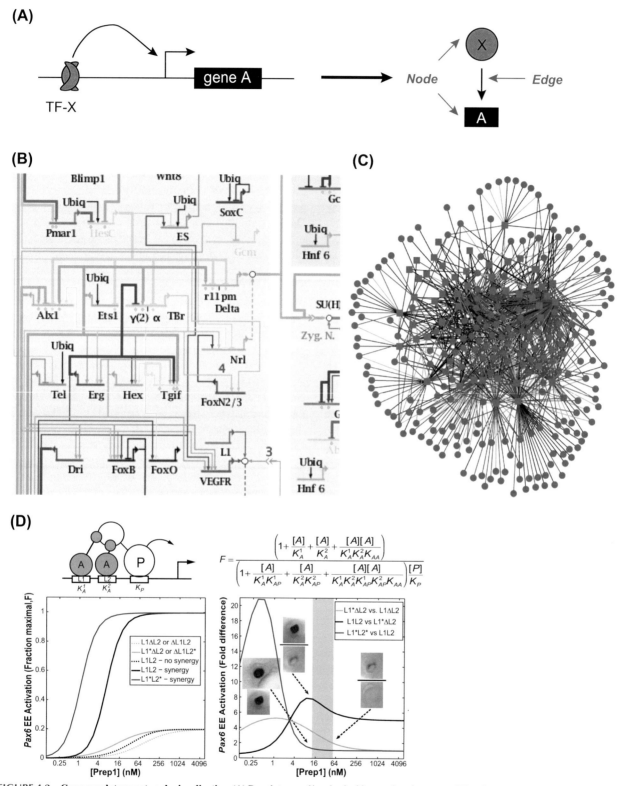

FIGURE 4.2 **Gene regulatory network visualization.** (A) Regulatory and/or physical interactions between a TF and a target gene can be visualized in GRN diagrams that contain nodes (the TFs and their target genes) and edges (physical and/or regulatory relationships between TFs and their target genes). (B) Section of a high-resolution wiring diagram of endomesoderm development in sea urchin (see Chapter 11). (C) Cytoscape-generated GRN of physical interactions between *C. elegans* TFs and metabolic gene promoters [28]. (D) Prep1 binding to two lower-affinity binding sites (L1 and L2) in the *Pax6* ectodermal enhancer (EE) drives expression of *Pax6* in the developing mouse lens. The Prep1 dependence of the EE was described using a biophysical model of Prep1 molecules (A in schematic) binding to sites L1 and L2 and synergistically recruiting a required co-factor (P in schematic). The model was

CRE can refer either to individual TF-binding sites, or to a collection of TF-binding sites clustered within a broader, contiguous stretch of sequence such as a CRM (Figure 4.1C). CRMs that activate gene expression are frequently referred to as transcriptional enhancers, while those that repress gene expression are referred to as transcriptional silencers. Classically, enhancers were defined as being able to activate gene expression regardless of distance from the transcription start site (TSS), location upstream or downstream of TSS, or orientation [30]. However, it has been found more recently that such activating elements often are not completely independent of location and orientation (indeed, such effects on CRM activity are often not even tested); instead, the term 'enhancer' is now more commonly used to refer to an activating CRM comprising a cluster of TF DNA-binding sites [30]. Other types of DNA elements, most notably gene promoters (see below) and insulators, also contribute to the regulation of gene expression. Insulators are DNA elements that, when placed between an enhancer and a promoter, can block the activating role of the enhancer on gene expression. In addition to such enhancer-blocking insulators, there are also barrier insulators, which block the spread of heterochromatinization and subsequent gene silencing [31].

In microbial genomes, such as *Escherichia coli* and *Saccharomyces cerevisiae*, CREs typically occur within a few hundred base pairs upstream of TSSs. Such upstream regions where regulatory elements occur are often loosely referred to as 'promoter regions'. These regions can be classified into: 1) basal promoters that harbor core elements that are bound by general TFs and RNA polymerase II; and 2) 'proximal promoters', which is a generic term that refers to the sequence around 1 kb upstream of the TSS. More complex eukaryotes additionally display regulation of promoter activity by distal regions, such as enhancers that can be located tens of kilobases upstream of TSSs. Various technologies, including chromatin immunoprecipitation (ChIP) (see Box 4.1) of RNA polymerase II [32], 5′ RACE (rapid amplification of cDNA ends) [33], 5′ CAGE (cap-analysis of gene expression) [34], and RNA-seq [35] have been used to identify 5′ ends of transcripts, and promoters are inferred as the sequence that is located immediately upstream of these 5′ ends.

TF-Binding Sites

The identification of the DNA-binding sites of TFs (typically ~6–15 bp in length) is of fundamental importance for the understanding of systems-level gene regulation. Various computational and experimental approaches (see below, and Box 4.1 and Box 4.2) have been developed to identify TF DNA-binding sites (either genomic sites or in vitro TF-DNA sequence specificities) and regulatory DNA motifs (here 'motif' refers to a computational model representing a set of similar sequences that putatively share a functional role, for instance by interacting with a particular TF). Computational (or in silico) approaches include searching non-coding genomic sequences upstream of co-expressed genes and genome-wide searches for over-represented phylogenetically conserved DNA sequence motifs [36–38]. The underlying hypothesis in such studies is that the non-coding regions' co-expressed genes are more likely to be bound and regulated by similar TF(s), and that such TF-binding sites have been conserved through evolution. Once TF-binding sites or putative CRMs have been identified, computational algorithms can be applied to search a genome, in particular non-coding regions, for matches to such sequences (Box 4.2). However, because these sites are short, many sequence matches occur by chance alone and may not actually serve a regulatory function. Furthermore, many such sites in eukaryotic genomes are not available for TF binding in a particular cell type or time point because they are occluded by nucleosomes. To address these challenges, more advanced computational algorithms have been developed that score genomic sequences in a weighted manner according to sequence accessibility [39].

Cis-Regulatory Modules (CRMs)

In the yeast *S. cerevisiae*, DNA-binding sites of regulatory TFs typically occur within ~600 bp upstream of genes and effective computational methods exist for mapping TFs and their associated DNA-binding site motifs to their target genes (Box 4.2) [40–42]. In contrast, metazoan CREs can be located far from the TSSs of the genes they regulate [43]. Moreover, in metazoans, regulatory motifs tend to co-occur

---

parameterized using protein-DNA-binding measurements and constrained by reporter gene assays in which the EE was used to drive β-galactosidase expression in the lens. (Top left) Schematic of the two-site model. (Top right) Biophysical function describing fraction of maximal EE activation (F), dependent on protein concentrations ([A], [P]), protein-DNA-binding constants (K_A, K_P), and protein-protein binding constants (K_{AP}, K_{AA}). (Lower left panel) Modeling of EE activation vs. Prep1 concentration (log 2 scale) for wild-type EE (i.e., L1L2) modeled with synergy between DNA-bound Prep1 molecules (black solid line) or without synergy (black dashed line), single-site mutations (ΔL1L2, L1ΔL2, gray dashed line) or high-affinity site mutations (ΔL1*L2*, L1ΔL2*, gray solid line; L1*L2*, blue solid line; * indicates high-affinity site mutant). (Lower right panel) Ratio of activation levels for reporter constructs modeled in the left panel to facilitate the comparison of model predictions and ratio of relative EE reporter levels. Estimates of the reporter level ratios (e.g. L1*ΔL2/L1ΔL2 > 2, L1L2/L1*ΔL2 > 4; L1*L2*/L1L2 < 1.3) provide constraints on the Prep1 concentration ([A] in model) and allowed estimation of physiological concentration of Prep1 (shaded red rectangle). Model correctly predicted greatest variation between native EE (L1L2) and mutant with high-affinity sites (L1*L2*) would occur at lower Prep1 concentrations (see red curve) that occur earlier in development. (*Adapted from [29] with permission from Cold Spring Harbor Laboratory Press*).

Box 4.1 Experimental Technologies for Identifying Protein−DNA Interactions

Chromatin immunoprecipitation (ChIP), coupled with either microarrays (ChIP-chip) or sequencing (ChIP-seq):
What it provides:
- In vivo occupancies of a protein at genomic loci

How it works:
- Cells are treated with a cross-linking agent (typically formaldehyde) and lyzed, chromatin is sheared into smaller DNA fragments; target protein is immunoprecipitated with an antibody specific for either the protein or a tag engineered onto the protein; cross-links are reversed to free the DNA bound by the protein; DNA is amplified and then either labeled for hybridization to microarrays or prepared for next-generation sequencing; statistical comparison to an 'input' sample (which indicates what was present before immunoprecipitation) permits identification of significantly 'bound' regions. Controls for antibody specificity can include a 'mock' ChIP (without antibody), ChIP in a sample that lacks the TF of interest (i.e., in a mutant or by RNAi), or ChIP with an independent antibody against the TF of interest

Throughput:
- Genome-scale
- Low at protein level because TFs need to be immunoprecipitated individually

Advantages:
- Condition-specific, in vivo binding data

Disadvantages:
- Difficulty in identifying individual TF−DNA-binding sites within a bound region
- Binding events can be due to direct or indirect TF-DNA association [39,44]
- For many TFs, specific ChIP-grade antibodies that do not cross-react with TF family members are not yet available
- Typically requires at least 5×10^7 cells
- Difficult to detect binding events that are rare (e.g., involve TFs that are expressed at low levels or in just a few cells within a population of cells)
- Does not provide data on all possible DNA-binding sites of TFs (i.e., condition-dependent binding)

DNA adenine methyltransferase (Dam) fusion and profiling (DamID):
What it provides:
- In vivo occupancies of a protein at genomic loci

How it works:
- Protein of interest is expressed as fusion with the *Escherichia coli* DNA adenine methyltransferase (Dam) protein; wherever the protein associates with DNA, Dam methylates adenines within nearby GATC sites; methylated sites are detected by digestion with a methyl-specific restriction enzyme followed by amplification, labeling, and hybridization to a DNA microarray; comparison to a 'Dam alone' control (i.e., no fusion) permits identification of significantly methylated, and thus 'bound', regions [45]

Throughput:
- Genome-scale (scale depends on genomic regions covered by the DNA microarrays)
- Proteins are fused to DamID and profiled individually

Advantages:
- Does not require ChIP-grade, specific antibodies
- In vivo assay
- Can be used for low-abundance TFs

Disadvantages:
- Spatial resolution of binding events is lower than that achieved by ChIP-seq
- Temporal resolution of binding events is lower than that achieved by ChIP-seq, since the Dam methylation event persists after TF binding
- Expression of Dam fusion protein needs to be optimized so that level of expression is low and does not result in potentially artifactual TF binding due to TF overexpression
- A higher-throughput version of this technology using sequencing instead of a microarray-based readout has not yet been described

Protein-binding microarrays (PBMs):
What it provides:
- High-resolution, comprehensive, in vitro DNA-binding specificity data for either individual TFs or TF complexes

How it works:
- Protein of interest is generated by in vitro translation, purified from native cells expressing the TF, or purified from a heterologous expression system, and then applied directly to a double-stranded DNA microarray (the DNA is naked, not chromatinized); the protein bound to the array is detected with an antibody specific for either the protein or a tag to which the protein is fused (by recombinant DNA cloning); if the primary antibody is not fluorophore-conjugated, then a fluorophore-conjugated secondary antibody is subsequently applied to the labeled, protein-bound array; data analysis of the antibody-derived fluorescence signal intensity compared to the DNA-derived fluorescence signal intensity permits identification of preferred DNA-binding sequences

Throughput:
- Can be either genome-scale covering all genomic TF-binding sites [46], or comprehensive, synthetic sequence design covering all possible k-mers (typically $k = 10$ or more bp) [47]
- Multiple proteins can be assayed simultaneously in a multiplex microarray format

Advantages:
- Provides data on DNA-binding specificity of particular protein(s) (e.g., TFs) under buffer conditions applied to microarray
- Multiple proteins can be assayed in parallel simultaneously
- Tens of thousands of different sequences can be assayed in a single PBM experiment

Disadvantages:
- Impact of in vivo interactions with other proteins or molecules is unknown unless specifically tested
- Naked DNA on microarrays does not provide information on protein binding to chromatinized DNA
- Suitable source of protein (full-length or DBD only) not always obtainable

Continued on next page

Box 4.1 Experimental Technologies for Identifying Protein—DNA Interactions—*Cont'd*

In vitro selection ('SELEX'):
What it provides:
- In vitro DNA-binding specificity data

How it works:
- A library of synthetic oligonucleotides, typically containing ~10–25 bp of degenerate sequence, is synthesized and incubated with a protein of interest under conditions of mild to moderate stringency; protein-bound DNA fragments are captured and amplified; typically several rounds of selection and elution are performed, sometimes entailing a mutagenic PCR step(s) to further increase DNA sequence diversity, in order to select for sequences bound with higher affinity; bound sequences are identified by sequencing [48]

Throughput:
- Typically one protein is assayed at a time
- Resolution and quantitative nature of data depend on complexity of initial oligonucleotide library, stringency of selection, and depth of DNA sequencing

Advantages:
- Complexity of initial oligonucleotide library can be quite high

Disadvantages:
- If not careful, one can over-select and obtain primarily high-affinity binding sequences and lose information on lower-affinity sequences
- Need relatively large amounts of active, purified protein

Yeast one-hybrid (Y1H):
What it provides:
- Interaction between a DNA fragment and a protein, which can be either a larger (~2 kb) genomic fragment or a short (6–20 bp, often multiple copies) DNA sequence such as a putative CRE [49–51]

How it works:
- DNA fragments of interest ('DNA baits') are cloned upstream of two reporter genes (typically *HIS3* and LacZ); the DNA bait::receptor constructs are integrated into the genome of a Y1H yeast strain into a mutant marker locus; ORFs encoding proteins of interest are fused to the Gal4 activation domain (AD) ('prey'); yeast with integrated DNA bait::receptor construct are transformed with a plasmid encoding the prey, or mated with a yeast strain carrying the prey construct; positive bait–prey interactions are identified from colonies where reporter gene activity is turned on: His3 expression is determined by growth on media lacking histidine and containing the competitive His3 inhibitor 3-aminotriazole and LacZ induction is measured by a colorimeteric (blue/white) assay

Throughput:
- When collections of TFs are used in an array format, the assay is proteome-scale. The coverage for TF collections for *C. elegans* is ~90% (834 out of 937 predicted TFs) [52], for human it is ~70% (988 out of 1434) [26], for *Drosophila* it is ~78% (588 out of 755) [53] and for *Arabidopsis* it is ~40% (645 out of 1500) [24]
- Multiple DNA baits can be processed simultaneously

Advantages:
- Interactions are tested in a eukaryotic organism (albeit in a heterologous context) independent of native conditions or native TF expression levels; as a result the assay is less condition dependent.
- Can be done at various levels of scale and throughput, from a single DNA fragment with one TF using standard molecular biology lab resources to multiple DNA fragments and large arrays of TFs with sophisticated robotic and computational tools

Disadvantages:
- DNA fragments are not tested in their native chromosomal context
- Not yet suitable for heterodimers
- Does not provide information on in vivo interactions with other proteins or molecules unless specifically tested
- TF collections are available for only some organisms and are not yet complete

---

as either homotypic or heterotypic TF DNA-binding site clusters within CRMs [54], whereas in *E. coli* and *S. cerevisiae* transcriptional regulation is often effected through single or pairs of TF DNA-binding sites [55,56]. Typical CRMs are ~200–1000 bp in length and contain one or more DNA-binding sites for one or more TFs that activate or repress the expression of target gene(s) [57]. Identification of tissue/cell-type-specific CRMs (either enhancers or silencers) remains a significant challenge and has been the focus of many computational and experimental studies (see below).

Computational methods for CRM prediction are currently based on the model that CRMs are essentially independent, functional regulatory units composed of clusters of TF-binding sites [58–60]. These methods typically favor the identification of evolutionarily conserved sequences [43,61–63]. Combinations of TF DNA-binding site motifs, or higher-order requirements on the arrangement (spacing, order, orientation) of the motifs relative to each other, that are associated with particular gene expression output patterns are often referred to as '*cis*-regulatory codes'.

One experimental strategy to identify enhancers has been to test conserved non-coding sequences for reporter activity in tissues of interest. For example, an enhancer trap of 1 Mb surrounding the gene Sonic Hedgehog (Shh), followed by testing of non-coding elements conserved between mouse and human, resulted in the identification of

Box 4.2 Computational Approaches for Predicting Transcriptional Regulatory Interactions

De novo motif finding:
What it provides:
- Sequence motif(s) over-represented within an input set of DNA sequences

How it works:
- Input sequences are searched, using one of multiple available search algorithms, for over-representation as compared to user-defined background sequences

Advantages:
- Does not require prior hypothesis of which TF DNA-binding sites might be over-represented

Disadvantages:
- Identified sequence motifs might not actually serve a regulatory role
- Identified motifs may not permit unambiguous mapping to a putative DNA-binding protein
- Depends on appropriate definition of background sequences

Scanning genomic sequence for TF binding site sequence matches:
What it provides:
- Locations and identities of candidate TF DNA-binding sites, according to user-defined search parameters, within user-defined input sequences

How it works:
- A user-defined input sequence (e.g., a promoter or candidate CRM) is scored for matches to TF binding site sequences using either a word-based or position weight matrix (PWM) model of the binding sequences [41]

Advantages:
- Can be automated to search many and/or lengthy sequences, for matches to binding sites for one or more TFs
- Systematic scan of acceptable DNA-binding site sequences with match scores (depending on choice of algorithm)
- Input DNA sequences can be either unbiased or filtered for likely regulatory regions

Disadvantages:
- Depends on quality and depth of TF–DNA-binding site data and corresponding binding model and user-defined scoring threshold
- Not all TF–DNA-binding site sequence matches will necessarily correspond to functional, regulatory sites

Gene regulatory network inference:
What it provides:
- Varies depending on the algorithm; can provide information on either regulatory DNA motifs or on regulatory proteins and their putative target genes

How it works:
- Lever algorithm identifies which DNA motifs (from a previously compiled set of motifs) or motif combinations are enriched within predicted CRMs associated with user-defined, input gene sets as compared to background genes [64]
- Module Networks algorithm applies a machine learning approach to user-input gene expression data to assign genes to sets of putatively co-regulated genes and to learn which regulatory proteins (from a previously compiled set of putative regulatory proteins) appear to regulate the gene sets [65]
- ARACNe algorithm applies a mutual information approach to identify interacting genes from user-input gene expression data [66]
- Bayesian analysis has been applied to gene expression data [67] and ChIP-chip data [68] to infer GRNs

Advantages:
- Can be automated to search many and/or lengthy sequences, for matches to binding sites for one or more TFs
- Input DNA sequences can be either unbiased or filtered for likely regulatory regions

Disadvantages:
- Enriched motifs or motif combinations might not permit unambiguous mapping to a putative DNA-binding protein
- Regulation of a gene set by a protein, as inferred from gene expression data, could be due to either direct or secondary effects
- Some regulatory proteins are not controlled at the level of gene expression, but rather post-transcriptionally

---

three novel enhancers that confer Shh-like expression [69]. Another example was the testing of 167 highly conserved non-coding elements that resulted in the identification of 61 enhancers that are active in the brain [70]. Another strategy has focused on non-coding regions containing multiple TF-binding sites identified by ChIP (see below and Box 4.1); seven of 13 regions bound by multiple sequence-specific cardiac TFs and the transcriptional co-activator p300 were found to drive cardiac expression in transient transgenic mouse embryos [71]. Finally, DNase I hypersensitive (DHS) sites [72–74] and regions detected in formaldehyde-assisted isolation of regulatory elements (FAIRE) assays [75] correspond to nucleosome-depleted regions and have been shown to be effective in predicting TF-binding locations that may function as enhancers [76]. Extensive data associated with regulatory elements, including DHS, FAIRE, and ChIP-seq on TFs, histone modifications, and CBP have been generated by the ENCODE project [77], the NIH Epigenomics Roadmap project, and other groups for many commonly studied human cell lines and cell types [78,79]. Similar data sets have been generated for cell lines and whole animals by the modENCODE Consortium for *D. melanogaster* [80] and *C. elegans* [81].

The functionality of enhancers can be studied in two complementary ways: either by using reporter constructs to determine enhancer activity, or by perturbing (deleting or mutating) enhancers in the genome. Perturbing CREs is relatively straightforward in *S. cerevisiae*, where highly efficient homologous recombination permits investigators to create deletions or mutant versions of CREs (or genes encoding the TFs that bind them) in the genome, and examining the effect of such deletions on gene expression or more global phenotypes. Such studies are much more challenging in metazoans, where homologous recombination is less efficient. Instead, heterologous reporter assays that measure enhancer activity are typically performed at non-native chromosomal loci or using plasmid-based assays [82,83]. *Drosophila* has been a major model organism for the study of enhancers in animals. In particular, significant efforts aimed at identifying transcriptional enhancers in *D. melanogaster* have focused on the blastoderm stage of development [84–86]. In addition, a GFP reporter assay has been employed to examine 44 DNA fragments, each ~3-kb long, for enhancer activity in the adult fly brain [87]. Other groups have taken a two-tiered approach whereby TF-occupied regions identified by ChIP were subsequently tested in reporter assays [88,89]. Altogether, it has been estimated that there are ~50 000 transcriptional enhancers in the *Drosophila* genome [87]. Despite all these efforts, at the time of writing there are still fewer than 900 known transcriptional enhancers in *D. melanogaster* [90]. Except for a handful of examples [82], CREs in *C. elegans* have been poorly characterized, and even basic trends about CRE complexity in *C. elegans*, such as the extent of combinatorial TF input and whether tissue-specific gene expression is regulated by complex transcriptional enhancers as in in *Drosophila* and mammals, remain unknown.

Testing of mammalian enhancers by in vivo knockout at a native chromosomal locus is a slow, laborious and costly approach that limits testing to just one or a few enhancers within a study [91]. Therefore, candidate mammalian enhancers are instead typically tested by reporter assays. Currently the largest-scale testing of mammalian enhancers can examine of the order of ~100 predicted enhancers in a single study. Such studies generally employ one of two methods: whole-mount *LacZ* reporter staining in mouse [70,92–94], or luciferase reporter assays in mammalian cell culture [95,96]. Mammalian enhancers can also be tested in zebrafish, which offers the advantage that it can be done at higher throughput than testing in mice [97]. However, a negative result can be more difficult to interpret, as the transcriptional regulatory networks in mammals might be substantially diverged from those in zebrafish, and because fish are not useful for assessing enhancer activity that may be specific to cell types found only in mammals or primates, such as certain neurons [98].

GRN Edges: Physical Interactions Between TFs and DNA

High-throughput experimental methods that can be used for the identification of physical TF–target gene interactions (Box 4.1) can be classified into two conceptually complementary categories [99]. The first are 'TF-centered' (protein-to-DNA) methods that identify DNA sequences or loci bound by individual TFs. Target genes are then usually attributed to a TF by their proximity to a TF-binding event. ChIP, combined with either microarrays (ChIP-chip) or high-throughput sequencing (ChIP-seq), is the most widely used method for the TF-centered identification of in vivo TF-DNA interactions, and has been used in a variety of model systems. For example, ChIP has been comprehensively used in *S. cerevisiae* to map physical interactions between most of the TFs and gene promoters both in rich media and under a set of different conditions [100,101].

An in vitro TF-centered approach is protein-binding microarray (PBM) technology, in which a TF is applied directly to double-stranded DNA microarrays covering a wide range of possible DNA-binding site sequences. PBMs have been used to determine the in vitro DNA-binding specificities of a variety of different classes of TFs from a wide range of organisms, including yeast [102], worms [18] and mice [103,104]. Based on the PBM results, position weight matrices (PWMs) can be constructed that capture the binding specificity of each TF. Such PWMs can then be used to identify potential binding sites for the TF in the relevant genome. Two additional TF-centered methods for TF–DNA interaction identification, SELEX and DamID, are described in Box 4.1.

The second category is comprised of 'gene-centered' (DNA-to-protein) methods that identify the repertoire of TFs that can interact with a regulatory genomic region of interest. Yeast one-hybrid (Y1H) assays provide such a gene-centered method, and have been mostly applied to the delineation of GRNs in the nematode *C. elegans* [28,52,105–107], though more recently Y1H resources have become available for other organisms as well [24,26,52].

TF- and gene-centered approaches are complementary because they enable a researcher to address different types of questions related to the systems biology of gene expression [44,99,108]. For instance, ChIP is highly useful when one is interested in identifying the in vivo DNA targets for one or a few TFs, e.g., involved in development or in a particular disease. Y1H assays, on the other hand, are particularly useful when one is interested in a single gene or a set of genes involved in a system or process of interest. For instance, when one would like to know the TFs that bind to a promoter of interest, it would not be practical to perform ChIP for all TFs; instead, Y1H assays could be used to screen a collection of TFs for binding to the promoter of interest. In addition to being conceptually complementary, the two approaches also have unique advantages and limitations (Box 4.1). For example, ChIP assays provide data on TF-binding specificity in particular cellular and environmental conditions, and can detect binding of TF complexes and modified TFs in vivo. Y1H assays can detect interactions with low-abundance TFs, identify multiple TFs in a single experiment, and detect direct interactions between DNA and TFs. PBM assays can provide a comprehensive survey of all possible DNA-binding site sequences recognized by TFs or TF complexes of interest, and they can be used as a biochemical tool to assess the impact of co-factors, such as interacting proteins, on DNA-binding specificity.

As mentioned above, a key advantage of ChIP is that it can be used to identify enhancers that are active in the cell or tissue used. In other words, in such experiments the identification of network *edges* can be used to annotate GRN *nodes*, although it may not be clear which gene(s) such putative enhancers regulate. For instance, genomic regions bound by the transcriptional co-activator p300/CBP and related histone acetyltransferases, sometimes together with chromatin signatures comprising particular histone modifications, can be used to accurately predict locations of active enhancers [93,109–111], even where cross-species sequence conservation is weak [92]. Based on such data, it has been estimated that there are of the order of 10^5-10^6 enhancers in the human genome [110].

Multiple experimental approaches have been developed to map *regulatory* interactions between TFs and their target genes, or the enhancers ascribed to these genes. In a series of classic, one-gene-at-a-time experiments, Davidson and colleagues delineated numerous regulatory relationships involved in endomesoderm development in the sea urchin (see Chapter 11). The compilation of this work has led to one of the first GRNs described in any metazoan system [112]. More recently, the development of a variety of high-throughput experimental technologies (Box 4.1) has enabled the large-scale identification of regulatory interactions, leading to the prospect of more complete GRNs for a variety of model organisms and humans. A widely used computational method is to infer GRNs from gene expression data (sometimes called 'reverse engineering') (Box 4.2). The underlying hypothesis is that sets of co-expressed genes ('gene modules' or 'gene batteries') are co-regulated by the same TF(s) (e.g., [64,65,113]). TFs are identified that exhibit a similar profile as a particular gene battery and are subsequently inferred to be responsible for, or involved in, generating the expression profile of the gene set [65,66]. Opposite expression profiles of TFs and target gene sets can also be informative because they may point to transcriptional repression. Regulatory network inference has been applied successfully to a variety of model systems, including yeast and mammals [66,114]. A major disadvantage of these approaches is that many of the interactions that are predicted are likely indirect, and many relevant interactions are missed because the TF that regulates a set of genes does not itself change in expression in the same way or at all [44] (see also Box 4.2).

GRNs: Visualization

When the interactions between TFs and their target genes are connected, complex and highly intricate networks emerge that are challenging to analyze by eye. Several tools are available for the visualization and interrogation of GRNs, including Cytoscape [115], Osprey [116], Nbrowse [117] and Biotapestry [118]. It is important to bear in mind that the graphs generated by these tools may reflect a compilation of molecular events that occur throughout the lifetime of the cell or organism, depending on the method that was used to delineate the network, and that not all interactions are necessarily direct or have a regulatory consequence.

GRNs: Data Quality

The quality of GRNs depends on the proportion of real interactions included and the number of included interactions that are real. False negatives refer to real interactions that are not detected with the experimental method used to delineate the network, and false positives refer to detected interactions that are not real. Issues related to false positives and negatives in different types of genomic data sets have been hotly debated over the past 15 years or so, particularly with regard to yeast two-hybrid and Y1H assays [44,119,120].

False negatives can arise because of limitations in each of the assays used to build a GRN. For example, in computationally inferred GRNs, false negatives occur when the expression of the regulatory TF itself does not change together with the gene module it regulates. Since many TFs are activated at the protein level by an outside signal that results in a post-translational modification (or removal thereof), the number of interactions missed by gene expression profiling is likely substantial. More technically, the degree of correlation between a TF and its target genes, together with the threshold applied, will determine which interactions are included and which are not.

The different types of physical interaction detection assay also each have their limitations. Gene-centered Y1H assays are tested in the heterologous environment of the yeast nucleus. This provides distinct advantages, but also results in limitations: Y1H assays are not effective at detecting interactions of DNA with heterodimeric TFs or TFs that need to be post-translationally modified prior to binding DNA. Further, TF binding in the context of the yeast nucleus to the DNA fragment assayed (which is integrated into the yeast genome) will depend on the assembly of nucleosomes in a similar manner as in the system from which the gene under study was selected. ChIP assays also have specific limitations: ChIP-grade antibodies are not available for most TFs, ChIP is challenging to perform for low-abundance TFs, or TFs expressed in only few cells in the organisms, and may detect indirect DNA−TF interactions due to the cross-linking of larger protein−DNA complexes such as enhancer−promoter loops (see Chapter 7). Finally, because of statistical thresholds chosen, true but weaker ChIP interactions may not have been included.

False positives are a much more contentious data quality issue than false negatives. First, with any assay, whether high- or low-throughput, there are two types of false positive: *technical false positives* that cannot be reproduced, even with the same assay, and *biological false positives* that can be reproduced technically, but that are not valid biologically. The first type of false positive is obviously highly detrimental to the quality of the resulting GRN, and many quality control steps have been introduced in the different assays to assure that such erroneous interactions are kept to a minimum (see Chapter 3). Examples for ChIP assays are the use of multiple antibodies in parallel, or comparing ChIP data in the test sample to data obtained in a sample in which the TF or its DNA-binding activity is lost due to a mutation or knockdown by RNAi. In Y1H assays interactions can be re-tested in fresh yeast cells, and, where possible, should confer a positive readout with both reporters. Biological false positives, or true negatives, are much more challenging to define and therefore to identify. Technically, it should be appreciated that the assay used to validate the regulatory or phenotypic consequence of a TF-DNA interaction will have its limitations and will therefore not detect all true interactions. Aside from technical issues, the question is how a biologically meaningful interaction is defined. Textbook views on transcriptional regulation (such as Figure 4.1C) draw a TF-binding event, link it to a single gene and assume that each binding event will have a regulatory consequence. It is becoming clear, however, that it is not that simple, as many TF-DNA-binding events appear to be silent (e.g., [121−123]. For instance, when ChIP and expression profiling data of the transcriptome were compared, only a small overlap in genes that are both bound by a TF and change in expression in its absence appears to be the rule rather than the exception [124].

There are both conceptual and technical reasons for the lack of overlap between physical and regulatory interactions [44]. First, regulatory relationships between TFs and genes that change in expression upon TF reduction or overexpression can be indirect. For instance, in linear cascades (Figure 4.3C), where one TF affects the expression of another, reduction of either can affect the expression of a downstream gene, but only one may directly bind its regulatory DNA region(s). Practically, it can also be that both TFs do physically associate with the gene in vivo, but only one can be detected with any of the physical interaction assays (Box 4.1). Conversely, several reasons can explain why not all physical interactions have a measurable regulatory consequence. For example, a physical interaction may affect gene expression only under particular conditions, for instance upon activation of the bound TF in response to an outside signal. Further, the functional role of a TF may be masked because of redundancy with other bound TFs. Also, the binding event may have been attributed to the wrong gene. This is a more likely issue in larger genomes where CREs can be located far from TSSs. Taken together, it is extremely challenging to identify which interactions in a GRN may never be functional. Overall, it is clear that there is no single method that will identify all TF−DNA interactions, and therefore there is a need for the development and application of a variety of methods, as well as the integration of the data obtained with other types of pertinent information such as gene expression patterns and phenotypes.

GRN Structure and Function

GRNs can be analyzed computationally using graph-theoretical algorithms and statistics, both for their overall structure or topology and for their local circuitry (see Chapter 9). Such analyses can provide insights into the design principles of gene regulation at different levels, which cannot be obtained from single-gene studies. The first level of GRN analysis is usually to determine the number of edges in which each node in the network participates. This is referred to as the degree of a node (Figure 4.3A). Because GRNs are both bipartite and directional, there are two types of degree: out-degree and in-degree. The out-degree (k_{out}) is defined as the number of genes bound or regulated by a particular TF, while the in-degree (k_{in}) is the number of TFs that bind or regulate a certain gene. To obtain a view of the overall wiring and hence the topology of a GRN, the degree for each node in the network is computed and the distribution of the degrees is plotted on a graph (Figure 4.3A). Interestingly, most real networks, whether biological or not, exhibit a scale-free degree distribution [125]. This indicates that the vast majority of nodes have relatively few connections, but that a small number of nodes are extremely highly connected. Such highly connected nodes are referred to as hubs. In GRNs, the out-degree is best fit by a scale-free degree distribution. The in-degree, however, is fit better by an exponential degree distribution [105]. Although the difference between these distributions is not yet understood, it is clear that most TFs bind or regulate small sets of genes,

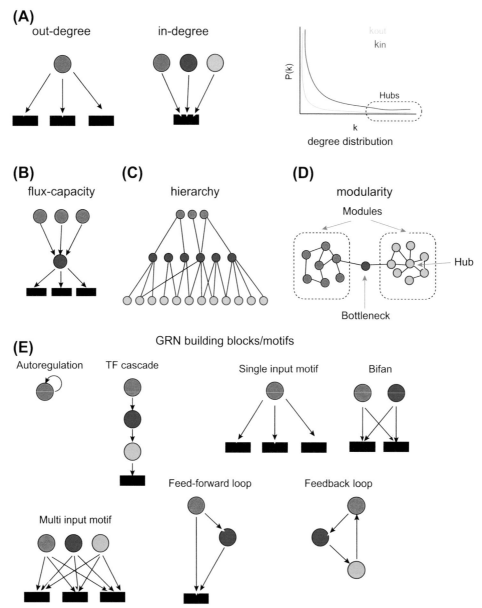

FIGURE 4.3 **GRN structure and topological features.** (A) The degree of a node refers to the number of connections it engages in. In GRNs, two types of degree can be distinguished: the out-degree (k_{out}) (left) refers to the number of genes targeted by a TF, and the in-degree (k_{in}) (middle) reflects the number of TFs that target a gene. All degrees in a GRN can be analyzed together to gain insight into the degree distribution (right). Nodes with a very high degree are referred to as hubs. (B) The product of the in- and out-degrees is referred to as the flux-capacity, and illustrates the number of information paths that pass through a node. (C) GRNs of *E. coli* and yeast are hierarchical, with different layers of TFs regulating one another. (D) GRNs are modular and bottleneck TFs with high between-ness (the number of paths it engages in that connect any two nodes in the network). For visualization purposes, the network is visualized with a single type of node (TF) and undirected edges (as in protein–protein interaction networks). (E) GRNs contain many different types of small circuits, or building blocks. When statistically overrepresented, these circuits are referred to as network motifs.

but that some TFs are broadly involved in gene control. Conversely, most genes are bound or regulated by a modest number of TFs, while some exhibit extensive regulatory input. Overall, the presence of hubs ensures network connectivity and integrity [125]. By combining the in- and the out-degree (k_{in} x k_{out}) the flux-capacity of a node can be computed, which indicates the number of information paths that pass directly through a node and a high flux-capacity may indicate the importance of that node to network functionality (Figure 4.3B) [106]. So far, the analysis of flux-capacity has been performed only on an integrated network that combines transcriptional regulation with post-transcriptional regulation by microRNAs in *C. elegans* [106]. In purely transcriptional GRNs (i.e., only involving TFs), flux-capacity implies a hierarchical structure, with TFs as both upstream regulators and downstream targets (Figure 4.3C).

Node betweenness is another measure of information flow through a node, and is also referred to as node centrality. This metric is defined as the number of times a path between any two nodes in the network passes through a node (i.e., not necessarily directly connecting to the node) (Figure 4.3D). Nodes with high between-ness are referred to as bottlenecks and often connect different network modules (see below).

Metabolic networks exhibit a hierarchical organization of modularity that is proposed to enable rapid reorganization and integration with other cellular functions [126]. A hierarchical organization has also been identified in GRNs, using a variety of computational and mathematical approaches. GRN hierarchy has so far been studied only in unicellular organisms, such as bacteria (*E. coli*) and yeast (*S. cerevisiae*), because sufficient amounts of high-quality data are not yet available for higher organisms. These unicellular GRNs were found to have a hierarchical organization with multiple layers of TFs that control the expression of other TFs. While some studies suggest a pyramidal structure for the network [127] (Figure 4.3C), others have identified an overall feedforward structure [128,129]. The different methods identified distinct numbers of hierarchical levels that can be further grouped into GRN layers. It has been proposed that a hierarchical GRN structure can lead to signal amplification and can confer robustness and adaptability [126,128]. Interestingly, the TFs that reside in different GRN layers have distinct attributes. For instance, TF hubs that regulate large numbers of downstream targets are rarely found at the top of the hierarchy. Similarly, bottleneck TFs are mostly found in the middle, or 'core' layers. There are several conflicting findings related to GRN hierarchy as well. Where some studies find that the TFs at the bottom are more essential [127], others find that it is the TFs at the top that are required most for viability [128]. Studies agree, however, that the TFs occurring in the top layer are most highly conserved. It will be interesting to determine whether the overall networks of metazoans and plants are also hierarchical, and whether there are different structures in subnetworks that control gene expression in the development of different tissues or under particular physiological or environmental conditions.

Network modules refer to 'neighborhoods' or clusters of nodes that are highly interconnected (Figure 4.3D). This is analogous to geographic locations such as countries in a continent, or neighborhoods in a city, that usually are inhabited by people with similar socioeconomic characteristics. Hence, network modules have been used to identify nodes with similar biological or biochemical attributes. Network modules can be identified using a variety of mathematical methods (see Chapter 9). A high degree of modularity has been proposed to facilitate a rapid response to external cues and has been observed in several metabolic networks and GRNs. In GRNs, two types of modularity can be studied: TF modules that are characterized by sharing target genes, and gene modules that share interacting TFs [27,28,107]. Redundant TFs are expected to reside in similar modules and may share target genes by binding highly similar DNA sequences. Such redundancies can be identified genetically, for instance in high-throughput genetic network screens in yeast (see Chapter 6) or by RNA interference in more complex metazoans (ref to Perrimon chapter).

GRNs can be further broken down into different types of small circuits or building blocks that provide different types of information flow and hence the logic of gene control (Figure 4.3E). A special type of such a building block is called a network motif, which is defined as a type of circuit that is overrepresented in real networks compared to randomized networks [130]. This is analogous to TF-binding motifs that can be enriched in the regulatory regions of their target genes compared to background sequence. Network motifs are identified by computationally randomizing GRNs. This can be done in different ways: by keeping the individual connectivity of each node but swapping different edges; by changing the connectivity of each node but keeping the overall degree distribution of the GRN the same; or by completely randomizing the GRN, which leads to a random degree distribution [106,131]. Preserving network architecture is usually the most informative, as most biological networks are scale free, and therefore complete randomization would distort the statistical significance because all nodes would have approximately the same degree.

The simplest GRN building block is autoregulation and involves only a single node: a TF that regulates its own expression. There are two types of autoregulation: one in which the TF represses its own expression (autorepression) and one in which the TF activates its own gene (autoactivation). Mathematical modeling has predicted possible functionalities for autoregulation. Negative autoregulation can confer an increase in the rate of reaching the steady state of the TF, and results in less variability [132,133]. This can be important to generate an appropriate response to a signal that induces the initial expression of the TF. Depending on the strength of its promoter and the input of other regulators, the steady-state level can differ. Further, the rate increase will depend on the strength of autorepression, which is a combination of the affinity of the TF for its own *cis*-regulatory regions and the mechanism by which it represses transcription. Finally, transcriptional autorepression is a more effective mechanism of gene regulation than protein degradation, which could result in the same net effect, but is more costly to the cell. Positive autoregulation can result in a slowing down in reaching steady-state TF levels and an increase in expression noise. Weak positive autoregulation, for instance caused by low-affinity binding of the TF to its own promoter, can result in large differences in TF concentration between different cells. Strong positive autoregulation can result in a bimodal distribution of TF expression in different cells where either high or low TF expression can be maintained. The latter can be useful to generate populations of cells with different properties during development or in response to different environmental (stress) cues. Strikingly, it was observed in early constructions of the *E. coli* GRN that ~40% of TFs negatively regulate their own expression [134]. However, only 6% autoactivate, suggesting that positive autoregulation is much less beneficial to the organism than autorepression. The *E. coli* GRN is still incomplete, and it is not clear which regulatory interactions correspond to physical interactions. Thus, it is possible that some autoregulation involves multiple TFs, rather than being caused by direct binding of TFs to their own promoter. Surprisingly, only 10 of 141 yeast TFs tested by ChIP bind their own promoter region (7%), and it is not clear how many of these events have a regulatory consequence [100,101]. This observation suggests that autoregulation may be more prevalent in prokaryotes, whereas more complex circuitry has evolved in eukaryotes. Comprehensive physical and regulatory datasets in more complex multicellular organisms will provide further insights into the prevalence of autoregulation in the different kingdoms of life.

TF cascades are the second type of building block that occurs in GRNs. They are defined as the sequential regulation of TFs by other TFs, followed by the regulation of downstream, non-TF target genes. Numerous TF cascades have been discovered in many systems, for instance in the context of animal development [135]. TF cascades function through a time delay because each step depends on the accumulation of sufficient amounts of the upstream TF. It is intuitive that this can be useful in development and differentiation, where new cellular states need to be sequentially acquired [136]. Whereas TF cascades do not occur frequently in *E. coli*, the yeast GRN contains 188 potential cascades with lengths between 3 and 10 regulators [101], suggesting that this gene regulatory mechanism is more prevalent in eukaryotes. The vast body of TF cascades that has been identified by more conventional studies suggests that TF cascades may occur even more frequently as complexity increases. Comprehensive data in more complex metazoans and plants will no doubt shed further light on TF cascades and their functions.

The third building block is called a single-input motif and is defined as a set of genes that are bound or regulated by the same TF (e.g., a gene module or gene battery). The bifan and multi-input motifs are variations of the single-input motif. A bifan is defined as two TFs binding or regulating the same two target genes. A multi-input motif corresponds to a gene battery that is co-regulated by multiple TFs. In the latter, not every TF necessarily controls each gene in the battery. This class of GRN circuits is highly useful in development, when sets of genes need to be turned on or off in cellular or organismal differentiation.

Feedforward loops are another important type of regulatory circuit in GRNs. These are defined by the regulation of a gene by two TFs, one of which also regulates the other [137]. Each of the edges in feedforward loops can be either activating or repressing. As a result, there are eight possible types of feedforward loop. When the overall outcome of the direct and indirect paths to the downstream target gene is the same (repressing or activating), the loops are called coherent; when the signs of direct and indirect edges are opposite, the loops are called incoherent. There are four loops of each type. However, each loop can behave differently, depending on the underlying logic structure. For instance, in an 'AND-logic gate' both TFs are required for expression of the downstream target gene, whereas in an 'OR-logic gate' one TF is sufficient. Each type of feedforward loop can confer a particular transcriptional response upon activation of the most upstream TF by an outside signal or by another TF [137]. For instance, in the simplest feedforward loop where all edges are activating, AND-logic provides a delay in expression of the target gene because sufficient amounts of both TFs need to accumulate, whereas OR-logic provides a delay in the reduction of target gene expression upon removal of the upstream signal because both TFs need to be inactivated. Although all types of feedforward loops can occur in GRNs, not all are a true network motif; i.e., only some occur more frequently than expected by chance [130]. Most frequently occurring in *E. coli* and *S. cerevisiae* are the coherent feedforward loop in which all edges are activating, and the incoherent feedforward loops where the upstream TF activates both the target gene and the downstream TF, and the latter represses the downstream target [101,138]. It remains an open question whether similar types of feedforward loop are enriched in the GRNs of complex multicellular systems and which type of logic (AND vs. OR) is most prevalent.

The final circuit that occurs in GRNs is the feedback loop. Feedback can involve a single TF (in the case of autoregulation discussed above), two TFs, or more complicated circuits comprising larger numbers of TFs. An example of a three-node feedback loop is shown in Figure 4.3E. Interestingly, although feedback was found in *E. coli* GRNs, it was not found to be enriched. Because feedback is so abundant throughout biology, this suggests that other types of biomolecule may be involved. Indeed, both protein–protein interactions and microRNAs were found to contribute [106,139]. However, since it has become clear that different types of network building blocks can be enriched in different species, transcriptional feedback loops might also occur in the GRNs of more complex systems such as animals and plants.

GRNs: Model Organisms

A variety of model organisms, ranging from relatively simple unicellular organisms to complex metazoans, have been employed in the study of GRNs. Each of the different major model organisms provides particular advantages and limitations, as discussed briefly here and in more detail in Box 4.3. Consequently, these organisms have been used to investigate different aspects of gene regulation in general and GRNs in particular. In the sections that follow, the major model organisms are discussed, starting with simpler organisms with compact genomes (i.e., microbes), and moving on to more complex organisms with larger genomes (i.e., animals and plants).

E. coli has been a long-standing bacterial model system for the investigation of basic phenomena in TF−DNA interactions, such as cooperativity in TF binding to multiple adjacent DNA recognition sites [140], and the roles of such interactions in the regulation of gene expression. However, a variety of features limit the general utility of *E. coli* as a model organism for understanding eukaryotic gene regulation. These include a compact genome, expression of polycistronic transcripts from operons, the lack of a nucleus, the lack of nucleosomes, and different inherent promoter activity in the absence of activators or repressors compared to eukaryotes [141]. Despite these limitations, studies in *E. coli* have provided important insights into basic biological mechanisms and GRNs, and *E. coli* is regaining attention with the growing number of microbiome studies and their importance for human health [142].

S. cerevisiae is one of the most useful eukaryotic model systems to study the systems biology of gene expression in eukaryotes because many basic biological processes, including transcriptional components such as histones and chromatin remodeling complexes, are conserved in yeast. In addition, it has a compact genome, a short doubling time, and can be easily genetically manipulated, and various genomic resources are available (see Chapters 6 and 8). *S. cerevisiae* served as a primary model system in the development of numerous genomic and proteomic technologies, including gene expression microarrays [143,144], large collections of bar-coded gene knockouts [145], ChIP-chip [146−148] (Box 4.1), genome-wide screens of protein−protein interactions by yeast two-hybrid assays [149,150] or by affinity purification coupled with mass spectrometry [151,152], large-scale genetic interaction screens [153,154], as well as computational approaches for predicting DNA regulatory elements (Box 4.2). The wealth of available genome sequence data and genomic data sets has permitted a number of pioneering, integrated approaches in GRN analysis in yeast [68,155]. Despite all the advantages of yeast, it has much simpler CREs and less combinatorial regulatory input by TFs than is typically found in higher eukaryotes.

A number of metazoans, including the nematode *C. elegans*, the fruit fly *D. melanogaster*, the sea urchin *Strongylocentrotus purpuratus*, and mouse, as well as human cell lines, have been utilized to investigate various aspects of gene regulation in animals. In *C. elegans*, gene-centered Y1H approaches have been used to delineate several medium-scale GRNs [28,105−107]. Although *C. elegans* has more protein-coding genes and TFs than *Drosophila*, intergenic regions are shorter, and it is not clear whether complex CRMs such as enhancers play a major role in gene regulation as they do in *Drosophila*, or if most regulation is conferred through the proximal gene promoter. Since *C. elegans* are transparent throughout their lifetime, green fluorescent protein (GFP) expression can be simply examined in living animals by light microscopy (see Chapter 19). For instance, by using transgenic animals that express GFP under the control of a variety of gene promoters it has been found that promoters often contain all elements required to correct tissue-specific expression [156−158]. This suggests that long-range gene control occurs much less frequently in *C. elegans* than in other complex metazoans.

Transcriptional enhancers have been studied extensively in *Drosophila*. In *Drosophila*, as in mammals, there are often multiple tissue-specific enhancers per gene, and enhancers can be found far from the gene(s) they regulate [159]. Detailed studies of *cis*-regulatory logic in transcriptional enhancers have been carried out in early blastoderm development [85] and embryonic mesoderm development [89,160]. Although dense clustering of TF-binding sites is often thought of as a central feature of transcriptional enhancers in *Drosophila*, recent analysis of 280 experimentally verified CRMs has revealed that binding-site clustering is typical of only a particular subclass of *Drosophila* enhancers [161]. Both *D. melanogaster* and *C. elegans* served as model organisms in efforts by the modENCODE Consortium to catalog the functional regulatory elements in those genomes [80,81], resulting in genomic data sets, in particular by ChIP, on in vivo TF-binding sites for a small number of TFs and hence the positions of putative DNA regulatory elements. Further, because only few TFs have been characterized by ChIP in these studies (22 of 937 for *C. elegans* and 41 of 755 for *Drosophila*), this has not yet led to extensive GRN models. Finally, testing the activity of putative CREs, such as tissue-specific transcriptional enhancers, is still only a moderate-throughput endeavor even in these model organisms.

The sea urchin *S. purpuratus* (see Chapter 11) offers several advantages. First, it has fewer gene duplications than vertebrates, with the practical consequence that there are fewer functionally redundant genes that need to be perturbed experimentally in genetic experiments of gene function. Second, it has a transparent embryo, allowing direct visualization of development and reporter gene

BOX 4.3 Model Organisms Commonly used in Studies of GRNs

Escherichia coli:
Advantages:
- Small, compact genome (K-12 strain [162]: 4288 protein-coding genes, including ~270 sequence-specific TFs [163]; 4.6 Mb; 11.2% non-coding)
- Short generation time (~20 min doubling time in optimal conditions)
- Relative ease of gene targeting and site-directed mutagenesis

Disadvantages:
- Unicellular

Saccharomyces cerevisiae (budding yeast):
Advantages:
- Small, compact genome (S288C strain: 5885 protein-coding genes [164], including ~200 sequence-specific TFs [12]; 12 Mb; ~30% non-coding)
- Short generation time (~90 min doubling time in optimal conditions)
- Facile gene targeting and site-directed mutagenesis (highly efficient homologous recombination)

Disadvantages:
- Unicellular, although there are different cell types according to mating type (a, alpha) and growth form (filamentous, non-filamentous)

Caenorhabditis elegans (nematode worm):
Advantages:
- Moderate-sized genome (~20000 protein-coding genes [165], including ~940 sequence-specific TFs [107,157]; 97 Mb; ~73% non-coding)
- Relatively short generation time (3 days in optimal conditions)
- RNAi knockdown of gene expression
- Moderate efficiency in genomic incorporation of transformed DNA

Disadvantages:
- More cumbersome gene targeting and site-directed mutagenesis (no homologous recombination)

Drosophila melanogaster (fruit fly):
Advantages:
- Moderate-sized genome (~13 600 protein-coding genes [166], including ~750 sequence-specific TFs [167]; ~180 Mb, of which 120 Mb is euchromatin; ~80% non-coding)
- Relatively short generation time (9 days in optimal conditions)
- RNAi knockdown of gene expression

Disadvantages:
- More cumbersome gene targeting and site-directed mutagenesis (inefficient homologous recombination)

Strongylocentrotus purpuratus (sea urchin):
Advantages:
- Highly efficient (although random) genomic incorporation of DNA injected into eggs
- Morpholino knockdown of transcripts [168]

Disadvantages:
- Large genome (~23 300 protein-coding genes, including somewhere between ~420 and ~620 predicted sequence-specific TFs; ~814 Mb [169]; ~93% non-coding; ~30 kb average intergenic distance)
- Slow generation time (~1 year) impedes genetic studies and development of transgenic lines
- More cumbersome gene targeting and site-directed mutagenesis (less efficient homologous recombination)

Mouse:
Advantages:
- Mammal; major non-primate model organism for human
- Many available inbred strains that have been bred selectively to carry human disease traits or other phenotypes of biomedical importance
- RNAi knockdown of gene expression

Disadvantages:
- Large genome (~22 000 protein-coding genes [170], including ~1600 sequence-specific TFs [171]; ~2.5 Gb euchromatic portion of the genome; ~97.9% non-coding)
- Slow generation time (~12 weeks)
- More cumbersome gene targeting and site-directed mutagenesis (less efficient homologous recombination)

Human cell lines:
Advantages:
- Relative ease of growth in lab conditions depending on cell type
- RNAi knockdown of gene expression

Disadvantages:
- Large genome (~22 300 protein-coding genes [172,173], including ~1400 sequence-specific TFs [11,26]; ~3.08 Gb, of which ~2.88 Gb is euchromatin (2.85 Gb of which is in the sequenced portion of the genome); ~98.1% non-coding)
- Transformed or otherwise immortalized cell lines are 'cancer-like' and not truly representative of the primary cells from which they have been derived
- More cumbersome gene targeting and site-directed mutagenesis (less efficient homologous recombination)

Arabidopsis thaliana (thale cress):
Advantages:
- Small genome (diploid) relative to other plants, many of which are polyploid (~25500 protein-coding genes, including ~1700 sequence-specific TFs; ~125 Mb, of which ~115 Mb is sequenced; ~70% non-coding [174])
- Relatively short generation time (~6 weeks in optimal conditions)
- Many inbred strains have been bred selectively to carry specific traits

Disadvantages:
- More cumbersome gene targeting and site-directed mutagenesis (plants preferentially utilize non-homologous end-joining instead of homologous recombination in DNA end repair)

expression. Third, reporter DNA constructs injected into the egg are subsequently stably incorporated with high efficiency, allowing accurate expression of these reporter constructs. This allows testing of the tissue-specific activity of putative CRMs. Thus, the connections between sequence-specific TFs and their target sites in the CRMs of their target genes have been dissected extensively in different tissues throughout the developmental course of *S. purpuratus* [54]. Comparisons of reporter activity between wild-type and mutant versions of CRMs, in which predicted TF-binding sites have been mutated, reveal functional TF-binding sites; analogously, performing such experiments in wild-type and mutant animals, in which the amount of a particular TF is altered genetically, indicates the contribution of the perturbed TF to reporter activity. Recently, a medium-throughput approach was developed to simultaneously inject and analyze multiple bar-coded reporter constructs [175]. There is an extensive understanding of GRNs in the sea urchin primarily from an abundance of data on the activities of CRMs and the effects of perturbations of TFs during development (notably, without genome-wide ChIP data on TF binding). As a result, sea urchin is perhaps best characterized in terms of the regulatory logic of the impact of TF input on target gene output, as well as how such individual connections are assembled into a GRN [54,176].

Progress in understanding GRNs in mouse and human has been far slower than in non-mammalian model organisms, with genome-scale progress in only a handful of systems, such as embryonic stem cells [177] and hematopoiesis [178,179]. Human and mouse have much larger genomes than many other model organisms. For instance, whereas the *C. elegans* genome is 100 Mb [165], the human is 30 times larger at 3 Gb [172], even though it harbors roughly the same number of protein-coding genes. One significant challenge is that CREs can be located far (hundreds of kilobases or even further away) from the genes they regulate in large genomes. In addition, many genes, including those encoding TFs, have undergone substantial gene duplication events, resulting in large families of at least partially functionally redundant genes, making it more difficult to use genetic approaches to unravel their function(s). There are also numerous practical limitations for experimental work in mouse, in particular the longer generation time than with simpler model organisms (see Box 4.3). Progress in the understanding of human GRNs is limited to experiments that can be performed using samples obtained from donors or on cell lines grown in cell culture, which are outside the organismal context and may not accurately reflect true physiology. Identification of CREs in these genomes has come largely from gene expression profiling studies combined with computational sequence motif analysis [64], cross-species sequence conservation analysis [38], ChIP studies including many as part of the human ENCODE project [77], and enhancer assays using reporter constructs tested in mice [70] or in human cell culture [96]. Using Y1H resources that were recently developed for the human genome, it will be possible to identify the TFs that can physically interact with identified CRMs [26].

The thale cress *Arabidopsis thaliana* has served as the primary model organism for plants (see Chapter 20). Aside from their importance in agriculture, plants have many of the same levels of complexity in gene regulation as animals, such as expansion of TF families and tissue-specific gene expression. Elegant studies on different tissues dissected from *A. thaliana* roots examined by gene expression profiling have uncovered gene expression patterns important in plant root development [180]. Such gene expression data, combined with other types of genomic, proteomic, and metabolite data, will be important for deciphering the GRNs in *A. thaliana* and other plants. Thus far, however, relatively few CREs and TF-binding sites have been mapped throughout the *A. thaliana* genome [181]. The mapping of GRNs has recently been facilitated by the development of a set of Y1H resources and their application to genes involved in the root [24,182].

A number of other model organisms has proven to be powerful in various ways for other types of studies such as those of patterning in vertebrate embryonic development, the generation of cellular extracts for in vitro investigations of cell biological and biochemical processes, or forward genetic screens [183–185]. These include: chicken (*Gallus gallus*), representing birds; African clawed frog (*Xenopus laevis*), representing amphibians; and zebrafish (*Danio reiro*), representing fish. However, we do not list these organisms in Box 4.3 because they have not yet been used extensively for large-scale studies of GRNs. We anticipate that the continued development of genomic resources in these and other model organisms [183,184,186] will continue to advance their use in such studies, and our understanding of how their gene regulatory and network properties may differ from those in the other major model organisms that thus far are better characterized.

FUTURE CHALLENGES

Although much effort has gone into using ChIP technologies to identify genomic sites occupied by TFs in vivo, a major question that persists is: which of the experimentally measured TF-binding events are direct and correspond to functional CREs that regulate gene expression? Similarly, with a few exceptions, the regulatory function of the vast majority of interactions detected by Y1H assays remains to be determined. While the functions of most ChIP-bound sites for any particular sequence-specific TF are as of yet unknown, most are not expected to correspond to active CREs. Other types of experiment will need to be performed to determine the functions and effects on gene regulation of individual TF-bound sites.

The importance of how CREs (either experimentally determined TF DNA-binding site motifs or computationally inferred regulatory motifs) are organized within more complex CRMs such as enhancers, and the importance of maintaining specific TF-binding site sequences, has only begun to be explored. Such effects will likely vary across different types of enhancers with some belonging to the 'enhanceosome' type for which a specific arrangement of TF DNA-binding sites is required, and others having more flexible organization or composition of DNA-binding sites as in 'billboard' type enhancers [187]. Improved computational methods and experimental testing of many CRMs will be required to decipher the 'grammar' of how the organization of TF-binding sites within CRMs confers the appropriate effect on gene expression in response to particular cellular and environmental contexts. Understanding the 'language' of how gene regulation is encoded in the genome will also permit more accurate prediction and understanding of variation within TF–DNA-binding sites, either within or between species. Accurate identification of non-coding regulatory variation and prediction of the associated effects on gene expression remain significant challenges and are important for a mechanistic understanding of the functional consequences of numerous disease- or other trait-associated non-coding variation identified in genome-wide association studies (GWAS) [188].

A variety of emerging model organisms is proving to be powerful in studies of the role of regulatory variation and the evolution of phenotypic variation. For example, multiple cases of changes in wing pigmentation patterns across insects (*Drosophila*, or more broadly, Diptera) have been found to be due to changes in CREs [189,190]. Partial or complete loss of pelvic structures in populations of three-spine stickleback fish (*Gasterosteus aculeatus*) show phenotypic variation across populations, with some populations exhibiting partial or complete loss of pelvic structures; it has recently been shown that pelvic loss in these populations is due to different deletions overlapping a tissue-specific enhancer of the Pituitary homeobox transcription factor 1 (Pitx1) gene [191].

GRNs evolve with the duplication and/or divergence of members of gene families, including TFs. Different DNA-binding site sequences and associated target genes can be recognized by different members of TF families [18,103,192–194]. Accurate models of GRNs will require an improved understanding of the redundant vs. the non-overlapping roles of paralogous regulatory factors.

Network models that capture the effects of different genetic, cellular, or environmental conditions will play important roles, as the complexity of GRNs is too high to permit reliable prediction from manual interpretation of the network. To this effect, it is important to move beyond qualitative network diagrams that are convenient for visual representation, to the development of quantitative and dynamic models that can be used to make accurate predictions of network behavior, in particular upon genetic, cellular, or environmental perturbation(s). Except for a handful of small, well-defined model systems, in general there is a lack of reliable experimental data on nuclear TF concentrations, TF–DNA-binding affinity constants (K_d values), protein localization data and measurements of protein activity (vs. just abundance), all of which are necessary to parameterize detailed computational models of transcriptional regulatory networks. Such data will be important for improved models of GRNs and a better mechanistic understanding of stochasticity ('noise') in gene expression [195–197]. In addition, there is a need for visual representations of networks that depict the condition-specific activity of portions of the networks, beyond simple graphs of edges connecting interacting nodes. Experimental testing of predictions arising from the inferred models will continue to be crucial for validation and refinement of network models.

This chapter has focused on the roles of TFs and CREs in GRNs. Numerous other regulatory mechanisms play important roles in regulatory networks. Some of these, including nucleosome occupancy and histone modifications, chromosomal conformations, non-coding regulatory RNAs such as lincRNAs and microRNAs, DNA methylation, RNA-binding proteins and untranslated regions (UTRs), affect transcript levels, whereas others, such as protein–protein interactions and small molecule ligands, affect protein levels or activity. Incorporation of these additional features into integrated models will be important for a more complete understanding and more accurate prediction of the effects of perturbations of GRNs.

ACKNOWLEDGEMENTS

The authors thank Trevor Siggers, Luis Barrera, Raluca Gordân, and Stephen Gisselbrecht from the Bulyk lab, and Alex Tamburino, Ashlyn Ritter and John Reece-Hoyes from the Walhout lab, for helpful comments in the preparation of this chapter. We are grateful to Trevor Siggers for creating the quantitative GRN display item shown in Figure 4.2D. Research in the Bulyk and Walhout labs is funded by grants from the National Institutes of Health [10].

REFERENCES

[1] Crick F. Central dogma of molecular biology. Nature 1970; 227:561–3.

[2] Halder G, Callaerts P, Gehring WJ. Induction of ectopic eyes by targeted expression of the eyeless gene in *Drosophila*. Science 1995;267:1788–92.

[3] Fukushige T, Hawkins MG, McGhee JD. The GATA-factor *elt-2* is essential for formation of the *Caenorhabditis elegans* intestine. Dev Biol 1998;198:286–302.

[4] Charoensawan V, Wilson D, Teichmann SA. Lineage-specific expansion of DNA-binding transcription factor families. Trends Genet 2010;26:388–93.

[5] Gamsjaeger R, Kong Liew C, Loughlin FE, Crossley M, Mackay JP. Sticky fingers: zinc-fingers as protein-recognition motifs. Trends Biochem Sci 2006;32:63–70.

[6] Tanaka Hall TM. Multiple modes of RNA recognition by zinc finger proteins. Curr Opin Struct Biol 2005;15:367–73.

[7] Bateman A, Coin L, Durbin R, Finn RD, Hollich V, Griffiths-Jones S, et al. The Pfam protein families database. Nucleic Acids Res 32 Database issue 2004:D138–41.

[8] Letunic I, Copley RR, Schmidt S, Ciccarelli FD, Doerks T, Schultz J, et al. SMART 4.0: towards genomic data integration. Nucleic Acids Res 2004;32:D142–4.

[9] Mulder NJ, Apweiler R, Attwood TK, Bairoch A, Barrell D, Bateman A, et al. The InterPro Database, 2003 brings increased coverage and new features. Nucleic Acids Res 2003;31:315–8.

[10] Reece-Hoyes JS, Deplancke B, Shingles J, Grove CA, Hope IA, Walhout AJM. A compendium of *C. elegans* regulatory transcription factors: a resource for mapping transcription regulatory networks. Genome Biol 2005;6:R110.

[11] Vaquerizas JM, Kummerfeld SK, Teichmann SA, Luscombe NM. A census of human transcription factors: function, expression and evolution. Nat Rev Genet 2009;10:252–63.

[12] Kummerfeld SK, Teichmann SA. DBD: a transcription factor prediction database. Nucleic Acids Res 2006;34:D74–81.

[13] Levine M, Tjian R. Transcription regulation and animal diversity. Nature 2003;424:147–51.

[14] Yen K, Narasimhan SD, Tissenbaum HA. DAF-16/Forkhead box O transcription factor: many paths to a single Fork(head) in the road. Antioxid Redox Signal 2011;14:623–34.

[15] Murre C, Bain G, van Dijk MA, Engel I, Furnari BA, Massari ME, et al. Structure and function of helix-loop-helix proteins. Biochim Biophys Acta 1994;1218:129–35.

[16] Glass CK. Differential recognition of target genes by nuclear receptor monomers, dimers and heterodimers. Endocri Rev 1994;15:391–407.

[17] Grove CA, Walhout AJ. Transcription factor functionality and transcription regulatory networks. Mol Biosyst 2008;4:309–14.

[18] Grove CA, deMasi F, Barrasa MI, Newburger D, Alkema MJ, Bulyk ML, et al. A multiparameter network reveals extensive divergence between *C. elegans* bHLH transcription factors. Cell 2009;138:314–27.

[19] Ravasi T, Suzuki H, Cannistraci CV, Katayama S, Bajic VB, Tan K, et al. An atlas of combinatorial transcriptional regulation in mouse and man. Cell 2010;140:744–52.

[20] Brasch MA, Hartley JL, Vidal M. ORFeome cloning and systems biology: standardized mass production of the parts from the parts-list. Genome Res 2004;14:2001–9.

[21] Walhout AJM, Temple GF, Brasch MA, Hartley JL, Lorson MA, van den Heuvel S, et al. GATEWAY recombinational cloning: application to the cloning of large numbers of open reading frames or ORFeomes. Methods Enzymol 2000;328:575–92.

[22] Reboul J, Vaglio P, Rual JF, Lamesch P, Martinez M, Armstrong CM, et al. *C. elegans* ORFeome version 1.1: experimental verification of the genome annotation and resource for proteome-scale protein expression. Nat Genet 2003;34:35–41.

[23] Rual J-F, Hirozane-Kishikawa T, Hao T, Bertin N, Li S, Dricot A, et al. Human ORFeome version 1.1: a platform for reverse proteomics. Genome Res 2004;14:2128–35.

[24] Gaudinier A, Zhang L, Reece-Hoyes JS, Taylor-Teeples M, Pu L, Liu Z, et al. An automated eY1H approach to elucidate *Arabidopsis* stele gene regulatory networks. Nat Methods 2011; 8:1053–5.

[25] Hu S, Xie Z, Onishi A, Jiang L, Lin J, Rho HS, et al. Profiling the human protein-DNA interactome reveals ERK2 as a transcriptional repressor of interferon signaling. Cell 2009;139:610–22.

[26] Reece-Hoyes JS, Barutcu AR, Patton McCord R, Jeong J, Jian L, MacWilliams A, et al. Yeast one-hybrid assays for high-throughput human gene regulatory network mapping. Nat Methods 2011;8:1050–2.

[27] Vermeirssen V, Deplancke B, Barrasa MI, Reece-Hoyes JS, Arda HE, Grove CA, et al. Matrix and Steiner-triple-system smart pooling assays for high-performance transcription regulatory network mapping. Nat Methods 2007;4:659–64.

[28] Arda HE, Taubert S, Conine C, Tsuda B, Van Gilst MR, Sequerra R, et al. Functional modularity of nuclear hormone receptors in a *C. elegans* gene regulatory network. Mol Syst Biol 2010;6:367.

[29] Rowan S, Siggers T, Lachke SA, Yue Y, Bulyk ML, Maas RL. Precise temporal control of the eye regulatory gene Pax6 via enhancer-binding site affinity. Genes Dev 2010;24:980–5.

[30] Bulger M, Groudine M. Functional and mechanistic diversity of distral transcriptional enhancers. Cell 2011;144:327–39.

[31] Wallace JA, Felsenfeld G. We gather together: insulators and genome organization. Curr Opin Genetic Dev 2007;17:400–7.

[32] Kim J, Bhinge AA, Morgan XC, Iyer VR. Mapping DNA-protein interactions in large genomes by sequence tag analysis of genomic enrichment. Nat Methods 2005;2:47–53.

[33] Denoeud F, Kapranov P, Ucla C, Frankish A, Castelo R, Drenkow J, et al. Prominent use of distal 5′ transcription start sites and discovery of a large number of additional exons in ENCODE regions. Genome Res 2007;17:746–59.

[34] Shiraki T, Kondo S, Katayama S, Waki K, Kasukawa T, Kawaji H, et al. Cap analysis gene expression for high-throughput analysis of transcriptional starting point and identification of promoter usage. Proc Natl Acad Sci U S A 2003;100:15776–81.

[35] Mortazavi A, Williams BA, McCue K, Schaeffer L, Wold B. Mapping and quantifying mammalian transcriptomes by RNA-seq. Nat. Methods 2008;5:621–8.

[36] Elemento O, Tavazoie S. Fast and systematic genome-wide discovery of conserved regulatory elements uisng a non-alignment based approach. Genome Biol 2005;6:R18.

[37] Huber BR, Bulyk ML. Meta-analysis discovery of tissue-specific DNA sequence motifs from mammalian expression data. BMC Bioinformatics 2006;7:229.

[38] Xie X, Lu J, Kulkobas EJ, Golub TR, Mootha V, Lindblad-Toh K, et al. Systematic discovery of regulatory motifs in human promoters and 3′UTRs by comparison of several mammals. Nature 2005;434:338–45.

[39] Gordan R, Hartemink AJ, Bulyk ML. Distinguhising direct versus indirect transcription factor-DNA interacitons. Genome Res 2009;19:2090–100.

[40] Hughes JD, Estep PW, Tavazoie S, Church GM. Computational identification of cis-regulatory elements associated with groups of

functionally related genes in *Saccharomyces cerevisiae*. J Mol Biol 2000;296:1205—14.

[41] Stormo GD. DNA-binding sites: representation and discovery. Bioinformatics 2000;16:16—23.

[42] Tavazoie S, Hughes JD, Campbell MJ, Cho RJ, Church GM. Systematic determination of genetic network architecture. Nat Genet 1999;22:281—5.

[43] Bulyk ML. Computational prediction of transcription factor binding site locations. Genome Biol 2003;5:201.

[44] Walhout AJM. What does biologically meaningful mean? A perspective on gene regulatory network validation. Genome Biol 2011;12:109.

[45] van Steensel B, Henikoff S. Identification of in vivo DNA targets of chromatin proteins using tethered Dam methyltransferase. Nat Biotechnol 2000;18:424—8.

[46] Mukherjee S, Berger MF, Jona G, Wang XS, Muzzey D, Snyder M, et al. Rapid analysis of the DNA-binding specificities of transcription factors with DNA microarrays. Nat Genet 2004;36:1331—9.

[47] Berger MF, Philippakis AA, Qureshi AM, He FS, Estep 3rd PW, Bulyk ML. Compact, universal DNA microarrays to comprehensively determine transcription-factor binding site specificities. Nat Biotechnol 2006;24:1429—35.

[48] Oliphant AR, Brandl CJ, Struhl K. Defining the sequence specificity of DNA-binding proteins by selecting binding sites from random-sequence oligonucleotides: analysis of yeast GCN4 protein. Mol Cell Biol 1989;9:2944—9.

[49] Deplancke B, Dupuy D, Vidal M, Walhout AJM. A Gateway-compatible yeast one-hybrid system. Genome Res 2004;14:2093—101.

[50] Reece-Hoyes JS, Deplancke B, Barrasa MI, Hatzold J, Smit RB, Arda HE, et al. The *C. elegans* Snail homolog CES-1 can activate gene expression in vivo and share targets with bHLH transcription factors. Nucleic Acids Res 2009;37:3689—98.

[51] Wang MM, Reed RR. Molecular cloning of the olfactory neuronal transcription factor Olf-1 by genetic selection in yeast. Nature 1993;364:121—6.

[52] Reece-Hoyes JS, Diallo A, Kent A, Shrestha S, Kadreppa S, Pesyna C, et al. Enhanced yeast one-hybrid (eY1H) assays for high-throughput gene-centered regulatory network mapping. Nat Methods 2011b;8:1059—64.

[53] Hens K, Feuz J-D, Iagovitina A, Massouras A, Bryois J, Callaerts P, et al. Automated protein-DNA interaction screening of *Drosophila* regulatory elements. Nat Methods 2011;8(12):1065—70.

[54] Davidson EH. Genomic Regulatory Systems: Development and Evolution. San Diego: Academic Press; 2001.

[55] Beer MA, Tavazoie S. Predicting gene expression from sequence. Cell 2004;117:185—98.

[56] Bulyk ML, McGuire AM, Masuda N, Church GM. A motif co-occurrence approach for genome-wide prediction of transcription factor binding sites in *Escherichia coli*. Genome Res 2004;14:201—8.

[57] Kulkarni MM, Arnosti DN. Cis-regulatory logic of short-range transcriptional repression in *Drosophila melanogaster*. Mol Cell Biol 2005;3411—20.

[58] Berman BP, Pfeiffer BD, Laverty TR, Salzberg SL, Rubin GM, Eisen MB, et al. Computational identification of developmental enhancers: conservation and function of transcription factor binding site clusters in *Drosophila melanogaster* and *Drosophila pseudoobscura*. Genome Biol 2004;5:R61.

[59] Wasserman W, Fickett J. Identification of regulatory regions which confer mucle-specific gene expression. J Mol Biol 1998;278:167—81.

[60] Zhou Q, Wong WH. CisModule: de novo discovery of cis-regulatory modules by hierarchical mixture modeling. Proc Natl Acad Sci U S A 2004;101:12114—9.

[61] Blanchette M, Bataille AR, Chen X, Poitras C, Laganiere J, Lefebvre D, et al. Genome-wide computational prediction of transcriptional regulatory modules reveals new insights into human gene expression. Genome Res 2006;16:656—68.

[62] Hallikas O, Palin K, Sinjushina N, Rautiainen R, Partanen J, Ukkonen E, et al. Genome-wide prediction of mammalian enhancers based on analysis of transcription-factor binding affinity. Cell 2006;124:47—59.

[63] Philippakis AA, He FS, Bulyk ML. Modulefinder: a tool for computational discovery of cis regulatory modules. Pac Symp Biocomput 2005:519—30.

[64] Warner JB, Philippakis AA, Jaeger SA, He FS, Lin J, Bulyk ML. Systematic identification of mammalian regulatory motifs' target genes and functions. Nat Methods 2008;5:347—53.

[65] Segal E, Shapira M, Regev A, Pe'er D, Boststein D, Koller D, et al. Module networks: identifying regulatory modules and their condition-specific regulators from gene expression data. Nat Genet 2003;34:166—76.

[66] Basso K, Margolin AA, Stolovitzky G, Klein U, Dalla-Favera R, Califano A. Reverse engineering of regulatory networks in human B cells. Nat Genet 2005;37:382—90.

[67] Hartemink AJ, Gifford DK, Jaakkola TS, Young RA. Using graphical models and genomic expression data to statistically validate models of genetic regulatory networks. Pac Symp Biocomput 2001:422—33.

[68] Hartemink AJ, Gifford DK, Jaakkola TS, Young RA. Combining location and expression data for principled discovery of genetic regulatory network models. Pac Symp Biocomput 2002:437—49.

[69] Jeong Y, El-Jaick K, Roessler E, Muenke M, Epstein DJ. A functional screen for sonic hedgehog regulatory elements across a 1 Mb interval identifies long-range ventral forebrain enhancers. Development 2006;133:761—72.

[70] Pennacchio LA, Ahituv N, Moses AM, Prabhakar S, Nobrega MA, Shoukry M, et al. In vivo enhancer analysis of human conserved non-coding sequences. Nature 2006;444:499—502.

[71] He A, Kong SW, Ma Q, Pu WT. Co-occupancy by multiple cardiac transcription factors identifies transcriptional enhancers active in heart. Proc Natl Acad Sci U S A 2011;108:5632—7.

[72] Crawford GE, Holt IE, Mullikin JC, Tai D, Blakesley R, Bouffard G, et al. Identifying gene regulatory elements by genome-wide recovery of DNase hypersensitive sites. Proc Natl Acad Sci U S A 2004;101:992—7.

[73] Hesselberth JR, Chen X, Zhang Z, Sabo PJ, Sandstrom R, Reynolds AP, et al. Global mapping of protein-DNA interactions in vivo by digital genomic footprinting. Nat Methods 2009;6:283—9.

[74] Sabo PJ, Hawrylycz M, Wallace JC, Humbert R, Yu M, Shafer A, et al. Discovery of functional noncoding elements by digital

analysis of chromatin structure. Proc Natl Acad Sci U S A 2004; 101:16837—42.
[75] Giresi PG, Kim J, McDaniell RM, Iyer VR, Lieb JD. FAIRE (Formaldehyde-Assisted Isolation of Regulatory Elements) isolates active regulatory elements from human chromatin. Genome Res 2007;17:877—85.
[76] Song L, Zhang Z, Grasfeder LL, Boyle AP, Giresi PG, Lee BK, et al. Open chromatin defined by DNaseI and FAIRE identifies regulatory elements that shape cell-type identity. Genome Res 2011;21:1757—67.
[77] The ENCODE Project Consortium. Identification and analysis of functional elements in 1% of the human genome by the ENCODE pilot project. Nature 2007;447:799—816.
[78] Ernst J, Kellis M. Discovery and characterization of chromatin states for systematic annotation of the human genome. Nat Biotechnol 2010;28:817—25.
[79] Ernst J, Kheradpour P, Mikkelsen TS, Shoresh N, Ward LD, Epstein CB, et al. Mapping and analysis of chromatin state dynamics in nine human cell types. Nature 2011;473:43—9.
[80] Roy S, Ernst J, Kharchenko PV, Kheradpour P, Negre N, Eaton ML, et al. Identification of functional elements and regulatory circuits by *Drosophila* modENCODE. Science 2010;330:1787—97.
[81] Gerstein MB, Lu ZJ, Van Nostrand EL, Cheng C, Arshinoff BI, Liu T, et al. Integrative analysis of the *Caenorhabditis elegans* genome by the modENCODE project. Science 2010;330: 1775—87.
[82] Ao W, Gaudet J, Kent WJ, Muttumu S, Mango SE. Environmentally induced foregut remodeling by PHA-4/FoxA and DAF-12/NHR. Science 2004;305:1743—6.
[83] Groth AC, Fish M, Nusse R, Calos MP. Construction of transgenic Drosophila by using the site-specific integrase from phage phiC31. Genetics 2004;166:1775—82.
[84] Berman BP, Nibu Y, Pfeiffer BD, Tomancak P, Celniker SE, Levine M, et al. Exploiting transcription factor binding site clustering to identify *cis*-regulatory modules involved in pattern formation in the *Drosophila* genome. Proc Natl Acad Sci USA 2002;99:757—62.
[85] Markstein M, Levine M. Decoding *cis*-regulatory DNAs in the *Drosophila* genome. Curr Opin Genet Dev 2002;12:601.
[86] Stathopoulos A, Van Drenth M, Erives A, Markstein M, Levine M. Whole-genome analysis of dorsal-ventral patterning in the *Drosophila* embryo. Cell 2002;111:687—701.
[87] Pfeiffer BD, Jenett A, Hammonds AS, Ngo TT, Misra S, Murphy C, et al. Tools for neuroanatomy and neurogenetics in *Drosophila*. Proc Natl Acad Sci U S A 2008;105:9715—20.
[88] Sandmann T, Girardot C, Brehme M, Tongprasit W, Stolc V, Furlong EE. A core transcriptional network for early mesoderm development in *Drosophila melanogaster*. Genes Dev 2007;21:436—49.
[89] Zinzen RP, Girardot C, Gagneur J, Braun M, Furlong EE. Combinatorial binding predicts spatio-temporal *cis*-regulatory activity. Nature 2009;462:65—70.
[90] Halfon MS, Gallo SM, Bergman CM. REDfly2.0: an integrated database of cis-regulatory modules and transcription factor binding sites in *Drosophila*. Nucleic Acids Res 2008;36: D594—8.
[91] Loots GG, Kneissel M, Keller H, Baptist M, Chang J, Collette NM, et al. Genomic deletion of a long-range cone enhancer misregulates sclerostin in Van Buchem disease. Genome Res 2005;15:928—35.
[92] Blow MJ, McCulley DJ, Li Z, Zhang T, Akiyama JA, Holt A, et al. ChIP-seq identification of weakly conserved heart enhancers. Nat Genet 2010;42:806—10.
[93] Visel A, Blow MJ, Li Z, Zhang T, Akiyama JA, Holt A, et al. ChIP-seq accurately predicts tissue-specific activity of enhancers. Nature 2009;457:854—8.
[94] Visel A, Prabhakar S, Akiyama JA, Shoukry M, Lewis KD, Holt A, et al. Ultraconservation identifies a small subset of extremely conserved developmental enhancers. Nat Genet 2008;40:158—60.
[95] Kumaki Y, Ukai-Tadenuma M, Uno KD, Nishio J, Masumoto KH, Nagano M, et al. Analysis and synthesis of high-amplitude *Cis*-elements in the mammalian circadian clock. Proc Natl Acad Sci U S A 2008;105:14946—51.
[96] Sun Q, Chen G, Streb JW, Long X, Yang Y, Stoeckert CJJ, et al. Defining the mammalian CArGome. Genome Res 2006;16: 197—207.
[97] Rada-Iglesias A, Bajpai R, Swigut T, Brugmann SA, Flynn RA, Wysocka J. A unique chromatin signature uncovers early developmental enhancers in humans. Nature 2011;470:279—83.
[98] Colombo JA, Reisin HD. Interlaminar astroglia of the cerebral cortex: a marker of the primate brain. Brain Res 2004;1006: 126—31.
[99] Arda HE, Walhout AJM. Gene-centered regulatory networks. Brief Funct Genomic Proteomic, http://dx.doi.org/10.1093/elp049 2009.
[100] Harbison CT, Gordon DB, Lee TI, Rinaldi NJ, Macisaac KD, Danford TW, et al. Transcriptional regulatory code of a eukaryotic genome. Nature 2004;431:99—104.
[101] Lee TI, Rinaldi NJ, Robert F, Odom DT, Bar-Joseph Z, Gerber GK, et al. Transcriptional regulatory networks in *Saccharomyces cerevisiae*. Science 2002;298:799—804.
[102] Zhu C, Byers K, McCord R, Shi Z, Berger M, Newburger D, et al. High-resolution DNA-binding specificity analysis of yeast transcription factors. Genome Res 2009;19:556—66.
[103] Badis G, Berger MF, Philippakis AA, Talukder S, Gehrke AR, Jaeger SA, et al. Diversity and complexity in DNA recognition by transcription factors. Science 2009;324:1720—3.
[104] Berger MF, Badis G, Gehrke AR, Talukder S, Philippakis AA, Pena-Castillo L, et al. Variation in homeodomain DNA-binding revealed by high-resolution analysis of sequence preferences. Cell 2008;133:1266—76.
[105] Deplancke B, Mukhopadhyay A, Ao W, Elewa AM, Grove CA, Martinez NJ, et al. A gene-centered *C. elegans* protein-DNA interaction network. Cell 2006;125:1193—205.
[106] Martinez NJ, Ow MC, Barrasa MI, Hammell M, Sequerra R, Doucette-Stamm L, et al. A *C. elegans* genome-scale microRNA network contains composite feedback motifs with high flux capacity. Genes Dev 2008;22:2535—49.
[107] Vermeirssen V, Barrasa MI, Hidalgo C, Babon JAB, Sequerra R, Doucette-Stam L, et al. Transcription factor modularity in a gene-centered *C. elegans* core neuronal protein-DNA interaction network. Genome Res 2007;17:1061—71.
[108] Walhout AJM. Unraveling transcription regulatory networks by protein-DNA and protein—protein interaction mapping. Genome Res 2006;16:1445—54.

[109] Chen X, Xu H, Yuan P, Fang F, Huss M, Vega VB, et al. Integration of external signaling pathways with the core transcriptional network in embryonic stem cells. Cell 2008;133:1106−17.

[110] Heintzman N, Hon GC, Hawkins RD, Kheradpour P, Stark A, Harp LF, et al. Histone modifications at human enhancers reflect global cell-type-specific gene expression. Nature 2009;459: 108−12.

[111] Heintzman ND, Stuart RK, Hon G, Fu Y, Ching CW, Hawkins RD, et al. Distinct and predictive chromatin signatures of transcriptional promoters and enhancers in the human genome. Nat Genet 2007;39:311−8.

[112] Davidson EH, Rast JP, Oliveri P, Ransick A, Calestani C, Yuh C-H, et al. A genomic regulatory network for development. Science 2002;295:1669−78.

[113] Wenick AS, Hobert O. Genomic cis-regulatory architecture and trans-acting regulators of a single interneuron-specific gene battery in *C. elegans*. Dev Cell 2004;6:757−70.

[114] Amit I, Garber M, Chevrier N, Leite AP, Donner Y, Eisenhaure T, et al. Unbiased reconstruction of a mammalian transcriptional network mediating pathogen responses. Science 2009;326: 257−63.

[115] Shannon P, Markiel A, Ozier O, Baliga NS, Wang JT, Ramage D, et al. Cytoscape: a software environment for integrated models of biomolecular interaction networks. Genome Res 2003;13:2498−504.

[116] Breitkreutz BJ, Stark C, Tyers M. Osprey: a network visualization system. Genome Biol 2003;4:R22.

[117] Kao HL, Gunsalus KC. Browsing multidimensional molecular networks with the generic network browser (N-browse). Curr Protoc Bioinformatics 2008;9. 9.11.

[118] Longabaugh WJR, Davidson EH, Bolouri H. Visualization, documentation, analysis, and communication of large-scale gene regulatory networks. Biochim Biophys Acta 2009;1789:363−74.

[119] Ge H, Walhout AJM, Vidal M. Integrating 'omic' information: a bridge between genomics and systems biology. Trends Genet 2003;19:551−60.

[120] von Mering C, Krause R, Snel B, Cornell M, Oliver SG, Fields S, et al. Comparative assessment of large-scale data sets of protein−protein interactions. Nature 2002;417:399−403.

[121] Li XY, MacArthur S, Bourgon R, Nix D, Pollard DA, Iyer VN, et al. Transcription factors bind thousands of active an inactive regions in the *Drosophila* blastoderm. PLoS Biol 2008;6:e27.

[122] MacQuarrie KL, Fong AP, Morse RH, Tapscott SJ. Genome-wide transcription factor binding: beyond direct target regulation. Trends Genet 2011;27:141−8.

[123] Tabuchi T, Deplancke B, Osato N, Zhu LJ, Barrasa MI, Harrison MM, et al. Chromosome-biased binding and gene regulation by the *Caenorhabditis elegans* DRM complex. PLoS Genet 2011;7:e1002074.

[124] Biggin MD. Animal transcription networks as highly connected, quantitative quanta. Dev Cell 2011;21:611−26.

[125] Albert R, Jeong H, Barabasi A-L. Error and attack tolerance of complex networks. Nature 2000;378:378−81.

[126] Ravasz E, Somera AL, Mongru DA, Oltvai ZN, Barabasi AL. Hierarchical organization of modularity in metabolic networks. Science 2002;297:1551−5.

[127] Yu H, Gerstein M. Genomic analysis of the hierarchical structure of regulatory networks. Proc Natl Acad Sci U S A 2006;103:14724−31.

[128] Jothi R, Balaji S, Wuster A, Grochow JA, Gsponer J, Przytycka TM, et al. Genomic analysis reveals a tight link between transcription factor dynamics and regulatory network architecture. Mol Syst Biol 2009;5:294.

[129] Ma HW, Buer J, Zeng AP. Hierarchical structure and modules in the *Escherichia coli* transcriptional regulatory network revealed by a new top-down approach. Nucleic Acids Res 2004;32: 6643−9.

[130] Milo R, Shen-Orr S, Itzkovitz S, Kashtan N, Chklovskii D, Alon U. Network motifs: simple building blocks of complex networks. Science 2002;298:824−7.

[131] Kashtan N, Itzkovitz S, Milo R, Alon U. Efficient sampling algorithm for estimating subgraph concentrations and detecting network motifs. Bioinformatics 2004;20:1746−58.

[132] Becksei A, Serrano L. Engineering stability in gene networks by autoregulation. Nature 2000;405:590−3.

[133] Rosenfeld N, Elowitz MB, Alon U. Negative autoregulation speeds the response times of transcription networks. J Mol Biol 2002;323:785−93.

[134] Thieffry D, Huerta AM, Perez-Rueda E, Collado-Vides J. From specific gene regulation to genomic networks: a global analysis of transcriptional regulation in *Escherichia coli*. Bioessays 1998;20:433−40.

[135] Davidson E, Levine M. Gene regulatory networks. Proc Natl Acad Sci U S A 2005;102:4935.

[136] Bolouri H, Davidson EH. Transcriptional regulatory cascades in development: initial rates, not steady state, determine network kinetics. Proc Natl Acad Sci U S A 2003;100:9371−6.

[137] Mangan S, Alon U. Structure and function of the feed-forward loop network motif. Proc Natl Acad Sci U S A 2003;100: 11980−5.

[138] Shen-Orr SS, Milo R, Mangan S, Alon U. Network motifs in the transcriptional regulation network of *Escherichia coli*. Nat Genet 2002;31:64−8.

[139] Yeger-Lotem E, Sattath S, Kashtan N, Itzkovitz S, Milo R, Pinter RY, et al. Network motifs in integrated cellular networks of transcription-regulation and protein−protein interaction. Proc Natl Acad Sci U S A 2004;101:5934−9.

[140] Ptashne M. A Genetic Switch. second ed. Cambridge, MA, Cell Press: Blackwell Scientific Publications; 1992.

[141] Struhl K. Fundamentally different logic of gene regulation in eukaryotes and prokaryotes. Cell 1999;98:1−4.

[142] Relman DA. New technologies, human-microbe interactions, and the search for previously unrecognized pathogens. J Infect Dis 2002;186(Suppl 2):S254−8.

[143] DeRisi JL, Iyer VR, Brown PO. Exploring the metabolic and genetic control of gene expression on a genomic scale. Science 1997;278:680−6.

[144] Wodicka L, Dong H, Mittmann M, Ho MH, Lockhart DJ. Genome-wide expression monitoring in *Saccharomyces cerevisiae*. Nat Biotechnol 1997:1359−67.

[145] Shoemaker DD, Lashkari DA, Morris D, Mittmann M, Davis RW. Quantitative phenotypic analysis of yeast deletion mutants using a highly parallel molecular bar-coding strategy. Nat Genet 1996;14:450−6.

[146] Iyer VR, Horak CE, Scafe CS, Botstein D, Snyder M, Brown PO. Genomic binding sites of the yeast cell-cycle transcription factors SBF and MBF. Nature 2001;409:533−8.

[147] Lieb JD, Liu X, Botstein D, Brown PO. Promoter-specific binding of Rap1 revealed by genome-wide maps of protein-DNA association. Nat Genet 2001;28:327–34.

[148] Ren B, Robert F, Wyrick JJ, Aparicio O, Jennings EG, Simon I, et al. Genome-wide location and function of DNA-binding proteins. Science 2000;290:2306–9.

[149] Ito T, Chiba T, Ozawa R, Yoshida M, Hattori M, Sakaki Y. A comprehensive two-hybrid analysis to explore the yeast protein interactome. Proc Natl Acad Sci USA 2001;98:4569–75.

[150] Yu H, Braun P, Yildirim MA, Lemmens I, Venkatesan K, Sahalie J, et al. High-quality binary protein interaction map of the yeast interactome network. Science 2008;322:104–10.

[151] Gavin A-C, Bosche M, Krause R, Grandi P, Marzloch M, Bauer A, et al. Functional organization of the yeast proteome by systematic analysis of protein complexes. Nature 2002;415:141–7.

[152] Ho Y, Gruhler A, Heilbut A, Bader GD, Moore L, Adams S-L, et al. Systematic identification of protein complexes in *Saccharomyces cerevisiae* by mass spectrometry. Nature 2002;415:180–3.

[153] Costanzo M, Baryshnikova A, Bellay J, Kim Y, Spear ED, Sevier CS, et al. The genetic landscape of a cell. Science 2010;327:425–31.

[154] Tong AHY, Evangelista M, Parsons AB, Xu H, Bader GD, Page N, et al. Systematic genetic analysis with ordered arrays of yeast deletion mutants. Science 2001;294:2364–8.

[155] Workman CT, Mak HC, McCuine S, Tagne JB, Agarwal M, Ozier O, et al. A systems approach to mapping DNA damage response pathways. Science 2006;312:1054–9.

[156] Dupuy D, Bertin N, Hidalgo CA, Venkatesan K, Tu D, Lee D, et al. Genome-scale analysis of in vivo spatiotemporal promoter activity in *Caenorhabditis elegans*. Nat Biotechnol 2007;25:663–8.

[157] Martinez NJ, Ow MC, Reece-Hoyes J, Ambros V, Walhout AJ. Genome-scale spatiotemporal analysis of *Caenorhabditis elegans* microRNA promoter activity. Genome Res 2008;18:2005–15.

[158] Reece-Hoyes JS, Shingles J, Dupuy D, Grove CA, Walhout AJ, Vidal M, et al. Insight into transcription factor gene duplication from *Caenorhabditis elegans* Promoterome-driven expression patterns. BMC Genomics 2007;8:27.

[159] Arnosti DN. Analysis and function of transcriptional regulatory elements: insights from Drosophila. Annu Rev Entomol 2003;48:579–602.

[160] Busser BW, Bulyk ML, Michelson AM. Towards a systems-level understanding of developmental regulatory networks. Curr Opin Genet Dev 2008;18:521–9.

[161] Li L, Zhu Q, He X, Sinha S, Halfon MS. Large-scale analysis of transcriptional *cis*-regulatory modules reveals both common features and distinct subclasses. Genome Biol 2007;8:R101.

[162] Blattner FR, Plunkett Gr, Bloch CA, Perna NT, Burland V, Riley M, et al. The complete genome sequence of *Escherichia coli* K-12. Science 1997;277:1453–74.

[163] Babu MM, Teichmann SA. Evolution of transcription factors and the gene regulatory network in *Escherichia coli*. Nucleic Acids Res 2003;31:1234–44.

[164] Goffeau A, Barrell BG, Bussey H, Davis RW, Dujon B, Feldmann H, et al. Life with 6000 genes. Science 1996;274:563–7.

[165] The *C. elegans* Sequencing Consortium. Genome sequence of the nematode *C. elegans*: a platform for investigating biology. Science 1998;282:2012–8.

[166] Adams MD, et al. The genome sequence of *Drosophila melanogaster*. Science 2000;287:2185–95.

[167] Pfreundt U, James DP, Tweedie S, Wilson D, Teichmann SA, Adryan B. FlyTF: improved annotation and enhanced functionality of the *Drosophila* transcription factor database. Nucleic Acids Res 2010;38:D443–7.

[168] McClay DR. Evolutionary crossroads in developmental biology: sea urchins. Development 2011;138:2639–48.

[169] Sodergren E, Weinstock GM, Davidson EH, Cameron RA, Gibbs RA, Angerer RC, et al. The genome of the sea urchin *Strongylocentrotus purpuratus*. Science 2006;314:941–52.

[170] Waterston RH, Lindblad-Toh K, Birney E, Rogers J, Abril JF, Agarwal P, et al. Initial sequencing and comparative analysis of the mouse genome. Nature 2002;420:520–62.

[171] Kanamori M, Konno H, Osato N, Kawai J, Hayashizaki Y, Suzuki H. A genome-wide and nonredundant mouse transcription factor database. Biochem Biophys Res Commun 2004;322:787–93.

[172] The International Human Genome Sequencing Consortium. Finishing the euchromatic sequence of the human genome. Nature 2004;431:931–45.

[173] Venter JC, et al. The sequence of the human genome. Science 2001;291:1304–51.

[174] initiative, T.A.G. Analysis of the genome sequence of the flowering plant *Arabidopsis thaliana*. Nature 2000;408:796–815.

[175] Nam J, Dong P, Tarpine R, Istrail S, Davidson EH. Functional *cis*-regulatory genomics for systems biology. Proc Natl Acad Sci U S A 2010;107:3930–5.

[176] Ben-Tabou de-Leon S, Davidson EH. Experimentally based sea urchin gene regulatory network and the causal explanation of developmental phenomology. Wiley Interdiscip Rev Syst Biol Med 2009;1:237–46.

[177] Young RA. Control of the embryonic stem cell state. Cell 2011;144:940–54.

[178] Novershtern N, Subramanian A, Lawton LN, Mak RH, Haining WN, McConkey ME, et al. Densely interconnected transcriptional circuits control cell states in human hematopoiesis. Cell 2011;144:296–309.

[179] Tijssen MR, Cvejic A, Joshi A, Hannah RL, Ferreira R, Forrai A, et al. Genome-wide analysis of simultaneous GATA1/2, RUNX1, FLI1, and SCL binding in megakaryocytes identifies hematopoietic regulators. Dev Cell 2011;20:597–609.

[180] Brady SM, Orlando DA, Lee JY, Wang JY, Koch J, Dinneny JR, et al. A high-resolution root spatiotemporal map reveals dominant expression patterns. Science 2007;318:801–6.

[181] Moreno-Risueno MA, Busch W, Benfey P. Omics meet networks — using systems approaches to infer regulatory networks in plants. Curr Opin Plant Biol 2010;13:126–31.

[182] Brady S, Zhang L, Megraw M, Martinez NJ, Jiang E, Yi CS, et al. A stele-enriched gene regulatory network in the *Arabidopsis* root. Mol Syst Biol 2011;7:459.

[183] Amaya E. Xenomics. Genome Res 2005;15:1683–91.

[184] Burt DW. Chicken genome: current status and future opportunities. Genome Res 2005;15:1692–8.

[185] Muller B, Grossniklaus U. Model organisms—a historical perspective. J Proteomics 2010;73:2054–63.

[186] Roest Crollius H, Weissenbach J. Fish genomics and biology. Genome Res 2005;15:1675−82.

[187] Arnosti DN, Kulkarni MM. Transcriptional enhancers: intelligent enhanceosomes or flexible billboards? J Cell Biochem 2005;94:890−8.

[188] Freedman ML, Monteiro AN, Gayther SA, Coetzee GA, Risch A, Plass C, et al. Principles for the post-GWAS functional characterization of cancer risk loci. Nat Genet 2011;43:513−8.

[189] Prud'homme B, Gompel N, Carroll SB. Emergin principles of Óregulatory evolution. Proc Natl Acad Sci U S A 2007;104(Suppl. 1):8605−12.

[190] Prud'homme B, Gompel N, Rokas A, Kassner VA, Williams TM, Yeh SD, et al. Repeated morphological evolution through *cis*-regulatory changes in a pleiotropic gene. Nature 2006;440:1050−3.

[191] Chan YF, Marks ME, Jones FC, Villarreal GJ, Shapiro MD, Brady SD, et al. Adaptive evolution of pevic reduction in sticklebacks by recurrent deletion of a Pitx1 enhancer. Science 2010;327:302−5.

[192] Fernandes L, Rodrigues-Pousada C, Struhl K. Yap, a novel family of eight bZIP proteins in *Saccharomyces cerevisiae* with distinct biological functions. Mol Cell Biol 1997;17:6982−93.

[193] Hollenhorst PC, Chandler KJ, Poulsen RL, Johnson WE, Speck NA, Graves BJ. DNA specificity determinants associate with distinct transcription factor functions. PLoS Genet 2009;5: e1000778.

[194] Tan K, Feizi H, Luo C, Fan SH, Ravasi T, Ideker TG. A systems approach to delineate functions of paralogous transcription factors: role of the Yap family in the DNA damage response. Proc Natl Acad Sci U S A 2008;105:2934−9.

[195] Elowitz MB, Levine AJ, Siggia ED, Swain PS. Stochastic gene expression in a single cell. Science 2002;297:1183−6.

[196] MacNeil L, Walhout AJM. Gene regulatory networks and the role of robustness and stochasticity in the control of gene expression. Genome Res 2011;21:645−57.

[197] Thattai M, van Oudenaarden A. Intrinsic noise in gene regulatory networks. Proc Natl Acad Sci U S A 2001;98:8614−9.

Chapter 5

Analyzing the Structure, Function and Information Flow in Signaling Networks using Quantitative Cellular Signatures

Meghana M. Kulkarni[1] and Norbert Perrimon[1,2]

[1]*Department of Genetics,* [2]*Howard Hughes Medical Institute, Harvard Medical School, 77 Avenue Louis Pasteur, Boston, MA 02115, USA*

Chapter Outline

The Concept of Linear Cassettes and Modularity in Signal Transduction	89
Genetic Dissection of Signal Transduction Pathways	92
Systems Approaches to Identify the 'Parts' of Cellular Signaling Networks	94
RNA Interference (RNAi)	94
Protein–Protein Interactions	97
Transcriptional Profiling	99
Quantitative RNAi Signatures or Phenoprints to Infer Context Dependent Information Flow Through Cellular Signaling Networks	100
Direction of Information Flow from Gene Expression Signatures	100
Direction of Information Flow from Phosphorylation Signatures	103
Concluding Remarks	106
References	106

THE CONCEPT OF LINEAR CASSETTES AND MODULARITY IN SIGNAL TRANSDUCTION

Cells communicate and respond to conditions in their local environment through signaling, a process consisting of a series of regulated steps that help propagate information across the external plasma membrane to the cell interior, and often to the nucleus, to regulate diverse cellular processes such as growth, proliferation, differentiation and apoptosis. The set of molecules recruited by a specific signal defines what is commonly referred to as a signal transduction pathway. Dissection of biological responses to similar families of ligands in various cell types and organisms revealed that these ligands regulate the activity of similar sets of downstream genes, a finding that led to the concept of 'evolutionarily conserved signal transduction cassettes' or modules [1]. A characteristic feature of these modules is the occurrence of a tight internal link between their individual components, and few, but well-defined connections to the rest of the system in which they operate. The concept of modules gained further acceptance when it became clear that these characteristic chains of events were iterated in the same pattern in different cellular and developmental situations. The extent of signaling modularity was underscored by studies of signal transduction in simple genetic model organisms, where signaling tends to be simpler than in more complex mammals (see below) [2]. The methodologies used to recognize and characterize these pathways relied mostly on the similarity in visible mutant phenotypes or screens in sensitized genetic backgrounds (see Box 5.1 for definition). The resounding conclusion from these studies was that of elegant simplicity: it is common for loss-of-function mutations (see Box 5.1 for definition) in different genes that constitute a specific pathway to result in identical phenotypes. For example, in the context of JAK/STAT signaling, mutations in the activating extracellular ligand (Unpaired/Upd), in the effector tyrosine kinase (Hopscotch/JAK), or in the effector transcription factor (STAT92E/Marelle) caused identical loss-of-function phenotypes, a feature consistent with linearity of the pathway [3] (Figure 5.1**A**). Furthermore, gain-of-function phenotypes (see Box 5.1 for definition) had strikingly opposite developmental phenotypes and could usually be reverted completely by removing the activity of a more downstream component of the pathway.

Box 5.1 Key Terms and Concepts

- **Sensitized genetic background**: A sensitized genetic background is a mutant state in which an allele of a gene leads to a weak phenotype in the biological process under study (for example eye development in *Drosophila*). Thus, a weak allele for gene X may lead to fruit flies with abnormal/small eyes, whereas a strong/null allele would produce flies with no eyes. In the sensitized background carrying the weak allele for gene X it is possible to screen for new mutants that either enhance the phenotype (small eyes to no eyes) or suppress the phenotype (small eyes to normal eyes). Such sensitized genetic screens can lead to the identification of genes that function in the same pathway or genes that act redundantly in parallel pathways.
- **Saturation screen**: A saturation screen is a genetic screen that is performed to discover all genes that are involved in a particular biological process. One of the first saturation screens was performed in *Drosophila* by Christiane Nüsslein-Volhard and Eric Wieschaus (1980) to uncover genes that were associated with embryonic lethality and changes in cuticle morphology. In such genetic screens, a mutagen such as a chemical or radiation is used to generate mutations in the organism's chromosomes. Individuals that exhibit the phenotype of interest are selected and the mutant alleles are mapped and cloned to identify every gene involved.
- **Loss-of-function mutation**: Changes in the DNA sequence of a gene that leads to reduced or abolished function of the gene product.
- **Gain-of-function mutation**: Changes in the DNA sequence of a gene that confers a new and/or abnormal function to the gene product.
- **Epistasis analysis**: Epistasis is the interaction between two or more genes where the effect of one gene on a particular phenotype is modified by other gene(s). The gene whose phenotype is manifested is called epistatic, while the gene whose phenotype is modified as the result of the epistatic gene is called hypostatic.
- **Synthetic lethality**: Mutations in two genes are said to be synthetically lethal when cells with either of the single mutations are viable but cells with both mutations are lethal. A synthetic lethal genetic screen, starts with a mutation in gene X that does not kill the cell, but may confer a weak phenotype (such as, slow growth). This genetic background is then used to systematically test mutations in other genes that may lead to lethality.
- **SDS-PAGE (sodium dodecyl-polyacrylamide gel electrophoresis)**: SDS-PAGE is a technique that is widely used in molecular biology and biochemistry to separate proteins as a function of their length and charge by application of an electric field.
- **iTRAQ (isobaric tags for relative and absolute quantification) and TMT (tandem mass tags)**: iTRAQ and TMT are used to identify differentially phosphorylated proteins between different samples. These amine-reactive molecules enable multiplexing of up to 4–8 samples: the small molecules, identical in structure and mass, differ in the isotopic substitution of atoms comprising their backbone. This altered distribution permits the unambiguous spectral identification of unique reporter ions generated from fragmentation of each tag during MS. The fragmentation of each tag during MS results in the release of a signature reporter ion that differs in mass from the other tags; the signature ions released from the six tag set differ successively by 1 Dalton and their relative levels can be considered to reflect differences in peptide levels between samples. iTRAQ and TMT enable the concurrent analysis of multiple samples, and the assessment of the relative levels of phosphopeptides.

Such properties have been instrumental in working out the epistatic (see Box 5.1 for definition) relationships between genes and in ordering components of a given pathway into linear, minimally branched cascades [4]. These pioneering studies have helped to reduce complex biological and developmental processes to a finite number of paradigms and clarified the identity and relationship of key components in evolutionary conserved pathways. Reflecting the implied linearity and independence of these pathways from each other, most are named after the activating ligand or a central effector protein, for example the Wnt/Wingless, Hedgehog (Hh), TGF-β, JAK/STAT, Toll, NFκB, Notch, receptor tyrosine kinase/extracellular regulated kinase (RTK/ERK), Akt/TOR, Jun Kinase (JNK), G protein-coupled receptor (GPCR) and steroid hormone pathways [5]. The analysis of signal transduction pathways in mammalian cells has presented a more complicated view, hinting at significant bridges or 'cross-talks' between various signal transduction modules/pathways. Cross-talk or interaction between pathways allows cells to respond in a coordinate manner to the combined extracellular and intracellular cues that they are exposed to in order to produce the appropriate response. Cross-talk between pathways accounts for many complex signaling behaviors, including signal integration, the ability to generate a variety of different responses to a signal, and/or to reuse proteins between pathways. For example, the SMAD proteins, which have been assigned to the TGF-β signaling cassette, can be phosphorylated by MAPK, which functions in a separate pathway in lower eukaryotes. Thus, rather than following a simple path, a signal received by a mammalian cell may be relayed through multiple channels. Similarly, other proteins such as Ras, protein kinase C (PKC), and protein kinase B (Akt) are also activated by multiple extracellular ligands. However, much of this knowledge has originated from studies in cell lines and in vitro

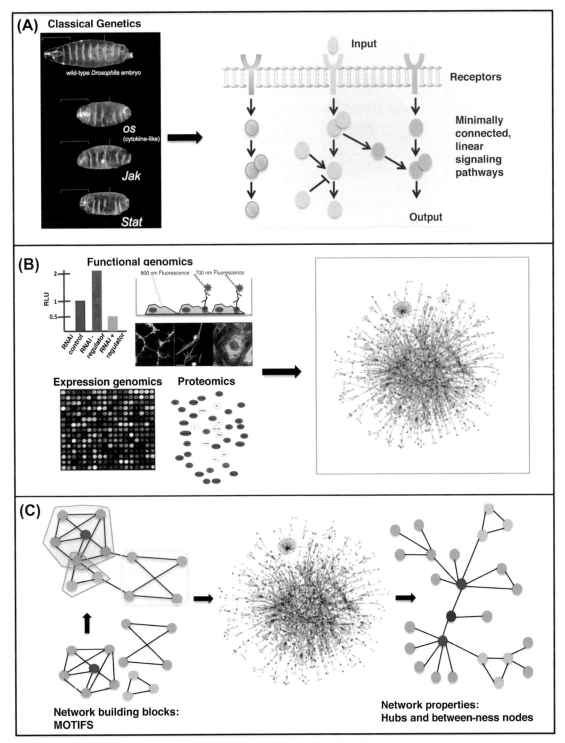

FIGURE 5.1 **Evolution of linear signal transduction pathways to highly organized signaling networks. A.** Mutant developmental phenotypes from classic genetics screens showed that mutations in the activating extracellular ligand (Unpaired/Upd), in the effector tyrosine kinase (Hopscotch/JAK), or in the effector transcription factor (STAT92E/Marelle) caused identical loss-of-function phenotypes, a feature consistent with linearity of signal transduction pathways. **B.** Systems-level, functional genomic, proteomic, and expression studies in the past few years have revealed that signaling is propagated within large networks consisting of hundreds or thousands of proteins. **C.** Structure–function analysis of signaling networks has led to the identification of network motifs, recurrent patterns of interconnections, that form the building blocks of networks as well as universal features such as Network Hubs (red circles) and between-ness nodes (blue circles) that are likely to be encoded by essential genes.

biochemical assays, and there is some question as to whether the same rules apply in vivo, or whether organisms have evolved tight controls to maintain modularity and prevent promiscuous cross talk.

More recently, newer findings from in vivo genetic studies have put the linear and simplistic model of signaling pathways increasingly at odds even in lower or simple metazoan model organisms. For example, during dorsal closure of the *Drosophila* embryo, the JNK, small GTPase and TGF-β pathways may act together or sequentially [6]. In addition, components that initially were thought to be unique to one pathway have now become implicated in others. For example, GSK3β and CK1α act as important regulators of both the Wingless and Hedgehog pathways in *Drosophila* [7]. Further, the Hippo pathway has recently been found to restrict Wingless/β-catenin signaling by promoting interaction between a canonical Hippo pathway target, the transcription factor TAZ/Yorkie, and Dishevelled, a canonical cytoplasmic component of the Wingless pathway [8]. In addition, the serine/threonine kinase Fused (Fu), a component of the canonical Hedgehog pathway, functions together with the E3 ligase Smurf to regulate the ubiquitylation and subsequent degradation of Thickveins (Tkv), a BMP receptor, during *Drosophila* oogenesis [9]. Altogether, an increasing number of examples escape the canonical view of linear signaling cassettes but rather argue in favor of more elaborate signaling mechanisms in which variations in both content and molecular interactions are a general feature of and between signaling pathways. In summary, our knowledge of the organization of signaling pathways is still rudimentary despite our extensive understanding of some of the players involved in signal transduction. Recognizing the flexible and interconnected nature of signaling cascades will promote a more systematic study of complex cellular signaling, which in turn may greatly improve our understanding of the origin of signaling versatility in development and pathology.

GENETIC DISSECTION OF SIGNAL TRANSDUCTION PATHWAYS

In 1958, G. Beadle and E. Tatum received the Nobel Prize in Physiology and Medicine for demonstrating that 'body substances are synthesized in the individual cell step-by-step in long chains of chemical reactions, and that genes control these processes by regulating definite steps in the synthesis chain (http: //www.nobel.se). Since the realization half a century ago that genes encode the building blocks that make up cells, identifying their functions has become a priority in the life sciences.

Historically, identifying gene function has relied on genetic approaches whereby the function(s) of a given gene is inferred from the phenotype(s) associated with a mutation in that gene. The systematic application of genetics has led to a wealth of knowledge in processes such as pattern formation during development, and signal transduction. For example, saturation screens (see Box 5.1 for definition) have led to a global understanding of pattern formation in the early *Drosophila* embryo, and to the identification of the key genes involved in the process [10]. A major result of these seminal studies was that genes exhibiting the same or similar morphological mutant phenotypes were often found to be part of the same signaling pathway.

An important consideration to keep in mind when taking a genetic approach to deduce gene function is that one studies the global response of the organism to a genetic perturbation. Thus, the endpoint phenotype may be telling us more about the way an organism responds to a genetic perturbation rather than about the wild-type function of the gene itself. A telling example is found in the context of Wingless (Wg/Wnt) signal transduction in *Drosophila*. There, the seven transmembrane protein DFz2 (*Drosophila* frizzled 2), which encodes the Wg receptor, regulates the activities of the Dishevelled (Dsh), Glycogen Synthase Kinase 3 (GSK3) and β-catenin proteins [11]. In the absence of DFz2, a related receptor encoded by Frizzled (Fz) can substitute for DFz2, suggesting that these two related receptors can act redundantly. However, in the presence of DFz2, Fz does not appear to regulate the activity of the Dsh/GSK3/β-catenin pathway, but instead is involved in the regulation of the Planar Cell Polarity (PCP) pathway. Although Fz does not transmit the Wg signal in the wild-type context, the structure of the signaling network allows the activity of Fz to be hijacked to compensate for the absence of DFz2. In this case, the analysis of mutations in *DFz2* failed to reveal the bona fide physiological function of DFz2 in Wingless signaling.

This simple example illustrates a critical but often overlooked concept: when interpreting the results of a genetic approach, there is the danger that our conclusions about the purported wild-type function of a single gene product might be obscured by our existing (but incomplete) knowledge of the signaling network of which it is a component. This is analogous to Plato's powerful Allegory of the Cave, which argues that our interpretation of the world around us is limited by observations made from our vantage point and current knowledge. Thus, in theory, the best way to fully evaluate the function of a single gene product would be to first have a global understanding of the cellular network in which they operate, then remove that component from the network and conclude about the function of that gene based on 'network knowledge'.

Access to the full repertoire of genes encoded by different genomes has made it possible to design new systems-level approaches based on the principles of 'reverse genetics' to construct such global cellular networks. Reverse genetics is an approach to discover the function of a gene by analyzing the effect of specific gene sequences on a phenotype, in contrast to forward genetics, which seeks to find the genetic basis of a particular phenotype. The availability of full genome sequences for all of the well-studied model organisms has given rise to the field of functional genomics, which attempts to describe gene sequences in terms of function (by perturbing the gene using, for example, RNA interference (RNAi) and examining the array of phenotypes associated with the perturbation), expression (by expression profiling using genome-wide microarrays) and protein—protein interactions (by proteomics studies, for example mass spectrometry (MS)). Functional genomic studies are generally carried out on a genome-wide scale using high-throughput methods rather than the more traditional 'gene-by-gene' approach, and can be implemented in both whole organisms and cell lines.

Importantly, functional genomic studies to analyze genotype—phenotype relationships using emerging technologies such as RNAi are based on quantitative (for example relative luminescence units from a luciferase reporter) rather than qualitative readouts (cuticle defects) to reflect effect on phenotype. The use of quantitative pathway reporters has led to the identification of hundreds of pathway components, each of which contributes to the measured phenotype, albeit in varying amounts. This can be best understood by ranking each gene by its effect (when knocked down by RNAi) on the quantitative phenotype being measured. In most cases this generates a continuous distribution from the strongest positive regulators to the strongest negative regulators [12]. In addition, large-scale interaction studies have shown that many proteins are involved in many interactions, both physical associations with other proteins, metabolites and nucleic acids as well as post-translational modifications, to regulate cellular and organismal functions. Thus, over the past decade or so, systems-level, functional, expression and proteomic studies have revealed that signaling is propagated within large networks consisting of hundreds or thousands of proteins (Figure 5.1B). This view is in direct contrast to the traditional reductive approaches discussed above, which focus on individual proteins and which had led to the consensus that signaling takes place largely within simple linear cascades. Furthermore, such analyses have led to the generation of network maps to represent cells as complex interconnected 'systems' rather than mere collections of individual molecules. In network representations of cells, 'nodes' represent proteins/metabolites/nucleic acids as 'parts' of the system [13]. The 'edges' represent relationships between the nodes. The generation of such network maps in the last decade has facilitated the identification of characteristic structural features or topologies inherent in most complex networks and has led to the realization that the structural organization of networks is key to their function. Topological analyses (Figure 5.1C) of different large-scale biological networks (metabolic networks [14], protein—protein interaction networks [15], and transcriptional regulatory networks [16]) have found that complex networks have a small-world property (that is, most nodes in the network can be reached from any other node by a small number of steps) [17] and are scale-free (that is, most nodes in the network have very few connections, whereas a few nodes have many connections) [18]. In at least three different eukaryotes — yeast, worm and fly [15, 19—21] — network hubs (nodes with a high degree of connectivity) are often encoded by essential genes. These shared network properties have been proposed to confer functional advantages such as robustness to fluctuations in the environment, tolerance to random mutations [22], and efficient information processing and flow [23]. Thus, the small-world and scale-free nature of biological networks is under positive evolutionary selection [24]. Another characteristic feature of biological networks are small, recurring patterns of interconnections called network motifs that are significantly more enriched in biological networks than in random networks [25,26]. These motifs form the basic building blocks of networks and are directly responsible for the dynamic information processing functions of biological systems (Figure 5.1C). For example, negative feedback loops increase the speed of response to incoming signals and help to reduce variations in protein levels across cells. On the other hand, positive feedback loops slow down the response time and can lead to a bi-stable (ON or OFF) switch-like behavior [27—30]. Another topological feature that has recently come into the limelight is the 'between-ness centrality' of nodes in a network [31,32]. Between-ness centrality measures the number of non-redundant shortest paths going through a given node [33,34]. Nodes with high between-ness are considered to represent bottlenecks within networks and are analogous to major intersections in a transportation network. Blockage of such intersections would cause a major traffic jam, leading to failure of the transportation system (Figure 5.1C), and so bottlenecks in signaling networks represent attractive targets for therapeutic intervention in the case of deregulated signaling. It has been shown that network bottlenecks are not necessarily hubs, but, like hubs, are more likely to be encoded by essential genes [19,35].

It is clear that to fully understand cellular responses to signaling will require approaches that go beyond the more classic genetic and biochemical studies. Indeed 'systems biology' approaches based on high-throughput, large-scale methods are needed to understand cellular responses to signals in toto. At the most basic level, a complete picture of signal transduction first requires a comprehensive 'parts list' of the components that participate in the cellular signaling network. The second step is to comprehensively identify the physical and functional relationships between the nodes and infer how information flows through the network. Importantly, since the same network is deployed to achieve distinct cellular functions in a context-dependent manner, it is essential to extend the systems biology approaches to construct network models that incorporate information regarding the dynamics of molecular interactions within cells. Inferring the flow of information through a network provides directionality to the edges between the nodes of the network. This can be achieved by systematically perturbing nodes, either by gene knockouts or by RNAi knockdown, followed by measurements of a wide variety of quantitative phenotypes. The phenotypes measured can include changes in gene expression, post-translational modifications such as phosphorylation and acetylation of key proteins, and/or in cellular morphology or behavior. Quantitative phenotypic signatures provide insight into the information processing function of signaling networks, which is key to achieving a mechanistic understanding of how various cellular processes are regulated in time and space.

SYSTEMS APPROACHES TO IDENTIFY THE 'PARTS' OF CELLULAR SIGNALING NETWORKS

RNA Interference (RNAi)

In recent years, as full genome sequences have become available for *Drosophila* [36], human [37,38] and other organisms, large-scale analyses of gene functions have given rise to the field of 'functional genomics'. RNAi has emerged as a unique and powerful functional genomics tool to effectively suppress gene expression in many animal systems [39]. In contrast to other genomic-based approaches, RNAi provides a direct link from gene to function. The development of genome-scale RNAi libraries that contain clones for most genes in a genome in multiple organisms from *Caenorhabditis elegans*, *Drosophila*, mouse and human cells, to the flatworm *Planaria* [40,41] and *Arabidopsis* [42], permits the rapid identification of all genes involved in a particular process [43–49]. Because RNAi is applicable to high-throughput genome-wide analyses it provides a tool to extract functional information globally and comprehensively.

The phenomenon of RNAi was first identified in plants and worms [50]. In *C. elegans*, the process of target gene suppression by RNAi can be triggered by injecting long dsRNAs (~500 nucleotides) into worms, by feeding them bacteria that express the dsRNA or by simply soaking them in solution containing the dsRNAs [51–53]. In *Drosophila*, dsRNAs can be delivered into embryos via injection or by generating transgenic animals that carry RNAi hairpin constructs for in vivo screens [54]. Importantly, the addition of long dsRNAs to *Drosophila* tissue culture cells (dsRNA bathing) can efficiently reduce the expression of target genes [55,56]. Using RNAi in cell lines has led to an explosion of genome-wide cell-based RNAi data for diverse biological processes, including signal transduction, host–pathogen interactions and oncogenesis [47].

When dsRNAs are introduced into cells, they are recognized and degraded by the conserved RNAse III family of nucleases known as Dicer [52,56–60]. Dicer enzymes process the dsRNA into 21–23 nucleotide (nt) short-interfering RNAs (siRNAs) that are incorporated into a multi-protein RNA-induced silencing complex (RISC). This complex directs the unwinding of the siRNAs contained within RISC, and guides RISC to the corresponding mRNA to eventually degrade the targeted transcript. Different types of RNAi reagent have been developed to knockdown target genes in different types of cells and organisms. The four most commonly used RNAi reagents include long dsRNAs (~500 nt), siRNAs (21–23 nt), short-hairpin RNAs (shRNAs; 70 nt) that can be produced exogenously or carried on an expression vector, and endoribonuclease-prepared siRNA (esiRNAs) [61–63]. Typically, RNAi reagents are delivered into cells by virus-mediated transduction for shRNAs, or by lipid-mediated transfection or electroporation for shRNAs, siRNAs, esiRNAs, and dsRNAs [44,46,64,65]. In the case of many *Drosophila* tissue culture cells, dsRNAs are directly taken up from the surrounding medium without the need for transfection [55,56].

The success of RNAi screening depends on the robustness of the cell-based assay, especially its suitability to high-throughput screening (HTS). Almost all HTS cell-based assays provide a quantitative readout for the biological process under study. Many assays use transcriptional reporters where a well-characterized transcriptional regulatory element that is known to respond to the signaling pathway under study is linked to a reporter such as luciferase, green fluorescence protein (GFP) or the *E. coli* β-galactosidase (*LacZ*). The overall output of the reporter can be rapidly measured using a standard plate reader [66–70]. Candidate RNAi hits are identified by their ability to affect the basal or induced expression of the reporter driven by the pathway responsive promoter. Transcriptional reporter-based assays have been used to identify regulators of individual transcription factors such as NFκB [71], E2F [72], and FoxO [73]. In addition, several transcriptional reporter-based RNAi screens have been conducted to identify the regulatory network surrounding cellular signaling pathways, including the Wnt pathway [69,74–76], the Hh pathway [7,70] and the JAK-STAT pathway [66,77]. Further screens based on Oct4 expression level or Oct4 driven GFP expression level have been used to identify regulators of stem cell identity [78,79]. Transcriptional reporter-based assays have several advantages, including easy adaptability to HTS, rapid and automated data collection, and the ability to identify both positive and negative regulators of the process under study. However, transcriptional reporter-based assays also suffer from a number of limitations that must be taken into careful consideration when interpreting the results from a given screen. For instance, the use of a single pathway-responsive promoter—reporter construct assumes that all signaling through the pathway converges on the single readout being assayed, and as a result components that do not converge on this readout will be missed. Furthermore, synthetic promoter constructs that are composed of a string of binding sites for a single downstream transcription factor do not reflect the endogenous context, as they lack sites for co-regulators or sequences that may be important for epigenetic regulation. These limitations contribute to the false negative rate (the number of true regulators missed) associated with the screen. Although it is not possible to completely eliminate false negatives in a genome-wide screen, one can estimate the false negative rate from benchmarking the data obtained with known pathway regulators. Perhaps a more serious problem is experimental variations due to non-specific factors affecting assay readout, leading to the accumulation of false positive hits in the screen. These include factors that affect the level of the reporter indirectly by having an effect on cell viability/proliferation, global transcription, protein translation and stability. Experimental variations are also introduced owing to differences in transfection efficiency in the case of screens where the reporter and/or RNAi reagents are transiently transfected into cells. Thus, appropriate normalization methods are required to account for false positives due to such non-specific factors. A commonly used normalization procedure is co-transfection of a control reporter (for example Renilla Luciferase) along with the experimental reporter (for example Firefly Luciferase). A good control reporter for assay normalization should include a constitutively active promoter that is inert to the pathway under study and achieves reporter expression significantly higher than background [80].

Transcriptional reporter assays have also been employed to identify transcription factor/signaling pathway regulators in a number of in vivo RNAi screens (i.e., in whole organisms rather than tissue culture cells). For instance, large collections of transgenic RNAi lines have been generated to conduct spatially and temporally defined in vivo screens in *Drosophila* [81,82]. Such resources can be used to systematically screen gene functions in specific tissues for phenotypes or effects on gene expression. For example, a *LacZ* reporter for Suppressor of Hairless (Su(H)) expression in the wing imaginal disc has been used as a readout in a screen for Notch pathway regulators [83]. In another case, serine proteases that are involved in the activation of the Toll pathway upon infection have been identified using a *drosomycin-LacZ* reporter assay [80].

Another quantitative cell-based assay makes use of specific antibodies that recognize protein modifications such as phosphoserine/tyrosine or methyl-lysine residues on key components of the signaling pathway itself or on proteins that form a part of the cell's response to activity through the pathway. For instance, genome-wide screens using phospho-specific antibodies have identified regulators of dually phosphorylated MAP kinase/ERK and phosphorylated Akt downstream of receptor tyrosine kinase (RTK) signaling [84–86]. The success of antibody-based assays is critically dependent on the availability of specific antibodies. Such assays can be performed using a simple plate reader to measure fluorescence emitted by the fluorescently coupled secondary antibody. Antibody selection and validation are critical for the development of any high-quality assay. The specificity of the antibody should be precisely evaluated by both Western blotting and immunocytochemistry (cell staining). An antibody that generates strong non-specific bands in a Western blot is not suitable for plate-based assays. Importantly, it must be determined beforehand that the antibody truly detects pathway activity in response to known stimuli and perturbations. A major difference between Western blots and plate-based assays is the context in which proteins are analyzed. In plate-based assays cultured cells are fixed to the bottom of a microplate, and therefore the immobilized antigens present a slightly different conformation than those that have been processed by SDS-PAGE (see Box 5.1 for definition) prior to Western blotting. Thus, it should be noted that although some primary antibodies perform well for Western blotting, they might exhibit poor binding characteristics on fixed antigens, resulting in low fluorescence signal in the plate-based format. As in the case for transcriptional reporter-based assays, the signal from the phospho-specific antibody must be normalized to account for variations in cell number across different wells in the plate. A wide array of fluorescent molecules, including DNA-binding dyes (DAPI, TO-PRO 3), actin-binding dyes (Phalloidin), antibodies to total protein and non-specific cytoplasmic protein stains, can be used for normalization. However, it must be first determined that the stain of choice does indeed provide a linear measure of cell number. These plate-based approaches provide an attractive alternative to high-throughput microscopy (also known as high-content screening, HCS) for assay readouts that are based on immunofluorescence detection. If high-resolution information is not central to the results of the screen, then plate reader assays [84–86] provide a significant advantage in terms of ease and speed of detection, as well as simplifying the downstream analysis to a single intensity measurement per well to report effect on pathway activity.

An alternative to the plate-based assays described above is the transfected cell microarrays that allow the miniaturization and simplification of high-throughput assays [87]. RNAi reagents are spotted on the surface of a standard glass microarray slide and are used to transfect cells. This generates a living cell microarray comprising locally transfected cells in a mixture of non-transfected cells [87,88]. Transfected cell microarrays facilitate large screens with many replicates, as they offer the advantage of using minimal amounts of antibody compared to traditional plate-screening formats. For example, Lindquist et al. performed a genome-scale RNAi screen on microarrays of *Drosophila* cells to identify novel regulators of mTOR (mammalian Target of Rapamycin) complex 1 (TORC1) signaling by immunofluorescence [89]. A total of 70 novel genes were identified as significant regulators of RPS6, a TORC1 effector. Cell microarrays facilitate large screens with many replicates, as they offer the advantage of using minimal amounts of antibody compared to the traditional plate-screening format. The amount of immunofluorescence from the phospho-specific antibody can be normalized to cytoplasmic area using standard image analysis software packages such as those developed for HCS image analysis [90].

Although plate reader assays have the advantage of speed and ease of performance, HCS is undeniably one of the most powerful HTS assays because it allows multiple cellular features/parameters, such as protein abundance as well as localization, to be measured simultaneously. Image-based screens typically use either fluorescently conjugated primary or secondary antibodies to visualize proteins or cellular structures of interest (for example anti-Fibrillarin antibody to visualize the nucleolus) or fluorescently labeled dyes and GFPs tagged with the appropriate localization signal (for example nuclei, mitochondria, Golgi, and actin filaments). HCS has been performed to identify targets of small molecules/drugs [91–95] and also in a number of RNAi screens to identify genes that affect diverse cellular functions, including cell morphology [96–98], cell cycle progression [99], mitosis [100,101], endocytosis [102, 103] and host–pathogen interactions [104–107]. Quantitative image analysis has also been used to identify genes required for growth and morphology of fluorescently labelled primary neurons/glia and muscle cells in response to RNAi-mediated gene knockdown [108,109]. Although multi-parametric, quantitative image analysis applied to large-scale functional genomic screens promises to generate systems-wide insights into many fundamental cellular processes, automated image acquisition and analysis, feature extraction, and data storage can be challenging and are still undergoing rapid development [110,111]. Other cell-based assays include the use of flow cytometry to measure response to RNAi treatments [112–114].

RNAi HTS in various cell lines using the different cell-based assays discussed above, have been conducted for a diverse array of biological processes, including cell viability [68], cell morphology [96,98], cell cycle [112], cytokinesis [91], susceptibility to DNA-damaging agents [115,116], RNA processing [117,118], general and specialized secretion [67], calcium stores [119–121], factors influencing polyQ aggregation and toxicity [122], mitochondrial dynamics [123], circadian clock [124], hypoxia [125], phagocytosis [113,126], innate immunity [127–129], cell susceptibility to infection by viruses or other intracellular pathogens [104–107] as well as most of the major signaling pathways [7,47,66,69,70,75,77,85,86, 130,131]. Results from these screens have not only identified new components of the process under consideration but have also provided insights into the complexity of signaling networks. RNAi screens in mammalian cells [47] have led to the identification of novel oncogenes and putative drug targets for the development of therapeutics [64,132–136].

Although HTS based on RNAi has transformed the field of systems biology in the identification of gene functions, it is important to keep in mind that inhibition of gene expression by RNAi is not the same as gene inactivation by mutation. RNAi acts at the level of the messenger RNA (mRNA), either by reducing mRNA levels or by blocking mRNA translation [137–139]. Thus, RNAi-based assays can suffer from high rates of false negatives due to incomplete knockdown of mRNA levels (or knockdown of only specific splice forms). Another significant issue associated with RNAi reagents is that they can lack specificity due to suppression of unintended genes, leading to false positives. False positives due to sequence-dependent off-target effects (OTEs) have been shown for RNAi reagents with ≥ 19 nt regions of homology with unintended targets [140]. It has also been demonstrated that sequence-dependent OTEs are particularly problematic when the RNAi reagents target gene regions containing CAN repeats (where N can be any nucleotide) [75] that are found in many fly genes. In addition, siRNAs can also interfere with mRNA stability and/or translation through the microRNA pathway [137–139]. MicroRNAs (miRNAs) are non-coding RNAs that are encoded by the organism's genome and help regulate gene expression. Mature miRNAs are 22 nt RNAs and are similar in structure to siRNAs that are produced from exogenously introduced long dsRNAs. miRNAs bind to complementary sites that are 7–8 nt long within 3′UTRs (untranslated region) of target genes, leading to cleavage or translational repression [141]. The siRNA and microRNA pathways converge downstream of initial processing steps and share some of the same silencing machinery [118]. The 5′ region of the siRNA can act like the seed region of a microRNA, which extends from position 2 to position 8 of the guide strand and is complementary to sequences in the 3′ untranslated region (3′UTR) of target genes. Since a perfect match of only 7–8 nt is required between the seed region and the target mRNA for repression, it is difficult to identify all of the many putative targets in a cell [142,143]. Thus, sequence-dependent OTEs of siRNAs seem to result at least in part via microRNA-like off-target activity, which may result from siRNAs entering the microRNA pathway and functioning as microRNAs on targets with matches to the seed region in their 3′UTRs.

Although the prevalence of OTEs was underestimated in early RNAi screens, a number of approaches have now been developed to minimize their effects [47,144]. These include the development of computational tools to design RNAi reagents with limited or no homology to genes other than the intended target; the use of multiple independent RNAi reagents targeting the same gene; and the rescue of the RNAi-induced phenotype by an RNAi-resistant version of the gene. Further, with the availability of the catalogue of expressed genes in a wide array of commonly used cell lines by RNA sequencing [145], false positives associated with a screen performed in *Drosophila* cells can be identified and filtered based on whether the targeted gene is expressed in the cell line being screened. One can also filter out potential false positives by removing genes that score in a large majority of RNAi screens.

Large-scale epistasis or synthetic lethality studies (see Box 5.1 for definition) using sensitized genetic backgrounds can also uncover new components of signaling pathways [146] because they tend to reveal genes that are involved in redundant or parallel pathways/complexes. Such screens are similar in concept to the synthetic genetic array (SGA) analysis in yeast, where the viability of a set of gene deletions has been tested in backgrounds where other genes have been similarly deleted (synthetic lethal) or overexpressed [147–149]. The results from these studies showed that RNAi of many individual genes does not affect growth, but that many genes do have a synthetic genetic growth phenotype in combination with other genes. These genetic interactions include both negative (aggravating) interactions as well as positive (alleviating) ones, where the phenotype of eliminating one gene is attenuated by the loss of a second one. Combinatorial RNAi experiments where dsRNAs are screened for their ability to suppress or enhance the effect caused by another dsRNA (or by small molecules) are also becoming increasingly common [150,151]. Examples of HTS for multiple genes by RNAi include 17 724 combinations that identified regulators of *Drosophila* JNK signaling [150], and combinatorial RNAi of disease relevant genes in *C. elegans*, which identified ~1750 novel functions for genes in signaling [152]. RNAi microarrays facilitate the miniaturization of combinatorial RNAi screens and provide an effective and economical way to conduct large-scale screens in tissue culture cells [89,153,154].

In addition to identifying new genes involved in a particular biological process, comprehensive and quantitative genetic interaction data can be used to shed light on the organizing principles of signaling networks and the ways in which distinct signaling modules are interconnected. Schuldiner and colleagues [155] developed a strategy for building large-scale genetic interaction maps called 'epistatic miniarray profiles' (E-MAPs) that allows one to group sets of genes based on their signature/patterns of genetic interactions. Using this strategy, an E-MAP of genes involved in the early secretory pathway (ESP) in the budding yeast was constructed which robustly identified known pathways and relationships, such as the effect of the unfolded protein response (UPR) pathway on secretory functions, and the hierarchical relationships of the different stages of vesicular trafficking. This study also identified a strong link between endoplasmic reticulum-associated degradation (ERAD) pathway and lipid biosynthesis, a connection that had been previously poorly characterized. The E-MAPs strategy has been successfully extended to study the networks of genes involved in creating, maintaining and remodeling the chromatin in response to various cues [156], and also to identify novel components of the RNAi machinery in the fission yeast *Schizosaccharomyces pombe* [157]. The success of these studies highlights the power of E-MAPs to provide a systems-level view of the functional topology of networks that cannot be obtained by other methods. Recently, the concept of E-MAPs has been successfully implemented in *Drosophila* cells using combinatorial RNAi screens [158]. In this study, pairwise interactions between 93 genes involved in signaling were evaluated using two independent RNAi reagents for each per target. This set of 93 genes included components of the three MAPK pathways (Ras-MAPK, JNK and p38 pathway) and all expressed protein and lipid phosphatases. The pairwise knockdowns were analyzed for their effects on cell number, mean nuclear area and nuclear fluorescence intensity and resulted in 73 728 measurements, from which interaction scores were estimated. The success of the strategy was reflected in the high frequency of interactions observed between known components of the Ras-MAPK signaling pathway and a clear separation from regulators of the JNK signaling pathway. In addition, the authors identified connector of kinase to AP-1 (*Cka*), a scaffold protein in the JNK signaling pathway [159], as a positive regulator of Ras-MAPK signaling, and thus a putative point of cross-talk between the two pathways was identified.

Functional genomic approaches at the level of whole systems are powerful because they can identify most genes that affect a given signaling network, and have revealed that, contrary to previous views, hundreds of genes may be a part of a signaling network. However, genetic studies do not distinguish between direct and indirect effects, and therefore it is not clear where in the network the different genes identified act. Understanding how they contribute to the overall structure of the cellular signaling network requires the integration of genetic data with other datasets such as protein–protein interaction networks.

Protein–Protein Interactions

Large-scale protein–protein interaction (PPI) mapping complements genetic studies by revealing physical associations and helps to define the physical signaling network. In PPI networks, nodes represent proteins and the edges represent a physical association between them. The methods most widely used to map PPI networks include the yeast two-hybrid (Y2H) system and its derivatives [160, 161], and affinity- or immunoprecipitation followed by mass spectrometry (AP/MS) [162–165]. PPI networks derived from Y2H methods are composed of binary (direct) interactions, whereas those derived from AP/MS techniques can be both direct and indirect, as they identify protein complexes.

Large-scale Y2H studies have been conducted with proteins from *Helicobacter pylori* [166], yeast [167–169], *C. elegans* [21,170,171], *Drosophila* [172–174], and human [175–179]. Strikingly, the three large-scale *Drosophila* Y2H mapping studies failed to fully recapitulate known signaling pathways. For example, querying the combination of these studies for Raf reveals only interactions with CG15422, Ras, Rhomboid, and Rap2L (http://itchy.med.wayne.edu/PIM2/PIMtool.html), neglecting to identify most known targets, scaffolds, and co-regulators of Raf activity. Thus, these 'proteome-scale' approaches, although they identified highly abundant or strongly interacting cellular components, failed to identify many interactors of signaling components — most likely because of the absence of endogenous signaling contexts [180]. For this reason, MS-based approaches have become more popular, especially as the difficulty of implementation and costs have dropped dramatically.

Comprehensive MS-based PPI mapping has been applied in yeast [181–185]. A global protein kinase and phosphatase interaction network identified 1844 interactions between 887 proteins [181]. The success of MS approaches has been aided by the increased sensitivity of MS technology and implementation of tandem affinity purification (TAP) of protein complexes [186]. Recently, tandem affinity purification followed by MS has been used to isolate protein complexes from *Drosophila* tissue culture cells and tissues (http: //flybase.org/). ~5000 *Drosophila* proteins were fused to a FLAG-HA tag so that the fusion proteins could be expressed and recovered with their interacting partners from cells, or from whole transgenic flies. In addition to proteome-scale AP/MS, a number of smaller studies have been conducted in human cells on signaling pathways such as TNF-α and Wnt [187,188], biological processes such as autophagy [189], protein families such as the de-ubiquitinating enzymes [190], and protein complexes such as the RNA–polymerase II and PP2A complexes [191,192].

Both Y2H and AP/MS PPI mapping methods have been applied to the characterization of cellular networks with disease relevance, such as virus–host interactions [193–197]. These proteomic studies have confirmed that cellular processes take place within large networks of interconnected proteins.

PPI approaches, as implemented thus far, have been incomplete for investigations of signal transduction because they (1) do not provide functional information, and (2) often take place outside the context of endogenous signaling. These issues can be addressed by combining proteomics with RNAi. For instance, Y2H was used to identify interactors of the DAF-7/TGF-β pathway in *C. elegans*, resulting in a network of 59 proteins, and RNAi was used to show that nine novel interactors functionally interact with the TGF-β pathway [198]. Another major study used a pathway-specific approach with liquid chromatography/tandem MS to characterize the interactors of 32 TNF-α/NFκB pathway components in mammalian cells under endogenous signaling conditions [187]. Interactors were identified at baseline and under TNF-α stimulus, revealing 221 interactions. RNAi was then used to determine their influence on signaling output. This study demonstrated the power of pathway-directed proteomics in endogenous signaling contexts. One limitation of this study, however, was the lack of rigorous quantitation of the assembly of signaling complexes. Most signaling complexes are highly dynamic, with components often held in inactive complexes that can change dramatically following stimulation. For example, Raf and KSR are held in separate inactive complexes bound to PP2A core components and 14-3-3 proteins; following stimulation, Ras induces the recruitment of PP2A regulatory subunits to Raf, dephosphorylation of 14-3-3 binding sites, release of 14-3-3 proteins, membrane recruitment, KSR and Raf co-localization, Raf phosphorylation of MEK, and MEK phosphorylation of MAPK [199,200].

RNAi and MS can also be combined by first starting with RNAi and then following up with MS, as has been demonstrated for RTK/ERK signaling at baseline and under insulin stimulation [84,85]. All of the major known components of the pathway were tagged. In addition, a control cell line was engineered to subtract common interactors/contaminants. Altogether, 54 339 peptides were identified representing 12 208 proteins, encompassing an unfiltered network of 5009 interactions among 1188 individual proteins. To provide a ranked list of novel pathway interactors, filtering out sticky proteins found in control preparations and providing a probability that the observed interactor is real, the significance analysis of interactome (SAINT) method was applied to the PPI dataset [181]. Using a SAINT cut-off of 0.83 and a false discovery rate (FDR) of 10%, a filtered PPI network of 386 interactions among 249 proteins surrounding the canonical components of the RTK/Ras/ERK signaling pathway was generated [84]. In this network canonical baits have multiple common interactors, as would be expected from a well-connected signaling pathway (as opposed to unbiased PPI mapping of random protein baits), as well as many unique interactors. Because the baits were purified under two conditions, baseline and insulin stimulation, the dynamics of the mini-proteome during signaling events was uncovered. As a measure of the sensitivity of the TAP/MS approach to network characterization, interactions among the canonical components and their known interactors were extracted. This canonical network recapitulated most of the known RTK-ERK signaling pathway.

Comparing the RTK-ERK PPI network to six unbiased genome-wide RNAi screens revealed that nearly half (119) of the proteins identified by PPI mapping scored in the RNAi screens, which is a significant enrichment relative to the entire genome (19%, $p < 7 \times 10^{-25}$) [187].

A major bottleneck in large-scale proteomics studies is the experimental validation of specific interactors or components of complexes. The combination of AP-MS with RNAi-mediated knockdown provides a way to directly validate specific PPIs. With differential labeling of the two proteomes to be compared (wild-type vs. RNAi knockdown) such analyses have the potential for accurate, highly quantitative results [201,202]. Currently, two major types of labeling technique are used for MS-based proteomics studies: metabolic labeling and chemical labeling [203]. Stable isotope labeling by amino acids in cell culture (SILAC) is considered to be the gold standard in the case of metabolic labeling. Here, isotopically labeled amino acids (for example arginine and lysine labeled with the stable ^{13}C and/or ^{15}N isotope) are incorporated into cellular proteins during normal protein biosynthesis [204]. Thus, the cells to be compared (for example wild-type vs. RNAi) are grown in media containing 'light' (normal) and 'heavy' (labeled) amino acids, respectively. After labeling, the two cell populations are mixed, fractionated, and subject to MS/MS analysis to quantify the differences between their two proteomes in a highly accurate manner [205,206]. Because the labels are carried by arginine and lysine residues tryptic digestion produces peptides that contain a labeled amino acid at the carboxy terminus. The heavy and light tryptic peptides elute together as pairs separated by a defined mass difference that allows the two proteomes to be distinguished in the MS/MS analyses. SILAC can be multiplexed to allow comparisons between three different proteomes simultaneously. SILAC-based differential labeling combined with RNAi, co-immunoprecipitation and quantitative MS analysis was used to detect and validate the cellular interaction partners of endogenous β-catenin and Cbl proteins in mammalian cells [202]. Alternatively, chemical labeling involves the use of isobaric tagging reagents such as iTRAQ (isobaric tags for relative and absolute quantification) [207] or TMT (tandem mass tags) [208] (see Box 5.1 for definitions) to label peptides after lysis and trypsinization. The peptides in samples to be compared are modified by covalent attachment of a unique tag or label, which enables the quantification of the same peptide across multiple samples. The uniquely labeled samples are combined and run through an MS analysis. Despite bearing distinct tags, the same peptides from the different samples are indistinguishable from each other in the first MS run, because the molecular weight of each tag is the same. However, during MS/MS each tag undergoes fragmentation, releasing a signature reporter ion. The signature reporter ions differ in mass between the tags and their relative levels serve as a measure of differences in the levels of a given peptide between samples. Labeling methods such as these provide a rapid means by which to quantitatively examine global proteome-level changes, and compare, for example, wild-type cells with those subjected to mutations, RNAi knockdown or small molecule treatments.

Transcriptional Profiling

Gene expression profiling using DNA microarrays, and more recently RNA-seq, has emerged as a valuable tool for broad correlation of gene activity with alterations in physiological or developmental states [209–211]. Transcriptional profiling experiments can be used to generate compendia of gene expression data across different cell types [212], diverse species [213], development times [214], and in response to distinct stimuli [215]. Such gene expression datasets have been commonly used to identify genes that function in common pathways or which encode components of the same complex. Studies in yeast have demonstrated that proteins that interact with each other show similar expression profiles to non-interacting proteins [216–219]. Gene expression profiling has been used to study signaling by wild-type and mutant receptor tyrosine kinases (RTKs) and has provided evidence for substantially overlapping immediate early transcriptional responses upon activation of PLCγ, PI3K, SHP2, and RasGAP proteins and their respective signaling pathways [220]. However, expression profiling studies do not provide details of how and where in this network pathways engage in cross-talk to specify the appropriate biological response.

Although expression profiling studies have become the gold standard for global responses to signaling, several recent studies have shown that correlation between transcriptome and proteome is only ~50% [221,222]. Proteomics approaches that directly measure the targets of signaling pathways — that is, the proteins — are more useful. For example, Yates and colleagues [223] compared protein abundance between wild-type *C. elegans* and those lacking the worm insulin receptor (InR) ortholog *daf-2*. This study revealed 86 proteins whose abundance changed following loss of InR, an important finding for a signaling pathway with known effects at the translational rather than the transcriptional level.

QUANTITATIVE RNAI SIGNATURES OR PHENOPRINTS TO INFER CONTEXT DEPENDENT INFORMATION FLOW THROUGH CELLULAR SIGNALING NETWORKS

The network maps discussed so far are primarily static and do not provide information regarding the direction of information flow through the nodes; instead, they provide the framework required to begin to dissect the functional, logical and dynamical nature of cellular signaling networks. The topological features of signaling networks reflect the need to process multiple input cues received by a cell, interpret them correctly and transmit the information to coordinate cellular activity and generate the proper phenotypic response [224–226]. Thus, signaling networks are highly dynamic, exist in distinct states, and are capable of deploying the same or overlapping set of signaling molecules in different ways, depending on the context and the input cues received by the cell. It has been demonstrated that, depending on the cumulative effects of the signals received by a cell, JNK activity, for instance, can be either pro- or anti-apoptotic [224, 226]. The challenge of future studies is to gain a mechanistic understanding of the direction of information flow, the dynamic nature of crosstalk between signaling pathways, and the hierarchical relationship between network components in response to a distinct set of stimuli. Such mechanistic insights will allow the generation of predictive (testable) models of how this information processing capacity of signaling networks is coopted in disease conditions to produce aberrant phenotypes, and will lead to the identification of novel drug targets and the development of more effective therapeutics.

In recent years it has been demonstrated, albeit on a small scale, that systematically perturbing the components of the network and simultaneously measuring multiple quantitative phenotypes in the presence or absence of specific input cues can be used to infer information flow through signaling networks. The phenotypes measured can include changes in gene expression, in phosphorylation of key signaling or target proteins and/or in cellular morphology. These quantitative phenotypes result from multiplexed assays and are therefore different from the cell-based assays used in RNAi HTS (described above), such that, instead of measuring a single transcriptional reporter or changes in the phosphorylation status of a single protein in response to the knockdown of a gene, RNAi signatures/phenoprints are composed of multiple measurements ranging anywhere from tens to hundreds of genes or proteins. A compilation of such quantitative phenotypes provides a unique, context/signal specific 'signature' or 'phenoprint' for each perturbed network component. Network components that are deployed in the same or similar manner in response to the incoming signal would tend to have similar phenoprints [227,228].

Direction of Information Flow from Gene Expression Signatures

Transcriptional signatures resulting from the loss/reduction of individual network components by RNAi can be used to infer the flow of information through proteins that are interconnected within a cellular network. This approach was used successfully in analyzing the response of *Drosophila* cells to microbial infection and lipopolysaccharides (LPS) [229]. In these studies, the topology of network connections was retrieved from experimentally measured global transcriptional responses to successive perturbations in pathway components. Genome-wide expression profiling and loss-of-function experiments using RNAi were used to determine the identity of the signaling pathways that control microbial challenge-induced cellular responses. Differential gene expression signatures appeared with discrete temporal patterns after LPS stimulation and septic injury, and could be assigned to the activation of distinct signaling pathways by impairing pathway-specific components using RNAi. Specifically, the results indicated that in addition to signaling through the Toll and Imd pathways, microbial agents induce signal transmission through the JNK and JAK/STAT pathways. Altogether, this demonstrated how data obtained from microarray expression profiling combined with the RNAi technology could be used to extract interconnections between different signaling pathways downstream of an extracellular stimulus.

Whole genome expression profiling [230] has identified gene expression signature-based analysis of signaling networks in a number of different model systems. One of the first successful applications of this approach was the generation of a compendium of gene expression profiles in yeast for 300 different mutations and small molecule treatments [231]. The 300 different mutations and chemical treatments specifically included 276 deletion mutants, 11 tetracycline-inducible essential genes and 13 small molecule inhibitors (data available from Rosetta Inpharmatics). The assumption was that the cellular state can be deduced from the global gene expression response, and the transcriptional profile of a gene in response to a change in cellular state (disease, cellular activity such as cell division, response to drugs or genetic perturbation) constitutes a unique quantitative molecular phenotype [210,232–239]. The study showed that genes known to be co-regulated could be easily detected, and mutations in genes or treatments with small molecules that regulate similar cellular processes displayed strikingly similar expression profiles. Most importantly, the study was able to assign eight unannotated genes to the regulation of pathways such as sterol metabolism, protein synthesis and mitochondrial function. Furthermore, the observation that gene expression profile in response to drug treatment phenocopies the loss-of-function profile of its target facilitated the identification of Erg2p, a sterol isomerase, as a novel target of the drug dyclonine [231].

Another study in *Saccharomyces cerevisiae* [240] analyzed the functional relationship between kinases and phosphatases by generating genome-wide expression signatures for 150 deletion mutants. Gene expression signatures were also generated for pairs of genes (kinases and phosphatases) that exhibit synthetic genetic interactions with the aim of investigating the mechanisms underlying the redundant relationships. The results of this study concluded that there are three types of redundant connection: (1) complete redundancy, where the two genes in a synthetic genetic interaction regulate the same set(s) of genes to an equal extent, such that the single mutants show no significant changes, but an effect on expression of regulated gene set(s) is seen only in the double mutant; (2) quantitative redundancy, where the two genes in a synthetic genetic interaction regulate the same set(s) of genes but to a quantitatively different extent. Here one of the single mutants shows no significant effect but the other does, and the effect on the expression of the regulated gene set(s) is amplified in the double mutant; and (3) mixed epistasis, where the two genes in a synthetic genetic interaction regulate some of the same gene set(s) via either complete or quantitative redundancy, while other gene set(s) behave in a completely different way. Mixed epistasis reflects only a partial overlap in function of the two genes in the synthetic genetic interaction. The authors concluded that such gene pairs share additional regulatory associations, such as inhibition of one by the other, and that mixed epistatic relationships provide the mechanisms to achieve signaling specificity in a context-dependent manner. Importantly, mixed epistasis was found to be the most common redundant relationship in signaling networks.

The Connectivity Map (Cmap) [241,242], identified functional connections between drugs, genes and diseases from expression profiles in a compilation of genome wide expression data from cultured human cells treated with either bioactive small molecules or genetic perturbations. Cmap incorporates pattern-matching algorithms that decode differential gene expression data into functional relationships between drugs, genes and diseases to generate testable hypotheses. Again, the underlying assumption of Cmap is that common gene expression changes reflect functional connectivity between the gene products targeted by either small molecule or various genetic perturbations. Functional connectivity is expected to reflect the role of gene products in a common biological process, in particular components of a specific signaling pathway. Cmap was used to identify the target of two previously uncharacterized natural products (celastrol and gedunin) that had inhibitory activity towards androgen receptor activity, with implications for the treatment of prostate cancer [243]. Gene expression signatures for each of the drugs were generated and used to search the Cmap database for similar gene expression patterns. The gene expression signatures of celastrol and gedunin were most similar to the signatures of inhibitors of the chaperone HSP90. This finding predicted that HSP90 was the most likely target of celastrol and gedunin activity, a hypothesis that was tested and validated experimentally. In another study, rapamycin was found to reverse the effects of resistance to the glucocorticoid dexamethasone in acute lymphoblastic leukemia (ALL) [244].

A number of publicly available compendia of gene expression profiles are available for data mining purposes, including the Global Cancer Map [245], Gene Expression Atlas [246,247], and Oncomine Cancer Profiling Database [248].

The overall logic of establishing connectivity based on gene expression signatures of RNAi treated cells is simple and schematically presented in Figure 5.2. The current challenge is to go beyond proof-of-principle studies and establish robust experimental protocols and computational tools that will allow large-scale implementation. This presents three main challenges: (1) the generation of gene expression signatures for every biological state of interest (disease, cellular activity such as cell division, response to drugs or genetic perturbation); (2) a cost-effective high-throughput platform for screening genetic perturbations or small molecule treatment using gene expression signatures; and (3) the development of computational tools for data analysis.

Gene expression signatures serve as molecular surrogates for biological states, are composed of tens to hundreds of genes, and are distinct for different biological states [235,249]. The ability to identify a gene expression signature that can serve as a quantitative molecular phenotype for a specific biological state holds great promise for the development of high-throughput small-molecule or RNAi screens using the signature of interest as the readout. The first gene expression signature-based screening (now called gene-expression-based high-throughput screening, GE-HTS) was conducted [249] to identify small molecules that induce the differentiation of acute myeloid leukemia cells. Using gene expression analysis of primary cells from patients and unaffected individuals identified a number of differentiation-correlated genes to generate a five-gene signature for leukemia cell differentiation. This signature was then used to screen a library of 1739 bioactive small molecules to identify those that induce the expression of the

FIGURE 5.2 **Inferring network structure from transcriptional signatures. A, B, C:** Given the structure and knowledge regarding the flow of information through a network of n components that transduce the input signal (IS) received by a cell, one can predict the effects of perturbing individual network components on the expression of known target genes (E). For example, given the linear structure and flow of information of the network shown in **A**, perturbing any of the network components will affect the expression of target genes E1–E6 in a similar manner as depicted in the schematized heat map. **D:** Given a set of gene expression signatures generated by perturbing the components of a network, it should be possible to reconstruct the structure of the network and deduce the flow of information through its resident components.

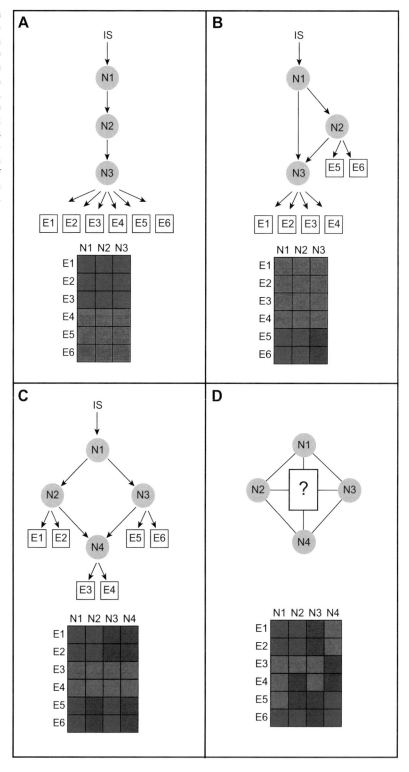

differentiation-correlated signature genes in leukemia cells. This study identified eight bioactives that were further validated as bona fide differentiation inducers.

The success of the first GE-HTS [249] led to the development of the Luminex xMAP technology suitable for HTS (see Box 5.2 for details on the Luminex technology). The technology combines multiplex ligation-mediated amplification [250–252] with optically tagged and barcoded microsphere and flow cytometric detection. The technology is currently capable of measuring up to 500 transcripts

Box 5.2 Luminex xMAP technology for multiplex gene expression analysis

The Luminex xMAP Technology (http: //www.luminexcorp.com) can be used to perform a wide variety of multiplex assays on the surface of 5.6 μm polystyrene microspheres. Each microsphere or bead is uniquely color-coded internally using precise concentrations of red and infrared fluorescent dyes, resulting in 500 spectrally distinct beads. This feature allows multiplexing of 1–500 analytes in a single sample. The surface chemistry of the microspheres allows capture reagents to be efficiently coupled to the beads to facilitate the measurement of different kinds of analyte in the sample. For example, capture reagents may include oligonucleotides, antibodies, peptides, enzyme substrates, or receptors, thus offering a wide range of applications, including gene expression analysis, detection of single nucleotide polymorphisms, protein expression analysis, detection of protein–protein interactions, quantification of antibody affinity and epitope mapping, serum analyte profiling and detection of enzyme/substrate or receptor–ligand reactions. After the analyte of interest (transcript, antibody, antigen, ligand, or substrate) is captured from the sample on the surface of the beads, the reactions are quantified in the Luminex analyzer, an instrument that combines high-tech fluidics based on the principles of flow cytometry and laser optics for signal detection and processing. In the analyzer, the microspheres pass through the detection chamber in a single file such that the reaction between the surface coated capture reagent and the analyte of interest can be quantitatively measured for each bead. In the detection chamber, a red laser or light-emitting diode (LED) is first used to classify each microsphere to one out of the 500 spectrally different sets. A second laser or LED excites the fluorescent dye associated with the reporter molecule that is used to detect the analyte of interest.

Peck and colleagues [253] developed the Luminex xMAP technology for gene expression signature analysis. Messenger RNA (mRNA) transcripts from each sample are captured on immobilized poly-dT in 384-well plates and are reverse transcribed to complementary DNA (cDNA). For each gene/transcript of interest two oligonucleotide probes are designed. The 5′ probe contains a 20 nt sequence complementary to the T7 primer site, a unique 24 nt sequence that serves as a barcode, and a 20 nt sequence complementary to the transcript of interest. Each 3′ probe is phosphorylated at its 5′ end and contains a 20 nt sequence contiguous with the gene-specific fragment of the 5′ probe followed by a 20 nt T3 primer site. Probe pairs for the transcripts of interest are mixed with cDNA from each sample, unbound probes are removed, and probe pairs annealed to contiguous regions of target mRNAs are ligated together to yield synthetic 104 nt templates for amplification. Universal T3 and 5′-biotinylated T7 primers are used to amplify the templates by PCR. The resulting biotinylated and bar-coded amplicons are hybridized to a pool of spectrally distinct microspheres. Each microsphere presents on its surface a distinct capture probe complementary to one of the barcodes. The hybridization reactions are finally reacted with streptavidin-phycoerythrin to fluorescently label biotin labels. Captured labeled transcripts of interest are quantified and beads decoded in the Luminex analyzer as described above. Luminex xMAP assays are carried out in a 96-well plate format, with up to 500 genes being measured in each well/sample.

within a single reaction in thousands of samples in a cost-effective manner [253]. (http: //www.luminexcorp.com)

GE-HTS has also been recently adopted to dissect the regulatory network controlling the transcriptional response of mouse primary dendritic cells to pathogens [215]. A 118-gene signature that defines the response of mouse primary dendritic cells to infection by pathogens was established using expression profiling of dendritic cells exposed to different pathogen-derived components (virus, Gram-positive and Gram-negative bacteria). The gene expression signature was then used to screen 125 transcription factors, including proteins that modify chromatin and proteins that bind RNA, for their role(s) in coordinating cellular response to pathogen infection. The reconstructed network model composed of the transcription factors and their cognate upstream signaling pathways helps to explain how pathogen-sensing pathways achieve specificity in their response to different microbial populations. The study used a screening platform called NanoString nCounter Gene Expression Assay (see Box 5.3 for details on the Nanostring technology). The nCounter Gene Expression Assay is a robust and highly reproducible method for detecting the expression of up to 800 genes in a single reaction. The biggest advantages of this platform are its high sensitivity, requirement for very small amounts of total RNA as starting material, and the lack of any enzymatic reactions to convert total RNA to cDNA and amplification of resulting cDNA by polymerase chain reaction (PCR).

Direction of Information Flow from Phosphorylation Signatures

Similar to gene expression signatures, the biological state of a cell can also be inferred from the phosphorylation profile of proteins that are themselves components of the cellular signaling network, as well as of proteins that form a part of the cellular response to signals impinging on the cell. Protein phosphorylation is a widespread post-translational modification and plays important roles in most biological processes in eukaryotic cells. The addition of phosphate groups on substrate proteins by kinases modulates the overall function of the substrates by directing their activity, localization and stability. Extensive protein-phosphorylation-mediated signaling networks direct the flow of

Box 5.3 Nanostring nCounter System for Mulplex Gene Expression Analysis

The nCounter assay can be used to detect several types of nucleic acid molecule, including mRNA, DNA and microRNAs. The nCounter assay is based on direct imaging of mRNA molecules of interest that are detected using target-specific, color-coded probe pairs [273]. It does not require the conversion of mRNA to cDNA via reverse transcription or the amplification of the resulting cDNA via PCR. A pair of sequence-specific probes — the capture and reporter probes — detects each target gene of interest. The capture probe contains, from 5' to 3', a 35—50-base sequence complementary to the target mRNA, a short sequence common to all capture probes, and a biotin affinity tag that provides a molecular handle for the attachment of target genes to facilitate detection. The reporter probe contains, from 3' to 5', a second 35—50-base sequence (complementary to the same target mRNA, near or contiguous with the target-specific sequence in the capture probe partner), a short sequence common to all reporter probes, and a color-coded molecular barcode. The common sequences included in all capture and reporter probes facilitate the removal of unbound excess probes during post-hybridization steps. The barcode contained in each reporter probe is composed of a linearized single-stranded M13 DNA molecule annealed to a series of six complementary RNA segments, each labeled with one of four spectrally non-overlapping fluorescent dyes. The arrangement of the differently colored RNA segments creates a unique color code for each target gene of interest. The different combinations of the four distinct colors at six contiguous positions allows for a large diversity of color-based barcodes, each designating a different gene transcript, that can be mixed together in a single reaction for hybridization and still be individually resolved and identified. The methodology offers the flexibility of multiplexing up to 800 reporter—capture probe pairs within a single reaction.

The target mRNA is mixed in solution with a large excess of the reporter and capture probe pairs, so that each targeted transcript finds its corresponding probe pair. After hybridization, excess unbound probes are washed away and the complexes, comprising target mRNA bound to specific reporter—capture probe pairs, are isolated. The biotin label at the 3' end of the capture probes is used to attach the complexes to streptavidin-coated slides. An electric field is applied to orient and extend the tripartite complexes on the surface of the slide to facilitate imaging and detection of the color-coded molecules. A microscope objective and a CCD camera are used to image the immobilized complexes. The number of molecules for a particular mRNA species is counted by decoding the unique pattern of the fluorescent colors encoded in each reporter probe. The protocol is performed from start to finish on the nCounter System, which is designed to provide hybridization, post-hybridization processing, and digital data acquisition capabilities in one simple workflow. The integrated system is composed of two instruments: the fully automated nCounter Prep Station for post-hybridization processing and the Digital Analyzer for imaging, data collection, and data processing.

---

information from cell surface receptors to effector molecules to regulate the response and functions of cells, tissues and organisms. Aberrant signaling due to misregulation of protein phosphorylation and dephosphorylation cascades is associated with many disease states, including most types of cancer.

A multiplex approach using phospho-specific antibodies and intracellular phospho-specific flow cytometry [254,255] to monitor changes in the level of phosphorylation of multiple key protein nodes in primary leukemic cells has been shown to have great potential in understanding how signaling through a network is co-opted in cancer cells to produce the aberrant phenotypes [256]. Phospho-specific antibodies to Stat1, Stat3, Stat5, Stat6, p38, and Erk1/2 were used to profile primary cells from patients with acute myeloid leukemia at basal state and following cytokine stimulation. Phosphorylation profiles of the six signaling proteins in acute myeloid leukemia cells were compared to those in normal blood cells to distinguish the leukemic signal transduction network from the healthy network. Using the same methodology, it is also possible to measure the effects of perturbations (genetic or small molecule) on these signaling events in cancer cells compared to normal cells with the aim of identifying potential drug targets.

Furthermore, one could determine the effect of such perturbations in either attenuating or enhancing the response of the cancer cells to other environmental cues.

Multicolor flow cytometry [257] has been used to measure 11 phosphoproteins and phospholipids simultaneously in response to stimulatory or inhibitory perturbations (small molecule inhibitors of key signaling components) to determine the effects of each condition on the cellular signaling networks in naive $CD4^+$ T primary cells. Bayesian network analysis was applied in order to infer causal connections between components of the network. Key to the success of this application was the use of the phosphorylation signatures of the 11 phosphoproteins and phospholipids in response to stimuli and perturbations.

Both studies discussed above use phospho-specific antibodies to key signaling or response proteins [256,257]. Such studies are limited by the availability of specific antibodies that recognize phosphorylated residues on proteins of interest. Proteome-scale MS-based studies provide one of the most comprehensive analyses of phosphorylation and do not depend on antibodies [258—260]. MS-based technologies provide highly quantitative and direct measurements that can detect the activities of many phosphorylation pathways simultaneously. The KAYAK (kinase activity assay for kinome profiling) method [261] can be used for the multiplexed measurement of phosphorylation events on 90 different peptides directly from cell lysates (see Box 5.4 for details on the KAYAK technology). Phosphorylated peptides are enriched using immobilized metal-ion affinity chromatography and analyzed by LC-MS techniques.

In addition to providing direct measurements on a proteome-wide scale, quantitative MS approaches facilitate the comparison of phosphoproteomes between wild-type cells and cells that have undergone manipulations of their signaling network components. In a recent study using mutant strains of *S. cerevisiae*, a label-free, quantitative phosphoproteomics approach was employed to determine the relationships between 97 kinases and 27 phosphatases and more than 1000 phosphoproteins [262]. Strikingly, inactivation of most (77%) kinases and phosphatases affects their immediate downstream targets as well as a large proportion of the overall signaling network. Owing to the inherent variation that exists between LC-MS experiments from run to run, label-free quantitation provides a relatively imprecise measurement of the differences in the phosphoproteome between wild-type and mutant yeast cells. Recent advances in techniques such as SILAC and chemical labeling, as well as improved sensitivity and dynamic range of peptide identification by current MS-based technologies, is enabling comprehensive and reproducible assessment of differences in phosphoproteomes [263]. Many groups have begun to apply this promising global approach to identify the effects of network perturbations on changes in the phosphoproteome [262,264—266]. iTRAQ labeling and phosphatase treatment was used to identify phosphorylation sites on the purified, auto-activated tyrosine kinase domain of fibroblast growth factor receptor 3 (FGFR3-KD) and to analyze complexes formed around the insulin receptor substrate homologue (chico) immunopurified from *Drosophila melanogaster* cells that were either stimulated with insulin or left untreated [267]. In two recent studies of the insulin signaling network in mammalian cells, Grb10 was identified as a mTORC1 substrate and was shown to be involved in feedback inhibition of the phosphatidylinositol 3-kinase (PI3K) and extracellular signal-regulated, mitogen-activated protein kinase (ERK-MAPK) pathways [268,269]. iTRAQ has also been used to compare the phosphoproteomes of cells treated with insulin to activate the pathway and cells that were pretreated with Torin 1 before insulin activation [268]. Torin 1 is a novel adenosine 5′-triphosphate (ATP)-competitive mTOR kinase domain inhibitor that blocks all known activities of both mTORC1 and mTORC2 complexes [270]. Yu and colleagues [269] used SILAC to quantify differences in the phosphoproteome of $TSC2^{-/-}$ MEFs in the presence and absence of rapamycin, as well as in the absence or presence of a drug

Box 5.4 KAYAK (Kinase Activity Assay for Kinome Profiling) Method for Multiplex Analysis of Kinase Activities

KAYAK is a multiplexed, MS-based kinase assay developed to measure the activity of multiple kinases from the same sample lysate [261,274]. A single MS run directly measures the phosphorylation of 90 synthetic peptides, thus providing a multiplexed assay to simultaneously monitor kinase activities from multiple signaling pathways. Key to the success of the KAYAK method is the design and synthesis of substrate peptides that can represent activity through the different core cellular signaling pathways. This set of substrate peptides also includes synthetic peptides containing phosphorylation sites with no associated kinase. Such peptides are identified from large-scale phosphoproteomics studies of cellular signaling networks and can be used to identify the responsible kinase via perturbation assays. The peptides are composed of 10—15 amino acid residues, with five residues upstream and four residues downstream of the phospho-acceptor site, and a C-terminal tripeptide of proline—phenylalanine—arginine to facilitate the incorporation and quantification of the stable isotope. The set of peptide substrates whose phosphorylation reflects activity through multiple signaling pathways is incubated together with a cell lysate to allow for phosphorylation by active kinases in the lysate. The in vitro kinase reactions are quenched, followed by the addition of stable isotope-labeled phosphopeptides of identical sequence (as internal standards), at a known concentration. Immobilized metal-ion affinity chromatography is used to enrich the phosphorylated substrate peptides and internal standard phosphopeptides, which are then analyzed by LC-MS techniques. The light (product) and heavy (internal standard) peptide pairs differ in mass by 6 daltons (Da) and although they co-elute, they can be quantified by the ratio of light-to-heavy areas under the curve from the raw spectra. Since the amount of each heavy phosphopeptide added is known, the ratio of the light to the heavy phosphopeptide provides the absolute amount of each product formed during the kinase reaction. The in vitro kinase reactions are carried out in a reaction volume of 50 μL and require only nanogram to microgram amounts of cell lysate. 5 μM of each substrate peptide is used in the reaction to reduce cross-phosphorylation of peptides by different kinases.

The KAYAK method was first applied to profile the activity of kinases in different cellular contexts, including mitogen-induced cell proliferation, inhibition of signaling pathways by known small molecule inhibitors, and a number of breast cancer cell lines [261]. This study also identified that a peptide derived from a PI3K regulatory subunit was a novel Src family kinase site in vivo.

(Ku-0063794, an ATP-competitive mTor inhibitor) leading to the identification of rapamycin-insensitive substrates of mTORC1 and mTORC2.

CONCLUDING REMARKS

Building on several decades of targeted classic genetics approaches, unbiased high-throughput technologies are beginning to generate a systems-level view of cellular signaling networks. In this chapter we have reviewed a number of experimental methods available to generate a comprehensive 'parts list' of cellular signaling networks. Further, we have described various approaches that can be used to construct network models based on the phenotypic signatures of each component. These techniques give us the unprecedented opportunity to evaluate globally and systematically the contribution of all genes to a specific biological process. However, the implementation of these methods is technically challenging and in some cases they are best used in combination, as integration of data sets increases the quality of the networks. Although the false positive and false negative rates for networks generated from high-throughput methods are currently relatively high, new experimental techniques and new methods for integrating multiple interacting data types will allow these networks to become powerful predictive tools.

A global view of cellular networks holds great promise in advancing our mechanistic understanding of how individual genetic alterations, as well as combinations of gene mutations, lead to a disease phenotype. For example, sequencing of cancer genomes [271] and genome-wide association studies [272] have identified hundreds of genetic aberrations that are linked to different cancers and complex diseases such as diabetes, obesity, hypertension and Crohn's disease. Comprehensive structure/function analysis of networks should help to understand the biological functions of many of the affected genes. Importantly, network analyses will facilitate the selection of protein targets for therapeutic intervention based on the underlying mechanisms of action. Furthermore, network maps will shed light on how certain drug–target interactions may lead to toxic effects. Such a mechanistic understanding is critical to the development of effective and safe treatments. Eventually, generation of comprehensive dynamic models of protein networks in response to signals over time will allow scientists to quantitatively predict the outcome of various perturbations.

REFERENCES

[1] Pawson T. Protein modules and signaling networks. Nature 1995;373:573–80.

[2] Noselli S, Perrimon N. Signal transduction. Are there close encounters between signaling pathways? Science 2000;290:68–9.

[3] Hou SX, Zheng Z, Chen X, Perrimon N. The Jak/STAT pathway in model organisms: emerging roles in cell movement. Dev Cell 2002;3:765–78.

[4] Thomas JH. Thinking about genetic redundancy. Trends Genet 1993;9:395–9.

[5] Perrimon N, Pitsouli C, Shilo B-Z. Signaling mechanisms controlling cell fate and patterning. Cold Spring Harb Perspect Biol 2012;4(8):pii: a005975.

[6] Noselli S, Agnes F. Roles of the JNK signaling pathway in Drosophila morphogenesis. Curr Opin Genet Dev 1999;9:466–72.

[7] Lum L, Yao S, Mozer B, Rovescalli A, Von Kessler D, Nirenberg M, et al. Identification of Hedgehog pathway components by RNAi in Drosophila cultured cells. Science 2003;299:2039–45.

[8] Varelas X, Miller BW, Sopko R, Song S, Gregorieff A, Fellouse FA, et al. The Hippo pathway regulates Wnt/beta-catenin signaling. Dev Cell 2010;18:579–91.

[9] Xia L, Jia S, Huang S, Wang H, Zhu Y, Mu Y, et al. The Fused/Smurf complex controls the fate of Drosophila germline stem cells by generating a gradient BMP response. Cell 2010;143:978–90.

[10] Nusslein-Volhard C, Wieschaus E. Mutations affecting segment number and polarity in Drosophila Nature 1980;287:795–801.

[11] Nusse R. Wnts and Hedgehogs: lipid-modified proteins and similarities in signaling mechanisms at the cell surface. Development 2003;130:5297–305.

[12] Friedman A, Perrimon N. Genetic screening for signal transduction in the era of network biology. Cell 2007;128:225–31.

[13] Barabasi AL, Oltvai ZN. Network biology: understanding the cell's functional organization. Nat Rev Genet 2004;5:101–13.

[14] Forster J, Famili I, Fu P, Palsson BO, Nielsen J. Genome-scale reconstruction of the Saccharomyces cerevisiae metabolic network. Genome Res 2003;13:244–53.

[15] Han JD, Bertin N, Hao T, Goldberg DS, Berriz GF, Zhang LV, et al. Evidence for dynamically organized modularity in the yeast protein–protein interaction network. Nature 2004;430:88–93.

[16] Deplancke B, Mukhopadhyay A, Ao W, Elewa AM, Grove CA, Martinez NJ, et al. A gene-centered C. elegans protein–DNA interaction network. Cell 2006;125:1193–205.

[17] Watts DJ, Strogatz SH. Collective dynamics of 'small-world' networks. Nature 1998;393:440–2.

[18] Barabasi AL, Albert R. Emergence of scaling in random networks. Science 1999;286:509–12.

[19] Hahn MW, Kern AD. Comparative genomics of centrality and essentiality in three eukaryotic protein-interaction networks. Mol Biol Evol 2005;22:803–6.

[20] Jeong H, Mason SP, Barabasi AL, Oltvai ZN. Lethality and centrality in protein networks. Nature 2001;411:41–2.

[21] Li S, Armstrong CM, Bertin N, Ge H, Milstein S, Boxem M, et al. A map of the interactome network of the metazoan C. elegans. Science 2004;303:540–3.

[22] Albert R, Jeong H, Barabasi AL. Error and attack tolerance of complex networks. Nature 2000;406:378–82.

[23] Toroczkai Z, Bassler KE. Network dynamics: jamming is limited in scale-free systems. Nature 2004;428:716.

[24] Aldana M, Cluzel P. A natural class of robust networks. Proc Natl Acad Sci U S A 2003;100:8710–4.

[25] Alon U. Network motifs: theory and experimental approaches. Nat Rev Genet 2007;8:450–61.

[26] Milo R, Shen-Orr S, Itzkovitz S, Kashtan N, Chklovskii D, Alon U. Network motifs: simple building blocks of complex networks. Science 2002;298:824–7.

[27] Davidson EH. Network design principles from the sea urchin embryo. Curr Opin Genet Dev 2009;19:535–40.

[28] Pomerening JR, Sontag ED, Ferrell Jr JE. Building a cell cycle oscillator: hysteresis and bistability in the activation of Cdc2. Nat Cell Biol 2003;5:346–51.

[29] Shen-Orr SS, Milo R, Mangan S, Alon U. Network motifs in the transcriptional regulation network of *Escherichia coli* Nat Genet 2002;31:64–8.

[30] Tyson JJ, Chen KC, Novak B. Sniffers, buzzers, toggles and blinkers: dynamics of regulatory and signaling pathways in the cell. Curr Opin Cell Biol 2003;15:221–31.

[31] Wang RS, Albert R. Elementary signaling modes predict the essentiality of signal transduction network components. BMC Syst Biol 2011;5:44.

[32] Yu H, Kim PM, Sprecher E, Trifonov V, Gerstein M. The importance of bottlenecks in protein networks: correlation with gene essentiality and expression dynamics. PLoS Comput Biol 2007;3:e59.

[33] Freeman LC. A set of measures of centrality based on betweenness. Sociometry 1977;40:35–41.

[34] Girvan M, Newman ME. Community structure in social and biological networks. Proc Natl Acad Sci U S A 2002;99:7821–6.

[35] Joy MP, Brock A, Ingber DE, Huang S. High-betweenness proteins in the yeast protein interaction network. J Biomed Biotechnol 2005;2005:96–103.

[36] Adams MD, Celniker SE, Holt RA, Evans CA, Gocayne JD, Amanatides PG, et al. The genome sequence of *Drosophila melanogaster* Science 2000;287:2185–95.

[37] Lander ES, Linton LM, Birren B, Nusbaum C, Zody MC, Baldwin J, et al. Initial sequencing and analysis of the human genome. Nature 2001;409:860–921.

[38] Venter JC, Adams MD, Myers EW, Li PW, Mural RJ, Sutton GG, et al. The sequence of the human genome. Science 2001;291:1304–51.

[39] Sharp PA. RNAi and double-strand RNA. Genes Dev 1999;13:139–41.

[40] Newmark PA. Opening a new can of worms: a large-scale RNAi screen in planarians. Dev Cell 2005;8:623–4.

[41] Newmark PA, Reddien PW, Cebria F, Sanchez Alvarado A. Ingestion of bacterially expressed double-stranded RNA inhibits gene expression in planarians. Proc Natl Acad Sci U S A 2003;100(Suppl. 1):11861–5.

[42] Brodersen P, Voinnet O. The diversity of RNA silencing pathways in plants. Trends Genet 2006;22:268–80.

[43] Boutros M, Ahringer J. The art and design of genetic screens: RNA interference. Nat Rev Genet 2008;9:554–66.

[44] Echeverri CJ, Perrimon N. High-throughput RNAi screening in cultured cells: a user's guide. Nat Rev Genet 2006;7:373–84.

[45] Martin SE, Caplen NJ. Applications of RNA interference in mammalian systems. Annu Rev Genomics Hum Genet 2007;8:81–108.

[46] Moffat J, Sabatini DM. Building mammalian signaling pathways with RNAi screens. Nat Rev Mol Cell Biol 2006;7:177–87.

[47] Mohr S, Bakal C, Perrimon N. Genomic screening with RNAi: results and challenges. Annu Rev Biochem 2010;79:37–64.

[48] Perrimon N, Mathey-Prevot B. Applications of high-throughput RNA interference screens to problems in cell and developmental biology. Genetics 2007;175:7–16.

[49] Root DE, Hacohen N, Hahn WC, Lander ES, Sabatini DM. Genome-scale loss-of-function screening with a lentiviral RNAi library. Nat Methods 2006;3:715–9.

[50] Fire A, Xu S, Montgomery MK, Kostas SA, Driver SE, Mello CC. Potent and specific genetic interference by double-stranded RNA in *Caenorhabditis elegans*. Nature 1998;391:806–11.

[51] Denli AM, Hannon GJ. RNAi: an ever-growing puzzle. Trends Biochem Sci 2003;28:196–201.

[52] Hannon GJ. RNA interference. Nature 2002;418:244–51.

[53] Paddison PJ, Hannon GJ. RNA interference: the new somatic cell genetics? Cancer Cell 2002;2:17–23.

[54] Perrimon N, Ni JQ, Perkins L. In vivo RNAi: today and tomorrow. Cold Spring Harb Perspect Biol 2010;2:a003640.

[55] Clemens JC, Worby CA, Simonson-Leff N, Muda M, Maehama T, Hemmings BA, et al. Use of double-stranded RNA interference in *Drosophila* cell lines to dissect signal transduction pathways. Proc Natl Acad Sci U S A 2000;97:6499–503.

[56] Hammond SM, Boettcher S, Caudy AA, Kobayashi R, Hannon GJ. Argonaute2, a link between genetic and biochemical analyses of RNAi. Science 2001;293:1146–50.

[57] Bernstein E, Caudy AA, Hammond SM, Hannon GJ. Role for a bidentate ribonuclease in the initiation step of RNA interference. Nature 2001;409:363–6.

[58] Bernstein E, Kim SY, Carmell MA, Murchison EP, Alcorn H, Li MZ, et al. Dicer is essential for mouse development. Nat Genet 2003;35:215–7.

[59] Carmell MA, Hannon GJ. RNase III enzymes and the initiation of gene silencing. Nat Struct Mol Biol 2004;11:214–8.

[60] Ketting RF, Fischer SE, Bernstein E, Sijen T, Hannon GJ, Plasterk RH. Dicer functions in RNA interference and in synthesis of small RNA involved in developmental timing in *C. elegans*. Genes Dev 2001;15:2654–9.

[61] Nowotny M, Yang W. Structural and functional modules in RNA interference. Curr Opin Struct Biol 2009;19:286–93.

[62] Rao DD, Vorhies JS, Senzer N, Nemunaitis J. siRNA vs. shRNA: similarities and differences. Adv Drug Deliv Rev 2009;61:746–59.

[63] Tilesi F, Fradiani P, Socci V, Willems D, Ascenzioni F. Design and validation of siRNAs and shRNAs. Curr Opin Mol Ther 2009;11:156–64.

[64] Iorns E, Lord CJ, Turner N, Ashworth A. Utilizing RNA interference to enhance cancer drug discovery. Nat Rev Drug Discov 2007;6:556–68.

[65] Lord CJ, Martin SA, Ashworth A. RNA interference screening demystified. J Clin Pathol 2009;62:195–200.

[66] Baeg GH, Zhou R, Perrimon N. Genome-wide RNAi analysis of JAK/STAT signaling components in *Drosophila*. Genes Dev 2005;19:1861–70.

[67] Bard F, Casano L, Mallabiabarrena A, Wallace E, Saito K, Kitayama H, et al. Functional genomics reveals genes involved in protein secretion and Golgi organization. Nature 2006;439:604–7.

[68] Boutros M, Kiger AA, Armknecht S, Kerr K, Hild M, Koch B, et al. Genome-wide RNAi analysis of growth and viability in *Drosophila* cells. Science 2004;303:832–5.

[69] DasGupta R, Kaykas A, Moon RT, Perrimon N. Functional genomic analysis of the Wnt-wingless signaling pathway. Science 2005;308:826–33.

[70] Nybakken K, Vokes SA, Lin TY, McMahon AP, Perrimon N. A genome-wide RNA interference screen in *Drosophila melanogaster* cells for new components of the Hh signaling pathway. Nat Genet 2005;37:1323–32.

[71] Li S, Wang L, Berman MA, Zhang Y, Dorf ME. RNAi screen in mouse astrocytes identifies phosphatases that regulate NF-kappaB signaling. Mol Cell 2006;24:497–509.

[72] Lu J, Ruhf ML, Perrimon N, Leder P. A genome-wide RNA interference screen identifies putative chromatin regulators essential for E2F repression. Proc Natl Acad Sci U S A 2007;104:9381–6.

[73] Mattila J, Kallijarvi J, Puig O. RNAi screening for kinases and phosphatases identifies FoxO regulators. Proc Natl Acad Sci U S A 2008;105:14873–8.

[74] Kategaya LS, Changkakoty B, Biechele T, Conrad WH, Kaykas A, Dasgupta R, et al. Bili inhibits Wnt/beta-catenin signaling by regulating the recruitment of axin to LRP6. PLoS One 2009;4:e6129.

[75] Ma Y, Creanga A, Lum L, Beachy PA. Prevalence of off-target effects in *Drosophila* RNA interference screens. Nature 2006;443:359–63.

[76] Tang W, Dodge M, Gundapaneni D, Michnoff C, Roth M, Lum L. A genome-wide RNAi screen for Wnt/beta-catenin pathway components identifies unexpected roles for TCF transcription factors in cancer. Proc Natl Acad Sci U S A 2008;105:9697–702.

[77] Muller P, Kuttenkeuler D, Gesellchen V, Zeidler MP, Boutros M. Identification of JAK/STAT signaling components by genome-wide RNA interference. Nature 2005;436:871–5.

[78] Ding L, Paszkowski-Rogacz M, Nitzsche A, Slabicki MM, Heninger AK, de Vries I, et al. A genome-scale RNAi screen for Oct4 modulators defines a role of the Paf1 complex for embryonic stem cell identity. Cell Stem Cell 2009;4:403–15.

[79] Hu G, Kim J, Xu Q, Leng Y, Orkin SH, Elledge SJ. A genome-wide RNAi screen identifies a new transcriptional module required for self-renewal. Genes Dev 2009;23:837–48.

[80] DasGupta R, Nybakken K, Booker M, Mathey-Prevot B, Gonsalves F, Changkakoty B, et al. A case study of the reproducibility of transcriptional reporter cell-based RNAi screens in *Drosophila*. Genome Biol 2007;8:R203.

[81] Dietzl G, Chen D, Schnorrer F, Su KC, Barinova Y, Fellner M, et al. A genome-wide transgenic RNAi library for conditional gene inactivation in *Drosophila*. Nature 2007;448:151–6.

[82] Ni JQ, Liu LP, Binari R, Hardy R, Shim HS, Cavallaro A, et al. A *Drosophila* resource of transgenic RNAi lines for neurogenetics. Genetics 2009;182:1089–100.

[83] Mummery-Widmer JL, Yamazaki M, Stoeger T, Novatchkova M, Bhalerao S, Chen D, et al. Genome-wide analysis of Notch signaling in *Drosophila* by transgenic RNAi. Nature 2009;458:987–92.

[84] Friedman AA, Tucker G, Singh R, Yan D, Vinayagam A, Hu Y, et al. Proteomic and functional genomic landscape of receptor tyrosine kinase and Ras to extracellular signal–regulated kinase signaling. Science Signaling 2011;4(196):rs10.

[85] Friedman A, Perrimon N. A functional RNAi screen for regulators of receptor tyrosine kinase and ERK signaling. Nature 2006;444:230–4.

[86] Kockel L, Kerr KS, Melnick M, Bruckner K, Hebrok M, Perrimon N. Dynamic switch of negative feedback regulation in *Drosophila* Akt-TOR signaling. PLoS Genet 2010;6:e1000990.

[87] Ziauddin J, Sabatini DM. Microarrays of cells expressing defined cDNAs. Nature 2001;411:107–10.

[88] Mishina YM, Wilson CJ, Bruett L, Smith JJ, Stoop-Myer C, Jong S, et al. Multiplex GPCR assay in reverse transfection cell microarrays. J Biomol Screen 2004;9:196–207.

[89] Lindquist RA, Ottina KA, Wheeler DB, Hsu PP, Thoreen CC, Guertin DA, et al. Genome-scale RNAi on living-cell microarrays identifies novel regulators of *Drosophila melanogaster* TORC1-S6K pathway signaling. Genome Res 2011;21:433–46.

[90] Carpenter AE, Jones TR, Lamprecht MR, Clarke C, Kang IH, Friman O, et al. CellProfiler: image analysis software for identifying and quantifying cell phenotypes. Genome Biol 2006;7:R100.

[91] Eggert US, Kiger AA, Richter C, Perlman ZE, Perrimon N, Mitchison TJ, et al. Parallel chemical genetic and genome-wide RNAi screens identify cytokinesis inhibitors and targets. PLoS Biol 2004;2:e379.

[92] Loo LH, Wu LF, Altschuler SJ. Image-based multivariate profiling of drug responses from single cells. Nat Methods 2007;4:445–53.

[93] Paran Y, Ilan M, Kashman Y, Goldstein S, Liron Y, Geiger B, et al. High-throughput screening of cellular features using high-resolution light-microscopy; application for profiling drug effects on cell adhesion. J Struct Biol 2007;158:233–43.

[94] Perlman ZE, Slack MD, Feng Y, Mitchison TJ, Wu LF, Altschuler SJ. Multidimensional drug profiling by automated microscopy. Science 2004;306:1194–8.

[95] Rickardson L, Wickstrom M, Larsson R, Lovborg H. Image-based screening for the identification of novel proteasome inhibitors. J Biomol Screen 2007;12:203–10.

[96] Bakal C, Aach J, Church G, Perrimon N. Quantitative morphological signatures define local signaling networks regulating cell morphology. Science 2007;316:1753–6.

[97] D'Ambrosio MV, Vale RD. A whole genome RNAi screen of *Drosophila* S2 cell spreading performed using automated computational image analysis. J Cell Biol 2010;191:471–8.

[98] Kiger AA, Baum B, Jones S, Jones MR, Coulson A, Echeverri C, et al. A functional genomic analysis of cell morphology using RNA interference. J Biol 2003;2:27.

[99] Moffat J, Grueneberg DA, Yang X, Kim SY, Kloepfer AM, Hinkle G, et al. A lentiviral RNAi library for human and mouse genes applied to an arrayed viral high-content screen. Cell 2006;124:1283–98.

[100] Goshima G, Wollman R, Goodwin SS, Zhang N, Scholey JM, Vale RD, et al. Genes required for mitotic spindle assembly in *Drosophila* S2 cells. Science 2007;316:417–21.

[101] Mukherji M, Bell R, Supekova L, Wang Y, Orth AP, Batalov S, et al. Genome-wide functional analysis of human cell-cycle regulators. Proc Natl Acad Sci U S A 2006;103:14819–24.

[102] Collinet C, Stoter M, Bradshaw CR, Samusik N, Rink JC, Kenski D, et al. Systems survey of endocytosis by multiparametric image analysis. Nature 2010;464:243–9.

[103] Pelkmans L, Fava E, Grabner H, Hannus M, Habermann B, Krausz E, et al. Genome-wide analysis of human kinases in clathrin- and caveolae/raft-mediated endocytosis. Nature 2005;436:78–86.

[104] Agaisse H, Burrack LS, Philips JA, Rubin EJ, Perrimon N, Higgins DE. Genome-wide RNAi screen for host factors required for intracellular bacterial infection. Science 2005;309:1248−51.

[105] Cherry S, Doukas T, Armknecht S, Whelan S, Wang H, Sarnow P, et al. Genome-wide RNAi screen reveals a specific sensitivity of IRES-containing RNA viruses to host translation inhibition. Genes Dev 2005;19:445−52.

[106] Cherry S, Kunte A, Wang H, Coyne C, Rawson RB, Perrimon N. COPI activity coupled with fatty acid biosynthesis is required for viral replication. PLoS Pathog 2006;2:e102.

[107] Philips JA, Rubin EJ, Perrimon N. *Drosophila* RNAi screen reveals CD36 family member required for mycobacterial infection. Science 2005;309:1251−3.

[108] Bai J, Binari R, Ni JQ, Vijayakanthan M, Li HS, Perrimon N. RNA interference screening in *Drosophila* primary cells for genes involved in muscle assembly and maintenance. Development 2008;135:1439−49.

[109] Sepp KJ, Hong P, Lizarraga SB, Liu JS, Mejia LA, Walsh CA, et al. Identification of neural outgrowth genes using genome-wide RNAi. PLoS Genet 2008;4:e1000111.

[110] Vizeacoumar FJ, Chong Y, Boone C, Andrews BJ. A picture is worth a thousand words: genomics to phenomics in the yeast *Saccharomyces cerevisiae*. FEBS Lett 2009;583:1656−61.

[111] Walter T, Shattuck DW, Baldock R, Bastin ME, Carpenter AE, Duce S, et al. Visualization of image data from cells to organisms. Nat Methods 2010;7:S26−41.

[112] Bjorklund M, Taipale M, Varjosalo M, Saharinen J, Lahdenpera J, Taipale J. Identification of pathways regulating cell size and cell-cycle progression by RNAi. Nature 2006;439:1009−13.

[113] Ramet M, Manfruelli P, Pearson A, Mathey-Prevot B, Ezekowitz RA. Functional genomic analysis of phagocytosis and identification of a *Drosophila* receptor for *E. coli*. Nature 2002;416:644−8.

[114] Ulvila J, Parikka M, Kleino A, Sormunen R, Ezekowitz RA, Kocks C, et al. Double-stranded RNA is internalized by scavenger receptor-mediated endocytosis in *Drosophila* S2 cells. J Biol Chem 2006;281:14370−5.

[115] Kondo S, Perrimon N. A genome-wide RNAi screen identifies core components of the G-M DNA damage checkpoint. Sci Signal 2011. 4, rs1.

[116] Ravi D, Wiles AM, Bhavani S, Ruan J, Leder P, Bishop AJ. A network of conserved damage survival pathways revealed by a genomic RNAi screen. PLoS Genet 2009;5:e1000527.

[117] Wagner EJ, Burch BD, Godfrey AC, Salzler HR, Duronio RJ, Marzluff WF. A genome-wide RNA interference screen reveals that variant histones are necessary for replication-dependent histone pre-mRNA processing. Mol Cell 2007;28:692−9.

[118] Zhou R, Hotta I, Denli AM, Hong P, Perrimon N, Hannon GJ. Comparative analysis of argonaute-dependent small RNA pathways in *Drosophila*. Mol Cell 2008;32:592−9.

[119] Feske S, Gwack Y, Prakriya M, Srikanth S, Puppel SH, Tanasa B, et al. A mutation in Orai1 causes immune deficiency by abrogating CRAC channel function. Nature 2006;441:179−85.

[120] Vig M, Peinelt C, Beck A, Koomoa DL, Rabah D, Koblan-Huberson M, et al. CRACM1 is a plasma membrane protein essential for store-operated Ca^{2+} entry. Science 2006;312:1220−3.

[121] Zhang SL, Yeromin AV, Zhang XH, Yu Y, Safrina O, Penna A, et al. Genome-wide RNAi screen of Ca(2+) influx identifies genes that regulate Ca(2+) release-activated Ca(2+) channel activity. Proc Natl Acad Sci U S A 2006;103:9357−62.

[122] Zhang S, Binari R, Zhou R, Perrimon N. A genomewide RNA interference screen for modifiers of aggregates formation by mutant Huntingtin in *Drosophila*. Genetics 2010;184:1165−79.

[123] Gandre-Babbe S, van der Bliek AM. The novel tail-anchored membrane protein Mff controls mitochondrial and peroxisomal fission in mammalian cells. Mol Biol Cell 2008;19:2402−12.

[124] Sathyanarayanan S, Zheng X, Kumar S, Chen CH, Chen D, Hay B, et al. Identification of novel genes involved in light-dependent CRY degradation through a genome-wide RNAi screen. Genes Dev 2008;22:1522−33.

[125] Dekanty A, Romero NM, Bertolin AP, Thomas MG, Leishman CC, Perez-Perri JI, et al. Drosophila genome-wide RNAi screen identifies multiple regulators of HIF-dependent transcription in hypoxia. PLoS Genet 2010;6:e1000994.

[126] Kocks C, Cho JH, Nehme N, Ulvila J, Pearson AM, Meister M, et al. Eater, a transmembrane protein mediating phagocytosis of bacterial pathogens in *Drosophila*. Cell 2005;123:335−46.

[127] Foley E, O'Farrell PH. Functional dissection of an innate immune response by a genome-wide RNAi screen. PLoS Biol 2004;2:E203.

[128] Gesellchen V, Kuttenkeuler D, Steckel M, Pelte N, Boutros M. An RNA interference screen identifies Inhibitor of Apoptosis Protein 2 as a regulator of innate immune signaling in *Drosophila*. EMBO Rep 2005;6:979−84.

[129] Kleino A, Valanne S, Ulvila J, Kallio J, Myllymaki H, Enwald H, et al. Inhibitor of apoptosis 2 and TAK1-binding protein are components of the *Drosophila* Imd pathway. EMBO J 2005;24:3423−34.

[130] Bartscherer K, Pelte N, Ingelfinger D, Boutros M. Secretion of Wnt ligands requires Evi, a conserved transmembrane protein. Cell 2006;125:523−33.

[131] Gwack Y, Sharma S, Nardone J, Tanasa B, Iuga A, Srikanth S, et al. A genome-wide *Drosophila* RNAi screen identifies DYRK-family kinases as regulators of NFAT. Nature 2006;441:646−50.

[132] Dasgupta R, Perrimon N. Using RNAi to catch *Drosophila* genes in a web of interactions: insights into cancer research. Oncogene 2004;23:8359−65.

[133] Gondi CS, Rao JS. Concepts in in vivo siRNA delivery for cancer therapy. J Cell Physiol 2009;220:285−91.

[134] Gondi CS, Rao JS. Therapeutic potential of siRNA-mediated targeting of urokinase plasminogen activator, its receptor, and matrix metalloproteinases. Methods Mol Biol 2009;487:267−81.

[135] Kim SY, Hahn WC. Cancer genomics: integrating form and function. Carcinogenesis 2007;28:1387−92.

[136] Wolters NM, MacKeigan JP. From sequence to function: using RNAi to elucidate mechanisms of human disease. Cell Death Differ 2008;15:809−19.

[137] Birmingham A, Anderson EM, Reynolds A, Ilsley-Tyree D, Leake D, Fedorov Y, et al. 3′ UTR seed matches, but not overall identity, are associated with RNAi off-targets. Nat Methods 2006;3:199−204.

[138] Jackson AL, Linsley PS. Recognizing and avoiding siRNA off-target effects for target identification and therapeutic application. Nat Rev Drug Discov 2010;9:57−67.

[139] Sigoillot FD, King RW. Vigilance and validation: Keys to success in RNAi screening. ACS Chem Biol 2011;6:47−60.

[140] Kulkarni MM, Booker M, Silver SJ, Friedman A, Hong P, Perrimon N, et al. Evidence of off-target effects associated with long dsRNAs in *Drosophila melanogaster* cell-based assays. Nat Methods 2006;3:833–8.

[141] Bartel DP. MicroRNAs: genomics, biogenesis, mechanism, and function. Cell 2004;116:281–97.

[142] Bartel DP. MicroRNAs: target recognition and regulatory functions. Cell 2009;136:215–33.

[143] Brennecke J, Stark A, Russell RB, Cohen SM. Principles of microRNA-target recognition. PLoS Biol 2005;3:e85.

[144] Echeverri CJ, Beachy PA, Baum B, Boutros M, Buchholz F, Chanda SK, et al. Minimizing the risk of reporting false positives in large-scale RNAi screens. Nat Methods 2006;3:777–9.

[145] Cherbas L, Willingham A, Zhang D, Yang L, Zou Y, Eads BD, et al. The transcriptional diversity of 25 *Drosophila* cell lines. Genome Res 2011;21:301–14.

[146] Boone C, Bussey H, Andrews BJ. Exploring genetic interactions and networks with yeast. Nat Rev Genet 2007;8:437–49.

[147] Pan X, Yuan DS, Xiang D, Wang X, Sookhai-Mahadeo S, Bader JS, et al. A robust toolkit for functional profiling of the yeast genome. Mol Cell 2004;16:487–96.

[148] Tong AH, Evangelista M, Parsons AB, Xu H, Bader GD, Page N, et al. Systematic genetic analysis with ordered arrays of yeast deletion mutants. Science 2001;294:2364–8.

[149] Tong AH, Lesage G, Bader GD, Ding H, Xu H, Xin X, et al. Global mapping of the yeast genetic interaction network. Science 2004;303:808–13.

[150] Bakal C, Linding R, Llense F, Heffern E, Martin-Blanco E, Pawson T, et al. Phosphorylation networks regulating JNK activity in diverse genetic backgrounds. Science 2008;322:453–6.

[151] Guertin DA, Guntur KV, Bell GW, Thoreen CC, Sabatini DM. Functional genomics identifies TOR-regulated genes that control growth and division. Curr Biol 2006;16:958–70.

[152] Lehner B, Tischler J, Fraser AG. RNAi screens in *Caenorhabditis elegans* in a 96-well liquid format and their application to the systematic identification of genetic interactions. Nat Protoc 2006;1:1617–20.

[153] Wheeler DB, Bailey SN, Guertin DA, Carpenter AE, Higgins CO, Sabatini DM. RNAi living-cell microarrays for loss-of-function screens in *Drosophila melanogaster* cells. Nat Methods 2004;1:127–32.

[154] Wheeler DB, Carpenter AE, Sabatini DM. Cell microarrays and RNA interference chip away at gene function. Nat Genet 2005;37(Suppl):S25–30.

[155] Schuldiner M, Collins SR, Thompson NJ, Denic V, Bhamidipati A, Punna T, et al. Exploration of the function and organization of the yeast early secretory pathway through an epistatic miniarray profile. Cell 2005;123:507–19.

[156] Schuldiner M, Collins SR, Weissman JS, Krogan NJ. Quantitative genetic analysis in *Saccharomyces cerevisiae* using epistatic miniarray profiles (E-MAPs) and its application to chromatin functions. Methods 2006;40:344–52.

[157] Roguev A, Bandyopadhyay S, Zofall M, Zhang K, Fischer T, Collins SR, et al. Conservation and rewiring of functional modules revealed by an epistasis map in fission yeast. Science 2008;322:405–10.

[158] Horn T, Sandmann T, Fischer B, Axelsson E, Huber W, Boutros M. Mapping of signaling networks through synthetic genetic interaction analysis by RNAi. Nat Methods 2011;8:341–6.

[159] Chen HW, Marinissen MJ, Oh SW, Chen X, Melnick M, Perrimon N, et al. CKA, a novel multidomain protein, regulates the JUN N-terminal kinase signal transduction pathway in *Drosophila*. Mol Cell Biol 2002;22:1792–803.

[160] Dreze M, Monachello D, Lurin C, Cusick ME, Hill DE, Vidal M, et al. High-quality binary interactome mapping. Methods Enzymol 2010;470:281–315.

[161] Fields S, Song O. A novel genetic system to detect protein–protein interactions. Nature 1989;340:245–6.

[162] Charbonnier S, Gallego O, Gavin AC. The social network of a cell: recent advances in interactome mapping. Biotechnol Annu Rev 2008;14:1–28.

[163] Gingras AC, Gstaiger M, Raught B, Aebersold R. Analysis of protein complexes using mass spectrometry. Nat Rev Mol Cell Biol 2007;8:645–54.

[164] Kocher T, Superti-Furga G. Mass spectrometry-based functional proteomics: from molecular machines to protein networks. Nat Methods 2007;4:807–15.

[165] Rigaut G, Shevchenko A, Rutz B, Wilm M, Mann M, Seraphin B. A generic protein purification method for protein complex characterization and proteome exploration. Nat Biotechnol 1999;17:1030–2.

[166] Rain JC, Selig L, De Reuse H, Battaglia V, Reverdy C, Simon S, et al. The protein–protein interaction map of *Helicobacter pylori*. Nature 2001;409:211–5.

[167] Ito T, Chiba T, Ozawa R, Yoshida M, Hattori M, Sakaki Y. A comprehensive two-hybrid analysis to explore the yeast protein interactome. Proc Natl Acad Sci U S A 2001;98:4569–74.

[168] Ito T, Tashiro K, Muta S, Ozawa R, Chiba T, Nishizawa M, et al. Toward a protein–protein interaction map of the budding yeast: A comprehensive system to examine two-hybrid interactions in all possible combinations between the yeast proteins. Proc Natl Acad Sci U S A 2000;97:1143–7.

[169] Uetz P, Giot L, Cagney G, Mansfield TA, Judson RS, Knight JR, et al. A comprehensive analysis of protein–protein interactions in *Saccharomyces cerevisiae*. Nature 2000;403:623–7.

[170] Reboul J, Vaglio P, Rual JF, Lamesch P, Martinez M, Armstrong CM, et al. *C. elegans* ORFeome version 1.1: experimental verification of the genome annotation and resource for proteome-scale protein expression. Nat Genet 2003;34:35–41.

[171] Walhout AJ, Sordella R, Lu X, Hartley JL, Temple GF, Brasch MA, et al. Protein interaction mapping in *C. elegans* using proteins involved in vulval development. Science 2000;287:116–22.

[172] Formstecher E, Aresta S, Collura V, Hamburger A, Meil A, Trehin A, et al. Protein interaction mapping: a *Drosophila* case study. Genome Res 2005;15:376–84.

[173] Giot L, Bader JS, Brouwer C, Chaudhuri A, Kuang B, Li Y, et al. A protein interaction map of *Drosophila melanogaster*. Science 2003;302:1727–36.

[174] Stanyon CA, Liu G, Mangiola BA, Patel N, Giot L, Kuang B, et al. A *Drosophila* protein-interaction map centered on cell-cycle regulators. Genome Biol 2004;5:R96.

[175] Colland F, Jacq X, Trouplin V, Mougin C, Groizeleau C, Hamburger A, et al. Functional proteomics mapping of a human signaling pathway. Genome Res 2004;14:1324–32.

[176] Lim J, Hao T, Shaw C, Patel AJ, Szabo G, Rual JF, et al. A protein–protein interaction network for human inherited ataxias and disorders of Purkinje cell degeneration. Cell 2006;125: 801–14.

[177] Rual JF, Venkatesan K, Hao T, Hirozane-Kishikawa T, Dricot A, Li N, et al. Towards a proteome-scale map of the human protein-protein interaction network. Nature 2005;437:1173–8.

[178] Stelzl U, Worm U, Lalowski M, Haenig C, Brembeck FH, Goehler H, et al. A human protein–protein interaction network: a resource for annotating the proteome. Cell 2005;122:957–68.

[179] Venkatesan K, Rual JF, Vazquez A, Stelzl U, Lemmens I, Hirozane-Kishikawa T, et al. An empirical framework for binary interactome mapping. Nat Methods 2009;6:83–90.

[180] von Mering C, Krause R, Snel B, Cornell M, Oliver SG, Fields S, et al. Comparative assessment of large-scale data sets of protein–protein interactions. Nature 2002;417:399–403.

[181] Breitkreutz A, Choi H, Sharom JR, Boucher L, Neduva V, Larsen B, et al. A global protein kinase and phosphatase interaction network in yeast. Science 2010;328:1043–6.

[182] Gavin AC, Aloy P, Grandi P, Krause R, Boesche M, Marzioch M, et al. Proteome survey reveals modularity of the yeast cell machinery. Nature 2006;440:631–6.

[183] Gavin AC, Bosche M, Krause R, Grandi P, Marzioch M, Bauer A, et al. Functional organization of the yeast proteome by systematic analysis of protein complexes. Nature 2002;415:141–7.

[184] Ho Y, Gruhler A, Heilbut A, Bader GD, Moore L, Adams SL, et al. Systematic identification of protein complexes in *Saccharomyces cerevisiae* by mass spectrometry. Nature 2002;415:180–3.

[185] Krogan NJ, Cagney G, Yu H, Zhong G, Guo X, Ignatchenko A, et al. Global landscape of protein complexes in the yeast *Saccharomyces cerevisiae*. Nature 2006;440:637–43.

[186] Puig O, Caspary F, Rigaut G, Rutz B, Bouveret E, Bragado-Nilsson E, et al. The tandem affinity purification (TAP) method: a general procedure of protein complex purification. Methods 2001;24:218–29.

[187] Bouwmeester T, Bauch A, Ruffner H, Angrand PO, Bergamini G, Croughton K, et al. A physical and functional map of the human TNF-alpha/NF-kappa B signal transduction pathway. Nat Cell Biol 2004;6:97–105.

[188] Major MB, Roberts BS, Berndt JD, Marine S, Anastas J, Chung N, et al. New regulators of Wnt/beta-catenin signaling revealed by integrative molecular screening. Sci Signal 2008. 1, ra12.

[189] Behrends C, Sowa ME, Gygi SP, Harper JW. Network organization of the human autophagy system. Nature 2010;466:68–76.

[190] Sowa ME, Bennett EJ, Gygi SP, Harper JW. Defining the human deubiquitinating enzyme interaction landscape. Cell 2009;138: 389–403.

[191] Cloutier P, Al-Khoury R, Lavallee-Adam M, Faubert D, Jiang H, Poitras C, et al. High-resolution mapping of the protein interaction network for the human transcription machinery and affinity purification of RNA polymerase II-associated complexes. Methods 2009;48:381–6.

[192] Glatter T, Wepf A, Aebersold R, Gstaiger M. An integrated workflow for charting the human interaction proteome: insights into the PP2A system. Mol Syst Biol 2009;5:237.

[193] Calderwood MA, Venkatesan K, Xing L, Chase MR, Vazquez A, Holthaus AM, et al. Epstein-Barr virus and virus human protein interaction maps. Proc Natl Acad Sci U S A 2007;104:7606–11.

[194] de Chassey B, Navratil V, Tafforeau L, Hiet MS, Aublin-Gex A, Agaugue S, et al. Hepatitis C virus infection protein network. Mol Syst Biol 2008;4:230.

[195] Jager S, Gulbahce N, Cimermancic P, Kane J, He N, Chou S, et al. Purification and characterization of HIV-human protein complexes. Methods 2011;53:13–9.

[196] Shapira SD, Gat-Viks I, Shum BO, Dricot A, de Grace MM, Wu L, et al. A physical and regulatory map of host-influenza interactions reveals pathways in H1N1 infection. Cell. 2009;139:1255–67.

[197] Uetz P, Dong YA, Zeretzke C, Atzler C, Baiker A, Berger B, et al. Herpesviral protein networks and their interaction with the human proteome. Science 2006;311:239–42.

[198] Tewari M, Hu PJ, Ahn JS, Ayivi-Guedehoussou N, Vidalain PO, Li S, et al. Systematic interactome mapping and genetic perturbation analysis of a *C. elegans* TGF-beta signaling network. Mol Cell 2004;13:469–82.

[199] Ory S, Zhou M, Conrads TP, Veenstra TD, Morrison DK. Protein phosphatase 2A positively regulates Ras signaling by dephosphorylating KSR1 and Raf-1 on critical 14–3–3 binding sites. Curr Biol 2003;13:1356–64.

[200] Ritt DA, Zhou M, Conrads TP, Veenstra TD, Copeland TD, Morrison DK. CK2 Is a component of the KSR1 scaffold complex that contributes to Raf kinase activation. Curr Biol 2007;17:179–84.

[201] Cuomo A, Bonaldi T. Systems biology 'on-the-fly': SILAC-based quantitative proteomics and RNAi approach in *Drosophila melanogaster*. Methods Mol Biol 2010;662:59–78.

[202] Selbach M, Mann M. Protein interaction screening by quantitative immunoprecipitation combined with knockdown (QUICK). Nat Methods 2006;3:981–3.

[203] Choudhary C, Mann M. Decoding signaling networks by mass spectrometry-based proteomics. Nat Rev Mol Cell Biol 2010;11:427–39.

[204] Ong SE, Mann M. Mass spectrometry-based proteomics turns quantitative. Nat Chem Biol 2005;1:252–62.

[205] Mann M. Functional and quantitative proteomics using SILAC. Nat Rev Mol Cell Biol 2006;7:952–8.

[206] Ong SE, Blagoev B, Kratchmarova I, Kristensen DB, Steen H, Pandey A, et al. Stable isotope labeling by amino acids in cell culture, SILAC, as a simple and accurate approach to expression proteomics. Mol Cell Proteomics 2002;1:376–86.

[207] Ross PL, Huang YN, Marchese JN, Williamson B, Parker K, Hattan S, et al. Multiplexed protein quantitation in *Saccharomyces cerevisiae* using amine-reactive isobaric tagging reagents. Mol Cell Proteomics 2004;3:1154–69.

[208] Thompson A, Schafer J, Kuhn K, Kienle S, Schwarz J, Schmidt G, et al. Tandem mass tags: a novel quantification strategy for comparative analysis of complex protein mixtures by MS/MS. Anal Chem 2003;75:1895–904.

[209] DeRisi J, Penland L, Brown PO, Bittner ML, Meltzer PS, Ray M, et al. Use of a cDNA microarray to analyse gene expression patterns in human cancer. Nat Genet 1996;14:457–60.

[210] Roberts CJ, Nelson B, Marton MJ, Stoughton R, Meyer MR, Bennett HA, et al. Signaling and circuitry of multiple MAPK pathways revealed by a matrix of global gene expression profiles. Science 2000;287:873–80.

[211] Spellman PT, Sherlock G, Zhang MQ, Iyer VR, Anders K, Eisen MB, et al. Comprehensive identification of cell cycle-regulated genes of

[211] ...the yeast *Saccharomyces cerevisiae* by microarray hybridization. Mol Biol Cell 1998;9:3273—97.

[212] Hsiao LL, Dangond F, Yoshida T, Hong R, Jensen RV, Misra J, et al. A compendium of gene expression in normal human tissues. Physiol Genomics 2001;7:97—104.

[213] Stuart JM, Segal E, Koller D, Kim SK. A gene-coexpression network for global discovery of conserved genetic modules. Science 2003;302:249—55.

[214] Stolc V, Gauhar Z, Mason C, Halasz G, van Batenburg MF, Rifkin SA, et al. A gene expression map for the euchromatic genome of *Drosophila melanogaster*. Science 2004;306:655—60.

[215] Amit I, Garber M, Chevrier N, Leite AP, Donner Y, Eisenhaure T, et al. Unbiased reconstruction of a mammalian transcriptional network mediating pathogen responses. Science 2009;326: 257—63.

[216] Ge H, Liu Z, Church GM, Vidal M. Correlation between transcriptome and interactome mapping data from *Saccharomyces cerevisiae*. Nat Genet 2001;29:482—6.

[217] Grigoriev A. A relationship between gene expression and protein interactions on the proteome scale: analysis of the bacteriophage T7 and the yeast Saccharomyces cerevisiae. Nucleic Acids Res 2001;29:3513—9.

[218] Jansen R, Greenbaum D, Gerstein M. Relating whole-genome expression data with protein—protein interactions. Genome Res 2002;12:37—46.

[219] Kemmeren P, van Berkum NL, Vilo J, Bijma T, Donders R, Brazma A, et al. Protein interaction verification and functional annotation by integrated analysis of genome-scale data. Mol Cell 2002;9:1133—43.

[220] Fambrough D, McClure K, Kazlauskas A, Lander ES. Diverse signaling pathways activated by growth factor receptors induce broadly overlapping, rather than independent, sets of genes. Cell 1999;97:727—41.

[221] Greenbaum D, Colangelo C, Williams K, Gerstein M. Comparing protein abundance and mRNA expression levels on a genomic scale. Genome Biol 2003;4:117.

[222] Nie L, Wu G, Culley DE, Scholten JC, Zhang W. Integrative analysis of transcriptomic and proteomic data: challenges, solutions and applications. Crit Rev Biotechnol 2007;27:63—75.

[223] Dong MQ, Venable JD, Au N, Xu T, Park SK, Cociorva D, et al. Quantitative mass spectrometry identifies insulin signaling targets in *C. elegans*. Science 2007;317:660—3.

[224] Janes KA, Albeck JG, Gaudet S, Sorger PK, Lauffenburger DA, Yaffe MB. A systems model of signaling identifies a molecular basis set for cytokine-induced apoptosis. Science 2005;310: 1646—53.

[225] Janes KA, Reinhardt HC, Yaffe MB. Cytokine-induced signaling networks prioritize dynamic range over signal strength. Cell 2008;135:343—54.

[226] Miller-Jensen K, Janes KA, Brugge JS, Lauffenburger DA. Common effector processing mediates cell-specific responses to stimuli. Nature 2007;448:604—8.

[227] Boulton SJ, Gartner A, Reboul J, Vaglio P, Dyson N, Hill DE, et al. Combined functional genomic maps of the *C. elegans* DNA damage response. Science 2002;295:127—31.

[228] Piano F, Schetter AJ, Morton DG, Gunsalus KC, Reinke V, Kim SK, et al. Gene clustering based on RNAi phenotypes of ovary-enriched genes in *C. elegans*. Curr Biol 2002;12:1959—64.

[229] Boutros M, Agaisse H, Perrimon N. Sequential activation of signaling pathways during innate immune responses in *Drosophila*. Dev Cell 2002;3:711—22.

[230] Schena M, Shalon D, Davis RW, Brown PO. Quantitative monitoring of gene expression patterns with a complementary DNA microarray. Science 1995;270:467—70.

[231] Hughes TR, Marton MJ, Jones AR, Roberts CJ, Stoughton R, Armour CD, et al. Functional discovery via a compendium of expression profiles. Cell. 2000;102:109—26.

[232] Alizadeh AA, Eisen MB, Davis RE, Ma C, Lossos IS, Rosenwald A, et al. Distinct types of diffuse large B-cell lymphoma identified by gene expression profiling. Nature 2000;403:503—11.

[233] DeRisi JL, Iyer VR, Brown PO. Exploring the metabolic and genetic control of gene expression on a genomic scale. Science 1997;278:680—6.

[234] Eisen MB, Spellman PT, Brown PO, Botstein D. Cluster analysis and display of genome-wide expression patterns. Proc Natl Acad Sci U S A 1998;95:14863—8.

[235] Golub TR, Slonim DK, Tamayo P, Huard C, Gaasenbeek M, Mesirov JP, et al. Molecular classification of cancer: class discovery and class prediction by gene expression monitoring. Science 1999;286:531—7.

[236] Gray NS, Wodicka L, Thunnissen AM, Norman TC, Kwon S, Espinoza FH, et al. Exploiting chemical libraries, structure, and genomics in the search for kinase inhibitors. Science 1998;281:533—8.

[237] Holstege FC, Jennings EG, Wyrick JJ, Lee TI, Hengartner CJ, Green MR, et al. Dissecting the regulatory circuitry of a eukaryotic genome. Cell 1998;95:717—28.

[238] Marton MJ, DeRisi JL, Bennett HA, Iyer VR, Meyer MR, Roberts CJ, et al. Drug target validation and identification of secondary drug target effects using DNA microarrays. Nat Med 1998;4:1293—301.

[239] Perou CM, Jeffrey SS, van de Rijn M, Rees CA, Eisen MB, Ross DT, et al. Distinctive gene expression patterns in human mammary epithelial cells and breast cancers. Proc Natl Acad Sci U S A 1999;96:9212—7.

[240] van Wageningen S, Kemmeren P, Lijnzaad P, Margaritis T, Benschop JJ, de Castro IJ, et al. Functional overlap and regulatory links shape genetic interactions between signaling pathways. Cell 2010;143:991—1004.

[241] Lamb J. The Connectivity Map: a new tool for biomedical research. Nat Rev Cancer 2007;7:54—60.

[242] Lamb J, Crawford ED, Peck D, Modell JW, Blat IC, Wrobel MJ, et al. The Connectivity Map: using gene-expression signatures to connect small molecules, genes, and disease. Science 2006;313:1929—35.

[243] Hieronymus H, Lamb J, Ross KN, Peng XP, Clement C, Rodina A, et al. Gene expression signature-based chemical genomic prediction identifies a novel class of HSP90 pathway modulators. Cancer Cell 2006;10:321—30.

[244] Wei G, Twomey D, Lamb J, Schlis K, Agarwal J, Stam RW, et al. Gene expression-based chemical genomics identifies rapamycin as a modulator of MCL1 and glucocorticoid resistance. Cancer Cell 2006;10:331—42.

[245] Ramaswamy S, Tamayo P, Rifkin R, Mukherjee S, Yeang CH, Angelo M, et al. Multiclass cancer diagnosis using tumor gene

[245] expression signatures. Proc Natl Acad Sci U S A 2001;98:15149−54.
[246] Su AI, Cooke MP, Ching KA, Hakak Y, Walker JR, Wiltshire T, et al. Large-scale analysis of the human and mouse transcriptomes. Proc Natl Acad Sci U S A 2002;99:4465−70.
[247] Su AI, Wiltshire T, Batalov S, Lapp H, Ching KA, Block D, et al. A gene atlas of the mouse and human protein-encoding transcriptomes. Proc Natl Acad Sci U S A 2004;101:6062−7.
[248] Rhodes DR, Yu J, Shanker K, Deshpande N, Varambally R, Ghosh D, et al. ONCOMINE: a cancer microarray database and integrated data-mining platform. Neoplasia 2004;6:1−6.
[249] Stegmaier K, Ross KN, Colavito SA, O'Malley S, Stockwell BR, Golub TR. Gene expression-based high-throughput screening(GE-HTS) and application to leukemia differentiation. Nat Genet 2004;36:257−63.
[250] Hsuih TC, Park YN, Zaretsky C, Wu F, Tyagi S, Kramer FR, et al. Novel, ligation-dependent PCR assay for detection of hepatitis C in serum. J Clin Microbiol 1996;34:501−7.
[251] Landegren U, Kaiser R, Sanders J, Hood L. A ligase-mediated gene detection technique. Science 1988;241:1077−80.
[252] Nilsson M, Barbany G, Antson DO, Gertow K, Landegren U. Enhanced detection and distinction of RNA by enzymatic probe ligation. Nat Biotechnol 2000;18:791−3.
[253] Peck D, Crawford ED, Ross KN, Stegmaier K, Golub TR, Lamb J. A method for high-throughput gene expression signature analysis. Genome Biol 2006;7:R61.
[254] Krutzik PO, Nolan GP. Intracellular phospho-protein staining techniques for flow cytometry: monitoring single cell signaling events. Cytometry A 2003;55:61−70.
[255] Perez OD, Nolan GP. Simultaneous measurement of multiple active kinase states using polychromatic flow cytometry. Nat Biotechnol 2002;20:155−62.
[256] Irish JM, Hovland R, Krutzik PO, Perez OD, Bruserud O, Gjertsen BT, et al. Single cell profiling of potentiated phospho-protein networks in cancer cells. Cell 2004;118:217−28.
[257] Sachs K, Perez O, Pe'er D, Lauffenburger DA, Nolan GP. Causal protein-signaling networks derived from multiparameter single-cell data. Science 2005;308:523−9.
[258] Ficarro SB, McCleland ML, Stukenberg PT, Burke DJ, Ross MM, Shabanowitz J, et al. Phosphoproteome analysis by mass spectrometry and its application to *Saccharomyces cerevisiae*. Nat Biotechnol 2002;20:301−5.
[259] Huttlin EL, Jedrychowski MP, Elias JE, Goswami T, Rad R, Beausoleil SA, et al. A tissue-specific atlas of mouse protein phosphorylation and expression. Cell 2010;143:1174−89.
[260] Wilson-Grady JT, Villen J, Gygi SP. Phosphoproteome analysis of fission yeast. J Proteome Res 2008;7:1088−97.
[261] Yu Y, Anjum R, Kubota K, Rush J, Villen J, Gygi SP. A site-specific, multiplexed kinase activity assay using stable-isotope dilution and high-resolution mass spectrometry. Proc Natl Acad Sci U S A 2009;106:11606−11.
[262] Bodenmiller B, Wanka S, Kraft C, Urban J, Campbell D, Pedrioli PG, et al. Phosphoproteomic analysis reveals interconnected system-wide responses to perturbations of kinases and phosphatases in yeast. Sci Signal 2010. 3, rs4.
[263] Ahrens CH, Brunner E, Qeli E, Basler K, Aebersold R. Generating and navigating proteome maps using mass spectrometry. Nat Rev Mol Cell Biol 2010;11:789−801.
[264] Chen SH, Albuquerque CP, Liang J, Suhandynata RT, Zhou H. A proteome-wide analysis of kinase-substrate network in the DNA damage response. J Biol Chem 2010;285:12803−12.
[265] Smolka MB, Albuquerque CP, Chen SH, Zhou H. Proteome-wide identification of in vivo targets of DNA damage checkpoint kinases. Proc Natl Acad Sci U S A 2007;104:10364−9.
[266] Tedford NC, Hall AB, Graham JR, Murphy CE, Gordon NF, Radding JA. Quantitative analysis of cell signaling and drug action via mass spectrometry-based systems level phosphoproteomics. Proteomics 2009;9:1469−87.
[267] Przybylski C, Junger MA, Aubertin J, Radvanyi F, Aebersold R, Pflieger D. Quantitative analysis of protein complex constituents and their phosphorylation states on a LTQ-Orbitrap instrument. J Proteome Res 2010;9:5118−32.
[268] Hsu PP, Kang SA, Rameseder J, Zhang Y, Ottina KA, Lim D, et al. The mTOR-regulated phosphoproteome reveals a mechanism of mTORC1-mediated inhibition of growth factor signaling. Science 2011;332:1317−22.
[269] Yu Y, Yoon SO, Poulogiannis G, Yang Q, Ma XM, Villen J, et al. Phosphoproteomic analysis identifies Grb10 as an mTORC1 substrate that negatively regulates insulin signaling. Science 2011;332:1322−6.
[270] Thoreen CC, Kang SA, Chang JW, Liu Q, Zhang J, Gao Y, et al. An ATP-competitive mammalian target of rapamycin inhibitor reveals rapamycin-resistant functions of mTORC1. J Biol Chem 2009;284:8023−32.
[271] Stratton MR, Campbell PJ, Futreal PA. The cancer genome. Nature 2009;458:719−24.
[272] Altshuler D, Daly MJ, Lander ES. Genetic mapping in human disease. Science 2008;322:881−8.
[273] Geiss GK, Bumgarner RE, Birditt B, Dahl T, Dowidar N, Dunaway DL, et al. Direct multiplexed measurement of gene expression with color-coded probe pairs. Nat Biotechnol 2008;26:317−25.
[274] Kubota K, Anjum R, Yu Y, Kunz RC, Andersen JN, Kraus M, et al. Sensitive multiplexed analysis of kinase activities and activity-based kinase identification. Nat Biotechnol 2009;27:933−40.

Chapter 6

Genetic Networks

Michael Costanzo[1], Anastasia Baryshnikova[1,2], Benjamin VanderSluis[3], Brenda Andrews[1,2], Chad L. Myers[3] and Charles Boone[1,2]

[1]*Banting and Best Department of Medical Research, The Donnelly Center for Cellular and Biomolecular Research, University of Toronto, Toronto, Ontario M5S 3E1, Canada,* [2]*Department of Molecular Genetics, University of Toronto, Toronto, Ontario M5S 3E1, Canada,* [3]*Department of Computer Science and Engineering, University of Minnesota, Minneapolis, MN 55455, USA*

Chapter Outline

Introduction	115
Defining Genetic Interactions	117
Negative Genetic Interactions	117
Positive Genetic Interactions	117
Experimental Approaches to Map Genetic Interaction Networks in Yeast	**118**
The Yeast Non-Essential Gene Deletion Collection	118
Genetic Interaction Mapping Technologies	118
Quantifying Genetic Interactions	121
Quantitative Genetic Interaction Profiles Reveal the Functional Organization of a Cell	121
Exploring Genetic Interaction Networks	**122**
Modular Network Structures Identify Functional Relationships between Pathways and Complexes	122
Genetic Networks Enable Functional Dissection of Pleiotropic Genes	125
Genetic Interactions as a Means of Studying the Evolution of Gene Duplicates	125
Integrating Genetic Interactions with Other Biological Networks	**127**
Genetic and Physical Interaction Networks	127
Genetic and Metabolic Networks	127
Mapping Genetic Interactions in Other Organisms	**128**
Genetic Interactions in Unicellular Organisms	128
Genetic Interactions in Metazoan Model Systems	128
Genetic Interactions in Mammalian Model Systems	128
Genetic Interactions and Cancer Therapeutics	129
Genetic Network Conservation	**129**
Conservation of Individual Interactions between Orthologous Gene Pairs	129
Conservation of Genetic Network Structure and Topology	129
Expanding Genetic Networks: Mutant Alleles, Conditions and Phenotypes	**130**
Conditional Alleles and Essential Genetic Interactions	130
Gain-of-Function Alleles	130
Condition-Specific Genetic Interactions	130
Quantitative Phenotypes to Measure Genetic Interactions	131
Genetic Interactions and Genome-Wide Association Studies	**131**
Acknowledgements	133
References	133

INTRODUCTION

The relation between an organism's genotype and its phenotype may be governed by an underlying set of complex genetic interactions [1]. To define the general principles of genetic networks, we have mapped genetic interactions systematically in the budding yeast *Saccharomyces cerevisiae* [2–4]. Like most eukaryotic organisms, the majority (~80%) of the ~6000 yeast genes are individually dispensable, with only a relatively small subset (~20%) required for viability [5,6]. The finding that most genes are non-essential may be indicative of the evolution of extensive buffering against genetic and environmental perturbations [7]. Genome-scale screens for genetic interactions provide a means of exploring this buffering capacity and mapping a functional wiring diagram of a cell. In particular, synthetic genetic array (SGA) methodology enables the systematic mapping of rare synthetic lethal genetic interactions amongst the set of ~5000 viable deletion mutants through an automated form of genetic analysis that produces high-density arrays of double mutants [4]. In addition to their functional information, the resultant genetic networks may provide fundamental insights into the genetic architecture underlying the genotype–phenotype relationship governing genetic diseases.

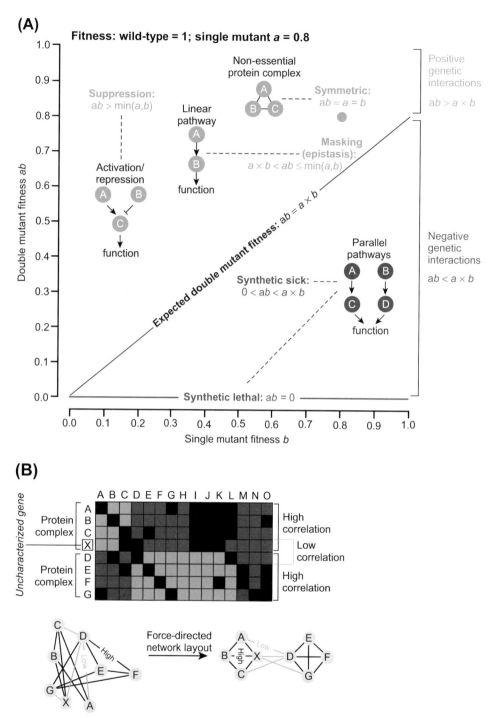

FIGURE 6.1 (A) A graphical representation of quantitative genetic interactions. In the illustrated scenario, the fitness of the wild-type and one of the single mutants (a) is assumed to be 1.0 and 0.8, respectively. The fitness of mutant b can vary between 0.0 and 1.0 and is plotted on the x-axis. The expected fitness of the resultant ab double mutant based on a multiplicative model ($a \times b$) is $0.8 \times b$ and is plotted as a solid black line. The observed double mutant fitness (ab) is plotted on the y-axis. **Negative genetic interactions.** Negative deviations from the expected fitness are scored as either synthetic sick or synthetic lethal interactions (pink area). Negative genetic interactions commonly occur between genes that function in parallel pathways to regulate the same essential function. **Symmetric positive interactions.** The measured fitness of the ab double mutant (0.8) is greater than the multiplicative expectation (0.64), indicating a positive genetic interaction. The interaction is classified as symmetric because the two single mutants (a and b) and the resultant double mutant (ab) exhibit the same fitness ($ab = a = b = 0.8$). Symmetric positive interactions are frequently observed between genes encoding members of the same non-essential protein complex. **Asymmetric positive interactions.** In this scenario, single mutants and double mutants differ in fitness. Positive deviations from expectation (green area) along with single mutant fitness comparisons allow the classification of asymmetric positive interactions into different subcategories, including masking interactions (light green area) and suppression (darker green area). **Masking** describes positive interactions where the double

DEFINING GENETIC INTERACTIONS

Geneticists have long recognized that genetic interactions are important for shaping the phenotypic landscape of a population. In 1909 William Bateson introduced the term 'epistasis' (i.e., 'standing upon') to describe a specific type of genetic interaction whereby one mutation masks the effects of another mutation on fur color in rabbits and mice [8] (Figure 6.1A). The same term was later adopted to describe genetic relationships that often occur among members of the same metabolic pathways, where the action of one enzyme depends on a substrate produced by another enzyme (Figure 6.1A). Ronald Fisher later expanded this term to include any multi-locus mutant effect that deviates from the additive combination of the corresponding individual loci [9]. Today, Fisher's epistasis is often used to generally define a genetic interaction as an unexpected phenotype that cannot be explained by the combined effects of the individual mutations (Figure 6.1A).

Adopting this definition, a genetic interaction between two genes can be experimentally defined based on measurement and comparison of three fundamental properties that include the single mutant phenotypes, an estimate of the expected double mutant phenotype, and a measurement of the observed double mutant phenotype (Figure 6.1A). Although these are relatively straightforward criteria, determining how mutations in different genes are expected to combine is not as obvious. The additive and multiplicative models of genetic interactions represent the two most common approaches for measuring non-independence of gene effects on phenotypic variance [10,11]. In the case of the additive model, the phenotypes associated with mutant alleles of individual genes are expected to combine, such that the double mutant phenotype should be equal to the sum of the two single mutant phenotypes. Alternatively, according to the multiplicative model, genes are expected to change a phenotype by a specific fraction, and thus a double mutant phenotype should be equal to the product of the two single mutants phenotypes. The choice of model is dependent on the particular phenotype and scale of measurement (e.g., linear vs. logarithmic) because genes may appear independent on one scale but show a genetic interaction if measured on a different scale [10,11]. In the case of yeast fitness, the expected double mutant phenotype is typically modeled as a multiplicative combination of single mutant phenotypes and genetic interactions are measured by the extent to which double mutants deviate from this multiplicative expectation [10,11] (Figure 6.1A).

Negative Genetic Interactions

Negative genetic interactions describe double mutants that exhibit a more severe phenotype than expected [10,11] (Figure 6.1A). Synthetic lethality represents an extreme negative interaction where two mutations, each causing little fitness defect on their own, result in an unviable phenotype when combined as double mutants. This phenomenon, initially observed among the progeny of intercrosses between natural variants of *Drosophila pseudoobscura*, provided the first insight into the degree of genetic variability concealed within natural populations [12]. Synthetic lethality has since been explored extensively in yeast and proven to be an extremely powerful genetic tool for identifying and characterizing genes in almost all biological processes [13–15]. Negative genetic interactions, such as synthetic sickness or synthetic lethality, are interesting because they often occur between genes that impinge on a common essential biological function [4]. Alternatively, genes functioning in the same essential pathway or complex may also share a negative genetic interaction if each mutation has adverse effects on pathway activity [16–18].

Positive Genetic Interactions

Positive interactions describe double mutants exhibiting a less severe phenotype than expected based on the product of the two single mutant phenotypes, and can be further sub-classified into a variety of categories (Figure 6.1A) [11, 19,20]. For example, the 'symmetric' category describes a type of positive interaction whereby the phenotypes associated with the single and the resultant double mutant are quantitatively indistinguishable [19,20]. Conversely, the 'asymmetric' class consists of those interactions in which the strength of the phenotypic effect varies between single and double mutants [19,20].

Importantly, these types of positive genetic interaction are often associated with different biological interpretations and, when measured quantitatively, offer the potential to infer biochemical relationships between gene products and elucidate how biological pathways and complexes relate to one another to modulate cellular functions [19–21]. For

---

mutant fitness is greater than the expected double mutant fitness but less than or equal to the fitness of the sickest single mutant. **Suppression** occurs when the observed double mutant fitness is greater than the fitness of the sickest single mutant. In the case of deletion mutants, suppression is often indicative of a negative regulatory relationship between the interacting genes. **(B)** The set of negative and interactions for a given mutant is referred to as a genetic interaction profile. Genes that share genetic interactions in common tend to be functionally related. Two-dimensional hierarchical clustering is an effective way to group genes together based on genetic interaction profile similarity. Arranging genes in this manner enables identification of highly correlated groups of genes that correspond to functionally related gene modules, including protein complexes and pathways, as well as novel functional predictions. In addition to clustering algorithms, genetic interactions can also be visualized as a correlation-based network connecting genes with similar genetic interaction profiles. Using a force-directed network layout, genes with highly similar genetic interaction profiles (black lines) are placed close to each other in the network, while genes with less similar interaction profiles (blue lines) are place further apart from one another.

example, genetic suppression represents an asymmetric positive interaction whereby the double mutant exhibits increased fitness relative to the sickest single mutant (Figure 6.1A). Genetic suppression between two interacting loss-of-function (LOF) deletion alleles often indicates that the suppressor gene functions as a negative regulator of a pathway associated with the interacting gene [17]. Our recent global analysis of the yeast genetic network revealed a large number of LOF suppression interactions, indicating that positive interactions between genes within different complexes or pathways are surprisingly prevalent [17].

On the other hand, genes encoding members of the same non-essential protein complex are frequently connected by symmetric genetic interactions, presumably because once the function of the complex is disrupted by the removal of the first component, the phenotype cannot be made worse by the removal of additional components (Figure 6.1A) [19–21]. The relationship between genetic and physical interactions is discussed in greater detail below (see Exploring Genetic Interaction Networks).

EXPERIMENTAL APPROACHES TO MAP GENETIC INTERACTION NETWORKS IN YEAST

The Yeast Non-Essential Gene Deletion Collection

With its elegant and straightforward genetics, *Saccharomyces cerevisiae* is a powerful model system to dissect the fundamental properties of eukaryotic cells at a molecular level and has served as a primary test bed for the development of most functional genomic methodologies. Large-scale genetic and phenotypic analyses have been made possible through the development and availability of a comprehensive collection of deletion mutants where each of the ~6000 yeast open reading frames was replaced with a dominant, drug-resistance marker flanked by unique synthetic 'barcode' sequences [5]. This systematic endeavour defined the set of ~1000 essential genes and generated a set of ~5000 viable deletion mutants [5,6]. The systematic mapping of genetic interactions was first made possible through the development of methodologies that take advantage of the yeast non-essential gene deletion collection.

Genetic Interaction Mapping Technologies

Synthetic genetic array (SGA) is an automated method that combines arrays of either non-essential gene deletion mutants, or conditional alleles of essential genes, with robotic manipulations for high-throughput construction of haploid yeast double mutants and identification of genetic interactions (Box 6.1) [3,4]. In its first large-scale application, SGA methodology was used to cross 132 query genes to the complete array of ~5000 viable haploid deletion mutants, resulting in a genetic interaction network consisting of ~1000 genes and ~4000 synthetic lethal/sick interactions [3]. This initial analysis provided important insight into fundamental genetic network properties and topology. For example, although generally rare, synthetic

BOX 6.1 Synthetic Genetic Array (SGA)

In a typical SGA screen a *MATα* mutant strain carries a 'query' mutation, marked with the dominant drug-resistance marker *natMX4* (closed black circle), is crossed to an array of ~5000 viable *MATa* deletion mutants or conditional alleles of essential genes, with each mutation marked with a *kanMX4* resistance cassette (closed red circle). The *CAN1* gene, which encodes an arginine permease, is replaced in the SGA query strain with a *MATa* haploid-specific reporter, *STE2pr-Sp_his5* such as *can1Δ:: STE2pr-Sp_his5*. The query strain also carries a deletion of the lysine permease gene *LYP1*. Following mating, diploid selection and sporulation, meiotic progeny are grown on media containing the G418 and nourseothricin and lacking histidine to select *MATa* haploid double mutants. In addition to positive selection, the media is also supplemented with canavanine and thialysine (toxic analogs of arginine and lysine, respectively) in order to counter-select against unsporulated diploid mutants that are heterozygous for *CAN1* and *LYP1* deletion mutations. The SGA procedure is briefly described below. For a more detailed protocol and media composition, refer to Baryshnikova et al. [119].

Mating: A *MATα* query deletion mutant is pinned onto the *MATa* non-essential gene deletion collection on rich media and incubated at 30°C.

Diploid selection: Resulting *MATa/α* zygotes are pinned onto rich media containing kanamycin and nourseothricin and incubated at 30°C to select heterozygous diploid double deletion mutants.

Sporulation: Diploids are transferred to sporulation medium and incubated for 5 days at room temperature.

MATa haploid selection: Spores are grown on synthetic medium lacking histidine and containing canavanine and thialysine (SD-his+canavanine+thialysine), which selects for *MATa* haploid cells expressing the *MATa*-specific reporter, *can1Δ:: STE2pr-Sp_his5* and selects against unsporulated diploids.

Single mutant selection: Colonies are transferred to SD-his+canavanine+thialysine growth medium supplemented with G418 to select for *MATa* haploids harboring the deletion array mutation.

Double mutant selection: Haploid double mutants are selected following growth on (SD-his+canavanine+thialysine) medium containing both G418 and nourseothricin.

BOX 6.1 Synthetic Genetic Array (SGA) – *Continued*

lethal interactions are rich in functional information often connecting genes with related biological functions and uncovering novel functional relationships [3]. Furthermore, this preliminary network highlighted the extensive genetic complexity encoded within the genome of a single unicellular organism. Indeed, estimates based on this small sampling of the yeast genetic network (~3%) suggested that a non-essential gene has, on average, approximately 30 interaction partners and, after considering the potential contribution of essential gene interactions, the complete *S. cerevisiae* was predicted to comprise ~200 000 synthetic sick or lethal interactions [3,22]. This finding implies that the potential for generating synthetic genetic interactions should increase with gene number, and the genomic load of synthetic effects may be even higher in more complex multicellular organisms, including humans [3,7,23].

BOX 6.1 Synthetic Genetic Array (SGA) — *Continued*

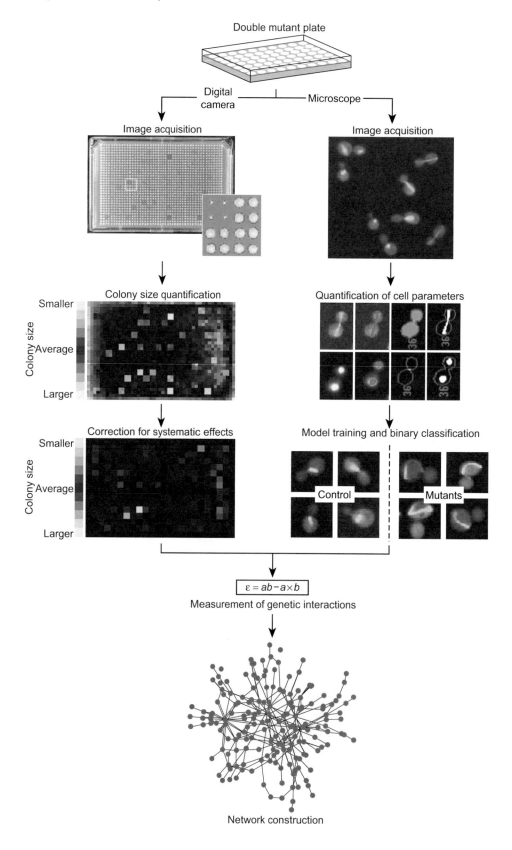

Methods complementary to SGA include dSLAM (diploid synthetic lethal analysis by microarray), which relies on the barcodes associated with each deletion mutant to enable quantification of double deletion strain abundance in a mixed population [24]. Briefly, a marked query mutation is introduced into a pooled set of heterozygote deletion strains containing an 'SGA marker' by mass transformation. Using the same selection steps as in SGA, double mutant haploids are selected and the barcode intensities of each strain in the pool (compared to a non-selected control pool) provide a measure of relative double mutant fitness. dSLAM has been applied extensively to map genetic interactions between genes involved in DNA integrity and histone modification [25,26]. Finally, a third method for genetic interaction discovery, called genetic interaction mapping (GIM), was used to examine interactions between genes involved in mRNA processing [27]. GIM represents a hybrid of SGA and dSLAM. Reminiscent of SGA, double mutants are generated by mating and sporulation, but as with dSLAM all steps are performed in a pooled format, which involves competitive growth of double mutant meiotic progeny, with interactions identified by comparison of barcode microarray hybridization intensities between double mutants and a reference population [27].

Quantifying Genetic Interactions

Early genetic interaction studies were predominantly based on binary (i.e., viable or unviable) assessment of cellular fitness [3,4]. Quantitative measurements enable identification of subtle negative and positive interactions and the construction of higher-resolution genetic networks. Most efforts to map quantitative genetic networks have thus far been based on the quantitative measurement of cell growth or fitness associated with yeast single and double deletion mutants. For example, a liquid growth profiling approach was used to quantitatively measure genetic interactions between a subset of genes involved in DNA replication and repair [19]. In another example, fitness was measured from fluorescence-labeled populations of wild-type cells mixed with either single or double mutant yeast strains to map a quantitative genetic interaction network for genes encoding components of the 26S proteasome [21]. A similar fluorescence-based assay was also applied to quantify genetic interactions between duplicated genes [28]. Despite providing high-resolution interaction measurements and illustrating the utility of quantitative genetic interaction analysis for functional analysis of pathways and protein complexes, these methods are not easily amenable to genome-scale studies [29].

The requirement for quantitative phenotypic measurements has imposed constraints on the scale and functional scope of quantitative genetic interaction studies. However, large-scale measurement of yeast colony size offers the potential to identify quantitative genetic interactions on a scale compatible with the throughput and capacity afforded by methods such as SGA [17,30]. In fact, correcting high-density yeast colony arrays for sources of systematic variability that plague most, if not all, array-based technologies resulted in quantitatively accurate single and double mutant colony size measurements (Box 6.2). These measurements overlap significantly with fitness measurements obtained using other high-resolution methods, indicating that colony size is a suitable proxy for yeast fitness [17]. The relative ease with which colony size-based fitness measurements can be obtained provides a reasonable compromise between experimental throughput and quantitative resolution. Indeed, a study combining SGA with a genome-scale colony size scoring methodology examined ~5.4 million gene pairs covering ~30% of the *S. cerevisiae* genome [2]. This large-scale endeavor measured single and double mutant yeast fitness to uncover ~170 000 genetic interactions (~113 000 negative and ~57 000 positive) and provide the first view of a quantitative, genome-scale genetic interaction network for a eukaryotic cell [2].

Quantitative Genetic Interaction Profiles Reveal the Functional Organization of a Cell

The set of synthetic lethal genetic interactions for a given gene, termed a genetic interaction profile, provides a rich

---

BOX 6.2 Computational Pipeline for Processing SGA Data

Single and double mutant array plates derived from SGA are photographed using a high-resolution digital camera. Mutant array plate images are then processed using custom-developed image processing software to identify colonies and measure their area in terms of pixels. To identify quantitative genetic interactions, yeast colony size pixel data are subjected to a series of normalization steps to correct for several systematic experimental effects, after which genetic interactions are measured by comparing corrected double mutant colony size to the colony size of the corresponding single mutants. This analysis generates a genetic profile for each array mutant that can be used to construct a correlation-based genetic interaction network (see Figure 6.1B and Figure 6.2). As an alternative to fitness-based genetic interactions, mutant array plates may carry a specific fluorescent reporter that can be analyzed using a high-content imaging system. Genetic interactions can then be identified and measured based on various cell biological parameters and phenotypes. Refer to Baryshnikova et al. [119] for detailed protocols and procedures pertaining to genetic interaction data acquisition, processing and analysis.

phenotypic signature indicative of gene function [3]. As a result, grouping genes according to their genetic interaction profiles, using standard clustering algorithms [31], is an effective and powerful way to precisely predict gene function [3] (Figure 6.1B).

Genetic interaction profile similarity has been expanded beyond synthetic lethal interaction profiles to include profiles derived from large-scale quantitative screens composed of both negative and positive interactions (Figure 6.1B). These quantitative genetic interaction profiles enabled the construction of a global network in which genes with similar interactions patterns are located next to one another while genes sharing less similar interaction profiles are further apart in the network [2]. The resulting network provides a multi-scale view of the functional organization within a cell (Figure 6.2). Globally, genes displaying tightly correlated profiles formed large and readily discernable clusters corresponding to distinct biological processes (Figure 6.2A). The relative distance between these clusters appeared to reflect shared functions highlighting the interdependencies of general cellular processes and the inherent functional organization of the cell. When observed in greater resolution, the genetic map enables dissection of broad biological processes into distinct yet interdependent gene functions (Figure 6.2B–C). In one region of the global network, separate gene clusters involved in various processes such as DNA replication, recombination and repair, microtubule biogenesis, RNA processing and RNA decay are readily distinguishable (Figure 6.2B). Finally, at its most detailed level and consistent with theoretical studies [32], the network is composed of highly organized modules corresponding to discrete biological pathways and/or protein complexes connected almost exclusively by a single type of genetic interaction (only negative or only positive interactions) and reveals a functional wiring diagram of the cell (Figure 6.2D–E). At this level of scrutiny, the modular organization of the genetic network also enables precise functional predictions for previously uncharacterized genes. Indeed, genetic analysis implicated three novel genes in the regulation of the general amino acid permease, *GAP1* [2] (Figure 6.2E).

EXPLORING GENETIC INTERACTION NETWORKS

Modular Network Structures Identify Functional Relationships between Pathways and Complexes

The ability of genetic interaction profiles to define functionally coherent clusters is possible because these networks are highly structured: interactions rarely occur in isolation but rather in larger 'blocks' that define specific functional modules (Figure 6.1B). For example, three genes, A, B and C, belonging to the same pathway or protein complex, should exhibit negative genetic interactions with a second pathway or complex composed of genes X, Y and Z if both pathways/complexes regulate a common essential function (Figure 6.3A). In general, this local network structure reveals genes that belong to the same functional module and the genetic relationships between separate modules that share partially redundant functions. Compromising the function of either non-essential module leaves the cell viable, but simultaneously compromising both results in cell death (Figure 6.3A). This relationship is aptly illustrated by genetic interactions between the Elongator (ELP) complex and the evolutionarily conserved URMylation pathway, revealing that these functional modules share distinct yet redundant roles in tRNA processing [33]. Several previous studies have highlighted the prevalence of this type of structure based on systematic mining of genetic interaction networks in *S. cerevisiae*, and it has been referred to as a 'between-pathway' network structure [18,34–36]. A recent survey of the yeast genetic network revealed that 58% of observed negative genetic interactions appear in structures involving at least nine total interactions (a minimum 3×3 gene matrix) (Figure 6.3B) and estimated that the true fraction may be as high as 75% if experimental false negatives and false positives are taken into account [36].

This modular structure suggests a fundamental principle of genetic redundancy: redundancy between genes appears to be more frequently the result of module-level compensation rather than single gene buffering, at least for yeast. The alternative situation is where redundancy is mainly encoded at the single gene level, which would instead produce mostly isolated negative interactions throughout the genome. It should be noted that while the latter case appears to be infrequent, at least in the context of the yeast genome [36], there are instances where single gene redundancy produces negative genetic interactions, particularly among pairs of gene duplicates (see Genetic Interactions as a Means of Studying the Evolution of Gene Duplicates). The degree to which module-level redundancy extends beyond yeast to genetic networks of higher eukaryotes remains unclear. Nonetheless, we anticipate that the functional relationships between modules may be conserved, and a detailed understanding of network structure may ultimately affect our ability to assess the conservation of genetic networks.

In addition to the 'between-pathway' structure, another type of genetic interaction network motif includes a set of negative interactions that connect a common set of genes (a 'clique-like' structure in graph theory terms). In this structure, known as 'within-pathway', a set of genes (D, E and F) all exhibit negative genetic interactions with each other (Figure 6.3A). Although the within-pathway structures

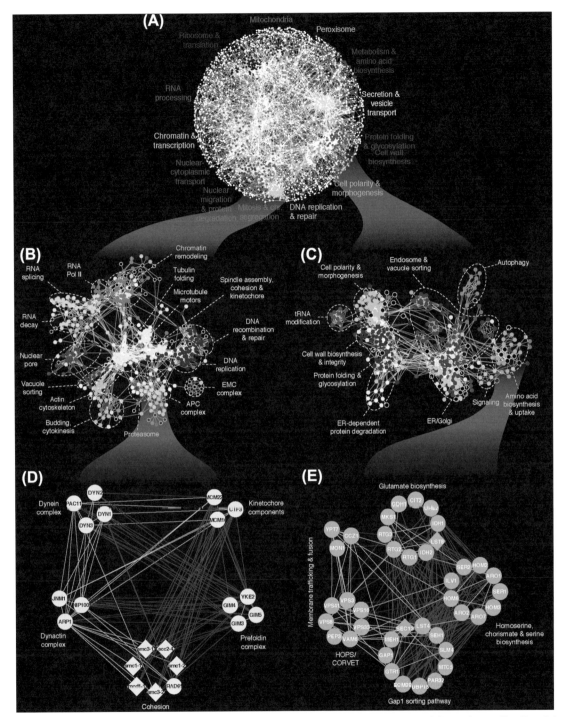

FIGURE 6.2 **Genetic network analysis.** (A) A correlation-based network connecting genes with similar genetic interaction profiles. Genetic interaction profile similarities were measured for all gene pairs by computing Pearson's correlation coefficients (PCC) and gene pairs whose profile similarity exceeded a PCC threshold >0.2 were connected in the network using the force-directed layout described in Figure 6.1B. (B,C) Magnification of the genetic network map resolves cellular processes with increased specificity. Sub-networks correspond to the indicated region of the global map described in (A). Node color corresponds to specific biological processes. Color schemes are unique to each panel in the figure. (D,E) Further magnification reveals modules corresponding to specific pathways and complexes connected by negative and positive genetic interactions. Subsets of genes belonging to spindle assembly, cohesion and kinetochore (D), as well as amino acid biosynthesis and uptake (E) regions of the network were selected. Genes are represented as nodes grouped according to profile similarity and edges represent negative (red) and positive (green) genetic interactions. Characterized genes are shown in blue (D) or green (E), while genes with previously unknown function are indicated in orange. *Modified from [2].*

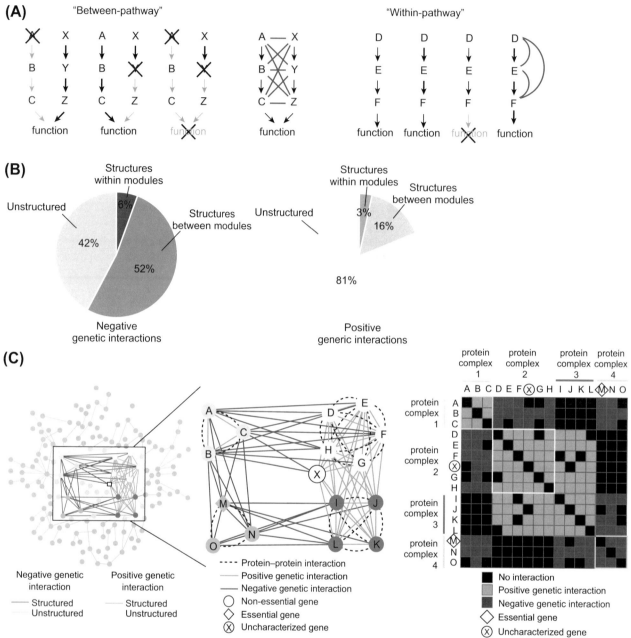

FIGURE 6.3 **Genetic interaction network structure: within- and between-pathway structures.** (A) Negative interactions (red lines) can arise from the disruption of parallel pathways converging on a common process ('between-pathway' genetic interactions), or can occur between pairs of genes within the same essential pathway or functional module ('within-pathway' genetic interactions). Semicircles depict conditional or hypomorphic (partially functional) alleles of essential genes. Thick black lines indicate wild-type function, while thin black lines indicate reduced function. Grey arrows indicate loss of function. (B) The fraction of negative and positive genetic interactions spanning within- and between-pathway structures, as well as unstructured genetic interactions. (C) **Properties of genetic interactions.** Schematic representation illustrating the complexity of within- and between-pathway structures. Both positive (green) and negative (red) interactions can exist as unstructured interactions, connect between two functional modules, or connect members within the same functional module. Circular and diamond shaped nodes represent non-essential and essential genes, respectively. The highly organized modular nature of the yeast genetic interaction network helps identify distinct groups of functionally related genes and how these different modules relate to one another. Although the vast majority (~98%) of genetic interactions (both negative and positive) tend to occur between pathways and protein complexes, negative interactions can connect members of the same essential protein complex while positive interactions can occur between members of non-essential complexes. Dashed lines indicate physical interactions. Similar to the network representation, two-dimensional clustering reveals distinct 'blocks' of genetic interactions that define functional modules. Interaction blocks highlighted along the diagonal axis of the matrix correspond to within-pathway structures while off-diagonal blocks represent between-pathway modules. *Modified from [120].*

appear to be much less abundant than between-pathway structures, they were frequently observed in systematic surveys of the *S. cerevisiae* interaction network [34,36]. For example, within-pathway structures were most recently reported to account for ~10% of all observed negative interaction modules [36] (Figure 6.3B). These modules often corresponded to essential protein [17,18,34] complexes or groups of co-regulated genes [36] (Figure 6.3C).

The interpretation of within-pathway negative genetic interactions appears to suggest instances where a functional module exhibits non-specific internal redundancy, where perturbation of one component can be tolerated but perturbation of any two components of the same functional module results in loss of function and cell death. Supporting this interpretation, protein complexes containing at least one essential gene tend to appear in these 'within-pathway' interaction structures [17,18] (Figure 6.3A, C). Presumably, these are instances where the function of the complex is essential, but the loss of a single non-essential component does not completely destabilize the complex. Perturbation of two or more of these non-essential components results in loss of complex function and cell viability. Alternatively, perturbation of two essential components (e.g., double mutants carrying conditional or hypomorphic alleles of essential genes) will also result in loss of protein complex function and lethality. It should be noted, however, that essential protein complexes explain only a minority (7%) of these within-pathway structures [36]. Interestingly, in yeast these negative genetic interaction structures are enriched for relatively specific functions, including genes involved in chromosome segregation and cell cycle [36].

Like negative genetic interactions, positive genetic interactions also exhibit modular structure, although to a lesser extent. In the same systematic survey based on the global yeast network, only 19% of positive interactions, compared to 58% of negative interactions, contribute to larger block structures (minimum 3 × 3 gene matrix) (Figure 6.3B), although it is possible that this difference may be partially related to decreased sensitivity and specificity in the experimental detection of positive genetic interactions [2,36]. Approximately 20% of positive interaction modules could be classified as within-pathway structures (Figure 6.3B) and, consistent with observations from smaller-scale studies [19, 37], these were enriched for non-essential protein complexes [36]. This is likely the result of a single gene deletion that is capable of completely disrupting protein complex function. In such a case, deleting genes encoding other components of the same complex will not have any additional effect on fitness due to symmetric positive interactions (Figure 6.1A) [19].

Most (~80%) of structured positive interactions reflected 'between-pathway' structures (Figure 6.3B). Although much more common, 'between-pathway' positive interaction structures are less well understood. Careful examination of these interactions identified several instances of genetic suppression that span all components of distinct protein complexes [17]. For example, deletion of any member of the conserved FAR complex was shown to suppress actin polarity and growth defects associated with mutant alleles of TORC2 [17]. Although the yeast FAR complex is largely uncharacterized, the orthologous complex in mammalian cells associates with protein phosphatase 2A (PP2A) [38]. An attractive hypothesis is that the FAR complex mediates its function by antagonizing TORC2 activity [17]. In general, mechanistic understanding of most between-pathway positive interaction structures remains an open question that deserves further study.

Genetic Networks Enable Functional Dissection of Pleiotropic Genes

The prevalence of modular structure in both negative and positive genetic interactions provides a rich basis for associating genes with a common function and broadly characterizing the functional organization of the genome. Each specific between-pathway set of interactions defines two specific functional modules and highlights their compensatory relationship. Interestingly, genes sometimes appear in several different between-pathway structures where each structure is composed of a unique set of genes, enriched for a distinct function. This observation highlighted the utility of genetic networks for identifying multifunctional or pleiotropic genes [36]. For example, *VIP1*, which encodes one of two yeast inositol pyrophosphate synthases required for synthesis of hexakisphosphate (IP6) and heptakisphosphate (IP7) [39], exhibited negative genetic interactions with 13 distinct modular structures [36]. Each of these structures reflected enrichment for a different function, including an unexpected one that associated *VIP1* with several genes involved in DNA replication and repair [36].

Genetic Interactions as a Means of Studying the Evolution of Gene Duplicates

One basis for redundancy within a genome is the presence of duplicated genes or paralogs. Although this has been a topic of interest for some time, large-scale genetic interaction networks provide a new tool for studying duplicate genes. Sequence studies of diverse organisms have revealed that a sizable fraction of many genomes consists of duplicate copies of existing genes. This is perhaps not surprising, given the long-held view that gene duplication is a primary mechanism for introducing novel gene function into a genome [40]. What is surprising, however, is the degree to which these duplicate pairs (or large gene families) can retain sequence similarity despite millions of years of opportunity for divergence [41,42]. Several studies have queried specific genetic interactions between duplicate gene pairs in *S. cerevisiae*, providing a direct measure of their functional redundancy. All of these studies found extensive evidence for redundancy, reporting that 25—35% of paralogs exhibit negative genetic interactions, which was between 10 and 20 times higher than expected by chance [28,43—45]. The high rate of negative genetic interaction observed among duplicates in yeast, many of which have been maintained for ~100 million years, presents an evolutionary paradox: why would functionally redundant copies of a gene be maintained with such high sequence similarity? Insight into this question was gained by looking beyond individual interactions connecting paralog pairs to examine broader genome-wide interaction profiles associated with these genes.

If duplicated genes share a high degree of redundancy, it was reasoned that deletion of a single duplicated gene should not yield many genetic interactions. However, functionally redundant duplicates often exhibit genetic interactions with other genes, suggesting that redundancy is at best incomplete [43]. Moreover, they also shared fewer interactions in common than other functionally related gene pairs, such as two distinct members of the same protein complex [43]. These observations may be reconciled by a simple model for how the functional redundancy shared between duplicate genes should affect their genetic interactions across the genome [46] (Figure 6.4). Immediately following a gene duplication event, all genetic interactions should be masked owing to complete functional redundancy of the duplicated genes, and only as the duplicated genes begin to diverge will genetic interactions begin to reappear. Thus, the observed genetic interaction profiles of duplicate gene pairs are generally incomplete, composed only of those interactions that reflect functions unique to the diverged duplicated gene (as through sub-/neo-functionalization), while interactions related to the common function(s) of the paralog pair cannot be detected in the absence of a single gene when one gene of the duplicated pair remains intact and functional. Analysis of the yeast genetic interaction map supported this hypothesis and suggested that genetic interactions are useful for dissecting the functional divergence of duplicate genes [46].

Additionally, we found that many duplicate pairs exhibited a highly asymmetric pattern of functional divergence where one gene retained or acquired a large number of genetic interactions while the other exhibited relatively few interactions [46]. Interestingly, the asymmetry in number of genetic interactions for duplicate gene pairs

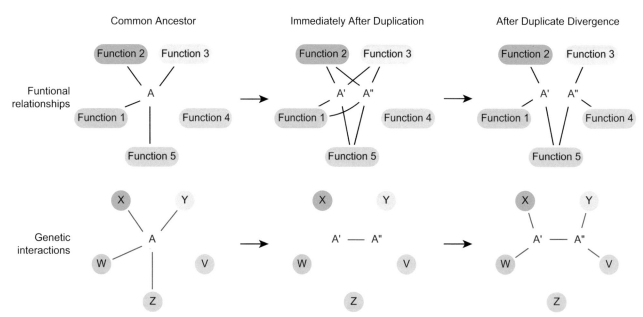

FIGURE 6.4 **Role of genetic interactions in the characterization of duplicate gene pairs.** A model for the buffering of genetic interactions by partially redundant genes. Gene A has no redundant partner and its set of functional relationships is revealed through negative genetic interactions. Immediately after duplication, genes A′ and A″ are fully redundant and their functional relationships are shared. Because each is capable of performing their common functions without the other, the deletion of A′ or A″ has negligible effects and does not exhibit negative interactions with any other genes. However, the simultaneous deletion of A′ and A″ reveals the original phenotype of their ancestor, and shows a negative genetic interaction. As A′ and A″ diverge, the redundancy becomes incomplete and unique deletion consequences emerge for each duplicate. Some of the negative genetic interactions observed for the ancestor gene A are not observed following duplication and divergence; for example, despite the functional relationship between A′ and A″ and Z, negative interactions are not observed with Z. A″ has evolved a new relationship with Function 4. A′ lacks this ability and we see a genetic interaction between A″ and V. *Modified from [46]*.

often corresponded with other asymmetric properties, including sequence evolution rates, protein—protein interaction degree, and single mutant fitness [41,42,47,48]. For example, the duplicate gene with the higher genetic interaction degree also often showed evidence of slower sequence evolution, higher protein—protein interaction degree, and lower single mutant fitness [46].

INTEGRATING GENETIC INTERACTIONS WITH OTHER BIOLOGICAL NETWORKS

Genetic and Physical Interaction Networks

One of the powerful aspects of genetic interaction networks is that they are highly complementary to other large-scale interaction networks that have been experimentally mapped in recent years. The main distinction is that genetic interactions capture functional consequences of (combined) genetic perturbations, whereas most other mapping efforts highlight physical interactions, such as protein—protein [49—52] or protein—DNA [53,54] interactions. Given this important difference, it is not surprising that there is little direct overlap between genetic interactions and physical networks. Only 10—20% of protein—protein interaction pairs were found to also share a negative or positive genetic interaction, which was well above the overlap expected by chance (~3%) but still quite modest. The overlap is even sparser when considering that only 0.4% of negative and 0.5% of positive genetic interaction pairs from a genome-wide network also showed a protein—protein interaction [2]. Thus, the large majority of either type of genetic interaction, negative or positive, do not reflect direct physical binding events among the corresponding gene products [2,17].

The relatively low overlap between genetic and physical interaction networks is closely related to the observation that most negative and positive genetic interactions occur in between-pathway network structures, which reflect two distinct modules of genes bridged by a large number of genetic interactions between them. In fact, protein complexes often appear within the sets of genes on either side of these between-pathway structures [34,36], suggesting that genetic and physical interactions are largely orthogonal to one another. Because genetic interactions are highly complementary to protein—protein interactions, a genetic interaction network should serve as a scaffold for unifying physical networks. Protein—protein interaction or protein—DNA interaction networks define connections that support local, molecular-level interactions, whereas genetic interactions capture more general information revealing how genes are broadly organized into modules that support cellular function. Indeed, several methods been developed to integrate genetic interactions with physical networks to globally define gene modules and their relationships [18,34,35].

Genetic and Metabolic Networks

The integrative analysis of genetic and metabolic networks has also been a fruitful area of study. Our relatively comprehensive understanding of the suite of metabolites, enzymes, and reactions that support life has enabled a number of advances in mechanistic modeling, particularly in prokaryotes and lower eukaryotes [55]. One focus in particular has been the use of techniques such as flux balance analysis (FBA) to model the phenotypic effect of single and combinatorial genetic perturbations on network flux and biomass yield [56,57]. In a landmark study, the FBA framework was used to measure yeast genetic interactions in silico based on predicted maximal rates of biomass production upon deletion of individual or pairs of genes [32]. Applying a multiplicative model for the expected combination of fitness defects, they estimated quantitative genetic interactions, both positive and negative, for all gene pairs involved in metabolism. Notably, in silico genetic interactions were highly modular in that genes involved in the same metabolic pathways (e.g., TCA cycle) tended to exhibit the same type of interaction (either positive or negative) both within and between pathways, a property coined as 'monochromaticity' [32]. Although this study preceded the results of large-scale mapping efforts that produced experimental measurements for negative and positive genetic interactions across a large fraction of the yeast genome, the predictions were in striking agreement with the experimental data. As discussed in the previous section, a large fraction of experimentally measured negative and positive genetic interactions fall into either monochromatic between- or within-pathway structures [2,36].

Methods based on mechanistic models of metabolism have also been used to predict specific genetic interactions. Szappanos et al. measured the ability of an FBA-based analysis of metabolic network models to specifically predict experimentally observed genetic interactions [58]. Based on a large collection of quantitative genetic interactions from ~200 000 double mutants in genes related to metabolism, they found that experimentally identified interactions, both positive and negative, were highly overrepresented among the interactions predicted by the in silico approach. However, in silico predictions suffered from poor sensitivity, covering only ~3% and ~12% of experimentally identified negative and positive interactions, respectively. The authors analyzed various factors to explain the low sensitivity of their modeling approach, and devised an iterative model refinement procedure that updated the underlying metabolic network model to optimize the agreement between predicted and measured genetic interactions. Interestingly, this refinement approach suggested removal of one of three NAD biosynthesis pathways in the reference yeast metabolic network. Indeed, further analysis confirmed that this particular pathway (a two-step biosynthetic route from aspartate to quinolinate) was inferred erroneously from *Escherichia coli* [58]. This study provides evidence for at least modest success of mechanistic models to predict genetic interactions in vivo, and perhaps more importantly, suggests a framework for using targeted experimental measurements to refine existing metabolic network models. Such an approach could have important applications as we begin to study genetic interactions in higher eukaryotes, where experimental technology for measuring genetic interactions is less scalable. In fact, similar metabolic network modeling approaches have recently been applied successfully to suggest new drug targets for specific cancers based on predictions of synthetic lethal interactions [59,60].

MAPPING GENETIC INTERACTIONS IN OTHER ORGANISMS

Genetic Interactions in Unicellular Organisms

Large-scale genetic interaction mapping techniques have been developed for the fission yeast *Schizosaccharomyces pombe* [61,62] and the Gram-negative bacterium *E. coli* [63,64]. These techniques are directly analogous to those employed in *S. cerevisiae*, in that they use genome-wide deletion collections and mating procedures to generate comprehensive sets of double mutants. Large-scale mapping of *S. pombe* genetic interactions identified similarities as well as significant differences in the wiring of its genetic network compared to the *S. cerevisiae* network [61, 62]. Similar to metazoans, the *S. pombe* genome encodes functional RNA interference (RNAi) machinery not found in typical *S. cerevisiae* laboratory strains, which likely account for at least some of the differences between the two fungal genetic networks [62].

Genetic Interactions in Metazoan Model Systems

Large-scale genetic interaction studies have also been described for the nematode worm *C. elegans* [65,66] and the fruit fly *D. melanogaster* [67]. These studies are made possible by organism-specific genome-wide RNAi libraries, which can be used to target individual genes and are highly compatible with high-throughput experimental approaches. A recent study combined a rigorous combinatorial RNAi experimental strategy with robust statistical modeling to systematically map pairwise genetic interactions between 93 genes encoding *Drosophila* signaling factors [67]. Rather than focusing on a single phenotype, identification of negative and positive genetic interactions was based on measurement of three distinct phenotypic features, including cell number, nuclear area, and the fluorescence intensity of stained nuclei [67]. Interestingly, these phenotypic measurements were highly complementary and uncovered rich and non-redundant sets of genetic interactions emphasizing the importance of diverse, quantitative phenotypic assays for mapping high-resolution genetic networks comprehensively [67] (see Expanding Genetic Networks: Mutant Alleles, Conditions and Phenotypes). Consistent with yeast studies, genes that shared similar function clustered together based on their genetic interaction profiles, independent of the phenotypes from which the profiles were generated. As a result, this compendium of negative and positive interactions can be used as a tool to predict *Drosophila* gene function. Indeed, comparison of genetic interaction profiles led to the discovery of a novel activator of RAS-MAPK pathway signaling, whose function is conserved from fruit flies to humans [67]. Importantly, although this study assayed genetic interactions on an intermediate scale, its potential for high-throughput application is obvious, suggesting that genome-scale metazoan genetic networks, analogous to those generated for yeast, are within grasp.

Genetic Interactions in Mammalian Model Systems

Currently, large-scale mammalian genetic interaction studies are based on genetic manipulation of cell cultures via RNAi-mediated gene inactivation. Approaches involving large RNAi libraries and comparative analyses of different cell lines have the potential to reveal genetic interactions specific to cell line mutations. Indeed, several efforts focused on specific cancer cell subtypes have discovered new synthetic lethal interactions with cancer-associated genes, each of which represents a new potential drug target. Two studies have reported new synthetic lethal interactions of KRAS, the serine/threonine kinase STK33 [68] and mitotic kinase PLK1 [69], which have obvious therapeutic relevance given the prevalence of KRAS mutations in various cancers. More recently, several groups have identified common and cancer cell-type-specific 'essential' genes in a panel of human cancer cell lines [70,71]. One of these studies assessed the essentiality of more than 11 000 genes in 102 human cancer cell lines [70]. This large-scale effort uncovered 54 genes that are specifically essential for the proliferation and viability of ovarian cells, including *PAX8*, which was identified as an ovarian lineage-specific dependency [70]. These cell-type specific interactions are presumably due to specific synthetic lethal interactions between the RNAi target and the genetic background of the tumor cell. Although the identification of relevant endogenous second site mutation(s) remains to be determined, a comprehensive understanding of the molecular vulnerabilities of different cancer cell types will undoubtedly provide a powerful roadmap to guide therapeutic approaches.

Genetic Interactions and Cancer Therapeutics

Insights derived from genetic interaction networks are directly relevant to human disease and, in particular, the development of targeted cancer therapies. The most commonly used cancer therapies involve administering high doses of radiation or toxic chemicals to the patient, which can help to suppress tumor growth but also cause substantial damage to normal cells. Synthetic lethal interactions suggest a much more targeted strategy for the treatment of cancer. The goal would be to identify a target gene that, when mutated or chemically inhibited, kills cells that harbor a second cancer-specific alteration due to a synthetic lethal interaction, but spares otherwise identical cells lacking the alteration, an idea originally proposed by Hartwell and colleagues [72]. In fact, this concept has recently been exploited in the development of poly (ADP-ribose) polymerase (PARP) inhibitors as novel chemotherapeutics for breast cancer [73,74]. Although PARP is not essential in normal cells, BRCA mutant cells are dependent on PARP for their survival. There are several PARP inhibitors in various stages of clinical trials, some with encouraging results [75].

GENETIC NETWORK CONSERVATION

Conservation of Individual Interactions between Orthologous Gene Pairs

Analysis of *S. cerevisiae* and *S. pombe* networks represents the most comprehensive comparison of genetic interaction conservation conducted to date. Although ~80% of *S. cerevisiae* essential gene orthologs are also indispensible for viability in *S. pombe* [76], two independent studies have found that most (~70%) of genetic interactions identified thus far appear to be species specific [61,62]. Significant rewiring, which may include more extensive buffering, of their genetic networks was not entirely unexpected, given that *S. pombe* and *S. cerevisiae* are separated from each other by close to 400 million years of evolution and exhibit substantial physiological differences. As a result, a ~30% overlap between the genetic networks of these two yeasts indicates that there is significant conservation of synthetic lethal genetic interactions over hundreds of years of evolution.

Analyses of individual genetic interactions between orthologous genes of yeast and worm provided weaker support for the conservation of specific interactions. Two studies found that less than 5% conservation of synthetic lethal genetic interactions identified by large-scale studies in *S. cerevisiae* were conserved in *C. elegans* [65,66]. On the other hand, a smaller study, employing very detailed quantification of worm mitotic spindle morphology, detected moderate but significant (~29%) conservation of genetic interactions between *S. cerevisiae* and *C. elegans*. There are a number of facts that may explain the low level of conservation estimated from the high- vs. low-throughput studies in *C. elegans* [77]. One possibility to account for the higher degree of conservation observed between different yeasts vs. the *S. cerevisiae*-metazoan comparison is that certain genetic interactions detectable in single-celled yeasts are likely to be conserved in multicellular organisms but difficult to detect due to cellular-level redundancy, tissue-specific or altered functions. It may be possible to detect these interactions using alternative phenotypic readouts that assay anatomical, developmental, or behavioral phenotypes. Indeed, a higher degree of genetic interaction conservation was observed between *S. cerevisiae* and *C. elegans*, at least with respect to genes involved in chromosome biology, when post-embryonic RNAi and the *C. elegans* vulval cell linage were used to measure somatic cell proliferation defects [78].

Although systematic comparative analyses involving mammalian genetic interaction networks are not yet feasible, evidence has shown that synthetic lethal interactions identified in lower eukaryotes, and involving highly conserved genes implicated in fundamental processes such as DNA synthesis and repair, can guide the selection of interactions that can be exploited to kill cancer cells in mammals. For example, a synthetic lethal genetic interaction between the yeast genes *RAD54* and *RAD27* was recapitulated in a RAD54B$^{-/-}$ human colorectal cancer cell line by shRNA-mediated targeting of the *RAD27* ortholog, *FEN1* [79].

Conservation of Genetic Network Structure and Topology

The studies described above suggest that genetic interactions can be conserved from yeast to higher organisms; however, the extent of this conservation remains unclear [80]. Despite this uncertainty, it is possible that genetic interactions are more generally conserved at the level of network structure and topology. Indeed, examination of the global yeast genetic network showed that, like other biological networks [81], most genes are sparsely connected, whereas a small number have many interactions and serve as network 'hubs' [2]. Whereas most genetic interactions occur between genes involved in the same biological process, network hubs tend to be pleiotropic and interact with many functionally diverse sets of genes [2]. Importantly, genes annotated to chromatin/transcription showed a significant number of genetic interactions with numerous different processes, indicating that genes involved in these functions are important for mediating cross-process connections in the genetic network [2]. Interestingly, chromatin and transcription-related genes were shown to have similar properties in the *C. elegans* genetic network [65]. The discovery of a central and hence highly pleiotropic role for chromatin- and transcription-related genes in both the *S. cerevisiae* and

C. elegans genetic networks provides evidence suggesting that network structure and topology may be conserved across organisms [2]. Therefore, these hubs may represent general buffers of phenotypic variation because they are capable of enhancing the phenotypic consequences associated with mutations in numerous different genes. This finding emphasizes the importance of identifying genes capable of multiple genetic interactions, since, in theory, genetic network hubs may have the potential to act as general modifiers of genetic diseases in humans [65].

Furthermore, comparative analysis of functional networks derived from *S. cerevisiae*, *S. pombe*, *C. elegans* and *D. melanogaster* demonstrated that conservation of biological interactions is maintained between species, but mainly at the module level rather than at the level of individual genes or proteins [82]. Another study noted similar module-level conservation when comparing chemical genetic interactions between *S. cerevisiae* and *S. pombe* [83], providing additional evidence to support conservation of general genetic network properties and suggesting that a complete yeast genetic network may serve as a template to guide experimental and computational analysis, as well as predicting genetic interaction hubs in complex organisms where genome-wide combinatorial perturbation analysis is more technically challenging.

EXPANDING GENETIC NETWORKS: MUTANT ALLELES, CONDITIONS AND PHENOTYPES

Conditional Alleles and Essential Genetic Interactions

Most yeast genetic interaction studies conducted thus far have focused on non-essential gene deletion alleles. However, a truly comprehensive and genome-wide genetic network must also include interactions involving essential genes. Mapping of essential genetic interactions is possible through the development of several strain collections where subsets of the ~1000 essential yeast genes are individually altered to produce either conditional alleles [84–86] or hypomorphic alleles that are compatible with viability [87]. Studies combining SGA analysis with a conditional allele collection showed that essential genes also act as hubs in the genetic network, engaging in synthetic lethal interactions five times more often than non-essential genes [22]. Another study emphasized the importance of essential genes for deciphering the complex relationship between genetic and physical interaction networks [17]. Essential genes tend to be highly conserved between species at the level of both sequence similarity [88] and fitness [76,80] and, as a result, continued analysis of essential gene interactions will provide a key resource for future comparative studies to assess the extent of genetic network conservation.

Gain-of-Function Alleles

Systematic analysis of deletion mutant alleles, both individually and in combination, have made an enormous impact on gene function discovery. However, while loss-of-function genetic analyses identify functionally coherent gene modules and the connections between them, they are not always sufficient for elucidating pathway architecture. Historically, dominant gain-of-function (GOF) mutations have provided an incredibly powerful means for determining gene position within a regulatory cascade ([89] for example). The availability of low- and high-copy plasmid libraries in which the expression of every yeast open reading frame (ORF) is controlled by the endogenous [90–92] or an inducible promoter [93–95] has enabled systematic and genome-wide investigation of GOF and gene dosage effects in wild-type and mutant strain backgrounds to complement previous genetic interaction studies [96,97].

Recently, a barcoded high-copy plasmid library was developed and screened for dosage suppression. Dosage suppressors of 41 temperature-sensitive different alleles of essential genes were identified [92]. An average of ~5 different genes were found to suppress the temperature-sensitive phenotype of each essential gene mutant, suggesting that, albeit more rare than LOF interactions such as synthetic lethality, dosage suppression is also a common genetic interaction. Furthermore, while they tend to connect functionally related genes, dosage suppression interactions often overlap with negative genetic interactions and physical interactions that occur within the essential pathway of the query gene; however, most interactions are novel, suggesting that dosage suppression represents a prevalent and distinct type of interaction that is rich in novel functional information [92]. Dosage suppression can identify pathway components that act downstream of the query gene, and therefore this type of genetic interaction offers the potential to decipher gene order and the directionality of biological circuits. Ultimately, we anticipate that a global dosage suppression map will be a major contributor to the construction of a complete and high-resolution cellular landscape comprising all types of genetic and physical interaction.

Condition-Specific Genetic Interactions

Although the majority of yeast genes are not required for viability under standard laboratory conditions, a large-scale survey of fitness demonstrated that nearly all yeast deletion mutants (97%) exhibit a growth defect in at least one chemical or environmental stress condition [98]. Furthermore, while the average gene shares a negative genetic interaction with ~2% of ORFs in the yeast genome, subsets of genes, such as those involved in metabolic processes, are statistically under-represented in the genetic interaction network [2]. These findings suggest that the lack of severe fitness defect associated with most genes cannot be explained entirely by functional redundancy between non-essential gene pairs. A comprehensive understanding of gene function and genetic interaction networks therefore requires further analysis in a variety of conditions that depend on the activity of otherwise dispensable genes.

One of the first studies to explore condition-specific genetic interactions in a quantitative manner examined all possible pairwise mutant combinations of a panel of 26 genes in the absence and presence of the DNA damaging agent methyl methanesulfonate (MMS) [19]. Although genes implicated in DNA replication and repair pathways tend to show many genetic interactions under standard laboratory conditions [2], exposure to MMS uncovered twice as many interactions, highlighting the potential of condition-specific genetic networks [19]. More recently, condition-specific genetic interaction screens have been extended to larger sets of DNA replication and repair genes [99]. Bandyopadhyay and co-authors analyzed all pairwise genetic interactions among 418 yeast genes, covering DNA repair proteins as well as a biased set of signaling molecules and transcriptional factors. Similarly to previous studies [19], a significant increase in genetic interaction density was observed in response to MMS treatment [99]. Substantial rewiring of the genetic network was also observed when cells were chemically challenged, indicating that, in addition to uncovering new interactions, condition-specific studies also reveal changes in network structure and topology [99]. Exploring genetic network dynamics in response to different experimental conditions will further the functional characterization of the cell as well as our understanding of how cells adapt to cope with environmental stress.

Quantitative Phenotypes to Measure Genetic Interactions

Owing to ease of measurement and amenability to high-throughput applications, most genome-scale studies conducted to date have focused on cell fitness as the phenotypic readout for genetic interactions [2,3,25,27]. However, it is becoming increasingly evident that more subtle and specific phenotypes harbor as much functional information as fitness. For example, nearly 50% of non-essential yeast deletion mutants, despite having no fitness defect, exhibit a number of morphological defects [100]. This finding suggests that investigation of complex phenotypes will uncover functional connections between genes that may not be apparent from growth data alone.

A variety of quantitative phenotypic assays have been developed to map the genetic networks that underlie specific biological processes, including yeast filamentous growth [20] as well as protein kinase A [101] and RAS signaling [67]. One study investigated the mechanisms of receptor endocytosis by measuring the efficiency of protein internalization in an array of single and double mutants [102]. Internalization efficiency was quantified using an enzymatic assay based on the cell surface localization of Snc1, the yeast vesicle-associated membrane protein (VAMP)/synaptobrevin homologue. Quantitative negative and positive genetic interactions derived from Snc1 localization signals were used to cluster genes into functional modules identifying 20 novel genes involved in Snc1 uptake at the plasma membrane [102]. Using SGA methodologies to introduce a green-fluorescent protein (GFP)-reporter gene into single and double deletion mutants, another group examined genetic interactions among genes implicated in the yeast unfolded protein response (UPR) [103]. Finally, fluorescent reporters have also been introduced into the yeast deletion collection using SGA to examine cell cycle-dependent transcription on a genome-wide scale [104,105].

The spectrum of phenotypic traits amenable to genome-scale mapping of genetic networks has expanded immeasurably with technological advances in high-throughput microscopy and imaging tools [106]. By combining cytological reporters with high-content screening it is now possible to classify and measure a variety of morphological and protein localization phenotypes in a large-scale manner [100,107]. For example, morphological data pertaining to the cell wall, actin cytoskeleton, and nuclear DNA have been systematically collected and analyzed for the entire set of *S. cerevisiae* non-essential gene deletion mutants [100]. Similar approaches have been applied in mammalian cells to characterize genes regulating various cellular processes, including cell morphology and cell cycle progression [108].

Combining high content screening with SGA (Box 6.2) enabled genome-wide screening of single and double mutants for phenotypic defects associated with mitotic spindle morphogenesis [109]. This endeavor generated a ~fourfold increase over the fitness-based genetic interaction network derived for the microtubule-binding protein Bim1 by identifying 100 novel genetic interactions that impinge on spindle morphology [109]. Thus, integrating high-throughput technologies for combining mutations, such as SGA, with diverse and quantitative phenotypic assays should lead to the construction of high-resolution networks that provide comprehensive genome coverage and the potential to incorporate temporal and environmental influences to accurately reflect global cellular functions.

GENETIC INTERACTIONS AND GENOME-WIDE ASSOCIATION STUDIES

In the past decade much progress has been made in our understanding of genetic interactions. Enabled by developments in experimental technology, it is possible to measure genetic interactions on a global scale in several model organisms. Genetic interaction networks have proved to be a powerful tool for dissecting gene function and understanding the systems-level organization of a genome. Beyond these important goals, large-scale genetic interaction maps also have the potential to reveal key insights into one of the most fundamental questions in modern biology: how genomes specify phenotypes, a major challenge towards understanding human disease.

Given increasingly scalable and affordable technology for mapping genotypes in the human population, genome-wide association studies (GWAS) for various disease traits have become commonplace. These efforts have quickly and dramatically increased our knowledge of genome variants associated with various human traits or disorders, to date producing nearly 1500 established associations (NIH GWAS Catalog, http://www.genome.gov/26525384). Despite this success, there are very few diseases, particularly among those that commonly afflict the population, for which the discovered variants are able to explain a substantial portion of the heritable variance [110]. For example, there are now 95 variants associated with LDL and HDL cholesterol levels, but the combination only explains 20–25% of the variation known to be heritable [110]. There have been a number of reasons proposed for this discrepancy, including the presence of rare variants, which may not be queried on current genotyping platforms, or the presence of numerous small effect variants that are below the level of detection given the sample size of a typical study [110,111]. Another potential explanation that we would like to consider is the influence of genetic interactions between variants [110,111].

Indeed, in a comparative study of closely related and interbreeding yeast strains, Dowell et al. [112] examined all genes for those that exhibit conditional essentiality, such that deletion of a specific gene is essential in one strain (individual) but not another. One possible explanation for conditional essentiality was that it results from a synthetic lethal interaction between two genes: the deleted gene and another containing natural variation that leads to a strain-specific LOF allele. However, in all cases of conditional essentiality examined, it appeared that multiple modifier loci were necessary to confer strain-specific essentiality [112]. Thus, genetic networks involving natural variation may be highly complex and often involve more than two genes.

The potential for genetic interactions to underlie a significant proportion of inherited phenotypes would cause an inflation in estimates of the total heritable variation, because current models used in such calculations assume a simple additive model [23]. Thus, the singly associated variants may represent the set of all causal variants, but interaction effects among them are also necessary to explain the missing variation [23]. This may suggest we have actually captured variants that explain more heritable variation than previously thought, but more work is required to define the model by which they influence phenotypic variation. Based on the large-scale reverse genetic screens in model organisms discussed in this review, genetic interactions are sufficiently common to support the plausibility of this scenario. For example, among the 5.4 million double mutants constructed in the latest yeast study, more than 3% exhibit detectable positive or negative genetic interactions, and the rate is substantially higher among genes that also exhibit fitness defects as single mutants [2].

Addressing the role of genetic interactions in the unexplained heritability of GWAS is challenging because these interactions are difficult to detect, given the parameters of most current association studies. In principle, each pair of variants genotyped in a particular study can be tested for interaction effects, and, in fact, there have been many recent studies that take this approach [113–115]. In general, the statistical power of such pairwise tests depends on several factors, including the allele frequency, population structure, and sample size, but under reasonable assumptions in a typical GWAS setting, recent estimates have placed the required sample size near 500 000 individuals to realistically detect such interactions [23], suggesting that this may not currently be a practical question without additional constraints.

One approach is to place additional constraints by only considering a subset of locus pairs meeting some statistical threshold, or more generally, requiring some other knowledge of association between the two genes. In some studies, the set of genomic loci tested for interaction effects was limited to the subset that showed significant association as individual factors [116]. Other studies have proposed to use literature-based or experimentally derived networks to filter the set of candidate pairs to be tested [117]. In a related study in yeast, Hannum et al. analyzed the structure of pairwise interactions derived from analysis of eQTL traits in a population of 112 segregants derived from an *S. cerevisiae* laboratory strain crossed with a wild isolate [118]. They leveraged the fact that they expected such interactions to exhibit modular structure relative to known protein complexes and pathways, as had been observed in genetic interactions derived from reverse genetic mapping approaches. They indeed found evidence that some pairwise interactions exhibited the same between-complex/between-pathway structure observed in reverse genetic interaction networks [118]. However, they observed little overlap between the genetic interactions derived from statistical analysis of eQTLs and those experimentally measured from constructing double mutants, which either suggests the types of interactions influencing expression variation are quite different from fitness-based reverse genetic screens or potentially reflects the statistical challenges associated with such an approach. The general strategy of leveraging genetic interactions derived from reverse genetic mapping approaches to narrow the search space for interactions in an appropriately matched GWAS context may be a fruitful approach, particularly as reverse genetic screens become more scalable in complex eukaryotes, including human cell lines.

ACKNOWLEDGEMENTS

CB, CLM and BA are supported by a grant from the National Institutes of Health (1R01HG005853). CB and BA are supported by grants from the Canadian Institutes of Health Research (MOP-102629)(MOP-97939) and the Ontario Research Fund (GL2-01-22). CB is supported by grants from the Canadian Institutes of Health Research (MOP-57830) and the Natural Sciences and Engineering Council of Canada (RGPIN 204899-06). BA is supported by a grant from the Canadian Institutes of Health Research (MOP-11206). CLM is supported by grants from the National Institutes of Health (1R01HG005084-01A1) and the National Science Foundation (DBI 0953881). CLM is partially supported by funding from the University of Minnesota Biomedical Informatics and Computational Biology program, and a seed grant from the Minnesota Supercomputing Institute.

REFERENCES

[1] Waddington CH. The Strategy of the Gene. Allen and Unwin, London 1957.
[2] Costanzo M, et al. The genetic landscape of a cell. Science 2010;327:425–31.
[3] Tong AH, et al. Global mapping of the yeast genetic interaction network. Science 2004;303:808–13.
[4] Tong AH, et al. Systematic genetic analysis with ordered arrays of yeast deletion mutants. Science 2001;294:2364–8.
[5] Giaever G, et al. Functional profiling of the *Saccharomyces cerevisiae* genome. Nature 2002;418:387–91.
[6] Winzeler EA, et al. Functional characterization of the *S. cerevisiae* genome by gene deletion and parallel analysis. Science 1999;285:901–6.
[7] Hartman JLt, Garvik B, Hartwell L, Principles for the buffering of genetic variation. Science 2001;291:1001–4.
[8] Bateson W. Mendel's Priciples of Heredity. Cambridge University Press, Cambridge;1909.
[9] Fisher RA. The correlation between relatives on the supposition of Mendelian inheritance. Proc R Soc Edinburgh 1918;52: 399–433.
[10] Phillips PC, Otto SP, Whitlock MC. Beyond the average: the evolutionary importance of gene interactions and variability of epistatic effects. In: Wolf JB, Brodie ED, Wade MJ, editors. Epistasis and the Evolutionary Process. New York: Oxford University Press Inc.; 2000. p. 20–40.
[11] Mani R, St Onge RP. J.L. t. Hartman, G. Giaever, F.P. Roth, Defining genetic interaction. Proc Natl Acad Sci U S A 2008;105:3461–6.
[12] Dobzhansky T. Genetics of natural poulations. Xiii. Recombination and variability in populations of *Drosophila pseudoobscura*. Genetics 1946;31:269–90.
[13] Novick P, Botstein D. Phenotypic analysis of temperature-sensitive yeast actin mutants. Cell 1985;40:405–16.
[14] Guarente L. Synthetic enhancement in gene interaction: a genetic tool come of age. Trends Genet 1993;9:362.

[15] Bender A, Pringle JR. Use of a screen for synthetic lethal and multicopy suppressor mutants to identify two new genes involved in morphogenesis in *Saccharomyces cerevisiae*. Mol Cell Biol 1991;11:1295–305.
[16] Boone C, Bussey H, Andrews BJ. Exploring genetic interactions and networks with yeast. Nat Rev Genet 2007;8:437–49.
[17] Baryshnikova A, et al. Quantitative analysis of fitness and genetic interactions in yeast on a genome-wide scale. Nat Methods 2010;7:1017–24.
[18] Bandyopadhyay S, Kelley R, Krogan NJ, Ideker T. Functional maps of protein complexes from quantitative genetic interaction data. PLoS Comput Biol 2008;4:e1000065.
[19] St. Onge RP, et al. Systematic pathway analysis using high-resolution fitness profiling of combinatorial gene deletions. Nat Genet 2007;39:199–206.
[20] Drees BL, et al. Derivation of genetic interaction networks from quantitative phenotype data. Genome Biology 2005;6:R38.
[21] Breslow DK, et al. A comprehensive strategy enabling high-resolution functional analysis of the yeast genome. Nat Methods 2008;5:711–8.
[22] Davierwala AP, et al. The synthetic genetic interaction spectrum of essential genes. Nat Genet 2005;37:1147–52.
[23] Zuk O, Hechter E, Sunyaev SR, Lander ES. The mystery of missing heritability: Genetic interactions create phantom heritability. P Natl Acad Sci USA 2012;109:1193.
[24] Pan X, et al. A robust toolkit for functional profiling of the yeast genome. Mol Cell 2004;16:487–96.
[25] Pan X, et al. A DNA integrity network in the yeast *Saccharomyces cerevisiae*. Cell 2006;124:1069–81.
[26] Lin YY, et al. Protein acetylation microarray reveals that NuA4 controls key metabolic target regulating gluconeogenesis. Cell 2009;136:1073–84.
[27] Decourty L, et al. Linking functionally related genes by sensitive and quantitative characterization of genetic interaction profiles. Proc Natl Acad Sci U S A 2008;105:5821.
[28] DeLuna A, et al. Exposing the fitness contribution of duplicated genes. Nat Genet 2008;40:676–81.
[29] Costanzo M, Baryshnikova A, Myers CL, Andrews B, Boone C. Charting the genetic interaction map of a cell. Curr Opin Biotechnol 2011;22:66–74.
[30] Collins SR, Schuldiner M, Krogan NJ, Weissman JS. A strategy for extracting and analyzing large-scale quantitative epistatic interaction data. Genome Biol 2006;7:R63.
[31] Eisen MB, Spellman PT, Brown PO, Botstein D. Cluster analysis and display of genome-wide expression patterns. Proc Natl Acad Sci U S A 1998;95:14863.
[32] Segre D, Deluna A, Church GM, Kishony R. Modular epistasis in yeast metabolism. Nat Genet 2005;37:77–83.
[33] Leidel S, et al. Ubiquitin-related modifier Urm1 acts as a sulphur carrier in thiolation of eukaryotic transfer RNA. Nature 2009; 458:228–32.
[34] Kelley R, Ideker T. Systematic interpretation of genetic interactions using protein networks. Nat Biotechnol 2005;23: 561–6.
[35] Ulitsky I, Shlomi T, Kupiec M, Shamir R. From E-MAPs to module maps: dissecting quantitative genetic interactions using physical interactions. Mol Syst Biol 2008;4:209.

[36] Bellay J, et al. Putting genetic interactions in context through a global modular decomposition. Genome Res 2011;21:1375–87.

[37] Collins SR, et al. Functional dissection of protein complexes involved in yeast chromosome biology using a genetic interaction map. Nature 2007;446:806–10.

[38] Goudreault M, et al. A PP2A phosphatase high density interaction network identifies a novel striatin-interacting phosphatase and kinase complex linked to the cerebral cavernous malformation 3 (CCM3) protein. Mol Cell Proteomics 2009;8:157–171.

[39] Mulugu S, et al. A conserved family of enzymes that phosphorylate inositol hexakisphosphate. Science 2007;316:106–9.

[40] Ohno S. Evolution by Gene Duplication. Springer-Verlag, Berlin 1970.

[41] Byrne KP, Wolfe KH. The Yeast Gene Order Browser: combining curated homology and syntenic context reveals gene fate in polyploid species. Genome Res 2005;15:1456–61.

[42] Kellis M, Birren BW, Lander ES. Proof and evolutionary analysis of ancient genome duplication in the yeast *Saccharomyces cerevisiae*. Nature 2004;428:617–24.

[43] Ihmels J, Collins SR, Schuldiner M, Krogan NJ, Weissman JS. Backup without redundancy: genetic interactions reveal the cost of duplicate gene loss. Mol Syst Biol 2007;3:86.

[44] Musso G, et al. The extensive and condition-dependent nature of epistasis among whole-genome duplicates in yeast. Genome Res 2008;18(7):1092–9.

[45] Dean EJ, Davis JC, Davis RW, Petrov DA. Pervasive and persistent redundancy among duplicated genes in yeast. PLoS Genet 2008;4:e1000113.

[46] VanderSluis B, et al. Genetic interactions reveal the evolutionary trajectories of duplicate genes. Mol Syst Biol 2010;6:429.

[47] Wagner A. Asymmetric functional divergence of duplicate genes in yeast. Mol Biol Evol 2002;19:1760–8.

[48] Conant GC, Wagner A. Asymmetric sequence divergence of duplicate genes. Genome Res 2003;13:2052–8.

[49] Gavin AC, et al. Proteome survey reveals modularity of the yeast cell machinery. Nature 2006;440:631–6.

[50] Krogan NJ, et al. Global landscape of protein complexes in the yeast *Saccharomyces cerevisiae*. Nature 2006;440:637–43.

[51] Yu H, et al. High-quality binary protein interaction map of the yeast interactome network. Science 2008;322:104–10.

[52] Tarassov K, et al. An in vivo map of the yeast protein interactome. Science 2008;320:1465–70.

[53] Harbison CT, et al. Transcriptional regulatory code of a eukaryotic genome. Nature 2004;431:99–104.

[54] Rhee HS, Pugh BF. Comprehensive genome-wide protein-DNA interactions detected at single-nucleotide resolution. Cell 2011; 147:1408–19.

[55] Feist AM, Herrgard MJ, Thiele I, Reed JL, Palsson BO. Reconstruction of biochemical networks in microorganisms. Nat Rev Microbiol 2009;7:129–43.

[56] Oberhardt MA, Palsson BO, Papin JA. Applications of genome-scale metabolic reconstructions. Mol Syst Biol 2009;5:320.

[57] Price ND, Reed JL, Palsson BO. Genome-scale models of microbial cells: evaluating the consequences of constraints. Nat Rev Microbiol 2004;2:886–97.

[58] Szappanos B, et al. An integrated approach to characterize genetic interaction networks in yeast metabolism. Nat Genet 2011;43: 656–62.

[59] Folger O, et al. Predicting selective drug targets in cancer through metabolic networks. Mol Syst Biol 2011;7:501.

[60] Frezza C, et al. Haem oxygenase is synthetically lethal with the tumour suppressor fumarate hydratase. Nature 2011;477:225–8.

[61] Dixon SJ, et al. Significant conservation of synthetic lethal genetic interaction networks between distantly related eukaryotes. Proc Natl Acad Sci U S A 2008;105:16653.

[62] Roguev A, et al. Conservation and rewiring of functional modules revealed by an epistasis map in fission yeast. Science 2008; 322:405–10.

[63] Butland G, et al. eSGA: *E. coli* synthetic genetic array analysis. Nat Methods 2008;5:789–95.

[64] Typas A, et al. High-throughput, quantitative analyses of genetic interactions in *E. coli*. Nat Methods 2008;5:781–7.

[65] Lehner B, Crombie C, Tischler J, Fortunato A, Fraser AG. Systematic mapping of genetic interactions in *Caenorhabditis elegans* identifies common modifiers of diverse signaling pathways. Nat Genet 2006;38:896–903.

[66] Byrne AB, et al. A global analysis of genetic interactions in *Caenorhabditis elegans*. J Biol 2007;6:8.

[67] Horn T, et al. Mapping of signaling networks through synthetic genetic interaction analysis by RNAi. Nat Methods 2011;8:341–6.

[68] Scholl C, et al. Synthetic lethal interaction between oncogenic KRAS dependency and STK33 suppression in human cancer cells. Cell 2009;137:821–34.

[69] Luo J, et al. A genome-wide RNAi screen identifies multiple synthetic lethal interactions with the Ras oncogene. Cell 2009; 137:835–48.

[70] Cheung HW, et al. Systematic investigation of genetic vulnerabilities across cancer cell lines reveals lineage-specific dependencies in ovarian cancer. P Natl Acad Sci USA 2011;108:12372.

[71] Marcotte R, et al. Essential gene profiles in breast, pancreatic, and ovarian cancer cells. Cancer Discovery 2012;2:172–89.

[72] Hartwell LH, Szankasi P, Roberts CJ, Murray AW, Friend SH. Integrating genetic approaches into the discovery of anticancer drugs. Science 1997;278:1064–8.

[73] Farmer H, et al. Targeting the DNA repair defect in BRCA mutant cells as a therapeutic strategy. Nature 2005;434:917–21.

[74] Bryant HE, et al. Specific killing of BRCA2-deficient tumours with inhibitors of poly(ADP-ribose) polymerase. Nature 2005; 434:913–7.

[75] O'Shaughnessy J, et al. Iniparib plus chemotherapy in metastatic triple-negative breast cancer. New Engl J Med 2011;364:205.

[76] Kim DU, et al. Analysis of a genome-wide set of gene deletions in the fission yeast *Schizosaccharomyces pombe*. Nat Biotechnol 2010;28:617–23.

[77] Tarailo M, Tarailo S, Rose AM. Synthetic lethal interactions identify phenotypic 'interologs' of the spindle assembly checkpoint components. Genetics 2007;177:2525–30.

[78] McLellan J, et al. Synthetic lethal genetic interactions that decrease somatic cell proliferation in *Caenorhabditis elegans* identify the alternative RFC CTF18 as a candidate cancer drug target. Mol Biol Cell 2009;20:5306–13.

[79] McManus KJ, Barrett IJ, Nouhi Y, Hieter P. Specific synthetic lethal killing of RAD54B-deficient human colorectal cancer cells by FEN1 silencing. Proc Natl Acad Sci U S A 2009;106:3276.

[80] Tischler J, Lehner B, Fraser AG. Evolutionary plasticity of genetic interaction networks. Nat Genet 2008;40:390–1.

[81] Barabasi AL, Oltvai ZN. Network biology: understanding the cell's functional organization. Nat Rev Genet 2004;5:101–13.

[82] Zinman GE, Zhong S, Bar-Joseph Z. Biological interaction networks are conserved at the module level. BMC Syst Biol 2011;5:134.

[83] Kapitzky L, et al. Cross-species chemogenomic profiling reveals evolutionarily conserved drug mode of action. Mol Syst Biol 2010;6:451.

[84] Ben-Aroya S, et al. Toward a comprehensive temperature-sensitive mutant repository of the essential genes of *Saccharomyces cerevisiae*. Mol Cell 2008;30:248–58.

[85] Mnaimneh S, et al. Exploration of essential gene functions via titratable promoter alleles. Cell 2004;118:31–44.

[86] Li Z, et al. Systematic exploration of essential yeast gene function with temperature-sensitive mutants. Nat Biotechnol 2011;29(4):361–7.

[87] Schuldiner M, et al. Exploration of the function and organization of the yeast early secretory pathway through an epistatic miniarray profile. Cell 2005;123:507–19.

[88] Pena-Castillo L, Hughes TR. Why are there still over 1000 uncharacterized yeast genes? Genetics 2007;176:7–14.

[89] Stevenson BJ, Rhodes N, Errede B, Sprague GF. Constitutive mutants of the protein kinase STE11 activate the yeast pheromone response pathway in the absence of the G protein. Gene Dev 1992;6:1293–304.

[90] Jones GM, et al. A systematic library for comprehensive overexpression screens in *Saccharomyces cerevisiae*. Nat Methods 2008;5:239–41.

[91] Ho CH, et al. A molecular barcoded yeast ORF library enables mode-of-action analysis of bioactive compounds. Nat Biotechnol 2009;27:369–77.

[92] Magtanong L, et al. Dosage suppression genetic interaction networks enhance functional wiring diagrams of the cell. Nat Biotechnol 2011;29:505.

[93] Gelperin DM, et al. Biochemical and genetic analysis of the yeast proteome with a movable ORF collection. Gene Dev 2005;19:2816–26.

[94] Hu Y, et al. Approaching a complete repository of sequence-verified protein-encoding clones for *Saccharomyces cerevisiae*. Genome Res 2007;17:536–43.

[95] Zhu H, et al. Global analysis of protein activities using proteome chips. Science 2001;293:2101–5.

[96] Sopko R, et al. Mapping pathways and phenotypes by systematic gene overexpression. Mol Cell 2006;21:319.

[97] Sharifpoor S, et al. Functional wiring of the yeast kinome revealed by global analysis of genetic network motifs. Genome Res 2012;22:791–801.

[98] Hillenmeyer ME, et al. The chemical genomic portrait of yeast: uncovering a phenotype for all genes. Science 2008;320:362–5.

[99] Bandyopadhyay S, et al. Rewiring of genetic networks in response to DNA damage. Science 2010;330:1385–9.

[100] Ohya Y, et al. High-dimensional and large-scale phenotyping of yeast mutants. PNAS 2005;102:19015.

[101] Van Driessche N, et al. Epistasis analysis with global transcriptional phenotypes. Nat Genet 2005;37:471–7.

[102] Burston HE, et al. Regulators of yeast endocytosis identified by systematic quantitative analysis. J Cell Biol 2009;185:1097–110.

[103] Jonikas MC, et al. Comprehensive characterization of genes required for protein folding in the endoplasmic reticulum. Science 2009;323:1693–7.

[104] Fillingham J, et al. Two-color cell array screen reveals interdependent roles for histone chaperones and a chromatin boundary regulator in histone gene repression. Mol Cell 2009;35:340–51.

[105] Costanzo M, et al. CDK activity antagonizes Whi5, an inhibitor of G1/S transcription in yeast. Cell 2004;117:899–913.

[106] Vizeacoumar FJ, Chong Y, Boone C, Andrews BJ. A picture is worth a thousand words: genomics to phenomics in the yeast *Saccharomyces cerevisiae*. FEBS Lett 2009;583:1656–61.

[107] Huh WK, et al. Global analysis of protein localization in budding yeast. Nature 2003;425:686–91.

[108] Moffat J, et al. A lentiviral RNAi library for human and mouse genes applied to an arrayed viral high-content screen. Cell 2006;124:1283–98.

[109] Vizeacoumar FJ, et al. Integrating high-throughput genetic interaction mapping and high-content screening to explore yeast spindle morphogenesis. J Cell Biol 2010;188:69–81.

[110] Lander ES. Initial impact of the sequencing of the human genome. Nature 2011;470:187–97.

[111] Manolio TA. Cohort studies and the genetics of complex disease. Nat. Genet. 2009;41:5–6.

[112] Dowell RD, et al. Genotype to phenotype: a complex problem. Science 2010;328:469.

[113] Hahn LW, Ritchie MD, Moore JH. Multifactor dimensionality reduction software for detecting gene–gene and gene–environment interactions. Bioinformatics 2003;19:376–82.

[114] Bush WS, Dudek SM, Ritchie MD. Parallel multifactor dimensionality reduction: a tool for the large-scale analysis of gene–gene interactions. Bioinformatics 2006;22:2173–4.

[115] Steffens M, et al. Feasible and successful: genome-wide interaction analysis involving all 1.9 x 10(11) pair-wise interaction tests. Hum Hered 2010;69:268–84.

[116] Evans DM, Marchini J, Morris AP, Cardon LR. Two-stage two-locus models in genome-wide association. PLoS Genet 2006;2:e157.

[117] Ritchie MD. Using biological knowledge to uncover the mystery in the search for epistasis in genome-wide association studies. Ann Hum Genet 2011;75:172–82.

[118] Hannum G, et al. Genome-wide association data reveal a global map of genetic interactions among protein complexes. PLoS Genet 2009;5:e1000782.

[119] Baryshnikova A, et al. Synthetic genetic array (SGA) analysis in *Saccharomyces cerevisiae* and *Schizosaccharomyces pombe*. Methods Enzymol 2010;470:146–80.

[120] Dixon SJ, Costanzo M, Baryshnikova A, Andrews B, Boone C. Systematic mapping of genetic interaction networks. Annu Rev Genet 2009;43:601–25.

Chapter 7

The Spatial Architecture of Chromosomes

Job Dekker[1] and Bas van Steensel[2]

[1]Program in Systems Biology and Program in Gene Function and Expression, Department of Biochemistry and Molecular Pharmacology, University of Massachusetts Medical School, Worcester, MA, USA, [2]Division of Gene Regulation, Netherlands Cancer Institute, Amsterdam, the Netherlands

Chapter Outline

Introduction	137
The Basic Material: the Chromatin Fiber	139
The Polymer Physics of Chromosomes	139
Persistence Length	140
Mass Density	140
Polymer States	140
The Ground State of an Unconstrained Chromosome	141
Polymer Conformation is Probabilistic	141
Nuclear Confinement and Formation of Chromosome Territories	141
Anchoring of the Genome to Fixed Scaffolds	142
Genome-Wide Techniques to Map Scaffold Interactions	142
Nuclear Lamina–Genome Interactions in Metazoans	143
Gene Attachment to Nuclear Pores	144
The Nucleolus as a Spatial Organizer	145
The Internal Organization of Chromosomes: Long-Range Interactions Along and Between Chromosomes	145
Molecular Techniques to Map Long-Range Chromatin Interactions	146
Determination of Polymer Parameters using 3C Chromatin Interaction Data	146
Co-association of Large Active and Inactive Chromosomal Domains leads to Chromosome Compartmentalization	147
Looping Interactions between Genomic Elements to Regulate Genes	147
A Dynamic Network View of Chromosome Folding	148
A Stochastic Interaction-Driven Model for Genome Folding and Nuclear Organization	148
Future Challenges	149
References	149

INTRODUCTION

A human diploid cell nucleus contains 46 chromosomes, each harboring one DNA double helix. If stretched out, these DNA molecules would each be 1.5–8 cm long, yet a typical human nucleus has a diameter of only 5–10 μm. Therefore, DNA must be extensively folded within the confines of the nucleus, resulting in roughly 10 000-fold compaction. At the same time, genes, regulatory elements and other genomic information are spread out across the 6 billion base pairs, but these elements must communicate appropriately, which often involves direct physical contacts, to ensure normal gene expression. Moreover, at every cell division all chromosomes are copied and the resulting duplicates are faithfully separated from one another. How is this remarkable topological feat accomplished?

Interphase chromatin and mitotic chromosomes were first described in the late 1800s. Using early light microscopes, Flemming was the first to observe compacted chromosomes and their dramatic changes in morphology during their formation in prophase and their apparent disappearance in interphase. Based on careful observation of the segregation patterns of chromosomes during meiosis, Sutton and Boveri subsequently concluded that chromosomes must be the carriers of the genetic information, several years before Morgan confirmed this by genetic studies in *Drosophila* and decades before the classic experiments by Avery that experimentally demonstrated that DNA is the genetic material.

For more than 100 years, microscopy remained the prime technique to study chromosome architecture and dynamics, but the dense packing of the chromatin fiber has made it difficult to resolve structural features. Electron microscopy allowed initial studies of chromosome folding, e.g., as described in early work by DuPraw [1]. A major breakthrough was the development of fluorescence in situ hybridization (FISH), which allowed the visualization of individual sequences or entire chromosomes inside the nucleus [2]. Such FISH studies have provided firm evidence that interphase chromosomes are not completely diffuse and non-randomly organized, as the early observations had suggested [3,4]. Two sets of key observations stand out. First, each chromosome occupies its own territory, with only limited intermingling (Figure 7.1A). Second, individual loci are often positioned non-randomly with respect to the nuclear periphery.

Over the past decade, several new technological developments have made it possible to study the molecular composition and the spatial architecture of chromosomes. On the one hand, new molecular mapping techniques have begun to yield genome-wide datasets that describe the localization of proteins and modifications along chromosomes as well as spatial folding of chromosomes in unprecedented detail. On the other hand, theoretical models and simulations can provide understanding of the

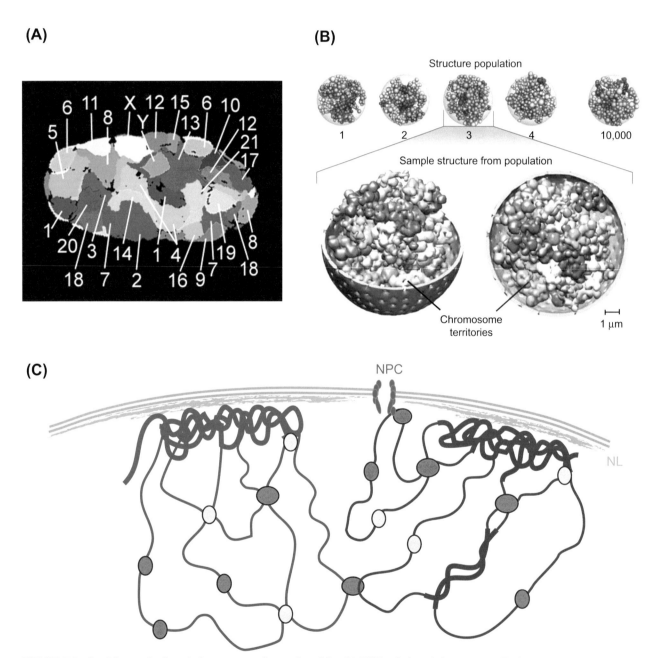

FIGURE 7.1 **Spatial organization of chromosomes: data and models.** (A) FISH painting of chromosomes. Each chromosome is labeled with a different color (image from [5], reproduced with permission). Note that the chromosomes each occupy a distinct territory. (B). Computational model of chromosome organization based on genome-wide chromatin interaction data [6]. Chromosomes are labeled in different colors. (C) Cartoon model of chromosome organization in a nucleus. Two chromosomes are indicated in purple and brown. Transcriptionally inactive chromatin (thick lines) tends to aggregate near the nuclear lamina, NL, (green); regulatory regions and active genes (thin lines) may be brought together in the nuclear interior by specific protein complexes (yellow and orange circles). NPC, nuclear pore complex. (Adapted from [7]).

fundamental physical principles that drive the folding of a chromosomal fiber [6,8,9]. With these techniques in place, it has now become possible to study the global characteristics and general mechanisms that drive chromosome folding. Thus, an approach combining biochemistry, genetics, imaging and biophysics promises to unravel one of the most fascinating topological phenomena in molecular biology. Integration of biophysical structures and dynamic spatial architectures of the genome with transcription profiles will provide a three-dimensional (3D) view of the systems biology of gene expression.

This chapter consists of several sections that each highlight a different, basic aspect of chromosome folding. First we describe the basic structure and protein composition of the chromatin fiber that makes up each chromosome. Then we discuss how fundamental principles of polymer physics drive key aspects of chromosome folding. Next, we consider how the anchoring of specific genomic loci to fixed nuclear scaffolds may constrain the topology of interphase chromosomes in relation to gene activity. Subsequently we discuss the formation of networks of intra- and inter-chromosomal contacts among linearly distant sequence elements, and their functional relevance. Finally, we will describe the three-dimensional architecture of chromosomes as self-organizing systems in which each of these processes contributes to build a functional cell nucleus.

THE BASIC MATERIAL: THE CHROMATIN FIBER

DNA and all associated proteins are collectively referred to as chromatin. In eukaryotes, most nuclear DNA is associated with specialized protein discs that consist of an octamer of histone proteins (two each of histones H2A, H2B, H3 and H4). About 145 bp of DNA is wrapped twice around a histone disc to form a nucleosome, and approximately 20–50 bp of 'linker' DNA separates two nucleosomes. This beads-on-a-string configuration, which has a thickness of about 10 nm, forms the first level of compaction of DNA. In vitro, without the help of other proteins, nucleosomal filaments can be induced to combine into a regularly shaped fiber of 30 nm diameter. Models of the 30 nm fiber are depicted in most standard molecular biology textbooks, but recent studies indicate that its occurrence in vivo may in fact be rare and limited to specialized cell types, such as sperm from sea urchin and erythrocytes from chicken [10,11].

Histones not only play a structural role, they also carry a range of post-translational modifications, such as methylation, acetylation, phosphorylation, ubiquitylation and several others. Some of these modifications alter the net charge of the histone particle, thereby altering the tightness of the histone–DNA interactions. Many others are recognized and bound by specific proteins. These histone modifications may therefore be regarded as signals that contribute to the local recruitment of proteins with particular structural or functional roles.

A rough estimate suggests that for each nucleosome there may be ~30 other protein molecules that make up chromatin [12]. One may therefore envisage the chromatin fiber as a nucleosomal array coated by a substantial layer of other proteins. It is likely that several thousands of distinct proteins contribute to this layer. Most of these proteins associate with only certain parts of the genome, their specificity being determined by direct binding to specific DNA sequences, by recognition of certain histone modifications, or by interaction with other proteins.

With dozens of histone modifications and thousands of proteins contributing to chromatin, it is theoretically possible that each small segment of the genome is bound by a different combination of proteins and histone marks. However, systematic mapping surveys in several multicellular organisms indicate that this is not the case. Rather, chromosomes appear to be organized into domains of relatively homogeneous protein composition, and the number of principal chromatin types appears to be limited to fewer than 10. For example, genome-wide mapping of the binding patterns of more than 50 chromatin proteins in *Drosophila* cells identified five major types of chromatin, defined by distinct, recurrent combinations of proteins [13]. These five types form domains that sometimes extend over more than 100 kb, often including multiple neighboring genes. Of these chromatin types, two are linked to gene repression while the other three appear conducive to transcription. Extensive mapping of many histone marks in various species also points to a limited number of chromatin states [12].

In summary, eukaryotic genomes are packaged into a nucleosomal fiber, which in turn is covered by a layer of many other proteins. A limited number of combinations of proteins and histone marks define distinct chromatin states that divide the genome into segments. How these different chromatin states may relate to the spatial organization of chromosomes will be discussed below.

THE POLYMER PHYSICS OF CHROMOSOMES

The chromatin fibers that make up chromosomes are very long, flexible polymers, and it is therefore not surprising that insights from the field of polymer physics have been instrumental in describing their three-dimensional folding and dynamics. In order to understand how chromosomes are organized inside the nucleus it is important to first outline the basic polymer principles of chromosomes that determine their conformation and shape in the absence of any biological and external structural constraints. We will briefly discuss the main physical parameters that describe properties of polymers and illustrate how they apply to chromosomes.

In the absence of any external constraints there are three critical physical properties that determine the shape of a polymer, or chromatin fiber, and the average volume that it will occupy: its persistence length, the mass density and the type of polymer state [14]. The book entitled *Giant Molecules, Here, There and Everywhere* by Grosberg and Khokhlov [15] is a good introduction to this field. These three parameters will be discussed below as they define the ground state of chromosomes on which other constraints, described in detail in subsequent sections, will act.

Persistence Length

The persistence length of a polymer is a measure of its flexibility, or bendability. Owing to the stiffness of a polymer, the bending angles of two positions located very close along the length of the polymer will be correlated, and at that length scale the polymer is considered rigid and straight. The persistence length is defined as the minimal distance between two positions on the polymer for which the bending angles are no longer correlated. Positions separated by a distance along the polymer that is several times the persistence length will display completely independent bending angles, and at that scale the polymer appears fully flexible. The persistence length of a polymer has a large impact on the average volume it will occupy. This volume is often expressed as the radius of gyration, which is related to the average end-to-end distance of the polymer. In general, the larger the persistence length the stiffer the polymer and the larger the radius of gyration of a given length of polymer becomes. This also directly implies that in the case of chromosomes modulation of persistence length, e.g., by chromatin modifications or DNA-binding proteins, will lead to shrinking or expanding of the chromatin fiber.

The persistence length of naked DNA is around 150 bp, which is around 50 nm [16]. This means that a segment of less than 150 bp is rather stiff and cannot easily form a loop or circle, whereas a DNA segment of several hundreds of bp can easily fold back on itself. The persistence length of chromatin fibers is less well established and wildly different values have been reported over the years, ranging from as small as a few kb (corresponding to tens of nm of chromatin in case of a 30 nm fiber, or around 100–150 nm for 10 nm fibers) to as large as tens of kb (corresponding to several hundred nm for a 30 nm fiber, or as long as a μm for a 10 nm fiber) [17]. Recent experiments and re-evaluations of earlier work are converging on a chromatin fiber that is less compact than a 30 nm fiber, with a persistence length of around 100 nm [18]. The value of the persistence length will directly influence the ability of chromatin fibers to form chromatin loops, as described below, and is therefore of significant biological interest.

Mass Density

The mass density of a polymer is a measure of the compaction of the polymer, which is typically expressed as the amount of mass per unit of contour length. In the case of DNA or chromatin fibers this is the contour length of a given number of base pairs. The mass density of a polymer is interesting because it relates its average size to the internal organization of the polymer itself. In the case of chromatin the mass density is related to how nucleosomes are packed together to form the chromatin fiber. For instance, B-DNA has a mass density of ~3 bp per nm. A 10 nm beads-on-a-string chromatin fiber has a mass density of ~15 bp per nm, whereas a more condensed 30 nm solenoid fiber has a mass density of ~ 90 bp per nm.

Intuitively it seems that increasing the mass density, e.g., through the formation of increasingly internally condensed chromatin fibers, will result in a polymer that occupies a smaller volume, i.e., has a smaller radius of gyration. However, this is complicated by the fact that increasing the mass density can also increase the persistence length, which in turn can lead to a larger radius of gyration. Thus, in order to understand the volume of a polymer one needs to know both the mass density and its persistence length. In addition, the specific polymer state will also affect the conformation and volume of a polymer, as we will explain below.

Polymer States

A chromatin fiber with a given persistence length and a given mass density can fold into several distinct types of spatial conformation that are related to the polymer state. The type of polymer state affects the overall volume occupied by a given polymer and the spatial distance between loci as a function of their separation along the length of the polymer. Intuitively, it can be understood that the longer the polymer is, the larger the volume it will occupy; and the farther apart two loci are spaced along the polymer, the larger their average spatial distance would be. However, the precise quantitative relationships between these parameters depend on the polymer state. A recent review by Fudenberg and Mirny provides a good introduction to these polymer states [19]. Here we describe them briefly.

The best-known polymer state is that of a random coil, also referred to as a random walk. In this state, the average spatial distance between two loci (R) scales with the square root of the distance along the polymer between them (s): $R(s) \sim s^{1/2}$. Thus, when the distance along the chromosome between two loci increases fourfold, the spatial distance between them increases twofold. For a random walk the probability that two loci will contact each other, e.g., forming a chromatin loop, decreases very rapidly with increasing distance along the polymer and scales as $P(s) \sim s^{-3/2}$.

When steric and excluded volume effects are dominant, e.g., when parts of the polymer repel each other, the polymer forms a so-called swollen globule. As one might expect, in this case the polymer will tend to occupy a larger volume than a random walk. For a swollen coil the spatial distance between segments along the polymer scales as $R(s) \sim s^{0.6}$, and the contact probability scales as $P(s) \sim s^{-1.8}$ [19].

The third polymer state is the fractal globule. Although this polymer state had been proposed more than two decades ago [20,21], it has only recently been explored in more detail because it has been found to describe experimental observations of chromosome folding [9,22]. It was originally proposed to be an intermediate or non-equilibrium state formed by collapse of a polymer upon itself. Such a collapse can be solvent induced, due to some intrinsic attraction between the segments that make up the polymer, or due to some externally imposed confinement. The fractal globule state leads to a more compact polymer than a random walk or a swollen coil. Direct computer simulation of a fractal globule polymer ensemble found that the spatial distance between loci scales as $R(s) \sim s^{1/3}$, reflecting a more compact conformation than a random walk polymer of the same length. Similarly, the contact probability of pairs of loci scales only as $P(s) \sim s^{-1}$, which means that contact probability decays less rapidly with distance along the polymer than a random walk or swollen coil. The fractal globule state has additional interesting characteristics. First, in contrast to the other polymer states, it has no knots or entanglements. Second, the fractal globule is not in equilibrium, and when left for a sufficient amount of time will convert to a random walk or swollen coil.

The Ground State of an Unconstrained Chromosome

Now that we have discussed the basic parameters that determine the shape of a polymer, we can estimate how they will affect the volume of a typical otherwise unconstrained interphase chromatin fiber. As an example, we will examine a chromatin fiber with the length of human chromosome 1 (247 Mb). When we assume that this fiber adopts a mass density that is typical for a 30 nm fiber (1 kb/11 nm), the contour length of this chromosome will be around 2.7 mm. The average end-to-end distance, or radius of gyration, of this chromosome will depend on its persistence length and the polymer state. We will assume a persistence length of 100 nm, corresponding to 9 kb, which is in accordance to a number of experimental observations. Then, when the chromosome folds according to a random walk, as would be expected in the absence of any confinement, the average end-to-end distance of the chromosome will only be around 0.02 mm, and the average volume of the chromosome will be around 1280 μm^3. This is about two to three times larger than the volume of a typical human cell nucleus (~500 μm^3). When the persistence length increases by a factor of 2, the chromatin fiber becomes less flexible and will occupy a much larger volume: around 9200 μm^3, or about 20 times the volume of the cell nucleus. A reduction in mass density, e.g., by chromatin decondensation, will lead to similar increases in occupied volume. This illustrates that folding of chromosomes inside cells must be highly constrained, at the least by confinement within the relatively small volume of the nucleus.

Polymer Conformation is Probabilistic

One fundamental difference between the structure of proteins and the folding of flexible chromatin polymers is that the latter do not form reproducible three-dimensional structures: each polymer molecule will follow its own unique three-dimensional path. Thus, the conformation of a polymer solution, or the folding of chromosomes across a cell population, should be viewed in statistical terms that describe a large ensemble of different conformations of otherwise identical polymer chains. Analysis of the spatial folding of chromosomes, e.g., by FISH, also reveals significant cell-to-cell variability, with distances between loci varying widely in a population of otherwise identical cells [23]. This commonly observed phenomenon, which can be wrongly interpreted to mean that the nucleus shows limited organization, is firmly rooted in the polymer properties of the chromatin fiber. As we will describe below, additional constraints on chromosomes will greatly reduce the number of possible conformations of the chromatin fiber, driving reproducible patterns of nuclear organization.

NUCLEAR CONFINEMENT AND FORMATION OF CHROMOSOME TERRITORIES

As outlined above, chromosomes are large, and unless constrained, occupy volumes that are significantly larger than the cell nucleus. Thus, the chromosomes must somehow be confined within the nucleus. Another striking feature of nuclear organization is that chromosomes do not readily mix: they occupy distinct territories, although there appears to be a degree of intermingling where two chromosomes touch each other [24,25]. Thus, each chromosome is in reality confined to a volume that is even smaller than the nucleus. Chromosome territories may be formed in early G1 cells when individual mitotic chromosomes decondense, and are then maintained because mixing of long polymers is an extremely slow process and only occurs at timescales that are significantly longer than the length of the cell cycle, or the lifetime of a differentiated cell [26].

A chromosome can fill up a volume that is as small as the volume of the chromatin fiber itself, which in the case of our example is only ~40 µm^3, much smaller than the size of the nucleus, but quite comparable to the size of chromosome territories. Within this small volume the chromatin can fold into a fractal globule state, as has been suggested for human chromosomes, but also as a confined random walk or swollen globule, or any mix of these. In any of the conformations the internal organization of the chromosome territory would be highly variable between cells, as the chromatin fiber can follow any random path within the constrained volume. However, recent studies, described below in more detail, show that the internal organization is not random but is further determined by the association of specific sub-chromosomal domains with nuclear structures such as the nuclear envelope, and by networks of long-range interactions between loci leading to formation of chromatin loops.

The nature of the constraint that can lead to highly compacted chromatin fibers can be extrinsic or intrinsic: an example of an extrinsic confinement would be the nuclear envelope, which can put chromosomes under pressure to occupy a limited volume. An example of an intrinsic constraint is the presence of long-range looping interactions between loci along the chromosome. In the following sections we will discuss the various constraints that act on the basic conformation of chromosomes dictated by their polymer characteristics and which modulate their conformation, their internal organization and their subnuclear positioning.

ANCHORING OF THE GENOME TO FIXED SCAFFOLDS

The polymer models described above assume that chromosomes are freely mobile within the spherical confines of the nucleus. However, there is considerable evidence that this assumption is not correct. Instead, specific regions of the genome appear to be attached to extrachromosomal scaffolds. This imposes substantial constraints onto the overall mobility, folding and positioning of chromosomes. The most prominent scaffold is the nuclear envelope. The interior surface of the nuclear envelope offers a very large and heterogeneous area for the anchoring of chromosomal regions. Indeed, various proteins on the inside of the nuclear envelope have been implicated in the tethering of specific chromosomal regions. In addition, the nucleolus may serve as a docking site.

Genome-Wide Techniques to Map Scaffold Interactions

Two complementary techniques are currently available for the detection of molecular contacts between the genome and specific scaffolds. These are chromatin immunoprecipitation (ChIP) and DNA adenine methyltransferase identification (DamID) (Box 7.1, Figure 7.2). Although both methods can also be used to map genomic binding sites of chromatin proteins and DNA-binding factors, we will only discuss their application for studies of chromosome anchoring.

ChIP starts by treatment of cells with formaldehyde, which rapidly enters the nucleus and cross-links DNA to its interacting proteins. Contacts between the genome and scaffold proteins are therefore fixed. After fragmentation of the nucleus by sonication, a scaffold protein of interest is purified by immunoprecipitation. DNA sequences attached to the protein are thus co-purified and can be identified and quantified using genomic tiling arrays, or by high-throughput sequencing.

DamID is based on a very different principle. It begins with the in vivo expression of a chimeric protein consisting of the scaffold protein of interest fused to DNA adenine methyltransferase (Dam). This fusion protein is incorporated into the scaffold. Thus, DNA sequences that contact the scaffold will also be in molecular proximity to the

---

BOX 7.1 DamID and ChIP Methods to Map Genome – Scaffold Interactions

Two complementary methods exist to identify genomic sequences that are in contact with a nuclear scaffold such as the nuclear lamina. DamID [27] is based on the integration of DNA adenine methyltransferase (Dam) into a scaffold, by in vivo expression of a chimearic protein consisting of Dam and a scaffold protein such as one of the lamins. As a consequence, any genomic DNA in contact with the scaffold will become adenine-methylated. This methylation tag, which is not endogenously present in DNA, can subsequently be mapped by a series of steps as outlined in Figure 7.2A. Adapted from [7].

ChIP [28] instead employs treatment of cells with a cross-linking agent such as formaldehyde to covalently attach scaffold proteins to contacting DNA elements. Next, chromatin (with attached scaffold fragments) is isolated, sheared, and subjected to immuno-purification with an antibody against a scaffold protein (such as a lamin). The thus obtained DNA fragments can be analyzed and mapped using genomic tiling microarrays or by high-throughput sequencing (Figure 7.2B).

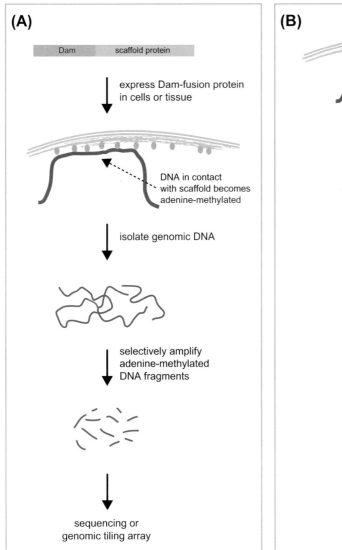
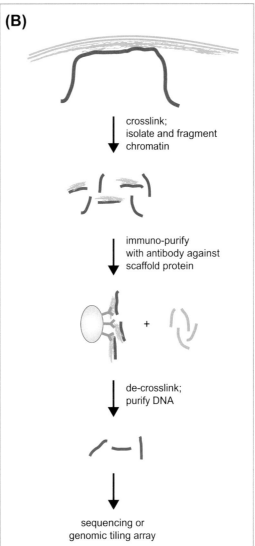

FIGURE 7.2 DamID and ChIP methods to map genome–scaffold interactions.

tethered Dam, and as a consequence are tagged by adenine methylation (^{m6}A). Because this covalent modification does not occur endogenously, contacts with the scaffold can be inferred from the ^{m6}A pattern, which can be mapped genome-wide using methods that employ methylation-sensitive restriction enzymes combined with a readout based on genomic tiling arrays or high-throughput sequencing.

Both methods yield very similar datasets that provide genome-wide views of the interactions of the genome with the scaffold protein of interest. It is important to note that neither method can discriminate direct protein–DNA contacts from indirect contacts (i.e., via another protein). Both methods currently require approximately 10 000 cells or more, hence the resulting maps represent population averages that must be interpreted accordingly: it is possible that a detected interaction occurs in only a fraction of the cells at any time. Nevertheless, the application of ChIP and DamID has yielded some remarkable insights into the interactions of the genome with components of the major scaffold, the nuclear envelope.

Nuclear Lamina–Genome Interactions in Metazoans

The inner nuclear membrane of virtually all metazoan cells is coated by the nuclear lamina (NL), a dense web of protein fibers. This network is primarily made of specialized intermediate filament proteins named lamins, and contains in addition a variety of other proteins, some of which are transmembrane proteins that are inserted into the inner nuclear membrane.

DamID and ChIP mapping studies in worms, fly, mouse and human cells have revealed remarkable patterns of genome—NL interaction [29—32]. In each species, hundreds of large genomic domains contact the NL. Mouse and human genomes harbor more than 1000 of these lamina-associated domains (LADs), distributed over all chromosomes. Mammalian LADs have a median size of about 0.5 Mb and together cover nearly 40% of the genome. A similar fraction of the genome was found to interact with the NL in worms and flies. How does so much DNA fit at the periphery? FISH microscopy of individual LADs suggests that the interactions of LADs with the NL do not take place in every cell at all times, but rather have a significant stochastic component. Thus, effectively only a subset of LADs may contact the NL at any given time.

This is not to say that LADs are 'softly' defined. On the contrary, genome-wide maps indicate that most mammalian LADs have remarkably sharp borders, which tend to be marked by specific sequence elements, such as binding sites for the insulator protein CTCF [30]. This indicates that LADs are at least in part 'hard-coded' in the genome itself. While some LADs are cell-type specific, the overall pattern of NL interactions is strikingly similar in a variety of cell types [31]. This raises the interesting possibility that a basal spatial organization of chromosomes is shared among all cell types.

It is not firmly established whether LADs are actively anchored to the NL, or instead are passively pushed to the nuclear periphery as a consequence of forces that drive inter-LAD regions (genomic regions not associated with the NL, and located inbetween LADs) to in the nuclear interior. Both mechanisms may contribute to the positioning of loci relative to the periphery. However, two lines of circumstantial evidence indicate that active anchoring of LADs occurs. First, lamins and other proteins of the NL are known to bind in vitro to DNA, nucleosomes, and various other chromatin components [33]. Second, the sharp demarcation of LAD borders argues against a passive 'brushing' of LADs against the NL. At present, molecular mechanisms that anchor LADs to the NL in vivo are not known.

Worms, fruit flies and mammals share the striking domain organization of NL interactions. Nevertheless, there are interesting differences. LADs in *D. melanogaster* are about five times smaller than their mammalian counterparts. This may be related to the smaller genome size and the much closer spacing of genes in the fly genome. Interestingly, the average number of genes per LAD is strikingly similar between human and fly, suggesting a role of LAD organization in the regulation of gene expression (see below). NL interactions in *C. elegans* occur primarily in the distal parts of the chromosomes, suggesting a spatial organization of chromosomes where central parts of the chromosomes are positioned in the nuclear interior and the chromosome ends at the periphery.

A common theme among the four investigated species is that NL interactions are tightly linked to gene repression.

Genes associated with the NL are almost invariably expressed at very low levels. The converse is not always true: not every inactive gene is associated with the NL. During differentiation of mouse embryonic stem cells, hundreds of genes shift their position relative to the NL. Detachment of genes from the NL coincides with activation of transcription, or with priming for activation at a later stage [31].

At least to some degree, positioning of genes at the NL plays a causal role in their repression. Artificial tethering of a reporter gene to the NL often leads to repression of this gene, and can also lead to downregulation of neighboring genes, although the magnitude of this repressive effect appears to depend on the reporter gene and its integration context [34—37]. Furthermore, knockout of lamin in worms and fruit flies caused relocation of NL-associated genes towards the nuclear interior, concomitant with increased expression (in *Drosophila*) or stochastic activation (in *Caenorhabditis*) of these genes [38,39]. These results do not rule out an inverse causal relationship, i.e., that repression of genes may cause their targeting to the NL. In fact, it is quite possible that a positive feedback loop exists between NL association and gene inactivation, which could contribute to stable gene silencing. In any case, these observations indicate that spatial organization of the genome is tightly linked to gene regulation.

Yeasts do not have an NL. Nevertheless, there are some interesting parallels with metazoans. In budding yeast the 32 telomeres cluster into 4—8 foci, and these foci are tethered to the periphery of the nucleus. This tethering is mediated by two nuclear envelope proteins which interact with components of telomeric chromatin. Like NL-associated chromatin in metazoans, telomeric chromatin in yeast represses transcription; moreover, detailed mechanistic studies have indicated that the peripheral positioning facilitates the silencing of telomeric sequences. Thus, very similar principles apply, even though the proteins involved differ.

Gene Attachment to Nuclear Pores

Nuclear pores (NPs) form another anchoring site for the genome at the nuclear envelope. NPs are large multi-subunit transport channels that perforate the nuclear membranes. They occupy positions in the nuclear envelope that are depleted of lamins. Electron microscopy indicates that the NL is generally in close contact with relatively condensed chromatin (consistent with the repressed state of LADs), whereas nuclear pores are surrounded by less condensed chromatin. Indeed, ChIP and DamID mapping studies in yeast and fly have revealed that NP proteins interact preferentially with a subset of active genes. There is a caveat, however: most NP proteins are not only located at NPs, but also roam the nucleoplasm, where they can interact with the genome. Thus, binding maps of most NP proteins cannot be easily interpreted in terms of spatial positioning of the genome. Nevertheless, FISH experiments and DamID studies with a NP protein that is exclusively located at the periphery indicate that some of the genes that interact with NP proteins are preferentially located at the nuclear envelope, suggesting that they indeed associate with NPs. In *S. cerevisiae*, short sequence elements have been identified that can target genes to NPs [40]. In *Drosophila* the NP-associated regions tend to be relatively short (compared to LADs), but a NP-targeting sequence motif has not been identified so far [41].

Both in yeast and in fruit flies the interaction of genes with NP proteins appears to enhance transcription. Interestingly, in yeast activation of the *INO1* gene triggers relocation to the nuclear periphery (presumably to an NP); this position is retained for several generations after the gene is shut off. During this period the gene can be turned on more rapidly, suggesting that NP association facilitates the swift activation of a gene in response to an external stimulus [42].

The Nucleolus as a Spatial Organizer

The nucleolus is a large solid structure inside the nucleus and hence another prime candidate to act as a scaffold for chromosome folding. Indeed, a diversity of DNA regions appear to be associated with the nucleolus. These regions were identified in human cells by microarray probing and high-throughput sequencing of DNA that co-purifies with nucleoli. Like LADs, nucleolus-associated domains (NADs) tend to be large (0.1–10Mb). Besides the 28S rRNA genes which are transcribed by RNA polymerase I inside nucleoli, NADs are enriched for specific gene sets, including olfactory genes and zinc-finger protein encoding genes. Furthermore, tRNA and 5S RNA genes are preferentially associated with nucleoli, which is of interest because these genes are transcribed by RNA polymerase III. Thus, the nucleolus acts as a docking platform for very specific sets of genes [43].

Interestingly, the nucleolus also appears to keep certain sequences spatially separated. This was observed for budding yeast chromosome XII, which harbors the rDNA gene cluster. Mapping by a 3C-derived method revealed that DNA sequences at either end of this cluster rarely contact one another, suggesting that the nucleolus acts as a physical barrier between the two chromosome segments [44].

THE INTERNAL ORGANIZATION OF CHROMOSOMES: LONG-RANGE INTERACTIONS ALONG AND BETWEEN CHROMOSOMES

The convoluted three-dimensional path of chromatin fibers is reflected, and in part determined, by multiple specific and non-specific long-range contacts between and along the chromosomes. These contact points can be experimentally detected by direct imaging, but more comprehensively and at higher resolution, i.e., at the scale of several kb for whole chromosomes or genomes, using a growing suite of molecular methodologies based on chromosome conformation capture (3C) technology [7,45,46] (Box 7.2). Applying such methods has led to the identification of widespread looping

BOX 7.2 Chromosome Conformation Capture-Based Technologies

A large variety of chromosome conformation capture (3C)-based technologies have been developed over the last decade [7,45,46]. All these methods rely on the same strategy of cross-linking interacting chromosomal loci, fragmenting chromatin followed by intra-molecular ligation to convert cross-linked pairs of loci into ligation products to obtain a genome-wide library of hybrid DNA molecules. Each hybrid molecule represents a pairwise interaction between the corresponding loci in a single cell. The different 3C-based methods mainly differ in the way the ligation products are being detected. The different 3C-based methods are outlined in Figure 7.3.

3C: In a classic 3C experiment the genome-wide ligation product library is interrogated by quantitative PCR to detect and quantify the presence and abundance of one ligation product (chromatin interaction) at the time [45]. This low-throughput approach is used for small-scale and locus-specific studies.

4C: the 4C method employs inverse PCR to amplify all DNA fragments ligated to a single fragment of interest [47,48]. The amplified DNA is then analyzed on a microarray or is directly sequenced. This method is used to obtain a genome-wide interaction profile of a single locus of interest.

5C: the 5C method employs highly multiplexed ligation-mediated amplification with pools of locus-specific primers to amplify large sets of targeted chromatin interactions [49]. For instance, all interactions among a set of promoters and a set of enhancers can be detected in parallel. Amplified interactions are directly sequenced.

Hi-C: Hi-C is a completely unbiased and genome-wide adaptation of 3C and differs from the other 3C-based methods by including a step prior to intra-molecular ligation during which the ends of the digested chromatin are filled in with biotinylated nucleotides [6,22]. As a result, ligation junctions will be marked by biotin, facilitating unbiased purification of all ligation junctions. Purified DNA is then directly sequenced to obtain a genome-wide chromatin interaction map.

ChIA-PET: The ChIA-PET method employs an immunoprecipitation step to enrich for chromatin interactions that involve loci bound by a protein of interest [50].

interactions between loci separated by large genomic distances, sometimes up to several Mb. Networks of such interactions may be involved in gene regulation and constrain the path of the chromatin fiber within chromosome territories in a cell type-specific manner.

Molecular Techniques to Map Long-Range Chromatin Interactions

3C-based technologies are used to detect pairs of loci located on the same chromosome or on different chromosomes that are in close spatial proximity [45]. Cells are cross-linked with formaldehyde to covalently link segments of chromatin fibers that are physically touching or interacting through co-binding proteins or subnuclear structures. This leads to linking of pairs of interacting loci throughout the genome. Chromatin is then solubilized and fragmented in small segments using a restriction enzyme to generate a complex mixture of linked pairs of restriction fragments. DNA is then intra-molecularly ligated so that pairs of interacting loci are converted into unique chimeric DNA molecules. Chimeric molecules, each representing a physical contact between a pair of chromatin segments, can then be detected and quantified by PCR, or more comprehensively by deep sequencing. The frequency with which a specific chimeric DNA molecule is detected is typically interpreted to represent the fraction of cells in the cell population in which a specific pair of loci is in close spatial proximity.

Over the years a large variety of 3C-based methods have been developed [7,46]. All of these rely on the basic concept of 3C, but differ in the way chimeric molecules are detected (Box 7.2; Figure 7.3). For instance, the Hi-C variant includes a step to mark ligation junctions with biotin to facilitate their purification, followed by deep sequencing to generate a long-range chromatin interaction map for a complete genome [22].

3C-based data provide insights into the probability, or frequency in the cell population, that any two loci are physically touching, and thus has become the method of choice to map the average spatial path of the chromatin fiber in a cell population.

Determination of Polymer Parameters using 3C Chromatin Interaction Data

As outlined above, the contact probability of a chromatin fiber is expected to decrease for loci located at increasingly large distances along the chromosome, with the precise function of this decay related to the polymer state of the chromosome. 3C-based interaction analyses have been used to study this relationship in detail. For yeast chromosomal arms it was found that the contact probability of loci scales as $P(s) \sim s^{-3/2}$, consistent with a random walk conformation [45]. Further studies that combined 3C assays and measurements of 3D distances between loci allowed estimation of the mass density and persistence length of yeast chromatin [18]. The mass density was found to be around 40 nm/kb, much longer than would be predicted for a canonical 30 nm fiber, and the persistence length was found to be around 95 nm, corresponding to 2.3 kb.

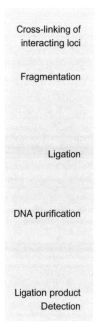

FIGURE 7.3 Chromosome conformation capture-based technologies. See box 7.2 for further description of these technologies. Figure reproduced with permission from [51].

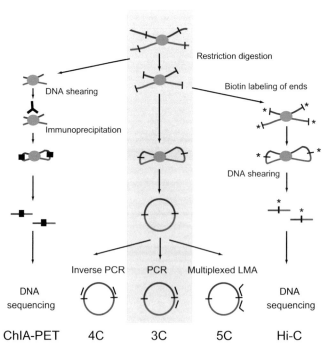

Analysis of contact probabilities along human chromosomes revealed a very different chromosome conformation than in yeast [22]. At a length scale of up to several hundred kb the contact probability decays much more slowly than any of the polymer states described above would predict. This indicates that at a scale of up to ~1 Mb the chromatin fiber adopts a rather compact conformation, with loci located far apart being located close in 3D space. One explanation for this phenomenon is the presence of abundant contacts between distal loci separated by up to 1 Mb, leading to a highly looped structure. This may be related to the presence of specific looping interactions between genes and their regulatory elements. Indeed, looping interactions can be readily detected between gene promoters and enhancer elements separated by hundreds of kb [52–54] as discussed in more detail below. At a length scale of several Mb the formation of specific loops between defined loci becomes increasingly unlikely and the contact probability decays as $P(s) \sim s^{-1}$, which is as predicted by a compact fractal globule conformation. As described above, this state is characterized by dense packing of the chromatin fiber in a largely unentangled state, and can be the result of confinement of the chromosome in a relatively small volume. Finally, at even longer length scales, e.g., ten to hundreds of Mb, the contact probability no longer decays as a function of genomic distance, indicating that the maximum spatial distance or volume of the chromosome has been reached. This plateau has also been observed by direct microscopic measurement of 3D distances between loci [55]. The plateau reflects the confined volume of the chromosome territory and determines the maximum distance by which any two loci on a chromosome can be separated.

Co-association of Large Active and Inactive Chromosomal Domains leads to Chromosome Compartmentalization

In complex genomes, such as those of human and mouse, large genomic regions enriched in genes are separated by regions that are relatively devoid of genes. 4C and Hi-C analyses showed that Mb-sized domains of active and gene dense chromatin preferentially interact with other large domains of active chromatin [22,47]. Similarly, inactive chromatin domains tend to associate. The molecular mechanisms underlying these co-associations remain to be identified. One contributing factor could be the tethering of inactive chromatin domains, often coinciding with LADs, to the nuclear lamina. Certain groups of inactive loci can also cluster together even when not tethered to the nuclear envelope. For example, loci bound by the repressive Polycomb complex can physically associate at subnuclear sites known as polycomb bodies [56,57]. Similarly, groups of active genes have been observed to co-localize at sites enriched in transcription and mRNA processing proteins, e.g., transcription factories and splicing speckles [58,59]. When combined, these various classes of associations lead to spatial separation of active and inactive chromatin within a chromosome territory. Notably, these co-associations are not limited to intra-chromosomal interactions, and can also be observed between two neighboring chromosome territories at their zone of limited intermingling [25]. Each of these interactions is of limited specificity: any two active or two inactive loci have a finite probability of interacting in a given cell, leading to significant cell-to-cell variation in the precise internal organization of a chromosome territory. However, in each cell active and inactive chromatin domains will tend to be spatially separate, and this compartmentalization is a major constraint on the path of the chromatin fiber in each cell. Further, in different cell types different groups of loci will be active, and thus the composition of the active and inactive compartment will change accordingly.

Looping Interactions between Genomic Elements to Regulate Genes

All types of interactions and constraints on chromosome conformation described so far are relatively non-specific. We have described general trends of co-association between active and inactive chromatin domains and with nuclear structures such as the nuclear lamina. These rather general interactions determine the positioning of chromosomes and their overall internal compartmentalization. However, at a smaller scale, within ~1 Mb, more specific interactions dominate. We consider chromatin interactions specific when they involve specific kb-sized or smaller loci and are mediated by specific proteins. Many of these interactions are involved in regulating gene expression.

Gene expression is controlled by a variety of regulatory elements, including enhancers, repressors and insulators. In small genomes such as that of yeast all elements that regulate a given gene are located within several hundred base pairs of the gene promoter, obviating the need for any higher-order folding of the chromatin fiber to bring them into direct physical contact. However, in complex genomes enhancer elements have been identified that are located up to 1 Mb from the gene they regulate, and often a number of unrelated genes can be found in between them. This raises two critical questions: how do they communicate with their target genes, and what determines the specificity of these interactions?

Two models propose mechanisms by which regulatory elements can contact distal genes over large genomic distances [60]. The looping model proposes that free passive diffusion of loci can bring regulatory elements into direct physical contact with their target promoter. Relatively stable association between the two loci will then lead to the formation of a chromatin loop. The tracking model also proposes that a chromatin loop is formed, but includes an active step in which protein complexes bound to the regulatory element 'track' along the chromatin fiber, pulling the regulatory element with them, until a target promoter is reached.

Over the last decade 3C-based studies have confirmed the formation of specific looping interactions between gene promoters and enhancers and insulators. One of the best-studied examples is the β-globin locus. This locus contains a cluster of related globin genes that encode the β-subunit of hemoglobin. A single complex gene regulatory element, the locus control region (LCR), is located 20–80 kb upstream of the globin genes and is required for high levels of gene expression. 3C assays have detected specific looping interactions between the LCR and the globin genes, but only in cells in which they are expressed [61]. Furthermore, specific transcription factors that bind the LCR and the globin promoters, e.g., GATA1 and EKLF1, were shown to mediate this interaction and to be required for globin gene activation [62,63].

Since these initial studies a large number of examples of chromatin looping interactions have been described that are involved in gene expression. These include other developmentally controlled gene clusters such as the α-globin [64,65], Th2 interleukin [66], and Hox gene clusters [67–69], but also more broadly expressed single gene loci such as BRCA1 [70], MYC [71] and CFTR [72,73]. A recent large-scale study of looping interactions that involve over 600 genes in the human genome performed in the laboratory of J. Dekker found that the majority of them in a given cell type are engaged in one or more chromatin looping interactions with putative gene regulatory elements located within several hundred kb [54]. Gene promoters can interact with more than one enhancer or insulator, but do not interact with all active enhancers located around the gene, indicating a significant level of specificity. How specificity is achieved is not known, but it seems highly plausible that combinations of specific DNA-binding proteins associating with the interacting loci, as well as other chromatin-bound complexes, will play a critical role.

A Dynamic Network View of Chromosome Folding

From these observations a picture of a highly looped and folded chromatin fiber emerges, at least at the sub-Mb scale. This structure is driven by a network of interactions, with varying specificities, between genes, gene regulatory elements, large chromosomal domains and between loci and nuclear substructures such as the NL and the nucleolus. This network is cell-type specific and determined by the set of active genes and regulatory elements present in that cell. At the scale of several hundreds of kb the path of the chromatin fiber is thus highly constrained by the combined action of specific associations between two or more loci. The fact that this class of chromatin interactions displays considerable specificity does not mean that the chromatin fiber will be folded in exactly the same 3D structure in each cell. The occurrence of each looping interaction is infrequent, across the cell population and perhaps over time within a single cell, and some interactions may be mutually exclusive. Thus, even at the sub-Mb scale folding of the chromatin fiber will be variable. Finally, the observation that chromatin loops are formed does not provide insights into the processes that bring the two interacting loci together. The tracking model would propose a directed and possibly active process, but at the length scale of hundreds of kb the chromatin fiber displays significant Brownian motion so that two loci can readily encounter one another by free diffusion. In addition, the persistence length of the chromatin fiber is relatively short, at several kb, compared to the relatively large chromatin loops typically observed (>10 kb), and thus the physical stiffness of chromatin is not limiting the occurrence of these interactions. Unraveling the molecular processes leading to loop formation, and their dynamic properties, is an exciting new challenge that may require the development of new or improved imaging and/or 3C-based technologies.

A network view of chromatin folding and nuclear organization also provides a new lens through which to view the regulation of genes through the combinatorial action of multiple, possibly cell-type-specific regulatory elements spread out over hundreds of kb. Current studies are aimed at generating detailed maps of these networks of gene-regulatory interactions. Over the next several years we will undoubtedly witness tremendous progress in genome-scale mapping of looping interactions, revealing new principles of gene regulation and the roles of the spatial organization of chromosomes in this process.

A STOCHASTIC INTERACTION-DRIVEN MODEL FOR GENOME FOLDING AND NUCLEAR ORGANIZATION

As outlined above, there are three main principles that drive genome organization: polymer physics, anchoring of chromosomes to nuclear scaffolds, and direct contacts among chromosomal loci. We propose that a hierarchical system of these defined, yet stochastic interactions along and between chromosomes, and between chromosomes and nuclear structures, determines the folding of chromosomes and the organization of the interphase nucleus in general. Below we describe this model in more detail and the predictions the model makes, and then outline some of the new questions that it invokes.

First, at a local level the physical properties such as mass density and persistence length of the chromatin fiber will be influenced by patterns of histone modification and by chromatin composition. Networks of long-range interactions between genes and regulatory elements will lead to complex looped configurations that directly affect gene expression. These conformations are between specific pairs of loci, but it is likely that not all interactions occur at the same time in a single cell. In some instances, such stochastic interactions between a gene and a distant strong enhancer can lead to 'jackpot' expression of the gene in a few cells [74]. Thus stochastic chromatin interactions may be related to the phenomenon of stochastic gene expression observed at the single cell level. At a higher level, groups of transcriptionally active loci located along the same chromosome will associate with each other, perhaps through association with subnuclear structures or bodies such as splicing speckles or transcription factories. Similarly, inactive domains tend to cluster and associate at the nuclear lamina. These interactions lead to compartmentalization of chromosomes, with inactive regions near the periphery of the nucleus and active chromatin located more centrally. Interactions between chromosomes are similarly related to co-association of active and inactive regions, and by anchoring of heterochromatic regions at the periphery or near nucleoli. Thus, the relative positioning of any locus in the genome will be dictated by its local looping interactions, by more global positioning of the larger chromosome domain it resides in near either active or inactive compartments, and finally by the anchoring of chromosome to subnuclear structures.

One critical aspect of the model is that the level of variability in interactions increases and the specificity decreases with the scale of chromosome organization: local interactions are highly specific and involve genes with their respective distal regulatory elements. At the next level, where groups of active or inactive loci cluster together, the level of specificity is lower, and which pairs of active loci are found together is more variable. Similarly, which inactive regions interact with each other and with the nuclear periphery can vary significantly between cells, although in all cells inactive regions are found near other inactive regions. Finally, the relative positions of chromosomes is predicted to be dependent on interactions between active and inactive regions located on different chromosomes, but the precise pattern of inter-chromosomal interactions will greatly differ from cell to cell.

In this model implementation of the same set of folding principles can lead to dramatic cell-to-cell differences in the spatial organization of the genome inside the nucleus. Considerable variability in organization is consistent with microscopic observations and 3C-based data indicating that any pair of loci can be in close spatial proximity in at least some cells. Another implication of the model is that each relatively large chromosomal domain is linked through interactions with many other loci and with subnuclear structure, making it relatively immobile. Further, the model predicts that movement of a genomic locus can only occur in coordination with the loci or structures it interacts with, or would require significant force to disrupt interactions. Many observations indeed suggest limited movement of chromosomes, even in response to stimuli that alter the expression of significant cohorts of genes [75,76]. Interestingly, more locally, within <1 μm, chromatin displays constrained Brownian motion [77,78]. At this length scale, which corresponds to up to several hundred kb, specific long-range interactions between genes and regulatory elements occur. Possibly the local movement facilitates the dynamic formation of these specific looping contacts between nearby genes and elements.

FUTURE CHALLENGES

The detailed view of genome organization provided by new technologies raises many new questions. How can a stochastic assembly of the nucleus be consistent with robust maintenance of gene expression programs and cell fate? Are the local and specific looping contacts between genes and nearby elements, e.g., within several hundred kb, sufficiently reproducible from cell to cell to yield correct gene expression? Does the local Brownian motion for these nearby loci ensure that these loci will encounter one another at an appropriate timescale in every cell? How can cells rearrange their chromosome structure, e.g., during differentiation or in response to signals? Does this require rewiring of all interactions, e.g., as could occur during cell division? Addressing these questions will require identification of the molecular mechanisms that mediate chromosomal anchoring and looping, and experimental manipulation of networks of chromosomal interactions combined with analysis of real-time dynamics of chromatin interactions and gene expression.

REFERENCES

[1] DuPraw EJ. Macromolecular organization of nuclei and chromosomes: a folded fibre model based on whole-mount electron microscopy. Nature 1965;206(982):338–43.

[2] Langer-Safer PR, Levine M, Ward DC. Immunological method for mapping genes on Drosophila polytene chromosomes. Proc Natl Acad Sci USA 1982;79(14):4381–5.

[3] Misteli T. Beyond the sequence: cellular organization of genome function. Cell 2007;128(4):787–800.

[4] Zhao R, Bodnar MS, Spector DL. Nuclear neighborhoods and gene expression. Curr Opin Genet Dev 2009;19(2):172–9.

[5] Bolzer A, et al. Three-dimensional maps of all chromosomes in human male fibroblast nuclei and prometaphase rosettes. PLoS Biol 2005;3(5):e157.

[6] Kalhor R, Tjong H, Jayathilaka N, Alber F, Chen L. Genome architectures revealed by tethered chromosome conformation capture and population-based modeling. Nat Biotechnol 2011;30(1):90–8.

[7] van Steensel B, Dekker J. Genomics tools for unraveling chromosome architecture. Nat Biotechnol 2010;28(10):1089–95.

[8] Umbarger MA, et al. The three-dimensional architecture of a bacterial genome and its alteration by genetic perturbation. Mol Cell 2011;44(2):252–64.

[9] Mirny LA. The fractal globule as a model of chromatin architecture in the cell. Chromosome Res 2011;19(1):37–51.

[10] Fussner E, Ching RW, Bazett-Jones DP. Living without 30nm chromatin fibers. Trends Biochem Sci 2011;36(1):1–6.

[11] Maeshima K, Hihara S, Eltsov M. Chromatin structure: does the 30-nm fibre exist in vivo? Curr Opin Cell Biol 2010;22(3):291–7.

[12] van Steensel B. Chromatin: constructing the big picture. EMBO J 2011;30(10):1885–95.

[13] Filion GJ, et al. Systematic protein location mapping reveals five principal chromatin types in *Drosophila* cells. Cell 2010;143(2):212–24.

[14] Tark-Dame M, van Driel R, Heermann DW. Chromatin folding –from biology to polymer models and back. J Cell Sci 2011;124(Pt 6):839–45.

[15] Grosberg AY, Khokhlov AR, de Gennes P-G. Giant Molecules, Here, There and Everywhere. World Scientific; 2011.

[16] Rippe K, von Hippel PH, Langowski J. Action at a distance: DNA-looping and initiation of transcription. Trends Biochem Sci 1995;20(12):500–6.

[17] Rippe K. Making contacts on a nucleic acid polymer. Trends Biochem Sci 2001;26(12):733–40.

[18] Dekker J. Mapping in vivo chromatin interactions in yeast suggests an extended chromatin fiber with regional variation in compaction. J Biol Chem 2008;283(50):34532–40.

[19] Fudenberg G, Mirny LA. Higher-order chromatin structure: bridging physics and biology. Curr Opin Genet Dev 2012;22(2):115–24.

[20] Grosberg AY, Nechaev SK, Shakhnovich EI. The role of topological constraints in the kinetics of collapse of macromolecules. J Phys 1988;49:2095–100.

[21] Grosberg AY, Rabin Y, Havlin S, Neer A. Crumpled globule model of the three-dimensional structure of DNA. Europhys Lett 1993;23:373.

[22] Lieberman-Aiden E, et al. Comprehensive mapping of long-range interactions reveals folding principles of the human genome. Science 2009;326(5950):289–93.

[23] Yokota H, van den Engh G, Hearst JE, Sachs RK, Trask BJ. Evidence for the organization of chromatin in megabase pair-sized loops arranged along a random walk path in the human G0/G1 interphase nucleus. J Cell Biol 1995;130(6):1239–49.

[24] Cremer T, Cremer M. Chromosome territories. Cold Spring Harb Perspect Biol 2010;2(3):a003889.

[25] Branco MR, Pombo A. Intermingling of chromosome territories in interphase suggests role in translocations and transcription-dependent associations. PLoS Biol 2006;4(5):e138.

[26] Rosa A, Everaers R. Structure and dynamics of interphase chromosomes. PLoS Comput Biol 2008;4(8):e1000153.

[27] Greil F, Moorman C, van Steensel B. DamID: mapping of in vivo protein–genome interactions using tethered DNA adenine methyltransferase. Methods Enzymol 2006;410:342–59.

[28] O'Geen H, Echipare L, Farnham PJ. Using ChIP-seq technology to generate high-resolution profiles of histone modifications. Methods Mol Biol 2011;791:265–86.

[29] Pickersgill H, Kalverda B, de Wit E, Talhout W, Fornerod M, van Steensel B. Characterization of the *Drosophila melanogaster* genome at the nuclear lamina. Nat Genet 2006;38(9):1005–14.

[30] Guelen L, et al. Domain organization of human chromosomes revealed by mapping of nuclear lamina interactions. Nature 2008;453(7197):948–51.

[31] Peric-Hupkes D, et al. Molecular maps of the reorganization of genome-nuclear lamina interactions during differentiation. Mol Cell 2010;38(4):603–13.

[32] Ikegami K, Egelhofer TA, Strome S, Lieb JD. *Caenorhabditis elegans* chromosome arms are anchored to the nuclear membrane via discontinuous association with LEM-2. Genome Biol 2010;11(12):R120.

[33] Dechat T, Pfleghaar K, Sengupta K, Shimi T, Shumaker DK, Solimando L, Goldman RD. Nuclear lamins: major factors in the structural organization and function of the nucleus and chromatin. Genes Dev 2008;22(7):832–53.

[34] Reddy KL, Zullo JM, Bertolino E, Singh H. Transcriptional repression mediated by repositioning of genes to the nuclear lamina. Nature 2008;452(7184):243–7.

[35] Finlan LE, Sproul D, Thomson I, Boyle S, Kerr E, Perry P, Ylstra B, Chubb JR, Bickmore WA. Recruitment to the nuclear periphery can alter expression of genes in human cells. PLoS Genet 2008;4(3):e1000039.

[36] Kumaran RI, Spector DL. A genetic locus targeted to the nuclear periphery in living cells maintains its transcriptional competence. J Cell Biol 2008;180(1):51–65.

[37] Dialynas G, Speese S, Budnik V, Geyer PK, Wallrath LL. The role of *Drosophila Lamin* C in muscle function and gene expression. Development 2010;137(18):3067–77.

[38] Shevelyov YY, Lavrov SA, Mikhaylova LM, Nurminsky ID, Kulathinal RJ, Egorova KS, Rozovsky YM, Nurminsky DI. The B-type lamin is required for somatic repression of testis-specific gene clusters. Proc Natl Acad Sci USA 2009;106(9):3282–7.

[39] Towbin BD, Meister P, Pike BL, Gasser SM. Repetitive transgenes in *C. elegans* accumulate heterochromatic marks and are sequestered at the nuclear envelope in a copy-number- and lamin-dependent manner. Cold Spring Harb Symp Quant Biol 2010;75:555–65.

[40] Ahmed S, Brickner DG, Light WH, Cajigas I, McDonough M, Froyshteter AB, Volpe T, Brickner JH. DNA zip codes control an ancient mechanism for gene targeting to the nuclear periphery. Nat Cell Biol 2010;12(2):111–8.

[41] Kalverda B, Pickersgill H, Shloma VV, Fornerod M. Nucleoporins directly stimulate expression of developmental and cell-cycle genes inside the nucleoplasm. Cell 2010;140(3):360–71.

[42] Light WH, Brickner DG, Brand VR, Brickner JH. Interaction of a DNA zip code with the nuclear pore complex promotes H2A.Z incorporation and INO1 transcriptional memory. Mol Cell 2010;40(1):112–25.

[43] Nemeth A, et al. Initial genomics of the human nucleolus. PLoS Genet 2010;6(2):e1000889.

[44] Duan Z, et al. A three-dimensional model of the yeast genome. Nature 2010;465(7296):363–7.

[45] Dekker J, Rippe K, Dekker M, Kleckner N. Capturing chromosome conformation. Science 2002;295(5558):1306–11.

[46] de Wit E, de Laat W. A decade of 3C technologies: insights into nuclear organization. Genes Dev 2012;26(1):11–24.

[47] Simonis M, Klous P, Splinter E, Moshkin Y, Willemsen R, de Wit E, van Steensel B, de Laat W. Nuclear organization of active and inactive chromatin domains uncovered by chromosome conformation capture-on-chip (4C). Nat Genet 2006;38(11):1348–54.

[48] Zhao Z, et al. Circular chromosome conformation capture (4C) uncovers extensive networks of epigenetically regulated intra- and interchromosomal interactions. Nat Genet 2006;38(11):1341–7.

[49] Dostie J, et al. Chromosome conformation capture carbon copy (5C): a massively parallel solution for mapping interactions between genomic elements. Genome Res 2006;16(10):1299–309.

[50] Fullwood MJ, et al. An oestrogen-receptor-alpha-bound human chromatin interactome. Nature 2009;462(7269):58–64.

[51] Sanyal A, Baù D, Martí-Renom MA, Dekker J. Chromatin globules: a common motif of higher order chromosome structure? Curr Opin Cell Biol 2013;23(3):325–31.

[52] Miele A, Dekker J. Long-range chromosomal interactions and gene regulation. Mol Biosyst 2008;4(11):1046–57.

[53] Dekker J. Gene regulation in the third dimension. Science 2008;319(5871):1793–4.

[54] Sanyal A, Lajoie BR, Jain G, Dekker J. The long-range interaction landscape of gene promoters. Nature 2012;489:109–13.

[55] Mateos-Langerak J, et al. Spatially confined folding of chromatin in the interphase nucleus. Proc Natl Acad Sci USA 2009; 106(10):3812–7.

[56] Bantignies F, Roure V, Comet I, Leblanc B, Schuettengruber B, Bonnet J, Tixier V, Mas A, Cavalli G. Polycomb-dependent regulatory contacts between distant Hox loci in *Drosophila*. Cell 2011;144(2):214–26.

[57] Tolhuis B, et al. Interactions among Polycomb domains are guided by chromosome architecture. PLoS Genet 2011;7(3):e1001343.

[58] Osborne CS, et al. Active genes dynamically colocalize to shared sites of ongoing transcription. Nat Genet 2004;36(10):1065–71.

[59] Brown JM, et al. Association between active genes occurs at nuclear speckles and is modulated by chromatin environment. J Cell Biol 2008;182(6):1083–97.

[60] Bulger M, Groudine M. Looping versus linking: toward a model for long-distance gene activation. Genes Dev 1999;13(19): 2465–77.

[61] Tolhuis B, Palstra RJ, Splinter E, Grosveld F, de Laat W. Looping and interaction between hypersensitive sites in the active beta-globin locus. Mol Cell 2002;10(6):1453–65.

[62] Drissen R, Palstra RJ, Gillemans N, Splinter E, Grosveld F, Philipsen S, de Laat W. The active spatial organization of the beta-globin locus requires the transcription factor EKLF. Genes Dev 2004;18(20):2485–90.

[63] Vakoc CR, Letting DL, Gheldof N, Sawado T, Bender MA, Groudine M, Weiss MJ, Dekker J, Blobel GA. Proximity among distant regulatory elements at the beta-globin locus requires GATA-1 and FOG-1. Mol Cell 2005;17(3):453–62.

[64] Vernimmen D, De Gobbi M, Sloane-Stanley JA, Wood WG, Higgs DR. Long-range chromosomal interactions regulate the timing of the transition between poised and active gene expression. EMBO J 2007;26(8):2041–51.

[65] Bau D, Sanyal A, Lajoie BR, Capriotti E, Byron M, Lawrence JB, Dekker J, Marti-Renom MA. The three-dimensional folding of the alpha-globin gene domain reveals formation of chromatin globules. Nat Struct Mol Biol 2011;18(1):107–14.

[66] Spilianakis CG, Flavell RA. Long-range intrachromosomal interactions in the T helper type 2 cytokine locus. Nat Immunol 2004;5(10):1017–27.

[67] Ferraiuolo MA, Rousseau M, Miyamoto C, Shenker S, Wang XQ, Nadler M, Blanchette M, Dostie J. The three-dimensional architecture of Hox cluster silencing. Nucleic Acids Res 2010;38(21):7472–84.

[68] Wang KC, et al. A long noncoding RNA maintains active chromatin to coordinate homeotic gene expression. Nature 2011;472(7341):120–4.

[69] Montavon T, Soshnikova N, Mascrez B, Joye E, Thevenet L, Splinter E, de Laat W, Spitz F, Duboule D. A regulatory archipelago controls Hox genes transcription in digits. Cell 2011;147(5):1132–45.

[70] Tan-Wong SM, French JD, Proudfoot NJ, Brown MA. Dynamic interactions between the promoter and terminator regions of the mammalian BRCA1 gene. Proc Natl Acad Sci USA 2008;105(13):5160–5.

[71] Wright JB, Brown SJ, Cole MD. Upregulation of c-MYC in cis through a large chromatin loop linked to a cancer risk-associated single-nucleotide polymorphism in colorectal cancer cells. Mol Cell Biol 2010;30(6):1411–20.

[72] Ott CJ, Blackledge NP, Kerschner JL, Leir SH, Crawford GE, Cotton CU, Harris A. Intronic enhancers coordinate epithelial-specific looping of the active CFTR locus. Proc Natl Acad Sci USA 2009;106(47):19934–9.

[73] Gheldof N, Smith EM, Tabuchi TM, Koch CM, Dunham I, Stamatoyannopoulos JA, Dekker J. Cell-type-specific long-range looping interactions identify distant regulatory elements of the CFTR gene. Nucleic Acids Res 2010;38(13):4325–36.

[74] Noordermeer D, et al. Variegated gene expression caused by cell-specific long-range DNA interactions. Nat Cell Biol 2011; 13(8):944–51.

[75] Kocanova S, Kerr EA, Rafique S, Boyle S, Katz E, Caze-Subra S, Bickmore WA, Bystricky K. Activation of estrogen-responsive genes does not require their nuclear co-localization. PLoS Genet 2010;6(4):e1000922.

[76] Hakim O, et al. Diverse gene reprogramming events occur in the same spatial clusters of distal regulatory elements. Genome Res 2011;21(5):697–706.

[77] Marshall WF, Straight A, Marko JF, Swedlow J, Dernburg A, Belmont A, Murray AW, Agard DA, Sedat JW. Interphase chromosomes undergo constrained diffusional motion in living cells. Curr Biol 1997;7(12):930–9.

[78] Chubb JR, Boyle S, Perry P, Bickmore WA. Chromatin motion is constrained by association with nuclear compartments in human cells. Curr Biol 2002;12(6):439–45.

Chapter 8

Chemogenomic Profiling: Understanding the Cellular Response to Drug

Anna Y. Lee[1], Gary D. Bader[1,2], Corey Nislow[1,2,3] and Guri Giaever[1,2,4]

[1]The Donnelly Centre for Cellular and Biomolecular Research, University of Toronto, Toronto, Ontario M5S 3E1, Canada, [2]Department of Molecular Genetics, University of Toronto, Toronto, Ontario M5S 3E1, Canada, [3]Banting and Best Department of Medical Research, University of Toronto, Toronto, Ontario M5S 3E1, Canada, [4]Department of Pharmaceutical Sciences, University of Toronto, Toronto, Ontario M5S 3M2, Canada

Chapter Outline

Impact of the Human Genome Project on Healthcare and Biology	**153**
Healthcare: Few New Therapies	153
Biology: The Cell is More Than the Sum of its Parts	154
Impact of Genomics: Shifting the Perceptions of Drug and Cellular Behavior	**154**
The Cell is a Highly Interactive Robust System	154
Yeast: A Pioneer and Driver of All Things 'Omic'	155
The Yeast Genome Project	155
The Yeast Deletion Project	155
Impact of Yeast in the Development of Genomic Technologies	157
Drug Behavior is Promiscuous	158
Modern Drug Discovery: A Historical Perspective	158
Drugs are Promiscuous; the Discovery of Polypharmacology	159
Phenotypic Screening	159
Chemical Probes; Like Drugs, but Designed to Serve as Tools to Study Gene Function	159
Chemogenomics	**160**
Target Identification/Mechanism of Drug Action	160
Compendium RNA Expression Approaches to Predict the Drug Target/Mechanism of Drug Action	161
Fitness-based Chemogenomic Profiling Approaches to Identify the Drug Target/Mechanism of Drug Action	162
HIPHOP	**162**
Chemical—Genetic Interactions Reveal Conditional Essentiality	163
Identification of Novel Chemical Probes	163
HIP Predicts the Yeast Druggable Genome is Double that of Current Estimates	164
Co-Inhibition Reflects Structure and Therapeutic Class	164
Co-Fitness Predicts Gene Function	166
The Multi-Drug Resistance Network	166
Global View of the Mechanisms Involved in Cellular Resistance to Small Molecules	166
Chemical Structures Associated with MDR Biological Processes	170
Conclusions and Future Challenges	**171**
References	**172**

IMPACT OF THE HUMAN GENOME PROJECT ON HEALTHCARE AND BIOLOGY

Healthcare: Few New Therapies

A decade after the completion of the human genome project (HGP), the much anticipated improvements in healthcare and the promised delivery of new therapies have yet to be realized. As the primary source of new therapies, the pharmaceutical industry has been in a position to exploit the hundreds of potential new targets revealed by the HGP. The HGP and the genetic variation studies that followed, however, revealed that most diseases are more complex than had been anticipated. As a result, drug discovery remains as challenging as ever. For example, despite enormous increases in research and development spending, the number of novel small-molecule drugs emerging from the pharmaceutical industry each year continues to decline. Combined with the crippling costs required to develop a new therapy (~1.8bn USD), most pharmaceutical executives describe the current state of drug discovery as unsustainable. Countless analyses have searched for underlying causes, yet none have gone beyond describing the symptoms; for example, high attrition rates during the most costly phases of clinical development, increased regulatory demands, and looming patent expirations for a large number of last-decade blockbuster drugs. Others question whether the decline is more fundamental in nature and challenge the validity of the modern drug discovery (MDD) paradigm. That is, will a small molecule with demonstrated ability to bind and/or alter the activity of the target protein in vitro with high affinity and/or high potency, translate into a successful and efficacious new therapy? Taken together, the causes for the decline in the productivity of the drug industry are complex. Struggling to survive, the industry has adopted a risk-averse approach, one reluctant to explore new target areas, focusing instead on 'druggable' or 'me-too' proteins that are already targeted by existing drugs, leaving little hope for diseases that currently lack therapeutic options. (http: //www.pharmalot.com/2012/03/pharma-execs-admit-our-model-is-broken/)

Biology: The Cell is More Than the Sum of its Parts

In contrast to the pharmaceutical industry, the HGP has had an enormous positive impact on the biological sciences, described by many as a 'genomic revolution'. The HGP set the stage for a new fearlessness, providing a much-needed reference from which a substantial number of large-scale consortia were launched, many ongoing (e.g., the 1000 Genomes Project, ENCODE, the Epigenomics Roadmap project [1−3]) and several successfully completed (e.g., International HapMap Project, The SNP Consortium [4,5]). The resulting catalogue of these vast datasets has provided a strong biological foundation and revealed a biological complexity far beyond that previously imagined. Combined with the established genomics concept that the ability of a gene to function is often dependent on interactions with other genes, these discoveries have made clear that the reductionist's one-by-one-approach of studying gene function in isolation has met its limit, and will be ineffective in improving our biological knowledge of the cell. In its place, a systems-level approach that incorporates the interaction-dependent functioning of genes will be required if we are to increase our understanding of cellular behavior. This new approach has taken shape as 'systems biology', and is based on the concept that by measuring the behavior of every 'part' of the cell, a coherent set of systems-level properties of cellular behavior will emerge. Systems-level approaches have been defined and shaped by genomics, whereby the cellular state is reported in a single system-wide profile comprised of individual readouts for every gene, protein, metabolite or other molecular element in the cell. These profiles are generated by 'omic' technologies that allow a cellular state to be measured from an increasing number of molecular perspectives. For example, these genomic technologies quantify (1) RNA expression levels, (2) genetic interactions, (3) protein abundance, modification and activity and (4) the cellular response to drug, the focus of this chapter. It is important to recognize that simple tabulation of the data from a single molecular perspective followed by a search for patterns in the data that may correlate with, for example, a particular disease, will not be sufficient to fully understand the systems-level behavior of the cell. Rather, data from all available perspectives should be integrated in order to account for the interactions and dependencies between molecular elements vital to cellular functions. By integrating these diverse systems datasets, it is the hope that the ultimate goal of systems biology — the ability to both understand and predict the cellular response to perturbation — will be realized.

In this chapter we discuss the contributions of chemogenomics to our understanding of the cellular response to small molecules. In order to appreciate how chemogenomics evolved to become a driver of biological discovery and modern drug discovery, we briefly review the genomic discoveries that have fundamentally altered our perception of how drugs act and cells behave. In addition, because this chapter is primarily based on chemogenomics in yeast, we provide a short history of the pioneering role of the budding yeast *Saccharomyces cerevisiae* in the development of genomic technologies. In the final two sections we discuss the biological discoveries that have been made using chemogenomic approaches, including their potential impact on the future of drug discovery.

IMPACT OF GENOMICS: SHIFTING THE PERCEPTIONS OF DRUG AND CELLULAR BEHAVIOR

The Cell is a Highly Interactive Robust System

Genomics studies have revealed that the cell is a system of interconnected, precisely regulated processes in continuous flux, where bioactive molecules in the cell interact with and respond to each other, while the cell continuously makes adjustments to the underlying biological architecture in an effort to maintain homeostasis. These studies have demonstrated that cellular behavior is inherently robust. For example, the deletion of a single gene often results in little or no phenotype owing to the ability of the cell to adjust to genetic perturbation through redundant or alternate pathways that allow homeostasis to be maintained. In support of this robust behavior, metabolic engineers have long realized that despite having the ability to modulate and control gene function, the cell stubbornly resists attempts to alter flux through a particular desired pathway [6]. It has also been recognized that the few genes that *do* have the ability to alter metabolic flux can often to alter the flux of many diverse cellular pathways [6]. Formally, a biological network can be described where 'nodes' represent genes, and two genes are connected by an 'edge' if both are, for example, members of the same metabolic pathway. 'Hubs' are nodes that are highly connected, i.e., are connected to many other 'nodes' or genes. If the network has a few hubs yet most nodes are connected to few other nodes, the topology is characteristic of a 'scale-free' network. Metabolic flux and several other biological networks (e.g., genetic or protein interaction networks) are examples of scale-free networks. This topology is responsible for the perturbational robustness of the cell as a system. Thus, perturbations of most genes (individually) may have surprisingly few effects, while those that target hub genes may dramatically alter the cellular response (see Chapter 9 for details).

Yeast: A Pioneer and Driver of All Things 'Omic'

At the end of every sequencing project, the daunting task of translating gene sequence into gene function presents itself. As the first eukaryotic organism sequenced, and equipped with a powerful arsenal of genetic and genomic tools, *S. cerevisiae* has served as a gold-standard model organism for functional annotation of gene sequences, and has provided essential templates for the development of nearly all things 'omic'.

The Yeast Genome Project

The bold idea to sequence the *S. cerevisiae* genome was formulated in 1986 by Andrew Goffeau (Université catholique de Louvain). At the time, the *S. cerevisiae* sequencing project (YGP) paralleled challenges similar in scope and scale to those faced by the HGP ~14 years later. In many key ways the YGP set the stage for the HGP; at the time, the *S. cerevisiae* genome was ~60 times the size of any previous sequencing effort, compared to the HGP which was ~10 times larger than all previous sequencing efforts combined. In addition, both projects were subject to public scrutiny, and many in the scientific community deemed the projects unjustifiable, owing to the enormous costs that many assumed would result in little biological value, and which would short-change hypothesis-driven science conducted in individual laboratories.

From the start, Goffeau, together with Steve Oliver (University of Manchester Institute of Science and Technology), appreciated that the magnitude of the YGP would require a consortium effort. Although this 'big science' approach had long been appreciated in the physical sciences, it was new to biology, and the YGP set an important model for future projects. In 1989, Goffeau built a European consortium of 35 laboratories; that, although not an easy task, was facilitated by the history of the tight-knit *S. cerevisiae* community. At the end of the YGP 92 laboratories worldwide were active participants, representing the largest collaborative sequencing effort to date. The YGP had a significant impact on the genome sequencing projects that followed, all having adopted similar combined-force strategies. The completion of the YGP in 1996 [7] revealed more than just DNA sequence: the most significant revelation was that, despite decades of effort, most of the protein-coding genes predicted from the DNA sequence did not correspond to any that had previously been encountered [8]. Moreover, 35% of the genes in the genome were designated orphan genes, that is homologs of these genes had also escaped detection in other organisms. These findings highlighted the importance of the full yeast genome sequence, particularly in light of the fact that by the end of the 1980s the gene discoveries were mostly rediscoveries, suggesting that the genome had been tapped out, when in fact the majority of the genes in the genome had not yet been identified.

The Yeast Deletion Project

Soon after the completion of the YGP the need to translate sequence into function loomed. One of the most effective approaches to elucidate gene function is to knock down or knock out the activity of a gene and observe the effect on the ability of the cell to function. Owing to the powerful and unique genetic tools available to yeast, constructing a gene deletion is straightforward and allows for a 'clean' site-specific start-to-stop replacement of the wild-type gene with a double-stranded laboratory-built DNA knockout cassette module [9]. The painstaking task of deleting and validating each of the ~6000 yeast genes one at a time inspired the creation of a second consortium, the Yeast Deletion Project (YDP) [10,11]. This effort, involving 16 laboratories worldwide and championed by the yeast geneticists Ron Davis (Stanford University) and Mark Johnson (Washington University), allowed the systematic construction of a genome-wide set of yeast deletion strains, indisputably the most powerful functional genomics tool available to date. Moreover, the results of the YGP had made clear that the standard genetic approaches were biased, yet few laboratories were interested in embarking on studies of orphan genes, because without a new toolkit they were unlikely to uncover any new biological insights. Thus the YDP served to remove this bias, by allowing all genes to be studied simultaneously with nearly the same effort as that required for the study of a single gene.

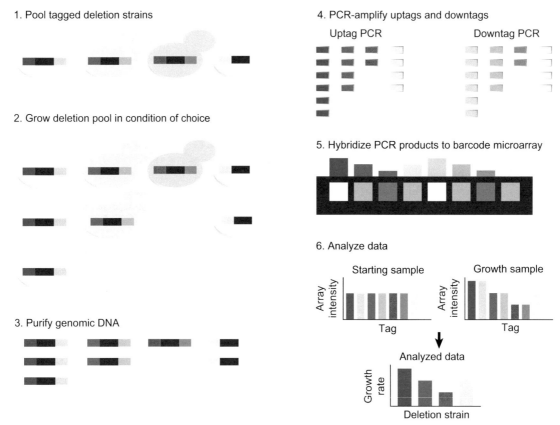

FIGURE 8.1 **Description of the competitive growth assay.** Fitness profiling of pooled deletion strains involves six main steps. 1. Strains are first pooled at approximately equal abundance. 2. The pool is grown competitively in the condition of choice. If a gene is required for growth under this condition, the strain carrying this deletion will grow more slowly and become underrepresented in the culture (red strain) over time. Similarly, resistant strains will grow faster and become overrepresented (blue strain). 3. Genomic DNA is isolated from cells harvested at the end of pooled growth. 4. Barcodes are amplified from the genomic DNA with universal primers in two PCR reactions, one for the uptags and one for the downtags. 5. Resulting PCR products are hybridized to a barcode microarray that quantitates the tag sequences, and therefore the relative abundance of each strain.. 6. Tag intensities for the treatment sample are compared to tag intensities for a control sample to determine the relative fitness of each strain. (*Figure originally published in [12]*).

The complete yeast deletion collection, often referred to as the Yeast KnockOut (YKO) collection, is comprised of four sets of strains: ~5000 haploids of mating type **a**, ~5000 haploids of mating type α, ~5000 homozygous deletion strains and ~6000 heterozygous strains. The additional ~1000 heterozygous strains represent the ~18% of genes that were determined to be essential (to unperturbed cells). To construct the yeast deletion collection, tens of thousands of oligonucleotides were necessary and the associated cost would have been prohibitive had they been purchased from commercial sources. Owing to the development of the first homemade 96-well oligonucleotide synthesizers at the Stanford Genome Technology Center, however, the oligonucleotides could be made at dramatically lower cost. Whereas most of the scientific community have realized the significance of the YDP, few appreciate that the project was contingent on the 96-well oligonucleotide synthesizer, a technology that had already been dismissed by the National Institute of Health (NIH) and by industry as a technology in search of an application. The 96-well synthesizer produced the eight unique primers required to construct and confirm each deletion strain. The construction strategy to make the knockout cassette modules used an elegant design devised by Davis (the proof-of-concept first demonstrated in bacteria [13]) that included the incorporation of two molecular tags or 'barcodes' (the second acting as a fail-safe) into each deletion strain. These molecular barcodes, comprising 20 bp DNA sequences that serve as unique strain identifiers, not only ensured against strain mix-up, but allowed the function of each gene to be interrogated in parallel by mixing them together. Specifically, in each competitive growth or 'fitness profiling' assay (Figure 8.1) a single culture is first inoculated with an approximately equal number of cells from each deletion strain. The culture is then grown in a condition of interest and samples are collected at several time points over a period of 5–20 generations. At the end of the experiment, genomic DNA is extracted from each sample and the molecular barcode tags are amplified in two PCR reactions (Figure 8.1). The abundance of each deletion strain is quantified by hybridizing the resulting PCR products to an oligonucleotide array carrying the complementary barcode sequences. The more important a gene is for growth in a given condition, the more rapidly the associated deletion strain will diminish relative to the other strains in the culture, reflected by the decreased intensity of the associated molecular barcodes on the 'tag' array. Thus, all genes required for growth are identified and quantitatively ranked in order of their relative importance for resistance to the condition of interest. This powerful assay has been applied in >10 000 individual studies, a large percentage of which have been driven by the HIPHOP chemogenomic screening platform, a variation on the fitness assay (Box 8.1), allowing the systematic interrogation of the relative requirement of each gene for resistance to the small molecule or drug tested (described in detail in the HIPHOP section).

Impact of Yeast in the Development of Genomic Technologies

The impact of the YGP and YKO projects is clear from the >3000 citations of the two publications describing the YKO collection [10,11]. Yeast moved on to become a highly respected innovator and test bed for virtually all 'omics' technologies, generating a vast number of comprehensive genome-wide datasets from these various platforms. Some of the genomic technologies that were first developed and tested in yeast include the first comprehensive gene expression microarray (following proof of concept [14]) and the molecular barcode microarray (the 'TAG series') designed for fitness profiling assays. In addition, yeast was used to test several generations of technology improvements of the microarray, many testing the performance of ever-increasing oligonucleotide densities. More recently, the first complete 4 bp resolution tiling array for both DNA strands of the yeast genome allowed the characterization of the complete transcriptome, leading to one of the first observations of pervasive transcription [15]. Other genomic technologies developed and tested using *S. cerevisiae* include, for example, platforms that measure protein modifications [16] and protein activity [17], protein–lipid interactions [18], genetic interactions (synthetic genetic array (SGA) technology) [19,20] and chemical–genetic interactions [21], among others. SGA technology, developed by Charlie Boone and Brenda Andrews at the University of Toronto, has attracted particular attention, allowing the identification of all (non-essential) genes that genetically interact with a chosen 'query' gene in a single experiment,

BOX 8.1 Fitness-Based Chemogenomic Dosage Assays

Chemogenomic profiles are generated by screening a library of yeast strains and measuring the sensitivity or resistance of each individual strain to small molecule of interest In *S. cerevisiae*, there are several full genome yeast collections available comprised of strains that each carry a gene whose copy number is either reduced or increased compared to a wild-type strain.

Here, we restrict our discussion to the yeast collections that carry DNA barcodes (20 bp in length) and therefore allowing parallel chemogenomic screens to be performed (as described in Fig. 8.1). These include:
- Yeast Knock Out (YKO) collection [10,11], each strain carries a start-to-stop deletion:
  - homozygous deletions strains; complete deletion of a nonessential gene in a diploid strain background
  - haploid deletion strains; complete deletion of a nonessential gene in a haploid strain background
  - heterozygous deletion strains; single-copy deletion of an essential or nonessential gene in a diploid strain background
- Decreased Abundance by mRNA Perturbation (DAmP) [22] collection [23]; each strain carrying a DAmP allele of an essential gene expressing ~10% of the wild-type gene dosage from its native promoter
- Molecular Barcoded Yeast Open Reading Frame (MoBY-ORF) collection; each gene carried on a high-copy 2μ plasmid expressed from its native promoter [24]
- MoBY-ORF 2.0 collection; each gene carried on a single-copy CEN plasmid expressed from its native promoter [25].

The non-competitive assays typically involve growing strains on plates where the fitness phenotype is quantified by colony size (e.g., [26]). In these assays, small molecules are added to solid media in an automation-friendly petri dish (known as an 'omni tray') to serve as the test plate; >600 such assays have been performed to date. Individual strains are arrayed onto the plates by 'pinning' from a 96- or 384-well formatted deletion collection. Sensitivity or resistance to small molecules is then measured by comparing the colony size of each strain to the same strain grown on a mock-treatment control plate, or to a control wild-type strain on the same plate, Colony size has often been assessed by visual inspection, although image analysis software can now be applied for better quantification. The collection of resulting relative growth measurements defines a chemogenomic profile specific to the tested small molecules, with the number of significant differences observed dependent on the sensitivity and precision of the growth metric used. The limitations of chemical screens performed in plate assays are that they are time consuming, costly in terms of the small molecule (requiring 10-100X the quantity required for a pooled liquid assay) and the cost/plate is 2-5 USD.

each packed with biological meaning. Once combined with an automated platform technology, SGA led to the first ever genome-scale di-genetic interaction map, providing a comprehensive view of the 'genetic landscape of the cell' [27] (see Chapter 6).

Drug Behavior is Promiscuous

Modern Drug Discovery: A Historical Perspective

The foundation of modern drug discovery rests on the paradigm of the early 1990s; that a small molecule with demonstrated ability to bind to/inhibit the activity of a target protein with high affinity/high potency in an in vitro biochemical assay will translate into an effective low-dose therapy. Traditionally, the potency of a small molecule measured in an in vitro binding or biochemical assay served as accepted evidence of the in vivo mechanism of action (MOA). In retrospect, given our current understanding of biology, the assumption that the in vitro behavior of a small molecule would exhibit the same behavior once in the in vivo context of the complex environment of a cell, let alone a human patient, seems improbable at best. How the MDD paradigm emerged and, moreover, was passionately supported from the early 1990s to the present day is thus a bit of a mystery, one best understood in its historical context.

Prior to ~1980 the discovery of new drugs was largely a result of careful pharmacology combined with serendipity and a heavy reliance on the isolation of natural products associated with long histories in traditional medicine. As advances in synthetic chemistry developed, the focus on developing improvements in compound isolation techniques shifted towards modifying existing drugs or natural products to improve their therapeutic value. Powerful bioactive molecules were also discovered in humans (e.g., hormones) during this period, and many were successfully modified to become drugs; thus the potent effects of small molecules had been realized. In the early 1990s, several advances in technology coalesced nearly overnight, and ushered in the era known as the high-throughput screening (HTS) revolution. Key to this revolution were (1) the discovery of recombinant DNA genetic engineering technologies, allowing large amounts of proteins to be produced in vitro; (2) improvements in chemical synthesis and combinatorial chemistry; and (3) the increased miniaturization, speed and liquid handling in robotic automation.

Together, these technological advances offered the pharmaceutical industry access to a vast number of novel compounds to explore new chemical space, and to increase the current screening throughput capacity by orders of magnitude (~1000 screens/day to 100 000+ screens/day). This opportunity proved too tempting to refuse and was quickly embraced. The advances in technology therefore led to (and in fact demanded) a new paradigm to justify the research and infrastructure investments that included large teams of chemists, automation engineers, and assay specialists to achieve this industrialization of drug discovery. The atmosphere was one of optimism and great promise, and from this frame of reference the modern discovery paradigm was born. In hindsight, the rush to embrace the target-oriented paradigm may have been premature, as little biological support could be provided to prove that the assumptions of the paradigm are valid and effective. At the same time, drug discovery invested heavily in genomics, and pursued many 'thinly' validated genomic targets, despite lacking a clear disease relevance. In a short time, mountains of data were produced by HTS, and the drive for ever faster and cheaper throughput became the central focus, fueled by the 'biotech boom' of the 1990s. By the end of the decade, however, despite the three orders of magnitude increase in HTS throughput, the return on the massive investment was a disappointment, as few new therapies had been discovered.

Thus the decade enthralled by technological advances ended with greater efforts being expended for diminishing returns, a result due in large part to incomplete knowledge of the underlying physiology and biology. One of the potential solutions to reversing this trend has been a renewed focus on academic discoveries and a fostering of public—private partnerships. In order for this approach to succeed, it is essential that (1) the successful strategies of pre-genomic drug discovery be reclaimed; (2) the mistakes of the HTS revolution are not repeated; and (3) the right technologies are used for the right applications. Today, new technologies continue to improve the efficiency and robustness of mining omic data for drug discovery opportunities, but to be effective they must be appropriately focused. To avoid the disconnect that occurred as a result of the assumption that increased data quantity = more new therapies, technologies must be actively and realistically evaluated for their ability to provide meaningful biological insight, and adjustments made in real time as needed.

Because of the dominating influence of target-oriented approaches to MDD over the last 20 years, it is not surprising that even well-established drugs are currently poorly characterized in vivo. The MOA of these older drugs must be revisited using in vivo approaches to understand their behavior. Indeed, several efforts that aim to repurpose older, already approved drugs require such approaches in order to illuminate new MOAs. Comprehensive characterization of MOA in vivo is difficult, and full understanding of in vivo MOA can, in theory, only be achieved by testing a small molecule for its ability to inhibit every protein activity in the cell. Such methods are currently unfeasible in mammalian cells, yet other approaches are available to address this issue (see the Chemogenomics section).

Drugs are Promiscuous; the Discovery of Polypharmacology

Molecular biology has long relied on the reductionist approach of intensive study of individual genes for a better understanding the cell. This approach has been tremendously successful, and is responsible for most of the major biological breakthroughs. However, as previously discussed, the reductionist approach does not effectively allow for the study of interaction-dependent gene function. Modern drug discovery adopted the molecular biologist's reductionist approach, revising the one gene—one protein—one function concept only slightly, to one gene—one protein—one disease—one drug. This 'forward' or 'target-oriented' drug discovery approach can be practically described as the concept that a small molecule with demonstrated ability to bind or inhibit a protein target with high potency in vitro will translate into a beneficial medical therapy in a human patient. As discussed above, this 'magic bullet' philosophy has rarely delivered, indicating that the effectiveness of a reductionist approach has reached its limits in MDD as well.

One explanation for the collective failure of target-oriented screening may lie in the recent 'discovery' that many drugs exhibit 'polypharmacology', defined as interacting with more than one protein target in the cell. Paralleling the interaction-pervasive behavior of genes, the discovery of 'promiscuous' drug behavior has proved with time to be much more prevalent than had been expected. Somewhat unexpectedly, polypharmacology is not limited to interactions with proteins in the same family as the intended target, but can involve proteins that seemingly share no structural or any other similarity with the intended target. Computational analyses that aim to predict polypharmacology have met with some success: specific scaffolds have been demonstrated to be promiscuous, and specific chemical properties are characteristic, though not definitive, of drugs that exhibit polypharmacology [28–30]. Interestingly, a recent analysis concluded that drugs exhibited a greater degree of polypharmacology than bioactive molecules, suggesting that polypharmacology is important for therapeutic efficacy [28].

Polypharmacology is now not only well established but is considered critical to the efficacy (and MOA) of approved drugs. Once again following molecular biology, systems-level approaches are required to increase our understanding of drug behavior. This approach is often referred to as network pharmacology, and although not yet fully accepted by the pharmaceutical industry, the need for such an approach has been highlighted in several recent studies [31,32]. Polypharmacology may be one answer to discovering new therapies; however, 'designing' polypharmacology will require screens that can accommodate drug effects on multiple targets, and allow the measurement and characterization of the cellular response, to guide our understanding of the net effect of multiple perturbations. More simply, designing polypharmacology will require a much deeper understanding of cellular biology and complex disease. The approach must first be tackled by understanding the cellular response to drug.

The development of chemogenomic technologies allows the cellular response to drug to be surveyed across the genome. These technologies are based on chemical—genetic strategies and allow comprehensive interrogation of compound-target relationships in the context of the living cell. Chemogenomic approaches have proved to be powerful tools that provide a genome-wide view of both the cellular targets and the biological networks or processes disrupted by the inhibition of the drug target.

Phenotypic Screening

Consistent with the accumulating failures of target-oriented screening, current trends in drug discovery suggest a re-emergence of phenotypic screens in drug discovery, bolstered by improved robotics, better imaging and image analysis algorithms. Phenotypic screens search for small molecules in whole cells or organisms that reverse a phenotype of interest, such as one that may be characteristic of a particular disease. A recent analysis [33] reported that a surprisingly high percentage (37% cell-based compared to 23% target-based) of the novel molecular entities (NMEs) discovered in the last decade resulted from cell-based phenotypic screening. This is despite the heavy bias on target-oriented screening campaigns during this period. Phenotypic screens are promising, but present new challenges. For example, because cellular phenotypes serve primarily as a cumulative readout of complex genetic and protein processes that may be coordinated by several pathways, mechanistic follow-up is required. However, tracking the cellular phenotypic effects to a specific target or mechanism is a challenge, and rapidly results in a new research bottleneck.

Chemical Probes; Like Drugs, but Designed to Serve as Tools to Study Gene Function

With more academic [34] and non-pharmaceutical groups (e.g., [35]) using small molecules to study biology, the term 'drug' as a catch-all to describe these molecules is inadequate, and in most cases wrong. By way of example, nocodazole is a microtubule depolymerizing molecule that has been instrumental in understanding many aspects of cytoskeletal and cellular biology. Nocodazole had its chance to become a drug, but failed in clinical trials. So rather than use the technically correct term 'failed drug' to describe nocodazole and similar molecules, a better definition was needed. The term 'chemical probe' was introduced for this purpose and has been widely accepted.

A chemical probe is a small molecule that targets a specific protein or other cellular structure and, by binding to its ligand, inhibits its activity. Nocodazole, and other chemical probes that can selectively inhibit specific cellular proteins, are powerful tools to elucidate gene function. Furthermore, when used in combination with other agents or together with genetic perturbations, they allow dissection of biological pathways at high resolution. Ultimately, once developed and validated, some chemical probes may graduate into lead candidates for new therapies, but this progression involves issues of chemical tractability that are often irrelevant for a probe to be a useful tool. Alternatively, a chemical probe can serve to demonstrate that a novel protein is druggable, thereby proving a potential for therapeutic intervention. When applied to a cell, the effects of chemical probes are complementary to genetic tools, yet they have distinct advantages. Chemical probes are rapid, reversible and tunable, and often directly affect the activity of a target protein. Unlike drugs, chemical probes are not constrained by pharmacokinetic and pharmacodynamic considerations, and therefore the biological space they can interrogate is potentially unlimited. Equally important, they are not constrained by pharmacoeconomic considerations, e.g., intellectual property is mostly irrelevant. An additional advantage of chemical probes is that, unlike genetic approaches to perturb function, they can often be readily transferred to other organisms (evolutionary conservation permitting) that may lack genetic tools. The ability to transfer chemical reagents across organisms increases their utility, and by making them readily available to researchers they promote interdisciplinary science. The concept of a community process where promising chemical probes are validated by community members as they use and modify them [36] has received much attention. The initial Molecular Libraries Initiative (MLI) component of the NIH roadmap [37] was motivated, in part, by 'the need to expand the availability, flexibility, and use of small-molecule chemical probes for basic research' [37]. Our own laboratory has adopted a strategy to provide resources for this community-driven effort to reduce costs and increase the efficiency of chemical probe identification and development [38].

The most important characteristic of a chemical probe (desirable but not necessary for a drug) is that the mechanism of action be well understood. Because the purpose of chemical probes is to act as molecular tools to elucidate gene function, if, for example, polypharmacological effects are not realized, the result may be mis-annotation of gene function. As already discussed, understanding in vivo MOA is challenging, particularly for compounds that act by polypharmacological mechanisms, where inhibiting one target may obscure the effects that result from the inhibition of is a secondary targets. The tools available to disentangle these different possibilities have been most extensively developed in model organisms, and several instructive examples are described in the following sections.

CHEMOGENOMICS

A chemical−genetic assay measures the response of a single gene to a small molecule, whereas a chemogenomic profile is a combined set of measurements of the response of each individual gene to a small molecule. Chemogenomic profiles can be generated by nearly all genomic technologies with the simple addition of a small molecule to the assay condition. Responses measured are platform dependent and can include, for example, RNA expression levels, protein abundance and/or modifications, metabolite abundance, and fitness in YKO and SGA screens.

Target Identification/Mechanism of Drug Action

Many large-scale chemogenomic datasets have been used to identify a small molecule's target. Knowledge of the target of a small molecule is increasingly important in drug development, providing the foundation for drug optimization. Often the need for target identification arises from a screen that identifies small molecules that induce a phenotype of interest, yet does not reveal the mechanism(s) that drive the phenotype. A common genomics approach to target prediction uses the chemogenomic profile of the small molecule of interest to query a large-scale reference set of profiles derived from characterized mutants (e.g., deletion strains) and/or small molecule with known and unknown targets/MOAs [26,39,40]. The assumption is that the small molecule of interest has a MOA similar to the MOA of the best match in the reference set, where matching is based on profile similarity. All such guilt-by-association approaches are inherently limited by the breadth and depth of the reference set.

Alternative approaches to target identification do not rely on guilt-by-association. Target-oriented (or forward) drug discovery dictates that, for lead small molecules that are highly potent, in vitro target will be the in vivo target. This approach (with important exceptions) has not generally been successful. A reverse drug discovery approach requires the measurement of the response of each gene to drug. HaploInsufficiency Profiling (HIP) in yeast provides such a reverse drug discovery approach (Box 8.1). Analogous to fitness profiling, HIP results in a list of the most likely drug targets, and often directly identifies the target.

Complementary assays that exploit modulations in gene dosage also allow direct drug target identification. These assays are discussed in further detail in the HIPHOP section below.

Compendium RNA Expression Approaches to Predict the Drug Target/Mechanism of Drug Action

The concept that large datasets could provide predictive value was first introduced to practice by small molecule screens performed at the National Cancer Institute (NCI), known as the 'NCI-60' project. The concept involved using profiles of cellular response to drug, measured across 60 cell lines, to classify small molecules by MOA and toxicity. Genome-wide expression studies borrowed many of the experimental design principles and analytics from the NCI-60 project [41].

In an innovative study [39] published soon after the introduction of RNA expression arrays, Hughes and colleagues used the NCI-60 concept to demonstrate that gene function and the mechanism of drug action could be predicted by querying a reference set to find the 'best match'. In this particular study [39], a 'compendium' of ~300 genome-wide RNA expression profiles, of yeast deletion strains and of a wild-type strain treated with small molecules, served as the reference set. The individual queries were expression profiles of either yeast strains deleted for a gene of unknown function, or a wild-type yeast strain treated with a small molecule of unknown mechanism. After computing the similarity between the query profile and the compendium profiles, in the example described here the best matches in the compendium for a query strain deleted for a gene with unknown function are strains deleted for genes involved in the ergosterol biosynthesis pathway, a common target of antifungal agents. These results suggested that the uncharacterized gene was also involved ergosterol biosynthesis, and following confirmation and an additional follow-up study, the gene was named *ERG28*. Eight other genes of unknown function were similarly annotated, thus proving the power of using large-scale datasets to predict gene function. In an additional experiment, a profile of wild-type yeast grown in the presence of dyclonine, a compound of unknown MOA, was used as the query. In this case, the query profile best matched the profile of the deletion strain *erg2Δ*, and it was inferred that Erg2 was therefore the drug target. This conclusion was based on an assumption that a decrease in protein activity due to inhibition by a small molecule would mimic the complete absence of gene activity. While genetic logic and the literature generally support the view that a decrease in protein activity due to inhibition by a drug would phenocopy a genetic decrease-of-function allele, it has not been demonstrated that the same holds true for a complete loss-of-function gene deletion. A later study demonstrated the drug target of dyclonine to be Erg24 [16], and importantly, when combined with an earlier study [42], provided compelling evidence that the dyclonine sensitivity observed in the *erg2Δ* strain was due to the downstream inhibition of Erg24. It is possible that the bona fide target of dyclonine might have been identified in this study had the reference set included the *ERG2* deletion strain; however, as *ERG24* is essential, a series of strains, each expressing a different level of Erg24, would have been required, and in addition, a similar series of strains representing all essential genes, making the construction of a comprehensive dataset impractical, and likely not even possible.

The discussion above strongly suggests that one of the conclusions of this study, that Erg2 is the protein target of dyclonine, was incorrect. The message provided by this re-examination of the compendium study clarifies the reasons why the incorrect conclusion was made: (1) the reference set was not sufficiently comprehensive; (2) the incorrect assumption that a gene deletion would mimic a decrease in protein activity due to inhibition by a small molecule; (3) the drug target was identified based on a downstream effect caused by the inhibition of the bona fide target. The take-home messages are: (1) a reference dataset can never be truly comprehensive; (2) the incorrect assumption made was based on current understanding of genetics and not on evidence; and (3) genomic technologies used for a particular application should, whenever possible, provide as direct a readout as possible, one best matched to the molecular level affected by the perturbagen. In this case, a transcriptional readout was used to measure the consequences of inhibiting a protein. The assumption that was made based on current understanding of genetics serves as a cautionary reminder that such assumptions can result in great cost, particular for large-scale genome-wide datasets. For example, a similar genetic-based assumption was made during the construction of the YKO collection. Specifically, it was assumed that the three auxotrophic markers included in the genetic background of the wild-type strain, from which all deletion strains were derived, were 'benign'. Recently, the auxotrophic markers have been shown to bias data in some cases and to limit the application of the YKO collection in others [43–45]. The details discussed here may seem trivial; however, these same concepts have provided the basis for much larger-scale reference datasets to be generated [46] so that, while proven powerful in inferring mechanism, future conclusions should be carefully considered to ensure that they are without bias. A second general impact of this study, one that came at great expense, was the futile pursuit of many genomic targets in drug discovery based on expression data alone. This highlighted the need to include, minimally, secondary evidence based on a complementary set of genomic data measured from a different molecular perspective before conclusions are drawn.

Fitness-based Chemogenomic Profiling Approaches to Identify the Drug Target/Mechanism of Drug Action

Non-competitive Fitness-based Chemogenomic Profiling Assays

Yeast-based chemogenomic platforms often take advantage of the YKO collection. The YKO collection and its variants have been used in > 600 individual chemical screens, measured by a simple count of the number of publications resulting from these chemical screens. Each publication presents the results from a genome-wide screen using a single, or sometimes a handful of small molecules, most often generated from plate-based assays that compare the colony sizes, that is, measurements of strain fitness, in the presence and absence of drug to identify the deletion strains that are sensitive or resistant to drug (Box 8.1). The sensitivity of each individual strain identifies a chemical–genetic interaction that can be of several types and often provides insight into the function of a gene. A deletion strain that is sensitive to drug indicates that the deleted gene function is required for resistance to drug. The function of the gene may, for example, be to buffer the drug target pathway, to transport the drug out of the cell and/or to detoxify the drug. The function of a gene, that when deleted results in resistance to drug, might be to import the drug into the cell, to activate the drug, or even might identify the drug target.

The fitness measurements for each individual strain, when combined, comprise a chemogenomic profile for the tested small molecule. The non-competitive assays rarely include heterozygous strains, and thus, they rarely test for drug-induced haploinsufficiency (see HIPHOP section). As a result, the target of the molecule is seldom identified (e.g., [21,26,47]). However, a profile often provides support for the identification of the general MOA of the small molecule. Gene Ontology (GO) analysis of a profile is useful for identifying the pathways associated with sensitivity/resistance to the tested molecule, towards inferring its MOA.

Competitive Fitness-Based Chemogenomic Profiling

Fitness-based chemogenomic assays that allow the pooling of strains require that the yeast library be barcoded. Five barcoded libraries exist: (1) the YKO homozygous and haploid non-essential gene deletion collection [10,11]; (2) the YKO heterozygous deletion collection [10,11]; (3) the Decreased Abundance by mRNA Perturbation (DAmP) [22] collection [23]; (4) the Molecular Barcoded Yeast Open Reading Frame (MoBY-ORF) collection, where each strain expresses a single gene from a single-copy CEN plasmid [25]; and (5) the MoBY-ORF 2.0 collection, where each strain expresses a single gene from a multi-copy 2 μm plasmid [48]. Importantly, only the assays using libraries with partial gene dosage or increased gene dosage allow the drug target to be directly identified.

In each fitness assay the strains are grown competitively (in a pool) in the presence and absence of a small molecule, as detailed in Figure 8.1 (e.g., [11,21,25]). In the assays using libraries with decreased gene dosage, the more important a gene is for resistance to drug, the more rapidly the abundance of a given strain, and hence the associated barcode sequences of the strain, diminishes from the culture. The same holds true for assays using libraries with increased gene dose; however, rather than the abundance of a strain diminishing in the pool, the more resistant a strain is to drug the greater the abundance will be in the culture. All genes can subsequently be identified and quantitatively ranked in order of their relative requirement for resistance or ability to confer resistance to drug.

The deletion or over-expression profiling assays described above, when used together, provide additional support for the MOA that each assay suggests or identifies. Again, GO analysis serves as an important tool to translate the profile from each assay into a general MOA for the tested small molecule. If many fitness-based profiles are collected in a compendium, an approach analogous to the compendium RNA expression approach described above, can be used to predict the MOA of a query molecule.

Compendiums of non-competitive fitness-based profiles can readily become much more powerful by adding only a few profiles of, usually related, small molecules, as they provide sufficient resolution to resolve slight differences between similar MOAs (examples include [26,49,50]). Mechanistic insights derived from existing chemogenomic profiles have also been proved to directly translate into mammalian cells and/or to be of clinical relevance in several cases [51–53]. Only a few chemogenomic studies have generated large-scale datasets to date [21,47]. Clearly, the more chemogenomic profiles in a dataset, the more powerful and valuable the dataset becomes in identifying general MOAs of small molecules while also providing insight into gene function [47,54].

HIPHOP

Although much of the discussion so far applies to the chemogenomic platform described by HIPHOP (because this platform has been explored so extensively) a section devoted to the discussion of HIP and HOP is warranted. The ability to pool yeast deletion strains in fitness assays (Figure 8.1, Box 8.1) enabled the development of a chemogenomic platform that allows genome-wide screens to be performed in an automated and quantitative manner; the platform includes Haplo Insufficiency Profiling (HIP) and HOmozygous Profiling (HOP). HIP involves measuring the relative requirement of each essential yeast protein for resistance to a particular small molecule in a single in vivo assay. Here, the requirement of a protein for resistance implies that the small molecule has inhibitory effects on the protein, and as such, HIP enables a systematic, unbiased exploration of the druggable fraction of the yeast genome. Designed to identify drug targets, HIP exploits drug-induced haploinsufficiency [55], a phenotype whereby a growth-inhibitory molecule induces a specific fitness defect in a heterozygous strain deleted for a single copy of the gene encoding the molecule's target [55,56]. To identify a drug target de novo, ~1100 heterozygous deletion strains (representing the essential genes) are tested in parallel, resulting in a fitness defect score for each strain/gene [21,56]. HIP has proved to be a highly successful approach to identifying the drug target and/or MOA and, significantly, in several instances the identified target or mechanisms have directly translated into mammalian systems and into clinical settings [51–53,55–59]. A complementary assay, HOP, allows parallel testing of all ~5000 homozygous deletion strains (Box 8.1), each carrying complete deletions of both copies of a non-essential gene, for sensitivity to a small molecule. The HOP assay provides rich information on genes that buffer the drug target pathway, including those involved in drug metabolism and detoxification [21,49]. HOP also results in a fitness defect score for each strain/gene tested, and a HIPHOP profile collates scores from the HIP and HOP screens of the same small molecule. In our most recent large-scale HIPHOP study, ~3500 genome-wide HIPHOP profiles were collected as a single set using a precisely defined protocol, representing the most comprehensive, systematic chemical–genetic screening effort to date. Selected results from this large-scale screen are highlighted below.

Chemical–Genetic Interactions Reveal Conditional Essentiality

The significant chemical–genetic interactions identified by HIPHOP provide insights into gene function that can be added to the existing functional annotation of a gene or, in some cases, serve as the only annotated characteristic of a gene. Several hundred genes are in fact annotated solely by a single chemical–genetic interaction, hinting at the underlying power of small molecules to induce measurable phenotypes. The potent ability of small molecules to induce a phenotype in a yeast deletion strain was demonstrated in a HIPHOP study that tested ~400 unique environmental and chemical conditions genome-wide [21]. Specifically, we demonstrated that 97% of the genes in the yeast genome are required for growth, defined by a significant fitness defect in at least one of the conditions tested. This was a remarkable finding, as despite decades of research, two-thirds of the genes in the genome was without annotation, an observation that had been attributed to genetic redundancy. This redundancy posed an evolutionary paradox as to why the cell would maintain a gene that could be deleted without consequence. Our results demonstrated that there was a cellular consequence; one that was obvious when the deletion strains were challenged by a small molecule introduced into their environment. This proved true for nearly all deletion strains, including those deleted for genes known to be duplicated during evolution, and those deleted for genes that overlap in function. For example, gene paralogs exhibited higher-than-average co-fitness values, indicating a shared but nonetheless distinct function. Co-fitness is a quantitative term that measures the similarity between the behavior of a pair of genes across all tested conditions [21]. Overall, this observation argues against the redundancy of duplicated genes because, if such genes were fully buffered, they would not be expected to exhibit a growth defect when individually deleted. Therefore, our finding supported a model [60,61] positing that such genes are partially redundant, with deletion of either duplicate resulting in a similar phenotype.

At the time, the definition of essentiality was controversial; however, the evidence provided by our study revised the definition of essential to be more complete by including the concept of 'conditional essentiality'. Genetic redundancy had thus been demonstrated to be limited, and is observed only in optimal growth conditions, i.e., standard laboratory conditions. While the conditional essentially observed in our study was due to an alteration in environmental conditions, the concept can be extended to other perturbations, for example those caused by introducing a genetic alteration into the cell. In addition, conditional essentiality has important implications in cancer therapeutic strategies that exploit local tumor environments to selectively target cancer cells using small molecules (e.g., [62]).

Identification of Novel Chemical Probes

To identify novel chemical probes and their cognate targets, we defined a 'HIP target' of a small molecule using the following criteria: (1) the fitness defect score of the heterozygous strain must pass a stringent significance threshold ($p \ll 0.01$); (2) the heterozygous deletion strain must exhibit the expected drug-induced haploinsufficiency phenotype in an isogenic culture; (3) the HIP profile of the small molecule inhibitor must identify only one or a few heterozygous strains (provided the identified strains pass the first two criteria). The final step ensures that the small molecule acts with a high degree of specificity. For example, the HIP profile of the chemical probe tunicamycin clearly shows that the probe exhibits a high degree of specificity, and identifies the heterozygous deletion strain $alg7\Delta/ALG7$ as having the greatest fitness defect, thereby successfully identifying the known yeast target of tunicamycin, $ALG7$ (Figure 8.2A). As a result of inhibiting $ALG7$, which encodes a protein required in the first step of N-linked glycosylation, protein folding in the endoplasmic reticulum is disrupted, which in turn induces the unfolded protein response (UPR). The UPR pathway has been extensively characterized and the primary cellular response is transmitted via the translational upregulation of the $HAC1$ transcription factor through an unconventional splicing mechanism by $IRE1$ (e.g., [63]). Both genes are 'classic' reporters of the UPR. As is readily apparent in the HOP profile of tunicamycin (Figure 8.2A), these genes and others exhibit highly significant chemical—genetic interactions with the probe. Moreover, GO-based functional enrichment analysis of all genes exhibiting chemical—genetic interactions with tunicamycin highlights the biological processes that are both directly (e.g., 'glycosylation') and indirectly related to the target of the probe (e.g., 'vacuolar transport'; Figure 8.2B). Taken together, the most significant interaction in the HIP profile identifies the drug target, while the HOP assay identifies the chemical—genetic interactions that buffer the target pathway, and when combined, the enrichment profile provides support for the MOA of tunicamycin, clearly characterizing the cellular response to the probe.

In addition to identifying the known targets of commonly used chemical probes (e.g., tunicamycin, rapamycin), we also rediscovered the targets of two compounds that were originally identified in our laboratory and published in separate studies (erodoxin and cantharidin [27,64]). To successfully identify targets with the above HIP target method, the small molecule must inhibit growth and the targets must have yeast homologs that are essential. Overall, we successfully identified the known targets of the screened molecules that meet these criteria, with few exceptions [65,66].

To date, our efforts to identify novel chemical probes, using the HIP assay and the criteria described above, have resulted in the identification of over 100 specific inhibitors that target 54 unique essential gene products. From the novel small molecule—HIP target interactions, six were validated by direct in vitro enzymatic (Ole1, Pma1, Sec14, Sec7) or polymerization/binding (Tub2, Arc19) assays and eight using in vivo measurements (Erg7, Sec13, Ssl2, Fas1, Erg1, Cdc12, Tor2, Erg11), including metabolomics, metabolite complementation, and fluorescence microscopy assays consistent with specific inhibition of protein activity [38]. Thus far we have successfully validated 26% of the novel HIP targets, demonstrating HIP as a robust platform for identifying both known and novel chemical probe and their cognate protein targets.

HIP Predicts the Yeast Druggable Genome is Double that of Current Estimates

From a global perspective, analysis of our HIP targets revealed a significant enrichment for 'druggable' proteins, i.e., those with precedence for inhibition by 'drug-like' small molecules, relative to the essential yeast genome. While the definition of 'druggable' is contentious, we observed similar enrichments using five different metrics that estimate druggability (χ^2 test, p < 0.001). Since the beginning of our ~3500 compound screen, the number of HIP targets (defined by our criteria) identified has increased approximately linearly with the number of compounds screened, and the rate of discovery of druggable compared to undruggable HIP targets is nearly identical. Assuming both discovery rates remain constant as more small molecules are screened has allowed an empirical prediction of the size of the yeast druggable genome by simple extrapolation. This analysis results in a prediction that 44% of the essential yeast genome is druggable by our HIP target criteria, nearly doubling current estimates [31]. By extension, using yeast—human orthology, a prediction of the human druggable genome is also twice that of current estimates. This result confirms the suspicion that the portion of the genome currently defined as undruggable is not likely to be due to inherent 3D chemical properties characteristic of the encoded proteins, but is more likely a designation assigned by default due to the focus of the pharmaceutical industry on proven druggable targets and a corresponding lack of exploration of other proteins in HTS campaigns [67]. In sum, our large HIPHOP dataset, collected in a systematic and unbiased manner, resulted in the first empirically based estimate of the yeast druggable genome, and by extension, the human druggable genome. We have identified lead chemical probes for 54 proteins so far, over half of which target undruggable proteins, representing a valuable resource to the yeast and chemical genetics communities, as well as to the systems biology community at large.

Co-Inhibition Reflects Structure and Therapeutic Class

A HIPHOP profile is a genome-wide in vivo snapshot of the cellular response to a small molecule, and as clearly illustrated in the tunicamycin profiles (Figure 8.2A), highlights

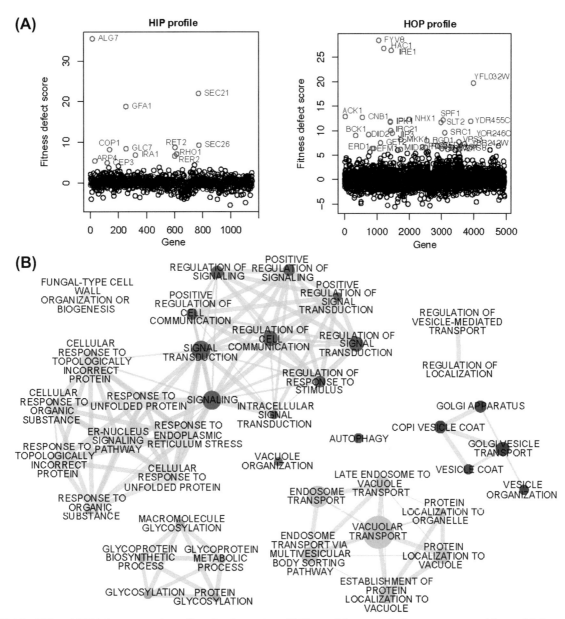

FIGURE 8.2 **HIP and HOP chemogenomic profiles of tunicamycin.** (A) Fitness defect scores for heterozygous essential gene deletion strains are shown in the left plot, i.e., the HIP profile, and those for the homozygous non-essential gene deletion strains are shown in the right plot, i.e., the HOP profile. In both plots, strains are arranged alphabetically by gene name on the x-axis. Significant chemical–genetic interactions are shown as red open circles ($p \leq 0.01$ for the HIP profile and $p \leq 10^{-4}$ for the HOP profile to improve visualization). Highly significant interactions in the HIP profile can predict the target of the tested small molecule. For example, *ALG7* is the most significantly sensitive and it is also a known yeast target of tunicamycin. Significant interactions in the HOP profile highlight genes that buffer the targeted pathway, including those involved in drug metabolism and detoxification. Tunicamycin is also known to induce the unfolded protein response and the HOP profile shows that regulators of this response, *HAC1* and *IRE1*, exhibit highly significant interactions with the small molecule. (B) Biological processes and protein complexes significantly enriched amongst all genes exhibiting chemical–genetic interactions with tunicamycin (FDR ≤ 0.1 by gene set enrichment analysis, [68]). Each node represents a gene ontology biological process or protein complex that is significantly enriched in the combined HIPHOP profile of tunicamycin. The size of a node is proportional to the significance with which the gene category is enriched (according to the FDR estimate), with larger nodes indicating greater significance. The width of an edge is proportional to the level of gene overlap between the two connected categories (i.e., gene sets). The color of a node shows its cluster membership, where clustering is based on the level of overlap between categories and thus groups together related categories. This network highlights biological processes that are both directly (e.g., glycosylation) and indirectly (e.g., vacuolar transport) related to the target of tunicamycin. The network was made using Cytoscape software [69].

the genes most required for resistance to the molecule. The biological 'richness' of each profile is readily appreciated in the GO enrichment maps generated for each small molecule screened (e.g., Figure 8.2B). Provided the profiled small molecules inhibit growth, HIPHOP profiles that are highly correlated indicate that the corresponding small molecules induce similar cellular responses, suggesting that they may have similar MOAs. We previously defined the quantitative term 'co-inhibition', a metric defined by the pairwise correlation between HIPHOP profiles [54]. From a global perspective, a weak but significant correlation between structural similarity of the profiled small molecules and co-inhibition was observed, suggesting that chemical structure influences patterns of inhibition, as has often been observed. Perhaps more interestingly, we observed a more significant association between co-inhibition and small molecules that belong to the same therapeutic class, compared to those that are structurally similar [54]. For example, clozapine and propiomazine clustered together based on co-inhibition and both are annotated as neuroleptics. Furthermore, of the pairs of small molecules that were co-inhibited *and* shared therapeutic class, more than 70% did not have significant structural similarity. This indicates that molecules with very different structures can result in similar biological responses. This finding was surprising, particularly as most of the therapeutic targets do not have a yeast homolog. However, the observed co-inhibition may arise from a variety of sources, for example structurally diverse molecules that inhibit different proteins within the same pathway, or different structures that inhibit the same target, such as those that may arise from 'me-too' targets, where different pharmaceutical companies search for molecules that inhibit the same target but must be structurally distinct (to avoid intellectual property issues), and strategic medicinal chemistry tactics may be in play. Whatever the cause, the observed strong co-inhibition of drugs that belong to the same therapeutic class suggests that yeast HIPHOP assays may be useful for classifying therapies and/or possibly for proposing alternatives.

Co-Fitness Predicts Gene Function

Many systems-level datasets can be used to define gene/protein signatures. For example, the genetic or protein–protein interactions of a gene/protein define a signature for that gene/protein. Pairs of genes (as opposed to compounds for co-inhibition) that are highly correlated in a HIPHOP dataset can be described as being co-fit. Pairs of co-fit genes often share function, and co-fitness predicts gene function on par with or better than other genomic platforms. Furthermore, when considering individual biological processes, co-fitness performed particularly well in predicting genes involved in certain processes, for example amino acid metabolism, compared to measures of similar gene signatures based on physical interactions, genetic interactions or gene expression [54]. In our more recent dataset of ~3500 HIPHOP profiles, hierarchical clustering of all genes by co-fitness defined a subcluster revealing the detailed pathway required for each step on the road to translation (Figure 8.3, discussed in further detail below; [70]).

We also found that essential genes are co-fit with other essential genes more frequently than expected [54]. This observation suggests that essential genes tend to work together in 'essential processes'. Indeed, if we define a complex as essential if more than 80% of its members are essential, a significantly greater number of complexes are essential in rich medium than expected by chance [71]. By the same token, we exploited our HOP results to identify conditionally essential complexes, that is, complexes where more than 80% of the (non-essential) members are significantly sensitive in a given condition. This analysis revealed that there are significantly more conditionally essential complexes than expected by chance, in 40% of the tested conditions [54]. In fact, vesicle transport genes involved in complexes are, in general, sensitive to a large number of diverse compounds, suggesting that these complexes are required for the cellular response to chemical stress. This finding supports and extends previous findings on multi-drug resistance (MDR) (e.g., [21,26]).

Significant co-fitness between two genes can be thought of as a functional linkage in the context of chemical stress, and as such, it is possible to construct a co-fitness network where nodes represent genes and edges represent co-fitness linkages. One could characterize such a network by, for example: (1) identifying sets of genes that are highly co-fit with one another to define functional modules relevant to chemical stress; (2) identifying relationships between the modules; and (3) defining global properties of the network, e.g., by examining distributions of gene centrality, or the shortest path length between genes. Importantly, the characterization of a co-fitness network may provide novel insights into the (re-)organization of genes, the dependencies between genes, and the importance of particular genes when the cell is under chemical stress conditions. Once integrated with other types of biological networks derived under chemical stress conditions as they become available, an even richer system-wide picture of a chemically perturbed cell should be possible.

The Multi-Drug Resistance Network

Global View of the Mechanisms Involved in Cellular Resistance to Small Molecules

Classically, MDR genes are defined by genes involved in xenobiotic metabolism, a process designed by the cell for the detoxification of foreign substances. The proteins

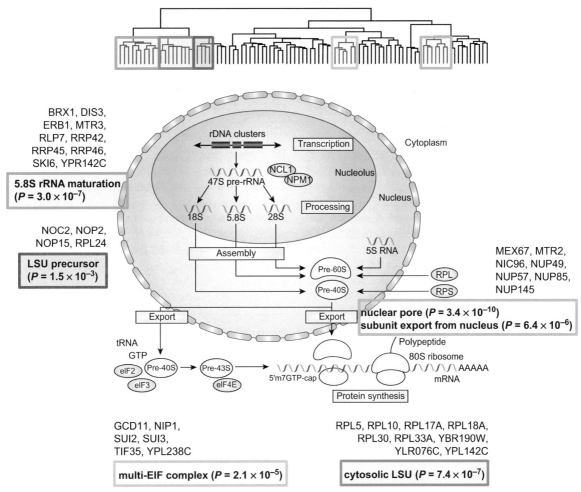

FIGURE 8.3 Co-fitness reveals the road to translation: translational regulation by genes involved in ribosomal biogenesis. Genes were clustered hierarchically based on co-fitness, i.e., similarity between their corresponding deletion strain signatures, where each signature contains HIPHOP measurements of the strain under ~3500 chemical perturbations conditions. Here we show a branch of the gene dendrogram that includes several clusters, highlighted in colored boxes, which are associated with ribosome biogenesis. For each highlighted cluster, select gene ontology biological processes and/or protein complexes that are significantly enriched amongst the member genes are shown in a box of the same color as the cluster box (corrected $p \leq 0.01$), next to a schematic of ribosome biogenesis in the cell. The multiple-test corrected p values measuring the significances of the enrichments are shown in parentheses. Further, the genes in the highlighted clusters are listed above the corresponding enrichment boxes. The biogenesis of the ribosome machinery is a highly coordinated process, which is composed of the synthesis and import of ribosomal proteins into the nucleus, synthesis and processing of ribosomal RNA (rRNA), assembly of ribosomal proteins and subsequent transport of the mature subunits into the cytoplasm. Most of these events take place in the nucleolus, except for 5S rRNA synthesis (which occurs in the nucleoplasm) and synthesis of ribosomal proteins (which occurs in the cytoplasm). The basic translation machinery is composed of ribosomal subunits, messenger RNAs (mRNAs), transfer RNAs (tRNAs), and translational initiation and elongation factors. First, the initiation factors eIF2, eIF3, tRNA and GTP are incorporated into a 40S ribosomal subunit to form a 43S complex. Second, eIF4e is recruited into the 43S complex to form a 48S complex with mRNA. Finally, a 60S ribosomal subunit and the 48S subunit form the final 80S complex. Abbreviations: eIF, eukaryotic initiation factor; LSU, large subunit of ribosome; rRNA, ribosomal RNA. *[Adapted by permission from Macmillan Publishers Ltd: Nature Reviews Cancer [72] copyright (2010)]*

encoded by MDR genes include cytochrome P450s that metabolize (~75% of all) drugs and to a much lesser extent multi-drug transporters, many belonging to the protein superfamily ATP-binding cassette transporters (ABC-transporter) that catalyze the active extrusion of a large number of unrelated small molecules out of the cell. From the few existing datasets that are large enough, yeast chemogenomic profiles identified a set of genes that, when deleted, are sensitive to multiple chemical perturbations [21,26,47]. Within this context, MDR genes have been generally defined by the corresponding deletion strains exhibiting a significant fitness defect to ~10–20% of the small molecules tested. Given the fact that yeast is notorious for its pleiotropic drug resistance (PDR) transporters and general ability to resist chemical stress, though many are redundant, we expected to identify PDR transporters as MDR genes by our criteria; instead, we identified surprisingly few, consistent with the findings of others [21,26,47].

The enormous size of our most recent dataset (an order of magnitude larger than those previously collected) allowed a more powerful analysis to be performed. Specifically, to better understand the cellular processes involved in the general response to chemical stress, we calculated the frequency of occurrence of significantly enriched GO biological processes and protein complexes across all small molecules tested. The result of this analysis revealed a coherent network of highly interconnected cellular processes that define a complex system of resistance mechanisms activated under conditions of chemical stress. The existence of such a drug resistance network is significant, as it implies that the cellular response to drug is in fact limited, and is defined and controlled by the regulation of processes identified in this network (Figure 8.4A).

Several of the enriched modules in the MDR network have been previously observed [21,26]; these include the ergosterol biosynthesis, vesicle-mediated trafficking and vacuolar biogenesis & regulation of pH modules. Many of the genes in the ergosterol biosynthetic pathway (highly homologous to the human cholesterol pathway), when deleted, are well known to increase the permeability of the cell to small molecules, likely due to effects on membrane composition. Several of these genes, particularly *ERG2* and *ERG6*, are often deleted specifically for the purpose of increasing the sensitivity of yeast to small molecules in drug studies. The requirement for the intracellular vesicle-mediated trafficking network for resistance to drug is not clear; however, perturbations in the cell wall and/or plasma membrane are the first cellular components that meet an incoming small molecule, and the response may require endocytosis for drug transport and sequestration. It may also include the activation of downstream signal transduction pathways and upregulation of proteins involved in remodeling the cell wall and/or plasma membrane, requiring the trafficking of these parts from the ER to the cell surface. Such processes require constant turnover in intracellular vesicular trafficking. This requirement for alterations in the activities of intracellular trafficking in response to drug is consistent with similar observations in mammalian cells [73–75].

The vacuolar biogenesis and regulation of pH *VB* module has also been previously recognized as required for MDR [21,26,47] in yeast. Our recent data provided an additional insight: many of the profile that were enriched in vacuolar biogenesis were profiles of cationic amphiphilic drugs (CADs), which often exhibited strong chemical—genetic interactions with *NEO1*. *NEO1* is a member of the evolutionarily conserved P-type flippases [76] and a major regulator of vacuolar pH. A temperature-sensitive *NEO1* strain exhibits hyper-acidified vacuoles [77], and because CADs are believed to accumulate in acidic vesicles due to their basic nitrogen [78], it follows that the *NEO1* chemical—genetic interaction might be the result of decreased vacuolar pH and enhanced trapping of CADs in the vacuole. Consistent with this proposed mechanism, raising vacuolar pH by treatment with bafilomycin A1, a specific inhibitor of vacuolar-type H^+-ATPase, conferred resistance to these drugs. In human patients, CADs are associated with drug-induced phospholipidosis (DIPL), a phospholipid storage disorder linked to clinical toxicities [78,79]. To explore the potential connection between the *VB* module and DIPL, we built a structural model based on small molecules associated with the *VB* module and demonstrated that the model performed well in predicting small molecules known to cause DIPL (based on the literature). Thus, yeast chemogenomic profiling is an effective predictor for small molecules that may cause DIPL. Moreover, our results identified potential genetic factors that might influence tolerance of CADs in patient populations.

Particularly striking in modules our MDR network involve ribosomal biogenesis and RNA processing, including core members of stress granules and P-bodies [80,81]. For example, the biological role of stress granules and P-bodies is an area of intensive research that, while well studied, has thus far escaped characterization, partly because the members that make up these stress granules and P-bodies are stress specific, and often include members that are seemingly involved in fundamental processes with widely different functional roles (e.g., [82,83]). Five modules that have *not* previously been associated with multidrug resistance include the nuclear pore complex (*NPC*), the cytoplasmic exosome, mRNA processing body and stress granule (*SG*), translation (*TL*), RNA polymerases I, II and III related functions (*RNAP*) and ribosome biogenesis (*RB*). Our data suggests that these modules form a coherent process that generally captures the cellular response to chemical stress. Upon close examination of these functionally linked processes, combined with hierarchical clustering of all genes based on co-fitness, a sub-cluster of ~400 co-fit genes emerged, revealing the processes required for each step on the road to translation at an exquisite level of detail (Figure 8.3). This road, defined experimentally, includes all of the steps required for ribosome biogenesis; a tightly regulated process that requires the coordination of several, sometimes simultaneously occurring processes, including the coordinated action of all three RNA polymerases. Ribosomal biogenesis starts with the synthesis of ribosomal proteins in the cytoplasm and import into the nucleus, followed by assembly of the subunits and synthesis of ribosomal RNA (rRNA) in the nucleolus. As the ribosome further matures, the subunits are transported through the nuclear pore complex and into the cytoplasm, where the final steps of ribosomal maturation occur. Co-fitness-based gene clusters are enriched for genes involved in discrete steps of ribosome biogenesis, and the

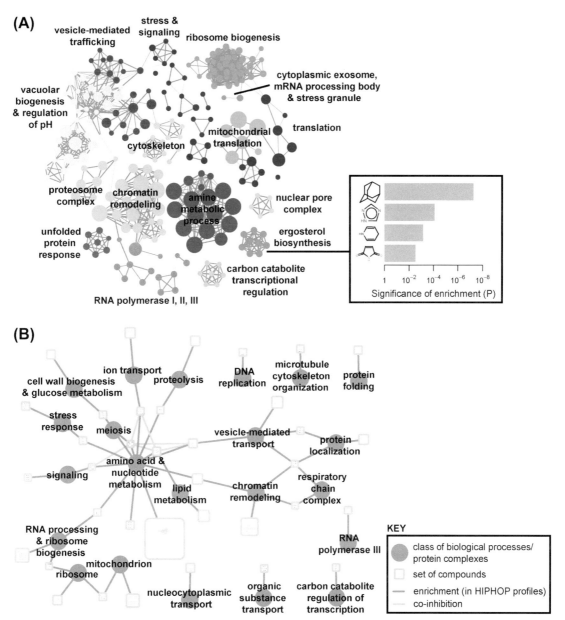

FIGURE 8.4 **The cellular response to chemical stress.** Biological processes and protein complexes frequently associated with chemical stress. (**A**) Each node represents a gene ontology (GO) biological process or protein complex that is significantly enriched in at least 1% of ~3500 HIPHOP profiles (FDR ≤ 0.1 by gene set enrichment analysis, [68]). The size of a node corresponds to the number of profiles in which the gene category is enriched, with larger nodes indicating more frequent enrichment. The width of an edge is proportional to the level of gene overlap between the two connected categories (i.e., gene sets). The color of a node shows its cluster membership, where clustering is based on the level of overlap between categories and thus groups together related categories. Each cluster is labeled with a general name that describes the member nodes. For the 'ergosterol biosynthesis' cluster, we show the top four most significantly enriched ring assemblies amongst the small molecules that are associated with the cluster, via enrichments in their profiles. (**B**) GO biological processes and protein complexes that are frequently associated with chemical stress, either alone or in combination. All considered processes and complexes were clustered based on the level of overlap between categories, and each resulting cluster is represented as a class of processes and complexes (pink nodes). A blue-bordered node is associated with a set of at least five small molecules with HIPHOP profiles that are significantly enriched (FDR ≤ 0.1 by gene set enrichment analysis, [68]) for processes/complexes only in the classes to which the node is connected (by pink edges). The size of a blue-bordered node is proportional to the number of small molecules associated with it, with larger nodes indicating more molecules. By extension, the sizes of these nodes indicate the prevalence of particular enrichment signatures. Two blue-bordered nodes are connected by a blue edge if at least 10% of the associated between-set small molecule pairs exhibit significant co-inhibition ($p \leq 0.01$). Interestingly, blue edges only occur between small molecule sets enriched for amino acid and nucleotide metabolism processes. Finally, each blue-bordered node contains a nested co-inhibition network of the associated small molecule set, where nodes represent molecules and edges represent significant co-inhibition ($p \leq 0.01$). The main network can be mined to identify chemical classes that target or at least affect multiple processes, and to potentially identify functional relationships between different processes. These functional relationships may only become apparent under specific perturbation conditions. The network was made using Cytoscape software [69].

relationships between these 'step' clusters in the hierarchy also reflect broader functional similarity (Figure 8.3). For example, the precursor and cytosolic ribosomal large subunit (LSU) clusters are closely related in the hierarchy.

A hallmark of stress is a rapid reprogramming of translation; translation is first sharply downregulated, while a few select transcripts are translationally upregulated, facilitating adaptation and promoting survival. The translation of mRNAs is a key regulatory point that is often controlled through the rate-limiting translational initiation step, which requires a complex interplay between translation initiation factors [84,85]. Ribosome profiling has also revealed that translation can be regulated through many distinct mechanisms, including, for example, specific translational uORF upregulation [86], stress granules that inhibit translation of specific transcripts [87], blocking of ribosomal subunit export at the nuclear pore complex [88,89], mRNA decay of specific transcripts, regulation mediated by the TOR pathway [21,61,64,90], control of ribosome biogenesis [89], alternate splicing [87], and co-regulation of transcription and translation [91]. The relative contributions of these diverse processes are often stress specific. In light of the many ways in which translational regulation can occur; the *NPC, SG, TL, RNAP* and *RB* module, our MDR network may not reflect a requirement. For example, ribosomal biogenesis may not be required for resistance to chemical stress per se, but its presence in the MDR network is instead is indicative that the translational reprogramming required during stress is mediated through these processes.

In light of the many translational regulatory mechanisms cited above, the novel modules in our MDR network may be connected by the following series of events that together define the cellular response to chemical stress: (1) global translational arrest, mediated by mechanisms that require genes represented in any or a combination of the *NPC, RB, RNAP* and *TL* modules; (2) mRNAs freed by aborted translation are sorted for different fates, either selective degradation (P-bodies) or storage/translational inhibition/trapping of translation initiation factors and 40S subunits in stress granules, requiring genes represented by the *SG* module; and (3) selected key genes or mRNAs are translationally upregulated (i.e., those not trapped by stress granules) by mechanisms that often include uORF and alternative splicing-enhanced translation [87], and may also involve many of the same mechanisms as those required for decreased translation. Finally, the select upregulated proteins act to adapt to stress and/or restore cellular homeostasis.

The literature cited above supports this general view of the cellular response to stress. Collectively, a new perspective of the cell has emerged, one where many genes are regulated by post-transcriptional and translational regulatory mechanisms [92–94] and fewer by transcriptional mechanisms. During stress, translation is the dominant regulatory mechanism [95] allowing for a rapid response, that can later be supplemented by increased or decreased transcription. This translational view of the cell is also supported by mammalian studies [96]. Moreover, many of the unique regulatory mechanisms, processes and even cellular components (e.g., P-bodies and *stress granules*) are activated only during conditions of stress, and the precise response (e.g., acting through one or a few genes) is specific to the particular stress condition. Several of these newly discovered post-transcriptionally regulated mechanisms have been recently linked to the progression of cancer [72,97], and may indicate potential mechanisms of cancer progression that enable cells to escape the many fail-safe mechanisms of this system [72]. This new research avenue has also been suggested as a rich source to mine for novel drug targets [98]. Albeit preliminary, the MDR network that emerged from our chemogenomic dataset pulls together many of the individual studies cited into a coherent framework that defines several novel mechanisms of resistance to stress, and thus the existence of this network represents an important biological discovery.

Further analyses combined with hypothesis-driven experiments are required to fully characterize and study the behavior of this network. Significantly, our first study of ~400 diverse environment and small molecule perturbations did not uncover the existence of this chemical stress response network, suggesting that the collection of a dataset an order of magnitude larger was essential to uncover this new biology, and supports a view that the dataset carries biologically rich information that can be explored at multiple levels of resolution. Overall, our MDR network exhibits many characteristics that are reminiscent of quiescence, rapamycin-induced inhibition of TOR and nitrogen starvation. Translational regulatory mechanisms may serve as a central control center; insulating the cell from chemical or other stress, activating stress-specific response mechanisms at precisely defined point in the network as needed, temporarily turning off translation, reducing growth, actively accumulating mRNAs in storage granules in the cytoplasm, and preparing the cell for rapidly restoration all systems when a favorable environment presents itself. In summary, although much analysis and further study lie ahead, the MDR network captures the first comprehensive genome-wide view of the cellular response to chemical stress, defining potentially novel resistance mechanisms that may be shared by other 'cellular states' and/or stress mechanisms.

Chemical Structures Associated with MDR Biological Processes

In addition to revealing the properties of the cellular response to chemical stress, we can refine our understanding of the relationship between specific biological processes and chemical stresses by examining the relevant chemical structures. For example, we identified ring assemblies that are significantly enriched among small molecules for which ergosterol biosynthesis is important for resistance (Figure 8.4). Ultimately, establishing such relationships between biological processes and chemical classes would permit the inference of biological effects based on chemical structure alone, and thus greatly simplify the characterization of the effects of specific small molecules and guide the development of drugs. To further exploit our dataset, we identified relationships between cellular processes in the context of chemical stress (Figure 8.4B). Specifically, we identified processes that are frequently enriched alone or together in the HIPHOP profiles of tested small molecules. The processes define a high-level signature of the cellular response to a subset of chemical stresses. As described before, it would be interesting to investigate whether small molecules inducing the same signatures have common structural features. For example, these signatures may reveal that a specific chemical class targets multiple processes. Similarly, processes that co-occur may be indicative of a functional relationship. For example, although several profiles are enriched for vesicle-mediated transport genes, a subset of these compounds are associated with both vesicle-mediated transport and protein localization (Figure 8.4B). As vesicle-mediated transport plays an important role in protein trafficking, a functional relationship between these main processes is expected. Additional study is required to characterize the identified relationships between cellular processes, and also their relationship back to specific chemical stresses; however, efforts in this direction are sorely needed for understanding the broad effects and/or polypharmacology of compounds, particularly for the development of therapeutics for complex diseases.

CONCLUSIONS AND FUTURE CHALLENGES

Over the last decade the HIPHOP platform has been developed, rigorously validated and applied in over 10 000 chemogenomic screens. From the beginning, the data generated in these screens have reported several themes in a consistent and recurring fashion. First, most small molecules do not act by inhibiting a single protein in the cell; however, there are exceptions. Defined by their highly specific HIP chemogenomic profiles, these exceptions, representing < 30 of the ~3500 small molecules screened, include: (1) FDA-approved drugs: methotrexate, statins, rapamycin, and the antifungal terbinafine; (2) established chemical probes: tunicamycin, cerulinin and aureobasidin; and (3) novel chemical probe, counting only those that we identified and verified in vitro or validated using other non-genetic assays: for example, sodium fluoride, cantharidin, and chemical probes targeting Tub2, Sec7, Pma1, Sec14, Sec13, Ero1, Ole1 and Erg7 [38]. In support of the validity of this 'theme', we previously demonstrated that a 'perfect' drug behaves 'perfectly' in the HIPHOP assay by using an engineered *CDC28* kinase inhibitor that specifically acts only on a *CDC28* allele designed to 'fit' the inhibitor. Ligand analogs, such as the engineered *CDC28* inhibitor, have been well established as highly potent and highly specific inhibitors of analog-sensitive alleles [99,100]. The resulting HIP chemogenomic profile of the *CDC28* inhibitor revealed an extraordinary degree of specificity for the Analog-sensitive allele, no sensitivity for the wild-type heterozygous *CDC28* deletion strain or any other strain [64]. In sum, while modern drug and chemical probe discovery aims to identify small molecules that inhibit a single protein with high specificity, and the HIPHOP platform can clearly identify such molecules, < 1% of the screened molecules to date demonstrate this property, suggesting that single-hit specificity rarely occurs in vivo.

Overall, from our large-scale discovery effort involving some 20+ million chemical−genetic interaction measurements, we have so far discovered ~100 novel inhibitors that hold promise for development into validated chemical probes. Over the course of the screening campaign the rate of HIP target discovery has been constant, that is, the number of HIP targets identified increased linearly with the number of small molecules screened. Moreover, the discovery rates are nearly equivalent for druggable compared to undruggable HIP targets. Significantly, nearly half of the HIP targets are considered 'undruggable'. This confirms the suspicions that the classification of a protein as undruggable is not due to an inherent inability of the protein to bind or be inhibited by a small molecule, but rather more likely due to the protein never having been explored in HTS campaigns and therefore, a lack of empirical evidence to support the definition of 'druggable'.

Based on our extensive experience, supported by the data provided in this chapter and the accumulating data in the literature, the in vitro MDD paradigm (with important exceptions) is unlikely to succeed as one that is viable, and many describe it as having been 'played out'. This also applies to the chemical probe initiative by the Molecular Libraries Screening Centers Network and others, because, while the focus is to develop molecular tools rather than drugs, the strategies in large part parallel those defined by MDD. Moreover, in order to qualify as a chemical probe, stringent criteria must be met. Many have expressed a view that these standards have been set too high, e.g., most approved drugs do not meet these criteria [101]. Indeed, a recent analysis found that oral drugs seldom possess nanomolar potency that is predicted by in vitro potencies, many have off-target effects, and moreover, in vitro potency does not correlate with therapeutic dose [102]. These findings, corroborated by our own, counteract the assumption of the MDD strategy that compounds with high in vitro potency will translate into low-dose therapeutics in vivo. There is no reason to believe that chemical probes would behave any differently; therefore, it follows that probes that meet these criteria are not likely to be superior as molecular tools for the elucidation of gene function.

The failures of the in vitro MDD paradigm naturally prompted a closer look at drugs that were successfully developed before this paradigm was adopted, using systems biology tools to gain a detailed yet comprehensive picture of the cellular response to each drug. Our chemogenomic data are consistent with other studies in demonstrating that most successful drugs exhibit polypharmacology. Meanwhile, studies mapping protein–protein, protein–DNA and genetic interactions on a genome-wide scale have given us a greater appreciation of the complexity of the cell, revealing many dependencies within as well as between functional modules. With this in mind, the polypharmacology of many drugs is now considered critical to therapeutic efficacy.

Chemogenomic profiling holds much promise towards an increased understanding of the mechanisms of drug action and the cellular response to chemical stress from the in vivo perspective of the cell. Combined with other systems-level approaches, these efforts are creating the foundation for a deeper understanding of disease, and ultimately, for a more effective polypharmacology-based drug discovery paradigm.

Table of Abbreviations

Abbreviation	Expansion
DAmP	Decreased Abundance by mRNA Perturbation
ENCODE	ENCyclopedia Of DNA Elements
GO	Gene Ontology
HGP	Human Genome Project
HIP	HapploInsufficency Profiling
HOP	HOmozygous Profiling
HTS	High-Throughput Screening
LSU	Large SubUnit of ribosome
MDD	Modern Drug Discovery
MDR	Multi-Drug Resistance
MOA	Mechanism Of Action
MoBY-ORF	Molecular Barcoded Yeast Open Reading Frame
NCI	National Cancer Institute (USA)
NIH	National Institute of Health (USA)
NME	New Molecular Entity
PDR	Pleiotropic Drug Resistance
SGA	Synthetic Genetic Array
UPR	Unfolded Protein Response
YDP	Yeast Deletion Project
YGP	Yeast Genome Project
YKO	Yeast KnockOut

REFERENCES

[1] Genomes Project C. A map of human genome variation from population-scale sequencing. Nature (7319):1061–73, http://dx.doi.org/10.1038/nature09534, 2010;467. PubMed PMID: 20981092; PubMed Central PMCID: PMC3042601.

[2] Consortium EP, Birney E, Stamatoyannopoulos JA, Dutta A, Guigo R, Gingeras TR, et al. Identification and analysis of functional elements in 1% of the human genome by the ENCODE pilot project. Nature (7146):799–816, http://dx.doi.org/10.1038/nature05874, 2007;447. PubMed PMID: 17571346; PubMed Central PMCID: PMC2212820.

[3] Bernstein BE, Stamatoyannopoulos JA, Costello JF, Ren B, Milosavljevic A, Meissner A, et al. The NIH roadmap epigenomics mapping consortium. Nat Biotechnol (10):1045–8, http://dx.doi.org/10.1038/nbt1010-1045, 2010;28. PubMed PMID: 20944595.

[4] International HapMap C, Altshuler DM, Gibbs RA, Peltonen L, Altshuler DM, Gibbs RA, et al. Integrating common and rare genetic variation in diverse human populations. Nature (7311):52–8, http://dx.doi.org/10.1038/nature09298, 2010;467. PubMed PMID: 20811451; PubMed Central PMCID: PMC3173859.

[5] Thorisson GA, Stein LD. The SNP consortium website: past, present and future. Nucleic Acids Res 2003;31(1):124–7. PubMed PMID: 12519964; PubMed Central PMCID: PMC165499.

[6] Hellerstein MK. A critique of the molecular target-based drug discovery paradigm based on principles of metabolic control: advantages of pathway-based discovery. metab eng (1):1–9, http://dx.doi.org/10.1016/j.ymben.2007.09.003, 2008;10. PubMed PMID: 17962055.

[7] Goffeau A, Barrell BG, Bussey H, Davis RW, Dujon B, Feldmann H, et al. Life with 6000 genes. Science 1996;274(5287). 546, 63–7.PubMed PMID: 8849441.

[8] Dujon B. The yeast genome project: what did we learn? trends genet: TIG 1996;12(7):263–70. Epub 1996/07/01. PubMed PMID: 8763498.

[9] Wach A, Brachat A, Pohlmann R, Philippsen P. New heterologous modules for classical or PCR-based gene disruptions in *Saccharomyces cerevisiae*. Yeast 1994;10(13):1793–808. PubMed PMID: 7747518.

[10] Winzeler EA, Shoemaker DD, Astromoff A, Liang H, Anderson K, Andre B, et al. Functional characterization of the S. cerevisiae genome by gene deletion and parallel analysis. Science

1999;285(5429):901–6. Epub 1999/08/07. PubMed PMID: 10436161.

[11] Giaever G, Chu AM, Ni L, Connelly C, Riles L, Veronneau S, et al. Functional profiling of the *Saccharomyces cerevisiae* genome. Nature (6896):387–91, http://dx.doi.org/10.1038/nature00935, 2002;418. PubMed PMID: 12140549.

[12] Pierce SE, Davis RW, Nislow C, Giaever G. Genome-wide analysis of barcoded *Saccharomyces cerevisiae* gene-deletion mutants in pooled cultures. Nat Protoc 2007;2(11):2958–74. PubMed PMID: 18007632.

[13] Hensel M, Shea JE, Gleeson C, Jones MD, Dalton E, Holden DW. Simultaneous identification of bacterial virulence genes by negative selection. Science 1995;269(5222):400–3. PubMed PMID: 7618105.

[14] Schena M, Shalon D, Heller R, Chai A, Brown PO, Davis RW. Parallel human genome analysis: microarray-based expression monitoring of 1000 genes. Proc Natl Acad Sci U S A 1996;93(20):10614–9. PubMed PMID: 8855227; PubMed Central PMCID: PMC38202.

[15] David L, Huber W, Granovskaia M, Toedling J, Palm CJ, Bofkin L, et al. A high-resolution map of transcription in the yeast genome. Proc Natl Acad Sci U S A (14):5320–5, http://dx.doi.org/10.1073/pnas.0601091103, 2006;103. PubMed PMID: 16569694; PubMed Central PMCID: PMC1414796.

[16] Ptacek J, Devgan G, Michaud G, Zhu H, Zhu X, Fasolo J, et al. Global analysis of protein phosphorylation in yeast. Nature (7068):679–84, http://dx.doi.org/10.1038/nature04187, 2005;438. PubMed PMID: 16319894.

[17] Zhu H, Bilgin M, Bangham R, Hall D, Casamayor A, Bertone P, et al. Global analysis of protein activities using proteome chips. Science (5537):2101–5, http://dx.doi.org/10.1126/science.1062191, 2001;293. PubMed PMID: 11474067.

[18] Gallego O, Betts MJ, Gvozdenovic-Jeremic J, Maeda K, Matetzki C, Aguilar-Gurrieri C, et al. A systematic screen for protein-lipid interactions in *Saccharomyces cerevisiae*. Mol Syst Biol: 430, http://dx.doi.org/10.1038/msb.2010.87, 2010;6. PubMed PMID: 21119626; PubMed Central PMCID: PMC3010107.

[19] Tong AH, Lesage G, Bader GD, Ding H, Xu H, Xin X, et al. Global mapping of the yeast genetic interaction network. Science (5659):808–13, http://dx.doi.org/10.1126/science.1091317, 2004;303. PubMed PMID: 14764870.

[20] Tong AH, Evangelista M, Parsons AB, Xu H, Bader GD, Page N, et al. Systematic genetic analysis with ordered arrays of yeast deletion mutants. Science (5550):2364–8, http://dx.doi.org/10.1126/science.1065810, 2001;294. PubMed PMID: 11743205.

[21] Hillenmeyer ME, Fung E, Wildenhain J, Pierce SE, Hoon S, Lee W, et al. The chemical genomic portrait of yeast: uncovering a phenotype for all genes. Science (5874):362–5, http://dx.doi.org/10.1126/science.1150021, 2008;320. PubMed PMID: 18420932; PubMed Central PMCID: PMC2794835.

[22] Breslow DK, Cameron DM, Collins SR, Schuldiner M, Stewart-Ornstein J, Newman HW, et al. A comprehensive strategy enabling high-resolution functional analysis of the yeast genome. Nat. Methods (8):711–8, http://dx.doi.org/10.1038/nmeth.1234, 2008;5. PubMed PMID: 18622397; PubMed Central PMCID: PMC2756093.

[23] Yan Z, Berbenetz NM, Giaever G, Nislow C. Precise gene-dose alleles for chemical genetics. Genetics (2):623–6, http://dx.doi.org/10.1534/genetics.109.103036, 2009;182. PubMed PMID: 19332878; PubMed Central PMCID: PMC2691769.

[24] Li Z, Vizeacoumar FJ, Bahr S, Li J, Warringer J, Vizeacoumar FS, et al. Systematic exploration of essential yeast gene function with temperature-sensitive mutants. Nat Biotechnol (4):361–7, http://dx.doi.org/10.1038/nbt.1832, 2011;29. PubMed PMID: 21441928; PubMed Central PMCID: PMC3286520.

[25] Ho CH, Magtanong L, Barker SL, Gresham D, Nishimura S, Natarajan P, et al. A molecular barcoded yeast ORF library enables mode-of-action analysis of bioactive compounds. Nat Biotechnol (4):369–77, http://dx.doi.org/10.1038/nbt.1534, 2009;27. PubMed PMID: 19349972.

[26] Parsons AB, Brost RL, Ding H, Li Z, Zhang C, Sheikh B, et al. Integration of chemical-genetic and genetic interaction data links bioactive compounds to cellular target pathways. Nat Biotechnol (1):62–9, http://dx.doi.org/10.1038/nbt919, 2004;22. PubMed PMID: 14661025.

[27] Costanzo M, Baryshnikova A, Bellay J, Kim Y, Spear ED, Sevier CS, et al. The genetic landscape of a cell. Science (5964):425–31, http://dx.doi.org/10.1126/science.1180823, 2010;327. PubMed PMID: 20093466.

[28] Hu Y, Bajorath J. Polypharmacology directed compound data mining: identification of promiscuous chemotypes with different activity profiles and comparison to approved drugs. J chem inf model (12):2112–8, http://dx.doi.org/10.1021/ci1003637, 2010;50. PubMed PMID: 21070069.

[29] Azzaoui K, Hamon J, Faller B, Whitebread S, Jacoby E, Bender A, et al. Modeling promiscuity based on in vitro safety pharmacology profiling data. Chem Med Chem (6):874–80, http://dx.doi.org/10.1002/cmdc.200700036, 2007;2. PubMed PMID: 17492703.

[30] Takigawa I, Tsuda K, Mamitsuka H. Mining significant substructure pairs for interpreting polypharmacology in drug-target network. PLoS One (2):e16999, http://dx.doi.org/10.1371/journal.pone.0016999, 2011;6. PubMed PMID: 21373195; PubMed Central PMCID: PMC3044142.

[31] Overington JP, Al-Lazikani B, Hopkins AL. How many drug targets are there? Nat Rev Drug Discov (12):993–6, http://dx.doi.org/10.1038/nrd2199, 2006;5. PubMed PMID: 17139284.

[32] Hopkins AL, Groom CR. The druggable genome. Nat Rev Drug Discov (9):727–30, http://dx.doi.org/10.1038/nrd892, 2002;1. PubMed PMID: 12209152.

[33] Swinney DC, Anthony J. How were new medicines discovered? Nat Rev Drug Discov (7):507–19, http://dx.doi.org/10.1038/nrd3480, 2011;10. PubMed PMID: 21701501.

[34] Oprea TI, Bologa CG, Boyer S, Curpan RF, Glen RC, Hopkins AL, et al. A crowdsourcing evaluation of the NIH chemical probes. Nat Chem Biol (7):441–7, http://dx.doi.org/10.1038/nchembio0709–441, 2009;5. PubMed PMID: 19536101.

[35] Colwill K. Renewable Protein Binder Working G, Graslund S. A roadmap to generate renewable protein binders to the human proteome. Nat Methods (7):551–8. Epub 2011/05/17, http://dx.doi.org/10.1038/nmeth.1607, 2011;8. PubMed PMID: 21572409.

[36] Frye SV. The art of the chemical probe. Nat Chem Biol 6(3):159–61. http://dx.doi.org/nchembio.296 [pii] 1038/nchembio.296. PubMed PMID: 20154659.

[37] Austin CP, Brady LS, Insel TR, Collins FS. NIH molecular libraries initiative. Science (5699):1138–9, http://dx.doi.org/10.1126/science.1105511, 2004;306. PubMed PMID: 15542455.

[38] Wallace IM, Urbanus ML, Luciani GM, Burns AR, Han MK, Wang H, et al. Compound prioritization methods increase rates of chemical probe discovery in model organisms. Chem Biol (10):1273−83, http://dx.doi.org/10.1016/j.chembiol.2011.07.018, 2011;18. PubMed PMID: 22035796.

[39] Hughes TR, Marton MJ, Jones AR, Roberts CJ, Stoughton R, Armour CD, et al. Functional discovery via a compendium of expression profiles. Cell. 2000;102(1):109−26. PubMed PMID: 10929718.

[40] Zhao R, Davey M, Hsu YC, Kaplanek P, Tong A, Parsons AB, et al. Navigating the chaperone network: an integrative map of physical and genetic interactions mediated by the hsp90 chaperone. Cell (5):715−27, http://dx.doi.org/10.1016/j.cell.2004.12.024, 2005;120. PubMed PMID: 15766533.

[41] Weinstein JN, Myers TG, O'Connor PM, Friend SH, Fornace Jr AJ, Kohn KW, et al. An information-intensive approach to the molecular pharmacology of cancer. Science 1997;275(5298):343−9. PubMed PMID: 8994024.

[42] Lai MH, Bard M, Pierson CA, Alexander JF, Goebl M, Carter GT, et al. The identification of a gene family in the *Saccharomyces cerevisiae* ergosterol biosynthesis pathway. Gene 1994;140(1):41−9. PubMed PMID: 8125337.

[43] Hueso G, Aparicio-Sanchis R, Montesinos C, Lorenz S, Murguia JR, Serrano R. A novel role for protein kinase Gcn2 in yeast tolerance to intracellular acid stress. biochem j (1):255−64, http://dx.doi.org/10.1042/BJ20111264, 2012;441. PubMed PMID: 21919885.

[44] Bauer BE, Rossington D, Mollapour M, Mamnun Y, Kuchler K, Piper PW. Weak organic acid stress inhibits aromatic amino acid uptake by yeast, causing a strong influence of amino acid auxotrophies on the phenotypes of membrane transporter mutants. eur j biochem/FEBS 2003;270(15):3189−95. PubMed PMID: 12869194.

[45] Klosinska MM, Crutchfield CA, Bradley PH, Rabinowitz JD, Broach JR. Yeast cells can access distinct quiescent states. genes dev (4):336−49, http://dx.doi.org/10.1101/gad.2011311, 2011;25. PubMed PMID: 21289062; PubMed Central PMCID: PMC3042157.

[46] Lamb J, Crawford ED, Peck D, Modell JW, Blat IC, Wrobel MJ, et al. The connectivity map: using gene-expression signatures to connect small molecules, genes, and disease. Science (5795):1929−35, http://dx.doi.org/10.1126/science.1132939, 2006;313. PubMed PMID: 17008526.

[47] Parsons AB, Lopez A, Givoni IE, Williams DE, Gray CA, Porter J, et al. Exploring the mode-of-action of bioactive compounds by chemical-genetic profiling in yeast. Cell. (3):611−25, http://dx.doi.org/10.1016/j.cell.2006.06.040, 2006;126. PubMed PMID: 16901791.

[48] Magtanong L, Ho CH, Barker SL, Jiao W, Baryshnikova A, Bahr S, et al. Dosage suppression genetic interaction networks enhance functional wiring diagrams of the cell. Nat Biotechnol (6):505−11, http://dx.doi.org/10.1038/nbt. 1855, 2011;29. PubMed PMID: 21572441.

[49] Lee W, St Onge RP, Proctor M, Flaherty P, Jordan MI, Arkin AP, et al. Genome-wide requirements for resistance to functionally distinct DNA-damaging agents. PLoS Genet (2):e24, http://dx.doi.org/10.1371/journal.pgen.0010024, 2005;1. PubMed PMID: 16121259; PubMed Central PMCID: PMC1189734.

[50] Tucker CL, Fields S. Quantitative genome-wide analysis of yeast deletion strain sensitivities to oxidative and chemical stress. Comparative and functional genomics (3):216−24, http://dx.doi.org/10.1002/cfg.391, 2004;5. PubMed PMID: 18629161; PubMed Central PMCID: PMC2447451.

[51] Blackman RK, Cheung-Ong K, Gebbia M, Proia DA, He S, Kepros J, et al. Mitochondrial electron transport is the cellular target of the oncology drug elesclomol. PLoS One (1):e29798, http://dx.doi.org/10.1371/journal.pone.0029798, 2012;7. PubMed PMID: 22253786; PubMed Central PMCID: PMC3256171.

[52] Skrtic M, Sriskanthadevan S, Jhas B, Gebbia M, Wang X, Wang Z, et al. Inhibition of mitochondrial translation as a therapeutic strategy for human acute myeloid leukemia. Cancer Cell (5):674−88, http://dx.doi.org/10.1016/j.ccr.2011.10.015, 2011;20. PubMed PMID: 22094260; PubMed Central PMCID: PMC3221282.

[53] Lain S, Hollick JJ, Campbell J, Staples OD, Higgins M, Aoubala M, et al. Discovery, in vivo activity, and mechanism of action of a small-molecule p53 activator. Cancer Cell. (5):454−63, http://dx.doi.org/10.1016/j.ccr.2008.03.004, 2008;13. PubMed PMID: 18455128; PubMed Central PMCID: PMC2742717.

[54] Hillenmeyer ME, Ericson E, Davis RW, Nislow C, Koller D, Giaever G. Systematic analysis of genome-wide fitness data in yeast reveals novel gene function and drug action. Genome Biol (3):R30, http://dx.doi.org/10.1186/gb-2010−11−3-r30, 2010;11. PubMed PMID: 20226027; PubMed Central PMCID: PMC2864570.

[55] Giaever G, Shoemaker DD, Jones TW, Liang H, Winzeler EA, Astromoff A, et al. Genomic profiling of drug sensitivities via induced haploinsufficiency. Nat Genet (3):278−83, http://dx.doi.org/10.1038/6791, 1999;21. PubMed PMID: 10080179.

[56] Giaever G, Flaherty P, Kumm J, Proctor M, Nislow C, Jaramillo DF, et al. Chemogenomic profiling: identifying the functional interactions of small molecules in yeast. Proc Natl Acad Sci U S A (3):793−8, http://dx.doi.org/10.1073/pnas.0307490100, 2004;101. PubMed PMID: 14718668; PubMed Central PMCID: PMC321760.

[57] Lum PY, Armour CD, Stepaniants SB, Cavet G, Wolf MK, Butler JS, et al. Discovering modes of action for therapeutic compounds using a genome-wide screen of yeast heterozygotes. Cell 2004;116(1):121−37. PubMed PMID: 14718172.

[58] Xu D, Jiang B, Ketela T, Lemieux S, Veillette K, Martel N, et al. Genome-wide fitness test and mechanism-of-action studies of inhibitory compounds in *Candida albicans*. PLoS Pathog. (6):e92, http://dx.doi.org/10.1371/journal.ppat.0030092, 2007;3. PubMed PMID: 17604452; PubMed Central PMCID: PMC1904411.

[59] Baetz K, McHardy L, Gable K, Tarling T, Reberioux D, Bryan J, et al. Yeast genome-wide drug-induced haploinsufficiency screen to determine drug mode of action. Proc Natl Acad Sci U S A (13):4525−30, http://dx.doi.org/10.1073/pnas.0307122101, 2004;101. PubMed PMID: 15070751; PubMed Central PMCID: PMC384780.

[60] Ihmels J, Collins SR, Schuldiner M, Krogan NJ, Weissman JS. Backup without redundancy: genetic interactions reveal the cost of duplicate gene loss. Mol Syst Biol :86, http://dx.doi.org/10.1038/msb4100127, 2007;3. PubMed PMID: 17389874; PubMed Central PMCID: PMC1847942.

[61] Musso G, Costanzo M, Huangfu M, Smith AM, Paw J, San Luis BJ, et al. The extensive and condition-dependent nature of epistasis among whole-genome duplicates in yeast. Genome Res. (7):1092−9, http://dx.doi.org/10.1101/gr.076174.108, 2008;18. PubMed PMID: 18463300; PubMed Central PMCID: PMC2493398.

[62] Jones NP, Schulze A. Targeting cancer metabolism — aiming at a tumour's sweet-spot. drug discov today (5–6):232–41, http://dx.doi.org/10.1016/j.drudis.2011.12.017, 2012;17. PubMed PMID: 22207221.

[63] Kaufman RJ. Stress signaling from the lumen of the endoplasmic reticulum: coordination of gene transcriptional and translational controls. genes dev 1999;13(10):1211–33. PubMed PMID: 10346810.

[64] Hoon S, Smith AM, Wallace IM, Suresh S, Miranda M, Fung E, et al. An integrated platform of genomic assays reveals small-molecule bioactivities. Nat Chem Biol (8):498–506, http://dx.doi.org/10.1038/nchembio.100, 2008;4. PubMed PMID: 18622389.

[65] Kim J, Tang JY, Gong R, Kim J, Lee JJ, Clemons KV, et al. Itraconazole, a commonly used antifungal that inhibits Hedgehog pathway activity and cancer growth. Cancer Cell (4):388–99, http://dx.doi.org/10.1016/j.ccr.2010.02.027, 2010;17. PubMed PMID: 20385363.

[66] Bendaha H, Yu L, Touzani R, Souane R, Giaever G, Nislow C, et al. New azole antifungal agents with novel modes of action: synthesis and biological studies of new tridentate ligands based on pyrazole and triazole. Eur J Med Chem (9):4117–24, http://dx.doi.org/10.1016/j.ejmech.2011.06.012, 2011;46. PubMed PMID: 21723647.

[67] Munos B. Lessons from 60 years of pharmaceutical innovation. Nat Rev Drug Discov (12):959–68, http://dx.doi.org/10.1038/nrd2961, 2009;8. PubMed PMID: 19949401.

[68] Subramanian A, Tamayo P, Mootha VK, Mukherjee S, Ebert BL, Gillette MA, et al. Gene set enrichment analysis: a knowledge-based approach for interpreting genome-wide expression profiles. Proc Natl Acad Sci U S A (43):15545–50, http://dx.doi.org/10.1073/pnas.0506580102, 2005;102. PubMed PMID: 16199517; PubMed Central PMCID: PMC1239896.

[69] Shannon P, Markiel A, Ozier O, Baliga NS, Wang JT, Ramage D, et al. Cytoscape: a software environment for integrated models of biomolecular interaction networks. Genome Res (11):2498–504, http://dx.doi.org/10.1101/gr.1239303, 2003;13. PubMed PMID: 14597658; PubMed Central PMCID: PMC403769.

[70] Tutucci E, Stutz F. Keeping mRNPs in check during assembly and nuclear export. nat rev mol cell biol (6):377–84. Epub 2011/05/24, http://dx.doi.org/10.1038/nrm3119, 2011;12. PubMed PMID: 21602906.

[71] Hart GT, Lee I, Marcotte ER. A high-accuracy consensus map of yeast protein complexes reveals modular nature of gene essentiality. BMC bioinformatics:236, http://dx.doi.org/10.1186/1471–2105–8-236, 2007;8. PubMed PMID: 17605818; PubMed Central PMCID: PMC1940025.

[72] van Riggelen J, Yetil A, Felsher DW. MYC as a regulator of ribosome biogenesis and protein synthesis. Nat Rev Cancer (4):301–9, http://dx.doi.org/10.1038/nrc2819, 2010;10. PubMed PMID: 20332779.

[73] Liang XJ, Mukherjee S, Shen DW, Maxfield FR, Gottesman MM. Endocytic recycling compartments altered in cisplatin-resistant cancer cells. cancer res (4):2346–53, http://dx.doi.org/10.1158/0008-5472.CAN-05–3436, 2006;66. PubMed PMID: 16489040; PubMed Central PMCID: PMC1382193.

[74] Rajagopal A, Simon SM. Subcellular localization and activity of multidrug resistance proteins. Mol Biol Cell (8):3389–99, http://dx.doi.org/10.1091/mbc.E02–11–0704, 2003;14. PubMed PMID: 12925771; PubMed Central PMCID: PMC181575.

[75] Liang XJ, Meng H, Wang Y, He H, Meng J, Lu J, et al. Metallofullerene nanoparticles circumvent tumor resistance to cisplatin by reactivating endocytosis. Proc Natl Acad Sci U S A (16):7449–54, http://dx.doi.org/10.1073/pnas.0909707107, 2010;107. PubMed PMID: 20368438; PubMed Central PMCID: PMC2867714.

[76] Hua Z, Fatheddin P, Graham TR. An essential subfamily of Drs2p-related P-type ATPases is required for protein trafficking between golgi complex and endosomal/vacuolar system. Mol Biol Cell (9):3162–77. Epub 2002/09/11, http://dx.doi.org/10.1091/mbc.E02–03–0172, 2002;13. PubMed PMID: 12221123; PubMed Central PMCID: PMC124150.

[77] Brett CL, Kallay L, Hua Z, Green R, Chyou A, Zhang Y, et al. Genome-wide analysis reveals the vacuolar pH-stat of *Saccharomyces cerevisiae*. PLoS One (3):e17619, http://dx.doi.org/10.1371/journal.pone.0017619, 2011;6. PubMed PMID: 21423800; PubMed Central PMCID: PMC3056714.

[78] Anderson N, Borlak J. Drug-induced phospholipidosis. FEBS lett (23):5533–40, http://dx.doi.org/10.1016/j.febslet.2006.08.061, 2006;580. PubMed PMID: 16979167.

[79] Reasor MJ, Kacew S. Drug-induced phospholipidosis: are there functional consequences? Exp Biol Med 2001;226(9):825–30. PubMed PMID: 11568304.

[80] Doostzadeh J, Davis RW, Giaever GN, Nislow C, Langston JW. Chemical genomic profiling for identifying intracellular targets of toxicants producing Parkinson's disease. Toxicol Sci 2007;95(1):182–7. PubMed PMID: 17043098.

[81] Costanzo M, Giaever G, Nislow C, Andrews B. Experimental approaches to identify genetic networks. Curr Opin Bio 2006;17(5):472–80. PubMed PMID: 16962766.

[82] Deutschbauer AM, Jaramillo DF, Proctor M, Kumm J, Hillenmeyer ME, Davis RW, et al. Mechanisms of haploinsufficiency revealed by genome-wide profiling in yeast. Genetics (4):1915–25, http://dx.doi.org/10.1534/genetics.104.036871, 2005;169. PubMed PMID: 15716499; PubMed Central PMCID: PMC1449596.

[83] Fraser HB, Hirsh AE, Giaever G, Kumm J, Eisen MB. Noise minimization in eukaryotic gene expression. PLoS Biol (6):e137, http://dx.doi.org/10.1371/journal.pbio.0020137, 2004;2. PubMed PMID: 15124029; PubMed Central PMCID: PMC400249.

[84] Lui J, Campbell SG, Ashe MP. Inhibition of translation initiation following glucose depletion in yeast facilitates a rationalization of mRNA content. Biochem Society Trans (4):1131–6, http://dx.doi.org/10.1042/BST0381131, 2010;38. PubMed PMID: 20659017.

[85] Altmann M, Linder P. Power of yeast for analysis of eukaryotic translation initiation. J Biol Chem (42):31907–12, http://dx.doi.org/10.1074/jbc.R110.144196, 2010;285. PubMed PMID: 20693283; PubMed Central PMCID: PMC2952190.

[86] Ingolia T, Ghaemmaghami S, Newman JR, Weissman JS. Genome-wide analysis in vivo of translation with nucleotide resolution using ribosome profiling. Science 2009;324(5924):218–23. PubMed PMID: 19213877.

[87] Kedersha N, Anderson P. Regulation of translation by stress granules and processing bodies. Prog Mol Biol Trans Sci :155–85, http://dx.doi.org/10.1016/S1877–1173(09)90004–7, 2009;90. PubMed PMID: 20374741.

[88] Lehar J, Stockwell BR, Giaever G, Nislow C. Combination chemical genetics. Nat Chem Biol (11):674–81, http://dx.doi.org/nchembio.120, 2008;4 [pii] 1038/nchembio.120. PubMed PMID: 18936752; PubMed Central PMCID: PMC2712875.

[89] Lopez A, Parsons AB, Nislow C, Giaever G, Boone C. Chemical-genetic approaches for exploring the mode of action of natural products. Prog Drug Res 2008;66(237):9–71. Epub 2008/04/18. PubMed PMID: 18416308.

[90] Yan Z, Costanzo M, Heisler LE, Paw J, Kaper F, Andrews BJ, et al. Yeast Barcoders: a chemogenomic application of a universal donor-strain collection carrying bar-code identifiers. Nat Methods (8):719–25, http://dx.doi.org/10.1038/nmeth 1231, 2008;5. PubMed PMID: 18622398.

[91] Smirnova JB, Selley JN, Sanchez-Cabo F, Carroll K, Eddy AA, McCarthy JE, et al. Global gene expression profiling reveals widespread yet distinctive translational responses to different eukaryotic translation initiation factor 2B-targeting stress pathways. Mol Cell Biol (21):9340–9, http://dx.doi.org/10.1128/MCB.25.21.9340–9349.2005, 2005;25. PubMed PMID: 16227585; PubMed Central PMCID: PMC1265828.

[92] Plotkin JB. Transcriptional regulation is only half the story. Mol Syst Biol :406, http://dx.doi.org/10.1038/msb.2010.63, 2010;6. PubMed PMID: 20739928; PubMed Central PMCID: PMC2950086.

[93] Vogel C. Translation's coming of age. Mol Syst Biol :498, http://dx.doi.org/10.1038/msb.2011.33, 2011;7. PubMed PMID: 21613985; PubMed Central PMCID: PMC3130562.

[94] Schwanhausser B, Busse D, Li N, Dittmar G, Schuchhardt J, Wolf J, et al. Global quantification of mammalian gene expression control. Nature (7347):337–42, http://dx.doi.org/10.1038/nature10098, 2011;473. PubMed PMID: 21593866.

[95] Spriggs KA, Bushell M, Willis AE. Translational regulation of gene expression during conditions of cell stress. Mol Cell (2):228–37, http://dx.doi.org/10.1016/j.molcel.2010.09.028, 2010;40. PubMed PMID: 20965418.

[96] Lander ES. Initial impact of the sequencing of the human genome. Nature (7333):187–97, http://dx.doi.org/10.1038/nature09792, 2011;470. PubMed PMID: 21307931.

[97] Siddiqui N, Borden KL. mRNA export and cancer. Wiley Interdiscip Rev RNA (1):13–25, http://dx.doi.org/10.1002/wrna.101, 2012;3. PubMed PMID: 21796793.

[98] Gassner NC, Tamble CM, Bock JE, Cotton N, White KN, Tenney K, et al. Accelerating the discovery of biologically active small molecules using a high-throughput yeast halo assay. J Nat Products 2007;70(3):383–90. PubMed PMID: 17291044.

[99] Knight ZA, Shokat KM. Features of selective kinase inhibitors. Chem Biol (6):621–37, http://dx.doi.org/10.1016/j.chembiol.2005.04.011, 2005;12. PubMed PMID: 15975507.

[100] Bishop AC, Buzko O, Shokat KM. Magic bullets for protein kinases. Trends Cell Biol 2001;11(4):167–72. Epub 2001/04/18. PubMed PMID: 11306297.

[101] Garcia-Serna R, Mestres J. Chemical probes for biological systems. Drug Discov Today (3–4):99–106. Epub 2010/11/26, http://dx.doi.org/10.1016/j.drudis.2010.11.004, 2011;16. PubMed PMID: 21093609.

[102] Gleeson MP, Hersey A, Montanari D, Overington J. Probing the links between in vitro potency, ADMET and physicochemical parameters. Nat Rev Drug Discov (3):197–208, http://dx.doi.org/10.1038/nrd3367, 2011;10. PubMed PMID: 21358739.

Chapter 9

Graph Theory Properties of Cellular Networks

Baruch Barzel,[1,2] Amitabh Sharma[1,2] and Albert-László Barabási[1,2]
[1]Center for Complex Network Research, Department of Physics, Northeastern University, 360 Huntington avenue, Boston, Massachusetts 02115, USA,
[2]Center for Cancer System Biology (CCSB) and Department of Cancer Biology, the Dana-Farber Cancer Institute and Department of Genetics, Harvard Medical School, 44 Binney street, Boston, Massachusetts, USA

Chapter Outline

Introduction	177
Biological Systems As Graphs	178
The Tools of Graph Theory	178
Erdős–Rényi – The Benchmark Network	178
Degrees and Degree Distribution	178
Network Paths and the Small World Phenomena	179
Clustering Coefficient	180
Successes and Failures of the Erdős–Rényi Model	180
Biological Small Worlds	180
Deviations from the Erdős–Rényi Model	180
Scale-Free Nature of Cellular Networks	181
The Scale-Free Property	181
Network Integrity and the Role of Hubs	182
The Origins of the Scale-Free Topology	183
Preferential Attachment in Biological Networks	183
Hierarchy and Modularity	184
Party vs. Date Hubs	184
Degree Correlations	184
Human Disease Network	185
The Building Blocks of Cellular Networks	186
Sub-graphs and Motifs	186
Randomized Networks	186
Autoregulation and the Feedforward Loop	187
Going Beyond Topology	187
Assigning the Weights	188
Characterizing the Weighted Topology	188
Topology Correlated Weights	188
Controllability	188
Differential Networks	189
From Structure to Dynamics	190
References	191

INTRODUCTION

From a conceptual point of view, the rise of systems biology can be described as the adoption of a broad-based perspective on biological systems. In that sense, the classical detailed biological analysis is complemented by a macroscopic description of the cell as a holistic unit [1–4]. This approach, aiming at a system-level understanding of biology, mandates a crude simplification of biological processes. In this light, the graph theoretic approach to biological systems focuses on the structural aspects of the interaction patterns, where the interacting species, be them genes, proteins or other biological components, are signified by nodes, and their interactions by the edges drawn between them. These network systems express the underlying architecture which enables the cellular functions to be carried out [5–9].

Undoubtedly, the functionality of the cell cannot be attributed to just one network, but rather to a set of interdependent networks ranging from the level of transcription to the processes of metabolism. It is common to divide the cellular functions into three distinct networks, the transcriptional network, the protein interaction network and the metabolic network [2]. Although we follow this division throughout this chapter, it should be acknowledged that the true functionality of the cellular unit is a result of the interdependence between these networks, and not merely the interactions in each of them alone. At the current state, the topology of these three fundamental cellular networks has been thoroughly mapped using high-throughput techniques. As a result, we now have reliable data on the interaction maps of many organisms. Some examples are protein–protein interaction networks, which have been constructed for organisms such as *Homo sapiens*, *Saccharomyces cerevisiae* (*S. cerevisiae*) *Helicobacter pylori* and others [10–20]. Regulatory and metabolic networks have also been successfully mapped for yeast, *Escherichia coli* and various other organisms [21–23]. However, we still lack data regarding other networks taking part in the cellular processes, such as sRNA and RNAi-mediated networks, about which we currently know little.

BIOLOGICAL SYSTEMS AS GRAPHS

Although the foundations of graph theory were laid purely out of mathematical curiosity, its applicability as a tool for the characterization of complex systems has long been appreciated. The notion is that the behavior of complex systems arises from the coordinated actions of many interacting components. The network abstraction can then be used to reveal the underlying structure of these interactions. The interacting components are signified by a series of nodes, and the interactions between selected pairs of these nodes are represented by the links (or edges) drawn between them. This abstract description eliminates some of the details associated with the specific nature of the system at hand. However, it allows one to utilize the well-established formalisms of graph theory, thus providing a powerful tool for the analysis and understanding of these complex systems. Moreover, this categorical representation applied to various systems provides the grounds for comparison between seemingly distinct networks. This process has proven highly beneficial, as one of the most important discoveries of recent years was that despite the diversity of cellular networks, several important universal properties are shared by them all [5].

In some cases the network description of a cellular system is straightforward and natural. Consider, for instance, the set of physical binding reactions between pairs of proteins or between proteins and other molecules, such as nucleic acids or metabolites. Here it seems natural to use a node for the representation of each molecular type, and an edge to denote each potential binding reaction. However, in other cases the network description is not unique, and may differ according to the motivation of the study. A simple example regards the transcriptional regulatory network. In this network the edges link between transcription factors and the genes that they regulate. Here the information flows from the regulating gene to the regulated one, so that the links are not symmetrical. The network is thus a *directed* network. Moreover, the relationship between a pair of interacting genes can be of an activating or of an inhibitory nature. Thus two different types of directed edges exist, which can be denoted by positive versus negative, or graphically by →versus ⊣. Nevertheless, in many contexts the directed nature of the interactions, or their sign, is not important, and it is sufficient to model the system using a regular undirected network. As a more complicated system we refer to metabolic networks. These systems can be conceptualized as networks on many levels of abstraction. For instance, one can visualize the molecular substrates as nodes, and the reactions transforming substrates to products as links. In this case the links are attributed by the enzyme catalyzing the reaction, and the graph is directed. However, there are contexts in which it is sufficient to use a simpler description, where the enzymes are ignored and the links are undirected. In this case the graph simply describes the interconnections between metabolites, leaving out the detailed chemistry that underlies these connections.

THE TOOLS OF GRAPH THEORY

In the basis of graph theory lies the insight that a complex system could be reduced into a series of abstract components tied together by a set of connections. The spark of this idea is commonly attributed to the 18th-century mathematician Leonhard Euler, who in 1735 used it to solve the problem of the Seven Bridges of Königsberg, a problem which back then confused the residents of the Prussian town. To show that one cannot visit all of the city's islands without crossing at least one of the city's seven bridges twice, Euler constructed an abstract map of the city in which the islands were represented by nodes and the bridges by edges. In doing so, Euler mapped a realistic problem, in all of its complexities, into a clean abstract mathematical representation, which allowed him to focus strictly on the structural crux of the problem. However, this spark remained dim and only re-emerged as an elaborate, formalized mathematical theory some 200 years later, in the 20th century, following the work of Paul Erdős and Alfréd Rényi.

Erdős–Rényi – The Benchmark Network

The most elementary network considered in graph theory, is the Erdős–Rényi random network, where each pair of nodes is connected with equal probability [24–26]. The properties of this prototypic network serve as a benchmark to which we later compare the more realistic networks of cellular biology. To construct an Erdős–Rényi network, we consider a set of N nodes. For each of the $\frac{N(N-1)}{2}$ pairs of nodes in the network we assign an edge with probability p, typically chosen so that $p \ll 1$. Simple as it may be, the Erdős–Rényi network features some surprising characteristics, commonly observed in many real-world biological networks.

Degrees and Degree Distribution

To analyze the components of the network we introduce some elementary network measures. For concreteness, we use the Erdős—Rényi network to exemplify them. The most basic characteristic of a node is its degree, k, defined as the number of links it has to other nodes in the network. In the Erdős—Rényi graph every node can potentially be linked to any of its $N-1$ counterparts with independent probability p. The average degree will thus be $\langle k \rangle = p(N-1) \approx pN$. The random nature of this network invokes some variability in the degree of the nodes, so that several nodes will have more links than the average, while others will have less. This variability can be described by the *degree distribution* of the graph. Denoted by $P(k)$, it is defined as the probability that a randomly selected node will have exactly k links. As we will see later, the degree distribution is one of the most fundamental characteristics of a network, carrying crucial information about its evolution and formation process. In the Erdős—Rényi network $P(k)$ follows a Poisson distribution, which indicates that most nodes are characterized by roughly the same degree, the probability to encounter a node with a degree significantly different than $\langle k \rangle$ being vanishingly small. The average degree is thus the *characteristic scale* of the degree distribution.

In directed networks, we distinguish between the *in-degree* of a node, denoting the number of incoming links, and the *out-degree*, denoting the number of outgoing links. For instance, if gene x regulates n other genes, it will have an out-degree of $k_{\text{out}} = n$, whereas if it is being regulated by m other genes, its in-degree will be $k_{\text{in}} = m$. Accordingly, the degree distribution in such networks is split into the incoming distribution $P(k_{\text{in}})$ and the outgoing distribution $P(k_{\text{out}})$. As an example, consider node number four in Figure 9.1, which in the undirected version of the graph (a) has a total degree of $k = 5$, while in the directed version (b) it is characterized by $k_{\text{in}} = 2$ and $k_{\text{out}} = 3$.

Network Paths and the Small World Phenomena

A crucial feature of any biological network is its ability to maintain a flow of information, mass or energy, between all of its nodes. From the graph theoretical perspective this requirement translates into the existence of a *network path* connecting all (or most) of the nodes in the network. By network path we refer to a route leading from one node to another by passing solely over existing links (Figure 9.1). Such a group of interconnected nodes constitutes a connected component, and if indeed a large fraction of the nodes in the graph can be reached from one another via these network paths, the graph is said to have a *giant connected component*. This seemingly remote feature appears rather frequently and does not require much high-level organization for it to be observed. In fact, for an Erdős—Rényi graph, a giant component will

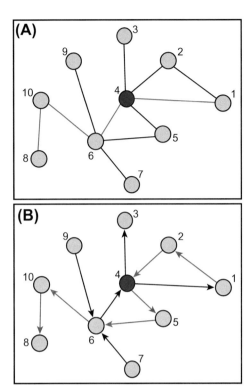

FIGURE 9.1 (A) An undirected network which includes 10 nodes. The network path between nodes 1 and 8 is emphasized (red). (B) In the directed version of this network, the path must advance in accordance with the direction of the edges.

emerge as long as $\langle k \rangle \geq 1$. Moreover, in the case where $1 \ll \langle k \rangle \ll \ln N$, this giant component is likely to encompass almost all of the nodes in the network [6,27].

Networks impose a unique metric by which the distance between nodes can be measured. Consider the network path between a pair of nodes. Its length is defined as the number of links it crosses. The length of the shortest path between a pair of nodes, x and y, is the network distance, l_{xy}. Figure 9.1(a) displays the shortest path between nodes 1 and 8. The network distance between these two nodes is $l_{1,8} = 4$. If no such path exists, then the pair of nodes cannot communicate with one another and their distance is set to be infinite. The distance between a pair of nodes parameterizes the potential to propagate information from one to the other, such that close nodes are more likely to affect one another than far-away nodes. Averaging over all pairs of nodes in the network gives the *average path length* of the network, $\langle l \rangle$, which offers a measure of the network's overall connectivity. If the network includes infinite paths, i.e., has isolated components, the averaging is commonly carried out over the nodes belonging to the giant connected component. For an Erdős—Rényi network (as well as almost all other random networks) we find that the average path length is strikingly small compared to the network size. To understand this we focus on the surrounding of a typical node in the network. It has $\langle k \rangle$ nearest neighbors, all at a distance of $l = 1$ from it; each of these neighbors has on average $\langle k \rangle$ neighbors of its own, so that there are roughly $\langle k \rangle^2$ nodes at a distance of $l = 2$. Following this logic, we find that the number of nodes within a given distance from any typical node inflates exponentially, leading to a logarithmic dependence of the average path length on the network size, N [27]

$$\langle l \rangle \approx \frac{\ln N}{\ln \langle k \rangle}. \qquad (1)$$

This simplified argument overlooks the possibility that some of the edges might be redundant, i.e., that some of the links emerging from nodes at a distance l might link to other nodes which are at the same distance. Still, for large N it captures the behavior of the network at the vicinity of a node, where the probability of linking to an already *used* node, and thereby creating a loop, is very small.

In a directed network, network paths are commonly defined to propagate only in the direction of the edges. This way the paths reflect the flow of information between the nodes. The network distance between a pair of nodes is no longer symmetrical, as displayed in Figure 9.1(b). The distance from node 1 to node 8 in the directed network is $l_{1,8} = 6$ (as opposed to $l_{1,8} = 4$ in the undirected version). However, no path exists in the opposite direction, rendering $l_{8,1}$ infinite.

Clustering Coefficient

Certain networks show a tendency to form clusters of interconnected nodes [28]. In such networks, if the nodes y and z are both connected to some other node, x, it is likely that there is also a direct link between y and z themselves. To quantify this we denote the number of links connecting x's nearest neighbors to one another by c_x. This number can take values ranging from zero, in the case that no such links exist, to $\frac{k_x(k_x - 1)}{2}$ in the case that all of the pairs among x's k_x nearest neighbors are connected. The clustering coefficient is thus defined as $C_x = \frac{2c_x}{k_x(k_x - 1)}$, taking values which are between zero and unity [29–30]. For instance, the clustering of node four in Figure 9.1 is $C_4 = \frac{2 \times 2}{5 \times 4} = 0.2$. Averaging over all the nodes in the network gives the average clustering coefficient $\langle C \rangle$. For an Erdős–Rényi graph the probability of all pairs of nodes being linked is uniform, regardless of whether they share a mutual neighbor or not. The average clustering coefficient is thus $\langle C \rangle_{ER} = p$, which, for a sparse graph, is typically very small.

SUCCESSES AND FAILURES OF THE ERDŐS–RÉNYI MODEL

Biological Small Worlds

The Erdős–Rényi network model is greatly oversimplified and therefore overlooks many important features observed in real biological networks. Nevertheless, this model does prove successful in predicting the overall connectivity observed is practically all analyzed biological network. These networks all feature a giant connected component, such that almost all pairs of nodes are connected by finite network paths. Moreover, the average path length is found to be consistent with Eq. (1), so that the interacting nodes in the network are typically just a few steps away from one another, meaning that cellular networks, like many other networks in nature, feature the small world effect. For instance, in metabolism it was found that most pairs of metabolites can be linked by paths averaging approximately three edges in length. These extremely short average path lengths are not unique to any specific species: they were found in as many as 43 different species, ranging from the evolutionary reduced metabolic network of parasitic bacteria to the highly developed networks of large multicellular organisms [31]. Similar, albeit less dramatic, results apply for protein and genetic interaction networks, where the average path length ranges from about four to eight edges [32–33].

Deviations from the Erdős–Rényi Model

In most cellular networks a tendency to form cliques is observed, where the neighbors of one node tend to be themselves connected. Thus the average clustering coefficient, $\langle C \rangle$, of most cellular networks is significantly larger than that of an equivalent Erdős–Rényi network. By an equivalent network, we refer to an Erdős–Rényi network with the same size and average degree. For instance, protein–protein interaction networks feature a clustering coefficient which is typically about an order of magnitude higher than that observed in their randomly rewired equivalents [32]. Similar findings also characterize metabolic networks [34–35].

The emergence of high clustering provides the first hint towards the recognition that the Erdős–Rényi model cannot account for the topological properties of realistic networks. However, the most significant indication in that direction comes from the degree distributions observed in actual networks. In contrast to the Poisson degree distribution, which is the fingerprint of the Erdős–Rényi graph, cellular networks consistently follow a power-law degree distribution [2,5,31,36–45], predicting that the probability for a randomly chosen node to have exactly k links is given by

$$P(k) \sim k^{-\gamma}, \qquad (2)$$

where γ takes values which are typically between 2 and 3. This finding has profound implications for the architecture of biological networks, as well as their evolution and functionality. These implications are discussed in the next section.

SCALE-FREE NATURE OF CELLULAR NETWORKS

As the structure of cellular networks was elucidated, it became evident that their topology does not obey the typical narrow distribution observed for many other quantities in nature. Instead of the commonly found Poisson, Gaussian and exponential distributions, cellular networks feature a power-law degree distribution. The first evidence for this came from metabolic networks, where we take the nodes to represent the metabolites and the directed links to represent the enzyme-catalyzed chemical reactions between them. The analysis of metabolic networks from as many as 43 different organisms revealed that they are all characterized by a power-law degree distribution [31]. Similar findings followed from the study of protein–protein interaction networks [32], and transcriptional regulatory networks [38,44].

The Scale-Free Property

In contrast to the Poisson (and other narrow) distributions, the power-law distribution is not concentrated around its mean. Networks characterized by such a degree distribution are thus highly non-uniform — most of the nodes have only a few links, whereas a few nodes have a disproportionately large number of links. These highly connected nodes, often called *hubs*, are the glue that binds the majority of low-degree nodes together. The presence of these hubs, which is strictly banned in narrow degree distributions, is observed in practically all the analyzed cellular networks, ranging from the ultra-reactive pyruvate and coenzyme A in metabolic networks, to the insulin receptor in protein–protein interaction networks [2,5]. This can be seen in Figure 9.2(a), where we display the protein–protein interaction network of *S. cerevisiae*. In this representation the node size is proportional to its degree, so that the clearly visible variability in the node sizes illustrates the heterogeneity in their degrees. While most proteins participate in one, two or three interactions, a few hubs participate in well above 10, and the degree of some even exceeds 100.

These highly heterogeneous topologies differ essentially from the classic Erdős–Rényi networks, in that they do not have a *typical node*. In an Erdős–Rényi network, the degrees of most nodes are in the vicinity of the average degree. The scarcity of nodes with any given degree can be estimated by comparing with the average degree of the network. In that sense the average degree provides a characteristic scale by which the rest of the nodes should be measured. In contrast, a power-law degree distribution, of the form of Eq. (2), allows for the coexistence of nodes with an extremely broad range of degrees, freeing the network of any typical scale. The cellular networks are thus *scale-free* (SF) [36]. Graphically, the power-law degree distribution forms a straight line when plotted on logarithmic axes, with the slope providing the scaling exponent, γ (Figure 9.3). This yields an intuitive illustration for the concept of the SF topology. It shows, graphically, that one cannot assign a typical scaling, since the graph duplicates itself regardless of the scaling used in the horizontal axis (denoting the degrees).

Of particular significance in the characterization of the degree distribution is the value of the scaling exponent, γ. The broadness of the distribution becomes larger as the

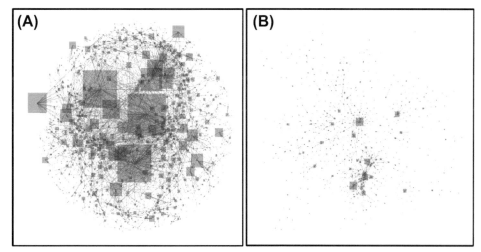

FIGURE 9.2 (A) The yeast protein–protein interaction network. The size of the nodes is proportional to their degree. The heterogeneity in the node sizes serves as a visual expression of the scale-free nature of the degree distribution. (B) A small portion of the network reveals that it is disassortative, i.e., that hubs are typically surrounded by low-degree nodes.

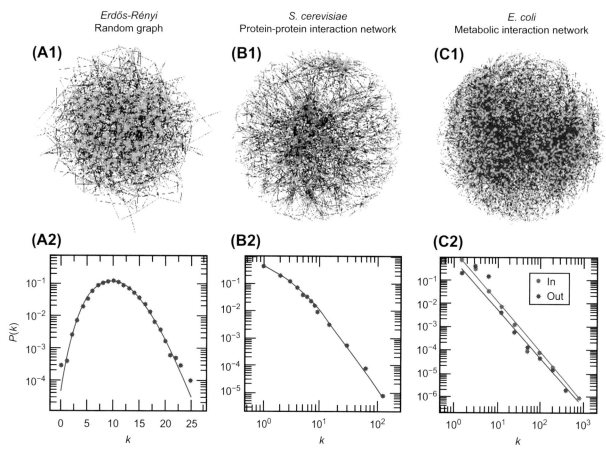

FIGURE 9.3 The degree distribution of the Erdős—Rényi network (A1) follows a Poisson (A2) as opposed to the cellular networks (B1) and (C1) for which it follows a power-law (B2) — (C2). While the Poisson distribution is concentrated around the mean, the degrees observed in the cellular networks range over several orders of magnitude. This non-uniformity of the cellular networks is expressed in the coexistence of dense and sparse patches visible in the network graph. For the metabolic network, where the edges are directed, both the in-degree and the out-degree distributions are plotted. In all graphs the dots represent the data, while the solid lines are fits to Poisson and to power-law distributions accordingly.

value of γ becomes smaller. This means that smaller γ values characterize more degree heterogeneous networks. More specifically, Eq. (2) features three different regimes as the value of γ is changed. To observe this we consider the value of the nth moment, $\langle k^n \rangle$, of the distribution, given by $\int k^n P(k) dk$. Note that for an infinite network, where k ranges from zero to infinity, this integral diverges if $n - \gamma \geq -1$. For $\gamma \leq 2$ the divergence is observed already at the level of the first moment. In such cases, the distribution is so broad that the average is undefined. In practice, when the network is finite, this will take form in a topology where almost all the nodes have degrees significantly lower than the average, and a small minority of nodes will have such a high degree that they connect directly to a significant fraction of the nodes in the network, i.e., their degree is of the order N. For $2 < \gamma \leq 3$ (typical of most cellular networks) the distribution has a finite average, but the second moment diverges. This means that the variance, $\sigma^2 = \langle k^2 \rangle - \langle k \rangle^2$, becomes undefined, capturing the high variability of the distribution. These mathematical pathologies are removed once $\gamma \geq 3$, as when the scaling exponent is above this threshold, for many practical purposes the scale-free nature of the distribution is no longer relevant. These three regimes are also expressed in the average path length of the network, as discussed in more detail below.

Network Integrity and the Role of Hubs

The SF topology allows for a disproportionate number of highly connected nodes. These nodes play a crucial role in the structural integrity of the network. To understand this we consider the majority of nodes in the network, which have only a few links. They are all likely to be connected to the hubs by very short paths (of one or two edges). In addition, any selected pair of hubs is also likely to be very close to one another, due to the large number of links that they have. The result is that in SF networks the path between nodes becomes even shorter than in Erdős—Rényi networks, the hubs playing the role of network *short-cuts*

[46–48]. In fact, in an SF network, where the scaling exponent is $2 < \gamma < 3$ the average path length satisfies $\langle l \rangle \sim \ln \ln N$, adding an additional logarithmic correction to the average path length characteristic of Erdős–Rényi networks (Eq. (1)). For $\gamma = 3$ it is found that $\langle l \rangle \sim \frac{\ln N}{\ln \ln N}$, and for $\gamma > 3$ the result is like that of the Erdős–Rényi network. Cellular networks, for which the scaling exponent is usually between 2 and 3, are thus *ultrasmall worlds* [49].

The analysis above, regarding the importance of the hubs as the structural backbone of SF networks, has some surprising implications on the robustness of cellular networks to random perturbations. Our intuition leads us to view complex systems as highly intricate structures, which depend strongly on the proper functionality of all of their components. When a significant fraction of their nodes fail, these systems are expected to become dysfunctional. In contrast, biological networks prove to be astoundingly resilient against component failure [50–54]. From the topological perspective this can be attributed to their SF topology and its hub-based backbone. Scale-free networks have been shown to maintain their structural integrity even under the deletion of as many as 80% of their nodes. The remaining 20% will still form a connected component [55–57]. This is although in an Erdős–Rényi network the removal of nodes beyond a certain fraction inevitably results in the network disintegrating into small isolated components [27]. The source of this topological resilience of SF networks is rooted in their inherent non-uniformity. The vast majority of nodes in SF networks have merely one or two links, playing a marginal role in maintaining the integrity of the network. Most random failures will occur on these unimportant nodes and thus will not significantly disrupt the network's functionality. The relative scarceness of the hubs, and, on the other hand, their central role in maintaining the network's structural integrity, ensures that random failures will rarely break down the network.

The robustness of cellular networks, which relies strongly on the hub nodes, is, however, a double-edged sword. Despite allowing the networks to withstand a large number of random failures, it makes them extremely vulnerable to intentional interventions. The removal of just a small number of key hubs will cause the SF network to break down into isolated dysfunctional clusters [56–57]. Supporting evidence for this comes from the small number of lethal genes found in many organisms, and, on the other hand, by the relatively large number of hubs found among these genes [39,41,58–65].

The Origins of the Scale-Free Topology

The SF topology is a universal feature of many real networks, both in the context of biology and in social and technological systems [5]. This ubiquitous topological feature not only characterizes the architecture of a given network, it also serves as an indicator for its formation process. This idea is captured by the Barabási–Albert model, which attributes the emergence of an SF topology to the presence of two fundamental formation processes: network growth and preferential attachment [36]. By growth we refer to the fact that networks are not static: they evolve in time by constantly adding new nodes and new links. By preferential attachment we mean that nodes are more likely to link to already highly connected nodes. For a more accurate definition, consider an evolving network, where at each time step a single new node is introduced, drawing m new links to any of the existing nodes. According to the preferential attachment mechanism the new node will choose to connect to an existing node, x, with a probability proportional to x's current degree, namely

$$P(x) = \frac{k_x}{\sum_i k_i}, \qquad (3)$$

where the sum in the denominator is over all nodes in the current state of the network. These two processes, growth and preferential attachment, give rise to the observed power-law degree distributions. It can be shown that any one of these processes alone is insufficient and does not yield the desired SF topology. Network growth is required, as otherwise the network reaches saturation and the degree distribution becomes nearly Gaussian. The preferential attachment mechanism is needed to support the formation of hubs [36]. By this mechanism, if a node has many links it is more likely to acquire new links, creating a state where the *rich get richer*. The result is that the more connected nodes gain new links at a higher rate, and eventually emerge as hubs. Eliminating the preferential attachment mechanism leads to an exponential distribution, much less broad than a power-law.

Preferential Attachment in Biological Networks

The realization of the Barabási–Albert model in the formation of cellular networks is rooted in the process of gene duplication [66–71]. This process is clearly responsible for network growth, as duplicated genes produce duplicate proteins and thus introduce new nodes into the network. The more delicate point is that gene duplication also adheres to the rules of preferential attachment. To understand this, consider an interaction network which grows via node duplication. At each time step a random node is chosen, say x, and an identical node, $\tilde{x}$, is created. This newly created *duplicate* node will have exactly the same interactions as the original node. This means that each of x's nearest neighbors will receive a new edge. Therefore, the distribution of new links in the network is biased towards the more connected nodes. Indeed, a node with many nearest neighbors is more likely to have one of its neighbors chosen for duplication. In fact, for a given node with degree k, the probability for a randomly chosen node to be linked to it is directly proportional to k. Thus its probability to gain a link in the growth process is also proportional to k, consistently with Eq. (3).

One of the predictions of the Barabási—Albert model is that nodes can become well connected by virtue of being older. A node that was introduced early in the history of the network will have more time to accumulate links, and, by the 'rich get richer' mechanism, enhance its chances of becoming a hub [36]. In metabolic networks, we find that the hubs do, indeed, tend to be older. Some examples are coenzyme A, NAD and GTP, remnants of the RNA world, which are among the most connected substrates of the metabolic network [34]. Similar findings rise from the analysis of protein—protein interaction networks, where, on average, the evolutionary ancient proteins are characterized by higher degrees [72—73]. This offers direct empirical evidence for the preferential attachment hypothesis.

HIERARCHY AND MODULARITY

The ability of complex systems to function properly and carry out vital tasks requires the cooperation of many independent components. In many artificial networks this is commonly achieved by relying on a hierarchical design. The network is layered, and nodes at one level orchestrate the behavior of their subordinates belonging to a level below. In that sense we tend to picture network hierarchy as a tree-like topology. However, the idea of having distinct hierarchical layers of nodes stands in sharp contrast to the scale-free nature of the cellular networks. The presence of hubs, which connect directly to a large fraction of the nodes in the network, will inevitably break down the layered topology. We therefore have to adopt a different notion of hierarchy to account for the functional design of biological networks.

The conceptual idea is that the functionality of these elaborate networks can be broken into distinct, relatively isolated tasks [35,74—78]. From a structural point of view, this will be expressed in networks composed of highly interconnected sub-graphs, or modules. The hierarchical disposition of a given node can be characterized by the number of such sub-graphs to which it belongs. This way, a node which is placed low in the hierarchy will participate in just one functional task, and hence belong to just one module. Higher in the hierarchy we find nodes that bridge between two or three different modules. Eventually, at the highest level of the hierarchy will reside the hubs, which do not belong to any specific sub-graph but rather connect many sub-graphs that would otherwise be isolated. The quantifiable fingerprint of such a hierarchical design can be found in $C(k)$, which describes the dependence of the clustering coefficient on the degree [35,45,79—80]. Low-degree nodes will tend to belong to a specific module and thus feature a high clustering coefficient — indeed, almost all their neighbors will themselves be part of the same module. The hubs, on the other hand, will be connected to many nodes from different modules, and accordingly will tend to have a low clustering coefficient.

The analysis of cellular networks shows clear evidence of hierarchical topology. The dependence of the clustering coefficient on the degree features a power-law scaling, $C(k) \sim k^{-\beta}$. This has been observed for metabolic networks [35], protein—protein interaction networks [32] and regulatory networks, with β taking values typically between 1 and 2.

PARTY VS. DATE HUBS

We have already acknowledged the crucial role that the hubs play in the integration of the network. In the above discussion we further emphasized their importance when the network has a modular structure, as the mediators between separate modules. In this context, an interesting distinction between two types of hubs has been proposed [81]. The first type, named *party* hubs, corresponds to our usual perception of hubs as nodes that interact with many other nodes simultaneously. The second type, *date* hubs, bind to their partners at different times or at different cellular locations. While the party hubs tend to interact within a module, it is the date hubs that typically connect between separate modules. So that it is mainly the latter that serve as the integrators of the network. In the analysis of the yeast protein—protein interaction network these two types of hubs were indeed identified [81]. When the date hubs were systematically removed, the network split into small disconnected modules. In contrast, the removal of party hubs, despite diluting the modules themselves, harmed the overall integrity of the network to a much lesser extent.

DEGREE CORRELATIONS

It is commonly observed in networks that similar nodes tend to connect to one another. This feature, termed assortative mixing, can be related to any characteristic of a node, and in particular to the node's degree. For instance, in social networks individuals with many friends tend to link to others who too have a high degree. However, as shown in Figure 9.2(b), in the featured protein—protein interaction network the opposite is true: the network is disassortative, which means that the hubs tend to avoid each other, leading to a network where highly connected nodes are surrounded by low-degree nodes [82]. This disassortativity is observed in most biological networks, including the metabolic and regulatory networks, and is, in fact, a property shared by technological networks, such as the power grid or the internet [83–84].

To classify a network as assortative or disassortative we first need to define our expectations of a *neutral* network. What we are aiming at is to characterize the expected correlations between the degrees of nearest neighbors in the absence of any assortative bias. To do this we consider the random selection of an edge in the network and calculate the probability that at one of its ends resides a node with degree of k and at the other a node with a degree of k'. Let us first calculate the probability for the first node: i.e., we are seeking the probability of finding a node with k links at the end of a randomly selected edge. This is essentially different from the direct selection of a random node, since it gives an advantage to nodes with a higher degree. The reason is simply because such nodes have a larger number of edges to which they are attached. For instance, to reach a node with a single edge through this procedure, one must pinpoint the one edge leading to it. On the other hand, there are k potential edges through which a k degree node can be reached, making this outcome k times more likely. Thus the desired probability is proportional to the abundance of k degree nodes, as well as to the degree itself, i.e., it is $q_k = \frac{kP(k)}{\langle k \rangle}$, where the denominator is used as a normalization constant. In a neutral network, the degree distribution of the nodes that lie at the other end of the selected edge is independent of q_k. Thus, in the absence of degree correlations, the probability that a randomly selected edge links between two nodes with a degree of k and k' is simply $Q_{kk'}^{\text{Neu}} = q_k q_{k'}$. To evaluate the assortativity of the network we compare the observed probability $Q_{kk'}$ to $Q_{kk'}^{\text{Neu}}$. For an assortative network, the observed probability will show a positive bias along the diagonal, where the value of k is close to that of k'. Disassortativity will be expressed as a negative bias along the diagonal, and a tendency to have more links where $k \neq k'$.

Another, more compact description of the degree correlations can be viewed by observing the average degree of a node's nearest neighbors. We denote this average by K_{nn}. We then average over all nodes with a given degree, k, to obtain $K_{\text{nn}}(k)$, namely the average degree of the neighbors surrounding a typical node with k links. In a neutral network, K_{nn} should not depend on k, but if degree correlations are present they will be expressed as a monotonic increase or decrease in $K_{\text{nn}}(k)$. In the protein–protein interaction network displayed in Figure 9.2, this dependency is clearly visible: the average degree of the hub's nearest neighbors is between 1 and 2, and yet the low-degree nodes are almost all connected to the hubs, so that for them K_{nn} is much greater. Indeed, the analysis shows that for this network $K_{\text{nn}}(k) \sim k^{-\alpha}$, where $\alpha \approx 0.24$ [85].

One can obtain an even more compact parameterization for a network's assortativity, by referring to the Pearson correlation coefficient measured between the degrees of pairs of connected nodes. This can be explicitly done by extracting the correlation coefficient for k and k' from the distribution given by $Q_{kk'}$. The result is [83]

$$r = \frac{\langle kk' \rangle - \langle k \rangle \langle k' \rangle}{\sigma^2}, \quad (4)$$

where σ^2 is the variance obtained from the distribution q_k. The parameter r takes values between 1, for a perfectly assortative network, and -1, when the network is perfectly disassortative. For the yeast protein–protein interaction network shown in Figure 9.2(a), it measures $r = -0.156$, confirming that the network is, indeed, disassortative.

The mechanism responsible for the disassortative nature of biological networks remains unclear. It cannot be accounted for by the Barabási–Albert mechanism, which does not yield any degree correlations [83]. From a functional point of view, it highlights the modular structure of biological networks, possibly strengthening even further the central role of the hubs. It was also shown that disassortativity harms the resilience of the network and makes it more vulnerable to the intentional removal of hubs, since in such networks the majority of low-degree nodes are connected solely to the hubs. On the other hand, disassortativity has a positive contribution to the integrity of the network when it is not under attack, as typically a disassortative network will feature a larger giant connected component than an assortative or a neutral one [83]. This once again emphasizes the resilience of cellular networks against random failure, compared to their vulnerability against selected node removal.

HUMAN DISEASE NETWORK

The applications of graph theory to systems biology can go beyond the mapping of the concrete network systems found within the cell. Graphs could also be used as a means of organizing biological information in a way that could potentially spark new insights. An innovative example is provided by the network approach to the study of human diseases [86]. In this approach two networks are constructed. The first is the human disease network. In this network the nodes represent genetic disorders and the edges link disorders which are associated with mutations in the same gene. The second network is the disease gene network. Here the nodes represent genes, and the edges link genes which are associated with the same disorder. Both networks are found to be highly clustered, showing that diseases, as well as disease genes, tend to divide into modules, or families, of related disorders. In the disease gene network genes that contribute to the same disorder tend to be correlated in many other ways as well. They have an increased tendency to be expressed together in specific tissues; they typically display high co-expression levels; and in many cases they share common cellular and functional characteristics, as annotated in the Gene Ontology [86]. This network is also in close relationship with the protein—protein interaction network, as disease-related genes are very likely to have their products interact together through physical binding. A surprising feature revealed by this analysis is the distinction between lethal genes and disease genes. In many cases, the products of lethal genes are highly connected nodes in the protein interaction network [41]. This emphasizes the importance of the hubs for the proper function of the network. In contrast, disease genes tend to avoid the hubs, and the vast majority of them are non-essential. It has been suggested that this is driven by natural selection, enabling the proliferation of mutations only if they harmed non-vital genes [86]. Confirming evidence comes from the fact that somatic mutations, which do not harm the organism's reproduction, are indeed more frequently related to hub genes. In a broader perspective, this network approach to human diseases offers a tool for the understanding of general patterns in genetic disorders, and could potentially reveal connections which are not apparent in the study of individual disorders [87—91].

THE BUILDING BLOCKS OF CELLULAR NETWORKS

In the previous section we discussed the macroscopic aspects of the hierarchical topology. We have shown that the hierarchy in cellular networks is closely intertwined with their modular structure. Indeed, from a functional point of view biology is full of examples of modularity, where a distinct group of proteins, genes or metabolites is responsible for the execution of some basic biological operations. Topologically, as discussed earlier, this is expressed in the emergence of various sub-graphs composed of highly interlinked groups of nodes. The high clustering typical of cellular networks provides the quantitative evidence for this modular network structure. In this section we focus on the typical recurring structures of these sub-graphs, and their meaning. In a sense, we are lowering the altitude of the bird's-eye perspective with which we viewed the networks until now, going from the macroscopic analysis to a more focused look at the building blocks of our complex systems.

Sub-graphs and Motifs

In order to conduct a fruitful analysis of network modules, we need first to develop a scheme by which we can identify what are meaningful modules. For instance, consider a tetrahedral sub-graph, which is a fully connected set of four nodes as shown in Figure 9.4(a). We can evaluate the abundance of this sub-graph in our network, but this will not be sufficient in order to tag it as a significant functional module. The randomness in the network topology makes it probable that such a module is due to appear in the network by chance. We thus consider a certain module to be a significant *motif* if it is over-represented in that network, that is, more abundant than expected by chance alone [92—93]. The idea is that if the network has the tendency to over-represent a certain module, there must be an evolutionary or functional need for it. Since natural selection discriminates on the basis of functional criteria, it will be these motifs that are likely to be capable of carrying important biological functions.

Randomized Networks

As stated above, for a certain module to qualify as a motif it must be more abundant than would be expected by chance. However, we have not accurately defined what we mean by

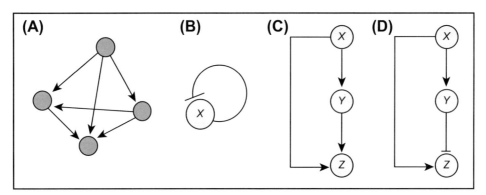

FIGURE 9.4 Network motifs: (A) The hypothetical tetrahedral motif; (B) the autoregulator; (C) the coherent feedforward loop and (D) the incoherent feedforward loop.

this criterion. It might, at first glance, seem intuitive to use the Erdős–Rényi networks as the grounds for comparison. However, a more careful look shows that this is not sufficient. The reason is that the expected frequency of a given sub-graph is dictated by the degree distribution of the graph as a whole. Consider, for instance, the tetrahedral module discussed above. This module consists of four nodes and six links. In order for such a module to emerge, first we need to have a node with a degree of at least 3. Then there must be at least three additional links among this node's nearest neighbors. The likelihood of the first condition is dictated by $P(k)$, and the likelihood of the second is determined by $C(k)$. In the broader sense what this means is that the macroscopic features of the network, given by $P(k)$ and $C(k)$, are in close relationship with the detailed structure of its modules. In the context of the current discussion, it states that the abundance of a given sub-graph is not independent of $P(k)$. Thus, in order to deem a certain module as over-represented in a particular network, we must compare its abundance to that of a randomized network with the same degree distribution [94]. Such a randomized network can be constructed by randomly rewiring all the links in the original network, preserving each node's degree, and hence $P(k)$, but deleting fine structure, such as the recurrence of motifs.

Autoregulation and the Feedforward Loop

We now briefly discuss two noted examples of highly recurring motifs found in transcriptional regulatory networks. The first motif is the negative autoregulator, which is one of the simplest and most abundant network motifs found in *E. coli* [95–96]. It includes a single transcription factor, which represses its own transcription. Graphically, this motif, shown in Figure 9.4(b), is simply a single node loop. It was shown to have two important functions. The first function is response acceleration. Compared to alternative regulating processes, such as protein degradation, the process of autoregulation allows for a faster response to signals. This was shown both theoretically and experimentally by employing synthetic gene circuits in *E. coli* [97]. The second advantage is that the motif increases the stability of the gene product concentration against stochastic noise. It therefore reduces the variations in protein levels between different cells [98–99].

Another motif frequently encountered in regulatory networks is the feedforward loop [100]. This motif consists of three nodes, x, y and z, where x is directly linked to both y and z, and in addition y is also directly linked to z (Figure 9.4(c),(d)). The direct links can symbolize the activation or the inhibition of the target gene, or any combination thereof. Thus eight different versions of this motif can be constructed, each with a different biological function [101]. To demonstrate the functional importance of this motif, we focus on two different versions of the motif. The first is a coherent feedforward loop, observed in the arabinose utilization and in the flagella systems of *E. coli* [102–103], and the second is an incoherent feedforward loop, which appears in the galactose system of *E. coli* [104].

In the coherent feedforward loop all the directed links represent the process of activation. Thus gene x activates both genes y and z, and yet gene y itself activates z once again. This might seem redundant, but can be shown to have important functional implications. Consider the case where the target gene, z, can only be activated if it receives a signal from both x and y. Using a computational analogy, we say that it serves as an *AND* gate, as it yields a *positive output* only when both of its *inputs* are positive. The motif will feature a time lag from when x is activated to when z responds. This is because z will be activated only after a sufficient concentration of y products has been produced. The result is that short sporadic expressions of the x gene will die off before z is ever activated. This motif, therefore functions as a filter, ignoring stochastic short-term perturbations and responding only to persistent ones. The complementary feature rises when the target gene serves as an *OR* gate. In this case, z is activated by either x or y. Here the delayed response will appear if x suddenly ceases to be expressed, in which case z will still remain active for some time, as long as a sufficient abundance of y's product persists. Thus the stability of z's expression is assured against sudden short-term drops in the production of x. This type of behavior is observed in the flagella system of *E. coli*, where a persistent activation of the flagella is maintained even under transient loss of the input signal [105].

A surprising, but nevertheless prevalent, version of the feedforward motif is the incoherent feedforward loop. Here, while x activates both y and z, the link between y and z is inhibitory. This seemingly contradictory wiring leads to an interesting functional feature. Consider a sudden activation of the gene x, due to, say, an external signal. As a result both y and z will be activated too. For a short time after x's activation, the expression levels of z will be constantly rising, owing to its activation by x. However, after a sufficient amount of y products has been produced, the expression of z will be suppressed, due to its inhibition by y. This version of the motif therefore translates a persistent signal induced by x into a spike of activation of the target gene, z.

GOING BEYOND TOPOLOGY

Despite its success, the purely topological approach possesses inherent limitations in the race to understand cellular networks. In focusing on topology alone, we have neglected the fact that *not all edges are created equal*.

In practice, the dynamical functionality of a complex network is probably affected not just by the binary pattern of who is connected to whom, but also by the nature of this connection and by its strength. Indeed, in a realistic biological network several reactions are more dominant than others — a feature that is overlooked by topology-based analyses. To obtain a more effective description we assign different *weights* to the edges, based on the intensity of the interaction [106−107]. This gives rise to weighted networks, where the link between a pair of nodes i, j, is no longer represented by the discrete state of present versus absent, but by a continuous number, w_{ij}, evaluating its importance.

Assigning the Weights

In metabolic networks the most natural measure for the weight of a given reaction is its flux, i.e., the rate by which a substrate is being converted to its product. The flux-balance approach has proved very successful in retrieving these fluxes [108−109]. In this approach one writes a set of equations for the metabolic fluxes, based on the assumption that all the metabolic reactions are balanced, that is, the concentrations of the reactants are at a steady state. This amounts to a set of linear algebraic equations for the fluxes. Typically these equations are underdetermined, as they include more variables than equations. To further narrow the solution space one imposes biological and chemical constraints. These constraints may emerge from experimental data, for instance, if some of the fluxes can be directly measured. Other constraints may be thermodynamic in nature, for instance if a certain reaction is known to be irreversible because the product has a much lower free energy than the substrates. Finally, after characterizing the diminished solution space to which the fluxes are constrained, the specific solution is chosen to be the one that optimizes the predefined biological function (e.g., maximal growth rate). For more detailed information regarding flux-balance methods see Chapter 12.

In transcriptional regulatory networks one can rely on microarray datasets to express the strength of a connection between a pair of genes. The co-expression of a pair of genes can be evaluated by measuring the correlations in their expression patterns. Alternatively, one can search for local similarities in the perturbed transcriptome profiles of the genes, and use those to infer the network connections and their weights [110−111].

Characterizing the Weighted Topology

Metabolic flux-balance analysis has been applied to the metabolic network of *E. coli*, and the complete weighted network has been obtained. Similarly to the topological findings, the weighted reactions were found to display strikingly high variability [112]. The reaction weights, based on the calculated fluxes, range over several orders of magnitude. The weight distribution, like the degree distribution, follows a power-law, $P(w) \sim w^{-\alpha}$, which, for *E. coli*, has a scaling exponent of $\alpha \approx 1.5$. While the specific fluxes for the different reactions depend on the environmental conditions, the aggregate behavior, captured by the weight distribution, remains unchanged under various environmental conditions. The fact that $\alpha < 2$ emphasizes the broadness of the observed weight distribution, as, mathematically, the mean value obtained for such a distribution diverges. In practice, one finds that almost all fluxes are below the average, and a few fluxes measure orders of magnitude above it, dominating the dynamics of the system. This provides an interesting illustration for the biochemical activity of metabolism. It suggests that under any given conditions metabolism is dominated by a small set of highly active reactions, embedded in a background of mostly dim chemical activity. A similar pattern of a highly uneven load distribution occurs in the regulatory network of *S. cerevisiae*. Also there most genes have weak correlations, while a few pairs show quite significant correlation coefficients [113].

Topology Correlated Weights

The connectedness of a node, characterized topologically, is captured by its degree. The weighted network analog is the node's *strength*. The strength is defined as the sum of all weights assigned to the node's set of links, namely $s_i = \sum_j w_{ij}$, where the absence of a link between a pair of nodes is denoted by setting the corresponding weight to zero. To characterize the relationship between the network topology and the link weights, we measure the dependence of the strength on the node degrees, namely $s(k)$. This function typically takes the form of $s(k) \sim k^\beta$ where a linear dependence, namely $\beta = 1$, reflects the absence of such degree-strength correlations, meaning that the weights are evenly distributed among the edges, so that a node acquires more strength simply because it has more links. However, in real systems it is commonly observed that $\beta > 1$. This implies that nodes with a higher degree tend to also have links with higher weight. This feature was explicitly observed in the *E. coli* metabolic network [114], where it was found that highly weighted links favored highly connected nodes. For each pair of connected nodes the weight, w_{ij}, is measured via flux balance analysis, finding that it features a power-law dependence on the degrees of the two linked nodes, i.e. $\langle w_{ij} \rangle \sim (k_i k_j)^\theta$, with $\theta \approx 0.5$.

Controllability

The functional state of a cellular network at any given time can be characterized by the concentrations of the reacting molecules, be them proteins, metabolites or any other biomolecules. This defines a vast state space that the network could explore. However, there are cases where some parts of this vast state space are restricted owing to the dynamics of the network. For instance, consider the simple case of just two interacting metabolites, X and Y, in which X produces Y, namely $X \rightarrow Y$. The state of this system is described by a point in the two-dimensional space given by $\vec{N} = (N_X, N_Y)$, where $N_{X(Y)}$ is the concentration of $X(Y)$. However, by assigning a certain value to N_X, the result for N_Y becomes predetermined via Y's production by X. The system is therefore confined to a small subspace of the complete two-dimensional state space. If one wished to steer the system to any desired state, the node X must be explicitly *driven* by an external input. We say [115–116] that this system can be *controlled* (that is, manipulated into any desired state) with a single *driver node*, X. In general, if we wish to control a complex network, we first need to identify the set of driver nodes that, if driven by different signals, offer full control over the network. We are particularly interested in identifying the minimum number of driver nodes, whose control is sufficient to fully control the system's dynamics [117]. An illustration for simple three-node networks is shown in Figure 9.5.

Applying the above concepts to transcriptional regulatory networks reveals that they are typically difficult to control. This is expressed by their relatively large number of driver nodes. As an example, for a typical transcriptional regulatory network approximately 80% of the nodes are driver nodes, indicating that in order to steer these networks, the majority of the nodes have to be explicitly controlled [117]. A higher extent of controllability is observed for metabolic networks, where the fraction of driver nodes is typically around one-third. From a topological point of view, dense and homogeneous networks are relatively easy to control, whereas sparse and inhomogeneous ones are hard to control. This implies that the degree distribution plays an important role in determining the controllability of a network. More specifically, scale-free networks, which are highly non-regular, will feature a large number of driver nodes and thus be difficult to control. Interestingly, controllability is not governed by the hubs, as the driver nodes tend to avoid the high-degree nodes.

The results presented above might seem to defy our intuitive perception of biological networks as systems which are expected to be firmly controlled. However, when examined once again, they might offer some deeper insight into the nature of control in biological systems. The fact that there are many driver nodes, and that they are typically the less central nodes in the network, is an expression of the highly constrained nature of these networks. It shows that the cellular networks are not *free* to explore the entire state space, but are rather confined to a restricted area of this space. Thus the only way in which these networks can be driven into a predefined final state is by explicitly driving almost each one of their nodes by an external signal — especially the low-degree nodes, whose state is otherwise governed by the hubs. In a sense, one can interpret these results as the strategy of cellular networks to circumvent external control and maintain their function, even if a large number of nodes are being influenced.

Differential Networks

A given network topology may give rise to a variety of dynamical behaviors under different dynamical rules or environmental conditions. External stimuli may trigger the activity of different parts of the network, further affecting its dynamical functionality. However, a much broader range of dynamical behaviors could be achieved under the rule of a non-static topology, where the structure of the network itself can react to environmental changes and external stimuli. Recently, it has been found that cellular networks indeed take advantage of this source of dynamical diversity,

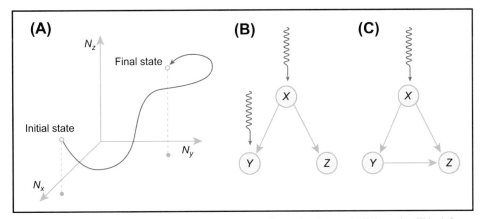

FIGURE 9.5 (A) The state of a three-node network is given by the concentration assigned to each of the nodes. This defines a point in the three-dimensional state-space. Controlling the network means steering it from any initial state to any desired final state. (B) In this network two of the three nodes must be explicitly controlled in order to manipulate the network. (C) Here it is sufficient to control just one node.

altering the architecture of the network itself under different conditions or biological states, such as tissue type, disease state or the surrounding environment [118].

FROM STRUCTURE TO DYNAMICS

The path taken throughout this chapter outlines, in some sense, the approach of network biology towards its future challenges. We begin by describing the network representation of cellular systems, examining their topological properties. We then follow with a discussion regarding motifs, weighted networks and controllability, which addresses the dynamics and function of these networks. In this spirit, we end this chapter with what is probably the most pressing challenge of this area of research — the *bridging between structure and dynamics*. We are currently at a stage where the topological aspects of cellular networks have been thoroughly elucidated and their evolutionary origins fairly understood. However, we still lack a complete theory which could interpret the topological findings into a set of dynamical predictions, from which the actual functionality of the networks could be inferred [5]. Below we stress, in a very broad fashion, the strategic path that could meet this challenge [119–120].

The most fundamental question we must address is whether the gap between structure and dynamical behavior could at all be bridged. We need to take into account that the topology is one actor in a highly detailed cast of network characteristics. In the most detailed description of these cellular systems, all the interactions can take on different reaction processes and different strengths. By reaction processes we refer to the types of interaction, e.g., chemical, regulatory, etc., and by strengths we mean that similar processes may occur at different rates. So whereas structurally we denote all the various interactions by network links, one should ask: is the process of genetic regulation really comparable to that of physical binding? Is it guaranteed that two structurally identical networks will express similar behavior even if they differ in some other details? Or perhaps these details, which are overlooked by graph theoretic analysis, are not important.

A more constructive approach towards the above questions is to ask how far *can* one actually progress with structure alone? It seems clear that a complete time-dependent dynamics of the system would require the incorporation of all of the details mentioned above, and is thus beyond the scope of network biology. On the other hand, what *could* be achievable based on a structural analysis is a macroscopic understanding of network dynamics. More specifically, network science is not expected to be successful in predicting the behavior of a specific set of nodes. It could, however, provide answers to general questions regarding the network as a whole. Questions such as: which are the most effective nodes in this network? Does this network support long-range interactions, or is the impact of nodes contained locally? Are the concentrations of the nodes in this network stable or governed by fluctuations? Will a small perturbation cause a macroscopic failure? At least some of these characteristics, and others like them, might be determined by the topology of the network, regardless of the other details of the interactions. And if that is the case, we should be able to address these important questions with the tools of network analysis.

In a broader perspective, applying graph theory to the study of complex systems is aimed at bringing about an intuitive, visual and mathematical toolkit for their understanding. In that sense, the challenge of this approach is to devise a set of intuitive dynamical interpretations to the already defined set of topological features. The idea is to assign a functional meaning to characterizations such as a *broad degree distribution*, *high clustering*, *small world-ness*, etc. Along this path, future research needs to challenge some of the common wisdoms regarding these attributions between structure and dynamics. For instance, the intuitive notion that in a small world topology all the nodes are affected by one another, since they are just a few reactions away; or the common perception that the hubs are the most influential nodes in the network. Once these statements, and others like them, are examined, they will bring forth a *new intuition* on the meaning of different structural attributes. Then, by analyzing the structure of a network, researchers will be able to make general assessments regarding its expected dynamics.

The rapidly improving experimental techniques in biology will hopefully enable the dynamical predictions derived from network analysis to be tested. However, even where the existing experimental procedures are insufficient, help might arrive from unexpected sources. Perhaps the greatest success of the network approach thus far is in revealing the universal nature of the topology of networks — cellular and others — providing a set of tools and criteria by which to classify and characterize the structure of these diverse systems [5]. A similar degree of universality in the dynamics of networks, if found, will provide us with a parallel set of unifying principles, allowing us to describe, using a common platform, various dynamical processes and make meaningful predictions on the behavior of networks from diverse fields. These universal dynamical aspects could then be inferred from one system to the other. Metaphorically, this expands the boundaries of the classic biology laboratories far beyond their traditional walls. As data are currently collected in vast amounts from biological, social and technological systems, the abilities that network science creates to learn from one system about the other provide a crucial source of empirical strength, a strength that may one day help make complex systems slightly more simple.

REFERENCES

[1] Alm E, Arkin A. Biological networks. Current Opinion in Structural Biology 2003;13:193−202.
[2] Barabási AL, Oltvai ZN. Network biology: understanding the cell's functional organization. Nature Reviews − Genetics 2004;5:101−13.
[3] Bray D. Molecular networks: the top-down view. Science 2003;301:1864−5.
[4] Vidal M, Cusick ME, Barabási AL. Interactome networks and human disease. Cell 2011;144:986−98.
[5] Albert R, Barabási AL. The statistical mechanics of complex networks. Review of Modern Physics 2002;74:47−97.
[6] Newman M. The structure and function of complex networks. SIAM Reviews 2003;45:167−256.
[7] Schadt EE, Lamb J, Yang X, Zhu J, Edwards S, Guhathakurta D, et al. An integrative genomics approach to infer causal associations between gene expression and disease. Nature Genetics 2005;37:710−7.
[8] Chuang HY, Lee E, Liu YT, Lee D, Ideker T. Network-based classification of breast cancer metastasis. Molecular Systems Biology 2007;3:140−50.
[9] François K. Biological Network. Complex Systems and Interdisciplinary Science − Volume 3. 1st ed. Hackensak, New-Jersey: World Scientific Publishing Company; 2007.
[10] Ihmels J, Friedlander G, Bergmann S, Sarig O, Ziv Y, Barkai N. Revealing modular organization in the yeast transcriptional network. Nature Genetics 2002;31:370−7.
[11] Stelzl U, Worm U, Lalowski M, Haenig C, Brembeck FH, Goehler H, et al. A human protein-protein interaction network: a resource for annotating the proteome. Cell 2005;122:957−68.
[12] Uetz P, Giot L, Cagney G, Mansfield TA, Judson RS, Knight JR, et al. A comprehensive analysis of protein-protein interactions in Saccharomyces cerevisiae. Nature 2000;403:623−30.
[13] Rain JC, Selig L, De-Reuse H, Battaglia V, Reverdy C, Simon S, et al. The protein-protein interaction map of Helicobacter pylori. Nature 2001;409:211−6.
[14] Giot L, Bader JS, Brouwer C, Chaudhuri A, Kuang B, Li Y, et al. A Protein Interaction Map of Drosophila melanogaster. Science 2003;302:1727−36.
[15] Ito T, Chiba T, Ozawa R, Yoshida M, Hattori M, Sakaki Y. A comprehensive two-hybrid analysis to explore the yeast protein interactome. Proc. Natl. Acad. Sci. USA 2001;98:4569−74.
[16] Costanzo MC, Crawford ME, Hirschman JE, Kranz JE, Olsen P, Robertson LS, et al. YPD, PombePD and WormPD: model organism volumes of the BioKnowledge library, an integrated resource for protein information. Nucleic Acids Res 2001;29:75−84.
[17] Li S, Armstrong CM, Bertin N, Ge H, Milstein S, Boxem M, et al. A map of the interactome network of the metazoan Caenorhabditis elegans. Science 2004;303:540−3.
[18] Costanzo M, Baryshnikova A, Bellay J, Kim Y, Spear ED, Sevier CS. The genetic landscape of a cell. Science 2010;327:425−56.
[19] De-Las Rivas J, Fontanillo C. Protein-protein interactions essentials: key concepts to building and analyzing interactome networks. PLoS Computational Biology 2010;6:e100807.
[20] Stumpf MP, Thorne T, de-Silva E, Stewart R, An HJ, Lappe M, et al. Estimating the size of the human interactome. Proc Natl Acad Sci 2008;105:6959−64.
[21] Tong AHY, Lesage G, Bader GD, Ding H, Xu H, Xin X, et al. Global mapping of the yeast genetic interaction network. Science 1999;286:509−21.
[22] Salgado H, Gama-Castro S, Peralta-Gil M, Díaz-Peredo E, Sánchez-Solano F, Santos-Zavaleta A, et al. RegulonDB (version 5.0): Escherichia coli K-12 transcriptional regulatory network, operon organization, and growth conditions. Nucleic Acids Research 2006;34:D394−7.
[23] Farkas I, Jeong H, Vicsek T, Barabási AL, Oltvai ZN. The topology of the transcription regulatory network in the yeast, Sacchromyces cervisiae. Physica A 2003;318:601−12.
[24] Erdős P, Rényi A. On Random Graphs I. Publicationes Mathematicae 1959;6:290−7.
[25] Erdős P, Rényi A. The Evolution of Random Graphs. Magyar Tud Akad Mat Kutató Int. Közl 1960;5:17−61.
[26] Gilbert EN. Random Graphs. Annals of Mathematical Statistics 1959;30:1141−4.
[27] Newman MEJ. Networks − an Introduction. 1st ed. New York: Oxford University Press; 2010.
[28] Holland PW, Leinhardt S. Transitivity in structural models of small groups. Comparative Group Studies 1971;2:107−24.
[29] Luce RD, Perry AD. A method of matrix analysis of group structure. Psychometrika 1949;14:95−116.
[30] Watts DJ, Strogatz S. Collective dynamics of small-world networks. Nature 1998;393:440−2.
[31] Jeong H, Tombor B, Albert R, Oltvai ZN, Barabási AL. The large-scale organization of metabolic networks. Nature 2000;407:651−5.
[32] Yook S, Oltvai ZN, Barabási AL. Functional and topological characterization of protein interaction networks. Proteomics 2004;4:928−42.
[33] Xu K, Bezakova I, Bunimovich L, Yi SV. Path lengths in protein-protein interaction networks and biological complexity. Proteomics 2011;11:1857−67.
[34] Wagner A, Fell DA. The small world inside large metabolic networks. Proc. Biological Science 2001;268:1803−10.
[35] Ravasz E, Somera AL, Mongru DA, Oltvai ZN, Barabási AL. Hierarchical organization of modularity in metabolic networks. Science 2002;297:1551−6.
[36] Barabási AL, Albert R. Emergence of scaling in random networks. Science 1999;286:509−21.
[37] Mason O, Verwoerd M. Graph theory and networks in Biology. IET Systems Biology 2007;1:89−119.
[38] Guelzim N, Bottani S, Bourgine P, Képès F. Topological and causal structure of the yeast transcriptional regulatory network. Nature Genetics 2002;31:60−3.
[39] Wagner A. The yeast protein interaction network evolves rapidly and contains few redundant duplicate genes. Molecular Biology and Evolution 2001;18:1283−92.
[40] Arita M. Scale-Freeness and Biological Networks. J. Biochemistry 2005;138:1−4.
[41] Jeong H, Mason SP, Barabási AL, Oltvai ZN. Lethality and centrality in protein networks. Nature 2001;411:41−3.
[42] Wuchty S. Scale-free behavior in protein domain networks. Molecular Biology and Evolution 2001;18. 1964−1702.
[43] Apic G, Gough J, Teichmann SA. An insight into domain combinations. Bioinformatics 2001;17:S83−9.
[44] Featherstone DE, Broadie K. Wrestling with pleiotropy: genomic and topological analysis of the yeast gene expression network. Bioessays 2002;24:267−74.

[45] Barabási AL, Dezső Z, Ravasz E, Yook SH, Oltvai Z. Scale-free and hierarchical structures in complex networks. AIP Conference Proceedings 2003;661:1–16.

[46] Chung F, Lu L. The average distance in random graphs with given expected degrees. Proc. Natl. Acad. Sci. USA 2002;99:15879–82.

[47] Bollobás B, Riordan O. The diameter of a scale-free random graph. Combintorica 2004;24:5–34.

[48] Chen F, Chen Z, Wang X, Yuan Z. The average path length of scale-free networks. Communications in Nonlinear Science and Numerical Simulation 2008;13:1405–10.

[49] Cohen R, Havlin S. Scale-free networks are ultra-small. Physical Review Letters 2003;90:058701:1–058701:4.

[50] Winzler EA, Shoemaker DD, Astromoff A, Liang H, Anderson K, Andre B, et al. Functional characterization of the Saccharomyces cerevisiae genome by gene deletion and parallel analysis. Science 1999;285:901–6.

[51] Giaever G, Chu AM, Ni L, Connelly C, Riles L, Véronneau S, et al. Functional profiling of the Saccharomyces cerevisiae genome. Nature 2002;418:387–91.

[52] Gerdes SY, Scholle MD, Campbell JW, Balázsi G, Ravasz E, Daugherty MD, et al. Experimental determination and system-level analysis of essential genes in Escherichia coli MG1655. J Bacteriology 2003;185:5673–84.

[53] Yu BJ, Sung BH, Koob MD, Lee CH, Lee JH, Lee WH, et al. Minimization of the Escherichia coli genome using a Tn5-targeted Cre/loxP excitation system. Nature Biotechnology 2002;20:1018–23.

[54] Kolysnychenko V, Plunkett G, Herring CD, Fehér T, Pósfai J, Blattner FR, et al. Engineering a reduced Escherichia coli genome. Genome Research 2002;12:640–7.

[55] Albert R, Jeong H, Barabási AL. Error and attack tolerance of complex networks. Nature 2000;406:378–82.

[56] He X, Zhang J. Why do hubs tend to be essential in protein networks? PLoS Genetics 2006;2:826–34.

[57] Havlin S, Cohen R. Complex Networks – Structure, Robustness and Function. 1st ed. New York: Cambridge University Press; 2010.

[58] Yu H, Greenbaum D, Xin Lu H, Zhu X, Gerstein M. Genomic analysis of essentiality within protein networks. Trends in Genetics 2004;20:227–31.

[59] Hahn MW, Kern AD. Comparative genomics of centrality and essentiality in three eukaryotic protein-interaction networks. Molecular Biology and Evolution 2005;22:803–9.

[60] Kamath RS, Fraser AG, Dong Y, Poulin G, Durbin R, Gotta M, et al. Systematic functional analysis of the Caenorhabditis elegans genome using RNAi. Nature 2003;421:231–8.

[61] Prachumwat A, Li WH. Protein function, connectivity, and duplicability in yeast. Molecular Biology and Evolution 2006;23:30–9.

[62] Yamada T, Bork P. Evolution of biomolecular networks: lessons from metabolic and protein interactions. Nature Reviews Molecular Cell Biology 2009;10:791–803.

[63] Liang H, Li WH. Gene essentiality, gene duplicability and protein connectivity in human and mouse. Trends in Genetics 2007;23:375–83.

[64] Wuchty S. Evolution and topology in the yeast protein interaction network. Genome Research 2004;14:1310–4.

[65] Fraser HB. Modularity and evolutionary constraint on proteins. Nature Genetics 2005;37:351–3.

[66] Rzhetsky A, Gomez SM. Birth of scale-free molecular networks and the number of distinct DNA and protein domains per genome. Bioinformatics 2001;17:988–96.

[67] Qian J, Luscombe NM, Gerstein M. Protein family and fold occurrence in genomes: power-law behavior and evolutionary model. J. Molecular Biology 2001;313:673–81.

[68] Bhan A, Galas DJ, Dewey TG. A duplication growth model of gene expression networks. Bioinformatics 2002;18:1486–93.

[69] Pastor-Satorras R, Smith E, Sole R. Evolving protein interaction networks through gene duplication. J Theoretical Biology 2003;222:199–210.

[70] Vazquez A, Flammini A, Maritan A, Vespignani A. Modeling of protein interaction networks. ComPlexUs 2003;1:38–44.

[71] Kim J, Krapivsky PL, Kahng B, Redner S. Infinite-order percolation and giant fluctuations in a protein interaction network. Physical Review E 2002;66:055101.

[72] Wagner A. How the global structure of protein interaction networks evolves. Proc Royal Society of London B 2003;270:457–66.

[73] Eisenberg E, Levanon EY. Preferential attachment in the protein network evolution. Physical Review Letters 2003;91:138701.

[74] Hartwell LH, Hopfield JJ, Leibler S, Murray AW. From molecular to modular cell biology. Nature 1999;402:C47–52.

[75] Wall ME, Hlavacek WS, Savageau MA. Design of gene circuits: lessons from bacteria. Nature Reviews Genetics 2004;5:34–42.

[76] Alon U. Biological networks: the tinkerer as an engineer. Science 2003;301:1866–7.

[77] Alberts B. The cell as a collection of protein machines: preparing the next generation of molecular biologists. Cell 1998;92:291–4.

[78] Ravasz E, Barabási AL. Hierarchical organization in complex networks. Physical Review E Statistical Nonlinear Soft Matter Physics 67 2003;026112.

[79] Deisboeck TS, Yasha-Kresh J, Kepler TB. Complex Systems in Biomedicine. 1st ed. New York, New York: Kluwer Academic Publishing; 2005.

[80] Dorogovtsev SN, Goltsev AV, Mendes JFF. Pseudofractal scale-free web. Physical Review E Statistical Nonlinear Soft Matter Physics 2002;65:066122.

[81] Han JD, Bertin N, Hao T, Goldberg DS, Berriz GF, Zhang LV, et al. Evidence for dynamically organized modularity in the yeast protein-protein interaction network. Nature 2004;430:88–93.

[82] Maslov S, Sneppen K. Specificity and stability in topology of protein networks. Science 2002;296:910–3.

[83] Newman MEJ. Assortative mixing in networks. Physical Review Letters 2002;89:208701.

[84] Vázquez A, Pastor-Satorras R, Vespignani A. Large-scale topological and dynamical properties of the Internet. Physical Review E 2002;65:066130–42.

[85] Colliza V, Flammini A, Maritan A, Vespignani A. Characterization and modeling of protein-protein interaction networks. Physica A 2005;352:1–27.

[86] Goh K, Cusick ME, Valle D, Childs B, Vidal M, Barabási AL. The human disease network. Proc Natl Acad Sci USA 2007;104:8685–90.

[87] Oti M, Snel B, Huynen MA, Brunner HG. Predicting disease genes using protein-protein interactions. J Medical Genetics 2006;43:691–8.

[88] Lage K, Karlberg EO, Størling ZM, Olason PI, Pedersen AG, Rigina O, et al. A human phenome interactome network of protein

complexes implicated in genetic disorders. Nature Biotechnology 2007;25:309–16.
[89] Franke L, Bakel H, Fokkens L, de-Jong ED, Egmont-Petersen M, Wijmenga C. Reconstruction of a functional human gene network, with an application for prioritizing positional candidate genes. The American Journal of Human Genetics 2006;78:1011–25.
[90] Sharma A, Chavali S, Tabassum R, Tandon N, Bharadwaj D. Gene prioritization in Type 2 Diabetes using domain interactions and network analysis. BioMed Central Genomics 2010;11:84–94.
[91] Köhler S, Bauer S, Horn D, Robinson PN. Walking the interactome for prioritization of candidate disease genes. American Journal of Human Genetics 2008;82:949–58.
[92] Shen-Orr SS, Milo R, Mangan S, Alon U. Network motifs in the transcriptional regulation network of Escherichia coli. Nature Genetics 2002;31:64–8.
[93] Milo R, Shen-Orr SS, Itzkovitz S, Kashtan N, Alon U. Network motifs: simple building blocks of complex networks. Science 2002;298:824–7.
[94] Vázquez A, Dobrin R, Sergi D, Eckmann JP, Oltvai ZN, Barabási AL. The topological relationship between the large-scale attributes and local interactions patterns of complex networks. Proc Natl Acad Sci USA 2004;101:17940–5.
[95] Savageau MA. Comparison of classical and autogenous systems of regulation in inducible operons. Nature 1974;252:546–9.
[96] Thieffry D, Huerta AM, Pérez-Rueda E, Collado-Vides J. From Specific gene regulation to genomic networks: a global analysis of transcriptional regulation in Escherichia coli. BioEssays 1998;20:433–40.
[97] Rosenfeld N, Elowitz MB, Alon U. Negative autoregulation speeds the response times of transcription networks. J Molecular Biology 2002;323:785–93.
[98] Becskei A, Serrano L. Engineering stability in gene networks by autoregulation. Nature 2000;405:590–3.
[99] Dublanche Y, Michalodimitrakis K, Kümmerer N, Foglierini M, Serrano L. Noise in transcription negative feedback loops: simulation and experimental analysis. Molecular Systems Biology 2006;2:41.
[100] Mangan S, Alon U. Structure and function of the feed-forward loop network motif. Proc Natl Acad Sci USA 2003;100:11980–5.
[101] Alon U. Network motifs: theory and experimental approaches. Nature Reviews Genetics 2007;8:450–61.
[102] Kalir S, Mangan S, Alon U. A coherent feed-forward loop with a SUM input function prolongs flagella expression in Escherichia coli. Molecular Systems Biology 2005;1. 2005.0006.
[103] Mangan S, Zaslaver A, Alon U. The coherent feed-forward loop serves as a sign-sensitive delay element in transcription networks. J Molecular Biology 2003;334:197–204.
[104] Mangan S, Itzkovitz S, Zaslaver A, Alon U. The incoherent feedforward loop accelerates the response-time of the gal system of Escherichia coli. J. Molecular Biology 2006;356:1073–81.
[105] Shen-Orr SS, Milo R, Mangan S, Alon U. Network motifs in the transcriptional regulation network of Escherichia coli. Nature Genetics 2002;31:64–8.
[106] Savageau M. Biochemical Systems Analysis: a Study of Function and Design in Molecular Biology; 1976. Addison-Wesley, Reading, 1976.
[107] Fell DA. Understanding the Control of Metabolism; 1997. Portland, London, 1997.
[108] Edwards JS, Ibarra RU, Palsson BO. In silico predictions of Escherichia coli metabolic capabilities are consistent with experimental data. Nature Biotechnology 2001;19:125–30.
[109] Ibarra RU, Edwards JS, Palsson BO. Escherichia coli K-12 undergoes adaptive evolution to achieve in silico predicted optimal growth. Nature 2002;420:186–9.
[110] Zhang B, Horvath S. A general framework for weighted gene coexpression network analysis. Statistical Applied Genetics and Molecular Biology 2005;4:17.
[111] de-la Fuente A, Brazhnik P, Mendes P. Linking the genes: inferring quantitative gene networks from microarray data. Trends in Genetics 2002;18:395–8.
[112] Almaas E, Kovács B, Vicsek T, Oltvai ZN, Barabási AL. Global organization of metabolic fluxes in Escherichia coli. Nature 2004;427:839–43.
[113] Balaji S, Babu MM, Iyer LM, Luscombe NM, Aravind L. Comprehensive analysis of combinatorial regulation using the transcriptional regulatory network of yeast. J Molecular Biology 2006;360:213–27.
[114] Macdonald PJ, Almaas E, Barabási AL. Minimum spanning trees of weighted scale-free networks. Europhysics Letters 2005;72:308–14.
[115] Kalman RE. Mathematical description of linear dynamical systems. Journal of the Society for Industrial and Applied Mathematics 1963;1:152–92.
[116] Slotine J-J, Li W. Applied Nonlinear Control; 1991. Prentice-Hall, 1991.
[117] Liu Y, Slotine J-J, Barabási AL. Controllability of complex networks. Nature 2010;473:167–73.
[118] Ideker T, Krogan NJ. Differential network biology. Molecular Systems Biology 2012;8:565.
[119] Barzel B, Biham O. Quantifying the connectivity of a network: the network correlation function method. Physical Review E 2009;80:046104–15.
[120] Barzel B, Biham O. Binomial moment equations for stochastic reaction systems. Physical Review Letters 2011;106:150602.

Section III

Dynamic and Logical Properties of Biological Systems

Chapter 10

Boolean Models of Cellular Signaling Networks

Zhongyao Sun[1] and Réka Albert[1,2]
[1]Department of Physics, The Pennsylvania State University, University Park, PA 16802, USA, [2]The Huck Institutes of the Life Sciences, The Pennsylvania State University, University Park, PA 16802, USA

Chapter Outline

Introduction	197
Boolean Networks and Biological Systems	198
Boolean Network Modeling	199
Reconstruct the Network Based on Biological Knowledge	199
Determine the Boolean Transfer Functions	200
Updating Schemes and Incorporating Time	201
Network Initialization	202
Steady-State Analysis	202
Model Validation: Reproduction of Known Results	205
Robustness against Disruptions and Useful Implications	206
Application Examples	207
T-LGL Leukemia Network Modeling	207
Pathogen–Immune Response Network	207
Conclusions and Future Directions	207
References	208

INTRODUCTION

Network structures permeate every sphere of cellular biology. The vast number of intra- and extracellular processes and interactions that form a complex web of mass, energy and signal transfer can intrinsically be described in network language. The nodes of a biological network are cellular components such as genes, RNAs, proteins and small molecules, and the edges are reactions, interactions, and regulatory or synthetic relationships among components. The process of transcribing coding DNA into mRNAs is either promoted or suppressed by transcription factors; the totality of these transcriptional processes can be integrated into gene regulatory networks [1]. Similarly, diverse interactions among proteins, or the biochemical reactions in cellular metabolism, can be readily depicted in network language [2–4]. In a network representation the elements of the system are represented by network nodes, and the interactions among elements as edges. The wealth of data and the affinity between network science and biology make network modeling of biological systems not only viable, but also powerful and uniquely useful (see Chapter 9).

In addition to the network structure (that is, the nodes and edges), it is often desired to be able to describe the dynamics of mass or information transfer through the network. Dynamic models characterize the nodes by states (e.g., concentration or activity), and the states of the nodes change in time according to the interactions encapsulated in the network. Continuous dynamic models use sets of differential equations to capture the detailed variation of concentrations of key substances [5] in the system (see Chapter 16). Despite the recent phenomenal growth of computing power, such approaches become practically impossible to implement when the number of nodes in a system reaches more than 100. Difficulties are also posed for parameter estimation in large-scale systems when there are insufficient temporal data. The aim to circumvent these difficulties makes discrete dynamic modeling such as Boolean networks [6–13] and Petri nets [14–16] particularly useful. Vast amounts of qualitative knowledge regarding regulatory relationships between cellular components have been either experimentally measured or inferred from biological evidence. The compilation of such knowledge forms the basis for network reconstruction, which then allows discrete modeling of a system, bypassing the obstacles posed by parameter estimation. The model can be tested against experimental evidence and refined iteratively. Discrete dynamic models generate insightful analysis about the interweaving of system components and the cascade of information flow, and enable in silico node knockout experiments that produce informative predictions about the system. Discrete dynamic models have been successfully implemented in numerous biological systems, facilitating the study and greater understanding of biological processes such as flower development [17,18], the yeast cell cycle [8,9,19], *Drosophila* embryonic development [20–24], hormone signaling in plants [10,25], the immune response [11], and T-cell signaling and differentiation [12,13, 26–28].

Boolean network modeling of biological systems is the simplest of the discrete dynamic models. Each node can have one of two discrete states, namely 0 and 1, instead of a continuously varying concentration. 0 or *OFF* means that the element represented by the node is inactive (e.g., an enzyme or transcription factor), or has a below-threshold concentration (e.g., a small molecule). 1 or *ON* represents the opposite, which is active or an above-threshold concentration [6,7]. The state of a system that has N nodes is therefore represented by an N-dimensional vector with each value being 0 or 1. The state space of such system contains a total of 2^N states. In order to carry out dynamic simulations, time is usually discretized into time steps. As time evolves, starting from one initial state vector, either predetermined or randomly generated, the system state vector can traverse the state space, reaching different parts of the space. It is not necessary that a dynamic trajectory traverse all 2^N states in the space; indeed, the observed trajectories converge into a stationary state or set of states after a much smaller number of state transitions.

In this chapter, we first explain the correspondence between Boolean networks and biological systems, then introduce the basic concepts of Boolean network modeling and elaborate in a step-by-step fashion the modeling procedure, starting from network reconstruction based on biological information compilation to model validation. We will present two successful applications of this methodology in cellular biology [11,12].

BOOLEAN NETWORKS AND BIOLOGICAL SYSTEMS

Here we introduce the basics of Boolean algebra and its connection with biological systems. In order to summarize all available information and graphically represent the system we are studying, the elements of the system, such as genes or proteins, become the nodes of the network. Nodes that represent a phenomenon or a certain biological result can also be included, for instance Stomatal Closure, or Apoptosis.

Edges are drawn between biologically or chemically related nodes to represent the relationship between them. Since the flow of signals and reaction fluxes is directional, the edges in the network will each have a direction that is consistent with biology; edges may also be characterized by a sign (+/−) that denotes the property of that edge: positive for activation and negative for inhibition. For example, a common representation of the synthesis reaction C + D → E connects the two reactants with the product of the reaction, as in Figure 10.1(A).

This representation does not, however, directly reflect the fact that C and D are both required for the reaction to take place. Thus the network must be complemented with rules that specify the ways in which all upstream components are combined. A natural and economical method is to use the Boolean operators *AND, OR, NOT*. A combination of these operators can describe most possible relationships or reactions between substances and components in a biological system. The logic dependence underlying a synthesis reaction can be described by the Boolean operator *AND*, so the reaction C + D → E becomes *C AND D = E** in Boolean language, where for simplicity the node names stand for the state of the node. Similarly, if a certain

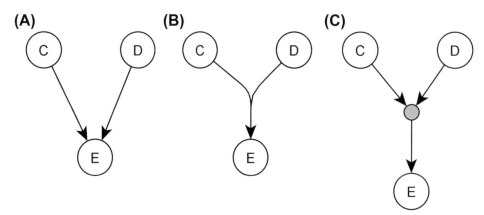

FIGURE 10.1 **Three different ways to represent the same synthesis reaction relationship between 3 nodes.** (A) Two separate edges directed from the reactants to the product. (B) Two edges from the reactants first merged together and then directed at the product. (C) An intermediate node (green circle) denoting the synergy between the two reactants is added. The reactants are connected to this intermediary node, which is in turn connected to the product.

result can stem from multiple causes, e.g., photosynthesis can be carried out under either blue or red light, then the Boolean equation denoting that will be Blue_light OR Red_light = Photosynthesis*. When a certain element or activity is negatively regulated by another, the *NOT* rule is used. The asterisks on the two example Boolean equations indicate that the specific processes to generate the product on the right-hand side take a certain amount of time to complete. We will return to this point later.

The additional information contained in the Boolean rules (e.g., the conditionality or independence of two edges incident on the same node) can be integrated into the network for a more complete representation. Merging edges may be used to represent the Boolean *AND* relation (Figure 10.1(B)), and separate edges for *OR* relations (Figure 10.1(A)) [29]. There is also a third choice, which involves an intermediate node, if the Boolean rule is *AND* (Figure 10.1(C)). Such a node would not exist if the rule were Boolean *OR* [21,30,31]. So far there is no universal 'standard' for how to map a system of complex relations onto a graph. The most appropriate choice is the one that best facilitates the analysis of the particular system under study.

BOOLEAN NETWORK MODELING

As mentioned in the introduction, a dynamic model characterizes a system's behavior over time. In Boolean dynamic modeling specifically, both time and the system's status are discretized. The state of a node can be 0 or 1, making the state of a system of N nodes an N-dimensional vector of 0s and 1s. A continuous time stream over a certain period (e.g., an experiment) is represented by a series of steps, which are an abstract representation of important time points at which biochemical events are taking place. The state of the system at a time step is determined by its predecessor state (or sometimes several predecessor states at earlier time steps) through what are called Boolean transfer functions. The calculation of a node state at a time step based on system state(s) at earlier step(s) is called updating. Depending on the updating scheme used, a number of nodes, ranging from 1 to N, are updated, thereby obtaining the system state at a new (future) time step. Given below is an example of a general expression of a Boolean transfer function of a certain node i. Suppose the state of node i at time step t is denoted as $V_{i,t}$. The transfer function through which $V_{i,t}$ is calculated is given as follows:

$$V_{i,t} = F_i\left(V_{k_1,\tau_{k_1}}, V_{k_2,\tau_{k_2}}, \ldots, V_{k_n,\tau_{k_n}}\right)$$

where $1 \leq i, k_1, \ldots, k_n \leq N$ are the node indices and $\tau_{k_1}, \ldots \tau_{k_n} \leq t$ denote the time step when the state of nodes $k_1, \ldots, k_n$ was last updated. We will revisit the details of time implementation later. As time evolves, the system state (N-dimensional vector) traverses the state space, and after a finite-duration transient behavior it settles into an attractor (stable dynamic behavior). Two types of attractors are possible: after hitting a certain state, any future updating results in the same state, hence the system reaches a 'fixed point' or steady state; or there exists a certain small set of states of the system G which the system keeps revisiting; that is, any updating of the system state will carry it to one of the states that belongs G. Depending on how updating is implemented, the second type of attractor of the system (non-steady state) can take on two possible forms, either an oscillation (a series of system states that repeat regularly) or a loose attractor (a random sequence of system states that is generated from a finite pool of states). See later in the chapter for further details.

Boolean dynamic modeling of a biological system is comprised of the following steps: reconstruct the network based on biological knowledge; determine the Boolean transfer functions; choose an updating scheme; determine the initial state of the model; analyze the model, including its attractors and state space; validate the model (reproduction of known results); and finally, study novel scenarios, e.g., robustness against disruptions, make useful predictions and inferences. We next look at each step in more detail and elaborate them through examples.

Reconstruct the Network Based on Biological Knowledge

The first step of Boolean dynamic modeling is to represent the system under study by a network, denoting the relevant elements of system by nodes and their pairwise relationships by edges. This is accomplished through extensive literature searches and compilation. Experimental databases available online, such as Transcription Factor Database (TRANSFAC, [32]) and Kyoto Encyclopedia of Genes and Genomes (KEGG, [33]), can be used for data mining to deduce causal relationships between components. Many, but not all, experiments indicate direct interactions or regulatory relationships of elements, such as transcription factor—gene interactions, enzymatic activities and protein—protein interactions. Genetic knockout or pharmaceutical evidence, such as exogenous application of a certain chemical, indicates regulatory relationships indirectly. With such relationships further inference and interpretation of experimental results might be needed to obtain the most proper regulatory relations to represent in the network [10]. For instance, experimental identification of the change in the activity level of a protein after a certain stimulus was applied implicates the protein as a potential downstream responder to that stimulus. Similarly, if the over-expression of a gene results in the downregulation of another gene or abnormally low activity of an enzyme, this implies that the gene negatively affects its target. A change in the concentration of a protein following a genetic knockout or over-expression can indicate the protein of interest as a potential regulation target of the gene being manipulated.

One can summarize the biological facts collected and list them in a table in order to synthesize and to present in a clear way all knowledge ready to be represented by a graph. Table 10.1 presents an example consisting of five nodes.

The network constructed based on Table 10.1 is shown in Figure 10.2.

Apart from straightforward evidences that indicate the regulatory relationship between two components as in Table 10.1, often experimental results can lead to complex inferences such as 'A promotes the process through which B activates C', the simplest case of which is that A catalyzes the reaction from B to C, or 'B induces the synthesis of C only in the absence of A'. Representation of such cases necessitates groups of three or more directed edges combining multiple nodes (as in Fig. 10.1(C)).

More often than not, even after careful synthesis the knowledge about a biological system is not sufficiently complete. Consider the hypothetical scenario of a signal that is known to function as an inhibitor of a certain output. Previous experiments indicate that the signal is activating node A, and the output is negatively regulated by node B, but no assay of the interaction between A and B has been carried out. Under such circumstances, one essentially needs to make reasonable and parsimonious assumptions to bridge the gap between existing evidences. In this case, a positive mass flow or regulatory relation oriented from A to B completes a feasible signal transduction pathway which is consistent with prior knowledge. In cases where contradicting observations are presented, one needs to critically examine and compare the methods by which the results are obtained, the environments under which the experiments were performed, and accept the better-supported relationship.

TABLE 10.1 A Listing of Biological Observations Describing the Causal Relationships Between Components of a System

Biological Evidence

Input activates both A and C.

A activates B, but is inhibited by B at the same time.

C is inhibited by A.

Output is activated by C.

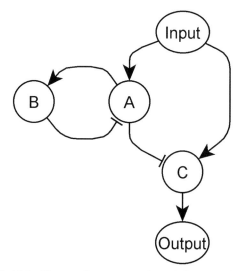

FIGURE 10.2 **The graph representation of the example system described in Table 10.1.** Input is the signal to the whole system and the sole source node (node with no incoming edges), and Output is the only sink node (node with no outgoing edges). '→' represents activation; '—|' represents inhibition.

Manual assembly and interpretation can become daunting for systems containing hundreds of nodes and abundant causal relationships. A computational methodology [34] has been developed into software packages (e.g., NET-SYNTHESIS, http://www.cs.uic.edu/~dasgupta/network-synthesis, [35]) aimed specifically at tackling the problem of network assembly at large scale. Taking a text file describing all causal relationships between system components as input, the software synthesizes and generates a network representation of the system, and outputs a file containing information of all the edges.

Determine the Boolean Transfer Functions

After the network backbone is assembled, the second crucial step towards dynamic simulation is to determine the Boolean transfer functions that govern the state transition of nodes through time. A node i might have one or more upstream regulators in the network. The Boolean transfer function expresses the way the states of these regulators are combined through the Boolean operators *AND*, *OR*, and *NOT*. The transfer functions are also referred to as Boolean rules. For clarity in the following examples we will denote the state of nodes by the node name and simplify the representation of time by only considering current and future time, the latter denoted by using an asterisk on the node name. When there is a single input, the state of the output at the next time step can take on one of the two possible states available, namely Output* = Input if there is a positive relationship between the output and the input, or

Output*=NOT Input if the input suppresses the output. When there are two or more input nodes, their states need to be combined via Boolean functions in a way that is consistent with available knowledge. A biochemical synthesis reaction A+B→C can naturally be represented by the Boolean rule: C* = A AND B. The statement 'B induces the synthesis of C only in the absence of A' can be represented by the Boolean rule C* = NOT A AND B, since both conditions (stimulus B as well as the absence of inhibitory factor A) need to be satisfied in order for C to be synthesized. On the contrary, if for instance two inputs function largely independently, do not exhibit synergy and can substitute one another, then the Boolean OR function should be used, e.g., C* = A OR B. The information of dependencies of input nodes can be obtained through literature search.

If in the system shown on Figure 10.2 both the presence of Input and the absence of B are required for the activation of A, then the Boolean function of A can hence be written as:

$$A^* = \text{Input } AND \text{ } (NOT \text{ } B)$$

Similarly, if either the absence of A or the presence of Input is sufficient to activate C, this leads to:

$$C^* = (NOT \text{ } A) OR \text{ Input}$$

The Boolean transfer functions of the system on Figure 10.2 are summarized in Table 10.2. These rules govern the state transitions of the nodes in the system.

The more inputs a node has, the more possible ways there are to combine all the input states. One needs to be particularly careful in such cases in order to obtain the rule that is able to generate dynamic simulation results that best fit the existing experimental data. It is likely that multiple 'trial and error' processes will have to take place before an optimal solution is found.

Instead of describing the state transition relationship with compact Boolean rules, one can also list all possible combinations of input states of a certain node i and assign output states to each of such combinations. This is called the truth table of a certain Boolean rule, which is equivalent to the rule itself. If node i has m input nodes, then the truth table will have 2^m rows.

Updating Schemes and Incorporating Time

We have now explained the network structure and the Boolean transfer functions that govern the state transitions of nodes. These two pieces comprise the major steps of the model. The next step, discretization of the continuous time stream into steps, is also non-trivial. There are different choices in terms of the number of nodes to be updated at each time step and the order in which the updating of those nodes is to be carried out, and these choices may affect the results.

Assume that the total number of nodes in the network is N. Starting from the general expression of Boolean rules established based on biological knowledge as described in the previous section, we will look at each updating scheme in detail.

Suppose a node i has n input nodes and the Boolean transfer function of node i can be written as follows:

$$V_{i,t} = F_i(V_{k_1,\tau_{k_1}}, V_{k_2,\tau_{k_2}}, ..., V_{k_n,\tau_{k_n}})$$

where $V_{i,t}$ is the state of node i at time step t, $1 \leq i, k_1, ..., k_n \leq N$ are the node indices, and $\tau_{k_1}, ... \tau_{k_n} \leq t$ denote the time step of the last update of the state of nodes $k_1, ..., k_n$. Two major types of updating exist, synchronous [36] and asynchronous [7].

For synchronous updating every node is updated once at each time step, using the states of the input nodes at earlier time steps as inputs to the transfer functions. In other words, $\tau_{k_j} \leq t - 1$, where $1 \leq j \leq n$. The fact that the input states to each and every transfer function are from earlier time steps makes the order in which all nodes are updated at each time step irrelevant, since all inputs have already been fixed before the updating.

Synchronous updating has a simple and straightforward formalism and leads to reproducible state changes. However, it overlooks the differences of timescales on which biological processes are taking place, which can range wildly from milliseconds for protein phosphorylation and post-translational modifications to hundreds of seconds for transcription and transcriptional regulations [37]. Asynchronous updating [7], which provides more detailed tracking of timescales and temporal orders [10–12], is devised in order to account for the diversity in duration of biological processes.

We will cover three types of asynchronous updating: general asynchronous updating [38], random order asynchronous updating [38,39], and deterministic asynchronous updating [40].

TABLE 10.2 Boolean Rules (Transfer Functions) of the System Shown in Figure 10.2, for Specific Assumptions on the Combination of Multiple Inputs.

Node	Boolean Rule (Transfer Function)
Input	—
A	A* = Input AND (NOT B)
B	B* = A
C	C* = (NOT A) OR Input
Output	Output* = C

In general asynchronous updating [38] only one node is randomly selected and updated at each time step. The states of the inputs to the node that is being updated are from earlier time steps, which means $\tau_{k_j} \leq t - 1$ for $1 \leq j \leq n$.

In random order asynchronous updating [38,39], at each time step a random permutation of the node labels (from 1 to N) is first generated, and the updating is carried out in that sequence in that time step. The random permutation of 1 to N is regenerated at each time step. Consider node i with n inputs and its updating:

$$V_{i,t} = F_i(V_{k_1,\tau_{k_1}}, V_{k_2,\tau_{k_2}}, ..., V_{k_n,\tau_{k_n}})$$

If node k_j comes before node i in the random sequence, meaning that by the time node i is being updated node k_j has already been updated at the current time step to its latest state, $\tau_{k_j} = t$ should be used. Otherwise, if node k_j comes after node i in the random sequence, meaning that at the time node i is being updated the state of node k_j has not been touched yet at the current time step, $\tau_{k_j} = t - 1$ should be used.

In deterministic asynchronous updating [40], every node is associated with an intrinsic timescale γ_i. The updating of node i only takes place if the current time t is a multiple of γ_i. One can clearly see that γ_i is the effective 'pace' of each reaction. In other words,

$$V_{i,t} = F_i(V_{k_1,\tau_{k_1}}, V_{k_2,\tau_{k_2}}, ..., V_{k_n,\tau_{k_n}}),$$

$t = c\gamma_i$, c is a positive integer.

When the updating of node i takes place, the states of its input nodes should be taken from the latest available time step, namely $\tau_{k_1}, ... \tau_{k_n} \leq t - 1$.

Several software packages can be used to carry out dynamic simulations based on Boolean network modeling on biological systems. An open-source Python package, BooleanNet [41], is available online at http://code.google.com/p/booleannet/. The input to the program is a text file containing all the Boolean transfer functions, and all of the updating schemes can be selected and readily implemented. Other software packages include CellNetAnalyzer (MATLAB package, [42]), GINsim [43], and BoolNet (R package, [44]).

Network Initialization

The system state is represented by an N-dimensional vector, for a system of N nodes (elements) (see earlier). To perform dynamic simulations of the system under the Boolean network framework, one needs to specify the first state in the state sequence (trajectory), namely the initial state of the system. The initial state of each node is determined such that it is consistent with known biological facts or experimental evidence. If an intracellular substance is known to be present under all conditions, it can be initialized to be in the 1, or *ON*, state. Initial states can also be assigned based on the question of interest. For example, one can implement a gene knockout by the initial and sustained *OFF* state of that particular gene. If there is insufficient experimental information concerning the concentration or activity level of an element, one can also randomize the initial state of that node. Therefore, a large number of runs of dynamic simulations can be carried out with each run randomizing the initial states of the nodes that cannot be predetermined [10]. The system will sample different potentially viable initial states and henceforth take varying routes in the state space.

Starting from the initial state, the system will evolve as time progresses and should eventually settle down into an attractor, that is, a time-invariant steady state or an (ordered or unordered) repetition of a certain finite set of states. These attractor states have been proposed as representations of cell fates going back to the 1940s work of C. Waddington [45,46] and later by S. Kauffman [6]. The complete set of states that can potentially reach a certain attractor through an updating scheme forms the basin of attraction of *that* attractor. Since synchronous updating is deterministic, the basins of attraction for different attractors will be distinct, whereas under asynchronous updating the basins of attraction for different attractors could share states and be partly overlapping with each other, as illustrated by the example in Figure 10.3 [39].

Steady-State Analysis

The system will have the same steady states (fixed points) under both synchronous and asynchronous updating, owing to the fact that a steady state repeats itself infinitely, making the order in which the nodes are updated irrelevant. BooleanNet [41] provides functions to detect the steady states of the system. CellNetAnalyzer [42] is also able to probe the steady states of the system, which are called 'logical steady states' in the package.

Before we illustrate in a step-by-step manner how to determine the steady states of a Boolean network, we note that there are two main focuses encountered in dynamic analysis of biological systems: determining the attractors of the whole system (of all nodes), and determining the attractors of a small set of designated *output nodes* of the Boolean network. The second focus is necessarily a subset of the first. The output-oriented analysis is often not only simpler, but also more relevant to signal transduction pathway-related research, where no more than several inputs (signals) are considered and the system's response is usually characterized by a single output node. In modeling gene regulatory networks, usually the attractors and the dynamic sequence of the

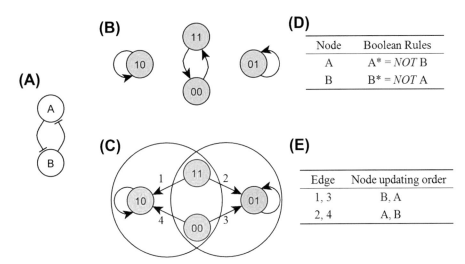

FIGURE 10.3 Ilustration of the basins of attraction of a system under different updating schemes. The system is composed of two mutually inhibiting nodes A and B (A). The Boolean functions are given in table (D). (B) and (C) show the state transition graphs of the system under synchronous update (B), and random order asynchronous update (C). The 2-digit vectors inside the circles represent the states of node A and B from left to right. In the case of random order update (C) the correspondence between the order in which A and B are updated and the particular state transitions, indicated by edge labels, is given in table (E). E.g., starting from an initial state of 11, if A is updated first and B is updated second, state 11 will evolve into state 01. The two unlabeled self-loops on states 01 and 10 in both (B) and (C) indicate that if the system is currently in one of those states, the future state will be the same as the current state, regardless of the manner of update, thus these are fixed points of the system. Under synchronous update (B), the system has 3 attractors: the two fixed points 01 and 10, and a cyclic attractor. Only the fixed points are preserved under asynchronous update (C). In (C), the states 11, 00, 10 form the basin of attraction of 10, and states 11, 00, 01 form the basin of attraction of 01. The states 11 and 00 belong to both basins of attraction.

whole system correspond to known biological events, such as certain phases of the cell cycle [8,47], apoptosis and cell differentiation [48], and hence the first focus, identifying attractors of the whole system, is most relevant.

The fixed points of a Boolean network model can be determined in several ways. One can analytically solve the Boolean equations. Since the system being in the steady state means that the state vector remains time-invariant, the future state of any node will be the same as its current state. Thus, the time-dependent features (indices or asterisks) of the Boolean equations can be removed and the set of resulting time-independent equations can readily be solved. One can also do repeated dynamic simulations of the system, updating the nodes' states according to their Boolean rules. One can also draw useful conclusions from the existence of particular interaction patterns (called 'network motifs') [49], such as feedforward or feedback loops (as first suggested by Thomas, [50]). Consider the following examples.

In the example of Figure 10.4, the system (hence also the Output) has a single attractor for any update method. For synchronous update, the attractor is a cycle of period 4 (Figure 10.5(A)). Under general asynchronous updating the attractor spans the whole state space. Every state in the latter attractor has multiple edges, both incoming and outgoing (Figure 10.5(B)). This is because a state can have different successor states if a different node is updated. The maximum number of possible outcomes from a state under the general asynchronous scheme equals the total number of nodes in the system. The system state will traverse all states in the attractor shown in Figure 10.5(B)), following the outgoing edges with a certain calculable probability. For both types of update, the average state level of the node Output (the last digit) in the attractor is 0.5. This could be interpreted as a 50% up time of a certain system activity.

The critical network motif [49] of the original network is the negative feedback on node A. It is the negative feedback, coupled with the Boolean AND rule for node A, that generates the oscillatory behavior of the system.

The next example network (Figure 10.6) differs from the previous one (Figure 10.4) in that it contains one more edge from B to the Output, and the state of the Output is determined by A and B independently, in a Boolean OR relation. We already know that the negative feedback on node A will generate an oscillatory behavior for itself as well as for node B, but does the new edge help improve the activation of the Output? The answer is yes. Under both updating schemes the average state level of the Output is ~0.75. If the Boolean relationship between A and B in the Output function is changed to AND, the Output level drops to ~0.25. It is indeed reasonable that a more stringent condition for activation will reduce the average Output level. Under synchronous update, the Output (and also the system) will have period-4 oscillations. In the state transition graphs of the system (Figure 10.7), (A) corresponds to Output* = A OR B, and (B) to Output* = A AND B.

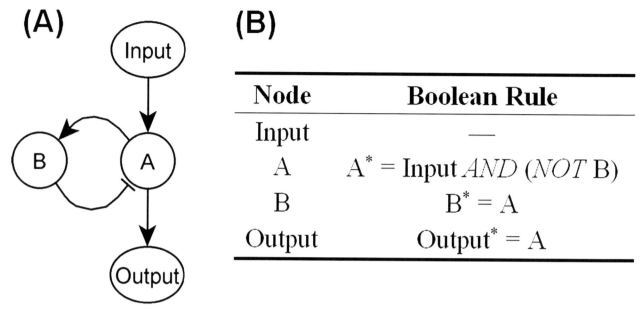

FIGURE 10.4 A four-node signal transduction network in which A is activated by Input, and its activation also *requires* the absence of B. (A) Network representation; (B) the list of Boolean rules.

Finally, consider yet another example (Figure 10.8). The characteristic feature of this network is two paths that both connect the Input and the Output, but functionally contradict each other. Given the *AND* rule for the Output, the steady state of Output is *OFF*, and more importantly, this outcome is independent of the state of the Input: the Output is decoupled from the Input in this case. If *OR* is used in the transfer function of the Output, its state will be at constant *ON* instead, but it will still be independent from the Input state.

Now let's apply the dynamic analysis to the network in Figure 10.2 and see whether a steady state of the Output is possible. Given that the Input node is *ON*, the negative feedback on node A will generate an oscillation of states for node A as well as node B. However, since the Boolean rule for node C is C* = (*NOT* A) *OR* Input, the oscillation coming from branch A will be combined by an *OR* rule with the *ON* state of the Input. The Output node will henceforth reach an *ON* steady state. The attractor for the state vector of the system, on the other hand, will be a limit cycle, since a subset of the nodes is oscillating.

In general, multiple network motifs can be present in the graph representation of a system under study, and each

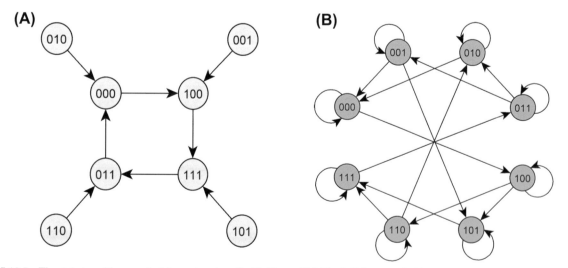

FIGURE 10.5 The state transition graph of the system described in Figure 10.4. The 3-digit vectors inside the circles represent the state of node A, B, and Output from left to right. The state of the Input is fixed at 1, indicating a constant ON signal to the system (not shown in the graph). A self-loop on a certain state indicates that if one of the nodes is updated, the system state remains the same. For 3 nodes, a total of 8 states exist. (A) Synchronous update. All 8 states are located in the same basin of attraction, of which 4 form the attractor. (B) General asynchronous update. The attractor is formed by all 8 states. These figures were generated by BooleanNet.

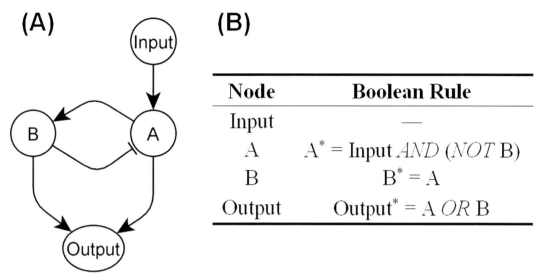

FIGURE 10.6 A four-node signal transduction network wherein A is activated by Input, and its activation also *requires* the absence of B. Either A or B is sufficient to activate the Output. (A) The network representation; (B) the list of Boolean rules.

has its own contribution to generate advantageous dynamic behaviors, stabilize the system function, prevent extreme behavior, or provide system redundancy and hence robustness [51–53]. Last but not least, the conditionality present in the Boolean transfer functions is often critical in determining the system behavior.

Model Validation: Reproduction of Known Results

An important step toward obtaining a successful model is to examine the dynamic sequence of system states in detail and compare with biological evidence. If, for instance, under reasonable assumptions, starting from a viable initial state, an oscillation of the output is observed, whereas it is biologically known that the output should exhibit a constant *OFF* state under such conditions, further modification of one or more Boolean transfer functions (or even the underlying network structure) is called for. Such discrepancies can also be a potential indication of components missing from the model. One needs to scrutinize all possibilities and make the most accurate, biologically truthful model revision by changing Boolean rules, rewiring edges between nodes, and/or adding/removing nodes to/from the network, so as to reproduce as many known

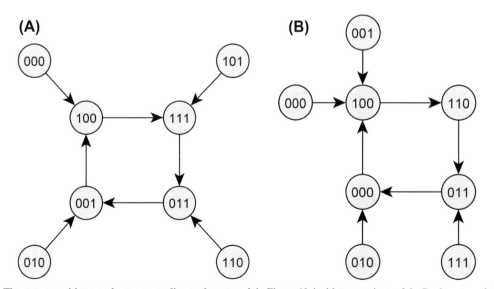

FIGURE 10.7 The state transition graphs corresponding to the network in Figure 10.4 with two variants of the Boolean transfer functions. The 3-digit vectors represent the state of node A, B, and Output from left to right. The state of the Input is fixed at 1. (A) Output* = A *OR* B. An average level of 0.75 of the node Output is observed in the attractor, which is a period-4 oscillation. (B) Output* = A *AND* B. An average level of 0.25 of the node Output is observed in the attractor.

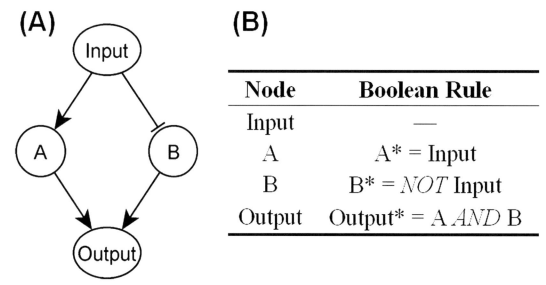

FIGURE 10.8 A four-node signal transduction network in which A is activated by Input, B is inhibited by Input, the activation of the Output requires the presence of both A and B. (A) The network representation; (B) the list of Boolean rules.

results as possible. In a noteworthy example, Samaga et al. performed a comparison between simulation results generated by a Boolean logic model based on the literature and data collected from cells on ErbB receptor phosphoprotein signaling [54]. They came to a set of 11 hypotheses regarding ErbB signaling resulting from the discrepancies between simulation results and experimental observations, among which five were supported by the literature, five led to further modifications of the model, and one implied the absence of specificity expected from a small molecular inhibitor.

Robustness against Disruptions and Useful Implications

An important additional assessment is whether the model is robust in terms of changes in interactions or Boolean transfer functions. Models that are extremely fragile to such changes may not be a good representation to biological systems, as the real systems exhibit substantial robustness to changes in concentrations, reaction rates or even mutations [55–59]. As currently comprehensive models of signal transduction systems are rare, the model should show reasonable robustness to changes in the network structure to instill confidence that its results will still stand after new components or interactions are discovered. The ability of the model to maintain the wild-type response under small topological perturbations can be tested by adding or deleting a randomly selected node or edge, rewiring edges in the network randomly (for example changing any pair of parallel edges to cross-edges) or making an inhibitory interaction into activation or vice versa. For example, if in Figure 10.1 node B is knocked out (constitutively *OFF*), the

Output of the system will *not* be affected, owing to the *OR* relation in the Boolean rule of node C. However, if the edge from Input to C becomes non-functional, a new oscillatory behavior of the system is exhibited. It is also possible that existing attractor basins get altered, or completely eliminated, and new ones arise from such perturbations of the system.

Validated models can be used to predict the outcomes of 'what if' scenarios — cases that have not yet been studied experimentally — and can generate testable predictions and significant insights. For example, Mendoza formulated a logic-based model of interactions among cytokines and transcription factors in helper T (Th) cells [28]. Dynamic simulation under all combinations of initial node states revealed four steady states: one corresponding to naïve Th cells, one corresponding to Th2 cells, and two corresponding to Th1 cells. The two Th1 cell attractors indicated two Th1 cell subpopulations with different levels of IFN-γ secretion, but the level of the IFN-γ receptor was the same in both attractors, a result supported by the literature. Mendoza studied in detail how node perturbations (knockout or over-expression) change the differentiation fate of Th cells. Several of the model results were supported by the experimental literature data, but numerous others are novel predictions [28].

We next take a closer look at Boolean models applied in two different contexts: survival signaling in T cells in T-LGL leukemia [12], and interactions between pathogens and a mammalian immune system [11]. The second example also demonstrates that Boolean modeling is a general approach that can be applied at different levels of biological organization, from the molecular to the population level [60].

APPLICATION EXAMPLES

T-LGL Leukemia Network Modeling

Cytotoxic T lymphocytes (CTLs) are a type of T cell that are activated in order to clear virus-infected cells. The activated population undergoes subsequent activation-induced cell death and eventually reaches a balance between proliferation, survival, and apoptosis. T-cell large granular lymphocyte (T-LGL) leukemia is characterized by an abnormal clonal proliferation of cytotoxic T lymphocytes that escaped activation-induced cell death [12]. The diseased T cells are insensitive to Fas-induced apoptosis and henceforth remain long-term competent. Zhang et al. constructed a T-LGL survival signaling network and a Boolean model of the network. Through a vast literature search and the use of NET-SYNTHESIS [35] to simplify the original network, a final network of 58 nodes and more than 100 edges was constructed. The input node to the system is 'Stimuli', representing antigen stimulus, and the output node of the system is 'Apoptosis', representing activation-induced cell death, which is the outcome of the system under normal conditions. The model [12] identified two proteins, IL-15 and PDGF, which are critical in inducing and maintaining the diseased cell behavior. The over-expression of the two is a sufficient condition to reproduce all known leukemic abnormalities. Furthermore, key survival mediators, such as NFκB, SPHK-1 and GAP were also proposed. These nodes settle down into ON or OFF state in T-LGL leukemia, and if the state of one of them is flipped artificially (ON$\rightarrow$OFF or OFF$\rightarrow$ON), apoptosis is induced. This offered potential therapeutic targets for curing T-LGL leukemia. Several of these predictions have been validated experimentally. The study also demonstrated that it is possible to integrate qualitative information regarding normal and diseased cellular signaling pathways into the same network, and discrete dynamic modeling of this network can be used to generate deep insight and validated predictions.

Pathogen—Immune Response Network

Interactions between an invading pathogen and the responses of the immune system form a complex signaling network [61—63]. Pathogens seek to evade or disrupt the host immune response to ensure access to nutrients and self-proliferation. For example, bacteria persist in hosts by interfering with antigen production or processing, subverting phagocytosis by immune cells, and/or by promoting host anti-inflammatory responses that under normal conditions would deactivate the protective effectors of the host. Pathogens therefore serve as the input to such a network, and the pathogen—host immune system interaction can result in either the clearance of the pathogen or persistent disease.

Thakar et al. [11] performed a comparative study between two species of the same genus *Bordetella* to better understand the different ways in which various virulence factors modulate immune responses and hence the adaptability of the pathogen and survivability of the hosts [11, 63]. Two separate network models for the interaction of these pathogens with their hosts were synthesized. The nodes of the networks were comprised of immune cells, cytokines, and antibodies, as well as conceptual nodes such as phagocytosis, among which 18 were common to both networks. 'Bacteria' is the input to the network, and phagocytosis can be considered the output, which in turn leads to the clearance of Bacteria, generating a negative feedback edge from the output to the input. The edges in the network represent cytokine production, cell recruitment and differentiation, and intercellular signaling processes.

Each network was then translated into a predictive dynamic model, using Boolean transfer functions reconstructed from the literature, and was validated by experimental observations. The model led to an understanding of puzzling differences between the two pathogens, e.g., that antibody transfer at the time of infection (similar to immunization) aided clearance in *Bordetella bronchiseptica* but not in *Bordetella pertussis*. On the contrary, the model predicted that for both pathogens a secondary infection would be cleared earlier than a first-time infection. Follow-up experiments were performed where a *B. pertussis* or *B. bronchiseptica* second challenge was given to convalescent hosts. The secondary infection was indeed cleared in 15 days, which is significantly less than the duration of a first-time infection. The model also predicts that certain cytokines are an equally effective prophylactic measure against both pathogens. The study identified three phases in *Bordetella* infection and offered a methodology of in silico evaluation of putative medical treatments. This methodology generates novel insights into the interplay between pathogen virulence factors and host immune response system, and can be readily adapted to similar systems.

CONCLUSIONS AND FUTURE DIRECTIONS

We have discussed Boolean modeling, a particular type of discrete dynamic modeling of cellular signaling networks. We detailed model construction and analysis of a model's temporal behavior and attractor structure. We illustrated through examples that Boolean modeling is a powerful tool for systems biology, especially when there is insufficient quantitative or kinetic information.

Boolean models can be readily tested against known biological knowledge, easily modified based on the latest progress in experimental data collection, and straightforwardly adapted to similar systems. They can be used to generate novel insights into a system that could not be otherwise obtained, to predict system behavior under various conditions, and to perform in silico experiments for node knockouts as well as gene over-expression.

The Boolean modeling framework can be readily expanded to include more quantitative details. Multilevel discrete dynamic models have more than two states [20, 22–24], and continuous -Boolean hybrid models characterize each component by two variables, a continuous variable akin to a concentration and a discrete variable akin to an activity [64]. The model can thus be adapted as new experiments are conducted and more evidence is accumulated. Such a process of trial and error, progress and adjustment plays a vital role in the advancement of system biology.

Table of System Biology Terms

Terms	Definitions
Node	A graphic representation of a component of a biological system, such as a protein, secondary messenger or small molecule
Edge	Any type of interaction that exists between nodes, e.g., chemical reaction, regulation or causal relationship
Node state	The value of the variable assigned to the node. It can mean concentration, activity, status. In Boolean modeling there are two node states: ON, meaning above threshold concentration or activity, and OFF, meaning below threshold concentration or activity
State of the system	An N-dimensional vector where N is the total number of nodes in the system. The ith dimension is the state of the ith node
Fixed point or steady state	A state vector of the system that is time-invariant. If the system reaches a steady state, it will stay in that state
State transition graph	A graph showing the evolution of the state of the system. The system will undergo state changes along the direction of the arrows
Attractor	A set of connected states of the system. Each state in the set has to be reachable by the system after a sufficiently long period of time. Fixed points (steady states) are a subtype of attractors
Basin of attraction	A set of connected states of the system. Given sufficiently long time, every state in the set *can* reach a specific attractor of the system. All such states that are able to reach the same attractor form the basin of attraction of *that* particular attractor

REFERENCES

[1] Lee TI, Rinaldi NJ, Robert F, Odom DT, Bar-Joseph Z, Gerber GK, Hannett NM, et al. Transcriptional regulatory networks in *Saccharomyces cerevisiae*. Science 2002;298:799–804.

[2] Schwikowski B, Uetz P, Fields S. A network of protein–protein interactions in yeast. Nat Biotechnol 2000;18:1257–61.

[3] Stelzl U, Worm U, Lalowski M, Haenig C, Brembeck FH, Goehler H, et al. A human protein–protein interaction network: a resource for annotating the proteome. Cell 2005; 122:957–68.

[4] Hatzimanikatis V, Li C, Ionita JA, Broadbelt LJ. Metabolic networks: enzyme function and metabolite structure. Curr Opin Struct Biol 2004;14:300–6.

[5] Chen KC, Csikasz-Nagy A, Gyorffy B, Val J, Novak B, Tyson JJ. Kinetic analysis of a molecular model of the budding yeast cell cycle. Mol Biol Cell 2000;11:369–91.

[6] Kauffman SA. Metabolic stability and epigenesis in randomly constructed genetic nets. J Theor Biol 1969;22:437–67.

[7] Thomas R. Boolean formalization of genetic control circuits. J Theor Biol 1973;42(3):563–85.

[8] Li F, Long T, Lu Y, Ouyang Q, Tang C. The yeast cell-cycle network is robustly designed. Proc Nat Acad Sci USA 2004;101:4781–6.

[9] Davidich MI, Bornholdt S. Boolean network model predicts cell cycle sequence of fission yeast. PLoS ONE 2008;3:e1672.

[10] Li S, Assmann SM, Albert R. Predicting essential components of signal transduction networks: a dynamic model of guard cell abscisic acid signaling. PLoS Biol 2006;4:e312.

[11] Thakar J, Pilione M, Kirimanjeswara G, Harvill ET, Albert R. Modeling Systems-Level Regulation of Host Immune Responses. PLoS Comput Biol 2007;3:e109.

[12] Zhang R, Shah MV, Yang J, Nyland SB, Liu X, Yun JK, et al. Network model of survival signaling in large granular lymphocyte leukemia. Proc Natl Acad Sci USA 2008; 105:16308–13.

[13] Saez-Rodriguez J, Simeoni L, Lindquist JA, Hemenway R, Bommhardt U, Arndt B, et al. A logical model provides insights into T cell receptor signaling. PLoS Comput Biol 2007;3:e163.

[14] Chaouiya C. Petri net modelling of biological networks. Brief Bioinform 2007;8:210–9.

[15] Peterson JL. Petri Net Theory and the Modeling of Systems, Prentice Hall PTR. NJ, USA: 1981.

[16] Sackmann A, Heiner M, Koch I. Application of Petri net based analysis techniques to signal transduction pathways. BMC Bioinformatics 2006;7:482.

[17] Mendoza L, Thieffry D, Alvarez-Buylla ER. Genetic control of flower morphogenesis in *Arabidopsis thaliana*: a logical analysis. Bioinformatics 1999;15:593–606.

[18] Espinosa-Soto C, Padilla-Longoria P, Alvarez-Buylla ER. A gene regulatory network model for cell-fate determination during *Arabidopsis thaliana* flower development that is robust and recovers experimental gene expression profiles. Plant Cell 2004;16: 2923–39.

[19] Faure A, Naldi A, Lopez F, Chaouiya C, Ciliberto A, Thieffry D. Modular logical modelling of the budding yeast cell cycle. Mol Biosyst 2009;5:1787–96.

[20] Sanchez L, Thieffry D. A logical analysis of the *Drosophila* gap-gene system. J Theor Biol 2001;211:115–41.

[21] Albert R, Othmer HG. The topology of the regulatory interactions predicts the expression pattern of the segment polarity genes in *Drosophila melanogaster*. J Theor Biol 2003;223:1–18.

[22] Sanchez L, Thieffry D. Segmenting the fly embryo: a logical analysis of the pair-rule cross-regulatory module. J Theor Biol 2003;224:517–37.

[23] Ghysen A, Thomas R. The formation of sense organs in *Drosophila*: a logical approach. Bioessays 2003;25:802–7.

[24] Gonzalez A, Chaouiya C, Thieffry D. Logical modelling of the role of the Hh pathway in the patterning of the *Drosophila* wing disc. Bioinformatics 2008;24:i234–40.

[25] Diaz J, Alvarez-Buylla ER. A model of the ethylene signaling pathway and its gene response in *Arabidopsis thaliana*: pathway cross-talk and noise-filtering properties. Chaos 2006;16:023112.

[26] Naldi A, Carneiro J, Chaouiya C, Thieffry D. Diversity and plasticity of Th cell types predicted from regulatory network modelling. PLoS Comput Biol 2010;6:e1000912.

[27] Naldi A, Berenguier D, Faure A, Lopez F, Thieffry D, Chaouiya C. Logical modelling of regulatory networks with GINsim 2.3. Biosystems 2009;97:134–9.

[28] Mendoza L. A network model for the control and differentiation process in Th cells. Biosystems 2006;84:101–14.

[29] Voit EO. Computational Analysis of Biochemical Systems—A Practical Guide for Biochemists and Molecular Biologists. Cambridge: Cambridge University Press; 2000.

[30] Wang R, Albert R. Elementary signaling modes predict the essentiality of signal transduction network components. BMC Syst Biol 2011;5:44.

[31] Saez-Rodriguez J, Alexopoulos LG, Epperlein J, Samaga R, Lauffenburger DA, Klamt S, et al. Discrete logic modeling as a means to link protein signaling networks with functional analysis of mammalian signal transduction. Molecular Systems Biol 2009;5:331.

[32] Wingender E, Dietze P, Karas H, Knuppel R. Transfac: a database on transcription factors and their DNA binding sites. Nucleic Acids Res 1996;24:238–41.

[33] Kanehisa M, Goto S. KEGG: Kyoto encyclopedia of genes and genomes. Nucleic Acids Res 2000;28:27–30.

[34] Albert R, DasGupta B, Dondi R, Kachalo S, Sontag E, Zelikovsky A, et al. A novel method for signal transduction network inference from indirect experimental evidence. J Comput Biol 2007;14:927–49.

[35] Kachalo S, Zhang R, Sontag E, Albert R, DasGupta B. NET-SYNTHESIS: a software for synthesis, inference and simplification of signal transduction networks. Bioinformatics 2008;24:293–5.

[36] Kauffman S. Origins of order: self-organization and selection in evolution. New York: Oxford Univ Press; 1993.

[37] Papin JA, Hunter T, Palsson BO, Subramanian S. Reconstruction of cellular signalling networks and analysis of their properties. Nat Rev Mol Cell Biol 2005;6:99–111.

[38] Harvey I, Bossomaier T. Time out of joint: Attractors in asynchronous random Boolean networks. In: Husbands P, Harvey I, editors. Proceedings of the Fourth European Conference on Artificial Life (ECAL97). Cambridge, MA: MIT Press; 1997. p. 67–75.

[39] Chaves M, Albert R, Sontag ED. Robustness and fragility of Boolean models for genetic regulatory networks. J Theor Biol 2005;235:431–49.

[40] Chaves M, Sontag ED, Albert R. Structure and timescale analysis in genetic regulatory networks. SystBiol (Stevenage) 2006;153:154–67.

[41] Albert I, Thakar J, Li S, Zhang R, Albert R. Boolean network simulations for life scientists. Source Code Biol Med 2008;3:16.

[42] Klamt S, Saez-Rodriguez J, Gilles ED. Structural and functional analysis of cellular networks with CellNetAnalyzer. BMC Syst Biol 2007;1:2.

[43] Gonzalez AG, Naldi A, Sánchez L, Thieffry D, Chaouiya C. GINsim: a software suite for the qualitative modeling, simulation and analysis of regulatory networks. Biosystems 2006;84:91–100.

[44] Mussel C, Hopfensitz M, Kestler HA. BoolNet—an R package for generation, reconstruction and analysis of Boolean networks. Bioinformatics 2010;26:1378–80.

[45] Waddington CH. Canalization of development and the inheritance of acquired characters. Nature 1942;150:563–5.

[46] Waddington CH. Epigenetics and evolution. Symp Soc Exp Biol 1953;7:186–99.

[47] Tyson J, Novak B. Regulation of the eukaryotic cell cycle: molecular antagonism, hysteresis, and irreversible transitions. J Theor Biol 2001;210:249–63.

[48] Huang AC, Hu L, Kauffman SA, Zhang W, Shmulevich I. Using cell fate attractors to uncover transcriptional regulation of HL60 neutrophil differentiation. BMC Syst Biol 2009;3:20.

[49] Shen-Orr S, Milo R, Mangan S, Alon U. Network motifs in the transcriptional regulation network of *Escherichia coli*. Nature Genetics 2002;31:64–8.

[50] Thomas R, D'Ari R. Biological Feedback. 1st ed. Florida: CRC Press; 1990.

[51] Shinar G, Milo R, MR M, Alon U. Input output robustness in simple bacterial signaling systems. Proc Natl Acad Sci USA 2007;104:19931–5.

[52] Tyson JJ, Borisuk MT, Chen K, Novak B. Analysis of complex dynamics in cell cycle regulation. In: Bower JM, Bolouri H, editors. Computational Modeling of Genetic and Biochemical Networks. Cambridge, MA: MIT Press; 2000. p. 287–305.

[53] Bhalla US, Iyengar R. Robustness of the bistable behavior of a biological signaling feedback loop. Chaos 2001;11:221–6.

[54] Samaga R, Saez-Rodriguez J, Alexopoulos LG, Sorger PK, Klamt S. The Logic of EGFR/ErbB Signaling: Theoretical Properties and Analysis of High-Throughput Data. PLoS Comput Biol 2009;5:e1000438.

[55] von Dassow G, Meir E, Munro EM, Odell GM. The segment polarity network is a robust developmental module. Nature 2000;406:188–92.

[56] Alon U, Surette MG, Barkai N, Leibler S. Robustness in bacterial chemotaxis. Nature 1999;397:168–71.

[57] Eldar A, Dorfman R, Weiss D, Ashe H, Shilo BZ, Barkai N. Robustness of the BMP morphogen gradient in *Drosophila* embryonic patterning. Nature 2002;419:304–8.

[58] Csete M, Doyle J. Bow ties, metabolism and disease. Trends Biotechnol 2004;22:446–50.

[59] Conant GC, Wagner A. Duplicate genes and robustness to transient gene knock-downs in *Caenorhabditis elegans*. Proc R Soc Lond B Biol Sci 2004;271:89–96.

[60] Campbell C, Yang S, Albert R, Shea K. A network model for plant–pollinator community assembly. PNAS 2011;108:197–202.

[61] Janeway Jr CA, Travers P, Walport M, Shlomchik MJ. Immunobiology. The Immune System in Health and Disease. 5th edition. New York: Garland Science; 2001.

[62] Mills KH, Boyd AP. Evasion of Immune Responses by Bacteria. In: Kaufmann SH, Steward M, editors. Topley and Wilson's microbiology and microbial infections. 10th edition. Lond: Edward Arnold; 2007.

[63] Thakar J, Albert R. Boolean models of within-host immune interactions. Curr Opin Microbiol 2010;13:377−81.

[64] Glass L. Classification of biological networks by their qualitative dynamics. J Theor Biol 1975;54:85−107.

Chapter 11

Transcriptional Network Logic: The Systems Biology of Development

Isabelle S. Peter and Eric H. Davidson
Division of Biology, MC 156—29, California Institute of Technology, Pasadena, CA 91125, USA

Chapter Outline

Development is a System-Wide Direct Output of the Genome 213
GRN Structural Components and Model Representation 214
The Regulatory State Concept 215
System-Wide and Deep Information Flow in Development 216
General Features of the Developmental Process 217
Developmental GRN Dynamics 221
Examples of GRN-Mediated Spatial Control in Development 222
    Endomesoderm Development in *S. purpuratus* Embryos 222
        Example 1: Use of Maternal Anisotropy to Define a Zygotic Regulatory Compartment 222

        Example 2: Double-Negative Gate Circuitry for Spatial Restriction of the Skeletogenic Cell Fate 222
        Example 3: Signaling and Cell Fate Specification 223
        Example 4: Parallel Mechanisms in Cell Fate Initiation 224
        Example 5: Mechanism of Irreversibility in the Developmental Process 224
        Example 6: GRN-Mediated Cell Fate Decision 224
Conclusion: The Explanation of Development 224
Note added in proof 226
Acknowledgement 226
References 226

How are the operations of development so coordinated as to give rise to a definitely ordered system? It is our scientific habit of thought to regard the operation of any specific system as determined primarily by its specific physical-chemical composition....This mechanistic assumption implies some specific structure or material configuration in the system, and since the organization of the egg is hereditary, the structure or configuration must be preserved by cell division without loss of its specific character....

(EB Wilson, The Cell in Development and Heredity, 3rd edn, 1924, p.1115).

The concepts and precepts of systems developmental biology are scarcely a recent invention. Once it was realized that the process of development is directly controlled by the genome (to use the modern word), the basic concepts of systems developmental biology were directly implied, just as the quote above illustrates. Wilson had no doubt whatsoever that the hereditary events of development are genomically determined, and his 'specific structure or material configuration in the system', his 'mechanistic assumptions', have in our time materialized in the form of the gene regulatory networks (GRNs) that lie at the heart of systems developmental biology. The idea that the zygote nucleus carries in its chromosomes the genetic information

required to direct development was already clearly formulated by the end of the 1880s, on the basis of a large amount of indirect evidence and logical deduction therefrom (reviewed in refs. [1,2]; for modern commentary see refs. [3—5]). Although there was no direct evidence, the concept of genomic control of development was sufficiently precise by 1889 that Boveri set out deliberately to provide an explicit direct test of it [4]. Though he failed in this initial attempt, he succeeded only a few years later with one of the classic systems biology experiments of the early 20th century. This was his famous 1902—1904 polyspermy experiment, done on sea urchin eggs [1,6—8]. Boveri made use of the unequal distribution of chromosomes resulting from fertilization with more than one sperm to study the developmental effects of aneuploidy. When the first four cells of the embryo were separated and individually cultured, only those containing a complete set of chromosomes produced normally developed larvae, whereas all other blastomeres experienced developmental failure. The analysis was quantitative and was replete with a predictive mathematical model, and it was entirely modern in its causal experimental logic. Boveri derived two essential conclusions: first, that the process of embryonic development requires the presence (and in some manner the action) of a complete set of chromosomes in every cell, and second, that each chromosome independently carries different hereditary information. Two decades later, when Wilson was writing the book from which the above quote comes, he still regarded Boveri's polyspermy experiment as the keystone of all the evidence by then accumulated for genomic control of development, even though, from 1910 on, Morganian genetics had provided multiple examples of specific gene mutations that produced specific (developmental) phenotypes. It was clear, and explicitly stated by Wilson [1], that the genome is an informational system resident in the many chromosomes of each nucleus; that indeed each chromosome contains a large number of distinct hereditary genetic components; and that development is an outcome of the operation of this whole system. The consequences of this most fundamental aspect of the developmental process were not really faced during the succeeding decades of 20th century single-gene developmental genetics and single-gene developmental molecular biology, and in many quarters have still not been absorbed. For it is necessary to consider and analyze the developmental process in system terms; that is, to understand all (or most) of the causal interactions that constitute the control system for each phase or element of the process. Only in this way can it be determined why development works as it does, what is the reason each event occurs where and when it does, and how this relates directly to the hereditary genomic information. Framework understanding of development can never be derived by adding together unrelated, single gene observations.

There is an even more deeply rooted origin to systems developmental biology, and that is the first feature of the developmental process to be seriously described in the earliest direct forerunners to the modern study of embryogenesis. This feature is the manifest increase in organismal complexity as embryonic development proceeds. Figure 11.1 is a reproduction of drawings of a chick embryo from the 17th century [9,10], designed explicitly to illustrate this point. The gist of an argument raging in the 18th and early 19th centuries was whether the observed increase in complexity during embryonic development simply reveals a preformed complexity resident from the beginning in the fertilized egg, or is actually progressively constructed de novo, a concept classically named 'epigenesis'. Wilson understood that the hereditary genomic information that organizes the events of development constitutes the solution to this venerable problem: '*...heredity is effected by the transmission of a nuclear preformation which in the course of development finds expression in a process of cytoplasmic epigenesis*' [1]. This

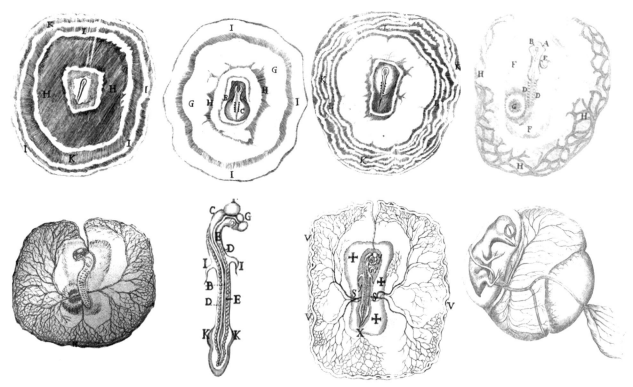

FIGURE 11.1 **The increasing complexity of the embryo.** This definitive and fundamental feature of embryogenesis was recognized as the basic challenge for understanding development even at the dawn of modern bioscience. The drawings of a developing chick embryo as seen through a microscope are reproduced from Marcello Malpighi's 1686 Opera Omnia [10], except for the dorsal view in color which was published by him in 1673 in Dissertatione Epistologica de Formatione Pulli in Ovo [9].

concept is essentially that of a hardwired genomic program for development. There is no concept more intrinsic to systems developmental biology than that of process control by a preformed genomic program, in exactly the same sense as the term 'program' is used for the code directing a complex stepwise computational operation.

We have two reasons to begin our chapter in this manner. First, we wish to make clear that our definition of systems developmental biology is primarily conceptual rather than one dependent on technology, throughput, or data mass. For us, systems developmental biology is the means of achieving a causal explanation of development, which begins with genomic regulatory information and extends vertically to all the different levels of biological organization directly affected by the developmental genomic control system. The explanation must encompass both the spatial allocation of gene expression and its temporal sequence. This is fundamentally a conceptual as well as an experimental discovery problem in information processing and regulatory logic, on a system-wide scale. Second, by alluding briefly to the deep roots of this conceptual problem, we can see that it was to a large extent formulated independently of the gigantic mass of current knowledge of molecular, biochemical, and cell biological detail. Indeed, the subject of this chapter is just how the 'material configuration' of the control system that Wilson deduced actually causes the process which, from then until now, has been termed 'development'.

DEVELOPMENT IS A SYSTEM-WIDE DIRECT OUTPUT OF THE GENOME

Biological organization underlies development at every level from genome to organism. If the question is how many genes are needed to form an adult organism, then the answer will be close to the total number of genes encoded in the genome. Numerous genetic screens have attempted to catalogue genes based on the developmental processes they are involved in. However, even though these studies have demonstrated a great many molecules involved in specific developmental processes, they are not designed to reveal how these individual components actually work together to form any given body part in a normally developing organism. It is the goal of systems biology to ask not only which molecular components are required for any given process, but also how these components operate on one another so as to promote function. Here we outline a system-level approach to development, addressing specifically how body plan organization is controlled, what types of molecule are required for this control and how they interact, and how both molecules and interactions are encoded in the genome so as to generate the developmental process. The developmental process is thus determined by the genomic control system — Wilson's 'material configuration' — and the outcome of this material configuration is the spatial allocation of cell types, tissues, organs and appendages that define the adult organism.

The organization of the body plan is determined by the assignment of cell fates in specific spatial arrangements oriented with respect to the body axes. The earliest assignment of spatial cell fate domains occurs at the transcriptional level, long before any morphogenetic functions are deployed. That is, embryonic organization depends directly on the transcription factors that come to be expressed in each cell and in each spatial domain. When and where transcription factors and signaling molecules are expressed during embryonic development are in turn determined by their transcriptional control apparatus. We can extend this principle from the earliest expression of the zygotic genome to the differentiated cell types of the adult organism. The formation of spatial regulatory domains depends on the genomically encoded structure of GRNs. These networks include both the regulatory genes and the regulatory interactions between them, such that each node of the network represents a regulatory gene and its *cis*-regulatory control system. Multiple regulatory interactions are required to control the expression of each regulatory gene, and each regulatory factor in turn provides inputs into multiple target genes; hence the network character of GRNs. GRNs control the spatial organization of the embryo, by determining the regulatory character of cell fate domains and by setting their boundaries.

How are GRNs encoded in the genome and how could they account for developmental complexity? Separate genomic entities encode the GRN nodes and linkages. Whereas the identity of the nodes is given by the protein-coding sequences of the regulatory genes, the GRN linkages are encoded in the *cis*-regulatory sequences associated with those genes. The transcriptional control system relies on multiple combinatorial information-processing functions: [1] each gene is regulated by several *cis*-regulatory modules, each with a defined spatial and temporal activity profile; [2] the activity of each *cis*-regulatory module is controlled by a specific combination of transcription factors, determined by the specificities of transcription factor-binding sites encoded in its regulatory sequences; [3] the *cis*-regulatory logic processing functions (AND, OR, NOT) define the relation between these inputs and the regulatory output of the module; [4] the transcriptional activity of any particular gene depends on the state of each *cis*-regulatory module at each time and place, and on the relative contributions of these regulatory modules.

In summary, developmental GRNs include a number of combinatorial logic processing functions which are essential for the increase in complexity during development. They explain why such diversity in developmental outcome can be achieved with a limited set of transcription factors.

Transcription factors, and even more so signaling molecules, are almost always used repeatedly during development. Their function, though, is always specific to the developmental context, and the sets of target genes they control depend on the presence of other transcription factors with which they combine to regulate gene expression. In the end, the overall developmental control system is directly dependent on regulatory genomic DNA sequence. From early embryogenesis to adult body plan, the regulatory genome defines the spatial and temporal organization of development.

GRN STRUCTURAL COMPONENTS AND MODEL REPRESENTATION

It is a great challenge for the mind to zoom out from transcriptional control of a single regulatory gene to the flow of regulatory information encoded in GRNs. To enable this shift in perception, adequate system level models are essential. Since the architecture of GRNs is directly encoded in genomic DNA, any model of a GRN has to be organized from the perspective of the genome. Figure 11.2 shows a GRN model produced in the BioTapestry platform [11,12]. The process modeled here is the specification of endomesoderm in the first 30 h of sea urchin embryo development [13–15]. The BioTapestry platform is designed to represent all biological entities that constitute GRNs. The components of GRNs and the symbolism used to represent them in BioTapestry are as follows:

Regulatory factors: At each node of the GRN the gene is indicated by a horizontal bar and an arrow, the arrow representing the protein-coding sequence and the ultimate protein product, i.e., the regulatory factor. These sequence-specific DNA-binding molecules are the moving parts (literally diffusing within the nucleus) of GRNs. The physical regulatory linkage is shown in the BioTapestry GRN model by a line emanating from the arrow at the gene encoding the regulatory factor.

Cis-*regulatory sequences*: The horizontal bar of each network node in the BioTapestry model specifically represents the *cis*-regulatory sequence of the respective

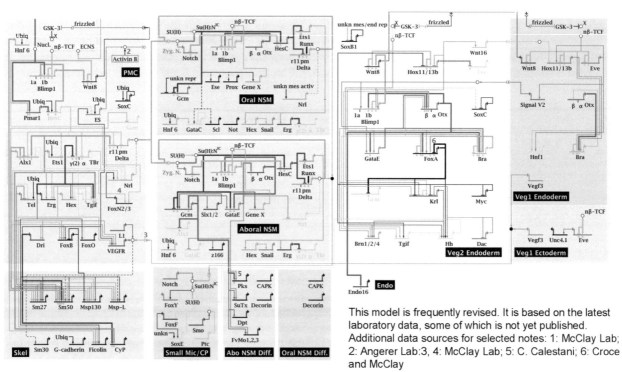

FIGURE 11.2 **The GRN underlying pregastrular specification of mesodermal and endodermal domains of the sea urchin embryo.** The network shows explicitly the regulatory interactions that underlie specification of the skeletogenic, mesodermal and endodermal domains of the sea urchin embryo (*Strongylocentrotus purpuratus*) up to the time when it gastrulates. By this point, entirely separate GRNs are in operation in each of these regions. All genes included in these GRNs encode transcription factors, or in a few cases signaling ligands or receptors, except for the samples of effector genes shown in the small rectangles at the bottom of the figure. The regulatory inputs into each gene, and the outputs from it to other genes, are indicated. Time runs approximately from top to bottom, and the GRNs portrayed include all linkages active at any point in the interval from very early cleavage to the onset of gastrulation. The large colored panels represent, from left to right, the GRN for specification of the skeletogenic lineage ('PMC'); the GRNs for specification of oral (top) and aboral (middle) mesoderm ('oral NSM' and 'aboral NSM'); the GRN for specification of anterior (future foregut) endoderm ('Veg2 endoderm'); and the GRN for specification of future posterior (hindgut) endoderm ('Veg1 endoderm'). A recent version of this continuously updated network model built in the BioTapesty computational platform is shown; for current status, see (http://sugp.caltech.edu/endomes/#BioTapestryViewer).

gene. Transcription factor-binding sites within the *cis*-regulatory sequences which affect the expression of their associated genes provide the endpoints of the regulatory linkages. If multiple *cis*-regulatory modules have been identified for a given gene, these modules can be represented by separate horizontal bars to reflect the regulatory logic specific for each module. In the example shown in Figure 11.2 this representation is used for the genes *delta* and *tbr*, for example.

Regulatory interactions: The physical interaction between transcription factors and DNA is directional in terms of biological information flow: the gene encoding the regulating input affects the expression of its target gene. The linkages between nodes in the BioTapestry GRN models therefore have a beginning and an end. This directionality is crucial for the causality of the entire process, which is also directional. Development never goes backwards. In the BioTapestry model, the linkage ends with an arrow if the transcription factor activates its target gene, whereas a small horizontal bar indicates repression.

Levels of experimental evidence: The BioTapestry model is based on several kinds of experimental evidence. First and foremost, every effort is made to include all regulatory genes specifically expressed at the relevant time and place that might perform a role in the control system (the total regulatory gene set is predicted from the genomic sequence). Information regarding the temporal course of regulatory gene expression and the exact spatial expression pattern, which is often dynamic, is essential. The key experimental evidence is obtained by perturbation analyses. *Trans*-perturbations are performed by blocking the expression of given genes followed by determination of the effects on all other genes in the system. This in itself is insufficient to distinguish between direct and indirect interactions. *Cis*-perturbations can accomplish this by demonstrating the specific import of *cis*-regulatory target sites for factors indicated by the *trans*-perturbations. Furthermore, it is to be noted that, as the analysis becomes more complete, multiple pieces of evidence bear on each node of the network. Details regarding the various kinds of evidence supporting the linkages in Figure 11.2 can be found at (http://sugp.caltech.edu/endomes/#BioTapestryViewer).

Spatial domains: The potential for all regulatory interactions is present in the genome of every cell of the organism. However, specific regulatory interactions are activated regionally, in specific spatial domains of the embryo. To represent regulatory interactions according to the process they drive, the network model in Figure 11.2 contains several colored rectangles labeled with the name of the respective embryonic domain in which the specification process occurs. Where the same regulatory gene is utilized in different spatial domains, it is shown separately in each network context in which it participates. This representation strategy preserves evidence of functionality which would be lost in a model in which each regulatory gene and its interactions would be represented only once, irrespective of the developmental process to which it pertains.

Temporal dynamics: Regulatory interactions occur in specific temporal windows. This information is not included in the two-dimensional GRN model in Figure 11.2, but is available online (http://sugp.caltech.edu/endomes/#BioTapestryViewer). In this continuously updated online version, regulatory interactions and regulatory genes are shown in gray as long as they are not active. A time-slider allows views of the GRN model at 3-hour intervals covering the entire process.

Signaling interactions: Signaling interactions occur between cells expressing a signaling ligand and cells expressing the cognate signaling receptor. Signaling is particularly important because it enables a specific GRN active in a given spatial domain to alter network function in another domain. Genes encoding signaling ligands and receptors are controlled by the inputs into them that are explicit in the GRNs, just as for other genes. However, the interactions between ligand and receptor occur in intercellular compartments. Reception of the signal activates an intracellular signaling cascade off the DNA, represented by a white circle in the BioTapestry model. Ultimately, the signal transduction biochemistry affects the function of an already present transcription factor (an immediate early response factor e.g., Suppressor of Hairless for Delta/Notch signaling or Tcf for Wnt signaling). Upon activation, the immediate early response factor induces novel regulatory gene expression. In the absence of the signaling interactions, this factor frequently associates with a co-repressor molecule and functions as a dominant repressor of the same target genes.

Predictive value: The BioTapestry model is designed to represent specific predictions of the genome-level interactions that constitute a whole developmental control system. At each individual node the requisite target site sequences are specifically implied, and these predictions are in turn amenable to direct experimental testing.

THE REGULATORY STATE CONCEPT

GRNs encode long chains of logic transactions. The result of these transactions is the developmental formulation of regulatory states. Regulatory states are defined by the combination of specifically expressed transcription factors and signaling molecules that discriminates these cells from others. But the individual components of regulatory states, that is, specific transcription factors, are rarely used in a uniquely specific developmental context but frequently recur in multiple regulatory states. All downstream functions in development are specified and driven by the regulatory state expressed at each time and place in the process.

Regulatory states are generated in a specific sequence as determined in the genome. Transitions of regulatory state are controlled between as well as within developmental domains, by the use of signaling interactions encoded in GRNs. These interactions ensure the coordinated activity of cell fate specification GRNs across the organism. Temporally, regulatory state transistion occurs continuously. However, and this is their main significance, regulatory states are finite in space, and their function is to discriminate cellular domains according to the cell fates each domain will give rise to. In space, regulatory states have clear borders. These borders are immediately apparent in any in situ hybridization experiment where gene expression is spatially non-uniform. The regulatory state in cells expressing given genes must be different from the regulatory state present in cells not expressing them.

Even though we can refer to the regulatory state as a single entity, it is actually a composite product of diverse regulatory instructions; some of these instructions depend on signaling interactions with adjacent domains, others on prior developmental inputs, others on temporal factors, others on cell lineage-specific regulatory factors, etc. All of these different upstream inputs together are required for the regulation of target regulatory genes. Here the combinatorial function of transcriptional control is the dominant feature. A subtle but essential consequence follows. This is that each *cis*-regulatory node controlling expression of the genes that constitute the regulatory state will be encoded differently, so that it will respond to different inputs. This can be clearly seen if we compare the control landscape underlying regulatory state to that underlying the differentiation gene battery, where all the genes respond to the same small set of regulators. The non-uniform use of regulatory information in the control of regulatory genes has its particular significance. An artist uses a palette of many colors which are kept separate to produce an infinite variety of different paintings, an impossibility were they all mixed together before application to the canvas.

The formation of regulatory states in embryonic development is directed and irreversible, and therefore also hierarchical. One intrinsic reason for the directedness of GRNs is the biochemical quality of regulatory interactions. The linkage between two nodes of a GRN represents the occupancy of a *cis*-regulatory module by a transcription factor, and results in a transcriptional control function which is unidirectional. The irreversibility is in addition ensured by specific GRN architectures. For example, signaling interactions are usually transient. Within a given time window signaling is required for the expression of regulatory genes, and perturbation of the signal leads to a failure in regulatory state propagation and cell fate acquisition. Later, the presence of the signaling ligand is no longer required for expression of these genes, and they are instead controlled by cross-regulation within the GRN; the new regulatory state is now irreversibly established. In general, GRN circuit design ensures the steady progression of regulatory states. The first zygotic developmental GRN(s) are activated by localized maternal inputs of regulatory significance. Thereafter, as embryonic domains are progressively subdivided, regulatory states propagate hierarchically. The end result is to ensure that the genes at the downstream periphery of GRNs, which encode the molecules that actually perform the functional and structural cellular processes, are expressed in the correct spatial location of the body plan.

SYSTEM-WIDE AND DEEP INFORMATION FLOW IN DEVELOPMENT

Novel developmentally functional assemblages, each consisting of many causally interrelated components, emerge at different levels. At the level of the terminal differentiation process, for instance, the terminal regulatory state causes more or less coordinate expression of multiple genes encoding functionally related proteins, and these interact in their own cellular domains to fulfill the requirement for specialized functionalities: think of genes encoding enzymes constituting pigment synthesis pathways [16], or the battery of genes producing eye lens crystallins [17,18], or the many other examples of specialized, dedicated, cell type-specific sets of effector gene batteries for which we now have some knowledge of transcriptional control [19]. In the strange-looking diagram in Figure 11.3 we see a cartoon that stresses the different levels of informational transaction in the hierarchical process of development. At the bottom is the genomic DNA, the unchanging source code, from which GRNs generate regulatory outputs in various parts of the embryonic organism. The GRNs are indicated in the gray platform representing the system of spatial specification of regional regulatory states. Progressively, these spatial domains are subdivided in each body part, as indicated by the red lines, and each new subdomain is defined by a new or partially new regulatory state. Black time arrows indicate developmental directionality. In the figure, vertical cylinders arising from the various spatial domains indicate that the local regulatory states (red vertical arrows) cause particular diverse morphogenetic gene cassettes (orange) and differentiation gene batteries (blue) to be activated. The impact of Figure 11.3 is that even though cell types can be compared at the effector gene level (horizontal planes), the causal explanation of their difference and specificity is rooted in the GRN control system (gray platform) and in the vertical flow of regional regulatory information (red arrows). Of course, few organisms operate in sequential lockstep as in this abstraction; that is, some differentiation genes are expressed in some parts before any morphogenetic genes are in other parts, and morphogenetic and differentiation gene expression may often temporally overlap. But Figure 11.3 is not

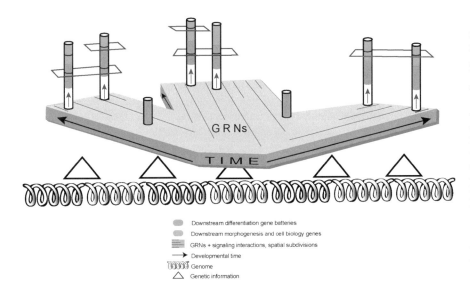

FIGURE 11.3 **An abstract image of the flow of genomic information for development of the body plan.** The regulatory genome is represented by the coil at the bottom of the diagram and the open triangles denote its developmental readout. Progressive spatial diversification of GRNs in diverse regions of the developing organism occurs on the grey platform as time proceeds. In each regional domain a different regulatory state is installed (white region of vertical cylinders with red arrows indicating the progression of development of each body part). These local regulatory states lead eventually to the regional activation of specific morphogenetic gene cassettes (orange region of cylinders) and of differentiation gene batteries (blue region of cylinders). See text for discussion and ramifications.

conceived as a temporal diagram: it is intended as an information flow diagram.

In each horizontally defined region of Figure 11.3 there is continuous flow of information from the genome vertically to every level of process. But conceptual residence on one of the horizontal planes provides no insight as to what comes immediately before, or before that. The issue is what constitutes explanation in development. If the explanation has to start with the ex cathedra 'givens' that provide the mountings for the upper-level planes, it can at best produce only isolated islands of causality. An explanation of development must be vertically related to the regulatory genome, the source code, the root of the explanatory chain of causality, or it will instead be rooted in a package of assumed phenomenology. An explanation of the spatially differential functions at any level must begin with the origins of the differential regulatory states in different domains. Therefore, what Figure 11.3 tells us is that explanation of development requires a system-wide and deep perception of causal relations: it requires a virtual framework of causality. This must originate in genomic information, and it must permit a seamless vertical transit from genomic information to any desired level of observation and analysis. Of course there will always be matters unknown, but an ultimate goal is a framework explanation that may retain a few islands of phenomenology floating within it. Developmental biology has largely forever been the reverse: a sea of phenomenology within which have floated a few poorly connected islands of causality.

GENERAL FEATURES OF THE DEVELOPMENTAL PROCESS

Development is the process by which the functional, morphological animal organism is built, beginning with the fertilized egg and terminating in the adult body plan. The major features of the body plan are established in exactly the same way in every individual of each species during embryogenesis, i.e., the positions, identity, and internal organization of organs, appendages and axial structures are the same per species. The individual divergence of characters within the species type that are due to genetic and occasionally environmental differences affects only the last phases of development, when finer-scale properties are established. On the other hand, the same suite of fundamental developmental mechanisms is used throughout the universe of animal forms to build the very diverse body plans that define different animal clades. Animal clades are hierarchically organized, indicating different levels of shared developmental programs: the upper-level characters that define phyla are shared by all members of that phylum (e.g., the shell glands of molluscs; the dorsal nervous systems of chordates; the jointed appendages of arthropods); fish, amphibians, reptiles, mammals are classes within the chordate subphylum vertebrates, and they all utilize similar developmental mechanisms to construct their axial skeletons, while within each of these classes the various orders, such as carnivores or rodents, share more detailed properties which, as is intuitively obvious, devolve from the shared developmental processes specific to each order, and so forth. In general, the genetic programs that operate development are organized hierarchically, function hierarchically, and display a hierarchical order in the breadth of their distributions when diverse animal species are compared [20–22].

The fact that all members of each species build their body plans in the same way, while members of different species do so in different ways, can only be explained if the control program for development is a heritable characteristic and the genomes of different species harbor different heritable control programs. Thus thinking about the readout of such control programs, i.e., the development of the body plan, leads to thinking about the origins of their different structures, i.e., evolution of the body plan, and vice versa. Here we can consider only the basic, common features of the mechanisms by which readout of the genomic control program is transformed into development of the animal body plan, as we have discussed elsewhere [22]. Although it lies outside our current province, deep understanding of the developmental control programs of animals will ultimately require understanding where they came from and how they were assembled in geologic deep time.

The most important thing to know about any complex progressive process is the principles by which its control system operates. As the previous sections illustrate, understanding the system-wide logic of GRN operation should provide a sufficient explanation for why development works as it does, in the ultimate terms of genomic regulatory information.

In the following is an outline of the essence of the developmental process as seen from the genomic perspective, deliberately devoid of detail and example, and designed simply to highlight the levers by which the genomic control program in principle produces development. This is intended for engineers rather than embryologists, logicians rather than cytologists, systems bioscientists rather than those focused on roles of specific genetic or biochemical functions that are required for given developmental events. GRNs are organized so as to control developmental process, and it is therefore essential to recall what this process consists of.

Before the GRNs operate: functional roles of egg and sperm: In the late 1870s and early 1880s it became clear from observations of fertilization in the eggs of sea urchins, plants, nematodes, and many other organisms that egg and sperm contributed equally to the diploid genome of the new organism that would develop from the fertilized egg [1]. But this basic pillar of understanding does not capture the asymmetry in functional roles of egg and sperm. In addition to its nuclear components the egg cytoplasm contributes much more. It contains a storehouse of all the housekeeping components, proteins and mRNAs encoding common structural and metabolic components, needed for any cell to exist [5]. The amount of these in the storehouse is, roughly speaking, enough to supply all the cells that the volume of egg cytoplasm will be divided into during embryogenesis. For until the embryo has become a larva able to feed, its cells must rely on maternal nutrients and most other bulk components needed for cellular life. We amniotes are an exception, since our embryos can absorb nutrient from the maternal environment.

The organization of the egg also determines the regulatory polarity of the embryo. Most regulatory factors are universally distributed in the egg, so that they will be present in all embryonic cells as cleavage divides it up.

However, there are always some factors of regulatory significance which by early cleavage are anisotropically distributed and are used to define the axes of the future embryo [19]. These may be transcription factors, cofactors thereof, or signaling components, which ultimately lead to activation of genes encoding transcription factors. When the cleavage planes divide up the egg, they generate a few specific domains that are distinct in their regulatory potentiality. Thus the founder cells of some specific cell lineages, or those populations of cells in given topological positions in the embryo, initiate different GRNs because they are endowed ab initio with different factors of regulatory significance, and these in turn operate to activate different early regulatory genes. The regulatory anisotropies of the egg provide the initial symmetry-breaking functions of the embryo, though these initial asymmetries account only for the very first crude regulatory distinctions in the embryo.

Embryonic specification: 'Specification' is an antique embryonic term with the fuzzy definition that it is the process by which cells of an embryo begin to acquire a particular fate. But for us it has a particular, sharp meaning: it is the process by which cells of an embryo acquire a given specific regulatory state. Specification is progressive: the initial inputs into a localized GRN lead toward the definitive regulatory state of that domain, but it always forms by a series of successive transcriptional steps. As just mentioned, owing to the anisotropic regulatory organization of the egg, different early embryonic cells initiate development by activating diverse regulatory genes [23]. But that is just the beginning. After very early development, spatial specification processes are all controlled by GRN circuit functions. Intercellular signaling is the main device used to define new spatial regulatory domains after the relatively simple initial spatial regulatory patterns are installed. As we stress in the following paragraphs, the generation and the interpretation of signals are both encoded properties of GRNs, and GRN circuitry contributes to spatial patterning of regulatory state in many additional ways. Below we consider several specific examples, and numerous particular GRN subcircuits that execute spatial patterning functions have been reviewed by us elsewhere [21,24]. These invariably involve transcriptional repression. Subcircuits that include spatial activation of genes encoding transcriptional repressors are used to set domain boundaries, to exclude alternative regulatory states in given domains and to define specification domains. The specificity of these repression functions is resident in the genomic *cis*-regulatory sequences of regulatory genes, as these determine the location of transcriptional repressor and activator target sites.

Signaling and regulatory state: For three-quarters of a century, ever since Spemann, developmental biology has been obsessed with signaling in embryogenesis. This was perhaps to be expected, given the dramatic and easily demonstrated phenotypic effects of perturbing signaling. This was initially done first by ectopic transplantation of signaling tissues, and later by numerous biochemical and molecular biological means. But, because of the structure of the developmental regulatory system, trying to understand development by focusing on signaling alone leads inexorably to a phenomenological cul de sac. It is like studying the output effects of flipping various switches on the panel of an unfamiliar electronic device. Signals per se have effects on development only indirectly, and only by causing alterations of regulatory state in the cells receiving the signal. The nature of the effects and the causal relations between the signal and the developmental consequence depend in detail on the structure of the GRN circuitry called into play by the change in regulatory state that has been effected by the signal. That is to say, the exact consequences of signals for development depend in any given case on the genomic regulatory wiring that mandates GRN circuitry. This must be true, because the same few signaling systems are used over and over in development. Developmentally important signals work in general as follows. Most signals utilized for developmental purposes are protein ligands encoded by genes activated as part of the function of regionally active GRNs. The signal ligands are externalized, and the receiving cells express transmembrane protein receptors (as part of the transcriptional function of their GRNs) that specifically recognize and bind the signaling ligands. The consequence of extracellular ligand—receptor binding is to alter molecular structure on the intracellular end of the receptor protein complex, which in turn affects intracellular biochemistry in such a way as to alter the regulatory activity of a dedicated target transcription factor present in the cytoplasm. Some signals cause the target transcription factor to acquire a cofactor it needs for activity, others cause it to be released from a cytoplasmic docking protein, and still others cause it to be chemically modified so as to activate it or stimulate its transit into the nucleus. Different ligand—receptor pairs affect intracellular biochemistry in different ways and utilize different target transcription factors, but the underlying logic is always the same: reception of the signal results in a specific, qualitative, gain of function change to the regulatory state. Thus signals indirectly result in new specification functions in those cells positioned to receive the signal. The key consequence is to cause new regulatory states to be set up in given spatial domains. This is what used to be called *inductive signaling*, defined phenomenologically by the signal-dependent appearance of a new state of specification in the receiving cells different from that of the sending cells. Since progressive installation of spatial regulatory states is causal for the whole developmental process, perturbation of signaling often produces dramatic phenotypic results.

There are three other aspects of developmental signaling that when viewed from the vantage point of regulatory state effects are general enough to require mention here. One is the 'Janus-like' behavior of many signal transduction systems [24,25]. If the transcription factor that serves as the early response factor for a given signal is in a cell receiving the signal, it produces a newly active regulatory factor as just discussed, but if not, the same factor will act as an obligate repressor. Therefore, were such a factor present in all cells of an embryo, it would act as a global control agent, binding the sites which it recognizes in its target genes in all cells and repressing these genes everywhere except where the signal is being received, while in these last cells it would contribute to their activation. A second often encountered aspect is that diffusible signals sometime function over distances as *vectorial patterning devices,* that is, different regulatory states are installed in response to the signal closer to its source where it is stronger, and others are installed farther away where the ligand concentration is lower. In some well-analyzed examples specific types of GRN circuitry can be shown to be responsible for discontinuous, hysteretic responses to different levels of signal input [24, 26—31], and this may be the general explanation for the often observed Boolean regulatory states set up in response to graded signaling ligand distributions. Thirdly, inductive signals are always transient. A predictable feature of developmental GRNs downstream of the primary signal response genes is circuitry that transfers dependence of the new regulatory state from the external signal to internal cross-regulatory functions encoded in the GRN architecture. For example, it is commonly found that the gene activated by a signal might activate other genes, which engage in a positive feedback loop (e.g., [14]). Once such a loop is set up it not only obviates further dependence on signal reception but also resets the levels of transcriptional output downstream [32].

Propagation of regulatory states within territorial domains: There is also another kind of signaling, in which cells of a territory all both send and receive the signal. Termed *community effect signaling* [33], this ensures that the regulatory states of these cells are homogeneous, and may be a fundamental mechanism by which the cells of multicellular tissues remain in lockstep at the level of developmental regulatory state. In several cases of community effect signaling a common circuitry feature is observed: the gene encoding the ligand responds to the very same early response transcription factor that is activated on reception of the signal. Its consequence is that all cells receiving the signal also send the signal, so that within the multicellular territory all the cells are locked into the mutual embrace of an intercellular feedback system. Any genes downstream which depend on the same activated transcription factor will now be expressed in these cells at similar rates [33]. Community effect signaling could be a very general mechanism underlying the remarkably similar functions of contiguous individual cells within common embryonic territories, and the GRN wiring underlying it one of the features that allows for multicellular territorial fate and function in development. An interesting difference from inductive signaling is that since it is essentially a maintenance device ensuring homogeneous function, in so far as regulatory state is concerned. Thus, in the cases we know about it is not transient as is inductive signaling, the fundamental role of which is to change spatial regulatory state rather than to maintain it.

Morphogenesis, differentiation, cell cycle control (growth): Thus far we have been concerned with the upper-level control system for development, the product of which is the plethora of unique regulatory states, dynamically defining in potential functional terms each component of the forming body plan. However, the vast majority of the genes in the genome, all but a few percent, and the vast majority of genes expressed in development, are not regulatory genes. They are the effector genes that do the cell biological work of development, expression of which is controlled transcriptionally by regional regulatory states as development proceeds. Eventually, the terminal regulatory states and the terminal cell type differentiations are installed. Cell cycle and morphogenetic effector genes are called into play early in embryogenesis, and progressively more differentiation genes are expressed as development proceeds. The *cell cycle* is driven by a series of enzymatic steps which affect the mitotic cytoskeletal and DNA replication apparatuses, homologous from yeast to man. In development, the deployment of the cell cycle is controlled transcriptionally, by preventing the expression of genes encoding cell cycle repressors or causing expression of genes activating the cycle at the appropriate times. These control functions are executed as an output of the regulatory state of each growing domain of the embryo [24]. A fascinating but largely unsolved aspect of development is exactly how *morphogenetic functions* are called into play, in a modular fashion, in given regulatory state domains. That is, to effect a morphogenetic function many separate cell biological functions have to be executed, and the cells that build multicellular structures have to have activated genes encoding the proteins that produce the right physical functions to generate these structures. For instance, in some animals the gastrular endoderm cells of an invaginating embryonic gut must have cell surface properties that enable them to slide over one another so as to accomplish the formation of an elongated tube, and they also have to have activated sets of genes encoding motility functions. No one yet knows how many separate genes are required to generate these phenotypes, but the number will not be small. Nor is it clear whether a majority of such genes in any given context have to be subject to direct transcriptional control by the local regulatory state, or alternatively, if most of these genes are widely expressed and only a minority of 'checkpoint' genes that trigger or nucleate the whole process have to be controlled directly by the GRN, though there is some evidence for the latter (for a review, see [24]). To solve morphogenetic control mechanisms, we will have to identify the genes required to make a tube, or an epithelium, or a bifurcation, or to control cycling in a growing tissue, etc., and relate the transcriptional control of those genes to the ambient regulatory state. *Differentiation* is a somewhat discrete regulatory problem, about which much more is known (for a review, see [24]). Differentiated cells express sometimes large sets of effector genes, the products of which are together used to produce the specialized functions mounted by these cells. Such differentiation gene batteries are expressed only in given cell types, while in contrast all cells sometimes execute cell division. Many different kinds of cell use overlapping morphogenetic functions, for instance those controlling cell motility, contractile behavior, and so forth. Differentiation effector genes in each battery are individually controlled at the transcription level by a small subset of the factors present in a cell type-specific regulatory state. Thus for differentiation gene batteries, as opposed to the higher levels of control in the GRN, wiring is largely parallel, in that many genes respond to the same few cell type-specific inputs.

The adult body plan: Development of the body plan terminates when there are no further body parts to be formulated. But of course this is not the end of development, broadly defined. There are in many animals extensive changes in scale and size from juvenile to adult. In mammals these changes are massive, and as in postnatal human brain development, they may be extremely complex but perhaps still use the same mechanistic features of embryonic development as do other body parts. In general, post-embryonic developmental processes also involve the peripheral aspects of developmental GRNs, such as the continuing use of morphogenetic, differentiation and cell cycle subprograms, in response to the terminal sets of spatial regulatory states. As has become a prominent realization lately, development in fact never terminates, as the differentiation functions of stem cells throughout adult life attest. But it is important to realize that stem cell re-creation of injured muscle or liver, or even diversification of immune cells from multipotential hematopoietic precursor cells, is a wholly different phenomenon from embryonic development of the body plan. Such adult stem cell differentiation is like terminal differentiation in embryogenesis: it requires in any given case relatively few specifically expressed regulatory factors, and it involves no novel spatial establishment of regulatory states, the fundamental function that underlies body plan development. Thus the genomic program for stem cell differentiation is fundamentally different from the genomic program for embryonic development. In some animals the regeneration of whole body parts can occur in the adult, but evidence indicates that the underlying processes are different from the spatial control processes of embryogenesis.

DEVELOPMENTAL GRN DYNAMICS

GRN dynamics denotes the changes in regulatory state encoded by GRN circuitry with time. Thus far the vast majority of dynamic regulatory processes that have been analyzed in developing animal systems are sequential changes in regulatory state that take place in the same cells. Over time, regulatory states change in these given cells, for example as a result of receiving signals, or in processes of stem cell differentiation, such as lineage choice functions in which given pluripotential cells respond to external conditions by activating one or another regulatory state. Such biological processes have been the subject of numerous ODE kinetic models (e.g., [26–31,34–37]). However, dynamic models of sequentially occurring changes in state within given cells or domains do not address the mechanisms causing changes in spatial regulatory state, with which we have so far been concerned.

Dynamic modeling of GRN outputs has been used to analyze several developmental mechanisms that execute spatial specification. The main function of the GRNs that control body plan formation is to set up progressively subdivided discrete regulatory state domains in space. They do this dynamically, of course, but the architecture of the GRNs that specify this process cannot be derived from the kinetic output of the network; rather, the kinetics can be derived from the GRN architecture if sufficient kinetic parameters are known. An example in which GRN architecture was used to construct a kinetic model concerns the dynamics with which spatial patterns of gap gene expression in the 14th cleavage cycle *Drosophila* embryo are generated [38,39]. Here previous studies had revealed much of the GRN topology, and this topology was used in the formulation of an ODE model. A large number of synthesis, decay and other kinetic parameters were extracted from a quantitative high-resolution imaging dataset and the observed process can be reconstructed when the model is run, with a few additional interactions indicated. In this system the constrained diffusion of transcription factors among nearby syncytial nuclei provides the spatial inputs to the genetically hard-wired transcriptional response system indicated by the GRN architecture. A second example is afforded by analyses of response to graded Hedgehog signaling in both the amniote neural tube [40,41], and in the *Drosophila* wing disc [42]. But in a large number of developmental contexts graded signal response is not an issue.

Specification in the earliest stages of embryogenesis depends at least in part on initial anisotropies, as above. Later spatial specification in embryos, and in body part subdivision, is usually accomplished through sequences of inductive signaling where there is generally only one response to the signal, and where signaling is often, and in some systems always, short range. One such system is the pregastrular sea urchin embryo, to which pertain the examples of GRN-mediated spatial specification processes discussed in the following section. Here we consider the dynamics of spatial specification in this embryo. Already by gastrulation (~ 600 cells), this embryo has constructed a mosaic of at least 10 distinct territorial spatial regulatory states, each giving rise to a different part of the completed post-gastrular larva.

What determines the tempo at which spatial regulatory state changes occur in this embryo? Detailed kinetic expression measurements for gene cascades in this system describe the dynamics of progression from one regulatory step to the next ('step time'). Although it had been noticed earlier in experimental GRN analyses [14,15], a consistently amazing result was the great regularity of the step time throughout many hours and many steps [45]. Gene to gene, the whole system runs on a similarly paced regulatory clock.

What determines the beat of the clock and why it operates the way it does is a particularly interesting problem, because despite the complexity of the process of embryonic development, it is temperature adjusted. Thus different species of sea urchin that live at different temperatures develop at different rates, but these developmental rates (e.g., time from fertilization to gastrulation) are predictable from one another because they closely follow the same Q10 rule that applies to basic molecular biology processes such as RNA or protein synthesis. That is, from cold water animals to flies to mammals, for each $10°$ C temperature change the real-time rate for these molecular processes changes by a factor of about 2 [5]. This turns out to provide a valuable clue to the clock-like behavior of at least some developmental systems. In 2003, a dynamic a priori analysis of step time in sea urchin embryos was carried out by Bolouri and Davidson [43], in which a model of *cis*-regulatory occupancy by transcription factors was constructed, based on thermodynamic principles [44], and *cis*-regulatory occupancy was related mathematically to the rate of transcriptional initiation and then combined with synthesis and turnover kinetics. This was essentially a first-principles estimation of what the step time should be for sea urchin embryos, assuming typical values of relative transcription factor–DNA equilibrium constants, and using canonical transcription, protein synthesis, and turnover rates that had been measured for these embryos at $15°$ C. There were two results that proved to be of immediate explanatory value for the problem under discussion here. First, the next step in the computed gene cascades occurs long before steady-state levels in mRNAs for the input transcription factors are ever achieved. Since the only way exact level control can be achieved is by balancing in and out rates — that is, synthesis and decay rates at steady state — this means that the regulatory system of the sea urchin embryo cares little about exact level and rate control. Indeed, comparisons among different individuals frequently show two- to threefold differences in transcription factor mRNA levels [45]. It is a forward drive, not a level-sensitive dynamic system. Therefore, it is only the kinetics with which newly activated genes produce transcription factor messages, and these mRNAs are translated to produce the factor proteins, which should count in determining the step time. But it is these same basic processes that obey the Q10 law, and the step time can be computed in terms of these very processes. Second, the Bolouri—Davidson model predicted a typical step time for the sea urchin system of about 3 hours. Thus the pace of the step time clock depends simply on basic biosynthetic dynamics, and this explains how a Q10 rule for development could apply, since in turn the dynamics of developmental specification processes depend directly on the rates of regulatory state changes, that is, on step time. A contributing factor, as yet unexplained biochemically, is the general quantitative similarity in regulatory gene transcription rates in this embryo, with only a few outliers (as indicated by the measurements in ref. [45]). In summary, the global dynamics of GRN function in embryogenesis may be more regular and less occult than meets the eye.

EXAMPLES OF GRN-MEDIATED SPATIAL CONTROL IN DEVELOPMENT

The GRNs driving the specification of endomesoderm in the first 30 hours of sea urchin embryo development have been analyzed systematically. Genes encoding transcription factors were identified in the entire genome based on their homology to known transcription factor families in other species [46—51]. Their spatial and temporal expression patterns served as the basis to generate a complete candidate list for the GRNs driving the formation of each specification domain in the early sea urchin embryo. Regulatory interactions between network components were subsequently identified in numerous morpholino perturbation and *cis*-regulatory analyses [14,15,52—56]. The resulting GRNs explain how individual cell fate specification processes occur and elucidate all the individual regulatory control mechanisms contributing to the organization of this developmental process. The level at which these mechanisms actually control process is not always apparent from a classic developmental biology perspective, as examples from the sea urchin endomesoderm GRNs demonstrate.

Endomesoderm Development in *S. purpuratus* Embryos

Very briefly, three cell lineages contribute to endodermal and mesodermal cell types in the sea urchin embryo [57, 58]. Their dispositions and signaling interrelations are indicated diagrammatically in Figure 11.4A,B. The central domain shown on the left produces the skeleton and the concentrically adjacent domains shown on the right (Figure 11.4B) produce the mesoderm and the endoderm. The skeletogenic cells are specified as discussed below, with an initial maternal input localized from the beginning in these cells. When they become specified they produce two essential signals. The immediately adjacent cells receive Signal1 (Figure 11.4A) and become mesoderm; the future endoderm receives Signal2 but not Signal1. In the following we discuss as examples of network-mediated fate decisions how these different states of specification arise.

Example 1: Use of Maternal Anisotropy to Define a Zygotic Regulatory Compartment

Spatially restricted zygotic gene expression is first observed in the skeletogenic precursors. The *pmar1* regulatory gene is expressed specifically in these cells and is the earliest specifically expressed zygotic transcription factor in the GRN driving skeletogenic cell specification [59]. *Pmar1* gene expression in these cells is controlled by two transcription factors which are present in the egg cytoplasm inherited by these cells as cleavage begins (maternal input, Figure 11.4A) [60—63]. The example shows how localized maternal regulatory factors result in the initiation of a specific localized zygotic regulatory state in the embryo.

Example 2: Double-Negative Gate Circuitry for Spatial Restriction of the Skeletogenic Cell Fate

Instead of simply activating other transcription factors in the skeletogenic GRN, Pmar1 functions as a repressor. Within the skeletogenic cells, Pmar1 transcriptionally represses the expression of the *hesc* gene which encodes another repressor. The *hesc* gene is under the control of ubiquitous activators, so it is expressed in all other cells of the embryo (Figure 11.4A, 'Spatial Circuit') [64]. Its target genes at this early developmental stage encode multiple transcription factors in the skeletogenic GRN (Fig. 11.4C, tan and pink areas). In the skeletogenic cells, because Pmar1 is expressed, *hesc* is repressed and these skeletogenic regulatory genes are expressed. The circuitry constitutes a double-negative logic gate. The use of this double repression circuitry not only drives gene

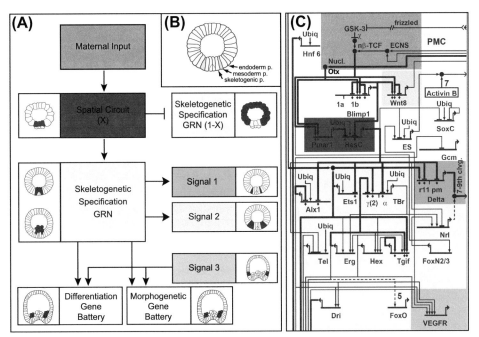

FIGURE 11.4 **Network circuitry underlying early spatial inputs in the sea urchin embryo.** (A) Color-coded diagram showing information flow that functions to provide spatial organization in the early embryo. Each colored box includes specific GRN circuitry, including emission and reception of signals. The location in the embryo where the respective GRN circuitries are active are shown by the red areas of the diagrammatic embryos. Where signal interactions are depicted, arrows lead from the source of the signal to the target cells. In the skeletogenic spatial circuit, X indicates the domain where the skeletogenic GRN is installed, 1-X the remainder of the embryo where it is repressed. (B) Lateral diagram of blastula stage embryo showing relative spatial disposition of skeletogenic, mesodermal and endodermal precursors (p.). (C) Skeletogenic GRN shown in detail with circuitry highlighted according to the color code of (A).

expression in a specific compartment but at the same time ensures that these same genes are repressed in all other cells. Especially in early developmental stages, when most cells are not yet specified, such binary regulatory outcome (X: where the skeletogenic genes are expressed, 1-X: the rest of the embryo) combined with repression provides a frequently used cell fate exclusion mechanism, which in one or the other of its states functions globally [21].

Example 3: Signaling and Cell Fate Specification

GRNs driving the specification of individual embryonic domains must communicate with each other to ensure correct spatial organization. Skeletogenic cells do not require any signaling inputs from the cells surrounding them to accomplish their own specification, but they do function as a signaling center for adjacent cells. That is, the skeletogenic GRN controls expression of genes encoding signal ligands. Downstream of the double-negative gate, under direct control of repressor HesC, is the gene encoding the signaling ligand Delta (Figure 11.4A,C 'Signal 1') [65–67]. Delta is a membrane-bound signaling ligand which activates the Notch receptor on immediately adjacent cells. The consequence is to activate a new network of transcriptional regulators in these adjacent cells which are thereby specified as mesoderm [54,68].

The specification of endodermal cell fates depends on another signaling interaction originating in the skeletogenic cells. The skeletogenic GRN causes expression of Wnt signaling ligands (Figure 11.4A, 'Signal 2') which diffuse to neighboring cells where, in consequence, an endodermal GRN is activated [15,53,69,70]. Multiple transcription factors in the GRN specifying endodermal fates are directly regulated by Tcf, which is the early response factor for Wnt signaling [52,53,55]. The Delta/Notch and Wnt signaling interactions globally control spatial expression of their respective endodermal and mesodermal target genes globally [53]. An interesting aspect of this circuitry is that both early response factors, Tcf (Wnt signaling) and Su(H) (Delta/Notch signaling) associate with the co-repressor Groucho to mediate repression of their target genes in cells not receiving the corresponding signaling ligand. Thereby, these target genes are activated in cells receiving the signal and repressed in all other cells, mediating an X/1-X regulatory logic similar to the one used to initiate the skeletogenic GRN though using different circuitry.

The skeletogenic cells are not dependent on any signaling input as long as they remain localized at the vegetal pole. But as soon as these cells start to migrate into the interior of the embryo, their position and further differentiation depend on signals emanating from their newly neighboring cells [71]. One of these signals (Figure 11.4A, 'Signal 3') is VEGF and the gene encoding its receptor (VEGFR) is transcribed under control of the skeletogenic GRN (Figure 11.4C).

Example 4: Parallel Mechanisms in Cell Fate Initiation

In the GRNs for the sea urchin endomesoderm we see repeatedly that the initial establishment of a spatial regulatory domain usually depends on a single regulatory input. Thus, mesoderm specification depends on the Delta/Notch signal from adjacent skeletogenic cells; endoderm specification requires Wnt signaling; and in the skeletogenic cells specification depends on transcription of *pmar1*. In all three domains, what begins with a unitary spatial input ends with the activation of a complex network of regulatory interactions. The progressive activation of a network of transcriptional interactions requires multiple regulatory steps and in the sea urchin embryo takes many hours to attain its terminal state. In all three cases the upstream input mediates a switch logic, such that the system is either repressed or activated in different locations. This 'bottleneck' aspect of the specification mechanism contributes in both evolution and development to a Boolean process of regional fate assignment.

Example 5: Mechanism of Irreversibility in the Developmental Process

Downstream of the early spatial activation system there predictably appears a network subcircuit which ensures the continued expression of regulatory genes even when the early inputs are no longer available. In the skeletogenic GRN, *pmar1* is expressed only transiently [64], but immediately downstream of the double-negative gate target genes there exists a triple positive feedback circuit [14]. These three genes maintain each other's expression, and once the system is activated it no longer requires the initial skeletogenic inputs. Similar positive feedback circuitry can be found in the mesoderm specification GRN (Figure 11.2, 'aboral NSM') downstream of the initial Delta/Notch signaling input [72]. Here again, the circuitry renders the subsequent regulatory state independent of continued signaling input. These positive feedbacks ensure that the regulatory state instructed by the signal or other transient initial input is maintained and cannot revert to the previous regulatory state.

Example 6: GRN-Mediated Cell Fate Decision

It is often found that in embryonic development a discrete progenitor cell population gives rise to multiple diverse cell fates. In the sea urchin embryo, the endoderm and mesoderm share a common progenitor cell lineage. Curiously, the mesodermal GRN and the endodermal GRN run simultaneously in these cells with no regulatory interactions between them (Figure 11.5A, red and blue networks) [53]. The activation of the GRNs occurs by using separate inputs, Delta/Notch signaling vs. Wnt signaling. Definitive separation of fates makes use of a canonical radial cleavage which separates an inner ring of cells destined to become mesoderm and an outer ring destined to become endoderm. The relation between regulatory state and these rings of cells is shown diagrammatically in Figure 11.5B, where endoderm fate is represented in red and mesoderm in green. Initially, both GRNs continue to operate in the inner ring of cells, but the circuitry shown in Figure 11.5C results in the extinction of the endodermal network in the inner ring of cells. Delta/Notch signaling, which is accessible only to the inner ring, because these cells are adjacent to the source of the Delta ligand (the skeletogenic cells, yellow in Figure 11.5B), is used to shut down endodermal regulatory genes [15,68]. Note that even before the definitive fate separation the progenitor cells are not 'bi-stable': which of their descendants are going to be mesodermal and which endodermal is hardwired.

The general import of these examples is that they illustrate in (minimum) detail exactly how GRN architecture results in direct and specific control of spatial regulatory state. In turn, viewing network circuitry in its developmental context illuminates the stepwise path of the developmental process.

CONCLUSION: THE EXPLANATION OF DEVELOPMENT

What would be the anatomy of an *explanation* of a developmental process of embryogenesis or body part formation? First, it must encompass all the parts of the apparatus that play a causal role, so that it is a system-wide and not an island-like explanation. Second, it should be predictive of the spatial innovations entailed in the process, so that the mechanism included in the explanation and the status at each given step should suffice to predict the next step. Third, it should be causally progressive, so that it captures the dynamics of the process. Fourth, it should be commutative, so that the mechanistic argument leads smoothly, without gaps, going either upstream from the observed developmental process back to the genome, or downstream from the genome to the observed developmental process.

Everything that happens in development depends causally on spatially defined regulatory states, and the process that embryologists used to call specification is neither more nor less than the installation of these regulatory states. Thus if we can *explain* the process of

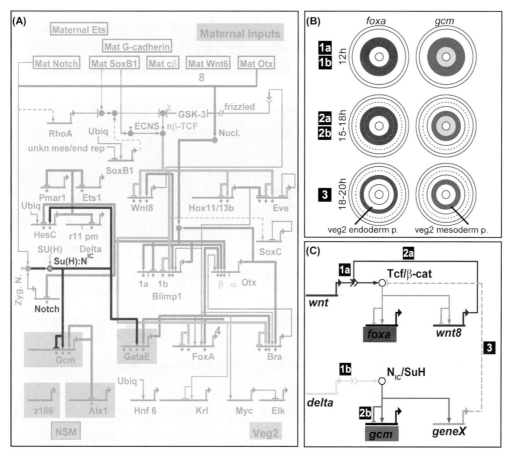

FIGURE 11.5 **Endoderm/mesoderm cell fate specification in the sea urchin embryo.** (A) Initiation of two non-interacting networks of regulatory gene expression within the same precursor cells early in the process of specification of endoderm and mesoderm [53]. A mesodermal GRN is initiated in response to Notch signaling (red GRN linkages); the endodermal GRN is initiated in response to Wnt signaling (blue GRN linkages). *Cis*-regulatory genomic hardwiring determines which regulatory gene responds to which signal transduction system. (B) Diagrammatic portrayal of progressive spatial separation of endodermal and mesodermal regulatory states. Red indicates expression of the endodermal GRN, here represented by the *foxA* regulatory gene. Green indicates expression of the mesodermal GRN here represented by the *gcm* regulatory gene. Yellow indicates the location of the skeletogenic cells and their expression of the Delta signaling ligand. At 12h a single ring of cells is precursor to both mesodermal and endodermal fate. At 15–18 h a radial cleavage separates this domain into two rings of cells, of which only the inner ring now expresses the mesodermal GRN. At 18–20h, mesodermal and endodermal fates are spatially separated with the terminal extinction of the endodermal GRN in the inner ring. (C) Network subcircuit responsible for activation of mesodermal and endodermal GRNs, here represented only by the *foxa* and *gcm* genes. The Notch dependent linkage which extinguishes Tcf function and hence endoderm gene expression is indicated.

specification for an episode of embryogenesis, or the formation of a body part, we will have taken a giant step toward explaining that aspect of development. We have comprehensive explanations couched in GRNs for pair rule and gap gene expression patterns [39,73], and for mesoderm diversification [74] in *Drosophila*; and for much of pregastrular specification in the sea urchin embryo [14,15,21,53]. Diverse specification GRNs have become available for embryonic processes in other model systems as well as for the development of at least some adult vertebrate and arthropod body parts (for reviews, see refs. [19,22,24]. As proof of principle, the best-known portions of the sea urchin GRNs summarized in Figure 11.2 generate explanations for specification that do indeed fulfill all the criteria for explanation listed above. Finally, the explanations we have for these specification functions provide a pathway causally linking the regulatory genome to regulatory states in embryonic space. Therefore we can understand both developmental specification in terms of the genomic program, and the regulatory genomic code in terms of its functional output, regional embryonic specification. There can be no satisfying explanation of development that is not rooted in the genomic information that encodes its control program, and in that sense specification GRNs provide the root for all downstream explanatory mechanisms.

But development scarcely ends with specification and the establishment of regulatory state patterns. By the same criteria, we lack any example of system-wide comprehensive explanations of developmental processes that carry all the way downstream to the deployment of morphogenesis cassettes and differentiation gene batteries, except for analyses of transcriptional control for various subsets of differentiation genes. In general, control of downstream gene expression remains unlinked to prior developmental specification functions. The way is open, however, to extend GRNs downstream so that they encompass the linkages between terminal local regulatory states and the effector genes that produce the observed morphogenetic and differentiation outcomes. This would be like traversing the vertical red arrows upwards in given domains of the cartoon of Figure 11.3. This is an enormous challenge, but the reward will be a true explanation of developmental processes that extends all the way from the genome, to the succession of regulatory states, to the control systems that animate differential effector gene expression.

NOTE ADDED IN PROOF

A recent computational model based on the GRN shown in Fig.11.2 demonstrates that in fact this network contains sufficient information to predict successfully almost all the observed spatial and temporal gene expression (Peter, I.S., Faure, E. and Davidson, E.H., Predictive computation of genomic logic processing functions in embryonic development. PNAS DOI 10.1073/pnas.1207852109)

ACKNOWLEDGEMENT

We are pleased to acknowledge support from NIH Grant HD037105, which provided support for ISP. We also acknowledge the support of the Lucille P. Markey Charitable Trust.

REFERENCES

[1] Wilson EB. The Cell in Development and Heredity. third Ed. New York, NY: Macmillan; 1924.
[2] Wilson EB. The Cell in Development and Heredity. first Ed. New York, NY: Macmillan; 1896.
[3] Gilbert SF. Embryological origins of the Gene Theory. J Hist Biol 1978;11:307–51.
[4] Laubichler MD, Davidson EH. Boveri's long experiment: sea urchin merogones and the establishment of the role of nuclear chromosomes in development. Dev Biol 2008;314:1–11.
[5] Davidson EH. Gene Activity in Early Development. third Ed. San Diego, CA: Academic Press/ Elsevier; 1986.
[6] Boveri T. Über mehrpolige mitosen als mittel zur analyse des zellkerns. Verhandlungen der Physikalische-medizinischen Gesellschaft zu Würzburg 1902;35:67–90.
[7] Boveri T. Zellenstudien V. Über die Abhängigkeit der Kerngrösse und Zellenzahl bei Seeigellarven von der Chromosomenzahl der Ausgangszellen. Jena: Gustav Fischer; 1905.
[8] Boveri T. Zellenstudien VI. Die Entwicklung dispermer Seeigeleier. Ein Beitrag zur Befruchtungslehre und zur Theorie des Kerns. Jena: Gustav Fischer; 1907.
[9] Malpighi M. Dissertatio epistolica de formatione pulli in ovo [Epistological dissertation on the formation of the chick in the egg]. London: Apud Joannem Martyn; 1673.
[10] Malpighi M. Opera Omnia. London: apud Robertum Scott; 1686.
[11] Longabaugh WJ, Davidson EH, Bolouri H. Computational representation of developmental genetic regulatory networks. Dev Biol 2005;283:1–16.
[12] Longabaugh WJ, Davidson EH, Bolouri H. Visualization, documentation, analysis, and communication of large-scale gene regulatory networks. Biochim Biophys Acta 2009;1789:363–74.
[13] Davidson EH, Rast JP, Oliveri P, Ransick A, Calestani C, Yuh CH, et al. A genomic regulatory network for development. Science 2002;295:1669–78.
[14] Oliveri P, Tu Q, Davidson EH. Global regulatory logic for specification of an embryonic cell lineage. Proc Natl Acad Sci USA 2008;105:5955–62.
[15] Peter IS, Davidson EH. A gene regulatory network controlling the embryonic specification of endoderm. Nature 2011;474:635–9.
[16] Braasch I, Schartl M, Volff JN. Evolution of pigment synthesis pathways by gene and genome duplication in fish. BMC Evol Biol 2007;7:74.
[17] Ilagan JG, Cvekl A, Kantorow M, Piatigorsky J, Sax CM. Regulation of alphaA-crystallin gene expression. Lens specificity achieved through the differential placement of similar transcriptional control elements in mouse and chicken. J Biol Chem 1999;274:19973–8.
[18] Cvekl A, Piatigorsky J. Lens development and crystallin gene expression: many roles for Pax-6. Bioessays 1996;18:621–30.
[19] Davidson EH. The Regulatory Genome. Gene Regulatory Networks in Development and Evolution. San Diego, CA: Academic Press/Elsevier; 2006.
[20] Davidson EH, Erwin DH. An integrated view of precambrian eumetazoan evolution. Cold Spring Harb Symp Quant Biol 2009; 74:65–80.
[21] Peter IS, Davidson EH. Modularity and design principles in the sea urchin embryo gene regulatory network. FEBS Letters 2009; 583:3948–58.
[22] Peter IS, Davidson EH. Evolution of gene regulatory networks controlling body plan development. Cell 2011;144:970–85.
[23] Davidson EH. How embryos work: a comparative view of diverse modes of cell fate specification. Development 1990;108:365–89.
[24] Davidson EH. Emerging properties of animal gene regulatory networks. Nature 2010;468:911–20.
[25] Barolo S, Posakony JW. Three habits of highly effective signaling pathways: principles of transcriptional control by developmental cell signaling. Genes Dev 2002;16:1167–81.
[26] Saka Y, Smith JC. A mechanism for the sharp transition of morphogen gradient interpretation in *Xenopus*. BMC Dev Biol 2007;7:47.
[27] Laslo P, Spooner CJ, Warmflash A, Lancki DW, Lee HJ, Sciammas R, et al. Multilineage transcriptional priming and determination of alternate hematopoietic cell fates. Cell 2006;126:755–66.
[28] Narula J, Smith AM, Gottgens B, Igoshin OA. Modeling reveals bistability and low-pass filtering in the network module determining blood stem cell fate. PLoS Comput Biol 2010;6:e1000771.
[29] Goentoro L, Shoval O, Kirschner MW, Alon U. The incoherent feedforward loop can provide fold-change detection in gene regulation. Mol Cell 2009;36:894–9.

[30] Spooner CJ, Cheng JX, Pujadas E, Laslo P, Singh H. A recurrent network involving the transcription factors PU.1 and Gfi1 orchestrates innate and adaptive immune cell fates. Immunity 2009;31:576–86.

[31] Chickarmane V, Enver T, Peterson C. Computational modeling of the hematopoietic erythroid-myeloid switch reveals insights into cooperativity, priming, and irreversibility. PLoS Comput Biol 2009;5:e1000268.

[32] Davidson EH. Evolutionary bioscience as regulatory systems biology. Dev Biol 2011;357:35–40.

[33] Bolouri H, Davidson EH. The gene regulatory network basis of the 'community effect,' and analysis of a sea urchin embryo example. Dev Biol 2010;340:170–8.

[34] Pimanda JE, Ottersbach K, Knezevic K, Kinston S, Chan WY, Wilson NK, et al. Gata2, Fli1, and Scl form a recursively wired gene-regulatory circuit during early hematopoietic development. Proc Natl Acad Sci USA 2007;104:17692–7.

[35] Bolouri H. Computational Modeling of Gene Regulatory Networks. London: Imperial College Press; 2008.

[36] Macarthur BD, Ma'ayan A, Lemischka IR. Systems biology of stem cell fate and cellular reprogramming. Nat Rev Mol Cell Biol 2009;10:672–81.

[37] Kirouac DC, Ito C, Csaszar E, Roch A, Yu M, Sykes EA, et al. Dynamic interaction networks in a hierarchically organized tissue. Mol Syst Biol 2010;6:417.

[38] Jaeger J, Blagov M, Kosman D, Kozlov KN, Manu, Myasnikova E, et al. Dynamical analysis of regulatory interactions in the gap gene system of *Drosophila melanogaster*. Genetics 2004;167:1721–37.

[39] Perkins TJ, Jaeger J, Reinitz J, Glass L. Reverse engineering the gap gene network of *Drosophila melanogaster*. PLoS Comput Biol 2006;2:e51.

[40] Dessaud E, Ribes V, Balaskas N, Yang LL, Pierani A, Kicheva A, et al. Dynamic assignment and maintenance of positional identity in the ventral neural tube by the morphogen sonic hedgehog. PLoS Biol 2010;8:e1000382.

[41] Ribes V, Balaskas N, Sasai N, Cruz C, Dessaud E, Cayuso J, et al. Distinct Sonic Hedgehog signaling dynamics specify floor plate and ventral neuronal progenitors in the vertebrate neural tube. Genes Dev 2010;24:1186–200.

[42] Nahmad M, Stathopoulos A. Dynamic interpretation of hedgehog signaling in the *Drosophila* wing disc. PLoS biology 2009;7:e1000202.

[43] Bolouri H, Davidson EH. Transcriptional regulatory cascades in development: initial rates, not steady state, determine network kinetics. Proc Natl Acad Sci USA 2003;100:9371–6.

[44] Shea MA, Ackers GK. The OR control system of bacteriophage lambda. A physical-chemical model for gene regulation. J Mol Biol 1985;181:211–30.

[45] Materna SC, Nam J, Davidson EH. High accuracy, high-resolution prevalence measurement for the majority of locally expressed regulatory genes in early sea urchin development. Gene Expr Patterns 2010;10:177–84.

[46] Howard-Ashby M, Materna SC, Brown CT, Chen L, Cameron RA, Davidson EH. Identification and characterization of homeobox transcription factor genes in *Strongylocentrotus purpuratus*, and their expression in embryonic development. Dev Biol 2006;300:74–89.

[47] Howard-Ashby M, Materna SC, Brown CT, Chen L, Cameron RA, Davidson EH. Gene families encoding transcription factors expressed in early development of *Strongylocentrotus purpuratus*. Dev Biol 2006;300:90–107.

[48] Howard-Ashby M, Materna SC, Brown CT, Tu Q, Oliveri P, Cameron RA, et al. High regulatory gene use in sea urchin embryogenesis: implications for bilaterian development and evolution. Dev Biol 2006;300:27–34.

[49] Materna SC, Howard-Ashby M, Gray RF, Davidson EH. The C2H2 zinc finger genes of *Strongylocentrotus purpuratus* and their expression in embryonic development. Dev Biol 2006;300: 108–20.

[50] Rizzo F, Fernandez-Serra M, Squarzoni P, Archimandritis A, Arnone MI. Identification and developmental expression of the ets gene family in the sea urchin (*Strongylocentrotus purpuratus*). Dev Biol 2006;300:35–48.

[51] Tu Q, Brown CT, Davidson EH, Oliveri P. Sea urchin Forkhead gene family: phylogeny and embryonic expression. Dev Biol 2006;300:49–62.

[52] Ben-Tabou de-Leon SB, Davidson EH. Information processing at the foxa node of the sea urchin endomesoderm specification network. Proc Natl Acad Sci USA 2010;107:10103–8.

[53] Peter IS, Davidson EH. The endoderm gene regulatory network in sea urchin embryos up to mid-blastula stage. Dev Biol 2010;340:188–99.

[54] Ransick A, Davidson EH. *Cis*-regulatory processing of Notch signaling input to the sea urchin glial cells missing gene during mesoderm specification. Dev Biol 2006;297:587–602.

[55] Smith J, Kraemer E, Liu H, Theodoris C, Davidson E. A spatially dynamic cohort of regulatory genes in the endomesodermal gene network of the sea urchin embryo. Dev Biol 2008;313:863–75.

[56] Wahl ME, Hahn J, Gora K, Davidson EH, Oliveri P. The *cis*-regulatory system of the tbrain gene: alternative use of multiple modules to promote skeletogenic expression in the sea urchin embryo. Dev Biol 2009;335:428–41.

[57] Cameron RA, Fraser SE, Britten RJ, Davidson EH. Macromere cell fates during sea urchin development. Development 1991;113: 1085–91.

[58] Ransick A, Davidson EH. Late specification of Veg1 lineages to endodermal fate in the sea urchin embryo. Dev Biol 1998;195: 38–48.

[59] Oliveri P, Davidson EH, McClay DR. Activation of pmar1 controls specification of micromeres in the sea urchin embryo. Dev Biol 2003;258:32–43.

[60] Leonard JD, Ettensohn CA. Analysis of dishevelled localization and function in the early sea urchin embryo. Dev Biol 2007;306:50–65.

[61] Weitzel HE, Illies MR, Byrum CA, Xu R, Wikramanayake AH, Ettensohn CA. Differential stability of beta-catenin along the animal-vegetal axis of the sea urchin embryo mediated by dishevelled. Development 2004;131:2947–56.

[62] Logan CY, Miller JR, Ferkowicz MJ, McClay DR. Nuclear beta-catenin is required to specify vegetal cell fates in the sea urchin embryo. Development 1999;126:345–57.

[63] Chuang CK, Wikramanayake AH, Mao CA, Li X, Klein WH. Transient appearance of *Strongylocentrotus purpuratus* Otx in micromere nuclei: cytoplasmic retention of SpOtx possibly mediated through an alpha-actinin interaction. Dev Genet 1996;19:231–7.

[64] Revilla-i-Domingo R, Oliveri P, Davidson EH. A missing link in the sea urchin embryo gene regulatory network: hesC and the double-negative specification of micromeres. Proc Natl Acad Sci USA 2007;104:12383–8.

[65] Revilla-i-Domingo R, Minokawa T, Davidson EH. R11: a cis-regulatory node of the sea urchin embryo gene network that controls early expression of SpDelta in micromeres. Dev Biol 2004;274:438–51.

[66] Smith J, Davidson EH. Gene regulatory network subcircuit controlling a dynamic spatial pattern of signaling in the sea urchin embryo. Proc Natl Acad Sci USA 2008;105:20089–94.

[67] Sweet HC, Gehring M, Ettensohn CA. LvDelta is a mesoderm-inducing signal in the sea urchin embryo and can endow blastomeres with organizer-like properties. Development 2002;129:1945–55.

[68] Croce JC, McClay DR. Dynamics of Delta/Notch signaling on endomesoderm segregation in the sea urchin embryo. Development 2010;137:83–91.

[69] Croce J, Range R, Wu SY, Miranda E, Lhomond G, Peng JC, et al. Wnt6 activates endoderm in the sea urchin gene regulatory network. Development 2011;138:3297–306.

[70] Wikramanayake AH, Peterson R, Chen J, Huang L, Bince JM, McClay DR, et al. Nuclear beta-catenin-dependent Wnt8 signaling in vegetal cells of the early sea urchin embryo regulates gastrulation and differentiation of endoderm and mesodermal cell lineages. Genesis 2004;39:194–205.

[71] Duloquin L, Lhomond G, Gache C. Localized VEGF signaling from ectoderm to mesenchyme cells controls morphogenesis of the sea urchin embryo skeleton. Development 2007;134:2293–302.

[72] Ransick A, Davidson EH. *Cis*-regulatory logic driving glial cells missing: self-sustaining circuitry in later embryogenesis. Dev Biol 2012;364:259–67.

[73] Sanchez L, Chaouiya C, Thieffry D. Segmenting the fly embryo: logical analysis of the role of the segment polarity cross-regulatory module. Int J Dev Biol 2008;52:1059–75.

[74] Bonn S, Furlong EE. *cis*-Regulatory networks during development: a view of Drosophila. Curr Opin Genet Dev 2008;18:513–20.

Chapter 12

Reconstruction of Genome-Scale Metabolic Networks

Hooman Hefzi,[1] Bernhard O. Palsson[1] and Nathan E. Lewis[1,2]
[1]*Bioengineering Department, University of California San Diego, La Jolla, CA, USA,* [2]*Department of Genetics, Harvard Medical School, Boston, MA 02115, USA*

Chapter Outline

Introduction	**229**
Cellular Metabolism	229
Metabolic Reconstructions	230
The Reconstruction Process	**230**
E. coli Core Metabolism	230
Stage 1: Creation of a Draft Reconstruction	230
Stage 2: Manual Curation	231
Additional Information and Verification	231
Incorporation of Spontaneous and Transport Reactions	232
The Biomass 'Reaction'	233
Stage 3: Converting a Network to a Mathematical Model	233
Stage 4: Network Evaluation	234
Topological Tests	234
Thermodynamic Tests	235
Phenotypic Tests	236
Quantitative Tests	236

Stage 5: Data Assembly, Dissemination, and Use	237
Reconstruction Standards	**237**
Applications	**237**
Metabolic Engineering	237
Biological Discovery	239
Phenotype Prediction and Evaluation	239
Fundamental Properties of Biological Networks	240
A Context for the Analysis of Large Data Sets	241
Multi-Cell and Microbial Community Metabolism	241
Future Directions	**242**
E-Matrix Reconstructions	242
O-Matrix Reconstructions	243
Toward a Whole Cell Model	244
Glossary	**244**
Acknowledgments	**245**
References	**245**

INTRODUCTION

Metabolism and enzymes are the classic topics of discussion in any undergraduate biochemistry course. Such textbooks are full of protein structures and descriptions of the topology and properties of short pathways. However, a living cell contains thousands of enzymes that do not act in isolation. In like manner, pathways in a cell often have complexity far beyond the simple organization shown in a textbook. Metabolic reconstructions aim to move beyond static pictures of protein structures and pathway concepts. This is done when they are used to model the functions of all enzymes in a cell, thus allowing us to understand how these parts and their interactions function together as a whole to make a living organism.

Cellular Metabolism

Cellular metabolism is organized around a large network of enzyme-catalyzed and spontaneous chemical reactions. These reactions involve a set of diverse metabolites, and all of the reactions function together to produce and recycle the materials for cell maintenance and growth (e.g., amino acids, lipids, ATP), signaling molecules (e.g., cAMP and H_2S), waste products, and molecules that affect the growth of surrounding organisms (e.g., antibiotics, quorum-sensing molecules, hormones, etc.). Metabolism is important in most cell and organism phenotypes. For example, it has been implicated as a contributing factor of human diseases such as diabetes [1] and cancer [2], making it an auspicious area for research. However, the size of these metabolic networks has made the study of their functions difficult, since they contain hundreds or thousands of reactions. Thus, systems-level methods have been developed to gain insight into how thousands of cellular components function together as a whole to provide the phenotypes we observe in living organisms.

Metabolic Reconstructions

Metabolic reconstructions are knowledge bases containing detailed and organized descriptive information for each enzyme in an organism. This information includes the stoichiometry of substrates and products of the reaction it catalyzes, reaction reversibility, and reaction localization [3]. This information can readily be found in detailed biochemical studies and inferred from annotated genomes. After it is organized in a knowledge base, it can be represented as a stoichiometric matrix (which we call here the M-matrix, for metabolism), in which each reaction corresponds to a column and each metabolite to a row.

Metabolic network reconstructions are a relatively modern development that arose in our efforts to understand the biochemical processes underlying cell functionality. Non-genome-scale reconstructions were first developed in the early 1980s [4,5]. A desire to understand these networks on a cell-wide level led to the first genome-scale metabolic reconstruction in 1999 [6]. Recent developments now allow the partial automation of the reconstruction process for genome-scale metabolic networks (e.g., via modelSEED [7] and the SuBliMinaL toolbox [8]), yielding an ever-growing number of metabolic reconstructions [3]. Here, we present an introduction to how these networks are constructed and their potential applications, as well as future directions in the field.

THE RECONSTRUCTION PROCESS

The process for constructing a metabolic reconstruction has been established (Figure 12.1), and a standardized protocol has been developed to assure the publication of high-quality genome-scale reconstructions [9]. Here, we highlight key steps and provide a simplified example of the reconstruction process for core *Escherichia coli* metabolism.

E. coli Core Metabolism

The core metabolic reconstruction of *E. coli* [10] consists of 95 reactions that catalyze transformations between 72 metabolites and accounts for glycolysis, the TCA cycle, the pentose phosphate shunt, a simplified version of oxidative phosphorylation and nitrogen metabolism, as well as secondary pathways such as gluconeogenesis and

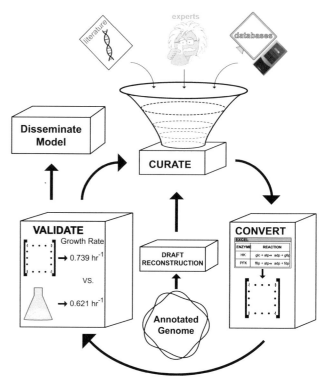

FIGURE 12.1 **Overview of the reconstruction process.** Curation, conversion, and validation are carried out iteratively until the model's predictive capabilities are sufficient for the scope of the reconstruction.

anapleurotic reactions (Figure 12.2). The network can be obtained from http://bigg.ucsd.edu

Stage 1: Creation of a Draft Reconstruction

The process begins with the creation of a preliminary draft reconstruction. An annotated genome for the organism of interest is obtained and queried to identify information such as genomic coordinates, names of loci, other accepted names, known functions of gene products, and known splice isoforms. Many of these items are catalogued in databases and tools such as Pathway tools [11], metaSHARK [12], KEGG [13], modelSEED [7], and others [14–16], as demonstrated in Box 12.1. Thus the acquisition of data for the draft reconstruction is amenable to automation [7,8].

For metabolic reconstructions, genes with metabolic functions need to be identified. Various approaches are commonly used, including searching for Enzyme Commission numbers [17], metabolic terms (e.g., dehydrogenase or kinase), or metabolic Gene Ontology terms [18]. Finally, reactions for each of the metabolic genes must be established. Again, multiple databases contain such information, including KEGG [13], Reactome [19], and BRENDA [20]. The metabolic genes and their associated reactions are collected and organized in preparation for stage 2, in which the network is manually curated.

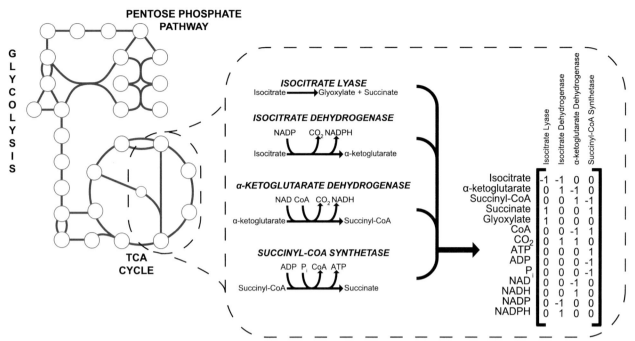

FIGURE 12.2 **Formulating the stoichiometric matrix.** A simplified schematic shows a few pathways present in the core metabolic reconstruction of E. coli. (Inset) The stoichiometric matrix is formulated for four reactions from the TCA cycle and glyoxylate shunt. Each reaction is represented as a column, each metabolite as a row. Each element in the matrix corresponds to a stoichiometric coefficient of a metabolite in a given reaction, with consumption denoted by a negative value and production by a positive value.

BOX 12.1 A Draft Reconstruction for *E. coli*'s Core Metabolism

For *E. coli*, specific databases such as EcoCyc [21] can be used to obtain a genome annotation. Some examples of the data gathered in this step are presented below in:

Gene alias	Locus name	EcoCyc function	EC number	Reaction
glk	b2388	Glucokinase	2.7.1.2	Glucose + ATP → Glucose-6-phosphate + ADP
pgi	b4025	Glucose-6-phosphate Isomerase	5.3.1.9	Glucose-6-phosphate ↔ Fructose-6-phosphate

Information for reactions in glycolysis, the TCA cycle, the pentose phosphate pathway is obtained from the annotation. When this was done for *E. coli* core metabolism, the reconstruction comprised 53 reactions.

Stage 2: Manual Curation

For most organisms, genome annotation is done primarily through homology methods. Therefore, reconstructions based solely on genome annotation may have many incorrect enzymatic activities, and will be missing reactions for which the associated enzymes were missed in the annotation process. Therefore, great care is taken to ensure that the reconstruction is accurate and complete for the organism of interest — i.e. efforts are made to verify that all reactions and genes included are actually present in the organism and that all known reactions and genes in the organism are included in the reconstruction. In addition, the cellular composition is determined. That is, the amounts of metabolites needed for cell growth and maintenance are determined. For example, the total amounts of proteins, mRNA, DNA, lipids, etc. are measured. Much of this information is organism specific. Thus the primary resources in this stage include either new experimental measurements or organism-specific databases (e.g., EcoCyc [21], AraCyc [22], SGD [23], etc.), textbooks, publications, and experts.

Additional Information and Verification

In the manual curation process, various pieces of information are determined and verified for each reaction and metabolite. For each reaction, the substrate and cofactor usage is verified using organism-specific literature when

BOX 12.2 Cofactor Specificity

Organism specificity is always a concern during manual curation. For example, the isocitrate dehydrogenase (IDH) reaction in *E. coli*'s core metabolism uses NADP/NADPH as cofactors rather than the NAD/NADH more commonly associated with the reaction [181], as a result of evolutionary adaption to permit growth on acetate. In fact, organisms with an isocitrate lyase gene (essential for growth on acetate), will always use NADP for the IDH reaction.

BOX 12.4 Transport Reactions

Experimentally, *E. coli* (among others) is known to have glucose and fructose import capabilities and export capabilities for metabolites such as lactate and ethanol. Thus, transport reactions are added for these metabolites. Twenty-one transport reactions (including diffusion processes for H_2O and CO_2) have been added to the *E. coli* core metabolic network, thereby increasing the total reaction count to 74.

possible, since cofactor specificity may vary between organisms (Box 12.2). If literature is unavailable, databases such as KEGG [13] and BRENDA [20] can be used. The localization of the reaction (i.e., the compartment in which each reaction occurs) can be determined using algorithms such as PSORT [24] or PASUB [25] if literature or experimental data is unavailable; if the algorithms are indeterminate, enzymes are often assumed to be in the cytosol. Care should be taken in assigning the correct localization for each reaction, since localization affects metabolic network functionality as in vivo metabolite availability may be limited due to transport costs. Accurate compartmentalization of reactions has been shown to increase the predictive power of metabolic reconstructions [26]. The subsystem (e.g., glycolysis or pyrimidine metabolism) for each reaction is also often established based on textbook or KEGG assignments [13]. Finally the gene–protein–reaction (GPR) association for each reaction (i.e., the flow from gene to enzyme to reaction catalyzed) should be found (Box 12.3). These are described as Boolean relationships between genes or proteins, and include information on isozymes, secondary or promiscuous enzyme activities, and protein complex participation. Primary literature and organism-specific databases are rich sources for this information. Moreover, careful curation of this information can greatly affect the reliability of simulations, such as gene deletion simulations.

In a similar manner, each metabolite must be carefully curated. For each metabolite the formula should be found.

BOX 12.3 GPR Associations

Data for GPR associations in *E. coli* are readily available. For example, *pykA* and *pykF* both encode proteins (PykA and PykF, respectively) that can catalyze the pyruvate kinase reaction (thus *pykA* OR *pykF* are required for the reaction). Similarly, *sdhA*, *sdhB*, *sdhC*, and *sdhD* are required for formation of the protein Sdh, which catalyzes the irreversible succinate dehydrogenase reaction in oxidative phosphorylation (thus *sdhA* AND *sdhB* AND *sdhC* AND *sdhD* are required for the reaction).

Multiple sources exist, e.g. KEGG [13], BRENDA [20], or PubChem [27]. It is preferable that multiple sources be used to ensure that the formula is accurate, since databases can vary in reported molecular formulae (e.g., protonation state). From these formulae the charged formula is usually determined, since physiological pH often results in non-neutral compounds (e.g., deprotonation of many acids). To aid in this, various software packages have been developed (e.g., Pipeline Pilot, ChemAxon's pK_a plugin, or ACD/Labs pK_a DB). Once all metabolite formulae are established, tests must be conducted to verify that all reactions are mass and charge balanced. Many reactions require the addition of water and protons to balance reaction oxygen and hydrogen. Any mistakes in mass balance (even with protons) can block normally functioning pathways or predict infinite ATP generation.

After reaction stoichiometry, gene association and metabolite information is verified, a confidence score is often provided as a semi-quantitative assessment of reaction quality. For example, reactions in many reconstructions are assigned a confidence score (from 0 to 4, 4 being the best) that reflects the amount of information available for the reaction. Guidelines for assigning confidence scores have been previously suggested [9].

Incorporation of Spontaneous and Transport Reactions

Not all reactions occurring in a cell are enzyme catalyzed, and some hydrophobic metabolites readily diffuse through cell membranes. Thus, spontaneous reactions should be added to the network to reflect these processes. Many such reactions are catalogued in databases such as KEGG [13] and BRENDA [20], or are characterized in the literature. Transport reactions (i.e., the transport of a metabolite across a membrane) can be either spontaneous or enzyme facilitated, depending on each individual metabolite. If evidence suggests that certain metabolites can be taken up from or secreted into the medium, or transported between cellular compartments, the associated transport reactions should be added accordingly.

Often biochemical and physiological data for transport reactions may not be available. In such cases, educated assumptions can be made based on similar transport reactions, and pathway-based algorithms can also be used to suggest potential transport reactions [28]. However, for any of these predictions, the lowest confidence score is assigned to the reaction as well as a note explaining why the transport reaction was added with a specific stoichiometry.

The Biomass 'Reaction'

The biomass objective function [29] or 'reaction' allows a metabolic network model to simulate growth. The biomass reaction is a mathematical representation of the molar amounts of various metabolites needed to make all cellular components for growth (Box 12.5). Therefore, in order to accurately model growth, it is necessary to know the overall composition (e.g., protein, lipids, etc.) of the cell and incorporate this information into the biomass reaction. Such information can be gathered from primary literature [30] or determined experimentally [31–34]. The amino acid and nucleic acid contributions can be measured or estimated from the genome, e.g., using the CMR database [35]. Lipid content is often inferred from experimental data by calculating the average molecular weight of fatty acid chains and using this to compute phospholipid contributions to biomass for the three major phospholipid groups [9]. Additionally, the content of the soluble pool (polyamines, vitamins, and cofactors) as well as the ion content of the cell needs to be determined and incorporated. Finally, the amount of energy the cell requires for replication, also known as the growth-associated maintenance (GAM), needs to be experimentally determined and incorporated into the biomass function as follows:

$$x\,ATP + x\,H_2O \rightarrow x\,ADP + x\,P_i + x\,H^+$$

Since GAM is the energy needed by the cell to replicate, x accounts for all the ATP requirements for macromolecule synthesis and other growth-related processes. The inclusion of GAM will result in a biomass function of the following general form:

$$dNTPs + AAs + lipid\ precursors + ions + soluble\ pool$$
$$+ ATP + H_2O \rightarrow ADP + P_i + H^+$$

with the stoichiometric coefficients being organism specific. Non-growth-associated maintenance (NGAM) is a measure of the energy needed for maintaining homeostasis (e.g., turgor pressure [36]) in the cell and is included as a separate reaction in the network with the form:

$$1\,ATP + 1\,H_2O \rightarrow 1\,ADP + 1\,P_i + 1\,H+$$

The flux through the NGAM reaction must be experimentally determined and constrained to the appropriate value when using the reconstruction for modeling.

Stage 3: Converting a Network to a Mathematical Model

To be used as a predictive tool, a reconstruction must be converted into a model. A reconstruction consists of a list of the reactions taking place within the organism and detailed information about each reaction and metabolite. However, a metabolic model is the computable form of the reconstruction. Therefore, to be a model, the network must be (1) described mathematically and (2) provide system boundaries with constraints on network inputs and outputs.

To describe the reconstruction mathematically, each reaction gathered during the reconstruction is included as a column in a stoichiometric matrix, **S**. Each metabolite in the network is included as a row. The elements of this matrix represent the stoichiometric coefficients of each metabolite in a given reaction. Metabolites that are consumed in a reaction have negative values and those produced have positive values. For example if in reaction j, metabolite i is consumed and has a stoichiometric coefficient of 1, then $S_{ij} = -1$ (Figure 12.2, inset).

After the system boundaries are set, extracellular metabolites that can be obtained from the media or secreted are determined and set as inputs or outputs of the model, respectively. For modeling purposes, the uptake or secretion of these extracellular metabolites is facilitated by using exchange reactions. These reactions allow the metabolites to enter or leave the system, thus, simulating their availability in the environment or their excretion (Box 12.6). Constraints can be subsequently placed on exchange reactions to provide upper limits on the uptake or secretion rates of their associate metabolites.

BOX 12.5 Biomass

Many important biomass components (e.g., nucleic acids) are not explicitly accounted for in core metabolism, so their precursors (e.g., pyruvate and L-glutamate for L-alanine) were substituted in the biomass reaction. The inclusion of the biomass reaction increases the reaction count of the reconstruction to 75.

BOX 12.6 Network Conversion and Exchange Reactions

A visualization of the conversion from reconstruction to stoichiometric matrix is seen in Figure 12.2, inset for a subset of the TCA cycle. An additional 20 reactions were added as exchange reactions to simulate known uptake and/or secretion capabilities of E. coli (e.g., glucose, ethanol, and lactate). This brings the final reaction count of the model to 95.

Additional pseudo-reactions can be specified as well. These include non-reversible demand reactions that remove a metabolite from the network. These are used for compounds that are known to be produced for processes beyond the scope of the model. In like manner, reversible sink reactions can be added for metabolites that are produced by reactions outside the scope of the reconstruction but are necessary for metabolic network function. Great care is taken with the addition of sink reactions, since their addition may allow the model to grow without using the metabolites in the medium. A visualization of these reactions is shown in Figure 12.3.

Once a network is converted into a mathematical model, the model can be used to simulate how different pathways are used under specific conditions. Classically, kinetic information has been used to model small-scale metabolic networks for which kinetic parameters are available. However, such information is lacking for most enzymes. Thus, large metabolic network models require simplifying assumptions and alternative approaches to be amenable to simulation. Flux-balance analysis (FBA) is one such method for computational modeling of these networks [37]. This technique assumes that the reaction rates (fluxes) within a cell can reach a steady state represented mathematically as $S \cdot v = 0$, where S is the stoichiometric matrix and v is the flux vector in which each reaction j has a flux value v_j. In this steady state, the requirements of mass-balance in chemical reactions, combined with constraints on specific fluxes (e.g., flux of input exchange reactions) allow for the prediction of reaction fluxes without requiring kinetic information.

Metabolic flux is often difficult to measure. This is particularly true for non-central metabolic pathways. Thus, the ability to predict flux for all pathways is particularly valuable. The S matrix for a genome-scale model is underdetermined (i.e., fewer metabolites than reactions). This means that genome-scale models will not provide a unique solution specifying the flux of each reaction in the network. However, they instead provide a range of feasible solutions, often referred to as the 'solution space'. This solution space contains all feasible steady-state flux values for each reaction, subject to the metabolic network stoichiometry and any user-inputted constraints (e.g., cellular metabolite consumption rates, product secretion rates, measured internal flux constraints, etc.). In FBA, linear programming is used to identify the fluxes for each reaction by optimizing some objective function (e.g., biomass precursor production). Figure 12.4 shows the procedure in more detail.

Stage 4: Network Evaluation

Once a reconstruction is converted to a model, the model is analyzed and used for simulation. Simulation results are validated against experimentally measured phenotypes. This is done to provide confidence that subsequent model predictions are accurate. A variety of tests should be used to increase model quality. Following each test, corrections to the reconstruction are made as needed through iterative rounds of manual curation, conversion to a model, and network evaluation. This process continues until the reconstruction is 'finished,' as determined by the ability of the model to carry out the purpose and scope desired for subsequent studies.

Topological Tests

The topology and stoichiometry of a metabolic network clearly affect model functions. Thus, the first validation done on a model involves verifying the mass balance of all reactions within the metabolic network (except for exchange, sink, demand, or biomass reactions). This is often followed by an assessment of metabolic dead ends, which are metabolites that are only produced or consumed. The identification of these can help identify gaps, which are model reactions that may be missing. Primary literature and genome annotation are used to identify potential genes and reactions present in the organism that would bridge identified gaps. Visualization of the metabolite's 'location' in the metabolic network can be helpful for identifying potential reactions that fill gaps in the network. Resources such as the KEGG database [13] or biochemical textbooks/maps can contain this information. Computational algorithms can aid in this process by suggesting missing reactions, which must be followed by experimental validation [38–40]. Care should be taken to ensure that gaps are not inappropriately

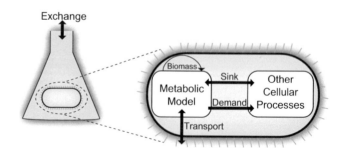

FIGURE 12.3 **A simplified representation of the various reaction types present in metabolic models.** Exchange reactions allow the flow of metabolites into or out of the medium, transport reactions allow the flow of metabolites into and out of the cell (or between intracellular compartments). Demand and sink reactions are for metabolites that are consumed or produced, respectively, by reactions outside the scope of the metabolic model.

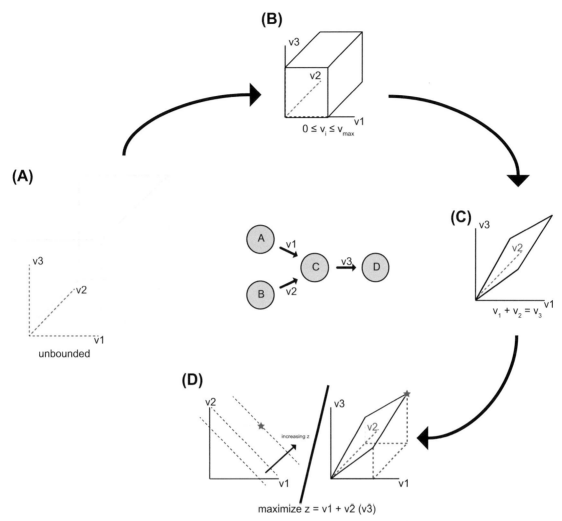

FIGURE 12.4 **The successive addition of constraints on a three reaction system.** By adding constraints, the size of the solution space decreases. In this solution space, methods such as flux balance analysis (FBA) can be used to find putative flux values for each reaction. (**A**) Without any bounds on the reaction fluxes (v1, v2, and v3 respectively), the solution space (blue) is unbounded. (**B**) Bounds are placed on the solution space by defining the limitations of each reaction flux based on measured substrate uptake and enzyme capacity limits. (**C**) The imposition of mass balance constraints limits the solution space even further. (**D**) By applying linear programming (LP) to optimize a cellular objective (in this case maximizing the sum of v1 and v2, which corresponds to v3), the optimal flux levels can be computed. In genome-scale models, optimization is often carried out for the biomass function.

filled (e.g., by adding in a gene that changes an auxotroph to an autotroph). However, reactions are occasionally added when detailed data are not available, only when it is necessary for network functionality (e.g., biomass precursor synthesis). These are subsequently flagged with a low confidence score. An appropriate treatment of model gaps can greatly improve the accuracy of model predictions.

Thermodynamic Tests

Once topological assessment is completed, efforts are turned toward making the model as physiologically accurate as possible. One concern with constraint-based models is that the simplification of thermodynamic constraints (i.e., reactions are only declared as reversible or irreversible) leads to the possibility of non-physiologically accurate cycles occurring in which no material is lost to other reactions outside the cycle. These are often called type III extreme pathways [41] or 'loops' and violate true thermodynamics [42]. Multiple methods have been developed to identify loops and remove them from a model [43–45]. If loops exist, reaction directionality can be constrained to break them if experimental data support the proposed changes to reaction directionality. Otherwise, iterative limitation of the directionality of pathway reactions can prove effective [42]. Alternative methods for removing thermodynamically infeasible loops have recently been developed [46–48]. However, when removing loops, care must be taken to avoid reaction modifications that introduce new pathway gaps or that affect the model's accuracy in replicating physiological data (e.g., growth rate or secretion products, as discussed below).

Phenotypic Tests

Once topological concerns and loops have been addressed, qualitative tests are conducted to assure that the model can make important metabolites and to identify which reactions cannot carry flux.

One important test of a network reconstruction is its ability to produce all essential biomass components and experimentally measured secretion products. When FBA predicts zero growth or no product secretion, non-producible metabolites are identified by using linear programming to maximize flux through demand reactions for each individual secretion product or reactant in the biomass function. Metabolites that cannot be synthesized according to the model are remedied through additional gap-filling, in which additional reactions are suggested that would allow synthesis of the metabolite to occur [9]. Since this process is sensitive to the choice of model growth medium, it is repeated multiple times for different in silico growth conditions.

Just as some biomass and secretion metabolites might not be produced, it is common for some reactions to be unable to carry flux. These reactions involve metabolites that are solely produced or consumed or that depend on other non-functional reactions. These can arise due either to knowledge gaps (when the connecting reactions are unknown) or to scope gaps if subsequent metabolic steps extend into a system outside the scope of the model. For knowledge gaps, gap-filling is done by adding the missing reaction, or a proposed reaction if none is known. However, scope gaps are better addressed with the use of sink and demand reactions, if doing so improves model accuracy for its intended purpose.

Once gap-filling is done, the comparison of phenotypic screens [49] with in silico gene deletion studies allow for a detailed assessment of the accuracy of functional reactions within the network. In this, single-gene deletion phenotypes (e.g., growth rate) are computed to validate the model in a holistic sense. When model-predicted growth differs from experimental growth phenotypes, these failures guide further curation. If the model predicts growth of a gene deletion mutant, whereas experimental data show the gene deletion to be lethal, then it is possible that a reaction has been improperly incorporated, that a gene associated with another reaction is silenced, or that the biomass reaction is incomplete. If, however, the model does not grow despite measured experimental growth, it is likely that reactions are missing from the reconstruction, and therefore gap-filling is needed [38]. It is important to note that the lack of regulatory rules (e.g., transcription factor activation/repression) in the model can also result in improper predictions. For example, growth in a medium with two carbon sources in it will not proceed in a diauxic [50] fashion as expected experimentally. Rather, the model will consume both carbon sources simultaneously and falsely predict a greater growth rate than would be expected experimentally.

Quantitative Tests

The ultimate goal of a genome-scale reconstruction is to be able to use it to make clear quantitative predictions that yield novel insight and knowledge. Therefore, it is critical to test the model to ensure its quantitative accuracy in predicting physiological parameters. Such quantitative tests may include predicting growth rate [51] or P/O ratio, which is a measure of ATP generated per electron pair used in oxidative phosphorylation [52].

For growth rate predictions, the biomass function is optimized under a variety of media conditions and subsequently compared with experimentally measured growth rates. Incorrect in silico predictions may include (1) no growth when growth is expected, (2) slower growth than physiologically expected, (3) faster growth than physiologically expected, or (4) growth when no growth is expected. These different scenarios have several possible explanations, and the appropriate steps to take for each scenario are discussed below.

In the first scenario, when the model predicts no growth while experimental assays demonstrate growth, it means that the network is incomplete (Box 12.7). Therefore, the reconstruction requires additional gap-filling and the accuracy of constraints (e.g., metabolite uptake or biomass function composition) needs to be verified.

> **BOX 12.7 Gap-Filling in the Core *E. coli* Model**
>
> Consider the in silico gene deletion of b3919, which codes for the triose phosphate isomerase protein that participates in glycolysis. Simulation of the knockout using the core metabolic model predicts that the deletion of this gene is lethal. However, experimentally, this mutant is known to exhibit growth. Investigation into the cause behind the false prediction shows that dihydroxyacetone phosphate (DHAP) can be produced but not consumed. Thus, the model predicts that the mutation is lethal because DHAP would accumulate. This fact suggests that there is possibly a missing reaction that would consume DHAP and prevent its accumulation in the knockout. In this case, the reactions necessary to account for DHAP utilization (e.g., methylglyoxal synthase) were not included in the reconstruction because they fell outside the scope of the reconstruction target of core metabolism. Various methods of correction exist, but as the pathway needed is outside the scope of the reconstruction, addition of a demand reaction for DHAP may be an appropriate course of action. Simulation of this modified version of core metabolism is able to make the proper qualitative prediction of growth for the triose phosphate isomerase mutant.

In the second scenario, in silico growth rates are slower than experimental measurements. This can occur if one or more components are not being synthesized at a high enough rate. Thus, the limiting components are identified by iteratively supplementing the model with each biomass component and measuring its effect on the predicted growth rate. This approach provides insight into which pathways require more careful curation.

The third scenario, in which model-predicted growth rates are higher than experimentally measured rates, can be explained by several means. One possibility is that the cell may have additional objectives beyond the synthesis of biomass [53–60]. Thus, when the model assumes the optimization of growth, it overestimates the amount of material and energy equivalents used for biomass production. Another possibility is that GAM or NGAM may have been improperly calculated for that growth condition. Overestimation of growth might also result from improperly included reactions. It is also recommended to constrain directionality of reactions that consume ATP or use quinones as electron acceptors to prevent these reactions from running in reverse and improperly increasing biomass reaction components. Model reactions that facilitate the erroneous high growth rate may be identified by testing the sensitivity the predicted growth rate to changes in flux for each reaction. Lastly, it is also possible that regulatory mechanisms suppress gene expression or enzyme activity in vivo. Thus, the experimental data may represent a non-ideal phenotype. Previous adaptive laboratory evolution (ALE) experiments have demonstrated that after hundreds of generations of exponential growth, bacterial strains can adapt towards the model-predicted growth rate [34,61–63].

In the fourth scenario, the model predicts that the organism can grow when experiments say otherwise. This may result from the improper inclusion of reactions, lack of accounting of regulatory mechanisms suppressing growth in vivo, or the presence of sink reactions that allow growth without consuming the metabolites in the medium. Careful curation, comparison with expression data, and iterative testing following sink adjustments can help reconcile such model–phenotype inconsistencies.

Stage 5: Data Assembly, Dissemination, and Use

Once a network reconstruction is completed, it is frequently made available to the research community. Often all of the relevant information, curator comments, and citations are presented in a spreadsheet, and the model is released in SBML format [64] for use with an array of modeling programs [65–69]. Important modeling parameters should be presented in an organized manner, thereby allowing for the replication of the validation tests. Careful efforts of model preparation and dissemination will allow iterative improvements on the scope of the reconstruction, as has been done for the genome-scale model for *E. coli* (improving coverage from 660 [70] to 1366 [71] genes) and *Saccharomyces cerevisiae* [72–74]. Moreover, the model can be used for a plethora of applications [75], such as the examples discussed later in this chapter.

RECONSTRUCTION STANDARDS

So far we have discussed the general concepts to be considered in a network reconstruction. However, it is important to follow all of the detailed steps in the published reconstruction protocol [9] to ensure a high-quality reconstruction and accurate model predictions. Failing to carefully curate a reconstruction can lead to a host of errors, such as the incorporation of incorrect reactions into the reconstruction, the loss of reactions that were missed in the draft reconstruction due to incomplete genome annotation, and inaccurate simulations due to incorrect stoichiometry (e.g., if protons or water are not added to balance reactions, processes such as oxidative phosphorylation will be affected). Thus, without curation through organism-specific literature or experimental data, mistakes of this sort will be pervasive. In addition, failing to validate the reconstruction computationally will result in a model with low predictive power that may not be able to simulate known phenomena.

APPLICATIONS

Since their development, genome-scale metabolic models have been deployed for a plethora of different applications and analyses. Many applications easily fit within one of six categories (Figure 12.5) [75,76]. Each category is described here and relevant examples are discussed.

Metabolic Engineering

The genetic engineering of organisms to produce commodity chemicals and pharmaceutical compounds more efficiently has been of interest in the life sciences for decades [77]. Prior to the advent of genome-scale models, metabolic engineering was approached on a small scale with targets based on intuitive changes in metabolic pathways closely connected with the product of interest [76], an approach that has been successful for certain applications [78]. However, this approach does not allow for the identification of many non-intuitive changes. Approaching metabolic engineering from a systems perspective (i.e., using genome-scale models) allows system-wide consequences of genetic manipulations to be predicted. In addition, these approaches can expand the list of potentially beneficial manipulations one can attempt.

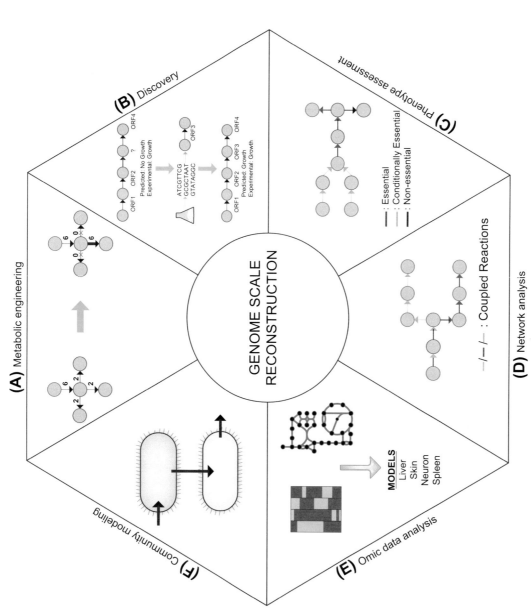

FIGURE 12.5 Applications of genome-scale metabolic reconstructions. (A) Computational algorithms can be used to determine combinatorial knockout effects on the production of a desired small-molecule product. Furthermore, some methods can make predictions of genetic and environmental perturbations that will help improve product yield. Blue numbers correspond to flux through a reaction, and the effect of a double KO to force flux to produce a desired metabolite is shown. (B) Biological discoveries are made as incorrect predictions are reconciled with experimental data. For example, missing reactions can be predicted, thus guiding detailed biochemical studies to identify genes for orphan and missing reactions. (C) Phenotypic behavior is used both as a predictive tool (e.g., gene essentiality or adaptive laboratory evolution) and as a method to guide model improvements. (D) Reconstructions can be used for network analysis, thereby generating hypotheses about network behavior (e.g., to predict coupled reactions or co-regulated genes). (E) High-throughput data can be used as a source for additional constraints or aid in the construction of tissue specific models, thereby adding value to omic data. (F) Multi-cell and community relationships can be modeled by integrating multiple metabolic models allowing for simulation of phenomena such as symbiosis, competition, or pathogenicity.

Many computer algorithms have been developed to predict genetic manipulations for metabolic engineering [79,80]. Two examples of interest are OptKnock [81] and OptStrain [82]. OptKnock attempts to maximize a user-determined objective function (e.g., production of a compound of interest) via iterative single-gene knockouts, while also subjecting constraints that maximize cellular growth (i.e., the biomass reaction). OptStrain is an algorithm that uses a large database of metabolic reactions drawn from multiple organisms to iteratively add heterologous reactions to a model in order to facilitate the production of a non-native compound of interest. Many successful strains have been predicted or engineered using these methods [83–92], and salient examples are highlighted below.

Lycopene production in *Escherichia coli* is accomplished by the introduction of heterologous genes into the metabolic network [93]. A computational search revealed single-gene knockouts (KO) that improved lycopene yield in *E. coli* over an already engineered parent strain while still maintaining a viable growth rate. Successful KOs were subjected to additional KOs, resulting in multi-gene deletion mutants with further improvements in lycopene yield, including a triple KO strain with 37% greater lycopene content than the original strain engineered for lycopene production. Of the three genes knocked out, none were a part of the recombinant lycopene synthesis pathway, with one gene (*fdhF*) improving yield only when part of a double or triple knockout strain. Additionally, another manipulation consisted of a gene (*talB*) that did not improve production as a single KO and reduced production as a double KO, but acted to improve production as part of a triple KO [84].

Another commodity chemical, 1,4-butanediol (BDO), is used to manufacture over 2.5 million tonnes annually of various polymers. Current production relies heavily on derivations of oil and natural gas, and hence a renewable approach is desirable. A strain of *E. coli* was engineered [92] to produce BDO by algorithmically identifying potential reactions that would produce BDO from various carbohydrate sources and introducing the corresponding heterologous proteins. The resultant strain was optimized for BDO production using OptKnock to identify a quadruple KO that would improve production. Genes essential for BDO production in this KO strain were identified and mutated or replaced to be functional under anaerobic conditions to facilitate the high degree of reduction necessary for BDO production. These modifications resulted in a strain that produced more than twice as much BDO as the parent strain. Additional modifications to remove a transcriptional repressor and enzymatic mutations to reduce inhibitory effects resulted in a final strain producing more than seven times as much BDO as the parent strain. Thus, as demonstrated in the examples mentioned here, genome-scale metabolic models provide a useful tool in metabolic engineering.

Biological Discovery

The reconstruction process can often lead to hypothesis-driven discoveries, especially during the network evaluation stage. Inconsistencies between model predictions and physiological data can necessitate investigation into possible explanations for the discrepancy, thereby possibly resulting in novel findings, such as those discussed below.

One such study [38] presented a process to discover new metabolic pathways in an organism, based on discrepancies between modeling and experimental data. By searching through all known metabolic reactions in other organisms, pathways were identified that would resolve the discrepancy. Experiments were subsequently designed to determine open reading frames (ORFs) that code the enzyme needed for the reaction. For example, a model of *E. coli* metabolism predicted that it could not grow on D-malate, whereas experimental data showed growth occurring. The algorithm identified transport and enzymatic reactions that were missing from the model for the catabolism of D-malate. Next, a set of genes was identified for these reactions. KO strains for each gene were grown on D-malate to see if growth was inhibited in these mutant strains. This resulted in the identification of three genes which, when incorporated into the model, reconciled the discrepancy. These genes were subjected to further biochemical testing, thereby demonstrating that two genes (*yeaU* and *yeaT*) were responsible for the catabolism of D-malate, and that another gene (*dctA*) acted as a transporter for the metabolite. In total, this study was able to expand annotation for eight genes in *E. coli*. While the extent of unknown information about organisms is not limited to metabolism, model-guided discovery can elucidate novel knowledge in metabolic gene product function and regulation.

Phenotype Prediction and Evaluation

A common use for genome-scale metabolic models is the examination and prediction of phenotypes under various states (e.g., wild-type vs. genetic manipulation). This process is cyclic, as failures guide experiments that result in iterative modifications to the model (e.g., gap-filling) that improve its accuracy for successive testing. Additional areas of focus in this category include the incorporation of additional data types (e.g., thermodynamic, metabolomic, and fluxomic) to provide bounds on constraints [36,94–101], to identify essential components (i.e., genes [36,102–105], metabolites [106,107], and reactions [95,108–110]), and to gain insight into evolution [61–63,111,112]. Studies in each area are explored in depth in the following sections.

Regulated reactions in *E. coli* were identified by using the Gibbs energies of formation for metabolites in each reaction [96]. Reactions that were far from equilibrium (having a large absolute value of the Gibbs energy for the reaction) were identified as likely being under some form of active regulation (due to high potential for wasted energy if left unregulated). This hypothesis was verified through quantitative metabolomic data, which in turn was used to further constrain the boundaries of reactions in the model.

Essential genes in *E. coli* have been predicted by the model by optimizing biomass formation for single-gene KOs at ~91% accuracy compared to experimental data [104] across the mutant strains in the Keio gene deletion collection [113]. In a similar approach, essential metabolites for *E. coli* have been predicted computationally [107] by blocking flux through all reactions producing a compound. In vivo experiments were used to test the prediction by knocking out multiple non-lethal reactions producing a compound and evaluating the growth rate of the organism. Compounds found by the model to be essential (e.g., tetrahydrofolate) exhibited at least a 50% decrease in growth rate following these knockouts, whereas non-essential compounds (e.g., 1-deoxy-D-xylulose 5-phosphate) showed little to no change following knockouts. This approach has subsequently proved useful for identifying new antibiotics that are analogs for essential metabolites [114].

It has been reported that some FBA-predicted in silico growth rates are higher than experimental values for wild-type [63], mutant [62], and engineered [62] *E. coli*. In these cases, it was theorized that the discrepancies were the result of suboptimal utilization of the metabolic network. To test this, adaptive laboratory evolution experiments (ALE) were conducted in which the cell cultures were maintained at exponential growth for hundreds of generations [61,115], allowing for selection of the fastest-growing strains. In these experiments, growth improved during ALE, often leading the phenotypes that were consistent with the model-predicted optimal phenotypes. By analyzing transcriptomic and proteomic data from evolved *E. coli* [111], it was found that the gene and protein expression of these evolved strains changed consistent with model-predicted usage. Essential and optimal genes for the model (coding for reactions carrying flux in the optimal solutions) showed significant increases in expression in the evolved strains, whereas genes encoding enzymes that were not used by the model showed significant decreases in expression.

Phenotypic evaluation of metabolic models thus represents a very broad area of study, with potential uses in increasing our understanding of cellular phenomena. Even though it has been one of the most popular uses for genome-scale metabolic models, the area of phenotype prediction and evaluation has many unexplored avenues of focus that will yield insights into how the gene content of a cell correlates with cell phenotypes.

Fundamental Properties of Biological Networks

The extent and careful curation of genome-scale metabolic networks (e.g., 92% of 1260 genes in the *i*AF1260 *E. coli* reconstruction [36] have experimentally validated functions) makes them ideal targets for holistic analysis of overarching network properties, including the identification of co-regulated genes using flux coupling analysis [116–118], network organization [119,120], and flux distribution logic [121].

Analysis of flux distribution across thousands of optimal growth simulation conditions and 50 000 non-optimal solutions showed that distribution of fluxes across *E. coli* followed a power-law distribution, with only a fraction of the available reactions in the network carrying high fluxes [121]. Albeit not noteworthy if the distribution was only observable under limited conditions, the fact that the phenomenon was observed across such a diverse range of conditions indicates that the distribution is likely physiologically advantageous in some way. This discovery holds importance in the area of synthetic biology (e.g., in building synthetic organisms [122] and efforts to create a minimal organism [123]) as well as for metabolic engineering, since perturbations in growth conditions showed redistribution of fluxes around the high-flux 'backbone' with little change in the reactions carrying low flux [75], indicating a degree of plasticity in the network.

Network analysis using in silico tools to identify co-regulated genes was carried out as early as 2002 [124] on a subsystem basis, and on a genome-scale in 2004 [116] by looking at flux ratios between reactions to categorize the coupling type. These results have been experimentally verified via metabolomic data of single-gene knockouts in yeast [117]. A similar approach looking at metabolite connectivity was used to calculate hard coupled reaction (HCR) sets (groups of reactions that are forced to operate in unison due to mass conservation and connectivity constraints [118]) in *Mycobacterium tuberculosis*. This analysis suggested several novel drug targets against the organism that could be used to bypass resistance or unwanted side-effect concerns of current therapeutics.

Thus, it has been demonstrated how the connectivity of a network and other properties contribute to the function of a biological system. Genome-scale metabolic reconstructions have aided in many such studies. Therefore, as these properties continue to be identified and characterized, they can potentially be used for many applications, such as improving our understanding of disease [125] and the development of treatments [126].

A Context for the Analysis of Large Data Sets

The interpretation of the vast amounts of data from various high-throughput (HT) techniques presents an ever-growing challenge for biological science [127]. Although statistical approaches in data analysis constantly reveal novel insights into cell properties [128], metabolic reconstructions provide a context-based approach toward HT data analysis [129]. Applications range from utilizing HT data to constrain the model [130,131], to the use of the model for HT data visualization [132], and even guiding tissue-specific reconstruction abstractions from genome-scale models [133–140].

Although the correlation between gene expression and protein expression is not exact [141,142], using microarray data to set constraints on in silico models has been shown to be effective in predicting metabolic behavior in yeast [133]. Alternative data types, e.g., fluxomic data, have also been used to predict internal fluxes in *E. coli* [130] by constraining extreme pathways of the model. Altogether, these applications function to provide an experimental basis for simulation constraints on models that further reduce the range of feasible phenotypes in the solution space for simulations, resulting in models that have greater predictive power.

Other studies utilize networks by overlaying HT data on them to visualize the effects of perturbations [132]. Albeit seemingly trivial, applications of this sort can facilitate the identification of metabolic hotspots or pathways that are significantly altered by providing a context for the data gathered. For example, one study [143] overlaid gene expression data after gastric bypass surgery on the human metabolic model and found that the expression levels of many reactions clustered together in metabolic pathways in a similar way as would be seen in the skeletal muscle of rhesus monkey when subjected to caloric restriction. Since the data compared gene expression data from patients before and a year after the surgery, this suggested that even after weight stabilization has occurred in these patients, their skeletal muscle metabolism is still showing the effects of caloric restriction.

A more recent application of gene expression data has been used in conjunction with a genome-scale model of human metabolism [143]. Construction of tissue-specific models of metabolism is facilitated by reducing the parent model based on the presence or absence of transcripts or proteins in HT data [133]. This approach was used to generate models for 10 different tissue types.

A later study [134] generated tissue-specific models for three neuron types and glial cells in the brain using similar techniques. The models were able to predict the resistance of GABAergic neurons to Alzheimer's damage and identify a mechanism underlying this response. This mechanism pointed to an enzyme, glutamate decarboxylase (GAD), that conferred the neuroprotective response. Microarray data from Alzheimer's disease patients provided additional support for the prediction of the role of GAD. It was found that GAD expression was highest in brain regions that were recalcitrant to neuron loss in AD, whereas the most severely affected regions showed the lowest expression levels in non-AD control patients. These models can thus be used for computational exploration of diseased states in human tissue in order to better understand the metabolic basis for a disease, as well as to potentially identify novel therapeutic targets.

Recently, another study used HT data to create a model for non-small cell lung cancer [136] that was used to identify novel drug targets. Using FBA, single-gene deletions and synthetic lethals (i.e., double deletions resulting in a lethal phenotype only when both genes are deleted) that reduced growth rate in the model were identified as potential drug targets. The set of 52 individual gene deletions included all but one of the metabolic anticancer drug targets approved by the FDA, as well as 13 genes targeted by non-cancer drugs currently being tested for cancer therapy applications. This left 31 potentially new targets for anticancer drugs. Additionally, novel drug targets were identified by using the synthetically lethal gene pairs and HT data on cancer and healthy cell gene expression data. By finding pairs where one gene was deleted in cancer cells but not in healthy cells, drugs targeting the second gene of a synthetically lethal pair would be lethal only to cancer cells, as healthy cells would have the first gene of the pair still active. For example, in healthy kidney cells the model predicted that haem oxygenase (*Hmox1*) and fumarate hydratase (*FH*) would function as a synthetic lethal gene pair. In hereditary leiomyomatosis and renal-cell cancer (HLRCC) there is often a germline mutation in the *FH* gene that prevents its expression, making *Hmox1* a potential target for targeted treatment. An experimental validation [144] showed that shRNA silencing *Hmox1* was lethal only to cells missing *FH*, thereby demonstrating that synthetic lethal gene pairings are a viable approach to developing targeted therapeutic approaches for cancer treatment.

Interpreting the vast quantities of data generated from high-throughput experiments has long been a challenge in the life sciences, with the ability to create data far outpacing the ability to analyze it. Metabolic models thus represent a potential context for data analysis, thus helping in the interpretation of this data in new biologically meaningful ways.

Multi-Cell and Microbial Community Metabolism

The analysis of metabolic models can provide insight into physiological phenomena for organisms in isolation. However, outside the laboratory cells often grow in diverse communities, giving rise to various beneficial interactions. These phenomena have been modeled for multistrain [145] and multispecies [135,146,147] networks, with applications in such diverse areas as biofuel engineering and diagnostics [79,135].

The general process in this area is to create metabolic models for each organism of interest and then integrate the models via reactions facilitating the transport of metabolites from one organism to the other [135,145−147]. For example, a host−pathogen model was recently reconstructed representing the interactions between the human alveolar macrophage and *M. tuberculosis* [135], thereby allowing the simulation of the metabolic interactions between the cells after *M. tuberculosis* invaded the macrophage. Evaluation of feasible steady-state solutions for the integrated model showed increases in flux through the glyoxalate shunt, decreased glycolysis, and increased gluconeogenesis. In addition, condition-specific models were created for various stages of infection and provided evidence that currently proposed drugs targeting the polyprenyl metabolic pathway [148,149] might only prove effective during latent stages of infection. This is because the pathway is not utilized during later infection stages.

In another study, intraspecies cooperation between auxotrophic strains of *E. coli* was investigated by finding synergistic growth pairs between conditionally lethal mutants [145]. Co-culture of the strains was found to result in synergistic growth when both strains received a benefit from the other via secreted products, and the secreted products provided a high benefit to the recipient strain while also not posing a large burden to the donor strain. Computationally, this corresponded to high shadow prices for received metabolites and low shadow prices for donated metabolites (i.e., the metabolites were highly beneficial to the recipient and did not significantly influence biomass production in the donor). Although the quantitative growth rate predictions were generally higher than experimental values, it was interesting that the prediction of cooperation could be computed purely from the stoichiometric matrices of the two strains, making the approach a valuable qualitative tool.

The development of multi-cell models is a relatively new area of research in constraint-based modeling. However, it is expected that it will continue to develop and play an important role, since community structure influences metabolism in the environment [150], human microbiomes [151], normal physiology [152], and pathophysiology [135].

FUTURE DIRECTIONS

Metabolic networks are central to almost all cellular processes. However, they represent only about one-third of genes in prokaryotes and an even smaller fraction in higher eukaryotes. Thus, for a more holistic view, goals exist to develop an integrated 'OME' model, consisting of the transcriptional regulatory network (O for operon), metabolism (M for metabolism), and transcription and translational processes behind enzymes (E for expression). It is expected that such a model would have greater predictive power than any of its components or combinations thereof. Reconstruction of metabolic networks is a well-established process and an E-matrix reconstruction has recently been developed for *E. coli* [153]. However, the reconstruction of transcriptional regulatory processes remains a relatively unrefined area of research.

E-Matrix Reconstructions

The procedure for E-matrix reconstruction is fairly similar to the protocol described earlier in this chapter, but the focus is on genes and reactions that are necessary for transcription and translation. The resulting model (Figure 12.6C) contains information describing the many processes [153], including transcription, mRNA degradation, translation, protein maturation and folding, protein complex formation, ribosome assembly, RNA processing, rRNA/tRNA modification, and tRNA charging.

These reconstructions represent another perspective on cellular processes beyond metabolism. Moreover, these can be computationally modeled in a fashion similar to metabolic reconstructions to investigate the effects of drugs, mutations, etc.

Efforts are being made to integrate the E-matrix with the metabolic network for an organism (ME-matrix) [154], which is accomplished fairly readily on a purely mathematical level by combining metabolites shared between the matrices into a single entry (row) and requiring that enzymes be synthesized by the transcription and translation machinery for a metabolic reaction to occur.

Specifically, the reaction fluxes of the E-matrix and the metabolic network are coupled together, such that if a metabolic reaction is needed, the flux through translation reactions for its associated enzyme must have a non-zero flux [153−155]. Thus reaction fluxes for transcription of the associated mRNA must also be non-zero, and reaction fluxes for ribosomal synthesis must also carry flux. Thus, these integrated models are able to more accurately predict phenotypes as they explicitly account for most cellular costs, including the costs of using the molecular machinery.

There are numerous potential applications of models that couple transcription, translation, and metabolism. These include (1) improvements in metabolic engineering, as the models explicitly account for recombinant protein synthesis cost in terms of metabolic burden, as well as (2) a network-level understanding of the effects of antibiotics, many of which [156−159] act by blocking the functionality of E-matrix components such as bacterial rRNA.

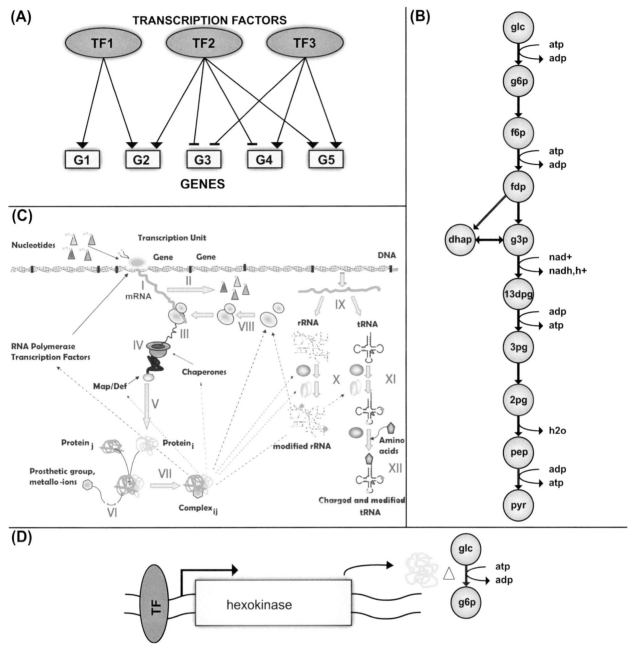

FIGURE 12.6 **The components of the OME-matrix.** (A) In a transcriptional regulatory network (TRN), transcription factors (TF) can modulate the expression of various genes. Activation is indicated by arrows, and repression is shown with horizontal bars. (B) Genes that are modulated by TFs often have gene products with important functions that depend on each other, such as with enzymes of the glycolytic metabolic pathway. (C) Transcription and translation link transcription regulation and enzyme function. An overview of transcriptional and translational machinery systems demonstrates the complexity of this link. Illustration used with permission from [153]. (D) The components each of these processes can be described mathematically, forming an OME-matrix including transcription regulation ('O' for operon), metabolism ('M'), and transcription/translation ('E' for expression). This will allow for the computation of how the processes work together as shown for a single GPR. The transcription factor binding to the ORF induces transcription of hexokinase mRNA, which is translated by E-matrix components. The hexokinase enzyme then catalyzes the conversion of glucose to glucose-6-phosphate.

O-Matrix Reconstructions

The process behind reconstructing the transcriptional regulatory network (TRN) is quite different from that of metabolism or transcription and translation, owing to the nature of the network. Where the latter networks can be described via a series of stoichiometric equations for component reactions, TRNs consist of a series of interactions (either activation or repression) between transcription factors (TFs) and their binding sites [3]. A common mode of determining connectivity between TFs and their binding sites is via chromatin immunoprecipitation, followed by hybridization to a microarray (ChIP-chip) [160] or sequencing (ChIP-seq) [161]. Targeted binding information for TFs can also be found using a yeast-1-hybrid system see Chapter 4). However, these approaches do not provide information about whether the TF activates or represses transcription of its target. Thus, gene expression analysis of TF KO strains is needed. Owing to the combinatorial interactions between TFs that act on some ORFs, gene expression assays in multiple-TF KO strains [162] or ChIP-seq assays for a TF in a different TF KO strain are needed to fully elucidate TF connectivity. If this information is available for major TFs in an organism under various conditions, a putative TRN can be constructed in a fairly automated fashion [3]. This is not to say that homology-based genomic data [163] or databases [164] cannot be useful in providing information for the reconstruction. However, experimental data are an important source of information and a method for algorithmic generation of such data has been described [165].

Although the generation of TRNs is better understood, it is more difficult to convert them into computable models that accurately recapitulate transcriptional regulation. However, a variety of methods are being attempted [166—169]. Generally, TRNs are represented as a network, with TF nodes connected to their ORF targets (Figure 12.6A). For small-scale networks, stochastic [170] and kinetic [171] models are readily computable, but they do not scale well for genome-scale models [3].

TRNs have been integrated with metabolism through a few approaches. In one approach called regulatory FBA (rFBA) the TRN rules were represented as a Boolean network that activated or repressed the flux through metabolic reactions associated with the genes the TFs regulated. This was applied to a simplified model of core metabolism and used for simulation [172]. Since the Boolean network needs to converge to a steady state, the simulations were conducted through an iterative series of calculations with the TRN and metabolic network. Each iteration started by first imposing the Boolean rules based on metabolite concentrations in the growth medium. FBA was then used to optimize growth with the metabolic network, and subsequent changes in extracellular metabolite concentration were computed based on the uptake and secretion profiles from the model. These concentrations were then used to set regulatory states of genes for the next iteration. Interestingly, this approach was successful in modeling phenomena controlled by regulatory mechanisms such as diauxies [50] and was later expanded to use ordinary differential equation-based modeling of TRN dynamics [173]. Current efforts to represent genome-scale TRNs in a computable fashion rely largely on Boolean network representation [167] and additive kinetic modeling approaches using concepts such as log-linear kinetics [3,174—176]. Integration with other network types using these representations (for non-genome-scale TRNs) [162,168,173,177] has been explored and was found to improve the predictive accuracy of the models.

Non-system-level understanding of regulatory networks has been used to guide metabolic engineering of yeast for increased galactose consumption [178]. Thus, a systems-based approach for metabolic engineering via TRN modifications may prove beneficial. Other applications include the use of integrated models to identify potentially novel regulatory interactions to explain discrepancies between experimental and predicted data, as was accomplished for *Halobacterium salinarum NRC-1*[179], *E. coli* [162], and *S. cerevisiae* [177].

Toward a Whole Cell Model

A major goal in systems biology is the attainment of a whole cell model that incorporates the function of every gene for an organism. Such a model might be amenable to interrogation of all biological processes of the organism to understand how all components are coordinated to yield the cell phenotype. Integration of the metabolic network (Figure 12.6B) with the O and E networks discussed above would theoretically result in the OME network (Figure 12.6D). However, the best mathematical representation of such a network on a genome-scale is an open question. Beyond this integration, the simulation of all processes within a cell would also require the inclusion of signaling networks, sRNA regulation [180], protein modifications, and account for cell division mechanisms. Ultimately, understanding the reactions behind each of these networks will allow representation in a stoichiometric matrix similar to those for the M and E matrices, making integration an easier task.

GLOSSARY

Biomass reaction A pseudo-reaction consisting of the summation of all biomass precursors in their appropriate fractional distribution.

Demand reaction Reactions added to the model to facilitate consumption of a metabolite when the actual consuming reaction is unknown or outside the scope of the model.

Exchange reaction Reactions added to the model to supply or remove metabolites from the in silico 'medium'.

Extreme pathway The boundary/edge vectors of the solution space to $S \cdot v = 0$. Each is linearly independent of the others and any point in the solution space can be described as a linear combination of extreme pathways.

Flux balance analysis A method of analysis for finding an optimal solution to $S \cdot v = 0$ using linear programming. This method relies on optimization for a user-specified reaction, such as the biomass reaction.

Gap A missing reaction in the network needed to connect one or more metabolic dead ends with the rest of the network.

Gene−protein−reaction (GPR) association A representation of the flow from gene to protein to reaction including logical operators (i.e., and/or) to describe protein complexes and isozymes.

Linear programming An optimization method where an objective function is maximized or minimized subject to a set of linear constraints.

Metabolic dead end Metabolites that can only be produced or only consumed in the network, and are therefore only associated with blocked reactions.

Metabolic model A computable form of a metabolic reconstruction where bounds are set on each reaction.

Metabolic reconstruction An organism specific list of reactions and their associated knowledge (e.g., associated genes, publications, stoichiometry) comprising the metabolic pathways present within the organism.

Objective function The reaction for which flux is maximized or minimized in an optimization problem.

Shadow price Value associated with each metabolite following optimization of the network via linear programming. The value corresponds to the amount that the objective function value would change with the incremental change of the exchange of that metabolite.

Sink reaction Reactions added to the model to facilitate supply of a metabolite when the relevant reaction is unknown or outside the scope of the model.

ACKNOWLEDGMENTS

This work was supported in part by a grant # R01GM057089 from the National Institutes of Health.

REFERENCES

[1] Kulkarni SS, Salehzadeh F, Fritz T, Zierath JR, Krook A, Osler ME. Mitochondrial regulators of fatty acid metabolism reflect metabolic dysfunction in type 2 diabetes mellitus. Metabolism 2012;61(2):175−85.

[2] Putluri N, Shojaie A, Vasu VT, Nalluri S, Vareed SK, Putluri V, et al. Metabolomic profiling reveals a role for androgen in activating amino acid metabolism and methylation in prostate cancer cells. PloS ONE 2011;6(7):e21417.

[3] Feist AM, Herrgard MJ, Thiele I, Reed JL, Palsson BO. Reconstruction of biochemical networks in microorganisms. Nat Rev Microbiol 2009;7:129−43.

[4] Fell DA, Small JR. Fat synthesis in adipose tissue. an examination of stoichiometric constraints. Biochem J 1986;238:781−6.

[5] Watson MR. A discrete model of bacterial metabolism. Comput Appl Biosci 1986;2:23−7.

[6] Edwards JS, Palsson BO. Systems properties of the haemophilus influenzae rd metabolic genotype. J Biol Chem 1999;274:17410−6.

[7] Henry CS, DeJongh M, Best AA, Frybarger PM, Linsay B, Stevens RL. High-throughput generation, optimization and analysis of genome-scale metabolic models. Nat Biotechnol 2010;28:977−82.

[8] Swainston N, Smallbone K, Mendes P, Kell D, Paton N. The SuBliMinaL toolbox: automating steps in the reconstruction of metabolic networks. J Integr Bioinform 2011;8(2):186.

[9] Thiele I, Palsson BO. A protocol for generating a high-quality genome-scale metabolic reconstruction. Nat Protoc 2010;5:93−121.

[10] Orth JD, Fleming RMT, Palsson BO. Reconstruction and use of microbial metabolic networks: the core *escherichia coli* metabolic model as an educational guide. In: Böck A, Curtiss III R, Kaper JB, Karp PD, Neidhardt FC, Nyström T, et al., editors. EcoSal—*Escherichia coli* and *Salmonella*: Cellular and Molecular Biology. Washington, DC: ASM Press; 2009 Chapter 10.2.1.

[11] Karp PD, Paley SM, Krummenacker M, Latendresse M, Dale JM, Lee TJ, et al. Pathway tools version 13.0: integrated software for pathway/genome informatics and systems biology. Brief Bioinform 2010;11:40−79.

[12] Pinney JW, Shirley MW, McConkey GA, Westhead DR. metaSHARK: Software for automated metabolic network prediction from DNA sequence and its application to the genomes of plasmodium falciparum and eimeria tenella. Nucleic Acids Res 2005;33:1399−409.

[13] Kanehisa M, Goto S, Hattori M, Aoki-Kinoshita KF, Itoh M, Kawashima S, et al. From genomics to chemical genomics: new developments in KEGG. Nucleic Acids Res 2006;34:D354−7.

[14] Notebaart RA, van Enckevort FH, Francke C, Siezen RJ, Teusink B. Accelerating the reconstruction of genome-scale metabolic networks. BMC Bioinformatics:296, http://dx.doi.org/10.1186/1471−2105−7-296, 2006;7.

[15] Borodina I, Nielsen J. From genomes to in silico cells via metabolic networks. Curr Opin Biotechnol 2005;16:350−5.

[16] Goesmann A, Haubrock M, Meyer F, Kalinowski J, Giegerich R. PathFinder: reconstruction and dynamic visualization of metabolic pathways. Bioinformatics 2002;18:124−9.

[17] Anonymous. Enzyme Nomenclature. San Diego, CA, USA: Academic Press; 1992.

[18] Gene ontology consortium. The gene ontology project in 2008. Nucleic Acids Res 2008;36:D440−4.

[19] Matthews L, Gopinath G, Gillespie M, Caudy M, Croft D, de Bono B, et al. Reactome knowledgebase of human biological pathways and processes. Nucleic Acids Res 2009;37:D619−22.

[20] Barthelmes J, Ebeling C, Chang A, Schomburg I, Schomburg D. BRENDA, AMENDA and FRENDA: the enzyme information system in 2007. Nucleic Acids Res 2007;35:D511−4.

[21] Karp PD, Riley M, Saier M, Paulsen IT, Collado-Vides J, Paley SM, et al. The EcoCyc database. Nucleic Acids Res 2002;30:56−8.

[22] Mueller LA, Zhang P, Rhee SY. AraCyc: A biochemical pathway database for arabidopsis. Plant Physiol 2003;132:453−60.

[23] Cherry JM, Adler C, Ball C, Chervitz SA, Dwight SS, Hester ET, et al. SGD: *Saccharomyces* genome database. Nucleic Acids Res 1998;26:73−9.

[24] Gardy JL, Laird MR, Chen F, Rey S, Walsh CJ, Ester M, et al. PSORTb v.2.0: Expanded prediction of bacterial protein subcellular localization and insights gained from comparative proteome analysis. Bioinformatics 2005;21:617−23.

[25] Lu Z, Szafron D, Greiner R, Lu P, Wishart DS, Poulin B, et al. Predicting subcellular localization of proteins using machine-learned classifiers. Bioinformatics 2004;20:547−56.

[26] Klitgord N, Segre D. The importance of compartmentalization in metabolic flux models: yeast as an ecosystem of organelles. Genome Inform 2010;22:41–55.

[27] Wheeler DL, Barrett T, Benson DA, Bryant SH, Canese K, Chetvernin V, et al. Database resources of the national center for biotechnology information. Nucleic Acids Res 2007;35:D5–12.

[28] Mintz-Oron S, Aharoni A, Ruppin E, Shlomi T. Network-based prediction of metabolic enzymes' subcellular localization. Bioinformatics 2009;25:i247–52.

[29] Feist AM, Palsson BO. The biomass objective function. Curr Opin Microbiol 2010;13(3):344–9.

[30] Neidhardt FC, Ingraham JL, Schaechter M. Physiology of the Bacterial Cell: A Molecular Approach. Sunderland, MA, USA: Sinauer Associates; 1990.

[31] Izard J, Limberger RJ. Rapid screening method for quantitation of bacterial cell lipids from whole cells. J Microbiol Methods 2003;55:411–8.

[32] Benthin S, Nielsen J, Villadsen J. A simple and reliable method for the determination of cellular RNA content. Biotechnol Tech 1991;5:39–42.

[33] Norris JR, Ribbons DW. Editors: Methods in microbiology. Volume 5. Academic Press, London; 1971

[34] Liao YC, Huang TW, Chen FC, Charusanti P, Hong JS, Chang HY, et al. An experimentally validated genome-scale metabolic reconstruction of klebsiella pneumoniae MGH 78578, iYL1228. J Bacteriol 2011;193:1710–7.

[35] Peterson JD, Umayam LA, Dickinson T, Hickey EK, White O. The comprehensive microbial resource. Nucleic Acids Res 2001; 29:123–5.

[36] Feist AM, Henry CS, Reed JL, Krummenacker M, Joyce AR, Karp PD, et al. A genome-scale metabolic reconstruction for *escherichia coli* K-12 MG1655 that accounts for 1260 ORFs and thermodynamic information. Mol Syst Biol 2007;3:121.

[37] Kauffman KJ, Prakash P, Edwards JS. Advances in flux balance analysis. Curr Opin Biotechnol 2003;14:491–6.

[38] Reed JL, Patel TR, Chen KH, Joyce AR, Applebee MK, Herring CD, et al. Systems approach to refining genome annotation. Proc Natl Acad Sci USA 2006;103:17480–4.

[39] Satish Kumar V, Dasika MS, Maranas CD. Optimization based automated curation of metabolic reconstructions. BMC Bioinformatics 2007;8:212.

[40] Orth JD, Palsson BO. Systematizing the generation of missing metabolic knowledge. Biotechnol Bioeng 2010;107:403–12.

[41] Schilling CH, Letscher D, Palsson BO. Theory for the systemic definition of metabolic pathways and their use in interpreting metabolic function from a pathway-oriented perspective. J Theor Biol 2000;203:229–48.

[42] Price ND, Thiele I, Palsson BO. Candidate states of *helicobacter pylori's* genome-scale metabolic network upon application of 'loop law' thermodynamic constraints. Biophys J 2006;90(11):3919–28.

[43] Bell SL, Palsson BO. Expa: A program for calculating extreme pathways in biochemical reaction networks. Bioinformatics 2005;21:1739–40.

[44] Yeung M, Thiele I, Palsson BO. Estimation of the number of extreme pathways for metabolic networks. BMC Bioinformatics 2007;8:363.

[45] Gudmundsson S, Thiele I. Computationally efficient flux variability analysis. BMC Bioinformatics 2010;11:489.

[46] Schellenberger J, Lewis NE, Palsson BO. Elimination of thermodynamically infeasible loops in steady-state metabolic models. Biophys J 2011;100:544–53.

[47] Henry CS, Broadbelt LJ, Hatzimanikatis V. Thermodynamics-based metabolic flux analysis. Biophys J 2007;92:1792–805.

[48] Hoppe A, Hoffmann S, Holzhutter HG. Including metabolite concentrations into flux balance analysis: thermodynamic realizability as a constraint on flux distributions in metabolic networks. BMC Syst Biol 2007;1:23.

[49] Bochner BR, Gadzinski P, Panomitros E. Phenotype microarrays for high-throughput phenotypic testing and assay of gene function. Genome Res 2001;11:1246–55.

[50] Loomis Jr WF, Magasanik B. Glucose-lactose diauxie in *escherichia coli*. J Bacteriol 1967;93:1397–401.

[51] Edwards JS, Ibarra RU, Palsson BO. In silico predictions of *escherichia coli* metabolic capabilities are consistent with experimental data. Nat Biotechnol 2001;19:125–30.

[52] Famili I, Forster J, Nielsen J, Palsson BO. *Saccharomyces cerevisiae* phenotypes can be predicted by using constraint-based analysis of a genome-scale reconstructed metabolic network. Proc Natl Acad Sci USA 2003;100:13134–9.

[53] Segre D, Vitkup D, Church GM. Analysis of optimality in natural and perturbed metabolic networks. Proc Natl Acad Sci USA 2002;99:15112–7.

[54] Burgard AP, Maranas CD. Optimization-based framework for inferring and testing hypothesized metabolic objective functions. Biotechnol Bioeng 2003;82:670–7.

[55] Schuetz R, Kuepfer L, Sauer U. Systematic evaluation of objective functions for predicting intracellular fluxes in *escherichia coli*. Mol Syst Biol 2007;3:119.

[56] Gianchandani EP, Oberhardt MA, Burgard AP, Maranas CD, Papin JA. Predicting biological system objectives de novo from internal state measurements. BMC Bioinformatics 2008; 9:43.

[57] Knorr AL, Jain R, Srivastava R. Bayesian-based selection of metabolic objective functions. Bioinformatics 2007;23:351–7.

[58] Holzhutter HG. The principle of flux minimization and its application to estimate stationary fluxes in metabolic networks. Eur J Biochem 2004;271:2905–22.

[59] Shlomi T, Berkman O, Ruppin E. Regulatory on/off minimization of metabolic flux changes after genetic perturbations. Proc Natl Acad Sci USA 2005;102:7695–700.

[60] Schuster S, Pfeiffer T, Fell DA. Is maximization of molar yield in metabolic networks favoured by evolution? J Theor Biol 2008; 252:497–504.

[61] Palsson B Ø. Adaptive laboratory evolution. Microbe 2011;6(2):69–74 (2011).

[62] Fong SS, Palsson BO. Metabolic gene-deletion strains of *escherichia coli* evolve to computationally predicted growth phenotypes. Nat Genet 2004;36:1056–8.

[63] Ibarra RU, Edwards JS, Palsson BO. *Escherichia coli* K-12 undergoes adaptive evolution to achieve in silico predicted optimal growth. Nature 2002;420:186–9.

[64] Hucka M, Finney A, Sauro HM, Bolouri H, Doyle JC, Kitano H, et al. The systems biology markup language (SBML): a medium for representation and exchange of biochemical network models. Bioinformatics 2003;19:524–31.

[65] Rocha I, Maia P, Evangelista P, Vilaca P, Soares S, Pinto JP, et al. OptFlux: An open-source software platform for in silico metabolic engineering. BMC Syst Biol 2010;4:45.

[66] Klamt S, Saez-Rodriguez J, Gilles ED. Structural and functional analysis of cellular networks with CellNetAnalyzer. BMC Syst Biol 2007;1:2.

[67] Becker SA, Feist AM, Mo ML, Hannum G, Palsson BO, Herrgard MJ. Quantitative prediction of cellular metabolism with constraint-based models: the COBRA toolbox. Nat Protoc 2007;2: 727–38.

[68] Gevorgyan A, Bushell ME, Avignone-Rossa C, Kierzek AM. SurreyFBA: a command line tool and graphics user interface for constraint-based modeling of genome-scale metabolic reaction networks. Bioinformatics 2011;27:433–4.

[69] Schellenberger J, Que R, Fleming RM, Thiele I, Orth JD, Feist AM, et al. Quantitative prediction of cellular metabolism with constraint-based models: the COBRA toolbox v2.0. Nat Protoc 2011;6: 1290–307.

[70] Edwards JS, Palsson BO. The *escherichia coli* MG1655 in silico metabolic genotype: its definition, characteristics, and capabilities. Proc Natl Acad Sci USA 2000;97:5528–33.

[71] Orth JD, Conrad TM, Na J, Lerman JA, Nam H, Feist AM, et al. A comprehensive genome-scale reconstruction of *escherichia coli* metabolism—2011. Mol Syst Biol 2011;7:535.

[72] Forster J, Famili I, Fu P, Palsson BO, Nielsen J. Genome-scale reconstruction of the *saccharomyces cerevisiae* metabolic network. Genome Res 2003;13:244–53.

[73] Duarte NC, Herrgard MJ, Palsson BO. Reconstruction and validation of *saccharomyces cerevisiae* iND750, a fully compartmentalized genome-scale metabolic model. Genome Res 2004;14:1298–309.

[74] Nookaew I, Jewett MC, Meechai A, Thammarongtham C, Laoteng K, Cheevadhanarak S, et al. The genome-scale metabolic model iIN800 of *saccharomyces cerevisiae* and its validation: a scaffold to query lipid metabolism. BMC Syst Biol 2008;2:71.

[75] Feist AM, Palsson BO. The growing scope of applications of genome-scale metabolic reconstructions using *escherichia coli*. Nat Biotechnol 2008;26:659–67.

[76] Oberhardt MA, Palsson BO, Papin JA. Applications of genome-scale metabolic reconstructions. Mol Syst Biol 2009;5:320.

[77] Windass JD, Worsey MJ, Pioli EM, Pioli D, Barth PT, Atherton KT, et al. Improved conversion of methanol to single-cell protein by methylophilus methylotrophus. Nature 1980;287:396–401.

[78] Bailey JE, Birnbaum S, Galazzo JL, Khosla C, Shanks JV. Strategies and challenges in metabolic engineering. Ann NY Acad Sci 1990;589:1–15.

[79] Lewis NE, Nagarajan H, Palsson BO. Constraining the metabolic genotype-phenotype relationship using a phylogeny of in silico methods. Nat Rev Microbiol 2012;10:291–305.

[80] Medema MH, van Raaphorst R, Takano E, Breitling R. Computational tools for the synthetic design of biochemical pathways. Nat Rev Microbiol 2012;10:191–202.

[81] Burgard AP, Pharkya P, Maranas CD. Optknock: a bilevel programming framework for identifying gene knockout strategies for microbial strain optimization. Biotechnol Bioeng 2003;84:647–57.

[82] Pharkya P, Burgard AP, Maranas CD. OptStrain: a computational framework for redesign of microbial production systems. Genome Res 2004;14:2367–76.

[83] Pharkya P, Burgard AP, Maranas CD. Exploring the overproduction of amino acids using the bilevel optimization framework OptKnock. Biotechnol Bioeng 2003;84:887–99.

[84] Alper H, Jin YS, Moxley JF, Stephanopoulos G. Identifying gene targets for the metabolic engineering of lycopene biosynthesis in *escherichia coli*. Metab Eng 2005;7:155–64.

[85] Alper H, Miyaoku K, Stephanopoulos G. Construction of lycopene-overproducing *E. coli* strains by combining systematic and combinatorial gene knockout targets. Nat Biotechnol 2005;23: 612–6.

[86] Fong SS, Burgard AP, Herring CD, Knight EM, Blattner FR, Maranas CD, et al. In silico design and adaptive evolution of *escherichia coli* for production of lactic acid. Biotechnol Bioeng 2005;91:643–8.

[87] Pharkya P, Maranas CD. An optimization framework for identifying reaction activation/inhibition or elimination candidates for overproduction in microbial systems. Metab Eng 2006;8:1–13.

[88] Lee SJ, Lee DY, Kim TY, Kim BH, Lee J, Lee SY. Metabolic engineering of *escherichia coli* for enhanced production of succinic acid, based on genome comparison and in silico gene knockout simulation. Appl Environ Microbiol 2005;71:7880–7.

[89] Wang Q, Chen X, Yang Y, Zhao X. Genome-scale in silico aided metabolic analysis and flux comparisons of *escherichia coli* to improve succinate production. Appl Microbiol Biotechnol 2006; 73:887–94.

[90] Park JH, Lee KH, et al. Metabolic engineering of *escherichia coli* for the production of L-valine based on transcriptome analysis and in silico gene knockout simulation. Proc Natl Acad Sci USA 2007;104:7797–802.

[91] Lee KH, Park JH, Kim TY, Kim HU, Lee SY. Systems metabolic engineering of *escherichia coli* for L-threonine production. Mol Syst Biol 2007;3:149.

[92] Yim H, Haselbeck R, Niu W, Pujol-Baxley C, Burgard A, Boldt J, et al. Metabolic engineering of *escherichia coli* for direct production of 1,4-butanediol. Nat Chem Biol 2011;7:445–52.

[93] Adam P, Hecht S, Eisenreich W, Kaiser J, Grawert T, Arigoni D, et al. Biosynthesis of terpenes: studies on 1-hydroxy-2-methyl-2-(E)-butenyl 4-diphosphate reductase. Proc Natl Acad Sci USA 2002;99: 12108–13.

[94] Beard DA, Liang SD, Qian H. Energy balance for analysis of complex metabolic networks. Biophys J 2002;83:79–86.

[95] Samal A, Singh S, Giri V, Krishna S, Raghuram N, Jain S. Low degree metabolites explain essential reactions and enhance modularity in biological networks. BMC Bioinformatics 2006;7:118.

[96] Kummel A, Panke S, Heinemann M. Putative regulatory sites unraveled by network-embedded thermodynamic analysis of metabolome data. Mol Syst Biol 2006;2:2006.0034.

[97] Kummel A, Panke S, Heinemann M. Systematic assignment of thermodynamic constraints in metabolic network models. BMC Bioinformatics 2006;7:512.

[98] Henry CS, Broadbelt LJ, Hatzimanikatis V. Thermodynamics-based metabolic flux analysis. Biophys J 2007;92:1792–805.

[99] Ederer M, Gilles ED. Thermodynamically feasible kinetic models of reaction networks. Biophys J 2007;92:1846–57.

[100] Hoppe A, Hoffmann S, Holzhutter HG. Including metabolite concentrations into flux balance analysis: thermodynamic realizability as a constraint on flux distributions in metabolic networks. BMC Syst Biol 2007;1:23.

[101] Warren PB, Jones JL. Duality, thermodynamics, and the linear programming problem in constraint-based models of metabolism. Phys Rev Lett 2007;99:108101.

[102] Ghim CM, Goh KI, Kahng B. Lethality and synthetic lethality in the genome-wide metabolic network of *escherichia coli*. J Theor Biol 2005;237:401—11.

[103] Gerdes S, Edwards R, Kubal M, Fonstein M, Stevens R, Osterman A. Essential genes on metabolic maps. Curr Opin Biotechnol 2006;17:448—56.

[104] Joyce AR, Reed JL, White A, Edwards R, Osterman A, Baba T, et al. Experimental and computational assessment of conditionally essential genes in *escherichia coli*. J Bacteriol 2006;188:8259—71.

[105] Motter AE, Gulbahce N, Almaas E, Barabasi AL. Predicting synthetic rescues in metabolic networks. Mol Syst Biol 2008; 4:168.

[106] Imielinski M, Belta C, Halasz A, Rubin H. Investigating metabolite essentiality through genome-scale analysis of *escherichia coli* production capabilities. Bioinformatics 2005;21:2008.

[107] Kim PJ, Lee DY, Kim TY, Lee KH, Jeong H, Lee SY, et al. Metabolite essentiality elucidates robustness of *escherichia coli* metabolism. Proc Natl Acad Sci USA 2007;104:13638—42.

[108] Burgard AP, Vaidyaraman S, Maranas CD. Minimal reaction sets for *escherichia coli* metabolism under different growth requirements and uptake environments. Biotechnol Prog 2001;17:791—7.

[109] Henry CS, Jankowski MD, Broadbelt LJ, Hatzimanikatis V. Genome-scale thermodynamic analysis of *escherichia coli* metabolism. Biophys J 2006;90:1453—61.

[110] Guimera R, Sales-Pardo M, Amaral LA. A network-based method for target selection in metabolic networks. Bioinformatics 2007; 23:1616—22.

[111] Lewis NE, Hixson KK, Conrad TM, Lerman JA, Charusanti P, Polpitiya AD, et al. Omic data from evolved E. coli are consistent with computed optimal growth from genome-scale models. Mol Syst Biol 2010;6:390.

[112] Nam H, Conrad TM, Lewis NE. The role of cellular objectives and selective pressures in metabolic pathway evolution. Curr Opin Biotechnol 2011;22:595—600.

[113] Baba T, Ara T, Hasegawa M, Takai Y, Okumura Y, Baba M, et al. Construction of *escherichia coli* K-12 in-frame, single-gene knockout mutants: the keio collection. Mol Syst Biol 2006;2: 2006.0008.

[114] Kim HU, Kim SY, Jeong H, Kim TY, Kim JJ, Choy HE, et al. Integrative genome-scale metabolic analysis of *vibrio vulnificus* for drug targeting and discovery. Mol Syst Biol 2011;7:460.

[115] Conrad TM, Lewis NE, Palsson BO. Microbial laboratory evolution in the era of genome-scale science. Mol Syst Biol 2011;7:509.

[116] Burgard AP, Nikolaev EV, Schilling CH, Maranas CD. Flux coupling analysis of genome-scale metabolic network reconstructions. Genome Res 2004;14:301—12.

[117] Bundy JG, Papp B, Harmston R, Browne RA, Clayson EM, Burton N, et al. Evaluation of predicted network modules in yeast metabolism using NMR-based metabolite profiling. Genome Res 2007;17:510—9.

[118] Jamshidi N, Palsson BO. Investigating the metabolic capabilities of *mycobacterium tuberculosis* H37Rv using the in silico strain iNJ661 and proposing alternative drug targets. BMC Syst Biol 2007;1:26.

[119] Gagneur J, Jackson DB, Casari G. Hierarchical analysis of dependency in metabolic networks. Bioinformatics 2003;19:1027—34.

[120] Sales-Pardo M, Guimera R, Moreira AA, Amaral LA. Extracting the hierarchical organization of complex systems. Proc Natl Acad Sci USA 2007;104:15224—9.

[121] Almaas E, Kovacs B, Vicsek T, Oltvai ZN, Barabasi AL. Global organization of metabolic fluxes in the bacterium *escherichia coli*. Nature 2004;427:839—43.

[122] Gibson DG, Glass JI, Lartigue C, Noskov VN, Chuang RY, Algire MA, et al. Creation of a bacterial cell controlled by a chemically synthesized genome. Science 2010;329:52—6.

[123] Forster AC, Church GM. Towards synthesis of a minimal cell. Mol Syst Biol 2006;2:45.

[124] Schilling CH, Covert MW, Famili I, Church GM, Edwards JS, Palsson BO. Genome-scale metabolic model of *helicobacter pylori* 26695. J Bacteriol 2002;184:4582—93.

[125] Barabasi AL, Gulbahce N, Loscalzo J. Network medicine: a network-based approach to human disease. Nat Rev Genet 2011;12:56—68.

[126] Kim HU, Kim SY, Jeong H, Kim TY, Kim JJ, Choy HE, et al. Integrative genome-scale metabolic analysis of *vibrio vulnificus* for drug targeting and discovery. Mol Syst Biol 2011;7:460.

[127] Palsson B, Zengler K. The challenges of integrating multi-omic data sets. Nat Chem Biol 2010;6:787—9.

[128] Boccazzi P, Zanzotto A, Szita N, Bhattacharya S, Jensen KF, Sinskey AJ. Gene expression analysis of *escherichia coli* grown in miniaturized bioreactor platforms for high-throughput analysis of growth and genomic data. Appl Microbiol Biotechnol 2005; 68:518—32.

[129] Lewis NE, Cho BK, Knight EM, Palsson BO. Gene expression profiling and the use of genome-scale in silico models of *escherichia coli* for analysis: providing context for content. J Bacteriol 2009;191:3437—44.

[130] Wiback SJ, Mahadevan R, Palsson BO. Using metabolic flux data to further constrain the metabolic solution space and predict internal flux patterns: the *escherichia coli* spectrum. Biotechnol Bioeng 2004;86:317—31.

[131] Colijn C, Brandes A, Zucker J, Lun DS, Weiner B, Farhat MR, et al. Interpreting expression data with metabolic flux models: predicting *mycobacterium tuberculosis* mycolic acid production. PLoS Comput Biol 2009;5:e1000489.

[132] Usaite R, Patil KR, Grotkjaer T, Nielsen J, Regenberg B. Global transcriptional and physiological responses of *saccharomyces cerevisiae* to ammonium, L-alanine, or L-glutamine limitation. Appl Environ Microbiol 2006;72:6194—203.

[133] Shlomi T, Cabili MN, Herrgard MJ, Palsson BO, Ruppin E. Network-based prediction of human tissue-specific metabolism. Nat Biotechnol 2008;26:1003—10.

[134] Lewis NE, Schramm G, Bordbar A, Schellenberger J, Andersen MP, Cheng JK, et al. Large-scale in silico modeling of metabolic interactions between cell types in the human brain. Nat Biotechnol 2010;28:1279—85.

[135] Bordbar A, Lewis NE, Schellenberger J, Palsson BO, Jamshidi N. Insight into human alveolar macrophage and *M. tuberculosis* interactions via metabolic reconstructions. Mol Syst Biol 2010;6:422.

[136] Folger O, Jerby L, Frezza C, Gottlieb E, Ruppin E, Shlomi T. Predicting selective drug targets in cancer through metabolic networks. Mol Syst Biol 2011;7:517.

[137] Jerby L, Shlomi T, Ruppin E. Computational reconstruction of tissue-specific metabolic models: application to human liver metabolism. Mol Syst Biol 2010;6:401.

[138] Chang RL, Xie L, Xie L, Bourne PE, Palsson BO. Drug off-target effects predicted using structural analysis in the context of a metabolic network model. PLoS Comput Biol 2010;6:e1000938.

[139] Becker SA, Palsson BO. Context-specific metabolic networks are consistent with experiments. PLoS Comput Biol 2008;4:e1000082.

[140] Vo TD, Paul Lee WN, Palsson BO. Systems analysis of energy metabolism elucidates the affected respiratory chain complex in leigh's syndrome. Mol Genet Metab 2007;91:15–22.

[141] Ideker T, Thorsson V, Ranish JA, Christmas R, Buhler J, Eng JK, et al. Integrated genomic and proteomic analyses of a systematically perturbed metabolic network. Science 2001;292:929–34.

[142] Chechik G, Oh E, Rando O, Weissman J, Regev A, Koller D. Activity motifs reveal principles of timing in transcriptional control of the yeast metabolic network. Nat Biotechnol 2008;26:1251–9.

[143] Duarte NC, Becker SA, Jamshidi N, Thiele I, Mo ML, Vo TD, et al. Global reconstruction of the human metabolic network based on genomic and bibliomic data. Proc Natl Acad Sci USA 2007;104:1777–82.

[144] Frezza C, Zheng L, Folger O, Rajagopalan KN, MacKenzie ED, Jerby L, et al. Haem oxygenase is synthetically lethal with the tumour suppressor fumarate hydratase. Nature 2011;477:225–8.

[145] Wintermute EH, Silver PA. Emergent cooperation in microbial metabolism. Mol Syst Biol 2010;6:407.

[146] Stolyar S, Van Dien S, Hillesland KL, Pinel N, Lie TJ, Leigh JA, et al. Metabolic modeling of a mutualistic microbial community. Mol Syst Biol 2007;3:92.

[147] Klitgord N, Segre D. Environments that induce synthetic microbial ecosystems. PLoS Comput Biol 2010;6:e1001002.

[148] Eoh H, Brown AC, Buetow L, Hunter WN, Parish T, Kaur D, et al. Characterization of the *mycobacterium tuberculosis* 4-diphosphocytidyl-2-C-methyl-D-erythritol synthase: potential for drug development. J Bacteriol 2007;189:8922–7.

[149] Boshoff HI, Xu X, Tahlan K, Dowd CS, Pethe K, Camacho LR, et al. Biosynthesis and recycling of nicotinamide cofactors in *mycobacterium tuberculosis*. an essential role for NAD in non-replicating *bacilli*. J Biol Chem 2008;283:19329–41.

[150] DeLong EF. The microbial ocean from genomes to biomes. Nature:200–6, http://dx.doi.org/10.1038/nature08059, 2009;459.

[151] Turnbaugh PJ, Gordon JI. The core gut microbiome, energy balance and obesity. J Physiol 2009;587:4153–8.

[152] Reichard GA, Moury NF, Hochella NJ, Patterson AL, Weinhouse S. Quantitative estimation of the cori cycle in the human. J Biol Chem 1963;238:495–501.

[153] Thiele I, Jamshidi N, Fleming RM, Palsson BO. Genome-scale reconstruction of *escherichia coli's* transcriptional and translational machinery: a knowledge base, its mathematical formulation, and its functional characterization. PLoS Comput Biol 2009;5:e1000312.

[154] Thiele I. Dissertation: A Stoichiometric Model of *Escherichia coli's* Macromolecular Synthesis Machinery and its Integration with Metabolism. San Diego, CA: University of California; 2009.

[155] Thiele I, Fleming RM, Bordbar A, Schellenberger J, Palsson BO. Functional characterization of alternate optimal solutions of *escherichia coli's* transcriptional and translational machinery. Biophys J 2010;98:2072–81.

[156] Connell SR, Tracz DM, Nierhaus KH, Taylor DE. Ribosomal protection proteins and their mechanism of tetracycline resistance. Antimicrob Agents Chemother 2003;47:3675–81.

[157] Shinabarger D. Mechanism of action of the oxazolidinone antibacterial agents. Expert Opin Investig Drugs 1999;8:1195–202.

[158] Gaynor M, Mankin AS. Macrolide antibiotics: binding site, mechanism of action, resistance. Curr Top Med Chem 2003;3:949–61.

[159] Davis BD. Mechanism of bactericidal action of aminoglycosides. Microbiol Rev 1987;51:341–50.

[160] Ren B, Robert F, Wyrick JJ, Aparicio O, Jennings EG, Simon I, et al. Genome-wide location and function of DNA binding proteins. Science 2000;290:2306–9.

[161] Johnson DS, Mortazavi A, Myers RM, Wold B. Genome-wide mapping of in vivo protein-DNA interactions. Science 2007;316:1497–502.

[162] Covert MW, Knight EM, Reed JL, Herrgard MJ, Palsson BO. Integrating high-throughput and computational data elucidates bacterial networks. Nature 2004;429:92–6.

[163] Perez-Rueda E, Collado-Vides J. The repertoire of DNA-binding transcriptional regulators in *escherichia coli* K-12. Nucleic Acids Res 2000;28:1838–47.

[164] Salgado H, Gama-Castro S, Peralta-Gil M, Diaz-Peredo E, Sanchez-Solano F, Santos-Zavaleta A, et al. RegulonDB (version 5.0): *Escherichia coli* K-12 transcriptional regulatory network, operon organization, and growth conditions. Nucleic Acids Res 2006;34:D394–7.

[165] Barrett CL, Palsson BO. Iterative reconstruction of transcriptional regulatory networks: an algorithmic approach. PLoS Comput Biol 2006;2:e52.

[166] de Jong H. Modeling and simulation of genetic regulatory systems: A literature review. J Comput Biol 2002;9:67–103.

[167] Gianchandani EP, Papin JA, Price ND, Joyce AR, Palsson BO. Matrix formalism to describe functional states of transcriptional regulatory systems. PLoS Comput Biol 2006;2:e101.

[168] Chandrasekaran S, Price ND. Probabilistic integrative modeling of genome-scale metabolic and regulatory networks in *escherichia coli* and *mycobacterium tuberculosis*. Proc Natl Acad Sci USA 2010;107:17845–50.

[169] Karlebach G, Shamir R. Modelling and analysis of gene regulatory networks. Nat Rev Mol Cell Biol 2008;9:770–80.

[170] Climescu-Haulica A, Quirk MD. A stochastic differential equation model for transcriptional regulatory networks. BMC Bioinformatics 2007;8(Suppl. 5):S4.

[171] Tyson JJ, Chen KC, Novak B. Sniffers, buzzers, toggles and blinkers: dynamics of regulatory and signaling pathways in the cell. Curr Opin Cell Biol 2003;15:221–31.

[172] Covert MW, Schilling CH, Palsson B. Regulation of gene expression in flux balance models of metabolism. J Theor Biol 2001;213:73–88.

[173] Covert MW, Xiao N, Chen TJ, Karr JR. Integrating metabolic, transcriptional regulatory and signal transduction models in *escherichia coli*. Bioinformatics 2008;24:2044–50.

[174] Workman CT, Mak HC, McCuine S, Tagne JB, Agarwal M, Ozier O, et al. a systems approach to mapping DNA damage response pathways. Science 2006;312:1054–9.

[175] Shmulevich I, Dougherty ER, Kim S, Zhang W. Probabilistic boolean networks: a rule-based uncertainty model for gene regulatory networks. Bioinformatics 2002;18:261−74.

[176] Segal E, Raveh-Sadka T, Schroeder M, Unnerstall U, Gaul U. Predicting expression patterns from regulatory sequence in *drosophila* segmentation. Nature 2008;451:535−40.

[177] Herrgard MJ, Lee BS, Portnoy V, Palsson BO. Integrated analysis of regulatory and metabolic networks reveals novel regulatory mechanisms in *saccharomyces cerevisiae*. Genome Res 2006;16:627−35.

[178] Ostergaard S, Olsson L, Johnston M, Nielsen J. Increasing galactose consumption by *saccharomyces cerevisiae* through metabolic engineering of the GAL gene regulatory network. Nat Biotechnol 2000;18:1283−6.

[179] Bonneau R, Facciotti MT, Reiss DJ, Schmid AK, Pan M, Kaur A, et al. A predictive model for transcriptional control of physiology in a free living cell. Cell 2007;131:1354−65.

[180] Shimoni Y, Friedlander G, Hetzroni G, Niv G, Altuvia S, Biham O, et al. Regulation of gene expression by small non-coding RNAs: a quantitative view. Mol Syst Biol 2007; 3:138.

[181] Zhu G, Golding GB, Dean AM. The selective cause of an ancient adaptation. Science 2005;307:1279−82.

Chapter 13

Genotype Networks and Evolutionary Innovations in Biological Systems

Andreas Wagner
University of Zurich, Institute of Evolutionary Biology and Environmental Studies, Y27-J-54, Winterthurerstrasse 190, CH-8057 Zurich, Switzerland

Chapter Outline

Introduction	251	The Diversity of Neighborhoods in Genotype Space	258
Metabolic Networks and Their Innovations	252	Genotype Networks and Their Diverse Neighborhoods Can Help Explain the Origin of New Phenotypes	259
Regulatory Circuits and Their Innovations	252		
Macromolecules and Their Innovations	253	Robustness, Genotype Networks and Environmental Change	260
Towards A Systematic Understanding of Innovation	254		
Genotype Spaces and the Phenotypes Therein	254	Conclusions and Future Challenges	261
Genotype Networks	256	References	261

INTRODUCTION

How new traits originate in life is a question that has occupied evolutionary biologists since Darwin's time. This holds especially for traits that are evolutionary innovations, i.e., qualitatively new features that benefit their carrier [1,2]. About this origin, the geneticist de Vries said in 1904 that Darwin's theory can explain the *survival* of the fittest but not its *arrival* [3]. Today, more than a century later, the biological literature contains many well-studied examples of innovations, fascinating case studies of natural history [1,4–10]. However, we still know little about any principles that might underlie the origin of innovations, other than the well-worn notion that a combination of mutation and natural selection may be necessary. We do not even know whether such principles exist. De Vries' statement makes clear that such principles would be principles of how biological systems bring forth novel and beneficial phenotypes. They would be principles of phenotypic variability.

To understand the origins of new phenotypes one needs to understand the relationship between genotype and phenotype. The genotype is the totality of an organism's genetic material. The phenotype is any other observable characteristic. It includes the morphology and behavior of complex organisms, the structure of cells, the expression pattern of genes and proteins, the biosynthetic abilities of an organism's metabolism, and the three-dimensional structure and function of its macromolecules, such as protein and RNA molecules.

New phenotypes often arise through mutations that alter an organism's genotype. Therefore, understanding phenotypic variability requires understanding how genotypic change translates into phenotypic change. Ideally, experimentation should provide this understanding [11,12]. However, a systematic understanding of the relationship between genotype and phenotype requires the analysis of thousands if not millions of different genotypes and their phenotypes. It is beyond reach of current experimental technologies for most systems. An alternative is to use existing comparative data about genotypes and their phenotypes. The necessary information is available only for a few kinds of system, for example proteins, where the structure and function of tens of thousands of proteins are available. In most other systems computational modeling of phenotypes will be essential for the foreseeable future. Fortunately, the tools of systems biology have allowed us to make great strides in such modeling. For example, within the last 15 years it has become possible to computationally predict the biosynthetic phenotypes of enormously complex metabolic networks comprising hundreds of enzymatic reactions [13,14]. The analyses reviewed in this chapter use such computational approaches, as well as comparative data and experimentation. Taken together, these three lines of evidence point to a series of surprisingly simple principles behind life's ability to produce novel and beneficial phenotypes, i.e., its innovability.

Here, I will first devote three short sections to three central classes of systems and their phenotypes. Changes in these systems are the foundations of most, if not all evolutionary innovations. These system classes are metabolic networks, regulatory circuits, and molecules such as proteins and RNA. Subsequently, I will suggest how one can study phenotypic variability systematically in these system classes. The next section explains two fundamental concepts, that of a genotype space and the phenotypes therein, for these system classes. The two sections after that summarize recent evidence that these system classes share two organizational features of genotype space that facilitate phenotypic variability and evolutionary innovation. These are the existence of genotype networks (to be defined further below) and of a great phenotypic diversity in different neighborhoods of genotype space. The next section explains how these concepts can help explain the origins of evolutionary innovations. A final section suggests why system classes as different as these can share such similarities, and especially the existence of genotype networks. The reason is that systems in these classes typically operate in changing environments, which endows them with robustness to environmental change, but also to genetic change. The existence of genotype networks is a consequence of such robustness.

I emphasize that the principles I discuss here by no means negate the importance of other factors, such as environmental change, phenotypic plasticity, multi-functionality of biological systems, epigenetic change, gene duplication, and gradual evolution from simple to complex systems, for the ability to bring forth variable phenotypes [15–20]. The principles I discuss are complementary to other factors, and may even help clarify the role these factors play in phenotypic variability. They do not only apply to qualitatively new phenotypes, but also to beneficial quantitative changes in existing phenotypes — evolutionary adaptations, in the jargon of evolutionary biologists. A more comprehensive treatment can be found elsewhere [10].

METABOLIC NETWORKS AND THEIR INNOVATIONS

Large-scale metabolic networks are systems of hundreds to more than 1000 chemical reactions that are at work in every organism [21]. Their most fundamental task is to transform sources of chemical elements and energy into a chemical form that is useful to the organism. Evolutionary innovations in metabolism fall into multiple categories. An especially prominent category concerns traits that allow organisms to survive on new sources of food. Prokaryotes are the undisputed masters of such innovations. They are able to survive on sources of carbon and energy that are bizarre and toxic (to us), including methane, hydrogen gas, crude oil, antibiotics, and xenobiotic chemicals [22–24].

The likely reason why many metabolic innovations occur in prokaryotes is the ability of prokaryotes to exchange genes through a variety of mechanism [25,26]. Horizontal gene transfer can transform the genome of a prokaryote on short evolutionary timescales, such that even different strains of the same bacterial species may differ in hundreds of genes. Such horizontal gene transfer is the cause of many evolutionary innovations. A candidate example involves the prokaryote *Sphingomonas chlorophenolica*, which is able to metabolize the toxic xenobiotic compound pentachlorophenol. It does so through a sequence of four chemical reactions [27], none of which is new to *S. chlorophenolica*. Two of them are involved in degrading naturally occurring chlorinated compounds in other organisms. Two others are involved in the metabolism of the common amino acid tyrosine [27]. The innovation in *S. chlorophenolica*'s metabolism is the combination of these reactions. Such new combinations of reactions can be easily achieved through horizontal transfer of enzyme coding genes.

Prokaryotes may be the most prolific metabolic innovators, but metabolic innovations also occur in the evolution of higher, multicellular organisms. An example is the urea cycle, an innovation that occurred during the evolution of land-living animals. It allows animals to dispose of ammonia, a waste product of their metabolism that is toxic to cells, by converting it into urea that is excreted in urine. The urea cycle consists of five metabolic reactions, none of which are new to their carrier. Individually, they are widespread in many organisms. Four of these reactions are involved in the biosynthesis of arginine, and the fifth is involved in the degradation of arginine [28]. What is new is the combination of these five reactions into a metabolic cycle, a major innovation of biological waste management.

REGULATORY CIRCUITS AND THEIR INNOVATIONS

Regulation is a process that changes the activity of genes and their products. It can affect transcription, translation, post-translational modification, transport, as well as several other aspects of gene and protein function. Among all the known modes of regulation, transcriptional regulation is perhaps the most prominent [29–33]. The reason is that most modes of regulation ultimately affect the regulation of transcription. Transcriptional regulation is thus a backbone of regulatory processes inside an organism. Transcriptional regulation involves specialized proteins called transcription factors that bind regulatory DNA near a gene. The binding of one or more transcription factors to such regulatory DNA can activate or repress the transcription of a gene through interactions with the RNA polymerase that is responsible for transcribing the gene.

Regulation is often mediated by complex regulatory circuits. Such circuits consist of multiple molecules that mutually influence each other's activity. Transcriptional regulation is no exception. Transcription factors form regulatory circuits that can comprise dozens of proteins. These proteins regulate the transcription of the genes encoding them, and of many other genes downstream of the circuit genes [34–39]. In doing so, the circuit's proteins produce a gene expression pattern in which specific genes are activated or repressed, a state that can vary in space and time. A gene expression pattern is a transcriptional regulation circuit's phenotype. Such phenotypes play central roles in physiology and in embryonic development, the process that creates a viable adult organism from a fertilized egg [34,35,40].

Regulatory circuits in general, and transcriptional regulation circuits in particular, are involved in the evolution of many new traits. One example involves the evolution of eyespots on the back of butterfly wings [41–43]. These traits may help butterflies deter predators [41–43]. Eyespots start to form during development in regions that are called eyespot foci. These foci express the protein *Distal-less*, which is causally involved in eyespot formation. The number of eyespots that form on a wing corresponds to the number of regions that express *Distal-less* during early wing development. What is more, grafts of *Distal-less* expressing cells to developing wing tissue can be sufficient for eyespot formation in the graft's recipient [44]. *Distal-less* is a transcription factor, a member of a complex regulatory circuit with other functions in the development of wings and legs [41–43].

Another example involves the evolution of dissected leaves in plants [45, 46]. The ancestral leaves of flowering plants were most likely simple leaves, which have an undivided leaf blade [47]. Dissected leaves evolved from such simple leaves. In a dissected leaf, the leaf blade is subdivided into multiple smaller leaflets. Leaf dissection is a trait that may facilitate heat dissipation in hot terrestrial environments and help increase CO_2 uptake in water [45, 46]. Dissected leaves may have originated multiple times in the evolution of flowering plants [47]. During the development of dissected leaves, transcription factors of the KNOX (KNOTTED1-like homeobox) family play a crucial role. They are expressed in leaf primordials, which form close to the growing tip of a plant's shoot. Increasing the expression of KNOX genes during leaf formation can increase the number of leaflets that are forming; conversely, reducing their expression can reduce this number of leaflets [48]. KNOX genes are part of a regulatory circuit [48].

These are just two examples where regulatory proteins and the regulatory circuits they form are critically involved in the formation of an organism and its parts, as well as in the formation of a structure that was an innovation when it first became fully formed. Other prominent examples include the role of Hox genes in the formation of axial structures such as limbs in vertebrates, or the role of MADS box genes in the formation and diversification of flowers [7, 41–43].

MACROMOLECULES AND THEIR INNOVATIONS

Individual proteins and RNA macromolecules are not usually considered the subject of systems biology, but they should be. They are systems whose parts are amino acid or nucleotide monomers. These parts are strung together to form a whole macromolecule that folds intricately in three-dimensional space. Such macromolecules are responsible for all enzyme-catalyzed reactions that take place in a cell. They serve numerous other functions in addition, including transport, structural support, and communication, and they are behind numerous if not all new molecular functions that originated in life's history. Some of these functions involve very little change in a macromolecule's genotype. (This genotype is the DNA string that encodes the molecule, but for many purposes, a protein's amino acid sequence or an RNA molecule's nucleotide sequence can be viewed as the genotype.) An example of a new function requiring little change involves the enzyme l-ribulose-5-phosphate 4-epimerase from the bacterium *Escherichia coli*, which is necessary for *E. coli* to grow on the sugar arabinose as a carbon and energy source. A single amino acid change from histidine to asparagine at position 97 of this enzyme suffices to create a new enzymatic function, an aldolase that joins one molecule of dihydroxyacetone phosphate and one of glycoaldehyde phosphate [49]. New functions in other molecules require greater amounts of amino acid change. Take as an example antifreeze proteins. These proteins occur in numerous organisms that have to survive cold conditions, such as Arctic and Antarctic fish, as well as overwintering insects and plants. Antifreeze proteins lower the freezing point of an organism's body fluids. They originated multiple times independently, in different organisms, and sometimes rapidly, through multiple amino acid changes in various ancestor proteins [50–52]. These are just two examples of myriad evolutionary innovations that occurred in biological macromolecules.

All these three kinds of change — in metabolism, in regulatory circuits, and in macromolecules — may be involved in any one innovation, and in ways that are difficult to disentangle. It is nonetheless useful to study these kinds of change separately, in order to find out whether any commonalities exist among them. Such commonalities would point to more general principles of phenotypic variability.

TOWARDS A SYSTEMATIC UNDERSTANDING OF INNOVATION

By themselves, examples of innovations like those just discussed may not help us answer whether broader principles of innovation exist. To this end, it may be necessary to study innovation more systematically. One way to do that is to study a 'space' of possible innovation in each of the three system classes. This space is vast, too vast to understand exhaustively. But even by examining small samples of this space it is possible to learn about the structure of the entire space, and thus about principles of innovability. I will next discuss such a more systematic approach for each system class. Before that, however, I want to highlight three goals that a systematic understanding of innovation — an innovability theory — should achieve (others are highlighted elsewhere [10]).

The first goal reflects perhaps the most difficult problem that the origin of new beneficial trait poses to our understanding of evolution. Most mutations that affect an organism's genotype are deleterious, that is, their effects harm rather than benefit their carrier. Such mutations produce inferior phenotypes. Thus, to find new and *superior* phenotypes organisms may have to explore many mutant genotypes. At the same time, however, organisms need to preserve existing, well-adapted phenotypes. In other words, organisms have to be conservative and explore many new phenotypes at the same time. How both objectives can be achieved simultaneously is a question that a theory of innovation would have to answer.

The second goal relates to the observation that many innovations in the history of life have occurred more than once [6]. Dissected leaves may have evolved multiple times; so did antifreeze proteins; and so did many metabolic innovations. For example, life has solved the metabolic problem of incorporating atmospheric carbon (CO_2) into biomass at least three times in different ways, that is, through the Calvin–Benson cycle, through the reductive citric acid cycle, and through the hydroxypropionate cycle [53].

A third goal relates to the observation that some innovations seem to combine existing parts of a system to create a new function. I mentioned a metabolic pathway that can degrade pentachlorophenol, as well as the urea cycle, both of which involve new combinations of existing enzymatic reactions — parts of a metabolic system. Is this combinatorial nature of innovations a peculiarity of some innovations, or is it a more general phenomenon?

The framework I will discuss here suggests an answer to all three questions.

GENOTYPE SPACES AND THE PHENOTYPES THEREIN

This section discusses genotypes and phenotypes separately for each of the three systems classes mentioned above.

The metabolic genotype of an organism is the totality of its DNA that encodes metabolic enzymes. It is thus fundamentally a subset of its genome. While one can think of this genotype as a DNA string, is often more expedient to represent the genotype more compactly. Here is a compact representation that is well suited to study innovation systematically [54]. Consider the known universe of biochemical reactions, that is, chemical reactions catalyzed by an enzyme that are known to occur somewhere in some organism. This known universe of biochemical reactions currently comprises more than 5000 such reactions [55]. One can write these reactions as a list, as shown in Figure 13.1a, which represents each reaction by its stoichiometric equation. Any one organism, such as a human or the bacterium *E. coli*, will have enzymes that catalyze some of these reactions but not others. For reactions that are catalyzed in any one organism, write 1 next to the stoichiometric equation shown in Figure 13.1a. For every reaction that does not occur, write 0. The result of this procedure is a binary string that indicates which reactions do or do not take place in the metabolism of an organism. It is a compact description of a metabolic genotype, comprising all the enzymatic reactions that take place in a metabolic network. With this definition in mind, the notions of metabolic genotype and metabolic network are used here synonymously.

The totality of all possible metabolic genotypes — the set of all the binary strings defined above — constitutes a metabolic *genotype space*, a collection of possible metabolic genotypes. This space is vast, containing more than 2^{5000} possible genotypes, more metabolisms than could ever be realized on earth (and many of them surely useless to life as we know it). It is the space of all possible metabolisms that can be realized with a given set of biochemical reactions. To understand metabolism and metabolic innovation systematically is to understand the structure of this space, and the metabolic phenotypes that exist in it.

A few further concepts are useful in discussing metabolic genotype space. The first is that of a *neighbor*. Two metabolic networks are neighbors in genotype space if they differ in a single chemical reaction. A *neighborhood* of a metabolic network comprises all its 1-neighbors, all metabolic genotypes that differ from it in a single reaction. These concepts can be extended to k-neighbors, networks that differ from a given network in k chemical reactions. The distance of two metabolic networks indicates the fraction of reactions in which they differ. Two metabolic networks have a distance of $D = 0$ if they contain the same reactions; a distance of $D = 0.5$ if 50% of reactions that are catalyzed by one network are not catalyzed by the other

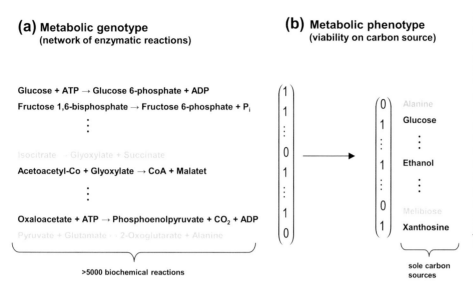

FIGURE 13.1 **Metabolic genotypes and phenotypes.** Panel a) shows the metabolic genotype of a genome-scale metabolic network. It can be represented in a simplified form as a binary string. The entries of this string correspond to one biochemical reaction in a 'universe' of known reactions. Panel b) shows one of many possible representations of a metabolic phenotype, a binary string representation whose entries correspond to individual carbon sources. This representation contains a one for every carbon source from which a metabolic network can synthesize all biomass precursors. *(Figure and caption adapted from [10]. Used with permission from Oxford University Press.)*

network or vice versa; and a distance of $D = 1$ if they differ in every single reaction [54,56,57].

There are as many ways to define a metabolic *phenotype* as there are tasks of metabolism. Metabolism detoxifies waste, synthesizes molecules for defense and communication, and manufactures all small precursor molecules for biomass synthesis. The latter task is the most fundamental, and I will therefore focus on metabolic phenotypes related to this task. For free-living organisms such as *E. coli* there are of the order of 60 small biomass precursors [58]. These include all proteinaceous amino acids, DNA nucleotide precursors, RNA nucleotide precursors, as well as multiple lipids and enzyme cofactors. A network's ability to synthesize all these molecules will depend on the nutrients that are available in an environment. Some organisms, such as *E. coli*, can survive in very simple, minimal chemical environments. These environments contain only one kind of molecule that provides each chemical element; at least one of these molecules also provides energy. Most free-living organisms can use multiple different sources of chemical elements and of energy.

These observations give rise to the following definition of a metabolic phenotype, which is focused on sources of carbon and energy but can be easily extended to sources of other chemical elements [54,57]. Consider a given number of molecules that could serve as sources of carbon energy to some organism. Write these molecules as a list, as shown in Figure 13.1b. If the metabolism of a given organism can synthesize biomass — that is, if it can sustain life on any one of these carbon sources (that is, the organism needs to be able to use this carbon source as its *only* carbon source) — write a 1 next to the carbon source. Otherwise, write a 0. In this way one can define a metabolic phenotype as a binary string that reflects an organism's viability on different

sources of carbon or other elements. Note that even for a modest number of 100 different potential carbon sources, the number of possible metabolic phenotypes is already 2^{100} or $>10^{30}$.

This representation of metabolic phenotypes lends itself to the systematic study of new metabolic phenotypes. Consider a genotypic change that causes the addition of chemical reactions to a metabolic network by horizontal gene transfer. If these new reactions allow an organism to survive on a carbon source that it had not been previously able to utilize, a metabolic innovation has arisen. In an environment where other carbon sources limit growth, or where they are absent, this ability can make a life-changing qualitative difference to its carrier.

I will now discuss analogous definitions of genotypes and phenotypes for regulatory circuits. The evolution of regulatory circuits, and especially of transcriptional regulation circuits, is difficult to study experimentally. Part of the reason is that regulatory DNA can occur far away from the genes it regulates; also, such DNA can change very rapidly on evolutionary timescales [36,59–66]. In addition, to understand the relationship between genotype and phenotype requires an analysis of many circuit genotypes and their phenotypes. For these reasons, computational models of regulatory circuits are still indispensable to understand genotype–phenotype relationships in such circuits. The evidence discussed below stems from well-studied models of transcriptional regulation circuits [67–71]. Variants of these models have been used successfully to understand the development of specific organisms, such as the early fruit fly embryo, and to predict the developmental changes in mutant embryos [67, 72–75]. In addition, they have helped us understand a variety of evolutionary phenomena, such as how regulatory circuits can evolve increased robustness to perturbations, and that cryptic variation — genotypic variation without phenotypic effects in a given environment — might facilitate evolutionary adaptation [76–80]. Note that circuits different from those discussed here, such as signaling circuits, can show properties similar to those highlighted below, which suggests that these properties may be generic features of regulatory circuits [71,81,82].

The genotype of a regulatory circuit comprises the genomic DNA that encodes all regulatory molecules, as well as the non-coding DNA that may help determine the interactions between them. For a transcriptional regulation circuit, this genotype typically includes the genes that encode the circuit's transcriptional regulators, as well as the regulatory DNA sequences that determine where a regulator binds, and which therefore determine who regulates who in the circuit. As in metabolism, there are more compact representations of a circuit's regulatory genotype than its DNA sequence. For example, one can represent the regulatory genotype simply through a square matrix $w = (w_{ij})$, whose entries w_{ij} reflect whether transcription factor j regulates the expression of transcription factor i in the circuit. In the simplest possible representation, this matrix contains only information about whether this interaction is activating ($w_{ij} = +1$), repressing ($w_{ij} = -1$), or absent. Even the simplest representation shows that the number of circuit genotypes will be very large, even for circuits with a modest number N of genes. That is, there are 3^{N^2} possible circuits. In more complicated representations of transcriptional regulation circuits, these interactions could assume a larger or a continuous range of values. Mutations in DNA that affect the regulatory interactions of circuit genes can change this circuit genotype. For example, a mutation in regulatory DNA that abolishes binding of a transcription factor to this DNA may also abolish regulation of a nearby gene by this transcription factor, and thus eliminate one of the regulatory interactions w_{ij} of this circuit ($w_{ij} \rightarrow 0$).

The mutual regulatory interactions of molecules in such a circuit will create a gene expression pattern. This expression pattern is a circuit's phenotype. It typically influences the expression of many genes downstream of the circuit, genes that influence physiological or developmental processes. Changes in such phenotypes caused by mutations of the circuit's regulatory genotype can help create new traits, some of which may become evolutionary innovations.

To study the origin of new gene expression phenotypes in such circuits systematically, one needs to think of any one circuit as being part of a much larger genotype space of circuits. This space contains all possible circuits of a given number N of genes. In this genotype space, two circuits are k-neighbors if they differ in k regulatory interaction. The k-neighborhood of a circuit comprises all circuits that differ from it in no more than k regulatory interaction. The distance D of two circuits can be defined as the fraction of regulatory interactions in which they differ. For example, two circuits would have a distance of $D = 0.2$, if they differed in 20% of their interactions. They would have a distance of $D = 1$ if they differed in every single interaction, that is, if no interaction that occurs in the first circuit also occurs in the second circuit [83,84].

The final class of systems to be discussed here are protein and RNA macromolecules. Their genotype spaces, also known as sequence spaces, have been studied for many years [85–87]. For protein strings of a given length N of amino acids, genotype space comprises all amino acids strings of length N, and thus a totality of 20^N such strings, because 20 different amino acids occur in most proteins. For RNA molecules of N nucleotides, it comprises 4^N possible RNA strings. As for metabolism and for regulatory circuits, the sizes of these genotype spaces can be astronomically large. Two protein and RNA molecules are k-neighbors in genotype space if they differ in k nucleotides or amino acids. The k-neighborhood of a molecule comprises all of its neighbors. The distance of two protein or RNA molecules can be defined in a variety of ways, one of them being the fraction of monomers in which they differ.

The phenotype of a protein or RNA molecule comprises its secondary structure, its tertiary structure — that is, its three-dimensional fold in space — as well as its biochemical function, be it catalytic, structural, or something else. Over the last 40 years the genotypes and phenotypes of tens of thousands of proteins have been characterized biochemically. They provide a rich source of information to study the relationship between genotype and phenotype [88]. Fewer RNA phenotypes are known, but for RNA secondary structures algorithms exist that can predict RNA phenotypes from genotypes [89,90]. Albeit not perfectly accurate for any one sequence, the relevant algorithms are sufficiently accurate (and also sufficiently fast) to characterize thousands to millions of different RNA genotypes and their phenotypes [87,91–93]. Because RNA secondary structure phenotypes are necessary for the functioning of many molecules, they are interesting study objects in their own right [94–96].

Genotype Networks

The genotype spaces of metabolic networks, regulatory circuits, and macromolecules are much too large to be characterized exhaustively. However, they can be characterized through (unbiased) sampling of genotypes or phenotypes, or through exhaustive enumeration of genotypes and phenotypes for small systems. These approaches can identify generic properties of such spaces, that is, properties that hold for typical genotypes and phenotypes.

Two such properties are discussed in this and the following section.

The first is that a given phenotype is typically not just formed by one or few genotypes, but by astronomically many genotypes [54,56,84,86,87]. In other words, vast sets of genotypes share the same phenotype. In some systems, such as metabolism, it is possible to characterize such sets of genotypes through Markov chain Monte Carlo sampling of genotype space [54,56]. This involves carefully designed random walks through genotype space. Briefly, one starts from a specific metabolic network (metabolic genotype) with a given number of reactions and a given phenotype. (This starting genotype can be viewed as a single point in metabolic genotype space.) Techniques such as flux-balance analysis allow one to compute this metabolic phenotype from the network's metabolic genotype [14,97]. One then either eliminates a specific, randomly chosen reaction from this starting genotype, or one adds a reaction chosen at random from the known universe of biochemical reactions. After this change to the network, one computes the phenotype of the changed network. If this phenotype is the same as before the change — that is, if this change has not altered the ability of the network to sustain life on a given spectrum of carbon sources — then the altered network is kept. Otherwise, the genotypic change is discarded and one reverts to the initial metabolic network. One then applies a second change (reaction deletion or addition), evaluates the phenotype, and keeps the altered network if it is unchanged, and so on, in a long sequence of $>10^5$ reaction changes, each of which has to keep the network's phenotype unchanged. This approach can not only sample sets of genotypes with a given phenotype uniformly, that is, in an unbiased manner, it also resembles the process by which metabolic networks evolve through the deletion and the addition of reactions to a network, for example through horizontal gene transfer.

Using this approach one can ask how different two metabolic genotypes that have the same phenotype can become? The answer is that they can become very different. For example, metabolic networks that have the same number of reactions as the *E. coli* network, and that can synthesize all *E. coli* biomass precursors in a minimal environment that contains glucose as the *sole* carbon source, can differ in more than 75% of their reactions. Moreover, any two such metabolic networks can typically be connected to one another. This means that sequences of reaction changes exist that can convert one metabolic network into the other, such that no individual change alters the phenotype [54,56]. In other words, metabolic genotypes with the same phenotype form extended networks — genotype networks — in metabolic genotype space. Note the distinction between a metabolic network and a genotype network: a metabolic network corresponds to a single point, a single genotype in genotype space; a genotype network is a network of such genotypes, and thus a network of metabolic networks. I will keep these two meanings of a network distinct.

Genotype networks exist for metabolic phenotypes that can sustain life on many different sole carbon sources, as well as on multiple carbon sources (when each source is provided as the sole carbon source). Even metabolic networks that can sustain life on up to 60 carbon sources form genotype networks that can still differ in 75% of their reactions. Extended genotype networks also exist for metabolic networks that contain different numbers of reactions, and for phenotypes that involve sources of chemical elements other than carbon [57]. Thus, the most basic properties of genotype networks are not highly sensitive to the phenotype one considers. Based on what we know, they appear to be generic features of metabolic genotype space.

To explore the genotype space of regulatory circuits one can use similar sampling approaches [83,84], and one finds a similar organization of this space. Two circuits with the same gene expression phenotype can have a genotype distance between $D=0.75$ and $D=1$, that is, they may differ in 75–100% of their regulatory interactions. What is more, circuits with very different genotypes can typically be connected through a sequence of steps, each of which changes a single regulatory interaction, but none of which alters the circuit's gene expression phenotype. These observations hold for broadly different gene expression phenotypes, and regardless of the number of genes or regulatory interactions in a circuit, except possibly for the smallest circuits [84,98].

Comparative data on the tertiary structure phenotypes of proteins demonstrates the existence of genotype networks here as well. Although exceptions exist [99], proteins with the same structure and/or function can differ in most of their amino acids [100–103]. Examples include oxygen-binding globins. These proteins occur both in animals and plants, probably share a common ancestor, but are extremely diverse in their genotypes. For example, no more than four of their more than 90 amino acids are absolutely conserved. Despite this genotypic divergence, globins have largely preserved their tertiary structure and their oxygen-binding ability [104–106].

Globins are not unusual in this regard. Other proteins with preserved phenotype are even more diverse. Take triose phosphate isomerase (TIM) barrel proteins. These proteins have preserved their tertiary structure but can differ in every single amino acid [107,108]. More generally, proteins with highly diverged genotype yet highly conserved phenotype are the rule rather than the exception [100–103]. Such proteins form vast genotype networks that extend far into genotype space. Phylogenetic analyses of related proteins from different organisms reveal a reflection of these networks in the tree of life [106].

Fewer RNA than protein phenotypes have been characterized experimentally, but computational analyses of the relationship between RNA sequence and secondary structure point to much the same phenomenon. RNA genotypes with the same secondary structure phenotype typically form large genotype networks that extend far into, and often all the way through, RNA genotype space [87,109].

In sum, metabolic networks, regulatory circuits, and macromolecules — very different kinds of systems — show a remarkable common property. Genotypes that have the same phenotype are typically organized in large genotype networks that reach far through genotype space. I will return to genotype networks later, when I discuss their significance for phenotypic variability.

The Diversity of Neighborhoods in Genotype Space

A second common property of the three system classes emerges from the analysis of genotypic neighborhoods. The neighborhood of a genotype is relevant for phenotypic variability, because it contains genotypes that can be easily reached from this genotype, that is, through one or few small genotypic changes. For an analysis of phenotypic variability, it is therefore useful to examine the spectrum of phenotypes $P1$ that occur in a given neighborhood of a genotype $G1$ that has some phenotype P. A simple question is whether the spectrum of phenotypes in this neighborhood depends on the genotype $G1$. More precisely, consider two genotypes $G1$ and $G2$ with the same phenotype P and a given distance D. Denote as $P1$ and $P2$ the sets of phenotypes (different from P) in their respective neighborhoods. How different is the set $P1$ from the set $P2$? That is, are most phenotypes in $P1$ also contained in $P2$? Or are most of these phenotypes unique to the neighborhood of $G1$, in the sense that they do not also occur in the neighborhood of $G2$?

In metabolism, one finds that the neighborhoods of two metabolic genotypes $G1$ and $G2$ sampled at random from the same genotype network contain mostly different novel phenotypes. In other words, the set $P1$ of new phenotypes in the neighborhood of $G1$ is very different from the set $P2$ of new phenotypes in the neighborhood of $G2$. This holds regardless of the specific genotypes $G1$ and $G2$, as well as regardless of the specific phenotype P that they have [54,57]. The situation in regulatory circuits is not much different. There, small neighborhoods around two circuits $G1$ and $G2$ may contain sets of phenotypes $P1$ and $P2$ that differ in the majority of their phenotypes, even for circuits whose genotypes differ little, that is, in no more than 20% of their regulatory interactions [83]. Much the same holds for protein and RNA molecules [87,92,110,111]. For example, a recent analysis studied more than 16 000 enzymes with known sequence, tertiary structure, and enzymatic function. It showed that small neighborhoods around two proteins $G1$ and $G2$ that differ at fewer than 25% of their amino acids can contain sets of $P1$ and $P2$ of new enzyme function phenotypes, such that the majority of enzymatic functions found in $P1$ are not contained in $P2$.

In sum, metabolic networks, regulatory circuits, and macromolecules show two common qualitative properties in the organization of their genotype space. Property 1 is the existence of genotype networks that reach far through genotype space. Property 2 is that small neighborhoods around different genotypes typically contain different phenotypes, even if the genotypes do not differ greatly. Figure 13.2 shows a schematic sketch of these properties. The large rectangle in the figure stands for a hypothetical genotype space. Each of the small circles stands for a single genotype. The open circles correspond to genotypes that share some hypothetical phenotype P (not shown). Two genotypes are connected by a straight line if they are

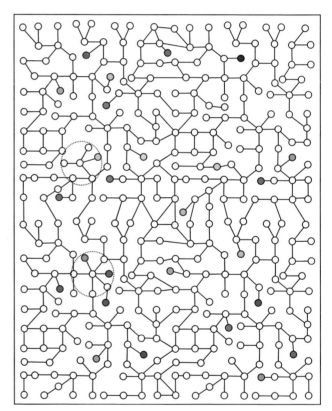

FIGURE 13.2 **A highly simplified schematic of the structure of a genotype network and the new phenotypes near it. See text for details.** Note that genotype networks are objects in a high-dimensional genotype space with counterintuitive geometric properties. Also, actual genotype networks contain an astronomical number of members. Individual genotypes may have hundreds to thousands of neighbors, only few of which can be shown. In addition, each of the genotypes shown in different colors is also part of a vast genotype network that is not shown. A figure like this can thus merely provide a modicum of intuition about the organization of genotype space. (*Adapted from [10]. Used with permission from Oxford University Press.*)

neighbors. The network of open circles stands for a large connected genotype network that traverses genotype space. The colored circles represent genotypes whose phenotype is different from P (one color per phenotype), and that are neighbors of the genotypes on the genotype network. Note that different regions of this hypothetical genotype space contain different colors. The two large dashed circles represent the neighborhoods of two genotypes on the genotype network. Note that the phenotypes (colors) in these two neighborhoods are different, a reflection of property 2. Note that Figure 13.2 represents a complex, vast, and high dimensional genotype space in a highly simplified, two-dimensional way. For example, actual genotype networks contain an astronomical number of members. Individual genotypes may have hundreds to thousands of neighbors, only few of which can be shown. In addition, each of the colored genotypes is also part of a vast genotype network that is not shown.

Genotype Networks and Their Diverse Neighborhoods Can Help Explain the Origin of New Phenotypes

I will now return to the three questions about the origin of new phenotypes posed earlier, and which a systematic understanding of phenotypic variability needs to address.

The first is that organisms need to preserve old, well-adapted phenotypes while exploring many new phenotypes. Properties 1 and 2 can jointly help answer this question. To see this, consider that all evolution takes place in populations of organisms, each with its own genotype. Envision a population of genotypes in any one of our three system classes. Individuals in this population have a phenotype that may be necessary for their survival, but somewhere in genotype space a superior phenotype may exist. The genotypes of individuals in this population suffer mutations that affect their genotype. Natural selection eliminates any mutants that have not preserved the old phenotype or replaced it with a superior phenotype. One can view such a population as a cloud of points [112] that diffuses on a genotype network through genotype space.

Genotype networks (property 1) allow the genotypes of individuals in such a population to change without affecting their phenotype. They allow the preservation of old phenotypes despite genotypic change. Over time, genotypes may change dramatically while preserving their phenotype. During this process, the population explores different regions of genotype space. Because of property 2, the diversity of genotypic neighborhoods, the neighborhoods of the population's genotypes will contain ever-changing sets of new phenotypes. This means that the population can explore different novel phenotypes in its neighborhood as its genotypes change. In sum, genotype networks and their neighborhoods allow populations to preserve old phenotypes while exploring many new phenotypes.

Neither property 1 nor property 2 alone would be sufficient for such exploration [10]. Without property 1 (no genotype networks), a population would have low genotypic diversity and could therefore not explore different neighborhoods in this space. The total number of phenotypes is much greater than the number of phenotypes in a neighborhood for any one of the three system classes [10]. Thus, the absence of genotype networks would mean that most novel phenotypes are off-limits to an evolving population. Conversely, in the absence of property 2, that is, if the neighborhoods of different genotypes contained mostly identical new phenotypes, the existence of genotype networks would be irrelevant to the exploration of novel phenotypes. The reason is that even though a population's genotypes could change during evolutionary exploration of a genotype network, the changing genotypes would have access to the same unchanging spectrum of novel phenotypes.

The second question posed earlier regards the multiple evolutionary origins of many evolutionary innovations [6, 53]. Such multiple origins may be difficult to understand, if one assumes that innovations are unique solutions to particular problems that life faces, and that they are unique because the underlying problems are difficult to solve. Viewing such solutions from the vantage point of a genotype space leads to a completely different perspective. There, a genotype with a specific phenotype can be viewed as a solution to a particular problem. The existence of vast genotype networks for typical phenotypes means that typical problems have not just one, but astronomically many solutions. Different genotypes on the same genotype network can be viewed as different solutions to the same problem. Populations of organisms that explore genotype space from different starting points may encounter different solutions. To be sure, most innovations may involve multiple changes in all three major system classes, but because genotype networks are ubiquitous in all three classes, so are multiple solutions to most problems. From this perspective, the multiple origins of many evolutionary innovations are not surprising but rather to be expected.

The third question is whether innovation is usually combinatorial, involving old parts that are combined to new purposes. Here again, the vantage point of a genotype space, which contains all possible innovations, suggests a very straightforward answer: all innovation is combinatorial. New functions of proteins emerge through new combinations of amino acids. New metabolic phenotypes emerge through new combinations of already existing biochemical reactions. And new gene expression patterns of regulatory circuits arise through new combinations of regulators and their interactions.

Innovations that involve new combinations of existing system parts have many guises. Students of embryonic development, for example, have coined the term co-option, the use of an existing regulator or an existing regulatory interaction for new purposes [113]. Examples include the regulator *Distal-less* mentioned earlier. It is involved in the development of insect legs and wings, but it has been co-opted to form eyespots [44,114]. *Distal-less* does not act alone in these processes. It is part of regulatory circuit that involves other molecules, some of which may also have changed their interactions and expression in helping form a new body structure. One can view the co-option of *Distal-less* as a special case of a more general principle, in which new combinations of regulators and their interactions specify new body parts.

In sum, evidence from three very different kinds of systems can help answer several related questions about the origins of new phenotypes. It can help us grasp how life can preserve old phenotypes while exploring many new phenotypes. It can help us understand how many evolutionary innovations have originated multiple times in the history of life. And it can help us appreciate that innovation will generally involve combinations of old parts to achieve new purposes. The fact that the properties described exist in very different kinds of system suggests that they apply to multiple different kinds of innovation. They are suitable to form the basis of a general innovability theory.

Robustness, Genotype Networks and Environmental Change

A question so far left open is why genotype networks exist in metabolism, regulatory circuits, and macromolecules. At first sight this may seem difficult to answer, because these system classes are so different. However, it can be shown that this commonality emerges from a very simple property that they share: the robustness of their phenotypes to mutational changes in individual system parts.

In the genotype space framework such robustness can be thought of as a property of individual genotypes. Mutations often change any one genotype into one of its neighbors. A loss-of-function mutation in an enzyme-coding gene may eliminate one reaction from a metabolic network and transform the network into one of its neighbors; a mutation-changing regulatory DNA may eliminate a transcription factor's binding to this DNA, and hence its regulatory interaction with a target gene, transforming the circuit into one of its neighbors; a nucleotide change in a protein-coding gene often transforms the protein into one of its neighbors. One way to quantify the robustness of a genotype is through the proportion of its neighbors that have the same phenotype as itself. Metabolic networks, regulatory circuits and macromolecules are all to some extent robust in this sense [39,115–123]. This robustness has been estimated experimentally in systems such as proteins through random mutagenesis experiments [115–118], in metabolic networks through knockout mutations of enzyme-coding genes, and in regulatory circuits through circuit rewiring [39,115–123]. Computer modeling confirms that such robustness is a generic feature of these three system classes [54,84,124]. Typically, between 10% and more than 50% of a genotype's neighbors have the same phenotype as itself, depending on the system and the individual genotype [10]. It can be shown mathematically that this property is both necessary and sufficient to bring forth genotype networks that are astronomically large, and that extend far through genotype space [10]. From this vantage point, one could argue that genotype networks are a consequence of robustness. (Their diverse phenotypic neighborhoods emerge from the fact that many more phenotypes exist than the neighborhood of any one genotype can contain [10].)

These observations raise a further question. What is the ultimate cause of this robustness? Although multiple answers have been proposed, the current best candidate emerges from the observation that living systems need to operate in different environments [119,125–129]. The notion of an environment should be broadly defined in this context, and include the biotic, chemical, and physical environment outside an organism, as well as inside its cells. For example, it includes the changing chemical environments that provide nutrients to a metabolic network, the different regions of a developing embryo in which a regulatory circuit is exposed to different chemical signals, and the intracellular chemical environment that macromolecules need to operate in.

The role of changing environments for robustness has been most thoroughly studied in the context of metabolic networks [56, 119,129–133]. A free-living organism such as *E. coli*, which encounters multiple different environment containing different nutrients, can sustain life on dozens of different nutrients. It also has a large metabolic network that comprises more than 900 reactions. In any one such environment, it is also robust to the removal of individual reactions [134]. For example, more than 70% of its reactions are dispensable in a minimal environment with glucose as the sole carbon source. Such robustness is not a peculiarity of the *E. coli* metabolic network: it is a general property of metabolic networks that can sustain life in multiple environments [57,119]. (Note that any reaction that is dispensable in one environment may be essential in a different environment [119].) If *E. coli* lived for many generations in an environment that did not vary in its nutrient composition, its robustness would slowly disappear. This is what happened in endosymbiotic organisms such as *Buchnera aphidicola*, a relative of *E. coli* that has lived for millions of years inside its host organism, an aphid

[135–137]. *Buchnera* has a much simpler metabolic network comprising only 263 reactions. During its long association with its host and the constant environment this host provides, *Buchnera* has lost the ability to survive on a broad spectrum of nutrients. What is more, it has also lost almost all robustness to the removal of chemical reactions from its metabolism [135].

A metabolism that is to sustain life in multiple different chemical environments needs specific enzymatic reactions to metabolize all the nutrients that these environments may contain. It therefore needs to be more complex than a metabolism highly specialized to one specific environment. This increased complexity endows metabolism with robustness to the removal of reactions in any one environment [10,57]. Although the relationship between environmental change, increased complexity, and robustness is not as well explored for proteins and regulatory circuits, similar arguments can be made for them [10].

In sum, the ability to cope with changing environments can require increased complexity of a biological system, which can cause robustness to genetic change in any one environment. Such robustness is responsible for the existence of genotype networks, which can facilitate the origin of new phenotypes. Minimally complex systems may not display some of the core properties discussed here [98], and may thus not be capable of exploring a broad spectrum of new phenotypes.

Conclusions and Future Challenges

Genotypic changes in metabolism, in regulatory circuits, as well as in protein and RNA molecules are involved in many if not all evolutionary innovations. These system classes are therefore important study objects to understand the phenotypic variability that brings forth evolutionary adaptations and innovations. I have discussed evidence that all three system classes have two common properties. The first is the existence of genotype networks, vast connected sets of genotypes with the same phenotype that reach far through genotype space. The second is the fact that the neighborhoods of different genotypes with the same phenotype typically contain very different new phenotypes. Together, these properties can help explain how living systems can preserve old phenotypes while exploring many new phenotypes, why many evolutionary innovations in life's history have occurred multiple times, and that evolutionary innovation has a fundamentally combinatorial nature. Robustness of a system to genetic change is both a necessary and sufficient criterion for the existence of genotype networks. A likely cause of such robustness is the fact that many biological systems need to operate in multiple environments.

The observations made here regard qualitative commonalities in the organization of genotype spaces. These spaces may harbor further, unrecognized similarities, but also differences among different kinds of systems. A combination of high-throughput genotyping with emerging technologies for high-throughput phenotyping [138], and sophisticated computer models of genotype–phenotype relationships may reveal many more such principles in the years to come. Given how vast genotype spaces are, myriad principles of phenotypic variability may still await discovery.

REFERENCES

[1] Muller GB, Wagner GP. Novelty in evolution – restructuring the concept. Annu Rev Ecol Evol Syst 1991;22:229–56.

[2] Pigliucci M. What, if anything, is an evolutionary novelty? In 20th Biennial meeting of the philosophy of science association. CANADA: Vancouver; 2006. p. 887–98.

[3] de Vries H. Species and Varieties, Their Origin by Mutation. Chicago, IL: The open court publishing company; 1905.

[4] Shubin N, Tabin C, Carroll S. Deep homology and the origins of evolutionary novelty. Nature 2009;457:818–23.

[5] Moczek AP. On the origins of novelty in development and evolution. Bioessays 2008;30:432–47.

[6] Vermeij GJ. Historical contingency and the purported uniqueness of evolutionary innovations. Proc Natl Acad Sci USA 2006;103: 1804–9.

[7] Irish VF. The evolution of floral homeotic gene function. Bioessays 2003;25:637–46.

[8] Shimeld SM, Holland PWH. Vertebrate innovations. Proc Natl Acad Sci USA 2000;97:4449–52.

[9] Gerhart J, Kirschner M. Cells, Embryos, and Evolution. Boston: Blackwell; 1998.

[10] Wagner A. The Origins of Evolutionary Innovations. A Theory of Transformative Change in Living Systems. Oxford, UK: Oxford University Press; 2011.

[11] Schultes E, Bartel D. One sequence, two ribozymes: implications for the emergence of new ribozyme folds. Science 2000;289: 448–52.

[12] Hayden E, Ferrada E, Wagner A. Cryptic genetic variation promotes rapid evolutionary adaptation in an RNA enzyme. Nature 2011;474:92–5.

[13] Feist AM, Herrgard MJ, Thiele I, Reed JL, Palsson BO. Reconstruction of biochemical networks in microorganisms. Nat Rev Microbiol 2009;7:129–43.

[14] Becker SA, Feist AM, Mo ML, Hannum G, Palsson BO, Herrgard MJ. Quantitative prediction of cellular metabolism with constraint-based models: the COBRA Toolbox. Nat Protocol 2007;2:727–38.

[15] West-Eberhard M. Developmental Plasticity and Evolution. New York, NY: Oxford University Press; 2003.

[16] Jeffery C. Moonlighting proteins. Trends Biochem Sci 1999;24:8–11.

[17] True HL, Berlin I, Lindquist SL. Epigenetic regulation of translation reveals hidden genetic variation to produce complex traits. Nature 2004;431:184–7.

[18] Masel J, Bergman A. The evolution of the evolvability properties of the yeast prion [PSI+]. Evolution 2003;57:1498–512.

[19] Jensen RA. Enzyme recruitment in evolution of new function. Annu Rev Microbiol 1976;30:409–25.

[20] Horowitz NH. On the evolution of biochemical syntheses. Proc Natl Acad Sci USA 1945;31:153–7.

[21] Palsson B. Metabolic systems biology. FEBS Lett 2009;583: 3900–4.

[22] Postgate JR. The Outer Reaches of Life. Cambridge, UK: Cambridge University Press; 1994.

[23] Dantas G, Sommer MOA, Oluwasegun RD, Church GM. Bacteria subsisting on antibiotics. Science 2008;320:100–3.

[24] Nohynek LJ, Suhonen EL, NurmiahoLassila EL, Hantula J, SalkinojaSalonen M. Description of four pentachlorophenol-degrading bacterial strains as *Sphingomonas chlorophenolica* sp nov. Syst Appl Microbiol 1996;18:527–38.

[25] Ochman H, Lawrence J, Groisman E. Lateral gene transfer and the nature of bacterial innovation. Nature 2000;405:299–304.

[26] Pal C, Papp B, Lercher MJ. Adaptive evolution of bacterial metabolic networks by horizontal gene transfer. Nat Genet 2005;37:1372–5.

[27] Copley SD. Evolution of a metabolic pathway for degradation of a toxic xenobiotic: the patchwork approach. Trends Biochem Sci 2000;25:261–5.

[28] Takiguchi M, Matsubasa T, Amaya Y, Mori M. Evolutionary aspects of urea cycle enzyme genes. Bioessays 1989;10:163–6.

[29] Alberts B, Johnson A, Lewis J, Raff M, Roberts K, Walter P. Molecular Biology of the Cell. New York, NY: Garland Science; 2008.

[30] Badis G, Berger MF, Philippakis AA, Talukder S, Gehrke AR, Jaeger SA, et al. Diversity and Complexity in DNA Recognition by Transcription Factors. Science 2009;324:1720–3.

[31] Walhout AJM. Unraveling transcription regulatory networks by protein-DNA and protein–protein interaction mapping. Genome Res 2006;16:1445–54.

[32] Davidson EH, Erwin DH. Gene regulatory networks and the evolution of animal body plans. Science 2006;311:796–800.

[33] Arnone MI, Davidson EH. The hardwiring of development: organization and function of genomic regulatory systems. Development 1997;124:1851–64.

[34] Hueber SD, Lohmann I. Shaping segments: Hox gene function in the genomic age. Bioessays 2008;30:965–79.

[35] Hughes CL, Kaufman TC. Hox genes and the evolution of the arthropod body plan. Evol Dev 2002;4:459–99.

[36] Tuch BB, Li H, Johnson AD. Evolution of eukaryotic transcription circuits. Science 2008;319:1797–9.

[37] Lee T, Rinaldi N, Robert F, Odom D, Bar-Joseph Z, Gerber G, et al. Transcriptional regulatory networks in *Saccharomyces cerevisiae*. Science 2002;298:799–804.

[38] Shen-Orr S, Milo R, Mangan S, Alon U. Network motifs in the transcriptional regulation network of *Escherichia coli*. Nat Genet 2002;31:64–8.

[39] Isalan M, Lemerle C, Michalodimitrakis K, Beltrao P, Horn C, Garriga-Canut M, et al. Evolvability and hierarchy in rewired bacterial gene networks. Nature 2008;452:840–5.

[40] Carroll SB, Grenier JK, Weatherbee SD. From DNA to Diversity. Molecular Genetics and the Evolution of Animal Design. Malden, MA: Blackwell; 2001.

[41] Stevens M, Stubbins CL, Hardman CJ. The anti-predator function of 'eyespots' on camouflaged and conspicuous prey. Behav Ecol Sociobiol 2008;62:1787–93.

[42] Stevens M, Hardman CJ, Stubbins CL. Conspicuousness, not eye mimicry, makes 'eyespots' effective antipredator signals. Behav Ecol 2008;19:525–31.

[43] Stevens M. The role of eyespots as anti-predator mechanisms, principally demonstrated in the Lepidoptera. Biol Rev 2005;80:573–88.

[44] Brakefield PM, Gates J, Keys D, Kesbeke F, Wijngaarden PJ, Monteiro A, et al. Development, plasticity and evolution of butterfly eyespot patterns. Nature 1996;384:236–42.

[45] Gurevitch J. Variation in leaf dissection and leaf energy budgets among populations of Achillea from an altitudinal gradient. Am J Bot 1988;75:1298–306.

[46] Givnish TJ. Comparative studies of leaf form – assessing the relative roles of selective pressures and phylogenetic constraints. New Phytol 1987;106:131–60.

[47] Bharathan G, Goliber TE, Moore C, Kessler S, Pham T, Sinha NR. Homologies in leaf form inferred from KNOXI gene expression during development. Science 2002;296:1858–60.

[48] Hay A, Tsiantis M. The genetic basis for differences in leaf form between Arabidopsis thaliana and its wild relative *Cardamine hirsuta*. Nat Genet 2006;38:942–7.

[49] Johnson AE, Tanner ME. Epimerization via carbon-carbon bond cleavage. L-ribulose-5-phosphate 4-epimerase as a masked class II aldolase. Biochemistry 1998;37:5746–54.

[50] Cheng CC-H. Evolution of the diverse antifreeze proteins. Curr Opin Genet Dev 1998;8:715–20.

[51] Shackleton NJ, Backman J, Zimmerman H, Kent DV, Hall MA, Roberts DG, et al. Oxygen isotope calibration of the onset of ice-rafting and history of glaciation in the North-Atlantic region. Nature 1984;307:620–3.

[52] Chen LB, DeVries AL, Cheng CHC. Convergent evolution of antifreeze glycoproteins in Antarctic notothenioid fish and Arctic cod. Proc Natl Acad Sci USA 1997;94:3817–22.

[53] Rothschild LJ. The evolution of photosynthesis… again? Philos Trans Ro Soc B Biol Sci 2008;363:2787–801.

[54] Rodrigues JF, Wagner A. Evolutionary plasticity and innovations in complex metabolic reaction networks. PLOS Comput Biol 2009;5:e1000613.

[55] Ogata H, Goto S, Sato K, Fujibuchi W, Bono H, Kanehisa M. KEGG: Kyoto encyclopedia of genes and genomes. Nucleic Acids Res 1999;27:29–34.

[56] Samal A, Rodrigues JFM, Jost J, Martin OC, Wagner A. Genotype networks in metabolic reaction spaces. BMC Syst Biol 2010;4:30.

[57] Rodrigues JF, Wagner A. Genotype networks in sulfur metabolism. BMC Syst Biol 2010;5:39.

[58] Feist AM, Henry CS, Reed JL, Krummenacker M, Joyce AR, Karp PD, et al. A genome-scale metabolic reconstruction for *Escherichia coli* K-12 MG1655 that accounts for 1260 ORFs and thermodynamic information. Mol Syst Biol 2007;3.

[59] Stone J, Wray G. Rapid evolution of *cis*-regulatory sequences via local point mutations. Mol Biol Evol 2001;18:1764–70.

[60] Martchenko M, Levitin A, Hogues H, Nantel A, Whiteway M. Transcriptional rewiring of fungal galactose-metabolism circuitry. Curr Biol 2007;17:1007–13.

[61] Tanay A, Regev A, Shamir R. Conservation and evolvability in regulatory networks: The evolution of ribosomal regulation in yeast. Proc Natl Acad Sci USA 2005;102:7203–8.

[62] Gasch AP, Moses AM, Chiang DY, Fraser HB, Berardini M, Eisen MB. Conservation and evolution of *cis*-regulatory systems in ascomycete fungi. PLOS Biol 2004;2:2202–19.

[63] Wray G, Hahn M, Abouheif E, Balhoff J, Pizer M, Rockman M, et al. The evolution of transcriptional regulation in eukaryotes. Mol Biol Evol 2003;20:1377–419.

[64] Ludwig MZ, Bergman C, Patel NH, Kreitman M. Evidence for stabilizing selection in a eukaryotic enhancer element. Nature 2000;403:564–7.

[65] Maduro M, Pilgrim D. Conservation of function and expression of unc-119 from two *Caenorhabditis* species despite divergence of non-coding DNA. Gene 1996;183:77–85.

[66] Romano L, Wray G. Conservation of Endo16 expression in sea urchins despite evolutionary divergence in both *cis* and *trans*-acting components of transcriptional regulation. Development 2003;130:4187–99.

[67] Jaeger J, Surkova S, Blagov M, Janssens H, Kosman D, Kozlov K, et al. Dynamic control of positional information in the early *Drosophila* embryo. Nature 2004;430:368–71.

[68] Sanchez L, Chaouiya C, Thieffry D. Segmenting the fly embryo: logical analysis of the role of the segment polarity cross-regulatory module. Int J Dev Biol 2008;52:1059–75.

[69] Albert R, Othmer HG. The topology of the regulatory interactions predicts the expression pattern of the segment polarity genes in *Drosophila melanogaster*. Journal of Theoretical Biology 2003;223:1–18.

[70] Ingolia NT. Topology and robustness in the *Drosophila* segment polarity network. PLOS Biol 2004;2:805–15.

[71] MacCarthy T, Seymour R, Pomiankowski A. The evolutionary potential of the *Drosophila* sex determination gene network. J Theor Biol 2003;225:461–8.

[72] Mjolsness E, Sharp DH, Reinitz J. A connectionist model of development. J Theor Biol 1991;152:429–53.

[73] Reinitz J, Mjolsness E, Sharp DH. Model for cooperative control of positional information in *Drosophila* by bicoid and maternal hunchback. J Exp Zool 1995;271:47–56.

[74] Reinitz J. Gene circuits for eve stripes: reverse engineering the Drosophila segmentation gene network. Biophys J 1999;76:A272–A272.

[75] Sharp DH, Reinitz J. Prediction of mutant expression patterns using gene circuits. BioSystems 1998;47:79–90.

[76] Azevedo RBR, Lohaus R, Srinivasan S, Dang KK, Burch CL. Sexual reproduction selects for robustness and negative epistasis in artificial gene networks. Nature 2006;440:87–90.

[77] Bornholdt S, Sneppen K. Robustness as an evolutionary principle. Proc R Soc Lon B Biol Sci 2000;267:2281–6.

[78] Wagner A. Does evolutionary plasticity evolve? Evolution 1996;50:1008–23.

[79] Siegal M, Bergman A. Waddington's canalization revisited: developmental stability and evolution. Proc Natl Acad Sci USA 2000;99:10528–10532

[80] Bergman A, Siegal M. Evolutionary capacitance as a general feature of complex gene networks. Nature 2003;424:549–52.

[81] Nochomovitz YD, Li H. Highly designable phenotypes and mutational buffers emerge from a systematic mapping between network topology and dynamic output. Proc Natl Acad Sci USA 2006;103:4180–5.

[82] Raman K, Wagner A. Evolvability and robustness in a complex signaling circuit. Mol BioSyst 2011;7:1081–92.

[83] Ciliberti S, Martin OC, Wagner A. Innovation and robustness in complex regulatory gene networks. Proc Natl Acad Sci USA 2007;104:13591–13596

[84] Ciliberti S, Martin OC, Wagner A. Circuit topology and the evolution of robustness in complex regulatory gene networks. PLOS Comput Biol 2007;3(2):e15.

[85] Maynard-Smith J. Natural selection and the concept of a protein space. Nature 1970;255:563–4.

[86] Lipman D, Wilbur W. Modeling neutral and selective evolution of protein folding. Proc R Soc Lon B 1991;245:7–11.

[87] Schuster P, Fontana W, Stadler P, Hofacker I. From sequences to shapes and back – a case-study in RNA secondary structures. Proc R Soc Lon B 1994;255:279–84.

[88] Berman H, Battistuz T, Bhat T, Bluhm W, Bourne P, Burkhardt K, et al. The Protein Data Bank. Acta Crystallogr B Biol Crystallogr 2002;58:899–907.

[89] Hofacker I, Fontana W, Stadler P, Bonhoeffer L, Tacker M, Schuster P. Fast folding and comparison of RNA secondary structures. Monatshefte fuer Chemie 1994;125:167–88.

[90] Flamm C, Fontana W, Hofacker I, Schuster P. RNA folding at elementary step resolution. RNA 2000;6:325–38.

[91] Fontana W, Schuster P. Shaping space: the possible and the attainable in RNA genotype–phenotype mapping. J Theor Biol 1998;194:491–515.

[92] Sumedha, Martin OC, Wagner A. New structural variation in evolutionary searches of RNA neutral networks. BioSystems 2007;90:475–85.

[93] Wagner A. Robustness and evolvability: a paradox resolved. Proc R Soc Lon B Biol Sci 2008;275:91–100.

[94] Jackson R, Kaminski A. Internal initiation of translation in eukaryotes: The picornavirus paradigm and beyond. RNA 1995;1;985–1000.

[95] Mandl C, Holzmann H, Meixner T, Rauscher S, Stadler P, Allison S, et al. Spontaneous and engineered deletions in the 3′ noncoding region of tick-borne encephalitis virus: construction of highly attenuated mutants of a flavivirus. J Virol 1998;72:2132–40.

[96] Iserentant D, Fiers W. Secondary structure of messenger-RNA and efficiency of translation initiation. Gene 1980;9:1–12.

[97] Feist AM, Palsson BO. The growing scope of applications of genome-scale metabolic reconstructions using *Escherichia coli*. Nat Biotech 2008;26:659–67.

[98] Cotterell J, Sharpe J. An atlas of gene regulatory networks reveals multiple three-gene mechanisms for interpreting morphogen gradients. Mol Syst Biol 2010;6:425.

[99] Doolittle R. The origins and evolution of eukaryotic proteins. Philos Trans Ro Soc B Biol Sci 1995;349:235–40.

[100] Thornton J, Orengo C, Todd A, Pearl F. Protein folds, functions and evolution. J Mol Biol 1999;293:333–42.

[101] Todd A, Orengo C, Thornton J. Evolution of protein function, from a structural perspective. Curr Opin Chem Biol 1999;3:548–56.

[102] Bastolla U, Porto M, Roman HE, Vendruscolo M. Connectivity of neutral networks, overdispersion, and structural conservation in protein evolution. J Mol Evol 2003;56:243–54.

[103] Rost B. Enzyme function less conserved than anticipated. J Mol Biol 2002;318:595–608.

[104] Aronson H, Royer W, Hendrickson W. Quantification of tertiary structural conservation despite primary sequence drift in the globin fold. Prot Sci 1994;3:1706–11.

[105] Hardison RC. A brief history of hemoglobins: Plant, animal, protist, and bacteria. Proc Natl Acad Sci USA 1996;93:5675–9.

[106] Goodman M, Pedwaydon J, Czelusniak J, Suzuki T, Gotoh T, Moens L, et al. An evolutionary tree for invertebrate globin sequences. J Mol Evol 1988;27:236–49.

[107] Copley RR, Bork P. Homology among $(\beta\alpha)_8$ barrels: Implications for the evolution of metabolic pathways. J Mol Biol 2000;303:627–40.

[108] Wierenga RK. The TIM-barrel fold: a versatile framework for efficient enzymes. FEBS Lett 2001;492:193–8.

[109] Schuster P. Molecular insights into evolution of phenotypes. In: Crutchfield JP, Schuster P, editors. Evolutionary dynamics: Exploring the Interplay of Selection, Accident, Neutrality, and Function. New York, NY: Oxford University Press; 2003. p. 163–215.

[110] Ferrada E, Wagner A. Evolutionary innovation and the organization of protein functions in sequence space. PLoS ONE 2010;5(11):e14172.

[111] Huynen MA. Exploring phenotype space through neutral evolution. J Mol Evol 1996;43:165–9.

[112] Eigen M. Viral Quasi-species. Scientific American 1993;269:42–9.

[113] True JR, Carroll SB. Gene co-option in physiological and morphological evolution. Annu Rev Cell Dev Biol 2002;18:53–80.

[114] Panganiban G, Rubenstein JLR. Developmental functions of the Distal-less/Dlx homeobox genes. Development 2002;129:4371–86.

[115] Huang W, Petrosino J, Hirsch M, Shenkin P, Palzkill T. Amino acid sequence determinants of beta-lactamase structure and activity. J Mol Biol 1996;258:688–703.

[116] Rennell D, Bouvier S, Hardy L, Poteete A. Systematic mutation of bacteriophage T4 lysozyme. J Mol Biol 1991;222:67–87.

[117] Weatherall DJ, Clegg JB. Molecular genetics of human haemoglobin. Annu Rev Genet 1976;10:157–78.

[118] Kleina L, Miller J. Genetic studies of the lac repressor. 13. Extensive amino-acid replacements generated by the use of natural and synthetic nonsense suppressors. J Mol Biol 1990;212:295–318.

[119] Wang Z, Zhang J. Abundant indispensable redundancies in cellular metabolic networks. Genome Biol Evol 2009;1:23–33.

[120] Blank LM, Kuepfer L, Sauer U. Large-scale C-13-flux analysis reveals mechanistic principles of metabolic network robustness to null mutations in yeast. Genome Biol 2005;6:R49.

[121] Stelling J, Klamt S, Bettenbrock K, Schuster S, Gilles ED. Metabolic network structure determines key aspects of functionality and regulation. Nature 2002;420:190–3.

[122] Segre D, Vitkup D, Church G. Analysis of optimality in natural and perturbed metabolic networks. Proc Natl Acad Sci USA 2002;99:15112–7

[123] Edwards JS, Palsson BO. The *Escherichia coli* MG1655 in silico metabolic genotype: its definition, characteristics, and capabilities. Proc Natl Acad Sci USA 2000;97:5528–33.

[124] Bornberg-Bauer E, Chan H. Modeling evolutionary landscapes: mutational stability, topology, and superfunnels in sequence space. Proc Natl Acad Sci USA 1999;96:10689–94

[125] Meiklejohn C, Hartl D. A single mode of canalization. Trends Ecol Evol 2002;17:468–73.

[126] Wagner A. Robustness and Evolvability in Living Systems. Princeton, NJ: Princeton University Press; 2005.

[127] Wagner GP, Booth G, Bagherichaichian H. A population genetic theory of canalization. Evolution 1997;51:329–47.

[128] Papp B, Teusink B, Notebaart RA. A critical view of metabolic network adaptations. HFSP J 2009;3:24–35.

[129] Soyer OS, Pfeiffer T. Evolution under fluctuating environments explains observed robustness in metabolic networks. PLOS Comput Biol 2010;6:e1000907.

[130] Vitkup D, Kharchenko P, Wagner A. Influence of metabolic network structure and function on enzyme evolution. Genome Biol 2006;7:R39.

[131] Papp B, Pal C, Hurst LD. Metabolic network analysis of the causes and evolution of enzyme dispensability in yeast. Nature 2004;429:661–4.

[132] Nishikawa T, Gulbahce N, Motter A. E. Spontaneous Reaction Silencing in Metabolic Optimization. PLOS Comput Biol 2008;4:e1000236

[133] Freilich S, Kreimer A, Borenstein E, Gophna U, Sharan R, Ruppin E. Decoupling environment-dependent and independent genetic robustness across bacterial species. PLOS Comput Biol 2010;6:e1000690.

[134] Reed JL, Vo TD, Schilling CH, Palsson BO. An expanded genome-scale model of *Escherichia coli* K-12 (iJR904 GSM/GPR). Genome Biol 2003;4:R54.

[135] Thomas GH, Zucker J, MacDonald SJ, Sorokin A, Goryanin I, Douglas AE. A fragile metabolic network adapted for cooperation in the symbiotic bacterium *Buchnera aphidicola*. BMC Syst Biol 2009;3:24.

[136] Pal C, Papp B, Lercher MJ, Csermely P, Oliver SG, Hurst LD. Chance and necessity in the evolution of minimal metabolic networks. Nature 2006;440:667–70.

[137] Yus E, Maier T, Michalodimitrakis K, van Noort V, Yamada T, Chen WH, et al. Impact of genome reduction on bacterial metabolism and its regulation. Science 2009;326:1263–8.

[138] Benfey PN, Mitchell-Olds T. Perspective – from genotype to phenotype: systems biology meets natural variation. Science 2008;320:495–7.

Chapter 14

Irreversible Transitions, Bistability and Checkpoint Controls in the Eukaryotic Cell Cycle: A Systems-Level Understanding

John J. Tyson[1] and Béla Novák[2]

[1]Department of Biological Sciences, Virginia Polytechnic Institute & State University, Blacksburg, VA 24061, USA, [2]Centre for Integrative Systems Biology, Department of Biochemistry, Oxford University, Oxford OX1 3QU, UK

Chapter Outline

Introduction	265	Start	275
Physiology of the Cell Cycle	265	Mitotic Checkpoint	276
Molecular Biology of the Cell Cycle	268	**Irreversible Transitions in the Mammalian Cell Cycle**	277
Irreversibility and Bistability	270	**Additional Checkpoints**	279
Irreversible Transitions in the Budding Yeast Cell Cycle	272	**Conclusions**	282
Two Alternative States	273	**References**	282

INTRODUCTION

The repetitive cycle of cell growth and division is fundamental to all aspects of biological growth, development and reproduction, and defects in cell growth and division underlie many human health problems, most notably cancer. For these reasons, a driving ambition of molecular cell biologists has been to discover the molecular basis of cell cycle regulation. This goal was largely achieved in the glory years of molecular biology (1980–2000), and Nobel Prizes were duly awarded in 2001 [1–3]. The end result was an appealing vision of a 'universal' molecular mechanism controlling the eukaryotic cell cycle [4]. But the initial appeal was quickly dispelled by a bewildering array of interacting genes and proteins that constitute the control system in any particular organism. For examples, see the interaction maps of cell cycle controls in mammalian cells [5] and in budding yeast cells [6,7]. Looking closely at these maps we can see, in places, clear connections between some molecular interactions and certain aspects of cell cycle progression. But can we identify any general principles of cell cycle regulation embedded in the network? Can we see how the gene–protein interactions in any particular organism determine the unique characteristics of cell proliferation in that organism?

The cell cycle is a particularly striking example of the necessity of systems-level thinking in 21st century molecular cell biology [8]. The resolute reductionism of the last century, albeit necessary for identifying the molecular components of cellular control systems and their interactions with binding partners, has proved insufficient for achieving an integrative understanding of the molecular basis of cell physiology. Putting the pieces back together requires new ways of thinking about and doing molecular biology – an approach now known as molecular systems biology. In this chapter, we show how systems-level thinking reveals deep and unexpected principles of cell cycle regulation.

Table 14.1 provides a glossary of technical terms used in this review.

Physiology of the Cell Cycle

The cell cycle is the sequence of events whereby a growing cell replicates all of its components and divides them more or less evenly between two daughter cells, so that the daughters receive all of the information and machinery necessary to repeat the process [9–11]. The most important components that need to be replicated and partitioned to

TABLE 14.1 Glossary

α Factor	A budding yeast pheromone that induces mating of haploid strains of opposite mating types (*a* and α).
Aneuploid	Having an unbalanced genome, with too few or too many copies of some chromosomes or sub-chromosomal regions.
Bifurcation point	A point in parameter space where a dynamical system undergoes a qualitative change in long-term behavior (e.g., from monostability to bistability).
Bistability	A dynamical system that has two stable steady states for physically realistic values of its variables ($0 \leq x_i \leq x_{i,\max}$).
Centromere	The region of a replicated chromosome where the two sister chromatids are held together by cohesin rings in prometaphase.
Checkpoint	A molecular mechanism that detects problems in cell cycle processes and prevents progression to later events until the earlier processes can be successfully completed.
Coupled reactions	A chemical reaction (R1) can be made to proceed in the non-spontaneous direction ($\Delta G_1 > 0$) by coupling it to a spontaneous reaction (R2; $\Delta G_2 < 0$) in such a way that $\Delta G_{\text{coupled}} = \Delta G_1 + \Delta G_2 < 0$.
Dynamic system	A system of interacting components (e.g., genes and proteins) that undergoes changes in time (and possibly also in space); in a continuous dynamical system, those changes can be described by a set of coupled differential equations.
Irreversibility, thermodynamic	At constant temperature and pressure, a non-equilibrium chemical reaction will proceed spontaneously (irreversibly) in the direction for which $\Delta G < 0$.
Irreversibility, dynamic	A transition in a dynamical system from one stable solution to another that is induced by a parameter change through a bifurcation point and that cannot be reversed by a small change in the parameter value in the opposite direction.
Kinetochore	The docking site for microtubules in the centromeric region of a chromatid.
Monostability	A dynamical system that has only one stable steady state for physically realistic values of its variables ($0 \leq x_i \leq x_{i,\max}$).
Nullcline	The locus of points in the phase plane where $f(x,y) = 0$ or $g(x,y) = 0$.
One-parameter bifurcation diagram	Plot of the steady state value of a dynamical variable as a function of a parameter of the dynamical system.
Parameter	A constant used to define the rate expression on the right-hand-side of a differential equation, e.g., the rate constant of a chemical reaction.
Phase plane	The state space of a two-variable dynamical system; $dx/dt = f(x,y)$, $dy/dt = g(x,y)$.
Proteosome	A multiprotein complex that functions as the cell's garbage disposal unit, capturing polyubiquitinated proteins and hydrolyzing them.
Restriction point	A point-of-no-return in G1 phase of the mammalian cell cycle with respect to removal of growth factor.
Reversibility, equilibrium	All chemical reactions are intrinsically reversible, and at equilibrium the forward and reverse reactions are occurring at the same rate.
Saddle-node bifurcation	A bifurcation point where a stable node and an unstable saddle point coalesce and annihilate each other.
Steady state	A constant solution $\{x(t) = x_o, y(t) = y_o\}$ of a dynamical system with the property that $f(x_o, y_o) = 0$ and $g(x_o, y_o) = 0$.
Steady state, stable	A constant solution of a dynamical system with the property that any small perturbation away from the steady state grows smaller as time proceeds.
Steady state, unstable	A constant solution of a dynamical system with the property that some small perturbations away from the steady state grow larger as time proceeds.
Stochastic	Random fluctuations in the number of molecules of a chemical species due to the probabilistic nature of chemical reactions.

TABLE 14.1 Glossary—cont'd

Two-parameter bifurcation diagram	Plot of regions of distinct dynamical behaviors (e.g., monostability or bistability) in dependence on two parameters of a dynamical system.
Ubiquitin	A small polypeptide (76 amino acids) that can be covalently linked to proteins in order to label the protein for degradation by proteasomes.
Variable	A time-dependent component of a dynamical system, e.g., the activity of a cyclin-dependent kinase.

daughter cells are the chromosomes — the cell's genetic material. In eukaryotic cells, the processes of chromosome replication and partitioning are accomplished in separate phases of the cell cycle (Figure 14.1): S phase (DNA synthesis) and M phase (mitosis).

A eukaryotic chromosome consists of a linear, double-stranded DNA molecule in close association with many types of DNA-binding proteins. During S phase the DNA molecule is carefully copied to produce two identical double-stranded DNA molecules, which become similarly decorated with DNA-binding proteins, to produce a pair of 'sister chromatids'. (Slight differences in the nucleotide sequences of sister chromatids may arise from mistakes — mutations — in the copying process.) As sister chromatid pairs are created during S phase they are physically bound together by protein bands called 'cohesin rings' [12]. After DNA synthesis is completed there is a gap (G2 phase) during which the cell prepares for mitosis. G2 cells are defined by having fully replicated chromosomes that have not yet started the mitotic process.

The goal of mitosis is to separate the sister chromatids, delivering one (and only one) copy of each chromosome to each of the incipient daughter cells. Mitosis is a complex and delicate process [13]. In prophase, the nuclear membrane breaks down, the cytoskeleton rearranges to form a bipolar spindle apparatus, and the replicated chromosomes condense into highly compacted, X-shaped bodies [14]. During prometaphase, microtubules emanating from the poles of the mitotic spindle search for the condensed chromosomes and bind to docking sites (kinetochores) on the sister chromatids at the mid-zone of the X, where the chromatids are still held together by 'centromeric' cohesins [15]. The goal is to attach the sister chromatids, by their kinetochore microtubules, to opposite poles of the mitotic spindle. When properly attached, each X is pulled to the midzone of the spindle (the metaphase plate) where it resides, under tension [16], until every replicated chromosome is properly aligned on the spindle (metaphase). At this point in time, the centromeric cohesin rings are cleaved, allowing the kinetochore microtubules to pull the sister chromatids to opposite poles of the cell (anaphase). Nuclear membranes are then reassembled around each of the segregated masses of chromosomes, forming a binucleate cell (telophase), which then divides down the middle to form two daughter cells, each with a full complement of unreplicated chromosomes (G1 phase).

Progression through the mitotic cell cycle is characterized by four crucial features [17]. First of all, to maintain a constant number of chromosomes per cell from generation to generation, it is necessary that S phase and M phase alternate, i.e., that progress through the cell cycle be

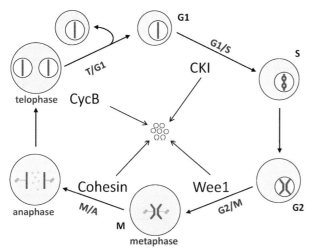

FIGURE 14.1 The eukaryotic cell cycle. Each 'cell' is bounded by a membrane enclosing cytoplasm (gray) with a nuclear compartment (pink) containing a representative chromosome (red bar). The chromosome goes through four distinct phases: G1, unreplicated; S, DNA synthesis; G2, replicated; M, mitosis. During G2 phase the sister chromatids are bound together by cohesin rings (yellow). Mitosis consists of distinct sub-phases: prometaphase (nuclear envelope breakdown, chromosome condensation, spindle assembly), prophase (alignment of replicated chromosomes on the spindle), metaphase (bi-orientation of all chromosomes on the central plate), anaphase (cleavage of cohesin rings and separation of sister chromatids to opposite poles of the spindle), telophase (reassembly of envelopes around the daughter nuclei), cell division. The four characteristic transitions of the cell cycle (G1/S, G2/M, M/A and T/G1) represent four natural stopping points ('checkpoints') where cell cycle progression can be halted if the cell detects any problems with completion of essential functions of the pre-transition state. As the cell passes each checkpoint it degrades a characteristic protein that had been inhibiting the transition (CKI, Wee1, cohesin, cyclin B). The seven small circles represent products of protein degradation.

unidirectional and irreversible: G1 → S → G2 → M → G1 →. Second, the cycle of DNA replication and division must be coordinated with the synthesis of all other cellular components (proteins, lipids, organelles, etc.). That is, the time required to complete the cell division cycle must be identical to the mass-doubling time of cellular growth processes. If cell growth and division are not balanced in this way, then cells will get either larger and larger or smaller and smaller each generation, and eventually they will die.

Third, although the cell cycle is a periodic process, it is not governed by a clock [17]. The time spent in each phase of the cell cycle is highly variable, because progression from one phase to the next depends not on time spent in the current phase but on successful completion of the essential tasks of this phase. These completion requirements are enforced by checkpoints [18–20] that guard the major transitions of the cell cycle: G1/S, G2/M, M/A (metaphase-to-anaphase) and T/G1 (telophase, cell division and return to G1, collectively known as 'exit from mitosis'). A checkpoint has three components [21]. Its surveillance mechanism looks for specific problems (incomplete replication of DNA, misalignment of chromosomes on the mitotic spindle, DNA damage). When a problem is detected, its error correction machinery is put into play (damage repair, reattachment of microtubules to kinetochores, etc.). In the meantime, the checkpoint proper blocks progression to the next stage of the cell cycle until the problem is resolved. These checkpoints ensure that the genome is passed down intact from generation to generation. When the checkpoints are compromised by mutations, daughter cells may inherit seriously damaged chromosomes (e.g., missing large pieces of the genome). Chromosomal abnormalities may trigger programmed cell death or malignant transformations of the damaged cell.

Fourth, the molecular mechanism controlling all the events of the cell cycle must be extremely robust: it must function perfectly under a wide variety of conditions and stresses, because mistakes can be lethal to the dividing cell and ultimately to the organism it supports. In particular, we must keep in mind that these molecular interactions are occurring within the small confines of a single cell, and the numbers of molecules participating in any aspect of the process may be extremely limiting. For example, in a haploid yeast cell there is (in general) only one copy of every gene, only a handful of copies of each specific mRNA, and only a few hundreds or thousands of molecules of specific regulatory proteins [22,23]. Basic laws of statistical physics demand that reactions among such small numbers of molecules must experience large stochastic fluctuations [24,25], yet cell cycle events are flawlessly orchestrated by the noisy molecular control system.

Any proposed explanation of the molecular basis of eukaryotic cell cycle controls must be consistent with these basic features of cell physiology. Informal, textbook explanations should not be accepted uncritically. Because the cell cycle is fundamentally a sequential process played out in time and space, the control system must be described in dynamic terms that provide insight into the general principles of temporal and spatial regulation and which account in quantitative detail for the idiosyncrasies of cell growth and division in particular organisms.

Molecular Biology of the Cell Cycle

In eukaryotes the basic events of the cell cycle — DNA synthesis and mitosis — are controlled by a family of cyclin-dependent protein kinases (CDKs). As their name implies, these enzymes, in conjunction with a suitable cyclin partner, phosphorylate protein targets [26,27] and thereby initiate processes such as DNA replication, nuclear envelope breakdown, chromosome condensation and mitotic spindle assembly. Hence the timing of cell cycle events depends on sequential waves of activation and inactivation of CDKs. During steady proliferation of most cell types the catalytic subunits (Cdk1, Cdk2, Cdk4 and Cdk6) are present in excess, and their activities are dependent on the availability of specific regulatory subunits (cyclin A, cyclin B, cyclin D, cyclin E) [28]. The abundance of each type of cyclin is controlled by its turnover (its rates of synthesis and degradation) (Figure 14.2). Cyclin synthesis rate is determined by the activity of specific transcription factors, and cyclin degradation rate is

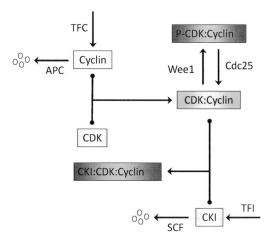

FIGURE 14.2 **Mechanisms for regulating the activity of a CDK: cyclin heterodimer.** The kinase subunit, CDK, is usually present in constant amount, in excess of cyclin subunits. The concentration of cyclin subunits is determined by the activities of its transcription factor (TFC) and its degradation machinery (APC or SCF). Active CDK:cyclin dimers can be inactivated by binding to a stoichiometric inhibitor (CKI), whose abundance is determined by the activities of its transcription factor (TFI) and its degradation machinery (SCF). In addition, the kinase subunit can be inactivated by phosphorylation (kinase = Wee1) and reactivated by dephosphorylation (phosphatase = Cdc25). The enzymes (TFC, APC, TFI, SCF, Wee1 and Cdc25) are all subject to their own regulatory interactions.

determined by the activity of specific E3 ubiquitin-ligating complexes (APC and SCF) [29]. Polyubiquitinated cyclin molecules are rapidly degraded by proteasomes in the cell.

Specific CDK:cyclin heterodimers are active at distinct phases of the cell cycle (Figure 14.3). In early G1 phase, cells are mostly devoid of cyclin molecules, except for minor amounts of cyclin D in combination with either Cdk4 or Cdk6 [28,30]. In late G1, cyclin E makes a brief appearance when, in combination with Cdk2, it turns on the transcription factor for cyclin A and turns off the ubiquitination of cyclin A [31]. Hence, cyclin A accumulates and, in combination with Cdk2, drives the cell through S phase. In G2 phase, cyclin A changes partners to Cdk1 and promotes the production of cyclin B. Cdk1:CycB heterodimers are essential for successful completion of mitosis. During prometaphase most cyclin A is degraded, but cyclin B persists at high levels right up to metaphase [32]. During metaphase and anaphase, cyclin B is rapidly cleared from the cell, leaving the daughter cells in G1 phase with only the remnant supply of cyclin D.

There are two other modes of CDK regulation that are crucial for cell cycle control (Figure 14.2). First, both Cdk1 and Cdk2 can be phosphorylated on neighboring threonine and tyrosine residues in the N-terminus of the polypeptide chain [33]. These phosphorylations, which significantly inhibit the activity of the CDK:cyclin heterodimer, are carried out by members of the Wee1 family of protein kinases. To regain catalytic activity, the heterodimer must be dephosphorylated by a member of the Cdc25 family of protein phosphatases [34]. How Wee1 and Cdc25 activities are regulated will be described later. Second, there exist families of cyclin-dependent kinase inhibitors (CKIs) that bind strongly to CDK:cyclin dimers to form inactive trimers [35]. The fraction of the CDK:cyclin pool that can be inhibited in this way depends on the abundance of CKI molecules. Like cyclins, the abundance of a CKI is determined by its rates of synthesis and degradation [36].

A few other molecular components deserve special attention (see Table 14.2 for a summary). Cyclin D, cyclin E and CKIs are ubiquitinated by an E3 ubiquitin ligase called SCF, which recognizes its substrates only when they are properly phosphorylated. Hence, the degradation of these regulatory proteins can be controlled by specific protein kinases. Cyclins A and B are ubiquitinated by the 'anaphase-promoting complex/cyclosome' (APC/C), which requires an auxiliary protein (Cdc20 or Cdh1) to target specific substrates to the APC. The phosphorylation states of the APC and its binding partners, Cdc20 and Cdh1, determine the activity of the complex [37]. In prometaphase APC:Cdc20 actively degrades cyclin A but not cyclin B. Cyclin B degradation is delayed until late metaphase [38], concurrently with securin (next paragraph).

Cohesin rings are cleaved by a protease called separase, which is kept inactive throughout most of the cell cycle by being bound to an inhibitor called securin [39]. During prometaphase, as the replicated chromosomes are being aligned on the mitotic spindle, the activity of APC:Cdc20 toward securin and CycB is blocked by an inhibitor, Mad2 [40]. When all chromosomes are properly aligned, the mitotic checkpoint is lifted and Mad2 is removed from APC:Cdc20, which then ubiquitinates securin, leading to its degradation by proteasomes. Free molecules of separase then cleave cohesin rings and promote anaphase.

The concurrent degradation of CycB during late metaphase and anaphase helps the cell to return to the G1 state. In budding yeast cells, as we describe later, the return to G1 is aided by the activation of Cdc14 (a Cdk counter-acting phosphatase) and Cdh1 (a Cdc20-homolog) [41].

From this brief description of the molecules that control cell cycle progression some specific features are incredibly obvious, such as the roles of CDKs in triggering DNA synthesis and mitosis, or the role of APC in promoting anaphase, but the subtle details of cell cycle control remain shrouded in mysteries. To understand exactly how the four fundamental properties of cell cycle progression are ensured by the underlying cell cycle machinery we must address the problem from a 'systems' point-of-view, asking two main questions. What are the basic principles of cell cycle regulation? And how are these principles implemented in molecular interactions? If we can answer these questions satisfactorily, then the whole welter of facts and

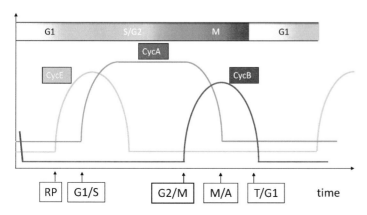

FIGURE 14.3 **The cell cycle of a 'generic' eukaryote.** We track fluctuations in three major classes of cyclins (A, B and E). In early G1 phase, all three classes of cyclins are absent. In mid-G1, CycE begins to rise at an event called the restriction point (RP) in mammalian cells (Start in yeast cells). CycA-dependent kinase is responsible for initiating DNA synthesis, so it rises at the G1/S transition. CycB-dependent kinase is responsible for mitosis, so its activity rises at the G2/M transition. CycA level falls in prometaphase, but CycB level falls later (after chromosome alignment on the metaphase plate).

TABLE 14.2 Molecular Components of the Cell Cycle Control System

Name	Yeast Ortholog	Description	Typical Reaction[a,b]
Cdk1, Cdk2, ...	Cdc28	Cylin-dependent kinase	Protein + ATP → Protein-P + ADP; Cdk$n$:CycX
CycA	Clb5, Clb6	S-phase promoting cyclin	ORC → ORC-P; Cdk2:CycA
CycB	Clb1, Clb2	M-phase promoting cyclin	Histone → Histone-P; Cdk1:CycB
CycD	Cln3	Growth-responding cyclin	RB → RB-P; Cdk4:CycD
CycE	Cln1, Cln2	Starter kinase	CKI → CKI-P; Cdk2:CycE
CAP		Counter-acting phosphatase	Protein-P + H_2O → Protein + P_i; CAP
CKI	Sic1	Cdk stoichiometric inhibitor	Cdk1:CycB + CKI → CKI: Cdk1:CycB
Cohesin	Scc1, etc.	Sister-chromatid cohesion	Xsome → Xtid: Cohesin: Xtid; S phase
Separase	Esp1	Cohesin degradation	Cohesin → DP; Separase
Securin	Pds1	Separase inhibitor	Separase + Securin ↔ Securin: Separase
APC	APC	Anaphase promoting complex	Securin: Separase → Separase + DP; APC:Cdc20
Cdc20	Cdc20	Targeting subunit of APC	APC + Cdc20 ↔ APC:Cdc20
Cdh1	Hct1	Homolog of Cdc20	Cdk1:CycB → Cdk1 + DP; APC:Cdh1
SCF	SCF	Skp-Cullin-Fbox complex	CKI-P → DP; SCF
Mad2	Mad2	Inhibitor of APC:Cdc20	APC:Cdc20 + Mad2$_A$ → MCC
	Cdc14	Exit phosphatase	APC:Cdh1-P → APC:Cdh1; Cdc14
	Net1	Inhibitor of Cdc14	Net1 + Cdc14 ↔ Net1:Cdc14
E2F	SBF	Transcription factor	*CYCE* promoter + E2F ↔ *CYCE*:E2F (active gene)
RB	Whi5	Inhibitor of E2F	RB + E2F ↔ RB:E2F
Wee1	Swe1	Tyrosine kinase	Cdk1:CycB → P-Cdk1:CycB; Wee1
Cdc25	Mih1	Tyrosine phosphatase	P-Cdk1:CycB → Cdk1:CycB; Cdc25

[a] A + B → C + D; E denotes a chemical reaction and its catalyzing enzyme. Reversible binding reactions are indicated by ↔.
[b] Orc, origin of replication complex; DP, degradation products; MCC, mitotic checkpoint complex.

speculations about the cell cycle will begin to make sense. And as we begin to understand the molecular logic of cell cycle control, we can expect to parlay this knowledge into significant advances in human health, agricultural productivity and biotechnological innovations.

IRREVERSIBILITY AND BISTABILITY

Irreversible progression through the cell cycle means that S and M phases always occur in strict alternation, separated by gaps — G1 and G2. When functioning properly, the cell cycle moves steadily forward like a ratchet device, not slipping backwards, say, to do two rounds of DNA synthesis without an intermediate mitosis. Of course, this happens sometimes, in cells under stress or in mutant cells [42], or even by design (in terminally differentiated polyploid cells) [43], but in general mitotic cell cycle transitions are irreversible. What sort of molecular interactions account for this directionality?

It is an alluring fact that at each irreversible transition of the cell cycle an important regulatory protein is degraded (Figure 14.1). At the G1/S transition, the CKI that was inhibiting CDK activity throughout G1 phase is rapidly degraded. At the G2/M transition, the Wee1 kinase that was inhibiting CDK activity throughout G2 phase is rapidly degraded [44]. At the M/A transition, the cohesin rings that were holding sister chromatids together prior to metaphase are cleaved by separase. As cells exit mitosis and return to G1, cyclin B subunits, which played essential roles in orchestrating mitosis, are degraded by the APC. These observations have led many commentators to suggest that proteolysis is the basis of irreversibility (Table 14.3). Is proteolysis one of the fundamental principles of cell cycle regulation that we are seeking?

TABLE 14.3 Proteolysis Provides Directionality to Cell Cycle Progression?

'The chemical irreversibility of proteolysis is exploited by the cell to provide directionality at critical steps of the cell cycle.' *Science* (1996)

'An obvious advantage of proteolysis for controlling passage through these critical points in the cell cycle is that protein degradation is an irreversible process, ensuring that cells proceed irreversibly in one direction through the cycle.' Textbook (2004)

'Importantly, the irreversible nature of proteolysis makes it an invaluable complement to the intrinsically reversible regulation through phosphorylation and other post-translational modifications.' Curr. Biol. (2004)

The hydrolysis of a peptide bond, $(AA)_N + H_2O \rightarrow (AA)_{N-1} + AA$, is a thermodynamically irreversible process, i.e., the free energy change of this chemical reaction is negative. But, in fact, every reaction that occurs in the cell has $\Delta G < 0$. If ΔG were positive, then the reaction would proceed in the opposite direction, as dictated by the second law of thermodynamics. Not only is protein degradation thermodynamically spontaneous but so is protein synthesis, $(AA)_N + AA \rightarrow (AA)_{N+1} + H_2O$, because the ribosome couples this reaction to the hydrolysis of four molecules of ATP, 4 ATP + 4 $H_2O \rightarrow$ 4 ADP + 4 P_i (P_i = inorganic phosphate ions). The overall reaction, polypeptide chain elongation + ATP hydrolysis, is an irreversible reaction ($\Delta G < 0$).

In the same way, protein phosphorylation and dephosphorylation are both intrinsically irreversible reactions:

Protein + ATP $\rightarrow$ Protein-P + ADP, $\Delta G < 0$.
Protein-P + $H_2O \rightarrow$ Protein + P_i, $\Delta G < 0$.

The net effect of a cycle of protein phosphorylation and dephosphorylation is the hydrolysis of one molecule of ATP to ADP + P_i. This is called a 'futile' cycle. Although futile from an energetic point of view, the cycle may be quite functional in the information economy of the cell, corresponding, in computer language, to flipping a bit from 0 to 1 and back to 0. ATP hydrolysis is the price the cell must pay for an elementary information-processing operation. A cycle of protein synthesis and degradation is also 'futile' in the same respect. The only difference is that it costs much more ATP to synthesize and then degrade a full protein — but then, there is much more 'information' in a complete protein than in a phosphorylated amino acid side chain.

It is incorrect to maintain that post-translational modifications of proteins are intrinsically reversible whereas proteolysis is intrinsically irreversible (Table 14.3). The reverse of protein degradation is protein synthesis in the same way that the reverse of protein dephosphorylation is protein phosphorylation [45]. The only differences are 'time and money': protein turnover (synthesis and degradation) takes longer and costs more than post-translational modifications. Indeed, it is common to find that protein synthesis and degradation are dynamically balanced, so that the total level of the protein is held at a constant value.

If protein turnover is rapid, then protein levels can be quickly ramped up or down, as changing circumstances dictate, simply by disturbing the balance between synthesis and degradation. In such cases, proteolysis is a 'dynamically reversible' process.

It is equally true of protein synthesis-and-degradation as of phosphorylation-and-dephosphorylation that the opposing processes are thermodynamically irreversible but dynamically reversible. Dynamic irreversibility — which is the type of irreversibility of relevance to cell cycle transitions — is not to be sought in the thermodynamic properties of individual reactions but in the dynamic behavior of sets of coupled reactions with feedback. Dynamic irreversibility is a systems-level property of molecular regulatory networks [45] (Figure 14.4).

Dynamic irreversibility is intimately connected to the bistability of chemical reaction networks. To explain the connection, consider the simple example of a transcription factor, e.g., E2F, that upregulates the expression of its own gene (Figure 14.4A). According to the basic principles of biochemical kinetics, we can describe the dynamic features of this little network by a pair of ordinary differential equations (ODEs):

$$\frac{dM}{dt} = k_{sm}H(P) - k_{dm}M$$
$$\frac{dP}{dt} = k_{sp}M - k_{dp}P \quad (1)$$

where M = [mRNA], P = [protein], k_{sm} etc. = rate constants for synthesis and degradation of mRNA and protein, and $H(P)$ = probability that the gene encoding the transcription factor is being actively transcribed. Suppose that this probability = 1 when the transcription factor is bound to the gene's promoter, and = ε when not bound. Then the function $H(P)$ is commonly taken to be a Hill function:

$$H(P) = \frac{\varepsilon K_p^n + P^n}{K_p^n + P^n} \quad (2)$$

where K_p is the equilibrium dissociation constant (units of concentration, say nM) of the transcription factor–promoter complex, and n is the Hill exponent, $n = 2$ or 4 depending on whether the transcription factor binds to the promoter as a homodimer or homotetramer.

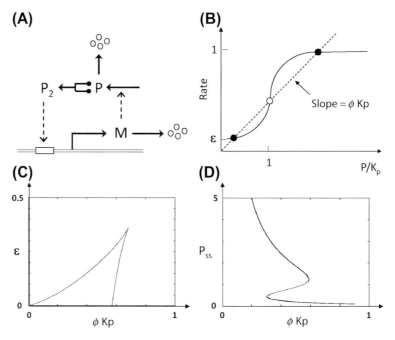

FIGURE 14.4 A simple bistable switch. (A) Wiring diagram. The synthesis of protein P is directed by mRNA M, which is transcribed from a gene controlled by a promoter (gray box on the double-stranded DNA molecule). The promoter is active when it is bound by dimers (or tetramers) of P. (B) Rates of synthesis (solid line) and degradation (dashed line) of P as functions of P/K_p. Black circles indicate stable steady states; white circle indicates an unstable steady state. Refer to Eqs. (1)–(3) in the text. (C) Domain of bistability in parameter space. (D) Signal-response curve. The steady state concentration of P is plotted as a function of $\phi \cdot K_p$ for $n = 4$ and $\varepsilon = 0.1$. The response is bistable for values of $\phi \cdot K_p$ in the range (0.301, 0.591).

At 'steady state' $dM/dt = dP/dt = 0$. Hence,

$$H(P) = \frac{k_{dm}}{k_{sm}}M = \frac{k_{dm}k_{dp}}{k_{sm}k_{sp}}P = \phi P \qquad (3)$$

where $\phi = k_{dm} k_{dp}/k_{sm} k_{sp}$ is a constant with units nM^{-1}. The steady-state concentrations of mRNA and protein are determined by solutions of the algebraic equation $H(P) = \phi P$, which is a cubic equation for $n = 2$ and a quintic equation for $n = 4$. From the graphs in Figure 14.4B it is easy to see that this algebraic equation may have one, two or three real positive roots (i.e., steady-state values of P), depending on the values of n, ε and $\phi \cdot K_p$. In Figure 14.4C we indicate how the number of steady states depends on ε and $\phi \cdot K_p$ for $n = 2$ and $n = 4$. In Figure 14.4D we show how the steady state values of P depend on $\phi \cdot K_p$ for $n = 4$ and $\varepsilon = 0.1$.

Figure 14.4D is known as a one-parameter bifurcation diagram (see Table 14.1 for definitions) for the steady-state solution, P_{ss}, of Eqs. (1) as a function of the dimensionless bifurcation parameter $\phi \cdot K_p$. For small values of $\phi \cdot K_p$ the gene is being actively transcribed (the gene is 'on') and P_{ss} is large. As $\phi \cdot K_p$ increases, P_{ss} steadily decreases until the gene abruptly turns off at $\phi \cdot K_p = 0.591$. At this value of $\phi \cdot K_p$ the dynamic system is said to undergo a 'saddle-node' bifurcation (see Table 14.1). The upper steady state (a 'stable node') coalesces with the intermediate steady state (an unstable 'saddle point') and they both disappear, leaving only one attractor of the dynamic system, namely the lower steady state (the alternative stable node). In the lower steady state gene expression is turned off, and it is not possible to turn it back on by a small decrease in the value of $\phi \cdot K_p$. The transition is 'irreversible'. In order to coax the gene to turn back on, $\phi \cdot K_p$ must be reduced below the other saddle-node bifurcation point at $\phi \cdot K_p = 0.301$. For $0.301 < \phi \cdot K_p < 0.591$, the reaction network is 'bistable', i.e., it can persist in one or another of two stable steady states (on or off). In the bistable zone, the two stable steady states are separated by an unstable steady state.

The bistable behavior we have illustrated with this simple model is completely representative of bistability in more complex networks. The irreversibility of transitions is related to bifurcation points in a bistable system, and bistability is a consequence of

Positive Feedback + Sufficient Non-linearity
+ Rate-constant Constraints.

In our example, the 'positive feedback' is obvious: the transcription factor P upregulates its own production. 'Sufficient non-linearity' is reflected in the sigmoidal Hill function with $n = 2$ or 4. (If $n = 1$, there can be no bistability.) 'Rate-constant constraints' are evident in Figure 14.4C: bistability is exhibited only within a limited range of rate constant values.

IRREVERSIBLE TRANSITIONS IN THE BUDDING YEAST CELL CYCLE

Figure 14.5 shows the molecular interactions underlying the irreversible transitions of the budding yeast cell cycle. Before describing these reaction networks in detail, we must introduce a few simplifications. In budding yeast,

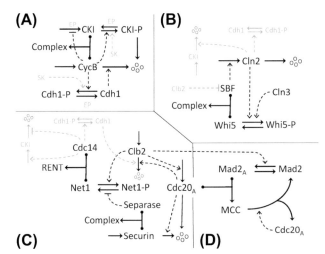

FIGURE 14.5 **The molecular control systems of the budding yeast cell cycle.** (A) The fundamental bistable switch, created by the antagonistic interactions between the B-type cyclins (CycB) and the G1-stabilizers (CKI and Cdh1). SK = starter kinase, EP = exit phosphatase. Four small circles are products of protein degradation. (B) The Start transition. Production of Cln2:Cdk1, the starter kinase of panel A, is controlled by the transcription factor, SBF, which is inactivated by binding to a stoichiometric inhibitor, Whi5. Phosphorylation of Whi5 releases SBF. Notice the double-negative feedback loop between Cln2 and Whi5. Cln3 is a 'growth indicator' that triggers the Cln2 switch. Notice also that Clb2-dependent kinase inactivates SBF, probably by direct phosphorylation. (C) Exit from mitosis. Active Cdc20:APC degrades securin and Clb2 (partially). Free separase degrades cohesin rings (anaphase) and helps to release Cdc14 from inhibition by Net1. Cdc14 is the exit phosphatase that activates Cdh1 and CKI. After Clb2 and other mitotic kinases are cleared by Cdh1:APC, Cdc20 is degraded, Cdc14 is inactivated, and the cell reverts to the G1 steady state. (D) Metaphase checkpoint. Clb2 and Cdc20 are components of both networks C and D. The checkpoint protein, Mad2, is activated by unaligned (tensionless) chromosomes, in a reaction that requires Clb2-kinase activity. Active Mad2 binds to and sequesters Cdc20:APC in the mitotic checkpoint complex (MCC). When all chromosomes have come into alignment on the metaphase plate, the rate of activation of Mad2 drops to zero, and the MCC starts to come apart in a reaction that is accelerated by free (active) Cdc20:APC.

many cell cycle genes come in pairs: *CLN1* and *CLN2*, *CLB1* and *CLB2*, *CLB5* and *CLB6* [46]. In our diagrams and models we lump these pairs together, i.e., 'Cln2' represents both Cln1 and Cln2 protein pools, 'Clb2' both Clb1 and Clb2, and 'Clb5' both Clb5 and Clb6. Sometimes we lump together Clb1, Clb2, Clb5 and Clb6 as 'Clb'. Also, both Sic1 and Cdc6 function as stoichiometric Cdk inhibitors in the mitotic cycle of budding yeast, so we refer to both of them together as 'CKI'. *CLN3* and *BCK2* encode proteins that are jointly responsible for growth sensitivity of the budding yeast cell cycle, and we usually are referring to these proteins jointly as 'Cln3' [7]. The expression of many cell cycle genes is controlled by regulated transcription factors, but we shall refer explicitly only to the regulation of *CLN1*, *CLN2*, *CLB5* and *CLB6* genes by SBF.

Two Alternative States

As suggested years ago by Nasmyth [47,48], we can think of the budding yeast cell cycle as an alternation between two self-maintaining states: G1 is a state characterized by unreplicated chromosomes and low activity of Clb-dependent kinases; and S-G2-M is a state characterized by high Clb-kinase activity and chromosomes in the process of being replicated and aligned on the mitotic spindle. The transition from G1 to S-G2-M involves commitment to a new round of DNA replication and preparation for division. The reverse transition involves partitioning the replicated chromosomes to daughter nuclei (anaphase and telophase) and cell division.

The G1 state is stabilized by inhibitors of Clb-kinase activity, namely CKI and Cdh1. In the S-G2-M state, the G1 stabilizers are neutralized and the Clb-dependent kinases are actively promoting DNA synthesis and mitosis [7,17]. As illustrated in Figure 14.5A, there is mutual antagonism between Clb-kinases and G1-stabilizers. The G1-stabilizers neutralize Clb-kinase activity (CKI binds to and inactivates Cdk1:Clb dimers, and Cdh1 promotes degradation of Clb subunits). On the other hand, active Cdk1:Clb dimers phosphorylate CKI and Cdh1, causing degradation of CKI and inactivation of Cdh1. These mutually antagonistic interactions create a basic bistable switch (see Figure 14.6 and Box 14.1). In the 'neutral' position (SK = 0 and EP = 0 in Figure 14.6), the CKI-Clb-Cdh1 dynamical network can persist in either of two stable steady states: Clb inactive, CKI and Cdh1 active (the G1 state), or Clb active, CKI and Cdh1 inactive (the S-G2-M state). This picture is the theoretical counterpart [20] of Nasmyth's intuitive notion of 'alternative self-maintaining states' [47].

In this theoretical framework the cell physiologist's G1/S transition corresponds to a saddle-node bifurcation (Figure 14.6, left), where the G1 branch of stable steady states ends, and the control system switches irreversibly to the S-G2-M branch of stable steady states. This transition is promoted by a 'starter kinase' (Figure 14.5A), which helps Clb-kinase to eliminate CKI and Cdh1. After the starter kinase has flipped the switch, it is no longer needed to maintain the system in the S-G2-M state. Similarly, the M/G1 transition ('exit from mitosis') corresponds to the reverse saddle-node bifurcation (Figure 14.6, right), where S-G2-M branch ends and the control system switches irreversibly back to G1. An exit phosphatase opposes Clb-kinase activity and helps the G1-stabilizers to reappear. Thereafter, it is no longer needed to maintain the system in the G1 state.

The theoretical picture in Figure 14.6 stands or falls on the presumed bistability of the molecular interactions in Figure 14.5A. Bistability at the G1/S transition has been confirmed experimentally by Cross [49] (see Figure 1 of that paper). That irreversible exit from mitosis is due to feedback loops rather than cyclin B degradation has been

BOX 14.1 Mutual Antagonism

The dynamic properties of the molecular regulatory system in Figure B1A can be described by a differential equation for $B = [\text{CycB:Cdk1}]$,

$$\frac{dB}{dt} = k_{sb} - k_{db}B - k_{dbc}C_AB, \quad (B1.1)$$

where k_{sb}, k_{db} and k_{dbc} are rate constants. The first subscript, 's' or 'd', refers to the type of reaction, 'synthesis' or 'degradation', the second subscript, 'b' in this case, refers to the chemical being synthesized or degraded, and the third subscript, 'c', refers to the enzyme catalyzing the reaction (whenever relevant). In Eq. (B1.1) C_A is the activity of the Cdh1:APC complex, which is given by the steady state solution of the multisite phosphorylation chain in Figure B1A. According to Kapuy et al. [101], C_A is given by the function

$$C_A = \frac{1 - (B/H)^{q+1}}{1 - (B/H)^{N+1}}. \quad (B1.2)$$

where q is the threshold number of phosphate groups above which Cdh1P$_i$ is inactive. In Eq. (B1.2), H = activity of the Cdk-counteracting phosphatase.

A representative set of parameter values for this dynamical system is given in Table B1. For this parameter set we plot C_A as a function of B in Figure B1B. In Figure B1C we plot, as functions of B, the two rate curves:

$$V_{\text{synthesis}} = k_{sb}, \text{ and } V_{\text{degradation}} = (k_{db} + k_{dbc}C_A)B. \quad (B1.3)$$

The points of intersection of these two curves (where $V_{\text{synthesis}} = V_{\text{degradation}}$) are steady state solutions of Eq. (B1.1). Clearly, the dynamical system may exhibit bistability, depending on the relative values of its parameters. In Figure B1D we indicate how the steady state values of B depend on H, with all other parameters fixed at their values in Table B1.

TABLE B1 Parameter values for the CycB-Cdh1 model

Parameter	Value	Parameter	Value
k_{sb}	0.1	q	2
k_{db}	0.1	N	9
k_{dbc}	1	H	0.25

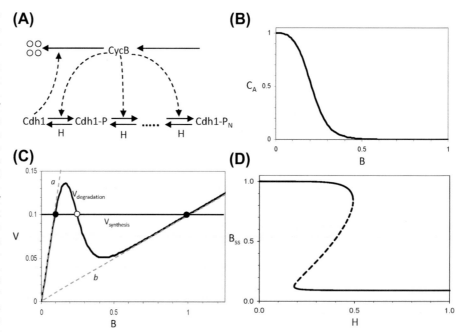

FIGURE B1 Bistability in a model of mutual antagonism. (A) Molecular regulatory network. 'CycB' represents the CycB:Cdk1 heterodimer (a protein kinase) and 'Cdh1' represents the Cdh1:APC complex (an E3 ubiquitin ligase). Cdh1 has multiple sites (N) of phosphorylation by CycB:Cdk1. The first q states of phosphorylation (Cdh1, Cdh1-P, ..., Cdh1-P$_q$) are assumed to be active, and states Cdh1-P$_{q+1}$, ..., Cdh1-P$_N$ inactive. H is a CDK-counteracting phosphatase. (B) Steady-state activity of Cdh1 as a function CycB-dependent kinase activity, from Eq. (B1.2). (C) Rates of synthesis and degradation of CycB, as functions of CycB-dependent kinase activity, from Eq. (B1.3). The intersection points correspond to two stable steady states of the dynamical system (black circles) and one unstable steady state (white circle). The slope of line a is $k_{db} + k_{dbc}C_T$, and the slope of line b is k_{db}. (D) Bifurcation diagram. The steady state values of B, from panel C, are plotted as functions of H, the activity of the counteracting phosphatase. For $0.184 < H < 0.492$, the regulatory network has three steady states, two stable (solid lines) and one unstable (dashed line). The turning points, at $H = 0.184$ and 0.492, are called saddle-node bifurcation points.

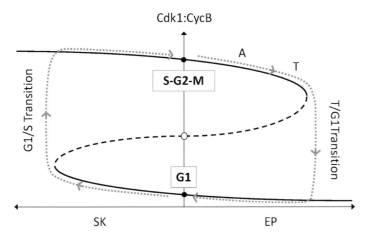

FIGURE 14.6 A dynamic view of cell cycle transitions. The network in Figure 14.5A creates a bistable switch between a stable G1 steady state (CycB-kinase activity low) and a stable S-G2-M steady state (CycB-kinase activity high). In early G1 phase, SK ≈ 0 and EP ≈ 0, and the cell is stuck in the stable G1 state. To exit G1 phase and begin DNA synthesis, the cell requires a starter kinase to drive the bistable switch past the saddle-node bifurcation point and to induce the G1/S transition (gray dotted line). After the transition is complete, SK activity drops back to zero, but the cell is now stuck in the stable S-G2-M state. To leave M phase and return to G1, via anaphase (A) and telophase (T), the cell requires an exit phosphatase to drive the bistable switch past a different saddle-node bifurcation point and to induce the T/G1 transition. After the transition is complete, EP activity drops back to zero, but the cell is now returned to the stable G1 state.

confirmed experimentally by Uhlmann's group [50]; see Figure 14.7 here.

If we accept this theoretical picture of the G1/S and M/G1 transitions, then the next logical issues concern regulation of the starter kinase and exit phosphatase.

Start

The G1/S transition in budding yeast is guarded by a checkpoint (called 'Start' by yeast physiologists) that controls production of the starter kinase, Cln2:Cdk1 [51]. As indicated in Figure 14.5B, Cln2 production is regulated by a transcription factor, SBF (a dimer of Swi4 and Swi6), which is kept inactive by binding to a stoichiometric inhibitor, Whi5 [52,53]. Phosphorylation of Whi5 and Swi6 leads to activation of SBF. In Figure 14.5B we simplify these interactions by assuming that Whi5 phosphorylation causes dissociation of the SBF:Whi5 complex. Because Cln2:Cdk1 is one of the kinases that can phosphorylate Whi5, Cln2 and Whi5 are involved in a classic double-negative feedback loop that creates a bistable switch for Cln2-kinase activity. If Cln2:Cdk1 activity is low, then Whi5 is unphosphorylated and SBF is retained in inactive complexes. But if Cln2:Cdk1 activity is high, then Whi5 is phosphorylated, SBF is active, and Cln2 is steadily synthesized.

At the Start transition, this switch is flipped from the Cln2-low state to the Cln2-high state [54]. The Start switch responds to two crucial physiological signals: cell growth and mating factor. Newborn daughter cells are too small to warrant a new round of DNA synthesis [55]. They must grow to a certain critical size before they can pass Start. In addition, budding yeast cells of mating type α respond to pheromone (α factor) by arresting in G1 phase before Start [51]. Hence, cell growth promotes the Start transition, whereas α factor inhibits it. Both signals appear to operate through the activity of Cln3-dependent kinase (and a second protein, Bck2 [56], that is still poorly characterized). Cell growth increases the net activity of Cln3:Cdk1 in the G1 nucleus of budding yeast cells, whereas α factor

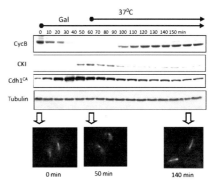

FIGURE 14.7 Reversible exit from mitosis in budding yeast. From Lopez-Aviles et al. [50]; used by permission. In this mutant strain of budding yeast, the *CDC20* gene has been placed under control of a methionine-repressible promoter (*MET-CDC20*), and a non-phosphorylable version of Cdh1 protein has been inserted, under control of a galactose-inducible promoter (*GAL-CDH1^{CA}*, 'CA' for 'constitutively active'). Finally, a temperature-sensitive allele (*cdc16^{ts}*) of an essential component of the anaphase promoting complex (APC) replaces the wildtype gene. This strain is perfectly normal when grown in glucose at 23°C (it has Cdc20, endogenous Cdh1, and active Cdc16 proteins). When grown in glucose + methionine at 23°C (time < 0), these cells arrest in metaphase, as indicated in the first column of the gel (lots of cyclin B, no CKI, a small amount of Cdh1^{CA} because the GAL promoter is slightly leaky). Furthermore, the nuclei have a metaphase morphology (micrograph at 0 min; red = spindle pole bodies, green = mitotic spindle, blue = DNA). At t = 0 the cells are transferred to galactose + methionine to induce the synthesis of non-phosphorylable Cdh1 protein, as witnessed by the third row of the gel. (The fourth row is a loading control.) Because Cdh1^{CA} protein cannot be phosphorylated by the high activity of CycB:Cdk1 in these cells, the cyclin B subunits are almost completely degraded by Cdh1^{CA}:APC over the course of 50 min (first row of the gel), and the G1-stabilizing cyclin-dependent kinase inhibitor (CKI) begins to appear (second row). Furthermore, the nuclei have adopted an interphase morphology after 50 min of treatment. At t = 50 min the cells are transferred to 37°C to inactivate APC. Despite the fact that the cells appear to have exited mitosis and returned to G1 phase (low CycB, high CKI), these cells return to mitosis, as evidenced by the facts that CycB returns, CKI is degraded, and the nuclei return to metaphase (micrograph at 140 min). If the treatment is continued for 60 min, and then the cells are transferred to 37°C, the cells proceed into G1 phase (not shown; see original paper). This behavior is clear evidence of a separatrix between two stable steady states: after 50 min treatment, the cells are still in the domain of attraction of the stable M-phase steady state, but after 60 min treatment, the cells have moved into the domain of attraction of the stable G1-phase steady state.

stimulates production of a CKI (Far1) that specifically inhibits Cln1,2,3-dependent kinase activities.

In the presence of α factor, the Start switch is permanently arrested in the Cln2-low position. In the absence of α factor, a small cell is arrested in the Cln2-low position; but as the cell grows, Cln3:Cdk1 activity increases. By phosphorylating Whi5, Cln3-kinase partially inactivates Whi5 and helps the Cln2 self-activation loop to engage [57]. The Cln2-low state is lost by a saddle-node bifurcation (Figure 14.8, left), and Cln2 begins to accumulate in a self-accelerating manner [20]. The Start switch is moving toward the Cln2-high position, but before it reaches this state, Cln2:Cdk1 flips the G1/S switch to the Clb2-high position, and Clb2:Cdk1 inactivates SBF (see Figure 14.5B). The Start switch never reaches the stable Cln2-high position, but instead drops back to the Cln2-low state (Figure 14.8, left). After Cln2 has done its job as a starter kinase to activate the B-type cyclins (DNA synthesis and mitosis), it drops back to the Cln2-low state in order not to interfere with mitotic exit [58].

A word about the interpretation of Figure 14.8. The left and right sides, considered separately, should be thought of as 'pseudo-phase planes'. The control system in Figure 14.5 has dozens of dynamical variables, and its vector field cannot be represented on a two-dimensional phase plane. Instead, we have plotted one-parameter bifurcation diagrams for subsets of the reaction network. For example, consider the left side of Figure 14.8 (similar reasoning applies to the right side). The black curve is the one-parameter bifurcation diagram for the bistable network in Figure 14.5A, treating Cln2-kinase as the bifurcation parameter and setting Cdc20 ≈ 0 and Cdc14 ≈ 0. The red curve is the one-parameter bifurcation diagram for the bistable network in Figure 14.5B, with Cln2-kinase as the variable and Clb2-kinase as the bifurcation parameter. Strictly speaking, these bifurcation curves are not nullclines on a phase plane, but they can be thought of as pseudo-nullclines. Their intersection points are certainly steady states of the full system (Figure 14.5A+B, with EP ≈ 0), and one can readily guess the stability of each steady state and the temporal evolution of the dynamical system when the steady states are removed by saddle-node bifurcations.

Mitotic Checkpoint

The M/G1 transition is guarded by the mitotic checkpoint, which controls activation of APC:Cdc20 and the exit phosphatase, Cdc14 [59]. Figure 14.5D diagrams the mitotic checkpoint proper, which controls activation of APC:Cdc20, and Figure 14.5C indicates how active Cdc20 initiates the activation of Cdc14 and the re-establishment of cells in G1.

The role of the mitotic checkpoint (the 'anaphase switch') is to prevent cells from cleaving cohesin rings until all replicated chromosomes are properly aligned on the mitotic spindle [60]. Alignment is judged, it seems, by tension within the centromeric region of bi-oriented chromosomes (one kinetochore attached by microtubules to one pole of the spindle, and the other kinetochore to the other pole). Unaligned, tensionless chromosomes generate a signal that activates Mad2, and active Mad2 binds to APC:Cdc20 [40], holding it in an inactive complex (the mitotic checkpoint complex, MCC, which is inactive with respect to degradation of separase and Clb2, but active on the degradation of Clb5). The signal from tensionless chromosomes to Mad2 depends on Clb-kinase activity, which is high in prometaphase. As soon as all chromosomes are properly aligned and under tension, the rate of Mad2 activation becomes 0, and the MCC begins to dissociate, releasing active APC:Cdc20.

The network in Figure 14.5C has two positive feedback loops. (1) Clb2:Cdk1 (via tensionless chromosomes) activates Mad2, which inactivates APC:Cdc20; whereas APC:Cdc20 degrades Clb2, thereby inactivating the kinase. This double-negative feedback loop creates a bistable switch at

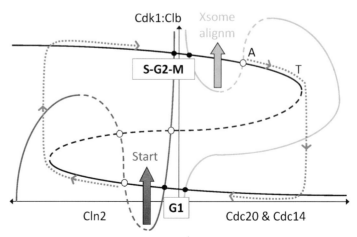

FIGURE 14.8 **Checkpoints in budding yeast.** We identify SK with Cln2:Cdk1 and EP with a combination of activities of Cdc20 and Cdc14. Left: the G1/S transition is guarded by a checkpoint called Start (Figure 14.5B), which creates a stable steady state of low Cln2 abundance. As the cell grows, the checkpoint is lifted (red arrow) and the stable G1 state is lost by a saddle-node bifurcation. Subsequently, Cln2 production induces the G1/S transition as in Figure 14.6. Right: the M/A transition is guarded by the mitotic checkpoint. When all chromosomes are properly aligned on the metaphase plate, the checkpoint is lifted (green arrow) and the stable metaphase-arrested state is lost by a saddle-node bifurcation. Subsequently, Cdc20 and Cdc14 are activated and they drive exit from mitosis and return to G1 phase.

the metaphase checkpoint [61]. (2) Active APC:Cdc20 promotes dissociation of the MCC. This self-activation loop accelerates the release from the checkpoint, so that anaphase follows soon after full chromosome alignment.

The anaphase switch (Box 14.2) has two stable steady states: (1) Clb2 level high, Mad2 active, Cdc20 inactive; and (2) Mad2 inactive, Cdc20 active, Clb2 level low. At the metaphase/anaphase transition, the switch is flipped from the Cdc20-inactive state to the Cdc20-active state. Consequently, APC:Cdc20 degrades securin, releasing separase to cleave cohesin rings and trigger anaphase (separation of sister chromatids). In addition, separase has a non-catalytic role [62], inhibiting the phosphatase that has been keeping Net1 active throughout the early stages of mitosis [63]. Inhibition of the phosphatase allows Clb2-kinase and other mitotic kinases (notably Polo kinase) to phosphorylate Net1 and release Cdc14 [63–65]. Meanwhile, APC:Cdc20 degrades Clb2, and the combination of low Clb2-kinase activity and high Cdc14-phosphatase activity silences the mitotic checkpoint [66]. (Cohesin cleavage at anaphase creates tensionless chromosomes, but they do not reactivate Mad2 because now Clb2:Cdk1 activity is low and Cdc14 activity is high.) Cdc14 promotes activation of the G1-stabilizers (CKI and Cdh1). CKI inhibits any remaining Clb-dependent kinase activity, and Cdh1 destroys Polo kinase and Cdc20.

In Figure 14.8 (right), we show how chromosome alignment removes the mitotic-arrest state by a saddle-node bifurcation, allowing the anaphase switch to flip on. Cdc20 and Cdc14 activities rise, but before they can reach the upper steady state, they induce degradation of Clb proteins and activation of the G1-stabilizers. The transition to G1 phase removes Cdc20 and Cdc14, restoring the newborn cells to the beginning of the cycle.

IRREVERSIBLE TRANSITIONS IN THE MAMMALIAN CELL CYCLE

The molecular machinery regulating progression through the mammalian cell cycle is very similar in principle to the yeast cell cycle. In the following paragraphs we will highlight the most important similarities and differences.

In mammalian cells, as in yeast, the G1 phase of the cell cycle is stabilized by three types of interaction that keep low the activities of S-phase promoting factor (SPF = Cdk2:CycA) and M-phase promoting factor (MPF = Cdk1:CycB): (1) high activity of APC:Cdh1, which promotes degradation of both CycA and CycB [67]; (2) high abundance of CKIs that inhibit SPF and MPF heterodimers [68]; (3) high abundance of an inhibitor (retinoblastoma protein, RB) of the transcription factors (E2F family) that promote synthesis of early cyclins (CycE and CycA) [69].

Each of these G1 stabilizers can be phosphorylated and neutralized by SPF and MPF, creating a fundamental bistable switch between steady states for G1 and S-G2-M. Progression through the mammalian cycle may also be envisioned as flipping this switch on by a starter kinase (Cdk2:CycE) and off by an exit pathway (Cdc20 and CAPs).

In particular, the Start transition in budding yeast is analogous to the 'restriction point' (RP) in the mammalian cell cycle [70]. The molecular mechanisms of the two checkpoints are almost identical under the identification of SBF with E2F, Whi5 with RB, Cln2 with CycE, Cln3 with CycD, and Clb5 with CycA. Both checkpoints are responsive to extracellular signals. In budding yeast cells α factor activates a MAP kinase pathway that upregulates a stoichiometric inhibitor of Cln3 and blocks cells in pre-Start. In mammalian cells, growth factor (GF) activates a MAP kinase pathway that upregulates a transcription factor for CycD and promotes passage through the RP. The logic differs because, for yeast cells, the default state of the cell cycle is vegetative growth and division, and yeast cells need a definite signal (α factor) to block progression through the cell cycle and start the mating process. For mammalian somatic cells, on the other hand, the default state is G1 arrest: only special cells under special circumstances are permitted to grow and divide. The permission is granted by specific GFs that promote passage through the RP.

Bistability at the mammalian RP has been demonstrated experimentally in elegant experiments by Yao et al. [71] (see Figure 3 in that paper). In a later paper [72], the same authors showed experimentally that RP bistability is due to the double-negative feedback loop (RB –| E2F → CycE –| RB).

Bistability in the mitotic exit mechanism of mammalian cells is still a matter of some disagreement [73,74], although in our opinion the evidence is definitely in favor of a bistable switch [45,75]. Our description of a bistable anaphase switch in budding yeast, based on experiments by Uhlmann and colleagues [66], is confirmed by similar experiments with fruit flies [76] and mammalian cells [77].

Budding yeast cells differ from most other types of organism (including fission yeast cells, plant cells, fruit fly embryos, frog embryos and mammalian cells) in lacking a checkpoint at the G2/M transition. Budding yeast cells are unique in having many small chromosomes that need not undergo much condensation during mitosis; hence they can go almost directly from S phase into mitosis [78]. Other organisms, on the contrary, need a gap phase (G2) between the end of S and the onset of M, during which chromosomes are replicated and available for transcription. When these cells enter mitosis, their chromosomes become so highly condensed that all transcription ceases. The duration of G2 is determined by the G2/M checkpoint. During G2 phase, mitotic cyclins accumulate in complexes that are

BOX 14.2 Anaphase Switch

The dynamic properties of the molecular regulatory system in Figure B2A can be described by a set of differential equations for $B = [\text{CycB}]$, $M_{AT} = [\text{Mad2}_A]_{\text{free}} + [\text{Mad2}_A\text{:Cdc20}]$, and $X = [\text{Mad2}_A\text{:Cdc20}]$,

$$\frac{dB}{dt} = k_{sb} - k_{db}B - k_{dbc}C \cdot B, \quad (B2.1A)$$

$$\frac{dM_{AT}}{dt} = k_{amb}L \cdot B \cdot M_I - k_{imh}H \cdot M - k_{break}C \cdot X, \quad (B2.1B)$$

$$\frac{dX}{dt} = k_{assoc}M \cdot C - k_{dissoc}X - k_{break}C \cdot X, \quad (B2.1C)$$

where $C = [\text{Cdc20}]_{\text{active}} = C_T - X$, $M_I = [\text{Mad2}]_{\text{inactive}} = M_T - M_{AT}$, and $M = [\text{Mad2}_A]_{\text{free}} = M_{AT} - X$. As usual, $C_T = [\text{Cdc20}]_{\text{total}}$ and $M_T = [\text{Mad2}]_{\text{total}}$. The rate constants k_{amb} and k_{imh} refer to the activation and inactivation of Mad2, k_{assoc} and k_{dissoc} refer to the association and dissociation of the Mad2$_A$:Cdc20 complex, and k_{break} refers to the breakdown of the Mad2$_A$:Cdc20 complex induced by active Cdc20 itself. The parameter $L =$ fraction of tensionless chromosomes (i.e., replicated chromosomes that are not properly aligned on the metaphase spindle), and the $H =$ activity of a counteracting phosphatase.

TABLE B2 Parameter values for the Mad2-Cdc20 model

Parameter	Value	Parameter	Value
k_{amb}	1	$K_{diss} = k_{diss}/k_{assoc}$	0.01
k_{imh}	0.01	$\kappa_{break} = k_{break}/k_{assoc}$	0.001
k_{break}	1	H	1
$B_T = k_{sb}/k_{db}$	1	M_T	2
$C_0 = k_{db}/k_{dbc}$	0.01	C_T	1

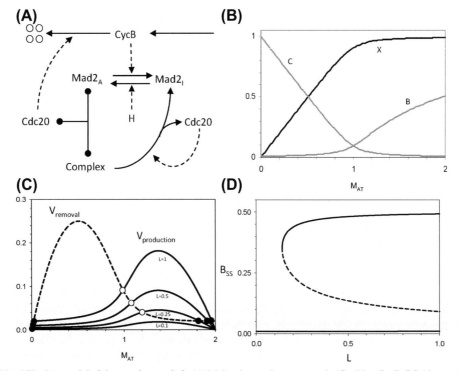

FIGURE B2 Bistability in a model of the anaphase switch. (A) Molecular regulatory network. 'CycB' = CycB:Cdk1 heterodimer, 'Cdc20' = Cdc20:APC complex (an E3 ubiquitin ligase), 'Mad2' = stoichiometric inhibitor of Cdc20:APC. Active Cdc20:APC primes CycB for degradation. CycB-dependent kinase activates Mad2, in conjunction with 'tensionless' chromosomes, L. H = Clb2-counteracting phosphatase. Notice that active Cdc20:APC promotes dissociation of the Mad2:Cdc20:APC complex. (B) The concentration, X, of the Mad2$_A$:Cdc20 complex is plotted as a function of $M_{AT} = [\text{Mad2}_A]_{\text{free}} + [\text{Mad2}_A\text{:Cdc20}]$. In addition, we plot $C = [\text{Cdc20}]_{\text{active}} = C_T - X$ and $B = [\text{CycB}]_{ss} = B_T C_0/(C_0+C)$ as functions of M_{AT}. (C) Rates of production and removal of active Mad2 as functions of M_{AT} from Eq. (B2.4). The production rate curve is drawn for several values of $L =$ fraction of tensionless chromosomes. The intersection points correspond to stable (black circles) and unstable (white circle) steady states of the dynamical system. (D) Bifurcation diagram. The steady state values of M_{AT} are plotted as functions of L. For $0.13 < L < 1$, the regulatory network has three steady states, two stable (solid lines) and one unstable (dashed line).

Continued on next page

BOX 14.2 Anaphase Switch—Cont'd

The steady state solution of Eq. (B2.1A) is $B = B_T C_0 / (C_0 + C)$, where $B_T = k_{sb}/k_{db}$ and $C_0 = k_{db}/k_{dbc}$. The steady state solution of Eq. (B2.1C) satisfies the quadratic equation

$$\kappa_{diss} X + \kappa_{break} (C_T - X) \cdot X = (M_{AT} - X) \cdot (C_T - X) \quad \text{(B2.2)}$$

where $\kappa_{diss} = k_{dissoc}/k_{assoc}$ and $\kappa_{break} = k_{break}/k_{assoc}$. Solving this quadratic equation for X, we obtain

$$F = M_{AT} + C_T + K_{diss} + K_{break} C_T$$

$$X = \frac{2 M_{AT} C_T}{F + \sqrt{F^2 - 4(1 + k_{break}) M_{AT} C_T}} \quad \text{(B2.3)}$$

In Figure B2B we plot X as a function of M_{AT}, along with $C = C_T - X$ and $B = B_T C_0/(C_0+C)$.

Substituting the results of the previous paragraph into Eq. (B2.1B), we obtain

$$\frac{dM_{AT}}{dt} = k_{amb} L \cdot \frac{B_T C_0}{C_0 + C_T - X} \cdot (M_T - M_{AT})$$
$$- [k_{imh} H \cdot (M_{AT} - X) + k_{break}(C_T - X) \cdot X] \quad \text{(B2.4)}$$
$$= V_{production} - V_{removal}$$

In Figure B2C we plot the two rate laws, $V_{production}$ and $V_{removal}$, as functions of M_{AT}, for a representative set of parameter values given in Table B2. The points of intersection of the two rate curves are steady state solutions of Eq. (B2.4). Clearly, the dynamical system may exhibit bistability, and in Figure B2D we indicate how the steady state values of M_{AT} depend on L, with all other parameters fixed at their values in Table B2. Notice that Mad2 remains active until $1 - L$, the fraction of aligned chromosomes, gets very close to 1, and then Mad2 is abruptly inactivated as the last chromosome aligns on the metaphase plate and the cell proceeds to anaphase.

---

inactivated by tyrosine phosphorylation of the Cdk subunit (Figure 14.2). To enter mitosis, these phosphate groups must be removed. The relevant phosphorylation and dephosphorylation reactions are:

Reaction	Enzyme
Cdk1:CycB → P-Cdk1:CycB (less active);	Wee1
P-Cdk1:CycB → Cdk1:CycB (more active);	Cdc25-P
Wee1 → Wee1-P (less active);	Cdk1:CycB
Wee1-P → Wee1 (more active);	CAP
Cdc25 → Cdc25-P (more active);	Cdk1:CycB
Cdc25-P → Cdc25 (less active);	CAP

Clearly, Cdk1:CycB and Cdc25 are involved in a positive feedback loop (mutual activation) [79,80], and Cdk1:CycB and Wee1 are involved in a double-negative feedback loop (mutual antagonism) [81,82]. This network controlling the G2/M transition is strongly bistable (Box 14.3), as first pointed out by Novak and Tyson [83]. The theoretical predictions of that paper were confirmed 10 years later by two groups independently and simultaneously [84,85] (Figure 14.9).

ADDITIONAL CHECKPOINTS

It should be obvious now that additional checkpoints can be created by realigning the curves in Figure 14.8. For instance (Figure 14.10, left), the cell can create a new checkpoint in late G1, with a high level — but low activity — of starter kinase (CycE), by synthesizing a stoichiometric inhibitor (CKI) of CycE:Cdk2. This is exactly the strategy used by mammalian cells to block entry into S phase if DNA damage is detected in G1. A surveillance mechanism upregulates a master transcription factor, p53, which induces synthesis of repair enzymes and of a CKI (p21[WAF1]) that inhibits CycE:Cdk2 and blocks cell cycle progression in late G1. If the damage can be repaired, then the CKI is removed and the cell can proceed into S phase, as usual. If the damage cannot be repaired, then p53 induces synthesis of pro-apoptotic proteins and the damaged cell commits suicide.

Similarly (Figure 14.10, right), the cell can create a checkpoint in telophase (T), with partially degraded CycB and incompletely released Cdc14, if the mitotic exit pathway (Figure 14.5C) is compromised. This strategy is employed by budding yeast cells to implement a 'spindle alignment' checkpoint [59]. Budding yeast cells determine the location of the cell division plane early in the cell cycle, at the point of bud emergence [78]. At the end of the cycle, the bud must separate from the mother cell at the neck between the two. Cell division will be successful only if the mitotic spindle is properly aligned with one pole in the mother-half of the cell and the other pole in the daughter-half. If this is the case, then, as the spindle elongates in late anaphase and pushes one mass of chromosomes into the bud, the spindle pole comes into contact with the bud cortex. This physical proximity brings together a G-protein (Tem1) and its GEF (Lte1), and the activated form of Tem1 activates a kinase that provides additional phosphorylation of Net1 and complete release of Cdc14 [86]. If the spindle is not properly aligned, then Tem1 does not become activated and Cdc14 is not fully released. Cell cycle progression blocks in telophase to give the cell time to reorient the spindle.

BOX 14.3 G2/M Checkpoint

The dynamic properties of the molecular regulatory system in Figure B3A can be described by three differential equations for $B = $ [MPF] = [CycB:Cdk1], $W = $ [Wee1] and $D_P = $ [Cdc25-P]:

$$\frac{dB}{dt} = (k_{ab} + k_{abd}D_P) \cdot (B_T - B) - (k_{ib} + k_{ib}W) \cdot B, \quad (B3.1A)$$

$$\frac{dD_P}{dt} = k_{adb}B \cdot (D_T - D_P) - k_{idh}H \cdot D_P, \quad (B3.1B)$$

$$\frac{dW}{dt} = k_{awh}H \cdot (W_T - W) - k_{iwb}B \cdot W, \quad (B3.1C)$$

where $k_{a\ldots}$ and $k_{i\ldots}$ are rate constants for activation and inactivation of the corresponding proteins, and the parameter H is the activity of the MPF-counteracting phosphatase. Without loss of generality, we can choose the units of H so that $k_{idh} = k_{adb}$.

The steady state solutions of Eqs. (B3.1B,C) are $D_P = D_T B/(H+B)$ and $W = W_T \varepsilon H/(\varepsilon H + B)$, where D_T and W_T are the total concentrations of Cdc25 and Wee1, as usual, and $\varepsilon = k_{awh}/k_{iwb}$. D_P and W as functions of B are plotted in Figure B3B, for the representative parameter values in Table B3. The rate curves,

$$\frac{dB}{dt} = \left(k_{ab} + k_{abd}D_T\frac{B}{H+B}\right) \cdot (B_T - B)$$
$$- \left(k_{ib} + k_{ibw}W_T\frac{\varepsilon H}{\varepsilon H + B}\right) \cdot B \quad (B3.2)$$
$$= V_{activation} - V_{inactivation}$$

are plotted in Figure B3C. The points of intersection of these two curves are steady state solutions of Eq. (B3.2). Clearly, the dynamical system may exhibit bistability, and in Figure B3D we indicate how the steady state values of B depend on B_T, with all other parameters fixed at their values in Table B3.

TABLE B3 Parameter values for the Wee1-MPF-Cdc25 model

Parameter	Value	Parameter	Value
k_{ab}	0	H	1
k_{abd}	1	B_T	0.3
k_{ib}	0	D_T	1
k_{ibw}	1	W_T	1
k_{awh}	0.02	k_{iwb}	1

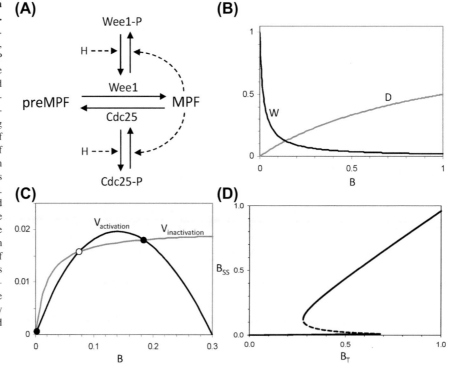

FIGURE B3 Bistability in a model of the G2/M transition. (A) Molecular regulatory network. 'MPF' = active CycB:Cdk1 kinase, 'preMPF' = inactive CycB:Cdk1-P dimer, 'Wee1' = tyrosine kinase (less active in the phosphorylated form), 'Cdc25' = tyrosine phosphatase (more active in the phosphorylated form). $H = $ Clb2-counteracting phosphatase. (B) Active forms of Wee1 and Cdc25 as functions of active MPF. (C) Rates of activation and inactivation of MPF as functions of MPF activity, from Eq. (B3.2). The intersection points correspond to stable (black circles) and unstable (white circle) steady states of the dynamical system. (D) Bifurcation diagram. The steady state values of active MPF are plotted as functions of total cyclin B, $B_T = $ [MPF] + [preMPF]. For $0.28 < B_T < 0.69$, the regulatory network has three steady states, two stable (solid lines) and one unstable (dashed line).

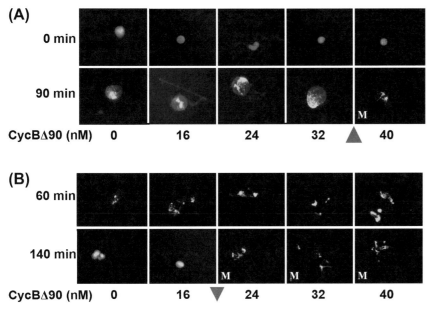

FIGURE 14.9 Bistability at the G2/M transition in frog egg extracts. From Sha et al. [85]; used by permission. Cytoplasmic extracts of frog eggs are used to measure the amount of cyclin B necessary to induce or sustain the kinase activity of CycB:Cdk1 heterodimers (referred to in this paper as MPF, mitosis promoting factor). Sperm nuclei in the extract (stained blue for chromatin) are photographed to stage the extracts. Interphase: round nuclei, dispersed chromatin, intact nuclear membrane, low activity of MPF (confirmed in separate experiments, not shown). Mitosis: highly condensed chromatin, no nuclear membrane, high activity of MPF. (A) Cyclin threshold for activation of MPF. Extracts are prepared in interphase ($t = 0$) in the presence of cycloheximide, to block all protein synthesis, including the synthesis of endogenous cyclin B. Samples of the extract are injected with increasing amounts of non-degradable cyclin B (CycBΔ90) at $t = 0$ and photographed at intervals thereafter ($t = 90$ min time point shown here). Extracts containing 0–32 nM CycBΔ90 have insufficient MPF activity to enter mitosis, but 40 nM CycBΔ90 or larger is enough to activate MPF and drive the nuclei into mitosis. (B) Cyclin threshold for inactivation of MPF. Extracts are prepared in interphase ($t = 0$) in the absence of cycloheximide and injected with increasing amounts of CycBΔ90. Whether the injected amount is small or large, synthesis of endogenous cyclin B drives the extract into mitosis by $t = 60$ min. At $t = 60$ min the extracts are treated with cycloheximide to prevent any further synthesis of endogenous cyclin B. As the extracts try to exit from mitosis, they activate Cdc20 and degrade the endogenous cyclin B proteins but leave behind the non-degradable CycBΔ90 molecules. Extracts containing 24 nM CycBΔ90 or higher retain MPF in the active form and block the nuclei in mitosis. Extracts containing 16 nM CycBΔ90 or lower have inactive MPF (confirmed in separate experiments, not shown) and return to interphase. For CycBΔ90 concentrations in the range 24–32 nM, the MPF control system is bistable: it can persist in an MPF-inactive state (panel A) or in an MPF-active state (panel B) under identical conditions on cyclin B concentration in the extract. Which state the extract adopts depends on whether it was initially in the MPF-inactive state (above) or in the MPF-active state (below).

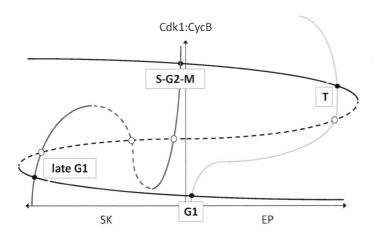

FIGURE 14.10 Additional checkpoints in late G1 and telophase. Left: a cell that has passed the restriction point (aka, Start) can still be blocked in late G1, with high levels of starter kinase, if a new checkpoint is created by a saddle-node bifurcation at the far left. Right: similarly, a cell that has completed the M/A transition can still be blocked in telophase (T), if a new checkpoint is created by a saddle-node bifurcation at the far right.

These checkpoints are imposed and lifted by saddle-node bifurcations; hence, just like the restriction point and exit from mitosis, they are dynamically irreversible transitions. On the other hand, unlike RP and exit, they are not universal. For instance, in contrast to budding yeast, mammalian cells do not have a telophase checkpoint. In mammalian cells, as in most cells, the division plane is determined late in the cell cycle, by the location of the mid-zone of the late anaphase spindle (which is the position where the metaphase plate assembled). Hence, as long as the M/A transition has been successfully completed, then cell division will automatically separate the two new daughter nuclei. There is no need for a telophase checkpoint, as in budding yeast, where exit from mitosis occurs in two stages.

Furthermore, the DNA damage checkpoint in budding yeast works completely differently than in mammalian cells. Damage in G1 phase causes a specific phosphorylation of Swi6 that blocks transcription of Cln2, the starter kinase [87]. Hence, young yeast cells with damage block at the early G1 checkpoint rather than the late G1 checkpoint. DNA damage received later in the cycle causes a block at the M/A transition in budding yeast cells [88], whereas most other cell types block at the G2/M transition.

Unicellular eukaryotes also differ from metazoan cells in the strength, duration and consequences of checkpoints. The purpose of checkpoints is to block progression through the cell cycle if problems arise that compromise successful replication of the cell and its genome. If the problem can be repaired, the checkpoint can be lifted and the cell can proceed with the replication—division cycle. But what should the cell do if the problem cannot be repaired? For a unicellular organism, the best strategy is to bypass the checkpoint after some time and proceed to cell division. The worst thing that can happen is that the problem is lethal and the cell dies. But in many cases the problem is not lethal, the cell survives and reproduces, and the daughter cells 'get on with life'. Maybe they carry some new mutations, maybe they are aneuploid or polyploid, but at least they are alive. For metazoans, on the contrary, these damaged cells inhabit a larger organism, and the mutations they carry may prove advantageous for the cell but fatal for the whole organism (think of malignant cancer cells). Hence, mammalian cells (which have been most thoroughly studied in this regard) tend to have stronger checkpoints, and if the damage cannot be repaired then the surveillance mechanism redirects the cell toward programmed cell death [89]. Better that the damaged cell be destroyed than that its progeny destroy the whole organism.

CONCLUSIONS

If our vision of the eukaryotic cell cycle control system is correct, then it should account naturally for the four characteristic features of mitotic cell division enumerated in the introduction. The first feature, that cell cycle progression is unidirectional and irreversible, and the third feature, that checkpoints guard the major transitions of the cell cycle, are the fundamental ideas behind our theory. We have shown how these features are based on the dynamics of the interacting genes and proteins that govern cell cycle progression.

The second feature, balanced growth and division, is a special case of the checkpoint paradigm. In unicellular eukaryotes, growth to a minimal size is a requirement for passing one of the checkpoints of the cell cycle. In budding yeast the size requirement is enforced at Start [55,90]. In fission yeast, it is enforced at the G2/M transition [91], as is also true of the acellular slime mold *Physarum polycephalum* [92]. Strict size control is also evident in the physiology of *Stentor* [93] and *Amoeba* [94,95], although the molecular details have never been worked out. By mutations, the size checkpoint can be moved to a different transition point; for example, $wee1\Delta$ mutants of fission yeast are size-regulated at the G1/S transition. In budding yeast, it should be noted, size control is strong in small daughter cells but weak in large mother cells [90]. It matters little that mother cells get progressively larger from generation to generation, because they eventually senesce and die.

Size control is less evident in metazoans than in yeast [96,97]. Unlike yeast cells, which maintain a stable size distribution over many generations of growth and division [98], cell lineages in metazoans are more variable in size, especially lineages of restricted proliferative potential. We would expect size control to be more evident in cells with high proliferative capacity (germline cells and stem cells).

Eggs and early embryos are interesting cases. Eggs grow very large and arrest in meiosis II with a single, haploid nucleus. After fertilization, the diploid zygote undergoes a series of rapid mitotic cycles (without growth) to create a ball of small, mononucleated cells (the blastula). Although details of these early embryonic divisions vary from one organism to another, in general the cell cycle checkpoints are not operating as usual. Around the mid-blastula stage the mitotic division cycles change dramatically, acquiring G1 and G2 phases and checkpoint controls, including growth controls. At this point the embryo starts to grow and develop, using resources stored in the egg or provided by the placenta.

The fourth feature, that the control system must be robust in the face of unavoidable molecular noise in the small environs of a cell, is beyond the scope of this chapter. Suffice it to say that in recent publications we have studied realistic stochastic models of bistability and irreversibility in the cell cycle engine and found that the control system shows exactly the same sort of robustness—variability exhibited by proliferating yeast cells [99—101].

If we have done our job well, then everything we have said should seem natural and intuitively appealing. If everything is so obvious, the skeptical reader might ask, 'Why do we need mathematical models and bifurcation diagrams? It's all right there in the reaction networks!' We hope most readers will not be so jaded. These ideas were not so obvious to us until we started thinking about cell cycle regulation in terms of mathematical models. We are convinced that a mathematical, systems-level approach to regulatory networks is absolutely essential to a correct understanding of physiological control systems like the eukaryotic cell cycle.

REFERENCES

[1] Hartwell LH. Nobel Lecture. Yeast and cancer. Biosci Rep 2002;22:373—94.

[2] Hunt T. Nobel Lecture. Protein synthesis, proteolysis, and cell cycle transitions. Biosci Rep 2002;22:465—86.

[3] Nurse PM. Nobel Lecture. Cyclin dependent kinases and cell cycle control. Biosci Rep 2002;22:487—99.

[4] Nurse P. Universal control mechanism regulating onset of M-phase. Nature 1990;344:503—8.

[5] Kohn KW. Molecular interaction map of the mammalian cell cycle control and DNA repair systems. Mol Biol Cell 1999;10:2703—34.

[6] Chen KC, Calzone L, Csikasz-Nagy A, Cross FR, Novak B, Tyson JJ. Integrative analysis of cell cycle control in budding yeast. Mol Biol Cell 2004;15:3841—62.

[7] Chen KC, Csikasz-Nagy A, Gyorffy B, Val J, Novak B, Tyson JJ. Kinetic analysis of a molecular model of the budding yeast cell cycle. Mol Biol Cell 2000;11:369—91.

[8] Tyson JJ, Chen K, Novak B. Network dynamics and cell physiology. Nat Rev Mol Cell Biol 2001;2:908—16.

[9] Mitchison JM. The Biology of the Cell Cycle. Cambridge: Cambridge University Press; 1971.

[10] Murray A, Hunt T. The Cell Cycle. An Intoduction. New York: W.H. Freeman and Company; 1993.

[11] Morgan DO. The Cell Cycle: Principles of Control. London. New Science Press; 2007.

[12] Nasmyth K, Haering CH. Cohesin: its roles and mechanisms. Annu Rev Genet 2009;43:525—58.

[13] Pines J, Rieder CL. Re-staging mitosis: a contemporary view of mitotic progression. Nat Cell Biol 2001;3:E3—6.

[14] Mitchison TJ, Salmon ED. Mitosis: a history of division. Nat Cell Biol 2001;3:E17—21.

[15] Peters JM, Tedeschi A, Schmitz J. The cohesin complex and its roles in chromosome biology. Genes Dev 2008;22:3089—114.

[16] Nezi L, Musacchio A. Sister chromatid tension and the spindle assembly checkpoint. Curr Opin Cell Biol 2009;21:785—95.

[17] Tyson JJ, Novak B. Temporal organization of the cell cycle. Curr Biol 2008;18:R759—68.

[18] Elledge SJ. Cell cycle checkpoints: preventing an identity crisis. Science 1996;274:1664—72.

[19] Tyson JJ, Novak B. Regulation of the eukaryotic cell cycle: molecular antagonism, hysteresis, and irreversible transitions. J Theor Biol 2001;210:249—63.

[20] Tyson JJ, Novak B, Chen K, Val J. Checkpoints in the cell cycle from a modeler's perspective. Prog Cell Cycle Res 1995;1:1—8.

[21] Murray AW. Creative blocks: cell-cycle checkpoints and feedback controls. Nature 1992;359:599—604.

[22] Ghaemmaghami S, Huh WK, Bower K, Howson RW, Belle A, Dephoure N, et al. Global analysis of protein expression in yeast. Nature 2003;425:737—41.

[23] Zenklusen D, Larson DR, Singer RH. Single-RNA counting reveals alternative modes of gene expression in yeast. Nat Struct Mol Biol 2008;15:1263—71.

[24] Pedraza JM, Paulsson J. Effects of molecular memory and bursting on fluctuations in gene expression. Science 2008;319:339—43.

[25] Swain PS, Elowitz MB, Siggia ED. Intrinsic and extrinsic contributions to stochasticity in gene expression. Proc Natl Acad Sci U S A 2002;99:12795—800.

[26] Murray AW. Recycling the cell cycle: cyclins revisited. Cell 2004;116:221—34.

[27] Morgan DO. Principles of CDK regulation. Nature 1995;374:131—4.

[28] Sherr CJ. Cancer cell cycles. Science 1996;274:1672—7.

[29] Vodermaier HC. APC/C and SCF: controlling each other and the cell cycle. Curr Biol 2004;14:R787—796.

[30] Satyanarayana A, Kaldis P. Mammalian cell-cycle regulation: several Cdks, numerous cyclins and diverse compensatory mechanisms. Oncogene 2009;28:2925—39.

[31] Hsu JY, Reimann JD, Sorensen CS, Lukas J, Jackson PK. E2F-dependent accumulation of hEmi1 regulates S phase entry by inhibiting APC(Cdh1). Nat Cell Biol 2002;4:358—66.

[32] Pines J. Mitosis: a matter of getting rid of the right protein at the right time. Trends Cell Biol 2006;16:55—63.

[33] Coleman TR, Dunphy WG. Cdc2 regulatory factors. Curr Opin Cell Biol 1994;6:877—82.

[34] Nilsson I, Hoffmann I. Cell cycle regulation by the Cdc25 phosphatase family. Prog Cell Cycle Res 2000;4:107—14.

[35] Martin-Castellanos C. Moreno: Recent advances on cyclins, CDKs and CDK inhibitors. Trends Cell Biol 1997;7:95—8.

[36] Besson A, Dowdy SF, Roberts JM. CDK inhibitors: cell cycle regulators and beyond. Dev Cell 2008;14:159—69.

[37] Peters JM. The anaphase promoting complex/cyclosome: a machine designed to destroy. Nat Rev Mol Cell Biol 2006;7:644—56.

[38] Clute P, Pines J. Temporal and spatial control of cyclin B1 destruction in metaphase. Nat Cell Biol 1999;1:82—7.

[39] Nasmyth K, Peters JM, Uhlmann F. Splitting the chromosome: cutting the ties that bind sister chromatids. Science 2000;288:1379—85.

[40] Musacchio A, Salmon ED. The spindle-assembly checkpoint in space and time. Nat Rev Mol Cell Biol 2007;8:379—93.

[41] Sullivan M, Morgan DO. Finishing mitosis, one step at a time. Nat Rev Mol Cell Biol 2007;8:894—903.

[42] Hayles J, Fisher D, Woollard A, Nurse P. Temporal order of S phase and mitosis in fission yeast is determined by the state of the p34cdc2-mitotic B cyclin complex. Cell 1994;78:813—22.

[43] Lee HO, Davidson JM, Duronio RJ. Endoreplication: polyploidy with purpose. Genes Dev 2009;23:2461—77.

[44] Michael WM, Newport J. Coupling of mitosis to the completion of S phase through Cdc34-mediated degradation of Wee1. Science 1998;282:1886—9.

[45] Novak B, Tyson JJ, Gyorffy B, Csikasz-Nagy A. Irreversible cell-cycle transitions are due to systems-level feedback. Nat Cell Biol 2007;9:724—8.

[46] Bloom J, Cross FR. Multiple levels of cyclin specificity in cell-cycle control. Nat Rev Mol Cell Biol 2007;8:149—60.

[47] Nasmyth K. At the heart of the budding yeast cell cycle. Trends Genet 1996;12:405—12.

[48] Piatti S, Bohm T, Cocker JH, Diffley JF, Nasmyth K. Activation of S-phase-promoting CDKs in late G1 defines a 'point of no return' after which Cdc6 synthesis cannot promote DNA replication in yeast. Genes Dev 1996;10:1516—31.

[49] Cross FR, Archambault V, Miller M, Klovstad M. Testing a mathematical model of the yeast cell cycle. Mol Biol Cell 2002;13:52—70.

[50] Lopez-Aviles S, Kapuy O, Novak B, Uhlmann F. Irreversibility of mitotic exit is the consequence of systems-level feedback. Nature 2009;459:592—5.

[51] Cross FR. Starting the cell cycle: what's the point? Curr Opin Cell Biol 1995;7:790—7.

[52] Costanzo M, Nishikawa JL, Tang X, Millman JS, Schub O, Breitkreuz K, et al. CDK activity antagonizes Whi5, an inhibitor of G1/S transcription in yeast. Cell 2004;117:899—913.

[53] Wittenberg C, Reed SI. Cell cycle-dependent transcription in yeast: promoters, transcription factors, and transcriptomes. Oncogene 2005;24:2746−55.

[54] Charvin G, Oikonomou C, Siggia ED, Cross FR. Origin of irreversibility of cell cycle start in budding yeast. PLoS Biol 2010;8:e1000284.

[55] Hartwell LH, Unger MW. Unequal division in *Saccharomyces cerevisiae* and its implications for the control of cell division. J Cell Biol 1977;75:422−35.

[56] Di Como CJ, Chang H, Arndt KT. Activation of CLN1 and CLN2 G1 cyclin gene expression by BCK2. Mol Cell Biol 1995;15:1835−46.

[57] Skotheim JM, Di Talia S, Siggia ED, Cross FR. Positive feedback of G1 cyclins ensures coherent cell cycle entry. Nature 2008;454:291−6.

[58] Kapuy O, He E, Lopez-Aviles S, Uhlmann F, Tyson JJ, Novak B. System-level feedbacks control cell cycle progression. FEBS Lett 2009;583:3992−8.

[59] Stegmeier F, Amon A. Closing mitosis: the functions of the Cdc14 phosphatase and its regulation. Annu Rev Genet 2004; 38:203−32.

[60] Vazquez-Novelle MD, Mirchenko L, Uhlmann F, Petronczki M. The 'anaphase problem': how to disable the mitotic checkpoint when sisters split. Biochem Soc Trans 2010;38:1660−6.

[61] He E, Kapuy O, Oliveira RA, Uhlmann F, Tyson JJ, Novak B. System-level feedbacks make the anaphase switch irreversible. Proc Natl Acad Sci U S A 2011;108:10016−21.

[62] Sullivan M, Uhlmann F. A non-proteolytic function of separase links the onset of anaphase to mitotic exit. Nat Cell Biol 2003;5:249−54.

[63] Queralt E, Lehane C, Novak B, Uhlmann F. Downregulation of PP2A(Cdc55) phosphatase by separase initiates mitotic exit in budding yeast. Cell 2006;125:719−32.

[64] Visintin C, Tomson BN, Rahal R, Paulson J, Cohen M, Taunton J, et al. APC/C-Cdh1-mediated degradation of the Polo kinase Cdc5 promotes the return of Cdc14 into the nucleolus. Genes Dev 2008;22:79−90.

[65] Vinod PK, Freire P, Rattani A, Ciliberto A, Uhlmann F, Novak B. Computational modelling of mitotic exit in budding yeast: the role of separase and Cdc14 endocycles. J R Soc Interface 2011;8: 1128−1141.

[66] Mirchenko L, Uhlmann F. Sli15(INCENP) dephosphorylation prevents mitotic checkpoint reengagement due to loss of tension at anaphase onset. Curr Biol 2010;20:1396−401.

[67] Rape M, Kirschner MW. Autonomous regulation of the anaphase-promoting complex couples mitosis to S-phase entry. Nature 2004;432:588−95.

[68] Sherr CJ, Roberts JM. CDK inhibitors: positive and negative regulators of G1-phase progression. Genes Dev 1999;13: 1501−12.

[69] Bartek J, Bartkova J, Lukas J. The retinoblastoma protein pathway and the restriction point. Curr Opin Cell Biol 1996; 8:805−14.

[70] Sherr CJ, Roberts JM. Living with or without cyclins and cyclin-dependent kinases. Genes Dev 2004;18:2699−711.

[71] Yao G, Lee TJ, Mori S, Nevins JR, You L. A bistable Rb-E2F switch underlies the restriction point. Nat Cell Biol 2008; 10:476−82.

[72] Yao G, Tan C, West M, Nevins JR, You L. Origin of bistability underlying mammalian cell cycle entry. Mol Syst Biol 2011; 7:485.

[73] Potapova TA, Daum JR, Byrd KS, Gorbsky GJ. Fine tuning the cell cycle: activation of the Cdk1 inhibitory phosphorylation pathway during mitotic exit. Mol Biol Cell 2009;20:1737−48.

[74] Potapova TA, Daum JR, Pittman BD, Hudson JR, Jones TN, Satinover DL, et al. The reversibility of mitotic exit in vertebrate cells. Nature 2006;440:954−8.

[75] Kapuy O, He E, Uhlmann F, Novak B. Mitotic exit in mammalian cells. Mol Syst Biol 2009;5:324.

[76] Oliveira RA, Hamilton RS, Pauli A, Davis I, Nasmyth K. Cohesin cleavage and Cdk inhibition trigger formation of daughter nuclei. Nat Cell Biol 2010;12:185−92.

[77] Potapova TA, Sivakumar S, Flynn JN, Li R, Gorbsky GJ. Mitotic progression becomes irreversible in prometaphase and collapses when Wee1 and Cdc25 are inhibited. Mol Biol Cell 2011;22: 1191−206.

[78] Nurse P. Cell cycle control genes in yeast. Trends Genet 1985;1:51−5.

[79] Kumagai A, Dunphy WG. Regulation of the cdc25 protein during the cell cycle in *Xenopus* extracts. Cell 1992;70:139−51.

[80] Trunnell NB, Poon AC, Kim SY, Ferrell Jr JE. Ultrasensitivity in the Regulation of Cdc25C by Cdk1. Mol Cell 2011;41:263−74.

[81] Kim SY, Ferrell Jr JE. Substrate competition as a source of ultrasensitivity in the inactivation of Wee1. Cell 2007;128: 1133−45.

[82] Tang Z, Coleman TR, Dunphy WG. Two distinct mechanisms for negative regulation of the Wee1 protein kinase. Embo J 1993;12: 3427−36.

[83] Novak B, Tyson JJ. Numerical analysis of a comprehensive model of M-phase control in *Xenopus* oocyte extracts and intact embryos. J Cell Sci 1993;106(Pt 4):1153−68.

[84] Pomerening JR, Sontag ED, Ferrell Jr JE. Building a cell cycle oscillator: hysteresis and bistability in the activation of Cdc2. Nat Cell Biol 2003;5:346−51.

[85] Sha W, Moore J, Chen K, Lassaletta AD, Yi CS, Tyson JJ, et al. Hysteresis drives cell-cycle transitions in *Xenopus laevis* egg extracts. Proc Natl Acad Sci U S A 2003;100:975−80.

[86] Bardin AJ, Visintin R, Amon A. A mechanism for coupling exit from mitosis to partitioning of the nucleus. Cell 2000;102:21−31.

[87] Sidorova JM, Breeden LL. Rad53-dependent phosphorylation of Swi6 and down-regulation of CLN1 and CLN2 transcription occur in response to DNA damage in *Saccharomyces cerevisiae*. Genes Dev 1997;11:3032−45.

[88] Cohen-Fix O, Koshland D. The anaphase inhibitor of *Saccharomyces cerevisiae* Pds1p is a target of the DNA damage checkpoint pathway. Proc Natl Acad Sci U S A 1997;94:14361−6.

[89] Menendez D, Inga A, Resnick MA. The expanding universe of p53 targets. Nat Rev Cancer 2009;9:724−37.

[90] Di Talia S, Skotheim JM, Bean JM, Siggia ED, Cross FR. The effects of molecular noise and size control on variability in the budding yeast cell cycle. Nature 2007;448:947−51.

[91] Fantes P, Nurse P. Control of cell size at division in fission yeast by a growth-modulated size control over nuclear division. Exp Cell Res 1977;107:377−86.

[92] Tyson J, Garcia-Herdugo G, Sachsenmaier W. Control of nuclear division in *Physarum polycephalum*: Comparison of

cycloheximide pulse treatment, uv irradiation, and heat shock. Exp Cell Res 1979;119:87—98.
[93] Frazier EA. DNA synthesis following gross alterations of the nucleocytoplasmic ratio in the ciliate *Stentor coeruleus*. Dev Biol 1973;34:77—92.
[94] Hartmann M. Über experimentelle unsterblichkeit von protozoen-individuen. Ersatz der Fortpflanzung von *Amoeba proteus* durch fortgesetzte Regenerationen. Zool Jahrb 1929;45.
[95] Prescott DM. Relation between cell growth and cell division. III. Changes in nuclear volume and growth rate and prevention of cell division in *Amoeba proteus* resulting from cytoplasmic amputations. Exp Cell Res 1956;11:94—8.
[96] Baserga R. Growth in size and cell DNA replication. Exp Cell Res 1984;151:1—5.
[97] Conlon I, Raff M. Differences in the way a mammalian cell and yeast cells coordinate cell growth and cell-cycle progression. J Biol 2003;2:7.
[98] Jorgensen P, Tyers M. How cells coordinate growth and division. Curr Biol 2004;14:R1014—1027.
[99] Barik D, Baumann WT, Paul MR, Novak B, Tyson JJ. A model of yeast cell-cycle regulation based on multisite phosphorylation. Mol Syst Biol 2010;6:405.
[100] Kar S, Baumann WT, Paul MR, Tyson JJ. Exploring the roles of noise in the eukaryotic cell cycle. Proc Natl Acad Sci U S A 2009;106:6471—6.
[101] Kapuy O, Barik D, Sananes MR, Tyson JJ, Novak B. Bistability by multiple phosphorylation of regulatory proteins. Prog Biophys Mol Biol 2009;100:47—56.

Chapter 15

Phenotypes and Design Principles in System Design Space

Michael A. Savageau
Biomedical Engineering Department and Microbiology Graduate Group, University of California, One Shields Avenue, Davis, CA 95616 USA

Chapter Outline

Background	**288**
Systems	288
Wholeness and Open Systems	288
Modules	290
Modules as Elements of Random Change	290
Modules as Designed Products of Selection	290
Interface	291
Interface and Function	291
Interface and Context	292
Design	292
Phenotypes	**293**
Generic Concept of Phenotype	294
Phenotypes from the Analytical Solution	294
Phenotypes from the Differential Equation	295
Phenotypes in System Design Space	295
Enumeration of Qualitatively Distinct Phenotypes	296
Robustness	296
Characterizing Performance	296
Logarithmic Gain	297
Parameter Sensitivity	297
Response Time	297
Global Tolerance	297
Comparison of Phenotypes	297
Criteria for Functional Effectiveness	297
Local Performance	297
Global Tolerance	298

Design Principles	**298**
Alternative Growth Modes of Phage Lambda	298
Developmental Decisions	299
Kinetic Model	299
Estimation of Parameter Values	300
Recast Equations	300
Number of Qualitatively Distinct Phenotypes in Design Space	301
Example of a Valid Phenotype (Case 11)	302
Dominance Conditions	302
Boundaries in System Design Space	302
Example of an Invalid Phenotype (Case 6)	303
Dominance Conditions	303
Evaluation of Local Performance	303
Quantitative Criteria	303
Analysis of Local Performance	303
Evaluation of Global Performance	305
Quantitative Criteria	305
Analysis of Global Performance	305
Biological Design Principles	305
Avoiding Inappropriate Switching	306
Maintaining the Temperate Lifestyle	307
Conclusions and Future Challenges	**307**
Acknowledgements	**308**
References	**308**

Throughout the pre-genomic era there was sustained interest in the relationship between genotype and phenotype. However, the announcement of the draft sequence of the human genome in 2000 revealed the true magnitude of the challenge. As Sydney Brenner [1] has stated, 'The problems faced by pre- and post-genomic genetics are much the same — they all involve bridging the chasm between genotype and phenotype'. The difficulty in relating these two levels of biological organization and function is hard to over-estimate. Moreover, between the levels of genotype and phenotype of the organism there are many intervening levels that form a rich hierarchy of molecular subsystems. Although we now have a generic *concept* of 'genotype' provided by the detailed DNA sequence, and next-generation sequencing is likely to produce complete manifestations of this concept in the near future, there is no corresponding generic *concept* of 'phenotype'. We have only some intuitive notions of what is meant by phenotype at the level of the organism, hair color of cats, shape and size of flowers, height and weight of livestock, not to mention disease states in humans. However, without a generic concept of phenotype there can be no context for a deep, predictive understanding of the relationship between genotype and phenotype. This chapter will address this issue and its importance for revealing the system design principles arrived at by the operation of natural selection. But first some background will be provided to set the stage.

BACKGROUND

The experimental study of specific systems by molecular biologists has revealed an immense variety of molecular mechanisms that are combined into complex networks, and the patterns of gene expression observed in response to environmental and developmental signals are equally diverse. Despite this impressive progress we are at a loss to understand the integrated behavior of most cell and molecular systems. Even in the best-studied organisms we are unable to predict the response to a specific change in its genome or its response to a novel compound in the environment. In short, our knowledge is still fragmented and descriptive: we have almost no understanding of the 'design principles' that govern cell and molecular systems. Uncovering these design principles will be important not only for understanding the normal function of cell and molecular systems but also for developing judicious methods to redirect normal expression for biotechnological purposes or to correct pathological expression for therapeutic purposes.

If there is to be an understanding based on design principles, then it must integrate the principles of physics, chemistry and biological organization. Although organisms must deal with many forms of energy, the basic unit of exchange is chemical. All other types of energy are interconvertible with the chemical form by means of specialized energy-transduction processes. Electromagnetic energy is converted to chemical energy in photosynthesis, whereas chemical energy is converted to mechanical energy in muscle contraction. Aside from such specialized energy conversions, the overwhelming majority of cellular functions are of a strictly chemical nature.

Although chemical change can be described in a variety of ways, the most fruitful in dealing with large systems are quantitative: to what extent a reaction normally takes place, and how fast it proceeds. These thermodynamic and kinetic aspects have had extensive development. The conventional representation in dealing with chemical kinetics is the mass-action formalism, whereas that in dealing with biochemical kinetics is the rational-function formalism, which results from mass action with constraints. Both of these formalisms can be considered special cases of the generalized mass-action formalism that will be discussed in detail later in this chapter.

Nothing in biology makes sense except in the light of systems.
with apologies to Dobzhansky

Systems

There is no universally accepted, unambiguous definition for what we mean by a 'system'. Any dictionary typically has dozens of definitions. I have previously formulated the following [2], which will be helpful for my purposes here:

A system can be defined as a collection of interacting parts that in some sense constitutes a whole. Everything excluded from the collection is considered the environment of the system. It is immediately obvious that if the parts mentioned in the above definition are also made up of interacting parts, then they too fit the definition and could therefore be considered as subsystems.

In this way a description of nature in terms of a hierarchy of systems is produced. For example, cells could be considered the subsystems of a tissue, tissues the subsystems of an organ, and organs the subsystems of an organism. This hierarchy could be greatly extended in both directions.

One does not study the universe but only some portion or level within its total hierarchical representation. In the selection of a system as a subject for investigation two conflicting demands are encountered: (a) the need to maintain wholeness, and (b) the need to limit the complexity of the problem.

Wholeness and Open Systems

The idea of wholeness connotes a complete or closed system, an idealization that is approached to varying extents in different cases. Many examples of close approximation to a closed system can be found among technological systems in which the concept of modularity originated. However, one of the most characteristic features of biological systems is that they are open, i.e., they are characterized by a high degree of interaction with their environment.

Biological systems seldom present themselves as well defined or closed: the investigator must make choices. This is an art, not a mechanical process, and the importance of intimate familiarity with the system should not be overlooked. Nevertheless, the choice for delineation of a system must be made according to some criterion such as the following: *Let systems be selected so as to maximize the numerical ratio of internal interactions to external interactions.* In this way the resulting system will tend to have a minimal number of interactions with its environment, but at the same time a maximum number of interactions will be included within the system. This choice will tend to preserve the integrity of critical functional groupings, which is important because certain phenomena cannot be adequately understood by simply analyzing the component parts of a system. At certain levels of organization new properties emerge that can be understood only at these levels; such phenomena must be treated as a whole.

Interactions between the system chosen in this way and its environment will remain, but the number will have been minimized. Thus, the identification of environmental interactions is facilitated and, when these are disrupted to study the system in isolation, the effects of the normal environment on the system can be most easily approximated by experimental means.

By this criterion for the delineation of a system, a cell constitutes a well-delineated system because the semipermeable plasma membrane tends to minimize the interactions between the cytoplasm and the extracellular milieu. On the other hand, a cell with a disrupted membrane would have greatly increased numbers of interactions between its cytoplasm and the extracellular milieu, and the system no longer would retain the properties characteristic of a whole cell.

Feasibility and the Limitation of Complexity

While the requirement of wholeness urges us to consider more and more interactions in describing a given system, time, resources, insight, and the precision of instruments place limits on the number of interactions that can be considered. The amount of detailed consideration that can be given to the environment on the one hand, and to the subsystems on the other, is limited. Fortunately, there are at least three different types of simplification — spatial, temporal, and functional — that arise naturally to limit the complexity of systems and make their analysis feasible.

Spatial Simplifications

Spatial or topological constraints are abundant in natural systems. For example, the relative specificity of enzymes for their reactants tends to limit the possible interactions among the molecules of a biochemical system. Thus, the enzymes of biochemical systems are natural subsystems at this level of organization. The formation of multi-enzyme complexes among the consecutive enzymes of a metabolic pathway, which results in 'channeling' of the intermediates, is another mechanism for spatially restricting the interaction among molecules. This leads to subsystems at a higher level of cellular organization. Channeling may be considered one of the simplest examples of compartmentalization [3], which is common among eukaryotic organisms. The mitochondria in such organisms provide one of the best-known illustrations of a spatially separate compartment in which certain metabolites have only limited access to the remainder of the cytoplasm. Cellular compartments form subsystems on a still higher level of organization. Such compartmentalization often permits these systems to be described by ordinary differential equations, whereas large distributed systems would require partial differential equations for their description.

Temporal Simplifications

There is generally a vast difference in the relaxation times or time constants that characterize the dynamic response of a system. Early on, biologists distinguished three levels of activity in time: evolutionary, developmental, and biochemical [4]. With the refinement of instrumentation, a fourth level was added: biomolecular. The time constants for evolutionary change in a given type of organism would be measured in terms of generations. Developmental change takes place within an organism's lifetime. Biochemical change has a relaxation time typically in the range of seconds to minutes, whereas biomolecular transitions can occur within millisecond and faster time constants.

This temporal separation of phenomena can simplify greatly the analysis of complex systems. The variables in such a system that respond much faster than the phenomena of interest can be assumed to be at their steady-state values; those variables responding more slowly than the phenomena of interest can be assumed to be constants or slowly varying parameters. For instance, if one is interested in the temporal behavior of a simple biochemical pathway responding to changes in the concentration of a specific ligand, then the molecular transitions of an enzyme molecule and the binding of ligand will typically occur much more rapidly than the time course of the overall reaction. Under these conditions the two processes could be assumed to be in steady state with respect to the rate of the overall reaction. At the other end of the temporal spectrum, changes in the total concentration of an enzyme that might occur during the course of development can be ignored, because such changes are so slight on the timescale of the overall reaction. Simplifications such as these have a long history of use in the physical sciences and are probably best illustrated quantitatively by the circuit theory approximation to time-varying electromagnetic phenomena [5].

Although the separation of timescales greatly simplifies the analysis of complex systems, the phenomena operating at the different scales remain integrated through constraints that are implicit in the parameters that characterize the system at any particular scale. For example, it is well known that the parameters that characterize the rational functions of steady-state enzyme kinetics on the biochemical timescale (Michaelis constants, maximal velocities, inhibition constants, Haldane relationships, etc.) are functions of the parameters that characterize the power-law functions of the underlying chemical kinetics on the biomolecular time scale (rate constants, kinetic orders, detailed balance, conservation constraints, etc.). Indeed, the systems approach is concerned specifically with understanding relationships that span interconnected scales by revealing the constraints between them and exploiting these constraints to simplify the analysis of the integrated system.

Functional Simplifications

This is a more abstract notion that is best illustrated with a specific example. Feedback control mechanisms (Figure 15.1), which are ubiquitous in living organisms, provide one of the best examples of functional simplification associated with subsystems on a level of organization encompassing not only material flow but also information flow for regulatory purposes. These mechanisms regulate certain variables in the system but also modify the mathematical form of the functional relationships between variables. Both of these properties can lead to a simplification of the system under consideration.

Feedback inhibition effectively maintains certain metabolic pools at relatively fixed concentrations, thereby reducing these variables of the system to constants for many practical purposes. Feedback inhibition also can simplify the effective mathematical complexity of rate laws. The kinetics of the regulatory enzyme in these systems is highly non-linear when characterized in vitro. However, in situ, the negative feedback tends to simplify the mathematical form of these kinetics. Therefore, the use of such simplified relationships in place of the original, highly non-linear ones can be justified, and it greatly simplifies the description of the functional relationship of velocity to reactant and modifier concentrations. This is a well-known property of negative feedback in electronic circuits [6]. Adding negative feedback around a transistor, which has a highly non-linear characteristic, is used to construct operational amplifiers with linear characteristics. In this way the non-linear characteristics of many technological components are purposely 'straightened' to the point of becoming linear in order to facilitate their analysis and make their behavior more predictable. However, in biological systems 'we have to deal with the system we are given, not the one we want' in most cases, although this may change somewhat as the field of synthetic biology matures [7]. The non-linear characteristics in biological systems are typically not 'straightened' all the way to linear; if they were, life as we know it would not exist. However, they are very often 'straightened' to power-law functions, a simplification that offers considerable advantages.

There are four conclusions to be drawn from the argument in this section. First, systems, as I have defined them, are naturally modular and pervasive in nature. Indeed, one might well use the terms system (or subsystem), component and module interchangeably. Second, systems function within the context provided by their environment. Third, systems interact with their environment across boundaries that have a specific material realization (this critically important point will be the focus of a later section). Fourth, to the extent that we can identify this modularity and its environmental context, our analysis of phenotypic function and system design, and ultimately our understanding of biological systems, is greatly facilitated.

Modules

Modules or subsystems are physical entities that can be viewed from two different perspectives: one with a focus on generating diversity for the evolution of new system functions, the other with a focus on understanding function and design of operational systems.

Modules as Elements of Random Change

Familiar examples of modules at the level of primary structure of macromolecules include amino acids in polypeptides and nucleotide bases in DNA and RNA. Modules at a higher level might include common folds and domains in proteins, and recognition sequences and exchangeable cassettes in nucleotide sequences. Such examples are often invoked to explain the advantages of modularity in creating the combinatorial expansion of diversity available for evolutionary innovation.

The shuffling of modules is involved at all levels in the process of mutation: base changes, deletions, insertions, inversions, duplications, recombinations, frame-shifts, truncations, and fusions. Most results are deleterious; a few are advantageous. Modularity clearly facilitates all these events. However, these are largely accidents that, at least initially, may not follow any clear rule *at the level of the nascent system*. The 'repurposing' of modules in this context is random, and only during the process of selection is the emerging system likely to exhibit rule-like behavior.

Once a system has been refined by the sculpting power of natural selection, one can begin to ask questions regarding how well it performs its functions, not only in an absolute sense but also in relation to alternatives that have evolved to perform similar functions. In this way one begins to identify design principles that apply to an entire class of systems.

Modules as Designed Products of Selection

The focus here is on modularity at a higher level of system organization and on the second of the perspectives mentioned in the introduction of this section. Historically, biologists have always identified modules for study. Biochemists have long focused on isolated enzymes as important modules. At the next level of metabolic organization one thinks of pathways, branch points and cycles as

FIGURE 15.1 Schematic diagram of a biosynthetic pathway with end-product inhibition.

modules that were recognized very early in the study of metabolism. Beginning in the late 1950s, it was recognized by Umbarger and Pardee and their colleagues that biosynthetic pathways involving end-product inhibition represent a prevalent module [8,9], and they and others noted that these modules were very analogous to technological devices based on feedback control. At the level of genetic encoding, Jacob and Monod [10] were quick to point out the implications of the operon module as a building block for metabolic and other functions.

The overall architecture of metabolism noted by Davis [11] provides numerous examples of the types of modules I have in mind. There is a large 'fan in' of catabolic pathways that converge toward the central region of intermediary metabolism, which is primarily concerned with energy production and the formation and exchange of small molecular weight intermediates. The products of intermediary metabolism then lead to a large 'fan out' of biosynthetic pathways that synthesize all the building blocks for the construction of a cell (Figure 15.2).

This overall organization greatly facilitated the study of metabolism and its genetic regulation. The pathways on the periphery tend to be modular, specialized, minimally connected to the main body of the metabolic network, and highly regulated. Classic examples of such systems include the lactose, arabinose and maltose catabolic systems, and the tryptophan, arginine and histidine biosynthetic systems. Their boundaries are relatively well defined and their physiological function is at least qualitatively clear, conditional mutants are easily selected, experimental studies can be performed without major perturbation of the other cellular subsystems, and the extent of regulation is large and easily detected. The modularity of these systems greatly facilitated the allocation of experimental effort, which led to rapid advances in understanding with separate modules being studied in parallel in separate laboratories around the world.

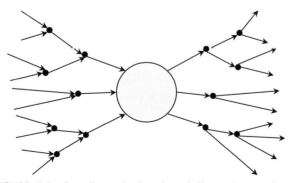

FIGURE 15.2 **Overall organization of metabolism.** A converging set of inducible catabolic pathways leads into intermediary metabolism, and a diverging set of repressible biosynthetic pathways leads out. See text for further discussion.

These studies also showed that the control of gene expression for these systems occurs largely at the level of transcription initiation, and that these modular systems are under the control of a small number of transcriptional regulators, mostly a combination of two or three specific and global regulators. It was found that control of effector gene expression in functionally similar systems nevertheless shows considerable diversity: changes in expression could be signaled by substrates, intermediates or products; controlled by activators or repressors; and coupled or uncoupled from regulator gene expression [12].

Modules as elements of change facilitate the random, combinatorial exploration of design space that creates the variation upon which natural selection acts. Modules as functional subsystems are the provisional fruits of natural selection that exhibit design principles. Their modularity facilitates the quantitative analysis and controlled comparison that lead to a deeper understanding of alternative designs.

Interface

The modules of technological systems are designed to have a common interface, e.g., conducting wires in electrical systems. Such modules also are more concerned with the transformation of signals or information than with the transformation of material. However, modules of biological systems are designed by natural selection to have fairly specific interfaces involving physical–chemical recognition of the molecules involved. Such modules also are often concerned necessarily with the transformation of material as well as with signals or information.

Interface and Function

The interface through which one biological module interacts with another is typically related intimately to its function. The term 'function' is used throughout this chapter, and is used throughout biology in many different ways. A metal in the reaction center of an enzyme might be said to have the function of coordinating various ligands in the protein. An enzyme in a signal transduction cascade might be said to have a kinase function. An operon in a metabolic system might be said to have an inducible catabolic function for supplying carbon in the production of biomass. When speaking of the function of a *system*, I mean this in the broadest terms to include not only such descriptive or qualitative characteristics, but also the quantitative aspects of performance such as specificity and activity of an enzyme, the behavior in time, such as how fast the system responds and the duty cycle specifying how often it may be required to respond, as well as more complex dynamic behavior including homeostasis, robustness, tolerance, oscillations, synchronization, hysteretic switching, digital counting, signal amplification, and noise-generated excitation.

Interface and Context

In this broad use of the term, function always occurs within particular contexts that require specific interfaces. Realizing a particular function in one context may require a design that differs from that for realizing the same function in a different context. For this reason, a class of mechanisms that performs the same function will often exhibit a common theme, but with important variations on the theme that undoubtedly reflect differences in context. This is clear from the examples of inducible catabolic and repressible biosynthetic systems referred to earlier. For example, the lactose and maltose systems are similar in certain aspects of function: both represent inducible catabolic systems and their principal effector function is a disaccharidase activity. However, the interface of these systems with their environment is materially different. Moreover, a repressor controls the lactose system, whereas an activator controls the maltose system. For some time many molecular biologists considered these different modes of control to be an inconsequential accident of evolution because they performed essentially the same 'function'. However, a more satisfying explanation for the difference in control — indeed a design principle — was eventually found in relation to the different environmental contexts in which these systems operate. The performance of these systems differs in the broader sense of system robustness and response time as a consequence of their different modes of control. Moreover, the selection of these alternative modes is related to the duty cycle or demand for their expression in different microenvironments [13–18]. For example, the substrate of the lactose system is rarely present in the natural environment of *Escherichia coli*, which selects for a negative mode of control, whereas the substrate of the maltose system is frequently present, which selects for a positive mode of control [13].

Many such biological design principles involve subtle variations on a theme, and their identification requires a comparative approach and quantitative criteria for comparison [2,12]. In some cases one starts with a well-characterized mechanism and attempts to identify which performance criteria it best fulfills. Good examples here include numerous two-component signal transduction systems that have been characterized biochemically and genetically and yet which are believed to be responding to environmental signals that in many cases are unknown. The design of prototype two-component systems exhibits a common variation in which the unphosphorylated sensor exhibits a second function (in addition to its primary function of phosphorylating the cognate response regulator) by acting as a phosphatase that dephosphorylates the response regulator, as in the case of the EnvZ–OmpR system [19]. In these cases it is called a bifunctional sensor, whereas when this second function is lacking, as in the case of the CheA–CheY system [19], it is called a monofunctional sensor [20]. Careful analysis predicts that an advantage of bifunctional sensors is the amplification of cognate vs. non-cognate signals that reduces the impact of noisy cross-talk. This is an irreducible consequence of bifunctional over monofunctional mechanisms, independent of the numerical values of the parameters [20]. However, the magnitude of the effect will depend on the specific parameter values. Recent experimental studies have found evidence that supports the role of bifunctional sensors in suppressing cross-talk [21,22]. Similarly, there are numerous toxin–antitoxin systems that have been characterized biochemically and genetically and yet their cellular role is unclear [23–25].

In other cases one starts from clear performance requirements and attempts to understand which mechanisms are best suited to meet those requirements. For example, one might know from the ecology of an organism that it must express a particular subsystem in a specific context, and yet the detailed mechanism by which the expression of the subsystem is controlled is still unknown.

Any design principle that can be established for a general class of such systems can be useful as a guide to an experimental program aimed at discovering the environmental context when the molecular mechanism is established, or at discovering the molecular mechanism when the environmental context is known [15].

Design

For a long time the issue of design principles in biology was not a legitimate consideration, and in some quarters it remains a taboo subject. This is a legacy of its misuse in arguments against evolution and in the 'just so' stories used to explain the evolution of biological systems. However, the term design has a rich and well-established meaning, and when used in the context of rigorous analysis and objective performance criteria, provides a deep understanding of the function and evolution of biological systems. This is now more widely accepted, as is evident in books being published on biological design principles and conferences being devoted to the subject.

Still, the question might be asked, are there design principles or rules that govern the patterns observed among biological systems? The answer depends upon whom one asks. There are some biologists who would answer: 'Of course there are rules, and it is the business of science to discover them!' This view has a long tradition embedded in positivist philosophy — the collection of empirical data, induction of rules, synthesis of general laws. Brahe, Kepler, Newton provide the paradigm. On the other hand, there are some biologists who would answer: 'No, there are no rules! Anything is possible. There is only what exists to be discovered and history'. This historical view is part of the Darwin legacy and, according to some, it has become the dominant view in modern biology. Webster and Goodwin [26] provide a fuller account of these two philosophies in the context of developmental biology.

In the realm of molecular genetics, leaders in the field have often expressed the latter view explicitly. To paraphrase a few: The rich variety of mechanisms governing gene expression is the result of historical accident; Nature is a tinkerer who haphazardly draws upon what already exists, not an engineer seeking optimal performance; The only rule is that there are no rules; Why are there positive and negative regulators? God only knows; There is no design — what works, works, what does not is dead.

I have presented elsewhere simple rules governing patterns of gene regulation that suggest how these two philosophies or views can be reconciled in specific cases [15]. A recently acquired function may not agree with some proposed rule. Such discrepancies can reflect historical contingencies associated with the origins of a mechanism. Although such discrepancies may be evident initially, they are not expected to survive long-term selective pressures that might enforce a rule. Differences may also be seen in the detailed molecular mechanisms by which a given type of system is realized. Such differences might be the result of historical accidents that are functionally neutral, or they might be governed by additional rules that have yet to be determined. One can always assume that certain differences are the result of historical accident, but such an explanation has no predictive power and tends to stifle the search for alternative hypotheses. It generally tends to be more productive if one starts with the working hypothesis that there are rules. One may end up attributing differences to historical accident, but in my opinion it is a mistake to start there.

Thus, accident and rule both play a role, but generally at different levels and during different periods. The result is plasticity in biological organization, but within bounds. Nature is optimizing, but subject to constraints (as is true of all optimization when properly understood). Mutation creates the diversity over which optimization acts; selection is the optimizer. Both the processes of mutation and selection are subject to constraints. External constraints are provided by origins and historical contingencies, whereas internal constraints are imposed by the laws of physics and the logic and dynamics of system organization.

More specific examples of such constraints include the limited time available to achieve some result, limitations on the amount of material or energy available, robustness in the face of uncertainty, etc. These are all obvious evolutionary constraints that lead to the emergence of systems with particular properties. The requirement for a fast temporal response leads to small pool sizes and negative feedback controls [2,27–30], and in some cases to cycles that can respond quickly [31,32]. The requirement for robustness also leads to the emergence of negative feedback controls at many levels, including metabolism [2,27]. Pathways that are physically or kinetically short favor dynamic stability [2,27,33]. The need for a commitment step in differentiation selects for hysteretic all-or-none switches, but there are many different realizations of such switches [12,34,35]. The need to reduce material and energy costs implies the emergence of cellular allocation schemes that prioritize the deployment of cellular resources according to environmental conditions such as the spectrum of available carbon sources. Think catabolite repression as a mechanism producing diauxic growth in the presence of mixed carbon sources such as lactose and glucose [36]. In Dobzhansky's well-known words, 'Nothing in biology makes sense except in the light of evolution.' This is certainly true of accident and design in biology. Both make use of modularity in evolution, but in very different ways.

PHENOTYPES

The second of the goals for this chapter is to provide a generic definition for what we mean by 'phenotype'. To put this in a larger context consider the grand challenge of relating genotype to phenotype [1]. It involves the fundamental unsolved problem of relating the 'genotype' — which has a well-defined, generic, digital representation — to the 'phenotype,' which has a poorly defined ad hoc analog representation. Without a rigorous generic definition of 'phenotype' to provide the context for a deep understanding of the relation between genotype and phenotype we are at a loss to know how many qualitatively distinct phenotypes are in an organism's repertoire, or the relative fitness of the phenotypes in different environments. These are practical challenges for clinicians attempting to develop therapeutic strategies to treat pathology and bioengineers wishing to redirect normal cellular functions for biotechnological purposes.

There are two fundamental problems that need to be solved in relating genotype to phenotype in a given environment: (a) the task of going from genome sequence to a model of the system, and (b) the task of going from the model to the phenotypic repertoire of the system in a given environment. The first task is the ongoing effort of experimental biology, and, whether considered from the bottom-up approach of molecular biology or the top-down approach of high-throughput technologies and computation, the magnitude of the problem is enormous [37]. An innovative approach to this first task is provided by the work of Cotterell and Sharpe [38], which suggests that it may be possible to start with a large class of models capable of capturing well-established features for a developmental system and systematically reduce this class to a significantly smaller set of models that improve the identification of candidate hypotheses for testing. The second task, assuming that an explicit model or suite of models is in hand, is also a daunting problem. Exploring the full phenotypic potential of even a relatively small non-linear model, say with 15 parameters, by analytical methods is intractable. Furthermore, an empirical sampling of alternatives suffers from a combinatorial explosion. (If one were to sample just 10 values of each parameter in all combinations it would require 10^{15} simulations, and one might still have missed some important behavior beyond the sampled range, or between sampled points.) We have recently provided a generic definition of 'phenotype' based on combinations of dominant processes operating within a system [39]. We showed how this definition partitions a 'system design space' into qualitatively distinct phenotypes with rigorously defined boundaries. With this approach, phenotypes are identified and enumerated, their relative fitness is compared, and their tolerance to phenotypic change measured.

Generic Concept of Phenotype

We define *a qualitatively distinct phenotype* as the set of concentrations and fluxes corresponding to a valid combination of dominant processes functioning within a system. Each of the terms in this definition requires further explanation for this definition to be useful. We do this first in the context of a very simple system and then show how this definition applies to more complex systems. With this approach in mind, we start with a two-step series of elementary chemical reactions as shown in Figure 15.3. For example, this could represent the reactions involving the acyclic form of D-glucose that exists during transitions between the alternative cyclic forms [40].

What is the repertoire of qualitatively distinct phenotypes for this simple system? If we could analytically solve for its behavior we might be able to identify distinct operating regimes from the structure of the solution.

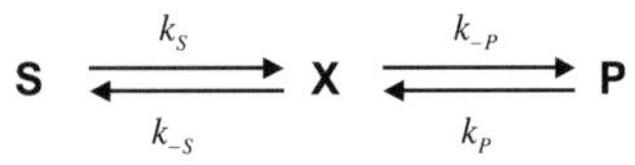

$$S \underset{k_{-s}}{\overset{k_s}{\rightleftarrows}} X \underset{k_P}{\overset{k_{-P}}{\rightleftarrows}} P$$

FIGURE 15.3 Two-step pathway of elementary first-order chemical reactions. The concentrations of the initial substrate, S, and the final product, P, are independent variables representing a fixed environment. The concentration of the intermediate, X, is the single dependent variable in this system. The rate constants k_i and k_{-i} are constrained by the equilibrium constants of the reversible reactions.

Phenotypes from the Analytical Solution

In fact, we can solve for its behavior, since it is a simple linear system. The differential equation governing this system is the following.

$$\frac{dX}{dt} = [k_s S - k_{-s} X] - [k_{-P} X - k_P P] \quad (1)$$

Setting the derivative to zero and solving for the dependent variable yields the steady-state solution

$$X = \frac{k_S S + k_P P}{k_{-S} + k_{-P}} \quad (2)$$

or, since detailed balance requires

$$\frac{k_S}{k_{-S}} = K_{eq}^S, \quad \frac{k_{-P}}{k_P} = K_{eq}^P \quad \text{and} \quad \frac{k_S k_{-P}}{k_{-S} k_P} = K_{eq}^S K_{eq}^P = K_{eq}$$

Equation (2) can be rewritten as

$$X = \frac{(k_{-S}/k_{-P}) K_{eq}^S S + P/K_{eq}^P}{(k_{-S}/k_{-P}) + 1} \quad (3)$$

Thus, the solution is characterized by two equilibrium constants (which are fixed thermodynamic quantities), two kinetic parameters (which are subject to change with the design of a catalyst), and the two independent concentration variables (which are subject to direct experimental manipulation of the environment).

Examination of the solution in Eq. (3) suggests four qualitatively distinct phenotypes based on the dominance of terms in its *numerator* and *denominator*:

Case 1: $X \approx K_{eq}^S S$

when $\dfrac{k_{-S}}{k_{-P}} > \dfrac{1}{K_{eq}\Gamma}$ and $\dfrac{k_{-S}}{k_{-P}} > 1$

Case 2: $X \approx (k_{-S}/k_{-P}) K_{eq}^S S$

when $\dfrac{k_{-S}}{k_{-P}} > \dfrac{1}{K_{eq}\Gamma}$ and $\dfrac{k_{-S}}{k_{-P}} < 1$

Case 3: $X \approx (k_{-P}/k_{-S}) P/K_{eq}^P$

when $\dfrac{k_{-S}}{k_{-P}} < \dfrac{1}{K_{eq}\Gamma}$ and $\dfrac{k_{-S}}{k_{-P}} > 1$

Case 4: $X \approx P/K_{eq}^P$

when $\dfrac{k_{-S}}{k_{-P}} < \dfrac{1}{K_{eq}\Gamma}$ and $\dfrac{k_{-S}}{k_{-P}} < 1$

where the conditions can all be expressed in terms of the genetically influenced and independent kinetic parameters for the mechanism $((k_{-S}/k_{-P}))$, the environmental influenced and independent variables ($\Gamma = S/P$), and the overall thermodynamic constant (K_{eq}).

This provides an attractive approach to the definition of phenotypes when an analytical solution is available. Unfortunately, such a solution is seldom available for the more complex non-linear systems that are characteristic of nearly all biological systems of interest.

Phenotypes from the Differential Equation

Might the same approach, based on the identification of dominant terms, be applied to the differential equation without having to obtain its analytical solution? The answer is yes! Examination of Eq. (1) suggests four qualitatively distinct phenotypes based on the dominance among its *positive* and *negative* terms:

Case 1: $\dfrac{dX}{dt} \approx [k_S S - k_{-S} X]$

when $\dfrac{k_{-S}}{k_{-P}} > \dfrac{1}{K_{eq}\Gamma}$ and $\dfrac{k_{-S}}{k_{-P}} > 1$

Case 2: $\dfrac{dX}{dt} \approx [k_S S - k_{-P} X]$

when $\dfrac{k_{-S}}{k_{-P}} > \dfrac{1}{K_{eq}\Gamma}$ and $\dfrac{k_{-S}}{k_{-P}} < 1$

Case 3: $\dfrac{dX}{dt} \approx [k_P P - k_{-S} X]$

when $\dfrac{k_{-S}}{k_{-P}} < \dfrac{1}{K_{eq}\Gamma}$ and $\dfrac{k_{-S}}{k_{-P}} > 1$

Case 4: $\dfrac{dX}{dt} \approx [k_P P - k_{-P} X]$

when $\dfrac{k_{-S}}{k_{-P}} < \dfrac{1}{K_{eq}\Gamma}$ and $\dfrac{k_{-S}}{k_{-P}} < 1$

This method of selecting one dominant positive term and one dominant negative term in general generates a set of nonlinear equations known as an S-system, which in steady state reduces to a linear problem for which one can obtain an explicit steady-state solution [2,12]. The steady-state solution of the dominant differential equations is obtained by setting the derivatives to zero and solving for the dependent variable, and the results are seen to be exactly the same as those obtained from the dominant terms of the analytical solution.

This method of selecting dominant positive and negative terms from the differential equation has several advantages over the method of selecting dominant terms from an analytical solution. First, the steady-state solution of the dominant differential equations, which is a linear problem, is much simpler than finding an analytical solution in the general case. Second, having the differential equations based on dominant positive and negative terms means that we also have access to the local dynamic behavior for each of the phenotypes. However, will this method tell us how many distinct phenotypes the system is capable of exhibiting? By integrating information from all the steady-state solutions and their corresponding boundary conditions, we can address this question in the context of the system design space.

Phenotypes in System Design Space

The phenotypes corresponding to the steady-state solutions only make sense if the solutions also satisfy the set of inequalities required to justify the assumption of dominance. This is a well-known linear programming problem involving solution of the steady-state equations, which are linear equations in logarithmic coordinates, along with the corresponding set of dominance conditions, which are linear inequalities in logarithmic coordinates [41,42]. The result is a set of boundaries delimiting valid regions, which define *qualitatively-distinct phenotypes* that can be visualized graphically in the *system design space*, as shown in Figure 15.4.

It should be emphasized that the phenotypes in this design space are 'generic' phenotypes for the entire class ('species') of such two-step reactions. The phenotype of an 'individual' member of this class will be located within this space when the parameters of the individual and its environment are specified. If an individual experiences a change in its genotype (genetically influenced parameters) or its environment (environmentally influenced variables), then the individuals location in design space will move accordingly and, if boundaries are crossed, there will be a qualitative change in phenotype. In any population of individuals there will undoubtedly be some heterogeneity in their genotypes and environments, and this will lead to

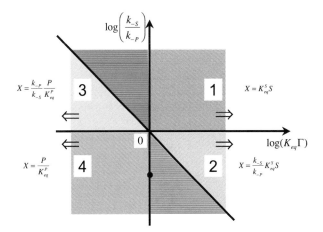

FIGURE 15.4 **System design space for the model in Figure 15.3.** The arrows show the direction of net flux in each quadrant. Case 1 (Blue): the first reaction is operating in quasi-equilibrium. Case 2 (Green): both reactions are operating in a quasi-irreversible forward direction. Case 3 (Cyan): both reactions are operating in a quasi-irreversible reverse direction. Case 4 (Red): the second reaction is operating in quasi-equilibrium. The black dot is the equilibrium state, $\log_{10}(k_{-S}/k_{-P}) \approx -0.25$, for the mutarotation of glucose in solution [40]. See text for discussion.

phenotypic variation in the population, as is commonly seen among individual cells in culture.

Enumeration of Qualitatively Distinct Phenotypes

There are seven qualitatively distinct phenotypes for this simple system, in agreement with our intuition (see also Figure 15.5). (a) The overall pathway operates with the first reaction in quasi-equilibrium and the net flux in the forward direction to the right (Case 1: upper right-hand quadrant). (b) The overall pathway operates with the first reaction in quasi-equilibrium and the net flux in the reverse direction to the left (Case 1: upper half of the upper left-hand quadrant). (c) The pathway operates quasi-irreversibly in the forward direction (Case 2: upper half of the lower right-hand quadrant). (d) It operates quasi-irreversibly in the reverse direction (Case 3: lower half of the upper left-hand quadrant). (e) The overall pathway operates with the second reaction in quasi-equilibrium and the net flux in the reverse direction to the left (Case 4: lower left-hand quadrant). (f) The overall pathway operates with the second reaction in quasi-equilibrium and the net flux in the forward direction to the right (Case 4: lower half of the lower right-hand quadrant). (g) Finally, the overall pathway is at thermodynamic equilibrium with zero net flux (vertical axis). In the case of more complex systems, the phenotypes typically will not be intuitively obvious, and not every assumption of dominance will yield a valid phenotype, as will be shown below.

Robustness

Robustness of organisms is widely observed although difficult to precisely characterize. Performance can remain nearly constant within some neighborhood of the normal operating regime, leading to homeostasis, but then abruptly break down with pathological consequences beyond this neighborhood. At the level of specific phenotypic function, the concept of robustness deals with the relationship between the physiological behavior and the underlying parameters of mechanistic models identified or hypothesized. Most approaches at this level have dealt with the local behavior as characterized by small (infinitesimal) changes. Robustness according to these approaches corresponds to parameter insensitivity — logarithmic sensitivities [43,44], linear sensitivities [45], or second-order sensitivities [46—48]. All of these approaches have shown what has been long known from experimental studies, that there is a spectrum of sensitivities with many parameters having very little influence and a smaller number having the major impact.

Although local approaches are useful for addressing the consequence of small changes, we need a global equivalent of *local robustness* (defined as insensitivity to small changes) in order to characterize the response to large changes. We call this concept *global tolerance*, and the boundaries between phenotypes in system design space provide a natural way of quantifying this concept [49].

Characterizing Performance

The system representation within each phenotypic region is always a simple S-system for which determination of local non-linear behavior reduces to conventional linear analysis [2,12]. Thus, the phenotypes involving local (small) variations are completely determined, and their relative fitness can be compared on the basis of relevant performance criteria. These criteria can be quantified using logarithmic gain, parameter sensitivity (local robustness), and response time. The boundaries that delineate a given phenotype can

FIGURE 15.5 **Phenotypes in the system design space for the model in Figure 15.3.** In each phenotypic region the pools associated with the intermediate concentration X, and the environmental concentrations S and P are represented graphically. The relative thickness of the arrows suggests the dominant fluxes. See legend for Figure 15.4 and text for discussion.

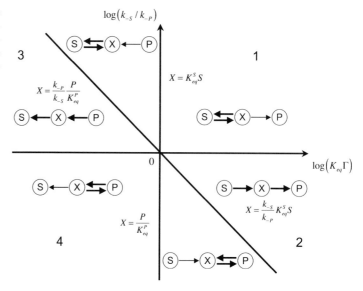

be used to quantify the global tolerance to large changes in genotype and environment as discussed below.

Logarithmic Gain

Signal transfer functions are used in various disciplines to characterize the response of dependent variables (such as concentrations and fluxes) to changes in the value of an independent variable. These amplification or gain functions actually represent logarithmic gain when they are defined in terms of a relative derivative [12,43]. For example, using the intermediate concentration X, pathway flux V and independent substrate concentration S, representative logarithmic gains for the pathway in Figure 15.3 are:

$$L(X,S) = \frac{\partial \text{Log } X}{\partial \text{Log } S} = \frac{\partial X}{\partial S} \frac{S}{X}$$

$$L(V,S) = \frac{\partial \text{Log } V}{\partial \text{Log } S} = \frac{\partial V}{\partial S} \frac{S}{V}$$

Values >1 signify amplification of the input signal, whereas values <1 signify attenuation of the signal; a positive sign indicates changes in the same direction, whereas a negative sign indicates changes in the opposite direction.

Parameter Sensitivity

The response of dependent concentrations and fluxes to a change in the value of the parameters that define the structure of the system (e.g., the rate constants here) are defined by the relative derivative of the explicit steady-state solution [12,50]. For example,

$$S(X, k_{-S}) = \frac{\partial \text{Log } X}{\partial \text{Log } k_{-S}} = \frac{\partial X}{\partial k_{-S}} \frac{k_{-S}}{X}$$

$$S(V, k_{-P}) = \frac{\partial \text{Log } V}{\partial \text{Log } k_{-P}} = \frac{\partial V}{\partial k_{-P}} \frac{k_{-P}}{V}$$

where the interpretation of magnitudes and signs is the same as for logarithmic gains. Small magnitudes (parameter 'insensitivity') imply that the system is robust to small (local) perturbations in the genotypically influenced parameters. Because parameter sensitivity and logarithmic gain factors have the same mathematical form, small magnitudes for logarithmic gains imply that the system is robust to small (local) perturbations in the environmentally influenced variables.

Response Time

The time it takes the system to respond to a local perturbation is determined by the inverse of the real part of the dominant eigenvalue [2,12].

Global Tolerance

Our generic definition of qualitatively distinct phenotypes and their boundaries in system design space provides a natural approach to the characterization of the system's response to large changes in its parameters [49]. We define global tolerance to such changes as the value of a parameter at the boundary between adjacent phenotypic regions relative to the normal operating value within a region (or the inverse if the normal value is greater than the value at the boundary). We will use the expression '$[T_D, T_I](\bullet, P_j)$' to describe the global tolerances of a given phenotype to changes in the parameter P_j, where $T_D = $ tolerance to a fold decrease and $T_I = $ tolerance to a fold increase (since boundaries can be crossed either by decreasing or increasing the value of a parameter).

Thus, systems located well within a physiological ('good') phenotypic region will be tolerant to large (global) changes in the values of their genotypically influenced parameters and environmentally influenced variables, because only large changes will be sufficient to move the system's location in system design space across a boundary into an adjacent region representing a qualitatively distinct (potentially pathological) phenotype.

Comparison of Phenotypes

Each phenotype has a characteristic set of values for its logarithmic gains, parameter sensitivities, response times and global tolerances. By making use of these values, the performance of the various phenotypes can be readily compared on the basis of relevant quantitative criteria.

Criteria for Functional Effectiveness

For the purposes of this example, assume the criteria for functional effectiveness of the pathway operating in the *forward* direction are those listed in Table 15.1.

Local Performance

A comparison of the local behavior for the relevant phenotypes is shown in Table 15.2. Based on this hypothetical scenario, we would conclude that the best overall phenotypic region is on the boundary between Region 2 and the right side of Region 4. (In this situation Regions 2 and 4 are essentially identical, and Region 1 has a greater intermediate concentration and slower response time than the other two regions; they trade advantages with respect to criteria 4 and 5, and all are the same with respect to criteria 1 and 2.) The equilibrium state for the mutarotation of glucose in solution is located on the positive y-axis ($\log_{10}(k_{-S}/k_{-P}) \approx -0.25$ [40]) in Figure 15.4. If the reaction were to be driven in the forward direction (β-D-glucose $\rightarrow$ α-D-glucose), then its operation would improve until it reaches the boundary with Region 2,

TABLE 15.1 Criteria for the Local Performance of the Pathway in Figure 15.3 Operating in the Forward Direction

Criterion number	Definition	Calculation
1	Maximum gain of pathway flux V in response to changes in substrate concentration S	$Max\ L(V, S)$
2	Minimum gain of pathway flux V in response to changes in product concentration P	$Min|L(V, P)|$
3	Minimum absolute value of intermediate X	$Min\ X$
4	Maximum robustness of flux V to variations in the rate constants of the reaction with S	$Min|S(V, k_S)|$
5	Maximum robustness of flux V to variations in the rate constants of the reaction with P	$Min|S(V, k_{-P})|$
6	Minimum response time	$Max\ \lambda\ (Min\ \tau_{1/2})$

TABLE 15.2 Comparison of Local Performance in the Three Relevant Phenotypic Regions of Figure 15.4 according to the Criteria In Table 15.1

Criterion	Preference[†]	Phenotypic Region*		
		1 (Right)	2	4 (Right)
1	⇑	1	1	1
2	⇓	0	0	0
3	⇓	SK_{eq}^S	Sk_S/k_{-p}	P/K_{eq}^P
4	⇓	0	1	1
5	⇓	1	0	0
6	⇑	k_{-S}	$k_{-P}^\S$	k_{-P}

[†]To improve performance, one must have either a high (⇑) or a low (⇓) value for the associated criterion.
*Right refers to the right-most region of Cases 1 and 4 in Figure 15.4.
§ Decreasing the response time (Criterion 6) by increasing k_{-P} implies a decrease in the absolute concentration of the intermediate X (Criterion 3).

and driving it further would move the system into Region 2 with no additional improvement in performance. On the other hand, if it were driven in the reverse direction, then its operation would move into the left side of Region 4, which is not the overall best phenotypic region when functioning in the reverse direction. Taken together, these results suggest that operation in the forward direction has some advantages. This suggestion would be consistent with the data showing preferential utilization of α-D-glucose by hexokinase in the case of human erythrocyte exposed to D-glucose at low concentrations; however, it would be inconsistent with the data showing preferential utilization of β-D-glucose at high concentrations [51].

Global Tolerance

For purposes of illustration, let us assume operating points in each of the four regions depicted in Figure 15.4. In Region 1 we shall assume the nominal values are $\log(K_{eq}\Gamma) = +2$ and $\log(k_{-S}/k_{-P}) = +1$. Although the global tolerances for each of the parameters and independent variables will require a calculation in general, the geometry of the system design space in Figure 15.4 is so simple that one can determine the global tolerances from the boundaries by inspection. In this case, the results are shown on the first line of entries in Table 15.3; similar results are shown on subsequent lines for the other regions and assumed operating points. These results make evident a number of symmetries in this system. Since the assumed operating points were all well removed from the boundaries, it is not surprising that the global tolerances are all large *fold* changes, which is in contrast to the values for local robustness that are often manifested as *percentage* changes.

DESIGN PRINCIPLES

To address the third goal of the Introduction, in this section I will show how the generic definition of phenotype applies in a more complex system, generates a more interesting system design space, and suggests non-intuitive biological design principles. The system selected for this purpose is the well-characterized model for the *cI* gene circuit involved in controlling lysogeny of bacteriophage lambda.

Alternative Growth Modes of Phage Lambda

Bacteriophage lambda can reproduce in two alternative modes of growth. In the *lytic* mode, the phage infects a bacterial cell, reproduces many copies of itself, lyses the host cell, and circulates through the environment to infect another host cell. In the *lysogenic* mode, the phage infects a bacterial cell, incorporates its DNA into the chromosome of the host cell, and remains quiescent, with its DNA being replicated along with that of the host. Lysogeny is typically a stable state unless the host cell is compromised in some fashion (e.g., DNA damaged by UV radiation), and then the phage undergoes an *induction* process by which it excises

TABLE 15.3 Global Tolerances to Change in Parameters (and Independent Variables)$[T_D, T_I](\bullet, P_j)$ with Assumed Values for Operating Points in Each of the Regions in System Design Space (see Figure 15.4)

Region	Log (Operating Point)	Parameters		Independent Variables	
		k_{-S}	k_{-P}	S	P
1	(+2,+1)	$[10, \infty]$	$[\infty, 10]$	$[1000, \infty]$	$[\infty, 1000]$
2	(+2,−1)	$[10, 10]$	$[10, 10]$	$[10, \infty]$	$[\infty, 10]$ *
3	(−2,+1)	$[10, 10]$	$[10, 10]$	$[\infty, 10]$	$[10, \infty]$
4	(−2,−1)	$[\infty, 10]$	$[10, \infty]$	$[\infty, 1000]$	$[1000, \infty]$

*Notation: [fold decrease, fold increase].

its DNA from the host chromosome and initiates lytic growth. The key regulatory interactions involved in maintaining the alternative fates of lambda are shown schematically in Figure 15.6.

Developmental Decisions

The lytic/lysogenic decision of phage lambda has long been studied as a model of cell-fate decisions in higher organisms, and has also been explored by mathematical modeling [52,53]. The regulator CII (not shown) initially is a critical agent in this decision, but once the developmental path to lysogeny is established it no longer plays a role [54]. The lysogeny/induction decision has received less attention as a model system, but it has also been well studied. The classic view held that CRO and CI form a double-negative bistable switch that plays an essential role in initiation of the induction process [55]. However, there is now evidence suggesting that CRO is not required for initiation of induction [56–58], although it can influence the threshold level of DNA damage that initiates induction [57] and the subsequent progress of lytic development [58].

The core regulator of the lysogenic state is the CI protein, which as a dimer represses the early lytic transcripts from promoters *pL* and *pR* by binding to operator sites *OL* and *OR* while activating its own transcription from promoter *pRM* at low concentrations and repressing it at high concentrations. Following DNA damage, which leads to an SOS response, RecA protein is activated and stimulates CI monomer auto-cleavage. The lowering of CI concentration, and hence the dimer concentration, results in deactivation of *cI* transcription and de-repression of lytic functions. A diagram emphasizing the core features of CI regulation is shown in Figure 15.7, along with a version having symbols that simplify the mathematical notation.

Kinetic Model

Early mathematical models of this system have been reviewed by Santillan and Mackey [59]. Well-established

Lytic growth

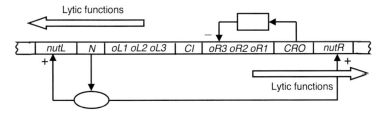

Lysogenic growth

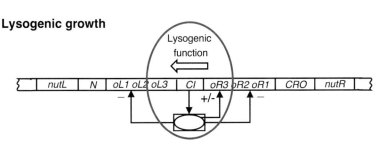

FIGURE 15.6 **Key regulatory circuits maintaining the alternative fates of bacteriophage lambda.** Two regulators maintain the lytic mode of replication; the N-gene product is a positive regulator required for activating transcription of the lytic-specific genes and CRO is a negative regulator required for repressing transcription of the lysogenic-specific genes. One regulator maintains the lysogenic mode of replication; the CI gene product is a negative regulator of the lytic transcripts and a positive regulator of its own transcription.

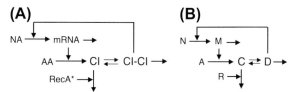

FIGURE 15.7 (A) Kinetic model of the *cI* gene circuit for phage lambda. Symbols: NA, nucleotide precursors; mRNA, transcript of *cI* gene; AA, activated amino acid precursors; CI, monomeric form of the CI protein; CI − CI, dimeric form of the CI protein; RecA*, activated form of the RecA protein. (B) Simplified symbolic version of the kinetic model in (A). Symbols: N, nucleotide precursors; M, transcript of *cI* gene; A, activated amino acid precursors; C Monomeric form of the CI protein; D Dimeric form of the CI protein; R, activated form of the RecA protein.

features of these models include first-order loss of CI mRNA, first-order translation of the CI mRNA, dimerization of CI monomers, and cooperative activation of *pRM* transcription by low concentrations of CI dimers and repression at high concentrations. These well-established features of the *cI* gene circuit are included in a number of detailed kinetic [59–63] and stochastic [64–66] models that incorporate the recent experimental results of Dodd et al. [67].

The following is a kinetic model that includes these conventional features of the system. It differs from existing models by the assumption of a phenomenological Hill-like function for the rate of *pRM* transcription and the potential for cooperative RecA* catalysis of CI monomer degradation under substrate-limited conditions. The first assumption appears to represent adequately the experimental data of Dodd et al. [67], as shown below. The second assumption addresses the detailed kinetics of CI degradation, which is unknown because it has been difficult to study for technical reasons (J.W. Little, personal communication). Nevertheless, fitting experimental data with this model yields an estimate of the capacity for regulation of CI degradation that is consistent with the limited experimental data available [68], as discussed below.

$$\frac{dM}{dt} = \left[\frac{\gamma_M K_D^P + D^p(\gamma_{MMax} + \gamma_M K_I^{-n} D^n)}{K_D^P + D^p(1 + k_I^{-n} D^n)}\right] - \delta_M M \quad (4)$$

$$\frac{dC}{dt} = \gamma_C M + 2\beta_D D - 2\gamma_D C^2 - \left[\frac{\delta_C K_R^a + \delta_{CMax} R^a}{K_R^a + R^a}\right]C \quad (5)$$

$$\frac{dD}{dt} = \gamma_D C^2 - \beta_D D - \delta_D D \quad (6)$$

The behavior described by these equations, which represent both mass-action and rational-function mechanisms, can be described as follows.

At very low concentrations of *D* there is a basal rate of transcription given by γ_M. As the concentration of *D* increases there is a cooperative activation of the rate, which is characterized by the Hill number *p*. When the concentration of *D* exceeds the intermediate value needed for half-maximal activation, the activation constant K_D, the increase in rate diminishes as it approaches the maximum velocity given by γ_{MMax}. As the concentration of *D* approaches and then exceeds the inhibition constant K_I, there is a cooperative repression of the rate, which is characterized by the Hill number *n*. The loss of mRNA, *M*, is first-order with rate constant δ_M.

The rate of translation is first-order in the concentration of *M* with rate constant γ_C. At very low concentrations of *R* there is a basal rate of loss of *C* given by δ_C. As the concentration of *R* increases there is a cooperative activation of the rate, which is characterized by the Hill number *a*. When the concentration of *R* exceeds the intermediate value needed for half-maximal activation, the activation constant K_R, the increase in rate diminishes as it approaches the maximum velocity given by δ_{CMax}. The reversible dimerization of CI is characterized by the second-order rate constant γ_D and the first-order rate constant β_D.

Estimation of Parameter Values

Dodd et al. [67] produced two sets of data critical for understanding the mechanism that controls *cI* gene transcription in lambda lysogens. In the first case, the *pRM* promoter is associated with wild-type operator sequences that allow binding of CI dimers at low concentration to the operator sites *OR1*−*OR2* and *OL1*−*OL2* to activate transcription, and that allow binding of CI dimers at high concentrations to the operator sites *OR3* and *OL3* to repress transcription. In the second case, the *pRM* promoter is associated with a disabled operator site *OL3* that fails to bind CI dimers, thereby eliminating repression at high concentrations. We fit our model to these data and obtained the values shown in Table 15.4 (for details see [69]).

Recast Equations

Equations (4) to (6) can be recast [12,70] exactly into a generalized mass action (GMA) system of equations within the power-law formalism by introducing two new variables

$$x_2 = K_D^p + x_1^p + K_I^{-n} x_1^{p+n} \quad \text{and} \quad x_3 = K_R^a + x_R^a.$$

$$\frac{dM}{dt} = \gamma_M K_D^p x_2^{-1} + \gamma_{MMax} D^p x_2^{-1} + \gamma_M K_I^{-n} D^{p+n} x_2^{-1} - \delta_M M \quad (7)$$

$$\frac{dC}{dt} = \gamma_C M + 2\beta_D D - 2\gamma_D C^2 - \delta_C K_R^a C x_3^{-1} - \delta_{CMax} R^a C x_3^{-1} \quad (8)$$

TABLE 15.4 Values of the Parameters for the CI Gene Circuit in Figure 15.7 Estimated from Experimental Data for *E. coli* Host Cells Grown in Rich Media with a Doubling Time of ~ 20 min [69]

Parameters	Values	Units
K_D	130	nM
K_I	320	nM
K_R	205*	nM
γ_{MMax}	355	LacZ units
γ_M	50	LacZ units
γ_C	0.0173	nM LacZ units^{-1} min^{-2}
γ_D	$1/\varepsilon^\dagger$	nM^{-1} min^{-1}
β_D	$32.6/\varepsilon$	min^{-1}
δ_M	0.116	min^{-1}
δ_{CMax}	7.49	min^{-1}
δ_C	0.0347	min^{-1}
δ_D	0.0347	min^{-1}
a	1	–
p	3	–
n	1.5	–

The concentration of RecA (R) is normalized to the geometric mean of the values for the hysteretic thresholds at the nominal value of K_D.
$^\dagger \varepsilon \ll 1$ for a dimerization rate that is much greater than rate of loss by dilution.

$$\frac{dD}{dt} = \gamma_D C^2 - \beta_D D - \delta_D D \quad (9)$$

$$x_2 = K_D^p + D^p + K_I^{-n} D^{p+n} \quad (10)$$

$$x_3 = K_R^a + R^a \quad (11)$$

In steady state the derivatives of Eqs. (7) to (9) are equal to zero, and the equations can be combined into the following system of non-linear algebraic equations:

$$0 = \gamma_M K_D^p x_2^{-1} + \gamma_{MMax} x_1^p x_2^{-1} + \gamma_M K_I^{-n} x_1^{p+n} x_2^{-1}$$
$$- 2\delta_M \delta_D \gamma_C^{-1} x_1 - \delta_M \delta_C \gamma_C^{-1} K_C^{1/2} K_R^a x_1^{1/2} x_3^{-1}$$
$$- \delta_M \delta_{CMax} \gamma_C^{-1} K_C^{1/2} x_R^a x_1^{1/2} X_3^{-1} \quad (12)$$

$$0 = K_D^p + x_1^p + K_I^{-n} x_1^{p+n} - x_2 \quad (13)$$

$$0 = K_R^a + x_R^a - x_3 \quad (14)$$

where $K_C = (\beta_D + \delta_D)\gamma_D^{-1}$, $x_1 = D$ and $x_R = R$.

Each of these equations is a sum of several positive and/or negative terms. When one term of each sign in each of the equations is dominant (i.e., is the term with the greatest magnitude among those of the same sign), the system of equations can be represented locally by a second system of equations, the S-system representation, which has a single analytical solution that is linear in the logarithms of the concentration variables and rate constants [71,2]. The boundaries between such dominant systems according to this method are not arbitrary, but are determined by the parameters of the original system.

Number of Qualitatively Distinct Phenotypes in Design Space

Since each term of a given sign in Eqs. (12) to (14) is potentially dominant, there are as many potential solutions as there are combinations of dominant terms; hence a bound on the total number (T) is provided by

$$T = \prod_{i=1}^{m} P_i * N_i$$

where m is the number of equations and P_i and N_i are the number of positive and negative terms in the ith equation. (In the case of Eqs. (12) to (14), the bound is $T = (3 * 3) * (3 * 1) * (2 * 1) = 54$.)

However, not all potential solutions are necessarily valid. The conditions for any given term to be dominant are provided by a set of linear inequalities in log space. A test of each potential solution against the inequalities necessary for its validity will determine whether or not a potential solution is in fact a valid solution. Substituting the valid solution into the corresponding dominance conditions yields a set of linear inequalities that defines the boundaries of the region in which the solution is valid. By following this strategy for the lambda *cI* gene circuit, and selecting the parameters R and K_D for the axes of a 2D plot that represents a slice through the design space, we obtain the result depicted in Figure 15.8. Note that only 10 of the 54 possibilities are valid solutions for this particular design space. Regions on the top of this design space [1,37,38] correspond to stable steady-state operating points with low values for the rate of transcription from the promoter *pRM*; we will henceforth refer to the qualitative nature of these as *lytic-like* or simply lytic states. Regions on the bottom of this design space [11,47,45,46] correspond to stable steady-state operating points with high values; we will henceforth refer to the qualitative nature of these as *lysogen-like* or simply lysogenic states. The regions in the middle [7,43,44] correspond to steady-state operating points with intermediate values that exhibit saddle-point instability. The overlapping regions thus represent hysteresis, which also is revealed by conventional bifurcation analysis (for a comparison of the methods see [72]). The system design spaces of some models have boundaries that correspond to other conventional bifurcations, such as the Hopf bifurcation for oscillations. However, most boundaries in system design space do not correspond to conventional

FIGURE 15.8 **System design space for the model in Figure 15.7.** The regions of different color represent qualitatively different phenotypes. The regions are numbered arbitrarily. The white dot represents the location of the wild-type system in the lysogenic state; the black dot represents the location of the system in the lytic state. Induction, caused by an increase in R, corresponds to rightward movement along the horizontal black line. Regions with a single number represent stable steady states (regions 1, 37 and 38 correspond to lytic phenotypes; regions 11, 47, 45 and 46 to lysogenic phenotypes). The band separating these regions represents hysteretic behavior with three steady states, one unstable and two stable, corresponding to the two neighboring stable regions. The color bar can be used to identify the three phenotypes in each overlapping region (regions 7, 43 and 44 correspond to the unstable steady-state phenotypes). Wild-type induction occurs across the critical hysteretic region (horizontal arrow). The value of K_D must lie between the dashed lines (vertical arrow) in order to maintain the temperate life cycle operating across the critical hysteretic region. See text for discussion.

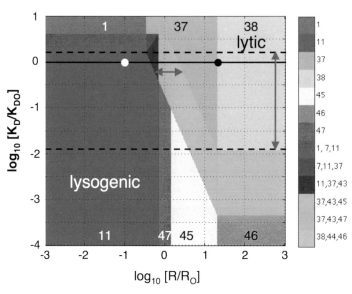

bifurcations; nevertheless, they do represent transitions between qualitatively-distinct phenotypes.

The qualitatively distinct phenotypes represented by each of the regions in Figure 15.8 can be readily characterized following the procedures illustrated in the simple case of a reversible pathway. However, it will be sufficient for our purposes here to characterize a couple of representative cases.

Example of a Valid Phenotype (Case 11)

If we pick the dominant terms to be the second positive term in Eq. (12), the first negative term in Eq. (12), the third positive term in Eq. (13) and the first positive term in Eq. (14), then the resulting S-system in steady state is

$$0 = \gamma_{MMax} x_1^p x_2^{-1} - 2\delta_M \delta_D \gamma_C^{-1} x_1$$
$$0 = K_I^{-n} x_1^{p+n} - x_2$$
$$0 = K_R^a - x_3$$

Solution

The steady-state solution for the dimer concentration D is readily obtained as

$$D = x_1 = \left(\frac{\gamma_{MMax} \gamma_C K_I^n}{2\delta_M \delta_D} \right)^{\frac{1}{n+1}}$$

and the corresponding values for the concentration of mRNA M and CI monomer C follow accordingly

$$M = \frac{\gamma_{MMax} K_I^n}{\delta_M x_1^n} = \left[\frac{2\delta_D K_I}{\gamma_C} \left(\frac{\gamma_{MMax}}{\delta_M} \right)^{\frac{1}{n}} \right]^{\frac{n}{n+1}}$$

$$C = \sqrt{\frac{(\beta_D + \delta_D)}{\gamma_D}} x_1 = \left(\frac{\beta_D + \delta_D}{\gamma_D} \right)^{0.5} \left(\frac{\gamma_{MMax} \gamma_C K_I^n}{2\delta_M \delta_D} \right)^{\frac{0.5}{n+1}}$$

For this solution to be valid it must satisfy the dominance conditions assumed above.

Dominance Conditions

Assuming the second positive term in Eq. (12) to be dominant over the other positive terms implies two dominance conditions:

$$\gamma_M K_D^p < \gamma_{MMax} x_1^p \qquad \gamma_{MMax} > \gamma_M K_I^{-n} x_1^n$$

Similarly, the assumption of dominant terms for the remaining cases in Eqs. (12) to (14) implies the remaining dominance conditions:

$$2\delta_D x_1^{1/2} > \delta_C K_R^a K_C^{1/2} x_3^{-1} \qquad 2\delta_D x_1^{1/2} > \delta_{CMax} K_C^{1/2} x_R^a x_3^{-1}$$

$$K_D^p < K_I^{-n} x_1^{p+n} \qquad 1 < K_I^{-n} x_1^n$$

$$K_R^a > x_R^a$$

Boundaries in System Design Space

Inserting the steady-state solution into the dominance conditions yields the boundaries within which the solution is valid (see Figure 15.8).

$$\log K_D < \frac{n(p-1)}{p(n+1)} \log K_I + \frac{n+p}{p(n+1)} \log \left(\frac{\gamma_{Mmax} \gamma_C}{2\delta_M \delta_D} \right)$$

$$\log x_R < \log K_R + \frac{1}{2a(n+1)} \log \left[\frac{\gamma_{Mmax} \gamma_C K_I^n}{\delta_M} \right]$$

$$+ \frac{2n+1}{2a(n+1)} \log [2\delta_D]$$

$$+ \frac{(n+1)}{2a(n+1)} \log \left[\frac{\gamma_D}{\delta_{CMax}^2 (\beta_D + \delta_D)} \right]$$

Example of an Invalid Phenotype (Case 6)

If we pick the dominant terms to be the first positive term in Eq. (12), the first negative term in Eq. (12), the third positive term in Eq. (13), and the second positive term in Eq. (14). The resulting S-system in steady state is

$$0 = \gamma_M K_D^p x_2^{-1} - 2\delta_M \delta_D \gamma_C^{-1} x_1$$

$$0 = K_I^{-n} x_1^{p+n} - x_2$$

$$0 = x_R^a - x_3$$

Solution

The steady-state solution for the dimer concentration D is

$$D = x_1 = \left(\frac{\gamma_M \gamma_C K_D^p K_I^n}{2\delta_M \delta_D}\right)^{\frac{1}{p+n+1}}$$

and the corresponding values for the concentration of mRNA M and CI monomer C follow accordingly

$$M = \frac{\gamma_M K_D^p K_I^n}{\delta_M x_1^{p+n}} = \left[\frac{2\delta_D}{\gamma_C}\left(\frac{\gamma_M K_D^p K_I^n}{\delta_M}\right)^{\frac{1}{p+n}}\right]^{\frac{p+n}{p+n+1}}$$

$$C = \sqrt{\frac{(\beta_D + \delta_D)}{\gamma_D} x_1} = \left(\frac{\beta_D + \delta_D}{\gamma_D}\right)^{0.5}\left(\frac{\gamma_M \gamma_C K_D^p K_I^n}{2\delta_M \delta_D}\right)^{\frac{0.5}{p+n+1}}$$

For this solution to be valid it must satisfy the dominance conditions assumed above.

Dominance Conditions

Assuming the first positive term in Eq. (12) to be dominant over the other positive terms implies two dominance conditions:

$$\gamma_M K_D^p < \gamma_{MMax} x_1^p \qquad K_D^p > K_I^{-n} x_1^{p+n}$$

Similarly, the assumption of dominant terms for the remaining cases in Eqs. (12) to (14) implies the remaining dominance conditions:

$$2\delta_D x_1^{1/2} > \delta_C K_R^a K_C^{1/2} x_3^{-1} \qquad 2\delta_D x_1^{1/2} > \delta_{CMax} K_C^{1/2} x_R^a x_3^{-1}$$

$$K_D^p < K_I^{-n} x_1^{p+n} \qquad 1 < K_I^{-n} x_1^n$$

$$1 > x_R^a$$

Note that the conditions $K_D^p > K_I^{-n} x_1^{p+n}$ and $K_D^p < K_I^{-n} x_1^{p+n}$ are mutually exclusive. Therefore, this particular assumption of dominance conditions cannot be satisfied and there is no corresponding valid phenotype.

Evaluation of Local Performance

In each region of Figure 15.8 the steady-state equations for the concentration of monomer CI (C), dimer CI (D) and mRNA (M) follow analytically from the form of the original non-linear equations, independent of the particular numerical values for the parameters. Thus, one can in principle characterize the various regions and compare their behaviors analytically without knowing the values of the parameters. This has been done for simple systems [39,49], but in practice it becomes intractable for large systems. The region of design space in which the system operates and the numerical positioning of the boundaries are determined when specific values for the parameters of the model are given.

Quantitative Criteria

Two criteria often found to correlate with the natural selection of a particular system design are the maximization (minimization) of a particular steady-state function and the response time following change. Moreover, when these values are optimal they often are robustly so (e.g., see [49]). Several criteria for effective local performance that reflect these expectations can be summarized as follows.

The lysogenic state should be locally robust to perturbations in the values of the parameters that define the system, and this robustness can be quantified by the parameter sensitivities, such as $S(D, p_j)$. It should not be influenced by fluctuations in the input signal (the level of DNA damage), which can be quantified by the logarithmic gain with respect to RecA* activity, such as $L(D, R)$. The robustness of the logarithmic gain in the face of perturbations in parameter values is also expected to be small, and this robustness can be quantified by parameter sensitivities such as $S[L(D, R), p_j]$. The dimer form of CI (D) is responsible for the primary regulatory actions, but similar criteria hold for the concentration of the monomer form of CI (C) and the mRNA (M) as well.

It is important to minimize the response time for restoring the system to its nominal steady state following small changes in the variables of the system, and this can be quantified in terms of the dominant eigenvalue $\lambda_{dominant}$. One could of course consider other criteria, but these will suffice for our purposes here.

Analysis of Local Performance

The local robustness of the system in each of the phenotypic regions of system design space can be calculated analytically for the model in Figure 15.7. From these results, one can predict maximum local robustness in most cases when the Hill numbers p and n are large, although trade-offs are involved in determining the optimum values, whereas local robustness is maximal when a is small (data not shown). By inserting the estimated values for the parameter (Table 15.4) into these analytical results one can convert them into numerical results. Small values imply superior performance for each of the criteria. The local robustness in each of the phenotypic regions reveals that on average the influence of perturbations in parameters is attenuated (magnitude of sensitivities less than one). The results for Case 11, shown in Table 15.5, range from a low of 0.1 to a high of 0.6 with an overall average of 0.37 if one omits all the cases with zero value.

The logarithmic gains of the system in response to the environmental input signal are all equal to zero,

$$L(M,R) = L(C,R) = L(D,R) = 0$$

i.e., fluctuations in the level of DNA damage (RecA* levels) are subthreshold and have no influence on the Case 11 phenotype of the system. It should be emphasized that this is robustness to local (small) changes. The system will respond to global (large) changes, which is the subject of global tolerance as discussed below.

The local response times in each of the phenotypic regions of system design space are readily calculated from the dominant S-system equations. In the Case 11 region (again assuming rapid dimerization)

$$\frac{dM}{dt} = \left[\frac{(\beta_D + \delta_D)K_I \gamma_{MMax}^{1/n}}{\gamma_D}\right]^n C^{-2n} - \delta_M M$$

$$\frac{dC}{dt} = \gamma_C M - \left[\frac{2\delta_D \gamma_D}{(\beta_D + \delta_D)}\right] C^2$$

which for local responses reduces to the following linear system [2,12]:

$$\frac{du_1}{dt} = \delta_M\{-u_1 - 2nu_2\}$$

$$\frac{du_2}{dt} = F_2\{u_1 - 2u_2\}$$

with $F_2 = (2\delta_D)^{\frac{n+0.5}{n+1}} \left[\frac{\gamma_D}{(\beta_D + \delta_D)}\right]^{0.5} \left(\frac{\gamma_{Mmax}\gamma_C K_I^n}{\delta_M}\right)^{\frac{0.5}{n+1}}$

With the values of the parameters in Table 15.4, the real part of the dominant eigenvalue is -0.317 min^{-1}, which corresponds to a half-time of 2.2 min.

TABLE 15.5 Summary of Analytical and Numerical Values for Local Robustness of the Lysogenic Phenotype Corresponding to Region Case 11 of the System Design Space in Figure 15.8. Numerical Results are Calculated on the Basis of the Parameter Values in Table 15.4, which were Determined from Experimental Data

Parameters	$S(M,p_i)$		$S(C,p_i)$		$S(D,p_i)$	
K_D	0	0	0	0	0	0
K_I	$n/(n+1)$	0.6	$0.5n/(n+1)$	0.3	$n/(n+1)$	0.6
K_R	0	0	0	0	0	0
γ_{MMax}	$1/(n+1)$	0.4	$0.5/(n+1)$	0.2	$1/(n+1)$	0.4
γ_M	0	0	0	0	0	0
γ_C	$-n/(n+1)$	-0.6	$0.5/(n+1)$	0.2	$1/(n+1)$	0.4
γ_D	0	0	-0.5	-0.5	0	0
β_D	0	0	$0.5\beta_D/(\beta_D+\delta_D)$	0.5	0	0
δ_M	$-1/(n+1)$	-0.4	$-0.5/(n+1)$	-0.2	$-1/(n+1)$	-0.4
δ_{CMax}	0	0	0	0	0	0
δ_C	0	0	0	0	0	0
δ_D	$n/(n+1)$	0.6	$\frac{0.5(n\delta_D - \beta_D)}{(\beta_D+\delta_D)(n+1)}$	-0.2	$-1/(n+1)$	-0.4
a	0	0	0	0	0	0
p	0	0	0	0	0	0
n	$\frac{n}{(n+1)^2}\ln\left(\frac{2\delta_M\delta_D K_I}{\gamma_{MMax}\gamma_C}\right)$	-0.21	$\frac{0.5n}{(n+1)^2}\ln\left(\frac{2\delta_M\delta_D K_I}{\gamma_{MMax}\gamma_C}\right)$	-0.10	$\frac{n}{(n+1)^2}\ln\left(\frac{2\delta_M\delta_D K_I}{\gamma_{MMax}\gamma_C}\right)$	-0.21

Evaluation of Global Performance

Characterizing local (small) changes in performance is important for characterizing the qualitatively distinct phenotypes in each region, but it does not address the overall behavior when phenotypes change as a result of large environmental stimuli or mutation in components of the system. The partitioning of system design space into qualitatively different phenotypic regions provides a means, based on the boundaries, of calculating tolerances to global (large) changes in parameter values [49]. These tolerances are defined for each parameter as the ratio of its value at the nominal steady state (normal operating point for the system) to its value on the boundary to an adjacent phenotypic region (or the inverse, depending on which value is the larger). The geometry of system design space in Figure 15.8 itself suggests additional criteria that are relevant for assessing global behavior of the system.

Quantitative Criteria

The criterion particularly important for the natural selection of a particular system design is large tolerances to a change in phenotype when the change produces dysfunction. Moreover, when these values are optimal they often are robustly so (again, see [49]). Several criteria for effective global performance that reflect this objective can be summarized as follows.

The hysteretic region (corresponding to the three overlapping Cases of 43, 37 and 47 or 45) in the design space of Figure 15.8 provides a type of 'safety factor'. Once a switch from lysogenic to lytic growth has been initiated, the phage becomes committed in the following sense: if there is a decrease in the SOS signal, it must be large enough to move the operation of the system completely back across the hysteretic region: any lesser decrease is insufficient to prevent completion of the switch. The hysteretic region also acts as a buffer to prevent inappropriate switching when there is only a transient small SOS signal, one that is insufficient to cause the system to cross the hysteretic region. This function is augmented by the slow response time noted above, which tends to filter out fast local transients.

These functions of the hysteretic region are enhanced by a large horizontal distance between the inclined lines in Figure 15.8, which can be quantified by the value of Δ_H, which is given by

$$\Delta_H = \left(\frac{\gamma_{Mmax}}{\gamma_M}\right)^{\frac{(2P-1)}{2pa}}$$

This safety factor should be robust to small parameter variations, which can be quantified by the parameter sensitivities $S(\Delta_H, P_j)$, and tolerant to large parameter changes, which can be quantified by the global tolerances $[T_D, T_I](\Delta_H, P_j)$.

For long-term survival of the temperate life cycle, the phage must be capable of repeatedly switching from lysogenic to lytic growth and back again across the critical hysteretic region. This is facilitated by having a design that places the operating point of the system between the dashed lines in Figure 15.8 and by having a large distance between these limits, i.e., above the lower limit

$$K_D^{lower} = \left(\frac{\gamma_D}{\beta_D + \delta_D}\right)\left(\frac{\gamma_{MMax}\gamma_C}{\delta_M\delta_{CMax}}\right)^2$$

and below the upper limit

$$K_D^{Upper} = \left(\frac{\gamma_M\gamma_C}{2\delta_M\delta_D}\right)\left(\frac{\gamma_{MMax}\gamma_C}{\delta_M\delta_{CMax}}\right)^{1/p}$$

Furthermore, this zone of operation should be robust to small variations in parameter values, which can be quantified by the parameter sensitivities $S(K_D^{lower}, p_j)$ and $S(K_D^{Upper}, p_j)$, and tolerant to large changes in parameter values, which can be quantified by the global tolerances $[T_D, T_I](K_D^{lower}, p_j)$ and $[T_D, T_I](K_D^{Upper}, p_j)$.

Analysis of Global Performance

The size of the hysteretic buffer and its global tolerance to large changes in parameter values can be calculated analytically; the numerical results, using the values of the parameters in Table 15.4, show that the breadth of this buffer is represented by a 5.1-fold change in RecA* activity (R). The global tolerances of the hysteretic buffer Δ_H to large variations in parameters $[T_D, T_I](\Delta_H, p_j)$ for γ_{Mmax}, γ_M, p and a are $[7.1, \infty]$, $[\infty, 7.1]$, $[6.0, \infty]$ and $[\infty, \infty]$, respectively (Table 15.6). The corresponding values for robustness to small variations in parameters of the hysteretic buffer $S(\Delta_H, P_i)$ are 0.83, −0.83, 0.85 and −4.25, respectively. All other parameters have no influence on the size of the hysteretic buffer, and thus their global tolerances and local robustness are effectively infinite.

The aspects of global behavior pertaining to the maintenance of the temperate lifecycle requires large tolerances to both the lower bound $[T_D, T_I](K_D^{lower}, p_j)$ and to the upper bound $[T_D, T_I](K_D^{Upper}, p_j)$. These tolerances can be calculated analytically, and the numerical results, using the values of the parameters in Table 15.4, are shown for Case 11 in Table 15.6. Most of the tolerances for change in parameters are essentially infinite; but if these are excluded, the resulting values range from a low of 1.6-fold to a high of 83-fold, with an overall average of 15-fold. The local robustness for these lower and upper bounds can be calculated as well (data not shown).

Biological Design Principles

The phenotypes of a system involve the interaction of the environmental (independent) variables and the

TABLE 15.6 Global Tolerances Relevant for the Case 11 Phenotype, Including Maintenance of the Phenotype, Proper Positioning of a Wide Hysteretic Buffer, and Survival of the Temperate Lifestyle $[T_D, T_I](\Delta H, P_j)$ are Best Ensured by having a Large Net Global Tolerance for all the Parameters. See Figure 15.8 and the text for discussion

Parameters	Case 9	Case 17	Δ_H	Δ_H^{Left}	Δ_H^{Right}	K_D^{Lower}	K_D^{Upper}	Net
K_D	[∞, ∞]	[∞, ∞]	[∞, ∞]	[∞, ∞]	[∞, ∞]	[83.2, ∞]	[∞, 1.60]	[83.2, 1.60]
K_I	[∞, 2.40]	[10.9, ∞]	[∞, ∞]	[∞, ∞]	[∞, ∞]	[∞, ∞]	[∞, ∞]	[10.9, 2.40]
K_R	[∞, ∞]	[∞, ∞]	[∞, ∞]	[4.30, ∞]	[∞, 8.00]	[∞, ∞]	[∞, ∞]	[4.30, 8.00]
γ_{MMax}	[2.40, ∞]	[∞, ∞]	[7.1, ∞]	[∞, ∞]	[∞, 8.00]	[∞, ∞]	[∞, ∞]	[2.40, 8.00]
γ_M	[∞, ∞]	[∞, 4.20]	[∞, 7.1]	[∞, ∞]	[∞, ∞]	[∞, ∞]	[2.02, ∞]	[2.02, 4.20]
γ_C	[∞, ∞]	[∞, ∞]	[∞, ∞]	[∞, ∞]	[∞, 8.00]	[∞, ∞]	[1.60, ∞]	[1.60, 8.00]
γ_D	[∞, ∞]	[∞, ∞]	[∞, ∞]	[∞, ∞]	[∞, 65.0]	[∞, ∞]	[20.0, ∞]	[20.0, 65.0]
β_D	[∞, ∞]	[∞, ∞]	[∞, ∞]	[∞, ∞]	[65.0, ∞]	[∞, ∞]	[∞, 20.0]	[65.0, 20.0]
δ_M	[∞, ∞]	[∞, ∞]	[∞, ∞]	[∞, ∞]	[8.00, ∞]	[∞, ∞]	[∞, 1.60]	[8.00, 1.60]
δ_{CMax}	[∞, ∞]	[∞, ∞]	[∞, ∞]	[∞, 4.20]	[8.00, ∞]	[∞, ∞]	[∞, ∞]	[8.00, 4.20]
δ_C	[∞, ∞]	[∞, ∞]	[∞, ∞]	[∞, ∞]	[∞, ∞]	[∞, ∞]	[∞, 4.60]	[∞, 4.60]
δ_D	[∞, ∞]	[10.9, ∞]	[∞, ∞]	[∞, ∞]	[∞, ∞]	[∞, ∞]	[∞, 1.60]	[10.9, 1.60]
a	[∞, ∞]	[∞, ∞]	[∞, ∞]	[1.39, ∞]	[∞, 50.0]	[∞, ∞]	[∞, ∞]	[1.39, 50.0]
p	[∞, 1.07]	[1.34, ∞]	[6.0, ∞]	[∞, ∞]	[∞, ∞]	[∞, ∞]	[∞, ∞]	[1.34, 1.07]
n	[∞, ∞]	[∞, ∞]	[∞, ∞]	[∞, ∞]	[∞, ∞]	[∞, ∞]	[∞, ∞]	[∞, ∞]

constellation of parameters that constitute the design of the system arrived at by natural selection. For this reason we choose to represent an important environmental variable of interest on the horizontal axis and an important structural parameter that is most revealing on the vertical axis. In the case of phage lambda, the environmental variable is the insult traditionally involving UV radiation (with RecA* activity as a proxy), and the structural parameter that proves revealing is the constant representing half-maximal activation of cI transcription.

The system passes through a progression of qualitatively distinct phenotypes – corresponding to Cases 11, 47, 45, 37 and 38 – as the level of RecA* (R) is steadily increased from its basal level in Region 11. In Case 11, the level of RecA* is subthreshold and the system, which has an intermediate level of CI mRNA (M) that sustains a high level of CI dimer (D), is unresponsive to fluctuations in RecA* (D is independent of R). In Case 47, the system responds to the elevated levels of RecA* by derepressing cI transcription thereby elevating M ($M \propto R^{2an/(2n+1)} \approx R^{3/4}$), which provides a negative feedback that tends to minimize the drop in D ($D \propto 1/R^{2a/(2n+1)} \approx 1/\sqrt{R}$, as opposed to $1/R$) and preserve the lysogenic state. In Case 45, the de-repression of cI transcription is maximal ($M = \gamma_{MMax}/\delta_M$), thereby eliminating the negative feedback component, the level of CI dimer drops more precipitously ($D \propto 1/R^{2a} \approx 1/R^2$, as opposed to $1/\sqrt{R}$), but the lysogenic state is still maintained. In these two phenotypes, the system is still resistant to the damage and is capable of fully recovering the basal lysogenic state if the environmental insult is removed. In Case 37, the system becomes committed to induction, the level of cI transcription drops to its basal level ($M = \gamma_M/\delta_M$), and the level of CI dimer continues to decrease ($D \propto 1/R^{2a} \approx 1/R^2$). This is the only transition in this progression of phenotypes that corresponds to a traditional bifurcation. Recovery from this commitment is only possible with a large decrease in RecA* activity (sufficient to retreat back across the hysteretic buffer zone). In Case 38, the level of cI transcription remains at its basal level ($M = \gamma_M/\delta_M$), the level of CI dimer no longer decreases as the level of RecA* activity becomes saturating (D is independent of R), and recovery is only possible with a large decrease in RecA* activity (sufficient to retreat back across the hysteretic buffer zone).

Avoiding Inappropriate Switching

One of the most important system design principles is made evident by these qualitatively distinct phenotypes in design space. Namely, the hysteretic buffer allows the system to maintain the lysogenic state to which it committed when favored by the prevailing environmental conditions. This state is maintained for a period, even in the face of deteriorating environmental conditions — presumably an advantage if the host is capable of recovering from sublethal damage. However, once committed to induction, there is a penalty for reversing the cell-fate decision, presumably because such vacillation would reduce the titer of phages ultimately produced to initiate a new round of infection if the host is damaged beyond repair. The determinants of the size of this buffer zone and its tolerance to large changes in the structural parameters may not be obvious, but they are readily quantified by means of the generic concept of phenotype and its manifestation in the system design space.

Near the hysteretic region fluctuations will result in some phages undergoing induction while others will not. This produces heterogeneity in the population that hedges bets in an unpredictable environment.

Maintaining the Temperate Lifestyle

Perhaps even more important are the system design principles that ensure long-term survival of the temperate lifecycle itself. If the constellation of structural parameters that determine the upper bound on K_D is violated, the phage becomes locked into the lytic mode of growth. The same result is obtained if the parameters experience variation that would shift the lower boundary of the hysteretic region to values lower than the basal value of the environmental stress. On the other hand, if the constellation of structural parameters that determine the lower bound on K_D is violated, the phage will reside permanently in the host's chromosome as a lysogen. Again, a similar result is obtained if the parameters experience variation that would shift the upper boundary of the hysteretic region to values higher than the maximal value of the environmental stress.

Maintenance of the phenotype (large tolerances to the adjacent phenotypes, e.g., Case 9 and Case 17 relative to Case 11), proper positioning of a wide hysteretic buffer (left margin greater than the basal value and right margin less than the maximal value of environmental stress), and proper positioning of K_M within its acceptable band of values (less than the upper bound and greater than the lower bound) are best ensured by having large global tolerances for all the parameters; thus, one can consider a minimum net tolerance for each of the parameters to be most relevant, as shown in Table 15.6. It should be noted that the tolerances mentioned above for the hysteretic buffer are all larger than these minimum net tolerances. The most restrictive net tolerances are for changes in the exponents a and p ([1.39,50.0] and [1.34,1.07], respectively, although most of the other tolerances for change in these exponents are essentially infinite). Most of the tolerances for change in the other parameters are essentially infinite; but if these are excluded, the resulting values for these parameters range from a low of 1.6-fold to a high of 83-fold, with an overall average of 22-fold.

These results suggest that the system design of the cI gene circuit is not only robust to local (small) changes in parameters, but it is also remarkably tolerant to global (large) changes in parameters and environmental stimuli. This is consistent with the work of John Little and colleagues [73–75], in which they have constructed variants with major changes in the binding constants, and indeed the identity of the circuit elements, and yet many of their constructs retain the essential qualitative features of the wild-type phage.

The relationships among these various elements of design, and their tolerances to large changes in parameter values, are not at all obvious, but again they are readily quantified by means of the generic concept of phenotype and its manifestation in system design space.

There are obvious analogies to other cell-fate decisions that are well known in bacteria as well as higher organisms. However, the detailed mechanisms involved can be quite different. For example, the decision in the case of lambda induction involves feedback regulation of a transcription initiation mechanism. That, for the case of spore formation in *Bacillus subtilis,* involves feedback regulation in a post-translational partner-switching mechanism [34]. The decision to switch from a proliferating eukaryotic cell to apoptosis [76] has clear similarities to the decision to switch from a stable lysogen to induction. Although the mechanisms that govern these systems are very different and most are not fully characterized, once an appropriate model has been formulated the procedure to develop the corresponding system design space is conceptually straightforward.

CONCLUSIONS AND FUTURE CHALLENGES

While the phenotypes and design principles in the case of very simple systems, such as the reversible pathway of the earlier example, may be obvious (or at least may become obvious in hindsight), this is seldom the case with more complex systems, as the example of lambda illustrates. We should have no illusions about the extreme challenges of relating genotype and environment to phenotype for complex organisms. However, without a generic concept of phenotype there can be no deep, predictive understanding of these important relationships. What then are the major challenges to developing this type of understanding?

The path forward will require major advances in moving from the digital information in the genome sequence to an analog model that captures the essential mechanistic underpinnings of system behavior in a given environmental context. Many of the approaches in this book, and others that have yet to be developed, will undoubtedly play a critical role in overcoming this challenge. However, progress is unlikely to involve a dramatic solution that starts with a DNA sequence and infers the underlying mechanism; indeed, it can be argued that there is insufficient information and knowledge of the constraints required to reduce the degrees of freedom necessary to accomplish such a feat. Rather, progress will be made, as in the past, by top-down high-throughput measurements and statistical analysis combined with bottom-up molecular characterization of subsystems involving hard physical and chemical constraints. This is the first of the two fundamental problems in bridging the chasm between genotype and phenotype that was mentioned in the introduction to the Phenotype Section. The second fundamental problem, the task of going from the model of a system to the elucidation of its phenotypic repertoire, is the long-term challenge being addressed by the system design space approach.

The system design space approach described in this chapter is applicable currently to any relatively small subsystem for which the molecular mechanisms have been identified. To deal with the size and subtleties of larger systems, future development of this approach will likely advance on three fronts. First, it should be possible to develop rules for producing the design space of a composite system by augmenting an existing design space with a newly acquired design space for an additional subsystem. In this way one could in principle move incrementally to larger system design spaces by building on the knowledge of constraints and design principles acquired earlier in the study of the constituent subsystems. Second, it should be possible to develop more efficient means of visualizing the enormous amount of data generated. In the lambda example, we have simply provided representative results for a single phenotype (Case 11); similar results are available for all of the realizable qualitatively distinct phenotypes. However, at this point more of the same types of figures and tables are overwhelming. Third, it should be possible to develop more efficient algorithms for identifying and analyzing phenotypes in system design space based on well-established theory. In general, the boundaries between phenotypes in the system design space are always straight lines (hyper-planes) in logarithmic space and the slopes are rational functions of the exponents in generalized mass action models. Algebraic geometry provides the theoretical foundation for this type of analysis, and there are opportunities for automating much of the analysis based on this linear theory.

As an example of what can be done with regard to this third opportunity is the set of algorithms already developed and assembled into a Matlab application called Design Space Toolbox [77], and there is the potential for much further automation. Establishing dominance is a standard mathematical problem in linear algebra that is solved everyday for problems of enormous size (e.g., scheduling in the airline industry), so solving each case is not a problem. Nevertheless, scale-up becomes an issue because of the large number of potential phenotypes, which is determined by the number of combinations of terms in the original equations, many of which will be invalid but still must be checked. This is an obviously parallizable problem yet to be implemented. As should be clear from these conclusions, the analysis of system design space is in its infancy and there are still abundant challenges and opportunities for further development.

ACKNOWLEDGEMENTS

I thank Rick Fasani for assistance in constructing the system design spaces, and Pedro Coelho, Dean Tolla, and Jason Lomnitz for fruitful discussions. This work was supported in part by US Public Health Service Grant R01-GM30054 and by a Stanislaw Ulam Distinguished Scholar Award from the Center for Nonlinear Studies of the Los Alamos National Laboratory.

REFERENCES

[1] Brenner S. Genomics: the end of the beginning. Science 2000;287:2173—4.

[2] Savageau M.A. Biochemical Systems Analysis: A Study of Function and Design in Molecular Biology. Reading, Mass: Addison-Wesley; 1996, [40th Anniversary reprinting (2009)] http: //www.amazon.com/Biochemical-Systems-Analysis-Function-Molecular/dp/1449590764/

[3] Srere PA, Mosbach K. Metabolic compartmentation: symbiotic organellar, multienzymic, and microenvironmental. Ann Rev Microbiol 1974;28:61—83.

[4] Waddington CH. The Strategy of the Genes. New York: Macmillan; 1957.

[5] Clement PR, Johnson WC. Electrical Engineering Science. New York: McGraw-Hill; 1960.

[6] Truxal JG. Automatic Feedback Control System Synthesis. New York: McGraw-Hill; 1955.

[7] Purnick PEM, Weiss R. The second wave of synthetic biology: form modules to systems. Nat Rev Mol Cell Biol 2009;10:410—22.

[8] Umbarger HE. Evidence for a negative-feedback mechanism in the biosynthesis of isoleucine. Science 1956;123.

[9] Yates RA, Pardee AB. Control of pyrimidine biosynthesis in *Escherichia coli* by a feedback mechanism. J Biol Chem 1956;221:757—70.

[10] Jacob F, Monod J. Genetic regulatory mechanisms in the synthesis of proteins. J Mol Biol 1961;3:318—56.

[11] Davis BD. The teleonomic significance of biosynthetic control mechanisms. Cold Spring Harb Symp Quant Biol 1961;26:1—10.

[12] Savageau MA. Design principles for elementary gene circuits: elements, methods, and examples. Chaos 2001;11:142—59.

[13] Savageau MA. Genetic regulatory mechanisms and the ecological niche of *Escherichia coli* Proc Nat Acad Sci USA 1974;71:2453—5.

[14] Savageau MA. Design of molecular control mechanisms and the demand for gene expression. Proc Nat Acad Sci USA 1977;74:5647—51.

[15] Savageau MA. Are there rules governing patterns of gene regulation? In: Goodwin BC, Saunders PT, editors. Theoretical biology — epigenetic and evolutionary order. Edinburgh: Edinburgh University Press; 1989. p. 42—66.
[16] Savageau MA. Demand theory of gene regulation. Genetics 1998;149:1665—91.
[17] Shinar G, Dekel E, Tlusty T, Alon U. Rules for biological regulation based on error minimization. Proc Nat Acad Sci USA 2006;103:3999—4004.
[18] Gerlanda U, Hwa T. Evolutionary selection between alternative modes of gene regulation. Proc Nat Acad Sci USA 2009;106:8841—6.
[19] Stock AM, Robinson VL, Goudreau PN. Two-component signal transduction. Annu Rev Biochem 2000;69:183—215.
[20] Alves R, Savageau MA. Comparative analysis of prototype two-component systems with either bifunctional or monofunctional sensors: differences in molecular structure and physiological function. Mol Microbiol 2003;48:25—51.
[21] Siryaporn A, Goulian M. Cross-talk suppression between the CpxA—CpxR and EnvZ—OmpR two-component systems in *E coli* Mol Microbiol 2008;70:494—506.
[22] Groban ES, Clarke EJ, Salis HM, Miller SM, Voigt CA. Kinetic buffering of cross talk between bacterial two-component sensors. J Mol Biol 2009;390:380—93.
[23] Gerdes K, Christensen SK, Lobner-Olesen A. Prokaryotic toxin-antitoxin stress response loci. Nat Rev Microbiol 2005;3:371—82.
[24] Magnuson RD. Hypothetical functions of toxin-antitoxin systems. J Bacteriol 2007;189:6089—92.
[25] Hayes CS, Low DA. Signals of growth regulation in bacteria. Curr Opin Microbiol 2009;12:667—73.
[26] Webster G, Goodwin BC. The origin of species: a structuralist approach. J Social Biol Struct 1982;5:15—47.
[27] Alves R, Savageau MA. Irreversibility in unbranched pathways: preferred positions based on regulatory considerations. Biophysical J 2001;80:1174—85.
[28] Rosenfeld N, Elowitz MB, Alon U. Negative autoregulation speeds the response times of transcription networks. J Mol Biol 2002;323:785—93.
[29] Wall ME, Hlavacek WS, Savageau MA. Design of gene circuits: lessons from bacteria. Nat Rev Genet 2004;5:34—42.
[30] Wong ML, Bolovan-Fritts C, Weinberger LS. Negative feedback speeds transcriptional response-time in human cytomegalovirus. Biophys J 2009;96:305a.
[31] Coelho PMBM, Salvador A, Savageau MA. Relating mutant genotype to phenotype via quantitative behavior of the NADPH redox cycle in human erythrocytes. PLoS One 2010;5(9):e13031.
[32] Tolla DA, Savageau MA. Phenotypic repertoire of the FNR regulatory network in *Escherichia coli*. Mol Microbiol 2011;79:149—65.
[33] Becskei A, Serrano L. Engineering stability in gene networks by autoregulation. Nature 2000;405:590—3.
[34] Igoshin OA, Price CW, Savageau MA. Signaling network with a bistable hysteretic switch controls developmental activation of the σ^F transcription factor in *Bacillus subtilis*. Mol Microbiol 2006;61:165—84.
[35] Igoshin OA, Alves R, Savageau MA. Hysteretic and graded responses in bacterial two-component signal transduction. Mol Microbiol 2008;68:1196—215.
[36] Brock TD. The Emergence of Bacterial Genetics. Cold Spring Harbor, New York: Cold Spring Harbor Laboratory Press; 1990.
[37] Hlavacek WS. How to deal with large models? Mol Sys Biol 2009;5:240—2.
[38] Cotterell J, Sharpe J. An atlas of gene regulatory networks reveals multiple three-gene mechanisms for interpreting morphogen gradients. Mol Syst Biol 2010;6:425.
[39] Savageau MA, Coelho PMBM, Fasani R, Tolla D, Salvador A. Phenotypes and tolerances in the design space of biochemical systems. Proc Natl Acad Sci USA 2009;106:6435—40.
[40] Pigman W, Isbell HS. Mutarotation of sugars in solution. Part 1. History, basic kinetics, and composition of sugar solutions. Adv Carbohydr Chem Biochem 1968;23:11—57.
[41] Dantzig GB. Linear Programming and Extensions. Princeton, NJ: Princeton University Press; 1963.
[42] Vanderbei RJ. Linear Programming: Foundations and Extensions. New York, NY: Springer; 2008.
[43] Savageau MA. Concepts relating the behavior of biochemical systems to their underlying molecular properties. Arch Biochem Biophys 1971;145:612—21.
[44] Kacser H, Burns JA. The control of flux. Symp Soc Exp Biol 1973;27:65—104.
[45] Heinrich R, Rapoport TA. A linear steady-state treatment of enzymatic chains. Critique of the crossover theorem and a general procedure to identify interaction sites with an effector. Euro J Biochem 1974;42:97—105.
[46] Salvador A. Synergism analysis of metabolic processes: I. Conceptual framework. Math Biosci 2000;163:105—29.
[47] Salvador A. Synergism analysis of metabolic processes: II. Tensor formulation and treatment of stoichiometric constraints. Math Biosci 2000;163:131—58.
[48] Gutenkunst RN, Waterfall JJ, Casey FP, Brown KS, Myers CR, et al. Universally sloppy parameter sensitivities in systems biology models. PLOS Comput Biol 2007;3:1871—8.
[49] Coelho PMBM, Salvador A, Savageau MA. Global tolerance of biochemical systems and the design of moiety-transfer cycles PLOS Comput Biol 2009;5(3):e1000319.
[50] Savageau MA. Parameter sensitivity as a criterion for evaluating and comparing the performance of biochemical systems. Nature 1971;229:542—4.
[51] Malaisse-Lagae F, Malaisse WJ. Anomeric specificity of D-glucose phosphorylation and oxidation in human erythrocytes. Int J Biochem 1987;19:733—6.
[52] Oppenheim AB, Kobiler O, Stavans J, Court DL, Adhya S. Switches in bacteriophage lambda development. Annu Rev Genet 2005;39:409—29.
[53] Tian T, Burrage K. Bistability and switching in the lysis/lysogeny genetic regulatory network of bacteriophage lambda. J Theor Biol 2004;227:229—37.
[54] Court DL, Oppenheim AB, Adhya S. A new look at bacteriophage λ genetic networks. J Bacteriol 2007;189:298—304.
[55] Ptashne M. A Genetic Switch. Cambridge, U.K: Blackwell Scientific Publications; 1992.
[56] Svenningsen SL, Costantino N, Court DL, Adhya S. On the role of Cro in lambda prophage induction. Proc Natl Acad Sci USA 2005;102:4465—9.
[57] Atsumi S, Little JW. Role of the lytic repressor in prophage induction of phage λ as analyzed by a module-replacement approach. Proc Natl Acad Sci USA 2006;103: 4558—63.

[58] Schubert RA, Dodd IB, Egan JB, Shearwin KE. CRO's role in the CI-CRO bistable switch is critical for λ's transition from lysogeny to lytic development. Genes Dev 2007;21:2461−72.

[59] Santillan M, Mackey MC. Why the lysogenic state of phage lambda is so stable: a mathematical modeling approach. Biophys J 2004;86:75−84.

[60] Vilar JM, Saiz L. DNA looping in gene regulation: from the assembly of macromolecular complexes to the control of transcriptional noise. Curr Opin Genet Dev 2005;15:136−44.

[61] Lou C, Yang X, Liu X, He B, Ouyang Q. A quantitative study of λ-phage switch and its components. Biophys J 2007;92:2685−93.

[62] Anderson LM, Yang H. DNA looping can enhance lysogenic CI transcription in phage lambda. Proc Natl Acad Sci USA 2008;105:5827−32.

[63] Anderson LM, Yang H. A simplified model for lysogenic regulation through DNA looping. 30th Annual International IEEE EMBS Conference, August 20−24. Canada: Vancouver, British Columbia; 2008; 607−610.

[64] Aurell E, Brown S, Johanson J, Sneppen K. Stability puzzles in phage lambda. Phys Rev E 2002;65:051914−9.

[65] Aurell E, Sneppen K. Epigenetics as a first exit problem. Phys Rev Lett 2002;88:048101.

[66] Zhu XM, Yin L, Hood L, Ao P. Calculating biological behaviors of epigenetic states in the phage lambda life cycle. Funct Integr Genomics 2004;4:188−95.

[67] Dodd IB, Perkins AJ, Tsemitsidis D, Egan JB. Octamerization of λ CI repressor is needed for effective repression of *pRM* and efficient switching from lysogeny. Genes Dev 2001;15:3013−22.

[68] Little JW. The SOS regulatory system: control of its state by the level of RecA protease. J Mol Biol 1983;167:791−808.

[69] Savageau MA, Fasani RA. Qualitatively distinct phenotypes in the design space of biochemical systems. FEBS Lett 2009;583:3914−22.

[70] Savageau MA, Voit EO. Recasting nonlinear differential equations as S-systems: a canonical nonlinear form. Math Biosci 1987;87:83−115.

[71] Savageau MA. Biochemical systems analysis II. The steady state solutions for an n-pool system using a power-law approximation. J Theoretical Biol 1969;25:370−9.

[72] Savageau MA. Design of the *lac* gene circuit revisited. Math Biosci 2011;231:19−38.

[73] Little JW, Shepley DP, Wert DW. Robustness of a gene regulatory circuit. EMBO J 1999;18:4299−307.

[74] Michalowski CB, Short MD, Little JW. Sequence tolerance of the phage lambda PRM promoter: implications for evolution of gene regulatory circuitry. J Bacteriol 2004;186:7988−99.

[75] Atsumi S, Little JW. A synthetic phage λ regulatory circuit. Proc Natl Acad Sci USA 2006;103:19045−50.

[76] Hanahan D, Weinberg RA. Hallmarks of cancer. Cell 2000;26:57−70.

[77] Fasani RA, Savageau MA. Automatic construction and analysis of the design space of biochemical systems. Bioinformatics 2010;26: 2601−9.

Chapter 16

System Biology of Cell Signaling

Chiara Mariottini and Ravi Iyengar

Department of Pharmacology and Systems Therapeutics and Systems Biology Center, New York Mount Sinai School of Medicine, New York, NY 10029, USA

Chapter Outline

Introduction	311
Cellular Signaling: Pathways to Networks	311
Isoforms of Signaling Molecules, Signaling Integration and Sorting	313
Scaffolding Proteins: Signaling Networks Form Signaling Complexes	314
Computational Analysis of Signaling Networks	316
Graph Theory-Based Models	316
Regulation By Network Motifs	316
Properties of a Cellular Signaling Network	317
Network Topology and Consolidation of Signals	318
Dynamical Models	319
Positive Feedback Loops can form Switches: The Concept of Bistability	319
Signaling Microdomains Within Cells	322
Conclusions and Future Challenges	323
Glossary	324
Acknowledgments	324
References	324

'Every object that biology studies is a system of systems.'

Francois Jacob (1974)

INTRODUCTION

It is intuitively obvious that a system is an entity made up of parts. This general definition is applicable to biological entities across all scales of organization. An approach to study systems is to consider them as networks of mutually dependent and thus interconnected components that generate a unified whole. Typically whole systems exhibit behaviors and have properties that cannot be attributed to an individual component. These system-level behaviors are called emergent properties [1–3]. Understanding how the different components and interactions drive the manifestation of emergent properties is one major goal of systems biology. Systems biology as a field is focused on the understanding of multicomponent functional units. It has begun to have enormous impact across biological sciences, including ecology, population biology, evolutionary biology as well as biochemistry, molecular and cell biology, physiology, development and genetics. Here we focus on systems biology at a cellular level.

Advances in experimental technologies that enable large-scale survey-type experiments, genome-wide studies such as whole genome sequencing [4,5], microarrays [6,7] and RNAseq profiling [8–10] and proteomics [11–13] have greatly increased in number during the past decade. Studies in molecular biology have increasingly adopted a systems-based approach to obtaining and processing high-throughput data and building computational models of cellular systems made up of multiple interacting components. Informatics research in systems biology is also developing approaches to catalog and process information in the extensive experimental biomedical research literature that has focused on studying individual cellular components in depth over the past five decades. Integrating detailed knowledge of individual components and interactions from earlier studies and contemporary experiments will allow us to obtain a mechanistic understanding of how components enable systems-level behaviors. Indeed, a grand challenge in systems biology is to develop detailed predictive models of mammalian cells.

CELLULAR SIGNALING: PATHWAYS TO NETWORKS

Intrinsic to cellular processes is the receipt and processing of information. The ability of a mammalian cell to live depends on its ability to receive information and respond to a constantly variable environment. Information is received through the cell's signaling pathways. Information transmittal through a pathway can often evoke a cellular response. Such responses in the context of tissues and organs can contribute to organismal behavior. A classic example is the fight or flight response. In response to an excitable threatening stimulus adrenaline is released into the bloodstream. Adrenaline can bind to its receptor, β-adrenergic receptor, which activates the signaling transducer Gs, which in turn activates the enzyme adenylyl cyclase which produces cAMP, a diffusible intracellular second messenger. Through the cAMP-dependent protein kinase (also known as PKA) cAMP, activates phosphorylase kinase (PhosK), which in turn activates glycogen phosphorylase (GlyPhos), which phosphorylates glycogen to catalyze the production of glucose-6-phosphatase, which is eventually converted to free glucose that can be used for energy metabolism in muscle. Thus this linear signaling pathway (Figure 16.1A, blue outline) produces energy required for the fight or flight response. In common parlance this response is called the 'adrenaline rush'.

Epinephrine (adrenaline) and norepinephrine have actions in many tissues and organs. In neurons β-adrenergic receptors, activated by the neurotransmitter norepinephrine through the second messenger cAMP regulate both PKA and Epac, leading to the final activation of MAP-kinase 1, 2 [14,15] (Figure 16.1A, red outline). AMP-GEF/Epac is a guanine nucleotide exchange factor (GEF). The GEF proteins promote the exchange of GDP for GTP on small GTPases. cAMP binds to Epac to activate the small GTPases RAP1A and RAP2A. Both PKA and MAPK in turn regulate the phosphorylation and activation of the transcription factor CREB [16,17] (Figure 16.1A). Many CREB-regulated genes have been shown to be involved in learning and memory processes [18,19].

The bifurcation of cAMP signals in neurons and other cell types where it can bind to and activate both PKA and Epac, shows the origins of a signaling network where a single signal (cAMP) can be routed to two different protein kinases (PKA and MAP-kinase 1, 2). These protein kinases have some common but also many different targets, and thus the cAMP signal can be routed to many targets within the cell.

Networks result from interconnections between signaling pathways. A certain signaling molecule can receive signals from multiple inputs. A small G protein such as Ras and a transcription factor, such as CREB, are examples of signaling proteins that receive signaling from multiple upstream pathways. Such signaling molecules can be considered as junctions that integrate signals. There are also molecules that route signals to multiple pathways. cAMP is one such molecule that can split signals to control multiple downstream pathways [20].

We can think of the mammalian cell as a complex network of signaling pathways resembling the network of different subway lines that form the New York City subway system (Figure 16.1B): focusing on a few subway lines or on a single station would give the observer only a partial idea of how the city works as people move from one place to another for work or other activities (Figure 16.1B). It would also not provide us with a realistic estimate of the number of people who take the subway every day. In order to get a more complete sense of the real number of people who ride the subway (estimated to be approximately 4.3 million every day, and >1 billion people per year!), one should consider the entire network of about 24 subway lines that form one of the largest subway systems in the world (Figure 16.1B, right panel). Similarly, in order to understand how the dynamics of signaling within the cell lead to coordination between multiple cellular functions, one needs to look at the entire network of signaling pathways and how they connect and regulate the various cellular machines in a synchronized manner. The multiplicity of signaling pathways and networking originates with the extracellular signals. Epinephrine and norepinephrine bind to three classes of receptors, the β-adrenergic, α_1-adenergic and α_2-adrenergic, which specifically couple to the Gs, Gq/11 and Gi/o pathways. Indeed, most natural ligands for G protein-coupled receptors bind to multiple classes of receptors which through different G protein-dependent pathways control many cellular processes that in turn regulate numerous physiological processes (Figure 16.1C).

Interactions between G protein pathways occur at multiple levels. Several G protein-coupled receptors interact with multiple heterotrimeric G proteins, and the downstream effectors can also serve as junctions between the different G protein pathways. The pathways are sufficiently interconnected that together they form a network. Such connectivity to form highly coupled networks has functional consequences. One of these is seen in drug—target interactions [21]. We can readily observe a visible cluster in the network between FDA-approved drugs and their targets (Figure 16.1D) as nearly 50% of all FDA approved drugs have G protein-coupled receptors as targets.

Although the description above has focused on G protein-coupled receptors, many other receptors, such as the growth factor receptors (receptor tyrosine kinases), ion channel receptors in the nervous systems, cytokine receptors and several other types of receptors, including nuclear receptors, are all interconnected.

One of the best examples where networking begins at the level of the receptor is the receptor tyrosine kinase [22—24] (Figure 16.2), which network can transmit growth factor signals through many different pathways. Such routing can regulate multiple independent cellular functions; also, signal routing through multiple pathways can

Chapter | 16 System Biology of Cell Signaling

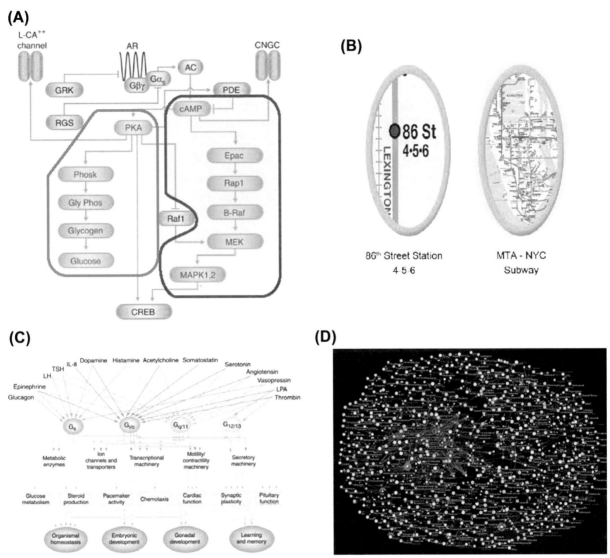

FIGURE 16.1 **The heterotrimeric G protein signaling network.** (A) The signaling network from the β-adrenergic receptor. The pathway on the left, outlined in blue, shows how adrenaline can induce through cAMP and PKA (c-AMP dependent kinase) glucose production for the fight or flight response. On the right, outlined in red, is a second pathway regulated by the same second messenger. cAMP can regulate Epac (the cAMP- and AMP-regulated exchange factor for RAP1) and lead to the activation of MAP-kinase 1,2 (mitogen-activated-protein kinase 1 and 2). Both PKA and MAPK can in turn phosphorylate and activate the transcription factor CREB (cAMP response element-binding protein) (adapted from [25]). (B) A common example with the complexity of cellular signaling systems. The mammalian cell can be compared to a complex network of signaling pathways resembling the many lines within the New York City subway system. (C) Many hormones and neurotransmitters interact with receptors coupled to different classes of G proteins. G protein-coupled receptors (GPCRs) often interact with several G proteins and regulate their downstream effectors to activate different intracellular responses (adapted from [25]). (D) The interconnection between the receptors and G proteins generate highly interconnected networks like the one seen in the center of the diagram representing interactions of FDA-approved drugs and their targets (*kindly provided by Dr Ma'ayan*).

lead to combinatorial signal specificity at the level of gene expression [22,26]. Such combinatorial specificity may be used as a mechanism to establish hierarchy among the regulated cellular processes. Such hierarchy can include cell motility preceding a secretion or proliferation response. In the cytoplasm such interconnectivity is largely mediated through several regulators guanine nucleotide exchange factors (GEFs) and GTPase-activating proteins (GAPs) of small G proteins and the three MAP-kinase pathways (MAPK 1, 2, p38 and JNK). In a few cases regulation of phosphatase activity is also used to obtain connectivity between pathways so as to form networks.

ISOFORMS OF SIGNALING MOLECULES, SIGNALING INTEGRATION AND SORTING

An important molecular characteristic that enables signal integration is the presence of isoforms of proteins, where each one has different inputs but all of which have common

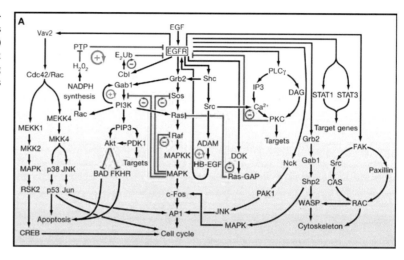

FIGURE 16.2 **The EGF—receptor signaling network.** Interactions of multiple effectors such as adaptors (Grb2), enzymes (PLC-γ) transcription factors (Stat3) lead to a complex intracellular signaling network that can regulate multiple cellular processes including cytoskeletal dynamics, cell cycle and apoptosis (*reprinted with permission from [23]*).

outputs. The GEFs for Ras display the value of having multiple isoforms. The Ras-GEF called Sos communicates signals from growth factor receptor pathways [27]. In contrast DAG-stimulated GEFs can activate Ras in response to signals from Gq/11 or Gi/o coupled pathways. [28].

An early example where isoforms enable integration is seen with mammalian adenylyl cyclases. There are 10 adenylyl cyclase isoforms that produce cAMP in response to signals from Gs-coupled receptors in response to Ca^{2+} from voltage gated calcium channels or glutamate-activated channels such as the NMDA receptors. Additionally, the production of cAMP can be inhibited by signals from Gi/o-coupled receptors and activated through PKC by signals from Gq-coupled receptors, depending on which isoform of adenylyl cyclase is present. This is summarized in Figure 16.3, adapted from a review published in 1993 by Pieroni et al. [20,29]. Isoform-dependent integration serves as a gauge to balance signals coming from many different receptors. In this way signaling information coming from junctions are routed towards different downstream cascades in such a way that they can in turn regulate numerous physiological events, as in the case of protein kinase A (Figure 16.3).

SCAFFOLDING PROTEINS: SIGNALING NETWORKS FORM SIGNALING COMPLEXES

Signaling networks are organized as complexes of signaling proteins. The organization of these complexes is dynamic and the complexes can often be assembled in response to incoming signals by proteins named scaffolding proteins. One of the best examples of these complexes is postsynaptic density (PSD), a protein-dense specialization attached to the synaptic junction in neurons. It lies adjacent to the cytoplasmic face of the postsynaptic plasma membrane, in close apposition to the active zone of the synapse and the docked synaptic vesicles in the presynaptic terminal. This location places the PSD directly in the path of the ionic fluxes and second-messenger cascades generated by neurotransmitters. The PSD functions as a structural matrix, which clusters ion channels in the postsynaptic membrane [30—32] and anchors signaling molecules such as protein kinases and phosphatases at the synapse [33]. Such anchoring has been shown to be important for the signaling events that underlie synaptic plasticity [34,35].

A well-known class of proteins called AKAPs (A-kinase anchoring proteins) functions as a scaffold for signaling networks [36—38] and assembles protein kinases and phosphatases to interact with receptors and channels. AKAPs provide a spatial dimension to signaling by forming signaling complexes that have the intrinsic capability to both consolidate and dissipate biochemical signals [20]. AKAPs themselves are targeted to distinct regions of the cell.

Another major class of scaffolding proteins is the PDZ domain-containing proteins that are central organizers of protein complexes at the plasma membrane of the glutamatergic synapses [39,40]. PDZ domain proteins often anchor channels and receptors to the postsynaptic density and interact with other scaffold proteins. AKAPs bind to PDZ domain-containing proteins PSD-95 and SAP-97, a critical step for the targeting of protein kinase A to AMPA receptors [41].

Typically scaffold proteins possess bidirectional specificity: they specifically recognize one or a group of signaling components and can interact simultaneously with a location within the cell, thereby enabling spatial organization of signaling pathways. Bidirectional specificity itself is a general mechanism for routing signals. An example of this bidirectional behavior is the mechanism of protein kinase A anchoring that ensures that PKA is exposed to localized changes in cAMP and is compartmentalized with

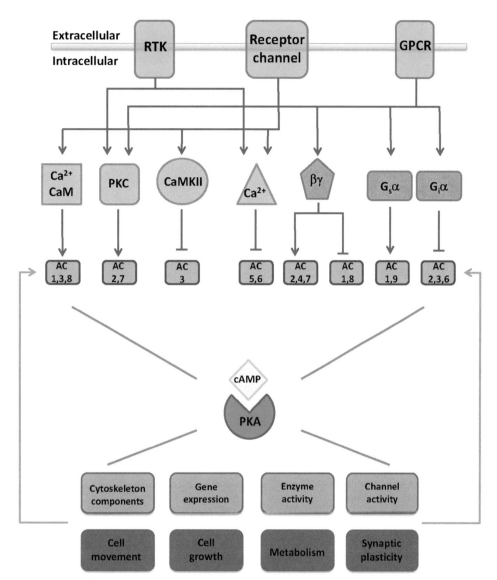

FIGURE 16.3 **Signal routing to cAMP involves different isoforms of adenylyl cyclase.** Different isoforms of adenylyl cyclase produce cAMP in response to differing signals received from G protein-coupled receptors (GPCRs), receptor channels (e.g., NMDA receptors) as well as receptors tyrosine kinase (RTKs). This ability to integrate signals coming from many pathways to regulate the levels of cAMP-dependent protein kinase (PKA) activity results in control of many physiological functions (*figure redrawn from [29] and [20]*).

respect to its substrates. Although cAMP is the only activator of PKA, scaffold regulatory proteins can control the temporal and spatial dynamics of PKA activation in response to specific stimuli. The coordinated actions of adenylyl cyclases and phosphodiesterases generate gradients and compartmentalized pools of cAMP, while AKAPs maintain the PKA holoenzyme at specific intracellular sites [42]. The AKAPs have two types of protein sequences that regulate PKA function: a conserved amphipathic helix that binds the R subunit dimer of the PKA holoenzyme, and a specialized targeting region that enables individual PKA−AKAP complexes to bind specific subcellular structures [36,43]. As a consequence, PKA is held by the AKAP in an inactive state at a defined intracellular location, where it is positioned to respond to cAMP by the local release of active C subunit. Thus, with appropriate scaffolding, a protein kinase with broad substrate specificity can rapidly phosphorylate specific targets in specific locations in response to a defined signal. Overall scaffolds serve as building blocks for molecular complexes that integrate and split signals. Such an assembly provides a natural mechanism to achieve selective separation of signaling components and thereby achieve specificity of signal routing. The scaffolds also provide a mechanism by which signals can be spatially resolved within the cell and thus provide the spatial dimension to signaling networks.

COMPUTATIONAL ANALYSIS OF SIGNALING NETWORKS

Owing to the large number of components, their varied molecular functions and their many interactions, it is impossible to intuit how such a complex system will function. To obtain such understanding computational analyses are essential. The goals of computational analyses are (1) to understand the organization of the signaling network, and (2) to understand how this organization leads to the processing of information as signal flows through the network. This processing of information leads to changes in the input—output relationships. Thus cell signaling networks are different from engineered communication networks. In the latter system the overriding goal is to maintain the fidelity of the input information so that output faithfully matches the input. In contrast, in cell signaling networks the system is configured to process information such that it can evaluate the information even as signal flows through the network and modulates output and responses in accordance with current capabilities of the cell. Such information processing can lead to many different kinds of outcome. Sometimes short-duration input signals are converted into longer-duration outputs so as to engage multiple cellular machines and change cell physiological functions. At other times sub-threshold signals may be dissipated, such that even though we may observe biochemical signaling reactions, no cell physiological responses are elicited. Two types of computational model have been widely used to study cell signaling networks, and these are discussed in the following sections.

Graph Theory-Based Models

The large numbers of components and interactions have made network analysis a very useful approach for understanding the organization of cell signaling networks. Substantial information about the topology can be gleaned from both undirected and directed graphs. Characteristics such as scale-free properties [44] and clustering coefficient [45] provide insight into the global topology of networks. At a more microscopic scale network motifs such as feedback loops, as well as feedforward and bifan motifs, serve as building blocks of networks as they occur within networks in a recurrent manner [46].

REGULATION BY NETWORK MOTIFS

A motif can be defined as a group of interacting components within the network which is capable of signal processing. Examples include **positive feedback loops** that enable propagation of signals across timescales, whereas a negative feedback loop limits the signal propagation through the network. **Negative feedback loops** are common in metabolic pathways which are frequently involved in maintaining homeostasis through end-product inhibition. Connectivity enables the formation of both positive and negative feedback loops, depending on the isoforms of the components present. A typical example in the neuron is the isoforms of the phosphodiesterase 4D that are activated by protein kinase A. The presence of these isoforms enables the formation of a negative feedback loop that controls the level of cAMP, thereby regulating signal flow to multiple downstream effectors. Network motifs display a range of signal processing capabilities, including the ability to delay or speed up signal flow, to enable the flow of information between different timescales, to filter noise, to sort or synchronize signals, to function as switches and to generate oscillations [47]. **Incoherent feedback loops** can function as sensors of fold change in both the Wnt [48] and EGF signaling [49] pathways. The topology of feedback and forward motifs is shown in Figure 16.4A.

One important feature of feedforward motifs is that they allow the transmission of signals across time scales. This 'temporal bridge' between the fast and slow timescales can have important effects in controlling several biological phenomena, for example proliferation and differentiation. One well-studied example is the MAPK-1,2 signaling network. MAPK-1 and -2 are activated within minutes in response to a wide number of external stimuli. In response to MAPK activation, a set of immediate-early genes (IEGs) are turned on in order to convert the initial MAPK activation into different long-lasting biological effects, such as proliferation or differentiation [50—52]. One crucial factor in deciphering the signal and translating information into the proper biological responses is the duration of the signal itself. In the rat neuronal cell line PC12 short activation of MAPK-1,2 leads to proliferation, whereas prolonged MAPK-1,2 signaling induces differentiation [52—54]. Murphy and co-workers have described a feedforward motif within the signaling network that decodes the extent of MAPK activation and consequently affects the expression of IEGs [55,56]. A summary of their results is represented in Figure 16.4B. One of the IEGs that are turned on in response to MAPK-1,2 activation is c-Fos. Since the timescale that is necessary for c-Fos transcription, translation and nuclear transportation is much longer that that required for transient MAPK-1,2 activation, only an initial prolonged MAPK signaling promotes c-Fos phosphorylation, which protects c-Fos against degradation by the ubiquitin—proteasome pathway [57,58]. A stable c-Fos (c-Fos-P) promotes transcription of c-Fos target genes even when MAPK signaling is turned off through a negative feedback loop involving MAPK-phosphatase that requires about 30—60 minutes to become operational. Consequently, the feedforward regulatory motif is responsible for decoding the signal duration and can therefore bridge the

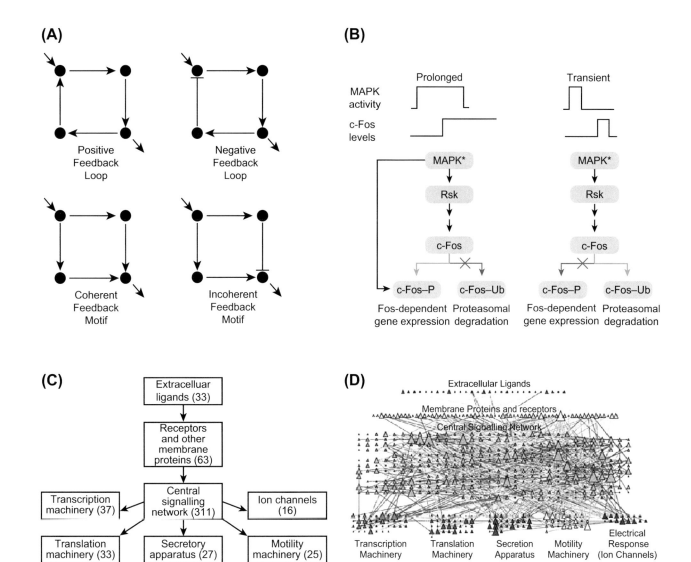

FIGURE 16.4 **Topology of cell signaling networks.** (A) The structure of regulatory motifs: positive and negative feedback loops (upper panel) and coherent and incoherent feedforward motifs (lower panel). Individual components are depicted as circles. Diagonal arrows depict signal flow into and out of the motif. Arrows indicate activation and plungers indicate inhibition. (B) A coherent feedforward motif can help filter short from long duration signals. The response of MAPK-1,2 activation involves the induction of immediate-early genes (IEGs), e.g., c-Fos. The presence of a feedforward motif within the network allows MAPK signals with longer duration to phosphorylate c-FOS and protect it against ubiquitin-mediated degradation, enabling bridging across timescales of minutes for MAPK activation and hours for c-Fos-regulated gene expression (*reprinted with permission from [59], based on experiments from [55]*). (C) The organization of a functional cell is shown as an interconnected network of a central signaling network and cellular machines. The block diagram shows how information generated from extracellular ligands is transmitted to different cellular machines and functional compartments. Numbers in parentheses represent the number of nodes. (D) Directed sign-specified network representation of signal flow: triangles represent nodes and their size is proportional to their level of connectivity (i.e., number of edges). Interactions that lead to activation are visualized as green arrows, while inhibitory ones are visualized as red ones and blue arrows represent neutral interactions (*figures (C) and (D) reprinted with permission from [60]*).

difference in the timescale activation for MAPK (minutes) and c-Fos-stimulated gene expression (hours).

PROPERTIES OF A CELLULAR SIGNALING NETWORK

A study of the minimal signaling network of a mammalian hippocampal CA1 neuron [60] provides an example of how signal processing may help control cellular decisions. The CA1 neuron was represented as a set of interacting components that make up a network of signaling pathways that connects to various cellular machines (Figure 16.4C). The fully connected network was built from known direct interactions, with functional effects documented in the experimental literature (Figure 16.4D).

To understand how signaling in the hippocampal neuron may lead to synaptic plasticity, the authors analyzed signal propagation from ligand—receptor interactions to their downstream effectors, such as the transcription factor CREB and the AMPA ion channel, for three key neurotransmitters involved in plasticity: glutamate (Glu), norephinephrine (NE) and the brain-derived neurotrophic factor (BDNF). Within the directed sign specified network the formation of regulatory motifs such as feedforward and feedback loops were identified. It was shown that norepinephrine, acting through the β-adrenergic receptor, evokes the formation of more positive feedback and coherent feedforward loops than negative feedback and incoherent feedforward loops. Since positive loops are associated with prolonging signals, and since long signals often drive physiological events, this network topology provides an explanation as to why norepinephrine, the cAMP pathway and CREB are closely associated with long-lasting neuronal plasticity. The study also identified the presence of different characteristics of the network as signal flows from the receptor to effectors: closer to the receptor the network contains a greater number of negative feedback loops than positive feedback loops. This indicates that the signaling network has a built-in control mechanism to limit signal propagation during the early steps. Such control systems, known in cell signaling research as desensitization mechanisms, are associated with almost all receptor pathways. As the signal flows further downstream, positive feedback loops tended to be more abundant than expected for all three ligands. This topology indicates that, because of abundance of negative feedback loops during the early steps, weak or short-lived signals may not penetrate deep into the network and thus may be unable to evoke a physiological response. However, as we penetrate deeper into the signaling network and the number of steps increases, the number of positive feedback loops appears to be more abundant in such a way that signals should persist and be able to evoke a biological response.

NETWORK TOPOLOGY AND CONSOLIDATION OF SIGNALS

Since interactions between components in the signaling complexes can be regulated by signal inputs, an important feature of biological signaling networks is that junctions and nodes can be assembled and disassembled in an activity-dependent manner, so that the system is able to assign a value to the signal which is either converted in the final biological response or safely dissipated within the network. This property sets biological signaling networks apart from physical networks, where network architecture is largely preset and generally is not reorganized by signal input. In biological systems the signaling network can estimate whether signals are of sufficient amplitude and duration to induce a physiological response and thus promote consolidation of signals.

The early stage of regulation of signal consolidation within signaling networks is located at the receptor level, where the process of desensitization can rapidly limit the signal flow. This type of regulation is seen in heterotrimeric G protein pathways, where receptor kinases (GRKs) rapidly uncouple the receptors from the G proteins [25,61]. To prevent the effect of consecutive and excessive stimulation of downstream components, the network operates through a mechanism that involves downregulation at the receptor level, i.e., receptors are removed from the cell surface membrane, the most common site of signal origination for intracellular networks. Many classes of receptors, including receptors tyrosine kinases, are subject to downregulation, thus controlling signal inflow upon repeated extracellular stimulation [62].

When signals of sufficient duration and amplitude get past the receptor, consolidation often occurs at the level of the transducer, such as G proteins. The extent of G protein activation reflects the duration, the amplitude and the strength of the signal. In disease states continual activation of G proteins is achieved by inhibiting the intrinsic GTPase. This can be achieved by inhibition of the GTPase by covalent modification, for example ADP ribosylation by cholera toxin, an enzyme (ADP-ribosyl transferase) secreted by the bacterium *Vibrio cholera*. The elevation of cAMP leads to inhibition of water reabsorption in the intestines, leading to diarrhea. Similarly, mutations that prevent GTP hydrolysis lead to a constitutively active Ras, thereby locking Ras in a permanently 'On' state. Activating mutations in Ras lead to elevated MAP-kinase activity and is frequently associated with many forms of human tumors [63]. Responsible for terminating the signaling event are a family of proteins that controls the GTPase activities of both small and large G proteins: these are named GTPase-activating proteins (or GAPs) for small G proteins, and regulators of G protein signaling (or RGS) for the heterotrimeric G proteins.

In this way the consolidation of the signaling to downstream components is affected both by the presence of the ligand at the receptor level and by the regulation of G protein activity. Additionally, interactions between members of the GAPs family may occur and result in junctions between different signaling pathways. Consequently, different branches of the same network communicate and transfer information, suggesting an intrinsic mechanism of control at the level of G protein regulators [64]. The third major locus of signal consolidation is at the level of protein kinases. Continuously activated protein kinases are capable of triggering intracellular signals leading to several physiological consequences. One of the most common examples is the persistently activated calcium-calmodulin kinase II (CAMKII) during long-term potentiation of synaptic responses [65], or the mutationally active forms of tyrosine kinases and MAP kinases associated with aberrant proliferation and neoplastic transformation [66]. Often the mechanism through which activation of these protein kinases is achieved is by phosphorylation in response to upstream signals. The duration of such activation is the result of the balance between input signaling and the activity of specific phosphatases. Alternatively, sequential inhibitory phosphorylation by protein kinases involved in other concurrent pathways can also regulate signal consolidation [67]. Scaffold proteins (such as AKAPs) can recruit protein kinases and phosphatases into dynamic complexes, thereby assembling the network topology required for regulation of signal consolidation necessary to evoke a physiological response.

In summary, the mechanism of signal consolidation depends on two fundamental capabilities of the network. First is the network's ability to propagate signaling information for various lengths of time across different cellular locations, in order to achieve physiological responses that depend on the integrated functioning of several cellular machines and which operate over different timescales. The second arises from differential connectivity of isoforms of a protein, resulting in both differential subcellular localization that depends on interactions with scaffolding proteins and the diverse ability to route signals through various pathways within a network, in response to proximal upstream and downstream components. Thus, only a proper spatial localization of the molecular entities involved in the network (i.e., the active form of MAP kinase or a second messenger such as cAMP) can result in a consolidated signal leading to cell physiological response. Such regulatory mechanisms are used by the cell to set local thresholds for the conversion of biochemical reactions into physiological responses [68]. These predicted system properties need to be tested by explicit experimentation.

DYNAMICAL MODELS

Although network models are very useful for understanding the relationship between the organization of components and their function, they are not fully predictive, as many of the capabilities, such as switching between different functional states, enabled by a particular organization such as a positive feedback loop, occur only when the concentration of components and reaction rates are in the appropriate range. It is reasonable to consider a mammalian cell as a complex chemical reactor, where reactions between different molecules, together with changes in their dynamic localization, give rise to the different mechanical, chemical and electrical properties of the cell. One of the major goals in system biology is to computationally represent this complex reactor using differential equations and mathematical modeling, to accurately predict the dynamics of such a cellular system. The mathematical models that are currently available to quantitatively describe the biological behavior of a mammalian cell unit can be divided into two main categories: deterministic and stochastic. **Deterministic models** are used when the number of molecules of each reactant is considered to be high enough such that the inherent biochemical noise can be averaged and each interaction (i.e., reaction) can be described by mean reaction rates. The change in the concentration of reactants with respect to time is completely determined by the initial concentrations of the reactants and the reaction rates between the reactants.

Stochastic models, by contrast, involve reactions between molecules that are present in small numbers and hence react with each other in a probabilistic manner. In a stochastic model, changes in the reactant concentrations with respect to time cannot be fully predicted from the initial conditions alone: in these systems, because the concentration of components is quite small, reactions can occur by chance. Such unregulated reactions generate what is termed biochemical 'noise'. Sometimes, depending on the trajectory and state of the system, such noise can be significant enough to switch the system from one biochemical state to another [69–72].

Generally dynamical models of biochemical systems, such as signaling networks, are based on solution kinetics, where the spatial heterogeneity of the cell is ignored, allowing for the pathway or network to be represented as a system of **ordinary differential equations (ODE)**. In order to better represent certain dynamic cellular features such as trafficking and transport, compartmental ODE models can be used with reactants confined to one or more compartments and a rate specified for movement between compartments. The movement of molecules across different compartments is modeled like a flux [73]. When explicit representation of the movement of each reactant in the system needs to be specified, then a system of **partial differential equations (PDE)** can be used. PDEs allow for representation of two or more variables, and usually changes in concentrations of reactants or products with respect to space and time are calculated. Both the compartmental and the reaction–diffusion models are a more realistic way to represent biochemical reactions within the cell, but they usually imply a high level of complexity in terms of great numbers of parameters to be considered, and therefore can be expensive in terms of required computational time. Regardless of this complexity, some stochastic models have been used to study specific systems such as neurotransmitter release, immunological synapses, endocytic vesicles or signaling in dendrites [74–77].

POSITIVE FEEDBACK LOOPS CAN FORM SWITCHES: THE CONCEPT OF BISTABILITY

One of the key findings from the use of dynamical models is that positive feedback loops can form switches, i.e., function as bistable systems. An early example of such bistable behavior can be found in the activation of protein kinase C (PKC) and mitogen-associated protein kinases (MAPK) 1 and 2, induced by stimulation of a receptor tyrosine kinase, such as the epidermal growth factor receptor (EGFR) [1]. The signaling pathways that are activated are represented in Figure 16.5A, left panel. Stimulation of EGFR triggers two interconnected pathways: the PLCγ-PKC pathway and the Ras-Raf-MAPK pathway (Figure 16.5A). These two pathways interact at two different levels: PKC activates Ras regulators [78] and possibly Raf [79,80], which in turn activates MAPK. MAPK can activate through phosphorylation the cytosolic phospholipase A-2 (cPLA2) [81,82] with production of arachidonic acid (AA). AA acts together with diacylglycerol (DAG) to activate PKC. This constitutes a positive feedback loop, and it is possible to computationally study the necessary conditions to achieve a sustained activation of the system in response to a brief stimulus. When the system is activated above a certain level (indicated in Figure 16.5A right panel by point T), both PKC and MAPK can reach a steady-state level of activation (point A); conversely, if the initial amount of stimulation is below a certain threshold (below point T), the system will go back to a basal level of activation (point B) after the ligand is removed. Therefore, the system exhibits bistable behavior.

Biological systems that behave in such a bistable fashion have the intrinsic property to store information, so that only external stimuli able to induce activation of PKC or MAPK above the intersection point T will move the system to switch from one state to another. Signals of sufficient amplitude and duration enable the network to switch from the 'low activity' state to the 'high activity' state, so under certain conditions the output signal can be sustained even when the input signal is removed [1,83]. The versatility and tissue level function of the PKC-MAPK-cPLA2 feedback loop has been demonstrated by Tanaka and Augustine [84], who showed that this positive feedback loop was necessary for cerebellar long-term depression, a form of synaptic plasticity. The PKC-MAPK-cPLA2 feedback loop functioned as a switch to couple brief Ca^{2+} input signals to protein synthesis that is required for long-term depression (LTD).

A recent example of how positive feedback loops function as memory storage devices to affect organismal behavior [85] focused on the potential role of synaptic plasticity during hunger. Yang and co-authors found that food deprivation results in persistent upregulation of excitatory synaptic inputs to AGRP neurons (Agouti-related protein, AgRP), which consequently display increased firing rates. AGRP neurons are a population of neurons localized in the hypothalamic arcuate nucleus that regulates feeding. These neurons are interconnected with a separate and functionally opposed population termed POMC neurons (so called because they express pro-opiomelanocortin), which indeed inhibit feeding [86]. The study showed the existence of a dynamic neural circuit with a reversible memory storage device that regulates food intake in response to energy deficit, thereby ensuring physiological homeostasis. The key control point is represented by excitatory synapses onto AGRP neurons activated by the hormone ghrelin. Ghrelin, released after food deprivation, binds to ghrelin receptors (Ghsr) located presynaptically, increasing glutamate release and so activating AGRP neurons through ionotropic glutamate receptors. 5′ Adenosine monophosphate-activated protein kinase (AMPK) is involved in the circuit and participates in a positive feedback loop that results in a switch-like behavior: indeed, the AMPK-mediated feedback upregulates activity at the presynaptic terminals for at least 5 hours. This results in a memory storage mechanism of the hormone ghrelin that would persistently promote synaptic activity onto AGRP neurons to stimulate feeding behavior (illustration of the model in Figure 16.5B, top panel). Simultaneously, those persistently activated synapses do require a separate inhibitory signal to switch off the upregulating inputs once the energy balance is restored. The hormone leptin, which is associated with long-term regulation of energy homeostasis [87], is well suited for this role. Yang et al. found that leptin is sufficient to reverse the elevated presynaptic activity onto AGRP neurons through an opioid receptor-dependent pathway. Indeed, leptin promotes POMC neuron activation [88] and the release of β-endorphin which represses elevated synaptic activity.

It is worth noting that this system presents similarities to a set/reset (SR) flip-flop memory storage circuit, which consists of two interconnected NOR logic gates (represented in Figure 16.5B, lower panel): the 'set' signal is the hormone ghrelin (S) and the 'reset' signal is leptin (R), and respectively they sense deficit or surfeit. This SR circuit is able to maintain high activity in response to S (even after S is turned off) until R becomes true. This bistable mechanism perfectly responds to set a range for energy balance (and not a set point) that is determined by respective thresholds for activation of S and R (Figure 16.5B, lower panel).

The physiological relevance of this circuit can be readily understood: the presence of a system that can set energy balance allows the organism to exhibit different behaviors when the energy component is not critical and does not represent a limiting factor. This switch controls the regulatory release of the two 'feeding control hormones', leptin and ghrelin, from the peripheral endocrine system in response to either deficit or surfeit energy levels, without necessarily reflecting the general energy state balance coming from the entire organism. Indeed, food deprivation usually leads to a rapid increase in ghrelin levels that are maintained only for 1−2 hours after refeeding; however, the time window of such hormonal activity, albeit transient in time, is enough to produce elevated firing rate at the level of AGRP neurons whose activation persists up to 24 hours

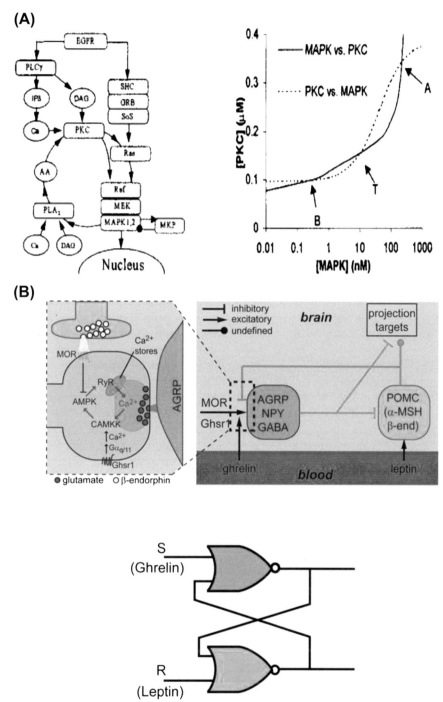

FIGURE 16.5 **A positive feedback loop as a bistable switch.** (A) (Left panel) Receptor tyrosine kinase (EGFR) signaling network that can induce activation of both phospholipase Cβ (PLCβ)-protein kinase C (PKC) pathway and the RAS-RAF-MAPK pathway. MAPK activates cytosolic phospholipase A-2 (cPLA2) leading to the production of arachidonic acid (AA), which together with diacylglycerol (DAG) activates PKC to enable a positive feedback loop. This can lead to prolonged activation of both PKC and MAPK1,2. (Right panel) Activity plots of PKC vs. MAPK show how this network can display bistable behavior. Activation of the receptor such that initial activity of PKC and MAPK is above the intersection point (point T) will move the system to a sustained activation state wherein both PKC and MAPK stay active even after the initial signal is withdrawn (point A). Conversely, if the initial stimulus is below point T, the system will be decay to the basal state B as soon as the stimulus is withdrawn (adapted from [1]). (B) A bistable switch in the memory circuit that controls hunger state and energy homeostasis. (Upper panel) AGRP neurons (named after the Agouti-related protein, AgRP) inhibit POMC neurons (expressing pro-opiomelanocortin) in response to circulating hormones released in the bloodstream (e.g., ghrelin) after food deprivation. The presence of an AMPK (5′ AMP-activated protein kinase)-dependent positive feedback loop persistently promotes synaptic activity onto AGRP neurons to stimulate feeding behavior until the energy balance is restored. (Lower panel) The system presents similarities to a set/reset (SR) flip-flop memory storage circuit where two hormones ('set': ghrelin and 'reset': leptin) separately signal deficit or surfeit. The model does not have a set point, but instead has a set range that is defined by the thresholds for activation of S and R *(reprinted with permission from [85])*.

after refeeding [89]. This memory device, using in this case an opioid receptor mechanism, triggers the signaling beyond the hormonal levels until the appropriate energy levels are restored.

SIGNALING MICRODOMAINS WITHIN CELLS

The ability of two cellular components to interact depends not only on their mutual chemical specificity but also on their co-localization in the various subcellular regions within the cell. To understand the organization and functional capabilities of cell signaling networks, we need to understand how spatial separation or co-localization is controlled to regulate the dynamic topology of cell signaling networks. Most signals that evoke physiological outcomes in a mammalian cell start at a cellular membrane level, where hormones and neurotransmitters bind their own receptors to activate the relevant intracellular signaling pathway so that signal starts to flow through the network. Often this involves local production near the cell surface membrane of important signaling molecules, such as second messengers (e.g., cAMP) as well as local activation of protein kinases, protein phosphatases and other signaling components. Such local production and transient elevation of activated signaling components ensures that sufficient signal flows through to evoke a local physiological response. Such regions are called **microdomains,** and can be described as dynamic regions of micrometer/submicrometer dimensions with increased concentration of one or more signaling components [90].

A microdomain has two identifying characteristics: a point at which the activated signaling component reaches its highest concentration, and a gradient that details how the concentration changes from the highest point to the surrounding neighboring regions. The slope of the gradient indicates the edges of the microdomain [91]. Typically these are not absolute values of the activated components; rather, they represent levels capable of transmitting signals to downstream components that can in turn evoke physiological responses. Combination of experiments together with PDE models has identified a number of variables that participate in controlling the nature and behavior of microdomains. These include the role of narrow regions with high surface to volume ratios that enable high local concentrations of GTPases [92,93]. Other interdependent factors include the topology of the signaling network, cell shape, diffusion coefficients and reaction kinetics (for an extended discussion see [91]). Advances in live cell imaging have enabled the demonstration and importance of microdomains for several signaling molecules, including cAMP [90,94–96], ions such as Ca^{2+} [97–99], GTPases [92,100] and protein kinases [101–103]. Hahn and co-workers have developed genetically encoded photoactivatable derivatives of GTPases and their regulators [104]. These molecular probes involve naturally occurring domains that undergo large conformational changes upon irradiation, and can be used as part of fusion constructs to control protein function. They produce reversible, repeatable, activation at visible wavelengths. These probes have been very useful in characterizing the role of the GTPases at the leading edge of the cell during cell motility [105].

To better understand the origins of microdomains from the control of the localization of intracellular signaling reactions, we describe a study from our laboratory by Neves and co-workers [106]. Using partial differential equations and real neuronal shapes obtained from microscopic images, Neves et al. modeled how signal from the β-adrenergic receptor activates MAPK-1,2, through the cAMP/PKA/b-Raf/MAPK-1,2 pathway. These studies demonstrated that when the signal originates at the plasma membrane and travels through the cytoplasm, the ratio of the surface area of the plasma membrane to the cytoplasmic volume (surface/volume (S/V) ratio) is one critical characteristic that determines microdomain dynamics. For a morphologically specialized cell such as a neuron, the S/V is greater in the dendrites than in the cell body. Since signals mostly start at the membrane and are dissipated in the cytoplasm through negative regulators such as phosphodiesterases and phosphatases, higher S/V ratios favor the formation of microdomains. Neves et al. conducted simulations using PDE models and showed that in neurons this would result in the selective formation of cAMP microdomains in the dendrites. These predictions were then experimentally verified.

Local activation of downstream components, such as MAPK and PKA, indicated that spatial information is propagated from cAMP microdomains to downstream components. Here factors other than surface to volume ratio also play a role. One obvious factor is diffusion coefficients of the signaling components involved. Simulations varying the diffusion coefficients of the different signaling components showed that the diffusion coefficients of most components of the signaling network did not affect microdomain characteristics when the signaling network was within thin dendrites such as those found in the dendritic arbors. The only exception was diffusion of negative regulators. These results suggested that in a reaction–diffusion model, using geometries obtained from real neurons, reactions are dominant over diffusion in the dendritic arbors. However, when the dendritic diameter increases, as seen for the primary dendrites or when the cell body is considered, the role of diffusion becomes critical for microdomain characteristics. Thus local shape and cellular geometry are important controllers of spatial signaling.

Simulations identified that a second important factor controlling microdomain dynamics is the negative regulator, such as a phosphatase. Experimental validation of the computational predictions in primary rat hippocampal neurons and hippocampal slices showed that transmission of the microdomain's characteristics from cAMP to MAPK is controlled by the phosphatase PTP. These simulations also revealed a new deep insight into information transfer during signal flow: information regarding activity state (i.e., fractional activation of a signaling component of interest) is different from spatial information (i.e., where within the cell is the component activated). In the system these authors studied this differential transfer of information is controlled by the phosphatase PTP, which inhibits MAPK [107,108] and which in turn is phosphorylated and inhibited by PKA. If inhibitory control of PTP by PKA is lost, the microdomains of activated MAPK are also lost, leading to a general activation of MAPK. These simulations were confirmed experimentally. Further simulations revealed that the transmission of spatial information is dependent on the kinetic parameters of PTP, especially its turnover rate (kcat), i.e., the number of times an enzyme molecule catalyzes its reaction per unit time.

If an upstream enzyme extensively activates a downstream enzyme (i.e., high kcat) then the microdomain characteristics of the upstream component will not be preserved in the downstream component. Appropriate tuning of kinetic parameters is critical for transmission of spatial information. Overall, three sets of characteristics — cell shape (i.e., surface-to-volume ratios), network topology and kinetic characteristics of the signaling components — are equally important for the formation and dissipation of microdomains. Each of these characteristics needs to be balanced with respect to the others. Therefore, microdomains of cell signaling components are truly an emergent systems phenomenon. Microdomains allow cells to produce selective responses. In neurons, cAMP microdomains in the dendrites allow for acute control of channel activity without producing gene expression-mediated long-lasting changes.

CONCLUSIONS AND FUTURE CHALLENGES

Over the past decade there has been substantial progress in understanding cellular functions at a systems level. We now know that:

1. The different signaling pathways interact to form extensive networks.
2. The presence of multiple isoforms of different signaling components with a mix-and-match input—output profile is essential for the formation of such networks that control multiple cellular machines.
3. The interconnections within the networks result in a topology that has an abundance of regulatory motifs, such as feedforward motifs and feedback loops that can process information to alter the relationship between signal inputs and cellular responses.
4. Positive feedforward loops can function as bistable switches and positive feedforward motifs function as time bridges wherein brief signals in the time scale of minutes can produce long-lasting effects such as gene expression changes which occur in the timescale of hours.
5. Cell shape, network topology and the kinetic characteristics of signaling components together regulate the formation of microdomains, which are localized regions of elevated signaling components. These microdomains are important for the cell's ability to produce specific responses using one cellular machine even when the signaling network is broadly connected to many cellular machines.

Questions that remain unanswered include how signaling networks vary between individuals depending on their genomic and epigenomic status. Answering these questions is likely to be critical for understanding individual variations in susceptibility to disease as well as the ability to respond to drug therapy. The availability of large-scale screening methods has allowed us to obtain genome-wide and cell-wide views of mRNAs and protein profiles, as well as to characterize the genome using a number of criteria. Mechanistic studies at the level of the genome and epigenomic changes have shown that there are multiple loci at which regulation of cellular components can be achieved. These include the role of polymorphisms and gene duplication or deletion, in changing the level of cellular components, in addition to mutations that alter the activity of components. Changes in DNA methylation, alterations in splicing, and changes in levels of microRNAs and regulation of translation, represent additional mechanisms by which the levels of cellular components may be controlled. How many of the epigenomic mechanisms are regulated by cell signaling networks is not fully understood and requires further study. How these various genomic and epigenomic mechanisms regulate the topology and activity of cell signaling networks is also not understood. It is likely that regulation in both directions is coupled and will have to be understood at multiple levels of organization: microscopic, mesoscopic and macroscopic. For such multiscale understanding we will need to understand both the topology and the dynamics of the systems. Computational approaches that explicitly integrate graph theory type analysis with explicit dynamical models, such as ODE and PDE models, are likely to be valuable. Similarly, experimental approaches that can measure the dynamics of many components and reactions will also be very useful. A personalized medicine study characterizing the dynamics of physiological responses in a single individual shows the power of correlating genomic status with the dynamics of physiological response [109]. As the systems we study become progressively larger, the tight, rapid and iterative coupling between experiments and modeling will become necessary. Such iterative approaches hold the promise of providing deep understanding of the mechanisms by which molecular interactions control organ, multiorgan and organismal level functions.

GLOSSARY

Bistability is an emergent property of a biological system that allows the system to exist stably in two states, such as active and inactive. The switch can by triggered by an incoming stimulus. Usually positive feedback is necessary for bistability, although it is not sufficient if the levels of the components of the system or reaction rates are not in the appropriate range.

Positive feedback occurs when a downstream protein within a pathway is able to activate its upstream regulator, resulting in the output signal being amplified with respect to the input. It is observed both at a cellular level and at tissue organ physiology level.

Negative feedback occurs when a downstream component inhibits the activity of an upstream regulator. Functionally a negative feedback loop will reduce output signal and can linearize the input/output relationship. A classic example is feedback inhibition of biosynthetic enzymes by metabolites such that metabolites control their levels within cells.

Coherent and incoherent feedforward motifs are organizational units (motifs) within networks. In a feedforward motif an upstream component has at least two paths to a downstream component. Often one path is longer than the other. A feedforward motif is called coherent if both paths result in activation of the downstream component. If one of the paths is inhibitory the feedforward loop is called incoherent.

Ordinary differential equations (ODEs) are used to represent systems where the rate of change in the variable of interest is determined with respect to one other variable, most often time.

Partial differential equations (PDEs) are used to represent systems in which the variable of interest changes with respect to two or more variables, most commonly time and space.

Deterministic models are computational models representing systems whose time evolution is entirely determined by the initial conditions. ODE and PDE based models are deterministic models.

Stochastic models are used to represent systems where the time evolution of the system has a probabilistic component. The trajectory of such systems cannot be predicted from initial conditions. Typically such systems involve interactions between cellular components that exist in very low copy numbers.

Microdomains are compartmentalized regions within a cell, often located in close proximity to the plasma membrane, that transiently contain high concentrations of activated signaling components. Microdomains may consist of anything from a short-lived cluster of several protein and lipid molecules to large organized domains tens or hundreds of nanometres in diameter.

ACKNOWLEDGMENTS

Research in our laboratory is supported by NIH grants GM54508 DK-087650 and System Biology Center Grant GM-071558.

REFERENCES

[1] Bhalla US, Iyengar R. Emergent properties of networks of biological signaling pathways. Science 1999;283:381–7.
[2] Kitano H. Computational systems biology. Nature 2002;420:206–10.
[3] Ideker T, Galitski T, Hood L. A new approach to decoding life: systems biology. Annu Rev Genomics Hum Genet 2001;2:343–72.
[4] Hoffmann S. Computational analysis of high throughput sequencing data. Methods Mol Biol 2011;719:199–217.
[5] Wheeler DA, Srinivasan M, Egholm M, Shen Y, Chen L, McGuire A, et al. The complete genome of an individual by massively parallel DNA sequencing. Nature 2008;452:872–6.
[6] Malone JH, Oliver B. Microarrays, deep sequencing and the true measure of the transcriptome. BMC Biol 2011;9:34.
[7] Szelinger S, Pearson JV, Craig DW. Microarray-based genome-wide association studies using pooled DNA. Methods Mol Biol 2011;700:49–60.
[8] Juhila J, Sipila T, Icay K, Nicorici D, Ellonen P, Kallio A, et al. MicroRNA expression profiling reveals miRNA families regulating specific biological pathways in mouse frontal cortex and hippocampus. PLoS One 2011;6:e21495.
[9] Islam S, Kjallquist U, Moliner A, Zajac P, Fan JB, Lonnerberg P, et al. Characterization of the single-cell transcriptional landscape by highly multiplex RNA-seq. Genome Res 2011;21:1160–7.
[10] Griffith M, Griffith OL, Mwenifumbo J, Goya R, Morrissy AS, Morin RD, et al. Alternative expression analysis by RNA sequencing. Nat Methods 2010;7:843–7.
[11] Sardiu ME, Washburn MP. Building protein–protein interaction networks with proteomics and informatics tools. J Biol Chem 2011;286:23645–51.
[12] Olsen JV, Blagoev B, Gnad F, Macek B, Kumar C, Mortensen P, et al. Global, in vivo, and site-specific phosphorylation dynamics in signaling networks. Cell 2006;127:635–48.
[13] Anderson NL, Anderson NG. Proteome and proteomics: new technologies, new concepts, and new words. Electrophoresis 1998;19:1853–61.
[14] Ross EM, Gilman AG. Resolution of some components of adenylate cyclase necessary for catalytic activity. J Biol Chem 1977;252:6966–9.

[15] Bos JL, de Rooij J, Reedquist KA. Rap1 signalling: adhering to new models. Nat Rev Mol Cell Biol 2001;2:369–77.

[16] Wu GY, Deisseroth K, Tsien RW. Activity-dependent CREB phosphorylation: convergence of a fast, sensitive calmodulin kinase pathway and a slow, less sensitive mitogen-activated protein kinase pathway. Proc Natl Acad Sci USA 2001;98:2808–13.

[17] Mayr B, Montminy M. Transcriptional regulation by the phosphorylation-dependent factor CREB. Nat Rev Mol Cell Biol 2001;2:599–609.

[18] Stevens CF. CREB and memory consolidation. Neuron 1994;13:769–70.

[19] Frank DA, Greenberg ME. CREB: a mediator of long-term memory from mollusks to mammals. Cell 1994;79:5–8.

[20] Jordan JD, Landau EM, Iyengar R. Signaling networks: the origins of cellular multitasking. Cell 2000;103:193–200.

[21] Ma'ayan A, Jenkins SL, Goldfarb J, Iyengar R. Network analysis of FDA approved drugs and their targets. Mt Sinai J Med 2007;74:27–32.

[22] Schlessinger J. Cell signaling by receptor tyrosine kinases. Cell 2000;103:211–25.

[23] Lemmon MA, Schlessinger J. Cell signaling by receptor tyrosine kinases. Cell 2010;141:1117–34.

[24] Schlessinger J, Lemmon MA. Nuclear signaling by receptor tyrosine kinases: the first robin of spring. Cell 2006;127:45–8.

[25] Neves SR, Ram PT, Iyengar R. G protein pathways. Science 2002;296:1636–9.

[26] Fambrough D, McClure K, Kazlauskas A, Lander ES. Diverse signaling pathways activated by growth factor receptors induce broadly overlapping, rather than independent, sets of genes. Cell 1999;97:727–41.

[27] Gureasko J, Galush WJ, Boykevisch S, Sondermann H, Bar-Sagi D, Groves JT, et al. Membrane-dependent signal integration by the Ras activator Son of sevenless. Nat Struct Mol Biol 2008;15:452–61.

[28] Feig LA, Cooper GM. Inhibition of NIH 3T3 cell proliferation by a mutant Ras protein with preferential affinity for GDP. Mol Cell Biol 1988;8:3235–43.

[29] Pieroni JP, Jacobowitz O, Chen J, Iyengar R. Signal recognition and integration by Gs-stimulated adenylyl cyclases. Curr Opin Neurobiol 1993;3:345–51.

[30] Kennedy MB. The postsynaptic density. Curr Opin Neurobiol 1993;3:732–7.

[31] Kennedy MB. The postsynaptic density at glutamatergic synapses. Trends Neurosci 1997;20:264–8.

[32] Ehlers MD, Mammen AL, Lau LF, Huganir RL. Synaptic targeting of glutamate receptors. Curr Opin Cell Biol 1996;8:484–9.

[33] Klauck TM, Scott JD. The postsynaptic density: a subcellular anchor for signal transduction enzymes. Cell Signal 1995;7:747–57.

[34] Kornau HC, Schenker LT, Kennedy MB, Seeburg PH. Domain interaction between NMDA receptor subunits and the postsynaptic density protein PSD-95. Science 1995;269:1737–40.

[35] Gardoni F, Schrama LH, Kamal A, Gispen WH, Cattabeni F, Di Luca M. Hippocampal synaptic plasticity involves competition between Ca2+/calmodulin-dependent protein kinase II and postsynaptic density 95 for binding to the NR2A subunit of the NMDA receptor. J Neurosci 2001;21:1501–9.

[36] Pawson T, Scott JD. Signaling through scaffold, anchoring, and adaptor proteins. Science 1997;278:2075–80.

[37] Edwards AS, Scott JD. A-kinase anchoring proteins: protein kinase A and beyond. Curr Opin Cell Biol 2000;12:217–21.

[38] Wong W, Scott JD. AKAP signalling complexes: focal points in space and time. Nat Rev Mol Cell Biol 2004;5:959–70.

[39] Ziff EB. Enlightening the postsynaptic density. Neuron 1997;19:1163–74.

[40] Garner CC, Nash J, Huganir RL. PDZ domains in synapse assembly and signalling. Trends Cell Biol 2000;10:274–80.

[41] Colledge M, Dean RA, Scott GK, Langeberg LK, Huganir RL, Scott JD. Targeting of PKA to glutamate receptors through a MAGUK-AKAP complex. Neuron 2000;27:107–19.

[42] Dell'Acqua ML, Smith KE, Gorski JA, Horne EA, Gibson ES, Gomez LL. Regulation of neuronal PKA signaling through AKAP targeting dynamics. Eur J Cell Biol 2006;85:627–33.

[43] Dell'Acqua ML, Scott JD. Protein kinase A anchoring. J Biol Chem 1997;272:12881–4.

[44] Barabasi AL, Albert R. Emergence of scaling in random networks. Science 1999;286:509–12.

[45] Watts DJ, Strogatz SH. Collective dynamics of 'small-world' networks. Nature 1998;393:440–2.

[46] Milo R, Shen-Orr S, Itzkovitz S, Kashtan N, Chklovskii D, Alon U. Network motifs: simple building blocks of complex networks. Science 2002;298:824–7.

[47] Lipshtat A, Purushothaman SP, Iyengar R, Ma'ayan A. Functions of bifans in context of multiple regulatory motifs in signaling networks. Biophys J 2008;94:2566–79.

[48] Goentoro L, Kirschner MW. Evidence that fold-change, and not absolute level, of beta-catenin dictates Wnt signaling. Mol Cell 2009;36:872–84.

[49] Cohen-Saidon C, Cohen AA, Sigal A, Liron Y, Alon U. Dynamics and variability of ERK2 response to EGF in individual living cells. Mol Cell 2009;36:885–93.

[50] Brightman FA, Fell DA. Differential feedback regulation of the MAPK cascade underlies the quantitative differences in EGF and NGF signalling in PC12 cells. FEBS Lett 2000;482:169–74.

[51] Yamada S, Taketomi T, Yoshimura A. Model analysis of difference between EGF pathway and FGF pathway. Biochem Biophys Res Commun 2004;314:1113–20.

[52] Schamel WW, Dick TP. Signal transduction: specificity of growth factors explained by parallel distributed processing. Med Hypotheses 1996;47:249–55.

[53] Traverse S, Seedorf K, Paterson H, Marshall CJ, Cohen P, Ullrich A. EGF triggers neuronal differentiation of PC12 cells that overexpress the EGF receptor. Curr Biol 1994;4:694–701.

[54] Mark MD, Liu Y, Wong ST, Hinds TR, Storm DR. Stimulation of neurite outgrowth in PC12 cells by EGF and KCl depolarization: a Ca(2+)-independent phenomenon. J Cell Biol 1995;130:701–10.

[55] Murphy LO, Smith S, Chen RH, Fingar DC, Blenis J. Molecular interpretation of ERK signal duration by immediate early gene products. Nat Cell Biol 2002;4:556–64.

[56] Murphy LO, MacKeigan JP, Blenis J. A network of immediate early gene products propagates subtle differences in mitogen-activated protein kinase signal amplitude and duration. Mol Cell Biol 2004;24:144–53.

[57] Okazaki K, Sagata N. The Mos/MAP kinase pathway stabilizes c-Fos by phosphorylation and augments its transforming activity in NIH 3T3 cells. EMBO J 1995;14:5048—59.

[58] Chen RH, Juo PC, Curran T, Blenis J. Phosphorylation of c-Fos at the C-terminus enhances its transforming activity. Oncogene 1996;12:1493—502.

[59] Eungdamrong NJ, Iyengar R. Computational approaches for modeling regulatory cellular networks. Trends Cell Biol 2004;14:661—9.

[60] Ma'ayan A, Jenkins SL, Neves S, Hasseldine A, Grace E, Dubin-Thaler B, et al. Formation of regulatory patterns during signal propagation in a mammalian cellular network. Science 2005;309: 1078—83.

[61] Pitcher JA, Freedman NJ, Lefkowitz RJ. G protein-coupled receptor kinases. Annu Rev Biochem 1998;67:653—92.

[62] Foveau B, Ancot F, Leroy C, Petrelli A, Reiss K, Vingtdeux V, et al. Down-regulation of the met receptor tyrosine kinase by presenilin-dependent regulated intramembrane proteolysis. Mol Biol Cell 2009;20:2495—507.

[63] Vakiani E, Solit DB. KRAS and BRAF: drug targets and predictive biomarkers. J Pathol 2011;223:219—29.

[64] Bar-Sagi D, Hall A. Ras and Rho GTPases: a family reunion. Cell 2000;103:227—38.

[65] Soderling TR. CaM-kinases: modulators of synaptic plasticity. Curr Opin Neurobiol 2000;10:375—80.

[66] Marshall CJ. Specificity of receptor tyrosine kinase signaling: transient versus sustained extracellular signal-regulated kinase activation. Cell 1995;80:179—85.

[67] Iyengar R. Gating by cyclic AMP: expanded role for an old signaling pathway. Science 1996;271:461—3.

[68] Teruel MN, Meyer T. Translocation and reversible localization of signaling proteins: a dynamic future for signal transduction. Cell 2000;103:181—4.

[69] Hasty J, Isaacs F, Dolnik M, McMillen D, Collins JJ. Designer gene networks: towards fundamental cellular control. Chaos 2001;11:207—20.

[70] Arkin A, Ross J, McAdams HH. Stochastic kinetic analysis of developmental pathway bifurcation in phage lambda-infected Escherichia coli cells. Genetics 1998;149:1633—48.

[71] McAdams HH, Arkin A. Stochastic mechanisms in gene expression. Proc Natl Acad Sci U S A 1997;94:814—9.

[72] Ozbudak EM, Thattai M, Lim HN, Shraiman BI, Van Oudenaarden A. Multistability in the lactose utilization network of Escherichia coli. Nature 2004;427:737—40.

[73] Dundr M, Hoffmann-Rohrer U, Hu Q, Grummt I, Rothblum LI, Phair RD, et al. A kinetic framework for a mammalian RNA polymerase in vivo. Science 2002;298:1623—6.

[74] Del Castillo J, Katz B. Statistical nature of facilitation at a single nerve-muscle junction. Nature 1953;171:1016—7.

[75] Resat H, Ewald JA, Dixon DA, Wiley HS. An integrated model of epidermal growth factor receptor trafficking and signal transduction. Biophys J 2003;85:730—43.

[76] Franks KM, Bartol Jr TM, Sejnowski TJ. A Monte Carlo model reveals independent signaling at central glutamatergic synapses. Biophys J 2002;83:2333—48.

[77] Lee KH, Dinner AR, Tu C, Campi G, Raychaudhuri S, Varma R, et al. The immunological synapse balances T cell receptor signaling and degradation. Science 2003;302:1218—22.

[78] Marais R, Light Y, Mason C, Paterson H, Olson MF, Marshall CJ. Requirement of Ras-GTP-Raf complexes for activation of Raf-1 by protein kinase C. Science 1998;280:109—12.

[79] Cacace AM, Ueffing M, Philipp A, Han EK, Kolch W, Weinstein IB. PKC epsilon functions as an oncogene by enhancing activation of the Raf kinase. Oncogene 1996;13:2517—26.

[80] Ueda Y, Hirai S, Osada S, Suzuki A, Mizuno K, Ohno S. Protein kinase C activates the MEK-ERK pathway in a manner independent of Ras and dependent on Raf. J Biol Chem 1996;271:23512—9.

[81] Lin LL, Wartmann M, Lin AY, Knopf JL, Seth A, Davis RJ. cPLA2 is phosphorylated and activated by MAP kinase. Cell 1993;72:269—78.

[82] Nemenoff RA, Winitz S, Qian NX, Van Putten V, Johnson GL, Heasley LE. Phosphorylation and activation of a high molecular weight form of phospholipase A2 by p42 microtubule-associated protein 2 kinase and protein kinase C. J Biol Chem 1993;268:1960—4.

[83] Bhalla US, Iyengar R. Robustness of the bistable behavior of a biological signaling feedback loop. Chaos 2001;11:221—6.

[84] Tanaka K, Augustine GJ. A positive feedback signal transduction loop determines timing of cerebellar long-term depression. Neuron 2008;59:608—20.

[85] Yang Y, Atasoy D, Su HH, Sternson SM. Hunger states switch a flip-flop memory circuit via a synaptic AMPK-dependent positive feedback loop. Cell 2011;146:992—1003.

[86] Aponte Y, Atasoy D, Sternson SM. AGRP neurons are sufficient to orchestrate feeding behavior rapidly and without training. Nat Neurosci 2011;14:351—5.

[87] Minokoshi Y, Alquier T, Furukawa N, Kim YB, Lee A, Xue B, et al. AMP-kinase regulates food intake by responding to hormonal and nutrient signals in the hypothalamus. Nature 2004;428:569—74.

[88] Cowley MA, Smart JL, Rubinstein M, Cerdan MG, Diano S, Horvath TL, et al. Leptin activates anorexigenic POMC neurons through a neural network in the arcuate nucleus. Nature 2001;411:480—4.

[89] Tschop M, Smiley DL, Heiman ML. Ghrelin induces adiposity in rodents. Nature 2000;407:908—13.

[90] Zaccolo M, Pozzan T. Discrete microdomains with high concentration of cAMP in stimulated rat neonatal cardiac myocytes. Science 2002;295:1711—5.

[91] Neves SR, Iyengar R. Models of spatially restricted biochemical reaction systems. J Biol Chem 2009;284:5445—9.

[92] Mochizuki N, Yamashita S, Kurokawa K, Ohba Y, Nagai T, Miyawaki A, et al. Spatio-temporal images of growth-factor-induced activation of Ras and Rap1. Nature 2001;411:1065—8.

[93] Nalbant P, Hodgson L, Kraynov V, Toutchkine A, Hahn KM. Activation of endogenous Cdc42 visualized in living cells. Science 2004;305:1615—9.

[94] Bacskai BJ, Hochner B, Mahaut-Smith M, Adams SR, Kaang BK, Kandel ER, et al. Spatially resolved dynamics of cAMP and protein kinase A subunits in Aplysia sensory neurons. Science 1993;260:222—6.

[95] Nikolaev VO, Bunemann M, Hein L, Hannawacker A, Lohse MJ. Novel single chain cAMP sensors for receptor-induced signal propagation. J Biol Chem 2004;279:37215—8.

[96] Zaccolo M, De Giorgi F, Cho CY, Feng L, Knapp T, Negulescu PA, et al. A genetically encoded, fluorescent indicator for cyclic AMP in living cells. Nat Cell Biol 2000;2:25—9.

[97] Cancela JM, Van Coppenolle F, Galione A, Tepikin AV, Petersen OH. Transformation of local Ca^{2+} spikes to global Ca^{2+} transients: the combinatorial roles of multiple Ca^{2+} releasing messengers. EMBO J 2002;21:909−19.

[98] Llinas R, Sugimori M, Silver RB. Microdomains of high calcium concentration in a presynaptic terminal. Science 1992;256:677−9.

[99] Marsault R, Murgia M, Pozzan T, Rizzuto R. Domains of high Ca^{2+} beneath the plasma membrane of living A7r5 cells. EMBO J 1997;16:1575−81.

[100] Janetopoulos C, Jin T, Devreotes P. Receptor-mediated activation of heterotrimeric G-proteins in living cells. Science 2001;291:2408−11.

[101] Nagai Y, Miyazaki M, Aoki R, Zama T, Inouye S, Hirose K, et al. A fluorescent indicator for visualizing cAMP-induced phosphorylation in vivo. Nat Biotechnol 2000;18:313−6.

[102] Ting AY, Kain KH, Klemke RL, Tsien RY. Genetically encoded fluorescent reporters of protein tyrosine kinase activities in living cells. Proc Natl Acad Sci U S A 2001;98:15003−8.

[103] Wang Y, Botvinick EL, Zhao Y, Berns MW, Usami S, Tsien RY, et al. Visualizing the mechanical activation of Src. Nature 2005; 434:1040−5.

[104] Gaits F, Hahn K. Shedding light on cell signaling: interpretation of FRET biosensors. Sci STKE 2003;2003:PE3.

[105] Wu YI, Frey D, Lungu OI, Jaehrig A, Schlichting I, Kuhlman B, et al. A genetically encoded photoactivatable Rac controls the motility of living cells. Nature 2009;461:104−8.

[106] Neves SR, Tsokas P, Sarkar A, Grace EA, Rangamani P, Taubenfeld SM, et al. Cell shape and negative links in regulatory motifs together control spatial information flow in signaling networks. Cell 2008;133:666−80.

[107] Blanco-Aparicio C, Torres J, Pulido R. A novel regulatory mechanism of MAP kinases activation and nuclear translocation mediated by PKA and the PTP-SL tyrosine phosphatase. J Cell Biol 1999;147:1129−36.

[108] Pulido R, Zuniga A, Ullrich A. PTP-SL and STEP protein tyrosine phosphatases regulate the activation of the extracellular signal-regulated kinases ERK1 and ERK2 by association through a kinase interaction motif. EMBO J 1998;17:7337−50.

[109] Chen R, Mias GI, Li-Pook-Than J, Jiang L, Lam HYK, Chen R, et al. Personal omics profiling reveals dynamic molecular and medical phenotypes. Cell 2012;148:1293−307.

Chapter 17

Spatial Organization of Subcellular Systems

Malte Schmick, Hernán E. Grecco and Philippe I.H. Bastiaens
Department of Systemic Cell Biology, Max Planck Institute of Molecular Physiology, Otto-Hahn-Str. 11, 44227 Dortmund, Germany

Chapter Outline

Motivation	329	Causality From Variation and Perturbation Analysis	336
Dimensionality Effects in Biochemical Reactions	330	Model-Driven Experimentation and Experimentally Driven Modeling	337
Towards a Systemic Understanding of Cellular Biology	330	Conclusions	339
Spatiotemporal Modeling of Cellular Processes	332	References	340
Spatiotemporal Quantification of Cellular Processes	334		

In this chapter we describe how spatially organized biochemical networks give rise to biological function on the cellular scale. We will first introduce self-organized reaction–diffusion systems, because many of them are used by living systems as platforms on which functionality is added. We then describe spatial patterning in biological systems, focusing on how intracellular compartmentalization enables function. Finally, we discuss how new insight into self-organization in reaction–diffusion systems can be obtained by combining microscopy and computational methods [1].

MOTIVATION

Biological systems strive against entropy by consuming energy and transforming it into order to reach a balance between influx of energy and the energy necessary to maintain a certain level of organization. This balance, in which the system is apparently 'resting,' can be defined as the state of the system. Take for example a mature multicellular organism that does not grow or develop although its cells keep dividing. A change in external conditions disturbs the balance that the organism maintains, so the organism needs to change as well. This adaptation requires the organism to sense and record the environmental and internal history, while computing and applying changes to its own conditions to re-acquire proper balance, i.e., a new state. Sensing, recording, computing and applying changes are therefore essential functions of all living biological systems. On this timescale, each cell encapsulates the biochemical reactions that evolved to perform these functions across the diffusively linked intracellular compartments.

In such a well-mixed reaction vessel, the functional possibilities are limited as all points perform the same operation simultaneously and in synchrony. Slowing down for example the diffusion of only one part of a reaction network constitutes a bridging of time and length scales, which introduces complexity to any system. Diffusion of reactants and products transmits information about a localized reaction beyond the area in which the reaction itself occurred. While each point in space is equivalent to any other, each senses a different environment and looks back to a different history. These reactions are limited only by the supply of energy and reactants, and thus Ångstrom-sized molecules can traverse meters (e.g., hormones such as adrenaline can travel large distances in an organism) and sub-second reactions can influence processes that occur at much longer timescales (e.g., seasonal cycles, such as mottling of fur). The richness of the involved reactions limits the scope, flexibility and robustness of the higher-order functions, such as a 'memory' of environmental changes, and hence the success of the organism: simple systems will be confined to simple functions, whereas complex ones might be able to resolve the challenges imposed by the ever-changing environment.

DIMENSIONALITY EFFECTS IN BIOCHEMICAL REACTIONS

The term 'cell', from the Latin for 'small room', describes the compartmentalization that enables life by confining reactions and thereby shielding them from the entropic effects of diffusion. The 'wall' of this 'small room' is the plasma membrane (PM), which in addition to defining the enclosed volume provides a reaction surface where local densities of reactants and products are higher. The PM comprises the organizational point of origin for bidirectional communication that integrates the intracellular state with the extracellular context via networks of interconnected interacting protein ensembles [2]. As a 2D surface, reactions and matter exchange function significantly differently from the enclosed cytosolic volume. A chemical example of this difference is the Beloussov–Zhabotinski reaction (BZ reaction), the prototype of a chemical oscillator, which consists of about 40 chemical reaction steps. In a well-stirred beaker, the color of the solution oscillates with a fixed period between two states (Figure 17.1). However, if 'spread' thin in a Petri dish, the reactants become an excitable medium [3]. Starting from a global excitable state, small hotspots of excitation (either triggered externally, or amplified from random fluctuation of the initial context) spread in concentric spiral waves, which can be reset by shaking the dish.

This example of self-organization in a lifeless/non-biological system has a striking impact on cellular biology and the way we try experimentally and theoretically to approach biological problems. The three main components of the BZ reaction, cerium sulfate, malonic acid and potassium bromate dissolved in sulfuric acid, are interacting far from equilibrium at the expense of energy. The interaction is sufficiently complex that the chemistry, which occurs in each point of the dish at a given time — even 60 years after its discovery — is not entirely understood. When this system is modeled, the large number of parameters yields a parameter space which is simply too large to analyze systematically. However, sensible simplifications yield an easy set of reactions that are closely related to the Lotka–Volterra equations, which govern predator–prey systems and form a cornerstone of biophysics and theoretical biology. These equations can be solved numerically in 2D and yield results that match the experiments closely with a manageable number of parameters. To introduce another layer of complexity, perturbing the system by adding methanol to the Petri dish hampers the periodic structures in such a way that they become frazzled and unstable. This chaotic behavior cannot be understood simply by looking at the waves, as they appear two-dimensional when seen from above, and necessitates expanding the model to include more parameters about the wave's 3D shape (Figure 17.1).

TOWARDS A SYSTEMIC UNDERSTANDING OF CELLULAR BIOLOGY

As humans, we are limited in our ability to fully understand any system in its totality without intense study. As a foundation of science we typically apply three fundamental techniques: reductionism, abstraction and generalization. Compartmentalization is an example of reductionism: we divide a problem into subsets with few critical components in each subset. For the critical components of a given subset, we collect experimental observations to formulate abstract rules that allow us to generalize the behavior of that subset to include the behavior of a different subset following similar rules. Herein lies the strength of the scientific principle and its danger: it allows us to deal with complicated systems by narrowing our focus until we lose

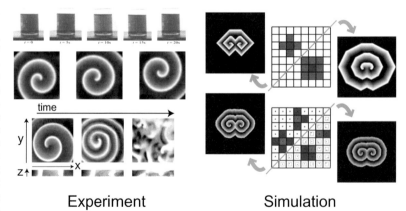

FIGURE 17.1 Beloussov–Zhabotinski reaction. Left panel: Experimental snapshot of the BZ reaction. Top row: color change signifying the cyclic reaction over a period of 20 s. Middle row: Progression of a spiral wave of the BZ reaction confined in a 2D dish as seen from above. Lower row: Adding methanol, the periodicity of the reaction becomes unstable and wave progression chaotic. This cannot be derived from the 2D projection. However, a cut in z-direction reveals that consecutive wave-fronts tilt increasingly, until the front of the following wave interferes with the back of the preceding one. This comprises an example for the type of Turing instability that leads to Turing patterns. Right panel: Cellular automaton implementation of the BZ reaction. The center column shows the local building blocks of the simulation and their connection. The displayed simulations (left and right column) are discretized to 300×300 of the small square cells depicted in the center column. The value of excitation in a given cell is distributed among its neighbors. Depending on the local definition of the neighborhood, varying spiral patterns result. Top row: Regular small or larger neighborhoods lead to too-regular large-scale patterns. Bottom row: Adding a stochastic position to the centers of each square, local neighborhoods become warped to a varying degree (left: 10% variation; right: 50% variation).

track of the bigger picture. In ethoecology, animal behavior at the single individual level is studied to deduce general principles of evolution that are also used to obtain insights into population dynamics. As a more pertinent example, although the amino acid sequences of many proteins and the interaction rules of amino acids are known, we are still unable to compute the folded structure of a protein from its amino acid sequence. The problem needs to be studied at its proper scale, hence the main tool in structural biology is measuring the structure of a folded protein with crystallography or NMR. From this structure we can derive putative interaction partners or correlate the structural differences of isomers with their respective functional differences. Conversely, complete knowledge of structure and functionality of, for example, a small G protein on a molecular level will never be sufficient to explain its role in the cellular context. To resolve this, additional layers of information, for example intracellular localization, the role of this protein in the different signaling networks, and their impact on inter- and intracellular communication, must be integrated. In this respect, systems biology is not supposed to be a new tool to gather detailed information on an isolated facet of biology. Instead, it strives to holistically collect data and paradigms from different disciplines into a more complete representation of the investigated problem at the correct scale. From this, the rules that govern systems behavior at different scales can then be extracted.

In cellular systems biology we are trying to piece together complex networks of protein interactions, which at first glance might be reduced to a few components that nonetheless yield an extraordinarily large diversity. This is inextricably linked to pattern formation, as can be seen from the fact that in vitro experiments rarely reproduce or quantify the in vivo functionality of proteins. The reason for this is the trivial seeming difference between a cell and a beaker: the beaker is well mixed, isotropic and large, while a cell is viscous, structured with shifting compartmentalization, but at least still large in comparison to the nanometer-size of a protein. In a real sense, cell biology boils down to pattern formation, because in its constant striving against entropy a cell must be continuously and consistently rebuilt. In this way, a cell depends on robust mechanisms of self-organization and thus poses the ideal platform to adapt these mechanisms to new functionality. On a larger organism scale, pattern formation is well established. In the 1960s the coloration of animals was linked to the action of so-called morphogens [4]. The interplay of apoptotic and proliferative networks via tissue spanning morphogenetic cues can result in embryonic changes. This is how tissue determines where to grow the fingers of a hand, and how to 'retract' a tadpole's tail. And the ability to differentiate tissue into veins on cue is essential for wound healing, and detrimental in case of tumor angiogenesis factor-induced tumor vascularization.

Similarly, the coupling of electrical and chemical stimuli across nerve cells with the intracellular protein state generates long-term memory in the brain. Furthermore, chemical synchronization by cyclic adenosine monophosphate (cAMP) waves analogous to the BZ reaction generates spatial organization in one stage of the lifecycle of *Dictyostelium discoideum*. When the resources of their environment diminish, single *Dictyostelium* cells exude cAMP. Neighboring cells not only can hydrolyze cAMP by a membrane-bound phosphodiesterase, but have receptors that are part of a system to sense shallow cAMP gradients with high precision and trigger a delayed but amplified cAMP release [5]. In close analogy with the BZ reaction, excitation centers are self-organized by spiral waves of cAMP. *Dictyostelium* uses this to organize a chemotactic gradient that leads to a multicellular organism. In reacting to this gradient, single *Dictyostelium* cells generate motive force by a very complex network of extracellular adhesions and intracellular cytoskeletal interactions. This last type of network is especially challenging, because correlation between spatial, temporal and compositional structure of the extracellular contacts is not clear. Does the context define the network motifs that organize the contacts, or vice versa?

For cell motility in tissue patterning the direction of cause and effect is unclear, which makes it an example of self-organization. With simultaneous upward and downward causation the large-scale patterns modulate the same local interactions from which they emerge. While the formation of the *Dictyostelium* slug on a larger scale obviously represents a self-organizational process, on the tissue level this type of intracellular pattern formation has been poorly represented in the literature, although it follows some similar principles, such as the emergence of patterning by coupling an autocatalytic amplification with a negative feedback. For example, these principles of self-organization are easily understood for cell polarity in the yeast *Saccharomyces cerevisiae* as a related phenomenon [6]. The essential component to establish cell polarity is CDC42, a membrane-bound small G protein of the Rho subfamily. Translocation of CDC42 to a subcellular spot triggers downstream cytoskeletal reorganization. Consider the interconnecting levels: a symmetric cell establishes an asymmetric protein distribution to derive an asymmetric cell shape. But how to spontaneously establish the necessary CDC42 localization from otherwise symmetric initial conditions? In contrast to the analogy of a parked car waiting to be started, proteins in cells are perpetually synthesized, modified, degraded and — most importantly — in motion. This refers to diffusion as well as a protein's 'state', as defined by its binding partners and post-translational modifications. In the case of the CDC42 protein these are different but interlinked processes: as a G protein it can exist in a GTP- or GDP-bound state, and as a protein with a hydrophobic part it exists in a steady state with a membrane-bound and a cytosolic fraction. It is important to stress that both kind of state are caused by dynamic and often cyclic processes, which are mediated by further binding partners. Guanine nucleotide binding is regulated by guanine nucleotide exchange factors (GEFs) and GTPase-activating proteins (GAPs), while scaffolds such as BEM1 increase the proximity of CDC42 and its GEF CDC24. Under 'normal' cellular circumstances these reactions are balanced out to remain stable. However, since BEM1-binding is CDC42-GTP dependent and the CDC42−BEM1−CDC24 complex favors CDC42-GTP binding, this positive feedback carries the potential to switch CDC42 activity and recruitment to a different mode. As Turing described for coupled reaction-diffusion systems [7], a shift in a parameter (e.g., expression level of CDC42) can destabilize a system, akin to a running car being shifted into gear. If the cell operates in this mode, a random fluctuation of CDC42-GTP can amplify into a sharp localization and can in the process deplete the cell of cytosolic CDC24 via diffusion. This is a method of communication, because it stops distant, less-pronounced spikes in CDC42-GTP concentration from growing further, delivering the 'message' that there exists a more successful competitor in the neighborhood. The result again is a stable dynamic cyclic process of binding/unbinding, albeit with a different localization pattern. The cytosol in this example serves two functions: in the inactive mode it smoothes the random fluctuations because of its rapid exchange of material. In the active mode its role is a medium for fast information transfer to suppress other sites of CDC42 localization if the fastest-growing site can, via CDC24 depletion, 'communicate' its success faster than the growth rate of its competitors. However, the cytosol can only achieve this in the presence of its complement: the plasma membrane with its reduced dimensionality which acts as a template for processes on a different timescale than in the cytosol. Another example for this principle is the generation of a cell-spanning gradient by anchoring the point source of activity at a specific site as in Fus3-phosphorylation in yeast, or pheromonally activated transmembrane receptors in the PM that diffuse at a timescale different from the spreading of their activity signal. In both cases perturbations of the steady state occur at the smallest relevant scale − post-translational modification in the form of, for example, phosphorylation of an amino-acidic residue. Furthermore, in both examples these minute changes can be responsible for a change in the cell's fate, be it differentiation, proliferation or death. Mitosis is an example where the spatially homogeneous nucleus of eukaryotic cells undergoes dramatic organization, which starts with mitotic spindle assembly and chromosome separation and leads to cytokinesis. In each of these processes multiple interaction networks govern the spatial restructuring, while at the same time the spatial organization triggers new downstream components. Specifically, the formation of microtubules into the mitotic spindle is a necessary step before microtubule-binding proteins can separate chromosomes, but timing does not imply causality, because the underlying processes can be interlinked. In all these examples the scientific goal to determine the exact configuration of the process, be it a snapshot (in time) or a zoom-in (in space) needs to be tempered with knowledge of the bigger picture to avoid postulating a standalone cause for a situation emerging from a complex interplay of spatiotemporal dynamics.

SPATIOTEMPORAL MODELING OF CELLULAR PROCESSES

Abstraction of a biological phenomenon means that its observation must be quantified and transferred into the common language of mathematics, independent of observer or instrument. With an equation that describes a hypothesis of the process and a set of numbers that describe a context, the outcome of a new experiment can be predicted and the actual results compared. This functions as a test of whether the abstraction of a process was valid. A misconception has long been perpetuated that complete knowledge of the parts of the process at its smallest scale can be extrapolated to the end result. In case of a simple chemical reaction it is impractical to consider the collision frequency of single molecules and their electrostatic interaction forces to derive the stoichiometry and speed of this reaction because of Avogadro's number of molecules. But the large number of interacting molecules in a small volume also saves the day, as an abstract value for the probability of a reaction to occur and the initial concentrations adequately describe the outcome of the reaction.

Thus, in understanding a dynamic process we choose a level of abstraction and formulate partial differential equations (PDEs), which describe the dynamics of the system at this level. Mathematically, a differential equation assumes that by dividing space and time into infinitesimally small elements, realistic behavior can be predicted if the current state is known. As is usually the case, if there are no analytical solutions numerical methods are employed. Here, the elements of time and space are chosen small enough to guarantee accuracy, but large enough so that the future state of the system to be modeled can be calculated in a reasonable timeframe. Non-linear dynamics and chaos theory (as used to describe turbulent flow, weather, etc.), however, have illustrated the limitation of this approach. If the system is sufficiently interconnected and has enough components, small computational errors propagate and accumulate so that the system may end up in different states depending on the initial conditions. In other words, local effects determine the global pattern. Unfortunately, biological systems tend to exhibit complex network motifs and a large number of components, making this kind of modeling extremely challenging.

With the increase in available computational power, Monte Carlo methods have gained in importance. Here, many stochastic snapshots of the system are evaluated to derive dynamic information. While this is robust to error propagation, the needed computational power is large in relation to the derived results. Even more computationally exhaustive is the method of molecular dynamics. Here, every component molecule and the forces acting upon it are tracked. This can give exceptional insight into the behavior of a small number of particles for short periods, but is very dependent on the assumptions and simplifications used for molecular shape, behavior and interaction.

As a phenomenological approach, based on the experiences with Turing patterns at a time when computational power was limited, the so-called cellular-automaton approach was developed. Similar to PDE, in this approach space and time are coarsely discretized, but so is the variable in question (e.g., concentration of protein). It thus becomes a state, in the most fundamental case binary (0s and 1s). The state at a certain position x at a later time-step $t+1$ is then determined by the state of x and its neighbors at time t by a fixed set of rules that are loosely related to the underlying PDEs. The Wolfram automaton is a simple 1D cellular automaton where time-evolution can be visualized as the second dimension in a 2D plot. Different rules lead to different behaviors that can, for example, be mapped to pigmentation patterns of mollusc shells [8].

As described previously, biological systems tend to exist in a steady state that is 'waiting' to be perturbed and thus to switch to a different mode of activity. This resembles a loose description of so-called excitable media, which are one example that has given prominence to the cellular automaton. All excitable media share a common 'rule set' that generates global patterns from extremely localized interactions. It can be reduced to the following for one observable substance:

1. Every point in space (called cell) is either in an excitable state with an excitation value of 0, or in the excited state with a value between 1 and n.
2. Excitation spreads via 'diffusion', i.e., the value of each cell is distributed among the cells in its neighborhood of a certain radius.
3. If a cell has state 0 and 'receives' excitation via diffusion, it becomes excited and its value is set to n; otherwise, its excitation value decays by 1 (a loss per cell additional to the redistribution by diffusion).

This phenomenological description of an excitable medium is the algorithm, which can be translated to a cellular automaton and renders BZ-like 2D patterns [9]. Here, rule 3 is an oversimplification of the complicated auto-catalytic process on a fast timescale, which is visible as a change of color in the system and can be triggered by the presence of minute traces of the activating component, but once this conversion is locally done, the system needs some time to recuperate (refractive time) before it again becomes excitable. This refractive time (e.g., the specific value of n) combined with the decay strength implemented in rule 3 and the diffusion coefficients determines the spatial distance between the spiral wave fronts.

The interaction can be extremely localized (here: activation within each cell and diffusion to its four neighboring cells) and still affect a global pattern. We can use this simple system to illustrate the effect of small differences in locality of this rule set being reflected in the global pattern. By changing the size of the diffusion radius or randomizing the shape of this neighborhood, the shape of the resulting large-scale spirals is determined (see Figure 17.1).

In this example, we minimize the system to one observable, which hides an interaction network of multiple agents. Expanding the above rule 3, one realizes that the topology of this kind of interaction network between the involved agents can also be a determining factor of a global pattern. By including a second substance that depends on the concentration of the first and at the same time influences it, one arrives at the classic activator–inhibitor type of system. As an example, two morphogens that act as activator and inhibitor of undifferentiated cells of a homogeneous tissue can form a spatially distinct pattern (fingerprint, retinal blood vessel network) of differentiated cells. These patterns are generated by the network topology of interacting substances. However, on the one hand its specific implementation follows stochastic fluctuations, which is why identical twins have distinct fingerprints and retinal patterns. On the other hand, global parameters (e.g., temperature, foreign substances or overlying gradients) can also affect the pattern formation and result, for example, in deformations in embryonal development, or can determine patterns of revascularization that are essential in wound healing but disastrous when initiated by tumors.

Similar pattern formation processes are also present at an intracellular scale. For example, a receptor tyrosine kinase (RTK) with a transmembrane domain to sense the presence of its extracellular growth factor ligand can be activated by ligand-induced di- or oligomerization, which results in autophosphorylation of the receptor by its intrinsic tyrosine kinase activity. As a balancing reaction to this activation, cytosolic protein tyrosine phosphatases (PTPs) inhibit this RTK activity by dephosphorylation of the receptors. RTK oligomerization and phosphorylation-induced increases in activity can be described as a positive feedback of RTK activity; this activity can also influence the inhibiting strength of PTPs either by inactivating PTPs via reactive oxygen species (ROS) or by translocating PTPs to the PM via adaptor modules that bind phosphorylated active receptors to in turn enhance the activity of the bound PTPs by phosphorylation. The system of RTKs interacting with different PTPs is an example where the inversion of an interaction from activating to inhibiting changes the global activity pattern from silenced but excitable to local activity hotspots. If, however, we track the localization of a PTP in response to its recruitment to the PM, the activity pattern becomes a time-dependent transient phenomenon (Figure 17.2). Before ligand binding, the system remains stable and activatable. Weak random activity fluctuations of the RTKs are countered by the cytosolic PTPs. Ligand binding constitutes a strong fluctuation as it continuously activates those RTKs to which the ligand is bound. As the system is a balance waiting to be tipped, this perturbation spreads more rapidly than diffusion of activated RTKs from this point of origin would allow. This is mediated by the inactivation of PTPs in close proximity to the PM via short-ranged action of ROS. The diffusion-limited translocation of a PTP to the PM by RTK activity-triggered binding to phosphotyrosines via an adaptor module again increases the strength of PTP activity in the globally activated state of RTK. As a result, in regions where fluctuations reduce RTK activity slightly, PTP activity further reduces RTK activity and hence ROS production. The system is still able to maintain hotspots of activity because the PM now has a pool of inactive, diffusing RTKs that can be locally reactivated. To test this hypothetical system and to put constraints on the physicochemical parameters of the modeling, experimental observations of the involved components are obviously crucial. However, reasonable, simple considerations about the biochemical properties of RTKs and PTPs and the interdependence of their activities allow the exploration of the possible manifestations of the system on the micrometer scale.

SPATIOTEMPORAL QUANTIFICATION OF CELLULAR PROCESSES

The macroscopic activity and localization patterns that give rise to cellular function emerge from the interaction and mobility of nanometer-sized molecules. These local molecular properties are themselves modulated by the global cellular patterns that they generate. Such simultaneous upward and downward causation is typical for self-organized systems. To understand the principles and mechanisms by which a cell operates, it is therefore necessary to quantify the progression of processes starting from a certain state with spatial and temporal resolution at the molecular level without losing sight of the cellular context.

Owing to its ability to resolve micrometer-sized structures, biological microscopy has been instrumental in the discovery and understanding of living systems. Microscopy as an extension of our eyes has quickly become a window into the subcellular world. At the dawn of microscopy, Robert Hooke, in 1665, published the first observation of the basic unit of life: the cell. One of the oldest preserved drawings is from the prolific microscopist Antonie van Leeuwenhoek (1632–1723), who observed a *lumen* in the red blood cells of salmon. The discovery of the nucleus showed that compartmentalization does not end at the cellular level but is a pattern that repeats within itself.

As staining agents with better contrast became available more than a century later, chromosomes, which strongly adsorbed basophilic dyes, were identified inside the nucleus by Walther Flemming and named by von Waldeyer-Hartz.

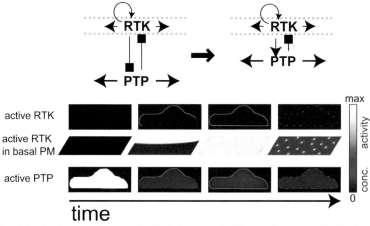

FIGURE 17.2 Cellular automaton simulation of an RTK-PTP-interaction in a 3D virtual cell, depicted in the simplified reaction schemes on top. Time progresses from left to right column, with a coloring table for individual pictures as last column. Top row: vertical section of the cell; normalized RTK activity; Middle row: top view of the basal membrane; Both: dark — no activity, bright — high activity. Bottom row: vertical section of the cell; normalized PTP concentration; dark — no concentration, bright — high concentration. Continuous PTP activity suppresses RTK activity, which remains below an activation threshold. A ligand-binding event perturbs the system above activation threshold. High RTK activity lowers PTP activity proximal to the PM, allowing the autocatalytic RTK activation to spread. Although initially localized, activation signal promptly spans the whole PM. High RTK activation also triggers PM binding of PTP via adaptor protein (lower row). This increases PTP concentration by sequestering PTP activity in close proximity to the RTK that raises RTK activation threshold. The PTP translocation transforms the double-negative feedback of RTK—PTP interaction to an activator—inhibitor topology. There, RTK activity in regions with slightly lower RTK activity is suppressed by the faster-diffusing PTP tethered to the membrane. A stable Turing pattern forms from this lateral inhibition. This is seen most easily in the middle row, where the RTK activity declines everywhere (white to yellow to blue) apart from the local hotspots. Here local RTK activity is sufficiently high to maintain its activity state, because the autocatalytic activation outperforms deactivation by PTP.

The term chromosome (colored body) that we still use today reveals that cellular structures were first discovered and named according to their appearance (phenotype), and only later were their functions studied and understood, mostly by other means. Such a merely descriptive usage of microscopy evolved to be the quantitative tool that it is today, suitable to understand the operation of spatially organized intracellular communication [10].

Fluorescence microscopy provides an unsurpassed way to observe the dynamics of molecules inside living systems and has therefore become a cornerstone of cell biology. In this technique molecules are specifically labeled with agents that absorb energy as photons of a given wavelength and re-emit it as photons of a different, usually red-shifted wavelength. In immunofluorescence and related methods, labeling is accomplished by using specific biomolecules such as antibodies to target fluorescent dyes to a protein of interest. As an excess of labeled material needs to be applied to the sample to ensure saturation of binding sites, washing of unbound material is essential to obtain contrast. While recent developments in chemical genomics have provided ways of accomplishing such labeling in living cells, most labeling methods are not suitable for live cells. Discovery and tooling of the green fluorescent protein (FP) was therefore a breakthrough in fluorescence microscopy. FPs can be fused to a protein of interest by genetic manipulation of cells to create a fluorescent chimeric version that can be expressed and imaged in living organisms.

With the ability to visualize molecules in living cells, the challenge then is to derive from the acquired fluorescence images information about molecular mobility, such as diffusion and transport, and molecular state such as interaction, conformation or post-translational modifications, to understand how biological patterns arise. Interactions, for example, could be investigated by simultaneous imaging of two or more labeled species. However, co-localization of different molecular species in the same spatially resolvable volume does not mean that interaction is occurring. A standard microscope is an optical low-pass filter that removes the fine details of an object, preventing the observation of sub-wavelength structures (Abbe diffraction limit). A point-like light source in the sample is reconstructed by a microscope into a larger image called point spread function (PSF), and its dimensions are related to the optics of the microscope and the wavelength of the light. For standard wide-field microscopy, the PSF is 250 nm in the lateral dimension (perpendicular to the beam propagation), and in the longitudinal dimension spans the height of a typical cell, resulting in a volume of about 5 fL. Such dimensions are much larger than the size of a typical protein (80 kDA equivalent to 10^{-7} fL), and therefore two apparently perfect co-localizing objects in an image can be far apart on the molecular scale.

Using a pinhole to reject out-of-focus light, confocal microscopy reduces the longitudinal dimension to <1 µm, enabling optical section of a typical cell into several slices. In spite of this achievement, the size of a typical confocal volume (1 fL) is still seven orders of magnitude larger than the size of a typical protein. Novel super-resolution techniques such as PALM/STORM or STED can routinely go down to a resolution of 20 nm. In PALM/STORM the ability to localize single fluorophores with nanometer accuracy is exploited to reconstruct a high-density super-resolution picture from a series of images of sparsely excited fluorophores [11,12]. In STED, the PSF of a confocal laser scanning microscope is sharpened by stimulated emission depletion of the fluorophores far away from the center of the PSF [13,14]. To identify and localize single molecules, both imaging methods benefit if the density of molecules to be detected is low. Longer acquisition times are required to attain the resolution needed to distinguish between interacting and nearby molecules (or a protein conformational change), thereby precluding its use to monitor fast cellular dynamics.

Although most of the aforementioned techniques have pushed the resolution well below the diffraction limit, a different subfamily of methods has been developed to directly assess functional observables such as diffusion, conformation and interacting populations of molecules. In a first group of functional techniques, the macroscopic fluorescent steady state is locally perturbed by photochemical means while monitoring its re-equilibration over the cell. Such recirculation of fluorescent species is a macroscopic reporter of the mediating intracellular processes: transport, binding and diffusion. The oldest and most common of these techniques is fluorescence recovery after photobleaching (FRAP), in which a region of the cell is first bleached using a strong laser beam for a short time; the recovery of the fluorescence intensity is monitored afterwards [15,16].

In a second group of functional techniques the macroscopic fluorescent steady state is dissected in space and time into the microscopic fluctuating parts that are averaged out on a larger scale. Such fluctuations are a product of the discrete molecular composition of the observed system and can be harvested for information about the underlying physicochemical processes that give rise to them. In an ergodic system, the observation in a limited region of space of a sufficient large number of events is equivalent to observing a statistical sample of all the ensemble's possible events. This principle is used in fluorescence correlation spectroscopy (FCS) to obtain absolute concentrations and diffusion times by measuring the motion of individual fluorescent species through a small (~ 0.5 fL) confocal volume [17,18]. With two-color FCS, the co-diffusion of proteins through the confocal volume will become coincidences in the intensity time traces [19].

Interaction (direct or mediated via a third protein) can be inferred from a high coincidence rate, as the co-diffusion of independent proteins is statistically rare.

While FRAP and FCS aim to obtain molecular information by observing the ensemble, they still cannot observe different conformational states directly, or distinguish between direct interactions and those mediated via a large protein complex. Foerster resonance energy transfer (FRET) measures molecular proximity by monitoring the far-field photophysical effects of the near-field dipole–dipole coupling between two fluorophores (usually called donor and acceptor). Such effects include the quenching of the donor fluorescence [20], the sensitized emission of the acceptor and the reduction in the donor fluorescence lifetime [21], and others [22]. As energy transfer only occurs when the distance between fluorophores is in the order of a few nanometers, FRET effectively senses proximity in a volume (10^{-5} fL) relevant to uncover interactions between proteins. In addition, FRET has been used as a basis for sensors that use conformational changes to relay information about activity, pH or concentration of molecular species [23].

In summary, functional fluorescence microscopy allows a cellular dynamic topographic map of proteins to be overlaid with topological information on the causality that determines protein state [24]. Such a state consists of mobility and population evolution of the different interacting or modified proteins. Altogether, this state is the molecular basis of the cell-spanning patterns that generate functionality on the micrometer scale [25].

CAUSALITY FROM VARIATION AND PERTURBATION ANALYSIS

Each cell is an individual entity that may respond to a signal differently from its neighbors, even in a clonal population exposed to the same environmental conditions. One of the sources of such variation are the stochastic properties of chemical reactions at low concentrations [26]. Many of the components involved in signaling are in such low numbers that actual reaction speeds can differ significantly from the average. The other source of variation arises from extrinsic factors such as its microenvironment. The accumulation of small differences in a series of events leads to an overall cell-to-cell variation in the internal state determining their response properties [27]. Cellular heterogeneity has been observed in a variety of cell types, ranging from bacteria to mammalian cells.

For example, in gene expression the process by which mRNAs and proteins are synthesized is inherently stochastic. This stochastic nature introduces fluctuations around the mean level of mRNA, causing identical copies of a given gene to express at different levels. This noise in gene expression is a substantial factor of population heterogeneity and a cause of variability in the cellular phenotype. Interestingly, the noise in the allele transcription in diploid cells is gene specific and not dependent on the regulatory pathway or absolute rate of expression. Moreover, mutations can alter the noise in gene expression, suggesting that it is an evolvable trait that can be optimized to balance fidelity and diversity in eukaryotic gene expression [28,29].

Methods that rely on the measurement of cell population averages provide a weighted summation over all possible states. In contrast, single cell techniques such as microscopy, flow cytometry and recently developed mass cytometry can assess such variation by providing cellular and, in the case of microscopy, subcellular information. The first clear advantage is that multimodal populations can be properly characterized. More importantly, single cell observations of the state of a set of intracellular components can provide information about the network connectivity between them.

A common feature of almost all signaling networks is the presence of feedback loops, and these are the basis for any regulated response [30,31]. An aspect of this network motif is that the response of any given protein within a feedback loop contains information about the dynamics of the network as a whole. Each protein can be considered as an embedded probe that relays the coherent response of the network. If the state of two proteins is monitored in a single cell, their relationship will be dictated by the connectivity in the network to which they belong. The paired observations in multiple cells will therefore represent a statistical sample of the landscape of possible relations compatible with the underlying network. While causal information cannot be derived from such correlations (correlation does not imply causation), the number of possible networks that can describe the system can be significantly reduced.

To obtain causal connections from observational data, extra information is needed. A priori knowledge, such as partial causal connectivity (or lack of), can be used to derive topological motifs. This has proved useful to detect positive feedback loops in tyrosine phosphorylation networks even in the case of the elusive autocatalytic loops [32]. Other means of deriving causality are by acquiring and analyzing richer datasets in which the time evolution of the system is followed [33]. In such time-dependent datasets the observation of subsequent events allows the analysis of flow of information and therefore enables the reconstruction of causality that is missing from static data.

While causality can be derived from correlative data in certain restricted cases, in general it must be derived from data after perturbation of the system. Here, one or more parameters of individual elements in a signaling network are perturbed, such as the activity level or concentration of a protein [34,35]. The effects of such interventions on the network components reflect the causal connectivity among them. A perturbation in a single node propagates throughout the network following its topology. By measuring the response of other nodes, this propagator can be determined. In perturbation analysis, the network components (or a linearly independent set of components) are sequentially perturbed to probe the network. By combining the propagators obtained from independent perturbations, the intrinsic structure of the system can be used to predict the system response to different input data.

Whereas in correlative studies the size of the dataset scales linearly with the network size, in perturbation studies this relationship is quadratic, as the number of possible connections scales in this way. Effective implementation of network reverse engineering methods to resolve network structure therefore demands a high-throughput automated system for data acquisition. Recently developed fluorescence lifetime imaging microscopy on cell arrays (CA-FLIM) is one such method [32]. Cells are plated on microscopy-compatible chambers where chemical (inhibitors) or genetic (siRNA, cDNA) perturbations have been spotted in a high-density grid (up to 25 spots/mm^2). Cells take up the material locally from the spot, creating an addressable array of perturbed samples. Fully automated frequency domain FLIM is used to traverse the array accurately determining in each spot the post-translational modification state with subcellular resolution. As was shown for the epidermal growth factor tyrosine phosphorylation network, CA-FLIM can provide correlative data on signal propagation. As FLIM measures the post-translational modification extent with spatial resolution, CA-FLIM is the basis to resolve the local structure of signaling networks in spatially regulated cellular processes. The current ability to measure 10^4 samples per day provides a way to derive causality in large networks (10^2 components) by perturbation analysis.

MODEL-DRIVEN EXPERIMENTATION AND EXPERIMENTALLY DRIVEN MODELING

From the causality maps, further insight into the spatio-temporal dynamics of biological processes can be obtained by iteratively integrating experiments with simulations. Partial information about the reactions involved in a biological process is used to simulate the biological process of interest and the resulting insights inspire new experimental observations. Biological parameters are obtained by fitting the data to hypothetical models, and new experiments that resolve the uncertainties/degeneracy of the models can be then performed.

As described earlier, information about causation and the agents involved can be extracted from the measurement of large populations and their variation. If the underlying agents and their interaction are known, a cellular automaton offers a rapid approach to recreate the observed behavior in silico on a single cell level. However, the prime challenge is currently the inverse process. This means, can we compute the network's causality if we know the topography of the systems and the players involved, even from single cell measurements? This would mean exploring the potential behaviors of all possible network motifs in simulation and then looking for distinguishing features that allow the exclusion of certain motifs. For the example of an RTK interacting with PTPs given above, that would mean deducing a double-negative feedback from an observed global activation pattern and excluding that for an observed formation of activity hotspots.

As a further example of deducing a network architecture from observed patterns, let us consider cell division in *Escherichia coli*. To measure the size of the cell and at the same time provide a spatial cue on where to divide, the MinD, MinC and MinE proteins self-organize into a cell-spanning oscillation of a MinD wave from pole to pole. A reconstituted system of MinD, MinE and ATP proved sufficient to generate 2D-patterns in vitro [36]. This biological example is akin to the BZ reaction and can be solved similarly to demonstrate the power of modeling in deriving a network topology from scant experimental knowledge. The reconstituted system in vitro exhibits waves of membrane-bound MinD/MinE with the following simplified characteristics:

- MinD can bind to the membrane, slowing its diffusion.
- MinD density increases slowly to a peak and drops more steeply as seen from the direction of the traveling wave.
- MinE density increases almost linearly towards the trailing edge of the wave and drops off very sharply at the end.

In this system we have two observables (MinD and MinE) in two states (free and bound). In trying to model this system, let us consider these two species in two states (free, bound; in case of MinD to the membrane, in case of MinE to MinD). This simple network therefore has four nodes pairwise-linked by binding reactions, resulting in four reaction rate constants, where each reaction rate can be influenced by each node in a positive or negative way (Figure 17.3). As long as no node influences its attached reaction rate, this system comprises linearly coupled differential equations, and a systematic check of these possibilities shows that no combination leads to traveling waves. Knowledge of the system further limits the amount of sensible connections: e.g., the ATPase MinD is known to bind to the PM in ATP-bound form and MinE mediates the MinD-ATPase activity, thereby increasing MinD-PM dissociation, and favorably binds to PM-bound MinD dimers. By this reasoning, MinE can be considered an antagonist to MinD-PM binding.

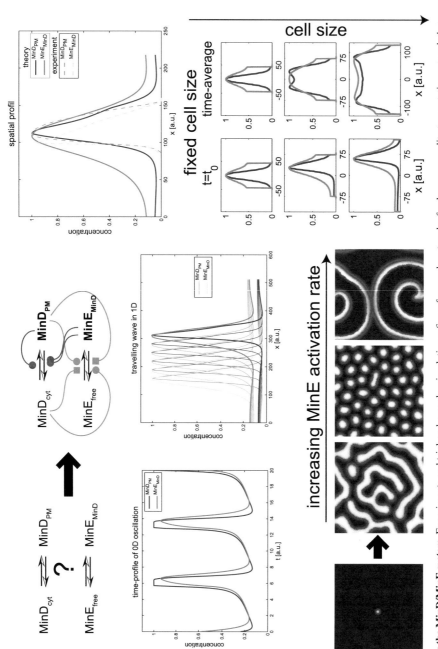

FIGURE 17.3 **Modeling the MinD/MinE system.** Experiments and trial-and-error by simulation refine a putative network that reproduces key features of the experiments. Main feature of the causality network are the red arrows: a positive feedback of MinD-dependent MinD recruitment to the PM and a more than linear increase of MinD–PM release dependent on MinE–MinD binding. Gray arrows depict the coupling of MinE activity to MinD–PM binding (mainly PM-bound MinD is activated by MinE, and MinE is released from MinD when not at PM). The network exhibits oscillations in '0D' of well-mixed reaction vessels, and wave progression in 1D that does not quite match the experimentally measured profiles. However, in 2D spiral waves and Turing patterns can be observed, and in a restricted 1D simulation the size of the cell sharply influences the time-average of MinDE peak concentration. Only in the case of medium size does the maximum of MinE coincide with a minimum of MinD. This can result in downstream activation of cell division localization in space and time.

To achieve a spontaneous increase in membrane binding of MinD in the presence of its antagonist, MinD binding must be self-referencing. This means that the presence of bound MinD favors further MinD binding, which can be caused by, for example, dimerization of MinD in the bound form. This introduces a first non-linearity in the reaction network. However, this needs to be balanced by a similar non-linearity on the antagonistic side, otherwise only high levels of membrane-bound MinD will be a stable solution. As postulated in the above paper, rapid rebinding of free MinE to MinD at high MinE concentrations is likely. This — partly by experimental knowledge, partly by exclusion via trial and error in simulations — reconstituted system is capable of oscillatory behavior in '0D' of a well-mixed reaction vessel and standing or travelling waves in 1D (Figure 17.3). When extended to 2D, these waves will self-propagate comparably to the reconstituted system described earlier [36]. Furthermore, a decrease in the MinE feedback strength results in localized Turing patterns. When restricting the 1D system to finite lengths, we realize that below a critical length wave propagation is impossible, whereas above a certain length multiple waves can form. Only at a certain length does the time-averaged maximum of MinE coincide with a minimum of MinD at the center of the cell, which could provide the spatial cue on where to divide, coincidentally with a cue on when the cell has grown large enough. This network operating with nm-sized molecules and reactions at microsecond timescales decides the fate of micrometer-sized bacteria after hours of growing. The above simulation hinges on the collision of scales in two instances: diffusion is about an order of magnitude slower in the membrane than in the cytosol, and recruitment of MinE to MinD is an order of magnitude slower than recruitment of MinD to the PM. The first lets a local reaction deplete a cytosolic pool fast enough to stay local, and the second is responsible for the delay of the MinE-peak in the wave with respect to the progressing MinD peak. These two assumptions need to be experimentally verified to validate the simulation. However, when considering the profile of the 1D travelling wave, a clear discrepancy with the experiment is obvious. The simplified model generates an almost symmetrical MinE profile of the 1D wave, whereas the experiment showed a linearly rising frontal edge with an abruptly decreasing tail edge. Even though the MinD profile is more similar, the MinE profile's only reproduced feature is its delay in respect to the MinD peak.

This points to the necessity for further experiments coupled with more detailed simulations to narrow down the reason for this discrepancy. For example, one could introduce further states to detail the order of MinD—MinE complex dissociation with regard to the state of PM binding and the ADP→ATP switch of MinD recently detached from the PM. Increasing the complexity of the model without increasing the number of observables at the same time runs not only the risk of over-fitting, but creates a false sense of security if the model can account for all previously observed experimental aspects. A more detailed model will create testable new features that need to be checked against experiments. In this way experiments and modeling must improve upon each other, and need to be interwoven more completely. The search for interaction partners of proteins (via screening or FRET approaches), their localization and transient translocation, the physicochemical properties that govern interaction and translocation timescales (reaction and association rates, diffusion coefficients) and their activity needs to continue. Equally essential are modeling approaches which can build on the existing experimental data to validate hypotheses and at the same time predict the missing experiments to complement current paradigms.

CONCLUSIONS

Biological systems manage to increase their own organization and maintain it at the expense of the order taken from the environment. Energy is consumed and disorder (entropy) is exported. It has been suggested that for a sufficiently complex network of chemical reactions, self-organization into autocatalytic loops will necessarily occur [37]. Such loops, as a form of closure, are a necessary topology for any system that needs to maintain itself, and hence are a prerequisite for life. Evolution shapes the path by favoring those structures adapted to the environment. In such self-organized systems, global coherence spontaneously emerges out of local interactions. This effect cannot be understood without the effect of simultaneous upward and downward causation of the involved scales. Closure is again present wrapping up the global and local scale.

Systems biology tries to describe living organisms beyond reductionism to create a coherent view of their operational principles and mechanisms. Our description starts at the scale which is most appropriate for the phenomenon that we are studying, being able to reach out to other levels when understanding increases. In the words of Sydney Brenner, we describe the system neither bottom-up nor top-down, but middle-out. Closure of biological systems makes the starting point logically irrelevant: 'There is no privileged level in biology that dictates the rest' [38]. And understanding how closure is achieved aids our understanding of living matter.

In cellular systems biology we work towards understanding how nanometer-sized molecules generate cell-spanning patterns from which biological function emerges. As we have described in this chapter, this requires bridging conceptually, theoretically and experimentally scales in space and time. A causal description of the molecular ecology that runs the cellular operation can aid in bridging the scale gap. Deriving such a causality map requires a tight iteration between experiments and simulations, directing the efforts of the former and increasing the predictive ability of the latter.

REFERENCES

[1] Kinkhabwala A, Bastiaens PI. Spatial aspects of intracellular information processing. Curr Opin Genet Dev 2010;20:31—40.
[2] Grecco HE, Schmick M, Bastiaens PI. Signaling from the living plasma membrane. Cell 2011;144:897—909.
[3] Zaikin AN, Zhabotin. Am. Concentration wave propagation in 2-dimensional liquid-phase self-oscillating system. Nature 1970;225:535.
[4] Wolpert L. Positional information and the spatial pattern of cellular-differentiation. Current Contents/Life Sciences 1986. 19—19.
[5] Manahan CL, Iglesias PA, Long Y, Devreotes PN. Chemoattractant signaling in *Dictyostelium discoideum*. Ann Rev Cell Dev Biol 2004;20:223—53.
[6] Chant J, Herskowitz I. Genetic-control of bud site selection in yeast by a set of gene-products that constitute a morphogenetic pathway. Cell 1991;65:1203—12.
[7] Turing AM. The chemical basis of morphogenesis. Philos Trans R Soc Lond B Biol Sci 1952;237:37—72.
[8] Kusch I, Markus M. Mollusc shell pigmentation: cellular automaton simulations and evidence for undecidability. J Theor Biol 1996;178:333—40.
[9] Markus M, Hess B. Isotropic cellular automaton for modeling excitable media. Nature 1990;347:56—8.
[10] Dehmelt L, Bastiaens PI. Spatial organization of intracellular communication: insights from imaging. Nat Rev Mol Cell Biol 2010;11:440—52.
[11] Kanchanawong P, Shtengel G, Pasapera AM, Ramko EB, Davidson MW, Hess HF, et al. Nanoscale architecture of integrin-based cell adhesions. Nature 2010;468:580—4.
[12] Betzig E, Patterson GH, Sougrat R, Lindwasser OW, Olenych S, Bonifacino JS, et al. Imaging intracellular fluorescent proteins at nanometer resolution. Science 2006;313:1642—5.
[13] Eggeling C, Ringemann C, Medda R, Schwarzmann G, Sandhoff K, Polyakova S, et al. Direct observation of the nanoscale dynamics of membrane lipids in a living cell. Nature 2009;457:1159—62.
[14] Hell SW, Wichmann J. Breaking the diffraction resolution limit by stimulated emission: stimulated-emission-depletion fluorescence microscopy. Opt Lett 1994;19:780—2.
[15] Axelrod D, Koppel DE, Schlessinger J, Elson E, Webb WW. Mobility measurement by analysis of fluorescence photobleaching recovery kinetics. Biophys Rev 1976;16:1055—69.
[16] Mueller F, Mazza D, Stasevich TJ, McNally JG. FRAP and kinetic modeling in the analysis of nuclear protein dynamics: what do we really know? Curr Opin Cell Biol 2010;22:403—11.
[17] Kim SA, Heinze KG, Schwille P. Fluorescence correlation spectroscopy in living cells. Nat Methods 2007;4:963—73.
[18] Magde D, Elson EL, Webb WW. Fluorescence correlation spectroscopy. II. An experimental realization. Biopolymers 1974;13:29—61.
[19] Maeder CI, Hink MA, Kinkhabwala A, Mayr R, Bastiaens PI, Knop M. Spatial regulation of Fus3 MAP kinase activity through a reaction-diffusion mechanism in yeast pheromone signalling. Nat Cell Biol 2007;9:1319—26.
[20] Wouters FS, Bastiaens PI, Wirtz KW, Jovin TM. FRET microscopy demonstrates molecular association of non-specific lipid transfer protein (nsL-TP) with fatty acid oxidation enzymes in peroxisomes. EMBO J 1998;17:7179—89.
[21] Wouters FS, Bastiaens PI. Fluorescence lifetime imaging of receptor tyrosine kinase activity in cells. Curr Biol 1999;9:1127—30.
[22] Jares-Erijman EA, Jovin TM. FRET imaging. Nat Biotechnol 2003;21:1387—95.
[23] Palmer AE, Qin Y, Park JG, McCombs JE. Design and application of genetically encoded biosensors. Trends Biotechnol 2011;29:144—52.
[24] Wouters FS, Verveer PJ, Bastiaens PI. Imaging biochemistry inside cells. Trends Cell Biol 2001;11:203—11.
[25] Zamir E, Bastiaens PI. Reverse engineering intracellular biochemical networks. Nat Chem Biol 2008;4:643—7.
[26] Elowitz MB, Levine AJ, Siggia ED, Swain PS. Stochastic gene expression in a single cell. Science 2002;297:1183—6.
[27] Snijder B, Pelkmans L. Origins of regulated cell-to-cell variability. Nat Rev Mol Cell Biol 2011;12:119—25.
[28] Raser JM, O'Shea EK. Control of stochasticity in eukaryotic gene expression. Science 2004;304:1811—4.
[29] Eldar A, Elowitz MB. Functional roles for noise in genetic circuits. Nature 2010;467:167—73.
[30] Csete ME, Doyle JC. Reverse engineering of biological complexity. Science 2002;295:1664—9.
[31] Zhu X, Gerstein M, Snyder M. Getting connected: analysis and principles of biological networks. Genes Dev 2007;21:1010—24.
[32] Grecco HE, Roda-Navarro P, Girod A, Hou J, Frahm T, Truxius DC, et al. In situ analysis of tyrosine phosphorylation networks by FLIM on cell arrays. Nat Methods 2010;7:467—72.
[33] Dunlop MJ, Cox 3rd RS, Levine JH, Murray RM, Elowitz MB. Regulatory activity revealed by dynamic correlations in gene expression noise. Nat Genet 2008;40:1493—8.
[34] Sachs K, Perez O, Pe'er D, Lauffenburger DA, Nolan GP. Causal protein-signaling networks derived from multiparameter single-cell data. Science 2005;308:523—9.
[35] Santos SD, Verveer PJ, Bastiaens PI. Growth factor-induced MAPK network topology shapes Erk response determining PC-12 cell fate. Nat Cell Biol 2007;9:324—30.
[36] Loose M, Fischer-Friedrich E, Herold C, Kruse K, Schwille P. Min protein patterns emerge from rapid rebinding and membrane interaction of MinE. Nat Struct Mol Biol 2011;18:577—83.
[37] Kauffman SA. Autocatalytic sets of proteins. J Theor Biol 1986;119:1—24.
[38] Noble D. The Music of Life: Biology Beyond Genes. Oxford University Press; 2008.

Section IV

Systems and Biology

Chapter 18

Yeast Systems Biology: Towards a Systems Understanding of Regulation of Eukaryotic Networks in Complex Diseases and Biotechnology

Juan I. Castrillo, Pinar Pir and Stephen G. Oliver

Cambridge Systems Biology Centre and Department of Biochemistry, University of Cambridge, Sanger Building, 80 Tennis Court Road, Cambridge CB2 1GA, UK

Chapter Outline

Introduction	343
Yeast for Systems Understanding of Eukaryotic Biology and Networks Applications	344
Yeast as a Model System for Comprehensive Systems Biology Studies	344
Data Analysis and Integration for Systems Biology Studies: State-of-the-art Towards Comprehensive Integration of 'Omics' Datasets	346
Need for a Clear Definition of Objectives and Experimental Design	346
Experimental Systems for Comprehensive Studies of Yeast Networks Dynamics: from Steady States to Time-Course Experiments through Perturbations	347
Comprehensive Data Analysis and Integration Methods: State-of-the-Art	348
Towards Comprehensive Integration of 'Omics' Datasets from Single Experiments	351
Yeast for Comprehensive Studies of Dynamics of Dysregulated Networks: Towards Rational Strategies and Applications in Biotechnology and Human Disease	353
Conclusions: Future Perspectives	357
Acknowledgements	357
References	357

INTRODUCTION

In February 2001, *Nature* and *Science* published two landmark papers which provided the first draft of the human genome sequence [1,2]. More than 10 years later, we are still far from understanding the exquisite complexity of human biology [3–8]. Many more initiatives to characterize human genetic and structural variations are being developed (e.g., the 1000 Genomes Project [5,9,10] and human metabolic networks are being carefully reconstructed and refined [11,12]. Moreover, it is clear that an increase in the limited coverage of human regulatory networks is required. As an example, current estimates indicate that actual protein–protein interaction maps may cover less than 10% of all potential human regulatory protein–protein interactions [13,14].

We are far from understanding not only the complexity of human biology, but also that of the eukaryotic cell and its basic mechanisms of adaptation and evolution [15–21]. 'Nothing makes sense in biology except in the light of evolution' [21]. The architecture and complexity of a biological system (e.g., single-celled, free-living prokaryote or eukaryote; parasite–host, or multicellular system) and its response to specific perturbations (e.g., disease) will always have to be contemplated in the light of evolution [16–19,22]. At this point, the humble yeast *Saccharomyces cerevisiae*, with its molecular mechanisms, biological networks and subcellular organization essentially conserved in all eukaryotes, is recognized as an optimal model for the study of the eukaryotic cell ([16,17] and references therein).

In many cases there are attempts to explain complex diseases in terms of the alteration of a single component (e.g., a 'gene'; see [23,24]); however, this is simplistic [4,7,16,25]. The application of progressively more affordable next generation-sequencing (NGS) technologies (e.g., exome sequencing and whole-genome sequencing (WGS; [26])) is already making an invaluable contribution [5,27,28] towards a more comprehensive holistic, 'systems-level' perspective (see below). First, it is important to recognize that, despite their limitations and the challenges ahead [29], NGS and advanced post-genomic technologies are already delivering clear examples of direct applications and improvements in human diagnostics and the treatment of diseases with a basic genomic component (see, for example: [30–34]). While this is opening the way towards a new era in the characterization and treatment of both Mendelian and, as yet, uncharacterized diseases [35–38], not all complex diseases will be resolved merely by the application of more powerful sequencing approaches.

The reality is that many complex traits and diseases are being revealed as multifactorial in nature [4,25,39] or involving a combination of genomic, epigenomic, and environmental factors (see next section).

Complex phenotypes may be more directly related to global 'systems' properties than they are to particular 'genes' or non-coding DNA sequences (Marc Vidal in [4]), with perturbations leading to changes in network interactions which, if not counteracted, will lead to global imbalances and complex diseases [7,16]. Essential steps towards clarifying the whole picture will be (1) the proper characterization and annotation of complex disease phenotypes, and (2) the definition of the dynamics of interactions within the networks underlying these complex diseases, towards early diagnosis and affordable treatment. Initiatives such as the definition of phenotype ontologies, in the OMIM and PhenOMIM databases are beginning to set the standards [7,40–44].

The idea that multi-scale dynamic complex systems formed by interacting macromolecules and metabolites, cells, organs, and organisms underlie some of the most fundamental aspects of life was proposed by a few visionaries half a century ago (see [45] and references therein) and systems biology is materializing as one of the major ideas of 'post-genomic biology' [4,45–47]. Systems biology is not so much concerned with inventories of working parts but, rather, with how those parts interact to produce units of biological organization whose properties are much greater than the sum of their parts [16,45–50]. For this purpose the use of model organisms in the elucidation of basic principles of eukaryotic biology and the dynamic behavior of cells subjected to specific perturbations appears essential. We submit that comprehensive experiments with *S. cerevisiae* and other model eukaryotes have the potential to unveil basic principles of internal organization at the cellular level [20], and the short- and long-term effects of perturbations and the dysregulation of networks that may illuminate the origin and sequence of events underlying complex phenotypes and diseases. These approaches will lead to direct applications in medicine and biotechnology (e.g., the early detection of imbalances; biomarker discovery [51]; see also below).

YEAST FOR SYSTEMS UNDERSTANDING OF EUKARYOTIC BIOLOGY AND NETWORKS APPLICATIONS

Yeast as a Model System for Comprehensive Systems Biology Studies

The new technologies of the post-genomic era are rediscovering the complexity of biological systems, with thousands of components (e.g., genes, transcripts, proteins, metabolites) interacting in finely tuned dynamic biological networks [16,52,53]. This is reflected in Figure 18.1, which shows the flow of genetic information (from DNA to RNA to proteins), together with the whole spectrum and exquisite choreography of biological interactions and networks in the eukaryotic cell: DNA–DNA; DNA–RNA; DNA–protein; DNA–metabolites; RNA–RNA; RNA–protein (RNP); RNA–metabolites; protein–protein; protein–metabolites and metabolite–metabolite interactions), with their dynamics and interaction with the environment being ultimately responsible for particular phenotypes. This 'ecosystem of organelles' which constitutes the eukarotic cell [54] is a direct consequence of millions of years of evolution [19,22].

A schematic representation of the eukaryotic cell, its basic architecture, the main levels of regulation at the (epi)-genome, transcriptome, proteome and metabolome, and its interactions with the environment (sensing natural fluctuations in external conditions together with those derived from interaction with other organisms, e.g., competition or cooperation) is presented in Figure 18.2. This intrinsically complex system involves the integration of mechanisms and networks in different compartments and at different levels of regulation, subjected to distributed, multilevel control [52,53,55]. The integration of comprehensive data from these levels into models with predictive and explanatory power constitutes one of the most exciting challenges of systems biology ([16] and references therein). In this context, the use of well-defined model systems, in properly designed experiments under controlled conditions, with proper data

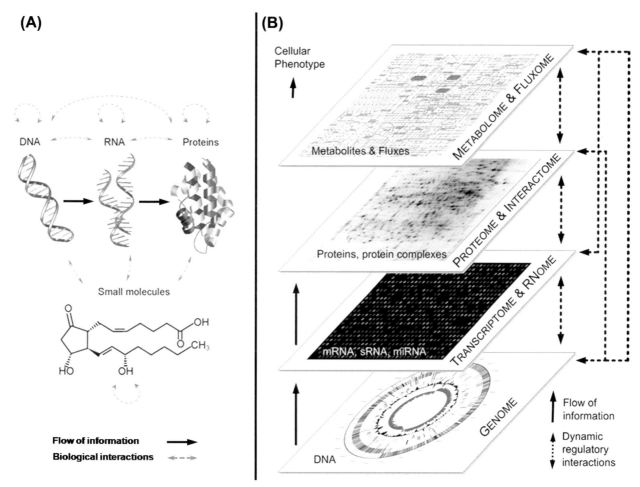

FIGURE 18.1 **Flow of genetic information, biological interactions and main molecular networks in the eukaryotic cell.** (A) Flow of genetic information (central dogma of molecular biology) from DNA to RNA to proteins [56–58] (solid arrows), and whole spectrum of biological interactions/networks at all different levels (from supramolecular to molecular level) in the eukaryotic cell: DNA–DNA; DNA–RNA; DNA–protein; DNA–metabolites; RNA–RNA; RNA–protein (RNP); RNA–metabolites; protein–protein; protein–metabolites and metabolite–metabolite interactions. *Adapted from [59] with permission from Faculty of 1000 Ltd (F1000 Ltd; http://f1000.com).* (B) Standard schematic representation of main biological entities and networks participating in the flow of genetic information (solid arrows), including whole spectrum of essential interactions/networks between them (dashed arrows), responsible for a specific phenotype in interaction with the environment. *Reproduced from [60] with permission from Springer.*

integration, using well-curated and up-to-date databases and data repositories, is of central importance [16,61].

S. cerevisiae is a species of budding yeast, a group of unicellular fungi belonging to the phylum *Ascomycota*. This yeast is being used as a model eukaryote because its basic mechanisms of DNA and chromosome replication, cell division, gene expression, translation, signal transduction and regulatory networks, metabolism and subcellular organization are essentially conserved between yeast and higher eukaryotes [61–64]. The essentials of eukaryotic biochemistry, as encompassed in a complete map of central metabolic pathways were first unveiled in *S. cerevisiae* [56,62,65,66], and a wide knowledge of the genetics, biochemistry, functional genomics and physiology of this yeast is now available [61–64,67,68]. Among the properties that make *S. cerevisiae* a particularly suitable organism for biological studies are that

it has Generally Regarded as Safe (GRAS) status and is a free-living organism with rapid growth, and simple methods of cultivation under defined conditions. It also has a well-defined genetic system with simple techniques of genetic manipulation. The yeast *S. cerevisiae* was the first eukaryotic organism for which the complete genome was sequenced [69], for which strategies for the proper annotation, curation and standards initiatives for maintenance of high quality curated databases and data repositories were implemented (e.g., *Saccharomyces* genome database; http://www.yeastgenome.org). Moreover, the majority of high-throughput postgenomic technologies (including NGS/RNA sequencing and the latest proteomics and metabolomics techniques) were first developed and validated in yeast (see [16,17] and references therein). The yeast *S. cerevisiae* is being used as a model organism to study cell growth; the cell cycle; checkpoints and

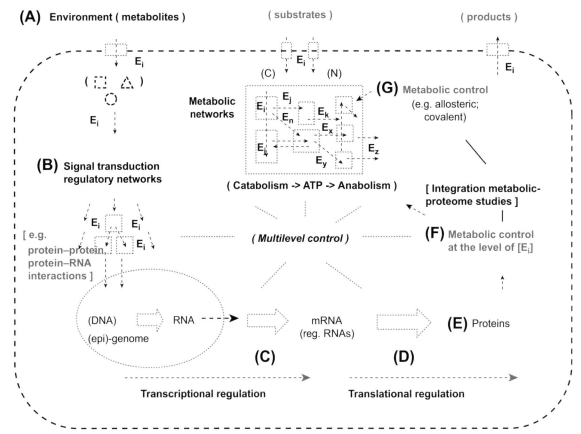

FIGURE 18.2 **Eukaryotic cell: hierarchy and levels of regulation.** Schematic description of hierarchy, main regulatory levels and networks in the eukaryotic cell, in direct interaction with the environment, subjected to distributed, multilevel control [53,55,70] (subcellular organelles other than the nucleus are omitted for clarity). (A) Environment: the eukaryotic cell is not isolated, sensing fluctuations in external conditions (e.g., concentrations of external compounds, pH, temperature) together with interactions with other organisms (e.g., via external signals; competition or cooperation). (B) Signal transduction regulatory networks (e.g., protein–protein networks; DNA–protein, protein–RNA (ribonucleoprotein) and metabolite–protein networks). (C) Gene expression at the transcriptional level, resulting in a pool of mRNAs and regulatory RNAs ('transcriptome', balance of synthesis and degradation). (D) Gene expression at the translational level, translational regulation. (E) Main pool of enzymes and regulatory proteins (i.e., 'proteome', balance of synthesis and degradation), responsible for central regulatory and metabolic networks. (F) and (G) Intracellular enzymes and metabolites concentrations (balance of uptake, intracellular synthesis and conversion) mainly responsible for in vivo regulation of metabolic fluxes and metabolic networks essential for balanced coupling between catabolism and anabolism. Abbreviations: $[E_i]$, pool of enzymes; (C) and (N), fluxes of assimilation of carbon and nitrogen sources, respectively.

their relation to cancer; cell polarity; control of pre-mRNA splicing; eukaryotic translation initiation; evolution; aging and extension of lifespan; protein folding and chaperone networks, and as a model to gain insight into the molecular pathology of neurodegenerative diseases ([16] and references therein; see also below). All these advantages are positioning *S. cerevisiae* at the forefront of the post-genomic era as a touchstone model for eukaryotic systems biology studies (see [16,61] and references therein). For an exhaustive selection of examples of advanced post-genomic technologies and the latest discoveries in eukaryotic biology using *S. cerevisiae* as a main reference model, at the (epi)-genome, genome organization, transcriptional, proteome and metabolome levels, as well as DNA–protein, RNA–protein, protein–protein, proteome–metabolome networks/interactions and internal compartmentalization levels the reader can refer to [16,17] and references therein. In the following sections, we aim to show the main strategies to convert into reality the huge potential of this reference model eukaryote for comprehensive studies of the dynamics of (dys-)regulated eukaryotic networks, with direct applications in biotechnology and human disease.

Data Analysis and Integration for Systems Biology Studies: State-of-the-art Towards Comprehensive Integration of 'Omics' Datasets

Need for a Clear Definition of Objectives and Experimental Design

A number of facts should be taken into account in all comprehensive systems biology studies (from yeast to human): (a) The need for a clear objective with clear-cut questions, balancing inductive (hypothesis-generating) approaches with hypothesis-driven experiments [71] in order to avoid becoming 'lost in high-throughput data' [53,72,73]; (b) The need for proper experimental design that minimizes the number of confounding variables and puts in place a bioinformatic and statistical strategy from the outset; (c) The need to acknowledge the uncertainties in biology. Even at the cellular level there are components and reactions still to be identified, localized, or properly annotated (e.g., ATP-producing and -consuming reactions; NAD(P) and NAD(P)H redox balances in different compartments; [3,15,56]; (d) The possibility of more than one function per gene, RNA, protein and/or metabolite, participating in different biological networks. As an example, studies on the dynamics of transcriptional regulatory networks of S. cerevisiae have revealed large topological changes depending on environmental conditions, with transcription factors altering their subcellular localization and interactions in response to stimuli, some of them serving as permanent hubs but most acting transiently in specific conditions only [74,75]; SGD database; www.yeastgenome.org; BioGRID, http://thebiogrid.org/; [16]; (e) The correct use of mutants (e.g., yeast auxotrophic and knockout mutants) and interpretation of the results obtained with them [76–79]; (f) The relevance of compartmentalization and distribution of functions between organelles in the light of evolution [22,54].

Experimental Systems for Comprehensive Studies of Yeast Networks Dynamics: from Steady States to Time-Course Experiments through Perturbations

Provided that external conditions such as pH, temperature, or the presence of toxic compounds have a negligible influence, most commonly encountered situations in nature can be summarized as nutrient starvation, nutrient limitation or nutrient excess. Free-living microbial eukaryotes have evolved in order to survive environmental perturbations, including those leading to changes in nutrient availability, with periods of starvation and nutrient limitation among the most common in nature [80].

A batch (flask) culture, most commonly used experimental system, is able to reproduce common transitions in nature (e.g., from starvation/quiescence to nutrient excess, nutrient limitation, and a new quiescence or maintenance state). Time-course experiments in batch, monitoring the different growth phases (quiescence; lag phase; acceleration; exponential growth; deceleration; stationary phase; maintenance or death), have the potential to characterize transitions in nature where the specific growth rate (μ) is continuously changing. However, their potential is hampered by operational and physiological limitations (including time-course limitations and the perturbations introduced by periodic sampling). Physiological studies in batch cultures are often limited to the analysis of dual transitions (e.g., from starvation to excess; from excess to nutrient limitation), or to the study of cells in exponential phase (a short interval in which all cells are growing at a constant growth rate, the maximum specific growth rate under the conditions tested, μ_{max}) in an essentially constant environment [81].

The main limitation of batch cultures, i.e., the difficulty of extracting proper biological conclusions from cultures in which growth rate and environmental conditions are changing, can be solved by using continuous (e.g., chemostat) cultures in steady state, in which the specific growth rate can be selected and fixed operationally, and the cells are growing (and can be long maintained) in a steady state at a constant growth rate, in a constant environment [81–83]. This characteristic has made the chemostat one of the preferred experimental systems for biochemical, physiological and functional genomic studies at the cellular level [81,82,84,85]. On the other hand, steady states alone are not able to realistically reproduce essential processes occurring during dynamic transitions (which are the most common events in nature, e.g., responses to environmental perturbations). Moreover, in a recent evolutionary study combining batch and chemostat cultivation, the majority of evolutionary mutations appeared to occur during dynamic transitions (batch culture) [86], which emphasizes their importance and the need for proper experimental designs for comprehensive studies of dynamic transitions.

The definition of an experimental system for comprehensive studies of eukaryotic network dynamics needs to satisfactorily address the issues explained above. Thus, the use of chemostat steady states combined with well-designed perturbations (transient or sustained), followed by time-course experiments and profiling (e.g., monitoring at different molecular and 'omics' levels), has advantages as a reference experimental system for next-generation comprehensive systems biology (NGSB) studies. Chemostat steady states combined with perturbations, monitoring short- and long-term responses have been applied to the elucidation of basic physiological responses and functional genomics studies [87–91]. Their huge potential will be clearly revealed with (a) the implementation of proper experimental designs with clear objectives, and (b) the integration of data from a range of 'omic' and molecular biological analyses [17,92–95] see also below). At the fundamental level, a first challenge will be the characterization of essential eukaryotic networks and their dynamics in comprehensive yeast systems biology studies, towards the elucidation of eukaryotic 'design' principles [17,53]. At the more applied level the challenge will be the characterization of early stages of dysregulation in eukaryotic networks responsible for the onset of complex diseases towards early diagnostics, and consequent informed decisions on therapies. This will also find application in the early detection and correction of imbalances in biotechnological processes (see below).

Comprehensive Data Analysis and Integration Methods: State-of-the-Art

Computational investigation of biological networks and their dysregulation involve the utilization and adaptation of concepts from network theory [96–100] and network biology ([101]; http://www.nrnb.org/). In this section we review bioinformatics methods and tools used in the construction and characterization of biological networks, and their integration with high-throughput 'omics' datasets in yeast systems biology research.

Networks datasets as a main source of primary (a priori) information and analysis

A considerable amount of information on yeast metabolic, regulatory and interactions networks is now available (e.g., *Saccharomyces* genome database; www.yeastgenome.org; [16,17,102] and references therein; [7]). Pioneering studies in yeast systems biology have exploited this information in new experimental designs, data integration and interpretation as a primary (a priori) source for the characterization and annotation of new network components. For example, based on the known components of the galactose utilization pathway and its regulation, and careful experimental design and proper integration of transcriptome and proteome profiles of the deletion mutants of the components of the pathway, Ideker and coworkers were able to identify new components and connections in the regulatory pathway [103].

New features from high-throughput data (e.g., transcriptome; proteome) can be extracted using unsupervised (i.e., no preconditions imposed) or supervised methods, followed by the use of networks for data interpretation. After properly selected pre-processing steps, specific for each technology (e.g., data normalization will be different if two-dyes or single-dye microarrays are used; [104,105]), the following methods are commonly used to summarize results from large datasets:

1. Principal components analysis (PCA) and partial least squares (PLS) methods. These are two singular value decomposition (SVD)-based methods that reduce the dimensionality of the datasets and relate them to each other [105]. A number of variations of these methods have been developed to extract biologically relevant information from high-dimensional and/or heterogeneous data. The datasets are transformed to fewer dimensions that capture the largest variation and eliminate noise, making it possible to visualize and interpret trends and patterns in large datasets [53].
2. Clustering methods [107] are used to identify subsets of cellular components (e.g., transcripts, proteins, metabolites) that exhibit similar trends across experiments. Also, the experimental conditions can be clustered to identify the conditions with similar profiles/responses. Clustering methods are often used for visualization of multidimensional data sets [53,108].
3. Analysis of variance (ANOVA) and analysis of covariance (ANCOVA) are often used to compare profiles from different states [108], when more than one factor/variable is included in the experimental design (e.g., effects of growth rate and/or limiting nutrients on gene expression). The analysis is usually done on a feature-by-feature basis (one gene/protein/metabolite analysed at a time) to decide whether the individual features have been affected by each specific change/perturbation during the experiment [53].
4. Comprehensive correlations between heterogeneous 'omics' datasets. These can reveal fundamental relationships between cellular components, for instance the existence of a limited correlation between proteome and transcriptome levels in yeast, revealing post-transcriptional regulation as a major phenomenon regulating protein levels in eukaryotes [53,110,111].

Computational tools (e.g., MATLAB; www.mathworks.co.uk; R project; http://www.r-project.org/) have many built-in functions, specialized toolboxes and packages to carry out the analyses referred to above. Specialized bioinformatics tools also provide a large range of such methods to both experts and less experienced users (e.g., GenePattern [112]; SIMCA-P+, www.umetrics.com; GeneSpring, Agilent Technologies; www.genespring.com/; Partek, www.partek.com/).

The functional relevance of genes, transcripts, proteins and/or metabolites with characteristic profiles/patterns is interpreted first, with the aid of the primary annotated networks. Metabolic and regulatory networks provide main causal relationships between interacting genes, transcripts, proteins and metabolites (e.g., see Figures 18.1 and 18.2). Changes in levels of a group of transcripts, proteins or metabolites as a response to specific perturbations can reveal the nature of the response by investigating how the responsive components are connected to each other in the biological networks. Some relevant tools are also often used to extract interesting features. Thus, for example, gene ontology (GO) hierarchical trees (http://www.geneontology.org/), make it possible to reveal, e.g., relevant groups of genes (with their specific function and/or participating in specific biological processes, GO categories or networks) with a significant biological role under the experimental conditions/perturbations tested. Networks, databases and tools to analyze and visualize them are available (e.g., *Saccharomyces* Genome Database, SGD; www.yeastgenome.org/, and GO Term Finder (http://www.yeastgenome.org/cgi-bin/GO/goTermFinder.pl); Cytoscape (open source bioinformatics software platform for visualizing molecular interaction networks www.cytoscape.org/ and BiNGO [113]). Main biological networks are available via general databases (e.g., STRING, http://string-db.org/) or specialized databases for a specific type of network (e.g., DryGIN; http://drygin.ccbr.utoronto.ca/).

In summary, annotated biological networks provide the first background information in order to interpret the results from high-throughput 'omics' studies using a combination of tools (see above). These have been used in, for example, basic functional annotation of unknown genes [114]; to investigate essential DNA damage repair mechanisms [115]; in systems biology studies on the control of cell growth [53,108,116,117] and the cell cycle [118,119]; and in comparisons of different strains used in integrative systems biology studies at the transcriptome, proteome and metabolome levels [120].

Regulatory networks (e.g., signaling and regulatory gene expression networks) (see also protein—protein and gene interaction networks)

Cellular systems regulate their activities via highly interconnected, dynamic, regulatory networks. Current knowledge is based on information generated by biochemical (bottom-up) approaches, although recent high-throughput studies together with advanced computational methods point to a larger number of new components (nodes) and connections between them (edges) [13,100].

The representation of a regulatory network is based on few features, e.g., nodes' activity levels (discrete/continuous); the relationships between nodes (directed/non-directed/mathematical function/relation) and the mathematical model implemented (stochastic/deterministic, or static/dynamic) [121]. Machine learning and validation methods have been developed for network inference in accordance with the above features [121—124]. Integration of gene expression profiles with DNA-binding profiles (e.g., ChIP-chip, [125]; and ChIP-Seq, [126]) or nucleosome positioning [127] and RNA-binding profiles [128] are increasing the accuracy of network inference methods considerably [129,130].

The topology of yeast regulatory networks has been investigated with different approaches that integrate high-throughput data combined with a wide range of methodologies. Construction of gene expression correlation networks as well as networks integrating gene expression, protein interactions, growth phenotype data, and transcription factor binding, revealed the modular organization of the yeast regulatory networks (e.g., gene transcriptional regulatory networks; [98], protein interactome networks [131]; and of the whole yeast 'system' [93,132]). At the transcriptional level, clustering of genome-wide gene expression profiles has been used to reveal the underlying topology, inference of regulatory networks, and for the functional annotation of unknown genes, based on the 'guilt-by-association' principle [133,134]. Dynamic transcript analysis has also been used to identify transcription factors (TFs) and their target genes regulating transient responses to stress [135] and specific nutritional perturbations [93].

Evidence of modularity of the yeast regulatory network, identifying regulatory modules from gene expression data, has also been demonstrated using Bayesian scoring and decision trees [133]. Jothi and co-workers, with a graph theoretical algorithm for hierarchical clustering of the yeast regulatory network, and integrating heterogeneous data sources with the network, revealed that TFs at the same hierarchical level show similar dynamics [136], and Youn and co-workers developed a clustering and maximization of likelihood-based learning algorithm from TF binding and gene expression, which can identify condition-specific regulation events [137]. As a final example, a study on network topology in combination with studies of noise in gene expression identified the tight transcriptional regulation of genes with noisy expression [138].

The main computational tools for inferring regulatory networks are often developed in platforms such as MATLAB (e.g., BayesNet; [139] or Bioinformatics (Mathworks, Inc)), and R programming language (e.g., LearnBayes; [140] and BoolNet [141]). Standalone tools are also available and, for example, yeast regulation networks and high-throughput datasets have been used to validate two of the publicly available computational tools for integration of data with regulation networks (bio-PIXIE, [142]; ChIP-Array, [143]; see below for a list of tools developed for computation and visualization of regulatory networks).

Inference and topological analysis of yeast regulation networks is a growing area in systems biology. As more experimental data from heterogeneous sources become available, new refined global models of regulation networks covering condition- and dynamics/time-dependence of the networks will be more feasible. The following section, focusing on physical and global genetic interaction networks in yeast, provides a broader perspective to inference and topological analysis of regulatory networks.

Protein—protein and gene interaction networks (PPI and GI networks)

Protein—protein interaction (PPI) networks describe physical interactions between proteins, taking place to mediate the assembly of proteins into protein complexes, or e.g., mediating signaling/regulation and transport events in the cell. Genetic interaction (GI) networks deal with pairs of proteins for which there is information that they interact functionally (i.e., the absence or presence of both proteins has a synergetic effect on the cell physiology/phenotype). The BioGRID Database (www.thebiogrid.org/; [144]) is the main repository of PPIs and GIs from a diverse selection of organisms, containing more than 60 000 unique PPIs and more than 120 000 unique GIs for *S. cerevisiae*, compiled and curated from more than 10 000 publications (as of October 2011), with a fivefold increase in number of PPIs, and a 15-fold increase in GIs since 2006 [145]. Topological investigations of the increasing number of PPIs and GIs are ongoing [145,146], together with studies providing new insights into their architecture, interrelationships and dynamics [131,147−151]. Clustering and machine learning algorithms dominate the computational methods developed for inference and analysis of PPIs and GIs [152], including hierarchical clustering, Bayesian networks, and decision trees.

Correlation studies between transcriptional (RNA) gene expression and PPI networks has revealed that co-expressed genes are more likely to encode proteins that interact [153], with combined scoring of co-expression and PPI networks revealing the order of interactions in a MAPK signaling pathway [154]. More importantly, the topology and modularity of yeast PPI networks ('interactome'), with prediction of localization of proteins, and their relations with other interactome networks in other organisms, from yeast to higher eukaryotes and humans, with an essentially conserved 'core' or essential proteins interactions, complexes and modules conserved in all eukaryotes, and their direct implications in molecular studies of human diseases, is a field of continuous investigation and constitutes one of the main challenges in the new era of systems biology ([7,16,17,45,102,155−157]). To address this challenge, different groups integrating PPIs datasets with additional levels of biological information are advancing the functional annotation of yeast genes [158,159], the identification of new functional modules in the PPI networks [160], and the capture of their dynamics during realistic physiological perturbations, namely, the dynamics of stress responses in yeast [161]. At the regulatory level, again, phosphorylation/de-phosphorylation of proteins are major regulatory events in the eukaryotic cell, and the yeast phospho-proteome (PhosphoGRID; www.phosphogrid.org/; [162]), i.e., a network of proteins that interact with protein kinases and phosphatases, was integrated with global PPI networks [163,164], revealing novel regulatory modules and patterns of interactions.

At the genetic interactions (GI) networks level, datasets obtained from three different experimental methods − synthetic genetic array (SGA) [165], epistatic miniarray profiling (E-MAP) [166], and genetic interaction mapping (GIM) [167] − have been compared and combined, with a consequent increase in the coverage and quality of GI maps [168].

Although not exempt from limitations, given that many other interactions are occurring in the cell (e.g., see Figure 18.1), a comprehensive 'compendium of interactions' with high predictive power was constructed by integrating heterogeneous data with PPI and GI networks in yeast. Park and co-workers integrated data from 30 different sources (or 'interaction' types) which included functional similarity, co-expression, synthetic lethality (GIs), PPI, phosphorylation and others [169]. In their study, 14 of the 20 novel genetic interactions proposed by their prediction algorithm could be verified experimentally. Studies of yeast interaction networks converted into Boolean networks have also shown that phenotypes can be predicted using the integrated interaction network RefRec (http://function.princeton.edu/bioweaver/; [170]).

Metabolic networks: Genome-wide reconstructed models

Metabolic networks describe the relationships between small biomolecules (metabolites) and the enzymes (proteins) that interact with them to catalyze a biochemical reaction. Metabolic networks, metabolic control and modeling of metabolic networks in genome-wide reconstructed models is a central area in systems biology [16,17,52,55,171,172].

Non-linear and dynamic models are often used to simulate metabolic and regulatory networks. However, such models usually focus on one or few pathway(s) of metabolism owing to technical limitations: e.g., the lack of empirical information on kinetic parameters and/or reaction mechanisms, high computational cost of non-linear parameter estimation, and lack of dynamic experimental data for verification of simulated results (for reviews of available methods see [121,173−176]). Stoichiometric models derived from metabolic networks are often genome wide and hence can be readily integrated with high-throughput data, Here, we will focus mainly on simulations with linear and steady-state stoichiometric models, with examples from yeast primary metabolism (i.e., catabolism and anabolism, with secondary metabolism not covered; [55]).

Metabolic networks essentially map the proteins/enzymes (often edges of the networks) interacting with metabolites (often nodes of the network), to catalyze metabolic reactions or to transport metabolites. Construction of such networks starts with the availability of the genome sequence of the organism (i.e., *S. cerevisiae*; www.yeastgenome.org; which provides the list of annotated enzymes) which is progressively refined with the aid of literature information [17,171,177,178]. A mathematical representation of metabolic networks can be used in simulations to predict the fluxes through the reactions, assuming that the culture is at a steady state (none of the metabolites accumulate in the cells and the extracellular growth medium is of a constant composition). The stoichiometry of the reactions in the model is usually known, hence these reactions can be listed as linear equations of mass conservation. Using metabolic flux analysis (MFA; or flux-balance analysis, FBA) [179,180], these equations are then used as constraints of a linear optimization problem, with selected objectives such as maximization of biomass production, minimization of oxygen or energy consumption, or other targets of interest. Experimental data on metabolite levels are also used as constraints to narrow down the space of possible solutions (each one a set of fluxes for all the reactions in the model). A variety of constraint-based methods have been developed for simulations with stoichiometric models [179,181,182].

The genome-wide metabolic network reconstructions have been distributed in non-standard formats until the SBML [183] format became available. The BioModels database [184] is one of the databases that store and distribute the models. The tools available for simulation of metabolic models are either toolboxes/packages developed for use with general computational tools (e.g., SBMLToolbox/COBRA/MATLAB; [185,186]), (SimBiology/MATLAB) or standalone platforms such as BioOpt (see BioMet online toolbox; www.sysbio.se/BioMet; [187]) and COPASI (www.copasi.org/; [188]).

Simulation of yeast metabolism via constraint-based methods has been integrated with information from viability of deletion mutants [189–191] and metabolome and transcriptome data (e.g., [192–195]). Constraint-based methods have also been used to predict protein localization [196].

The topology of yeast metabolic networks has been investigated using high-throughput data [197,198], as well as being compared and integrated with other networks [102,164,199,200].

Simulations of metabolic models are usually based on the assumption of steady-state cultures. On the other hand, industrial yeast cultures are traditionally dynamic batch or fed-batch cultures (e.g., wine production; recombinant protein production processes). Apart from other modeling strategies, the construction of a hybrid model of yeast culture based on the metabolic network and a non-linear kinetic description of nutrient uptake rates makes it possible to predict the behavior of dynamic yeast cultures [201].

Metabolic models are not limited to constraint-based modeling [202,203]. Thus, metabolic modeling from graph theory and logical modeling have been used as background knowledge for hypothesis generation by robot scientists, with experimentation and machine learning able to generate additional knowledge (e.g., identification of genes responsible for catalyzing specific reactions in the yeast metabolic network; [204,205]).

Metabolic models of *S. cerevisiae,* as a model eukaryote, can be extended with the integration of signaling/regulation events [102] and dynamics [201] as suggested by Bailey a decade ago [206], providing an optimum platform for the global investigation of metabolism [52,55]. Comparison and integration of yeast metabolic and regulation networks with PPI networks from an evolutionary point of view based on their topological properties has revealed the importance of temporal and spatial aspects [200], and large metabolic fluxes as a possible driving force [207]. Finally, the integration of kinetic data from quantitative proteomics and metabolomics, the two levels most directly involved in metabolic control, has been proposed as the most comprehensive way to integrate metabolomics in genome-wide systems biology models [55].

Network analysis/visualization tools

A search in the Nucleic Acids Research (NAR) directory of biology-related web servers databases and tools (http://bioinformatics.ca/links_directory/) using the keyword 'interaction' brought in 121 results (retrieved October 2011), the majority of these designed for storing protein–protein interaction (PPI) datasets, or predicting novel PPIs, most of them with visualization tools. A smaller subset of these has been designed to work with arbitrary networks and is able to handle more than one type of molecular interaction data. Table 18.1 shows a non-exhaustive list of the publicly available tools. As an example, GEOMI ([208]; http://www.systemsbiology.org.au/downloads_geomi.html) allows four-dimensional (4D; i.e., three dimensions + time-course dynamics) network visualization of protein interactions, and has been applied to the visualization and analysis of the *S. cerevisiae* complexome network [209]. International community efforts are progressively being developed to promote collaborations and the implementation of new open source tools for network-based visualization and networks analysis (e.g., Cambridge Networks Network, CNN, http://www.cnn.group.cam.ac.uk/; National Resource for Network Biology, http://www.nrnb.org).

Towards Comprehensive Integration of 'Omics' Datasets from Single Experiments

Data mining and processing of 'omics' datasets (e.g., transcriptome, proteome, metabolome) towards comprehensive integration (e.g., principal components analysis; proteome–transcriptome correlations) should ideally be performed with data extracted from the same experiment, using standardized, curated sampling protocols. Also, comprehensive integration will always require a prior

TABLE 18.1 Tools and Repositories for Analysis and Visualization of Biological Networks

Tool	Application	Details
Bind [210]	Repository of biomolecular interactions, 1500+ species	Static database (not updated)
BioGRID [144]	Repository of genetic and protein interactions, model organisms (~50)	Database
DryGin [211]	Repository of genetic interactions, yeast	Database
DIP [212]	Repository of protein–protein interactions from 436 organisms	Database
ConsensusPathDB [213]	Repository of biomolecular interactions, analysis of networks, 3 species	Database
WEbInterViewer [155]	Visualization and comparison of molecular networks	Web server
Torque [214]	Small-scale, blast-based multi-species PPI search tool	Web server
GraphWeb [215]	Retrival/analysis/visualization of molecular networks	Web server
String [216]	Retrival/analysis/visualization of molecular networks, 1100+ organisms	Database
Cytoscape [217]	Biological network analysis and visualization	Software
Cytoscape/CyClus3D [218]	Network clustering plug-on for Cytoscape	Plug-on
PathBlast [219]	Blast-based multispecies PPI comparison/alignment tool	Web server
NetworkBlast [220]	PPI analysis of protein complex identification	Web server
Neat [221]	Analysis of networks and clusters	Web server
BioMet [187]	Stoichiometric analysis and data integration tool for metabolic networks	Software/web server
Osprey [222]	Visualization of biological networks	Software (last update: 2004)
VisAnt [223]	Biological network analysis and visualization	Software/web server
Biological Networks [224]	Analysis and visualization of heterogeneous biological data	Software
Babelomics [225]	Integration and analysis of omics data	Web server
COBRA [185]	MATLAB toolbox for simulation and analysis of metabolic networks	MATLAB Toolbox
SMBL toolbox [186]	MATLAB toolbox for generating and modifying models in SMBL format	MATLAB Toolbox
RefRec [170]	Reconstruction and analysis of yeast biomolecular interaction network	Model distributed in SMBL format
Ondex [226]	Graph-based analysis of high throughput data	Software
PPISearch [227]	Blast-based multi-species analysis of pairs of interacting proteins	Web server
PathGuide [228]	A list 325 pathway-related resources	Web page
DEEP [229]	Integration of expression data with biological networks	Web server
GEOMI [208,209]	Four-dimensional (4D) visualisation and analysis of protein interaction networks.	Web server

knowledge of the characteristics and limitations of the techniques used. As an example, high-throughput microarrays and RNA sequencing (RNA-Seq) methods provide complete information on changes in transcriptional expression of all genes in an annotated genome (microarrays) [230] or in the whole genome (coding and noncoding regions; RNA-Seq) [231], whereas proteome quantitative methods often result in a much lower coverage ([232,233]) and frequently contain missing values. A careful integration of high-throughput 'omics' data will need to address not merely normalization methods and/or missing values, but the intrinsic limitations of the latest techniques, with the latest bioinformatics and statistical methods available. As a case example, high-throughput transcriptome, proteome and metabolome data were initially obtained from steady-state chemostat cultures in a yeast systems biology study of cell growth under different nutrient limitations [53]. Since the 'omic' data were gathered by different methodologies, specific data analysis procedures were used to minimize systematic bias and to be consistent in data processing prior to integration. Table 18.2 summarizes the pipeline and data analysis steps performed to handle each class of 'omic' data. The genome-scale integrative data analysis and modeling of dynamic processes from different 'omic' levels, obtained with the latest techniques, will be a recurrent challenge in systems biology [53,172].

Yeast for Comprehensive Studies of Dynamics of Dysregulated Networks: Towards Rational Strategies and Applications in Biotechnology and Human Disease

Advanced biotechnological processes and complex diseases, multifactorial in nature, constitute a major challenge, for which a rational strategy for comprehensive dynamic studies at the cellular level is currently lacking. Complex diseases, traditionally classified based, for example, on basic symptoms, organ localization and/or late phenotypes, are being progressively better characterized, with more insight into their primary molecular origin. However, the common trend to reduce the problem to a single causation (e.g., candidate 'disease gene'; 'risk gene' or 'risk mutation' [234–236]), overlooking the essential role of biological networks and interactions, their interplay and dynamics, needs to be addressed [7,16]. Until we understand diseases as altered states of human biological networks [237], with intrinsic dynamics and interplay, not only between them (Figures 18.1 and 18.2) but also with our microbiome ([238]; Human microbiome project; http://commonfund.nih.gov/hmp/), in constant relationship with the environment, and the reality of the recently proved human metabolic individuality [239], our vision will be incomplete. Progress is being made, and we are beginning to consider complex diseases to be more likely due to alterations of 'system' properties [7], starting with efforts directed towards better mapping and characterization of the interactomes of different eukaryotic organisms, with progressive incorporation of more layers of information [7,240]. Figure 18.1, and global maps of human binary protein–protein interactions (CCSB Interactome Database, http://interactome.dfci.harvard.edu/), provide a basis for seeking dysregulated pathways/networks [235,241], specific 'disease networks' or alteration in 'networks properties' [7,234,242]. Many of these studies and maps, however, need to incorporate more information on changes of topology, and the interplay and dynamics of networks characteristic of the specific biological problem/disease, initially at the cellular level, from comprehensive experiments under controlled conditions (see previous section).

With this perspective, we submit that complex imbalances and diseases should be viewed as shown in Figure 18.3. The genome and epigenome ('nature') define the essential networks, homeostatic states, and initial susceptibility to dysregulation, which will be subjected to a sequence of environmental perturbations ('nurture'). Perturbations will result in transient deactivation of redundant networks and activation of defense or stress responses, until a new homeostatic state is restored. Complex (e.g., multifactorial) and/or sustained perturbations that overcome the intrinsic defenses and stress responses may result in specific cascades of dysregulation, which result in highly complex acute imbalances and diseases. From here, periodic longitudinal monitoring in proper experimental systems (e.g., chemostat steady states subjected to carefully designed perturbations; see above) allow the characterization of homeostatic and perturbed networks in systems biology by comprehensive experiments in model organisms (e.g., yeast) [53,93,243]. These have the potential to unveil the origin, early stages, and dynamics of progression of complex imbalances and diseases towards early diagnosis (e.g., characterization of relevant, multiple biomarkers, at different 'omic' levels; [244]) and rational, truly affordable, strategies for prevention and therapy. At this point, it is relevant to note that, although the essential core of eukaryotic machinery is conserved in all eukaryotes [16,53,56], yeast models of human disease have shortcomings and there will always be doubts about how closely the conditions in yeast recapitulate conditions in differentiated human cells. However, the large number of molecular tools and the huge potential for high-throughput genetics and chemical screenings position yeast as a first-line approach to tackle complex human diseases. With careful experimental designs towards the characterization of biomarkers and networks dynamics, the new knowledge can be used to develop advanced experimental strategies in higher eukaryotes [245]. Based on this, the facts explained in previous sections, and its huge potential for integrative systems biology studies [17], we submit that *S. cerevisiae* is an optimum model organism for comprehensive studies of dynamics of dysregulated networks at the essential cellular level, to enable rational strategies and applications in biotechnology and disease.

Yeast will be invaluable in most challenges at the forefront of systems biology, such as (a) characterization of essential eukaryotic networks, common core to all eukaryotes: their architecture, hierarchy, interplay, dynamics, changes in topology under different conditions and perturbations, principles of adaptability, flexibility and robustness in eukaryotic networks [246]; (b) characterization of steady states and dynamics of activation of transient defense networks (e.g., stress, proteostasis networks), their interaction and interplay with essential networks, dynamics

TABLE 18.2 Pipeline and Data Analysis Steps of Transcriptome, Proteome and Metabolome Datasets Obtained from Same Experiments, Prior to Integration

Steps	Data analysis procedures	Transcriptome	Proteome	Internal metabolites	External metabolites
1	Analytical method	Affymetrix YG-S98 GeneChip	iTRAQ	GC-TOF-MS	GC-TOF-MS
	Growth rates (h-1)	0.07, 0.1, 0.2	0.1, 0.2	0.07, 0.1, 0.2	0.07, 0.1, 0.2
	Nutr. limitations	C, N, P, S	C, N, P, S	C, N, P, S	C, N, P, S
	Replicates	4	1	3	3
2	Identification and quantification	Affymetrix GCOS software [a]	Genome Annotating Proteomic Pipeline [h] (GAPP) system, Mascot score and link to the quantification data in a relational database	Mass spectral libraries, reverse match score >700 and analysis of pure standards	Mass spectral libraries, reverse match score >700, and analysis of pure standards
3	Heavily-tailed distribution	Data logged	Data logged	Data logged	Data logged
4	Experiment-wise normalization	—	Each run normalized to pooled standard	—	—
5	Sample-wise normalization	RMA [b–d] normalization	MAD (median absolute deviation) [e]	Normalized to internal standard	Normalized to internal standard
6	Missing value imputation (MVI)	Not required	MVI [f]	MVI [f]	MVI [f]
7	Measurement-wise normalization	MAD [e]	MAD [e]	MAD [e]	MAD [e]
8	Principal Components Analysis (PCA)	NIPALS [g]	NIPALS [g]	NIPALS [g]	NIPALS [g]

RMA, robust multichip average (RMA) quantile normalization; MVI, missing-value imputation. MAD (median absolute deviation) normalization; non-linear iterative partial least squares (NIPALS) algorithm.
[a] Affymetrix website [http://www.affymetrix.com]
[b] Bolstad BM, Irizarry RA, Astrand M, Speed, TP: A comparison of normalization methods for high density oligonucleotide array data based on variance and bias. Bioinformatics 2003, 19: 185–193.
[c] Irizarry RA, Bolstad BM, Collin F, Cope LM, Hobbs B, Speed TP: Summaries of Affymetrix GeneChip probe level data. Nucleic Acids Res 2003, 31: e15.
[d] RMAExpress software [http://rmaexpress.bmbolstad.com/]
[e] Yang YH, Dudoit S, Luu P, Lin DM, Peng V, Ngai J, Speed TP: Normalization for cDNA microarray data: a robust composite method addressing single and multiple slide systematic variation. Nucleic Acids Res 2002, 30: e15.
[f] Oba S, Sato MA, Takemasa I, Monden M, Matsubara K, Ishii S: A Bayesian missing value estimation method for gene expression profile data. Bioinformatics 2003, 19: 2088–2096.
[g] Wold H: Estimation of Principal Components and Related Models by Iterative Least Squares. In Multivariate Analysis. Edited by Krishnaiah PR. New York: Academic Press; 1966: 391–420.
[h] Genome Annotating Proteomic Pipeline (GAPP) [http://www.gapp.info]

([53]); Suppl. File 4. Supplementary Methods, with permission).

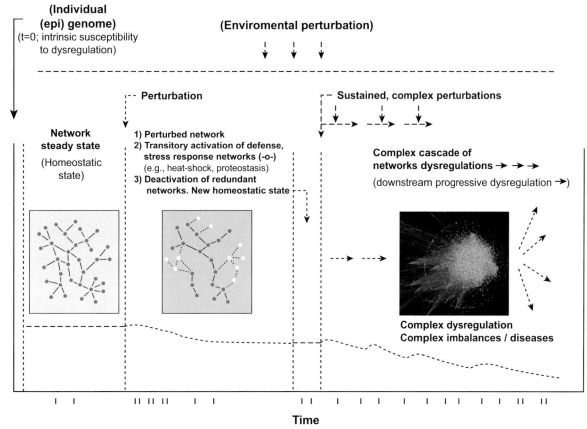

FIGURE 18.3 **Dynamics of dysregulation of biological networks towards recovery of homeostatic state and/or cascades of complex dysregulation/imbalances/diseases.** The genome and epigenome ('nature') define/underlie the essential networks, homeostatic states and initial susceptibility to dysregulation of an organism, which will be subjected to a specific sequence of environmental perturbations ('nurture') during its lifetime. Simple/mild perturbations result in transient deactivation of redundant networks and subnetworks (gray nodes and edges) and activation of defense, stress responses (e.g., heat shock, protein homeostasis, inflammatory and/or immunological networks; see new nodes and edges, -o-) until a new homeostatic state is restored. More importantly, complex (e.g., multifactorial) and/or sustained perturbations overcoming the intrinsic defenses and stress responses may result in specific sequences/cascades of dysregulations, which propagate towards intertwined essential biological networks, resulting in highly complex acute imbalances and diseases, with potential collapse of the whole networks system. Periodic, longitudinal monitoring towards characterization of homeostatic and perturbed networks in different genetic backgrounds in molecular and systems biology comprehensive experiments in model organisms (i.e., from yeast to human) [53,93,243] have the potential to unveil the origin, early stages, and dynamics of progression of complex imbalances and diseases well before the 'point of no return' (e.g., sequence of dysregulation of essential networks in neurological diseases and cancers), towards early diagnosis (e.g., characterization of relevant, multiple biomarkers, at different 'omic' levels) and rational, truly affordable, counteracting and therapeutic strategies. Human interactome network picture visualized by Cytoscape 2.5. Human microbiome networks [238] with direct interactions with the human interactome are omitted for clarity. Dataset created by Andrew Garrow at Unilever UK. Author: Keiono, reproduced under GNU Free Documentation License and Creative Commons (CC) licenses (http://en.wikipedia.org/wiki/File: Human_interactome.jpg).

of recovery of new steady states, or propagation of cascade effects towards more acute phenotypes; (c) the derivation of key 'design' principles, universal rules in dynamic networks, principles of biological circuits, modules, entities, networks and their interplay [247,248], e.g., 'bimodal gene expression activation in 'stress inducible genes' [249], systems-level circuitry principles of the eukaryotic cell cycle [250]; (d) 'humanized' yeast strains can be constructed, allowing the 'awesome power' of yeast systems biology to be brought to bear on the problems of human disease. As clear examples, such yeast systems biology studies would be of great interest to advance in the insight and progression of complex diseases using, among others: *S. cerevisiae* cells expressing the oncogene-like *RAS2* [251]; yeast cells expressing human mutant isocitrate dehydrogenase IDH, producing the oncometabolite 2-hydroxyglutarate (2HG), leading to impairment in histone demethylation, heterochromatin modifications in human gliomas and acute myeloid leukemia [252—254] for the study of interdependence of genetic and epigenetic alterations; yeasts expressing human disease alleles causing homocystinuria [255]; yeast strains with the lipid defect responsible for Niemann—Pick type C (NP-C) disease [256]; a yeast model of human protein aggregation in amyotrophic lateral sclerosis (ALS) [257]; a yeast model of Aβ toxicity in yeast [258]. In biotechnological processes, yeasts (e.g., *S. cerevisiae, Pichia pastoris*) expressing human wild-type and misfolded (variant) proteins [259] are being used for the comprehensive characterization of bottlenecks in recombinant protein production, and of protein quality control mechanisms and dynamics.

The application of comprehensive yeast systems biology strategies to the above and more advanced examples in multidisciplinary, collaborative efforts has the potential to accelerate the translation of fundamental research concepts and basic knowledge into direct applications for the benefit of society [16]. Together with this, fundamental and practical studies using yeast as a touchstone model in post-genomic research are progressively appearing, with direct applications in biotechnology and medicine. Relevant selected examples are presented in Table 18.3. For more complete information on the latest

TABLE 18.3 Yeast Systems Biology Advanced Studies and Applications in Human Disease and Biotechnology

Yeast advanced studies and applications	Reference
1) Yeast models for comprehensive Systems Biology studies of dynamics of dysregulated networks in biotechnology and disease, early detection of imbalances and networks dynamics	[51,240,243,260]
Yeast cell cycle model/design principles	[250].
Yeast model studies on eukaryotic cell growth (balanced cell growth)	[53,108,117,261,262]
Yeast breast cancer model (yeast with human mutations and increased cell proliferation) (uncontrolled growth)	[263]
Yeast expressing parasite genes and their human orthologs for protein quality control networks studies and identification of antiparasitic agents	[264, 265]
Yeast as a model for aging research studies	[266]
Yeast models of protein folding diseases and human neurodegenerative disorders	[245,257,258,267–272]
Yeast industrial strains expressing human wild-type and misfolded (variant) proteins	[259]
2) Yeast molecular, high-throughput Systems Biology studies and applications in biotechnology and human disease	[16,273]
Yeast functional screen to identify new candidate amyotrophic lateral sclerosis (ALS) genes	[274]
Yeast genetics and metabolic profiling for characterization of human disease alleles	[255]
Yeast-based assay for identification of drugs active against human mitochondrial disorders	[275]
Phenotypic screening for compounds that rescue TDP-43, α-synuclein, and polyglutamine proteotoxicity in yeast	[276]
Humanized yeast-based high-throughput screen to identify small activators for therapeutic intervention in neurodegenerative diseases	[277]
Yeast high-throughput screen for inhibitors of $A\beta_{42}$ oligomerization	[278]
Yeast genomics and drug target identification	[279]
Chemical genomic (chemogenomic) profiling. Screenings of targets genes and biomolecules using yeast as a model	[280,281]
Yeast systems biotechnology Development of cell factories for production of advanced biofuels	[282]
Systems biotechnology Recombinant protein production in yeasts	[283]
Yeast genome-scale metabolic networks in metabolic engineering and biotechnology processes From modules and network models to network responses	[284]
Metabolic engineering and systems biology systems-level approach for engineering of yeast cell factories	[285,286]

applications, the reader is referred to the latest reviews and books on the subject [16,17,61,273].

CONCLUSIONS: FUTURE PERSPECTIVES

Eukaryotic life is rich in complexity [4,8], and yeast can help us to rediscover and appreciate this: 'Do you not know that our soul is composed of harmony?' [287].

However, this exquisite complexity should not be a deterrent but, rather, a stimulus to overcome the challenges, advance progress towards specific objectives (e.g., better characterization of biological networks and imbalances responsible for a specific disease) and contribute to new applications for the benefit of society. Comprehensive yeast systems biology experiments under controlled conditions can uncover the complexity and interplay of biological networks with their dynamics, at the essential cellular level, towards the characterization of dysregulated networks in biotechnology and disease.

In order to realize the full potential of yeast as a reference model in systems biology and permit direct applications, the high level of international collaboration between yeast research groups seen in the yeast genome sequencing [68] (http://www.yeastgenome.org) and functional genomics projects [288–291] must be continued and even enhanced, with incorporation of experts from different scientific backgrounds, in comprehensive interdisciplinary efforts. To this end, the Yeast Systems Biology Network (YSBN) (http://www.gmm.gu.se/globalysbn/), a global consortium of researchers working in systems biology of the yeast *S. cerevisiae*, promotes interdisciplinary collaborations, projects and initiatives between experts in the different fields. A relevant example is the UNICELLSYS project (http://www.unicellsys.eu/), a systems biology initiative with the overall objective of a quantitative understanding of control of cell growth and proliferation. These joint initiatives and projects can be reproduced in systems biology studies at higher physiological levels [246,292,293], with more complex model organisms, higher eukaryotes (e.g., fly; mouse; mammalian systems; [294–299] and, ultimately, humans [300,301], towards progress and new applications in systems biotechnology [297], medicine and complex disease studies [240,292,293,302,303,304]. Thus, the Virtual Physiological Human (VPH) initiative represents a network of excellence [305,306]; http://www.vph-noe.eu/objectives, and VPH projects are directed towards modeling a wide range of human organs and systems in both healthy and diseased state ([307] and references therein; http://www.vph-noe.eu/vph-projects). In order to advance in their objectives, all these community efforts will always need to incorporate fundamental principles on eukaryotic networks from systems biology studies at the cellular level. In this journey, the humble yeast *Saccharomyces cerevisiae* can accompany us along the path towards new discoveries and direct applications [4,17,288,302,303].

ACKNOWLEDGEMENTS

This work was supported by BBSRC grants BB/C505140/2 and BB/F00446X/1 as well as by a contract from the European Commission under the FP7 Collaborative Programme, UNICELLSYS.

REFERENCES

[1] Lander ES, Linton LM, Birren B, International Human Genome Sequencing Consortium, et al. Initial sequencing and analysis of the human genome. Nature 2001;409: 860–921.

[2] Venter JC, Adams MD, Myers EW, et al. The sequence of the human genome. Science 2001;291:1304–51.

[3] Alberts B. Lessons from genomics. Science 2011;331:511.

[4] Heard E, Tishkoff S, Todd JA, Vidal M, Wagner GP, Wang J, et al. Ten years of genetics and genomics: what have we achieved and where are we heading? Nat Rev Genet 2010;11:723–33.

[5] Gonzaga-Jauregui C, Lupski JR, Gibbs RA. Human genome sequencing in health and disease. Annu Rev Med 2012;63:35–61.

[6] Marshall E. Human genome 10th anniversary. Waiting for the revolution. Science 2011;331:526–9.

[7] Vidal M, Cusick ME, Barabási AL. Interactome networks and human disease. Cell 2011;144:986–98.

[8] Hayden EC. Human genome at ten: life is complicated. Nature 2010;464:664–7.

[9] 1000 Genomes Project Consortium. A map of human genome variation from population-scale sequencing. Nature 2010;467:1061–73. PubMed PMID: 20981092 http://www.1000genomes.org/.

[10] Mills RE, Walter K, Stewart C. et al., 1000 Genomes Project. Mapping copy number variation by population-scale genome sequencing. Nature 2011;470(7332):59-65.

[11] Bordbar A, Palsson BO. Using the reconstructed genome-scale human metabolic network to study physiology and pathology. J Intern Med 2012;271:131–41.

[12] Stobbe MD, Houten SM, Jansen GA, et al. Critical assessment of human metabolic pathway databases: a stepping stone for future integration. BMC Syst Biol 2011;5:165.

[13] Barabási AL, Gulbahce N, Loscalzo J. Network medicine: a network-based approach to human disease. Nat Rev Genet 2011;12:56–68.

[14] Liu, Y. Y., Slotine, J. J., Barabási, A. L. (2011). Liu et al. reply to: Müller, F. J., Schuppert, A. (2011) Few inputs can reprogram biological networks. *Nature* 478, E4; discussion E4-E5.

[15] Alberts B. A grand challenge in biology. Science 2011;333:1200.

[16] Castrillo JI, Oliver SG. Yeast systems biology: the challenge of eukaryotic complexity. Methods Mol Biol 2011;759:3–28. PubMed PMID: 21863478.

[17] Castrillo JI, Oliver SG, editors. Yeast Systems Biology. Methods and Protocols. Methods in Molecular Biology vol. 759. (MiMB Series. Editor-in-chief. Prof John M. Walker). New York: Humana Press, Springer; 2011.

[18] Darwin C, Wallace AR. On the Tendency of Species to Form Varieties; and on the Perpetuation of Varieties and Species by

Natural Means of Selection. J of the Proceedings of the Linnean Society of London. Zoology 1858;3:46–50.

[19] Darwin C. On the Origin of Species by Means of Natural Selection, or the Preservation of Favoured Races in the Struggle for Life. first ed. London: John Murray; 1859.

[20] Nurse P, Hayles J. The cell in an era of systems biology. Cell 2011;144:850–4.

[21] Travis J. Mysteries of the cell. Science 2011;334:1046.

[22] Dobzhansky T. Biology, molecular and organismic. Am Zool 1964;4:443–52.

[23] Pearson H. Genetics: what is a gene? Nature 2006;441:398–401.

[24] Pennisi E. Genomics. DNA study forces rethink of what it means to be a gene. Science 2007;316:1556–7. PubMed PMID: 17569836.

[25] Pang CP, Baum L, Lam DS. Hunting for disease genes in multifunctional diseases. Clin Chem Lab Med 2000;38:819–25.

[26] DeFrancesco L. Life technologies promises $1,000 genome. Nat Biotechnol 2012;30:126. PubMed PMID: 22318022.

[27] Casals F, Idaghdour Y, Hussin J, et al. Next-generation sequencing approaches for genetic mapping of complex diseases. J Neuroimmunol 2012;248(1–2):10–22.

[28] Ku CS, Cooper DN, Polychronakos C, et al. Exome sequencing: dual role as a discovery and diagnostic tool. Ann Neurol 2012;71:5–14. PubMed PMID: 22275248.

[29] Calvo SE, Compton AG, Hershman SG, et al. Molecular diagnosis of infantile mitochondrial disease with targeted next-generation sequencing. Sci Transl Med 2012;4. 118ra10.

[30] Choi M, Scholl UI, Ji W, et al. Genetic diagnosis by whole exome capture and massively parallel DNA sequencing. Proc Natl Acad Sci U S A 2009;106:19096–101. PubMed PMID: 19861545.

[31] Davies K. Hugh Rienhoff's voyage round his daughter's DNA. Bio-IT World Sept–Oct 2010, http: //www.bio-itworld.com/BioIT_Article.aspx?id=101664; 2010.

[32] Maher B. Personal genomics: his daughter's DNA. Nature 2007;449:773–6.

[33] Mayer AN, Dimmock DP, Arca MJ, et al. A timely arrival for genomic medicine. Genet Med 2011;13:195–6.

[34] Worthey EA, Mayer AN, Syverson GD, et al. Making a definitive diagnosis: successful clinical application of whole exome sequencing in a child with intractable inflammatory bowel disease. Genet Med 2011;13:255–62.

[35] Auffray C, Caulfield T, Khoury MJ, et al. Looking back at genomic medicine in 2011. Genome Med 2012;4(1):9.

[36] Gilissen C, Hoischen A, Brunner HG, et al. Disease gene identification strategies for exome sequencing. Eur J Hum Genet 2012;20:490–7.

[37] Li MX, Gui HS, Kwan JS, et al. A comprehensive framework for prioritizing variants in exome sequencing studies of Mendelian diseases. Nucleic Acids Res 2012;40(7):e53.

[38] Puffenberger EG, Jinks RN, Sougnez C. Genetic mapping and exome sequencing identify variants associated with five novel diseases. PLoS One 2012;7:e28936.

[39] Monico CG, Milliner DS. Genetic determinants of urolithiasis. Nat Rev Nephrol. 2011;8:151–62.

[40] Ahmed N, Dawson M, Smith C, et al. Biology of Disease. New York: Garland Science. Taylor and Francis group; 2006.

[41] Amberger J, Bocchini C, Hamosh A. A new face and new challenges for Online Mendelian Inheritance in Man (OMIM®). Hum Mutat 2011;32:564–7.

[42] Mungall CJ, Gkoutos GV, Smith CL, et al. Integrating phenotype ontologies across multiple species. Genome Biol 2010;11:R2.

[43] van Triest HJ, Chen D, Ji X, et al. PhenOMIM: An OMIM-based secondary database purported for phenotypic comparison. Conf Proc IEEE Eng Med Biol Soc 2011;2011:3589–92.

[44] Washington NL, Haendel MA, Mungall CJ, et al. Linking human diseases to animal models using ontology-based phenotype annotation. PLoS Biol 2009;7:e1000247.

[45] Vidal M. A unifying view of 21st century systems biology. FEBS Lett. 2009;583:3891–4.

[46] Kitano H. Systems biology: a brief overview. Science 2002;295:1662–4.

[47] Nurse P. The great ideas of biology. Clin Med 2003;3:560–8.

[48] Arkin AP, Schaffer DV. Network news: innovations in 21st century systems biology. Cell 2011;144:844–9.

[49] Kruger RP. Systems biology. Cell 2011;144(827):829.

[50] Walhout AJM, Aebersold R, Meyer T, et al. Systems biology: what's the next challenge? Cell 2011;144(6):837–8.

[51] Villoslada P, Baranzini S. Data integration and systems biology approaches for biomarker discovery: challenges and opportunities for multiple sclerosis. J Neuroimmunol 2012. 2012 Jan 24. [Epub ahead of print].

[52] Castrillo JI, Oliver SG. Metabolomics and systems biology in *Saccharomyces cerevisiae*. In: Karl Esser K, editor. The Mycota. A Comprehensive Treatise on Fungi as Experimental Systems for Basic and Applied Research. Fungal Genomics, vol. XIII. Berlin: Springer; 2006. p. 3–18.

[53] Castrillo JI, Zeef LA, Hoyle DC, et al. Growth control of the eukaryote cell: a systems biology study in yeast. J Biol 2007;6:4.

[54] Klitgord N, Segrè D. The importance of compartmentalization in metabolic flux models: yeast as an ecosystem of organelles. Genome Inform. 2010;22:41–55.

[55] Castrillo JI, Oliver SG. Metabolic control in the eukaryotic cell, a systems biology perspective. In: Stansfield I, Stark MJR, editors. Methods in Microbiology. Yeast Gene Analysis. second ed., vol. 36. London: Academic Press; 2007. p. 527–54.

[56] Alberts B, Johnson A, Lewis J, Raff M, Roberts K, Walter P. Molecular Biology of the Cell. 5th edn. New York: Garland Science, Taylor and Francis Group; 2008.

[57] Crick FH. On protein synthesis. Symp Soc Exp Biol 1958;12:138–63.

[58] Crick F. Central dogma of molecular biology. Nature 1970;227:561–3.

[59] Antunes LC, Davies JE, Finlay BB. Chemical signaling in the gastrointestinal tract. F1000 Biol Rep 2011;3:4.

[60] Kohlstedt M, Becker J, Wittmann C. Metabolic fluxes and beyond-systems biology understanding and engineering of microbial metabolism. Appl Microbiol Biotechnol 2010;88: 1065–75.

[61] Castrillo JI, Oliver SG. Yeast as a touchstone in post-genomic research: strategies for integrative analysis in functional genomics. J Biochem Mol Biol. 2004;37:93–106.

[62] Rose AH, Harrison JS. The Yeasts, vols. 1–6. London, UK: Academic Press; 1987–1995.

[63] Sherman F. An introduction to the genetics and molecular biology of the yeast *Saccharomyces cerevisiae*. In: Meyers RA, editor. The

Encyclopedia of Molecular Biology and Molecular Medicine. vol. 6. Germany: VCH Publisher Weinheim; 1997. p. 302−25.

[64] Sherman F. Getting started with yeast. Modified from Methods Enzymol 2006;350:3−41. http://www.urmc.rochester.edu/labs/Sherman-Lab/publications/pdfs/Getting-Started-With-Yeast. pdf.

[65] Lehninger AL. Biochemistry. second ed. New York: Worth Publishers; 1975.

[66] Fell DA. Understanding the Control of Metabolism. London: Portland Press; 1997.

[67] Feldmann H. Yeast: Molecular and Cell Biology. Wiley-VCH; 2010.

[68] Stansfield I, Stark MJ. Yeast Gene Analysis. second ed. London: Academic Press; 2007.

[69] Goffeau A, Barrell BG, Bussey H, et al. Life with 6000 genes. Science 1996;274(546):563−7.

[70] Ramanathan A, Schreiber SL. Multilevel regulation of growth rate in yeast revealed using systems biology. J Biol 2007;6:3.

[71] Kell DB, Oliver SG. Here is the evidence, now what is the hypothesis? The complementary roles of inductive and hypothesis-driven science in the post-genomic era. Bioessays 2004;26:99−105.

[72] Cohen J, Medley G. Stop Working and Start Thinking. second ed. New York: Garland Science, Taylor and Francis Group; 2005.

[73] Teitelman R. On the genome, the markets and data. The Deal Economy 2012. The Deal Pipeline, http://www.thedeal.com/thedealeconomy/on-the-genome-the-markets-and-data.php 2012.

[74] Balaji S, et al. Comprehensive analysis of combinatorial regulation using the transcriptional regulatory network of yeast. J Mol Biol 2006;360:213−27.

[75] Luscombe NM, et al. Genomic analysis of regulatory network dynamics reveals large topological changes. Nature 2004;431:308−12.

[76] Davies J. Regulation, necessity, and the misinterpretation of knockouts. Bioessays 2009;31:826−30.

[77] DeLuna A, Springer M, Kirschner MW, Kishony R. Need-based up-regulation of protein levels in response to deletion of their duplicate genes. PLoS Biol 2010;8:e1000347.

[78] Gout JF, Kahn D, Duret L. Paramecium post-genomics consortium. The relationship among gene expression, the evolution of gene dosage, and the rate of protein evolution. PLoS Genet 2010;6:e1000944.

[79] Pronk JT. Auxotrophic yeast strains in fundamental and applied research. Appl Environ Microbiol 2002;68:2095−100.

[80] Ferenci T. Regulation by nutrient limitation. Curr Opin Microbiol 1999;2(2):208−13.

[81] Fiechter A, Kappeli O, Meussdorfer F. Batch and continuous culture. In: Rose A, Harrison JS, editors. The Yeasts, vol. 2. London: Academic Press; 1987. p. 99−129.

[82] Fiechter A. Continuous cultivation of yeasts. Methods Cell Biol 1975;11:97−130.

[83] Kubitschek HE. Introduction to Research with Continuous Cultures. Prentice hall biological techniques series. New Jersey: Prentice-Hall Englewood Cliffs; 1970.

[84] Hayes A, et al. Hybridization array technology coupled with chemostat culture: tools to interrogate gene expression in *Saccharomyces cerevisiae*. Methods 2002;26:281−90.

[85] Weusthuis RA, et al. Chemostat cultivation as a tool for studies on sugar transport in yeasts. Microbiol Rev 1994;58:616−30.

[86] Gresham D, et al. The repertoire and dynamics of evolutionary adaptations to controlled nutrient-limited environments in yeast. PLoS Genet 2008;4:e1000303.

[87] Käppeli O. Regulation of carbon metabolism in *Saccharomyces cerevisiae* and related yeasts. Adv Microb Physiol 1986;28:181−209.

[88] Käppeli O, Gschwend-Petrik M, Fiechter A. Transient responses of *Saccharomyces uvarum* to a change of the growth-limiting nutrient in continuous culture. J Gen Microbiol 1985;131:47−52.

[89] Petrik M, Käppeli O, Fiechter A. An expanded concept for the glucose effect in the yeast *Saccharomyces uvarum*: involvement of short- and long-term regulation. J Gen Microbiol 1983;129:43−9.

[90] van den Brink J, Daran-Lapujade P, Pronk JT. de Winde JH. New insights into the *Saccharomyces cerevisiae* fermentation switch: dynamic transcriptional response to anaerobicity and glucose-excess. BMC Genomics 2008;9:100.

[91] Rintala E, Jouhten P, Toivari M, Wiebe MG, Maaheimo H, Penttilä M, et al. Transcriptional responses of *Saccharomyces cerevisiae* to shift from respiratory and respirofermentative to fully fermentative metabolism. OMICS 2011;15 (7−8):461−76.

[92] Bull AT. The renaissance of continuous culture in the post-genomics age. J Ind Microbiol Biotechnol 2010;37:993−1021. PubMed PMID: 20835748.

[93] Dikicioglu D, et al. How yeast re-programmes its transcriptional profile in response to different nutrient impulses. BMC Syst Biol. 2011;5:148.

[94] Winder CL. Lanthaler K. The use of continuous culture in systems biology investigations. Methods Enzymol 2011;500:261−75.

[95] Jameson D, Verma M, Westerhoff HV. Methods Enzymol, 500. London: Academic Press; 2011.

[96] Almaas E. Biological impacts and context of network theory. J Exp Biol 2007;210(Pt 9):1548−58.

[97] Emmert-Streib F, Dehmer M. Networks for systems biology: conceptual connection of data and function. IET Syst Biol 2011;5(3):185−207.

[98] Kim H, Hu W, et al. Unraveling condition specific gene transcriptional regulatory networks in *Saccharomyces cerevisiae*. BMC Bioinformatics 2006;7:165.

[99] Kim TY, Kim HU, et al. Data integration and analysis of biological networks. Curr Opin Biotechnol 2010;21:78−84.

[100] Liu YY, Slotine JJ, Barabási AL. Controllability of complex networks. Nature 2011;473:167−73.

[101] Cagney, G, Emili, A, editors. Network Biology. Methods and Applications. Methods in Molecular Biology (MiMB Series. Editor-in-chief. Prof. John M. Walker). vol. 781. New York: Humana Press, Springer; 2011. ISBN 978−1-61779−275−5.

[102] Lee JM, Gianchandani EP, et al. Dynamic analysis of integrated signaling, metabolic, and regulatory networks. PLoS Comput Biol 2008;4(5):e1000086.

[103] Ideker T, Thorsson V, et al. Integrated genomic and proteomic analyses of a systematically perturbed metabolic network. Science 2001;292(5518):929−34.

[104] Ideker T, Thorsson V, et al. Testing for differentially-expressed genes by maximum-likelihood analysis of microarray data. J Comput Biol 2000;7(6):805−17.

[105] Irizarry RA, Hobbs B, et al. Exploration, normalization, and summaries of high density oligonucleotide array probe level data. Biostatistics 2003;4(2):249−64.

[106] Wold, et al. Principal component analysis. Chemometrics and Intelligent Laboratory Systems 1987;2:37–52.

[107] Datta S. Comparisons and validation of statistical clustering techniques for microarray gene expression data. Bioinformatics 2003;19(4):459–66.

[108] Gutteridge A, Pir P, et al. Nutrient control of eukaryote cell growth: a systems biology study in yeast. BMC Biol 2010; 8(1):68.

[109] Rutherford A. Introducing ANOVA and ANCOVA: a GLM Approach. London: Sage Publ; 2001.

[110] De Sousa Abreu R, Penalva LO, et al. Global signatures of protein and mRNA expression levels. Mol Biosyst 2009;5(12): 1512–26.

[111] Griffin TJ, Gygi SP, et al. Complementary profiling of gene expression at the transcriptome and proteome levels in *Saccharomyces cerevisiae*. Mol Cell Proteomics 2002;1(4):323–33.

[112] Reich M, Liefeld T, et al. Genepattern 2.0. Nat Genet 2006;38(5): 500–1.

[113] Maere S, Heymans K, et al. BiNGO: a Cytoscape plugin to assess overrepresentation of gene ontology categories in biological networks. Bioinformatics 2005;21(16):3448–9.

[114] Hughes TR, Marton MJ, et al. Functional discovery via a compendium of expression profiles. Cell 2000;102(1):109–26.

[115] Brown JA, Sherlock G, et al. Global analysis of gene function in yeast by quantitative phenotypic profiling. Mol Syst Biol 2006;2. 2006 0001.

[116] Pir P, Kirdar B, et al. Integrative investigation of metabolic and transcriptomic data. BMC Bioinformatics 2006;7:203.

[117] Pir P, et al. The genetic control of growth rate: a systems biology study in yeast. BMC Syst Biol 2012;6:4.

[118] Boer VM, Crutchfield CA, et al. Growth-limiting intracellular metabolites in yeast growing under diverse nutrient limitations. Mol Biol Cell 2010;21(1):198–211.

[119] Brauer MJ, Huttenhower C, et al. Coordination of growth rate, cell cycle, stress response, and metabolic activity in yeast. Mol Biol Cell 2008;19(1):352–67.

[120] Canelas AB, Harrison N, et al. Integrated multilaboratory systems biology reveals differences in protein metabolism between two reference yeast strains. Nat Commun 2010;1:145.

[121] Hecker M, Lambeck S, et al. Gene regulatory network inference: data integration in dynamic models – a review. Biosystems 2009;96(1):86–103.

[122] Cooke EJ, Savage RS, et al. Computational approaches to the integration of gene expression, ChIP-chip and sequence data in the inference of gene regulatory networks. Semin Cell Dev Biol 2009;20(7):863–8.

[123] De Smet R, Marchal K. Advantages and limitations of current network inference methods. Nat Rev Microbiol 2010; 8(10):717–29.

[124] Geurts P, Irrthum A, et al. Supervised learning with decision tree-based methods in computational and systems biology. Mol Biosyst 2009;5(12):1593–605.

[125] Horak CE, Snyder M. ChIP-chip: a genomic approach for identifying transcription factor binding sites. Methods Enzymol 2002;350:469–83.

[126] Jothi R, Cuddapah S, et al. Genome-wide identification of in vivo protein-DNA binding sites from ChIP-Seq data. Nucleic Acids Res 2008;36(16):5221–31.

[127] Huttenhower C, Mutungu KT, et al. Detailing regulatory networks through large scale data integration. Bioinformatics 2009;25(24): 3267–74.

[128] Joshi A, Van de Peer Y, et al. Structural and functional organization of RNA regulons in the post-transcriptional regulatory network of yeast. Nucleic Acids Res 2011;39:9108–17.

[129] Bonneau R. Learning biological networks: from modules to dynamics. Nat Chem Biol 2008;4(11):658–64.

[130] Karlebach G, Shamir R. Modelling and analysis of gene regulatory networks. Nat Rev Mol Cell Biol 2008;9(10):770–80.

[131] Han JD, et al. Evidence for dynamically organized modularity in the yeast protein–protein interaction network. Nature 2004;430: 88–93.

[132] Tanay A, Sharan R, et al. Revealing modularity and organization in the yeast molecular network by integrated analysis of highly heterogeneous genome wide data. Proc Natl Acad Sci U S A 2004;101(9):2981–6.

[133] Segal E, Shapira M, et al. Module networks: identifying regulatory modules and their condition-specific regulators from gene expression data. Nat Genet 2003;34(2):166–76.

[134] Tian W, Zhang LV, et al. Combining guilt-by-association and guilt-by-profiling to predict *Saccharomyces cerevisiae* gene function. Genome Biol 2008;9(Suppl. 1):S7.

[135] Miller C, Schwalb B, et al. Dynamic transcriptome analysis measures rates of mRNA synthesis and decay in yeast. Mol Syst Biol 2011;7:458.

[136] Jothi R, Balaji S, et al. Genomic analysis reveals a tight link between transcription factor dynamics and regulatory network architecture. Mol Syst Biol 2009;5:294.

[137] Youn A, Reiss DJ, et al. Learning transcriptional networks from the integration of ChIP-chip and expression data in a non-parametric model. Bioinformatics 2010;26:1879–86.

[138] Li C, Donizelli M, et al. BioModels Database: an enhanced, curated and annotated resource for published quantitative kinetic models. BMC Syst Biol 2010;4:92.

[139] Murphy K. Bayes Net toolbox for Matlab. Computing science and statistics (pages not specified), http://code.google.com/p/bnt/; 2001.

[140] Albert R. Bayesian computation with R. Springer Science+Business Media, LLC; 2009.

[141] Mussel C, Hopfensitz M, et al. BoolNet – an R package for generation, reconstruction and analysis of Boolean networks. Bioinformatics 2010;26(10):1378–80.

[142] Myers CL, Robson D, et al. Discovery of biological networks from diverse functional genomic data. Genome Biol 2005;6(13):R114.

[143] Qin J, Li MJ, et al. ChIP-Array: combinatory analysis of ChIP-seq/chip and microarray gene expression data to discover direct/indirect targets of a transcription factor. Nucleic Acids Res 2011;39(Web Server issue):W430–6.

[144] Stark C, Breitkreutz BJ, et al. The BioGRID Interaction Database: 2011 update. Nucleic Acids Res 2011;39(Database issue): D698–704.

[145] Reguly T, Breitkreutz A, et al. Comprehensive curation and analysis of global interaction networks in *Saccharomyces cerevisiae*. J Biol 2006;5(4):11.

[146] Michaut M, Baryshnikova A, et al. Protein complexes are central in the yeast genetic landscape. PLoS Comput Biol 2011;7(2): e1001092.

[147] Collins SR, Kemmeren P, et al. Toward a comprehensive atlas of the physical interactome of *Saccharomyces cerevisiae*. Mol Cell Proteomics 2007;6(3):439−50.
[148] Gentleman R, Huber W. Making the most of high-throughput protein-interaction data. Genome Biol 2007;8(10):112.
[149] Hakes L, Pinney JW, et al. Protein−protein interaction networks and biology−what's the connection? Nat Biotechnol 2008;26(1): 69−72.
[150] Przulj N. Protein−protein interactions: making sense of networks via graph-theoretic modeling. Bioessays 2011;33(2):115−23.
[151] Sharan R, Ideker T. Modeling cellular machinery through biological network comparison. Nat Biotechnol 2006;24(4): 427−33.
[152] Browne F, Zheng H, Wang H, Azuaje F. From Experimental Approaches to Computational Techniques: A Review on the Prediction of Protein−Protein Interactions. Advances in Artificial Intelligence 2010. Volume 2010, Article ID 924529, http://dx.doi.org/10.1155/2010/924529.
[153] Ge H, Liu Z, et al. Correlation between transcriptome and interactome mapping data from *Saccharomyces cerevisiae*. Nat Genet 2001;29(4):482−6.
[154] Liu Y, Zhao H. A computational approach for ordering signal transduction pathway components from genomics and proteomics Data. BMC Bioinformatics 2004;5:158.
[155] Han K, Ju BH, et al. WebInterViewer: visualizing and analyzing molecular interaction networks. Nucleic Acids Res 2004;32(Web Server issue):W89−95.
[156] Lee K, Chuang HY, et al. Protein networks markedly improve prediction of subcellular localization in multiple eukaryotic species. Nucleic Acids Res 2008;36(20):e136.
[157] Scott MS, Calafell SJ, et al. Refining protein subcellular localization. PLoS Comput Biol 2005;1(6):e66.
[158] Cho YR, Shi L, et al. A probabilistic framework to predict protein function from interaction data integrated with semantic knowledge. BMC Bioinformatics 2008;9:382.
[159] Zhao XM, Wang Y, et al. Gene function prediction using labeled and unlabeled data. BMC Bioinformatics 2008;9:57.
[160] Maraziotis IA, Dimitrakopoulou K, et al. Growing functional modules from a seed protein via integration of protein interaction and gene expression data. BMC Bioinformatics 2007;8:408.
[161] Wang YC, Chen BS. Integrated cellular network of transcription regulations and protein−protein interactions. BMC Syst Biol 2010;4:20.
[162] Stark C, Su TC, et al. PhosphoGRID: a database of experimentally verified in vivo protein phosphorylation sites from the budding yeast *Saccharomyces cerevisiae*. Database (Oxford) 2010. 2010: bap026.
[163] Ptacek J, Devgan G, et al. Global analysis of protein phosphorylation in yeast. Nature 2005;438(7068):679−84.
[164] Yachie N, Saito R, et al. Integrative features of the yeast phosphoproteome and protein−protein interaction map. PLoS Comput Biol 2011;7:e1001064.
[165] Tong AH, Evangelista M, et al. Systematic genetic analysis with ordered arrays of yeast deletion mutants. Science 2001; 294(5550):2364−8.
[166] Schuldiner M, Collins SR, et al. Exploration of the function and organization of the yeast early secretory pathway through an epistatic miniarray profile. Cell 2005;123(3):507−19.
[167] Decourty L, Saveanu C, et al. Linking functionally related genes by sensitive and quantitative characterization of genetic interaction profiles. Proc Natl Acad Sci U S A 2008;105(15): 5821−6.
[168] Linden RO, Eronen VP, et al. Quantitative maps of genetic interactions in yeast − comparative evaluation and integrative analysis. BMC Syst Biol 2011;5:45.
[169] Park CY, Hess DC, et al. Simultaneous genome-wide inference of physical, genetic, regulatory, and functional pathway components. PLoS Comput Biol 2010;6(11):e1001009.
[170] Aho T, Almusa H, et al. Reconstruction and validation of RefRec: a global model for the yeast molecular interaction network. PLoS One 2010;5(5):e10662.
[171] Nookaew I, et al. Genome-scale metabolic models of *Saccharomyces cerevisiae*. Methods Mol Biol 2011;759:445−63.
[172] Schwartz JM, Gaugain C. Genome-scale integrative data analysis and modeling of dynamic processes in yeast. Methods Mol Biol 2011;759:427−43. PubMed PMID: 21863501.
[173] Bruggeman FJ, Westerhoff HV. The nature of systems biology. Trends Microbiol 2007;15(1):45−50.
[174] Chou IC, Voit EO. Recent developments in parameter estimation and structure identification of biochemical and genomic systems. Math Biosci 2009;219(2):57−83.
[175] Klipp E. Modelling dynamic processes in yeast. Yeast 2007;24(11):943−59.
[176] Costa RS, Machado D, Rocha I, Ferreira EC. Critical perspective on the consequences of the limited availability of kinetic data in metabolic dynamic modelling. IET Syst Biol 2011;5(3):157−63.
[177] Feist AM, Herrgard MJ, et al. Reconstruction of biochemical networks in microorganisms. Nat Rev Microbiol 2009; 7(2):129−43.
[178] Forster J, Famili I, et al. Genome-scale reconstruction of the *Saccharomyces cerevisiae* metabolic network. Genome Res 2003;13(2):244−53.
[179] Lee JM, Gianchandani EP, et al. Flux balance analysis in the era of metabolomics. Brief Bioinform 2006;7(2):140−50.
[180] Varma A, Palsson BO. Metabolic flux balancing: basic concepts, scientific and practical use. Bio/Technology 1994;12:994−8.
[181] Orth JD, Thiele I, et al. What is flux balance analysis? Nat Biotechnol 2010;28(3):245−8.
[182] Reed JL, Senger RS, et al. Computational approaches in metabolic engineering. J Biomed Biotechnol 2010;2010:207414.
[183] Hucka M, Finney A, et al. The systems biology markup language (SBML): a medium for representation and exchange of biochemical network models. Bioinformatics 2003;19(4): 524−31.
[184] Li J, Min R, et al. Exploiting the determinants of stochastic gene expression in *Saccharomyces cerevisiae* for genome-wide prediction of expression noise. Proc Natl Acad Sci U S A 2010;107(23):10472−7.
[185] Becker SA, Feist AM, et al. Quantitative prediction of cellular metabolism with constraint-based models: the COBRA toolbox. Nat Protoc 2007;2(3):727−38.
[186] Keating SM, Bornstein BJ, et al. SBMLToolbox: an SBML toolbox for MATLAB users. Bioinformatics 2006;22(10):1275−7.
[187] Cvijovic M, Olivares-Hernandez R, et al. BioMet Toolbox: genome-wide analysis of metabolism. Nucleic Acids Res 2010;38(Web Server issue):W144−9.

[188] Mendes P, Hoops S, et al. Computational modeling of biochemical networks using COPASI. Methods Mol Biol 2009;500:17−59.

[189] Famili I, Forster J, et al. *Saccharomyces cerevisiae* phenotypes can be predicted by using constraint-based analysis of a genome-scale reconstructed metabolic network. Proc Natl Acad Sci U S A 2003;100(23):13134−9.

[190] McGary KL, Lee I, et al. Broad network-based predictability of *Saccharomyces cerevisiae* gene loss-of-function phenotypes. Genome Biol 2007;8(12):R258.

[191] Snitkin ES, Dudley AM, et al. Model-driven analysis of experimentally determined growth phenotypes for 465 yeast gene deletion mutants under 16 different conditions. Genome Biol 2008;9(9):R140.

[192] Cakir T, Patil KR, et al. Integration of metabolome data with metabolic networks reveals reporter reactions. Mol Syst Biol 2006;2:50.

[193] Daran-Lapujade P, Jansen ML, et al. Role of transcriptional regulation in controlling fluxes in central carbon metabolism of *Saccharomyces cerevisiae*. A chemostat culture study. J Biol Chem 2004;279(10):9125−38.

[194] Fazio A, Jewett MC, et al. Transcription factor control of growth rate dependent genes in *Saccharomyces cerevisiae:* a threefactor design. BMC Genomics 2008;9:341.

[195] Patil KR, Nielsen J. Uncovering transcriptional regulation of metabolism by using metabolic network topology. Proc Natl Acad Sci U S A 2005;102(8):2685−9.

[196] Mintz-Oron S, Aharoni A, et al. 'Network-based prediction of metabolic enzymes' subcellular localization. Bioinformatics 2009;25(12):i247−252.

[197] Becker SA, Price ND, et al. Metabolite coupling in genome-scale metabolic networks. BMC Bioinformatics 2006;7:111.

[198] Burgard AP, Nikolaev EV, et al. Flux coupling analysis of genome-scale metabolic network reconstructions. Genome Res 2004;14:301−12.

[199] Ruppin E, Papin JA, et al. Metabolic reconstruction, constraint-based analysis and game theory to probe genome-scale metabolic networks. Curr Opin Biotechnol 2010;21(4):502−10.

[200] Yamada T, Bork P. Evolution of biomolecular networks: lessons from metabolic and protein interactions. Nat Rev Mol Cell Biol 2009;10(11):791−803.

[201] Vargas FA, Pizarro F, et al. Expanding a dynamic flux balance model of yeast fermentation to genome-scale. BMC Syst Biol 2011;5(1):75.

[202] Oberhardt MA, Palsson BO, et al. Applications of genome-scale metabolic reconstructions. Mol Syst Biol 2009;5:320.

[203] Osterlund T, Nookaew I, Nielsen J. Fifteen years of large scale metabolic modeling of yeast: developments and impacts. Biotechnol Adv 2012;30(5):979−88.

[204] King RD, Rowland J, et al. The automation of science. Science 2009;324:85−9.

[205] Whelan K, et al. Representation, simulation, and hypothesis generation in graph and logical models of biological networks. Methods Mol Biol 2011;759:465−82. PubMed PMID: 21863503.

[206] Bailey JE. Complex biology with no parameters. Nat Biotechnol 2001;19(6):503−4.

[207] Durek P, Walther D. The integrated analysis of metabolic and protein interaction networks reveals novel molecular organizing principles. BMC Syst Biol 2008;2:100.

[208] Goel A, Li SS, Wilkins MR. Four-dimensional visualisation and analysis of protein−protein interaction networks. Proteomics 2011;11:2672−82.

[209] Li SS, Xu K, Wilkins MR. Visualization and analysis of the complexome network of *Saccharomyces cerevisiae*. J Proteome Res 2011;10:4744−56.

[210] Isserlin R, El-Badrawi RA, et al. The Biomolecular Interaction Network Database in PSI-MI 2.5. Database (Oxford) 2011. 2011: baq037. Print 2011.

[211] Koh JL, Ding H, et al. DRYGIN: a database of quantitative genetic interaction networks in yeast. Nucleic Acids Res 2010;38(Database issue):D502−507.

[212] Salwinski L, Miller CS, et al. The database of interacting proteins: 2004 update. Nucleic Acids Res 2004;32(Database issue):D449−51.

[213] Kamburov A, Pentchev K, et al. ConsensusPathDB: toward a more complete picture of cell biology. Nucleic Acids Res 2011;39(Database issue):D712−7.

[214] Bruckner S, Huffner F, et al. TORQUE: topology-free querying of protein interaction networks. Nucleic Acids Res 2009;37(Web Server issue):W106−108.

[215] Reimand J, Tooming L, et al. GraphWeb: mining heterogeneous biological networks for gene modules with functional significance. Nucleic Acids Res 2008;36(Web Server issue):W452−459.

[216] Szklarczyk D, Franceschini A, et al. The STRING database in 2011: functional interaction networks of proteins, globally integrated and scored. Nucleic Acids Res 2011;39(Database issue): D561−568.

[217] Smoot ME, Ono K, et al. Cytoscape 2.8: new features for data integration and network visualization. Bioinformatics 2011;27(3): 431−2.

[218] Audenaert P, Van Parys T, et al. CyClus3D: a Cytoscape plugin for clustering network motifs in integrated networks. Bioinformatics 2011;27(11):1587−8.

[219] Kelley BP, Yuan B, et al. PathBLAST: a tool for alignment of protein interaction networks. Nucleic Acids Res 2004;32(Web Server issue):W83−88.

[220] Kalaev M, Smoot M, et al. NetworkBLAST: comparative analysis of protein networks. Bioinformatics 2008;24:594−6.

[221] Brohee S, Faust K, et al. NeAT: a toolbox for the analysis of biological networks, clusters, classes and pathways. Nucleic Acids Res 2008;36(Web Server issue):W444−451.

[222] Breitkreutz BJ, Stark C, et al. Osprey: a network visualization system. Genome Biol 2003;4(3):R22.

[223] Hu Z, Hung JH, et al. VisANT 3.5: multi-scale network visualization, analysis and inference based on the gene ontology. Nucleic Acids Res 2009;37(Web Server issue):W115−121.

[224] Kozhenkov S, Dubinina Y, et al. BiologicalNetworks 2.0 − an integrative view of genome biology data. BMC Bioinformatics 2010;11:610.

[225] Medina I, Carbonell J, et al. Babelomics: an integrative platform for the analysis of transcriptomics, proteomics and genomic data with advanced functional profiling. Nucleic Acids Res 2010;38(Web Server issue):W210−213.

[226] Kohler J, Baumbach J, et al. Graph-based analysis and visualization of experimental results with ONDEX. Bioinformatics 2006;22(11):1383–90.

[227] Chen CC, et al. PPISearch: a web server for searching homologous protein–protein interactions across multiple species. Nucleic Acids Res 2009;37(Web Server issue): W369–75.

[228] Bader GD, Cary MP, et al. Pathguide: a pathway resource list. Nucleic Acids Res 2006;34(Database issue):D504–506.

[229] Degenhardt J, Haubrock M, et al. DEEP–a tool for differential expression effector prediction. Nucleic Acids Res 2007;35(Web Server issue):W619–624.

[230] Hayes A, Castrillo JI, Oliver SG, Brass A, Zeef LAH. Transcript analysis: a microarray approach. In: Stansfield, Stark MJR, editors. Methods in Microbiology. Yeast Gene Analysis. second ed., vol. 36. London: Academic Press; 2007. p. 189–219.

[231] Waern K, Nagalakshmi U, Snyder M. RNA sequencing. Methods Mol Biol 2011;759:125–32.

[232] Picotti P, Bodenmiller B, Mueller LN, Domon B, Aebersold R. Full dynamic range proteome analysis of *S. cerevisiae* by targeted proteomics. Cell 2009;138(4):795–806.

[233] Rees J, Lilley K. Enabling technologies for yeast proteome analysis. Methods Mol Biol 2011;759:149–78.

[234] Barrenas F, Chavali S, Holme P, et al. Network properties of complex human disease genes identified through genome-wide association studies. PLoS One 2009;4:e8090.

[235] Kim YA, et al. Identifying causal genes and dysregulated pathways in complex diseases. PLoS Comput Biol 2011;7: e1001095.

[236] Schäfer SA, Machicao F, Fritsche A, et al. New type 2 diabetes risk genes provide new insights in insulin secretion mechanisms. Diabetes Res Clin Pract 2011;93(Suppl. 1):S9–S24.

[237] Friend SH. The need for precompetitive integrative bionetwork disease model building. Clin Pharmacol Ther 2010;87:536–9.

[238] Cho I, Blaser MJ. The human microbiome: at the interface of health and disease. Nat Rev Genet 2012;13:260–70. PubMed PMID: 22411464.

[239] Suhre K, Shin SY, Petersen AK, et al. Human metabolic individuality in biomedical and pharmaceutical research. Nature 2011;477:54–60.

[240] Wang X, Wei X, Thijssen B, et al. Three-dimensional reconstruction of protein networks provides insight into human genetic disease. Nat Biotechnol 2012;30:159–64.

[241] Ulitsky I, Krishnamurthy A, Karp RM, et al. DEGAS: *de novo* discovery of dysregulated pathways in human diseases. PLoS One 2010;5:e13367.

[242] Agarwal S, Deane CM, Porter MA, et al. Revisiting date and party hubs: novel approaches to role assignment in protein interaction networks. PLoS Comput Biol 2010;6:e1000817.

[243] Chen R, et al. Personal omics profiling reveals dynamic molecular and medical phenotypes. Cell 2012;148:1293–307.

[244] Wang K, Lee I, Carlson G, et al. Systems biology and the discovery of diagnostic biomarkers. Dis Markers 2010; 28:199–207.

[245] Kryndushkin D, Shewmaker F. Modeling ALS and FTLD proteinopathies in yeast: an efficient approach for studying protein aggregation and toxicity. Prion 2011;5(4):250–57.

[246] Bashan A, Bartsch RP, Kantelhardt JW, et al. Network physiology reveals relations between network topology and physiological function. Nat Commun 2012;3:702. http://dx.doi.org/10.1038/ncomms1705.

[247] Alon U. An Introduction to Systems Biology. Design Principles of Biological Circuits. Chapman & Hall/CRC mathematical and computational and biology series. Boca Raton: Taylor & Francis Group; 2007.

[248] Lan Y, Mezić I. On the architecture of cell regulation networks. BMC Syst Biol 2011;5:37. PubMed PMID: 21362203.

[249] Pelet S, Rudolf F, Nadal-Ribelles M, et al. Transient activation of the HOG MAPK pathway regulates bimodal gene expression. Science 2011;332:732–5.

[250] Ferrell Jr JE. Simple rules for complex processes: new lessons from the budding yeast cell cycle. Mol Cell 2011; 43:497–500.

[251] Lee C, Raffaghello L, Brandhorst S, et al. Fasting cycles retard growth of tumors and sensitize a range of cancer cell types to chemotherapy. Sci Transl Med 2012;4:124ra27.

[252] Chowdhury R, Yeoh KK, Tian YM, et al. The oncometabolite 2-hydroxyglutarate inhibits histone lysine demethylases. EMBO Rep. 2011;12:463–9. PubMed PMID: 21460794.

[253] McCarthy N. Metabolism: unmasking an oncometabolite. Nat Rev Cancer 2012;12:229.

[254] Xu W, Yang H, Liu Y, et al. Oncometabolite 2-hydroxyglutarate is a competitive inhibitor of α-ketoglutarate-dependent dioxygenases. Cancer Cell 2011;19:17–30.

[255] Mayfield JA, Davies MW, Dimster-Denk D, et al. Surrogate genetics and metabolic profiling for characterization of human disease alleles. Genetics 2012;190(4):1309–23.

[256] Munkacsi AB, Chen FW, Brinkman MA, et al. An 'exacerbate-reverse' strategy in yeast identifies histone deacetylase inhibition as a correction for cholesterol and sphingolipid transport defects in human Niemann-Pick type C disease. J Biol Chem 2011;286:23842–51.

[257] Johnson BS, McCaffery JM, Lindquist S, et al. A yeast TDP-43 proteinopathy model: exploring the molecular determinants of TDP-43 aggregation and cellular toxicity. Proc Natl Acad Sci U S A 2008;105:6439–44.

[258] Treusch S, Hamamichi S, Goodman JL, et al. Functional links between Aβ toxicity, endocytic trafficking, and Alzheimer's disease risk factors in yeast. Science 2011;334:1241–5.

[259] Kumita JR, Johnson RJ, Alcocer MJ, et al. Impact of the native-state stability of human lysozyme variants on protein secretion by *Pichia pastoris*. FEBS J 2006;273:711–20.

[260] Gruson D, Bodovitz S. Rapid emergence of multimarker strategies in laboratory medicine. Biomarkers 2010;15:289–96.

[261] Jorgensen P, Tyers M, Warner JR. Forging the factory: ribosome synthesis and growth control in budding yeast. In: Hall MN, Raff M, Thomas G, editors. Cell Growth. Control of Cell Size. New York: Cold Spring Harbor Laboratory Press; 2004. p. 329–70.

[262] Przytycka TM, Andrews J. Systems-biology dissection of eukaryotic cell growth. BMC Biol 2010;8:62.

[263] Li XC, Schimenti JC, Tye BK. Aneuploidy and improved growth are coincident but not causal in a yeast cancer model. PLoS Biol 2009;7:e1000161.

[264] Bell SL, Chiang AN, Brodsky JL. Expression of a malarial Hsp70 improves defects in chaperone-dependent activities in ssa1 mutant yeast. PLoS One 2009;6:e20047.

[265] Bilsland E, Pir P, Gutteridge A, et al. Functional expression of parasite drug targets and their human orthologs in yeast. PLoS Negl Trop Dis 2011;5:e1320.

[266] Breitenbach M, Jazwinski SM, Laun. P, editors. Aging Research in Yeast. Subcellular Biochemistry, vol. 57. New York: Springer; 2012. p. 365.

[267] Bharadwaj P, Martins R, Macreadie I. Yeast as a model for studying Alzheimer's disease. FEMS Yeast Res 2010;10:961–9.

[268] Lindquist S. Using yeast to understand protein folding diseases: an interview with Susan Lindquist by Kristin Kain. Dis Model Mech 2008;1:17–9.

[269] De Vos A, Anandhakumar J, Van den Brande J, et al. Yeast as a model system to study tau biology. Int J Alzheimers Dis 2011 (2011), Article ID 428970.

[270] Mason RP, Giorgini F. Modeling Huntington disease in yeast: perspectives and future directions. Prion 2011;5(4):269–76.

[271] McGurk L, Bonini NM. Cell biology. Yeast informs Alzheimer's disease. Science 2011;334:1212–3.

[272] Ocampo A, Barrientos A. Developing yeast models of human neurodegenerative disorders. Methods Mol Biol 2011; 793:113–27.

[273] Zhang N, Bilsland E. Contributions of *Saccharomyces cerevisiae* to understanding mammalian gene function and therapy. Methods Mol Biol 2011;759:501–23.

[274] Couthouis J, Hart MP, Shorter J, et al. A yeast functional screen predicts new candidate ALS disease genes. Proc Natl Acad Sci U S A 2011;108:20881–90.

[275] Couplan E, Aiyar RS, Kucharczyk R, et al. A yeast-based assay identifies drugs active against human mitochondrial disorders. Proc Natl Acad Sci U S A. 2011;108:11989–94.

[276] Tardiff DF, Tucci ML, Caldwell KA, et al. Different 8-hydroxyquinolines protect models of TDP-43 protein, α-synuclein, and polyglutamine proteotoxicity through distinct mechanisms. J Biol Chem. 2012;287:4107–20.

[277] Neef DW, Turski ML, Thiele DJ. Modulation of heat shock transcription factor 1 as a therapeutic target for small molecule intervention in neurodegenerative disease. PLoS Biol 2010;8: e1000291.

[278] Park SK, Pegan SD, Mesecar AD, et al. Development and validation of a yeast high-throughput screen for inhibitors of Aβ oligomerization. Dis Model Mech 2011;4:822–31.

[279] Bharucha N, Kumar A. Yeast genomics and drug target identification. Comb Chem High Throughput Screen 2007;10:618–34.

[280] Andrusiak K, Piotrowski JS, Boone C. Chemical–genomic profiling: systematic analysis of the cellular targets of bioactive molecules. Bioorg Med Chem 2012;20:1952–60.

[281] Proctor M, Urbanus ML, Fung EL, et al. The automated cell: compound and environment screening system (ACCESS) for chemogenomic screening. Methods Mol Biol 2011; 759:239–69.

[282] De Jong B, Siewers V, Nielsen J. Systems biology of yeast: enabling technology for development of cell factories for production of advanced biofuels. Curr Opin Biotechnol 2012; 23(4):624–30.

[283] Mattanovich D, Branduardi P, Dato L, et al. Recombinant protein production in yeasts. Methods Mol Biol 2012; 824:329–58.

[284] Soh KC, Miskovic L, Hatzimanikatis V. From network models to network responses: integration of thermodynamic and kinetic properties of yeast genome-scale metabolic networks. FEMS Yeast Res 2012;12:129–43.

[285] Kim IK, Roldão A, Siewers V, et al. A systems-level approach for metabolic engineering of yeast cell factories. FEMS Yeast Res. 2012;12:228–48.

[286] Nielsen J, Pronk JT. Metabolic engineering, synthetic biology and systems biology. FEMS Yeast Res 2012;12:103.

[287] Da Vinci L. Trattato della pittura. Parte prima/23. Carabba Editore. Manuscript compiled from the originals by F. Melzi. 1540. Leonardo Digital Edition (LDE VU) page 14v. F. Fiorani digital project. University of Virginia, http://www.treatiseonpainting.org/exist/leonardo/pages/vu/0034#chap_top; 1947.

[288] Oliver SG. Yeast as a navigational aid in genome analysis. Microbiology 1997;143:1483–7.

[289] Oliver SG. Functional genomics: lessons from yeast. Philos Trans R Soc Lond B Biol Sci 2002;357:17–23.

[290] Winzeler EA, Shoemaker DD, Astromoff A, et al. Functional characterization of the *S. cerevisiae* genome by gene deletion and parallel analysis. Science 1999;285:901–6.

[291] Giaever G, Chu AM, Ni L, et al. Functional profiling of the *Saccharomyces cerevisiae* genome. Nature 2002;418:387–91.

[292] MacLellan WR, Wang Y, Lusis AJ. Systems-based approaches to cardiovascular disease. Nat Rev Cardiol 2012;9:172–84.

[293] Sperling SR. Systems biology approaches to heart development and congenital heart disease. Cardiovasc Res 2011;91: 269–78.

[294] Antony PM, Diederich NJ, Balling R. Parkinson's disease mouse models in translational research. Mamm Genome 2011;22:401–19.

[295] Cox B, Kotlyar M, Evangelou AI, et al. Comparative systems biology of human and mouse as a tool to guide the modeling of human placental pathology. Mol Syst Biol 2009;5:279.

[296] Mori MA, Liu M, Bezy O, et al. A systems biology approach identifies inflammatory abnormalities between mouse strains prior to development of metabolic disease. Diabetes 2010;59: 2960–71.

[297] O'Callaghan PM, James DC. Systems biotechnology of mammalian cell factories. Brief Funct Genomic Proteomics. 2008;7:95–110.

[298] Stuart LM, Boulais J, Charriere GM, et al. A systems biology analysis of the *Drosophila* phagosome. Nature 2007;445: 95–101.

[299] Yu D, Corbett B, Yan Y, et al. Early cerebrovascular inflammation in a transgenic mouse model of Alzheimer's disease. Neurobiol Aging 2012. 2012 Mar 20. [Epub ahead of print] PubMed PMID: 22440674.

[300] Goto JJ, Tanzi RE. The role of the low-density lipoprotein receptor-related protein (LRP1) in Alzheimer's A beta generation: development of a cell-based model system. J Mol Neurosci. 2002;19:37–41.

[301] Joyner MJ, Pedersen BK. Ten questions about systems biology. J Physiol 2011;589:1017–30.

[302] Antony PM, Balling R, Vlassis N. From systems biology to systems biomedicine. Curr Opin Biotechnol 2012;23(4):604–8.

[303] Clermont G, Auffray C, Moreau Y, et al. Bridging the gap between systems biology and medicine. Genome Med 2009;1:88. PubMed PMID: 19754960.

[304] McDonald JF. Integrated cancer systems biology: current progress and future promise. Future Oncol. 2011;7:599–601.

[305] Sansom C, Mendes M, Coveney P. Modelling health and disease. Ingenia 2011;47:27–32.

[306] Viceconti M, Clapworthy G, Van Sint Jan S. The virtual physiological human – a european initiative for *in silico* human modelling. J Physiol Sci. 2008;58:441–6.

[307] Sansom C, Mendes M, Coveney P. Modelling the virtual physiological human. BioTechnologia. J Biotech, Comput Biol and Bionanotechnol 2011;92(3):225–29.

Chapter 19

Systems Biology of *Caenorhabditis elegans*

Andrew Fraser[1] and Ben Lehner[2]

[1]The Donnelly Centre, University of Toronto, 160 College Street, Ontario M5S 3E1, Canada, [2]European Molecular Biology Laboratory (EMBL)-Centre for Genomic Regulation (CRG) Systems Biology Unit and Institució Catalana de Recerca i Estudis Avançats (ICREA), CRG, Universitat Pompeu Fabra (UPF), c / Dr Aiguader, 88, Barcelona 08003, Spain

Chapter Outline

Introduction: The System before the Sequence	367
Forward and Reverse Genetics in the Worm: How 97mb says 'Make a Worm'	369
From 'How Does This Work?' to 'What Does This Do?'	369
Classical Screens: Finding the Core Modules that Control Worm Development	370
Classical Screens with Next-Generation Sequencing – Genetics Leaps Forward	371
Persuading Worms to take Enough Drugs – Chemical Genomics in the Worm	372
Reverse Genetics and the Magic of RNAi	373
Expressing and Regulating an Animal Genome	376
Genome-Wide Maps of Normal and Perturbed Gene Expression	377
Single Cell-Resolution Analysis of Gene Expression	377
Global Maps of Transcription Factor-Binding Sites and Chromatin Organization	378
Variation in Gene Expression among Individuals and its Phenotypic Consequences	379
Proteomics in *C. elegans*: Global Maps of Protein Expression and Protein–Protein Interactions	380
Integrative and Dynamic Modeling to Link Genotype to Phenotype	380
Data Integration and Genome-Scale Networks to Connect Genes and Modules to Phenotypic Variation	380
Dynamic Models of Developmental Processes	382
Outlook	382
References	384

INTRODUCTION: THE SYSTEM BEFORE THE SEQUENCE

When it comes to systems biology, *Caenorhabditis elegans* is unique amongst the current major model organisms. Unlike in the fly, the yeasts, or the mouse, systems biology has not emerged as a later development of classical genetics studies but has been at the heart of the worm community philosophy from the very start. The goal of completeness (later refined to CAP – Complete, Accurate, and Permanent) was central in Sydney Brenner's initial proposal to the MRC – indeed, his vision of 'taming' the worm culminates with the modest idea that 'We intend to identify every cell in the worm and trace lineages'. The audacity of this idea of completeness, in effect the intellectual 'taming' of a complex biological system, set the tone of early worm research and it is an approach that persists.

The worm is the intellectual descendant of phage research and of the glimpses of the logic of genetic control seen through detailed analyses of prokaryotic information processors such as the *lac* operon. In this sense, the first key steps in the establishment of the worm have a fundamentally different flavor from those of the fly. Fly research, in its earliest days, was by necessity top-down. T.H. Morgan's fly room was established in the first decade of the 20th century at a time when almost all the basic principles of genetics were unknown. Although the Boveri–Sutton chromosome theory had proposed as early as 1903 that chromosomes were the fundamental carriers of heritable genetic material [1,2], the first two decades of the 20th century were awash with conflicting theories as to how evolution and inheritance operated. By 1915, however, the publication of *The Mechanism of Mendelian Heredity* by Morgan et al. [3] was a huge advance in genetics and provided a single accessible synthesis of the major concepts of, among others, Mendel, de Vries, Boveri, and Sutton, and shaped the next phase of genetics research. Thus the 'top-down' nature of early fly research (starting with phenotypic outcome and attempting to unravel the mechanistic basis) was born out of necessity

— the fly was established at a time when the world of molecular biology was in darkness, and to ask how a fly 'works' from the molecular perspective was abstract in the extreme.

In the early 1960s the world was completely different. If one compares the theorizing and abstraction of Schrodinger's *What is Life?* [4] with the concrete molecular biological knowledge crystallized in Crick's central dogma [5,6], it is clear that biology could be questioned from a new standpoint. The basic molecular processes underpinning life and heredity were known and there were at least the beginnings of an information theory approach to genetics, either through the pioneering concepts of Shannon [7,8] or (amongst other examples) through the experimental dissection of the logic underpinning the *lac* operon of Jacob and Monod [9–11]. It became possible to think from bottom-up, from the molecule to the system — to attempt to understand how the central molecules and fundamental processes of molecular biology are organized to direct the development and function of complex tissues. The idea to use the worm as a simple model system to elucidate the molecular mechanisms and genetic logic underpinning both development and nervous system function resulted from discussions between Brenner and Crick in the early 1960s. In effect, Brenner and Crick rejected the fly as ultimately 'untameable' — the cellular complexity of the fly is immense compared to that of the worm, and this, coupled to its more complex lifecycle, meant that the goal of anchoring phenotypic analysis in a complete cellular lineage was impossible.

The initial phase of worm-taming has been described extensively elsewhere, most obviously in the Nobel lectures of 2002, but ultimately rests on four major achievements: the construction of both the genetic map [12] and the physical genomic map [13–15]; the complete description of the lineage of worm development [16–18]; the comprehensive mapping of the physical and chemical network of the nervous system; and, finally, the complete genome sequence. This broad summary clearly hides huge amounts of painstaking work and groundbreaking technical innovation. Whether one considers the meticulous and intricate work of tracing the lineage and of reconstructing the network of cellular contacts of the neuromuscular system, or the myriad technical achievements necessary to allow the sequencing of a complex animal genome (including, among others, the development of PHRED [19] and PHRAP, improvements in clone handling and sequencing technology, and the development of computational tools such as ACeDB [20] to store, query, and analyze a previously unapproachable level of data), these steps were heroic.

By the end of 1998, with the publication of the complete genome sequence, these major early achievements meant that for the first (and still only) time for any animal, the genotype–phenotype problem is bounded: both the genotype and phenotype are completely known, every base pair, every cell. While the work of understanding how the genetic material dictates the phenotype of a worm from a single-cell fertilized egg through to the 959 cells of the adult hermaphrodite is immense and ongoing, at least the start and endpoint of this question are fixed — in Brenner's words, complete, accurate, and permanent. In this chapter, we first set out a brief historical perspective of the advances in the understanding of worm biology prior to the publication of the genome sequence, then illustrate how the genome sequence has driven the systematic studies to understand how the 20 000-odd coding genes encoded in six chromosomes coordinate development and function. We first discuss systematic analyses of gene function through genetic screens, then set out the key systems advances in understanding the coordinate regulation of gene expression at the level of RNA then protein, then end with computational approaches that attempt to integrate the vast datasets produced by genomics technologies to understand how genes are organized into the modules that ultimately read the genetic code and translate it to the phenotype.

Before moving into the post-genome world, it is worth looking again at the map of the neuromuscular system of the worm as a vignette illustrating the way that systems principles — first identify the individual components, then examine the connections between them, and finally understand how the systems-level organization explains the biological properties — underpin worm research. There is a large body of primary literature leading to the insights outlined here, but ultimately the best overviews can be found in White et al. [21] and in Richard Durbin's Ph.D. Thesis ([22] and online at http://www.wormatlas.org/ver1/durbinv1.2/durbinindex.html). First, the components, in this case the neurons, were identified — every adult hermaphrodite has 302 neurons and the series of divisions and specific cell deaths that gives rise to these 302 cells is essentially invariant. Second, the functions of many of these individual cells was studied through laser ablation — individual neurons were 'deleted' from the developing animal and the outcome both on the nervous system structure and on behavior could be studied, thus establishing the requirements for the correct development and function of the worm neuromuscular system (reviewed in [23]). Third, the connections either between neurons or between neurons and muscle cells were deduced from examining electron micrographs of serial sections through a number of individual worms. The architecture of the nervous system is highly reproducible between animals, and in total there are approximately 5000 chemical synapses, 2000 neuromuscular junctions, and 600 gap junctions. Finally, having identified the complete set of cells involved and the complete set of connections between these cells, researchers were at last able to ask systems-level questions, and specific network features could be identified. Neurons of the same functional class tended to make gap junctions with other members of the same class, but chemical synapses tend to be between classes. Patterns of triangular connectivity between neurons are highly over-represented, the circuitry of the nerve ring is highly directional (unlike the general organization in vertebrates for example), and processing depth is shallow — the number of connections between stimulus and output is very small. Ultimately, this systems-level view of the nervous system provides both a map to guide hypothesis-driven experiments (e.g., in experiments monitoring the real-time visualization of neuron firing following stimulation) and a rational framework for interpreting the outcomes of such studies. By layering newer experimental data onto this complete network of the neuromuscular system, researchers now understand phenomena such as the short- and long-term modulation of responses to odors [24], and the physical basis for pheromone attraction and social behavior in worms [25].

The 'first components, then connections, then system-level properties' approach has thus been central to *C. elegans* research from its inception. The lineage, the nervous system structure, the genome sequence are all complete, permanent, and accurate. For the remainder of this chapter we will examine how genetic approaches to understanding how genotype dictates phenotype have been combined with gene expression analyses and physical interaction mapping to ultimately generate integrated models for gene function in vivo. Finally, we end with a brief section regarding future directions in the field.

FORWARD AND REVERSE GENETICS IN THE WORM: HOW 97MB SAYS 'MAKE A WORM'

From 'How Does This Work?' to 'What Does This Do?'

The world beyond the concrete truths of genome sequence and cell lineage is comparatively confusing and ill-defined. The genome sequence of the N2 Bristol isolate is indeed a platform for investigating biology [26]; however, whereas the question 'What is the complete genome sequence of the N2 Bristol isolate?' has a unique and well-defined answer, the questions that follow become increasingly complex and poorly constructed. 'What are the genes encoded in the *C. elegans* genome?' sounds beguilingly simple, yet leads rapidly into the territory usually occupied by first-year genetics essays on definitions of what makes up a gene. Depending on the gene prediction method, and on whether one defines genic units as being only those encoding proteins or instead include the ever-increasing number of non-coding transcribed regions of the genome, predicted gene numbers can vary widely. But compared to gene number, it is once one enters the realm of gene function that things begin to become complicated.

A question such as 'What are the functions of all of the genes encoded in the *C. elegans* genome?' appears superficially simple. We immediately reach conceptually (for example) to analogies of cars, with parts lists and amateur mechanic exploded diagrams of gear boxes and carburetors, each engine 'module' made up of small number of intimately connected components and the modules linked together into units of increasing complexity until we can view the engine as a single unit of function. Unfortunately, gene function is seldom viewed in such conceptually simple terms, and it is very rare that two researchers agree on what a gene 'does'. For example, the function of the key cellular oncogene *c-myc* could alternately (and equally accurately) be described as a transcription factor of the basic helix—loop—helix leucine zipper class; as an oncogene whose over-expression leads to cancer; as a dimerization partner of the transcription factor MAX; as an activator of apoptosis and cell proliferation; or as a gene that is essential for murine embryonic development (reviewed in [27]). Clearly, attempts have been made to rationalize this kind of mish-mash of molecular, cellular, biochemical, and organismal function: the Gene Ontology project [28] is the most widely known and used, and organizes gene function into hierarchies within three broad domains, cellular component, biological process, and molecular function.

From the point of view of genetics, however, the simplest functional question to ask of a gene is 'What are the phenotypic consequences of inherited variation in the sequence of that gene?', and this is where we will focus in this section. We first lay out briefly the progress made through classical genetic screens in which mutagens have been used to generate genetic variation leading to mutant phenotypes, then examine how knowledge of the genome sequence, along with the recent advances in sequencing power, has greatly accelerated these approaches, opening the door to more sensitive and higher-coverage classical genetic screens.

Classical ('forward') genetics experiments all investigate the connection between genotype and phenotype in one direction only: mutants are first identified on the basis of phenotypic differences from the wild-type state, and the causative sequence variations are subsequently identified either through classical genetics mapping strategies or (more recently) through brute force sequencing. In essence, for any single mutant the question being asked is 'This worm looks mutant: what is the sequence change responsible for this?'. Identifying multiple independent mutants that have similar mutant phenotypes, widens the question to ask 'What is the set of genes that affect this biological process?' and this approach, forward genetics (from mutant phenotype to causative mutation), dominated experimental genetics for decades.

However, there is an alternative way to investigate connection between genotype and phenotype, asking instead 'If I change the activity of gene X, what happens?', and this is the area that has been revolutionized by knowledge of the genome sequence. Reverse genetics — 'Here is gene X, what does it do?' — is the natural direction of investigation for systems biology in the post-genomics era, when the 'parts list' is known. Any child given a simple alarm clock will first ask 'How does this work?' (forward genetics); However, once that same child that lifts off the back of the clock to see the mechanism, they will intuitively rephrase the question to 'What does this bit do?' (reverse genetics). The majority of this section will focus on the progress made in the post-genomics era in systematic approaches to understanding the in vivo functions of genes, and will cover both the ongoing attempts to make large-scale deletion collections and genome-scale RNA-mediated interference (RNAi) screens. Although we will set out the technologies, we will also focus on the insights into gene function, genome evolution, and the molecular machineries underpinning *C. elegans* development and behavior that arise from such systematic analyses.

Classical Screens: Finding the Core Modules that Control Worm Development

From the mid 1970s to the mid 1990s classical genetics in *C. elegans* opened up major areas of animal biology, providing the first view of the key components of the conserved pathways regulating cell death (reviewed in [29], controlling axon guidance (overview in [30]), and/or transducing a touch signal to a muscle response (summarized in [31]), among many others. Processes ranging from dosage compensation and gender determination to the development of tissues such as the vulva and neuromuscular system were all subjected to exhaustive genetic analysis, and this phase of *C. elegans* classical forward genetics screening identified many of the core molecular pathways that are key for metazoan development, including the CED-9 CED-3 CED-4 core apoptotic machinery, netrin signaling in axon guidance, and microRNAs. These major advances are all testament to the power of basic research to revolutionize our view of biology and spawn huge fields of cutting-edge therapeutics (for example microRNA and RNA-mediated interference-based therapies). Far from being ill-conceived meanderings that are often pejoratively termed 'fishing expeditions', they are clear models of hypothesis-driven research. The hypothesis is that knowledge is better than ignorance; or, more prosaically, that if you have no fish, a fishing expedition is a fine plan.

The initial 20 years of classical genetics took understanding of the molecular machineries underpinning worm development from literally nothing to a series of core modules whose components could be ordered by epistasis, and these modules still form the central foundation for what we know today. To any worm biologist, if I said 'vulval induction', this would immediately mean at least *lin-3, let-23, let-60, lin-45* (for review see [32]); phagocytosis would give *ced-1, ced-6, ced-7* and *ced-2, ced-5* and *ced-10* (reviewed in [33]); and regulation of embryonic polarity would immediately call up *par-1* and *par-2, par-3, par-6* and *pkc-3* (see [34] for a review). There are similar modules for many key processes, and this phase of module discovery through classical genetics screens is one of the great highlights of *C. elegans* research.

Despite the great advances made by classical genetics, like any technique it has its blind spots, several of which are important for systems-level analysis. One is the difficulty of identifying any multigenic effect de novo. One clear illustration of this is in the development of the *C. elegans* vulva. Multiple genetic screens have been carried out to identify genes that affect vulval development, and these identified key components of the EGF-ras-raf, Notch, and Wnt pathways (summarised in [32]). Each of these genes is required for normal vulval development — for example, loss of function mutations in *let-60*, the *C. elegans* ras ortholog, give vulvaless worms, whereas gain of function mutations give multivulval animals [35]. So far, so good. However, the molecular dissection of the *lin-15* locus revealed a more complex story [36]. A large deletion of the *lin-15* locus gives multivulval (Muv) animals, suggesting that *lin-15* encodes a repressor of vulval fate. Detailed dissection of this region of the genome revealed not one, but two genes in that deletion, *lin-15A* and *lin-15B*. These two are apparently (almost) completely redundant as regards vulval fate — deletion of either gene has no effect, but only a double deletion has the Muv phenotype. Subsequent upon this finding other such synthetic multivulval (*synmuv*) genes have been identified, some that are redundant with *lin-15A* (e.g., *lin-35*), and others that are redundant with *lin-15B* (e.g., *lin-8*). In all cases but one their identification through genetic screens relied entirely on secondary screens after the detection of an extremely rare event involving a double mutation between *lin-8* and *lin-9* [37]. How many such multigenic interactions are there that have hidden from classical screening? The indications from yeast are that such redundancies are extremely widespread. Data from systematic synthetic lethal screens indicate that a few percent of bigenic interactions are non-additive, and this number appears to be conserved from yeasts to worm, suggesting that the case of *lin-15A* and *lin-15B* is far from the exception and that the roles of many genes in key biological pathways have been hidden from classical screens.

Another class of genotype—phenotype connection that cannot be detected trivially using forward genetics screens is the early developmental role for genes that have a zygotic requirement but which also have a large maternal contribution. The early events of embryogenesis are almost completely reliant on the large amounts of mRNA and protein contributed by the maternal germline, and so embryos that are homozygous null for a gene that plays an essential role in early embryogenesis may nonetheless go through early embryogenesis absolutely normally if it was laid by a heterozygous mother. Since many of the genes that play key roles in early embryogenesis also function later in the life of the animal, one cannot derive embryos from homozygous null mothers (and thus eliminate maternal contribution). The roles that such genes play in the key initial events of embryogenesis are thus hidden from classical genetics, and developing models of the machineries underpinning embryogenesis is therefore impossible without alternative approaches.

A third issue, particularly from the point of view of systems biology, is the difficulty of estimating how near any mutagenesis screen is to saturation. Although approximate calculations can be done to estimate the coverage of a mutagenesis screen, it is clear that mutagenesis is far from random, and that even screens which isolated multiple independent alleles of the same gene that have the exact same mutation (e.g., more than one allele of the G13E mutation of the ras orthologue *let-60* were identified in genetic screens for vulval mutants [35]), which one might therefore imagine to have reached saturation, missed multiple genes that were subsequently identified to affect the same process. This raises clear issues: if it is hard to estimate coverage and false negative rates of any screen, it makes systems-level analysis of any identified pathway extremely precarious. In Rumsfeld-speak, while 'known unknowns' (measurable false negatives rates) are tolerable for systems biology, the 'unknown unknowns' are not. The under-saturation of many screens is obvious — many genes that were isolated through forward genetics screens have only had a single allele isolated, suggesting that there must be many potential 'hits' that fell just below this threshold and thus escaped identification.

Finally, many genes have weak loss-of-function phenotypes that may be hard to detect in an otherwise wild-type background; for this reason, modifier screens have often been used to identify additional components of specific pathways that eluded basic screens. In particular, screening in 'sensitized' backgrounds harboring weak mutations in a pathway of interest have been useful for identifying additional pathway genes whose loss of function phenotypes are weak in a wild-type background. However, while such enhancer screens can identify genes that increase the phenotypic penetrance of a specific mutation, there is often little or nothing known about the specificity of such interactions from classical screens. For example, mutations in *hmg-1.2*, a HMG-box transcriptional regulator, strongly enhance the effect of mutations in the Wnt pathway on male ray development [38]. The conclusion drawn could therefore be that *hmg-1.2* is a specific regulator of Wnt signaling in male ray development. However, systematic studies show that reductions in *hmg-1.2* activity enhance mutations in a wide variety of signaling pathways, including EGF, Wnt, Notch and Ephrin pathways [39]. Therefore, just as testing a single pairwise protein—protein interaction can give a positive result that may be misleading owing to the 'stickiness' of one of the proteins, so a genetic screen to identify enhancers can be misinterpreted due to the promiscuity of the enhancer and the lack of systematic pairwise testing. This creates problems for any downstream systems-level analysis, as without a measure of specificity the importance of the involvement of a gene in a given pathway may be overestimated.

Despite the caveats of classical screening in the worm, the components identified in such screens and their ordering in pathways through epistasis analysis have yielded key frameworks that served to direct the hypothesis-driven detailed molecular analyses that ultimately fleshed out the physical basis for the genetic pathways. For example, in the cell death pathway, genetic screens identified three loci that were critical for programmed cell death, *ced-3*, *ced-4* and *ced-9*; *ced-9* was then shown to be genetically upstream of *ced-3* and *ced-4*. Following the molecular cloning of these genes, and through biochemical and molecular analyses by several groups, we now know that in the 959 cells that survive, CED-9 physically interacts with CED-4, restraining it from binding the pro-apoptotic form of CED-3 [40–42]; release of CED-9 from CED-4 in the 131 cells destined for death, largely through interaction with the BH3-only containing protein EGL-1 [43], releases CED-4 and it binds CED-3, leading to CED-3 auto-proteolysis to yield the mature active cysteine protease which is the key apoptotic effector. This core apoptotic module is highly conserved, and classical screens were critical for the identification of all the components, which we now understand at atomic resolution. Classical genetics has been crucial in defining similar core molecular modules for processes ranging from mechanosensation to sex determination, and these genetically identified modules can serve as the starting point for rigorous modeling approaches such as those being pioneered in the vulva.

Classical Screens with Next-Generation Sequencing — Genetics Leaps Forward

In a classical genetics screen in the worm, generating and picking mutants is easy: identifying the causative mutation (and hence the genotype—phenotype inference) is not — or at least was not. In the pre-genome sequence era, the approach to identifying the causative mutation was limited by sequencing power, and proceeding in broad terms as follows (reviewed by Fay et al. in depth in WormBook [44]). During a screen multiple mutants are typically isolated; for each mutant, the locus responsible for the mutant phenotype of interest is first crudely mapped to linkage groups, then to large regions of a specific chromosome using classical genetics. Guided by this crude mapping, complementation tests are carried out to identify how many independent loci have been isolated in the screen. For any single locus, the next step is to rescue the mutant phenotype — whether as cosmids, YACs, fosmids or BACs, some large fragments of the genome covering the region identified by genetic mapping are injected into the mutant animal and the resultant transgenics tested for rescue. This is repeated using increasingly smaller test fragments until a fragment is isolated that is small enough to sequence using standard Sanger sequencing and traditionally, together with cDNA library probing and Northern blots the transcriptional unit that rescues the mutant phenotype can thus be identified. Finally, having identified the gene affected, PCR and sequencing can home in on the sequence changes in the mutant, and the story is therefore complete, one gene at a time.

The process outlined above is clearly laborious and extremely low throughput. Cloning a single gene, along with some basic molecular characterization, was typically the body of a 5-year PhD. As a result, there is often extensive triaging of the mutants to be mapped — mutants with low-penetrance phenotypes are often frozen away and neglected, and anything with complex phenotypes indicating extensive pleiotropy is often ignored. Researchers understandably initially focus on the clean and the strong, 100% penetrant phenotypes with no confounding additional defects. If one can only identify one or two genes a year, this makes sense, but it does lead to an ascertainment bias in the hits from these screens that is hard to quantify and hard to compensate for in network or systems-level analysis of the molecular pathways identified.

Why was the throughput of these screens so low? The problem was not usually that of screening enough worms, or picking enough mutants, but rather the slow process of mapping leading to the isolation of a small enough piece of rescuing DNA that was worth sequencing. The availability of the genome sequence and the huge recent advances in sequencing power have changed this landscape massively, however. For mapping, the identification of many SNPs between the canonical N2 Bristol isolate sequence [26] and the CB4856 (often called the Hawaiian isolate) has revolutionized the mapping of mutant loci in the worm [45]. Rather than rely on visually obvious mapping markers, which are only dense enough to typically map down a mutant locus to a region of around 100 genes, one can now do rapid SNP mapping of the locus to a handful of genes, a huge advance. The other critical advance is in sequencing power — using 'next-generation' sequencing technology of any type currently gives sufficient coverage of the 97 Mb worm genome to confidently identify any single-base mutation for around $100. Combining SNP-mapping with whole-genome sequencing therefore allows the extremely rapid identification of causative mutations [46,47], and this should open the way to deeper, more saturating forward genetics. This in turn should provide more accurate frameworks for the systems analysis of pathways leading to quantitative and predictive modeling, as has been attempted for the vulva (for example, by [48–51]) and which one hopes might one day approach the beautiful fusion of theoretical and experimental analysis of signaling in bacterial chemotaxis (reviewed in [52]).

Persuading Worms to take Enough Drugs — Chemical Genomics in the Worm

Although most of this section focuses on using genetic perturbations to examine the genotype–phenotype connection (whether through mutagenesis or using RNA-mediated interference), chemical genomics provides a completely different entry point to perturbation of genetic networks (reviewed, for example in [53]). Chemical screens can not only identify vital medical compounds, but on a research level provide many crucial tools — it is hard to imagine mammalian signaling research without the sets of compounds used to activate or repress specific pathways. However, chemical genomics in the worm is complicated by a specific feature of normal worm biology: worms live in dirt. To deal with this complex environment, they have evolved a defensive arsenal of cytochrome P450s and drug transporters to ensure that any potentially harmful molecules to which they are exposed are either rapidly modified or removed from their bodies [54]. In the standard laboratory environment this makes little difference: unchallenged by toxins, these defensive weapons lie unused. However, this raises clear difficulties for chemical genomics, as many of the chemicals in commercially available compound libraries are either rapidly modified, rendering them broadly ineffective, or pumped out of the worm so that insufficient intracellular concentrations build up and no effect is seen. The cuticle is also a major barrier to drug entry, and while some drugs penetrate relatively easily, others must be applied to the intact animal at hundreds to thousands times higher concentrations than their estimated target affinity, increasing cost and, in some cases, leading to solubility problems. Together, these issues have limited the number of de novo compound screens carried out in the worm (compared, say, with the extensive small molecule screens carried out either in yeast or in mammalian cells in culture), and so chemical genomics has yet to make a similar impact on systems-level understanding of worm biology.

Despite the difficulty of delivering many drugs efficiently and at effective concentrations into the worm, some pioneering work in small molecule screening has been carried out. Small molecule screens of several thousand commercially available compounds [55] have identified novel Ca-channel antagonists and a novel chemical inhibitor of DAF-9 a cytochrome P450 with a key role in dauer formation [56,57]. In addition to identifying drugs that affect the worm to produce biologically interesting phenotypes, through the use of downstream forward genetics screens, they were often able to identify the true target of the identified compound, taking the process of small molecule from lead compound to drug target in a rapid manner. Crucially, they also developed a set of rules that are predictive of the uptake and bioactivity of small compounds into worms [54]. This should open the way for more large-scale screening of pre-selected small molecule libraries, with high coverage of the chemical space that is predicted to be bioactive in worms. To date, however, these approaches are in their infancy, and while such screens are exciting possible sources both of novel anthelmintics and of key research tools such as pathway agonists and antagonists, this is currently a work in progress rather than a fully established area of systems research in the worm.

Reverse Genetics and the Magic of RNAi

The worm has ~20 000 predicted coding genes. For each gene, reverse genetics asks 'If I change the activity of this gene, what is the effect on the organism?'. The key for systems biology is to ask this comprehensively — to systematically perturb the activity of each and every predicted gene and examine the phenotypic effect. In *S. cerevisiae* and *S. pombe*, researchers have constructed collections of strains in which each predicted gene has been perturbed in a similar manner. For example, there are deletion collections, in which each predicted gene has been deleted by chromosome engineering [58,59], and over-expression collections in which each predicted open reading frame is placed under promoter giving high expression levels [60,61]. With such collections, rather than use random mutagenesis to attempt to identify the genes required for a specific process, one can scan the genome a gene at a time and ask whether perturbing each and every predicted gene individually affects the process of interest. In this section, we examine the progress made in the worm to establish analogous collections, and how the data generated from such systematic surveys of gene function have given key insights into the organization of biological processes and the evolution of gene function.

The publication of the genome sequence of the worm in 1998 opened the possibility of systematic reverse genetics: here are all the genes, what does each one do in vivo? To begin to answer this requires a means of perturbing each gene in a specific and targeted manner that can be carried out at sufficient throughput to analyze an entire genome. In yeast, the relative ease of chromosome engineering made the direction obvious: delete the genes one by one by targeted disruption and see what happens. In the worm, no such technology existed — targeted knockouts were a pipedream. By pure chance, the key enabling technology for systematic reverse genetics in the worm was discovered at a time that coincided nearly perfectly with the assembly of the genome sequence. The first paper reporting the potent and specific action of RNA interference (RNAi) was published in 1998 [62]; by 2000 around 25% of worm genes had been targeted by RNAi and the effects reported [63,64]; and by 2003 the results of the first genome-scale RNAi screens were published [65–69]. The speed with which RNAi was adopted and used by the community was remarkable: here was a magic switch that allowed a researcher to turn down the level of any gene of interest, specifically, and in vivo. Not only was it a technology perfectly married to the genome sequence, but it was also astoundingly easy and cheap. The discovery that dsRNAs could be delivered to worms not only by injection or soaking following in vitro synthesis, but that feeding dsRNA-expressing bacteria to worms could generate similar knockdown effects [70–72] paved the way for the development of a genome-scale RNAi library of dsRNA-expressing bacteria covering the great majority of predicted genes [66]. With this library, simply by feeding bacteria to worms, any researcher can target a specific gene in one worm or in a billion, examine all the members of a molecular pathway, target pairs of genes, or — most relevantly for system biology — carry out genome-scale RNAi screens.

RNAi has completely transformed worm genetics, and to date over 50 genome-scale screens have been carried out. Each screen walks through the genome, a gene at a time, and asks 'If I knock down expression of this gene by RNAi, does it affect my process of interest?'. An RNAi screen, then, is just like a classical genetics screen but without the need for positional cloning and without random mutagenesis and the resulting uncertainty concerning coverage — every gene can be targeted, (almost) every gene can be tested. Screens have been done for genes that affect crude phenotypes such as viability or brood size [66]; for genes that are involved in the development of specific tissues such as the vulva (e.g., [73]) or the neuromuscular system [74]; and for genes that affect individual molecular processes such as DNA damage response [68], fat metabolism [65], or even RNAi itself [75]. The initial output of any screen is a list of validated hits: this set of genes gives this specific phenotype when knocked down by RNAi. These hits are curated in public databases, most obviously in WormBase [76,77], and now any researcher interested in a specific gene can, within a few clicks, identify the set of known RNAi phenotypes for that gene as well as identifying all other genes that share similar RNAi phenotypes; for around 1000 genes one can even view detailed time-lapse movies showing the effect of removing that gene on early embryogenesis. This wealth of functional data is a source for many systems-level analyses, several of which are covered in subsequent sections.

Despite the power of RNAi, no technology is perfect and no screen is ideal — false positives and false negatives can confound many systems analyses if they are not measured. In a typical RNAi screen the false negative rate can be estimated based on recovery of previously known genes — typically false negative rates are ~50% [66]. For some screens this is higher, either because of the difficulty of the assay or because the process being examined is partially refractory to RNAi. Certain tissues, most specifically neurons, are partially protected from RNAi [66] owing to reduced efficiency in dsRNA uptake, and this leads to a greater false negative rate due to insufficient knockdown. These issues can be largely circumvented, however, either by using mutants that have increased RNAi efficiency such as *rrf-3* [78] or *lin-35* [74,79] or by engineering worms to increase dsRNA uptake in neuronal tissues [80]. The intrinsic false positive rate of RNAi in the worm appears very low — estimates from genome-scale RNAi screens place the level of false positives due to off-target effects that cannot be trivially excluded at well under 5% [66]. At first sight this appears puzzling, given the large confounding issues with off-target effects in mammalian RNAi screens. However, RNAi in mammals and RNAi in worms are fundamentally different. In mammals, typically a single siRNA (or at best a small pool) is used to target a transcript in any individual assay [81]; that siRNA affects more transcripts than the intended target and hence has off-target effects at the concentrations required to generate efficient knockdown. In the worm, however, the dsRNAs used to target a gene are large, typically in the range of 1–1.5 kb — these were processed in vivo by Dicer [82–84] to yield a complex mix of siRNAs. Each siRNA has a different set of off-targets; all share the same 'true' target. The key is that unlike in mammals, where an individual siRNA is delivered at concentrations sufficient to generate knockdown, in the worm any single siRNA is present at extremely low concentrations — knockdown is achieved by the additive effect of all siRNAs. In this way off-targets are at undetectable levels (since off-targets are siRNA-specific), whereas knockdown of the intended target is efficient (since all siRNAs target the same transcript).

As with any screening technology, hits need to be validated, which can either be done quickly by testing that two non-overlapping dsRNAs targeting the identified hit yield the same RNAi phenotype (thus excluding any possibility of off-target effects) or, more convincingly, by confirming any RNAi phenotype by using a genetic mutant. In this regard, the large-scale ongoing efforts to generate loss-of-function genetic alleles for all *C. elegans* genes are immensely useful. These methods either use random mutagenesis and PCR-based methods to identify substantial deletions within coding regions (and thus presumptive null alleles) (reviewed in [85]) or, more recently, efforts are under way to use the immense power of next-generation sequencing to identify nonsense mutations in each and every gene following high doses of mutagen treatment [86]. Together with pre-existing alleles isolated in forward genetic screens and freely available from the *C. elegans* stock center (CGC, https: //cgcdb.msi.umn.edu/), such mutant collections mean that for a large proportion of hits in any RNAi screen, one can simply order the corresponding deletion allele; combining this with Mos1-based methods, it is possible to extend this to follow up essentially all the hits from any RNAi screen with genetic mutants in a fairly rapid manner, a major advance for the field. Note that in this chapter we have not discussed in any depth the impact that these deletion collections have made on *C. elegans* systems biology, since to date there has been little systematic phenotypic characterization of these deletion lines beyond the identification of essential genes, but this is likely to come in the future. For now, being able to move rapidly from finding a hit in an RNAi screen to getting a genetic mutant in that gene simply by sending an email is already a huge impact.

RNAi screens provide a rapid way to systematically assess gene function in vivo, and it is worth returning briefly to compare their strengths with classical genetics screens. Classical genetics screens had several problems discussed above: the difficulty of examining the roles of essential genes in early embryogenesis due to maternal contribution; difficulty in assessing saturation; and complications arising from genetic interactions, both because of problems with de novo identification of synthetic genetic interactions and because of difficulties in assessing the specificity of genetic interactions. RNAi, in each of these areas, is highly complementary to classical genetics. First, RNAi targets all mRNAs whether derived maternally or zygotically. RNAi is thus an ideal tool to investigate the roles of all genes in the key processes of early embryogenesis. Several groups have carried out genome-scale screens to identify all genes required for early embryogenesis, and have gone on to examine in great detail the precise nature of their role through time-lapse observation of developing embryos [64,87–90]. In this way the exact defects leading to embryonic lethality can be pinpointed — polarity defects, cytokinesis failures, errors in spindle orientation and so on. This level of detail has allowed the construction of complex models of the machineries regulating early embryogenesis in the worm [87,89,91], and these are described in more detail below. Second, coverage is far easier to assess with RNAi than with random mutagenesis screens, with their unknown biases. The typical RNAi screen uses the Ahringer library [63,66] which covers ~80% of predicted genes; most screens have an estimated false negative rate of a little under 50% [66]. Thus coverage can be readily assessed, making systems analysis far easier.

Identifying synthetic genetic interactions by classical genetics is hard without a starting point: if two independent genes must both be mutated in order for a phenotype to become detectable, this is extremely improbable in any screen unless they are chromosomal neighbors, for example *lin-15A* and *lin-15B* [36] or (in an example from the fly) *rpr, hid* and *grim* [92–94]. By RNAi, however, this raises few problems aside from those of scale. Genetic interactions can be mapped de novo either by carrying out an RNAi screen in a mutant background and comparing the phenotypes observed with those seen in wild-type worms [39,95,96], or by carrying out combinatorial RNAi [97] in an RNAi hypersensitive strain like *rrf-3* [69] or *lin-35* [74,79]. This not only allows the direct and unbiased identification of synthetic genetic interactions, but can also address directly the issue of specificity of any genetic interaction detected. For example, we previously investigated genetic interactions between genes involved in *C. elegans* signal transduction and transcriptional networks and identified a set of ~350 genetic interactions out of ~65 000 tested pairwise interactions [39]. Unlike in classical genetic screens, we know immediately which interactions yielded synthetic phenotypes without any need for cloning, we know the number of tested interactions precisely (and thus can estimate the rate of non-additive genetic interactions in vivo), and we can also estimate the specificity of any interaction with a component or pathway tested. Whereas most genes show synthetic genetic interactions with a single signaling pathway, a handful of genes, all encoding chromatin regulators, have synthetic interactions with every signaling pathway tested. Similar 'hubs' in the genetic networks in humans may have key disease importance, as inherited variation in such hubs would be likely to affect the phenotypic outcome of a large proportion of other personal variant alleles.

Taken together, then, what have these RNAi screens taught us about gene function in the worm in a way that would have been inaccessible from classical genetics? First, it is possible to examine the relationship between the molecular functions of a gene and its in vivo role in a way that was never previously possible. Crude trends are immediately obvious — almost half the genes giving sterile phenotypes are involved in protein translation, whereas genes that only affect more complex post-embryonic tissue development are highly enriched for signaling molecules and transcription factors [63,66]. Second, one can also begin to see evidence of the role of domain innovation in the evolution of novel biological processes in a way that is statistically relevant and not anecdotal. For example, genes that have RNAi phenotypes that only affect complex multicellular processes such as tissue organization, body size regulation, or the coordination of movement are enriched for domains that are only present in animal genomes, suggesting that the evolution of complex behavior and multicellular architecture has in part been driven by the evolution of novel protein folds and functions, rather than simply through the rewiring of existing components and domains [66]. Third, systematically derived phenotypic datasets such as those from RNAi screens can be easily compared to analogous datasets in other organisms: for example, over 60% of the 1:1 orthologs of *S. cerevisiae* 'essential' genes have lethal or sterile RNAi phenotypes in the worm [98], showing that the core functions of the eukaryotic cell are highly conserved and carried out by identical machineries across long evolutionary periods. RNAi-based genetic interaction screens in the worm indicated that ~4% of genetic interactions in the worm are non-additive [39]; similar analysis of synthetic lethal interactions in yeast [99] gives a similar proportion of non-additive interactions, suggesting that the underlying architectures of genetic interaction networks are maintained over long evolutionary timescales.

One can also use the data deriving from systematic RNAi screens to re-examine some of the basic implications of the neutral theory of evolution and natural selection. Some genes have extremely strong RNAi phenotypes: worms with these genes knocked down either die or are completely sterile. Others have much weaker defects, such as mild deviations from normal vulval development or subtle body length changes. Finally, there are a large number of genes that have no detected phenotype in any of the 50 genome-scale screens to date. Both intuitively and based on theoretical predictions, it seems reasonable that the stronger the loss-of-function phenotype of a gene (that is, the stronger the effect of inheritance of a mutation in that gene), the stronger the strength of negative selection acting on that gene over evolutionary timescales. With the systematic datasets deriving from RNAi screens, one can begin to examine these predictions. At a crude level, there are indications that this is true: a higher proportion of genes with strong (i.e., lethal or sterile) RNAi phenotypes have a yeast ortholog than genes with weak or no detectable RNAi phenotypes [63,66]. Looking at more sensitive tests for strength of negative selection, one can indeed see a difference in ka/ks for genes with sterile phenotypes compared to other genes [100]. Most recently, assays were developed to examine the effects of targeting genes by RNAi on the overall reproductive fitness of a population of worms [101]. Strikingly, these studies indicate that the great majority of genes are required for wild-type fitness and thus have detectable RNAi phenotypes (unlike had been surmised from previous screens) and, furthermore, that there is a clear, albeit weak, correlation between the level of requirement of a gene for wild-type fitness and the level of negative selection acting on that gene as measured by ka/ks. The systematic and unbiased datasets deriving from RNAi screens in the worm thus allow one to confirm at last some of the basic predictions of molecular evolution and natural selection.

Finally, one can use systematic RNAi data to examine large-scale functional organization of chromosomes. At one extreme, one might imagine the organization of genes on chromosomes to be entirely random and genes with similar phenotypes to therefore be scattered evenly across the entire genome. The other extreme alternative would be a situation of high compartmentalization of the genome — all genes affecting neuromuscular development might reside in clusters, all genes required for normal vulval development in other clusters. Some clustering is immediately evident: of the genes with detectable phenotypes in a series of genome-scale RNAi screens, ~70% of those on autosomes have non-viable (lethal or sterile) phenotypes, whereas only ~30% of genes on the X chromosome have non-viable phenotypes [66]. Looking more closely, one can identify clear clustering of genes with specific phenotypes in broad chromosomal domains: the autosomal central clusters are highly enriched for non-viable genes and the autosomal arms are typically under-enriched for non-viable genes, whereas the X chromosome is enriched for genes with more subtle, post-embryonic phenotypes [108]. None of the above questions could have been examined without the systematic, unbiased datasets provided by RNAi screens.

Genome-scale RNAi screens have therefore been key for systems biology in the worm. They have fleshed out many of the pathways initially identified by classical genetics screens — the first genome-scale RNAi screens to be published multiplied the number of genes with identified in vivo loss-of-function phenotypes by a factor of 3 over everything identified in 30 years of forward genetics, as an example [66]. They have identified the great majority of genes that orchestrate the events of early embryogenesis, pinpointing exactly which specific process they control, and such high-resolution high-coverage screens have allowed both the prediction of molecular functions of previously unknown genes and the systems-level analysis of embryogenesis. Specifically, the unbiased nature of the screens has also allowed researchers to ask basic questions about genetics — how the molecular function of a gene relates to its organismal role, how new functions arise in biology, and how natural selection acts through perturbed phenotypes. RNAi emerged at a perfect time in the history of worm research, but it also helps fill in some of the dark spaces that are hard to illuminate with classical screens — combinatorial genetics, early roles for maternally expressed genes, and so on. Likewise, classical genetics provides tools in the form of genetic mutants, but also can attack areas that RNAi will always miss — hypermorphic alleles, for example, have been critical in dissecting signaling pathways (e.g., [35,102]), but RNAi cannot generate these, nor can it trivially generate loss-of-function phenotypes in screens for genes whose products have long half-lives. RNAi and classical genetics are likely to coexist for a long time to come, each providing different views of gene function and tools for further experiments.

The loci defined by classical genetics screens, and their ordering into genetic pathways by epistasis, provide an entry point for functional studies. However, without knowing what the identified genes encode, where and when they are expressed, and how their protein products interact, a genetic pathway remains an abstract informational entity. Going from the relationships between genotype and phenotype revealed by genetics (whether forward or reverse) to molecular models of the machineries underpinning biology requires many other data sources. In the following sections we describe systematic approaches to examine gene expression patterns, gene regulation, and physical interaction networks in the worm, and then the computational approaches developed to integrate genetic, gene expression, physical interaction, and other datasets into single networks that can in turn provide insight into the functional modules that coordinate and carry out worm development.

EXPRESSING AND REGULATING AN ANIMAL GENOME

One central challenge in systems biology is to understand — and to be able to predict — how genes are expressed in a reproducible spatial and temporal pattern. Such a grand challenge can be tackled at multiple levels, for example by detailed analysis of the *cis* and *trans* regulatory control of individual genes, by focusing on the regulatory interactions involved in specifying particular cell lineages or tissue types, and through the global mapping of gene expression patterns and regulatory interactions. Moreover, whereas many studies have focused primarily on understanding transcriptional control, post-transcriptional processes such as the regulation of mRNA stability, splicing and translation must also be considered. Ultimately, a predictive model of gene regulation during development will also need to incorporate the signaling interactions between cells and how these connect to gene expression, as well as biophysical influences on development, pattern formation and gene expression.

Genome-Wide Maps of Normal and Perturbed Gene Expression

The invention of microarrays, and more recently the implementation of massively parallel sequencing technology, allows the levels of essentially all mRNAs to be assayed in parallel (see chapter by Guigo). Initial gene expression profiling studies used microarrays to quantify the levels of mRNAs in RNA samples extracted from whole animals or whole embryos, for example to map how gene expression changes during development [103], as animals age [104–106], or in response to environmental perturbations such as heat shock [107]. These early experiments, together with expressed sequence tag (EST) cDNA sequencing, also served an important role in validating the existence of many only computationally predicted genes.

More recently, deep sequencing and tiling microarrays have been used to provide a more comprehensive view of the transcribed genome, including maps of 5′ ends [108], *trans*-splicing [109], alternative slicing [110], 3′ UTRs [111,112], and non-coding RNAs [113]. For example, hundreds of alternative splicing events were found to be developmentally regulated [110]. Moreover, expression has been compared between two different nematode species [114].

A major technical challenge when mapping gene expression in a rapidly developing multicellular animal such as *C. elegans* is the need to map gene expression to individual cells or tissues. Three main approaches have addressed this challenge: (1) using mutants with transformed cell fates so that the majority of the mRNA derives from a single cell lineage or type; (2) using dissection to isolate tissue samples; and (3) using cell sorting to isolate cells expressing a reporter gene of interest. Examples of the first approach include studies that have used well-characterized mutants that affect the fates of early blastomeres to map genes expressed in particular tissues and lineages, such as the pharynx [115], the C lineage [116], and the male and female germlines [117]. Examples of the second approach include the manual dissection of germline tissue [118], and examples of the third approach include the use of tissue-specifically expressed poly(A) binding protein to isolate genes expressed in the muscle [119] and intestine cells [120], and the use of tissue-specific GFP reporters and cell sorting to map genes expressed at particular time points [121] or in particular embryonic cells, such as the touch receptor neurons [122], and later in many different cell types as part of the mod-ENCODE project [123].

A few studies have also used genome-wide expression profiling to understand gene regulation, an elegant example being studies on the development of the pharynx. Here, genes expressed in the pharynx were first globally identified by comparing the expression profiles of mutant embryos with excess or missing pharyngeal cells. Most pharynx-expressed genes identified this way contained binding sites for the master regulator of pharyngeal development PHA-4 in their promoters, and interestingly, the strength of these binding sites correlated to some extent with the timing of transcription induction [115]. Further, by searching for enriched sequence motifs in the promoters of genes activated either early or late during development (information that was obtained from the NEXTDB in situ database, see below), motifs that act in combination with PHA-4 to specify early or late expression were also identified [124].

As a second example, microarray-based expression profiling has also been used in combination with genetic perturbations to map gene regulatory networks during the development of the posterior of the embryo [116,125]. Here, genetic mutants with fate conversions were used to identify genes expressed in the C lineage [116], and a combination of expression profiling in mutants, yeast one-hybrid analyses and reporter construct experiments was used to infer regulatory interactions among transcription factors [116,125]. Examining the interactions among transcription factors specifying muscle and epidermal cell fates allowed the authors to propose that these two fates cross-repress each other, and also that the two pathways differ in the topology of the interactions that lead to all-or-none fate conversions [125]. Two technical challenges that have possibly discouraged the wider adoption of this approach are, first, the need to obtain pure tissue samples in the perturbed animals and, second, the redundancy and compensation in transcription networks that mean inhibiting even master regulators sometimes only has a subtle effect on gene expression [126].

Single Cell-Resolution Analysis of Gene Expression

An alternative approach to using microarrays or RNA sequencing to globally gene expression is a 'gene-centric' approach that aims to map all of the cells in which a particular gene is expressed. Five important contributions have been made towards this goal in *C. elegans*.

First, the spatial and temporal expression of many genes has been studied at low resolution by in situ hybridization, with the resulting images available in a free-to-access database (NEXTDB, http://nematode.lab.nig.ac.jp/). These in situ data provide information on the expression of thousands of genes, but have the disadvantage that the expression patterns have not been systematically annotated in a way that makes them useful for computational analysis. Obviously, information is only provided on mRNA, not protein expression.

Second, a computational approach has been developed that can recognize 357 of the 558 nuclei in the first (L1) larval stage from their reproducible locations with 86%

accuracy [127]. This allows the expression patterns of fluorescent protein reporter constructs such as GFP to be mapped at cellular resolution in the L1 larva. To date, data have been released for the expression patterns of GFP reporters driven by the promoters of 93 different genes [128]. Mapping these expression patterns onto the lineage fates predicts when their expression first starts during development, and at which point cells diverge in fate [128].

Third, an approach has been developed that allows for the semi-automated tracking of all nuclei up to the ~350 cell stage of embryonic development, using time-lapse confocal microscopy and histone−GFP reporter constructs to visualize all nuclei [129]. The method can be used to map gene expression patterns using a strain in which histone−GFP reporters are expressed in all nuclei and a histone−mCherry reporter is expressed from the promoter of a gene of interest. Although relatively low throughput in capacity (for each gene a transgenic histone−mCherry reporter strain must be constructed, and each dataset requires long time-lapse recordings and a substantial effort to correct the tracked lineages), such an approach has the potential to map the expression patterns of genes of interest at cellular resolution, including following perturbation experiments [130].

Fourth, spatiotemporal expression profiles have been generated for ~900 different promoter−GFP constructs using an adapted flow-cytometer (a 'worm sorter') [131]. The advantages of this technique are the high-throughput generation of profiles, and the quantitative expression data. The main disadvantage of the technique is that the spatial expression information is compressed from three dimensions to one, with expression levels only quantified along the long axis of the worm.

Fifth, an approach has been developed that uses multiple (~50) fluorescently labelled probes to visualize individual mRNA molecules in developing embryos (single molecule fluorescence in situ hybridiazation, smFISH, [132]). To date this method has only been used to quantify expression levels in whole embryos, and in particular how this expression varies across isogenic individuals [133,134]. There is, however, the potential that smFISH could be quite widely used to provide cell-resolution information on gene expression patterns for many genes, although at the moment the cost of the probes would limit such a project. A disadvantage of smFISH is that it uses fixed samples, so dynamic changes in gene expression must be indirectly inferred.

Global Maps of Transcription Factor-Binding Sites and Chromatin Organization

One central aspect of gene regulation is the physical binding of transcription factors to specific regions of the genome. Recently, binding sites for transcription factors have been mapped genome-wide in multiple organisms. In *C. elegans* two principal approaches have been used. The first is a 'gene-centered' approach where the yeast one-hybrid system is used to test for interactions between a promoter of interest and each transcription factor from a library expressed in the yeast nucleus [135] (see Chapter 4). The second approach involves cross-linking proteins to DNA, shearing chromatin into small fragments, precipitating DNA bound to a protein of interest using antibodies, and then identifying this DNA using microarrays or deep sequencing (referred to as ChIP-CHIP or ChIP-seq, respectively; see Chapter 4).

C. elegans protein−DNA interactions have been mapped quite extensively using the yeast one-hybrid system and, when combined with perturbation experiments, have provided insights into the regulation of intestine-expressed genes [135], the regulation of fat metabolism [136], neuronal development [137], feedback interactions between transcription factors and miRNAs [138], and the functional divergence of transcription factor gene families during evolution [139].

The mapping of *C. elegans* protein−DNA interactions in vivo by chromatin immunoprecipitation has been most extensively undertaken as part of the modENCODE project. Here the strategy has been to express GFP-tagged transcription factors from large genomic constructs (to recapitulate as closely as possible the endogenous regulation of a gene) and to use ChIP-seq to determine sites of enriched binding along the genome [140]. Analyzing the enrichment sites of 22 transcription factors revealed a surprising lack of specificity in their binding, with many inferred binding sites detected in the upstream regions of highly expressed genes [140].

As for most of the gene expression datasets discussed above, the current maps of transcription factor-binding sites in *C. elegans* suffer from the major disadvantage that they were performed from whole embryo or whole worm extracts, and thus represent the average binding signals from complex mixtures of cell types. Thus transcription factor binding that is specific to only a subset of cells may not be detected, and signals from different cell types may be mixed. Moreover, information on how transcription factor binding changes in different cell types is lacking. In the future, approaches such as those employed in *Drosophila* to perform ChIP from individual tissues [141] may prove useful. Moreover, at least in *C. elegans*, there is still a big gap between the genome-wide maps of transcription factor binding and an understanding of how genes are differentially regulated in space and time: to date there has been little success in using genome-wide transcription factor-binding datasets to predict gene expression.

A second aim of the modENCODE project has been to produce genome-wide maps of chromatin organization [142]. Here immunoprecipitation is performed with antibodies against histone post-translational modifications [142] or histone variants [143] rather than transcription factors. Clustering the data from ChIP-CHIP experiments mapping 19 histone modifications and eight chromatin-associated proteins revealed five groups of marks associated with X chromosome silencing, gene activation and gene repression. Moreover at the chromosome scale, the pattern of histone modifications mirrored those of gene expression, with repressive marks concentrated in the regions of elevated meiotic recombination in the chromosome arms [142]. Similarly, genome-wide maps of chromosome binding by the dosage compensation complex responsible for downregulation of gene expression from both X chromosomes in hermaphrodites have been constructed, revealing extensive binding at transcription start sites [144] and sequence-independent propagation of the complex from X-linked recruitment sites [145].

Perhaps one of the most interesting insights obtained from these genome-wide maps of chromatin organization is evidence for the epigenetic transmission of information from the germline to the zygote [146]. Mapping the binding sites of the H3K36 methyltransferase MES-4 in embryos revealed a striking correlation between MES-4 binding and genes that were previously expressed in the maternal germline [146]. Together with the semi-stable transmission of RNAi-induced chromatin states [147] and phenotypes [148], this provides genome-wide evidence for the transmission of chromatin states across generations. As for transcription factors, a major issue with mapping chromatin states in *C. elegans* is the diversity of cell types present in a typical sample. To date, cell-type specific maps of chromatin modifications have not been produced. Also, chromatin confirmation capture-based approaches to map the three-dimensional organization of chromosomes in the nucleus (see Chapter 7) have not yet been applied to *C. elegans*.

Variation in Gene Expression among Individuals and its Phenotypic Consequences

Isogenic individuals often show substantial phenotypic variation, even when they share a common, controlled environment. This is true of species from microorganisms to mice, and identifying the causes and consequences of inter-individual variation presents an important challenge for systems biology. *C. elegans* represents an attractive model for the study of inter-individual variation: laboratory strains are isogenic and self-fertilizing, wild-type animals produce several hundred genetically identical offspring, the invariant cell lineage means that individual cells can be identified with confidence, and the small and transparent embryo allows behaviors and gene expression to be easily quantified in vivo. As such, a series of recent studies have examined different aspects of inter-individual phenotypic variation in *C. elegans*, focusing both on the causes of phenotypic variation in wild-type individuals [149–152] and on the causes of inter-individual variation in the outcome of mutations [133,134,149].

One of the most striking phenotypes with inter-individual variation in both humans and worms is lifespan. In laboratory conditions, *C. elegans* has a lifespan of about 2 weeks, but there is substantial variation in lifespan between isogenic individuals even on the same agar plate [152,153]. Several studies have now identified gene expression variation that can predict with reasonable accuracy the remaining lifespan of individuals part way through their life [150–152]. The first of these studies focused on variation in lifespan after the application of a lifespan-extending heat stress [151]. Animals experiencing a mild environmental stress such as a heat shock induce the expression of molecular chaperones, and display an extended lifespan. Using an *hsp-16.2–GFP* reporter construct and a 'worm sorter' to quantify expression in individual worms revealed substantial variation in the induction of chaperones after a heat stress. Moreover, separating individuals according to their levels of chaperone induction revealed that those with higher chaperone expression tended to live longer [151]. Variation in *hsp-16.2* induction correlates with inter-individual variation in the duration of stress signaling, and so likely reflects pre-existing molecular heterogeneity in a population [149]. This heterogeneity might reflect a fitness trade-off between reproductive development and stress resistance: although individuals with higher chaperone expression live longer and are more stress resistant, they develop more slowly and so will be outcompeted in benign conditions [149].

More recently, reporters have also been described that can predict the remaining lifespan of non-pre-stressed individuals [53,68,73,101,110,117,119,132,134,146,150–152,154,155]. These reporters include components and targets of the insulin-like/DAF-16 signaling pathway, and may reflect the differing extent to which worms are experiencing or detecting pathogenic infection [152]. Additional reporters that partially predict remaining lifespan quantify the expression of miRNAs previously suggested to influence lifespan. Indeed, quantifying the early-to-mid adult expression of three miRNAs in individuals predicted more than 40% of the variation in their remaining lifespan. Two of these miRNAs may act upstream of the insulin-like signaling pathway to influence lifespan variation [150].

The other context in which inter-individual variation has been studied in *C. elegans* is with respect to variation in the outcome of mutations. Many mutations are incompletely penetrant, affecting only a subset of isogenic individuals who carry them. For example, partial loss-of-function mutations in the maternal transcription factor SKN-1 only cause defects in intestine development in some worms. Using single molecule fluorescence in situ hybridization it was shown that the expression of a gene downstream of SKN-1 becomes highly variable in the mutants, with this variation propagating to an all-or-none induction of the master intestine regulator ELT-2. Thus it was argued that variation in gene expression underlies incomplete penetrance [134].

A second study addressed variation in the outcome of a deletion of the gene *tbx-9* [133]. In this study, live imaging of embryos expressing fluorescent protein reporter constructs was used to relate early variation in gene expression to later variation in the phenotypic outcome of the mutation. Variation in two buffering systems was found to partially predict the outcome of the mutation in each individual. First, the compensatory upregulation of a partially redundant gene, *tbx-8*, was found to vary among individuals, with very strong upregulation masking the effect of the inherited mutation in *tbx-9*. Second, a surprising amount of variation in the induction of molecular chaperones such as Hsp90 (DAF-21) was observed among isogenic embryos, and this variation also partially predicted the later outcome of the inherited mutation. By simultaneously quantifying the variation in both systems, more accurate predictions could be made about the phenotypic outcome of the mutation in each individual [133]. Chaperone activity is an important influence on the outcome of many mutations, so inter-individual variation in chaperone expression may be quite a promiscuous determinant of mutation outcome [149].

Proteomics in *C. elegans*: Global Maps of Protein Expression and Protein—Protein Interactions

Compared, for example, to the situation in yeast, proteomic approaches have not been very widely employed by the *C. elegans* community. However, shotgun proteomics (see Chapter 1) confirmed the expression of >10 000 proteins (more than half of genes) [156]. Not surprisingly, the set of identified proteins was biased against hydrophobic transmembrane proteins, short proteins and genes expressed in only a small number of neurons, highlighting further improvements that are needed in the approach. Comparing mRNA and protein expression levels between *C. elegans* and *Drosophila melanogaster* suggested that protein levels might be more conserved during evolution than mRNA expression [156]. Quantitative proteomics has also been used to validate predicted miRNA targets [157], and proteomic approaches have been used to systematically identify sperm chromatin components [158], targets of the insulin-like signaling pathway [159], and to identify several hundred proteins that aggregate in old animals [160]. This last study suggested that protein insolubility is an inherent part of normal aging in *C. elegans*. Efforts have also been initiated to systematically map protein post-translational modifications, with nearly 7000 phosphorylation sites identified on 2400 proteins [161]. The *C. elegans* genome encodes a large number of nematode-specific kinases, and indeed the phosphoproteome was found to be rather distinct from that of other higher eukaryotes, with different inferred kinase substrate motifs and low conservation of the phosphorylated sites in other higher eukaryotes [161].

C. elegans has been a leading model system for the comprehensive mapping of protein—protein interactions, primarily using the yeast two-hybrid system (see Chapter 3) In this approach, two proteins (or protein domains) are fused respectively to a DNA-binding domain and a transcription activation domain, and are expressed in the yeast nucleus. If the two proteins interact, then expression of a reporter construct is activated. Early pioneering studies systematically mapped protein interactions among proteins implicated in regulating a common developmental process — vulva development [162] — among subunits of a protein complex [163], or among genes expressed in the germline [164]. Additional studies used protein interaction mapping in combination with genetic perturbations to identify new regulators of the DNA damage response [165] and TGF-β signaling [166]. These efforts were extended to the scale of a few thousand screens against cDNA libraries, identifying >2000 high confidence two-hybrid interactions [167]. More recently, ~1800 yeast two-hybrid interactions were reported from an experiment testing for interactions among ~10 000 proteins [168]. Moreover, a strategy to identify interaction domains has allowed the minimally interacting regions to be identified for over 200 proteins involved in early embryogenesis [169]. These datasets provide a fantastic resource for the *C. elegans* community, for example allowing researchers to identify new proteins potentially involved in a process of interest. However, it is important to note that efforts to map protein—protein interactions in *C. elegans* (the 'interactome') are far from complete: to date, <5000 protein interactions have been described, from an estimated total of >100 000 interactions [168]. Moreover, no large-scale studies have been initiated to map protein interactions in *C. elegans* through the isolation of protein complexes, and more work is needed in this area.

INTEGRATIVE AND DYNAMIC MODELING TO LINK GENOTYPE TO PHENOTYPE

Data Integration and Genome-Scale Networks to Connect Genes and Modules to Phenotypic Variation

Aside from yeast, *C. elegans* is still the only eukaryote in which systematic and comprehensive reverse genetic screens have been performed for many different phenotypes. This has led to *C. elegans* being adopted as a model system for developing and testing computational approaches that aim to globally identify and to predict how proteins are organized into molecular machines and functional modules, and how genetic perturbations in these modules relate to phenotypic variation [170].

One of the first studies to use data integration to predict how genes interact focused on 600 transcripts expressed in the germline [164]. Three types of data were used to link genes in a network: protein–protein interactions from two-hybrid experiments, the sharing of common loss-of-function phenotypes reported in RNAi experiments, and correlated expression in microarray experiments. Evidence was presented that genes predicted to interact functionally by more than one type of evidence were indeed more likely to act in a common process [164].

This approach has been extended with the inclusion of much more detailed loss-of-function phenotypes curated from time-lapse recordings of the first two cell divisions of embryonic development [87], and from imaging the architecture of the germline [91]. In the first study, protein interaction, expression similarity, and phenotypic similarity were again used to connect genes in a network. By only considering interactions predicted by two or more pieces of evidence, it was suggested how 305 essential genes are organized into quite distinct functional modules, and protein localization experiments were used to provide further evidence to support functional predictions for ten previously uncharacterized proteins [87]. In the second study, phenotypic profiles alone were used to construct a network for >500 essential genes that had similar resolution to genetic interaction-based profiling in yeast (see Chapter 6).

These studies employed the intuitive approach of constructing networks of interactions by using either all interactions derived from different types of data or by taking the overlapping between different datasets. The disadvantage of the first approach is that both reliable and non-reliable interactions are combined. This can result in large networks, but these networks are potentially contaminated by many false positive (i.e. wrong) interactions. The disadvantage of the second approach is that, by only considering interactions supported by more than one piece of evidence, many real interactions are probably discarded, thereby dramatically reducing both the number of interactions and the number of genes included in a network. That is, the reliability of the network is increased, but at the expense of reducing its coverage.

A general strategy to avoid this loss of coverage is to use more sophisticated machine learning-based methods for data integration [171–175]. Many different datasets can be used to predict whether two genes functionally interact or act in a common molecular process, but these datasets can be of very different reliability. The trick is first to estimate how reliable each dataset is, and then, when integrating the datasets, to weight them according to this reliability score. Thus a dataset measured to have high reliability would contribute more to a final network than a dataset measured to have low reliability. In effect, each interaction in the final network has a confidence score assigned to it, and a high confidence interaction might derive from either a single piece of reliable evidence or from multiple less reliable sources. This approach therefore allows large networks to be constructed without sacrificing quality. Moreover, because each interaction in the final network is a probability, one can choose to analyze a very large network including both low and high confidence interactions, or a smaller network only considering more reliable interactions above a chosen threshold [171,174].

The first study to use a machine-learning approach to predict functionally related genes in *C. elegans* used logistic regression to combine interactions derived from protein interactions, gene expression, phenotypes, and functional annotations from three model organisms [175]. To benchmark the interactions, a gold standard positive set of physical and genetic interactions was used. This allowed a network of >17 000 interactions to be constructed. The utility of the network was demonstrated by identifying new modifiers of mutations in the *let-60 (Ras)*, and *itr-1* (1,4,5-trisphosphate (IP_3) receptor) genes [175].

A second integrated network, Wormnet, was constructed from even more diverse data sources, for example protein interactions and co-expression relationships from many species, the co-citation of genes in the scientific literature, and the co-inheritance of *C. elegans* genes across bacterial species [172]. The different datasets were benchmarked using the sharing of Gene Ontology annotations between genes, and integrated using a modified naïve Bayesian approach. No data from genome-wide RNAi screens were used to construct the network, allowing these gene–phenotype links to be used to evaluate how well the network predicts loss-of-function phenotypes. Strikingly, genes that share common RNAi phenotypes are highly clustered in the network. This demonstrates that genes linked to diverse phenotypes ranging from the subcellular to the physiological level (such as lifespan) can be successfully predicted in an animal using a single integrated network. Indeed, the network was used successfully to predict new suppressors of the Retinoblastoma/SynMuv tumor suppressor pathway and to connect the Dystrophin and Ras/MAPK signaling pathways. The second updated release of the network has improved predictive performance, and consists of close to 100 000 links between >15 000 genes (75% of the genome) [173].

Genome-wide integrated networks provide a global picture of how biological functions and phenotypes relate to the underlying molecular machinery. However, they also provide a rich resource for a research community. For example, the networks capture and summarize a huge amount of existing biological knowledge and so can be used to rapidly search for connections between a novel set of genes, such as those identified in a genetic screen or in an expression profiling experiment. Moreover, given a set of genes known to be important for a particular process, a network can be used to predict new genes that are likely to function in that process. Search tools such as that provided for Wormnet (www.functionalnet.org/wormnet) make this process of 'network-guided genetic screening' straightforward.

Dynamic Models of Developmental Processes

One ultimate goal of systems biology is to produce mechanistic and predictive models describing biological systems. Such mechanistic modeling necessarily involves considering the dynamics of a system, but dynamical modeling is, unfortunately, another area where systems research in *C. elegans* has rather lagged behind that in the other two main invertebrate systems of yeast and *D. melanogaster*. Indeed, only two developmental processes have received substantial interest from the modeling community: symmetry breaking in the first cell division of the embryo [176,177], and the early development of the hermaphrodite vulva [48,50,51,178–180].

With respect to the first cell division, two elegant studies have aimed to connect biophysical processes to the generation of asymmetry, modeling the physical mechanisms by which stable asymmetric distributions of proteins arise [176], and how cortex–microtubule interactions position the mitotic spindle in response to these polarity cues [177]. Crucially, in both cases the plausibility of the biophysical models was assessed using quantitative imaging. For example, in the first study it was shown that advective transport by the flowing cell cortex, when coupled to a PAR protein reaction–diffusion system, can be sufficient to trigger polarity in an otherwise stably unpolarized system [176]. This coupling of mechanical and biochemical modeling is an approach that may be crucial for building additionally mechanistically grounded predictive models of development. In general, a better understanding of biophysical processes during *C. elegans* development is required [181].

The second system that has been quite extensively modeled in *C. elegans* is the patterning of the precursor cells during the early development of the hermaphrodite vulva [48,50,51,178–180]. Here, six cells of initially equal developmental potential adopt three different developmental fates, a process that involves the interactions between at least three different signal transduction pathways. Vulva development has been modelled using a number of different approaches [48,50,51,178–180].

These models have highlighted the importance of a sequential activation of the epidermal growth factor receptor (EGFR) and LIN-12/Notch signaling pathways, and how the lack of a delay in signaling results in unstable fate patterns [48]. Moreover, they have also illustrated how quantitative changes in a common regulatory network may be sufficient to explain the differences in vulva patterning that are observed among different species [179].

As in many other multicellular systems, a key limitation when it comes to constructing dynamical models is the lack of data on important in vivo biochemical and biophysical parameters such as rate constants, protein concentrations, tensions and viscosities. As such, it is likely the combination of modeling and quantitative experimental validation will be most fruitful in the immediate future.

OUTLOOK

In this chapter we have attempted to outline the key systems-level approaches that have been employed to date by the *C. elegans* community, and where future directions might best be focused. However, our aim has also been to emphasize that *C. elegans* is a rather unique animal when it comes to systems biology. This is not primarily because it is a relatively 'simple' and highly experimentally tractable organism, although these are, of course, key advantages. Rather, it is because 'systems thinking' has existed in the *C. elegans* community from its very inception, and because the phenotype of *C. elegans* is a uniquely well-defined and reproducible problem at the cellular level. Only for this animal is the complete list of cell divisions, movements, differentiations and deaths known and quantifiable during development, and only in this animal is the complete anatomy of the adult, including the nervous system, described and quantifiable at cellular resolution. We and others [182] would therefore argue that, if our aim is to achieve a comprehensive and quantitative understanding of an animal life form, then that animal should be *C. elegans*.

Brenner famously quips that there are only three important questions in biology [183]: How does it work? How is it built? And how did it get that way? If we want to answer these questions in a comprehensive and predictive manner, then *C. elegans* was — and still is — the most logical system to study. How does a single cell develop into an animal? How does the organization of that animal determine its functions? And how did such an animal come into being in the first place through the process of evolution?

We will take it as read that there will be continued progress in using the worm to unravel the genetic basis for many complex animal behaviors and processes, from alternative splicing to the development of the neuromuscular system to the role of non-coding RNAs in animal development. Genetic screens, whether using the classical mapping—rescue—cloning approach for mutant alleles, exploiting the power of SNP-mapping and next-generation sequencing [46], or using RNAi, will always be powerful ways to open up new areas of biology. In addition to this, the resolution at which we can view the organismal state is ever increasing — from cellular level anatomy, we are already entering the world of high-resolution molecular anatomy [184], any cell type being defined not only by lineage, function, and morphology, but also by a genome-scale view of molecular expression markers, and soon metabolite markers are likely to follow. This increase in resolution will feed into computational modeling as outlined above, and our view of the wild-type N2 animal and how it got to be that way will always be increasing in resolution. However, we will focus on a small number of areas of potential future progress that address central issues of animal biology and genetics which can only now begin to be addressed thanks to the current state of knowledge and technology.

The first is the move outside the cozy, contained world of the N2 reference and into the far murkier reality of natural variation. Perhaps social scientists will simply see the guiding hand of postmodern relativism at work (inviolate axiomatic frameworks shattered, a world of slippery inconstancy), but ultimately this is a vital area of genetics in which the worm can make a tremendous impact. A central problem in human genetics is to understand the genetics underlying natural variation in quantitative traits, made particularly acute by the advent of personal sequencing. At the heart of this is a major issue of statistical power: given an unlimited number of individuals and crosses, one could map all the possible variants that affect any traits. However, there are only 7 billion humans on earth — an ecologically massive figure, but for the examination of the key issue regarding whether hundreds of relatively rare alleles of low effect size may cumulatively explain a large component of the variation in any trait [185], this is not vast by any means. Every human is unique, and a human geneticist ultimately has to work at best with the 14 billion currently existing human genomes. However, any worm can be cloned indefinitely, and rational mating schemas can be drawn up to generate collections of intercrossed lines which allow the mapping of natural variant alleles that affect any traits of interest [155,186]. Statistical power — an unlimited number of individuals and crosses, and even an unlimited number of individuals of any genotype — is not an issue (except in practice; we choose to ignore this confining issue for now).

To begin to address natural variation in quantitative traits, the worm is now well placed. Thanks to pioneering efforts of the Felix and Kruglyak groups among others, in addition to the reference N2 genome, there is an analog of the human HapMap [187] delineating the genetic relationships among the major subgroups of the *C. elegans* global population. There are collections of recombinant inbred lines (and advanced RIL collections with increased mapping precision) [155,186], introgression lines (ILs) [188], and expression data for many of these which can allow for eQTL mapping in a whole animal. Crucially, these datasets and collections mean that we have an accurate overview on the total genetic variation in the entire *C. elegans* species and, given any specific quantitative trait that varies between any two naturally occurring isolates, we have the means to map the genetic variants that explain the phenotypic variation. Thus, one might ultimately envision a comprehensive analysis of the dissection of the genetic basis for all phenotypic variation in *C. elegans* (or in a related species such as *C. briggsae*) — the insights into the genetics of natural variation in another animal, man, would be huge. We anticipate that in the coming decade or two attempts to understanding the genetic architecture of natural variation in the worm, and the computational predictive methods developed, will play a major role in understanding natural variation in humans and in increasing our ability to predict our phenotypes in health and disease from our personal genome sequences.

To study natural variation in quantitative traits not only requires knowledge of genetic variation, but also requires the accurate measurement of quantitative traits and we anticipate that this will also be a major area of growth. While many of the phenotypes studied in classical genetic screens have been characterized in a quantitative manner — for example the number of vulval precursor cells induced to a vulval fate [35], or the number of corpses in the anterior pharynx of the developing embryo (for example, in [189]) — this quantitation has been almost entirely manual to date. However, an increasing number of groups are pushing ahead in the development of automated phenotyping platforms on a variety of fronts. Analysis of static images [154,190,191], much like the high-content high-throughput style of mammalian cell-based assays, is clearly one option, and software such as CellProfiler [192], an open source image analysis package, may prove to be key tools for the worm. Researchers have also tried to use automated analysis of time-resolved image series, either analyzing films of worms swimming or crawling [193—197], or using automated imaging for the automated lineaging of developing embryos [129,130,198]. Finally, broad population phenotypes such as fitness have also been examined quantitatively using either a commercial worm sorter or even food utilization rates [101,199]. The increased sensitivity and automated statistical definition of mutant phenotypes allows for far more complex genetic analyses than would be possible for crude quantitative phenotypes. Finally, the myriad possibilities opened up by microfluidics to trap, sort, perturb, and image worms are also beginning to bear fruit, and this will also clearly expand greatly as a field in the coming years [200—202]. The well-defined body morphology, small size, near-invariant lineage and transparency, together with the wealth of known mutant phenotypes, make the worm an ideal test case for the development of automated images, and thus we believe that improvements in automated imaging in the worm will be an area of progress that will have an impact well beyond the worm field.

Thirdly, we believe that research into the mind of the worm — how information, in the form of electrical and chemical signals, flows through the neuromuscular system and integrates multiple inputs and stimuli into robust behavioral outcomes — is about to be transformed by a number of technologies. The physical map of the connections between all neurons and between neurons and muscle cells is in some way analogous to the genome. The genome encodes a huge number of possible cellular states, and the precise sets of genes turned on and off determine a large part of the functional state of the cell. Similarly, the nervous system map shows the possible ways in which information could flow and, while laser ablations show key requirements for specific neurons, cross-talk and feedback can complicate the interpretation of these results — visualizing the actual temporal and spatial flow of information from stimulus to response is the crucial missing piece to understanding the worm brain. Great advances in imaging and fluorescent reporters, together with advances in microfluidics (e.g., in accurately controlling the worm microenvironment [203]), mean it is now possible to carry out real-time imaging of neurons firing in vivo (e.g., in [204–207]). Using the genetic toolbox available to study the nervous system — large mutant collections that have perturbed neuronal lineages, multiple cell-specific fluorescent markers etc. — together with the guiding map of the physical map of the nervous system of the worm, real-time imaging will yield key insights into how information is processed by an intact and complete animal brain.

Finally, we also anticipate an ever-deepening analysis of other related nematodes. *C. briggsae* was Sydney Brenner's first identified worm to 'tame', and comparisons between *C. elegans* and *C. briggsae* can illuminate many aspects of genomic and biological evolution. Many new tools have been developed recently [208], and a sister species has been identified for *C. briggsae* which should open many new doors here. Further, intriguing studies have begun to refocus on examining the evolutionary plasticity of development. For example, analysis of vulval development in various nematodes illustrates the intricate evolutionary rewirings of EGF, Notch and Wnt pathways in vulval induction (reviewed in [209]), and examination of the genetic basis for hermaphroditism in *C. elegans* and *C. briggsae* finds that this has been derived separately and arises through different mechanisms [210]. Even at the simple morphological level, plasticity is everywhere: the events of early embryogenesis have been studied in a wide variety of nematodes, showing the dizzying number of ways in which a near-identical four-cell embryo can be derived [211]. Many of the earliest analyses of embryogenesis in any nematode were undertaken in other nematodes such as *Ascaris* [212] and, while *C. elegans* has clearly outpaced research on these other worms, it will be intriguing to see how the re-application of new tools and technologies, together with full genomic sequences, can open up these fields. The approaches pioneered in *C. elegans* are likely to be key also in the medically vital studies of various pathogenic nematodes. Many of these, whether plant pathogens or animal pathogens, are competent for RNAi, and this is likely to identify key potential drug targets and pathways. Together, the systematic examination of the genetics of development in a wide range of related nematodes is likely to shed key light on the constraints and mechanisms of evolution, and this will be an intriguing area to follow.

Since Sydney Brenner's first concerted drive to establish *C. elegans* as a simple animal model to understand the genetic basis for development and the function of the neuromuscular system, giant strides have been taken. However, the simplicity of the biology of the worm and the awesome power of worm genetics will continue to shed light on central problems in genetics and animal biology in the coming decades as new technologies and new approaches open up untouched areas.

REFERENCES

[1] Boveri. Ergebnisse über die Konstitution der chromatischen Substanz des Zelkerns. Fisher; 1904.
[2] Sutton WS. The chromosomes in heredity. Biol Bull 1903;4:231–51.
[3] Morgan TH, S.A.H., Muller HJ, Bridges CB. The Mechanism of Mendelian Heredity. New York: Henry Holt and Company; 1915.
[4] Schrodinger E. What Is Life? Cambridge University Press; 1944.
[5] Crick F. Central dogma of molecular biology. Nature 1970;227:561–3.
[6] Crick FH. On protein synthesis. Symp Soc Exp Biol 1958;12:138–63.
[7] Shannon C. An algebra for theoretical genetics. In: Massachussetts Institute of Technology. Cambridge: M.I.T.; 1943.
[8] Shannon CaWW. The Mathematical Theory of Communication. Univ. Of Illinois Press; 1949.
[9] Jacob F, Monod J. Genetic regulatory mechanisms in the synthesis of proteins. J Mol Biol 1961;3:318–56.
[10] Jacob F, Monod J. Biochemical and genetic mechanisms of regulation in the bacterial cell. Bull Soc Chim Biol (Paris) 1964;46:1499–532.
[11] Jacob F, Ullman A, Monod J. [The promotor, a genetic element necessary to the expression of an operon]. C R Hebd Seances Acad Sci 1964;258:3125–8.

[12] Brenner S. The genetics of *Caenorhabditis elegans*. Genetics 1974;77:71–94.

[13] Coulson A, Kozono Y, Lutterbach B, Shownkeen R, Sulston J, Waterston R. YACs and the *C. elegans* genome. Bioessays 1991;13:413–7.

[14] Coulson A, Sulston J, Brenner S, Karn J. Toward a physical map of the genome of the nematode *Caenorhabditis elegans*. Proc Natl Acad Sci U S A 1986;83:7821–5.

[15] Coulson A, Waterston R, Kiff J, Sulston J, Kohara Y. Genome linking with yeast artificial chromosomes. Nature 1988;335:184–6.

[16] Kimble J, Hirsh D. The postembryonic cell lineages of the hermaphrodite and male gonads in *Caenorhabditis elegans*. Dev Biol 1979;70:396–417.

[17] Sulston JE, Horvitz HR. Post-embryonic cell lineages of the nematode, *Caenorhabditis elegans*. Dev Biol 1977;56:110–56.

[18] Sulston JE, Schierenberg E, White JG, Thomson JN. The embryonic cell lineage of the nematode *Caenorhabditis elegans*. Dev Biol 1983;100:64–119.

[19] Ewing B, Hillier L, Wendl MC, Green P. Base-calling of automated sequencer traces using phred. I. Accuracy assessment. Genome Res 1998;8:175–85.

[20] Eeckman FH, Durbin R. ACeDB and macace. Methods Cell Biol 1995;48:583–605.

[21] White JG, S.E., Thopmson JN, Brenner S. The structure of the nervous system of the nematode Caenorhabditis elegans. Phil Trans Royal Soc London 1986;314:1–340.

[22] Durbin RM. Studies on the Development and Organisation of the Nervous System of the Nervous System of *Caenorhabditis elegans*. Cambridge: University of Cambridge; 1987.

[23] Fang-Yen C, Gabel CV, Samuel AD, Bargmann CI, Avery L. Laser microsurgery in *caenorhabditis elegans*. Methods Cell Biol 107:177–206.

[24] Chalasani SH, Kato S, Albrecht DR, Nakagawa T, Abbott LF, Bargmann CI. Neuropeptide feedback modifies odor-evoked dynamics in *Caenorhabditis elegans* olfactory neurons. Nat Neurosci 2010;13:615–21.

[25] Macosko EZ, Pokala N, Feinberg EH, Chalasani SH, Butcher RA, Clardy J, et al. A hub-and-spoke circuit drives pheromone attraction and social behaviour in *C. elegans*. Nature 2009;458:1171–5.

[26] The *C. elegans* Sequencing Consortium. Genome sequence of the nematode *C. elegans*: a platform for investigating biology. Science 1998;282:2012–8.

[27] Evan GI, Littlewood TD. The role of c-myc in cell growth. Curr Opin Genet Dev 1993;3:44–9.

[28] Ashburner M, Ball CA, Blake JA, Botstein D, Butler H, Cherry JM, et al. Gene ontology: tool for the unification of biology. The Gene Ontology Consortium. Nat Genet 2000;25:25–9.

[29] Hengartner MO. The biochemistry of apoptosis. Nature 2000;407:770–6.

[30] Merz DC, Culotti JG. Genetic analysis of growth cone migrations in *Caenorhabditis elegans*. J Neurobiol 2000;44:281–8.

[31] O'Hagan R, Chalfie M. Mechanosensation in *Caenorhabditis elegans*. Int Rev Neurobiol 2006;69:169–203.

[32] Sternberg PW, Vulval development (June, 25 2005), WormBook, ed. The *C. elegans* Research Community, WormBook, doi/10.1895/wormbook.1.6.1, http://www.wormbook.org.

[33] Hengartner MO. Apoptosis: corralling the corpses. Cell 2001;104:325–8.

[34] Gonczy P, Rose LS. Asymmetric cell division and axis formation in the embryo. WormBook 2005:1–20.

[35] Beitel GJ, Clark SG, Horvitz HR. *Caenorhabditis elegans* ras gene let-60 acts as a switch in the pathway of vulval induction. Nature 1990;348:503–9.

[36] Clark SG, Lu X, Horvitz HR. The *Caenorhabditis elegans* locus lin-15, a negative regulator of a tyrosine kinase signaling pathway, encodes two different proteins. Genetics 1994;137:987–97.

[37] Ferguson EL, Horvitz HR. The multivulva phenotype of certain *Caenorhabditis elegans* mutants results from defects in two functionally redundant pathways. Genetics 1989;123:109–21.

[38] Jiang LI, Sternberg PW. An HMG1-like protein facilitates Wnt signaling in *Caenorhabditis elegans*. Genes Dev 1999;13:877–89.

[39] Lehner B, Crombie C, Tischler J, Fortunato A, Fraser AG. Systematic mapping of genetic interactions in *Caenorhabditis elegans* identifies common modifiers of diverse signaling pathways. Nat Genet 2006;38:896–903.

[40] Chinnaiyan AM, Chaudhary D, O'Rourke K, Koonin EV, Dixit VM. Role of CED-4 in the activation of CED-3. Nature 1997;388:728–9.

[41] Chinnaiyan AM, O'Rourke K, Lane BR, Dixit VM. Interaction of CED-4 with CED-3 and CED-9: a molecular framework for cell death. Science 1997;275:1122–6.

[42] James C, Gschmeissner S, Fraser A, Evan GI. CED-4 induces chromatin condensation in *Schizosaccharomyces pombe* and is inhibited by direct physical association with CED-9. Curr Biol 1997;7:246–52.

[43] Conradt B, Horvitz HR. The *C. elegans* protein EGL-1 is required for programmed cell death and interacts with the Bcl-2-like protein CED-9. Cell 1998;93:519–29.

[44] Fay D. Genetic mapping and manipulation: chapter 1–Introduction and basics (February 17, 2006), WormBook, ed. The *C. elegans* Research Community. WormBook, doi/10.1895/wormbook.1.90.1, http://www.wormbook.org.

[45] Wicks SR, Yeh RT, Gish WR, Waterston RH, Plasterk RH. Rapid gene mapping in *Caenorhabditis elegans* using a high density polymorphism map. Nat Genet 2001;28:160–4.

[46] Doitsidou M, Poole RJ, Sarin S, Bigelow H, Hobert O. *C. elegans* mutant identification with a one-step whole-genome-sequencing and SNP mapping strategy. PLoS One 2010;5:e15435.

[47] Hobert O. The impact of whole genome sequencing on model system genetics: get ready for the ride. Genetics 2010;184:317–9.

[48] Fisher J, Piterman N, Hajnal A, Henzinger TA. Predictive modeling of signaling crosstalk during *C. elegans* vulval development. PLoS Comput Biol 2007;3:e92.

[49] Fisher J, Piterman N, Hubbard EJ, Stern MJ, Harel D. Computational insights into *Caenorhabditis elegans* vulval development. Proc Natl Acad Sci U S A 2005;102:1951–6.

[50] Kam N, Kugler H, Marelly R, Appleby L, Fisher J, Pnueli A, et al. A scenario-based approach to modeling development: a prototype model of *C. elegans* vulval fate specification. Dev biol 2008;323:1–5.

[51] Li C, Nagasaki M, Ueno K, Miyano S. Simulation-based model checking approach to cell fate specification during *Caenorhabditis elegans* vulval development by hybrid functional Petri net with extension. BMC Syst Biol 2009;3:42.

[52] Shimizu TS, Bray D. Modelling the bacterial chemotaxis receptor complex. Novartis Found Symp 2002;247:162–77. discussion 177–181, 198–206, 244–152.

[53] Roemer T, Davies J, Giaever G, Nislow C. Bugs, drugs and chemical genomics. Nat Chem Biol 2011;8:46–56.

[54] Burns AR, Wallace IM, Wildenhain J, Tyers M, Giaever G, Bader GD, et al. A predictive model for drug bioaccumulation and bioactivity in *Caenorhabditis elegans*. Nat Chem Biol 2010; 6:549–57.

[55] Burns AR, Kwok TC, Howard A, Houston E, Johanson K, Chan A, et al. High-throughput screening of small molecules for bioactivity and target identification in *Caenorhabditis elegans*. Nat Protoc 2006;1:1906–14.

[56] Kwok TC, Ricker N, Fraser R, Chan AW, Burns A, Stanley EF, et al. A small-molecule screen in *C. elegans* yields a new calcium channel antagonist. Nature 2006;441:91–5.

[57] Luciani GM, Magomedova L, Puckrin R, Urbanus ML, Wallace IM, Giaever G, et al. Dafadine inhibits DAF-9 to promote dauer formation and longevity of *Caenorhabditis elegans*. Nat Chem Biol 2012;8:318.

[58] Kim DU, Hayles J, Kim D, Wood V, Park HO, Won M, et al. Analysis of a genome-wide set of gene deletions in the fission yeast *Schizosaccharomyces pombe*. Nat Biotechnol 2010;28:617–23.

[59] Winzeler EA, Shoemaker DD, Astromoff A, Liang H, Anderson K, Andre B, et al. Functional characterization of the *S. cerevisiae* genome by gene deletion and parallel analysis. Science 1999;285:901–6.

[60] Jones GM, Stalker J, Humphray S, West A, Cox T, Rogers J, et al. A systematic library for comprehensive overexpression screens in *Saccharomyces cerevisiae*. Nat Methods 2008;5:239–41.

[61] Sopko R, Huang D, Preston N, Chua G, Papp B, Kafadar K, et al. Mapping pathways and phenotypes by systematic gene overexpression. Mol Cell 2006;21:319–30.

[62] Fire A, Xu S, Montgomery MK, Kostas SA, Driver SE, Mello CC. Potent and specific genetic interference by double-stranded RNA in *Caenorhabditis elegans*. Nature 1998;391:806–11.

[63] Fraser AG, Kamath RS, Zipperlen P, Martinez-Campos M, Sohrmann M, Ahringer J. Functional genomic analysis of *C. elegans* chromosome I by systematic RNA interference. Nature 2000;408:325–30.

[64] Gonczy P, Echeverri C, Oegema K, Coulson A, Jones SJ, Copley RR, et al. Functional genomic analysis of cell division in *C. elegans* using RNAi of genes on chromosome III. Nature 2000;408:331–6.

[65] Ashrafi K, Chang FY, Watts JL, Fraser AG, Kamath RS, Ahringer J, et al. Genome-wide RNAi analysis of *Caenorhabditis elegans* fat regulatory genes. Nature 2003;421:268–72.

[66] Kamath RS, Fraser AG, Dong Y, Poulin G, Durbin R, Gotta M, et al. Systematic functional analysis of the *Caenorhabditis elegans* genome using RNAi. Nature 2003;421:231–7.

[67] Lee SS, Lee RY, Fraser AG, Kamath RS, Ahringer J, Ruvkun G. A systematic RNAi screen identifies a critical role for mitochondria in *C. elegans* longevity. Nat Genet 2003;33:40–8.

[68] Pothof J, van Haaften G, Thijssen K, Kamath RS, Fraser AG, Ahringer J, et al. Identification of genes that protect the *C. elegans* genome against mutations by genome-wide RNAi. Genes Dev 2003;17:443–8.

[69] Simmer F, Moorman C, van der Linden AM, Kuijk E, van den Berghe PV, Kamath RS, et al. Genome-wide RNAi of *C. elegans* using the hypersensitive rrf-3 strain reveals novel gene functions. PLoS Biol 2003;1:E12.

[70] Kamath RS, Martinez-Campos M, Zipperlen P, Fraser AG, Ahringer J. Effectiveness of specific RNA-mediated interference through ingested double-stranded RNA in *Caenorhabditis elegans*. Genome Biol 2001;2:RESEARCH0002.

[71] Timmons L, Court DL, Fire A. Ingestion of bacterially expressed dsRNAs can produce specific and potent genetic interference in *Caenorhabditis elegans*. Gene 2001;263:103–12.

[72] Timmons L, Fire A. Specific interference by ingested dsRNA. Nature 1998;395:854.

[73] Poulin G, Dong Y, Fraser AG, Hopper NA, Ahringer J. Chromatin regulation and sumoylation in the inhibition of Ras-induced vulval development in *Caenorhabditis elegans*. EMBO J 2005;24:2613–23.

[74] Sieburth D, Ch'ng Q, Dybbs M, Tavazoie M, Kennedy S, Wang D, et al. Systematic analysis of genes required for synapse structure and function. Nature 2005;436:510–7.

[75] Kim JK, Gabel HW, Kamath RS, Tewari M, Pasquinelli A, Rual JF, et al. Functional genomic analysis of RNA interference in *C. elegans*. Science 2005;308:1164–7.

[76] Harris TW, Antoshechkin I, Bieri T, Blasiar D, Chan J, Chen WJ, et al. WormBase: a comprehensive resource for nematode research. Nucleic Acids Res 2010;38:D463–467.

[77] Yook K, Harris TW, Bieri T, Cabunoc A, Chan J, Chen WJ, et al. WormBase 2012: more genomes, more data, new website. Nucleic Acids Res 2012;40:D735–741.

[78] Simmer F, Tijsterman M, Parrish S, Koushika SP, Nonet ML, Fire A, et al. Loss of the putative RNA-directed RNA polymerase RRF-3 makes *C. elegans* hypersensitive to RNAi. Curr Biol 2002;12:1317–9.

[79] Lehner B, Calixto A, Crombie C, Tischler J, Fortunato A, Chalfie M, et al. Loss of LIN-35, the *Caenorhabditis elegans* ortholog of the tumor suppressor p105Rb, results in enhanced RNA interference. Genome Biol 2006;7:R4.

[80] Calixto A, Chelur D, Topalidou I, Chen X, Chalfie M. Enhanced neuronal RNAi in *C. elegans* using SID-1. Nat Methods 2010;7:554–9.

[81] Elbashir SM, Harborth J, Lendeckel W, Yalcin A, Weber K, Tuschl T. Duplexes of 21-nucleotide RNAs mediate RNA interference in cultured mammalian cells. Nature 2001;411: 494–8.

[82] Grishok A, Pasquinelli AE, Conte D, Li N, Parrish S, Ha I, et al. Genes and mechanisms related to RNA interference regulate expression of the small temporal RNAs that control *C. elegans* developmental timing. Cell 2001;106:23–34.

[83] Ketting RF, Fischer SE, Bernstein E, Sijen T, Hannon GJ, Plasterk RH. Dicer functions in RNA interference and in synthesis of small RNA involved in developmental timing in *C. elegans*. Genes Dev 2001;15:2654–9.

[84] Knight SW, Bass BL. A role for the RNase III enzyme DCR-1 in RNA interference and germ line development in *Caenorhabditis elegans*. Science 2001;293:2269–71.

[85] Moerman DG, Barstead RJ. Towards a mutation in every gene in *Caenorhabditis elegans*. Briefings Funct Genomics Proteomics 2008;7:195–204.

[86] Flibotte S, Edgley ML, Maydan J, Taylor J, Zapf R, Waterston R, et al. Rapid high resolution single nucleotide polymorphism-comparative genome hybridization mapping in *Caenorhabditis elegans*. Genetics 2009;181:33–7.

[87] Gunsalus KC, Ge H, Schetter AJ, Goldberg DS, Han JD, Hao T, et al. Predictive models of molecular machines involved in *Caenorhabditis elegans* early embryogenesis. Nature 2005;436:861–5.

[88] Piano F, Schetter AJ, Mangone M, Stein L, Kemphues KJ. RNAi analysis of genes expressed in the ovary of *Caenorhabditis elegans*. Curr Biol 2000;10:1619–22.

[89] Sonnichsen B, Koski LB, Walsh A, Marschall P, Neumann B, Brehm M, et al. Full-genome RNAi profiling of early embryogenesis in *Caenorhabditis elegans*. Nature 2005;434:462–9.

[90] Zipperlen P, Fraser AG, Kamath RS, Martinez-Campos M, Ahringer J. Roles for 147 embryonic lethal genes on *C. elegans* chromosome I identified by RNA interference and video microscopy. EMBO J 2001;20:3984–92.

[91] Green RA, Kao HL, Audhya A, Arur S, Mayers JR, Fridolfsson HN, et al. A high-resolution *C. elegans* essential gene network based on phenotypic profiling of a complex tissue. Cell 2011;145:470–82.

[92] Abrams JM, White K, Fessler LI, Steller H. Programmed cell death during *Drosophila* embryogenesis. Development 1993;117:29–43.

[93] Steller H, Abrams JM, Grether ME, White K. Programmed cell death in *Drosophila*. Philosophical transactions of the Royal Society of London Series B. Biological sciences 1994;345:247–50.

[94] White K, Grether ME, Abrams JM, Young L, Farrell K, Steller H. Genetic control of programmed cell death in *Drosophila*. Science 1994;264:677–83.

[95] Byrne AB, Weirauch MT, Wong V, Koeva M, Dixon SJ, Stuart JM, et al. A global analysis of genetic interactions in *Caenorhabditis elegans*. J Biol 2007;6:8.

[96] Lehner B, Tischler J, Fraser AG. RNAi screens in *Caenorhabditis elegans* in a 96-well liquid format and their application to the systematic identification of genetic interactions. Nat Protoc 2006;1:1617–20.

[97] Tischler J, Lehner B, Chen N, Fraser AG. Combinatorial RNA interference in *Caenorhabditis elegans* reveals that redundancy between gene duplicates can be maintained for more than 80 million years of evolution. Genome Biol 2006;7:R69.

[98] Tischler J, Lehner B, Fraser AG. Evolutionary plasticity of genetic interaction networks. Nat Genet 2008;40:390–1.

[99] Costanzo M, Baryshnikova A, Bellay J, Kim Y, Spear ED, Sevier CS, et al. The genetic landscape of a cell. Science 2010;327:425–31.

[100] Cutter AD, Payseur BA, Salcedo T, Estes AM, Good JM, Wood E, et al. Molecular correlates of genes exhibiting RNAi phenotypes in *Caenorhabditis elegans*. Genome Res 2003;13:2651–7.

[101] Ramani AK, Chuluunbaatar T, Verster AJ, Na H, Vu V, Pelte N, et al. The majority of animal genes are required for wild-type fitness. Cell 2012;148:792–802.

[102] Greenwald I, Seydoux G. Analysis of gain-of-function mutations of the lin-12 gene of *Caenorhabditis elegans*. Nature 1990;346:197–9.

[103] Baugh LR, Hill AA, Slonim DK, Brown EL, Hunter CP. Composition and dynamics of the *Caenorhabditis elegans* early embryonic transcriptome. Development 2003;130:889–900.

[104] Budovskaya YV, Wu K, Southworth LK, Jiang M, Tedesco P, Johnson TE, et al. An elt-3/elt-5/elt-6 GATA transcription circuit guides aging in *C. elegans*. Cell 2008;134:291–303.

[105] Golden TR, Melov S. Microarray analysis of gene expression with age in individual nematodes. Aging Cell 2004;3:111–24.

[106] Lund J, Tedesco P, Duke K, Wang J, Kim SK, Johnson TE. Transcriptional profile of aging in *C. elegans*. Curr Biol 2002;12:1566–73.

[107] GuhaThakurta D, Palomar L, Stormo GD, Tedesco P, Johnson TE, Walker DW, et al. Identification of a novel cis-regulatory element involved in the heat shock response in *Caenorhabditis elegans* using microarray gene expression and computational methods. Genome Res 2002;12:701–12.

[108] Lamm AT, Stadler MR, Zhang H, Gent JI, Fire AZ. Multimodal RNA-seq using single-strand, double-strand, and CircLigase-based capture yields a refined and extended description of the *C. elegans* transcriptome. Genome Res 2011;21:265–75.

[109] Allen MA, Hillier LW, Waterston RH, Blumenthal T. A global analysis of *C. elegans* trans-splicing. Genome Res 2011;21:255–64.

[110] Ramani AK, Calarco JA, Pan Q, Mavandadi S, Wang Y, Nelson AC, et al. Genome-wide analysis of alternative splicing in *Caenorhabditis elegans*. Genome Res 2011;21:342–8.

[111] Jan CH, Friedman RC, Ruby JG, Bartel DP. Formation, regulation and evolution of *Caenorhabditis elegans* 3′;UTRs. Nature 2011;469:97–101.

[112] Mangone M, Manoharan AP, Thierry-Mieg D, Thierry-Mieg J, Han T, Mackowiak SD, et al. The landscape of *C. elegans* 3′UTRs. Science 2010;329:432–5.

[113] Lu ZJ, Yip KY, Wang G, Shou C, Hillier LW, Khurana E, et al. Prediction and characterization of noncoding RNAs in *C. elegans* by integrating conservation, secondary structure, and high-throughput sequencing and array data. Genome Res 2011;21:276–85.

[114] Yanai I, Hunter CP. Comparison of diverse developmental transcriptomes reveals that coexpression of gene neighbors is not evolutionarily conserved. Genome Res 2009;19:2214–20.

[115] Gaudet J, Mango SE. Regulation of organogenesis by the *Caenorhabditis elegans* FoxA protein PHA-4. Science 2002;295:821–5.

[116] Baugh LR, Hill AA, Claggett JM, Hill-Harfe K, Wen JC, Slonim DK, et al. The homeodomain protein PAL-1 specifies a lineage-specific regulatory network in the *C. elegans* embryo. Development 2005;132:1843–54.

[117] Reinke V, Gil IS, Ward S, Kazmer K. Genome-wide germline-enriched and sex-biased expression profiles in *Caenorhabditis elegans*. Development 2004;131:311–23.

[118] Jiang M, Ryu J, Kiraly M, Duke K, Reinke V, Kim SK. Genome-wide analysis of developmental and sex-regulated gene expression profiles in *Caenorhabditis elegans*. Proc Natl Acad Sci U S A 2001;98:218–23.

[119] Roy PJ, Stuart JM, Lund J, Kim SK. Chromosomal clustering of muscle-expressed genes in *Caenorhabditis elegans*. Nature 2002;418:975–9.

[120] Pauli F, Liu Y, Kim YA, Chen PJ, Kim SK. Chromosomal clustering and GATA transcriptional regulation of intestine-expressed genes in *C. elegans*. Development 2006;133:287–95.

[121] Stoeckius M, Maaskola J, Colombo T, Rahn HP, Friedlander MR, Li N, et al. Large-scale sorting of *C. elegans* embryos reveals the dynamics of small RNA expression. Nat Methods 2009;6:745–51.

[122] Zhang Y, Ma C, Delohery T, Nasipak B, Foat BC, Bounoutas A, et al. Identification of genes expressed in *C. elegans* touch receptor neurons. Nature 2002;418:331−5.

[123] Spencer WC, Zeller G, Watson JD, Henz SR, Watkins KL, McWhirter RD, et al. A spatial and temporal map of *C. elegans* gene expression. Genome Res 2011;21:325−41.

[124] Gaudet J, Muttumu S, Horner M, Mango SE. Whole-genome analysis of temporal gene expression during foregut development. PLoS Biol 2004;2:e352.

[125] Yanai I, Baugh LR, Smith JJ, Roehrig C, Shen-Orr SS, Claggett JM, et al. Pairing of competitive and topologically distinct regulatory modules enhances patterned gene expression. Mol Syst Biol 2008;4:163.

[126] Baugh LR, Hunter CP. MyoD, modularity, and myogenesis: conservation of regulators and redundancy in *C. elegans*. Genes Dev 2006;20:3342−6.

[127] Long F, Peng H, Liu X, Kim SK, Myers E. A 3D digital atlas of *C. elegans* and its application to single-cell analyses. Nat Methods 2009;6:667−72.

[128] Liu X, Long F, Peng H, Aerni SJ, Jiang M, Sanchez-Blanco A, et al. Analysis of cell fate from single-cell gene expression profiles in *C. elegans*. Cell 2009;139:623−33.

[129] Bao Z, Murray JI, Boyle T, Ooi SL, Sandel MJ, Waterston RH. Automated cell lineage tracing in *Caenorhabditis elegans*. Proc Natl Acad Sci U S A 2006;103:2707−12.

[130] Murray JI, Bao Z, Boyle TJ, Boeck ME, Mericle BL, Nicholas TJ, et al. Automated analysis of embryonic gene expression with cellular resolution in *C. elegans*. Nat Methods 2008;5:703−9.

[131] Dupuy D, Bertin N, Hidalgo CA, Venkatesan K, Tu D, Lee D, et al. Genome-scale analysis of in vivo spatiotemporal promoter activity in *Caenorhabditis elegans*. Nat biotechnol 2007;25:663−8.

[132] Raj A, van den Bogaard P, Rifkin SA, van Oudenaarden A, Tyagi S. Imaging individual mRNA molecules using multiple singly labeled probes. Nat Methods 2008;5:877−9.

[133] Burga A, Casanueva MO, Lehner B. Predicting mutation outcome from early stochastic variation in genetic interaction partners. Nature 2011;480:250−3.

[134] Raj A, Rifkin SA, Andersen E, van Oudenaarden A. Variability in gene expression underlies incomplete penetrance. Nature 2010;463:913−8.

[135] Deplancke B, Mukhopadhyay A, Ao W, Elewa AM, Grove CA, Martinez NJ, et al. A gene-centered *C. elegans* protein-DNA interaction network. Cell 2006;125:1193−205.

[136] Arda HE, Taubert S, MacNeil LT, Conine CC, Tsuda B, Van Gilst M, et al. Functional modularity of nuclear hormone receptors in a *Caenorhabditis elegans* metabolic gene regulatory network. Mol syst biol 2010;6:367.

[137] Vermeirssen V, Barrasa MI, Hidalgo CA, Babon JA, Sequerra R, Doucette-Stamm L, et al. Transcription factor modularity in a gene-centered *C. elegans* core neuronal protein−DNA interaction network. Genome Res 2007;17:1061−71.

[138] Martinez NJ, Ow MC, Barrasa MI, Hammell M, Sequerra R, Doucette-Stamm L, et al. A *C. elegans* genome-scale microRNA network contains composite feedback motifs with high flux capacity. Genes Dev 2008;22:2535−49.

[139] Grove CA, De Masi F, Barrasa MI, Newburger DE, Alkema MJ, Bulyk ML, et al. A multiparameter network reveals extensive divergence between *C. elegans* bHLH transcription factors. Cell 2009;138:314−27.

[140] Niu W, Lu ZJ, Zhong M, Sarov M, Murray JI, Brdlik CM, et al. Diverse transcription factor binding features revealed by genome-wide ChIP-seq in *C. elegans*. Genome Res 2011;21:245−54.

[141] Bonn S, Zinzen RP, Girardot C, Gustafson EH, Perez-Gonzalez A, Delhomme N, et al. Tissue-specific analysis of chromatin state identifies temporal signatures of enhancer activity during embryonic development. Nat Genet 2012;44:148−56.

[142] Liu T, Rechtsteiner A, Egelhofer TA, Vielle A, Latorre I, Cheung MS, et al. Broad chromosomal domains of histone modification patterns in *C. elegans*. Genome Res 2011;21:227−36.

[143] Whittle CM, McClinic KN, Ercan S, Zhang X, Green RD, Kelly WG, et al. The genomic distribution and function of histone variant HTZ-1 during *C. elegans* embryogenesis. PLoS Genet 2008;4:e1000187.

[144] Ercan S, Giresi PG, Whittle CM, Zhang X, Green RD, Lieb JD. X chromosome repression by localization of the *C. elegans* dosage compensation machinery to sites of transcription initiation. Nat Genet 2007;39:403−8.

[145] Ercan S, Dick LL, Lieb JD. The *C. elegans* dosage compensation complex propagates dynamically and independently of X chromosome sequence. Curr Biol 2009;19:1777−87.

[146] Rechtsteiner A, Ercan S, Takasaki T, Phippen TM, Egelhofer TA, Wang W, et al. The histone H3K36 methyltransferase MES-4 acts epigenetically to transmit the memory of germline gene expression to progeny. PLoS Genet 2010;6.

[147] Gu SG, Pak J, Guang S, Maniar JM, Kennedy S, Fire A. Amplification of siRNA in *Caenorhabditis elegans* generates a transgenerational sequence-targeted histone H3 lysine 9 methylation footprint. Nat Genet 2012;44:157−64.

[148] Greer EL, Maures TJ, Ucar D, Hauswirth AG, Mancini E, Lim JP, et al. Transgenerational epigenetic inheritance of longevity in *Caenorhabditis elegans*. Nature 2011;479:365−71.

[149] Casanueva MO, Burga A, Lehner B. Fitness trade-offs and environmentally induced mutation buffering in isogenic *C. elegans*. Science 2012;335:82−5.

[150] Pincus Z, Smith-Vikos T, Slack FJ. MicroRNA predictors of longevity in *Caenorhabditis elegans*. PLoS Genet 2011;7:e1002306.

[151] Rea SL, Wu D, Cypser JR, Vaupel JW, Johnson TE. A stress-sensitive reporter predicts longevity in isogenic populations of *Caenorhabditis elegans*. Nat Genet 2005;37:894−8.

[152] Sanchez-Blanco A, Kim SK. Variable pathogenicity determines individual lifespan in *Caenorhabditis elegans*. PLoS Genet 2011;7:e1002047.

[153] Herndon LA, Schmeissner PJ, Dudaronek JM, Brown PA, Listner KM, Sakano Y, et al. Stochastic and genetic factors influence tissue-specific decline in ageing *C. elegans*. Nature 2002;419:808−14.

[154] Raviv TR, Ljosa V, Conery AL, Ausubel FM, Carpenter AE, Golland P, et al. Morphology-guided graph search for untangling objects: *C. elegans* analysis. Medical image computing and computer-assisted intervention: MICCAI International Conference on Medical Image Computing and Computer-Assisted Intervention 2010;13:634−41.

[155] Rockman MV, Kruglyak L. Recombinational landscape and population genomics of *Caenorhabditis elegans*. PLoS Genet 2009;5:e1000419.

[156] Schrimpf SP, Weiss M, Reiter L, Ahrens CH, Jovanovic M, Malmstrom J, et al. Comparative functional analysis of the *Caenorhabditis elegans* and *Drosophila melanogaster* proteomes. PLoS Biol 2009;7:e48.

[157] Jovanovic M, Reiter L, Picotti P, Lange V, Bogan E, Hurschler BA, et al. A quantitative targeted proteomics approach to validate predicted microRNA targets in *C. elegans*. Nat Methods 2010;7:837–42.

[158] Chu DS, Liu H, Nix P, Wu TF, Ralston EJ, Yates 3rd JR, et al. Sperm chromatin proteomics identifies evolutionarily conserved fertility factors. Nature 2006;443:101–5.

[159] Dong MQ, Venable JD, Au N, Xu T, Park SK, Cociorva D, et al. Quantitative mass spectrometry identifies insulin signaling targets in *C. elegans*. Science 2007;317:660–3.

[160] David DC, Ollikainen N, Trinidad JC, Cary MP, Burlingame AL, Kenyon C. Widespread protein aggregation as an inherent part of aging in *C. elegans*. PLoS Biol 2010;8:e1000450.

[161] Zielinska DF, Gnad F, Jedrusik-Bode M, Wisniewski JR, Mann M. *Caenorhabditis elegans* has a phosphoproteome atypical for metazoans that is enriched in developmental and sex determination proteins. J Proteome Res 2009;8:4039–49.

[162] Walhout AJ, Sordella R, Lu X, Hartley JL, Temple GF, Brasch MA, et al. Protein interaction mapping in *C. elegans* using proteins involved in vulval development. Science 2000;287:116–22.

[163] Davy A, Bello P, Thierry-Mieg N, Vaglio P, Hitti J, Doucette-Stamm L, et al. A protein–protein interaction map of the *Caenorhabditis elegans* 26S proteasome. EMBO Rep 2001;2:821–8.

[164] Walhout AJ, Reboul J, Shtanko O, Bertin N, Vaglio P, Ge H, et al. Integrating interactome, phenome, and transcriptome mapping data for the *C. elegans* germline. Curr Biol 2002;12:1952–8.

[165] Boulton SJ, Gartner A, Reboul J, Vaglio P, Dyson N, Hill DE, et al. Combined functional genomic maps of the *C. elegans* DNA damage response. Science 2002;295:127–31.

[166] Tewari M, Hu PJ, Ahn JS, Ayivi-Guedehoussou N, Vidalain PO, Li S, et al. Systematic interactome mapping and genetic perturbation analysis of a *C. elegans* TGF-beta signaling network. Mol Cell 2004;13:469–82.

[167] Li S, Armstrong CM, Bertin N, Ge H, Milstein S, Boxem M, et al. A map of the interactome network of the metazoan *C. elegans*. Science 2004;303:540–3.

[168] Simonis N, Rual JF, Carvunis AR, Tasan M, Lemmens I, Hirozane-Kishikawa T, et al. Empirically controlled mapping of the *Caenorhabditis elegans* protein–protein interactome network. Nat Methods 2009;6:47–54.

[169] Boxem M, Maliga Z, Klitgord N, Li N, Lemmens I, Mana M, et al. A protein domain-based interactome network for *C. elegans* early embryogenesis. Cell 2008;134:534–45.

[170] Lehner B. Modelling genotype-phenotype relationships and human disease with genetic interaction networks. J Exp Biol 2007;210:1559–66.

[171] Fraser AG, Marcotte EM. A probabilistic view of gene function. Nat Genet 2004;36:559–64.

[172] Lee I, Lehner B, Crombie C, Wong W, Fraser AG, Marcotte EM. A single gene network accurately predicts phenotypic effects of gene perturbation in *Caenorhabditis elegans*. Nat Genet 2008;40:181–8.

[173] Lee I, Lehner B, Vavouri T, Shin J, Fraser AG, Marcotte EM. Predicting genetic modifier loci using functional gene networks. Genome Res 2010;20:1143–53.

[174] Lehner B, Lee I. Network-guided genetic screening: building, testing and using gene networks to predict gene function. Briefings Funct Genomics Proteomics 2008;7:217–27.

[175] Zhong W, Sternberg PW. Genome-wide prediction of *C. elegans* genetic interactions. Science 2006;311:1481–4.

[176] Goehring NW, Trong PK, Bois JS, Chowdhury D, Nicola EM, Hyman AA, et al. Polarization of PAR proteins by advective triggering of a pattern-forming system. Science 2011;334:1137–41.

[177] Kozlowski C, Srayko M, Nedelec F. Cortical microtubule contacts position the spindle in *C. elegans* embryos. Cell 2007;129:499–510.

[178] Giurumescu CA, Sternberg PW, Asthagiri AR. Intercellular coupling amplifies fate segregation during *Caenorhabditis elegans* vulval development. Proc Natl Acad Sci U S A 2006;103:1331–6.

[179] Giurumescu CA, Sternberg PW, Asthagiri AR. Predicting phenotypic diversity and the underlying quantitative molecular transitions. PLoS Comput Biol 2009;5:e1000354.

[180] Sun X, Hong P. Computational modeling of *Caenorhabditis elegans* vulval induction. Bioinformatics 2007;23:i499–507.

[181] Mayer M, Depken M, Bois JS, Julicher F, Grill SW. Anisotropies in cortical tension reveal the physical basis of polarizing cortical flows. Nature 2010;467:617–21.

[182] Kitano H, Hamahashi S, Luke S. The perfect C. elegans project: an initial report. Artificial life 1998;4:141–56.

[183] Brenner S. Turing centenary: Life's code script. Nature 2012;482:461.

[184] Gerstein MB, Lu ZJ, Van Nostrand EL, Cheng C, Arshinoff BI, Liu T, et al. Integrative analysis of the *Caenorhabditis elegans* genome by the modENCODE project. Science 2010;330:1775–87.

[185] Manolio TA, Collins FS, Cox NJ, Goldstein DB, Hindorff LA, Hunter DJ, et al. Finding the missing heritability of complex diseases. Nature 2009;461:747–53.

[186] Seidel HS, Rockman MV, Kruglyak L. Widespread genetic incompatibility in *C. elegans* maintained by balancing selection. Science 2008;319:589–94.

[187] Andersen EC, Gerke JP, Shapiro JA, Crissman JR, Ghosh R, Bloom JS, et al. Chromosome-scale selective sweeps shape *Caenorhabditis elegans* genomic diversity. Nat Genet 2012;44:285–90.

[188] Doroszuk A, Snoek LB, Fradin E, Riksen J, Kammenga J. A genome-wide library of CB4856/N2 introgression lines of *Caenorhabditis elegans*. Nucleic Acids Res 2009;37:e110.

[189] Shaham S, Reddien PW, Davies B, Horvitz HR. Mutational analysis of the *Caenorhabditis elegans* cell-death gene ced-3. Genetics 1999;153:1655–71.

[190] Moy TI, Conery AL, Larkins-Ford J, Wu G, Mazitschek R, Casadei G, et al. High-throughput screen for novel antimicrobials using a whole animal infection model. ACS Chem Biol 2009;4:527–33.

[191] Wahlby C, Riklin-Raviv T, Ljosa V, Conery AL, Golland P, Ausubel FM, et al. Resolving Clustered Worms Via Probabilistic Shape Models. Proceedings / IEEE International Symposium on Biomedical Imaging: from nano to macro IEEE International Symposium on Biomedical Imaging 2010;2010:552–5.

[192] Carpenter AE, Jones TR, Lamprecht MR, Clarke C, Kang IH, Friman O, et al. CellProfiler: image analysis software for identifying and quantifying cell phenotypes. Genome Biol 2006;7:R100.

[193] Cronin CJ, Feng Z, Schafer WR. Automated imaging of *C. elegans* behavior. Methods Mol Biol 2006;351:241–51.

[194] Feng Z, Cronin CJ, Wittig Jr JH, Sternberg PW, Schafer WR. An imaging system for standardized quantitative analysis of *C. elegans* behavior. BMC bioinformatics 2004;5:115.

[195] Stephens GJ, Bueno de Mesquita M, Ryu WS, Bialek W. Emergence of long timescales and stereotyped behaviors in *Caenorhabditis elegans*. Proc Natl Acad Sci U S A 2011;108:7286–9.

[196] Stephens GJ, Johnson-Kerner B, Bialek W, Ryu WS. Dimensionality and dynamics in the behavior of *C. elegans*. PLoS Comput Biol 2008;4:e1000028.

[197] Stephens GJ, Johnson-Kerner B, Bialek W, Ryu WS. From modes to movement in the behavior of *Caenorhabditis elegans*. PLoS One 2010;5:e13914.

[198] Murray JI, Bao Z, Boyle TJ, Waterston RH. The lineaging of fluorescently-labeled *Caenorhabditis elegans* embryos with StarryNite and AceTree. Nat Protoc 2006;1:1468–76.

[199] Elvin M, Snoek LB, Frejno M, Klemstein U, Kammenga JE, Poulin GB. A fitness assay for comparing RNAi effects across multiple *C. elegans* genotypes. BMC Genomics 2011;12:510.

[200] Chronis N. Worm chips: microtools for *C. elegans* biology. Lab chip 2010;10:432–7.

[201] Chung K, Crane MM, Lu H. Automated on-chip rapid microscopy, phenotyping and sorting of *C. elegans*. Nat Methods 2008;5:637–43.

[202] Heng X, Erickson D, Baugh LR, Yaqoob Z, Sternberg PW, Psaltis D, et al. Optofluidic microscopy –a method for implementing a high resolution optical microscope on a chip. Lab chip 2006;6:1274–6.

[203] Albrecht DR, Bargmann CI. High-content behavioral analysis of *Caenorhabditis elegans* in precise spatiotemporal chemical environments. Nat Methods 2011;8:599–605.

[204] Faumont S, Rondeau G, Thiele TR, Lawton KJ, McCormick KE, Sottile M, et al. An image-free opto-mechanical system for creating virtual environments and imaging neuronal activity in freely moving *Caenorhabditis elegans*. PLoS One 2011;6:e24666.

[205] Haspel G, O'Donovan MJ, Hart AC. Motoneurons dedicated to either forward or backward locomotion in the nematode *Caenorhabditis elegans*. J Neurosci 2010;30:11151–6.

[206] Kerr RA, Schafer WR. Intracellular Ca2+ imaging in *C. elegans*. Methods Mol Biol 2006;351:253–64.

[207] Santos SI, Mathew M, Loza-Alvarez P. Real time imaging of femtosecond laser induced nano-neurosurgery dynamics in *C. elegans*. Optics express 2010;18:364–77.

[208] Zhao Z, Flibotte S, Murray JI, Blick D, Boyle TJ, Gupta B, et al. New tools for investigating the comparative biology of *Caenorhabditis briggsae* and *C. elegans*. Genetics 2010;184:853–63.

[209] Felix MA, Barkoulas M. Robustness and flexibility in nematode vulva development. Genetics 2012;28(4):185–95.

[210] Hill RC, de Carvalho CE, Salogiannis J, Schlager B, Pilgrim D, Haag ES. Genetic flexibility in the convergent evolution of hermaphroditism in *Caenorhabditis nematodes*. Dev Cell 2006;10:531–8.

[211] Brauchle M, Kiontke K, MacMenamin P, Fitch DH, Piano F. Evolution of early embryogenesis in rhabditid nematodes. Dev biol 2009;335:253–62.

[212] Boveri T. Die Entwicklung von Ascaris megalocephala mit besonderer Rucksicht auf die Kernverhaltnisse. Gustav Fischer; 1899.

Chapter 20

Arabidopsis as a Model for Systems Biology

Philip N. Benfey[1] and Ben Scheres[2]

[1]*Department of Biology and Duke Center for Systems Biology, Duke University, Durham, NC 27708, USA,* [2]*Department of Biology, Utrecht University, Padualaan 8, 3584 CH Utrecht, The Netherlands*

Chapter Outline

Introduction	391
Systems Analysis of *Arabidopsis* Development	**392**
Background	392
Analysis of Gene Activity in Space and Time	393
The Use of Spatiotemporal Specific Expression Data to Analyze Development	393
Toward an Understanding of the Dynamics of Asymmetric Cell Division	397
Computational Modeling of Root Development	398
Auxin Flow and Stem Cell Specification	398
Mutual Support for Root Hair Patterning	399
Integrated Analysis of Shoot Development	399
Gene Activity in Space and Time	399
Coupled Gene Regulatory Networks in the Shoot	400
Modeling Regulatory Networks in the Shoot	400
Development and the Response to Environmental Stress	**401**
Tissue-Specific Responses to Environment	401
A Major Challenge: The Transition to Flowering	402
Conclusions and Perspective	**403**
References	**404**

INTRODUCTION

In this age of rapidly decreasing costs for genome-wide assays, the question often arises, 'Are model systems still all that useful?' The question is particularly raised for plants, for which there are relatively few models and lots of non-model plants that are valuable crops. At least in the area of systems biology, the answer is still, 'Yes, model systems are very useful.' Their utility arises from the very nature of systems biology, which aims to understand the dynamic connections among cellular components. This is a very difficult problem to solve owing to the number of components and the complexity of their interactions. For the near future, at least, it is only in model organisms that large quantities of information of sufficient quality about cellular components can be obtained. It is also of great benefit to have available a broad and deep literature on a model organism to both inform and potentially validate hypotheses generated from systems level analyses.

Among plants, the leading model system for systems biology work is *Arabidopsis thaliana*. It was originally chosen as a favored model system for molecular genetic analysis because of its rapid generation time, large number of progeny and relatively small diploid genome. It was the first plant to have its genome sequenced, and a large number of genome-wide expression datasets have been generated. An extensive library of insertional mutants has facilitated the identification and characterization of many genes. In the process, *Arabidopsis* research has contributed to a vastly enhanced understanding of many basic plant processes, including hormone production and response, circadian rhythms, and plant growth and development. For development, plants in general, and *Arabidopsis* in particular, have features that greatly simplify analysis. For example, root and stem tissues are organized as concentric cylinders of different types of cells (Figure 20.1). Because there is no cell movement, cellular positions are fixed with respect to their neighbors.

In this chapter, we will not try to give an exhaustive overview of systems-level analysis in *Arabidopsis*, but instead discuss a few informative examples. We will focus on the application of systems biology to development and to the response to stress, in part because other productive areas, such as circadian rhythm, are covered in other chapters.

FIGURE 20.1 Schematic of *Arabidopsis* embryo and seedling. Embryonic stages are depicted with dotted lines indicating the provenance of different cellular populations. Magnified regions of the shoot and root apical meristems are shown with cell types indicated in the root meristem and expression domains for different genes color coded in the shoot meristem.

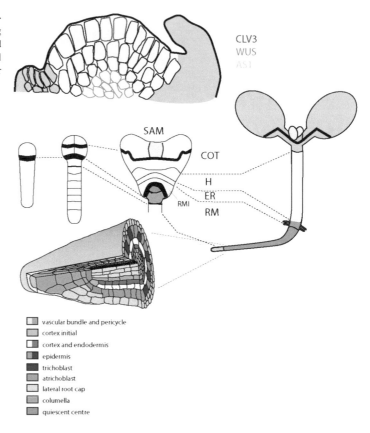

SYSTEMS ANALYSIS OF *ARABIDOPSIS* DEVELOPMENT

Background

Because plants cannot move, their developmental strategies necessarily differ from those of animals. Whereas animals generally have all of their organs formed during embryogenesis, plant embryos can be thought of as two sets of stem cells, one that will form the aerial tissues and the other that will form the roots (Figure 20.1). These stem cell populations or 'meristems' are the source of all post-embryonic growth and development, and it is nearly impossible to predict the ultimate form and structure of a plant from its embryo. For instance, the embryo of an oak tree does not look that different from the embryo of a corn plant. How meristems are able to generate all of the different types of organs in a plant is a fascinating question, which is beginning to be addressed through systems-level analysis. We will discuss work that addresses both root and shoot development.

One essential component of systems analysis is the ability to dissect gene activity in space and time. The root of *Arabidopsis* has become a favored organ for such analysis, owing primarily to the simplicity of its organization. The stem cell niche is located at the tip of the root (Figure 20.1) and it produces its progeny through a series of highly stereotyped divisions. As one moves from the stem cell niche up the root toward the shoot, the cells of each type become older. Thus, to a first approximation any root cell can be specified along two axes, its place on the radial axis identifying the type of cell and its position on the longitudinal axis determining where along the developmental timeline it is. This simplification effectively reduces what is normally a four-dimensional problem (three spatial dimensions plus time) to a two-dimensional problem. Being able to specify cell type and developmental stage along two axes has greatly facilitated approaches that treat this organ as a heterogeneous set of closely coordinated cell types, rather than assaying the organ in its entirety. Similar approaches are currently being used in shoot development, despite the higher complexity of this organ system.

A second component of systems analysis is the generation of models from data. When regulatory networks are not linear, for example when there are feedbacks, it is no longer trivial to understand their workings, as not only the quality of interactions (positive or negative) but also their strength determine the outcome. Mathematical or computational analysis is then required to calculate how regulatory circuits work. This approach has been increasingly adopted over recent years to study how patterns are generated within roots and shoots [1].

Here, we review how both components of systems biology have been adapted to the study of plant development and sketch how this is increasing our understanding of the factors and the logic that drive plant development.

Analysis of Gene Activity in Space and Time

An early effort to assay gene expression in individual cell types within a plant organ involved the use of five transgenic lines in which individual cell populations were marked with green fluorescent protein (GFP) [2]. Enzymes were used to dissociate the cells of the *Arabidopsis* root, which were subsequently sorted using a fluorescence-activated cell sorter (FACS). From the cells enriched for GFP expression, RNA was extracted and used for microarray analysis. To profile expression along the developmental timeline, roots were dissected into sections corresponding to developmental zones and RNA was extracted from the isolated sections and used for microarray analysis. A more recent analysis extended this Root Map to include nearly all of the known cell types in the root [3]. Because root growth is not precisely synchronous, to gain a more precise knowledge of developmental stages, 13 sections along the longitudinal axis (Figure 20.2) were cut from individual roots and mRNA was extracted and used for microarray analysis.

The microarray data sets were first analyzed separately for the cell types and developmental stages. In each case clusters of co-regulated genes were identified that were enriched for biological functions. In the case of the developmental stage-specific data a surprising finding was that a large number of the co-regulated clusters exhibited fluctuating behavior such that their expression peaked at two or more different stages of development [3]. This was unexpected because development is normally thought of as a unidirectional, progressive process. It is an example of the type of finding that could not have been foreseen prior to performing this type of discovery-driven experiment. More recently, the same approach has been used to profile small RNAs in specific cell types and developmental stages. At least 70 novel microRNAs (see Chapter 2) were identified using this approach, and many previously characterized microRNAs were shown to be expressed in a cell type- or developmental stage-specific manner.

Two different approaches have been used to combine the radial and longitudinal datasets to infer expression for any cell within the root. The first approach used the developmental stage data to divide expression for each cell type proportionally along the longitudinal axis [2]. For example, if expression for gene X was 80 units in the epidermis and its expression was distributed along the longitudinal axis with 50% just above the stem cell niche, then 40 units of expression would be assigned to the epidermal cells in this location. A more sophisticated approach made use of an expectation maximization algorithm, which treated the longitudinal and radial datasets as a matrix in which the sums for each row and column must be constant [4]. Validation efforts have generally been consistent with predictions of the expectation maximization approach. However, for the long term it will be important to determine directly the expression levels in cells at different developmental stages along each of the cell files in the root.

Several alternative approaches have been used to assay genome-wide cell type-specific expression in plants. These include the use of laser capture microdissection and tagged ribosomal proteins coupled with immunoprecipitation. Despite being tested on *Arabidopsis* [5], laser capture microdissection combined with microarray analysis has been used most extensively for rice [6]. Immunoprecipitation of polysomes involves fusing an epitope tag such as the antigenic portion of the Myc protein to a ubiquitously expressed ribosomal protein expressed behind a tissue-specific promoter. The construct is introduced into plants, allowing polysomes to be immunoprecipitated with antibodies specific to the epitope. This was performed on *Arabidopsis* root tissues with several tissue-specific promoters driving the epitope-tagged ribosomal protein [7]. The ability to assay polysomes makes this approach well suited to identifying the mRNAs that are actively undergoing translation. A more recently developed technique involves isolating nuclei that have been tagged using tissue-specific promoters [8]. To date, this approach has been primarily used for chromatin immunoprecipitation assays (see Chapter 4), although it would be interesting to perform a side-by-side comparison of RNAs found in the nucleus and those located in the cytoplasm.

The Use of Spatiotemporal Specific Expression Data to Analyze Development

The primary utility of genome-wide expression datasets is to generate hypotheses that will shed light on biological processes. A straightforward hypothesis is that a gene expressed specifically in a particular cell type and/or developmental stage is involved in a process performed specifically in that cell or stage. An example of testing this type of hypothesis is the identification of a family of genes required for Casparian strip formation, a terminal differentiation property of endodermal cells, based on their specific expression in maturing endodermal cells [9].

Another example is an effort to identify transcriptional regulators of the transition from cellular proliferation to differentiation. The immediate progeny of stem cells

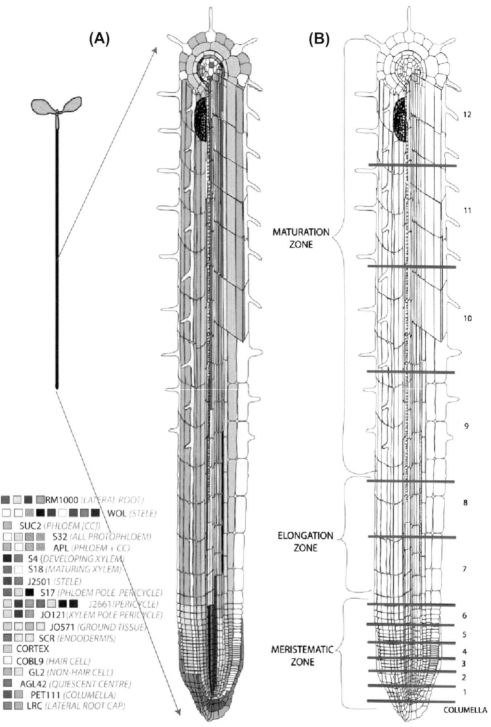

FIGURE 20.2 **Profiling mRNAs at cell type-specific and developmental stage-specific resolution.** (A) Schematic of cell types and (B) developmental zones profiled. (C) Dominant expression patterns identified among cell populations. *(Adapted from [3].)*

usually pass through a 'transit amplifying' stage during which there is rapid cellular proliferation, prior to the onset of differentiation. In the root, the onset of differentiation occurs when cells stop dividing and begin to elongate, a process that can increase their volume over 100-fold.

From the developmental stage-specific expression data, transcription factors were identified whose expression peaked at the transition from cell division to elongation [10]. Insertional mutations were tested in 50 genes, resulting in the identification of one mutant with enhanced

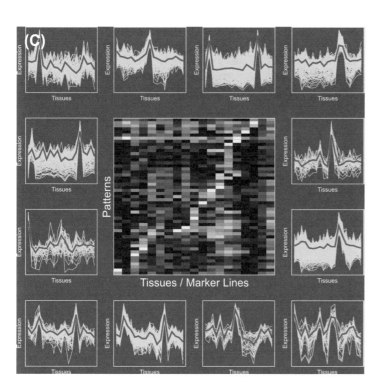

FIGURE 20.2 (*continued*).

root growth. The growth enhancement was due to an increase in the number of cells undergoing division as well as cells elongating to a larger size. When ectopically expressed, this gene, named UPBEAT1 (UPB1), caused the root to have a reduced growth rate with fewer dividing cells and shorter cells in the elongation zone. To identify likely targets of UPB1, which encodes a bHLH transcription factor, microarray studies were performed on the mutant and the ectopic expressor. This led to the identification of a set of peroxidase genes as being repressed by UPB1 specifically in the elongation zone. Chromatin immunoprecipitation followed by microarray analysis (ChIP-chip) revealed that the promoters of these peroxidase genes were bound directly by UPB1. Because peroxidases can control reactive oxygen species (ROS), the ROS status in the root was investigated using reagents that respond to different ROS molecules. One type of ROS, hydrogen peroxide, was found to be localized primarily in the elongation zone, while another type, superoxide, was localized primarily to the meristematic zone. The amount and localization of these ROS changed dramatically in the *upb1* mutant and ectopic expressor. Manipulation of the different ROS with small molecule agents provided evidence that they play a central role in regulating the transition from cellular proliferation to differentiation. Thus, a hypothesis based on tissue-specific expression data led to the discovery of a key transcription factor that acts through peroxidases to regulate the levels of reactive oxygen species, which in turn control root growth rates [10]. Because UPB1 expression is responsive to levels of hydrogen peroxide, there is a feedback loop that provides a means by which this subnetwork could control ROS homeostasis.

A different type of hypothesis was generated from the cell type-specific data set. The transcriptional regulation that results in tissue-specific expression could be due to the activity of transcription factors whose expression is enriched in that tissue. Alternatively, tissue-specific expression might be primarily controlled by transcription factors whose expression overlaps in specific tissues, but the expression of the factors themselves is not specifically in that tissue. To determine which of these alternative hypotheses was more likely to be correct, an experiment was designed to identify the regulators for a set of genes expressed in the vascular tissue and surrounding pericycle (collectively known as the 'stele').

The yeast one-hybrid assay (see Chapter 4) was used on full-length promoters of a set of 172 genes whose expression was enriched in the stele [11]. These 172 promoters actually belonged to the set of tissue-enriched transcription factors that were tested for their ability to bind to these promoters. In this way, both binding to other factor genes as well as binding to their own promoters would be tested in the same assay. The results were that approximately 25% of the genes were placed into a network (Figure 20.3A), suggesting that a substantial amount of tissue-specific expression results from regulation by transcription factors enriched in that tissue. The surprise came when the effects of mutating upstream factors on the expression of downstream genes was determined. This was

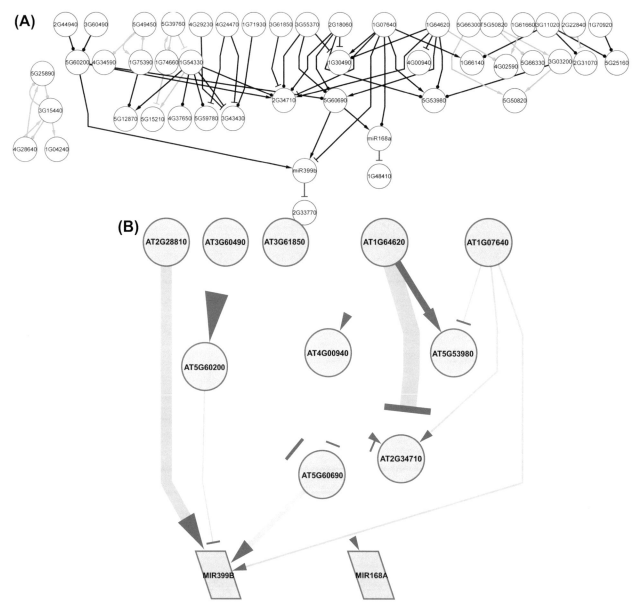

FIGURE 20.3 Gene regulatory network of the root vascular tissue. (A) The gene regulatory network generated through yeast one-hybrid analyses. (B) Modeling strength of interactions using Bayesian inference based on the effect of mutations in upstream regulators on expression of downstream targets. *(Adapted from [11].)*

done by performing quantitative PCR on the transcripts of target genes in the background of insertional mutants for the upstream regulators. In *Arabidopsis* there is a low rate of return on standard reverse genetics screens, which has been attributed to a large amount of functional redundancy. In the stele-specific gene regulatory network there were many cases of promoters being bound by more than one upstream regulator. A potential reason for the observed high level of functional redundancy was that transcription factors played redundant roles in binding promoters. However, in over 60% of the tested cases there was a reproducible change in expression of the downstream gene in the presence of a mutated upstream factor [11]. This suggested that the reported redundancy is more likely to be a function of the topology of the regulatory networks rather than the fact that multiple transcription factors bind to the same promoter. The direction of change in transcript level (up- or downregulated) of the downstream target enabled the relationship to be designated as that of either an activator or a repressor. In addition to specifying the type of interaction, this analysis quantified the extent of transcriptional change, which was used to formulate a model that specified which regulators played the most important roles in controlling downstream expression (Figure 20.3B) [11].

Toward an Understanding of the Dynamics of Asymmetric Cell Division

Although all of the developmental stages are present along the longitudinal axis of the root, there are dynamic aspects of development that can only be explored with the addition of real-time analysis. An example is the cell division that generates the cortex and endodermis from a single stem cell. In fact, there are two asymmetric divisions that must occur: the first regenerates the stem cell and a daughter cell on its shootward side (Figure 20.4A); the second divides the daughter cell into the first cells of the endodermis and cortex cell lineages (Figure 20.4A). The plant-specific transcription factors, SHORTROOT (SHR) and SCARECROW (SCR) are required for the second asymmetric division to occur [12,13]. In mutants of either factor, there is a single cell layer between the epidermis and pericycle. However, the identity of this cell layer differs in the two mutants. In *shr* the endodermis is missing and there is only cortex, whereas in *scr* there are characteristics of both cortex and endodermis in the mutant cell layer. These phenotypic differences were the basis for a model in which SHR acts through SCR to effect the asymmetric division that gives rise to cortex and endodermis. Further characterization of the two genes resulted in a more nuanced hypothesis in which SHR interacts directly with SCR, and together they regulate downstream genes required for the asymmetric division [14]. An interesting feature of the SHR protein is that it is initially synthesized in the vascular tissue and then physically moves to the adjacent cell layer, where it interacts with SCR, which appears to be immobile [15]. It is this interaction that prevents SHR from moving to the outer layers of the root [14].

To address the question of how SHR and SCR regulate the asymmetric cell division, an inducible version of SHR was constructed by fusing the glucocorticoid receptor to the SHR protein. In the absence of steroid hormone, the glucocorticoid receptor causes SHR to be retained in the cytoplasm, preventing it from entering the nucleus and controlling downstream genes. When the synthetic steroid hormone, dexamethasone, is added to the growth medium, SHR is released and enters the nucleus. This construct was transformed into the *shr* mutant. With the addition of dexamethasone, complete rescue of the mutant phenotype occurred [16]. A second experiment was then performed in which dexamethasone was added and roots of the transgenic plants were imaged over time [17]. The first divisions of the mutant layer occurred around 6 hours after the addition of dexamethasone, followed by rapid division within the next hour (Figure 20.4B). What was surprising was that the first asymmetric division did not occur in the

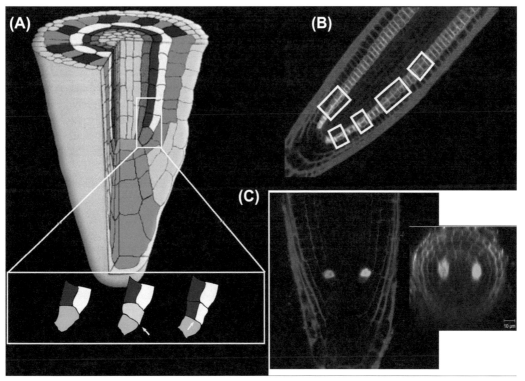

FIGURE 20.4 **(A) Asymmetric division of the Cortex/Endodermal Initial (CEI) depicted in a schematic.** (B) Asymmetric divisions of the mutant layer after induction of SHR activity. (C) Expression of cyclin D6 in the CEI daughter cell. *(Adapted from [17].)*

daughter of the stem cell, but rather took place in cells further up the root (Figure 20.4B). The onset of asymmetric divisions at 6 hours suggested that genes responsible for these divisions should be activated in that timeframe. Therefore, a time course experiment was performed, taking samples every 3 hours after the addition of dexamethasone. To focus on genes activated in the cell layer undergoing the asymmetric cell divisions, a marker was introduced into the line that expresses GFP specifically in the cell layer prior to division and in both cell layers after the division (Figure 20.4B). This was used for cell sorting, followed by hybridization of the RNA from the sorted cells to microarrays.

Analysis of the time course data identified a cluster of co-regulated genes whose expression went up markedly at 6 hours after the addition of dexamethasone [17]. Among the genes in the cluster were some that had been previously identified as direct targets of SHR, including SCR. A surprising finding was that a cell cycle component, a D-type cyclin, was also in the cluster. Because there are many cells undergoing division in the root, there was no reason to believe that a protein involved in regulating the cell cycle would be specifically regulated during a particular asymmetric division. Further characterization showed that both SHR and SCR could bind to the promoter region of the D-type cyclin, and that it was not expressed in every dividing cell but only in cells undergoing asymmetric divisions dependent on SHR and SCR (Figure 20.4C). Mutation of the D-type cyclin resulted in a delay in the asymmetric division, while mis-expression in the mutant layer of *shr* plants could partially rescue the loss of the asymmetric division. Taken together these results showed how a systems-level analysis with cell type-specific resolution over time could lead to the identification of a dedicated circuit regulating a specific asymmetric cell division critical for patterning of the root [17].

Computational Modeling of Root Development

Detailed knowledge on the transcriptome of an organ identifies new factors that contribute to development and allows us to elaborate the essential gene regulatory networks. However, the logic of developmental regulation is often not immediately apparent from a description of the dynamics of the molecules that guide it. In the *Arabidopsis* root, two examples illustrate this point well. First, we discuss the relationship between the major driving signal for stem cell specification, auxin, and the transcription factors that are required for stem cell specification. Second, we describe a lateral inhibition process that leads to a regular pattern of hair-bearing and hairless epidermal cells.

Auxin Flow and Stem Cell Specification

Auxins are a class of small molecules with indole acetic acid as a major representative. They serve as plant growth regulators whose distribution conveys patterning information and have been identified as key players in many developmental processes. In roots, high levels of auxin are correlated with stem cell specification, and membrane proteins of the PIN-FORMED (PIN) family that facilitate auxin transport are key factors for the accumulation of auxin in the stem cell niche [18,19]. The auxin transport system plays an important role in the ability of stem cells to regenerate after removal from the root [20]. How is this system of stem cells maintained in such a way that it can be rebuilt during regeneration? Answers to this question do not follow from the molecular connections between the main players in stem cell specification by themselves, but require an analysis of these links at the cellular level. Construction of a 'digital root', where cells were represented using a formalism that allows them to grow and divide, allowed an understanding of how auxin transport and auxin diffusion in roots translate polar localization of PIN proteins into a stable gradient of auxin with a peak at the stem cell position [21]. Simulation of growth and cell division, both under the control of auxin, revealed that such an auxin gradient is stable at the multicellular level while cells grow, divide and expand. Simulation of surgical manipulations and laser ablations, known to regenerate stem cells, revealed that the model had the capacity to emulate regeneration. This model represented an early step towards understanding dynamic pattern formation in roots. Important future additions are needed, which explicitly explain the coordinated polarization of the PIN proteins. This polarization is expected to be in part self-organizing, but the molecular components that govern polarization are only partially understood, which makes it difficult to formalize their activity and derive models that can be tested by experiment. Next-generation models should explicitly include self-organizing mechanisms that explain formation of the root stem cell system.

How does auxin program the stem cell state? Auxin-responsive transcription factors are required in the embryo for the expression of transcription factors of the PLETHORA (PLT) family, which maintain root identity and the stem cell state in the root apex [22,23]. Some PLT members can ectopically induce roots when expressed in shoot primordia, indicating that they are necessary and sufficient for root stem cell specification and maintenance [23]. While auxin accumulation is necessary to activate PLT expression, PLT function is required for the subsequent expression of genes encoding PIN auxin transporters [23]. Therefore, transcription factors expressed under the influence of auxin signal transduction in turn regulate auxin flow. Finally, induction of PLT expression is auxin dependent, but other factors are involved in maintenance of the PLT gradient [24,25]. The relationships between auxin accumulation and transcription factor action are not so much complicated in the sense of having many components, but are complex in the sense of being connected in a way that does not intuitively reveal the properties of the network. Here, formal descriptions and computer simulations of the relationships between components will be necessary to understand the developmental logic of root initiation and stem cell programming.

Mutual Support for Root Hair Patterning

Epidermal cell patterns are highly accessible, and the ease by which mutants in such patterns can be selected have made them favorites for the genetic dissection of pattern-forming mechanisms in animals and plants alike. The *Arabidopsis* root epidermis is one of the examples where molecular genetic analyses have yielded a wealth of components involved in the binary decision of whether or not to make hairs [26,27]. The primary role of root hairs, which protrude from the base of certain epidermal cells, is to expand the absorptive surface of the root. Genetic screens for roots with increased or decreased numbers of hairs resulted in the identification of two variants of transcription factor complexes in hair cells and hairless cells. In hairless cells, the active WEREWOLF (WER) complex promotes transcription of downstream targets that suppress the hair fate [28]. In hair cells, this complex is prevented from forming and replaced by a CAPRICE (CPC) complex through a combination of signaling from subepidermal cell layers through the SCRAMBLED (SCM) receptor kinase and by the lateral cell-to-cell movement of inhibitory transcription factors such as CPC originating from the hairless cells [29,30]. Several working models had been proposed for the way in which this regulatory network leads to correct pattern formation, relying on conventional notions of local activation and lateral inhibition, but formal analysis of the components indicated that a novel mechanism, where neighboring cells support the alternative cell type, best fitted all results [31]. This work indicates once more that complex networks are rarely amenable to intuitive explanations but need downstream analysis for deeper understanding. Interestingly, a study on regulators of SHR and SCR activity which function in the ground tissue to restrict the range of SHR action [32], revealed that these have a cell-autonomous role in the ground tissue for the production of signals that guide root hair patterning [33]. This work indicates that movement of SHR from stele to ground tissue is connected to a biasing system that impinges on epidermal patterning. This illustrates how interconnected the gene regulatory networks involved in root development are. One of the lessons to be learned is that it may not be possible to reconstruct organ development as a simple addition of modular gene regulatory networks. Rather, 'modules' only appear as such after crude analysis, and they may collapse into larger networks upon closer inspection. This posits a considerable challenge for next-generation computational models that summarize regulatory interactions and loops in the context of an entire organ.

Integrated Analysis of Shoot Development

Plant shoots contain tip regions that are sites of continuous cell division, growth and pattern formation. In the slowly dividing center of shoots, stem cells reside within a 'central zone' (CZ) (Figure 20.1). The underlying 'rib meristem' (RZ) zone contains transiting cells that dynamically form a spatially stable organizing center required to maintain the overlying stem cells. Organs such as leaves initiate when stem cell daughters emanating from the CZ reach the peripheral zone (PZ) (Figure 20.1) [34,35]. Roots branch by creating new stem cells through dedifferentiation of cells from one internal cell layer well outside the growth zones. Shoot branching and differentiation events, on the contrary, are scattered over the growing apex as new organs repeatedly form in close proximity to the stem cell region. These organs in turn develop new stem cell regions, creating nested patterns of differentiating organs and stem cell groups. While performing this iterative developmental act shoots can undergo major phase changes, dramatically illustrated in the switch from vegetative development, characterized by the production of leaves, to the formation of flowers, which is brought about by extensive modulation of developmental programs.

Gene Activity in Space and Time

The mix of patterning and differentiation zones in the growth regions of shoots has not precluded the use of dissection methods similar to those described for root transcriptome mapping. Expression profiles corresponding to developmental zones have been collected by enzymatically digesting the cells walls of the shoot apex followed by FACS using markers for CZ, RZ and PZ in a mutant background with a highly enriched shoot apical meristem tissue fraction (Figure 20.2) [36]. As observed in root samples, this approach significantly increased the sensitivity with which rare transcripts could be detected, and yielded about 1000 genes that were unique to cell population-specific data. In addition, mining and validation of differentially expressed genes revealed novel markers for the stem cell niche, and indicated an enrichment in DNA repair as well as DNA- and histone-modifying enzymes, in line with important roles for epigenetic control and control of genome stability in stem cells. The study also revealed novel expression patterns in the shoot apex, indicating that spatiotemporal patterns of gene expression in the shoot meristem have more complexity than previously anticipated. It will be important to extend this data set with more markers, which can be chosen from the existing digital expression atlas, to obtain a more complete picture of the shoot transcriptome. One important addition to the current system will be to find ways to extend this analysis from the inflorescence meristem to the vegetative meristem, which is deeply buried in leaves and therefore relatively inaccessible. Furthermore, it will be important to include differentiating cells in shoot organs in the gene expression maps. Only then will it be possible to compare shoot and root transcriptomes with equal sensitivity and accurately define genes as being root- or shoot-specific.

Coupled Gene Regulatory Networks in the Shoot

The intertwined nature of shoot development has also been advantageous. The repetitive nature of development allows initially subtle developmental distortions to be amplified. This has facilitated the identification of numerous genes which specifically affect shoot development (for recent reviews see [37–40]). Traditionally, these pathways were studied in relative isolation, leading to the description of major subnetworks. Although the separation of the shoot gene regulatory network has several characterizations, for the purpose of this chapter it is sufficient to define four major subnetworks. First, the WUSCHEL (WUS) – CLAVATA3 (CLV3) feedback loop controls the size of the stem cell domain by positive regulation of WUS transcription factor-expressing organizing cells in the RZ on CLV3-signal peptide-expressing stem cells, and repression of WUS expression by CLV3 action (Figure 20.1) [41–43]. Second, the CZ and PZ express the SHOOT MERISTEMLESS (STM) transcription factor, which maintains a mutual repression loop between STM family members and the AS1 transcription factor specific to organ primordia (Figure 20. 1) [44,45]. Third, a PIN–auxin network regulates auxin flow such that it converges on organ initiation points (Figure 20.1) [46–48] Fourth, an abaxial–adaxial specific transcription factor network operates through mutual inhibition and miRNA control to establish polar domains on organ primordia and maintain stem cells (Figure 20.1) [49–55].

The genetic studies that identified the components of the shoot regulatory network indicated early on that connections between the various subnetworks exist. One example is the discovery that the plant hormone cytokinin is not only an important player in the STM pathway [56,57], but at the same time the cytokinin response is repressed by WUSCHEL [58]. In addition, cytokinins suppress CLV1 levels as well as cytokinin sensitivity [59].

More recently, in line with the intertwined nature of shoot development, multiple connections between the various subnetworks have emerged through analysis of the target genes of some of the key transcription factors. A genome-wide identification of WUS response genes and WUS-binding sites indicated that WUS mainly represses transcripts within its expression domain in the RZ, and indicated that among these repressed genes are positive regulators for auxin signaling and negative regulators for cytokinin signaling [60]. While this study reveals that WUS likely acts through many target genes, the limited overlap between WUS-binding sites and WUS-dependent expression indicates that caution is needed in attempts to identify target genes of factors that act in very few cells.

Modeling Regulatory Networks in the Shoot

The many connections between regulatory subnetworks in shoot development that have either been identified by genetics or assumed from target gene lists indicate that the control of shoot development is complex. In addition, the processes that control shoot development operate at different levels. For example, diffusible peptide signals are coupled to transcriptional responses in different regions; auxin transport, biosynthesis and perception are regulated in distinct zones and must be considered in multicellular context; and growth that is controlled by gene regulatory networks leads to morphological changes that can subsequently influence signal flow and gene regulation. For these reasons, a formal representation of our knowledge on regulatory networks in the shoot and the use of computer simulations to study network properties is necessary. Without such analysis, it is not possible to understand the morphogenetic properties of the system.

An example of the insights that modeling can offer comes from attempts to understand the observed self-organization of WUS expression after laser ablations [61]. By implementing a reaction–diffusion model on an extracted cell template of the shoot apex, it was shown that local activation of WUS yielded a central WUS domain that was able to regenerate [62]. The CLV–WUS loop was modeled to explain how the WUS-expressing organizer cells and the CLV3-expressing stem cells maintain each other [63]. In addition, alternative feedback loops between WUS expression and cytokinin signaling were modeled and compared to experiment, supporting the hypothesis that the WUS domain can be established in a stable zone through which cells are traversing when cytokinin signaling influences WUS levels, cytokinin response regulators and the CLV pathway [59]. Collectively, these models increasingly pinpoint the function of feedback motifs in the regulation of dynamic expression domains.

Insightful modeling has also been performed on the PIN–auxin network that is associated with the patterned positioning of organ primordia at the shoot apex known as phyllotaxis. A mapping of experimentally determined PIN protein polarity and abundance on realistic cellular templates of the shoot apex and computation of the resultant fluxes showed that the convergent PIN polarization that accompanies the formation of new organ primordia is sufficient to initiate patterned auxin maxima [64]. The use of different scenarios as to how auxin concentrations might feedback on PIN polarization and abundance revealed how auxin maxima could self-organize in spatial patterns at the shoot apex, forming regular patterns of outgrowth [65,66]. Changes in parameters of the implemented feedback could be tuned to resemble existing phyllotactic patterns [66]. These findings bring to the fore two equally important messages. First, the work beautifully illustrates that feedback between auxin levels and polar auxin transport molecules can generate patterns de novo. Second, different assumptions on feedback, none of which are completely supported by biochemical mechanisms, yield self-organizing systems, which tells us that models do not always provide insight into mechanisms. Another issue that has been raised is that these phyllotaxis models are essentially two-dimensional and do not explicitly take into account the relation of auxin maxima at the surface and the internal auxin transport away from them in underlying tissues. A revised model was proposed in which cells at the surface transport auxin up the concentration gradient, whereas inner cells transport auxin down the flux gradient [67]. Again, it has yet to be determined whether the proposed switch function between the two mechanisms can be underpinned by biochemical mechanisms. Based on the observation that PIN polarity and microtubule arrangement respond coordinately to stresses and strains, a phyllotaxis model has been proposed that posits cell wall strains as an intermediate factor in the feedback mechanism between auxin and PIN polarity [68]. Although the biochemical nature of such a feedback remains to be clarified, the model is attractive because it can unify the work on gene regulatory networks and shoot development summarized above with a hitherto unconnected body of interesting observations indicating direct influence of cell wall composition on organ outgrowth at the shoot apex [69,70].

Finally, models for trichome patterning, which shares many of the factors acting in root hair patterning [27], have been constructed which indicate that transcriptional control of negative regulators and trapping of moving proteins contribute to pattern formation [71,72].

The above examples illustrate that important progress has been made in the identification and analysis of regulatory subnetworks in shoot development. However, the numerous connections between the various subnetworks have not yet been sufficiently dealt with, and it is very likely that these are key to some of the fundamental properties of the shoot system as a self-organizing unit. For example, the sizes of central and peripheral zones, now modeled mainly by focusing on the WUS–CLV–cytokinin network, have profound influence on the pattern of organ initiation, currently modeled with a primary focus on auxin distribution patterns. It is clear that genes that influence meristem size and cytokinin signaling influence phyllotaxis [73]. In addition, PLT genes expressed in the shoot apex control phyllotactic patterns, in part through regulation of PIN levels in primordia but also through changes in the shape and size of the meristem [74]. In addition, the interplay between adaxial and abaxial (equivalent to dorsal and ventral) polarity genes and the interaction between STM and its homologs, along with primordium-expressed genes, influence meristem maintenance. For a true understanding of shoot development all of these factors will have to be taken into account, which will require multilevel models that reconstitute shoot morphogenesis as well as the observed expression domains of key regulators in all of the subnetworks. Moreover, predictions of these models will have to be tested rigorously in order to arrive at the proper description of the most relevant interactions and components. For this, classic genetic analysis will not be sufficient, as it is more likely that models will vary in dynamic responses to transient perturbations. In *Arabidopsis*, the stage has been set for dynamic experiments that test existing models with the development of methods for sophisticated visualization and transient gene manipulation within the shoot apex [75–77].

DEVELOPMENT AND THE RESPONSE TO ENVIRONMENTAL STRESS

Because plants are sessile organisms, they have evolved different strategies from animals to deal with environmental stimuli. Many of their responses involve modifying their development. The aerial parts of plants grow toward light and respond to complex cues when deciding to flower; roots grow toward water and nutrients. Therefore, plant developmental responses can be thought of as akin to animal behavior.

Tissue-Specific Responses to Environment

Most of the initial work on responses to environmental stimuli was performed at the level of whole plants or entire organs. More recently the question has been asked, 'Are there differential responses among different cell types to environmental stimuli?' Put another way, the question could be, 'Do cells respond to stimuli as individual entities or as part of a coordinated response within the organism?' Addressing this question requires the analysis of the response to stimuli of individual cell types. One effort in this direction involved growing plants on high levels of salt

[78]. Since excess salinity is a common problem in soils around the world, the experimental results could have real-world applications to agriculture. To define a timeframe in which plants should be treated, the first step was to run a time course experiment with microarray analyses performed at multiple time points after exposure to high salt media. It was found that a substantial change in the expression of a large number of transcription factors occurred after 1 hour of exposure to high salt. This time point was then selected for the analysis of six individual cell types and four developmental stages isolated through FACS sorting and microdissection. The remarkable finding was that far more genes were differentially regulated within individual cell types and/or developmental stages than across all cell types and developmental stages. Moreover, with the heightened resolution this approach afforded, there were about eight times as many genes identified as being responsive to salt stress than had been previously identified using entire organs or whole plants [78]. This strongly suggested that even though salt is equally toxic to all cells, differential responses have evolved that are beneficial to the organism as a whole.

So what are the genes that respond to salt doing, and why is their expression regulated in a tissue-specific manner? Some answers are emerging. A striking morphological change that occurs upon encountering a high-salinity environment is swelling of the cells in the cortex. This could be an effort by the root to lower the salt concentration through dilution in the cortex cells. Within the cortex, a set of genes that regulates cell expansion were specifically downregulated upon salt stress [79]. Similarly, root hair elongation is aborted upon exposure to salt, which could be an effort to reduce salt uptake. Genes involved in root hair outgrowth are downregulated specifically in the root epidermis [79].

Similar approaches have been used to analyze the response at the level of individual cell types and developmental stages of growth in low iron or after nitrogen stimulation [80,81]. In the former, a gene named POPEYE whose expression is enriched in the pericycle when grown under iron deficiency conditions was shown to be required for root growth in low iron conditions. It encodes a transcription factor of the bHLH family, and through ChIP-chip analysis it was found to bind the promoters of a set of genes involved in metal ion homeostasis [81]. It also is co-regulated with a second gene named BRUTUS, which when mutated results in faster-growing roots under iron deficiency conditions. Both POPEYE and BRUTUS are able to interact with other bHLH proteins, suggesting that they act as a complex [81]. Thus, a gene regulatory network active in the response to low iron has begun to emerge with a transcription factor complex upstream of a set of genes involved in iron homeostasis.

A Major Challenge: The Transition to Flowering

The transition from vegetative growth to reproductive development imposes dramatic changes upon the developmental programs within the shoot apex. The fertilization process takes place in flowers and has to be exquisitely timed with respect to the right season and circumstances, and the production of flowers and seeds that result from successful fertilization instead of photosynthetic leaves imposes physiological demands on the plant. In perennial plants, this decision is made on a subpopulation of shoot nodes, but annuals such as *Arabidopsis* undergo an all-or-none transition to flowering, heightening the importance of proper timing of this process. For these reasons, the floral transition is controlled by the integration of many endogenous and environmental cues. Genes controlling flowering time were among the first described by *Arabidopsis* geneticists [82], and the collective effort of numerous groups has led to a molecular genetic dissection of flowering time control, where separate genetic pathways were shown to mediate the effect of day length, light quality and exposure to cold periods, all converging on the transcriptional activity of a few 'floral pathway integrator' genes encoding transcription factors [83]. The pathway integrators mediate the transition to flowering and later activate floral meristem identity genes, which change the developmental program of shoot meristems to allow the production of patterned floral organs — which are modified leaves [84].

Over the last decade many molecular details on how the perception of environmental inputs is transferred to the activity of floral pathway integrator genes have been elucidated, and for an updated account of this work we refer the reader to recent reviews [85,86]. This work has clarified the structure of five separate pathways that control flowering. (1) The photoperiod pathway utilizes rhythmic stimulation of expression conveyed by the circadian clock (see Chapter 21) in combination with direct light inputs to activate, in long days, the CONSTANS (CO) transcription factor in leaves. CO contributes to the activation of the floral pathway integrator FT, constituting a mobile signal that can move to the shoot apex and induce the floral transition. (2) FT is also regulated by the transcriptional repressor FLC, whose abundance decreases in prolonged cold periods due to local changes in chromatin structure. (3) FLC abundance is also regulated by general factors involved in chromatin modification and/or RNA processing, which fall into the 'autonomous pathway' that could monitor developmental age. (4) Increases in ambient temperature promote flowering and feed into FT abundance. (5) The plant hormone gibberellic acid and high sugar levels facilitate the induction of pathway integrators in the meristem. These factors may read out a variety of endogenous cues related to aging and carbohydrate status.

What can systems biology contribute to the understanding of the elaborate regulatory network that guides the transition to flowering? First, more complete descriptions of gene activity networks can help map out how different pathways are connected. Recently, high-confidence targets of the APETALA1 (AP1) transcription factor were identified by a combination of gene expression analyses after AP1 induction and deep sequencing of AP1-bound nuclear DNA (ChIP-Seq). From these studies a dual role emerges for AP1, in which it acts first to downregulate repressors of the floral transition and then as an activator for floral organ formation genes, including its own dimerization partners [87]. A similar dual role was extracted from target gene analysis and chromatin-binding studies for the APETALA2 (AP2) transcription factor. AP2 targets also fall into two broad classes, flowering time genes that repress the floral transition, and floral identity genes that regulate regional identity in the flower [88]. These dual roles for two important factors in the floral transition pathway indicate once more that subnetworks become more intimately connected when more information becomes available. It will be very informative to obtain target gene analyses of each of the major transcription factors that guide the transition from vegetative to generative meristems in the shoot. It will be important to see whether the separation into subnetworks derived from decades of genetic analysis will ultimately need to be replaced by a description of a linked 'network of networks'.

To determine whether large, complicated networks can be decomposed into subnetworks or whether they need to be considered in their fully integrated state, a description of network structures is not sufficient. Again, modeling the effect of network connections will be needed to substantiate whether our current, essentially linear view on how environmental inputs initiate flowering holds true. When genome-wide data indicate that such inputs are intertwined, surprising behavior may be expected when the networks are simulated. A workable approach to tackle large-scale information processing networks is to perform Boolean network analysis, which simplifies quantitative inputs in gene regulatory networks (see Chapter 10). Such an analysis has been performed for floral pattern formation [89,90], and it seems worthwhile to integrate similar models with an interlinked flowering time network. At the same time, it will be necessary to model these networks in space, as all factors are operating in a spatial context and their activity modifies the spatial context. An early example of such models generates, from a hypothesis at the molecular level, understanding at the level of whole plant architecture [91]. Here, a simple network for the floral transition is modeled in a spatial context, which provides tantalizing insights into the architecture of floral branching structures as well as in to the constraints that guide their evolution. Ultimately, a much more fine-grained integration will be necessary to understand, for example, how the floral transition changes not only the identity of organs formed on the shoot apex but also their relative positioning, which is phyllotaxis. Finally, theoretical analysis is not only helpful to ascend to higher levels of integration, but can also provide key insights into basic molecular mechanisms that underlie many of the processes captured more abstractly in global networks. Very recently, computational modeling has been adapted to understand how the accumulation of cold can lead to stable repression of the FLC transcription factor [92]. The observed two-step accumulation of epigenetic marks at the key flowering regulator FLC was used to inspire a model in which nucleation of repressive marks and a bias in histone dynamics towards the repressive mark were simulated on in silico loci. The model predicted stochastic switching of individual cells to the repressed FLC state, where the amount of switched cells depended on the length of the cold induction, and this stochastic behavior was experimentally validated.

CONCLUSIONS AND PERSPECTIVE

One of the goals of systems biology is to map the relationships between genotype and phenotype. The gene products encoded in the genome interact in complicated and complex networks, and from those interactions the biological processes that underlie phenotype are enabled. In this view, phenotype is an emergent property of the networks whose components are directly or indirectly defined by genotype. An added measure of complexity is that networks respond differentially to environmental stimuli.

Plants are tractable systems in which advances have been made on this very big problem. Their sessile nature allows environmental perturbations to be controlled and monitored. Aspects of their growth and development help to simplify the analysis of responses, particularly at the resolution of individual cell types.

We have provided several examples in which high-throughput expression data were used as a starting point to probe biological processes. The results have been the identification of genes and subnetworks that regulate the process under study. The primary focus to date has been on transcriptional regulation, in large part because the tools available for genome-wide transcriptional analysis are far more robust than for analysis of other cellular components. With the advent of the next generation of sequencing platforms an area that is rapidly opening is post-transcriptional regulation by non-coding RNAs. To achieve anything like the same depth and breadth of analysis for proteins and metabolites will require advances in current technologies.

The next challenge is to move from the identification of networks to an understanding of how they function in real time. We have given a number of examples that show how computational modeling is necessary to address this issue in any depth. While these models already provide much information with crude parameter settings, and are in a number of cases very robust to parameters, it will become necessary to measure at least the range of a number of parameters in specific interaction networks. For example, the difference between signal damping and oscillation can reside in how much delay there is in a negative feedback loop. A near-term goal is to be able to monitor various components of a network as they interact. This is likely to involve live imaging facilitated by microfluidics. In the longer term it may be necessary to design new mathematical analysis tools that can indicate which parameter measurements in a complex regulatory network are minimally required to test it aagainst alternatives. For each of these future directions, it appears that plant models provide fertile testing grounds.

REFERENCES

[1] Roeder AH, Tarr PT, Tobin C, Zhang X, Chickarmane V, Cunha A, et al. Computational morphodynamics of plants: integrating development over space and time. Nat Rev Mol Cell Biol 2011;12:265–73.

[2] Birnbaum K, Shasha DE, Wang JY, Jung JW, Lambert GM, Galbraith DW, et al. A gene expression map of the *Arabidopsis* root. Science 2003;302:1956–60.

[3] Brady SM, Orlando DA, Lee JY, Wang JY, Koch J, Dinneny JR, et al. A high-resolution root spatiotemporal map reveals dominant expression patterns. Science 2007;318:801–6.

[4] Cartwright DA, Brady SM, Orlando DA, Sturmfels B, Benfey PN. Reconstructing spatiotemporal gene expression data from partial observations. Bioinformatics 2009;25:2581–7.

[5] Kerk NM, Ceserani T, Tausta SL, Sussex IM, Nelson TM. Laser capture microdissection of cells from plant tissues. Plant Pathol 2003;132:27–35.

[6] Jiao Y, Tausta SL, Gandotra N, Sun N, Liu T, Clay NK, et al. A transcriptome atlas of rice cell types uncovers cellular, functional and developmental hierarchies. Nat Gene 2009;41:258–63.

[7] Mustroph A, Zanetti ME, Jang CJ, Holtan HE, Repetti PP, Galbraith DW, et al. Profiling translatomes of discrete cell populations resolves altered cellular priorities during hypoxia in *Arabidopsis*. Proc Natl Acad Sci U S A 2009;106(44):18843–8.

[8] Deal RB, Henikoff S. A simple method for gene expression and chromatin profiling of individual cell types within a tissue. Dev Cell 2010;18(6):1030–40.

[9] Roppolo D, De Rybel B, Tendon VD, Pfister A, Alassimone J, Vermeer JE, et al. A novel protein family mediates Casparian strip formation in the endodermis. Nature 2011;473:380–3.

[10] Tsukagoshi H, Busch W, Benfey PN. Transcriptional regulation of ROS controls transition from proliferation to differentiation in the root. Cel 2010;143:606–16.

[11] Brady SM, Zhang L, Megraw M, Martinez NJ, Jiang E, Yi CS, et al. A stele-enriched gene regulatory network in the *Arabidopsis* root. Mol Syst Biol 2011;7:459.

[12] Di Laurenzio L, Wysocka-Diller J, Malamy JE, Pysh L, Helariutta Y, Freshour G, et al. The *SCARECROW* gene regulates an asymmetric cell division that is essential for generating the radial organization of the *arabidopsis* root. Cell 1996;86:423–33.

[13] Helariutta Y, Fukaki H, Wysocka-Diller JW, Nakajima K, Jung J, Sena G, et al. The SHORT-ROOT gene controls radial patterning of the *Arabidopsis* root through radial signaling. Cell 2000;101:555–67.

[14] Cui H, Levesque MP, Vernoux T, Jung JW, Paquette AJ, Gallagher KL, et al. An evolutionarily conserved mechanism delimiting SHR movement defines a single layer of endodermis in plants. Science 2007;316:421–5.

[15] Nakajima K, Sena G, Nawy T, Benfey PN. Intercellular movement of the putative transcription factor SHR in root patterning. Nature 2001;413:307–11.

[16] Levesque MP, Vernoux T, Busch W, Cui H, Wang JY, Blilou I, et al. Whole-genome analysis of the SHORT-ROOT developmental pathway in *Arabidopsis*. PloS Biology 2006;4:e143.

[17] Sozzani R, Cui H, Moreno-Risueno MA, Busch W, Van Norman JM, Vernoux T, et al. Spatiotemporal regulation of cell-cycle genes by SHORTROOT links patterning and growth. Nature 2010;466:128–32.

[18] Friml J, Benkova E, Blilou I, Wisniewska J, Hamann T, Ljung K, et al. AtPIN4 mediates sink-driven auxin gradients and root patterning in *Arabidopsis*. Cell 2002;108:661–73.

[19] Sabatini S, Beis D, Wolkenfelt H, Murfett J, Guilfoyle T, Malamy J, et al. An auxin-dependent distal organizer of pattern and polarity in the *Arabidopsis* root. Cell 1999;99:463–72.

[20] Xu J, Hofhuis H, Heidstra R, Sauer M, Friml J, Scheres B. A molecular framework for plant regeneration. Science 2006;311:385–8.

[21] Grieneisen VA, Xu J, Maree AF, Hogeweg P, Scheres B. Auxin transport is sufficient to generate a maximum and gradient guiding root growth. Nature 2007;449:1008–13.

[22] Aida M, Beis D, Heidstra R, Willemsen V, Blilou I, Galinha C, et al. The PLETHORA genes mediate patterning of the *Arabidopsis* root stem cell niche. Cell 2004;119:109–20.

[23] Galinha C, Hofhuis H, Luijten M, Willemsen V, Blilou I, Heidstra R, et al. PLETHORA proteins as dose-dependent master regulators of *Arabidopsis* root development. Nature 2007;449:1053–7.

[24] Kornet N, Scheres B. Members of the GCN5 histone acetyltransferase complex regulate PLETHORA-mediated root stem cell niche maintenance and transit amplifying cell proliferation in *Arabidopsis*. Plant Cell 2009;21:1070–9.

[25] Matsuzaki Y, Ogawa-Ohnishi M, Mori A, Matsubayashi Y. Secreted peptide signals required for maintenance of root stem cell niche in *Arabidopsis*. Science 2010;329:1065–7.

[26] Schellmann S, Hulskamp M. Epidermal differentiation: trichomes in *Arabidopsis* as a model system. Int J Dev Biol 2005;49:579–84.

[27] Schiefelbein J. Cell-fate specification in the epidermis: a common patterning mechanism in the root and shoot. Curr Opin Plant Biol 2003;6:74–8.

[28] Lee MM, Schiefelbein J. WEREWOLF, a MYB-related protein in *Arabidopsis*, is a position-dependent regulator of epidermal cell patterning. Cell 1999;99:473–83.

[29] Kurata T, Ishida T, Kawabata-Awai C, Noguchi M, Hattori S, Sano R, et al. Cell-to-cell movement of the CAPRICE protein in *Arabidopsis* root epidermal cell differentiation. Development 2005;132:5387–98.

[30] Kwak SH, Shen R, Schiefelbein J. Positional signaling mediated by a receptor-like kinase in *Arabidopsis*. Science 2005;307:1111–3.

[31] Savage NS, Walker T, Wieckowski Y, Schiefelbein J, Dolan L, Monk NA. A mutual support mechanism through intercellular movement of CAPRICE and GLABRA3 can pattern the *Arabidopsis* root epidermis. PLoS Biol 2008;6:e235.

[32] Welch D, Hassan H, Blilou I, Immink R, Heidstra R, Scheres B. *Arabidopsis* JACKDAW and MAGPIE zinc finger proteins delimit asymmetric cell division and stabilize tissue boundaries by restricting SHORT-ROOT action. Genes Dev 2007;21:2196–204.

[33] Hassan H, Scheres B, Blilou I. JACKDAW controls epidermal patterning in the *Arabidopsis* root meristem through a non-cell-autonomous mechanism. Development 2010;137:1523–9.

[34] Meyerowitz EM. Genetic control of cell division patterns in developing plants. Cell 1997;88:299–308.

[35] Steeves TA, Sussex IM. Patterns in Plant Development. Cambridge: Cambridge University Press; 1989.

[36] Yadav RK, Girke T, Pasala S, Xie M, Reddy GV. Gene expression map of the *Arabidopsis* shoot apical meristem stem cell niche. Proc Natl Acad Sci U S A 2009;106:4941–6.

[37] Aida M, Tasaka M. Genetic control of shoot organ boundaries. Curr Opin Plant Biol 2006;9:72–7.

[38] Hay A, Tsiantis M. KNOX genes: versatile regulators of plant development and diversity. Development 2010;137:3153–65.

[39] Rieu I, Laux T. Signaling pathways maintaining stem cells at the plant shoot apex. Semin Cell Dev Biol 2009;20:1083–8.

[40] Sablowski R. The dynamic plant stem cell niches. Curr Opin. Plant Biol 2007;10:639–44.

[41] Fletcher JC, Brand U, Running MP, Simon R, Meyerowitz EM. Signaling of cell fate decisions by CLAVATA3 in *Arabidopsis* shoot meristems. Science 1999;283:1911–4.

[42] Mayer KF, Schoof H, Haecker A, Lenhard M, Jurgens G, Laux T. Role of WUSCHEL in regulating stem cell fate in the *Arabidopsis* shoot meristem. Cell 1998;95:805–15.

[43] Schoof H, Lenhard M, Haecker A, Mayer KF, Jurgens G, Laux T. The stem cell population of *Arabidopsis* shoot meristems is maintained by a regulatory loop between the CLAVATA and WUSCHEL genes. Cell 2000;100:635–44.

[44] Byrne ME, Barley R, Curtis M, Arroyo JM, Dunham M, Hudson A, et al. Asymmetric leaves1 mediates leaf patterning and stem cell function in *Arabidopsis*. Nature 2000;408:967–71.

[45] Long JA, Moan EI, Medford JI, Barton MK. A member of the KNOTTED class of homeodomain proteins encoded by the STM gene of *Arabidopsis*. Nature 1996;379:66–9.

[46] Benkova E, Michniewicz M, Sauer M, Teichmann T, Seifertova D, Jurgens G, et al. Local, efflux-dependent auxin gradients as a common module for plant organ formation. Cell 2003;115:591–602.

[47] Reinhardt D, Frenz M, Mandel T, Kuhlemeier C. Microsurgical and laser ablation analysis of interactions between the zones and layers of the tomato shoot apical meristem. Development 2003;130:4073–83.

[48] Reinhardt D, Mandel T, Kuhlemeier C. Auxin regulates the initiation and radial position of plant lateral organs. Plant Cell 2000;12:507–18.

[49] Eshed Y, Baum SF, Bowman JL. Distinct mechanisms promote polarity establishment in carpels of *Arabidopsis*. Cell 1999;99:199–209.

[50] Juarez MT, Kui JS, Thomas J, Heller BA, Timmermans MC. microRNA-mediated repression of rolled leaf1 specifies maize leaf polarity. Nature 2004;428:84–8.

[51] Kerstetter RA, Bollman K, Taylor RA, Bomblies K, Poethig RS. KANADI regulates organ polarity in *Arabidopsis*. Nature 2001;411:706–9.

[52] Mallory AC, Reinhart BJ, Jones-Rhoades MW, Tang G, Zamore PD, Barton MK, et al. MicroRNA control of PHABULOSA in leaf development: importance of pairing to the microRNA 5' region. EMBO J 2004;23:3356–64.

[53] McConnell JR, Emery J, Eshed Y, Bao N, Bowman J, Barton MK. Role of PHABULOSA and PHAVOLUTA in determining radial patterning in shoots. Nature 2001;411:709–13.

[54] Nogueira FT, Madi S, Chitwood DH, Juarez MT, Timmermans MC. Two small regulatory RNAs establish opposing fates of a developmental axis. Genes Dev 2007;21:750–5.

[55] Siegfried KR, Eshed Y, Baum SF, Otsuga D, Drews GN, Bowman JL. Members of the YABBY gene family specify abaxial cell fate in *Arabidopsis*. Development 1999;126:4117–28.

[56] Jasinski S, Piazza P, Craft J, Hay A, Woolley L, Rieu I, et al. KNOX action in *Arabidopsis* is mediated by coordinate regulation of cytokinin and gibberellin activities. Curr Biol 2005;15:1560–5.

[57] Yanai O, Shani E, Dolezal K, Tarkowski P, Sablowski R, Sandberg G, et al. *Arabidopsis* KNOXI proteins activate cytokinin biosynthesis. Curr Biol 2005;15:1566–71.

[58] Leibfried A, To JP, Busch W, Stehling S, Kehle A, Demar M, et al. WUSCHEL controls meristem function by direct regulation of cytokinin-inducible response regulators. Nature 2005;438:1172–5.

[59] Gordon SP, Chickarmane VS, Ohno C, Meyerowitz EM. Multiple feedback loops through cytokinin signaling control stem cell number within the *Arabidopsis* shoot meristem. Proc Natl Acad Sci U S A 2009;106:16529–34.

[60] Busch W, Miotk A, Ariel FD, Zhao Z, Forner J, Daum G, et al. Transcriptional control of a plant stem cell niche. Dev Cell 2010;18:849–61.

[61] Reinhardt D, Pesce ER, Stieger P, Mandel T, Baltensperger K, Bennett M, et al. Regulation of phyllotaxis by polar auxin transport. Nature 2003;426:255–60.

[62] Jonsson H, Heisler M, Reddy GV, Agrawal V, Gor V, Shapiro BE, et al. Modeling the organization of the WUSCHEL expression domain in the shoot apical meristem. Bioinformatics 2005;21(Suppl 1):i232–40.

[63] Geier F, Lohmann JU, Gerstung M, Maier AT, Timmer J, Fleck C. A quantitative and dynamic model for plant stem cell regulation. PLoS One 2008;3:e3553.

[64] de Reuille PB, Bohn-Courseau I, Ljung K, Morin H, Carraro N, Godin C, et al. Computer simulations reveal properties of the cell–cell signaling network at the shoot apex in *Arabidopsis*. Proc Natl Acad Sci U S A 2006;103:1627–32.

[65] Jonsson H, Heisler MG, Shapiro BE, Meyerowitz EM, Mjolsness E. An auxin-driven polarized transport model for phyllotaxis. Proc Natl Acad Sci U S A 2006;103:1633–8.

[66] Smith RS, Guyomarc'h S, Mandel T, Reinhardt D, Kuhlemeier C, Prusinkiewicz P. A plausible model of phyllotaxis. Proc Natl Acad Sci U S A 2006;103:1301–6.

[67] Bayer EM, Smith RS, Mandel T, Nakayama N, Sauer M, Prusinkiewicz P, et al. Integration of transport-based models for phyllotaxis and midvein formation. Genes Dev 2009;23:373–84.

[68] Heisler MG, Hamant O, Krupinski P, Uyttewaal M, Ohno C, Jonsson H, et al. Alignment between PIN1 polarity and microtubule orientation in the shoot apical meristem reveals a tight coupling between morphogenesis and auxin transport. PLoS Biol 2010;8:e1000516.

[69] Fleming AJ, McQueen-Mason S, Mandel T, Kuhlemeier C. Induction of leaf primordia by the cell wall protein expansin. Science. 1997;276:1415–8.

[70] Peaucelle A, Louvet R, Johansen JN, Hofte H, Laufs P, Pelloux J, et al. *Arabidopsis* phyllotaxis is controlled by the methyl-esterification status of cell-wall pectins. Curr Biol 2008;18:1943–8.

[71] Bouyer D, Geier F, Kragler F, Schnittger A, Pesch M, Wester K, et al. Two-dimensional patterning by a trapping/depletion mechanism: the role of TTG1 and GL3 in *Arabidopsis* trichome formation. PLoS Biol 2008;6:e141.

[72] Digiuni S, Schellmann S, Geier F, Greese B, Pesch M, Wester K, et al. A competitive complex formation mechanism underlies trichome patterning on *Arabidopsis* leaves. Mol Syst Biol 2008;4:217.

[73] Giulini A, Wang J, Jackson D. Control of phyllotaxy by the cytokinin-inducible response regulator homologue ABPHYL1. Nature 2004;430:1031–4.

[74] Prasad K, Grigg SP, Barkoulas M, Yadav RK, Sanchez-Perez GF, Pinon V, et al. *Arabidopsis* PLETHORA transcription factors control phyllotaxis. Curr Biol 2011;21:1123–8.

[75] Reddy GV, Meyerowitz EM. Stem-cell homeostasis and growth dynamics can be uncoupled in the *Arabidopsis* shoot apex. Science 2005;310:663–7.

[76] Vernoux T, Brunoud G, Farcot E, Morin V, Van den Daele H, Legrand J, et al. The auxin signalling network translates dynamic input into robust patterning at the shoot apex. Mol Syst Biol 2011;7:508.

[77] Yadav RK, Tavakkoli M, Reddy GV. WUSCHEL mediates stem cell homeostasis by regulating stem cell number and patterns of cell division and differentiation of stem cell progenitors. Development 2010;137:3581–9.

[78] Dinneny JR, Long TA, Wang JY, Jung JW, Mace D, Pointer S, et al. Cell identity mediates the response of *Arabidopsis* roots to abiotic stress. Science 2008;320:942–5.

[79] Dinneny JR. Analysis of the salt-stress response at cell-type resolution. Plant Cell Environ 2010;33(4):543–51.

[80] Gifford ML, Dean A, Gutierrez RA, Coruzzi GM, Birnbaum KD. Cell-specific nitrogen responses mediate developmental plasticity. Proc Natl Acad Sci U S A 2008;105:803–8.

[81] Long TA, Tsukagoshi H, Busch W, Lahner B, Salt DE, Benfey PN. The bHLH transcription factor POPEYE regulates response to iron deficiency in *Arabidopsis* roots. Plant Cell 2010;22:2219–36.

[82] Redei GP. Supervital mutants of *Arabidopsis*. Genetics 1962; 47:443–60.

[83] Simpson GG, Dean C. *Arabidopsis*, the Rosetta stone of flowering time? Science 2002;296:285–9.

[84] Weigel D, Meyerowitz EM. The ABCs of floral homeotic genes. Cell 1994;78:203–9.

[85] Farrona S, Coupland G, Turck F. The impact of chromatin regulation on the floral transition. Semin Cell Dev Biol 2008;19:560–73.

[86] Srikanth A, Schmid M. Regulation of flowering time: all roads lead to Rome. Cell Mol Life Sci 2011;68:2013–37.

[87] Kaufmann K, Wellmer F, Muino JM, Ferrier T, Wuest SE, Kumar V, et al. Orchestration of floral initiation by APETALA1. Science 2010;328:85–9.

[88] Yant L, Mathieu J, Dinh TT, Ott F, Lanz C, Wollmann H, et al. Orchestration of the floral transition and floral development in *Arabidopsis* by the bifunctional transcription factor APETALA2. Plant Cell 2010;22:2156–70.

[89] Espinosa-Soto C, Padilla-Longoria P, Alvarez-Buylla ER. A gene regulatory network model for cell-fate determination during *Arabidopsis thaliana* flower development that is robust and recovers experimental gene expression profiles. Plant Cell 2004;16: 2923–39.

[90] Mendoza L, Alvarez-Buylla ER. Dynamics of the genetic regulatory network for *Arabidopsis thaliana* flower morphogenesis. J Theor Biol 1998;193:307–19.

[91] Prusinkiewicz P, Erasmus Y, Lane B, Harder LD, Coen E. Evolution and development of inflorescence architectures. Science 2007;316:1452–6.

[92] Angel A, Song J, Dean C, Howard M. A Polycomb-based switch underlying quantitative epigenetic memory. Nature. 2011;476: 105–8.

Chapter 21

The Role of the Circadian System in Homeostasis

Anand Venkataraman, Heather Ballance and John B. Hogenesch

Department of Pharmacology, Penn Center for Bioinformatics, Penn Genome Frontiers Institute, Institute for Translational Medicine and Therapeutics, University of Pennsylvania Perelman School of Medicine, Philadelphia, PA 19104, USA

Chapter Outline

Introduction	407
Evolution of Clocks	407
Role of the clock in neurological functions	408
Sleep	409
Homeostasis: You Notice It When It's Broken	412
Cognition (Learning and Memory)	412
Neuropsychiatric Disorders	414
Clocks in Energy and Metabolic Homeostasis	415
What are Peripheral Clocks?	415
Understanding Energy Homeostasis with Animal Models of the Clock	417
Molecular Integrators of Clock and Metabolism	418
Conclusion	419
References	419

INTRODUCTION

In the early 1700s, Jean-Jacques d'Ortous de Mairan reported that the daily opening and closing of the mimosa plant's leaves persisted in the complete absence of all external cues [1]. de Mairan inferred that plants have an endogenous biological clock, and the field of chronobiology was officially born. Since this observation, scientists have investigated and identified biological rhythms in almost all living organisms. The term circadian (Latin *circa* = about, *dies* = day) was coined by Franz Halberg to describe the near-24-hour period of these biological rhythms [2,3]. Remarkably, the fundamental molecular machinery driving these rhythms is conserved between unicellular prokaryotic cyanobacteria and mammals [4].

Using reductionist approaches, research endeavors from de Marian's time have increasingly focused on understanding the molecular nuts and bolts of the circadian machine. This approach led researchers to understand the basic mechanisms of clock function across species, but by itself this approach could not explain all the complex properties of circadian clocks. One of the key questions is why we need clocks at all. Another is how clocks act in the homeostatic process. Over the past decade integrative or systems biology approaches have once again allowed us to step back and appreciate the influence of the circadian machinery on the overall regulation of our body's homeostasis.

In the late 19th and 20th centuries the concept of 'homeostasis' was formalized by the likes of Claude Bernard [5] and Walter Cannon [6]. The word homeostasis, derived from Greek meaning 'similar' and 'stability', refers to the properties by which biological systems keep their physiology at steady-state levels [7]. For example, mammals maintain their body temperature within a precise and narrow range throughout the day. When this maintenance fails, e.g., during a fever and recovery, homeostatic mechanisms are in place to return temperature to a normal working range. Importantly, these mechanisms need to work in an open system and constantly adapt to environmental changes. In this chapter, we will detail ways in which the biological clock helps maintain homeostasis amidst daily changes in the environment, and what can happen to the organism when clock-enabled homeostasis is disrupted.

EVOLUTION OF CLOCKS

Early life was faced with a volatile environment, with wild fluctuations in levels and intensity of light as well as temperature. To explain how clocks could help these primitive organisms maintain their homeostasis, let us begin by asking what evolutionary selection pressures could have enabled the clock to arise. Among many possibilities, there is some empirical evidence for the 'escape from light' hypothesis [8]. This hypothesis, framed by Dr Collin Pittendrigh, father of modern chronobiology, suggests that the selection pressure for clock evolution was a cellular-coping mechanism against the ionizing radiation of the sun [8]. Certain physiological processes, such as DNA replication, can be compromised by photodamage. Mechanisms that restricted or 'scheduled' such photosensitive processes to the night could have provided the earliest push for the evolution of circadian clocks. To test this hypothesis, Carl Johnson's group tested survival curves of the unicellular algae *Chlamydomonas reinhardtii* after exposure to ultraviolet (UV) radiation at different times of day [9]. These algal cells showed increased susceptibility to UV radiation at night. An extension of this hypothesis is the temporal compartmentalization of somewhat antagonistic processes. For example, in unicellular cyanobacteria, biochemically incompatible reactions such as oxidation/reduction, photosynthesis/nitrogen fixation, etc., are scheduled for different times of the day, an elegant solution [8,10].

Next let us ask, how did the clock machinery evolve to its current state? There are two models to describe the evolution of clock mechanisms: (1) the classic model — complex clocks evolved by refinements of a primitive clock, and (2) the convergent model — repeated but independent selection pressure led to the generation of clocks across the entire biosphere [8,11,12]. The classic model refers to the prevalence of the PAS domain (named after the three founding members of the PAS domain superfamily, *Drosophila* PER, mammalian ARNT and *Drosophila* SIM) in clock-related proteins throughout the evolutionary tree of life [13]. PAS domains were originally identified as protein—protein interaction domains, but were subsequently found to also bind small ligands (e.g., in the aryl hydrocarbon receptor [14]). However, reports suggest that PAS domains have evolved from ancient environment-sensing proteins. Based on homology, PAS domains could have evolved from LOV (light, oxygen, voltage sensing; [13,15]) and/or bacterial blue-light receptor, photoactive yellow protein (PYP; 13) domains. Incorporation of these environmental sensing domains in an oscillatory feedback loop with a 24-hour period could have initiated clock evolution. Evolutionary fitness provided by the circadian system ensured its selection over time, albeit with growing complexity.

In contrast, the convergent model draws attention to the differences in the oscillators. The most compelling argument here has been the complete lack of homology of clock genes in cyanobacteria, plants, and animals, suggesting that circadian rhythms have evolved at least three times [12].

Additionally, finer differences are observed, such as non-overlapping clock gene expression patterns between the silk moth, *Drosophila* and mice [16—19]. Thus the convergent model proposes that evolutionarily distinct pacemakers converged to a feedback activation—repression loop mechanism, presumably because of their robustness in generating 24-hour rhythmic oscillations. Although strong empirical evidence is still lacking, as it often is for evolutionary arguments, clocks could have evolved by a process that included both the above-mentioned models.

Why do we and most other organisms on Earth still have a clock? This is particularly interesting for nocturnal rodents, who are largely shielded from the radiation of the sun. For example, the blind subterranean mole rat, *Spalax ehrenbergi*, lives in total darkness but still has a circadian clock [20]. A salient feature of the circadian pacemaker is its ability to maintain robust oscillation with conserved amplitude, period length and phase even when faced with acute environmental perturbations. Our circadian pacemakers are resistant to acute environmental perturbations such as diurnal changes in temperature (morning vs. afternoon), light (cloudy vs. clear sky), humidity etc., allowing us to keep track of time. However, clocks are also adaptable to prolonged perturbations such as seasonal changes in day length or temperature [8,21,22]. Thus the most fundamental advantage of an internal clock is to synchronize one's behavior/physiology to the state of the prevailing environment.

Some of the adaptive advantages of biological clocks have now been identified and related to various facets of survival in the wild. Examples include avoiding predators (animals with lesioned central clock do not survive as well as non-lesioned animals in the wild; [23,24]), preventing desiccation (an example is the timed eclosion rhythms of the fruitfly larvae to the morning, when humidity is relatively high; [25]), finding mates (temporal separation by speciation; [26]), synchronizing feeding with food availability, and tuning metabolic pathways to allow seasonal features such as hibernation.

In summary, it is important to appreciate that evolution favors mechanisms that work — anticipating and reacting to changes in the environment such as light, temperature and humidity, is an integral part of this survival strategy. For the rest of this chapter we will focus on our current understanding of the adaptive advantages of the mammalian biological clock, with an emphasis on its role in behavior, physiology, and molecular homeostasis.

ROLE OF THE CLOCK IN NEUROLOGICAL FUNCTIONS

Before we begin our discussion on neurophysiological functions of the clock, let us review the organization of the

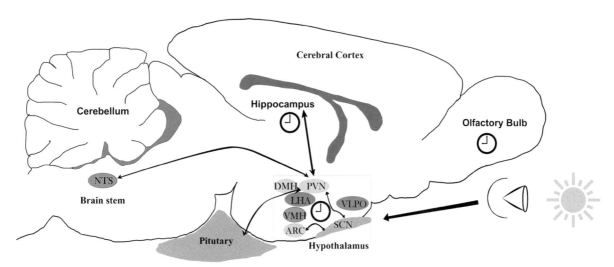

FIGURE 21.1 **Inter-clock communication within the brain.** A schematic sagittal view of the mouse brain with the master clock, the suprachiasmatic nucleus (SCN) positioned in the anterior hypothalamus. This small cluster of ~50 000 neurons in the SCN receive photic input from the retina and are able to autonomously generate circadian oscillation in phase with the environment. Efferent processes from the SCN innervate neighboring nuclei such as the arcuate nucleus (ARC), paraventricular nucleus (PVN) and lateral hypothalamus (LHA) which regulate a plethora of homeostatic processes. Furthermore, the PVN functions as a relay center and connects with numerous other regions, such as the hippocampus of the temporal lobe, the nucleus of the solitary tract (NTS) in the brainstem and the pituitary gland — the basecamp for all major regulatory hormones. In addition to the SCN, various other oscillators have been discovered within the brain, e.g., the olfactory bulb in rodents, the ARC nucleus and more recently, the hippocampus, the center for learning and memory. DMH, dorsomedial hypothalamus; VMH, ventromedial hypothalamus; VLPO, ventrolateral preoptic nucleus. Figure not drawn to scale.

circadian pacemaker in mammals. The master clock in mammals is a cluster of ~50 000 neurons placed anteriorly to the optic chiasm and hence referred to as the suprachiasmatic nucleus (SCN; Figure 21.1). The SCN receives sensory light inputs from the retina, and in turn the SCN connects with structures within the brain and in peripheral organs (Figure 21.1; also discussed further below). The SCN is able to generate autonomous and sustained circadian oscillations that can be observed at the electrophysiological, behavioral and molecular levels [5,27].

Sleep

The influence of the biological clock is most obvious in the timing of sleep. In diurnal organisms, behavioral and cognitive performance peak during the subjective day, while the opposite occurs in nocturnal animals. However, in both diurnal and nocturnal animals performance is also associated anecdotally with sleep/wake history, i.e., a sleep-deprived individual will perform poorly compared to a fully rested individual [28]. Such observations suggest an obvious link between the circadian system and the sleep—wake cycle. But why do we sleep? Numerous hypotheses have been proposed, and this topic remains contentious. A recent hypothesis by Tononi and colleagues has gained much popularity [29]. They suggest that sleep provides a mechanism to reboot neuronal activity and return it to baseline (energetically) after a day of interaction with the physical world [29].

First, let us define sleep, which differs depending on the organism studied and how it is evaluated [30,31]. Sleep can be scored qualitatively based on behavioral parameters such as a period of inactivity or poor responsiveness to external stimuli, and quantitatively by physiological parameters such as brain electrical activity (at least in higher vertebrates such as birds and mammals). In the absence of electrophysiological signs, reptiles, amphibians, fish, and invertebrates are said to undergo 'rest' resembling that of higher vertebrates (with some exceptions: 30,31). Classically, studies of human sleep disorders were carried out using electroencephalographic (EEG) recordings that identified two major components in sleep architecture. These are named after their phenotypic correlates in eye movement, rapid eye movement (REM) sleep and non-REM (NREM) sleep [32]. During waking, brain waves are observed to be high-frequency low-amplitudes spikes that are believed to arise from the internal desynchrony of the active cortical neurons [33]. With progression of sleep into REM and subsequently to the more stable and deeper NREM sleep, EEG recordings tend towards lower-frequency (<2 Hz) high-amplitude waves (>75 μV; also termed δ waves) potentially arising from residual but synchronized neuronal activity [33,34]. In humans NREM is further subdivided into the N1, N2 and N3 stages. The N3 stage (also called slow-wave sleep (SWS)) represents the deepest form of sleep, where 20% or more of all recorded signals observed within a 30 s window (an epoch) are δ waves [32,35]. The prevalence and amplitude of EEG recordings in SWS, referred to as 'δ power' are hypothesized to be a direct readout of the 'sleep pressure', or our intrinsic drive to get sleep (more on the 'sleep homeostat' later). Hence, when sleep-deprived one experiences an increased 'need' for sleep which correlates with a corresponding increase in δ power for that individual (also referred to as 'rebound sleep'; [36]).

As discussed above, our δ power reflects our 'sleep pressure', but what gates the onset of sleep? In answering this question we arrive at the relationship between the circadian system and the sleep–wake pathways. In 1982 Borbély proposed the two-process model, which in simple terms proposes the presence of a circadian component or 'Process-C' that gates the onset of sleep/wake to an ecologically relevant time, while a 'Process-S' acts as a sleep homeostat to determine the duration, quality, and depth of sleep [37]. It is known that under a normal sleep–wake cycle the homeostatic drive or Process-S builds up in phase with the circadian-active phase or Process-C. But during sleep Process-S is dissipated in an exponential manner, with mammals spending about one-third of a given 24-hour day resting or inactive. In corollary, Process-C enables the consolidation of sleep–wake episodes by counteracting the increasing homeostatic pressure of Process-S during the rest and wake phases, respectively [38–40].

The existence of the two oscillating processes, Process-C and Process-S, is now widely accepted. However, the extent of the interdependence or independence of these two processes remains contentious. Behavioral, surgical, and genetic perturbation protocols have been utilized in an effort to disentangle the contributions of these two oscillating systems. Behavioral disentanglement protocols, for example the 'forced desynchrony' and 'spontaneous internal desynchrony' protocols, involve the complete removal of the subject from the environmental and social cues required for the normal circadian entrainment to the 24-hour day. Under the 'spontaneous internal desynchrony' protocol, the subjects self-select their light–dark cycles; under these 'free running' conditions, the homeostatic drive and consolidation of sleep were found to remain intact. However, dramatic differences in the phase relationship between the two processes were observed: the onset of the NREM sleep phase advanced by 6 hours to overlap the nadir of core-body temperature rhythms, which oscillates in a circadian fashion [41,42]. In 'forced desynchrony', instead of a normal 24-hour day, the subjects are exposed to a sleep–wake schedule imitating either very short (20 hr) or long (28 hr) days. Such schedules are beyond the entrainment limits, causing the biological clock to 'free-run' and desynchronize with the sleep–wake cycle. Thus, under forced desynchrony, sleep episodes occur at all phases of the endogenous circadian cycle. But unlike sleep-deprivation protocols, only minor changes in slow-wave sleep were observed, indicating relative independence of NREM sleep from Process-C [39,43]. However, these studies and others [44] have also shown a strong correlation of the circadian phase with the onset/density of REM and wake-promoting signals.

Another facet uncovered in these studies is the involvement of melatonin, a hormone secreted rhythmically by the pineal gland, in modulating sleep efficiency [43]. In the absence of photic entrainment, completely blind patients often exhibit drastic circadian-rhythm irregularities and poor sleep–wake rhythms [45]. Circadian re-entrainment and enhanced sleep consolidation are observed after scheduled, long-term administration of melatonin in these blind patients [46,47]. Melatonin is observed to phase-shift the circadian sleep–wake cycle, and although the mode of action is not completely understood, studies suggest that melatonin inhibits neuronal firing by acting through its cognate receptors in the master-clock, the SCN [48–51]. Taken together, behavioral disentanglement protocols have not only helped identify Process-C-independent variables but also have definitively validated the presence of a cross-talk between Process-C and Process-S.

Although the anatomical location of the sleep homeostat is not known, the central oscillator governing the body clock is located in the suprachiasmatic nuclei (SCN) of the hypothalamus. Thus surgical removal of the SCN followed by perturbation of sleep–wake states has been performed to disentangle the physiological role of sleep from circadian functions. A classic study involving lesions of the SCN in diurnal squirrel monkeys found a dramatic 4-hour overall increase in the total sleep time in the lesioned monkeys compared to their control counterparts [52]. Therefore, this study suggested that the circadian pathways were involved mainly in wake-promoting pathways. However, subsequent studies using nocturnal rats, mice, hamsters and others indicated a more complex role of SCN involving both sleep- and wake-promoting functions ([53,54]; also see review by Ralph Mistelberger [55] for extensive discussion on this topic). Pleiotropic roles of any anatomical loci in the brain, along with technical challenges of these studies, have been a major limitation, making it difficult to reproduce and compare the different studies. Two brain regions, the ventrolateral preoptic nucleus (VLPO) and the lateral hypothalamic area (LHA; see Figure 21.1), have been unequivocally associated with sleep- and wake-promoting centers, respectively. However, a detailed study on the circadian output following VLPO or LHA lesions remains to be performed.

Mechanistic understanding of the clock has proved useful in uncovering its relationship to sleep. Molecular, cellular, and behavioral studies have revealed a basic mechanism of clock function in insects and mammals. The molecular clock is made of interlocking feedback loops governed by two families of transactivators (Bmal1/2 and

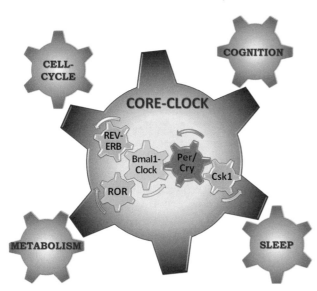

FIGURE 21.2 **The circadian clock is a gear within a much larger machine.** The core clock comprises interlocking feedback loops of transactivators (Bmal1/Clock) and transrepressors (Per/Cry) that are regulated at the transcriptional level by nuclear hormone receptors — RORs and REV-ERBs — and at the post-translational level by casein kinase 1 (CSK1) to produce rhythmic output with a 24-hour period. This core clock is represented as a cog among many that together comprise the physiological machinery of an organism.

Clock/Npas2) and two families of transrepressors (Per1/2/3 and Cry1/2) (see Figure 21.2; [56,57]). *Drosophila* and mouse models have been made for all of these components, and sleep has been characterized in many of them. $Cry1/Cry2^{-/-}$ double-knockout mice display a longer and, more consolidated NREM sleep with higher δ power than wild-type littermates [58]. In other words, sleep analyzed in $Cry1/Cry2^{-/-}$ double-knockout mice, which are behaviorally arrhythmic, resembled EEG recordings from wild-type mice following sleep deprivation. Similar disruption of sleep homeostasis has been reported in other clock-mutant mice (Table 21.1), emphasizing the strong relationship between these two critical processes.

Mammalian Per genes seems to be the exception to this rule. $Per1^{-/-}$, $Per2^{-/-}$ and $Per1/Per2^{-/-}$ mice display normal sleep architecture and rebound following sleep deprivation [69,70]. Interestingly, in rodents and humans the Per3 isoform, a 'redundant' component of the clock, has been discovered to have a stronger role in sleep homeostasis than its more celebrated siblings. Humans homozygous for a naturally occurring polymorphism in the *Per3* locus ($Per3^{5/5}$) had comparatively higher sleep pressure and were more susceptible to the effects of sleep loss [71]. Follow-up studies in *Per3*-mutant mice recapitulated some of these phenotypes [72]. Additionally, carriers of spontaneously occurring point mutations identified predominantly for their circadian deficits were subsequently also found to have sleep–wake disorders. For example, in 1999, the Ptáček group first published the identification of a circadian disorder in humans called familial advanced sleep phase syndrome (FASPS; [73]). As the name suggests, the sleep–wake rhythms in FASPS patients are phase advanced [73]. Subsequently, Ptáček and Fu's group identified that the syndrome is caused by a point mutation in hPer2 [74] or by a T44A missense mutation in the enzyme casein-kinase 1 δ (CK1δ; [75]). CK1δ and CK1ϵ are already known in the field of circadian biology to be involved in turnover of the PER proteins and to drastically affect the circadian clock in behavioral assays [76–79].

Another example, Dec2, a basic helix–loop–helix (bHLH) protein that is thought to repress Clock/Bmal1, is conserved between mice and humans and a related ortholog is present in flies [80,81]. Although neither Dec1 nor Dec2 knockout mice have circadian locomotor activity deficits, they do respond poorly to light pulses [82]. Recently Fu's group identified a point mutation in Dec2 (P385R) in humans associated with short sleep. Most humans need about 8 hours of sleep per night, two individuals carrying this Dec2-point mutation need only 6 hours per night [83]. Although many people may feel as if they only need 6 hours of sleep, the majority will suffer attention deficits after only one or a few days of 6-hour sleep [28]. In mice,

TABLE 21.1 Status of Sleep Homeostasis in Clock-Mutant Mice

Knockout(gene)	Wheel running	Sleep homeostasis effect	Reference
Clock	Rhythmic	↓NREM time and consolidation; ↓REM after sleep deprivation	[59–61]
Bmal1	Arrhythmic	↑Total sleep; ↑fragmentation; ↓SWS; ↓rebound sleep after sleep deprivation	[62,63]
Npas2	Rhythmic	↓NREM time in dark-phase;	[64–66]
Dbp	Rhythmic	↓NREM time and consolidation; ↓REM after sleep deprivation	[67,68]
Per1-Per2	Arrhythmic	No change in Waking, NREM, REM; No difference in rebound sleep	[69,70]
Per3	Rhythmic	↓Wakefulness, ↑REM, ↑NREM — in light phase; ↓REM, ↓NREM time in dark-phase; ↓Wakefulness, ↑REM — after sleep deprivation	[70,72]

knockout of Dec2 did not result in a decrease in sleep, but transgenic mice carrying the specific point mutation exhibited decreased sleep time without affecting circadian period, suggesting a dominant effect [83].

One caveat of the genetic studies mentioned above is the possibility of a tissue-specific non-circadian role of core clock genes, i.e., a given gene might regulate the core clock in SCN but homeostatic sleep drive in another brain region such as VLPO. Under such conditions global deletion of a clock gene will presumably produce circadian and sleep—wake phenotypic defects. Hence to elucidate clearly whether Process-C directly impinges on homeostatic sleep, sleep analysis may need to be conducted in animals with SCN-specific gene ablation of the core clock genes [84].

In summary, the sleep homeostat and the circadian clock are highly interconnected at the molecular and behavioral levels. The clock, which is thought to help maintain adaptive homeostasis, and sleep, which is a critical homeostatic need, work together.

HOMEOSTASIS: YOU NOTICE IT WHEN IT'S BROKEN

In vertebrates, much of what we know about the role of the clock in physiology and behavior is driven by observational studies in normal animals or human subjects, or in mouse models of core clock genes. Conceptually, it is easy to understand how the clock and sleep are related: after all, each of us experiences this relationship almost every day. Less well understood, however, is the way that the clock influences other less obvious systems such as cognition and metabolism. Finally, how it does so, largely through a cascade of signaling between the hypothalamus, other nuclei in the CNS, and the periphery, remains largely a mystery.

Cognition (Learning and Memory)

You are likely familiar with the phrase 'I'm too tired to think'. Disruption of the circadian clock disrupts not only the sleep—wake cycle but many other dependent processes, notably cognition. Mammals are able to learn the characteristics of their environment, remember them, anticipate their recurrence, and adapt their behavior accordingly. Put differently, animals schedule their feeding patterns, sleep—wake cycles, and associated homeostatic processes based on experience. However, as an added layer of complexity, the circadian homeostat impinges on our ability to perform cognitive tasks and we show time-of-day difference in tasks such as learning, registering and recalling memories. Maybe we are too tired to think because we are not supposed to think at certain times of day (or night).

But how are memories made? The hippocampus, located deep within the mediotemporal lobe, is considered a basecamp for cognitive processes (see Figure 21.1). Having paid attention to an event (learning), memory of that event can be acquired, retained, and eventually recalled. Consolidation is a process by which short-term memory (STM), which lasts on a scale of seconds to minutes, is stabilized (by association with cortical circuitry) to long-term memory (LTM) that can last for hours, months, and even years [85,86]. Although the molecular distinction of these two processes has been difficult to dissect, the sensitivity of these two processes seems to differ across phylogeny. STM in invertebrates has been shown to be sensitive to circadian perturbations [87,88]. Interestingly, in vertebrates STM remains unaffected, but the consolidation and recall of memory was found to be profoundly affected by circadian perturbation [89].

Over the last few decades, studies have explored the role of circadian rhythms in cognition using behavioral, neurophysiological, and genetic perturbation. However, given our discussion on the influence of sleep or behavior state, we can now appreciate the complexity of understanding the circadian component in cognition. To overcome the influence of the master oscillator, early studies included electrolytic lesioning of the SCN followed by behavioral analysis of hippocampal-specific tasks such as inhibitory avoidance conditioning [90]. These studies found clear time-of-day memory retention deficits ([90]; but also see; [91,92]). Others have approached this question differently, using behavioral perturbations similar to those discussed above in the section on sleep. In a recent publication by the Colwell group [93], circadian perturbations were achieved by acute phase-shifts in the light—dark cycle that mimicked 'jet-lag'. These mice displayed no change in overall sleep-time but showed clear deficits in memory recall in a fear-conditioning paradigm (see also [94]). This work neatly adds to previous observations that the peak-time in memory acquisition and recall exhibits a circadian profile in mice ([95]; but also see [96,97]). Next, studies exploring adult hippocampal neurogenesis have provided an unexpected glimpse at the influence of the circadian homeostat in learning and memory [98]. Adult hippocampal neurogenesis has been shown to increase after two circadian-driven events, namely a hippocampal-dependent learning task [98,99] and exercise, i.e., wheel running activity [100,101]. Indeed $Per2^{Brdm}$ (non-functional $Per2$) mice were shown to have increased neuronal stem cell proliferation, suggesting the involvement of core clock proteins in the cell cycle [102–104], and more specifically adult neurogenesis [105]. This view has been strengthened by studies wherein experimental jet-lag (a 6-hour phase advance every 3 days for 25 days) in adult hamsters led to pronounced deficits in hippocampal neurogenesis and concurrent deficit in hippocampal learning and memory

[106]. Furthermore, these deficits persisted even after the animals were returned to normal LD cycles, suggesting a long-term effect of circadian disruption on cognitive functions.

Our discussion thus far suggests a causal role of the SCN master clock in learning and memory processes in the hippocampus. However, over the last decade studies have enabled us to appreciate the hippocampus as a circadian pacemaker in its own right. First, circadian rhythms observed at the molecular and neurophysiological levels persisted in hippocampal tissue culture, i.e., in the absence of SCN connections [107]. Importantly, time-of-harvest did not seem to influence the hippocampus clock, giving further credence for the presence of an endogenous oscillator within the hippocampus. Second, circadian rhythms observed in at least some forms of learned behavior do not seem to require an intact SCN [91, 92]. So what drives the clock in the hippocampus? The same set of core clock genes as seen in the SCN seems to be responsible for driving rhythms in the hippocampus. Bmal1 [108,109], Clock [108], Cry(1&2) [109] and Per(1&2) [105,107] are all known to oscillate robustly in the hippocampus, albeit often peaking out of phase with the SCN and even at different phases within the sub-regions of the hippocampus. For example, Per2 expression in the CA1−3 regions is anti-phase to the SCN, with peaks in the former at ZT2 and in the latter at ZT14 [107]. Also, Bmal1 expression peaks around ZT10 in the CA3 region of the hippocampus but at ZT2 in the CA1 region [108]. One speculation is that by peaking at different phases, circadian genes guide the diverse biological and physiological functions within these sub-regions. Within the hippocampus molecular rhythms of clock genes most likely drive the rhythms observed in its synaptic plasticity, a neurophysiological correlate of learning and memory. Long-term potentiation (LTP) is a synaptic plasticity event that is often regarded as an onset marker for learning and memory. Following repeated stimulation, a neuronal circuit can be said to have undergone LTP if it is rendered more sensitive to subsequent stimulations. Studies spanning three decades have consistently reported a time-of-day effect on the incidence and magnitude of LTP in electrophysiological recordings from mouse hippocampal slices [110−112]. Thus the circadian homeostat enables metaplasticity [113] wherein one level of plasticity, time-of-day, layers atop another, synaptic plasticity. Behaviorally, hippocampal-dependent tasks such as spatial/novel object recognition and classic fear conditioning are thought to elicit LTP-like responses and concurrently exhibit time-of-day differences in such tests. Taken together, the hippocampus utilizes the same set of core-clock genes as found in the SCN to drive circadian rhythms found in hippocampal-related molecular, physiological and behavioral assays.

But why do we need a functional clock in the hippocampus? In 1949 Donald Hebb posited that repeated feedback loops of actively firing neurons are required for memory consolidation (also referred to as the Hebbian or the dual-trace theory) [114,115]. We also know that repeated and persistent kinase activation, protein synthesis, and gene expression are needed for new memory formation. Such reverberating loops of activity/inactivity are best achieved by the oscillating components of the hippocampal clock [106]. In summary, the hippocampal clock can be postulated to perform three main functions: (a) it provides a time-stamp for certain forms of memory; (b) it sub-serves pleiotropic function using the same network of core clock genes; and (c) it provides a framework for 'reverberating circuits' enabling memory consolidation. In support of this hypothesis, mice with global deletion of clock genes such as $Npas2^{-/-}$ [117], $Cry(1,2)^{-/-}$ [118], $Per2^{-/-}$ [107], $Bmal1^{-/-}$ [119] and $VIP^{-/-}$ [120] exhibit cognitive defects. However, global deletion of clock genes is confounded by the potential role of clock genes in sleep and other behavioral states of these animals (as discussed above). Hence, although a strong precedent exists for the role of clock genes in cognition, there is a clear need for a re-evaluation of cognitive abilities in mice with tissue-specific ablation of clock genes. Various genetic methods are now available for tissue-specific ablation or silencing of genes (see reviews [121−124]).

The cross-talk between circadian timing, synaptic plasticity mechanisms and memory formation has been evolutionarily conserved between invertebrates (insects (*Drosophila*), gastropods (*Aplysia*)) and vertebrates (zebrafish, mouse, and human) [125]. The extracellular signal-regulated kinase (ERK) isoforms participate in the MAPK/cAMP cascade and have long been known to be indispensable for memory consolidation [126−128] as well as light-dependent phase resetting of the SCN [129−131]. In recent landmark studies, circadian rhythms of ERK and other signaling components of the MAPK/cAMP cascade were shown to influence the time-of-day differences in cognition [96,132]. Eckel-Mahan et al. [96] provided the first molecular and mechanistic details of this cross-talk in a mammalian system. They demonstrated that within the CA1 and CA3 sub-regions of the hippocampus, (a) ERK phosphorylation occurred in a circadian fashion, and (b) these rhythms were indispensable for the consolidation of hippocampal-dependent contextual fear-conditioned memory. Furthermore, the authors showed that the circadian peak in Erk phosphorylation coincided with the animal's peak capacity in forming new long-term hippocampal-dependent memory. A follow-up study from the same group confirmed that SCN-lesioned mice lose hippocampal rhythms in the signaling components of the cAMP-MAPK-CREB pathway along with deficits in fear-conditioned memory [133]. Interestingly, this circadian-dependent suppression was alleviated by pharmacological facilitation of the MAPK pathway in *Aplysia* [132], raising the interesting possibility of circadian-phase specific de-repression of cognitive abilities.

In summary, a picture emerges wherein the hippocampus uses its own clock to enable cognitive homeostasis. Just as light entrains the SCN, a tantalizing suggestion is that cognitive processes could function as *zeitgebers* (time cues) for the hippocampal clock [134]. If true, this would indicate a reciprocal relationship between cognitive performance and the circadian phase of the hippocampal clock. Expounding on this hypothesis, it would be easy to reason that cognitive tasks need to be performed in sync with the master clock. Indeed, various studies have found direct and significant correlation between an individual's chronotype (preference for activity in either morning or evening) and cognitive performance [135,136]. Further, however impractical, this would argue against the rigid design of timetables in schools and other educational institutions [137]. Another implication involves professions that have high demands for cognitive performance often atop a compromised circadian homeostat, such as air traffic controllers, pilots, shift workers or doctors. With improving knowledge of signaling pathways it might become feasible pharmaceutically to relieve the impact of the clock in at least some aspects of cognitive performance.

Neuropsychiatric Disorders

The clinical manifestations of many central nervous system (CNS) disorders are deregulated circadian rhythms (core body temperature, cortisol/melatonin levels) and/or sleep cycles. Although it is difficult to directly resolve causality, these studies indicate that disruption of circadian oscillators can affect CNS disease severity.

In the mammalian brain, in addition to the hippocampus, various extra-SCN or slave oscillators have been discovered in the last decade (see Figure 21.1; [138]). Synchronizing factors such as hormones, neuronal connections, paracrine signals, metabolites and body temperature can cause changes in phase/expression levels in a tissue-specific (and often sub-structural) level without affecting the SCN (see below for more discussion on this topic). For example, glucocorticoids such as corticosterone can modulate expression of clock proteins such as Per1 in the hippocampus but not in the SCN [139,140]. Thus, rhythmic signals orchestrated by the SCN synchronize peripheral clocks, including those within the brain. Various aspects of modern life, including abnormal light exposure via artificial lightning, sleep deprivation induced by shift work, jet-lag due to travel across time zones, side effects of medications, and abnormal social environments, ultimately lead to desynchrony of these internal clocks [141,142]. This is because the SCN resets over a period of days, whereas peripheral tissues may take shorter or longer. In the CNS, a loss in communication between the SCN and other brain nuclei can aggravate or even result in neuropsychiatric disorders. A comprehensive review of all neuropsychiatric disorders influenced by the circadian system is beyond the scope of this discussion and is covered elsewhere [141–143]. Instead, we will briefly discuss a few well-known links between the circadian system and some neurological disorders followed by our current understandings/hypotheses regarding the mode of this action.

There is a strong link between disruption of circadian homeostasis and bipolar disorder. Bipolar disorder (BPD) is associated with rapid switching between a spectrum of symptoms that range from depression to mood elevation (mania and hypomania) and from low-grade mood cycling to full psychosis. Disruption of the 24-hour sleep–wake cycle is one of the strongest etiological triggers for a relapse into mania, and therapies for BPD often involve regimens of stable and adequate sleep [141]. Transgenic mice carrying the dominant negative Clock ($Clk^{\Delta 19}$) allele recapitulate hallmarks of BPD, namely a decreased need for sleep, increased motor activity, lower anxiety, and an increased susceptibility to drugs of abuse such as cocaine [144]. Similar results are obtained by RNAi-dependent knockdown of the Clock gene (*Clk*) in the ventral tegmental area (VTA), a brain region involved in the 'reward circuitry' (discussed below; [145]). Furthermore, treatment of $Clk^{\Delta 19}$ mice with lithium, a mood-stabilizing drug used in human patients, restores normal behavior [146]. SNP analysis and gene-association studies have confirmed association of *Clock* with BPD and have extended the association of other clock genes, including *Bmal1* and *Per3* [147].

Normal seasonal variation is enough to disrupt the homeostasis of certain patients with seasonal affective disorder (SAD). For example, in many parts of the world, day length and light exposure change with the seasons. SAD manifests circa-annually, with individuals having normal mental health throughout most of the year but experiencing depressive symptoms seasonally (usually in the winter; [148,149]). In fact, SAD is considered to be a type of major depressive disorder (MDD) according to international classification schemes of psychiatric disorders (the Diagnostic and Statistical Manual of Mental Disorders (DSM-IV) of the American Psychiatric Association and the International Classification of Disorders (ICD-10) of the World Health Organization). Restoring light, i.e., light therapy, is currently the most effective treatment for SAD [150]. Blue-light-photosensitive retinal ganglion cells (pRGCs) use the photopigment melanopsin (Opn4) and project from the retina into the SCN and VLPO. pRGCs can trigger a cascade of neuronal activities that are critical for the entrainment of the molecular clock to the local light cycle and in regulating sleep–wake cycles. Interestingly, polymorphisms in the gene encoding Opn4 are linked to SAD [151]. Furthermore, gene-association studies have also found strong correlation in polymorphisms within the clock genes Bmal1, *Npas2* and *Per2* with susceptibility to SAD [152]. Similar roles of the circadian system have been identified in other diseases such as schizophrenia [153,154], addiction [144,146,155,156] and anxiety/hyperactivity-based disorders [119,157,158]. Taken together these studies emphasize the strong relationship between circadian system-driven homeostasis and our neuropsychological state.

How does the circadian state influence such a diverse range of neuropsychiatric disorders? Although recent studies provide some insight, research is required to provide in-depth mechanistic details for the cross-talk between the biological clock and CNS processes. As discussed above, lithium, used clinically to treat BPD and other mood disorders, is also known to be a potent inhibitor of glycogen synthase kinase 3β (GSK3β) [6,7]. Interestingly, GSK3β can influence the nuclear localization and stability of circadian clock proteins REV-ERBα [159] and PER2 [160,161]. In the nucleus, REV-ERBα functions as a potent repressor of Bmal1 transcription [162]. However, lithium has many more targets than just GSK3β, and it remains to be seen whether its effect on depression is mediated through the clock.

The dopamine system provides another potential mechanism. Dopamine-mediated neurotransmission in the midbrain, specifically in the ventral tegmental area (VTA), is a crucial component of the brain's reward circuit. Deregulation of dopamine-based neurocircuitry is often associated with susceptibility to drugs of abuse such as cocaine and methamphetamines and neuropsychiatric disorders such as depression, attention deficit hyperactivity disorder (ADHD) and schizophrenia [141]. As discussed above, $Clk^{\Delta 19}$ mice were found to exhibit BPD-like behavior [144]. The same study also found that the activity of tyrosine hydroxylase, the principal enzyme involved in dopamine synthesis, is elevated in the VTA of these mice. Restoration of the wild-type clock gene in the VTA rescued the mice from some of the mood-related abnormalities. A more recent study identified monoamine oxidase A (MAOA), a key enzyme involved in the breakdown of dopamine under circadian control [163]. This study shows that transcription of MAOA both in vitro and in vivo is under the control of clock genes Npas2, Bmal1, and Per2. Additionally, the *Per2*-mutant mice (*Per2^{Brdm}*) display reduced MAOA levels with increased dopamine levels, and interestingly were found to be more resilient than wild-type mice in behavioral tests that induced a depression-like phenotype [163]. Finally, disruption in the circadian secretion of neurotransmitters such as neuropeptide Y (NPY) or vasoactive intestinal peptide (VIP) could also lead to anxiety-like behavior and aggression [164,165].

In hindsight, time-of-day effects on our emotional and mental faculties seem obvious, but the molecular relevance of these effects is only beginning to be appreciated. The causal roles of 'nature vs. nurture' in an individual's cognitive/behavioral output have been a long-standing debate. We know that genetic make-up can predispose one to neurological and behavioral imbalances; however, circadian research has provided insight into how these genetically driven imbalances can at times be compensated or exacerbated by environmental perturbations. For example, the influence of social pressures such as 'social jet-lag', regimented/restricted sleep or activity, or light/noise exposure at night (especially for city dwellers) could have neurological consequences during development [166–171].

CLOCKS IN ENERGY AND METABOLIC HOMEOSTASIS

What are Peripheral Clocks?

The need to metabolize and generate energy for sustenance, growth and reproduction is an inescapable fact of all forms of life. It is therefore not surprising that this primeval requirement is intricately connected with our biological clock. Soon after identifying the first clock genes [172], ground-breaking studies from the Kay and Schibler groups changed our perception that clocks were restricted to the brain. Kay's group showed that body parts from *Drosophila* were able to oscillate in culture [173], and Schibler's group next demonstrated robust oscillations in immortalized Rat1 fibroblasts [174]. We now know that robust gene-expression rhythms can be detected in tissue explants from almost all peripheral organs [175,176]. (Exceptions include thymus and testis — organs populated largely by undifferentiated cells [177–179]). Subsequently, using genome-wide gene-expression profiling of peripheral tissues such as liver [180–183], skeletal muscle [184,185], heart [186,187] and adrenal glands [188], our laboratory and others have shown that ~10% of the expressed transcripts in each tissue exhibit circadian rhythms in expression. Intriguingly, by increasing the time resolution of tissue collection we have also identified rhythmic genes with ultradian periods in the second or third harmonics (12 h and 8 h) of the circadian rhythms [181]. Taken together, we can definitively state that the clock seems to be ubiquitously present — within the brain (referred to as the SCN and extra-SCN clocks) and in most peripheral organs (referred to as peripheral clocks). We can follow this statement by three obvious questions: Are peripheral clocks similar to the central clock?, How does one clock 'talk' with another? and Why do we need so many clocks? Let us try to address these questions in the order listed.

To better understand the following discussion, let us refresh our memories with the most simplified model of the core clock. In this model, interlocking feedback loops of transactivators (Bmal1; Clock/Npas2) and transrepressors [Per(1,2); Cry(1,2)] are regulated at the transcriptional level by nuclear hormone receptors — RORs (Rora, Rorb, Rorc) and REV-ERBs (Nr1d1 and Nr1d2) and at the post-translational level by casein kinase 1-δ/ε (CSK1-δ/ε) to produce rhythmic output with a 24-hour period (see Figure 21.2). Peripheral clocks harbor this core-clock machinery of the SCN but with some interesting (and many speculative) differences. In the SCN, the *Clock* gene is functionally compensated by *Npas2* [189], but is indispensable in peripheral tissues [190]. This suggests tissue-specific differences in the molecular make-up of the clock network. This hypothesis is strengthened by recent studies wherein tissue-specific disruption of clock in the liver [191] and pancreas [192] gave mirror-opposite phenotypes.

In addition to these intracellular differences, strong intercellular coupling is a unique property of the SCN. Three well-defined modes of cell—cell communication exist in the SCN: gap junctions, peptidergic signaling using the neuropeptide vasointestinal peptide (VIP) and its cognate VPAC2 receptor, and GABA signaling [27]. Rhythms of peripheral organs ex vivo dampen in the absence of the intercellular coupled architecture seen in the SCN. Hence, we can envision that in vivo peripheral oscillators require synchronization cues potentially from the SCN (and/or other yet unrecognized master oscillator(s) [193]). This hypothesis is demonstrable as peripheral clocks are susceptible to entrainment by a wide variety of environmental cues, such as temperature and fasting/feeding schedules. The 24-hour period of the clock is found to be resistant to temperature fluctuations within the biological range (i.e., clocks are temperature compensated) [194], but unlike in the SCN, changes in the ambient temperature cause re-entrainment or phase resetting of the peripheral clocks [195]. Also, gene-expression rhythms in the peripheral clocks of $Cry1^{-/-}/Cry2^{-/-}$ and SCN lesioned mice are restored in a restricted feeding paradigm, indicating that even in the absence of the central clock, feeding schedule is functioning as a *zeitgeber* (time cue) for the peripheral clocks [196—198]. In summary, all clocks use the same building blocks, but the network properties and mechanistic details of peripheral clocks are probably different from that of the SCN [199].

Causality in inter-clock communication is the subject of intense investigation. The SCN can communicate with the other clock-containing regions (directly or indirectly) by neuronal innervation or by long-range humoral signals such as glucocorticoid hormones (Figure 21.3; [27,200]). Retrograde and anterograde labeling have been used to map direct neuronal innervation from the SCN to the cell bodies in the neighboring hypothalamic nuclei, for example the arcuate nucleus (ARC), periventricular nucleus (PVN), lateral hypothalamic area (LHA) and dorsomedial hypothalamic area (DMH) (see Figure 21.1; [201,202]). The ARC is a well-known center for the synthesis and secretion of neuropeptides that can either increase appetite (orexigenic), namely neuropeptide Y/Agouti-related protein

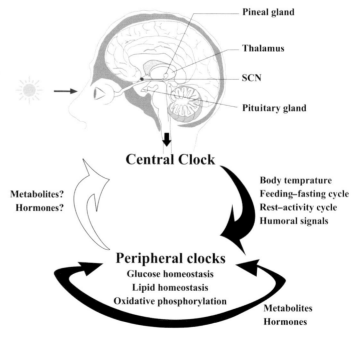

FIGURE 21.3 **A schematic representation of cross-talk between the central and peripheral clocks.** These communications are known to be governed by environmental, physiological and metabolic cues. The nature and extent of this inter-clock communication are subjects of intense discussion (see text). An increasing pool of research is identifying the causal role of peripheral clocks in regulating various metabolic processes, and the dysregulation of this interplay is found to be an etiological factor for human diseases such as obesity and diabetes.

(NPY/AgRP) or suppress appetite (anorexigenic) namely pro-opiomelanocortin/cocaine and amphetamine-regulated transcript (POMC/CART; [203]). Ex vivo monitoring of 27 brain regions from transgenic rats expressing a luciferase reporter expressed under the control of the Per1 promoter (Per1:LUC) demonstrated that about half of them are rhythmic in the explant culture [204]. From this study it was not surprising to discover an ARC clock, given that the ARC reciprocally innervates the SCN and regulates two robust circadian events, namely appetite and feeding. Also, orexin/hypocretin-producing neurons in the LHA are known to be important in long-term weight homeostasis [205]. Connecting via relay centers such as the PVN, the SCN can communicate with the nucleus of the solitary tract (NTS), a satiety center in the brainstem and dorsally placed pituitary gland, the basecamp of all major regulatory hormones (see Figure 21.1; [205,206]). However, the mechanistic relevance of the ARC clock, or SCN innervations into the LHA, for appetite and feeding remains to be fully characterized. The extensive breadth of the SCN's innervation is beyond the scope of this discussion, but the basic message here is that SCN carries the advantage of prime real estate: it resides in the premier homeostatic neighborhood in the brain, the hypothalamus, allowing association with clocks that also happen to be major homeostatic regulators of energy and metabolism in peripheral organs [207–209].

To recap, we have clocks in peripheral organs that are entrained by various environmental cues but eventually are reset by the master oscillator in the SCN. This brings us to the most pertinent question of this discussion: Why do we need so many clocks? One of the first things we do after waking up is break our fast, i.e., breakfast. While eating our gut motility increases, nutrition from food is absorbed, metabolized in the liver, and the energy generated is stored as fat or used by muscles for activity. As our energy needs are met, the CNS registers satiety to inhibit the feeding drive. Just as we show behavioral rhythms in sleep/activity, many aspects of our metabolism also show circadian variation, such as serum levels of insulin [210–212], glucagon [213], adiponectin [192,214], corticosterone [215–217], leptin and ghrelin [218,219]. The circadian alignment of our behavioral and metabolic rhythms is critical for maintaining homeostasis. Using the forced desynchrony protocol, Scheer et al. recently investigated the impact of a misalignment between metabolic rhythms and the central clock in human subjects [220]. Participants in this study slept 12 hours out of phase from their habitual times and were found to have decreased leptin (predictive of increased appetite), increased blood glucose levels despite increased insulin (also termed insulin resistance), inverted cortisol rhythm, and increased blood pressure, and some were symptomatic of a pre-diabetic state.

Furthermore, genome-wide association studies have suggested a connection between clock-gene polymorphisms and obesity and diabetic disorders [221,222]. Concurrently, narcoleptic patients who display extreme daytime sleepiness due to loss of hypocretin-producing neurons are associated with an increased incidence of obesity [223–225]. Similarly, patients with night-time eating syndrome (NES) show disrupted patterns of sleep and metabolic rhythms [226], consume significantly more food at night, and are prone to obesity with a high risk for diabetes (although their total food-intake is similar to that of control subjects; [227]). Similar results have been obtained from animal studies. Scheduling an isocaloric high-fat diet [228] or even a normal diet [229] to the wrong time of day in rodents leads to increased weight gain. Thus the interplay of lighting and feeding schedules is probably required to maintain the homeostatic control of body weight [230,231]. Interestingly, when genetically obese strains of mice that exhibit disrupted diurnal feeding rhythms were fed exclusively at night (active phase), they displayed an amelioration in their metabolic and obesity status [232]. Thus a reciprocal relationship between clock disruption and metabolic pathologies seems to create a reinforcing loop leading to rapid progression of the metabolic disease.

In summary, our discussion suggests that we need peripheral clocks to provide time stamps for the various organs involved in food processing and metabolism for maximal efficiency. Also, within any given peripheral organ, clocks compartmentalize biochemically incompatible reactions to different times of the day, a functionality conserved throughout evolution (see above). For example, metabolic enzymes involved in glycogenolysis (energy producing) and glycogenesis (energy storage) need to be synthesized in temporally restricted timeframes corresponding with food absorption and nutrient metabolism. Finally, liver enzymes such as cytochrome P450 monooxidase detoxify xenobiotics at the cost of producing harmful reactive oxygen species, a worthwhile endeavor only after consumption of food. It therefore makes sense to schedule such reactions to times when they are needed. So, taken together, we need peripheral clocks for three main functions: circadian alignment of different organ systems, compartmentalization of incompatible reactions, and restricted expression of reactions with adverse side effects to time-points when they are needed (see Figure 21.3; reviewed in [233]).

Understanding Energy Homeostasis with Animal Models of the Clock

Genetic animal models with global and tissue-specific clock deficiencies have indicated a major role for clocks in lipid and carbohydrate metabolism. Mice with a global deletion of *Bmal1* (Bmal1$^{-/-}$) display complete loss of locomotor rhythms with poor body weight but increased adiposity, arthropathy (diseased joints) and myopathy (weak muscles) at an early age [191,234–237]. A study by McDearmon et al. [235] suggests that Bmal1 expression in the brain is necessary for rhythmic activity and its expression in the skeletal muscle is required for maintaining body weight. Whereas *Per2*$^{-/-}$ mice gain more weight on a high-fat diet than controls [238], mice deficient in the circadian deadenylase nocturin remain lean and resistant to weight gain compared to control mice [239]. *Clock*^{Δ19} mutant mice are obese and hyperphagic [240]. Further investigation revealed that clocks in enterocytes (cells lining the small-intestinal wall) schedule absorption of triglycerides for night-time [241]. These rhythms are lost in the enterocytes of *Clock*^{Δ19}-mutant mice, leading to hypertriglyceridemia (a causal factor for obesity) in these mice [242]. Next, in a restricted feeding paradigm (food made available only for 4 hours), *Npas2*$^{-/-}$ mice lost weight even though the total amount of food consumed was comparable to that of control mice [243]. This suggests a potential role of Npas2 in synchronizing feeding behavior with food availability. However, results from the global knockout mice in the aforementioned studies need to be interpreted with caution. As discussed earlier, core clock genes such as *Bmal1*, *Clock* and *Npas2* are known to have pleiotropic functions, and their roles in energy and metabolism are only beginning to be addressed using tissue-specific knockouts. The gravity of these concerns is highlighted by two recent studies involving tissue-specific ablation of Bmal1 in the liver and pancreas. While deletion of *Bmal1* in liver causes hypoglycemia (restricted to the fasting phase i.e., during the day in nocturnal mice kept under LD cycle) [191], deletion in the pancreas causes hyperglycemia and hypoinsulinemia [192], indicating antagonistic functions of Bmal1 in the liver and pancreas. Maintaining steady-state serum levels of glucose is critical as it acts as a fuel source for the brain, and these studies clearly indicate the importance of peripheral clocks in glucose homeostasis. Further investigation revealed that mice with liver-specific knockout of Bmal1 had constitutively low levels of the glucose exporter GLUT2 that led to the hypoglycemia phenotype. The clock in the pancreas regulates insulin secretion. In contrast to liver, the absence of Bmal1 expression in the pancreas led to hyperglycemia as the normal glucose-stimulated insulin (GIS) secretion response of the pancreas was lost [210].

Molecular Integrators of Clock and Metabolism

At the molecular level we can envision proteins with overlapping functions in circadian and metabolic pathways that help integrate the two systems. Studies indicate that nuclear hormone receptors (NR) are one such class of molecular integrators. NRs such as REV–ERBα provide robustness to the central clock [244] and *REV–ERBα*$^{-/-}$ mice exhibit shorter period in wheel running activity [162]. REV–ERBα binds to RORE elements and achieves transcriptional repression by recruiting nuclear receptor corepressor 1 (NCoR1) and histone deacetylase 3 (HDAC3) to its target genes, such as Bmal1 [162,245]). Inhibiting the complex of REV–ERBα and NCoR1-HDAC3 leads to circadian and metabolic deficits [246]. It is interesting to note that REV–ERBα was originally identified as a regulator of lipid metabolism and adipogenesis [247]. More recently, the discovery of heme as REV–ERB's ligand [248] has cast REV–ERB as a major nodal point for the integration of the clock and metabolism [249]. Heme is a cofactor for enzymes such as catalases, peroxidases and cytochrome p450 enzymes, playing a role in oxygen and drug metabolism [250,251]. Heme is now also known to improve thermal stability of REV–ERB and hence enhances its interaction with co-repressor complex NCoR1-HDAC3 [248]. Furthermore, the rate-limiting enzyme in heme biosynthesis, aminolevulinate synthase 1 (ALAS1), is regulated by NPAS2 and expressed in a circadian manner [250]. Heme in turn binds to NPAS2 and inhibits its transactivation activity [252]. Thus heme and heme-binding proteins are beginning to be appreciated as integrators of the transcriptional feedback loop of the circadian system and enzymatic reactions in metabolism. Other nuclear hormone receptors, such as estrogen-related receptor-α (ERRα; [253]), the family of peroxisome proliferator-activated receptors (PPAR; [254–256]), and glucocorticoid receptors [257] are being recognized as integrators and are being incorporated in wire diagrams depicting the metabolic clock. In fact, an exhaustive circadian profiling identified the rhythmic expression of 25 of 49 NR in various peripheral tissues [258], indicative of many more prospective NR integrators of clocks with metabolism.

In addition to NR, recent studies indicate that nutrient sensors for energy (AMP/ATP ratio) and redox state (NAD$^+$/NADH ratio) can also function as couplers of the circadian and metabolic systems [259]. NAD$^+$ functions as a coenzyme for oxidoreductases that connect the glycolysis/citric-acid cycles to mitochondrial oxidative phosphorylation leading to ATP production, the energy currency in cells [260]. Additionally, NAD$^+$ also functions as coenzyme for SIRT1, a mammalian ortholog of the yeast sirtuin family [261]. SIRT1 is a nutrient sensor that is stimulated under conditions such as fasting or calorie restriction, and restores energy homeostasis by deacetylating key proteins in carbohydrate and lipid catabolism, such as PPARγ-coactivator-α (PGC1α; [262–265]) and liver X receptor (LXR; [266]). Recently SIRT1 was identified to rhythmically bind with CLOCK and counteract

CLOCK's intrinsic acetyltransferase activity for target proteins like BMAL1 [267,268]. Taken together, SIRT1 is a NAD^+ dependent nutrient sensor, and the rhythmic binding of CLOCK-SIRT1 potentially leads to a counterbalancing interplay of acetylation/deacetylation of target genes. Subsequent studies have identified that NAMPT (nicotinamide and 5′-phosphoribosyl 1-pyrophosphatase), a rate-limiting enzyme in the NAD^+ biosynthesis pathway, is regulated by the SIRT1-CLOCK-BMAL complex driving rhythms in NAMPT expression [269,270]. Thus in addition to the transcriptional and translational feedback loops, an enzymatic feedback loop involving clock and NAD^+-dependent sensors such as SIRT1 is emerging as molecular integrator of the clock and metabolic systems. Another integrator identified recently is an enzyme, AMP-activated protein kinase (AMPK), which is activated by high AMP/ATP ratio (indicative again of a low energy state; [271]). AMPK levels were found to be rhythmic in mouse liver, hypothalamus and isolated fibroblasts; once activated, AMPK can relieve the negative feedback arm of the core clock by destabilizing Per2 and Cry1 proteins [272,273]. Interestingly, AMPK is also known to regulate NAD^+ levels [259,274], and $AMPK^{-/-}$ mice lose rhythms in *Nampt* expression and core-body temperature [275].

CONCLUSION

In conclusion, if homeostasis is the seesaw act of keeping physiology and behavior constant, then the circadian clock is its counterweight, helping to offset predictable daily changes to this balance imposed by the environment. What might have started billions of years ago as a way to keep chemically incompatible reactions separated in time, has evolved into a handy mechanism to orchestrate the increasingly complex physiology and behavior of multicellular and multiorgan organisms. Of a litany of examples, we have focused on three: sleep, other cognitive processes, and metabolism. Sleep, the most obvious of these examples, is affected by its timing, which is set by the circadian clock, and a sleep homeostat, which measures how much we have against how much we need. Circadian clock-deficient mice and flies have overt problems with their sleeping patterns (among many other phenotypes). The same holds true for certain clock genes in humans: get too little sleep or wake up at the wrong time, and performance suffers. The converse is also true: a litany of CNS diseases from Alzheimer's to schizophrenia to depression are all comorbid with sleep and clock disorders. These links continue to metabolism, where sleep disruption can cause metabolic problems, and metabolic perturbations can affect sleep and the clock. For homeostasis, timing may not be the only thing, but it is indispensable.

REFERENCES

[1] De Mairan J-J d O. Observation Botanique. Histoire de l'Academie Royale des Sciences 1729:35–6.
[2] Halberg F. Physiologic 24-hour periodicity; general and procedural considerations with reference to the adrenal cycle. Int Z Vitaminforsch Beih 1959;10:225–96.
[3] Halberg F. Chronobiology. Annu Rev Physiol 1969;31:675–725.
[4] Bell-Pedersen D, Cassone VM, Earnest DJ, Golden SS, Hardin PE, Thomas TL, Zoran MJ. Circadian rhythms from multiple oscillators: lessons from diverse organisms. Nat Rev Genet 2005;6:544–56.
[5] Ko CH, Takahashi JS. Molecular components of the mammalian circadian clock. Hum Mol Genet 2006;15(Spec No 2):R271–7.
[6] Ryves WJ, Harwood AJ. Lithium inhibits glycogen synthase kinase-3 by competition for magnesium. Biochem Biophys Res Commun 2001;280:720–5.
[7] Stambolic V, Ruel L, Woodgett JR. Lithium inhibits glycogen synthase kinase-3 activity and mimics wingless signalling in intact cells. Curr Biol 1996;6:1664–8.
[8] Pittendrigh CS. Temporal organization: reflections of a Darwinian clock-watcher. Annu Rev Physiol 1993;55:16–54.
[9] Ouyang Y, Andersson CR, Kondo T, Golden SS, Johnson CH. Resonating circadian clocks enhance fitness in cyanobacteria. Proc Natl Acad Sci U S A 1998;95:8660–4.
[10] Edery I. Circadian rhythms in a nutshell. Physiol Genomics 2000; 3:59–74.
[11] Marques MD, Waterhouse JM. Masking and the evolution of circadian rhythmicity. Chronobiol Int 1994;11:146–55.
[12] Horton TH. Handbook of behavioral neurobiology. In: Takahashi JS, Turek FW, Moore RY, editors. Conceptual issues in the ecology and evolution of circadian rhythms. New York: Kluwer Academic/Plenum Publishers; 2001.
[13] Kay SA. PAS, present, and future: clues to the origins of circadian clocks. Science 1997;276:753–4.
[14] McIntosh BE, Hogenesch JB, Bradfield CA. Mammalian Per-- Arnt-Sim proteins in environmental adaptation. Annu Rev Physiol 2010;72:625–45.
[15] Taylor BL, Zhulin IB. PAS domains: internal sensors of oxygen, redox potential, and light. Microbiol Mol Biol Rev 1999; 63:479–506.
[16] Sauman I, Reppert SM. Circadian clock neurons in the silkmoth *Antheraea pernyi*: novel mechanisms of period protein regulation. Neuron 1996;17:889–900.
[17] Sassone-Corsi P. Circadian rhythms. Same clock, different works. Nature 1996;384:613–4.
[18] Shearman LP, Zylka MJ, Weaver DR, Kolakowski Jr LF, Reppert SM. Two period homologs: circadian expression and photic regulation in the suprachiasmatic nuclei. Neuron 1997;19:1261–9.
[19] Albrecht U, Sun ZS, Eichele G, Lee CC. A differential response of two putative mammalian circadian regulators, mper1 and mper2, to light. Cell 1997;91:1055–64.
[20] Avivi A, Albrecht U, Oster H, Joel A, Beiles A, Nevo E. Biological clock in total darkness: the Clock/MOP3 circadian system of the blind subterranean mole rat. Proc Natl Acad Sci U S A 2001;98:13751–6.
[21] Paranjpe DA, Sharma VK. Evolution of temporal order in living organisms. J Circadian Rhythms 2005;3:7.
[22] Cloudsley-Thompson JL. Adaptive functions of circadian rhythms. Cold Spring Harb Symp Quant Biol 1960;25:345–55.

[23] DeCoursey PJ, Krulas JR, Mele G, Holley DC. Circadian performance of suprachiasmatic nuclei (SCN)-lesioned antelope ground squirrels in a desert enclosure. Physiol Behav 1997;62: 1099−108.

[24] DeCoursey PJ, Krulas JR. Behavior of SCN-lesioned chipmunks in natural habitat: a pilot study. J Biol Rhythms 1998;13:229−44.

[25] Pittendrigh CS. On temperature independence in the clock System controlling emergence time in Drosophila. Proc Natl Acad Sci U S A 1954;40:1018−29.

[26] Miyatake T, Matsumoto A, Matsuyama T, Ueda HR, Toyosato T, Tanimura T. The period gene and allochronic reproductive isolation in *Bactrocera cucurbitae*. Proc Biol Sci 2002;269:2467−72.

[27] Welsh DK, Takahashi JS, Kay SA. Suprachiasmatic nucleus: cell autonomy and network properties. Annu Rev Physiol 2010; 72:551−77.

[28] Van Dongen HP, Maislin G, Mullington JM, Dinges DF. The cumulative cost of additional wakefulness: dose-response effects on neurobehavioral functions and sleep physiology from chronic sleep restriction and total sleep deprivation. Sleep 2003; 26:117−26.

[29] Tononi G, Cirelli C. Sleep function and synaptic homeostasis. Sleep Med Rev 2006;10:49−62.

[30] Siegel JM. Sleep viewed as a state of adaptive inactivity. Nat Rev Neurosci 2009;10:747−53.

[31] Cirelli C. The genetic and molecular regulation of sleep: from fruit flies to humans. Nat Rev Neurosci 2009;10:549−60.

[32] Iber C, American Academy of Sleep Medicine. The AASM Manual for the Scoring of Sleep and Associated Events: Rules, Terminology and Technical Specifications. Westchester, IL: American Academy of Sleep Medicine; 2007.

[33] Steriade M, McCormick DA, Sejnowski TJ. Thalamocortical oscillations in the sleeping and aroused brain. Science 1993; 262:679−85.

[34] Davis H, Davis PA, Loomis AL, Harvey EN, Hobart G. Changes in human brain potentials during the onset of sleep. Science 1937; 86:448−50.

[35] Dijk DJ. Regulation and functional correlates of slow wave sleep. J Clin Sleep Med 2009;5:S6−15.

[36] Franken P, Dijk DJ. Circadian clock genes and sleep homeostasis. Eur J Neurosci 2009;29:1820−9.

[37] Borbely AA. A two process model of sleep regulation. Hum Neurobiol 1982;1:195−204.

[38] Dijk DJ, Czeisler CA. Paradoxical timing of the circadian rhythm of sleep propensity serves to consolidate sleep and wakefulness in humans. Neurosci Lett 1994;166:63−8.

[39] Dijk DJ, Czeisler CA. Contribution of the circadian pacemaker and the sleep homeostat to sleep propensity, sleep structure, electroencephalographic slow waves, and sleep spindle activity in humans. J Neurosci 1995;15:3526−38.

[40] Dijk DJ, von Schantz M. Timing and consolidation of human sleep, wakefulness, and performance by a symphony of oscillators. J Biol Rhythms 2005;20:279−90.

[41] Phillips AJ, Czeisler CA, Klerman EB. Revisiting spontaneous internal desynchrony using a quantitative model of sleep physiology. J Biol Rhythms 2011;26:441−53.

[42] Weitzman ED, Czeisler CA, Zimmerman JC, Ronda JM. Timing of REM and stages 3 + 4 sleep during temporal isolation in man. Sleep 1980;2:391−407.

[43] Dijk DJ, Shanahan TL, Duffy JF, Ronda JM, Czeisler CA. Variation of electroencephalographic activity during non-rapid eye movement and rapid eye movement sleep with phase of circadian melatonin rhythm in humans. J Physiol 1997;505:851−8.

[44] Cajochen C, Wyatt JK, Czeisler CA, Dijk DJ. Separation of circadian and wake duration-dependent modulation of EEG activation during wakefulness. Neuroscience 2002;114:1047−60.

[45] Skene DJ, Arendt J. Circadian rhythm sleep disorders in the blind and their treatment with melatonin. Sleep Med 2007;8:651−5.

[46] Palm L, Blennow G, Wetterberg L. Long-term melatonin treatment in blind children and young adults with circadian sleep--wake disturbances. Dev Med Child Neurol 1997;39:319−25.

[47] Sack RL, Brandes RW, Kendall AR, Lewy AJ. Entrainment of free-running circadian rhythms by melatonin in blind people. N Engl J Med 2000;343:1070−7.

[48] Dubocovich ML. Melatonin receptors: role on sleep and circadian rhythm regulation. Sleep Med 2007;8(Suppl 3):34−42.

[49] Liu C, Weaver DR, Jin X, Shearman LP, Pieschl RL, Gribkoff VK, Reppert SM. Molecular dissection of two distinct actions of melatonin on the suprachiasmatic circadian clock. Neuron 1997;19:91−102.

[50] Lewy AJ, Ahmed S, Jackson JM, Sack RL. Melatonin shifts human circadian rhythms according to a phase-response curve. Chronobiol Int 1992;9:380−92.

[51] Lewy AJ, Bauer VK, Ahmed S, Thomas KH, Cutler NL, Singer CM, Moffit MT, Sack RL. The human phase response curve (PRC) to melatonin is about 12 hours out of phase with the PRC to light. Chronobiol Int 1998;15:71−83.

[52] Edgar DM, Dement WC, Fuller CA. Effect of SCN lesions on sleep in squirrel monkeys: evidence for opponent processes in sleep−wake regulation. J Neurosci 1993;13:1065−79.

[53] Mistlberger RE, Bergmann BM, Rechtschaffen A. Relationships among wake episode lengths, contiguous sleep episode lengths, and electroencephalographic δ waves in rats with suprachiasmatic nuclei lesions. Sleep 1987;10:12−24.

[54] Wurts SW, Edgar DM. Circadian and homeostatic control of rapid eye movement (REM) sleep: promotion of REM tendency by the suprachiasmatic nucleus. J Neurosci 2000;20:4300−10.

[55] Mistlberger RE. Circadian regulation of sleep in mammals: role of the suprachiasmatic nucleus. Brain Res Brain Res Rev 2005; 49:429−54.

[56] Hogenesch JB, Herzog ED. Intracellular and intercellular processes determine robustness of the circadian clock. FEBS Lett 2011;585:1427−34.

[57] Panda S, Hogenesch JB, Kay SA. Circadian rhythms from flies to human. Nature 2002;417:329−35.

[58] Wisor JP, O'Hara BF, Terao A, Selby CP, Kilduff TS, Sancar A, Edgar DM, Franken P. A role for cryptochromes in sleep regulation. BMC Neurosci 2002;3:20.

[59] Naylor E, Bergmann BM, Krauski K, Zee PC, Takahashi JS, Vitaterna MH, Turek FW. The circadian clock mutation alters sleep homeostasis in the mouse. J Neurosci 2000;20:8138−43.

[60] Vitaterna MH, King DP, Chang AM, Kornhauser JM, Lowrey PL, McDonald JD, et al. Mutagenesis and mapping of a mouse gene, Clock, essential for circadian behavior. Science 1994;264:719−25.

[61] Jin X, Shearman LP, Weaver DR, Zylka MJ, de Vries GJ, Reppert SM. A molecular mechanism regulating rhythmic output from the suprachiasmatic circadian clock. Cell 1999;96:57−68.

[62] Laposky A, Easton A, Dugovic C, Walisser J, Bradfield C, Turek F. Deletion of the mammalian circadian clock gene BMAL1/Mop3 alters baseline sleep architecture and the response to sleep deprivation. Sleep 2005;28:395–409.

[63] Bunger MK, Wilsbacher LD, Moran SM, Clendenin C, Radcliffe LA, Hogenesch JB, et al. Mop3 is an essential component of the master circadian pacemaker in mammals. Cell 2000;103:1009–17.

[64] Franken P, Dudley CA, Estill SJ, Barakat M, Thomason R, O'Hara BF, McKnight SLC. NPAS2 as a transcriptional regulator of non-rapid eye movement sleep: genotype and sex interactions. Proc Natl Acad Sci U S A 2006;103:7118–23.

[65] Dudley CA, Erbel-Sieler C, Estill SJ, Reick M, Franken P, Pitts S, McKnight SL. Altered patterns of sleep and behavioral adaptability in NPAS2-deficient mice. Science 2003;301:379–83.

[66] Reick M, Garcia JA, Dudley C, McKnight SL. NPAS2: an analog of clock operative in the mammalian forebrain. Science 2001;293:506–9.

[67] Ripperger JA, Shearman LP, Reppert SM, Schibler UC. CLOCK, an essential pacemaker component, controls expression of the circadian transcription factor DBP. Genes Dev 2000;14:679–89.

[68] Franken P, Lopez-Molina L, Marcacci L, Schibler U, Tafti M. The transcription factor DBP affects circadian sleep consolidation and rhythmic EEG activity. J Neurosci 2000;20:617–25.

[69] Shiromani PJ, Xu M, Winston EM, Shiromani SN, Gerashchenko D, Weaver DR. Sleep rhythmicity and homeostasis in mice with targeted disruption of mPeriod genes. Am J Physiol Regul Integr Comp Physiol 2004;287:R47–57.

[70] Bae K, Jin X, Maywood ES, Hastings MH, Reppert SM, Weaver DR. Differential functions of mPer1, mPer2, and mPer3 in the SCN circadian clock. Neuron 2001;30:525–36.

[71] Viola AU, Archer SN, James LM, Groeger JA, Lo JC, Skene DJ, von Schantz M, Dijk DJ. PER3 polymorphism predicts sleep structure and waking performance. Curr Biol 2007;17:613–8.

[72] Hasan S, van der Veen DR, Winsky-Sommerer R, Dijk DJ, Archer SN. Altered sleep and behavioral activity phenotypes in PER3-deficient mice. Am J Physiol Regul Integr Comp Physiol 2011;301:R1821–30.

[73] Jones CR, Campbell SS, Zone SE, Cooper F, DeSano A, Murphy PJ, et al. Familial advanced sleep-phase syndrome: a short-period circadian rhythm variant in humans. Nat Med 1999;5:1062–5.

[74] Toh KL, Jones CR, He Y, Eide EJ, Hinz WA, Virshup DM, et al. An hPer2 phosphorylation site mutation in familial advanced sleep phase syndrome. Science 2001;291:1040–3.

[75] Xu Y, Padiath QS, Shapiro RE, Jones CR, Wu SC, Saigoh N, et al. Functional consequences of a CKIdelta mutation causing familial advanced sleep phase syndrome. Nature 2005;434:640–4.

[76] Lowrey PL, Shimomura K, Antoch MP, Yamazaki S, Zemenides PD, Ralph MR, et al. Positional syntenic cloning and functional characterization of the mammalian circadian mutation tau. Science 2000;288:483–92.

[77] Ralph MR, Menaker M. A mutation of the circadian system in golden hamsters. Science 1988;241:1225–7.

[78] Meng QJ, Logunova L, Maywood ES, Gallego M, Lebiecki J, Brown TM, et al. Setting clock speed in mammals: the CK1 epsilon tau mutation in mice accelerates circadian pacemakers by selectively destabilizing PERIOD proteins. Neuron 2008;58:78–88.

[79] Lee C, Etchegaray JP, Cagampang FR, Loudon AS, Reppert SM. Posttranslational mechanisms regulate the mammalian circadian clock. Cell 2001;107:855–67.

[80] Honma S, Kawamoto T, Takagi Y, Fujimoto K, Sato F, Noshiro M, et al. Dec1 and Dec2 are regulators of the mammalian molecular clock. Nature 2002;419:841–4.

[81] Lim C, Chung BY, Pitman JL, McGill JJ, Pradhan S, Lee J, et al. Clockwork orange encodes a transcriptional repressor important for circadian-clock amplitude in Drosophila. Curr Biol 2007;17:1082–9.

[82] Rossner MJ, Oster H, Wichert SP, Reinecke L, Wehr MC, Reinecke J, et al. Disturbed clockwork resetting in Sharp-1 and Sharp-2 single and double mutant mice. PLoS One 2008;3:e2762.

[83] He Y, Jones CR, Fujiki N, Xu Y, Guo B, Holder Jr JL, et al. The transcriptional repressor DEC2 regulates sleep length in mammals. Science 2009;325:866–70.

[84] Husse J, Zhou X, Shostak A, Oster H, Eichele G. Synaptotagmin10-Cre, a driver to disrupt clock genes in the SCN. J Biol Rhythms 2011;26:379–89.

[85] Morgado-Bernal I. Learning and memory consolidation: linking molecular and behavioral data. Neuroscience 2011;176:12–9.

[86] McGaugh JL. Memory—a century of consolidation. Science 2000;287:248–51.

[87] Decker S, McConnaughey S, Page TL. Circadian regulation of insect olfactory learning. Proc Natl Acad Sci U S A 2007;104:15905–10.

[88] Lyons LC, Roman G. Circadian modulation of short-term memory in Drosophila. Learn Mem 2009;16:19–27.

[89] Devan BD, Goad EH, Petri HL, Antoniadis EA, Hong NS, Ko CH, et al. Circadian phase-shifted rats show normal acquisition but impaired long-term retention of place information in the water task. Neurobiol Learn Mem 2001;75:51–62.

[90] Stephan FK, Kovacevic NS. Multiple retention deficit in passive avoidance in rats is eliminated by suprachiasmatic lesions. Behav Biol 1978;22:456–62.

[91] Mistlberger RE, de Groot MH, Bossert JM, Marchant EG. Discrimination of circadian phase in intact and suprachiasmatic nuclei-ablated rats. Brain Res 1996;739:12–8.

[92] Cain SW, Ralph MR. Circadian modulation of conditioned place avoidance in hamsters does not require the suprachiasmatic nucleus. Neurobiol Learn Mem 2009;91:81–4.

[93] Loh DH, Navarro J, Hagopian A, Wang LM, Deboer T, Colwell CS. Rapid changes in the light/dark cycle disrupt memory of conditioned fear in mice. PLoS One 2010;5.

[94] Craig LA, McDonald RJ. Chronic disruption of circadian rhythms impairs hippocampal memory in the rat. Brain Res Bull 2008;76:141–51.

[95] Chaudhury D, Colwell CS. Circadian modulation of learning and memory in fear-conditioned mice. Behav Brain Res 2002;133:95–108.

[96] Eckel-Mahan KL, Phan T, Han S, Wang H, Chan GC, Scheiner ZS, Storm DR. Circadian oscillation of hippocampal MAPK activity and cAmp: implications for memory persistence. Nat Neurosci 2008;11:1074–82.

[97] Gerstner JR, Yin JC. Circadian rhythms and memory formation. Nat Rev Neurosci 2010;11:577–88.

[98] Deng W, Aimone JB, Gage FH. New neurons and new memories: how does adult hippocampal neurogenesis affect learning and memory? Nat Rev Neurosci 2010;11:339–50.

[99] Aimone JB, Wiles J, Gage FH. Potential role for adult neurogenesis in the encoding of time in new memories. Nat Neurosci 2006;9:723–7.

[100] Holmes MM, Galea LA, Mistlberger RE, Kempermann G. Adult hippocampal neurogenesis and voluntary running activity: circadian and dose-dependent effects. J Neurosci Res 2004;76:216–22.

[101] van Praag H, Kempermann G, Gage FH. Running increases cell proliferation and neurogenesis in the adult mouse dentate gyrus. Nat Neurosci 1999;2:266–70.

[102] Rana S, Mahmood S. Circadian rhythm and its role in malignancy. J Circadian Rhythms 2010;8:3.

[103] Pando BF, van Oudenaarden A. Coupling cellular oscillators—circadian and cell division cycles in cyanobacteria. Curr Opin Genet Dev 2010;20:613–8.

[104] Merrow M, Roenneberg T. Cellular clocks: coupled circadian and cell division cycles. Curr Biol 2004;14:R25–6.

[105] Borgs L, Beukelaers P, Vandenbosch R, Nguyen L, Moonen G, Maquet P, et al. Period 2 regulates neural stem/progenitor cell proliferation in the adult hippocampus. BMC Neurosci 2009; 10:30.

[106] Gibson EM, Wang C, Tjho S, Khattar N, Kriegsfeld LJ. Experimental 'jet lag' inhibits adult neurogenesis and produces long-term cognitive deficits in female hamsters. PLoS One 2010; 5:e15267.

[107] Wang LM, Dragich JM, Kudo T, Odom IH, Welsh DK, O'Dell TJ, Colwell CS. Expression of the circadian clock gene Period2 in the hippocampus: possible implications for synaptic plasticity and learned behaviour. ASN Neuro 2009;1.

[108] Wyse CA, Coogan AN. Impact of aging on diurnal expression patterns of CLOCK and BMAL1 in the mouse brain. Brain Res 2010;1337:21–31.

[109] Jilg A, Lesny S, Peruzki N, Schwegler H, Selbach O, Dehghani F, Stehle JH. Temporal dynamics of mouse hippocampal clock gene expression support memory processing. Hippocampus 2010;20: 377–88.

[110] Chaudhury D, Wang LM, Colwell CS. Circadian regulation of hippocampal long-term potentiation. J Biol Rhythms 2005;20: 225–36.

[111] Barnes CA, McNaughton BL, Goddard GV, Douglas RM, Adamec R. Circadian rhythm of synaptic excitability in rat and monkey central nervous system. Science 1977;197:91–2.

[112] Harris KM, Teyler TJ. Age differences in a circadian influence on hippocampal LTP. Brain Res 1983;261:69–73.

[113] Abraham WC, Bear MF. Metaplasticity: the plasticity of synaptic plasticity. Trends Neurosci 1996;19:126–30.

[114] Morris RG. D.O. Hebb: The organization of behavior, Wiley: New York; 1949. Brain Res Bull 1999;50:437.

[115] Hebb DO, Martinez JL, Glickman SE. The organization of behavior – a neuropsychological theory – Hebb, Do. Contemp Psychoanal 1994;39:1018–20.

[116] Roth TL, Sweatt JD. Rhythms of memory. Nat Neurosci 2008; 11:993–4.

[117] Garcia JA, Zhang D, Estill SJ, Michnoff C, Rutter J, Reick M, et al. Impaired cued and contextual memory in NPAS2-deficient mice. Science 2000;288:2226–30.

[118] Van der Zee EA, Havekes R, Barf RP, Hut RA, Nijholt IM, Jacobs EH, Gerkema MP. Circadian time-place learning in mice depends on Cry genes. Curr Biol 2008;18:844–8.

[119] Kondratova AA, Dubrovsky YV, Antoch MP, Kondratov RV. Circadian clock proteins control adaptation to novel environment and memory formation. Aging 2010;2:285–97.

[120] Chaudhury D, Loh DH, Dragich JM, Hagopian A, Colwell CS. Select cognitive deficits in vasoactive intestinal peptide deficient mice. BMC Neurosci 2008;9:63.

[121] Smedley D, Salimova E, Rosenthal N. Cre recombinase resources for conditional mouse mutagenesis. Methods 2011;53:411–6.

[122] Guan C, Ye C, Yang X, Gao J. A review of current large-scale mouse knockout efforts. Genesis 2010;48:73–85.

[123] Birling MC, Gofflot F, Warot X. Site-specific recombinases for manipulation of the mouse genome. Methods Mol Biol 2009; 561:245–63.

[124] Premsrirut PK, Dow LE, Kim SY, Camiolo M, Malone CD, Miething C, et al. A rapid and scalable system for studying gene function in mice using conditional RNA interference. Cell 2011; 145:145–58.

[125] Gerstner JR, Lyons LC, Wright Jr KP, Loh DH, Rawashdeh O, Eckel-Mahan KL, Roman GW. Cycling behavior and memory formation. J Neurosci 2009;29:12824–30.

[126] Atkins CM, Selcher JC, Petraitis JJ, Trzaskos JM, Sweatt JD. The MAPK cascade is required for mammalian associative learning. Nat Neurosci 1998;1:602–9.

[127] Sweatt JD. Mitogen-activated protein kinases in synaptic plasticity and memory. Curr Opin Neurobiol 2004;14:311–7.

[128] Kelly A, Laroche S, Davis S. Activation of mitogen-activated protein kinase/extracellular signal-regulated kinase in hippocampal circuitry is required for consolidation and reconsolidation of recognition memory. J Neurosci 2003;23:5354–60.

[129] O'Neill JS, Maywood ES, Chesham JE, Takahashi JS, Hastings MH. cAMP-dependent signaling as a core component of the mammalian circadian pacemaker. Science 2008;320:949–53.

[130] Prosser RA, McArthur AJ, Gillette MU. cGMP induces phase shifts of a mammalian circadian pacemaker at night, in antiphase to cAMP effects. Proc Natl Acad Sci U S A 1989;86:6812–5.

[131] Obrietan K, Impey S, Smith D, Athos J, Storm DR. Circadian regulation of cAMP response element-mediated gene expression in the suprachiasmatic nuclei. J Biol Chem 1999;274: 17748–56.

[132] Lyons LC, Collado MS, Khabour O, Green CL, Eskin A. The circadian clock modulates core steps in long-term memory formation in *Aplysia*. J Neurosci 2006;26:8662–71.

[133] Phan TX, Chan GC, Sindreu CB, Eckel-Mahan KL, Storm DR. The diurnal oscillation of MAP (mitogen-activated protein) kinase and adenylyl cyclase activities in the hippocampus depends on the suprachiasmatic nucleus. J Neurosci 2011;31: 10640–7.

[134] Eckel-Mahan KL, Storm DR. Circadian rhythms and memory: not so simple as cogs and gears. EMBO Rep 2009;10:584–91.

[135] Besoluk S, Onder I, Deveci I. Morningness–eveningness preferences and academic achievement of university students. Chronobiol Int 2011;28:118–25.

[136] Schmidt C, Collette F, Cajochen C, Peigneux P. A time to think: circadian rhythms in human cognition. Cogn Neuropsychol 2007; 24:755–89.

[137] Lufi D, Tzischinsky O, Hadar S. Delaying school starting time by one hour: some effects on attention levels in adolescents. J Clin Sleep Med 2011;7:137−43.

[138] Guilding C, Piggins HD. Challenging the omnipotence of the suprachiasmatic timekeeper: are circadian oscillators present throughout the mammalian brain? Eur J Neurosci 2007;25: 3195−216.

[139] Conway-Campbell BL, Sarabdjitsingh RA, McKenna MA, Pooley JR, Kershaw YM, Meijer OC, et al. Glucocorticoid ultradian rhythmicity directs cyclical gene pulsing of the clock gene period 1 in rat hippocampus. J Neuroendocrinol 2010; 22:1093−100.

[140] Gilhooley MJ, Pinnock SB, Herbert J. Rhythmic expression of per1 in the dentate gyrus is suppressed by corticosterone: implications for neurogenesis. Neurosci Lett 2011;489:177−81.

[141] McClung CA. Circadian genes, rhythms and the biology of mood disorders. Pharmacol Ther 2007;114:222−32.

[142] Wulff K, Gatti S, Wettstein JG, Foster RG. Sleep and circadian rhythm disruption in psychiatric and neurodegenerative disease. Nat Rev Neurosci 2010;11:589−99.

[143] Menet JS, Rosbash M. When brain clocks lose track of time: cause or consequence of neuropsychiatric disorders. Curr Opin Neurobiol 2011;21(6):849−57.

[144] McClung CA, Sidiropoulou K, Vitaterna M, Takahashi JS, White FJ, Cooper DC, Nestler EJ. Regulation of dopaminergic transmission and cocaine reward by the Clock gene. Proc Natl Acad Sci U S A 2005;102:9377−81.

[145] Mukherjee S, Coque L, Cao JL, Kumar J, Chakravarty S, Asaithamby A, et al. Knockdown of Clock in the ventral tegmental area through RNA interference results in a mixed state of mania and depression-like behavior. Biol Psychiatry 2010; 68:503−11.

[146] Roybal K, Theobold D, Graham A, DiNieri JA, Russo SJ, Krishnan V, et al. Mania-like behavior induced by disruption of CLOCK. Proc Natl Acad Sci U S A 2007;104:6406−11.

[147] Lamont EW, Coutu DL, Cermakian N, Boivin DB. Circadian rhythms and clock genes in psychotic disorders. Isr J Psychiatry Relat Sci 2010;47:27−35.

[148] Magnusson A, Boivin D. Seasonal affective disorder: an overview. Chronobiol Int 2003;20:189−207.

[149] Roecklein KA, Rohan KJ. Seasonal affective disorder: an overview and update. Psychiatry (Edgmont) 2005;2:20−6.

[150] Benedetti F, Barbini B, Colombo C, Smeraldi E. Chronotherapeutics in a psychiatric ward. Sleep Med Rev 2007;11:509−22.

[151] Roecklein KA, Rohan KJ, Duncan WC, Rollag MD, Rosenthal NE, Lipsky RH, Provencio I. A missense variant (P10L) of the melanopsin (OPN4) gene in seasonal affective disorder. J Affect Disord 2009;114:279−85.

[152] Partonen T, Treutlein J, Alpman A, Frank J, Johansson C, Depner M, et al. Three circadian clock genes Per2, Arntl, and Npas2 contribute to winter depression. Ann Med 2007; 39:229−38.

[153] Zhang J, Liao G, Liu C, Sun L, Liu Y, Wang Y, et al. The association of CLOCK gene T3111C polymorphism and hPER3 gene 54-nucleotide repeat polymorphism with Chinese Han people schizophrenics. Mol Biol Rep 2011;38:349−54.

[154] Takao T, Tachikawa H, Kawanishi Y, Mizukami K, Asada T. CLOCK gene T3111C polymorphism is associated with Japanese schizophrenics: a preliminary study. Eur Neuropsychopharmacol 2007;17:273−6.

[155] Spanagel R, Pendyala G, Abarca C, Zghoul T, Sanchis-Segura C, Magnone MC, et al. The clock gene Per2 influences the glutamatergic system and modulates alcohol consumption. Nat Med 2005;11:35−42.

[156] Abarca C, Albrecht U, Spanagel R. Cocaine sensitization and reward are under the influence of circadian genes and rhythm. Proc Natl Acad Sci U S A 2002;99:9026−30.

[157] Brophy K, Hawi Z, Kirley A, Fitzgerald M, Gill M. Synaptosomal-associated protein 25 (SNAP-25) and attention deficit hyperactivity disorder (ADHD): evidence of linkage and association in the Irish population. Mol Psychiatry 2002;7:913−7.

[158] Zhang H, Zhu S, Zhu Y, Chen J, Zhang G, Chang H. An association study between SNAP-25 gene and attention-deficit hyperactivity disorder. Eur J Paediatr Neurol 2011;15:48−52.

[159] Yin L, Wang J, Klein PS, Lazar MA. Nuclear receptor Rev-erbalpha is a critical lithium-sensitive component of the circadian clock. Science 2006;311:1002−5.

[160] Iitaka C, Miyazaki K, Akaike T, Ishida N. A role for glycogen synthase kinase-3beta in the mammalian circadian clock. J Biol Chem 2005;280:29397−402.

[161] Kaladchibachi SA, Doble B, Anthopoulos N, Woodgett JR, Manoukian AS. Glycogen synthase kinase 3, circadian rhythms, and bipolar disorder: a molecular link in the therapeutic action of lithium. J Circadian Rhythms 2007;5:3.

[162] Preitner N, Damiola F, Lopez-Molina L, Zakany J, Duboule D, Albrecht U, Schibler U. The orphan nuclear receptor REV-ERBalpha controls circadian transcription within the positive limb of the mammalian circadian oscillator. Cell 2002; 110:251−60.

[163] Hampp G, Ripperger JA, Houben T, Schmutz I, Blex C, Perreau-Lenz S, et al. Regulation of monoamine oxidase A by circadian-clock components implies clock influence on mood. Curr Biol 2008;18:678−83.

[164] Karl T, Burne TH, Herzog H. Effect of Y1 receptor deficiency on motor activity, exploration, and anxiety. Behav Brain Res 2006; 167:87−93.

[165] Wersinger SR, Caldwell HK, Christiansen M, Young 3rd WS. Disruption of the vasopressin 1b receptor gene impairs the attack component of aggressive behavior in mice. Genes Brain Behav 2007;6:653−60.

[166] Hagenauer MH, Perryman JI, Lee TM, Carskadon MA. Adolescent changes in the homeostatic and circadian regulation of sleep. Dev Neurosci 2009;31:276−84.

[167] Kohyama J. Sleep health and asynchronization. Brain Dev 2011; 33:252−9.

[168] Wittmann M, Dinich J, Merrow M, Roenneberg T. Social jetlag: misalignment of biological and social time. Chronobiol Int 2006; 23:497−509.

[169] Randler C, Bilger S. Associations among sleep, chronotype, parental monitoring, and pubertal development among German adolescents. J Psychol 2009;143:509−20.

[170] Biggs SN, Lushington K, van den Heuvel CJ, Martin AJ, Kennedy JD. Inconsistent sleep schedules and daytime behavioral difficulties in school-aged children. Sleep Med 2011;12:780−6.

[171] Colrain IM, Baker FC. Changes in sleep as a function of adolescent development. Neuropsychol Rev 2011;21:5−21.

[172] King DP, Zhao Y, Sangoram AM, Wilsbacher LD, Tanaka M, et al. Positional cloning of the mouse circadian clock gene. Cell 1997;89:641–53.

[173] Plautz JD, Kaneko M, Hall JC, Kay SA. Independent photoreceptive circadian clocks throughout Drosophila. Science 1997; 278:1632–5.

[174] Balsalobre A, Damiola F, Schibler U. A serum shock induces circadian gene expression in mammalian tissue culture cells. Cell 1998;93:929–37.

[175] Yamazaki S, Numano R, Abe M, Hida A, Takahashi R, Ueda M, et al. Resetting central and peripheral circadian oscillators in transgenic rats. Science 2000;288:682–5.

[176] Yoo SH, Yamazaki S, Lowrey PL, Shimomura K, Ko CH, Buhr ED, et al. PERIOD2:: LUCIFERASE real-time reporting of circadian dynamics reveals persistent circadian oscillations in mouse peripheral tissues. Proc Natl Acad Sci U S A 2004; 101:5339–46.

[177] Yagita K, Horie K, Koinuma S, Nakamura W, Yamanaka I, Urasaki A, et al. Development of the circadian oscillator during differentiation of mouse embryonic stem cells in vitro. Proc Natl Acad Sci U S A 2010;107:3846–51.

[178] Liu S, Cai Y, Sothern RB, Guan Y, Chan P. Chronobiological analysis of circadian patterns in transcription of seven key clock genes in six peripheral tissues in mice. Chronobiol Int 2007;24:793–820.

[179] Alvarez JD, Sehgal A. The thymus is similar to the testis in its pattern of circadian clock gene expression. J Biol Rhythms 2005; 20:111–21.

[180] Panda S, Antoch MP, Miller BH, Su AI, Schook AB, Straume M, et al. Coordinated transcription of key pathways in the mouse by the circadian clock. Cell 2002;109:307–20.

[181] Hughes ME, DiTacchio L, Hayes KR, Vollmers C, Pulivarthy S, Baggs JE, et al. Harmonics of circadian gene transcription in mammals. PLoS Genet 2009;5. e1000442.

[182] Akhtar RA, Reddy AB, Maywood ES, Clayton JD, King VM, Smith AG, et al. Circadian cycling of the mouse liver transcriptome, as revealed by cDNA microarray, is driven by the suprachiasmatic nucleus. Curr Biol 2002;12:540–50.

[183] Oishi K, Miyazaki K, Kadota K, Kikuno R, Nagase T, Atsumi G, et al. Genome-wide expression analysis of mouse liver reveals CLOCK-regulated circadian output genes. J Biol Chem 2003; 278:41519–27.

[184] Zambon AC, McDearmon EL, Salomonis N, Vranizan KM, Johansen KL, Adey D, et al. Time- and exercise-dependent gene regulation in human skeletal muscle. Genome Biol 2003;4:R61.

[185] McCarthy JJ, Andrews JL, McDearmon EL, Campbell KS, Barber BK, Miller BH, et al. Identification of the circadian transcriptome in adult mouse skeletal muscle. Physiol Genomics 2007;31:86–95.

[186] Storch KF, Lipan O, Leykin I, Viswanathan N, Davis FC, Wong WH, Weitz CJ. Extensive and divergent circadian gene expression in liver and heart. Nature 2002;417:78–83.

[187] Martino T, Arab S, Straume M, Belsham DD, Tata N, Cai F, et al. Day/night rhythms in gene expression of the normal murine heart. J Mol Med 2004;82:256–64.

[188] Oishi K, Amagai N, Shirai H, Kadota K, Ohkura N, Ishida N. Genome-wide expression analysis reveals 100 adrenal gland-dependent circadian genes in the mouse liver. DNA Res 2005; 12:191–202.

[189] DeBruyne JP, Weaver DR, Reppert SM. CLOCK and NPAS2 have overlapping roles in the suprachiasmatic circadian clock. Nat Neurosci 2007;10:543–5.

[190] DeBruyne JP, Weaver DR, Reppert SM. Peripheral circadian oscillators require CLOCK. Curr Biol 2007;17:R538–9.

[191] Lamia KA, Storch K-F, Weitz CJ. Physiological significance of a peripheral tissue circadian clock. Proc Natl Acad Sci U S A 2008;105:15172–7.

[192] Marcheva B, Ramsey KM, Buhr ED, Kobayashi Y, Su H, Ko CH, et al. Disruption of the clock components CLOCK and BMAL1 leads to hypoinsulinaemia and diabetes. Nature 2010;466:627–31.

[193] Dibner C, Schibler U, Albrecht U. The mammalian circadian timing system: organization and coordination of central and peripheral clocks. Annu Rev Physiol 2010;72:517–49.

[194] Brown SA, Zumbrunn G, Fleury-Olela F, Preitner N, Schibler U. Rhythms of mammalian body temperature can sustain peripheral circadian clocks. Curr Biol 2002;12:1574–83.

[195] Buhr ED, Yoo SH, Takahashi JS. Temperature as a universal resetting cue for mammalian circadian oscillators. Science 2010; 330:379–85.

[196] Damiola F, Le Minh N, Preitner N, Kornmann B, Fleury-Olela F, Schibler U. Restricted feeding uncouples circadian oscillators in peripheral tissues from the central pacemaker in the suprachiasmatic nucleus. Genes Dev 2000;14:2950–61.

[197] Vollmers C, Gill S, DiTacchio L, Pulivarthy SR, Le HD, Panda S. Time of feeding and the intrinsic circadian clock drive rhythms in hepatic gene expression. Proc Natl Acad Sci U S A 2009;106: 21453–8.

[198] Stephan FK, Swann JM, Sisk CL. Entrainment of circadian rhythms by feeding schedules in rats with suprachiasmatic lesions. Behav Neural Biol 1979;25:545–54.

[199] Gomez-Abellan P, Gomez-Santos C, Madrid JA, Milagro FI, Campion J, Martinez JA, et al. Circadian expression of adiponectin and its receptors in human adipose tissue. Endocrinology 2010;151:115–22.

[200] Abrahamson EE, Moore RY. Suprachiasmatic nucleus in the mouse: retinal innervation, intrinsic organization and efferent projections. Brain Res 2001;916:172–91.

[201] Saeb-Parsy K, Lombardelli S, Khan FZ, McDowall K, Au-Yong IT, Dyball RE. Neural connections of hypothalamic neuroendocrine nuclei in the rat. J Neuroendocrinol 2000;12:635–48.

[202] Yi C-X, van der Vliet J, Dai J, Yin G, Ru L, Buijs RM. Ventromedial arcuate nucleus communicates peripheral metabolic information to the suprachiasmatic nucleus. Endocrinology 2006;147:283–94.

[203] Blevins JE, Baskin DG. Hypothalamic-brainstem circuits controlling eating. Forum Nutr 2010;63:133–40.

[204] Abe M, Herzog ED, Yamazaki S, Straume M, Tei H, Sakaki Y, et al. Circadian rhythms in isolated brain regions. Clin EEG Neurosci 2002;22:350–6.

[205] Morton GJ, Cummings DE, Baskin DG, Barsh GS, Schwartz MW. Central nervous system control of food intake and body weight. Nature 2006;443:289–95.

[206] Tonsfeldt KJ, Chappell PE. Clocks on top: the role of the circadian clock in the hypothalamic and pituitary regulation of endocrine physiology. Mol Cell Endocrinol 2012;349:3–12.

[207] Nader N, Chrousos GP, Kino T. Interactions of the circadian CLOCK system and the HPA axis. Trends Endocrinol Metab 2010;21:277–86.

[208] Kalsbeek A, Palm IF, La Fleur SE, Scheer FAJL, Perreau-Lenz S, Ruiter M, et al. SCN outputs and the hypothalamic balance of life. J Biol Rhythms 2006;21:458–69.

[209] Saper CB, Lu J, Chou TC, Gooley J. The hypothalamic integrator for circadian rhythms. Trends Neurosci 2005;28:152–7.

[210] Marcheva B, Ramsey KM, Bass J. Circadian genes and insulin exocytosis. Cell Logist 2011;1:32–6.

[211] Zhang EE, Liu AC, Hirota T, Miraglia LJ, Welch G, Pongsawakul PY, et al. A genome-wide RNAi screen for modifiers of the circadian clock in human cells. Cell 2009; 139:199–210.

[212] La Fleur SE, Kalsbeek A, Wortel J, Buijs RM. A suprachiasmatic nucleus generated rhythm in basal glucose concentrations. J Neuroendocrinol 1999;11:643–52.

[213] Ruiter M, La Fleur SE, van Heijningen C, van der Vliet J, Kalsbeek A, Buijs RM. The daily rhythm in plasma glucagon concentrations in the rat is modulated by the biological clock and by feeding behavior. Diabetes 2003;52:1709–15.

[214] Ando H, Yanagihara H, Hayashi Y, Obi Y, Tsuruoka S, Takamura T, et al. Rhythmic messenger ribonucleic acid expression of clock genes and adipocytokines in mouse visceral adipose tissue. Endocrinology 2005;146:5631–6.

[215] De Boer SF, Van der Gugten J. Daily variations in plasma noradrenaline, adrenaline and corticosterone concentrations in rats. Physiol Behav 1987;40:323–8.

[216] Moore RY, Eichler VB. Loss of a circadian adrenal corticosterone rhythm following suprachiasmatic lesions in the rat. Brain Res 1972;42:201–6.

[217] Thanos PK, Cavigelli SA, Michaelides M, Olvet DM, Patel U, Diep MN, Volkow ND. A non-invasive method for detecting the metabolic stress response in rodents: characterization and disruption of the circadian corticosterone rhythm. Physiol Res 2009;58:219–28.

[218] Ahima RS, Prabakaran D, Flier JS. Postnatal leptin surge and regulation of circadian rhythm of leptin by feeding. Implications for energy homeostasis and neuroendocrine function. J Clin Invest 1998;101:1020–7.

[219] Bodosi B, Gardi J, Hajdu I, Szentirmai E, Obal Jr F, Krueger JM. Rhythms of ghrelin, leptin, and sleep in rats: effects of the normal diurnal cycle, restricted feeding, and sleep deprivation. Am J Physiol Regul Integr Comp Physiol 2004;287:R1071–9.

[220] Scheer FA, Hilton MF, Mantzoros CS, Shea SA. Adverse metabolic and cardiovascular consequences of circadian misalignment. Proc Natl Acad Sci U S A 2009;106:4453–8.

[221] Liu YJ, Guo YF, Zhang LS, Pei YF, Yu N, Yu P, et al. Biological pathway-based genome-wide association analysis identified the vasoactive intestinal peptide (VIP) pathway important for obesity. Obesity 2010;18:2339–46.

[222] Sookoian S, Gemma C, Gianotti TF, Burgueno A, Castano G, Pirola CJ. Genetic variants of Clock transcription factor are associated with individual susceptibility to obesity. Am J Clin Nutr 2008;87:1606–15.

[223] Froy O. Metabolism and circadian rhythms—implications for obesity. Endocr Rev 2010;31:1–24.

[224] Kok SW, Overeem S, Visscher TL, Lammers GJ, Seidell JC, Pijl H, Meinders AE. Hypocretin deficiency in narcoleptic humans is associated with abdominal obesity. Obes Res 2003; 11:1147–54.

[225] Dahmen N, Bierbrauer J, Kasten M. Increased prevalence of obesity in narcoleptic patients and relatives. Eur Arch Psychiatry Clin Neurosci 2001;251:85–9.

[226] Goel N, Stunkard AJ, Rogers NL, Van Dongen HP, Allison KC, O'Reardon JP, et al. Circadian rhythm profiles in women with night eating syndrome. J Biol Rhythms 2009;24:85–94.

[227] Gallant AR, Lundgren J, Drapeau V. The night-eating syndrome and obesity. Obes Rev 2012;13(6):528–36.

[228] Arble DM, Bass J, Laposky AD, Vitaterna MH, Turek FW. Circadian timing of food intake contributes to weight gain. Obesity 2009;17:2100–2.

[229] Salgado-Delgado R, Angeles-Castellanos M, Saderi N, Buijs RM, Escobar C. Food intake during the normal activity phase prevents obesity and circadian desynchrony in a rat model of night work. Endocrinology 2010;151:1019–29.

[230] Reiter RJ, Tan DX, Korkmaz A, Ma S. Obesity and metabolic syndrome: Association with chronodisruption, sleep deprivation, and melatonin suppression. Ann Med 2012;44(6): 564–77.

[231] Fonken LK, Workman JL, Walton JC, Weil ZM, Morris JS, Haim A, Nelson RJ. Light at night increases body mass by shifting the time of food intake. Proc Natl Acad Sci U S A 2010; 107:18664–9.

[232] Masaki T, Chiba S, Yasuda T, Noguchi H, Kakuma T, Watanabe T, et al. Involvement of hypothalamic histamine H1 receptor in the regulation of feeding rhythm and obesity. Diabetes 2004;53:2250–60.

[233] Schibler U. The daily timing of gene expression and physiology in mammals. Dialogues Clin Neurosci 2007;9:257–72.

[234] Rudic RD, McNamara P, Curtis AM, Boston RC, Panda S, Hogenesch JB, Fitzgerald GA. BMAL1 and CLOCK, two essential components of the circadian clock, are involved in glucose homeostasis. PLoS Biol 2004;2:e377.

[235] McDearmon EL, Patel KN, Ko CH, Walisser JA, Schook AC, Chong JL, et al. Dissecting the functions of the mammalian clock protein BMAL1 by tissue-specific rescue in mice. Science 2006; 314:1304–8.

[236] Bunger MK, Walisser JA, Sullivan R, Manley PA, Moran SM, Kalscheur VL, et al. Progressive arthropathy in mice with a targeted disruption of the Mop3/Bmal-1 locus. Genesis 2005; 41:122–32.

[237] Kondratov RV, Kondratova AA, Gorbacheva VY, Vykhovanets OV, Antoch MP. Early aging and age-related pathologies in mice deficient in BMAL1, the core component of the circadian clock. Genes Dev 2006;20:1868–73.

[238] Yang S, Liu A, Weidenhammer A, Cooksey RC, McClain D, Kim MK, et al. The role of mPer2 clock gene in glucocorticoid and feeding rhythms. Endocrinology 2009;150:2153–60.

[239] Green CB, Douris N, Kojima S, Strayer CA, Fogerty J, Lourim D, et al. Loss of Nocturnin, a circadian deadenylase, confers resistance to hepatic steatosis and diet-induced obesity. Proc Natl Acad Sci U S A 2007;104:9888–93.

[240] Turek FW, Joshu C, Kohsaka A, Lin E, Ivanova G, McDearmon E, et al. Obesity and metabolic syndrome in circadian Clock mutant mice. Science 2005;308:1043–5.

[241] Hussain MM, Pan X. Clock genes, intestinal transport and plasma lipid homeostasis. Trends Endocrinol Metab 2009; 20:177–85.

[242] Pan X, Hussain MM. Clock is important for food and circadian regulation of macronutrient absorption in mice. J Lipid Res 2009; 50:1800–13.

[243] Wu X, Wiater MF, Ritter S. NPAS2 deletion impairs responses to restricted feeding but not to metabolic challenges. Physiol Behav 2010;99:466–71.

[244] Liu AC, Tran HG, Zhang EE, Priest AA, Welsh DK, Kay SA. Redundant function of REV-ERBalpha and beta and non-essential role for Bmal1 cycling in transcriptional regulation of intracellular circadian rhythms. PLoS Genet 2008;4:e1000023.

[245] Yin L, Lazar MA. The orphan nuclear receptor Rev-erbalpha recruits the N-CoR/histone deacetylase 3 corepressor to regulate the circadian Bmal1 gene. Mol Endocrinol 2005;19:1452–9.

[246] Alenghat T, Meyers K, Mullican SE, Leitner K, Adeniji-Adele A, Avila J, et al. Nuclear receptor corepressor and histone deacetylase 3 govern circadian metabolic physiology. Nature 2008;456: 997–1000.

[247] Yin L, Wu N, Lazar MA. Nuclear receptor Rev-erbalpha: a heme receptor that coordinates circadian rhythm and metabolism. Nucl Recept Signal 2010;8:e001.

[248] Raghuram S, Stayrook KR, Huang P, Rogers PM, Nosie AK, McClure DB, et al. Identification of heme as the ligand for the orphan nuclear receptors REV-ERBalpha and REV-ERBbeta. Nat Struct Mol Biol 2007;14:1207–13.

[249] Yin L, Wu N, Curtin JC, Qatanani M, Szwergold NR, Reid RA, et al. Rev-erbalpha, a heme sensor that coordinates metabolic and circadian pathways. Science 2007;318:1786–9.

[250] Furuyama K, Kaneko K, Vargas PD. Heme as a magnificent molecule with multiple missions: heme determines its own fate and governs cellular homeostasis. Tohoku J Exp Med 2007;213:1–16.

[251] Ponka P. Cell biology of heme. Am J Med Sci 1999;318:241–56.

[252] Dioum EM, Rutter J, Tuckerman JR, Gonzalez G, Gilles-Gonzalez MA, McKnight SL. NPAS2: a gas-responsive transcription factor. Science 2002;298:2385–7.

[253] Dufour CR, Levasseur MP, Pham NH, Eichner LJ, Wilson BJ, Charest-Marcotte A, et al. Genomic convergence among ERRalpha, PROX1, and BMAL1 in the control of metabolic clock outputs. PLoS Genet 2011;7:e1002143.

[254] Gutman R, Barnea M, Haviv L, Chapnik N, Froy O. Peroxisome proliferator-activated receptor alpha (PPARalpha) activation advances locomotor activity and feeding daily rhythms in mice. Int J Obes 2011;36:1131–4.

[255] Oishi K, Shirai H, Ishida N. CLOCK is involved in the circadian transactivation of peroxisome-proliferator-activated receptor alpha (PPARalpha) in mice. Biochem J 2005;386:575–81.

[256] Schmutz I, Ripperger JA, Baeriswyl-Aebischer S, Albrecht U. The mammalian clock component PERIOD2 coordinates circadian output by interaction with nuclear receptors. Genes Dev 2010;24:345–57.

[257] Lamia KA, Papp SJ, Yu RT, Barish GD, Uhlenhaut NH, Jonker JW, et al. Cryptochromes mediate rhythmic repression of the glucocorticoid receptor. Nature 2011;480:552–6.

[258] Yang X, Downes M, Yu RT, Bookout AL, He W, Straume M, et al. Nuclear receptor expression links the circadian clock to metabolism. Cell 2006;126:801–10.

[259] Canto C, Auwerx J. PGC-1alpha, SIRT1 and AMPK, an energy sensing network that controls energy expenditure. Curr Opin Lipidol 2009;20:98–105.

[260] Imai S. 'Clocks' in the NAD World: NAD as a metabolic oscillator for the regulation of metabolism and aging. Biochim Biophys Acta 2010;1804:1584–90.

[261] Schwer B, Verdin E. Conserved metabolic regulatory functions of sirtuins. Cell Metab 2008;7:104–12.

[262] Rodgers JT, Lerin C, Haas W, Gygi SP, Spiegelman BM, Puigserver P. Nutrient control of glucose homeostasis through a complex of PGC-1alpha and SIRT1. Nature 2005;434:113–8.

[263] Li S, Chen XW, Yu L, Saltiel AR, Lin JD. Circadian metabolic regulation through crosstalk between casein kinase 1delta and transcriptional coactivator PGC-1alpha. Mol Endocrinol 2011; 25:2084–93.

[264] Sonoda J, Mehl IR, Chong LW, Nofsinger RR, Evans RM. PGC-1beta controls mitochondrial metabolism to modulate circadian activity, adaptive thermogenesis, and hepatic steatosis. Proc Natl Acad Sci U S A 2007;104:5223–8.

[265] Nemoto S, Fergusson MM, Finkel T. SIRT1 functionally interacts with the metabolic regulator and transcriptional coactivator PGC-1{alpha}. J Biol Chem 2005;280:16456–60.

[266] Li X, Zhang S, Blander G, Tse JG, Krieger M, Guarente L. SIRT1 deacetylates and positively regulates the nuclear receptor LXR. Mol Cell 2007;28:91–106.

[267] Asher G, Gatfield D, Stratmann M, Reinke H, Dibner C, Kreppel F, et al. SIRT1 regulates circadian clock gene expression through PER2 deacetylation. Cell 2008;134:317–28.

[268] Nakahata Y, Kaluzova M, Grimaldi B, Sahar S, Hirayama J, Chen D, et al. The NAD+-dependent deacetylase SIRT1 modulates CLOCK-mediated chromatin remodeling and circadian control. Cell 2008;134:329–40.

[269] Ramsey KM, Yoshino J, Brace CS, Abrassart D, Kobayashi Y, Marcheva B, et al. Circadian clock feedback cycle through NAMPT-mediated NAD+ biosynthesis. Science 2009;324:651–4.

[270] Nakahata Y, Sahar S, Astarita G, Kaluzova M, Sassone-Corsi P. Circadian control of the NAD+ salvage pathway by CLOCK-SIRT1. Science 2009;324:654–7.

[271] Kahn BB, Alquier T, Carling D, Hardie DG. AMP-activated protein kinase: ancient energy gauge provides clues to modern understanding of metabolism. Cell Metab 2005;1:15–25.

[272] Lamia KA, Sachdeva UM, DiTacchio L, Williams EC, Alvarez JG, Egan DF, et al. AMPK regulates the circadian clock by cryptochrome phosphorylation and degradation. Science 2009;326:437–40.

[273] Um JH, Yang S, Yamazaki S, Kang H, Viollet B, Foretz M, Chung JH. Activation of 5'-AMP-activated kinase with diabetes drug metformin induces casein kinase Iepsilon (CKIepsilon)-dependent degradation of clock protein mPer2. J Biol Chem 2007;282:20794–8.

[274] Canto C, Gerhart-Hines Z, Feige JN, Lagouge M, Noriega L, Milne JC, et al. AMPK regulates energy expenditure by modulating NAD+ metabolism and SIRT1 activity. Nature 2009;458:1056–60.

[275] Um JH, Pendergast JS, Springer DA, Foretz M, Viollet B, Brown A, et al. AMPK regulates circadian rhythms in a tissue- and isoform-specific manner. PLoS One 2011;6:e18450.

Chapter 22

Biological and Quantitative Models for Stem Cell Self-Renewal and Differentiation

Huilei Xu[1,3], Dmitri Papatsenko[2,3], Avi Ma'ayan[1,3] and Ihor Lemischka[2,3*]

[1]*Department of Pharmacology and System Therapeutics, Mount Sinai School of Medicine, Systems Biology Center New York, One Gustave L. Levy Place, New York, NY 10029, USA,* [2]*Department of Regenerative and Developmental Biology, Mount Sinai School of Medicine, One Gustave L. Levy Place, New York, NY 10029, USA,* [3]*Black Family Stem Cell Institute, Mount Sinai School of Medicine, One Gustave L. Levy Place, New York, NY 10029, USA*

Chapter Outline

Current Problems and Paradigms in Stem Cell Research	427
Empirical vs. Systems Studies, Quantitative Biological Models	427
Deterministic and Stochastic View of Self-Renewal and Differentiation	428
Examples of Quantitative Models Explaining Stem Cell Behavior	430
Models for Embryonic Stem Cells	430
Models for Hematopoietic Stem Cells	431
Strategies for Model Construction and Validation	431
Data Integration and Network Construction	431
Simple Binary Models for Complex Gene Networks	432
Modular Network Design: from Kernels to Integrated Models	433
Role of Transcription Regulatory Signals and Transcriptional Gene Networks	434
Dynamic Biological Reaction Model for the Core Pluripotency Network	435
Model Validation and Overfitting	437
Information Flow and Epigenetic Landscapes in Differentiation	437
Information Flow and Epigenetic Memory	437
Waddington Landscapes and Attractor States	438
References	438

CURRENT PROBLEMS AND PARADIGMS IN STEM CELL RESEARCH

Empirical vs. Systems Studies, Quantitative Biological Models

Recent studies have revealed the potential of reprogramming or transforming somatic cells into induced pluripotent stem cells (iPS) [1–3]. However, the efficiency of reprogramming remains low and the quality of the obtained cells is often questionable [4]. The original retroviral reprogramming method can transform only about 0.01% of fibroblasts into iPS; more relevant to medicine, 'cleaner' adenoviral reprogramming or direct delivery of reprogramming proteins into cells are even less efficient (0.0001–0.001%) [5]. The original reprogramming 'cocktail' of four factors (Oct3/4, Sox2, c-Myc, and Klf4) and its consequent modifications [6–9] were determined in rather empirical studies, supported by bioinformatics and genome-wide explorations [10,11], but not much in part of predictive biological models. Despite a constantly growing reprogramming field, the low efficiency and the associated expenses still impose barriers to therapeutic applications [4,12,13]. Systems approaches, supported by quantitative models, can bring new solutions and, possibly, solve the current limits.

In general, the first step in model construction includes broad integration of data, involving genome-wide expression or epigenetic studies [14,15]. Typically, these studies identify the most prominent candidate genes selectively expressed in self-renewing or differentiating stem cells. Linking such candidate genes into networks is based on their co-expression or the presence of similar binding patterns for transcriptional regulators in the gene control regions. However, the resulting gene regulatory networks (GRNs) are often too complex and require filtering and partition steps before the model can be constructed. Largely, models in the stem cell field are focused on a single kernel and involve few transcriptional regulators [16–22].

Quantitative dynamical models are biological hypotheses expressed in a formal way. Quantitative models provide room for better capturing diversity and complexity of biological systems, but require formalization and/or quantification of biological data. Current quantitative predictions still require experimental validation, since the models contain many unknown or ambiguous parameters.

Roughly, all quantitative models are focused either on biochemical (BRN — biochemical reaction networks) or statistical interactions observed between genes in gene networks (SIN — statistical influence networks) [23]. Typically, BRN-based models describe gene product dynamics and rely on systems of ordinary differential equations (ODE) or stochastic differentiation equations (SDE) [21,24], where the reaction constants are often represented by unknown parameters. Certain BRN-based models are limited by consideration of the reaction's steady states (or gene response potentials) and rely on systems of logistic functions describing, for instance, fractional occupancy of binding sites for transcription factors (BSTF) in the gene control regions [25–27]. Since BRN-based models attempt to capture many biological details, they are better to handle smaller gene networks including typically about a dozen of genes.

A second broad class of models, based on SIN, requires no specific knowledge regarding the already known biology and may rely on genome-wide readings of gene expression patterns, the results of statistical analysis of gene associations in large databases, or both. SIN-based models include random and probabilistic Boolean models (RBN and PBN [24]). In the late 1960s, Kauffman [28] first proposed the use of random Boolean networks to model gene regulatory networks. More recently, probabilistic Boolean networks (PBNs) have been developed as the stochastic extension of classic Boolean networks [29,30]. In a PBN, each node could potentially have more than one possible Boolean function to be randomly selected at each time step, and the resultant output of each function carries a probability. Subsequently, the long-run steady-state behavior can be studied in the context of Monte Carlo Markov chains. It is assumed that concentration levels in GRNs can be approximated by the Hill function, which in extreme cases (high cooperativity) approaches a step function [31], therefore the Boolean formalism can capture many dynamic properties of GRNs.

As it appears today, exploration of a complex system, such as the self-renewal gene network in embryonic stem cells, with largely unknown gene interactions, may be successfully modeled by starting from SINs, such as Boolean networks. Critical network domains and kernels, crystallized from these preliminary studies, may be analyzed in detail, using steady-state or dynamic BRNs, as long as the required biological knowledge is available for the interacting genes.

Deterministic and Stochastic View of Self-Renewal and Differentiation

One of the general key features shared by both embryonic and adult stem cell types [32] is their ability to self-renew. It is believed that self-renewal and pluripotency depend on a relatively small set of gene network circuits (kernels) connecting few transcriptional regulators. It is assumed that these core transcriptional circuits occupy the top position in the regulatory hierarchy within a stem cell of any kind and regulate expression of thousands genes, which actually define the self-renewal state [5,33,34]. In the case of embryonic stem cells (ESCs), the regulatory kernel involves the transcription factors Nanog, Oct4 and Sox2 [35,36]. In the case of adult bone marrow hematopoietic stem cells (HSCs), the potential candidate regulators include Fli1, Scl, Gata2 and few other transcription factors [37]. Figure 22.1 shows a suggested design for the two core stem cell gene regulatory networks from vertebrates and a retina determination network from flies [38–40]; all three

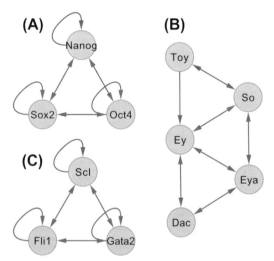

FIGURE 22.1 Functional redundancy in the core regulatory networks. (A) Shows a minimal gene network for mouse embryonic stem cell. (B) Retina determination core network from flies. A network with identical elements: Six1 (So)–Eya2 (Eya)–Dach2 (Dac) is involved in the determination of muscle cells in vertebrates. (C) The suggested minimal core network for hematopoietic stem cells. All three networks share the same network motif, a fully connected triad. Such a design can provide substantial functional redundancy to the master regulator networks in general. While the actual gene regulatory networks underlying pluripotency or hematopoiesis are much larger, the core elements appear to represent central, relatively independent kernels.

master regulator networks share a common network motif — a fully connected triad, suggesting functional redundancy across species and cell types. Despite variations in gene expression in individual cells [20,41—43], the topology and molecular interactions in the core networks ensure the presence of stable states or *attractors* encompassing the self-renewal state. In this context, attractors are sets of concentrations of the core transcription factors, towards which the concentrations evolve over time. In the case of embryonic stem cells, two or even more attractors apparently correspond to the pluripotent state [20,21], since pluripotent cells expressing both high and low concentrations of major pluripotency regulators, such as Nanog, have been detected [20,21,44]. Theoretical studies of network motifs such as the fully connected triad shown in Figure 22.1 [45], with topology similar to the core pluripotency network and few published quantitative models, largely support the presence of more than one 'self-renewal attractor' in ESCs [16,21,22].

The described studies allow the narrowing down of the general question, 'What is self-renewal and pluripotency?' to specific questions of the reconstruction of stem cell regulatory gene networks and identifying and explaining attractor states for the most critical network domains.

The core regulatory networks are placed in the context of much broader arrays of genes, rendering vital stem cell functions. One such function is susceptibility and response to external signals. Figure 22.2 summarizes major pathways of signaling found in the embryonic pluripotency gene network. LIF signaling is mediated by members of the Klf family of transcription factors (Klf4 is essential for maintenance of induced pluripotent stem cells, iPSC) [46,47]. In turn, the expression levels of Klfs are positively regulated by the core pluripotency factors Nanog, Oct4 and Sox2 [48]. The transcriptional repressor Tcf3 represents another interesting example of interlocking between extrinsic signaling pathways and the core pluripotency transcriptional network. A target and negative regulator of the core pluripotency factors, Tcf3 can also mediate the Wnt signaling pathway, which has been shown to promote pluripotency [49—53]. In this sense, the position of Tcf3 in the hierarchy of pluripotency networks resembles that of Klf4, but it remains to be determined whether Klf4 and Tcf3 are 'true' integral parts of the (extended) core pluripotency circuit. Embryonic stem cells are able to maintain their pluripotent state even in the absence of external signals, required for culturing (LIF, BMP) but for a limited time only [53]; external signaling becomes completely dispensable if cell proliferation suppressors Erk1/2 and Gsk are inhibited [22,54]; in this sense the network is self-sustainable. Functional redundancy and robustness of the network are supported by the observation that certain mouse ESCs can 'adapt' in culture to remain pluripotent

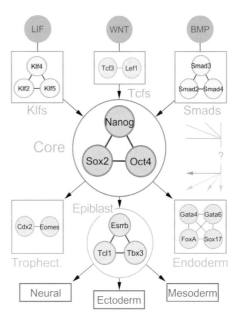

FIGURE 22.2 **Signaling pathways in mouse embryonic stem cells.** Links between the major signaling pathways and the core pluripotency network are shown. Domains with green nodes correspond to 'sensory' elements, mediating extrinsic signals; domains with yellow nodes are 'executive' elements. Arrows on the right side of the core circuit show possible feedbacks, bypasses and parallel processing. Direction of links displays information flow rather than direction of actual interactions between the genes. Circuits mediating signaling (such as Klfs or Smads, Gatas) have connectivity features apparently similar to those of the core pluripotency circuit, which may also serve to achieve functional redundancy at each level of the signaling cascade.

even in the absence of some of the major core transcription factors, such as Nanog, which highlights the flexibility and robustness of the pluripotency network [54].

Many network components mediate signaling and differentiation pathways (Figure 22.2) and affect the dynamic properties of the core network circuit; it is believed that these inputs, along with the core factors Nanog, Oct4 and Sox2, contribute to the semi-stable states described via dynamic attractors (see above). How to integrate all these inputs, given the high complexity of the extended pluripotency gene network, is an important emerging problem.

During the early differentiation process the semi-stable self-renewal states undergo transition and the stem cells differentiate. One current view is that variation in gene expression and the consequent existence of attractor states may be the reason behind this transition. The hypothesis emphasizes the role of stochastic gene expression in cell fate determination, so many current models explaining stem cell behavior include both deterministic and stochastic components [18,20—22,42]. Indeed, it has been found that concentrations of the core factors measured in individual pluripotent ES cells form statistical distributions featuring local peaks (bimodality), possibly corresponding to dynamical semi-stable states or pluripotent attractors [20,21,43,55]. In extreme cases, stochastic variations in gene expression may lead to a loss of one or more of the key self-renewal factors. The remaining skewed and/or incomplete combinations of the core factors may result in commitment to a specific cell lineage [10,55]. For instance, alternative expression of Oct4 or Sox2 in mouse ESCs may result in commitment of cells to either mesendodermal (ME, Oct4) or neuroectodermal (NE, Sox2) fates [22, 56, 57].

Major efforts in the stem cell field are currently focused on finding optimal factor combinations or 'cocktails' that will ensure stem cell differentiation towards desired cell types or tissues [1,58–60]. Hopefully, future quantitative biological models explaining stem cell behavior will help to carry out these studies in a more systematic and predictive manner.

EXAMPLES OF QUANTITATIVE MODELS EXPLAINING STEM CELL BEHAVIOR

Models for Embryonic Stem Cells

Embryonic stem cells are becoming a major subject for quantitative modeling owing to the relatively high level of understanding of this particular system. Early models focused on the structure of the core pluripotency network and transitions between alternative pluripotent states, corresponding to dynamic attractors (see Figure 22.1A). Three potential scenarios were revealed, all producing bimodal statistical distributions for Nanog concentrations, observed in ESC culture in vivo [43]. In the first scenario, highly cooperative interactions between the core components produced a classic bistable switch with two attractors, the first corresponding to the high and the second corresponding to the low concentration of Nanog (see Figure 22.3A) [20,21]. It has been assumed that stochastic variations in gene expression may lead to transition between the states without loss of pluripotency [18]. Addition of a transcriptional repressor to the core pluripotency network produced a second scenario, oscillation (Figure 22.3B) [21]. Interestingly, in the context of the oscillator model, no stochastic noise is actually required to achieve transition between the 'high' and the 'low' states. The addition of inhibition to the core circuit may reflect known antagonistic interactions between the transcriptional regulators Cdx2 and Oct4 [61], Zfp281 and Nanog [62] or Tcf3 and Nanog [63]. Some of the core factors, such as Oct4, may be involved in both activation and repression [3]; the dual functions of these pluripotency factors lead to a third possible scenario, involving mutual repression (Figure 22.3C)[64]. While the unidirectional repressive interactions considered above may produce oscillatory behavior of the system, mutual repression results in a scenario similar to the classic phage lambda switch [65], but rather different from the cooperative switch considered above.

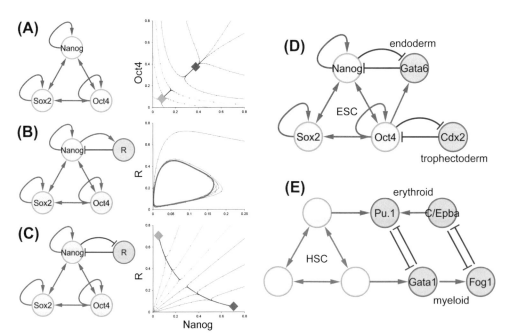

FIGURE 22.3 **Known models and scenarios for embryonic and hematopoietic stem cells.** (A–C) Models for pluripotency in ESC. Dynamic solutions (phase space) are shown on the right side of each panel. Cooperative interactions between the core factors (A) may produce a bistable system with two point attractors; unilateral repression, combined with self activation produces oscillatory behavior (limit cycle attractor, B); mutual repression results in bistability again (C), but with different distribution of the point attractors. (D) Model describing alternative cell fate commitment in ESC. Two loosely linked bistable switches ensure the fate choice between endoderm (Gata6 expression) and trophectoderm (Cdx2 expression). (E) Two parallel bistable switches reinforce each other and govern alternative commitment of hematopoietic progenitors (here HSC) to erythroid or myeloid cell fates.

Differentiation of embryonic stem cells became a subject of quantitative exploration as well [16]. Mutual repression between Cdx2 and Oct4 has been suggested to regulate differentiation to the trophectoderm lineage, while mutual repression between Gata6 and Nanog regulates differentiation to the endoderm lineage (see Figure 22.3D). One can see that the model integrates a pair of bistable switches, each combining a single pluripotency factor and a single lineage commitment regulator [19] along with bimodal distributions for the core factor concentrations. The model predicted a quite interesting concentration-dependent ('bell-shaped') response of Gata6 to the concentration of Oct4, resulting from direct activation and indirect repression of Gata6 by Oct4. Analogous concentration-dependent responses are known in other well-studied systems, such as *Drosophila* embryogenesis [66]. Mutual exclusion between the alternative cell fates may be achieved, in part, via Oct4: activation of Cdx2 (trophectoderm primed cells) causes Oct4 repression and prevents Oct4-mediated activation of Gata6, thus blocking endodermal cell fate. Interestingly, the explored network topology and the model suggested that Nanog overexpression, not suppression of the lineage-specific factor Gata6, is the optimal way of reprogramming endodermic cells into iPS. Inferring differentiation and reprogramming conditions from future quantitative models may potentially supplement or even replace empirical studies in the reprogramming field.

Models for Hematopoietic Stem Cells

Hematopoietic stem cells (HSCs) are undifferentiated cells that give rise to all types of blood lineage progenitors and terminally differentiated blood cells. HSCs are found in myeloid tissues such as red bone marrow with frequency $\sim 10^{-4}$, relative to other cell types. Unlike with embryonic stem cells, HSC culturing is not yet attainable, and purification for medical applications is inefficient and very expensive.

A large proportion of HSCs in mouse are present in a non-proliferating or quiescent (dormant) state. Under native conditions with a very low frequency HSCs may enter the cell cycle, which typically leads to irreversible differentiation of HSCs to blood lineage progenitors. In this respect it is still not quite clear whether the dormant HSC state is analogous to the pluripotent ESC state. Recently, an architecture for the core HSC network, presumably responsible for transition between the dormant and differentiated states, has been suggested (see Figure 22.1B) [37]. Models describing this transition are not yet available, with the exception of quantitative analysis of HSC pool exhaustion and cellular aging due to proliferation [67,68].

One well-studied model for hematopoiesis focuses on a switch governing differentiation of HSCs into erythroid or myeloid lineages [69]. The switch is mediated by mutual repression between the transcription factors Pu.1 and Gata1 (see Figure 22.3E) [70]. In this case, however, the bistable switch is placed directly between these factors, responsible for the alternative cell fate commitment, and it is also reinforced by yet another parallel switch downstream (C/Epbα—Fog1).

One can see that in both considered cases (see differentiation of ESC above) differentiation events are typically associated with mutual repression; however, the bistable switches occupy quite different positions relative to each other and to the core network. In the case of ESCs, their communication is indirect and much 'softer'.

A simple three-gene model of hematopoiesis utilizing Boolean networks has also been proposed [71]. In Boolean terms, the stem cell genes should be 'ON' and the differentiation genes 'OFF' in the dormant state; the genes switch to opposite states when the cells are fully committed. Under the Boolean network framework, a hematopoietic stem cell represented by three genes can fall into either a 'rest' or a 'cycling' attractor state according to a set of rules governing the three genes. Interestingly, this may correspond to switching between the bistable and the oscillatory scenarios, described for ESCs above (see Figure 22.3A and B).

STRATEGIES FOR MODEL CONSTRUCTION AND VALIDATION

Data Integration and Network Construction

The functional properties of various biological systems can be uncovered by inferring the architecture of biological networks and modeling GRN dynamics using either reaction-based (BRN) or statistical-based (SIN) models (Figure 22.4). With the advent of high-throughput biotechnologies, efforts have been made to map and reverse-engineer the GRNs of stem cells. The past few years have witnessed the development of resources for stem-cell-centered networks and databases for the broad stem cell community. A number of regulatory interactions are deposited in network-based repositories such as PluriNetWork [72], Plurinet [73] and iScMiD [36]. Additionally, WikiPathways provides the CIRW portal that highlights pathways contributed to and maintained by the stem cell community [74]. In the context of transcriptome data, databases such as FunGenES and StemBase contain gene expression studies and others with interactive query tools and a web interface [75,76]. In addition to gene expression data, genome-wide protein—gene-binding ChIP-seq/chip data in stem cells are collected into the public domain in NCBI's GEO [77] database, while databases such as ChEA [78] and ESCDb [79] collect a number of processed ChIP-seq/chip results and allow users to query for genes of

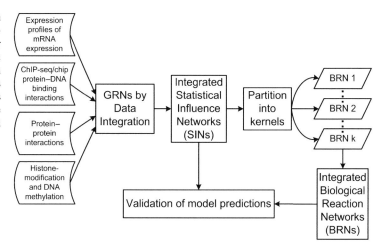

FIGURE 22.4 **Data integration and model construction strategies.** Flow charts show possible ways of data integration, construction of SIN and BRN-based models and their relationships. While the construction of SINs, such as Boolean networks and models, requires data integration, construction of BRNs requires preliminary analysis of the networks. BRNs display network modularity; simplified to the level of nodes with known input/output characteristics, BRNs may be consequently 'integrated back' into larger networks and models.

interest. Other miscellaneous databases include StemDB (http://www.stemdb.org), which holds stem cell-related information (mRNA expression profiles, antibodies, primers and protocols).

New provisional gene networks for mouse ESCs are constantly being generated and updated [14,15,20,34,36,72]. However, these networks are often focused on various specific aspects of mESC biology, constructed from data obtained under different conditions (e.g., different knockdowns or differentiation conditions) and carry substantial disagreements. Similarly designed genome-wide studies of the core pluripotency factors protein—protein interactions (interactomes) [80,81] may produce only 20—40% of true positives that are consistent between the studies [82]. Comparisons of large numbers of genome-wide datasets may produce even weaker agreements.

On account of the enormous amount, diversity and ambiguity of currently available data, one prominent goal is to integrate various kinds of genome-wide data across multiple regulatory layers in order to build more consistent GRNs and predictive quantitative models.

Simple Binary Models for Complex Gene Networks

Abstract information regarding regulatory interactions between genes can be approximated by rather simple Boolean on/off switches; Boolean networks are among the common computational approaches to model the dynamics of the GRNs [28,31,83,84]. A Boolean network can be represented by a directed graph $G(V, F)$, as in Figure 22.5A. G is defined as a set of vertices (nodes) $V = \{x_1, ..., x_n\}$ connected by a set of Boolean functions $F = \{f_1, ..., f_n\}$. Under the Boolean framework, each variable can take binary values of 1 representing the 'ON' or active state, or 0 representing the 'OFF' or inactive state. The state of each node at time t is determined by the value of other nodes in the previous time step according to a list of Boolean functions $F = \{f_1, ..., f_n\}$. A Boolean function f_k is a logical function operating on the values of upstream nodes regulating the activity of node k via Boolean operators such as 'AND', 'OR' and 'NOT'. At each time step the nodes can be updated in a synchronous or asynchronous manner. Synchronous updating simply assumes that the reactions in the network G have similar timescales. Therefore, a node is updated after all rules have been applied and all nodes are updated simultaneously; alternatively, asynchronous updating takes temporal ordering into account and can be categorized as either undeterministic, stochastic asynchronous, such as randoming order updates [85], or deterministic updating [86]. The total number of states of a network is finite (2^N for a network G consisting of N nodes). For synchronous and deterministic asynchronous updating, the system typically arrives at steady states relatively quickly characterized by either a limited cycle or a fixed-point attractor. In contrast, states outside the attractors are transient and unstable. For each attractor, the set of all transient states leading to that attractor constitute the basin of attraction similar to the behavior of continuous variables dynamical systems (see Figure 22.5C,D).

Boolean network modeling can capture the collective behavior of sophisticated regulatory networks and has been applied to explore several complex biological systems. A Boolean network has been developed to simulate the yeast cell cycle and to predict cell cycle events [87]. A probabilistic Boolean network has been successfully applied to analyze the dynamical behavior of a subnetwork consisting of 15 genes in human glioma [88]. Based on the analysis of joint steady-state probabilities for Tie-2, NFκB and TGF-β3, NFκB, the model predicted function for Tie-2, a receptor tyrosine kinase involved in tumor development. Frequently, Boolean networks are used to infer the underlying structure of GRNs from high-throughput time series of microarray data [89—92]. The REVEAL algorithm developed by Liang was among the first formalism to infer Boolean model structure. The algorithm uses mutual information to determine the

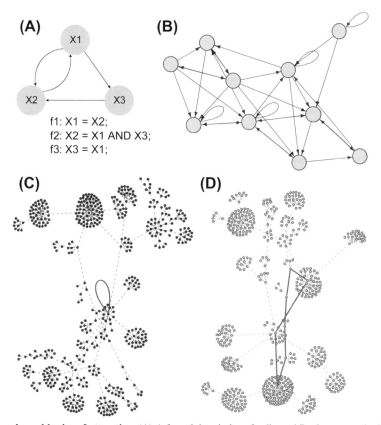

FIGURE 22.5 **Boolean networks and basins of attraction.** (A) A formal description of a directed Boolean network; three Boolean functions (*f*1-*f*3) link the network nodes. (B–D) Examples of Boolean networks and their dynamical solutions. (C) A basin for a single point attractor. The network nodes correspond to vectors of on/off node states. Red edge (loop) corresponds to a transition between the attractor states (the same state). (D) A basin for a cycle attractor. The red edges show transition between the consecutive attractor states. These cyclic attractors were observed in networks describing periodic biological processes, such as the cell cycle.

dependency and regulation among nodes based on state transition tables corresponding to time series of gene expression patterns. As a result, a minimal set of inputs is identified that can uniquely determine the output at the next time point for each variable in the network. In practice, a semi-quantitative approach has been developed to infer the regulatory interactions in differentiating embryonic stem cells from gene expression data. The method combines the Boolean updating framework with internal continuous expression levels [93]. The resulting optimized network revealed a hierarchical structure with Oct4, Nodal and E-cadherin on top and regulatory flow to Foxa2 through Oct4.

To facilitate the construction of BNs, there are several software packages available for (re)constructing, simulating and visualizing Boolean networks (Table 22.1). For example, general toolboxes such as the RBN toolbox, the PBN toolbox [30] and the CellNetOptimizer [94] are publicly available and can be used in programming under the Matlab environment. Additionally, CellNetOptimizer can be downloaded as a Bioconductor package in R. Other interactive graphics tools such as NetBuilder [95] and DDlab [96] allow users to create, visualize and simulate genetic regulatory networks, including discrete Boolean networks. Additionally, BooleanNet [97] is a tool used to simulate gene regulatory networks in a Boolean formalism; similarly, the R package BoolNet [98] can generate, simulate and reconstruct Boolean networks with support for three types of BN: synchronous, asynchronous and probabilistic. It also integrates with existing visualization tools such as Pajek [99] and BioTapestry [100].

Modular Network Design: from Kernels to Integrated Models

Self-renewal or differentiating cell states are characterized by specific molecular signatures involving protein, mRNA and miRNA concentrations as well as numerous specific genetic and epigenetic interactions [15,59,101,102]. In turn, these genome-wide changes are believed to be under the control of small genetic networks (kernels) at the top of a gene regulatory hierarchy. Known or suggested kernels are shown in Figure 22.1 for ESC and HSC. In the case of pluripotent stem cells, the emerging core circuit (kernel) may include three to six transcription factors (Nanog, Oct4, Sox2, Klf4, Myc, Tcf3) [20,34,35,103,104]. This particular

TABLE 22.1 Software for Network Construction and Network Modeling

Name	Source	Implementation environment
BNM	www.rustyspigot.com/software/BooleanNetwork/?url=/software/BooleanNetwork	Standalone
BooleanNet	code.google.com/p/booleannet/	Python
BoolNet	cran.r-project.org/web/packages/BoolNet/	R/Bioconductor
CellNetAnalyzer	www.mpi-magdeburg.mpg.de/projects/cna/cna.html	Matlab
CellNetOptimizer	sites.google.com/site/saezrodriguez/software/cellnetoptimizer	Matlab; R/Bioconductor
DDlab	www.ddlab.com/	Standalone
DVD v1	dvd.vbi.vt.edu/cgi-bin/git/dvd.pl	Web-based
Learnboo	www.maayanlab.net/ESCAPE	Matlab
NetBuilder	strc.herts.ac.uk/bio/maria/NetBuilder/	Standalone
Odefy	www.helmholtz-muenchen.de/cmb/odefy	Matlab
PBN toolbox	code.google.com/p/pbn-matlab-toolbox/	Matlab
RBN toolbox	www.teuscher.ch/rbntoolbox/	Matlab
RBNLab	sourceforge.net/projects/rbn/	Java

network kernel, however, is a part of a larger (extended) regulatory network, which includes a broader range of regulators (such as Dax1, Rex1, Tbx3, Esrrb, Klf5) [34,50,105,106] (shown in Figure 22.2).

Analyses of the published studies and available networks suggest the presence of at least two types of network domain linked to the core pluripotency circuit: (1) upstream 'sensory' network elements related to interpretation of extrinsic signals and communication with the core pluripotency circuit; and (2) downstream 'executive' elements that interpret states of the core pluripotency network and transmit this information to epigenetic modifiers and downstream lineage commitment genes. Interestingly, 'sensory' and 'executive' domains as well as the core networks are represented by several transcription factors with redundant properties, such as Smads or Klfs. This network shows an emerging layout for partitioning the large network into smaller units, kernels and plug-ins, which can be utilized for the construction of independent biochemical reaction network (BRN) and models. Consequently, such robust elements with known input/output characteristics may be integrated into a large model explaining the properties of the entire pluripotency network, in a way similar to the assembly of computer parts on a motherboard (see Figure 22.4). Straightforward construction of BRN networks from integrated data (such as co-expression networks) is difficult, as BRNs involve multiple unknown parameters corresponding to binding constants, synthesis and degradation rates. Accounting for all these parameters typically results in high levels of ambiguity, so the preliminary Boolean network analysis and partition into independent kernels seems to be a necessary step. In addition, kernels may correspond to different molecular mechanisms, such as transcriptional regulation or signal transduction, and combining such diverse chemical reactions with different kinds of chemical constants is yet another problem.

A future integrated model, created from the small robust kernels, should predict diverse outcomes, such as alternative differentiation to endoderm/trophectoderm lineages in response to Gata6/Cdx2 balance (see Figure 22.3D) or alternative differentiation to mesendoderm/neuroectoderm fate in response to Oct4/Sox2 balance [22,56,57] or suppression of commitment to a neural fate in response to BMP signaling [6,107,108].

Role of Transcription Regulatory Signals and Transcriptional Gene Networks

Sometimes, molecular mechanisms of gene interactions are difficult to capture by the Boolean network models described above. For instance, high or low concentrations of Oct4 trigger differentiation, whereas moderate Oct4 levels promote pluripotency [109,110]. This rich response by Oct4 target genes resembles the concentration-dependent response of Hunchback (Hb) target genes in *Drosophila* development [66,111–113]. Strikingly, in the fly development, graded responses to upstream regulators are the key feature, important for embryo development and morphological specification [114,115]. Boolean models are not capable of capturing concentration-dependent responses; therefore, certain biological systems often require biological reaction networks (BRN) and models.

Core networks, determining key self-renewal and differentiation decisions, are transcriptional; for this reason, BRNs describing stem cell behavior often rely on transcriptional regulation. Steady-state models for transcriptional regulation typically take into account a single step in gene regulation, binding of transcription factors to their sites in gene regulatory regions, promoters and enhancers [116]. Such site occupancy models, or fractional occupancy models (FOM), were successfully used in many biological systems, including phages, bacteria, yeast and flies [26,65,117,118]; however, their distribution in the stem cell field is currently limited by insufficient knowledge regarding the pluripotency gene's control regions. Thus, binding sites for the core pluripotency factors (Oct4, Nanog, Sox) were observed in the upstream regions of many potential targets, but only a single structured element, the Oct4-Sox2 sequence element, has been identified so far. Binding sites for Oct4 and Sox2 are found in close proximity in many genes, suggesting the formation of heterodimers between these two critical pluripotency factors [56,119]. Potential cooperativity between interacting factors may change their binding properties and the responses of their target genes, thereby producing new dynamic solutions or cell states (cell lineages). It is quite possible that such signal integration contributes to the alternative lineage commitment in the case of alternative expression of Oct4 vs. Sox2 or concentration-dependent response of the genes to Oct4.

Mapping transcription regulatory signals and binding sites in gene control regions essential in self-renewal or differentiation serves (1) to establish directed and signed GRNs, and (2) to provide additional data (along with gene expression) for model validation. Steady-state FOMs or dynamic models based on ordinary and stochastic differential equations (ODEs or SDEs, respectively) require many rate constants, such as affinities of binding sites, to be known or at least estimated in order to reduce the space of potential solutions and the level of model ambiguity. Integration of several data types may produce highly valuable maps for distribution of binding motifs and the binding motif combinations in promoters and enhancers. Most important data sources include (1) chromatin immunoprecipitation followed by deep sequencing or microarrays (ChIP-chip or ChIP-seq); (2) evolutionary conservation of DNA in the gene control regions; (3) bioinformatics analyses of the control regions for the presence of binding motif matches; and (4) finding enhancers specifically activated in stem cells. ChIP data relevant to stem cells are widely available in dedicated databases, such as CHEA [78], but the method itself suffers from high false positive rates and its current resolution is not sufficient for mapping single binding sites [120]. Preliminary filtering of ChIP-X data based on data regarding chromatin structure [121,122] may be a solution. Methods based on genome alignments, such as the PhastCons program [123,124], help to narrow down the regions indicated by ChIP-seq. Mapping binding motif matches using bioinformatics analyses of promoter and enhancer DNA sequences is useful, but also sensitive to the quality of the binding motif models (typically alignments of known binding sites for a given transcription factor) [125]. Finally, enhancer and promoter DNA sequences, specifically activated in stem cells, may be identified in independent genome-wide studies in vivo [126]. These types of data have been proved to be extremely valuable in the analysis of transcription regulatory signals [127,128], but are not widely available in the stem cell field.

Dynamic Biological Reaction Model for the Core Pluripotency Network

A *deterministic* dynamical model for the minimal core pluripotency network, shown in Figure 22.6A, has been expressed using a system of three ordinary differential equations (ODE) involving fractional site occupancy (BRN) under the synthesis terms (x=Nanog, Oct4, Sox2):

$$\frac{d[x]}{dt} = \alpha(1 - (1 - P_{Nanog})(1 - P_{Oct-Sox})) - \beta[x] \quad (1)$$

Nanog may bind its target genes as a homodimer, which can be adequately described in terms of the fractional site occupancy (FOM model) describing cooperativity between Nanog monomers C_{N-N} and Nanog-DNA binding K_N. The binding constants in FOMs are sometimes expressed via true thermodynamic terms ($K = \exp(-\Delta G/RT)$). The same framework may serve to describe the formation of Oct4-Sox2 heterodimers, where the heterodimers are virtually present only when bound to bipartite sites and heterotypic cooperative interactions are expressed using cooperativity C_{O-S} between the Sox2 (S) and Oct4 (O) monomers:

$$P_{Oct-Sox} = \frac{K_s[S] + K_o[O] + C_{o-s}K_s[S]K_o[O]}{1 + K_s[S] + K_o[O] + C_{o-s}K_s[S]K_o[O]} \quad (2)$$

Expansion of the model for the core pluripotency network may be necessary for several reasons. For instance, it has been suggested that negative feedback plays a critical role in the function of the core circuit. Indeed, some core factor target genes, such as Tcf3, encode transcriptional repressors, which in turn target the core factors. Second, it has been found that different concentrations (threshold levels) of some core pluripotency factors (Oct4) may induce alternative differentiation events. Such threshold responses are quite common in developmental and differentiation pathways. Sometimes they are mediated by

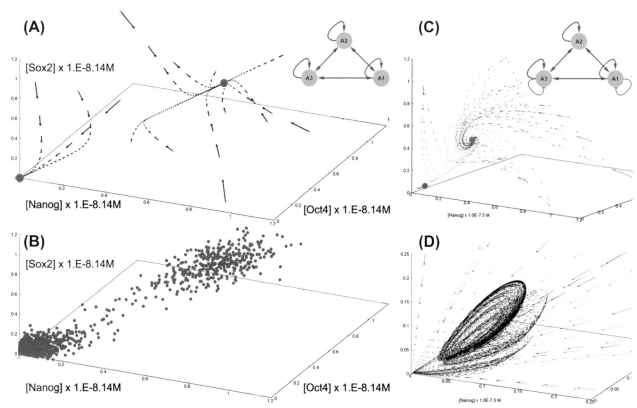

FIGURE 22.6 **Bistable and oscillating scenarios for the core pluripotency networks.** (A) Phase space of the model for three core pluripotency regulators linked into fully connected triad (see Eqs 1–3), the system is bistable at the selected parameter ranges; the two dynamic attractors are shown as red dots. (B) The same phase space after incorporation of stochastic noise into the model. Dots correspond to cell states with respect to Nanog/Oct4/Sox2 concentrations. (C, D) Dynamic solutions for the same triad network, but with additional repressor links, corresponding, for instance, to repression of Oct4 by Tcf3, and repression of Nanog by Zfp281. The presence of repressors cause oscillations of Nanog, Oct4 and Sox2 concentrations. Depending on the parameter setting, the system can either maintain its bistability (C, damped oscillator) or become a true oscillator with a limit cycle attractor type (D). Red dots on D (finite simulation states) display the attractor states.

concentration-dependent action (dual regulation) of dual or alternative transcriptional regulators.

Another important expansion of the model is the incorporation of stochastic terms, necessary to describe stochastic switching between metastable system states or dynamic attractors. In our previous work we employed a model for Nanog based on stochastic differential equations (SDE). Accordingly, the stochastic term is included in the model:

$$\frac{d[x]}{dt} = \alpha(1 - (1 - P_{Nanog})(1 - P_{Oct-Sox})) - \beta[x] + g(x)\xi(0, \sigma) \quad (3)$$

A system of three SDEs for the core network components (x=Nanog, Oct4, Sox2), such as in Eq. 3 (representative of one factor) incorporates both *deterministic and stochastic* terms and will serve as the initial point in the exploration of stochastic switches between metastable states. The stochastic term in Eq. 3 incorporates a concentration-dependent function $g(x)$, which is often approximated as $\sqrt{x}$ and the random variable ξ, which is the zero mean Gaussian 'white noise'. Such a stochastic term is general, as it specifies no particular source of noise in the system (Figure 22.6B). At the same time, at least two potential sources of noise in transcriptional networks are well known. First, the most obvious source is mRNA degradation, which follows exponential decay. At the level of the cell population this will produce a Gaussian distribution for the number of mRNA molecules per cell. Other sources of noise are related to mRNA synthesis because the initiation of mRNA synthesis is inherently stochastic owing to variations in the times required for the assembly of large eukaryotic transcription initiation complexes. Initiation complexes may also need to interact with upstream enhancers, which is yet another stochastic process.

The stochastic distributions for gene activity levels (such as shown in the Figure 22.6B) fit well with the actual experimental data obtained from cell sorting or cell imaging. Bistable models can be fitted to the intensity distribution data by adjusting three to four parameters, including global binding affinities and noise levels.

Extensions of the simple cooperativity-based bistable model described by Eqs 1 and 3 may include a unilateral repression of the core circuit's members by one or more repressors (see scenario in Figure 22.3B). In the context of the core network such unilateral repression results in oscillatory behavior. In the case of two or more repressors their oscillations may or may not synchronize, depending on the relative strength of linkage between the core factors involved in the unilateral repression (see Figure 22.6C, D). This supports the possibility for strange attractor-like steady states and/or limit cycles for the core pluripotency network.

Model Validation and Overfitting

A simple way to validate quantitative models is to perform a series of deletion or over-expression in silico to test the consistency between predictive behavior from the quantitative simulations and biological outcomes from experimental gene perturbations. For example, in the Boolean formalism the predictive results could be steady states/attractors of a network after in-silico simulations of gene knockdown by forcing that gene to be constantly 'off'. Gene expression pattern of the steady states after such perturbation can then be examined by RNA interference-mediated gene knockdown followed by real-time PCR of the network genes. Approximate distribution of dynamic attractors and the presence of limit cycles is a more challenging problem. Mapping phase spaces such as those shown in Figure 22.6 may be explored based on single-cell gene expression data. A major experimental challenge here is to obtain reliable protein level readouts in single cells for as many network components as possible. Tools available for such measurements include transgenic cell lines expressing GFP under the control of regulatory elements for the core pluripotency factors Nanog or Oct4 [10]. These cell lines provide the power to measure gene expression in live cells and record gene expression dynamics. However, these measurements can be done in a single channel only for either Nanog or Oct4. Routine immunostaining in combination with flow cytometry will allow an increase in the number of channels to three or four (1 − transgenic Nanog + 2−3 channels for immunostaning) [129]. This number of channels will entirely satisfy elementary models, such as the core circuit model (see preliminary results), but will be insufficient for larger models. Methods based on high-throughput real-time PCR assays (BioMark System, Fluidigm Corporation) are able to deliver readings for up to 96 genes (mRNA) in a single cell, but the number of cells (96) in this method requires the use of multiple chips to achieve statistically sound cell numbers. Recently emerging mass cytometry-based methods [130] read the stable lanthanide isotope tags attached to antibodies using metal-chelating labeling reagents, and allow reading of 30 channels (proteins) and more in a large number of cells.

Improper balance between the number of open (unknown) parameters in the system and the amount of available experimental data (such as single cell readings) may be a serious issue. If few data are available, overwhelming numbers of parameters may result in data overfitting, when the available data fit almost any model. The presence of overfitting may be displayed by deletion or over-expression in silico, as described above. In the case of BRN-based models, the number of parameters may be reduced by learning the relative values of binding constants from promoter and enhancer DNA sequences (see above) or using global parameters for most of the reaction components. For instance, degradation and synthesis constants as well as constants for gene expression noise may be considered identical for all network components regulated in a similar manner. Indeed, biological networks typically have sufficient robustness to tolerate significant changes in component concentrations. The described strategies may help to minimize the number of open parameters to three to five in the case of models for single kernels (such as the core circuit alone), and 10−20 in the case of larger, integrated networks [113].

INFORMATION FLOW AND EPIGENETIC LANDSCAPES IN DIFFERENTIATION

Information Flow and Epigenetic Memory

One interesting problem emerging from quantitative analysis of stem cell gene networks is informational flow in the system during self-renewal and differentiation. The information content of any pluripotent cell state and each transient or terminally differentiated state can be characterized, for instance, by a genome-wide set of gene expression levels. This rather trivial representation is supported by frequently observed specific 'molecular signatures' characterizing pluripotent and differentiated cell types in microarray studies [14]. During transition between two (rather arbitrary defined) neighboring cell states n and $n+1$, genes specific to $n+1$ are activated, whereas some genes specific to n are shut down and some genes specific to n maintain or slightly change their activity levels (Figure 22.7A). One can see that the new information, required to achieve state $n+1$ from state n, has been read from the genome. Intuitively, the amount of new information may not exceed the amount of binary information contained in the DNA encoding the newly activated genes, with the account of their transcription regulatory regions as well. In the case of differentiation, the genome plays a role similar to that of a storage device on a computer, such as a CD or 'read only memory' (ROM). Interestingly, epigenetic information (genes which are still active, DNA methylation, chromatin modification states and so on) from state n to state $n+1$ may be considered as temporary memory or, by analogy with computers, 'random access memory – RAM'. The epigenetic memory provides means for self-renewal, where a given number of active network components fluctuates between different semi-stable states.

Finding cell-specific markers or tissue-specific transcriptional regulators is, in fact, tracing the emergence of new information in differentiating cells. Separation of the new, specific information from epigenetic memory may help focus on the right candidate genes and reduce the complexity of gene networks and the network-based quantitative models.

Waddington Landscapes and Attractor States

Stem cells are capable of self-renewal and of differentiating into terminal cell types. Waddington [131] described the cells in a metaphorical way as marbles rolling down an epigenetic landscape containing 'hills' and 'valleys' during the development process: the 'valleys' represent cell types separated by the 'hills'. From the network biology perspective, cells settled in the 'valleys' are in the stable states, i.e., fixed-point attractors. Alternatively, cells oscillating around multiple 'valleys' are unstable. Different stable or transitive states are defined by gene expression pattern and epigenetic status. Huang et al. provided the first experimental evidence that cell types can be represented by high-dimensional attractor states [132]. Specifically, they differentiated human leukemia cells (HL60) into neutrophils by two different stimuli. Using gene expression microarray, they showed that HL60 adopted two different trajectories of differentiation which finally converged into the same 'end program' attractor state under the two conditions. Random Boolean networks (RBNs) were applied as a simple but generic model to mimic cell fate commitment back in the 1960s [84]. When each node in a random network had fixed number, K, of input nodes (parents), the network displayed interesting dynamics. For $K=2$, the network exhibited 'critical' dynamics, which is the borderline between chaotic and ordered states. Apparently, the model captured the generic properties of biological gene regulatory networks corresponding to 'critical' dynamics, which seems to reflect an equilibrium between adaptiveness and robustness. Therefore, instead of stable attractors, stem cells would be expected to reside in semi-stable states in order to prepare themselves for environmental changes, such as external differentiation signals. Sometimes, such semi-stable states on the borderline of chaos are called 'primed' states, poised for differentiation [42,133,134] (see Figure 22.7B). Apparently, cells may

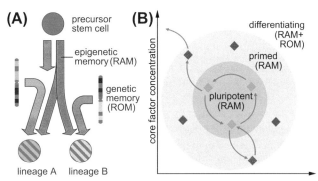

FIGURE 22.7 Information flow in pluripotency and differentiation. (A) Epigenetic memory is a set of active genes and gene products in a cell. This information is analogous to random access memory of computers (RAM). Upon differentiation, new information is added to the system with newly activated genes, this new information is retrieved from genome. The discrete genome information has analogy to read only memory of computer devices (ROM). Gene activation is shown by the green → red color gradient; shutdown of genes is shown by red → green gradient. (B) Pluripotent and primed pluripotent states, attractors (shown by the green and the red diamonds correspondingly) utilize epigenetic memory (RAM), though the already present genes may by changing their expression levels. Exit to differentiating state (transitions between states are shown by arrows) typically results in massive activation of new genes, addition of new information (ROM) to the system. The diagram on (B) may be considered as a projection of multidimensional space (such as that shown in Figure 22.5) to a surface corresponding to concentrations of just two core pluripotency factors.

travel from one semi-stable state to another by means of molecular noise in gene expression (see Figure 22.6B), or expose cyclic behavior, described via limit cycles, damping oscillations or even strange attractors.

REFERENCES

[1] Takahashi K, Yamanaka S. Induction of pluripotent stem cells from mouse embryonic and adult fibroblast cultures by defined factors. Cell 2006;126(4):663–76.

[2] Carvajal-Vergara X, et al. Patient-specific induced pluripotent stem-cell-derived models of LEOPARD syndrome. Nature 2010;465(7299):808–12.

[3] Marikawa Y, et al. Dual roles of Oct4 in the maintenance of mouse P19 embryonal carcinoma cells: as negative regulator of Wnt/beta-catenin signaling and competence provider for Brachyury induction. Stem 2011;20(4):621–33.

[4] Hayden EC. Stem cells: the growing pains of pluripotency. Nature 2011;473(7347):272–4.

[5] Kim HD, et al. Transcriptional regulatory circuits: predicting numbers from alphabets. Science 2009;325(5939):429–32.

[6] Chiang PM, Wong PC. Differentiation of an embryonic stem cell to hemogenic endothelium by defined factors: essential role of bone morphogenetic protein 4. Development. 2011;138(13): 2833–43.

[7] Son EY, et al. Conversion of mouse and human fibroblasts into functional spinal motor neurons. Cell Stem Cell 2011; 9(3):205–18.

[8] Green MD, et al. Generation of anterior foregut endoderm from human embryonic and induced pluripotent stem cells. Nat Biotechnol 2011;29(3):267–72.

[9] Tavernier G, et al. Activation of pluripotency-associated genes in mouse embryonic fibroblasts by non-viral transfection with in vitro-derived mRNAs encoding Oct4, Sox2, Klf4 and cMyc. Biomaterials. 2012;33(2):412–7.

[10] Ivanova N, et al. Dissecting self-renewal in stem cells with RNA interference. Nature 2006;442(7102):533–8.

[11] Guan Y, et al. A genomewide functional network for the laboratory mouse. PLoS Comput Biol 2008;4(9):e1000165.

[12] Tang C, Drukker M. Potential barriers to therapeutics utilizing pluripotent cell derivatives: intrinsic immunogenicity of in vitro maintained and matured populations. Semin Immunopathol 2011;33(6):563–72.

[13] Leeb C, et al. New perspectives in stem cell research: beyond embryonic stem cells. Cell Prolif 2011;44(Suppl. 1):9–14.

[14] Lu R, et al. Systems-level dynamic analyses of fate change in murine embryonic stem cells. Nature 2009;462(7271):358–62.

[15] Markowetz F, et al. Mapping dynamic histone acetylation patterns to gene expression in nanog-depleted murine embryonic stem cells. PLoS Comput Biol 2010;6(12):e1001034.

[16] Chickarmane V, et al. Transcriptional dynamics of the embryonic stem cell switch. PLoS Comput Biol 2006;2(9):e123.

[17] Hoffmann M, et al. Noise-driven stem cell and progenitor population dynamics. PLoS One 2008;3(8):e2922.

[18] MacArthur BD, Please CP, Oreffo RO. Stochasticity and the molecular mechanisms of induced pluripotency. PLoS One 2008;3(8):e3086.

[19] Chickarmane V, Peterson C. A computational model for understanding stem cell, trophectoderm and endoderm lineage determination. PLoS One 2008;3(10):e3478.

[20] MacArthur BD, Ma'ayan A, Lemischka IR. Systems biology of stem cell fate and cellular reprogramming. Nat Rev Mol Cell Biol 2009;10(10):672–81.

[21] Glauche I, Herberg M, Roeder I. Nanog variability and pluripotency regulation of embryonic stem cells—insights from a mathematical model analysis. PLoS One 2010;5(6):e11238.

[22] Thomson M, et al. Pluripotency factors in embryonic stem cells regulate differentiation into germ layers. Cell 2011;145(6):875–89.

[23] Price ND, Shmulevich I. Biochemical and statistical network models for systems biology. Curr Opin Biotechnol 2007;18(4):365–70.

[24] Shmulevich I, Aitchison JD. Deterministic and stochastic models of genetic regulatory networks. Methods Enzymol 2009;467:335–56.

[25] Zinzen RP, et al. Computational models for neurogenic gene expression in the drosophila embryo. Curr Biol 2006;16(13):1358–65.

[26] Janssens H, et al. Quantitative and predictive model of transcriptional control of the *Drosophila melanogaster* even skipped gene. Nat Genet 2006;38(10):1159–65.

[27] Gertz J, Siggia ED, Cohen BA. Analysis of combinatorial cis-regulation in synthetic and genomic promoters. Nature 2009;457(7226):215–8.

[28] Kauffman SA. Metabolic stability and epigenesis in randomly constructed genetic nets. J Theor Biol 1969;22(3):437–67.

[29] Gershenson C. Classification of random boolean networks. In: Artificial life VIII, editor. Proceedings of the eight international conference on artificial life. Sydney, Australia: MIT Press; 2002.

[30] Shmulevich I, et al. Probabilistic boolean networks: a rule-based uncertainty model for gene regulatory networks. Bioinformatics 2002;18(2):261–74.

[31] de Jong H. Modeling and simulation of genetic regulatory systems: a literature review. J Comput Biol 2002;9(1):67–103.

[32] Armstrong L, et al. Editorial: our top 10 developments in stem cell biology over the last 30 years. Stem Cells 2012;30(1):2–9.

[33] Orkin SH, Zon LI. Hematopoiesis: an evolving paradigm for stem cell biology. Cell 2008;132(4):631–44.

[34] Kim J, et al. An extended transcriptional network for pluripotency of embryonic stem cells. Cell 2008;132(6):1049–61.

[35] Loh YH, et al. The Oct4 and Nanog transcription network regulates pluripotency in mouse embryonic stem cells. Nat Genet 2006;38(4):431–40.

[36] Xu H, et al. Toward a complete in silico, multi-layered embryonic stem cell regulatory network. Wiley Interdiscip Rev Syst Biol Med 2010;2(6):708–33.

[37] Pimanda JE, et al. Gata2, Fli1, and Scl form a recursively wired gene-regulatory circuit during early hematopoietic development. Proc Natl Acad Sci U S A. 2007;104(45):17692–7.

[38] Gehring WJ. New perspectives on eye development and the evolution of eyes and photoreceptors. J Hered. 2005;96(3):171–84.

[39] Silver SJ, Rebay I. Signaling circuitries in development: insights from the retinal determination gene network. Development 2005;132(1):3–13.

[40] Punch VG, Jones AE, Rudnicki MA. Transcriptional networks that regulate muscle stem cell function. Wiley Interdiscip Rev Syst Biol Med 2009;1(1):128–40.

[41] Cai L, Friedman N, Xie XS. Stochastic protein expression in individual cells at the single molecule level. Nature 2006;440(7082):358–62.

[42] Hanna J, et al. Direct cell reprogramming is a stochastic process amenable to acceleration. Nature 2009;462(7273):595–601.

[43] Kalmar T, et al. Regulated fluctuations in nanog expression mediate cell fate decisions in embryonic stem cells. PLoS Biol 2009;7(7):e1000149.

[44] Niwa H. How is pluripotency determined and maintained? Development 2007;134(4):635–46.

[45] Milo R, et al. Network motifs: simple building blocks of complex networks. Science 2002;298(5594):824–7.

[46] Jiang J, et al. A core Klf circuitry regulates self-renewal of embryonic stem cells. Nat Cell Biol 2008;10(3):353–60.

[47] Zhang P, et al. Kruppel-like factor 4 (Klf4) prevents embryonic stem (ES) cell differentiation by regulating Nanog gene expression. J Biol Chem 2010;285(12):9180–9.

[48] Niwa H, et al. A parallel circuit of LIF signalling pathways maintains pluripotency of mouse ES cells. Nature 2009;460(7251):118–22.

[49] Nguyen H, Rendl M, Fuchs E. Tcf3 governs stem cell features and represses cell fate determination in skin. Cell 2006;127(1):171–83.

[50] Yi F, et al. Opposing effects of Tcf3 and Tcf1 control Wnt stimulation of embryonic stem cell self-renewal. Nat Cell Biol 2011;13(7):762–70.

[51] Wend P, et al. Wnt signaling in stem and cancer stem cells. Semin 2010;21(8):855–63.

[52] Sokol SY. Maintaining embryonic stem cell pluripotency with Wnt signaling. Development 2011;2011:8.

[53] Berge DT, et al. Embryonic stem cells require Wnt proteins to prevent differentiation to epiblast stem cells. Nat Cell Biol 2011;13(9):1070–5.

[54] Nichols J, Smith A. The origin and identity of embryonic stem cells. Development 2011;138(1):3–8.

[55] Chambers I, et al. Nanog safeguards pluripotency and mediates germline development. Nature 2007;450(7173):1230–4.

[56] Rizzino A. Sox2 and Oct-3/4: a versatile pair of master regulators that orchestrate the self-renewal and pluripotency of embryonic stem cells. Wiley Interdiscip Rev Syst Biol Med 2009; 1(2):228–36.

[57] Plachta N, et al. Oct4 kinetics predict cell lineage patterning in the early mammalian embryo. Nat Cell Biol 2011;13(2):117–23.

[58] Yu J, et al. Induced pluripotent stem cell lines derived from human somatic cells. Science 2007;318(5858):1917–20.

[59] Ang YS, et al. Stem cells and reprogramming: breaking the epigenetic barrier? Trends Pharmacol Sci 2011;32(7):394–401.

[60] Maekawa M, et al. Direct reprogramming of somatic cells is promoted by maternal transcription factor Glis1. Nature 2011; 474(7350):225–9.

[61] Strumpf D, et al. Cdx2 is required for correct cell fate specification and differentiation of trophectoderm in the mouse blastocyst. Development 2005;132(9):2093–102.

[62] Fidalgo M, et al. Zfp281 functions as a transcriptional repressor for pluripotency of mouse embryonic stem cells. Stem Cells 2011;2011(13):736.

[63] Ombrato L, Lluis F, Cosma MP. Regulation of self-renewal and reprogramming by TCF factors. Cell 2012;11(1):39–47.

[64] Andrecut M, et al. A general model for binary cell fate decision gene circuits with degeneracy: indeterminacy and switch behavior in the absence of cooperativity. PLoS One 2011;6(5):e19358.

[65] Ptashne M. Principles of a switch. Nat Chem Biol 2011; 7(8):484–7.

[66] Papatsenko D, Levine MS. Dual regulation by the Hunchback gradient in the *Drosophila* embryo. Proc Natl Acad Sci U S A 2008;105(8):2901–6.

[67] Glauche I, et al. Stem cell proliferation and quiescence – two sides of the same coin. PLoS Comput Biol 2009;5(7):e1000447.

[68] Glauche I, Thielecke L, Roeder I. Cellular aging leads to functional heterogeneity of hematopoietic stem cells: a modeling perspective. Aging Cell 2011;10(3):457–65.

[69] Chickarmane V, Enver T, Peterson C. Computational modeling of the hematopoietic erythroid-myeloid switch reveals insights into cooperativity, priming, irreversibility. PLoS Comput Biol 2009;5(1):e1000268.

[70] Loose M, Swiers G, Patient R. Transcriptional networks regulating hematopoietic cell fate decisions. Curr Opin Hematol 2007;14(4):307–14.

[71] Preisler HD, Kauffman S. A proposal regarding the mechanism which underlies lineage choice during hematopoietic differentiation. Leuk Res 1999;23(8):685–94.

[72] Som A, et al. The PluriNetWork: an electronic representation of the network underlying pluripotency in mouse, and its applications. PLoS One 2010;5(12):e15165.

[73] Muller FJ, et al. Regulatory networks define phenotypic classes of human stem cell lines. Nature 2008;455(7211):401–5.

[74] Kelder T, et al. WikiPathways: building research communities on biological pathways. Nucleic Acids Res 2011;40(Database issue):D1301–7.

[75] Porter CJ, et al. StemBase: a resource for the analysis of stem cell gene expression data. Methods Mol Biol 2007;407:137–48.

[76] Schulz H, et al. The FunGenES database: a genomics resource for mouse embryonic stem cell differentiation. PLoS One 2009; 4(9):e6804.

[77] Barrett T, et al. NCBI GEO: mining tens of millions of expression profiles – database and tools update. Nucleic Acids Res 2007;35:D760–5. Database issue.

[78] Lachmann A, et al. ChEA: transcription factor regulation inferred from integrating genome-wide ChIP-X experiments. Bioinformatics. 2010;26(19):2438–44.

[79] Jung M, et al. A data integration approach to mapping OCT4 gene regulatory networks operative in embryonic stem cells and embryonal carcinoma cells. PLoS One 2010;5(5):e10709.

[80] van den Berg DL, et al. An Oct4-centered protein interaction network in embryonic stem cells. Cell 2010;6(4):369–81.

[81] Pardo M, et al. An expanded Oct4 interaction network: implications for stem cell biology, development, and disease. Cell 2010;6(4):382–95.

[82] Lemischka IR. Hooking up with Oct4. Cell 2010;6(4):291–2.

[83] Kauffman S. Self-Organization and Selection in Evolution, The Origins of Order. Oxford: Oxford University Press; 1993.

[84] Kauffman SA. Sequential DNA replication and the control of differences in gene activity between sister chromatids – a possible factor in cell differentiation. J Theor Biol 1967; 17(3):483–97.

[85] Chaves M, Albert R, Sontag ED. Robustness and fragility of Boolean models for genetic regulatory networks. J Theor Biol 2005;235(3):431–49.

[86] Aracena J, et al. On the robustness of update schedules in Boolean networks. Biosystems 2009;97(1):1–8.

[87] Davidich MI, Bornholdt S. Boolean network model predicts cell cycle sequence of fission yeast. PLoS One 2008;3(2):e1672.

[88] Shmulevich I, et al. Steady-state analysis of genetic regulatory networks modelled by probabilistic boolean networks. Comp Funct Genomics 2003;4(6):601–8.

[89] Liang S, Fuhrman S, Somogyi R. Reveal, a general reverse engineering algorithm for inference of genetic network architectures. Pac Symp Biocomput 1998:18–29.

[90] Akutsu T, Miyano S, Kuhara S. Identification of genetic networks from a small number of gene expression patterns under the Boolean network model. Pac Symp Biocomput 1999: 17–28.

[91] Akutsu T, Miyano S, Kuhara S. Algorithms for inferring qualitative models of biological networks. Pac Symp Biocomput 2000:293–304.

[92] Ideker TE, Thorsson V, Karp RM. Discovery of regulatory interactions through perturbation: inference and experimental design. Pac Symp Biocomput 2000:305–16.

[93] Lutter D, Bruns P, Theis FJ. An ensemble approach for inferring semi-quantitative regulatory dynamics for the differentiation of mouse embryonic dtem cells using prior knowledge. Adv Exp Med Biol 2012;736:247–60.

[94] Saez-Rodriguez J, et al. Discrete logic modelling as a means to link protein signalling networks with functional analysis of mammalian signal transduction. Mol Syst Biol 2009:5.

[95] Brown CT, et al. New computational approaches for analysis of cis-regulatory networks. Dev Biol 2002;246(1):86–102.

[96] Wuensche A. Discrete dynamics lab: tools for investigating cellular automata and discrete dynamical networks. In: Komosinski M, Adamatzky A, editors. Artificial life models in software 2009;Chapter 8. 2nd Ed. London: Springer; 2009. p. 215–58.

[97] Albert I, et al. Boolean network simulations for life scientists. Source Code Biol Med 2008;3:16.

[98] Mussel C, Hopfensitz M, Kestler HA. BoolNet — an R package for generation, reconstruction and analysis of Boolean networks. Bioinformatics 2010;26(10):1378–80.

[99] Batagelj V, Mrvar A. Pajek — program for large network analysis. Connections 1998;21:45–57.

[100] Longabaugh WJ, Davidson EH, Bolouri H. Computational representation of developmental genetic regulatory networks. Dev Biol 2005;283(1):1–16.

[101] Hirai H, Karian P, Kikyo N. Regulation of embryonic stem cell self-renewal and pluripotency by leukaemia inhibitory factor. Biochem J 2011;438(1):11–23.

[102] Zuccotti M, et al. Gatekeeper of pluripotency: a common Oct4 transcriptional network operates in mouse eggs and embryonic stem cells. BMC Genomics 2011;12:345.

[103] Hagos EG, et al. Expression profiling and pathway analysis of Kruppel-like factor 4 in mouse embryonic fibroblasts. Am J Cancer Res 2011;1(1):85–97.

[104] Hishida T, et al. Indefinite self-renewal of ESCs through Myc/Max transcriptional complex-independent mechanisms. Cell Stem Cell 2011;9(1):37–49.

[105] Sun C, et al. Dax1 binds to Oct3/4 and inhibits its transcriptional activity in embryonic stem cells. Mol Cell Biol 2009; 29(16):4574–83.

[106] Bourillot PY, Savatier P. Kruppel-like transcription factors and control of pluripotency. BMC Biol 2010;8:125.

[107] Watabe T, Miyazono K. Roles of TGF-beta family signaling in stem cell renewal and differentiation. Cell Res 2009;19(1):103–15.

[108] Zhou J, et al. High-efficiency induction of neural conversion in human ESCs and human induced pluripotent stem cells with a single chemical inhibitor of transforming growth factor beta superfamily receptors. Stem Cells 2010;28(10):1741–50.

[109] Niwa H, Miyazaki J, Smith AG. Quantitative expression of Oct-3/4 defines differentiation, dedifferentiation or self-renewal of ES cells. Nat Genet. 2000;24(4):372–6.

[110] Matoba R, et al. Dissecting Oct3/4-regulated gene networks in embryonic stem cells by expression profiling. PLoS One 2006;1:e26.

[111] Schulz C, Tautz D. Autonomous concentration-dependent activation and repression of Kruppel by hunchback in the *Drosophila* embryo. Development 1994;120(10):3043–9.

[112] Zuo P, et al. Activation and repression of transcription by the gap proteins hunchback and Kruppel in cultured *Drosophila* cells. Genes Dev 1991;5(2):254–64.

[113] Papatsenko D, Levine M. The Drosophila gap gene network is composed of two parallel toggle switches. PLoS One 2011; 6(7):e21145.

[114] Davidson EH, Levine MS. Properties of developmental gene regulatory networks. Proc Natl Acad Sci U S A 2008; 105(51):20063–6.

[115] Papatsenko D. Stripe formation in the early fly embryo: principles, models, and networks. Bioessays 2009;31(11):1172–80.

[116] Davidson EH, et al. A genomic regulatory network for development. Science 2002;295(5560):1669–78.

[117] Zinzen RP, Papatsenko D. Enhancer responses to similarly distributed antagonistic gradients in development. PLoS Comput Biol 2007;3(5):e84.

[118] He X, et al. Thermodynamics-based models of transcriptional regulation by enhancers: the roles of synergistic activation, cooperative binding and short-range repression. PLoS 2010; 6(9):e1000935.

[119] Kuroda T, et al. Octamer and Sox elements are required for transcriptional cis regulation of Nanog gene expression. Mol Cell Biol 2005;25(6):2475–85.

[120] MacArthur S, et al. Developmental roles of 21 Drosophila transcription factors are determined by quantitative differences in binding to an overlapping set of thousands of genomic regions. Genome Biol 2009;10(7):R80.

[121] Wu H, et al. Genome-wide analysis of 5-hydroxymethylcytosine distribution reveals its dual function in transcriptional regulation in mouse embryonic stem cells. Genes 2011;25(7):679–84.

[122] You JS, et al. OCT4 establishes and maintains nucleosome-depleted regions that provide additional layers of epigenetic regulation of its target genes. Proc Natl Acad Sci U S A 2011;108(35):14497–502.

[123] Hubisz MJ, Pollard KS, Siepel A. PHAST and RPHAST: phylogenetic analysis with space/time models. Brief 2011;12(1): 41–51.

[124] Pollard KS, et al. Detection of nonneutral substitution rates on mammalian phylogenies. Genome Res 2009;20(1):110–21.

[125] Robasky K, Bulyk ML. UniPROBE, update 2011: expanded content and search tools in the online database of protein-binding microarray data on protein–DNA interactions. Nucleic Acid Res 2011;39(Database issue):D124–8.

[126] Yaragatti M, Basilico C, Dailey L. Identification of active transcriptional regulatory modules by the functional assay of DNA from nucleosome-free regions. Genome Res 2008;18(6):930–8.

[127] Lifanov AP, et al. Homotypic regulatory clusters in Drosophila. Genome Res 2003;13(4):579–88.

[128] Schroeder MD, et al. Transcriptional control in the segmentation gene network of Drosophila. PLoS Biol 2004;2(9):E271.

[129] McKenna BK, et al. A parallel microfluidic flow cytometer for high-content screening. Nature 2011;8(5):401–3.

[130] Ornatsky O, et al. Highly multiparametric analysis by mass cytometry. J Immunol Methods 2010;361(1–2):1–20.

[131] Bolhuis J, Hogan J. The Development of Animal Behavior: A Reader. Oxford: Wiley-Blackwell; 1999.

[132] Huang S, et al. Cell fates as high-dimensional attractor states of a complex gene regulatory network. Phys Rev Lett 2005; 94(12):128701.

[133] Hanna JH, Saha K, Jaenisch R. Pluripotency and cellular reprogramming: facts, hypotheses, unresolved issues. Cell 2010;143(4):508–25.

[134] Enver T, et al. Stem cell states, fates, and the rules of attraction. Cell Stem Cell 2009;4(5):387–97.

Section V

Multi-Scale Biological Systems, Health and Ecology

Chapter 23

Systems Medicine and the Emergence of Proactive P4 Medicine: Predictive, Preventive, Personalized and Participatory

Leroy Hood[1], Mauricio A. Flores[2], Kristin R. Brogaard[1], and Nathan D. Price[1]

[1]*Institute for Systems Biology, 401 N. Terry Ave, Seattle, WA 98121, USA*
[2]*P4 Medicine Institute, 401 N. Terry Ave, Seattle, WA 98121, USA*

Chapter Outline

Introduction	445	Two Big Challenges: Education and Information	
Systems Medicine	447	Technology for Healthcare	464
Five Systems' Strategies for Dealing with Biological Complexity	449	Impact of P4 Medicine on Society	464
		How to Bring P4 Medicine to Patients	465
P4 Medicine	460	Acknowledgments	465
		References	466

INTRODUCTION

Medicine is undergoing a revolution that will transform the practice of healthcare in virtually every way. This revolution is emerging from the convergence of systems biology — a holistic approach to biology (and medicine) — and the digital revolution. Systems biology is opening what has historically been the black box of our individual biological systems as they change over time. The digital revolution is vastly expanding our capacity to generate and analyze 'big data' sets, deploy this information in business and social networks and create digital consumer devices to measuring personal information. Both of these new capabilities will be deployed in systems medicine (see below).

Systems biology emerged at least partly in response to the incredible complexity of biological systems, both normal and diseased [1,2]. This complexity arises from the natural process of Darwinian evolution — random mutations followed by natural selection generated in large part by the environment. Darwinian evolution is a random and chaotic process — building new complexities on top of previously evolved complexities. Indeed, biological systems resemble Rube—Goldberg devices. Consider a Rube—Goldberg device that attaches 14 different gadgets together to cool the temperature of soup. Deciphering the complexity of this 'soup-cooling system' requires (1) defining the components of the soup-cooling system, (2) determining how these components interact with one another, and (3) delineating the dynamics of these components in space and time that are necessary for carrying out their function of soup cooling. These are precisely the elements that systems biology attempts to define when deciphering the complexity of biological systems. To achieve actionable understandings of biological complexity, the analyses must be global (comprehensive), integrative and dynamic.

Systems medicine, the healthcare-focused derivative of systems biology, is beginning to alter the face of healthcare through (1) a systems approach to disease, (2) driving the emergence of technologies that permit the exploration of new dimensions of patient data space (e.g., sequencing the individual human genome), (3) the analyses of the quantized units of biological information (single genes, single molecules, single cells, single organs to provide disease-relevant information on health or disease for the individual), and (4) the resulting explosion of patient data that is transforming traditional biology and medicine into an information science [1–11].

The digital revolution has already transformed communications, finance, retail and information technology by harnessing big datasets through computational analyses and by creating powerful new business and social networks. The digital revolution is contributing to individualize healthcare in several important ways: (1) by providing tools and strategies for managing and analyzing large biological and environmental datasets; (2) by catalyzing the invention of personal monitoring devices that can digitalize biological and social information, thus enabling an assessment of wellness and disease for the individual (e.g., the 'quantified self'); and (3) by providing models for the creation of consumer (patient)-driven and consumer (patient)-participating social networks that focus on optimizing wellness and/or dealing with disease (Figure 23.1) [8—10,12].

The convergence of the digital revolution and systems approaches to wellness and disease is beginning to lead a proactive P4 medicine that is predictive, preventive, personalized and participatory [1,13—15]. Thus 'P4 medicine' is the clinical application of the tools and strategies of systems medicine to quantify wellness and demystify disease (Figure 23.2) for the wellbeing of the individual. The digital revolution has given scientists the ability to generate and analyze previously inconceivably large quantities of digital data. Using these new capabilities and employing the domain expertise of biology to direct the development of software, systems biologists have developed powerful new suites of tools for mining, integrating and modeling 'big data' sets of heterogeneous biological data to generate predictive and actionable models of health and disease for each patient [10,16—19]. 'Actionable' means that the data provide information that is useful for improving the health of the individual patient. Thus, systems biologists have transitioned from the reductive studies of traditional biology that focus on a few genes or proteins to the new holistic and comprehensive analyses of systems biology, analyzing how all of the components of biological systems interact.

Unlike the reactive, limited-data population-based hierarchical approach of contemporary evidence-based healthcare [20], P4 medicine is not confined to clinics and hospitals. It will be practiced in the home, as informed and networked consumers use new information, tools and resources such as wellness and navigation coaches and digital health information devices to better manage their health. Below we provide a brief picture of systems medicine and its role in the emergence of this proactive P4 medicine. We then will explore the multi-dimensionality of P4 medicine, with its many implications for the individual consumer (patient) and for society.

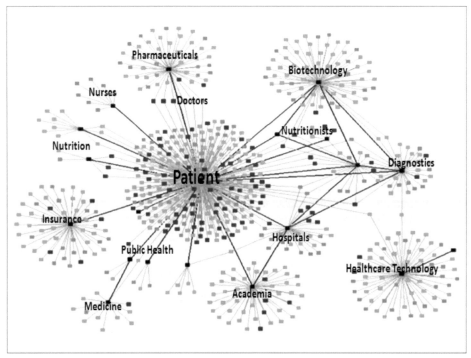

FIGURE 23.1 **A network depicting the interacting components of the healthcare system indicating the dominant role patients will have in advancing P4 medicine through their consumer-driven social networks.** Networks allow one to organize and model data and are important in dealing with the signal to noise problem of large data sets.

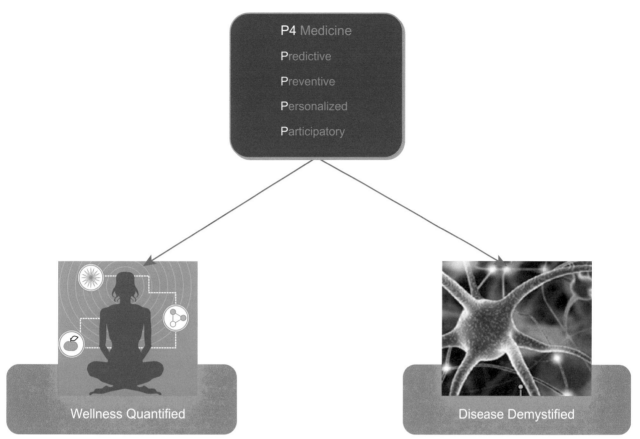

FIGURE 23.2 A schematic representation of the two major objectives of P4 medicine: quantizing wellness and demystifying disease.

SYSTEMS MEDICINE

We predict that within 10 years every healthcare consumer will be surrounded by a virtual cloud of billions of data points (Figure 23.3). These will range from molecular and cellular data, to conventional medical data, to enormous amounts of imaging, demographic and environmental data. Big data sets are essential to decipher 'signal' from the noise generated by the complexities of disease and wellness. As noted above, the complexity of biological systems arises from the random and chaotic processes of Darwinian evolution. The principal data analyzed by systems medicine relate to the following central concerns about human biological systems: the identification all of the system components, establishing their interactions, assessing the dynamics of those interactions — both temporal and spatial — and then attempting to understand how the system as a whole executes its biological functions and exhibits specific phenotypes.

A systems approach to disease takes into account the myriad social and environmental factors that compound the innate complexity of human biology and which are crucial determinants of health. It does so by treating disease as a consequence of genetic and/or environmental perturbations of biological networks. These disease-perturbed networks all express altered information that changes dynamically across time (and often space) and arises from the perturbations. The amalgamation of this distorted information explains the pathophysiology of the disease, as discussed below in the context of a specific systems-disease example, neurodegeneration.

The above description is not intended to suggest that disease-perturbed networks are of a single type: there is a 'network of networks' that reflects the multi-dimensionality of both biological processes and disease processes (Figure 23.4). Thus we can describe genetic networks arising from the genome; molecular networks arising from protein and DNA interactions; cellular networks arising from cell interactions; organ networks arising from organ–organ interactions; and finally, social networks arising largely in part from the nature of an individual's environmental interactions [2,17,21]. The integration of all of these networks is necessary to understand their functioning in the context of the individual.

The ascertainment of these networks requires enormous amounts of data. For some of these measurements, the tools are just now being developed. Big data sets pose two

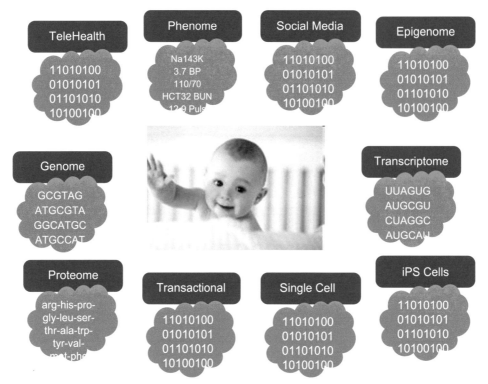

FIGURE 23.3 **In 10 years a virtual cloud of billions of data points will surround each patient.** These data will be of many different types and, accordingly, multistage. The challenge will be to convert these data into simple hypotheses about health and disease for the individual.

FIGURE 23.4 **A figure depicting the 'network of networks' that specifies the nature of some of the integrated networks that specify normal biology and disease.** The genetic, molecular, cellular, organ and individual networks are represented, and represent a fully integrated network of networks. Networks are powerful tools for integrating and modeling biological data. Networks also provide a powerful means for dealing with signal to noise problems. (*Figure adapted from [60].*)

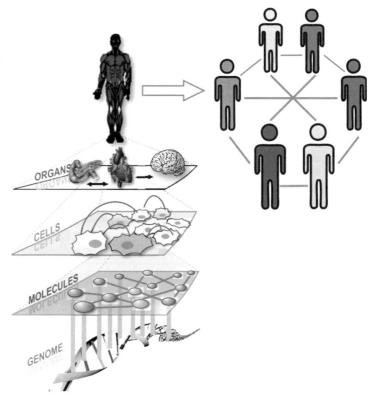

significant problems. First, how to deal with the enormous signal-to-noise challenges intrinsic to all large data sets, and second, how to convert data into knowledge. Solving these problems is the role of systems medicine. The key for systems medicine in the future will be to ascertain and deconvolute the 'network of networks' for each individual and to be able to follow its dynamics in response to various types of biological information, providing fundamental insights into wellness and disease.

Looking back in history there were four paradigm changes that led directly to systems medicine: (1) automated and high-throughput biological technology, (2) the Human Genome Project, (3) the creation of cross-disciplinary biology institutes, and (4) the creation of systems biology as an area of research [22,23] We will spend some time focusing on one of these, the Human Genome Project, to describe the revolutionary effect that it has had on biology.

The first meeting on the Human Genome Project was held in the spring of 1985. Twelve 'experts' had been invited to Santa Cruz to consider whether sequencing the human genome was advisable. This committee came to two conclusions: (1) the project was feasible albeit technically difficult; and (2) the group was split 6 to 6 on whether it was a good idea. In the mid and late 1980s perhaps 90% of biologists were opposed to the project, as was the National Institutes of Health (for reasons such as: big science is bad; the genome is mostly junk, so why sequence it? no good scientists would participate in such a mindless endeavor, etc.). A committee of the National Academy of Sciences with both opponents and proponents was convened to consider this possibility, and their unanimously favorable report turned the tide. The project was initiated in 1990 and finished in 2004 — under budget and ahead of schedule.

Many have argued that the genome project did not fulfill its promise, but the truth is quite the opposite. The Human Genome Project has transformed both biology and medicine in several important ways:

1. It made systems biology possible by providing a complete parts list of all (most) of the genes (and by inference their corresponding proteins) in the human and several model organisms. This parts list was essential for global and integrative systems approaches, not to mention all subsequent human biology research.
2. It 'democratized' all genes (and indeed any region of the genome) by making all genes available to all biologists.
3. It pushed high-throughput biology to the next stage by driving technologies to increase the speed and accuracy of DNA sequencing to pioneer other genomic technologies, such as DNA arrays and parallel sequencing.
4. It made mass spectrometry-based proteomics possible by providing the sequences of proteins and their corresponding tryptic peptides.
5. It brought computer scientists, theoretical physicists, software engineers and mathematicians into biology to deal with the exponentially increasing data sets, thereby providing new software and mathematical tools for converting data into knowledge.
6. By requiring that genomic data be made public as soon as it was determined, it pioneered the idea of open data, rapidly published and available to all.
7. It also pioneered the importance of assessing data quality and the software necessary for this assessment.
8. It made the genomes of microbes, plants, and animals accessible to all biologists, transforming many fields of biology (e.g., microbiology, virology, immunology, etc.).
9. It revolutionized our understanding of molecular evolution.
10. For medicine, it made possible new approaches to genomic diagnoses; it enabled personalized medicine; and it is forging new approaches to assessing proper drugs (therapies) for differing kinds of cancer.
11. It brought biology into the realm of big science and initiated a big science/small science debate that continues even today. Big and small science can be beautifully integrated, and each plays an important role in deciphering biological complexity [1].

The Battelle Memorial Institute has recently estimated that the Human Genome Project has led to almost $800 billion in benefits for an initial investment of about $3.5 billion. It is clear from skimming though these benefits that the project enabled and enriched systems approaches to biology and medicine in many different ways that go far beyond genomics itself.

Many people use the term 'genomic medicine' to denote the medicine of the future, yet in principle genomic medicine is one-dimensional in nature, only encompassing nucleic acid information. Systems medicine, in contrast, is holistic and utilizes all types of biological information, including DNA, RNA, protein, metabolites, small molecules, interactions, cells, organs, individuals, social networks and external environmental signals — integrating them so as to lead to predictive and actionable models for health and disease.

FIVE SYSTEMS' STRATEGIES FOR DEALING WITH BIOLOGICAL COMPLEXITY

To develop predictive and actionable models we need to tease apart the complexity of health and disease. Systems medicine employs five strategies to deal with biological complexity: (1) viewing medicine as an informational science, (2) creating a cross-disciplinary infrastructure in which to implement systems medicine, (3) employing experimental systems approach to disease that are holistic and integrative, (4) driving the development of new technologies that permit the exploration of new dimensions of patient data space, and (5) developing new analytical tools, both computational and mathematical, for capturing, validating, storing, mining, integrating and finally modeling data so as to convert data into knowledge.

These five pillars of systems medicine permit biological complexity to be deciphered by providing a path forward for both generating large amounts of data, integrating and modeling these data in ways that reduce noise and delineate biological mechanisms. They create the conceptual framework for converting data into knowledge.

1. Systems medicine views medicine as an informational science, providing an intellectual framework for dealing with complexity. Fundamentally, there are two types of biological information: the digital information of the genome and the environmental signals that come from outside the genome. Together these two types of information are integrated in the individual organism (e.g., a human) to produce its phenotype, healthy or diseased. These two types of information and the phenotypes they produce are connected through biological networks that capture, transmit, integrate signals and then pass the information to molecular machines that execute the functions of life. It is the dynamics of networks and molecular machines that constitute a major focus of systems studies. The 'network of networks' adds yet another multiscale challenge to organizing and integrating information (Figure 23.4).

    As noted above, systems medicine postulates that disease arises from disease-perturbed networks (perturbed by genetic changes and/or environmental signals). Altered molecular machinery encoded by the disease-perturbed networks leads to the pathophysiology of the disease. Thus following the dynamics of the disease-perturbed networks gives deep insights into disease mechanisms and provides a powerful tool for dealing with the signal to noise challenges of big data sets. The utility of this approach has been demonstrated in two mouse models — mouse neurodegeneration (prion infection) [16] and glioblastoma — from mice genetically engineered in a combinatorial manner with oncogenes and tumor suppressors). The prion model of neurodegenerative is discussed later in this chapter.

    To obtain the necessary information for systems medicine it is also critical to integrate and model the many diverse data types, including from animal models, that follow disease progression. The reasons for this are obvious. One cannot follow the disease from initiation to the end in humans; one cannot usually easily sample the diseased tissue at multiple different time points; nor can one experimentally perturb the system with environmental signals.

    The need for systems dynamics data to deal with noise and create models emphasizes the importance of experimental animal disease models where the starting point of the disease process can be known (e.g., by genetic activation of the disease process or the experimental initiation of disease such as an infection) and the dynamics followed until death. The key point is that animal models must closely mimic their human counterpart diseases. Scientists must clearly identify those aspects of the disease-perturbed systems that are orthologous to human disease and those that are unique to the animal — and use the former for gaining dynamical insights into human disease. When the disease process is translated into network dynamics, determining orthology between the animal model and human disease becomes much simpler. Indeed, one can draw inferences from model organism disease-perturbed networks that are orthologous to their human counterparts and ignore the disease-perturbed networks that are not orthologous. This approach enables animal studies to be powerfully informative about human disease.

2. Our belief is that a special infrastructure is required for practicing systems medicine. This belief is driven by the conviction that leading-edge biology must drive the development of new high-throughput technologies to explore new dimensions of patient data space. The data arising from these technologies in turn require the pioneering of new analytical tools for the integration and modeling of diverse data types. We have termed this the 'holy trinity' of biology — biology drives technology drives analytical tools — and integrated them together to revolutionize our understanding of medicine (Figure 23.5).

    This approach requires a cross-disciplinary environment where biologists, chemists, computer scientists, engineers, mathematicians, physicists and physicians all learn to speak the languages of the other disciplines and work together in biology-driven teams to achieve this holy trinity. To be effective, this cross-disciplinary environment requires the 'democratization' of data generation and data analysis tools; that is, it is essential to make these tools accessible to all individual scientists so that they may carry out either big science or small science projects.

    Thus the infrastructure of systems medicine consists of both the instrumentation to generate data for the diverse 'omic' technologies (genomics, proteomics, metabolomics, interactomics, cellomics, etc.) and a culture that encourages scientists to learn to speak the languages of

Chapter | 23 Systems Medicine and the Emergence of Proactive P4 Medicine

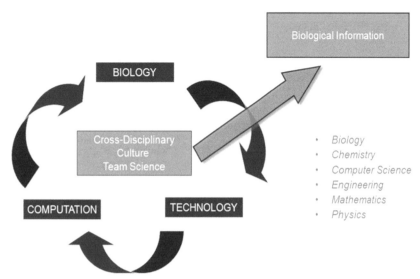

FIGURE 23.5 **The 'holy trinity of biology' where biology drives technology drives computational/mathematical tools.** Practicing this ideal requires a cross-disciplinary environment where scientists of many different disciplines (see lower right-hand side of figure) learn to speak the languages of the other scientists and to work together in teams. When the holy trinity is practiced effectively enormous amounts of biological information can be generated rapidly.

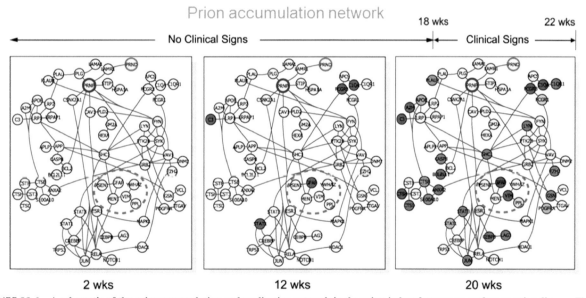

FIGURE 23.6 **A schematic of the prion accumulation and replication network in the prion-induced mouse neurodegenerative disease.** The red indicates transcript levels that have been increased in the brains from prion-infected animals compared with normal control brains. The yellow indicates transcripts that are the same in control and diseased animals. The three panels represent the network at 2, 12 and 20 weeks in animals that live about 22 weeks with this disease. The disease-perturbed networks appear about eight weeks before the clinical signs appear in these animals.

multiple scientific disciplines and how to work together in biology-driven teams practicing the holy trinity in the context of specific big or small science problems. This systems-driven infrastructure is what the Institute for Systems Biology (ISB) has spent the first 10 years of its existence creating [24].

3. Experimental systems approaches to disease and wellness are holistic. To decipher biological complexity, systems medicine depends on generating global and comprehensive data sets, following the dynamics of disease-perturbed networks across disease initiation and progression.

Ultimately integrating diverse data types together to create predictive and actionable models (Figure 23.6) [2,24,25]. Thus systems medicine will give fundamental new insights into disease mechanisms — and open new opportunities for diagnosis, therapy and prevention.

The idea of systems biology can be illustrated by the example of how a radio converts radio waves into sound waves [26]. It is clear that merely understanding the function of each individual radio component would not provide us with an understanding of how radio waves are converted into sound waves. The same principle applies to complex biological systems: understanding what individual genes or proteins do does not tell us how the biological systems in which they operate function. An engineer would connect the radio parts into their circuits and come to understand how the circuits worked individually and then collectively to convert radio into sound waves. So it is with living organisms: they employ biological circuits or networks to manage biological information and convert it into phenotype and function, and to understand these we must understand the dynamics of biological networks in handing information.

To generate holistic data systems biology (and therefore systems medicine) has three central elements. (1) It is hypothesis driven, where a model (which is a formally structured, precise and potentially complex) is formulated from existing data. Hypotheses from model predictions are then tested with systems perturbations and the high-throughput acquisition of data. The data (and metadata) are then reintegrated back into the model with appropriate modifications, and this process is repeated iteratively until new predictions from theory and experimental data are in agreement. (2) It is based on high-throughput data that should be (i) global (comprehensive), (ii) generated from different multi-scale data types (e.g., DNA, RNA, protein, metabolites, interactions, etc.), (iii) used to monitor networks dynamically, (iv) employed to provide deep insight into biology, and (v) integrated using proper statistics and bioinformatics to handle the enormous signal-to-noise problems. (3) Models may be descriptive, graphical or mathematical as dictated by the amount of available data, but they must be predictive. For medical use, predictions made must be actionable and useful for treating patients.

Boosting signal-to-noise in complex biology is essential for deciphering complexity. To reduce noise and to enhance statistical power, biologists have leveraged two fundamental ideas: filters and integrators [27]. Filters are used to winnow down the number of candidates based on the biological assumptions about complexity (e.g., modularity, hierarchical organization, complexity arising from evolution and inheritance). Integrators leverage the availability of complementary data of genome, transcriptome, miRNAome, proteome, metabolome, and interactome. Successful application of these strategies in disease will lead to a transformational understanding of disease and therapeutics.

The framework for approaching these studies in a holistic way is a systems approach to disease. As discussed above, the key idea is that disease arises as a consequence of the perturbation of one or more biological networks in the relevant organ. This perturbation alters the information the network encodes in a dynamic manner that changes during the progression of the disease (e.g., changing levels of mRNAs, miRNAs, or even proteins) — and these altered levels explain the pathophysiology of the disease and provide new insights into diagnosis and therapy.

A systems approach to a neurodegenerative disease in mice. We will illustrate this holistic systems approach as it applies to neurodegenerative disease (prion disease) in mice. This disease is initiated by the injection of 'infectious prion proteins' into the brains of mice. An important point is that we know precisely when the disease is initiated (at injection), allowing us to follow the dynamics of the disease process from initiation to termination. We analyzed the brain transcriptomes of the infected mice at 10 time points across the approximately 22 weeks of disease progression, in addition to the transcriptome of their healthy littermates. This procedure identified 7400 differentially expressed genes (DEGs), which represented a staggering signal-to-noise problem.

There are two types of noise: *technical noise*, which comes from the instrumentation/procedures for handling the data, and *biological noise*, which arises from biological processes other than neurodegeneration contributing to the phenotypic measurements. When you measure any aspect of phenotype in an organism, often those phenotypes are the sum of a number of different biologies. Hence one must use a deep understanding of biology to subtract from the biology of interest (neurodegeneration) the signals resulting from the other biologies.

To overcome the noise we carried out this study in eight different inbred-strain/prion-strain combinations of infected mice. With more data and a deep biological understanding of the disease process we were able to subtract away noise. For example, in the double-knockout mouse for the prion gene, after injection with infectious prions the animals never develop the disease. Thus, any changes in the brain transcriptomes of these animals were irrelevant to the prion neurodegeneration response and could be subtracted away. With seven additional subtractions, we identified a core of about 333 differentially expressed genes that encoded the basic prion neurodegeneration process. We mapped these DEGs on to four major biological networks of the prion disease process that had been defined by serial histopathology of the diseased brains. We then integrated the transcriptome data with (1) serial brain histopathological analyses of these animals, (2) serial sagittal brain sections stained for infectious prions, (3) clinical signs of the disease and (4) blood biomarker analyses. Figure 23.6 illustrates one of the major dynamically changing networks (prion replication and accumulation).

We drew the following conclusions from this study: (1) The disease starts with one or a few networks being

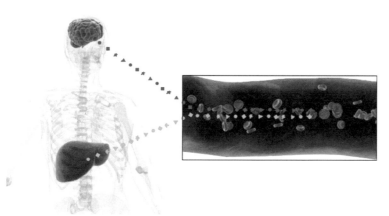

FIGURE 23.7 **A diagram of organ-specific blood fingerprints (collections of organ-specific proteins) from the brain and the liver.** For example, in a normal brain, each of the proteins in the brain-specific blood fingerprint will have one set of levels. In a diseased brain, the proteins whose cognate networks have become disease perturbed will change their levels. Since each disease leads to distinct combinations of disease-perturbed networks an analysis of the brain-specific protein fingerprints can distinguish healthy from diseased brains, and if diseased can stratify (e.g., distinguish from one another) the distinct types of brain disease. Thus organ-specific brain fingerprints can provide early detection, a stratification of different types of disease and the ability to follow the progression of the disease (not shown).

disease perturbed, and as the disease progresses more and more networks are recruited to become disease perturbed. (2) Two-thirds of the DEGs mapped into the four major networks — and the dynamics of these four networks explained virtually every aspect of known prion disease. (3) These four major networks were disease perturbed in a serial manner: first, prion replication and accumulation; second, glial activation; third, the degeneration of neuronal axons and dendrites; and finally neuron apoptosis. The importance of this observation is that if one is interested in new approaches to diagnostics and therapeutics the initial disease-perturbed network is a logical place to start. (4) The remaining one-third of the DEGs identified six new networks that were heretofore unknown to prion disease — the so-called 'dark genes and networks of prion disease' as identified by the global analyses of normal and diseased brain transcriptomes. These insights emphasize the importance of global analyses of the transcriptomes. (5) These studies suggested new approaches to blood diagnostics that are discussed below. (6) For therapy it is obvious that the first and most proximal prion-specific network should be re-engineered with drugs to make it behave in a more normal manner and hopefully abrogate the downstream consequences of this pathological progression. It is clear that multiple drugs will be required to re-engineer biological networks. We are now learning how to re-engineer biological networks with drugs in microbes, so that we can come to understand the basic logic of the approach and then extend it to higher model organisms (mice) before applying it to humans.

A holistic approach to disease will use blood as a window for monitoring health (wellness) and disease. A systems approach to blood diagnostics emerged from two ideas arising from the prion studies. First, some transcripts are expressed in their disease-perturbed networks 8 weeks or more before the first clinical signs (e.g., 10 weeks and 18 weeks, respectively). We were able to demonstrate that several of these differentially expressed gene (DEG) transcripts encoded proteins expressed in the blood, and we could see the altered protein levels in the blood. This was an example of presymptomatic diagnosis, a long-sought keystone of early disease detection. However, these DEGs were expressed in several different organs, so we could not be certain of the location of the disease-perturbed process directly from observing protein concentration changes in the blood. Second, in order to obtain blood markers with organ-specific addresses, we identified transcripts that were organ specific by deep comparative transcriptome analyses across 40+ different organs in humans and mice (Figure 23.7). From these analyses, through an examination of the human and mouse blood protein databases, and experimental mass spectrometry analyses, we were able to identify about 100 brain-specific proteins in humans and mouse. Of these, about 95% were orthologous between the two species (the presumption is that they will reflect similar activities in the two species), and these proteins collectively constituted a brain-specific blood fingerprint. We were able to show that some of these brain-specific proteins could also be used for presymptomatic diagnosis of prion disease in mice. Additionally, brain-specific blood proteins encoded by each of the four distinct networks exhibit concentration changes in the blood in a serial manner consistent with the order of disease perturbation of their cognate transcriptional networks. These data demonstrate that we will be able to assess both early disease detection and disease progression from the blood.

In the organ-specific blood protein fingerprints each individual protein assesses the behavior of its cognate biological network, distinguishing normal functioning from disease-perturbed functioning by changes in their blood concentration levels. Because each disease perturbs different combinations of networks, the brain-specific blood fingerprints will be able to distinguish normal from disease and, if diseased, identify the disease. This will enable the five holy grails of blood disease diagnosis: (1) presymptomatic diagnosis; (2) stratification of a disease into its different subtypes; (3) assessment of the progression of the disease; (4) following patient response to therapy; and (5) identifying recurrences. We are now applying this strategy to identify human organ-specific blood biomarkers for several cancer types. In addition to blood proteins as tumor biomarkers, circulating DNAs, mRNAs, and microRNAs, as well as circulating tumor cells have also been studied which can serve as surrogate disease biomarkers and for monitoring cancer recurrence [28–32]. The organ-specific blood fingerprints are powerful aids to disease diagnostics, assessing drug toxicities, validating the orthologies between human disease and animal models of that disease, assessing multiorgan responses to diseases (beginning to define the organ/organ communicating networks (Figure 23.4) and, as we shall discuss later, assessing longitudinally across time the wellness of individual patients).

4. The systems approach to disease mandates the need to develop new or emerging technologies that can explore new dimensions of patient data space as reflected in part by the dynamics of the network of networks. These technologies include new approaches to genomics, proteomics, metabolomics, interactomics, cellomics, organomics, in vitro and in vivo imaging, and other high-throughput phenotypic measurements [33–38]. Microfluidic and nanotechnology approaches are moving many of these assays towards further miniaturization, parallelization, automation and integration of complex chemical procedures [39–41]. The areas of in vitro imaging and high-throughput phenotypics assays are going to contribute enormously to expanding the data repertoires of a systems approach to disease. But these new technologies must be driven by the real needs of biology or medicine. The outcome is an exponentially increasing ability to generate enormous amounts of digitalized personal data — big data — that necessitates a mandate to translate these data into knowledge. Let us briefly consider several areas in which emerging technologies are or will transform a systems approach to medicine.

DNA sequencing. Automated DNA sequencing was the cornerstone technology for sequencing the genome [42]. Since the initial completion of the human genome sequence [43] there has emerged a series of 'next-generation sequencing (NGS)' technologies that have exponentially increased the throughput of DNA sequencing while bringing down the costs dramatically (through parallelization and miniaturization of the process). NGS generates short sequence reads (e.g., 50–100 base pair reads) which have enabled the rapid sequencing of human genomes, the direct sequence analyses of transcriptomes and miRNAomes, as well as the analyses of some epigenetic features such as methylation. DNA arrays remain useful for looking at genetic variation and at the quantification of RNA populations (transcriptomes and miRNAomes) — although it is clear that in time DNA sequencing will replace these DNA array analyses. There is now emerging a third generation of DNA sequencing instruments that employ nanopores for threading single DNA molecules through the pores to enable the electronic analysis of single-stranded DNA molecules [44]. These new techniques have the potential for extremely long sequence reads (50–100 kb or more) and the ability for such extensive parallelization of the sequencing runs (through threading many DNA molecules simultaneously through many nanopores) that one may imagine doing a sequence run for a complete human genome sequence in a fraction of an hour, rather than the day to week or more that is required by current NGS instruments.

NGS has enabled striking new strategies for generating data (e.g., RNAseq — the quantification of complete transcriptomes, exon sequencing — the sequence analyses of all (most) of the exons in a genome and family genome sequencing — determining the complete human genome sequences of all the members of a family).

Complete family genome sequencing integrates genetics with genomics and in doing so raises fascinating possibilities for delineating diverse chromosomal features. For example, in the sequence analysis of a family of four where the mother and father were healthy and the two children each had two recessive genetic diseases, we had hoped to identify a modest number of gene candidates to explain the two genetic diseases (Figure 23.8) [33]. To our surprise, family genome sequences enabled far more in the way of family analysis. First, we were able to correct about 70% of the DNA sequencing errors merely by a consideration of the principles of Mendelian genetics. Second, we were able to identify rare variants merely by asking whether two or more members of the family exhibited the variant (thus eliminating the possibility of DNA sequencing errors). Third, we were able to determine the sites of chromosomal recombination, and accordingly could determine complete chromosomal haplotypes for each member of the family. This turned out to be important, as it reduced the chromosomal space in which disease genes might reside (by asking which haplotype regions were shared by the diseased children compared to their healthy parents). In a family of four the two affected genes must reside within a defined quarter of

Chapter | 23 Systems Medicine and the Emergence of Proactive P4 Medicine

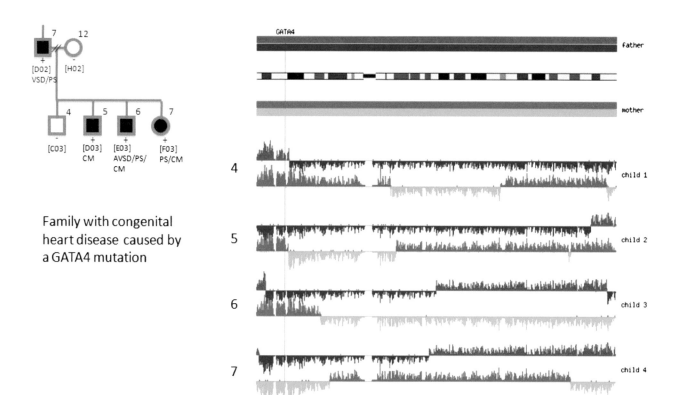

FIGURE 23.8 A schematic depicting the haplotypes of the members of a family of six. The family tree is indicated at the left. The four parental haplotypes (two for each parent) are indicated by four different colors. The portions of the parental haplotypes that are passed on to each child are indicated by the same colors. Each color change denotes a site of chromosomal recombination. Family genome sequencing permits one to determine these recombinational sites with great precision. The important point is that the genes that cause particular diseases must reside in areas of shared haplotype by those individuals in the family exhibiting the disease.

the genome, independent of any genetic models for the two diseases. Fourth, we were able to determine the intergenerational mutation rate for the two children (about 35 mutations per child). In this regard, it is interesting to note that because of intergenerational mutations there is no such thing as genetically identical twins. Finally, we were able to reduce the candidate gene list for the two diseases to just four possibilities. The correct defects could readily be associated with the diseases by using other genetic analyses of these defects. Thus whole-family genome sequencing is a powerful approach to enriching the signal-to-noise intrinsic to human genetic studies. Family genome studies constitute a powerful new approach to identifying the genome elements that are responsible for health or disease.

The technologies of genomics are becoming increasingly mature. This means that companies will be able to perform genomic analyses far more effectively than most academics, and these analyses will be increasingly outsourced to highly efficient vendors.

Mass spectrometry and the identification and quantification of proteins in proteomes. The analysis of proteins differs from that of their genomic counterparts in several ways. First, DNA is basically digital in nature, i.e., sequences and functions are specified by a digital four-letter language. In contrast, proteins are synthesized as linear structure, but fold into three-dimensional structures to execute their functions. We can, however, digitize the identification and quantification of proteins through analysis by mass spectrometry (see below). Second, proteins exhibit enormous complexity in structures due to many different procedures associated with their synthesis, including RNA editing, RNA splicing, protein modification, protein processing, etc. Indeed, some have estimated that the human genome may produce a million or more proteins [45]. Third, proteins are dynamic, changing their structures as they execute their functions and as they interact with other small and large molecules [46]. Fourth, proteins cannot be analyzed in a global or comprehensive manner (unlike genomic features) because of the enormous dynamic range of protein expression (10^6 in tissues and 10^{10} in blood), which exceeds the dynamic range of detection by the mass spectrometer. Finally, there is no protein amplification method equivalent to the polymerase chain reaction (PCR) of nucleic acids to enable the amplification and analysis of rare proteins.

A proteome is the collection of proteins that are present in a biological entity — a cell, an organ, the blood or an individual. The Human Genome Project has given us the sequences of most of the human proteins (and their tryptic peptides) and this has enable mass-spectrometry-based proteomics. Mass spectrometry can identify (and quantify) tryptic peptides of the proteins (those generated by the proteolytic enzyme trypsin cleaving either at the amino acid residues lysine or arginine). The mass spectrometer has the ability to separate and determine the mass to charge ratios of the tryptic peptides (and thus identify them). Quantification is achieved by determining the frequency of a particular tryptic peptide in a proteome mixture and averaging the frequencies of all the identifiable tryptic peptides shared by a particular protein.

Mass spectrometry has been used in two ways. A shotgun proteome analysis attempts to quantify all the tryptic peptides present in the given proteome. This procedure permits only the more dominant proteins to be quantified accurately because of the limited dynamic range and the fact that the peptides of rare proteins will be seen rarely, if at all. A targeted proteome analysis permits 100 proteins to be identified in a complex mixture by synthesizing isotopically labeled 'standard' peptides to be compared against their counterpart peptides in the proteome mixture. The triple quadrapole mass spectrometer has the ability to search out the peptides from the proteins that will be quantified by these assays. The targeted proteomic approach is called selective or multiple reaction monitoring (SRM or MRM) mass spectrometry. Recently, standard assays have been developed for most of the 20 000 or so human proteins [47]: just as the Human Genome Project 'democratized' all human genes by making them accessible to every biologist, so this proteome project has 'democratized' human proteins. Mass spectrometry can also be used to analyze the proteins present in organelles and can analyze those proteins interacting with one another (after pulling down and purifying the interacting protein complexes with specific antibodies). Mass spectrometry has also been used to look at translational chemical modifications and the protein forms arising from alternative RNA splicing.

The SRM assays have been used in assaying brain and liver organ-specific blood proteins, both in the organs and in the blood. As noted above in the section discussing 'blood as a window', one would like to contemplate the ability to analyze, say, 50 organ-specific proteins from each of 50 different human organs, and to follow for each patient's protein footprint across time to assess health vs. disease. As one contemplates the possibility of analyzing organ-specific blood proteins several times a year in the blood of the 340 million patients in the US, the mass spectrometer does not have the extendable throughput to manage 680 million samples. In the future these analyses will be done by microfluidic chips with ELISA (antibody) assays for each protein. A recent microfluidic protein chip has been designed with 50 ELISA assays for blood proteins that can be analyzed from 200 nL of plasma in about 5 minutes across a dynamic range of close to 10^5 and a sensitivity in the mid-atomole range [39]. We believe that the potential for this chip can be extended to thousands of protein measurements. Generating pairs of antibodies for this many proteins would be extremely expensive and time-consuming, and, moreover, antibodies are not very stable. Accordingly, one will have to develop more effective, stable and scalable protein-capture agents. Aptamers (for example, 60-mers of DNA) and trimeric or tetrameric peptide fragments (e.g., D amino acid 6-mers) appear to be interesting candidates as new protein-capture agents [48,49].

The mass spectrometer is one of the most powerful approaches for analyzing metabolites [35,50]. The challenges for metabolite analysis are generally similar to those of proteins (distinguishing the enormous number of different metabolites, facilitating their identification dealing with their broad dynamic ranges of expression, etc.).

Single-cell analyses. Most of our understanding of development, physiological responses and the initiation and progression of disease comes from studies that assess populations of cells. In the future the analyses of large numbers of *individual* cells will become important and standard practice — analyses that allow the genomes, epigenomes, transcriptomes, miRNAomes, proteomes, metabolomes, and interactomes within single cells to be determined. A variety of microfluidic and nanotechnologic approaches are being applied to these problems. Single-cell analyses will allow us to answer two fundamental questions. First, do discrete, quantized populations of cells exist within given organs? Preliminary results suggest that the answer to this question is yes. The fundamental issue is: what are the biological roles of these quantized populations? For example, they may represent a series of transition intermediates on their way to a final end stage. Alternatively, they may represent discrete populations of cells each with a separate function within the organ. Second, the expression of some of the information molecules within individual cells may behave in a stochastic manner. Single-cell analyses, properly executed, will permit us to distinguish between quantized cell populations and stochastic variability. We believe that single-cell analyses will be a critical tool in the future for deciphering biological complexity.

As indicated in the introduction to this section, many additional technologies are emerging that will open up the exploration of new dimensions of patient data space.

5. The 'data explosion' requires that new analytic tools be created for capturing, validating, storing, mining, integrating and finally modeling all of these biological data sets, thus helping to convert them into knowledge. A critical point is that these software solutions must be driven by the needs of leading-edge biology and medicine — and by biological domain expertise. One big revolution in medicine is that we will create massive amounts of digital data for the 'quantified self' of each individual that will transform our ability to monitor and optimize our own wellness. The following sections discuss in detail the transformation of big data sets to medically relevant information

Computational integration of 'quantified self' data will revolutionize health. Information on the quantified self provides enormous potential for the future of P4 medicine, as we are able to harness this information productively through powerful data analysis and large-scale computation (Figure 23.9). The key issue is how such large repositories of data can be turned into actionable knowledge. The potential of this endeavor is enormous, as we will gain unprecedented detail about how our bodies work, what brings about disease, and how wellness can be maintained. Interpreting multifaceted biological data deeply for each individual — and integrating it broadly across populations — will open new vistas of biological knowledge and clinical power. The pace of the technological changes will be quick, driven by exponentially rising computational power to take advantage of the exponentially rising amounts of high-throughput biological data. This is P4 medicine's heritage from the digital revolution.

Four factors will be important for dealing with the striking signal-to-noise issues of large data sets: (1) the integration of similar data types from different laboratories to enormously enlarge the data sets analyzed (see below); (2) the integration of data of different types — including molecular, cellular, conventional medical and phenotypic data; (3) the transformation of these data into the 'network of networks' for each individual patient — and following the dynamics of the 'network of networks'; and (4) the use of subtractive biological analyses to eliminate various forms of biological noise, as described in the prion discussion (see above). Each of these approaches represents significant computational/mathematical, technical and biological challenges, some of which are illustrated in the following discussion.

Realizing the power of high-dimensional diagnostics requires overcoming very significant data analysis challenges. This exciting future will only be realized as we address very significant computational and data analysis challenges. The human body is an enormously complex dynamic system interacting with an ever-changing and diverse environment. As the capacity to make molecular measurements continues to increase in scope and precision, the challenge of finding the relevant signals amidst the sea of observations can be daunting. As with any complex system, causality is often difficult to find and there are many ways that systems can break down and result in disease. Our bodies have enormously intricate and beautiful approaches for dealing with disease, for example via the immune system, and thus the residual medical problems we must solve must consider the consequences of the highly adaptive protective immune responses: both those that have been successful and those that were unsuccessful (such as the failure to check malignant cancers). Thus, these problems are often highly challenging, including from an informational point of view.

The primary challenge of big data in biology is to separate relevant signal from noise, including both technical noise (from measurements) and biological noise (from other biological factors besides those of interest). The number of measurements that come from the quantified self present significant hazards for proper interpretation. Having increasing ability to make precise measurements is exciting because so much new information is available, but care must be taken not to build overly complex models that appear very good on initial data assessment, but which fail when moved forward towards potential clinical use. Using an overly complex model that fits the already observed data really well, but then does not maintain accuracy when applied to new data for the same phenomenon (i.e., the model is fitting noise rather than the true underlying relationship), is called overfitting. In biological and clinical studies with 'omics' data, we are typically in what statisticians refer to as the small samples size regime. That is, we have very many more variables than we do observations. For example, there are tens of thousands of different transcripts measured in a human transcriptome (the variables), but generally only of the order of 100 or so samples (the observations) in a given study. This is exactly the opposite situation of what is desired to reliably use measurements to distinguish classes and establish reliable relationships among the variables (e.g., transcripts): one would like to have very many observations relative to the number of variables that are being used. Because the number of variables is

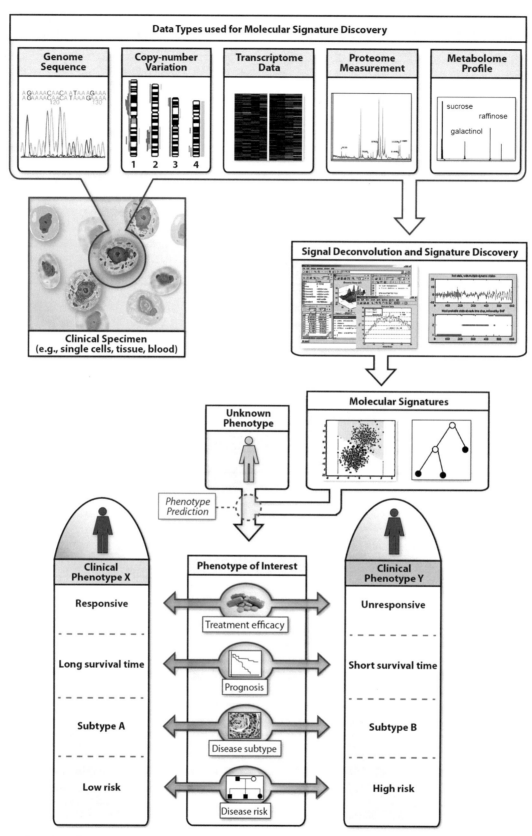

FIGURE 23.9 Overview of the discovery and application of molecular signatures for phenotypic assessment, including disease diagnosis, treatment selection, prognosis, and risk assessment. *(Figure adapted from [51].)*

so high, this increases the chances that overfitting to the observed data will become an significant issue, leading to results that appear promising but that in reality will not hold up for clinical use. For the quantified self, transcriptomics will be only one of the data types, and thus the number of measurements means that statistical approaches to control for overfitting will be essential [36,51]. Fortunately, the continued dramatic reduction in cost for many 'omics' technologies will help to mitigate some of these challenges by making it possible to analyze larger sample numbers, but for what is contemplated in the near future for P4 medicine we are still very much in the small sample regime, and will be for quite some time to come.

This high dimensionality of data exacerbates issues related to data reproducibility, which can be difficult to reproduce in detail from study to study, or even from batch to batch [52]. For example, in order to statistically expect overlap of 50% in identified differentially expressed genes between two different studies comparing breast cancer to normal tissue, one would need of the order of thousands of samples [53]. Almost all individual studies today have fewer than that, and so even if executed with the highest possible experimental rigor one would not expect differentially expressed gene lists between studies to be very similar — even if everything is done correctly by both laboratories doing the studies. The same is true for identified molecular signatures, where a number of molecular measurements are coupled with a computational algorithm to differentiate phenotypes (e.g., make a disease diagnosis) [54].

Importantly, when one study is used to train a molecular signature to differentiate between different phenotypes (e.g., cancer vs. control) and then tested on a separate study of the same phenotypes, very often the classifier will fail, or at best the signature performance degrades severely. A primary reason for this drop in performance comes from heterogeneity between studies, due both to underlying variance in the biology of the patients studied and to technical variations in precisely how the data were measured, normalized, and analyzed. Whenever two individual studies are compared it is very often the case that the differences that we will refer to here as laboratory effects are greater than the differences due to phenotype (e.g., cancer vs. normal). One powerful means for making signatures much more robust is to integrate their identification across multiple studies at multiple sites [55]. In such integrated studies, the signal associated with the phenotype difference is amplified, while the laboratory effects are damped out. Signatures learned across multiple different studies from multiple different laboratories perform much better on average on yet additional studies than do signatures learned from one study alone. This fact argues strongly for the need to [1] build consortiums that enable the integration of large amounts of data from multiple sites [14] and [2] make data publicly available so that they can be aggregated together in meta-analyses. Such data integration is essential to enable P4 medicine.

Computational challenges of blood as a window. The computational challenges associated with maintenance of wellness and the pre-symptomatic diagnosis and prevention of disease are particularly challenging. For example, the envisioned blood diagnostics of the future must be able to distinguish not just one disease from normal but rather must differentiate any possible disease against the background of normal that can be affected by many conditions — including even mundane changes such as diet, exercise, time of day, sleep cycle and so forth. As is well appreciated in machine learning, accuracies tend to degrade quickly as more potential phenotypes need to be separated simultaneously. Deciphering signal from noise against such a dynamic and multifaceted background is a daunting challenge indeed. A number of strategies will therefore be important to harness the information content of the blood as a window to health and disease. 1) It is unlikely that any one data platform will be sufficient to achieve the accuracies that will be needed for clinical practice across the wide range of possible disease states. Therefore, to achieve the predictive and preventive aspects of P4 medicine will require multifaceted data analysis, including multiple sources of molecular data from the blood. As is described here, the blood contains enormous numbers of different molecular information sources, including not just the proteins, but also metabolites, miRNAs, mRNAs, circulating cells, antibodies and so forth. It will also be important to link these molecular data with clinical data as well as input from activated and digitally networked patients (such as changes in lifestyle, environmental exposures). Patient activation refers to a person's willingness and ability to manage their health and healthcare, as measured by the Patient Activation Measure [56]. 2) Another key to address this challenge is to build coarse-to-fine hierarchies, where coarse overall assessments are made initially and then followed by tests of increasingly finer levels of specificity, for disease diagnosis and wellness monitoring. For example, organ-specific blood proteins can be used to first answer the question of what organ system is being perturbed. Following this assessment, more specific molecular markers of finer resolution that differentiate different diseases of the organ system can then be used to narrow down the disease possibilities further. Once a disease is distinguished, yet finer resolution molecular signatures can be used to help determine perturbed networks and hence the best therapy options. 3) Considering the plethora of computational challenges above, there are many instances where it is hard to imagine amassing sufficient statistical power to address all the relevant states of wellness and disease. The key in these prevalent cases will be to leverage deep biology and knowledge of mechanisms. In this case, the mapping of molecular networks through systems biology approaches — and the interplay between genomics and the environment — will be crucial to deciphering what signals of the quantified self really matter to disease treatment and health maintenance.

Health information systems of the future. Looking forward, the development of P4 medicine based on 'omics' data will require extremely large repositories of data in minable health information systems where signatures are constantly evaluated, updated, locked down and then re-evaluated for efficacy. Such systems will be based largely on data in the 'real world' of patient treatment and clinical outcomes — since this is where the vast majority of medically relevant data will come from in the future. Learning what factors most affect patient outcomes by broadly measuring new data sources and linking these back to activated patient communities will serve as a powerful paradigm for developing new tests to move forward iteratively through clinical evaluation. Importantly, these systems will be unbiased in the sense that they will record both positive and negative outcomes as seen in the clinic equally (a key problem with current literature practices, where essentially only good outcomes are widely published and reported in the development phases, and bad outcomes are selected against). Such systems will need to be very expansive in terms of numbers of samples and measurements to be integrated — and there will thus be institutional barriers to sharing data that will need to be overcome through collaborative models. Whether the data will be formed as raw data or processed into metadata for storage is a critical question, and almost certainly we will move toward the storage of metadata to reduce the data dimensionality. As an example, the 6 billion nucleotides of the human genome can be compared against a reference genome — and since humans differ by about 0.1% of their genomes we could store only the differences, reducing the data dimensionality by three orders of magnitude. There will be a strong financial incentive to collaborate, as the clear winners in this area will come only from those who can achieve sufficient sample numbers across a sufficient breadth of the population to identify the most robust signatures. Thus, there will be enormous commercial opportunities that will form the basis for emerging health information companies that will mine this data and produce content that is directly usable by consumers (patients) as well as physicians. While the number of signatures that are translated currently is small relative to the number that have been reported in papers, reasons for this from a statistical point of view — given that we are deeply in the small sample regime — are clear. There is every reason to believe that as the data are integrated at the scale that is necessary, with the rigor that is necessary, and with the connection to biological networks and mechanisms that is necessary, these approaches will indeed transform the practice of medicine. And they will enable all of the ideal features of diagnostics — early detection, assessment of the stage of disease progression, stratification of disease, following the response to therapy and detecting reoccurrences of disease.

Systems medicine provides powerful approaches for dealing with signal-to-noise issues and biological complexity. Systems medicine allows us to reduce enormously the dimensionality of the search space for accurate and robust biomarkers. For example, systems approaches have led to organ-specific, cell type-specific and organelle-specific biomarkers reflecting the functioning of key disease-perturbed networks. Such approaches have also been used to identify biomarkers in the blood, using secreted proteins, proteins cleaved from the membranes of the disease-perturbed cells, cytoplasmic and nuclear proteins reflecting the death of cells, etc. The important point going forward is to start with a narrow and targeted set of biomarkers to search for molecular fingerprints that can reflect the disease, including early diagnosis, stratification of the disease types, and assessing the progression of the disease. Accordingly much smaller populations of patients can be used to identify valid biomarker signatures with focused studies that leverage biological knowledge. Exactly this approach has been applied successfully to several mouse model diseases.

These five pillars of systems medicine together with the digital revolution have given rise to the medical opportunities embodied in P4 medicine — prediction, prevention, personalization and participation.

P4 MEDICINE

Systems medicine is focused on developing biological, technical and computational tools to decipher the complexities of disease. P4 medicine employs the strategies and tools of systems medicine for quantifying wellness and demystifying disease for the benefit of the individual, as well as dealing with the societal opportunities and challenges created by this revolution in medicine.

A key component of P4 medicine is the 'activated' (continually informed patients possessing the knowledge, skill and confidence necessary to manage their health and healthcare and that of their families) and 'networked' (patients connected with other patients and other members of their community, working together to enhance personal and community wellbeing) patients [56]. The convergence of systems approaches to wellness and disease with activated and networked patients will result in a P4 medicine that integrates discovery science with clinical practice and health management by networked and activated patients. This integration will generate the virtual clouds of billions of data points for each patient (Figure 23.3). Analyses of this data cloud will decipher the 'network of networks' for each patient and lead to discoveries and the optimization of wellness and disease emerging from relevant patient social networks. P4 medicine is now pioneering something that never existed before — actionable understandings of disease and wellness as a continuum of network states unique in time and space to each individual human being that can be perturbed by various means, including but by no means limited to drugs, to restore and maintain health.

These databases of patient information and social networks will provide physicians and other healthcare providers with the information they need to deliver care tailored to the circumstances of each individual. We are beginning to see this in the case of certain cancers, where the DNA sequencing of the tumors provides insights into the mutations in signal transduction pathways that inform the choice of therapy for the individual patient [57, 58]. In addition, data from clinical encounters, and from networked patients and consumers actively involved in managing their health and that of their families, will provide millions of data points that will enable systems biologists to decipher signal from noise in complex biological networks. These data will be funneled to scientific research centers and emerging health information companies to fuel large-scale studies. Eventually 100 000s or even millions of patients' data on physiological, cellular, and molecular markers will be used to increasingly demystify disease and quantify wellness by augmenting the relevant biological signals.

An exciting new cycle of accelerating biomedical innovation will emerge as systems-medicine discoveries are routed back to patients and consumers, thereby generating more data to fuel further advances in actionable insights into individual biological systems. P4 medicine includes novel techniques and paradigms for facilitating new relationships between scientists, care providers, patients and consumers, as well as for dealing with the social opportunities and challenges that inevitably will arise from these new relationships. Here is a 10-year glimpse into the envisioned future of the 4 Ps.

Predictive. In 10 years nearly everyone will have his or her genome sequenced. 'Actionable genetic variants' — those whose identification opens the door to a course of action that will improve physical health or relieve anxiety for the individual — will drive forward the acceptance of the complete genome sequence as a part of the individual's medical record. While most of the medically relevant variation catalogued to date occurs in the coding regions of genes, it is becoming clear that variations in non-coding regions, copy number, structural variations, and other features of chromosomal architecture play a role in disease etiology. We believe that complete genome sequencing should be done in families. Family sequencing enables the correction of a significant fraction of the DNA sequencing errors, thus generating very accurate sequences. It also provides a deep understanding of the one-dimensional organization of the many genetic variants in the chromosomes of each individual (this is denoted haplotype determination) (Figure 23.8), thereby enormously facilitating the discovery of disease genes or loci. We have identified more than 300 of these actionable variants and many more are being identified with every passing year. Indeed, every person's genome will be reviewed yearly for new actionable variants — and these will provide powerful insights to optimize the wellness of the individual. The genomes of individuals will provide an investment in individual information that will permit a yearly optimization of wellness for the rest of the person's life.

P4 medicine is making blood a window for assessing health and disease. Organ-specific proteins found in the blood will be analyzed in a longitudinal manner across the individual's life (Figure 23.7). Within 10 years, we envision a hand-held device that can prick your finger, take a fraction of a droplet of blood and quantify several thousand organ-specific proteins in 5 minutes. This device will permit the health or impairment status for each of your major 50 organ systems to be followed in a longitudinal manner over time [41]. Moreover, these 'microfluidic protein assays' will permit hundreds of millions of patients to be analyzed routinely, e.g., biannually. Thus transitions from health to disease may immediately be identified and acted upon. The organ-specific blood fingerprints will allow us to stratify diseases into their distinct subtypes (for an impedance match against proper therapies) and to follow the progression of a disease. In the future different drugs will be effective against different stages of a disease. In addition, organ-specific blood protein fingerprints will enable one to analyze the responses of multiple organs to a given disease — and thereby extend the analysis of the 'networks of networks' from DNA, molecules and cells to organs (Figure 23.4). Finally, organ-specific blood protein will enable one to rapidly and precisely identify off-target reactions of drugs, thereby providing a powerful approach to assessing drug toxicities.

Preventive. Systems analyses provide insights into the dynamics of disease-perturbed networks. A new 'network-centric' rather than 'gene-centric' approach to choosing drug targets will employ multiple drugs to 're-engineer' a disease-perturbed network to make it behave in a more normal manner. We are now exploring this possibility in microorganisms to learn the fundamental principles of network re-engineering before applying this strategy to the more complex requirements of human disease therapies [59]. Our conviction is that the treatment of disease in the future will often require the combination of two or more drugs. Tools are now being developed to explore the combinatorial analyses of the hundreds of drugs that have met the safety requirement — both those that were efficacious for a particular disease and those that were not. Hence for the effective and non-useful drugs perhaps combinations could be identified to more effectively attack a wide variety of diseases. This could make drugs more effective and far less expensive, because there is a rationale for the choice of drug targets and their perturbations by drugs. Through longitudinal multi-biomarker (blood protein, miRNA, mRNA, metabolites etc.) analyses, P4 medicine will be able to predict the potential future emergence of disease-perturbed networks in patients and then design 'preventive drugs' that will block the emergence of these disease-perturbed networks and their cognate diseases. A systems approach to the immune response will, in time, give us a deeper understanding of how to create effective cellular as well as humoral immune responses, permitting us to create effective vaccines for scourges such as AIDS. Clearly, stem cells will provide powerful possibilities in the future for replacing damaged cellular and even organ components (as well as being powerful tools for understanding disease mechanism and stratifying disease). Finally, the digitized data defining the quantified self will provide powerful new insights into optimizing wellness for the individual. Indeed, the focus of P4 medicine will increasingly move from disease to wellness.

Personalized. On average, humans differ from one another by about 6 million nucleotides in their genomes — hence we, individually, are genetically unique. Even identical twins each may exhibit 35 different 'intergenerational mutational nucleotide differences' from each parent and from one another [33]. Each person must be treated as a unique individual and not as a statistical average. We must take account of the fact individuals vary in ways that significantly affect effective treatment. Individuals should each serve as their own controls to determine when their own data reflect transitions from health to disease. Moreover, there is a growing sentiment that observations on single individuals may collectively provide fundamental new insights into the disease (or wellness) process. The so-called experiments where patient N equals 1 may open up powerful new approaches to more effectively dealing with the individual patient and aggregating the useful data they generate. Indeed, the first molecularly detailed tracking of two individuals (the 'quantified self') is already yielding fascinating insights [12,25]. Imagine, in 10 years, 340 million Americans each with billions of data points: this will potentially create a powerful aggregated data source from which to infer the predictive medicine of the future.

In this regard, we believe that it will be critical to champion the idea that it is essential that all patients' data (with appropriate privacy measures including anonymization) be made available through an appropriately constituted entity for qualified researchers and physicians to mine for the predictive medicine of the future. After all, this contribution will enable us to revolutionize the healthcare for our children and grandchildren — a point to which most patients are responsive. Moreover, laws must be passed to protect the individual against the exploitation of their medical data by other elements of society, such as employers or insurance companies. It is interesting to note that all of us make available our entire financial histories to three credit agencies for the convenience of having a credit card. Surely patients will recognize the benefits of being able to mine their collective data to pioneer the future of P4 medicine for the benefit of their families.

Participatory. P4 medicine relies greatly on the positive contributions of activated patients and consumers. Our existing healthcare system is not well adapted to exploit the new capabilities of P4 medicine. Physicians as well as pharmaceutical and medical device companies are compensated solely for the delivery of specific procedures and products, hence they have limited financial incentive to deploy new innovations to predict or prevent disease or to maintain wellness. Moreover, the healthcare industry is locked into financial and regulatory models based on large-scale population studies that ignore crucial genetic and environmental exposure differences among individuals. Pressure for change is beginning to be felt as the medical profession faces the looming challenge of increasingly being compensated for outcomes as opposed to service delivery. However, the most important source of pressure for change will be newly activated and networked patients and consumers. Collectively, they will constitute a vital new stakeholder in P4 medicine very different from the passive recipients (patients) of expert advice characteristic of pre-digital medicine. Activated and networked consumers will do more than demand more effective healthcare — they will help direct the changes to achieve it.

Activated and networked consumers are beginning to push for healthcare that is adapted to their own particular circumstances, including their individual genome (which is static and need be sequenced only once) and dynamic measurements such as from blood (which change over time, must be measured longitudinally in time, and can track changing wellness and disease states) [10]. They are also beginning to push for new ways in which to engage with our science-based healthcare system to maintain wellness and achieve life goals as well treating disease. Because of the reactive nature of the existing healthcare system, which is more accurately described as the 'disease management industry', the vast bulk of health management in areas such as nutrition, exercise and sleep takes place in the home, without the assistance of physicians or other professionally trained care providers. The writers of bestsellers about the latest diet and purveyors of unregulated health products operate largely outside the constraints of science-based healthcare. Today's educated consumers are increasingly conscious of this fact and are beginning to demand that science-based healthcare address their need for assistance in managing these other areas of their own health. They are stimulating the growth of a new market for devices that deliver increasingly real-time digital data about every aspect of their health (quantifying their wellness), ranging from activity levels to vital signs, and many of them are starting to come to their physicians for help in interpreting these data. The digital revolution is beginning to help fashion a new dimension for healthcare — the wellness arising from the quantified self.

P4 medicine responds to these growing demands by providing patients and consumers with actionable information that they can use to improve their health. Clinical institutions using wellness coaches, genetic counselors and physicians can provide this information cost-effectively. It will be conveyed largely through digitally linked social networks, the most important of which will be family networks. One effective strategy may be to identify family members who are the most active in setting familial health-related standards and in caring for members with health problems, and then to work with those individuals to help them do a better job. Medicine today systemically approaches patients as statistical abstractions, relying on the efforts of time-pressed physicians to achieve some degree of personalized care. Working with family and social networks will allow P4 medicine to systemically and more effectively deal with the reality of the social context in which patients and consumers are embedded and which largely determines how they eat, exercise and sleep. Activated patents and consumers working effectively within their family and social networks to utilize the increasingly real-time flow of personalized health-related data will be able to reduce the incidence of and to better manage complex diseases such as type II diabetes, which account for a huge percentage of total healthcare costs and, of course, to optimize their wellness.

To achieve this goal we need to develop and continually update a 'gold standard' of reliable data, information and explanations of disease and health that will meet the needs of both physicians and patients. Developing and maintaining this gold standard will require a close working relationship between clinical, systems-based scientists and patients. In addition, we need to develop ways to actively counter misinformation that might begin to spread through social networks and to correct misinformation often found on the current medical websites and in other sources of medical information. While significant, these challenges are outweighed by benefits to be gained by wide dissemination of actionable 'gold standard' of personalized health data reviewed for consistency with the standards of our science-based healthcare system.

Ultimately patients will be recognized as not only a source of disease problems to be solved but as a source of disease and wellness solutions as revealed by their data. Creative new forms of engagement with networked and activated consumers as active participants in healthcare, as opposed to passive recipients of expert advice, will become a major source of value tapped by the healthcare revolution. Networked and activated participants will find new ways to adjust diet and exercise to move their biomarkers in the direction of better health. Crowd sourcing these problems will yield many benefits. For example, researchers will be able to correlate their behavior changes with biomarker fluctuations, their genome, their medical histories and other key parameters. Such data from millions or even tens of thousands of patients would provide researchers with deep insights into the effects of nutrition and exercise that have never before been possible. These large-scale personalized data sets would be the basis for the quantification of wellness, providing society with a far more effective understanding of the effects of diet, exercise and sleep on highly stratified population sectors.

The digitization of P4 medicine enables its distribution to all citizens of the world — both developed and less developed. For example, we remember those who thought the initial large brick-like cell phones of the early 1990s were ridiculous and could not imagine their widespread acceptance (there are now more than 4 billion cell phones worldwide). Today, a woman in a rural village in India can make a living for her family with a cell phone thanks to the digitization of communications, with its potential for transforming the economic conditions of even the poor. So we will see a 'democratization of P4 medicine' throughout the world as inexpensive and digitized P4 medicine becomes available (see below).

To summarize, we argue that P4 medicine has two major objectives for each participant: to quantify their wellness and to demystify their disease (Figure 23.2). Our feeling is that the quantification of wellness will become increasingly important over time, ultimately dominating as the concern of most individuals. Table 23.1 provides a striking comparison of proactive P4 medicine with contemporary, reactive evidence-based medicine.

TWO BIG CHALLENGES: EDUCATION AND INFORMATION TECHNOLOGY FOR HEALTHCARE

One big challenge for P4 medicine is the education of consumers, patients, physicians, and the members of the broader medical community, including the principal stakeholders in the healthcare industry. This education will present an enormous challenge and will ultimately require the effective exploitation of social networks for education integrated with new effective information technology teaching strategies. Many individuals will initially want to remain 'old-fashioned patients' letting the doctor tell them what is best. However, once individuals see the power of consumer-driven medicine to improve individual health, that will change (just as skepticism has disappeared with the widespread acceptance of cell phones).

The Institute for Systems Biology has successfully developed modules for teaching systems biology to insert leading-edge biology into high-school biology courses, and is currently developing similar modules for P4 medicine. Early education of consumers/patients is key. Another interesting idea is the suggestion that there be a commercial TV program, hopefully with very broad coverage, along the lines of the forensic CSI TV program to explore solving the problems of P4 medicine in a well-written and compelling manner that brings a knowledge of P4 medicine to the average viewer, just as CSI has brought insights in crime forensics to a broad audience. Another possibility is that one could use computer-game-like strategies to bring the principles of P4 medicine to patients, physicians and members of the healthcare community — at least for those who are comfortable with the digital revolution.

Another challenge is how to produce an information technology (IT) for healthcare that can handle the enormous multi-scale data dimensionality that will arise from P4 medicine — for in the end P4 medicine is defined by the interconnected 'network of networks' — genetic networks connected to molecular networks, to cellular networks, to organ networks, to the networks of individuals in society, for each provide unique insights into the complexities of disease (Figure 23.4). We must understand the individual in the context of all of these integrated networks, as this is the only way to capture both the digital information of the genome and all of the diverse environmental signals impinging on the individual from many different sources. This requirement places enormous demands on the need to develop an effective IT for healthcare. Healthcare IT must be comprehensive, interoperable, data-driven (e.g., bottom-up), biology-driven and, we believe, fundamentally open source. It is probably beyond the capacity of any single organization to fashion a comprehensive IT for healthcare that goes beyond medical records to encompass the collection and distribution of the entire heterogeneous data cloud at the heart of P4 medicine. Yet that is what is required, and an effectively orchestrated open-source approach could transform IT for healthcare. We must be able to capture the deep insights that will come from various patient social networks.

IMPACT OF P4 MEDICINE ON SOCIETY

P4 medicine will have an enormous impact on society and healthcare.

1. P4 medicine will transform the practice of healthcare in virtually every way. Table 23.2 provides a summary of some of the powerful new strategies and technologies P4 medicine will create and employ.

2. P4 medicine will require that all healthcare companies rewrite their business plans in the next 10 years or so. Many will not be able to do so and will become 'industrial dinosaurs'. There will be enormous economic opportunities for the emergence of new companies tailored to the needs and opportunities of P4 medicine.

3. P4 medicine will at some time in the future turn around the ever-escalating costs of healthcare and will in fact reduce these costs to the point where P4 medicine can be exported to the developing world, enabling a 'democratization of healthcare' unimaginable even 5 years ago. These savings will arise from many of the features described in Table 23.2: the early diagnosis and hence more effective treatment of disease; the stratification of each major disease into its major subtypes to achieve a proper impedance match for each individual against a drug effective for a particular subtype of disease; the ability to identify genetic variants that cause drugs to be metabolized in a manner dangerous to the patient (this is termed pharmacogenomics, and more than 50 such variants have been identified to date); the ability to 're-engineer' disease-perturbed networks with drugs to generate a powerful and less-expensive rationale for drug-target selection; an increasing focus on wellness for each individual; and the emergence of striking near-term advances in modern medicine. These include an increasing ability to deal effectively with cancer, and to use stem cells for replacement therapy. Additionally, these advances will lead to new approaches to diagnostics and understanding disease mechanisms, an understanding of aging that will allow individuals to optimize and extend their effective mental and physical health routinely into their 80s and 90s, an understanding of the metagenome (e.g., population of microbes) of the gut and other body surfaces that will provide deep insights into one incredibly important manner in which the microbes of our environment influence our health, and finally the emergence

of a deep understanding of neurodegeneration to avoid the personal and societal tragedies of diseases such as Alzheimer's and Parkinson's.

4. P4 medicine, through its driving of the emergence of new technologies and computational techniques, is pioneering the digitization of medicine. The 'quantified self' will provide the data that will enable each individual to optimize his or her own health. This will also provide the data to empower P4 medicine to revolutionize healthcare through consumer-driven social networks (Figure 23.1). The digitization of medicine through the generation of big data sets for each individual — allowing one to sculpt with exquisite specificity wellness and appropriate responses to emerging diseases — is one of the transforming aspects of P4 medicine. Another important implication arising from the digitization of medicine is the fact that personal data will become incredibly inexpensive (e.g., it is estimated the first human genome sequence finished in 2003 cost about $1 billion; today it costs a few thousand dollars, and in a few years genome sequences will cost perhaps $100) — thus digital technologies and their exponentially declining costs will contributing significantly to reversing the escalating costs of healthcare.

5. P4 medicine will bring increased wealth to the healthcare systems, communities and nations that practice it. The decreasing costs of healthcare have been mentioned above. Many economic opportunities will evolve from the knowledge of P4 medicine through the transformation of the healthcare industry. We predict that there will be a 'wellness industry' that will emerge over the next 10–15 years that will in time far exceed the size of the healthcare industry. P4 medicine is an area replete with economic opportunities for those who are practicing it at the leading edge.

6. The patient (consumer), through social networks, will drive the emergence of P4 medicine. Because of intrinsic conservatism and sclerotic bureaucratic systems, physicians, healthcare specialists and the healthcare industry will take a back seat to the power of patient-driven social networks in bringing change to the healthcare system. Indeed, patients may be the only driving force capable of truly changing our contemporary healthcare system to the proactive P4 mode.

HOW TO BRING P4 MEDICINE TO PATIENTS

The challenges of bringing P4 medicine to patients and consumers have two critical dimensions, technological and societal. The latter is far more complex. A variety of efforts are focused on bringing personalized medicine to healthcare and to dealing with some of the societal issues of this new medicine (including the Personalized Medicine Coalition, established to advance the future of personalized medicine by connecting the scientists, clinicians, the media and the general community in a common goal). We will discuss below only our own efforts to bring P4 medicine to patients.

At the Institute for Systems Biology, in conjunction with Ohio State Medical School and PeaceHealth clinics, we have created the P4 Medicine Institute (P4MI), a non-profit organization that is committed to creating a network of six or so clinical centers with the ISB to employ the strategies and tools of P4 medicine in pilot projects to prove the power of P4 medicine. Pilot project success will be critical in convincing conservative physicians, a skeptical medical community and an often bureaucratic and herd-driven healthcare industry as to the potential of P4 medicine. P4MI is also interested in bringing relevant industrial partners to this clinical network. In addition, P4MI has established a Fellows Program to begin delving into some of the societal challenges of P4 medicine, eventually expressing these issues through 'white papers' on economics, the 'gold standard of healthcare information', ethics, regulations, etc. Strategic partnerships are a critical component for bringing P4 to patients worldwide and gaining its widespread acceptance in the national and international medical communities.

As the P4MI pilot programs become successful and begin demonstrating the power of P4 medicine, we would like next to persuade a small country to build a P4 medicine/healthcare system. This country could play a key leadership role in pioneering the new medicine of the 21st century, just as Johns Hopkins Medical School in the early 1900s adopted some recommendations of the Flexner Report on the future of medicine (sponsored by the Carnegie Foundation), which included the recommendation that medicine should integrate basic research and clinical medicine. Thus Johns Hopkins propelled itself from a mediocre medical trade school into a world leader in US and world medicine. Now there is now a similar opportunity for institutions (and countries) that pioneer P4 medicine/healthcare to become world leaders. It will take courage, leadership, resources and an effective communication of the vision to the stakeholders to catalyze the revolution in P4 medicine/healthcare. It goes without saying that any nation that assumes a leading-edge leadership role in catalyzing the emergence of P4 healthcare will be in a unique position to transform the healthcare of its citizens, to revolutionize its medical research agenda and to take advantage of the associated economic opportunities associated with the newly emerging world of P4 medicine.

ACKNOWLEDGMENTS

Thanks to Lee Rowen for advice and counsel in preparing this paper and Jaeyun Sung for help with Figure 23.9. This paper was in part adapted from a paper now in press — L Hood and M Flores; Systems Medicine and the Emergence of Proactive P4 Medicine: Predictive, Preventive, Personalized and Participatory; New Biotechnology, in press. LH would like to acknowledge the support of the Luxembourg Centre for Systems Biomedicine and the University of Luxembourg, the General Medical Sciences Center for Systems Biology GM076547 and a Department of Defense contract on Liver Toxicity W911SR-09-C-0062. NDP acknowledges funding from an NIH-NCI Howard Temin Pathway to Independence Award in Cancer Research, a Roy J. Carver Young Investigator Grant, and the Camille Dreyfus Teacher–Scholar program.

REFERENCES

[1] Hood L. Deciphering Complexity: A personal view of systems biology and the coming of 'Big' science. Genet Eng News 2011;31:131.

[2] Barabasi AL, Oltvai ZN. Network biology: understanding the cell's functional organization. Nat Rev Genet 2004;5:101–13.

[3] Hood L, Heath JR, Phelps ME, Lin B. Systems biology and new technologies enable predictive and preventative medicine. Science 2004;306:640–3.

[4] Patrinos GP, Brookes AJ. DNA, diseases and databases: disastrously deficient. Trends Genet 2005;21:333–8.

[5] Loscalzo J, Kohane I, Barabasi AL. Human disease classification in the postgenomic era: a complex systems approach to human pathobiology. Mol Syst Biol 2007;3:124.

[6] Auffray C, Chen Z, Hood L. Systems medicine: the future of medical genomics and healthcare. Genome Med 2009;1:2.

[7] Price N, Edelman L, Lee I, Yoo H, Hwang D, Carlson G, et al. Systems biology and the emergence of systems medicine. Genomic and Personalized Medicine: From Principles to Practice 2009;1:131–41.

[8] Buell J. The Digital Medicine Revolution in Healthcare. American College of Healthcare Executives, Chicago 2011.

[9] Garg V, Arora S, Gupta C. Cloud computing approaches to accelerate drug discovery value chain. Comb Chem High Throughput Screen 2011;14:861–71.

[10] Topol E. The Creative Destruction of Medicine: How the Digital Revolution Will Create Better Health Care, a Member of the Perseus Books Group. New York: Basic Books; 2012.

[11] Collino S, Martin FP, Rezzi S. Clinical metabolomics paves the way towards future healthcare strategies. Br J Clin Pharmacol 2012. doi: 10.1111/j.1365-2125.2012.04216.x.

[12] Smarr L. Quantified Health: A 10-year detective story of the digitally enabled genomic medicine. Strateg News Lett 2011;14:1–33.

[13] Weston AD, Hood L. Systems biology, proteomics, and the future of health care: toward predictive, preventative, and personalized medicine. J Proteome Res 2004;3:179–96.

[14] Hood L, Friend SH. Predictive, personalized, preventive, participatory (P4) cancer medicine. Nat Rev Clin Oncol 2011;8:184–7.

[15] Tian Q, Price ND, Hood L. Systems cancer medicine: towards realization of predictive, preventive, personalized and participatory (P4) medicine. J Intern Med 2012;271:111–21.

[16] Hwang D, Lee IY, Yoo H, Gehlenborg N, Cho JH, Petritis B, et al. A systems approach to prion disease. Mol Syst Biol 2009;5:252.

[17] Hidalgo CA, Blumm N, Barabasi AL, Christakis NA. A dynamic network approach for the study of human phenotypes. PLoS Comput Biol 2009;5:e1000353.

[18] Kinross JM, Darzi AW, Nicholson JK. Gut microbiome–host interactions in health and disease. Genome Med 2011;3:14.

[19] Williams A, Smith JR, Allaway D, Harris P, Liddell S, Mobasheri A. Applications of proteomics in cartilage biology and osteoarthritis research. Front Biosci 2012;17:2622–44.

[20] Janecka I. Is U.S. health care an appropriate system? A strategic perspective from systems science. Health Res Policy Syst 2009;7.

[21] Christakis NA, Fowler JH. The spread of obesity in a large social network over 32 years. N Engl J Med 2007;357:370–9.

[22] Hood L. A personal journey of discovery: developing technology and changing biology. Annu Rev Anal Chem (Palo Alto Calif) 2008;1:1–43.

[23] Hood L. Acceptance remarks for Fritz J. and Delores H. Russ Prize. NAE Journal The Bridge 2011;41:46–9.

[24] Hood L, Rowen L, Galas DJ, Aitchison JD. Systems biology at the institute for systems biology. Brief Funct Genomic Proteomic 2008;7:239–48.

[25] Chen R, Mias GI, Li-Pook-Than J, Jiang L, Lam HY, Chen R, et al. Personal omics profiling reveals dynamic molecular and medical phenotypes. Cell 2012;148:1293–307.

[26] Lazebnik Y. Can a biologist fix a radio? Or, what I learned while studying apoptosis. Biochemistry 2004;69:1403–6.

[27] Ravasi T, Suzuki H, Cannistraci CV, Katayama S, Bajic VB, Tan K, et al. An atlas of combinatorial transcriptional regulation in mouse and man. Cell 2010;140:744–52.

[28] Keller A, Backes C, Leidinger P, Kefer N, Boisguerin V, Barbacioru C, et al. Next-generation sequencing identifies novel microRNAs in peripheral blood of lung cancer patients. Mol Biosyst 2011;7:3187–99.

[29] Pinzani P, Salvianti F, Zaccara S, Massi D, De Giorgi V, Pazzagli M, et al. Circulating cell-free DNA in plasma of melanoma patients: qualitative and quantitative considerations. Clin Chim Acta 2011;412:2141–5.

[30] Schwarzenbach H, Hoon DS, Pantel K. Cell-free nucleic acids as biomarkers in cancer patients. Nat Rev Cancer 2011;11:426–37.

[31] Chapman MH, Sandanayake NS, Andreola F, Dhar DK, Webster GJ, Dooley JS, Pereira SP. Circulating CYFRA 21–1 is a specific diagnostic and prognostic biomarker in biliary tract cancer. J Clin Exp Hepatol 2011;1:6–12.

[32] Matsusaka S, Suenaga M, Mishima Y, Kuniyoshi R, Takagi K, Terui Y, et al. Circulating tumor cells as a surrogate marker for determining response to chemotherapy in Japanese patients with metastatic colorectal cancer. Cancer Sci 2011;102: 1188–92.

[33] Roach JC, Glusman G, Smit AF, Huff CD, Hubley R, Shannon PT, et al. Analysis of genetic inheritance in a family quartet by whole-genome sequencing. Science 2010;328:636–9.

[34] Chiu CL, Randall S, Molloy MP. Recent progress in selected reaction monitoring MS-driven plasma protein biomarker analysis. Bioanalysis 2009;1:847–55.

[35] Yao M, Ma L, Duchoslav E, Zhu M. Rapid screening and characterization of drug metabolites using multiple ion monitoring dependent product ion scan and postacquisition data mining on a hybrid triple quadrupole-linear ion trap mass spectrometer. Rapid Commun Mass Spectrom 2009;23:1683–93.

[36] Ma S, Funk CC, Price ND. Systems approaches to molecular cancer diagnostics. Discov Med 2010;10:531–42.

[37] Bartfai T, Buckley PT, Eberwine J. Drug targets: single-cell transcriptomics hastens unbiased discovery. Trends Pharmacol Sci 2012;33:9—16.

[38] Liu L, Ye Q, Wu Y, Hsieh WY, Chen CL, Shen HH, et al. Tracking T-cells in vivo with a new nano-sized MRI contrast agent. Nanomedicine 2012; http://dx.doi.org/10.1016/j.nano.2012.02.017.

[39] Heath JR, Davis ME, Hood L. Nanomedicine targets cancer. Sci Am 2009;300:44—51.

[40] Chakraborty M, Jain S, Rani V. Nanotechnology: emerging tool for diagnostics and therapeutics. Appl Biochem Biotechnol 2011;165:1178—87.

[41] Shi Q, Qin L, Wei W, Geng F, Fan R, Shin YS, et al. Single-cell proteomic chip for profiling intracellular signaling pathways in single tumor cells. Proc Natl Acad Sci U S A 2012;109:419—24.

[42] Hodgson J. Gene sequencing's industrial revolution. Spectrum 2000;37:36—42.

[43] International human genome sequencing consortium. Finishing the euchromatic sequence of the human genome. Nature 2004;431:931—45.

[44] Oxford Nanopore Press release. DNA 'Strand Sequencing' on the High-Throughput GridION Platform and Presents MinION, a Sequencer the Size of a USB Memory Stick. Oxford: Oxford Nanopore; 2012.

[45] American Medical Association (AMA) Proteomics, http://www.ama-assn.org; 2012.

[46] Juritz EI, Alberti SF, Parisi GD. PCDB: a database of protein conformational diversity. Nucleic Acids Res 2011;39:D475—479.

[47] Moritz R. Institute for systems biology. Pers Commun 2012.

[48] Connor AC, McGown LB. Aptamer stationary phase for protein capture in affinity capillary chromatography. J Chromatogr A 2006;1111:115—9.

[49] Zichel R, Chearwae W, Pandey GS, Golding B, Sauna ZE. Aptamers as a sensitive tool to detect subtle modifications in therapeutic proteins. PLoS One 2012;7:e31948.

[50] Rakhila H, Rozek T, Hopkins A, Proudman S, Cleland L, James M, et al. Quantitation of total and free teriflunomide (A77 1726) in human plasma by LC-MS/MS. J Pharm Biomed Anal 2011;55:325—31.

[51] Sung J. Molecular signatures from omics data: from chaos to consensus. Biotechnol J 2012;7(8)946—57.

[52] Leek JT, Scharpf RB, Bravo HC, Simcha D, Langmead B, Johnson WE, et al. Tackling the widespread and critical impact of batch effects in high-throughput data. Nat Rev Genet 2010;11:733—9.

[53] Ein-Dor L, Zuk O, Domany E. Thousands of samples are needed to generate a robust gene list for predicting outcome in cancer. Proc Natl Acad Sci U S A 2006;103:5923—8.

[54] Quackenbush J. Microarray analysis and tumor classification. N Engl J Med 2006;354:2463—72.

[55] Sirota M, Dudley JT, Kim J, Chiang AP, Morgan AA, Sweet-Cordero A, et al. Discovery and preclinical validation of drug indications using compendia of public gene expression data. Sci Transl Med 2011;3:96ra77.

[56] Hibbard JH, Stockard J, Mahoney ER, Tusler M. Development of the Patient Activation Measure (PAM): conceptualizing and measuring activation in patients and consumers. Health Serv Res 2004;39:1005—26.

[57] Pleasance ED, Cheetham RK, Stephens PJ, McBride DJ, Humphray SJ, Greenman CD, et al. A comprehensive catalogue of somatic mutations from a human cancer genome. Nature 2010;463:191—6.

[58] Dancey JE, Bedard PL, Onetto N, Hudson TJ. The genetic basis for cancer treatment decisions. Cell 2012;148:409—20.

[59] Bonneau R, Facciotti MT, Reiss DJ, Schmid AK, Pan M, Kaur A, et al. A predictive model for transcriptional control of physiology in a free living cell. Cell 2007;131:1354—65.

[60] Barabási A-L. N Engl J Med 2007;357(4):404—7

Chapter 24

Cancer Systems Biology: A Robustness-Based Approach

Hiroaki Kitano
The Systems Biology Institute, Tokyo, Japan; Okinawa Institute of Science and Technology; Sony Computer Science Laboratories, Inc., Tokyo, Japan; Department of Cancer Systems Biology, The Cancer Institute, Tokyo, Japan

Chapter Outline

Introduction	469	Mechanisms for Cancer Robustness	471
Cancer as a Robust System	469	Robustness Trade-offs	473
What is Robustness?	469	Theoretically Motivated Therapy Strategies	473
Robustness and Homeostasis	470	A Proper Index of Treatment Efficacy	475
Mechanisms for Robustness	471	Long-Tail Drug	476
Systems Control	471	Open Pharma	477
Fault Tolerance	471	Conclusion	478
Modularity	471	Acknowledgements	478
Decoupling	471	References	478

INTRODUCTION

Cancer as a Robust System

Cancer is a heterogeneous and highly robust disease that represents the worst-case scenario of entire system failure: a fail-on fault where malfunction components are protected by mechanisms that support robustness in normal physiology [1,2]. It involves hijacking the robustness mechanisms of the host. The survival and proliferation capability of tumor cells are robustly maintained against a range of therapies, due to intratumoral genetic diversity, feedback loops for multidrug resistance, tumor–host interactions, etc. This chapter examines why cancer is robust against therapeutic interventions and tries to elucidate possible options for us to control it.

What is Robustness?

Robustness is a property of the system that maintains a certain function despite external and internal perturbations [3]. It is distinctively a system-level property that cannot be observed by just looking at components. Robustness is observed ubiquitously within and among biological systems; from bacteria to human beings. For example, bacteria robustly maintain their chemotaxis circuit's capability against a broad range of perturbations in external chemical changes and internal fluctuations of enzyme dosages [4]. Robustness also applies to engineering systems. For instance, a modern airplane (system) maintains its flight path (function) against atmospheric turbulence (perturbations). Specific aspects of the system, the functions to be maintained, and the types of perturbations that the system is robust against, must be well defined in order to make solid deductions.

Robustness is not necessarily identical to homeostasis or stability. Homeostasis is a term coined by Walter Cannon, meaning an identical (homo) state (stasis). In 'Wisdom of the body', Cannon describes it as follows:

'The everlasting state maintained in the body may be called an equilibrium state. But, this word has a fairly precise meaning now since it has been used for a comparatively simple physicochemical state in which known forces are balanced, namely used for closed systems. The interrelated physiologic action that maintains the main portions of stable states in a living body, which contains the brain, the nerves, the heart, the lungs, the kidneys and the spleen to fulfill their functions collaboratively, is so complex and unique that I have been proposing to use a special word of 'homeostasis'

for such a state. This word does not indicate something that is fixed and does not move, or a stasis. It means a certain state that may change but is relatively constant.'[5]

Stability is a concept similar to homeostasis because it is basically judged by the degree of maintenance of a state. Robustness, on the other hand, is focused on the maintenance of functions as a criterion, not of the maintenance of a state. That is to say, if a state changes substantially in order to maintain the functions, it can be considered as a kind of robustness, but not as homeostasis or stability.

With mathematical abstraction, the state of the system can be expressed in N-dimensional phase space. Figure 24.1 represents a simplified two-dimensional phase space. When perturbations are imposed, the state of the system drifts in phase space. If the degree of perturbation is small and the system tends to be stable, then the phase space trajectory of the system's state orbits around the 'basin' of the attractor and gradually returns to the initial state. Bacterial chemotaxis machinery is an example of such a case.

However, there are cases where perturbations trigger a transition of the state from one basin of the attractor to the other. Systems may be robust in this case by switching to a new state that maintains functions, rather than returning to the original state. Nature provides a satisfactory example: the tardigrade [6]. This creature crawls around in its normal state. When it enters in a high-salt environment or a very dry environment, it becomes dehydrated, stops metabolic activity and enters into a kind of freeze-dry state. It survives in this state for years and can start crawling again when water is available. In this case, it gave up the homeostasis of metabolism and moved into the state of survival without water, to cope with a major external perturbation. This creature is robust against dehydration stress, but this robustness is generated by the sacrifice of the homeostasis of metabolism.

Robustness and Homeostasis

Attention should be paid to the essential difference between robustness and homeostasis. There is a major impact on selection in evolution from the functions associated with higher probabilities of survival and reproduction, not directly from the states. However, the functions related to maintaining the state of an individual are subject to selection only when they are linked to the possibility of survival and reproduction. The system's behaviour of increasing the probability of survival by moving into a completely different state to cope with perturbations, just like the response of the tardigrade against dehydration, may be considered robust, but it cannot be identified as the maintenance of homeostasis. When the continuation of a certain state is effective in maintaining the functions of a system, robustness and homeostasis (or stability) are equivalent, but when the transition of the system into a new state is an effective response to a disturbance, a robust system will abandon homeostasis.

In addition, there are some cases in which robustness is maintained because a system is unstable. HIV has a very high mutation rate and part of its gene sequence changes frequently. Because of this, HIV escapes from immune system control. In the case of cancer, the chromosome becomes more unstable during advanced stages and groups

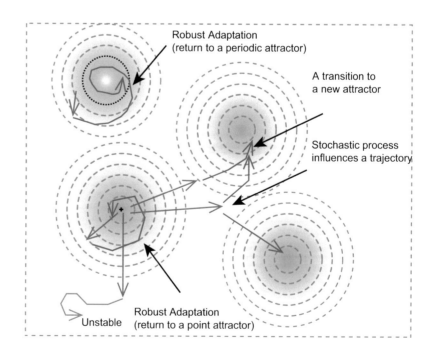

FIGURE 24.1 **Robustness and stability.**

with a variety of abnormalities in chromosomes, which coexist even in a solid cancer. Therefore, even if a part of a tumor cluster is eliminated by an anticancer drug, subsets of tumor cells with genetic properties tolerant to anticancer drugs will survive and continue to proliferate. This strategy is made possible by the instability of chromosomes.

MECHANISMS FOR ROBUSTNESS

Here we investigate basic mechanisms for making systems robust. There are at least four basic mechanisms: systems control, fault tolerance, modularity, and decoupling.

Systems Control

Extensive systems control is used in biological systems. Negative feedback, positive feedback, and feedforward are used to make a system dynamically stable around the specific state of the system, or to form bistable and multistable switches that drive transition of the state into different attractors. Bacteria chemotaxis is an example of how negative feedback enables robust, but fragile control against a wide range of fluctuations in chemical concentration [4,7,8]. Owing to integral feedback, bacteria can sense changes in chemoattractant and chemorepellent environments, independent of absolute concentration, so that proper chemotaxis behaviour is maintained over a wide range of ligand concentrations. In addition, the same mechanism makes it insensitive to changes in rate constants involved in the circuit. This is an example of how systems control is used to maintain a system's state within an attractor so that proper function can be maintained.

Positive feedbacks are often used to create bistability in signal transduction and the cell cycle, so that the system is tolerant against minor perturbations from stimuli. Details are available elsewhere [9–11].

Fault Tolerance

Fault-tolerant mechanisms increase tolerance against component failure and environmental changes by providing alternative components or methods to ultimately maintain a function of the system. Sometimes there are multiple components that are similar to each other and are redundant. In other cases, multiple components or circuits with overlapping functions may exist to compensate for any component insufficiency. This is called diversity. The difference between redundancy and diversity is clear. For example, multiple phone lines are 'redundant', but alternative access to internet, phone, fax, and other means of communication is 'diversity'. Redundancy and diversity are often considered as opposites, but it is more consistent to view them as different ways of providing alternative fail-tolerant mechanisms.

Modularity

Modularity provides isolation of a perturbation from the rest of the system. The cell is the most significant example. More subtle examples are modules of biochemical and gene regulatory networks. Modules that buffer perturbations also play an important role during developmental processes, so that proper pattern formation can be accomplished [12–14]. The definition of modules and the methods for their detection are still controversial, but the general consensus is that modules do exist and that they play an important role [15].

Decoupling

Decoupling isolates low-level noise and fluctuations from functional-level structures and dynamics. One example is genetic buffering by Hsp90, in which misfolding of proteins due to environmental stresses is repaired. Thus the effects of such perturbations are isolated from the functions of circuits. This mechanism also applies to genetic variations, where genetic changes in coding regions that may affect protein structures are masked because protein folding is fixed by Hsp90, unless such masking is removed by extreme stress [16–18]. Emergent behaviour of complex networks also exhibits such buffering properties [19]. These effects may constitute canalization, as proposed by Waddington [20].

The airplane as an example of a sophisticated engineering system clearly illustrates how these mechanisms work as a whole system (Figure 24.2). An airplane is supposed to maintain the pilot's desired flight path against atmospheric perturbations and various internal perturbations, including changes in the center of gravity due to fuel consumption and movement of passengers, as well as mechanical inaccuracies. This function is carried out by an automatic flight control system (AFCS) using movable flight control surfaces (rudder, flaps, elevators, etc.) and a propulsion system (engines). Extensive negative feedback control is used to correct deviations from the desired flight path. The reliability of the AFCS is critical for stable flight. To increase reliability, the AFCS is composed of three independently implemented modules (a triple redundancy system) that all meet the same functional specifications. Most parts of the AFCS are digitized, so that low-level noise from voltage fluctuations are effectively decoupled from digital signals that define the function of the system. Because of these mechanisms, modern airplanes are highly robust against various perturbations.

MECHANISMS FOR CANCER ROBUSTNESS

Cancer is robust against various therapeutic interventions. There are three groups of mechanisms that make cancer

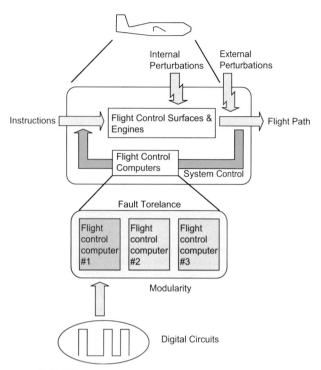

FIGURE 24.2 Robustness mechanisms in an airplane.

robust: intratumoral heterogeneity, intracellular feedback loops, and host—tumor entrainment.

Intratumoral genetic heterogeneity is a major source of robustness in cancer. Chromosome instability facilitates the generation of intratumoral genetic heterogeneity through gene amplification, chromosomal translocation, point mutations, aneuploidy, etc. [21—24] Intratumoral genetic heterogeneity is one of the most important features of cancer that provides fault tolerance for the tumor to survive and grow again despite various therapies, because some tumor cells may have genetic profiles that are resistant to these therapies. Whereas there are only a few studies on intertumor genetic heterogeneity, available observations in certain types of solid tumor indicate that there are multiple sub-clusters of tumor cells within one tumor cluster in which each sub-cluster has different chromosomal aberrations [25—29]. This implies that each sub-cluster is developed as a clonal expansion of a single mutant cell. The creation of a new sub-cluster depends on the emergence of a new mutant that is viable for clonal expansion. A computational study demonstrates that the spatial distribution within a tumor cluster enables the coexistence of multiple sub-clusters [30]. The issue of heterogeneity also applies for cancer stem cells. The heterogeneity of cancer stem cells provides heterogeneity at genetic and epigenetic levels.

Multidrug resistance is a cellular-level mechanism that provides robustness of viable tumor cells against toxic anticancer drugs. In general, this mechanism involves over-expression of genes such as MDR1 that encode an ATP-dependent efflux pump, P-glycoprotein (P-gp), which effectively pumps out a broad range of cytotoxins [31,32]. Trials to mitigate the function of P-gp using verapamil and ciclosporin its derivative PSC833 have been disappointing [33].

Host—tumor entrainment includes various tumor—host interactions as well as remodeling and control of the tumor microenvironment. Tumor—host interactions play major roles in tumor growth and metastasis [34]. When tumor growth is not balanced by vascular growth, a hypoxic condition emerges in tumor clusters [35]. This triggers HIF-1 upregulation, which induces a series of reactions that maintain normal physiological conditions [36]. Upregulation of HIF-1 induces upregulation of VEGF, which facilitates angiogenesis, and uPAR and other genes that enhance cell motility [35]. These responses solve the restriction of hypoxia for tumor cells, either by providing oxygen to the tumor cluster or by moving tumor cells to a new environment — resulting in further tumor growth or metastasis. Interestingly, macrophages are found to move via chemotaxis into a tumor cluster. Such a macrophage is called a tumor associated macrophage (TAM), and is found to over-express HIF-1[37]. This means that a macrophage that is supposed to remove tumor cells may instead be built-in to feedback loops to facilitate tumor growth and metastasis.

In addition, tumor cells may evolve through self-extending symbiosis [38]. In this case, tumor cells can enhance their robustness against various perturbations through horizontal gene transfer and uptake of chromosomes, leading to symbiosis with other cells in the form of cell fusion, and the formation of symbiotic relationships with surrounding environments [39—42]. This implies that tumor cells may be considered as a group of cells that have become somewhat detached from the host system and begin evolving independently, so that a wide range of phenomena, such as self-extending symbiosis, occur in tumor cells, thereby enhancing their robustness against perturbations. There are arguments that cancer cells can also carry out cell fusion with macrophages to change their character to support metastasis [39,40]. Furthermore, macrophages accumulate around a cancer and over-express genes that may contribute to the growth of a tumor. For patients who suffer from an immune deficit due to AIDS, Kaposi sarcomas resulting from AIDS appear with a high percentage, but the incidence rate of other cancers, such as breast cancer and lung cancer, are substantially lower than usual. This seems to indicate that the innate immune systems are hijacked by cancer cells so as to contribute to their proliferation and metastasis. If these observations are correct, the cancer itself seems to evolve to capture the surrounding host systems as extended cancer systems. So far, such phenomena have only been reported independently, and not placed in a unified perspective.

Reorganizing these findings under the coherent view of cancer robustness will provide us with a guideline for further research.

ROBUSTNESS TRADE-OFFS

Systems that acquire robustness against certain perturbations through either design or evolution have intrinsic trade-offs between robustness, fragility, performance, and resource demands. Carlson and Doyle, using simple examples from physics and forest fires, argued that systems which are optimized for specific perturbations are extremely fragile against unexpected perturbations [43, 44]. Ceste and Doyle further argued that robustness is a conserved quantity [45]. This means when robustness is enhanced against a range of perturbations, it must then be compensated for by fragility elsewhere, as well as compromised performance and increased resource demands.

Robust yet fragile trade-offs can be understood intuitively using the airplane example. Modern commercial airplanes are by a great magnitude more robust against atmospheric perturbations than the Wright flyer, which is attributable to a sophisticated flight control system. However, such a flight control system relies completely on electricity. In the unthinkable event of total power failure in which all electricity in the airplane is lost, it cannot be controlled at all. Obviously, airplane manufacturers are well aware of this and take all possible measures to minimize such a risk. On the other hand, despite its vulnerability against atmospheric perturbations, the Wright Flyer will never be affected by power failure, because there is no reliance on electricity. This extreme example illustrates systems that are optimized for certain perturbations could be extremely fragile against unusual perturbations. Trade-offs are expected to exist not only between robustness and fragility, but also between robustness and performance, and robustness and resource demands (Figure 24.3). At the same time, it should be noted that such trade-offs may hold only under a certain condition that a system is sufficiently optimized and has no room for optimization without trade-offs [46].

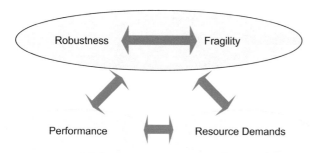

FIGURE 24.3 Robustness trade-offs.

THEORETICALLY MOTIVATED THERAPY STRATEGIES

Given the highly complex control systems and genetic heterogeneity of tumors, random trials of potential targets are not as effective as one would wish. There is a need for theoretically motivated approaches that guide us to therapies that best counter the disease. The implication of cancer robustness is that there are specific patterns of behavior and weakness in robust systems as well as a rational way of controlling and fixing systems, and such general principles also apply to cancer. Thus, there must be theoretically motivated approaches for the prevention and treatment of cancer.

Strategies for cancer therapy may depend upon the level of robustness of tumors in specific patients. When robustness is low and genetic heterogeneity is low, then there is a good chance that the use of drugs with specific molecular targets may be effective by causing a common mode failure: a type of failure where all redundant subsystems fail for the same reason. The example of chronic myeloid leukemia (CML) therapy using imatinib metylate may provide us with some insights [47,48]. Although this is speculative, the dramatic effect of imatinib metylate on early-stage CML may be due to a common mode failure, but resistance in advanced stages may be due to heterogeneity. For this strategy to be effective, there must be proper means of diagnosing the degree of intratumoral genetic variations. Also, the most effective molecule for a target needs to be recognized in ways that lead to identification and optimization processes.

However, for patients with an advanced-stage cancer, intratumoral genetic heterogeneity may be already high and various feedback controls may be significantly upregulated. In these cases, drugs that are effective in the early stage may not work as expected, owing to the heterogeneous response of tumor cells and feedbacks to compensate for perturbations. For these cases, therapy and drug design need a drastic shift from a single target-oriented approach to a systems-oriented approach. Then, the question is what approach shall be taken to target the system? There seem to be only a few theoretically motivated countermeasures.

First, the robustness/fragility trade-off implies that cancer cells that have gained increased robustness against various therapies may have a point of extreme fragility. Targeting such a point of fragility may provide dramatic effects against the disease. The major challenge is to find such a point of fragility. When it turns out that a cancer is tolerant to a specific drug after its provision but is fragile to another drug, the therapeutic strategy may be to use a series of regimens that continue to attack the point of fragility. For example, cancer cells that survived via evolution to increased robustness to the first drug may have fragility to another drug. Eventually, the cancer cell clusters that establish robustness to the second drug will appear, for which a third drug should be prepared. This is therefore the method for exploring the fragility of cancer by changing the treatments consecutively. Further exploration of the above approach would be via the development of a method for controlling and creating/inducing fragility artificially.

Some readers may wonder if such a convenient set of circumstances could really happen. In fact, the phenomenon whereby a cancer that obtained tolerance to a certain type of anticancer drug became fragile to other anticancer drugs was already known in the 1970s as 'collateral sensitivity'. Skipper et al. investigated into when a cancer becomes tolerant to one of 30 drugs, and to which other drugs that cancer becomes fragile [49]. Figure 24.4 shows the tolerance to 10 anticancer drugs out of 30. This research was not pursued, but it shows the trade-offs of robustness and so needs to be studied anew.

Second, there are possibilities for deploying more 'ecological' approaches. When cancer cells with multiple kinds of genetic diversity coexist, they may be in competitive and symbiotic situations. There is an interesting example in the treatment of prostate cancers. For prostate cancer, androgen-dependent cancer cells proliferate to grow a tumor. Blocking androgen to destroy those cells is the main purpose of treatment. However, it is known that after the shrinkage of tumor primarily consisting of androgen-dependent cells, androgen-independent cancer cells begin to proliferate. The androgen-independent cancer cells are more malignant and harder to cure. In the situation where the androgen-dependent cells and the androgen-independent cells coexist, it is observed that there is a possibility of slight reductions in androgen-independent cancer cell clusters. In this case, it will be possible to establish a treatment to control the proliferation and reduction of the tumor by blocking androgen intermittently. The actual clinical results show the effectiveness of such an intermittent androgen control method [50−52]. Aihara and his team at the University of Tokyo conducted mathematical analyses of these dynamics to derive an optimum treatment strategy [53]. This is a method of controlling the ecology of cancer cell clusters having diversity. Through the further generalization of this method and the systematic use of multiple treatments, the method of controlling the ecology of cancer will become an important approach in the future.

Third, approaches that avoid increases in robustness are another possibility. Since genetic heterogeneity is enhanced, at least in part, by somatic recombination, selectively inducing cell cycle arrest in tumor cells can effectively control the robustness. There is a theoretical possibility that such subtle control can be achieved by careful combination of multiple drugs that specifically perturb biochemical interactions. A computational study indicates that removal or attenuation of specific feedback loops involved in the cell cycle reduces robustness of the cell cycle against changes in rate constants [54]. The challenge is to find appropriate combinations of drugs that can effectively induce cell cycle arrest only in tumor cells, but not in other cells. While this approach uses a combination of multiple drugs, the hope is to find a set of drugs that can be administered at minimum dosage and toxicity. This approach results in dormancy of tumors. Cancer dormancy has already been proposed [55,56] and several reports indicate that induced dormancy has been observed in the mouse [57,58]. However, these studies report cases where tumor cell proliferation is offset by increases in apoptosis. Since heterogeneity may increase by cell proliferation, this type of dormancy, which can be called 'pseudo dormancy,' does not prevent an increase of heterogeneity, hence robustness is not controlled. Genuine dormancy needs to induce cell cycle arrest as selectively as possible.

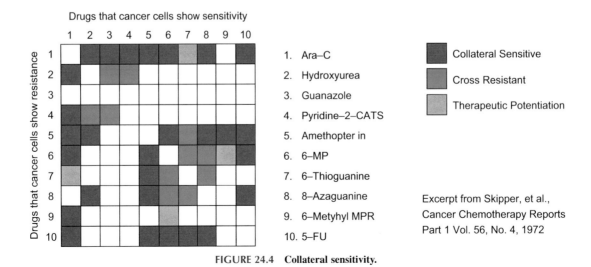

FIGURE 24.4 Collateral sensitivity.

Fourth, an approach to actively reduce intratumoral genetic heterogeneity, followed by therapy by molecular targeted drugs, may be a viable option. If we can design an initial therapy to impose a specific selection pressure on tumors, in which only cells with specific genetic variations survive the therapy, then a reduction in genetic heterogeneity may be achieved. Then, if a tumor cell population is sufficiently homogeneous, a drug that specifically targets a certain molecule may have significant impact on the remaining tumor cell population. An important point here is that the drugs used should not enhance mutation and chromosomal instability. If mutations and chromosomal instability are enhanced, particularly by the initial therapy, heterogeneity may quickly increase so that the second-line therapy will be ineffective. The drawback of this approach is that it does not eliminate the fundamental chromosome instability that continues to generate tumor cells with diverse genetic backgrounds.

Alternatively, a method to enhance chromosomal instability selectively in cells that already have unstable chromosomes could be one candidate. The point here is whether or not such effects can be achieved with sufficient selectivity. A non-selective approach to increase chromosomal instability has been proposed [59], but it may enhance chromosome instability of cells that are relatively stable, so it potentially promotes malignancy.

Fifth, one may wish to retake control of feedback loops that give rise to robustness in epidemic states. Since the robustness of a tumor is often caused by host—tumor feedback controls, robustness of a tumor can be seriously mitigated if such feedback loops can be controlled. One possible approach is to introduce a decoy that effectively disrupts feedback control or invasive mechanisms of the epidemic state. Such an approach is proposed in AIDS therapy, so that a conditionally replicating HIV-1 (crHIV-1) vector which has only a *cis* region but not a *trans* region is introduced [60,61]. This decoy virus dominates the replication machinery, so that the HIV-1 viruses are pushed into latency, instead of eradication. In a solid tumor, an interesting idea has been proposed to use a tumor-associated macrophage (TAM) as delivery vehicle for the vector [37, 62]. TAM migrates into a solid tumor cluster and upregulates HIF-1, which facilitates angiogenesis and metastasis. If TAM can be used to retake control, robustness may be well controlled and self-extending symbiosis in cancer evolution may be aborted.

Sixth, it is critically important that therapeutic interventions specifically target tumor cells, but not normal cells. Simply identifying causative genes is not enough, as most disease-causative genes also bear important functions for the operation of normal cells. The difference between target and off-target cells has to be identified. Recently, the author's team created a novel biological assay method called 'gTOW' — genetic Tug-of-War — that enables us to measure the quantitative upper bound of genes that can be over-expressed without perturbing cellular functions such as proliferation [63]. This method, when extended to mammalian cell systems, may help us discover differences of robustness between target and off-target cells against perturbation. The same perturbation may hit the point of fragility in target cells but be tolerable for normal cells. As such differences may stem from the properties of networks, focusing on such differences between cells may provide us a broader margin of therapeutic windows than from current dosage-based methods. At the same time, the use of multiple components to explore differences of robustness has been proposed. The spread spectrum control and the Long-tail problem have been proposed as mathematical problems that may provide us the means to design drugs with large numbers of components [64].

A PROPER INDEX OF TREATMENT EFFICACY

It is important to recognize that, in the light of cancer robustness theory, tumor mass reduction is not an appropriate index for therapy and drug efficacy judgment. As already discussed, reduction of tumor mass does not mean that the proliferation potential of a tumor is generally decreased. It merely means that a sub-population of tumor cells that respond to the therapy were eradicated or significantly reduced. The problem is that remaining tumor cells may be more malignant and aggressive, so that therapies for the relapsed tumor could be extremely ineffective. This is particularly the case when drugs used to reduce tumor mass are toxic and potentially promote mutations and chromosomal instability in non-specific ways. It may even enhance malignancy by imposing selective pressures favoring resistant phenotypes and enhancing genetic diversity, as well as providing niches for growth by eradicating a fragile sub-population of tumor cells.

A proper index should rather be based on control of robustness, either to minimize increased robustness or to reduce robustness. This can be achieved by inducing dormancy, actively imposing selective pressure to reduce heterogeneity, or exposing fragility that can be the target of therapies to follow, thereby retaking control of feedback regulation. The outcome of controlling the robustness may vary from moderate growth of the tumor, to dormancy where there is no tumor mass growth or significant reduction in tumor mass. It should be noted that robustness control does not exclude the possibility of significant tumor mass reduction. If we can target a point of fragility of the tumor, it may trigger a common mode failure and result in significant tumor mass reduction. However, this is a result of controlling robustness, and should not be confused with therapies aimed at tumor mass reduction, because robustness has to be controlled first, to actively exploit a point of fragility.

However, this criterion poses a problem for drug design, because the current efficacy index by which antitumor drugs are measured is based on tumor mass reduction. Drugs that induce dormancy will not satisfy this efficacy criterion. Whether such an approach can be taken may depend on a perception change in practitioners, drug industries, and regulatory authorities.

LONG-TAIL DRUG

One of the points clarified in the discussions so far is that the living body is a robust and evolving system, that its control cannot be effective if an effort is made to simply identify one factor and suppress it, and that it may rather bring about undesirable side effects. Actually, it has been reported that the genes involved in the initiation and maintenance of cancers are quite often the hubs of networks [65,66]. The fact that the causative gene of a cancer is mostly found in the hub of a network means that, when the causative gene is set as a target of a drug, there is an inherent risk of severe side effects. Statistical studies indicate that the average number of interactions involving targeted molecules of anticancer drugs is significantly higher than those of non-cancer drugs [67]. This difference seems to reflect the degree of side effects between anticancer drugs and other drugs. In order to avoid the side effects, it is necessary to adapt the strategy to avoid perturbing major hubs in molecular interaction networks. In view of the above, cancer treatment requires strategic and logical approaches.

When we look at the distribution of the number of interactions that each molecule has, it seems to exhibit a power law distribution. That is, the distribution indicates that only a few proteins have a large number of interacting partners, while a majority of proteins interact with only small numbers of other molecules. The distribution becomes a straight declining line on a double logarithmic chart, while on a single logarithmic chart (vertical axis: logarithm), it becomes a curved line with a very long tail in the direction to the right (Figure 24.5). To show the distribution with a double logarithm, the appearance frequency is plotted on the vertical axis and the number of interactions on the horizontal axis. With a single logarithm, the number of interactions is plotted on the vertical axis and the appearance frequency or the order on the horizontal axis in most cases. Because of the shape of this single logarithmic graph, it is sometimes called the long-tail distribution.

This long-tail distribution is well known in the world of internet business. This is because the long-tail distribution has become the theoretical background of the profit structure of online shops. It started from the study of the sales of Amazon.com by a team from the MIT business school [68]. Amazon.com sells more than 2.5 million books online, and the sales amount and ranking of each book form the long-tail distribution. This means a few bestsellers and a great many bad sellers form the long-tail distribution. It turns out that half of the total sales came from the books with sales ranking No.1 to about No. 40 000. However, the other half (estimations differs in the range of 57–30%, depending on studies) resulted from poorer sellers at No. 40 000 or lower. In other words, even the bad sellers have a big impact on total sales. It means that a concentration on bestsellers may miss the business opportunity available in the thin but long tail.

What does this imply for the interactions within cells that show the same distribution pattern statistically? Is it not possible to generate the effects similar to those by the drugs that suppress very important genes and proteins by applying weak suppression widely to the genes and proteins that do not seem to be so important? This leads to the conjecture that an effective synergy can be obtained by targeting multiple non-hub genes and proteins and intervening in them weakly rather than strongly. Let's call it a 'Long-tail Drug'. If it goes well, treatments can be provided without great disturbances to the hub molecules, hence with fewer side effects.

Researches that aim to create effective drugs by disturbing multiple target molecules with multiple chemical

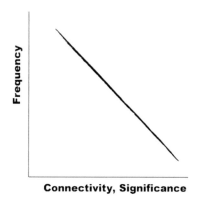
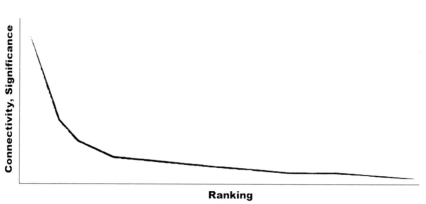

FIGURE 24.5 Long-tail distribution.

compounds have already started. CombinatoRX, a Boston-based drug development venture, has been exploring a drug that can generate useful synergy by combining two out of 1200 kinds of drugs [68]. As a result, it found some interesting combinations. For example, there is data that, if chlorpromazine (commercial name Contmin), a psychotropic drug, and pentamizin (commercial name: Benambax), an antibiotic substance of *Pneumonia carinii* therapeutic agent, are given at the same time, the cytostatic effect increases and an effect similar to or stronger than that of paclitaxel (commercial name Taxol) used as an anticancer drug can be obtained according to the experiments of cultured cells. It is the important experimental data to indicate that unexpected synergy is obtained by multiple drugs, but no methodology has been established that can estimate 'what kinds of effects can be obtained when which molecules are set as targets.' At present, screening to examine possible combinations one after another is mainly done. This approach will be practical for selecting the combination of two chemical substances and a few targets from the molecules whose mechanisms are known to some extent, but it will not be realistic for identifying high-order and effective combination efficiently.

A group of Hungarian researchers, using the metabolism model of budding yeasts and bacteria, calculated how much the metabolic rate decreases when the largest hub is eliminated. Then they computed how many interactions are necessary to realize a similarly decreased metabolic rate using the method of reducing the number of interactions by half [69,70]. As a result, it was found that 10−50 interactions need to be suppressed. Theoretically this shows the possibility of having a big impact on the cells with the use of the combination of weak interactions.

These results imply that the approach of intervening in multiple targets with the use of network structures and multiple chemical substances, when applied to the development of new drugs and the study of combination therapy using existing drugs, becomes an important option. The ultimate goal is a long-tail drug combination that generates the synergy of weak interactions with the simultaneous use of very many substances. The procedure is to avoid as much as possible the important targets — the so-called hubs — and try to get the synergy effect on many targets that are not thought to be important but which exist abundantly in the long tail of the distribution of the number of interactions of proteins and genes [64].

Since there are enormous numbers of possibilities in the selection of target molecules and in the combinations of drugs in this case, it is necessary to establish technology to efficiently select effective combinations with the fewest side effects. The author considers that the approach of measuring the differences of robustness against the altered expression of candidate genes and the altered activity of candidate proteins between the target cells and the off-target cells will become a workable option.

OPEN PHARMA

The generation of synergy effect from the parallel use of multiple drugs indicates the possibility of producing new effects through the combination of off-patent drugs. For example, given that a certain chemical substance was developed and patented, the effect similar to that of new chemical substance may be realized by the combination of off-patent chemical substances. In this case, the value of patented chemical substance may be significantly reduced. When a combination of off-patent chemical substances is packaged as a drug, clinical trials need to be done with the package, which will cost almost the same amount of money as a normal new drug. If the simple prescription of multiple drugs will do, it could be a less costly solution. For example, Toxol cost about 40 000JPY per 100 mg in Japan, Contmin and Benambax cost about 3000JPY in total. If the prescription of these drugs generates a certain effect, the drug cost decreases dramatically.

At the same time, to discover the effects of the combinations of drugs it is necessary to tackle enormous numbers of combinations. The screening of the combinations of only two drugs cost a lot, and so the screening of the combinations of three or more drugs is too much for one company. Disclosing the combination information and searching for new combinations in open collaboration will be effective. At the same time, the combination of off-patent chemical substances should be aggressively examined. That is to say, it is the opening of medicine manufacturing processes to the public. In this paradigm, part of the pharmaceutical industry changes from a manufacturing industry to a service industry. That is, the companies will provide the public with the information on the combination of drugs and their usage instead of developing new drugs. The same thing has already happened in the computer industry. IBM used to sell patent-protected computers, but it has become a solution provider to support positively open sources now (but their mainframe is doing well). Similar things may happen to the pharmaceutical industry.

This indicates a very interesting possibility. If the combinations of inexpensive drugs produce effects similar to those of expensive ones, they will contribute to the substantial decrease in drug cost in medical expenses, and at the same time they will be able to respond to the substantial needs for the drugs that can be used for the people who cannot afford to use these expensive drugs economically.

When this technology advances further, it will become possible to design drugs optimized to individual genetic and epigenetic characteristics by replacing some ingredients of multiple usable drugs with others. In Silicon Valley now, ventures that analyze the genomes of individuals and offer advice on their disease risks have emerged. The combination of these technologies and the long-tail drugs would open the paradigm of the drugs for individual medical care. This technology will be expensive in the initial stages, and so it will be used mainly by wealthy people. When it makes various combinations openly available, it will become less costly and be more widely used.

CONCLUSION

This chapter has discussed basic ideas behind biological robustness and its implications for cancer research and treatment. Biological robustness is one of the essential features of living systems that is argued to be tightly coupled with evolution. It may also shape the basic architectural feature of biological systems that are robust and evolvable. This chapter provides overall perspectives and presents the framework of thoughts that may turn into more solid theories of biological robustness. One of the major consequences is to identify trade-offs between robustness, fragility, resource demands, and performance. Fragility is particularly relevant to disease. At the same time, cancer establishes its own robustness as it evolves in a patient. Its success may be a result of hijacking the robustness intrinsic to the host system. Understanding of this complex nature of biological systems may have profound implications for future biomedical research. Several theoretically motivated strategies can be derived. Some of them may not be practical, but if even one such idea becomes reality it will result in ground-breaking progress in cancer therapeutics.

ACKNOWLEDGEMENTS

This research is supported in part by ERATO-SORST Program (Japan Science and Technology Agency: JST), the Genome Network Project (Ministry of Education, Culture, Sports, Science, and Technology), BBSRC-JST Strategic Collaboration Program of JST to the Systems Biology Institute.

REFERENCES

[1] Kitano H. Cancer robustness: tumor tactics. Nature 2003; 426(6963):125.
[2] Kitano H. Cancer as a robust system: implications for anticancer therapy. Nat Rev Cancer 2004;4(3):227−35.
[3] Kitano H. Biological robustness. Nat Rev Genet 2004;5:826−37.
[4] Alon U, et al. Robustness in bacterial chemotaxis. Nature 1999;397(6715):168−71.
[5] Cannon W. The Wisdom of the Body. New York: Norton; 1932.
[6] Crowe JH, Crowe LM. Preservation of mammalian cells-learning nature's tricks. Nat Biotechnol 2000;18(2):145−6.
[7] Barkai N, Leibler S. Robustness in simple biochemical networks. Nature 1997;387(6636):913−7.
[8] Yi TM, et al. Robust perfect adaptation in bacterial chemotaxis through integral feedback control. Proc Natl Acad Sci U S A 2000;97(9):4649−53.
[9] Tyson JJ, Chen K, Novak B. Network dynamics and cell physiology. Nat Rev Mol Cell Biol 2001;2(12):908−16.
[10] Ferrell Jr JE. Self-perpetuating states in signal transduction: positive feedback, double-negative feedback and bistability. Curr Opin Cell Biol 2002;14(2):140−8.
[11] Chen KC, et al. Integrative analysis of cell cycle control in budding yeast. Mol Biol Cell 2004;15(8):3841−62.
[12] Eldar A, et al. Robustness of the BMP morphogen gradient in *Drosophila* embryonic patterning. Nature 2002;419(6904):304−8.
[13] Meir E, et al. Robustness, flexibility, and the role of lateral inhibition in the neurogenic network. Curr Biol 2002;12(10):778−86.
[14] von Dassow G, et al. The segment polarity network is a robust developmental module. Nature 2000;406(6792):188−92.
[15] Schlosser G, Wagner G. Modularity in Development and Evolution. Chicago, IL: The University of Chicago Press; 2004.
[16] Rutherford SL, Lindquist S. Hsp90 as a capacitor for morphological evolution. Nature 1998;396(6709):336−42.
[17] Queitsch C, Sangster TA, Lindquist S. Hsp90 as a capacitor of phenotypic variation. Nature 2002;417(6889):618−24.
[18] Rutherford SL. Between genotype and phenotype: protein chaperones and evolvability. Nat Rev Genet 2003;4(4):263−74.
[19] Siegal ML, Bergman A. Waddington's canalization revisited: developmental stability and evolution. Proc Natl Acad Sci U S A 2002;99(16):10528−32.
[20] Waddington CH. The Strategy of the Genes: A Discussion of Some Aspects of Theoretical Biology. New York, NY: Macmillan; 1957.
[21] Lengauer C, Kinzler KW, Vogelstein B. Genetic instabilities in human cancers. Nature 1998;396(6712):643−9.
[22] Li R, et al. Aneuploidy vs. gene mutation hypothesis of cancer: recent study claims mutation but is found to support aneuploidy. Proc Natl Acad Sci U S A 2000;97(7):3236−41.
[23] Tischfield JA, Shao C. Somatic recombination redux. Nat Genet 2003;33(1):5−6.
[24] Rasnick D. Aneuploidy theory explains tumor formation, the absence of immune surveillance, and the failure of chemotherapy. Cancer Genet Cytogenet 2002;136(1):66−72.
[25] Baisse B, et al. Intratumor genetic heterogeneity in advanced human colorectal adenocarcinoma. Int J Cancer 2001;93(3): 346−52.
[26] Fujii H, et al. Frequent genetic heterogeneity in the clonal evolution of gynecological carcinosarcoma and its influence on phenotypic diversity. Cancer Res 2000;60(1):114−20.
[27] Gorunova L, et al. Cytogenetic analysis of pancreatic carcinomas: intratumor heterogeneity and nonrandom pattern of chromosome aberrations. Genes Chromosomes Cancer 1998;23(2):81−99.
[28] Gorunova L, et al. Extensive cytogenetic heterogeneity in a benign retroperitoneal schwannoma. Cancer Genet Cytogenet 2001;127(2): 148−54.
[29] Frigyesi A, et al. Power law distribution of chromosome aberrations in cancer. Cancer Res 2003;63(21):7094−7.
[30] Gonzalez-Garcia I, Sole RV, Costa J. Metapopulation dynamics and spatial heterogeneity in cancer. Proc Natl Acad Sci U S A 2002;99(20):13085−9.

[31] Juliano RL, Ling V. A surface glycoprotein modulating drug permeability in Chinese hamster ovary cell mutants. Biochim Biophys Acta 1976;455(1):152–62.
[32] Nooter K, Herweijer H. Multidrug resistance (mdr) genes in human cancer. Br J Cancer 1991;63(5):663–9.
[33] Tsuruo T, et al. Overcoming of vincristine resistance in P388 leukemia in vivo and in vitro through enhanced cytotoxicity of vincristine and vinblastine by verapamil. Cancer Res 1981;41(5):1967–72.
[34] Bissell MJ, Radisky D. Putting tumors in context. Nat Rev Cancer 2001;1(1):46–54.
[35] Harris AL. Hypoxia – a key regulatory factor in tumor growth. Nat Rev Cancer 2002;2(1):38–47.
[36] Sharp FR, Bernaudin M. HIF1 and oxygen sensing in the brain. Nat Rev Neurosci 2004;5(6):437–48.
[37] Bingle L, Brown NJ, Lewis CE. The role of tumor-associated macrophages in tumor progression: implications for new anticancer therapies. J Pathol 2002;196(3):254–65.
[38] Kitano H, Oda K. Self-extending symbiosis: a mechanism for increasing robustness through evolution. Biol Theory 2006;1(1):61–6.
[39] Pawelek JM. Tumor-cell fusion as a source of myeloid traits in cancer. Lancet Oncol 2005;6(12):988–93.
[40] Pawelek J, et al. Co-opting macrophage traits in cancer progression: a consequence of tumor cell fusion? Contrib Microbiol 2006;13:138–55.
[41] Vignery A. Macrophage fusion: are somatic and cancer cells possible partners? Trends Cell Biol 2005;15(4):188–93.
[42] Ogle BM, Cascalho M, Platt JL. Biological implications of cell fusion. Nat Rev Mol Cell Biol 2005;6(7):567–75.
[43] Carlson JM, Doyle J. Highly optimized tolerance: a mechanism for power laws in designed systems. Phys Rev E Stat Phys Plasmas Fluids Relat Interdiscip Topics 1999;60(2 Pt A):1412–27.
[44] Carlson JM, Doyle J. Complexity and robustness. Proc Natl Acad Sci U S A 2002;99(Suppl. 1):2538–45.
[45] Csete ME, Doyle JC. Reverse engineering of biological complexity. Science 2002;295(5560):1664–9.
[46] Kitano H. Violations of robustness trade-offs. Mol Syst Biol 2010;6:384.
[47] Hochhaus A. Cytogenetic and molecular mechanisms of resistance to imatinib. Semin Hematol 2003;40(2 Suppl. 3):69–79.
[48] Hochhaus A, et al. Roots of clinical resistance to STI-571 cancer therapy. Science 2001;293(5538):2163.
[49] Skipper HE, et al. A quick reference chart on cross resistance between anticancer patients. Cancer Chemother Rep 1972;56(4):493–8.
[50] Gleave M, et al. Intermittent androgen suppression for prostate cancer: rationale and clinical experience. Eur Urol 1998;34(Suppl. 3.):37–41.
[51] Gleave ME, et al. Neoadjuvant androgen withdrawal therapy decreases local recurrence rates following tumor excision in the Shionogi tumor model. J Urol 1997;157(5):1727–30.
[52] Bruchovsky N, et al. Intermittent androgen suppression for prostate cancer: canadian prospective trial and related observations. Mol Urol 2000;4(3):191–9. discussion 201.
[53] Shimada T, Aihara K. A nonlinear model with competition between prostate tumor cells and its application to intermittent androgen suppression therapy of prostate cancer. Math Biosci 2008;214(1–2):134–9.
[54] Morohashi M, et al. Robustness as a measure of plausibility in models of biochemical networks. J Theor Biol 2002;216(1):19–30.
[55] Takahashi Y, Nishioka K. Survival without tumor shrinkage: re-evaluation of survival gain by cytostatic effect of chemotherapy. J Natl Cancer Inst 1995;87(16):1262–3.
[56] Uhr JW, et al. Cancer dormancy: opportunities for new therapeutic approaches. Nat Med 1997;3(5):505–9.
[57] Holmgren L, O'Reilly MS, Folkman J. Dormancy of micrometastases: balanced proliferation and apoptosis in the presence of angiogenesis suppression. Nat Med 1995;1(2):149–53.
[58] Murray C. Tumor dormancy: not so sleepy after all. Nat Med 1995;1(2):117–8.
[59] Sole RV. Phase transitions in unstable cancer cell populations. Eur Phys J 2003;B(35):117–23.
[60] Dropulic B, Hermankova M, Pitha PM. A conditionally replicating HIV-1 vector interferes with wild-type HIV-1 replication and spread. Proc Natl Acad Sci U S A 1996;93(20):11103–8.
[61] Weinberger LS, Schaffer DV, Arkin AP. Theoretical design of a gene therapy to prevent AIDS but not human immunodeficiency virus type 1 infection. J Virol 2003;77(18):10028–36.
[62] Owen MR, Byrne HM, Lewis CE. Mathematical modelling of the use of macrophages as vehicles for drug delivery to hypoxic tumor sites. J Theor Biol 2004;226(4):377–91.
[63] Moriya H, Shimizu-Yoshida Y, Kitano H. In vivo robustness analysis of cell division cycle genes in *Saccharomyces cerevisiae*. PLoS Genet 2006;2(7):e111.
[64] Kitano H. A robustness-based approach to systems-oriented drug design. Nat Rev Drug Discov 2007;6(3):202–10.
[65] Yao L, Rzhetsky A. Quantitative systems-level determinants of human genes targeted by successful drugs. Genome Res 2008;18(2):206–13.
[66] Goh KI, et al. The human disease network. Proc Natl Acad Sci U S A 2007;104(21):8685–90.
[67] Hase T, et al. Structure of protein interaction networks and their implications on drug design. PLoS Comput Biol 2009;5(10):e1000550.
[68] Brynjolfsson E, Smith MD, Hu Y. Consumer Surplus in the Digital Economy: Estimating the Value of Increased Product Variety at Online Booksellers. Cambridge, MA: MIT Sloan School of Management; 2003.
[69] Agoston V, Csermely P, Pongor S. Multiple weak hits confuse complex systems: a transcriptional regulatory network as an example. Phys Rev E Stat Nonlin Soft Matter Phys 2005;71(5 Pt 1):051909.
[70] Csermely P, Agoston V, Pongor S. The efficiency of multi-target drugs: the network approach might help drug design. Trends Pharmacol Sci 2005;26(4):178–82.

Chapter 25

Systems Immunology: From Cells and Molecules to a Dynamic Multi-Scale System

Mark M. Davis[1] and Shai S. Shen-Orr[2]

[1]Department of Microbiology & Immunology, Howard Hughes Medical Institute, Beckman Center, Room B221, Stanford University School of Medicine, Stanford, CA 94305, USA, [2]Department of Immunology, Faculty of Medicine, Technion, 1 Efron St. Haifa, 31096, Israel

Chapter Outline

Introduction	481
Cell-Focused Systems Immunology	484
Reconstructing Cellular Networks	484
Exploring Cellular Diversity	486
Limitations of a Cell-Focused Approach	487
System-Focused Systems Immunology	488
Human Immune Monitoring	488
Antibody and TCR Repertoire Diversity	489
Limitations of a System-Focused Approach	491
Multi-Scale-Focused Systems Immunology	491
System-Wide Meso-Scale Cellular Automata Models of Immunity	491
Statecharts as a Rich Framework to Model Immunity	493
Limitations of a Multi-Scale-Focused Approach	494
Conclusions and Future Challenges	495
References	495

INTRODUCTION

Immunity to infectious diseases is orchestrated by a highly complex and diffuse system of specialized cells and organs that flourishes on diversity and is in a constant interplay with its environment. Leukocytes, or white blood cells, comprise a large number of the specialized cell types that constitute the immune system. They reside in different tissues and organs as well as circulate throughout the blood and lymphatic system. Leukocytes are derived from a single type of hematopoietic stem cell in the bone marrow but go through a differentiation process that ultimately yields several basic cell types, such as lymphocytes, monocytes and granulocytes (see Figure 25.1A for an overview of the major cell types). Immune responses in vertebrates are divided into two major types: innate and adaptive, whereas invertebrates have just the innate types of responses. The innate system is specific for broad categories of molecules, such as double-stranded RNAs or lipopolysaccharides and other molecules that are characteristic of pathogens, and is comprised of cells and mechanisms that form the host's first line of defense. In contrast, the cells of the adaptive immune system are slower to respond initially yet have molecules such as antibodies or T-cell receptors that are specific for particular molecules, as well as a 'memory'-stage cell that can be maintained for decades if not throughout an organism's lifetime. The cells of the innate immune system (dendritic cells, macrophages, among others) serve as a first line of defense, as well as trainers and prompters to cells of the adaptive immune system (lymphocytes), making stimulatory cytokines and also presenting peptides derived from foreign antigens that are bound specifically to molecules of the major histocompatibility complex (MHC) (peptide-MHC) on the cell surface of antigen-presenting cells for recognition by T-cell receptors. Whole antigens are also presented to B cells, which use cell surface immunoglobulins, which upon binding a specific antigen stimulate those cells to initiate a response. As research tools have become more sophisticated, new immune cell subtypes (i.e., cells with differing functionality at a given condition) are continuously being discovered such that the estimates for the total number of distinct cells now numbers in the hundreds. These cells are the 'quanta' of the immune system, each a largely independent agent capable of determining its next action based on environmental input. The importance of the immune system to human health is difficult to overstate. The

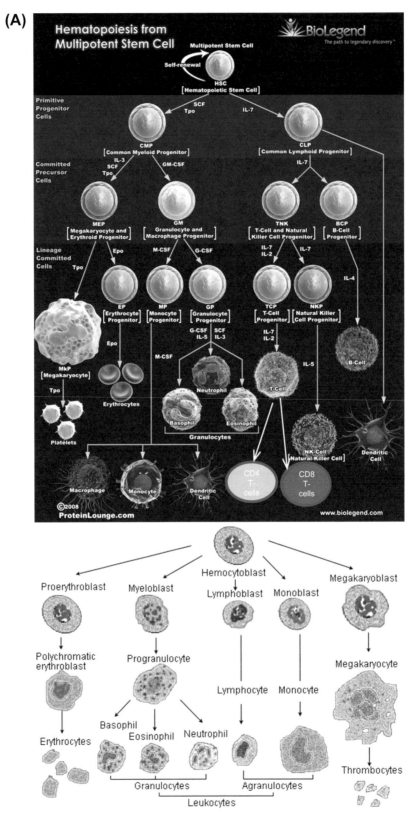

FIGURE 25.1 **The immune system is a complex diffuse system containing many cell types that requires the application of multiple, complementary approaches.** (A) The cell lineage of hematopoiesis. All immune cells are derived from a single precursor but differentiate through a series of developmental and signaling events to many different cell types, each with its own state and functionality. Shown are only the basic cell types, yet many more cell-type subsets exist, and more are increasingly being discovered. (B) Systems immunology aims to 'put the pieces together' such that we can develop a true understanding of how immunity functions, why has it developed as it is, and what the forces are that shape it. We delineate in the discussion three approaches researchers are using to probe the immune system. Each has its advantages and disadvantages, yet all are needed for a systems-view of immunology to emerge. As human health is highly dependent on the immune system, a direct benefit of improved systems-level understanding of immunology would be expected in the clinic.

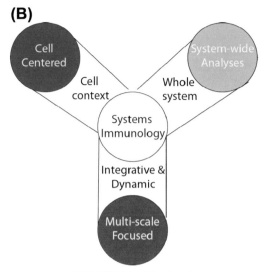

FIGURE 25.1 *(Continued)*

development of vaccines for infectious diseases such as smallpox, measles etc., has had an enormous impact over the last century, and remains the most efficacious medical intervention to date in terms of lives saved. However, many challenges remain to be resolved, such as developing effective vaccines for HIV, malaria and tuberculosis, among others, and while there are promising developments we have still a long way to go in using the immune system to combat cancer and many other diseases in which it is involved. Lastly, there are dozens of syndromes in which the immune system attacks the body's own tissues or organs, such as type 1 diabetes, rheumatoid arthritis, multiple sclerosis and many others for which we have an inadequate understanding and thus far no ability to cure.

As in other fields, a reductionist approach has been the predominant strategy in immunology research for many years. Research is often focused on understanding the role of a single cell type or compartment (e.g., B cells). Although fruitful, such studies often disregard the cooperativity and cross-talk between the different components of the immune system. Similarly, traditional studies of individual genes have often been difficult owing to the large number of different cell types in which each immune gene is expressed, often with different functions. The effect of cytokines on cells often differs according to the cell type and the stage at which the interaction occurs. Immunologists in each such subfield have made huge advances toward understanding particular segments of the immune system; however, assembling the whole puzzle from its pieces is both necessary and now more possible owing to the fact that many key parts and principles of the immune system have become well understood and can be measured simultaneously. Systems immunology has recently emerged as a distinct discipline and is the topic of discussion in this chapter. In fact, a systems approach has deep roots in immunology itself, as 'systems-wide' questions and reasoning have always occupied an important place in some of the earliest work in the field; however, in more recent decades difficulties encountered in the more theoretical work has forced a retreat to more reductionist approaches.

We define systems immunology as a new start at 'putting the pieces together' with the goal of directly studying how immunity functions as a system. That is, how changes in one compartment of the immune system affect other parts, and how together these components orchestrate a response that is greater than the sum of its parts. The experimental approach taken is quantitative and requires extensive computation to incorporate the results into increasingly comprehensive and accurate models. We will distinguish measurements on an 'omics' scale (referred to here as 'high-bandwidth biology') that often aims to detect biomarkers, from systems immunology, which focuses more on interactions than on individual components, although these two directions of research often overlap. In other biological research areas, such as cancer and genetics, systems biology approaches have been around for a decade or more. In immunology, however, these approaches are much more recent owing to the high system complexity and its diffuse nature, and for some immune components, such as cell types, it is only very recent developments that enable us to enumerate their many different subsets.

In systems immunology there been no single starting point, nor a single research focus. Rather, researchers are converging from a variety of fields: from immunologists and clinicians who measure the immune system at an 'omics' scale to systems biologists, computer scientists and mathematical modelers who are challenged by the 'next frontier' and drawn by the clinical relevance of the field. The former bring immune know-how and measurement techniques, the latter bring methodology and approaches originally developed for other domains and now being adapted to systems immunology. Together these researchers are discovering a remarkable level of complexity, such as additional specialized cell types [1,2], non-uniform sampling of variable diverse joining (VDJ) recombination (see later, [3,4]), and complex temporal regulatory programs for intercellular communication [5,6].

Here we review the development of this nascent field and the initial findings and concepts that have emerged. We structure our discussion into three main sections: cell-focused, system-focused, and multi-scale focused. This delineation emphasizes both the principal differences in the way researchers are probing the immune system as well as the insights obtained (Figure 25.1B). In the first, cell-focused section, the emphasis is on understanding intracellular components and mechanisms in detail, which does not consider interactions with the rest of the immune system. In the second, system-focused section, the immune system is discussed as a whole, how it is studied using high-throughput approaches. These approaches usually do not consider cellular context, which can make results difficult to interpret. The last, multi-scale focused section explores integrative models that encode information immune system biology at multiple levels (e.g., genes, proteins, cells, cell populations, processes etc.). The aim is to identify emergent phenomena and knowledge gaps. It should be noted, however, that the specification of the models is an enormous task and only small segments of the immune system can be studied at present. The premise for pursuing a systems immunology approach is that developing a quantitative understanding of how the many components of the immune system interact would yield a true understanding of the immune system as well as generate predictions about the immune response of a given individual, outcomes expected to be of high clinical relevance.

CELL-FOCUSED SYSTEMS IMMUNOLOGY

Immunologists use proteins that are expressed on the cell surface (cell-surface markers) to quantify and isolate heterogeneous cell populations (such as those obtained from peripheral blood) via flow cytometry. As cells of the immune system are highly specialized, much of the work in immunology has been to understand the specific immune processes mediated by each cell, and how they relate to overall immune protection. Cells can be studied in culture, and can be harvested then manipulated and reintroduced into the organism. Beyond research as to their role in immunity, their high accessibility and ease of manipulation have made immune cells a top choice for systems biologists interested in eukaryotic cellular regulatory networks [5,7–13].

Reconstructing Cellular Networks

Immune cells are a useful model for developing network reconstruction methodologies because they are highly accessible and can therefore serve as a general model for eukaryotic mammalian cells. The pathways or networks related to innate and adaptive immunity are ultimately involved in the determination of successful immune responses as well as in disease. One methodology for eukaryotic network reconstruction was originally developed in yeast [14] and includes four steps: (1) global profiling of gene expression, for instance using microarrays; (2) computational inference to identify regulators and their putative target genes based on co-expression; (3) repeated measures with perturbations to obtain increasingly accurate interactions; and (4) selected validation of predicted regulatory interactions (see also Chapter 4). With the development of novel technologies, molecular resources for perturbation as well as progress in automation, the same approach can now be carried out in mammals, yet still requires herculean efforts and extensive validation.

Regulatory network reconstruction has been applied to several different hematopoietic-derived immune cell types, including B-cell transcriptional and post-transcriptional regulatory networks [9–11], as well as $CD4^+$ T cells signaling pathways that were derived from single-cell intracellular measurements of phosphorylated protein abundance obtained from thousands of cells via flow cytometry [7]. This impressive proof-of-concept can now be expanded significantly using the recently developed mass cytometry approach (Box 25.1, Figure 25.2). In the innate system, systems biology techniques have been employed to map out the regulatory circuitry (transcriptional and signaling) in macrophage and dendritic cells [5,8,12,13]. For these, the focus has been on mapping the circuitry downstream of the Toll-like receptor (TLR) family, a class of pattern recognition receptors each capable of recognizing

---

BOX 25.1 Exploring cells in high dimension — the use of mass cytometry to quantify cell identity and function

Cells are the quanta of the immune system and their identity and function can be understood by the degree to which they express proteins on the cell surface or intracellularly. Traditionally, the workhorse tool of immunology has been the flow cytometer, which optically measures the return fluorescence from cells stained with fluorophore-labeled antibodies bound to cell-surface and intracellular proteins. A difficulty for using multiple fluorophores simultaneously in flow cytometry is that the fluorescence emission spectrum spills over from its characteristic wavelength to interfere with the reading from other channel. One can use 10–15 different labels before the overlapping emissions spectra become too complex to accurately separate them.

Cytometry by time-of-flight (CyTOF) is a recently introduced technology that measures the abundance of metal isotope labels on antibodies and other tags (such as peptide-MHC tetramers for labeling specific T cells) on single cells using mass spectroscopy. The advantage of this approach is that it allows many more molecules to be used in combination to assay a single sample (blood or single-cell suspension of tissues) than with fluorescence label-based cytometry [1–4]. This allows much more information to be obtained from each cell. An illustration of this is that whereas 10 labels gives approximately 1000 possible combinations, 30 give over 1 000 000 (using the formula $x = 2^{n-1}$, where x is the number of combinations and n is the number of labels). Thus there is much greater resolving power with the number of labels possible with this new technology.

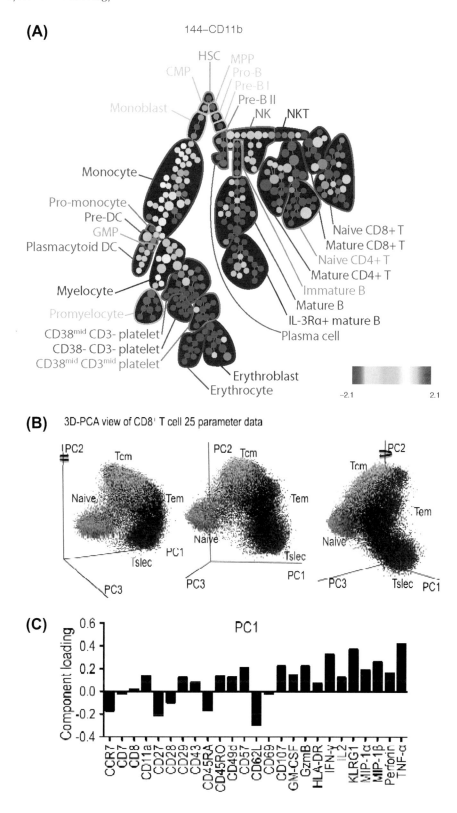

FIGURE 25.2 **Mass cytometry enables single-cell highly dimensional analysis of immunity and demands new algorithmic and visualization techniques.** (A) Immunophenotypic progression in healthy human bone marrow measured using mass cytometry profiling and visualized by SPADE in a tree plot. Tree was constructed by using 13 cell-surface antigens. The size of each circle in the tree indicates the relative frequency of cells that fall within the 13-dimensional confines of the node boundaries. Node color is scaled to the median intensity of marker expression of the cells within each node, expressed as a percentage of the maximum value in the data set (CD11b is shown). Putative cell populations were annotated manually and are represented by colored lines encircling sets of nodes that have CD marker expression emblematic of the indicated subset designations. (*Taken from* [2].) (B) One mass cytometry data set is plotted on the first three principal component axes (a representative PMA$^+$ ionomycin-stimulated CD8$^+$ T cell sample) and shown from three different perspectives (rotated around the PC2 axis). After gating by surface marker phenotype, naive (green), central memory (Tcm, yellow), effector memory (Tem, blue), and short-lived effector (Tslec, red) cell populations are overlaid to identify the main phenotypic clusters. (C) The principal component analysis parameter loadings (weighting coefficients) for the first components are plotted. (*Taken from* [1]) (Still need to seek permission).

a set of molecules that characterize pathogens but not the host. Multiple genes that were previously not known to function in the immune system were discovered to play a role in the immune response [12]. For example, Aderem and colleagues found that the transcription factor ATF3, which was previously implicated in the regulation of stress response, cell cycle and apoptosis, but not in immunity, is a regulator of the TLR response. More recently, Hacohen, Regev, Amit and collaborators, discovered a host of cell-cycle regulators that have been co-opted for antiviral transcriptional responses in non-dividing dendritic cells [5,8]. Thus, one insight that emerged from these studies is that the set of 'immunologically important genes' is still poorly defined. This is even reflected among already discovered associations, as there is inconsistency between the existing databases for immune-related genes [15]. Because new immune genes are still being discovered, it is challenging to test all genes and pre-filter them by known function or in another unbiased fashion to reduce the number of hypotheses to test further [16]. Studies of eukaryotic regulatory networks are revealing network motifs [17] used by eukaryotic cells to elicit specific regulatory functions (see Chapter 4). For example, TLR-4, NFκB and ATF generate a pulse response of downstream targets to lipopolysaccharide stimulation through an incoherent feedforward loop (TLR4 activates NFκB and ATF, with NFκB activating downstream targets such as IL-6 and ATF repressing the same targets) [12,18]. Similarly, Amit et al. discovered a large number of coherent feed forward loops in a pathogen-sensing network [5], which at least in some cases appears to protect the system from activating a complex regulatory program in the face of a transient, rather than persistent signal [18]. As additional layers of regulation are incorporated into the network reconstruction [7,8] new motifs incorporating multiple layers of regulation and timescales are likely to be discovered.

Exploring Cellular Diversity

Each of the cell types of the immune system has specific, highly specialized functions. For example, cytotoxic T cells (also known as $CD8^+$ T cells) can kill other cells, whereas T-helper cells (also known as $CD4^+$ T cells) generally provide stimulatory signals to other cells to enable them to perform their particular tasks. Many immune-related disorders are directly associated with a malfunction of a specific cell type. For instance, the inability to create T (H)-17 helper T-cells in Job's syndrome patients is the causative agent for high susceptibility to fungal infections [19], whereas a reduction in lifespan of $CD8^+$ naïve and memory cells can lead to severe cutaneous viral infections [20].

Changes in cell type abundance and relative frequency may have important clinical implications. The complete blood count (CBC) is a clinical assay that enumerates five major leukocyte classes in blood based on cell shape and size (automatic cell counters rely on cell size and isoelectric focusing). First used in 1957, it is one of the tests most commonly prescribed by physicians because it can be indicative of a recent infection or disease. Yet, as the number of immune cell subtypes discovered has grown beyond those basic cell types assayed in the CBC, no standard clinical test measuring cell-type subset abundance followed (e.g., neither the loss of T (H)-17 cells or the reduction in $CD8^+$ naïve and memory cells, described above, would have been identified by a CBC). Rather, currently in the clinic, and at the risk of missing prognostic disease-relevant information, the abundance of immune cell subsets is either not tested, tested at a low resolution, or tested for specific cell subsets based on other patient phenotypic information for disease differential diagnosis.

As different cell types correspond to different functions, an important question is how many different cell types there are in the immune system, and by which markers they can be identified. In many cases different cell subtypes have vastly differing expression profiles, although the reusability of the gene modules and regulatory programs in a different cellular context is evident [21]. For example, the respective gene expression and microRNA profile of 38 and 27 classic hematopoiesis cell types were recently profiled in humans [21,22], whereas the ImmGen project, a consortium effort that aims to profile the gene expression pattern of every known mouse immune cell type, has already profiled over 200 different cell type subsets [23], though for the majority of these cell type subsets their clinical effects are still not sufficiently well known.

Perhaps an even more relevant question is, to what degree do cells occupy discrete 'cell-states', as defined by their gene expression, protein and functional responses? So far, scientists have placed artificial 'cell-state' barriers, although it is possible that there is a continuum of functionality. Recent technological breakthroughs now enable the measurement of immune cell subsets at the single-cell level and at a high dimension, that is, measurement of multiple proteins or genes on the same single cell (Boxes 1 and 2), and are beginning to shed light on this question.

For example, recently Nabel, Sékaly and colleagues showed that immunization-specific gene expression signatures were detectable at a single-cell $CD8^+$ T-cell level following vaccination, and define novel T-cell subsets whose relative frequency in alternative vaccine strains is variable [24]. Thus, current 'standards' for cell-type subsetting are in flux, which makes it more difficult to design and evaluate $CD8^+$ T cells vaccines, and more generally the relationship between cell-type subset and the immune response.

Novel cytometry-based studies evaluate differences at the single-cell level and at high dimensionality for thousands of cells [1,2,25]: Bendall et al. used mass cytometry to profile the entire hematopoietic lineage of an individual human bone marrow at single-cell resolution for a combined total of 52 cell surface and intracellular proteins [2]. Such studies are essentially ex vivo, as they are done on non-separated primary cells that have been stimulated and fixated such that the effects of the local milieu on cells is, to a large extent, maintained. Cells were measured not only at rest, but also when stimulated with a variety of stimuli and inhibitors, and the abundance of phosphorylated proteins was measured at a single-cell level. The increased dimensionality into the immune system is so high that standard methods for manual cell subset identification (also known as 'gating') are inadequate, and novel automatic gating and visualization strategies had to be developed [1,26] (see Figure 25.2A,B). Furthermore, the ex vivo profiling methodology used and the orders-of-magnitude higher resolution attained are assured to advance the clinical goals of improved diagnostics and personalized healthcare [27–31]. Simultaneously, they are expected to reveal important new insights into immune system heterogeneity.

Recently, Newell et al. analyzed antigen-specific $CD8^+$ T cells to map the expression landscape of 17 cell surface proteins and nine intracellular cytokines known to be variably expressed in $CD8^+$ T cells. Through the use of principal component analysis, a mathematical transformation which converts a set of possibly correlated variables (i.e., 17 cell surface proteins) into a set of linearly uncorrelated ones (i.e., a combination of said proteins), this large phenotypic landscape was projected onto a set of vectors that best explain the variation observed in the data. Plotting the first three principal components on a three-dimensional axis showed a continuum of cells, reproducible between individuals, with 'hotspots' in established 'phenotypic clusters' such as naïve, central memory and effector memory (Figure 25.2B). In total, the first three components explained 60% of the observed variation. Additionally, the data showed more than 200 functional phenotypes (different cytokine signatures) just within the $CD8^+$ T-cell subsets, showing a much greater complexity than previously appreciated. How this diversity is generated is illustrated by data showing that different stimulations of these cells yield different responses, and that T cells specific for different viruses express both distinct and overlapping sets of cytokines [1]. This suggests that the immune system can take a combinatorial approach to its functional responses in addition to the genetic diversity encoded in its antigen receptors. Alternatively, this high amount of variation may be quiescent with respect to cellular output or yield a uniform response across cells. Single cells studies of NFκB response to TNF-α stimulation indicate that cells respond heterogeneously and in a binary manner (i.e., either respond or not respond) as a function of stimulus concentration [32], suggesting that initial conditions would play a significant role in determining which cells respond and the nature of the response.

The large degree of functional diversity, both in cells of the same specificity as well as across the entire hematopoietic lineage [2], provides the immune system with remarkable flexibility and indicates a need for mechanisms that enable intercellular communication and convergence of responses between cells. Supporting this, Love and colleagues recently showed the existence of an intricate regulatory program for cellular cytokine release [6]. Using a novel microfluidic device that allows for serial, time-dependent single-cell analysis [33] (see Box 25.2), they showed that polyfunctional human T cells (i.e., T cells releasing multiple cytokines) release their cytokines predominantly in a sequential manner, which begins asynchronously from cell to cell [6]. As virtually all previous measurements were made at a specific single endpoint, this temporal complexity was previously masked and just integrated sum over time of cytokine secretion noted.

Limitations of a Cell-Focused Approach

With advances in automation, a scaling-up of methodologies is likely to occur such that ultimately we will be able to delineate regulatory networks for a large variety of cell types. Immune cells are highly affected by the complex local microenvironment they experience. Thus, how much of the

---

BOX 25.2 Empowering single-cell measurements through microfluidics devices

Much of our knowledge in immunology comes from bulk measurements of many cells together. Owing to problems of averaging and noise, the behavior of cells as inferred from average measurements often drowns cell-to-cell differences and may not reflect the behavior of any single cell. Microfluidic devices can control the flow of minute amounts of liquids or cells and may be designed for sensitive detection of analytes at a single-cell level. Detection (often light) is usually set up in an automated manner such that many cells may be measured. For example, cells may be moved through the device by capillary flow, measured and imaged, or placed stationary into an array of serially micro-engraved wells, each of which is imaged. A recent newcomer to immunology, microfluidic devices are already being utilized for single-cell gene expression studies run in multiplex (across multiple cells and genes) [5], the measurement of single cells cytokine secretion [6] and cell counting [7,8], to name but a few. The miniaturization of these devices and their relative low cost and transportability are promising for the future development of microfluidics-based diagnostics.

inferred network reflects in vivo regulation remains an open question. Further, non-combinatorial perturbations may not accurately reflect overall functionality [8], as partial redundancy is abundant in many biological processes.

As technologies able to perform multiple single cell measurements increase in scale (currently an upper limit of 100 proteins can be simultaneously profiled by mass cytometry, and 96 genes can be assayed by microfluidics) they could serve an important role in whole network reconstruction based on cell-to-cell variance [7,34,35]. However, recent work showing large functional cell-to-cell variation in an individual sample between what were previously considered the same cell types, complicates the interpretation of inferred networks for any single cell [1,32]. Hence a prime challenge is to understand the origin of the newly unmasked variation, its function, and how it relates ultimately to variation in immune response.

SYSTEM-FOCUSED SYSTEMS IMMUNOLOGY

Protective immunity is not the end outcome of any single cell, but rather relies on processes that span multiple cells or cell populations and which are greatly affected by the microenvironment. In contrast to the above cell-focused approach, a growing number of medical researchers and immunologists are embracing high-bandwidth technologies to assay the entire immune system. A strong driving force for these efforts is clinical, as no good metrics of immunological health currently exist [36]. Early efforts used gene expression microarrays to profile immune system state; more recently other multiplex immunoassays technologies probing different layers of the immune system have been introduced: soluble proteins by bead and electrochemiluminesce based assays from blood-derived serum or plasma; multicolor flow and mass-based cytometry for enumerating protein markers for different cell types; and the abundance of intracellular phosphoproteins, to name a few. These pan-cell system-wide analyses are not only yielding blood-based biomarkers, but also provide paradigm-shifting new insights into immunology. In the first part of this section, we examine the insights from recent efforts to comprehensively profile human subjects to understand the immunological mechanisms of vaccination [37–39] (the approach is sometimes referred to as 'systems vaccinology'). We note that these represent only a subset of efforts that are applied to 'system-level' profiling of immune-related diseases, infections and inflammation [39–41]. We will highlight common themes and challenges that have arisen in the first part of this section. In the second part, we will focus on the application of next-generation DNA sequencing to the analysis of the B-cell antibody repertoire and TCR sequences.

Human Immune Monitoring

Vaccination marks one of the greatest achievements of science and is by far the medical intervention that has saved the most lives. However, infectious disease and immune-related diseases contribute to over 25% of deaths in developed countries and more than 50% of deaths in developing countries. Our understanding of the immunological mechanisms by which vaccines work is limited, and we do not yet know why vaccines fail to protect a certain percentage of people in specific subpopulations. For example, roughly only 40% of adults over the age of 65 mount an effective response to the standard yearly influenza vaccine [42], and only 50% of children in developing countries exhibit neutralizing antibodies to oral polio vaccination, vs. 90% in the developed world [43]. Studies in cancer research have shown that disease-specific signatures can be detected in gene expression, often earlier than a definite diagnosis can be made clinically, and that subgroups of patients may be identified with specific gene expression signatures that are prognostic of treatment outcome [44,45]. Phenotypically, it has long been noted that there is considerable variation in the immune responses of individuals. Thus, it is tempting to speculate that there are specific immune states, even among those considered healthy, and that these states affect vaccine responses as well as disease susceptibility and outcome.

Vaccination presents a safe and controlled perturbation in which to study immune system response. Vaccine efficacy varies as a function of many variables, including the type of vaccine (e.g., killed or attenuated virus), the adjuvant introduced with the vaccine to stimulate innate response, and, as mentioned above, it is affected by demographic and geographic attributes. Recent studies are combining multiple high-bandwidth measurements of the immune system in humans and animal models measured pre and post vaccination. The outcome variable in these studies is the degree of protection elicited by a given vaccine in a subject. For most vaccines, this is currently measured by the magnitude of the antigen (vaccine)-specific antibody titers in blood after vaccination. However, it is well known that in some cases this correlate does not actually correspond to protection, or at least corresponds to only a part of the requirements for protection. This exemplifies our limited understanding of the biological mechanisms involved in the response to a given vaccine. Thus, one goal is to identify improved correlates of protection (biomarkers) that can be used to identify individuals who respond suboptimally to vaccination, and the mechanism of immunity involved in eliciting response, whose study may help devise improved vaccines and vaccination protocols [37,38].

Two independent studies employing a systems approach to vaccination were performed on the yellow fever YF-17D vaccine, a good choice for a first detailed study of vaccination, as the vaccine is extremely effective (over 90% of vaccinees are protected), and because it is known to elicit neutralizing antibodies that last more than three decades. Blood samples were taken from cohorts of individuals on the day of the vaccination (time 0) and at several time points afterwards, until 60 days after vaccination and including days 3–7, which covers the initial innate response. Both studies showed a strong change in gene expression following vaccination, which could be detected as early as day 3 and which peaked by day 7. Gene set functional enrichment analysis showed that the changes in gene expression corresponded not only to factors directly related to antibody production, but also to other effectors of the immune response, most prominently activation of virus-dependent innate immunity [46,47] and the proliferation of antigen-specific cytotoxic T cells, whose numbers varied significantly between individuals [47]. The strong innate response upon vaccination was unexpected because the formation of neutralizing antibodies is primarily an adaptive immunity-dependent process. This suggests that there is a strong dependence of the adaptive arm of the immune response on the innate arm. Using gene expression data from the earlier time points (days 0–7), Pulendran and colleagues could predict the magnitude of the later adaptive response of antigen-specific cytotoxic T-cell proliferation. Close examination of the set of predictor genes implicated that the stress response shuts down protein translation as part of the immune mechanisms of vaccination unappreciated [47]. Thus, such studies can serve both for the identification of predictive biomarkers and as a stepping stone for the exploration of immune mechanisms involved in vaccination.

In a follow-up publication, Nakaya et al., in a similar longitudinal experimental design, analyzed influenza response over a 3-year period. As noted above, the effect of influenza vaccination is suboptimal, and individuals are exposed to a variant of virus and/or receive vaccinations on an annual basis. All adults have been previously exposed to at least some flu strains, which greatly confounds analysis because prior exposure has primed memory responses which may activate in some individuals based on virus strain similarity, and because it is unclear what the base response is. Furthermore, for influenza, two different vaccines are in common use: a live attenuated virus given to individuals under the age of 65 (LAIV) and a killed virus (TIV) usually given to at-risk groups. The two vaccines do not elicit an antibody response of the same magnitude, and although a common gene signature is observed, the genes induced by the two vaccines are for the most part quite different. Comparing the gene expression signature of YF-17D-vaccinated individuals to those vaccinated with influenza showed a distinct vaccine-specific signature for both vaccines, although a common interferon-related gene was commonly induced by both YF-17D and LAIV but not TIV. Thus, different vaccines and different vaccine types appear to activate distinct immune mechanisms.

The new insights into the mechanisms of vaccine response can help to develop improved vaccines. For instance, Pulendran and colleagues vaccinated mice with synthetic nanoparticles that contained antigens plus ligands that signal through TLR4 and TLR7, and showed that they induced synergistic increases in antigen-specific neutralizing antibodies, compared to immunization with single TLR ligands [48]. Such coupling between systems analyses and targeted combinatorial vaccine design heralds a new era of highly targeted and personalized vaccination strategies.

Antibody and TCR Repertoire Diversity

The adaptive arm of immunity tailors responses for any encountered antigen. To do so, B and T cells need to generate an enormous repertoire of structural diversity in antigen-recognizing proteins, including antibodies (also known as immunoglobulins) and T-cell receptors (TCR), so that they may be able to recognize and eliminate the large range of possible foreign invaders and cancerous cells. The immune system generates structural diversity through the recombination of three separate and highly variable gene segments loci, termed variable (V), diversity (D) and joining (J) to form a very large number of polypeptides (heavy and light chains of antibodies and α, β, γ and δ of T-cell receptors) that can combine to form heterodimeric antigen-recognition domains. An allelic exclusion mechanism generally allows only a single VDJ combination to be expressed in a given cell, despite the additional chromosomal copies, and a separate mechanism activated following antigen recognition assures high specificity of the receptor/antibody to the antigen through hyper-mutation and selection.

VDJ recombination can create as many as 10^8 different combinations. With an estimated cell count of 10^{11} different B and T cells in an individual human being, it is assumed that this mechanism generates a sufficiently large repertoire for immune system antigen recognition. However, until recently, surveying even a small fraction of an individual's repertoire was considered an impossible task. As a consequence, even basic questions related to the structure and dynamics of the repertoire had gone unanswered for decades. Next-generation DNA sequencing now offers an opportunity to start exploring the basic principles of repertoire selection, as well as its relation to disease.

An individual zebrafish maintains at any one time an average of 300 000 B cells, five orders of magnitudes less than a human, and up to 975 unique VDJ combinations. Seizing on this lower complexity, Quake and colleagues used next-generation DNA sequencing to sequence heavy chain V regions (V_H), from 14 different animals to near-saturation [4]. These VH sequences can be captured in a single sequence read and contain the majority of antibody diversity. Analysis of each fish's repertoire strengthened early observations [49] that VDJ recombination is biased and does not occur with equal probability for each combination. Rather, the antibody repertoire was found to be unevenly distributed, with a high number of combinations expressed at a low abundance and a small number of combinations appearing at very high abundance to make up a rough total of 9% of combinations in any single fish, with significant overlap between fish. A specific VDJ combination may account for more than one antibody due to the aforementioned processes of somatic hyper-mutation and junctional changes. By shifting sequence alignment parameters the researchers were able to estimate which distinct antibodies were produced. The lower and upper bounds on the estimated total number of antibodies per fish were 1200 and 6000, respectively, whereas the antibody abundance frequency followed a power-law distribution which has also been observed in many other complex systems (see Chapter 9). Analysis of Vβ diversity from the TCR of human $CD8^+$ naïve and memory T cells yielded similar results: an uneven repertoire distribution, a large number of distinct combinations expressed in a single individual at a single point in time [50] ($>10^6$, though still negligible compared with the upper theoretical limit of 10^{11}), and a significant overlap of sequence repertoires between any two individuals, particularly in naïve cells [51].

The expressed repertoire is thought to be dependent on VDJ recombination and mutation, phylogenetic history and the pathogen environments. It is possible that the source of the observed structure in the repertoire is simply the result of convergence. That is, that the environment individuals experience yields correlated VDJ usage. Alternatively, the uneven distribution of VDJ choice may reflect a bias in the VDJ recombination mechanisms, either due to long-term (over generations) selection factors or to previously unappreciated inherent limitations in the recombination and mutation mechanisms that yield antibody and TCR diversity. This would have important implications, as it is thought that the diversity of repertoire defines the range of pathogens to which the organism can respond effectively. Using the above-described zebrafish dataset [4], Mora et al. borrowed from statistical physics to measure the D-region diversity using entropy and test the degree to which the diversity could be explained by structural constraints [52]. The genomic origins of the D region is 11–14 nucleotides long, whereas in the zebrafish antibody data it was the most variable sequenced region, with sequences ranging from 3 to 18 nucleotides in length, which is often difficult to align to the original genomic copy. Modeling repertoire sequences in protein space, they assumed that every possible sequence is possible (a space as large as 20^L, where L is D-region peptide length and $L_{max} = 8$), and assessed the measure of diversity in sequence space, or entropy). Next they analyzed to what extent the entropy was reduced by observed amino acid correlations in the data. Whereas simply taking the observed biases in the use of single amino acids into account did not reduce the entropy significantly, accounting for next-most-simple correlations between 'nearest neighbor' and the 'next nearest neighbors' amino acid captured between two-thirds and 90% of all correlated structure in the distribution of sequences [52]. Although the model used is somewhat naïve, in that it gives equal probability to all protein sequences irrespective of the original genomic sequence of the antibody, the correlations between amino acid distributions suggests constraints in somatic hyper-mutation and junctional diversity mechanisms.

Additional basic insights into repertoire dynamics are emerging from studying how the antibody repertoire changes as a function of development and disease. Jian et al. [3] compared the repertoire of antibody diversity between young and older fish and found restricted segment usage in young fish compared to older animals. Diversity increases with age, but primarily through less biased somatic hyper-mutation and random clonal expansion effects, whereas a common deterministic program, with as-yet unknown mechanisms, biases VDJ use [3]. Boyd et al. quantified B-cell IgH antibody repertoire in 150 healthy humans and in blood cancer patients. From healthy individuals they were able to estimate the normal B-cell repertoire complexity, whereas disease-specific signatures were observed in cancer patients, which allowed quantification of the number and identity of the dominant cancerous B-cell clonal receptors [53].

Apparent lessons on repertoire dynamics can be learned not only from organism development, but also from cellular differentiation itself. By separating $CD8^+$ naïve and memory cell types prior to sequencing TCR repertoire, Robins et al. showed that repertoire sequences with high relative frequency in naïve cells are more likely to be observed in the memory compartment [50]. No correlation was observed with respect to repertoire size between the two cells subsets, suggesting that the mechanism to determine the clone size in the two cell subsets is different. Wang et al. sequenced the TCR Vβ and Vα repertoires from one human individual but separately for each of eight different cell subsets. When examining the TCRs whose sequence was most abundant (dominant clones), they detected significant overlap in TCR repertoire between the different $CD8^+$ cell subsets. There has been a long-standing debate on the role of individual TCRs in determining cell fate. Two theories prevail: that a cell's fate is determined in a 'stochastic' fashion with survival dependent on a later selective TCR signal, or in an 'instructive' manner by the affinity of the interaction of the TCR with a cognate peptide—MHC complex. Although the observed overlap in TCR between cell subsets may be explained by independent expansion of cell subsets due to antigenic challenge, it also suggests that the determination of cell fate is a random process independent of TCR affinity, which is in support of the stochastic model. Ultimately, the outcome of having shared or closely similar TCR repertoire between different cell subsets is yet another layer of increased coordination between compartments of the immune system.

Limitations of a System-Focused Approach

Blood is a complex tissue, particularly with respect to its many white blood cell subsets, each of which has its own molecular signature. The system-focused approaches discussed here (both for human immune monitoring and for antibody repertoire) use novel technologies to measure an enormous number of immune system components. Yet with few exceptions they do so while losing the cellular context of the data gathered. This results in several limitations. For example, the frequency of a given cell subset in the blood can vary markedly (2—10-fold difference) between individuals or over time, and thus differences detected from blood between disease and control may to a large extent reflect the fluctuations in cell type subset frequency between samples [54,55], rather than true molecular-level changes associated with disease.

For human immune monitoring-focused studies it is currently impractical to capture whole genome information on each and every cell type or single cell, and for most diseases it is unclear whether there is a predominant cell type that can be sorted and studied (i.e., in a cell-focused manner as described above) associated with a disease. Experimental designs that include time courses elevate this problem to some extent, as an individual's samples are compared to one another rather than across individuals, for which there is usually much more variation. Recently, we and others have developed computational solutions for post-hoc extraction of cell type-specific information from heterogeneous tissue data (Box 25.3) [56,57]. Such techniques often increase the signal-to-noise ratio in system-wide studies by orders of magnitude and greatly facilitate data interpretation, while also allowing the gathering of both cell-specific and system-wide information, thereby effectively creating a bridge between these two approaches.

MULTI-SCALE-FOCUSED SYSTEMS IMMUNOLOGY

Protective immunity is not the final outcome of any single cell, but rather requires the execution of multiple processes that draw on functionality elicited by many cell types communicating between one another, and sometimes even between larger cell populations. In dissecting molecular and cellular data (i.e., cell-focused and system-focused) we remain far from a true understanding the integrated system. One main problem is that it becomes difficult for humans to maintain an integrated and correct picture of knowledge that extends beyond a few interacting entities. Computational modeling of immune systems aims to provide a solution to this problem by incorporating different data types.

Ideally, any computational model of the immune system would seek to include representation for each of the many involved cell types and their intercellular communications. A first choice when generating a computational model is to define the exact phenomenon that will be modeled as well as to decide the level of detail required. Model building iterates between two steps: specification and simulation. Specification can be broken into three parts: (1) defining model assumptions, which may be as general as how two entities (e.g., proteins) interact, and how unknown information is treated; (2) explicit inputting of all known information on the studied phenomenon, which is usually a tedious task that requires collection of data from scientific papers and measurements, and their translation into a well-defined executable specification; and (3) setting of boundary conditions, e.g., the interface between the studied phenomenon and other phenomena external to it.

Immune system modeling has been around for several decades, but the computational infrastructure capabilities only recently allow us to go beyond toy and educational models to models that provide research with novel insights and hypotheses [58]. Many groups have used mathematical and computational models to gain a better understanding of immunological data. As the topic of this section is specifically building an integrated understanding of the immune response, we focus on models that span a large number of cells and in which the phenomena are being modeled at the whole organ or system level, rather than a smaller cell—cell interaction level. We will describe here two levels of models: cellular automata-based models and Statecharts-based models. These two 'model specification languages' differ greatly in their richness to express complicated constructs, and from this stem large differences in the specification efforts required by the researcher and ultimately the breadth of the immune system that the model covers. Common to both is that through simulation coupled to graphical visualizations, both model classes are beginning to unravel the complex cross-talk between cells, and are providing a framework to study emergent system properties.

System-Wide Meso-Scale Cellular Automata Models of Immunity

The simplest models rely on the ideas of cellular automata, discrete models consisting of a grid of cells, each in one of

BOX 25.3 Cleaning the mess — computational approaches for obtaining cell type specific information from heterogeneous tissues

For many measurements performed on tissue samples consisting of more than one cell type, current experimental techniques and cost limitations allow measurement of a few biological species across many different cell types (e.g., as in flow or mass-based cytometry) or many biological species across a heterogeneous tissue that is devoid of cell-type information (a single isolated cell type being a specific subset of the latter). Ideally, a researcher may desire to measure many biological species, as they are affected by their tissue microenvironment but at a cell type-specific resolution, and for all of the different cell types in a tissue.

Statistical deconvolution techniques incorporate tissue heterogeneity information to capture both cell type-specific and system-wide information from the same samples. These techniques are performed in silico after data generation and are based on the idea of exploiting sample to sample variation. First developed in yeast [9,10], they have been enhanced and tested for the analysis of human blood gene expression samples [11,12], though the basic concept can be expanded to other complex tissues [13] or measurement types. By tracking how measured gene expression fluctuates between samples in relation to cell-frequency changes, it is possible to accurately estimate the average type-specific expression of each cell type. This estimated cell type-specific expression may then be used to identify cell type-specific expression differences between groups of samples for each cell type present in the tissue above a certain (empirically determined) frequency, or reconstruct biological samples having removed the effects of one or more of the cell types. Conversely, through tracking the transcript abundance of combinations of cell-specific markers it is possible to accurately estimate the frequency of each cell subset. The sensitivity of cell type specific expression performed in this manner is often orders of magnitude higher than that obtained by analyzing heterogeneous tissue samples and provides a cellular context for each of the detected differences between case and control.

References

[1] Bendall SC, Simonds EF, Qiu P, Amir el AD, Krutzik PO, Finck R, et al. Single-cell mass cytometry of differential immune and drug responses across a human hematopoietic continuum. Science 2011;332(6030):687–96.

[2] Newell EW, Sigal N, Bendall SC, Nolan GP, Davis MM. Cytometry by Time-of-Flight Shows Combinatorial Cytokine Expression and Virus-Specific Cell Niches within a Continuum of CD8 ($^+$) T Cell Phenotypes. Immunity 2012;36(1):142–52.

[3] Ornatsky OI, Kinach R, Bandura DR, Lou X, Tanner SD, Baranov VI, et al. Development of analytical methods for multiplex bio-assay with inductively coupled plasma mass spectrometry. J Anal At Spectrom 2008;23(4):463–9.

[4] Bandura DR, Baranov VI, Ornatsky OI, Antonov A, Kinach R, Lou X, et al. Mass cytometry: technique for real time single cell multitarget immunoassay based on inductively coupled plasma time-of-flight mass spectrometry. Anal Chem 2009;81(16):6813–22.

[5] Flatz L, Roychoudhuri R, Honda M, Filali-Mouhim A, Goulet JP, Kettaf N, et al. Single-cell gene-expression profiling reveals qualitatively distinct CD8T cells elicited by different gene-based vaccines. Proc Natl Acad Sci USA 2011;108(14):5724–9.

[6] Han Q, Bradshaw EM, Nilsson B, Hafler DA, Love JC. Multidimensional analysis of the frequencies and rates of cytokine secretion from single cells by quantitative microengraving. Lab Chip 2010;10(11):1391–400.

[7] Garcia D, Ghansah I, Leblanc J, Butte MJ. Counting cells with a low-cost integrated microfluidics-waveguide sensor. Biomicrofluidics 2012;6(1):14115–141154.

[8] Leblanc J, Mueller AJ, Prinz A, Butte MJ. Optical planar waveguide for cell counting. Appl Phys Lett 2012;100(4): 43701–437015.

[9] Lu P, Nakorchevskiy A, Marcotte EM. Expression deconvolution: a reinterpretation of DNA microarray data reveals dynamic changes in cell populations. Proc Natl Acad Sci USA 2003;100(18):10370–5.

[10] Stuart RO, Wachsman W, Berry CC, Wang-Rodriguez J, Wasserman L, Klacansky I, et al. In silico dissection of cell type-associated patterns of gene expression in prostate cancer. Proc Natl Acad Sci USA 2004;101(2):615–20.

[11] Abbas AR, Wolslegel K, Seshasayee D, Modrusan Z, Clark HF. Deconvolution of blood microarray data identifies cellular activation patterns in systemic lupus erythematosus. PLoS One 2009;4(7):e6098.

[12] Shen-Orr SS, Tibshirani R, Khatri P, Bodian DL, Staedtler F, Perry NM, et al. Cell type-specific gene expression differences in complex tissues. Nat Methods 2010;7(4):287–9.

[13] Kuhn A, Thu D, Waldvogel HJ, Faull RL, Luthi-Carter R. Population-specific expression analysis (PSEA) reveals molecular changes in diseased brain. Nat Methods 2011;8(11):945–7.

---

a finite of states and a set of rules executed temporally defining neighborhood cell interactions. Models in this class make many simplifying assumptions and usually lack all but the rudimentary principles of the biological processes being modeled. To develop large-scale models, researchers usually specify a small number of cell types and molecular entities and the set of rules that describe cell and molecular interaction behavior. They then create many replicate cellular entities, such that the total number of cells may be very large, even though the actual number of cells of the same cell type and state may be limited. Cell-to-cell variation may be generated by external input (e.g., a stimulation which may be spatially or temporally restricted) which would define each cell's initial conditions and ultimately cellular behavior.

cImmSim is a stochastic cellular automata system based on an the original ImmSim model proposed by Seiden and Celada [59]. It incorporates a general textbook model of innate and adaptive immunity, including several cell types, states they can adopt with probabilistic transition rules, antibodies and antigen. Cells may have unique properties, such as their receptor specificity, specified by a bit string (e.g., 001110), and an interaction between two such entities occurs with a probability that is a function of bit string similarity (modeling affinity). All the feasible interactions among cells and molecules take place within a lattice with cell-to-cell interactions occurring between neighboring cells ordered on the lattice. Time is discretized, and at every single time step entities (cells and molecules) may diffuse to adjoin neighborhoods on the lattice. For example, a B-cell entity in an 'exposing' state implies (but is not directly modeled) that the cell has phagocytosed one antigen and has already processed it. If the bind with the MHCII molecule is successful then the cell is exposing the MHCII molecule bond with one antigen peptide. If an interaction with a T-helper cell in an 'Active' state occurs (as determined by a stochastic event and position on the lattice), the action is to update both the T-helper and B cells instances to a 'Stimulated' state [60]. Millions of cells can be simulated simultaneously, and their states computed in parallel and asynchronously [60]. Simulating viral infections (e.g., HIV and Epstein—Barr) using this model, Castiglione and colleagues showed that they could reproduce disease natural history phenomena such as time from infection to AIDS development in the model [61], and then use it to suggest the best time to administer antiretroviral therapy (HAART) [62].

Statecharts as a Rich Framework to Model Immunity

Cellular automata-based models lack detailed biological information on the modeled interactions. As such, they are limited in the amount of insight they can yield, and an observed disagreement between them and what is observed in the actual system is common and often goes unexplained. Other models in use to model immune-related phenomena are much richer in their specification of the details of the biological process. These require gathering large amounts of information from the literature and encoding information gathered from text, tables and figures into a machine-understandable format.

The modeling language of Statecharts is a visual formalism invented to aid the design of complex man-made reactive systems [63]. It is mathematically well defined, which makes it amenable to execution by computers through software packages implementing the formalism. It has been adapted for use in modeling biological systems [64]. Behavior in Statecharts is described using states and events that cause transitions between states. States may contain substates that are suitable for multilayered systems (e.g., molecules within cells). In addition, Statecharts allow for 'orthogonal states' where the same object (i.e., a cell) may exist in different states. In modeling the immune system, this functionality is co-opted to describe the differentiation of a cell from one subtype to another (e.g., from a naïve cell expressing CD4 to a memory cell when co-expressing the cell surface protein CD45RO). Transitions in Statecharts take the system from one state to another; the biological equivalent is the result of an interaction either between two cells or between a cell and various molecules. Although models here may be as simple as the one defined using cellular automata, the high expressivity of the programming language allows one to define much more complicated models that include many states and many substates, as well as spatial information without loss of reliability (Figure 25.3A). A novel visualization layer sits on top of these models and displays a visual image of cells, molecules and their interactions as well as enabling user manipulation of the simulation parameters (Figure 25.3B). The end result is a mathematically precise presentation of the underlying simulation that is executable and is akin to watching a video recording of a cell or a tissue through a microscope.

Using Statecharts, Efroni, Harel and Cohen modeled first the biological process of T-cell maturation in the thymus [65,66]. The migration of cells in the thymus depends on many factors, including their receptor, the chemokine gradients, epithelial cells, cell proliferation and survival. The thymic environment, loaded with molecules and cells, presents a challenge to many researchers from different fields who have detailed knowledge of some of its parts, but yet wish to comprehend either system-level effects of a molecular level change or to identify the origin of a system-level property. Efroni et al. integrated data generated from reductionist biology from hundreds of scientific publications into a specification model encoded in Statecharts.

In silico testing the effects of loss-of-function mutations in the CXCR4 and CCR9 chemokine receptors on cell migration matched those observed in histological samples, with the added advantage of being able to observe in real time the dynamics of cell migration in thousands of cells. A developing thymocyte must commit to becoming either a $CD4^+$ or a $CD8^+$ cell in the thymus. The decision-making process is obscure and the ratio of mature $CD4^+$ to $CD8^+$ is unequal (roughly 2:1). Where information is not known, the Statecharts framework enables the simultaneous specification of alternative theories. With respect to lineage

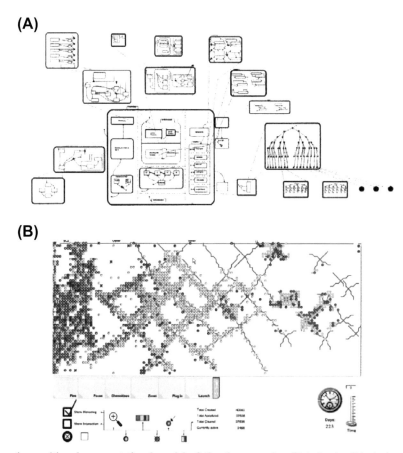

FIGURE 25.3 **An integrative multi-scale computational model of the thymus using Statecharts.** (A) A theory of interactions between thymic epithelial cells and thymocytes presented as a Statechart. The visual multilayered nature of the language enables specification of rich models emulating real biology. (B) A snapshot of the simulated thymus at run time. The visualization overlay on top of the executable program enables intuitive understanding of the dynamics of the model and an interface for the modeler to alter parameters in the simulation. (*Taken from* [65])

commitment, the model showed that the thymus is packed full of cells and competition exists between cells for interaction space. Suffice it to say that the rate of disassociation of $CD8^+$ cells from epithelial cells is lower than that of CD4, to create the observed unequal ratio. This is in contrast to the two dominant molecular-based models described above (see Systems-focused section for more information on the stochastic and instructive lineage commitment models). The existence of competition has not been previously discussed, and gaining the insight from in vivo experiments that cell niche competition could be responsible for the observed cell lineage ratio would have been difficult to deduce. Thus, the framework provides an integrative multi-scale view that enables us to capture emergent phenomena and matches the way we reason in biology.

Limitations of a Multi-Scale-Focused Approach

Owing to the high complexity of the system, its component-rich multi-scale nature and the many unknowns, a full-scale and accurate in-detail model of the immune system is likely impractical in the foreseeable future. A big bottleneck lies in the effort for accurate specification, which is currently an extensive manual process requiring expert curators. Thus the current models are either limited in the richness of the biology they express (e.g., cellular automata) or rich in their expression but narrow in scope (e.g., Statecharts). In an era where gigabytes of new immune-related data are being generated every day, this is a serious concern as the former does not capture the reality being measured, whereas the latter does not capture the scale. Solutions to these concerns must advance on two fronts: first, as best as possible, specification from the literature must become an automated process. This will require advanced natural language processing, automated ontology building, and comprehensive high-quality standardized annotation of genome-scale data, all fields still in their infancy. Beginning with the cell—cytokine network, we have been advancing this effort through immuneXpresso, an automated engine aimed at constructing an immune-related knowledge base [67]. A second front is the incorporation of information generated by 'omics' technologies. Machine learning techniques, such as Bayesian network analyses, are the state of the art for reasoning over such data. However, the underlying specification for the models being used by these statistical methodologies is usually abstract and rarely incorporates detailed knowledge in the field. Hybrid models combining the rich knowledge gathered to date with the novel high-bandwidth data are needed to advance towards full-scale detailed immune system simulation.

CONCLUSIONS AND FUTURE CHALLENGES

We have described a new and upcoming direction for systems-levels analysis, that of systems immunology. The aim of systems immunology is to 'put the pieces together' such that we develop a true understanding of how immunity functions, why has it developed as it is, and what the forces are that shape it. Likely more so than other fields, immunology bridges basic and clinical research. As a derivative of that unique position, and due to the high complexity of the human immune system, systems immunology is also leading the charge to enhance our understanding of immunity and its critical role in disease to a level that allows predictive personalized medicine.

We grouped the many research efforts in systems immunology based on the principal approaches taken. One is no more important than another: rather, they are all complementary and synergistic. For brevity, we did not go into detail on the numerous features of the immune system that would require a combined quantitative-experimental effort to understand. For example, the sensory organlike — sensitivity and specific recognition properties of the T-cell receptor for a particular MHC-peptide complex derived from a pathogen [68]. Cell-focused, systems-focused or multi-scale focused, it is our view that all are needed to advance on this momentous task and that true understanding of immunity, the end goal of systems immunology and likely an emergent phenomenon in its own right, will only emerge if it becomes seamless to transition between these different approaches.

For this to happen, system-wide measurements of molecules must find their way back into to the cell, where they naturally reside, and this genome-wide information integrated into dynamic multi-scale models of cell populations. To do so will require both a strong computational effort as well as the development of technologies to assay for molecules other than proteins, which, like mass-cytometry, are highly dimensional yet work at a single cell level and on all cells of the immune system. We believe that this may require training a new breed of scientists (perhaps those reading this chapter), adept in immunology in all of its many layers and complexity, yet also highly quantitative and able to develop the computational and technological advances needed.

Despite these challenges, we cannot emphasize enough what a transformative time this is for immunology. As an illustration of this, few immune-related citations in this chapter are older than 2–3 years. We do not think this is because of bias on our part, but rather because little of systems immunology and systems-like approaches to immunology existed before then. It is only now that we have reached a time where we have the ability to peek into this fascinating system, to discover how little we actually understood to date, and try to understand the whole. A direct benefit of this should be a better understanding of the human immune system and some of its many diseases, offering us new opportunities for drug target development and even in vitro clinical testing for diagnostics and treatment decision support [69].

REFERENCES

[1] Newell EW, Sigal N, Bendall SC, Nolan GP, Davis MM. Cytometry by time-of-flight shows combinatorial cytokine expression and virus-specific cell niches within a continuum of CD8 (+) T cell phenotypes. Immunity 2012;36(1):142–52.

[2] Bendall SC, Simonds EF, Qiu P, Amir el AD, Krutzik PO, Finck R, et al. Single-cell mass cytometry of differential immune and drug responses across a human hematopoietic continuum. Science 2011;332(6030):687–96.

[3] Jiang N, Weinstein JA, Penland L, White 3rd RA, Fisher DS, Quake SR. Determinism and stochasticity during maturation of the zebrafish antibody repertoire. Proc Natl Acad Sci USA 2011; 108(13):5348–53.

[4] Weinstein JA, Jiang N, White 3rd RA, Fisher DS, Quake SR. High-throughput sequencing of the zebrafish antibody repertoire. Science 2009;324(5928):807–10.

[5] Amit I, Garber M, Chevrier N, Leite AP, Donner Y, Eisenhaure T, et al. Unbiased reconstruction of a mammalian transcriptional network mediating pathogen responses. Science 2009;326(5950): 257–63.

[6] Han Q, Bagheri N, Bradshaw EM, Hafler DA, Lauffenburger DA, Love JC. Polyfunctional responses by human T cells result from sequential release of cytokines. Proc Natl Acad Sci USA 2012;109(5):1607–12.

[7] Sachs K, Perez O, Pe'er D, Lauffenburger DA, Nolan GP. Causal protein-signaling networks derived from multiparameter single-cell data. Science 2005;308(5721):523–9.

[8] Chevrier N, Mertins P, Artyomov MN, Shalek AK, Iannacone M, Ciaccio MF, et al. Systematic discovery of TLR signaling components delineates viral-sensing circuits. Cell 2011;147(4): 853–67.

[9] Basso K, Margolin AA, Stolovitzky G, Klein U, Dalla-Favera R, Califano A. Reverse engineering of regulatory networks in human B cells. Nat Genet 2005;37(4):382–90.

[10] Mani KM, Lefebvre C, Wang K, Lim WK, Basso K, Dalla-Favera R, et al. A systems biology approach to prediction of oncogenes and molecular perturbation targets in B-cell lymphomas. Mol Syst Biol 2008;4:169.

[11] Wang K, Saito M, Bisikirska BC, Alvarez MJ, Lim WK, Rajbhandari P, et al. Genome-wide identification of post-translational modulators of transcription factor activity in human B cells. Nat Biotechnol 2009;27(9):829–39.

[12] Gilchrist M, Thorsson V, Li B, Rust AG, Korb M, Roach JC, et al. Systems biology approaches identify ATF3 as a negative regulator of Toll-like receptor 4. Nature 2006;441(7090):173–8.

[13] Ramsey SA, Klemm SL, Zak DE, Kennedy KA, Thorsson V, Li B, Gilchrist M, et al. Uncovering a macrophage transcriptional program by integrating evidence from motif scanning and expression dynamics. PLoS Comput Biol 2008;4(3):e1000021.

[14] Segal E, Shapira M, Regev A, Pe'er D, Botstein D, Koller D, et al. Module networks: identifying regulatory modules and their condition-specific regulators from gene expression data. Nat Genet 2003;34(2):166–76.

[15] Clancy T, Pedicini M, Castiglione F, Santoni D, Nygaard V, Lavelle TJ, et al. Immunological network signatures of cancer progression and survival. BMC Med Genomics 2011;4:28.

[16] Bourgon R, Gentleman R, Huber W. Independent filtering increases detection power for high-throughput experiments. Proc Natl Acad Sci USA 2010;107(21):9546–51.

[17] Alon U. Network motifs: theory and experimental approaches. Nat Rev Genet 2007;8(6):450–61.

[18] Litvak V, Ramsey SA, Rust AG, Zak DE, Kennedy KA, Lampano AE, et al. Function of C/EBPdelta in a regulatory circuit that discriminates between transient and persistent TLR4-induced signals. Nat Immunol 2009;10(4):437–43.

[19] Milner JD, Brenchley JM, Laurence A, Freeman AF, Hill BJ, Elias KM, et al. Impaired T (H)17 cell differentiation in subjects with autosomal dominant hyper-IgE syndrome. Nature 2008;452(7188):773–6.

[20] Randall KL, Chan SS, Ma CS, Fung I, Mei Y, Yabas M, et al. DOCK8 deficiency impairs CD8 T cell survival and function in humans and mice. J Exp Med 2011;208(11):2305–20.

[21] Novershtern N, Subramanian A, Lawton LN, Mak RH, Haining WN, McConkey ME, et al. Densely interconnected transcriptional circuits control cell states in human hematopoiesis. Cell 2011;144(2):296–309.

[22] Petriv OI, Kuchenbauer F, Delaney AD, Lecault V, White A, Kent D, et al. Comprehensive microRNA expression profiling of the hematopoietic hierarchy. Proc Natl Acad Sci USA 2010;107(35):15443–8.

[23] Heng TS, Painter MW. The immunological genome project: networks of gene expression in immune cells. Nat Immunol 2008;9(10):1091–4.

[24] Flatz L, Roychoudhuri R, Honda M, Filali-Mouhim A, Goulet JP, Kettaf N, et al. Single-cell gene-expression profiling reveals qualitatively distinct CD8 T cells elicited by different gene-based vaccines. Proc Natl Acad Sci USA 2011;108(14):5724–9.

[25] Gibbs Jr KD, Gilbert PM, Sachs K, Zhao F, Blau HM, Weissman IL, et al. Single-cell phospho-specific flow cytometric analysis demonstrates biochemical and functional heterogeneity in human hematopoietic stem and progenitor compartments. Blood 2011;117(16):4226–33.

[26] Qiu P, Simonds EF, Bendall SC, Gibbs Jr KD, Bruggner RV, Linderman MD, et al. Extracting a cellular hierarchy from high-dimensional cytometry data with SPADE. Nat Biotechnol 2011;29(10):886–91.

[27] Irish JM, Myklebust JH, Alizadeh AA, Houot R, Sharman JP, Czerwinski DK, et al. B-cell signaling networks reveal a negative prognostic human lymphoma cell subset that emerges during tumor progression. Proc Natl Acad Sci USA 2010;107(29):12747–54.

[28] Kotecha N, Flores NJ, Irish JM, Simonds EF, Sakai DS, Archambeault S, et al. Single-cell profiling identifies aberrant STAT5 activation in myeloid malignancies with specific clinical and biologic correlates. Cancer Cell 2008;14(4):335–43.

[29] Irish JM, Czerwinski DK, Nolan GP, Levy R. Altered B-cell receptor signaling kinetics distinguish human follicular lymphoma B cells from tumor-infiltrating nonmalignant B cells. Blood 2006;108(9):3135–42.

[30] Krutzik PO, Irish JM, Nolan GP, Perez OD. Analysis of protein phosphorylation and cellular signaling events by flow cytometry: techniques and clinical applications. Clin Immunol 2004;110(3):206–21.

[31] Irish JM, Hovland R, Krutzik PO, Perez OD, Bruserud O, Gjertsen BT, et al. Single cell profiling of potentiated phospho-protein networks in cancer cells. Cell 2004;118(2):217–28.

[32] Tay S, Hughey JJ, Lee TK, Lipniacki T, Quake SR, Covert MW. Single-cell NF-kappaB dynamics reveal digital activation and analogue information processing. Nature 2010;466(7303):267–71.

[33] Han Q, Bradshaw EM, Nilsson B, Hafler DA, Love JC. Multidimensional analysis of the frequencies and rates of cytokine secretion from single cells by quantitative microengraving. Lab Chip 2010;10(11):1391–400.

[34] Sachs K, Itani S, Fitzgerald J, Wille L, Schoeberl B, Dahleh MA, et al. Learning cyclic signaling pathway structures while minimizing data requirements. Pac Symp Biocomput 2009:63–74.

[35] Sachs K, Itani S, Carlisle J, Nolan GP, Pe'er D, Lauffenburger DA. Learning signaling network structures with sparsely distributed data. J Comput Biol 2009;16(2):201–12.

[36] Davis MM. A prescription for human immunology. Immunity 2008;29(6):835–8.

[37] Pulendran B, Li S, Nakaya HI. Systems vaccinology. Immunity 2010;33(4):516–29.

[38] Zak DE, Aderem A. Systems biology of innate immunity. Immunol Rev 2009;227(1):264–82.

[39] Rappuoli R, Aderem A. A 2020 vision for vaccines against HIV, tuberculosis and malaria. Nature 2011;473(7348):463–9.

[40] Chaussabel D, Pascual V, Banchereau J. Assessing the human immune system through blood transcriptomics. BMC Biol 2010;8:84.

[41] Nakaya HI, Li S, Pulendran B. Systems vaccinology: learning to compute the behavior of vaccine induced immunity. Wiley Interdiscip Rev Syst Biol Med 2012;4(2):193–205.

[42] Goodwin K, Viboud C, Simonsen L. Antibody response to influenza vaccination in the elderly: a quantitative review. Vaccine 2006;24(8):1159–69.

[43] Paul Y. Polio eradication in India: have we reached the dead end? Vaccine 2010;28(7):1661–2.

[44] Ebert BL, Golub TR. Genomic approaches to hematologic malignancies. Blood 2004;104(4):923–32.

[45] Segal E, Friedman N, Kaminski N, Regev A, Koller D. From signatures to models: understanding cancer using microarrays. Nat Genet 2005;37(Suppl):S38–45.

[46] Gaucher D, Therrien R, Kettaf N, Angermann BR, Boucher G, Filali-Mouhim A, et al. Yellow fever vaccine induces integrated multilineage and polyfunctional immune responses. J Exp Med 2008;205(13):3119−31.

[47] Querec TD, Akondy RS, Lee EK, Cao W, Nakaya HI, Teuwen D, et al. Systems biology approach predicts immunogenicity of the yellow fever vaccine in humans. Nat Immunol 2009;10(1):116−25.

[48] Kasturi SP, Skountzou I, Albrecht RA, Koutsonanos D, Hua T, Nakaya HI, et al. Programming the magnitude and persistence of antibody responses with innate immunity. Nature 2011;470(7335):543−7.

[49] Alt FW, Ferrier P, Malynn B, Lutzker S, Rothman P, Berman J, et al. Control of recombination events during lymphocyte differentiation. Heavy chain variable region gene assembly and heavy chain class switching. Ann N Y Acad Sci 1988;546:9−24.

[50] Robins HS, Campregher PV, Srivastava SK, Wacher A, Turtle CJ, Kahsai O, et al. Comprehensive assessment of T-cell receptor beta-chain diversity in alphabeta T cells. Blood 2009;114(19):4099−107.

[51] Robins HS, Srivastava SK, Campregher PV, Turtle CJ, Andriesen J, Riddell SR, et al. Overlap and effective size of the human $CD8^+$ T cell receptor repertoire. Sci Transl Med 2010;2(47):47ra64.

[52] Mora T, Walczak AM, Bialek W, Callan Jr CG. Maximum entropy models for antibody diversity. Proc Natl Acad Sci USA 2010;107(12):5405−10.

[53] Boyd SD, Marshall EL, Merker JD, Maniar JM, Zhang LN, Sahaf B, et al. Measurement and clinical monitoring of human lymphocyte clonality by massively parallel VDJ pyrosequencing. Sci Transl Med 2009;1(12):12ra23.

[54] Whitney AR, Diehn M, Popper SJ, Alizadeh AA, Boldrick JC, Relman DA, et al. Individuality and variation in gene expression patterns in human blood. Proc Natl Acad Sci USA 2003;100(4):1896−901.

[55] Cobb JP, Mindrinos MN, Miller-Graziano C, Calvano SE, Baker HV, Xiao W, et al. Application of genome-wide expression analysis to human health and disease. Proc Natl Acad Sci USA 2005;102(13):4801−6.

[56] Shen-Orr SS, Tibshirani R, Khatri P, Bodian DL, Staedtler F, Perry NM, et al. Cell type-specific gene expression differences in complex tissues. Nat Methods 2010;7(4):287−9.

[57] Kuhn A, Thu D, Waldvogel HJ, Faull RL, Luthi-Carter R. Population-specific expression analysis (PSEA) reveals molecular changes in diseased brain. Nat Methods 2011;8(11):945−7.

[58] Halling-Brown M, Pappalardo F, Rapin N, Zhang P, Alemani D, Emerson A, et al. ImmunoGrid: towards agent-based simulations of the human immune system at a natural scale. Philos Transact A Math Phys Eng Sci 2010;368(1920):2799−815.

[59] Seiden PE, Celada F. A model for simulating cognate recognition and response in the immune system. J Theor Biol 1992;158(3):329−57.

[60] Bernaschi M, Castiglione F. Design and implementation of an immune system simulator. Comput Biol Med 2001;31(5):303−31.

[61] Castiglione F, Poccia F, D'Offizi G, Bernaschi M. Mutation, fitness, viral diversity, and predictive markers of disease progression in a computational model of HIV type 1 infection. AIDS Res Hum Retroviruses 2004;20(12):1314−23.

[62] Paci P, Martini F, Bernaschi M, D'Offizi G, Castiglione F. Timely HAART initiation may pave the way for a better viral control. BMC Infect Dis 2011;11:56.

[63] Harel D. Statecharts − a visual formalism for complex-systems. Sci Comput Program 1987;8:231−2281.

[64] Kam N, Cohen I R, Harel D. The immune system as a reactive system: modeling T cell activation with statecharts. To appear in Bull Math Bio. An extended abstract of this paper appeared in Proc. Visual Languages and Formal Methods (VLFM01), part of IEEE Symposia on Human-Centric Computing Languages and Environments (HCC01), pp. 15−22 (2001).

[65] Efroni S, Harel D, Cohen IR. Toward rigorous comprehension of biological complexity: modeling, execution, and visualization of thymic T-cell maturation. Genome Res 2003;13(11):2485−97.

[66] Efroni S, Harel D, Cohen IR. Emergent dynamics of thymocyte development and lineage determination. PLoS Comput Biol 2007;3(1):e13.

[67] Shen-Orr SS, Goldberger O, Garten Y, Rosenberg-Hasson Y, Lovelace PA, Hirschberg DL, et al. Towards a cytokine-cell interaction knowledgebase of the adaptive immune system. Pac Symp Biocomput 2009:439−50.

[68] Davis MM, Krogsgaard M, Huse M, Huppa J, Lillemeier BF, Li QJ. T cells as a self-referential, sensory organ. Annu Rev Immunol 2007;25:681−95.

[69] Amit I, Regev A, Hacohen N. Strategies to discover regulatory circuits of the mammalian immune system. Nat Rev Immunol 2011 Nov 18;11(12):873−80.

Chapter 26

Causal Inference and the Construction of Predictive Network Models in Biology

Eric E. Schadt
Department of Genetics and Genomic Sciences, Mount Sinai School of Medicine, New York City, NY 10029, USA

Chapter Outline

Introduction	499
A Movie Analogy for Modeling Biological Systems	502
Causality as A Statistical Inference	503
From Assessing Causal Relationships Among Trait Pairs to Predictive Gene Networks	505
Building from the Bottom Up or Top Down?	505
An Integrative Genomics Approach to Constructive Predictive Network Models	506
Integrating Genetic Data as a Structure prior to Enhancing Causal Inference in the Bayesian Network Reconstruction Process	507
Incorporating other 'Omics' Data as Network Priors in the Bayesian Network Reconstruction Process	507
Illustrating the Construction of Predictive Bayesian Networks with an Example	508
Step 1: Identification of the URA3-Centered de novo Biosynthesis of Pyrimidine Ribonucleotides Subnetwork	509
Step 2: Reconstructing Networks using only Expression and Metabolite Traits (Excluding DNA Variation Data)	509
Step 3: Constructing Priors Using eQTL Data	510
Step 4: Constructing Priors using KEGG Data	510
Step 5: Constructing Networks using Expression Data, Metabolite Data, and the Genetic and Canonical Pathway Priors defined in Steps 3 and 4	510
Step 6: Comparing the Networks Constructed in Steps 2 and 5	510
Networks Constructed from Human and Animal Data Elucidate the Complexity of Disease	510
Conclusion and Future Directions	511
Unifying Bottom-Up and Top-Down Modeling Approaches	512
A Need to Review How We Work Together	512
References	513

INTRODUCTION

The big data revolution is all around us, permeating nearly every aspect of our lives. Electronic devices that consume much of our attention on a daily basis enable rapid transactions among individuals on an unprecedented scale, where all of the information involved in these daily transactions can be seamlessly stored in digital form, whether the transactions involve cell phone calls, text messages, credit card purchases, emails, or visits to the doctor's office, in which all tests carried out are digitized and entered into your electronic medical record (Figure 26.1).

In 1984 IBM introduced the PC Junior as among the first attempts to bring computer technology into the home. Basic models of this first PC came with 64 kilobytes (kb) of random access memory and a 5.25 inch floppy drive capable of storing 360 kb of data, all at a cost of roughly $1500 (at 2010 prices). Today, for roughly the same price, one can purchase a laptop computer with more than a terabyte of disk storage (a seven orders of magnitude increase over the PC Junior) and a couple of gigabytes (Gb) of random access memory (not to mention high-end graphical processing units and high-resolution displays). The digital universe more generally now far exceeds 1 zetabyte (that is, 21 zeros or one billion terabytes). Thus, our ability to store and access unimaginable amounts of data has been revolutionized by technological innovations that are often observed to operate at super-Moore's Law rates.

The life and biomedical sciences have not stood on the sidelines of this revolution. There has been an incredible wave of new technologies in genomics — such as next-generation sequencing technologies [1], sophisticated imaging systems, and mass spectrometry-based flow cytometry [2] — enabling data to be generated at very large scales. As a result we can monitor the expression of tens of

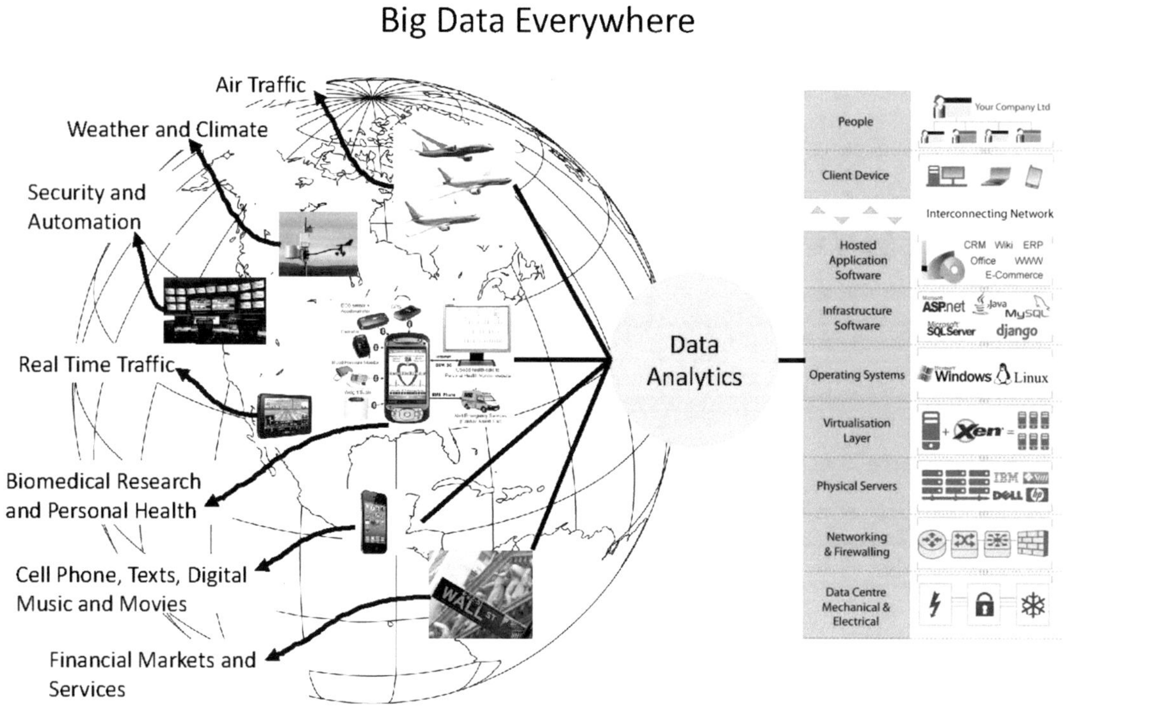

Big Data Everywhere

FIGURE 26.1 **Big data are all around us, enabled by technological advances in micro- and nano-electronics, nano materials, interconnectivity provided by sophisticated telecommunication infrastructure, massive network-attached storage capabilities, and commodity-based high-performance computing infrastructures.** The ability to store all credit card transactions, all cell phone traffic, all email traffic, video from extensive networks of surveillance devices, satellite and ground sensing data informing on all aspects of the weather and overall climate, and now generate and store massive data informing on our personal health, including whole genome sequencing data and extensive imagining data, is driving a revolution in high-end data analytics to make sense of the big data, drive more accurate descriptive and predictive models that inform decision making on every level, whether identifying the next big security threat or making the best diagnosis and treatment choice for a given patient.

thousands of protein- and non-coding genes simultaneously [3,4], score hundreds of thousands of SNPs (single nucleotide polymorphisms) in individual samples [5], sequence entire human genomes now for less than $5000 [6], and relate all of these data patterns to a great diversity of other biologically relevant information (clinical data, biochemical data, social networking data etc.). Given technologies on the horizon such as the IBM DNA transistor with theoretical sequencing limits in the hundreds of millions of bases per second per transistor (imagine millions of these transistors packed together in a single handheld device) [7], we will not be talking in the future about Google rolling through neighborhoods with Wi-Fi sniffing equipment [8], but rather about DNA sniffing equipment rolling through neighborhoods sequencing everything they encounter in real time, and then pumping such data into big data clouds to link with all other available information in the digital universe.

Keeping pace with these life sciences technology advances are information technology advances, in which now more 'classic' information-savvy companies such as Microsoft, Amazon, Google, Facebook, Ebay, and Yahoo, as well as a new breed of emerging big data mining companies such as Recorded Future, Factual, Locu, and Palantir, have led the way in becoming masters of petabyte- and exabyte-scale datasets, linking pieces of data distributed over massively parallel architectures in response to user requests and presenting them to the user in a matter of seconds. Following these advances in other disciplines, we are on track to access the same types of tools to tackle the big data problems now being faced by the life and biomedical sciences. But large-scale data generation and big computer infrastructures are just two legs of the stool regarding what is needed to revolutionize our understanding of living systems. While the data revolution is driven by technologies that provide insights into how living systems operate, understanding living systems will require that we master the information the high-throughput technologies are generating.

If we want to achieve understanding from big data, organize it, compute on it, build predictive models from it, then we must employ statistical reasoning beyond the more classic hypothesis testing of yesteryear. We have moved well beyond the idea that we can simply repeat experiments to validate findings generated in populations. In fact, while first instances of the Central Dogma of Biology looked something like the simple graph depicted in Figure 26.2a, today, given the complex interplay of

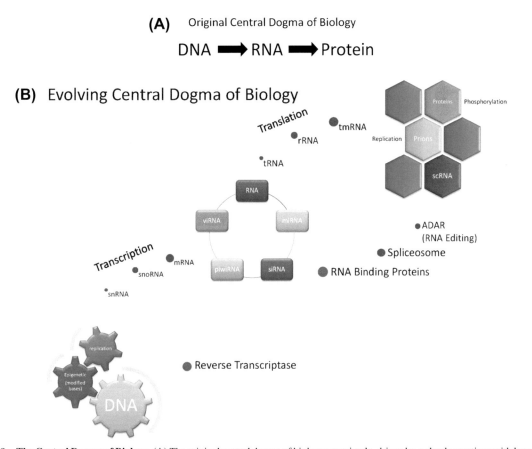

FIGURE 26.2 **The Central Dogma of Biology. (A)** The original central dogma of biology was simple, driven by early observations with low-resolution tools that uncovered a central relationship between DNA, RNA and proteins, namely that RNA is transcribed from DNA and RNA in turn is translated into proteins. **(B)** New higher-resolution technologies have enabled a far more complex view of the central dogma to emerge, with epigenetic changes to DNA that are transgenerational, leading to non-Mendelian patterns of inheritance, a complex array of RNA molecules such as microRNA, viRNA, piwiRNA, and siRNA that do not code for proteins but carry out complex regulatory functions, and sophisticated protein complexes involved in splicing, RNA editing, and RNA binding all feeding back on transcription, leading to a more network-oriented view of the central dogma.

multiple dimensions of data (DNA, RNA, protein, metabolite, cellular, physiologic, ecologic and social structures more generally) demands a more holistic view be taken in which we embrace complexity in its entirety, the central dogma is evolving to look something more like the graph depicted in Figure 26.2b. Our emerging view of complex biological systems is one of a dynamic, fluid system that is able to reconfigure itself as conditions demand [9–13]. Despite these transformative advances in technology and the need to embrace complexity, it remains difficult to assess where we are at with respect to our understanding of living systems, relative to a complete comprehension of such systems. One of the primary difficulties in our making such an assessment is that the suite of research tools available to us seldom provides insights into aspects of the overall picture of the system that are not directly measured.

In this chapter I discuss one class of modeling approaches that can integrate across diverse types of data and on broad scales in ways that enable others to interpret their data in a more informative context, to derive predictions that inform decision making on multiple levels, whether deciding on the next set of genes to validate experimentally, or the best treatment for a given individual given detailed molecular and higher-order data on their condition. Central to these models will be inferring causality among molecular traits and between molecular and higher-order traits by leveraging DNA as a systematic source of perturbation. In contrast to the more qualitative approaches biological researchers have employed in the past, getting the most from these new types of high-dimensional large-scale data requires constructing more complex, predictive models from them, refining the ability of such models to assess disease risk, progression, and best treatment strategies, and ultimately translating these complex models into a clinical setting where doctors can employ them as tools to understand most optimally your current condition and how best to improve it. Such solutions require a robust engineering approach, where integrating the new breed of large-scale datasets streaming out of the biological sciences and constructing predictive models from them will require approaches more akin to those employed by physicists, climatologists, and other strongly quantitative disciplines that have mastered the collection and predictive modeling of high-dimensional data.

A MOVIE ANALOGY FOR MODELING BIOLOGICAL SYSTEMS

The full suite of interacting parts in living systems, from the molecular to the ecological level, if they could be viewed collectively over time, would enable us to achieve a more complete understanding of cellular, organ, and organism-level processes, in much the same way as we achieve understanding by watching a movie. The continuous flow of information in a movie enables our minds to exercise an array of priors that provide the appropriate context and which constrain the possible relationships (structures) not only within a given frame or scene, but over the entire course of the movie. As our senses take in all of the streaming audio and visual information, our internal network reconstruction engine (centered at the brain) pieces the information together to represent highly complex and non-linear relationships depicted in the movie, so that in the end we are able to achieve an understanding of what the movie intends to convey at a hierarchy of levels.

If, instead of viewing a movie as a continuous stream of frames of coherent pixels and sound, we viewed single dimensions of the information independently, understanding would be difficult if not impossible to achieve. For example, consider a 1.5-hour feature-length film comprising 162 000 frames (30 frames per second), where each frame consists of 1280×720 (roughly one million) pixels. One way to view the film would be as a single frame in which the intensity value for each pixel across all 162 000 frames was averaged. This gross aggregate average would provide very little, if any, information regarding the movie, not unlike our attempts to understand complex living systems by examining single snapshots of a subset of molecular traits in a single cell type and in a single context at a single point in time. If we viewed our movie as independent one-dimensional slices through its frames, where each slice was viewed as pixel intensities across that one dimension changing over time (like a dynamic mass spec trace), this view would provide significantly more information, but it would still be very difficult to understand the meaning of the movie by looking at all of the one-dimensional traces independently, unless more sophisticated mathematical algorithms were employed to link the information together.

Despite the complexity of biological systems, even at the cellular level, research in the context of large-scale high-dimensional 'omics' data has tended to focus on single data dimensions, whether constructing co-expression networks based on gene expression data, carrying out genome-wide association analyses based on DNA variation information, or constructing protein interaction networks based on protein—protein interaction data. Although we achieve some understanding in this way, progress is limited because none of the dimensions on their own provide a complete enough context within which to interpret results fully. This type of limitation has become apparent in genome-wide association studies (GWAS) or whole-exome or genome sequencing studies, where thousands of highly replicated loci have been identified and highly replicated as associated with disease, but our understanding of disease is still limited because the genetic loci do not necessarily inform on the gene affected, on how gene function is altered, or more generally, how the biological processes involving a given gene are altered at particular points of time or in particular contexts [3—5,14]. It is apparent that if different biological data dimensions could be formally considered simultaneously, we would achieve a more complete understanding of biological systems [3,4,15—17]. (See the documentary film 'The New Biology' at http://www.youtube.com/watch?v=sjTQD6E3lH4.)

To obtain a more complete understanding of biological systems, we must not only evolve technologies to sample systems at ever higher rates and with ever greater breadth, we must also innovate methods that consider many different dimensions of information to produce more descriptive models (movies) of the system. There are of course many different types of modeling approaches that have been and continue to be explored. Descriptive models quantify relationships among variables in data that can in turn enable classification of systems under study into different meaningful groups, whether stratifying disease populations into disease subtypes to assign patients to the most appropriate treatment, or categorizing customers by product preference, descriptive models are useful for classifying, but cannot necessarily be used to predict how any given variable will respond to another at the individual level. For example, whereas patterns of gene expression such as those identified for breast cancer and now in play at companies such as Genomic Health, can very well distinguish good from poor prognoses [18,19], such models are not generally as useful for understanding how genes in such patterns are causally related, or for distinguishing key driver genes from passenger genes.

Predictive models, on the other hand, incorporate historic and current data to predict how one variable may respond to another in a particular context, or predict response or future states of components of a system at the individual level. In the biological context, predictive models aim to accurately predict (*in silico*) molecule expression-level changes, cell state dynamics, and phenotype transitions in response to specific perturbation events.

For example, understanding how the breast cancer patterns that predict prognosis are actually related to one another in probabilistic causal ways can lead to an understanding of how perturbing a given gene (say, for treatment) will affect that pattern or outcome, or understanding for a given patient how a given variant may affect their response to a given treatment. Key to constructing predictive models is elucidating causal relationships between traits of interest. Resolving causal relationships requires a systematic source of perturbation, and here I discuss the use of DNA variation as a systematic perturbation source to infer causal relationships among molecular traits and between molecular traits and higher-order traits such as disease [3,4,20–26].

CAUSALITY AS A STATISTICAL INFERENCE

In the life sciences, most researchers are accustomed to thinking about causality from the standpoint of physical interactions. In the molecular biology or biochemistry setting, when two molecular traits are indicated as causally related, we typically mean that one of the molecular entities (e.g., a small molecule compound) has been determined experimentally to physically interact with or to induce processes that directly affect the other molecular entity (e.g., the target protein of the small molecule) and consequently leads to a phenotypic change of interest (e.g., lower LDL cholesterol levels). In this case we have an understanding of the causal factors relevant to the activity of interest, so that careful experimental manipulation of these factors allows for the identification of genuine causal relationships. However, in the context of many thousands of variables related in unknown ways, the aim is to examine the behavior of those variables across populations in ways that facilitate statistically inferring causal relationships. For example, statistical associations between changes in DNA, changes in molecular phenotypes, and changes in higher-order phenotypes such as disease, can be examined for patterns of conditional dependency among the variables that can allow directionality to be inferred among them. In this case we can employ indirect measures of processes that mediate changes in one trait conditional upon another to make a statistically inferred causal link. This is not unlike the types of statistical inferences that are leveraged in other disciplines to make new discoveries. For example, <5% of known extrasolar planets have been directly observed, so that most are observed indirectly. One method for detecting planets that cannot be observed directly considers that when a planet is orbiting a star, the gravitational pull of the planet on the star will place the star into a subtle orbit, which from our vantage point will appear as the star moving in a cyclical fashion closer to and further away from the earth. Such movement can be measured as displacements in the star's spectral lines due to the Doppler effect [27], and so the presence of the planet acting on the star can be statistically inferred.

Critical to identifying causal relationships is distinguishing between correlation and causation. The old adage 'correlation does not imply causation' is familiar to most. This is among the first fallacies one learns about in beginning logic courses: *post hoc ergo propter hoc* (Latin for 'after this, therefore because of this'). Measurements taken on independent variables over time can be correlated because trends reflected by such variables are coincidentally similar, or changes in each variable are independently caused by a common source, in addition to being correlated as a result of a cause-and-effect relationship. It is also interesting to note that while correlation and causation are related, our intuitive notation that causation implies correlation is not always correct either. For example, suppose U and V are random variables with the same distribution, and suppose $X = U + V$ and $Y = U - V$. In this case the covariance between X and Y (defined as $E(XY) - E(X)E(Y)$, where E represents the expectation function) is 0 and so the correlation is 0, even though there is a direct functional dependence between the variables [28]. Only when two variables are linearly dependent (which is often the case in research) is our intuitive notion of functional dependence implying perfect correlation met.

Structure learning approaches that seek to infer causal relationships among correlated variables often employ conditional dependency arguments or mutual information measures to resolve causality by introducing a third correlated variable. By conditioning each of the variables on the third and examining the residual correlation between them in each case, a decision can be made as to the direction of the flow of information between the variables. However, this type of reasoning has generally failed to result in predictive causal inference, because in the absence of systematic perturbations the number of graphs that can be represented between just three traits is large (125 graphs representing directed and undirected relationships between three correlated variables are possible), and many of these possible relationships between the traits are not statistically distinguishable [29]. For example, if variables X, Y, and Z are observed to be correlated and the true relationship between the variables is $X \rightarrow Z \leftarrow Y$, this relationship cannot be statistically distinguished from $X \rightarrow Y \leftarrow Z$ and $Z \rightarrow X \leftarrow Y$, even though these relationships give rise to contradictory causal relationships.

To break this type of statistical symmetry a source of perturbation is required. Classically in biology we have introduced artificial perturbations by knocking a gene out, over-expressing a gene, or chemically perturbing a given protein to assess the consequences on a given trait of interest. If experimentally controlled artificial perturbations on a given gene cause a change in a trait of interest, then we infer a causal relationship between that gene and trait.

However, DNA variation in the germline provides an excellent systematic perturbation source that can also be used to resolve causal relationships in biological systems. Because variations in DNA cause variations in RNA, proteins, metabolites and subsequently higher-order phenotypes, this source of variation can be leveraged to infer causality. Unlike artificial perturbations such as gene knockouts, transgenics, and chemical perturbations that may induce artificial correlations that are not observed in more natural settings, naturally occurring genetic variation defines those perturbations that give rise to the broad array of phenotypic variations (such as disease and drug response) that we are precisely interested in elucidating. The past 7 years have demonstrated that causal links between DNA variations and molecular and higher-order phenotypes can provide information on causal relationships between those traits [3,4,21–24,26,30–34]. Causality in this instance can be inferred because there is random segregation of the chromosomes during gametogenesis, thereby providing the appropriate randomization mechanism to protect against confounding, similar to what is achieved in randomized clinical trials by randomly assigning patients to treatments to test the causal effects of a drug of interest [35,36]. However, quantifying the uncertainty in making such causal calls has been challenging. For example, causal effect estimates often considered in Mendelian randomization approaches can be confounded by pleiotropic effects and reverse causation, limiting the utility of such approaches for problems that involve the reconstruction of regulatory networks, in which pleiotropy is common and there may be little prior information regarding the structure of the causal relationships between the traits of interest [21].

Recently, though, formal statistical tests for inferring causal relationships between quantitative traits mediated by a common genetic locus have been developed [21]. To understand how such a test works, consider marker genotypes at a given DNA locus L that are correlated with a given molecular phenotype, G, and a higher-order phenotype T (Figure 26.3). The causal relationship G → T is implied if three conditions are satisfied under the assumption that L is sufficiently randomized: (1) L and G are associated; (2) L and T are associated; and (3) L is independent of T given G (T|G) [30]. If L is independent of G|T this is consistent with T → G, and if L is associated with G|T then this is consistent with G → T. We can boil all of these observations down to four conditions from which a statistical test can be formed to test for causality: (1) L and T are associated; (2) L is associated with G|T; (3) G is associated with T|L; and (4) L is independent of T|G. Each of these conditions can be assessed with a corresponding statistical test. For example, if we assume the marker corresponding to locus L is biallelic, where L_1 and L_2 represent indicator variables for the two alleles in a co-dominant

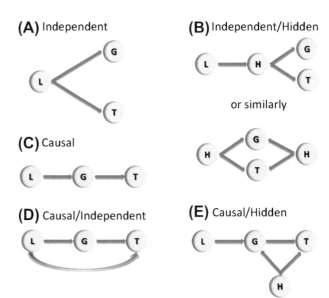

FIGURE 26.3 Given that two traits G and T are correlated in a given population with changes in DNA at locus L, there are five basic causal models to consider in testing the hypothesis that variations in trait G cause variations in trait T. Here H denotes an unmeasured molecular or higher-order trait.

coding scheme, then the four conditions above can be tested in the parameters of the following three regression models:

$$T_i = \alpha_1 + \beta_1 L_{1i} + \beta_2 L_{2i} + \varepsilon_{1i} \quad (1.1)$$

$$G_i = \alpha_2 + \beta_3 T_i + \beta_4 L_{2i} + \varepsilon_{2i} \quad (1.2)$$

$$T_i = \alpha_3 + \beta_6 G_i + \beta_7 L_{1i} + \beta_8 L_{2i} + \varepsilon_{3i} \quad (1.3)$$

where the ε_{ij} represent independently distributed random noise variables with variance σ_i^2 [30]. Given these models, the four component tests of interest are:

$$H_0 : \{\beta_1, \beta_2 = 0\}, H_1 : \{\beta_1, \beta_2\} \neq 0 \quad (1.4)$$

$$H_0 : \{\beta_4, \beta_5 = 0\}, H_1 : \{\beta_4, \beta_5\} \neq 0 \quad (1.5)$$

$$H_0 : \beta_6 = 0, H_1 = \beta_6 \neq 0 \quad (1.6)$$

$$H_0 : \{\beta_7, \beta_8 \neq 0\}, H_1 : \{\beta_7, \beta_8\} = 0 \quad (1.7)$$

The four conditions of interest can be tested using standard F tests for linear model coefficients (conditions 1–3) and a slightly more involved test for the last condition, given that it is an equivalence testing problem [21]. Given these individual statistical tests on the different regression parameters, a causal inference test can then be carried out by testing the strength of the chain of mathematical conditions that collectively are consistent with causal mediation (i.e., the strength of the chain is only as strong as its weakest link, so that the intersection of the rejection regions of the component tests provides for the causality test we seek). For a series of statistical tests of size α_r and rejection region R_r, the 'intersection-union' test with rejection region is a level $\cap R_r \sup(\alpha_r)$ test, so that the p value for the causal inference test corresponds to the p value for an intersection-union test, or, simply, the supremum of the four p values for the component tests [30]. This test has been implemented as the CIT package in the R statistical programming language and is freely available.

Applications of this type of test can be applied to resolve the types of causal relationships depicted in Figure 26.3. Application of these ideas in segregating mouse populations have led to the identification and validation of many genes causal for a number of metabolic traits, including obesity, diabetes and heart disease. In one such population constructed between the B6 and DBA inbred strains of mouse, 111 F2 intercross animals were placed on a high-fat atherogenic diet for 4 months at 12 months of age. All animals were genotyped using a genome-wide panel of markers, clinically characterized with respect to a number of metabolic traits, and the livers were expression profiled using a comprehensive gene expression microarray. Given the pattern of genetic association between the metabolic and gene expression traits, causal inference testing was carried out to identify the genes in this population best supported as causal of obesity-related traits [22,34]. Of the top nine genes identified in this study supported as causative of obesity-related traits, eight were ultimately experimentally validated [23]. The only gene that failed to validate was an X-linked gene that was lethal if completely knocked out, and so represented a more complicated example that the appropriate tools could not be constructed to validate.

FROM ASSESSING CAUSAL RELATIONSHIPS AMONG TRAIT PAIRS TO PREDICTIVE GENE NETWORKS

Leveraging DNA variation as a systematic perturbation source to resolve the causal relationships among traits is necessary but not sufficient for understanding the complexity of living systems. Cells are comprised of many tens of thousands of proteins, metabolites, RNA, and DNA, all interacting in complex ways. Complex biological systems are comprised of many different types of cells operating within and between many different types of tissues that make up different organ systems, all of which interact in complex ways to give rise to a vast array of phenotypes that manifest themselves in living systems. Modeling the extent of such relationships between molecular entities, between cells, and between organ systems is a daunting task. Networks are a convenient framework for representing the relationships among these different variables. In the context of biological systems, a network can be viewed as a graphical model that represents relationships among DNA, RNA, protein, metabolites, and higher-order phenotypes such as disease state. In this way, networks provide a way to visualize extremely large-scale and complex relationships among molecular and higher-order phenotypes such as disease in any given context.

Building from the Bottom Up or Top Down?

Two fundamental approaches to the reconstruction of molecular networks dominate computational biology today. The first is referred to as the bottom-up approach, in which fundamental relationships between small sets of genes that may comprise a given pathway are established, thus providing the fundamental building blocks of higher-order processes that are then constructed from the bottom up. This approach typically assumes that we have more complete knowledge regarding the fundamental topology (connectivity structure) of pathways, and given this knowledge, models are constructed that precisely detail how changes to any component of the pathway affect other components, as well as the known functions carried out by the pathway (i.e., bottom-up approaches are hypothesis driven). The second approach is referred to as a top-down approach in which we take into account all data and our existing understanding of systems and construct a model that reflects whole system behavior, and from there tease apart the fundamental components from the top down. This approach typically assumes that our understanding of how the network is actually wired is sufficiently incomplete, that our knowledge is sufficiently incomplete, that we must objectively infer the relationships by considering large-scale high-dimensional data that informs on all relationships of interest (i.e., top-down approaches are data driven).

Given our incomplete understanding of more general networks and pathways in living systems, this chapter focuses on a top-down approach to reconstructing predictive networks, given this type of structure learning from data is critical to derive hypotheses that cannot otherwise be efficiently proposed in the context of what is known (from the literature, pathway databases, or other such sources). However, top-down and bottom-up approaches are complementary to one another, although these approaches have largely been pursued as separate disciplines, with, interestingly, little cross-talk between them. One of the future directions discussed in the conclusion is the need to mathematically unify these two classes of predictive modeling to produce probabilistic causal networks that more maximally leverage all available data and knowledge.

In the context of integrating genetic, molecular profiling and higher-order phenotypic data, biological networks are comprised of nodes that represent molecular entities that are observed to vary in a given population under study (e.g., DNA variations, RNA levels, protein states, or metabolite levels). Edges between the nodes represent relationships between the molecular entities, and these edges can either be directed, indicating a cause-and-effect relationship, or undirected, indicating an association or interaction. For example, a DNA node in the network representing a given locus that varies in a population of interest may be connected to a transcript abundance trait, indicating that changes at the particular DNA locus induce changes in the levels of the transcript. The potentially millions of such relationships represented in a network defines the overall connectivity structure of the network, or what is otherwise known as the topology of the network. Any realistic network topology will be necessarily complicated and nonlinear from the standpoint of the more classic biochemical pathway diagrams represented in textbooks and pathway databases such as KEGG [37]. The more classic pathway view represents molecular processes on an individual level, whereas networks represent global (population level) metrics that describe variation between individuals in a population of interest that in turn defines coherent biological processes in the tissue or cells associated with the network.

An Integrative Genomics Approach to Constructive Predictive Network Models

Systematically integrating different types of data into probabilistic networks using Bayesian networks has been proposed and applied for the purpose of predicting protein–protein interactions [38] and protein function [39]. However, these Bayesian networks are still based on association between nodes in the network, as opposed to causal relationships. As discussed above for the simple case of two traits, from these types of networks we cannot infer whether a specific perturbation will affect a complex disease trait. To make such predictions we need networks capable of representing causal relationships. Probabilistic causal networks are one way to model such relationships from the top down, where causality again in this context reflects a probabilistic belief that one node in the network affects the behavior of another. Bayesian networks [40] are one type of probabilistic causal network that provide a natural framework for integrating highly dissimilar types of data.

Bayesian networks are directed acyclic graphs in which the edges of the graph are defined by conditional probabilities that characterize the distribution of states of each node given the state of its parents [40]. The network topology defines a partitioned joint probability distribution over all nodes in a network, such that the probability distribution of states of a node depends only on the states of its parent nodes: formally, a joint probability distribution $p(X)$ on a set of nodes X can be decomposed as $p(X) = \prod p(X^i | \text{Pa}(X^i))$, where $\text{Pa}(X^i)$ represents the parent set of X^i. The biological networks of interest we wish to construct are comprised of nodes that represent a quantitative trait such as the transcript abundance of a given gene or levels of a given metabolite. The conditional probabilities reflect not only relationships between genes, but also the stochastic nature of these relationships, as well as noise in the data used to reconstruct the network.

Bayes' formula allows us to determine the likelihood of a network model M given observed data D as a function of our prior belief that the model is correct and the probability of the observed data given the model: $P(M|D) \sim P(D|M)P(M)$. The number of possible network structures grows super-exponentially with the number of nodes, so an exhaustive search of all possible structures to find the one best supported by the data is not feasible, even for a relatively small number of nodes. A number of algorithms exist to find the optimal network without searching exhaustively, e.g., Monte Carlo Markov Chain (MCMC) [41] simulation. With the MCMC algorithm, optimal networks are constructed from a set of starting conditions. This algorithm is run thousands of times to identify different plausible networks, each time beginning with different starting conditions. These most plausible networks can then be combined to obtain a consensus network. For each of the reconstructions using the MCMC algorithm, the starting point is a null network. Small random changes are made to the network by flipping, adding, or deleting individual edges, ultimately accepting those changes that lead to an overall improvement in the fit of the network to the data. To assess whether a change improves the network model or not, information measures such as the Bayesian information criterion (BIC) [42] are employed, which reduces overfitting by imposing a cost on the addition of new parameters. This is equivalent to imposing a lower prior probability $P(M)$ on models with larger numbers of parameters.

Even though edges in Bayesian networks are directed, we cannot in general infer causal relationships from the structure directly, just as discussed above in relation to the causal inference test. For a network with three nodes, X_1, X_2, and X_3, there are multiple groups of structures that are mathematically equivalent. For example, the three models $M1 : X_1 \rightarrow X_2 X_2 \rightarrow X_3$; $M2 : X_2 \rightarrow X_1 X_2 \rightarrow X_3$; and $M3 : X_2 \rightarrow X_1 X_3 \rightarrow X_2$, are all Markov equivalent, meaning that they all encode for the same conditional independence relationship: $X_1 \perp X_3 | X_2$, X_1 and X_3 are independent conditional on X_2. In addition, these models are mathematically equivalent:

$$\begin{aligned} p(X) &= p(M1|D) = p(X_2|X_1)p(X_1)p(X_3|X_2) \\ &= p(M2|D) = p(X_1|X_2)p(X_2)p(X_3|X_2) \\ &= p(M3|D) = p(X_2|X_3)p(X_3)p(X_1|X_2) \end{aligned}$$

Thus, from correlation data alone we cannot infer whether X_1 is causal for X_2 or vice versa from these types of structures. It is worth noting, however, that there is a class of structures, V-shaped structures (e.g., Mv : $X_1 \rightarrow X_2, X_3 \rightarrow X_2$), that have no Markov equivalent. In such cases it is possible to infer causal relationships based on correlation data alone. Because there are more parameters to estimate in the Mv model than in the M1, M2, or M3 models, there is a large penalty in the BIC score for the Mv model. Therefore, in practice, a large sample size is needed to differentiate the Mv model from the M1, M2, or M3 models.

Integrating Genetic Data as a Structure prior to Enhancing Causal Inference in the Bayesian Network Reconstruction Process

In general, Bayesian networks can only be solved to Markov equivalent structures, so that it is often not possible to determine the causal direction of a link between two nodes even though Bayesian networks are directed graphs. However, the Bayesian network reconstruction algorithm can take advantage of genetic data to break the symmetry among nodes in the network that lead to Markov equivalent structures, thereby providing a way to infer causal directions in the network in an unambiguous fashion [24]. The reconstruction algorithm can be modified to incorporate genetic data as prior evidence that two quantitative traits may be causally related based on previously described causality test [24]. The genetic priors can be constructed from three basic sources. First, gene expression traits associated with DNA variants that are coincident with the gene's physical location (referred to as *cis* acting expression quantitative trait loci or *cis* eQTL) [43] are allowed to be parent nodes of genes with coincident *trans* eQTLs (the gene in this case does not physically reside at the genetic locus of interest), $p(cis \rightarrow trans) = 1$, but genes with *trans* eQTLs are not allowed to be parents of genes with *cis* eQTLs, $p(trans \rightarrow cis) = 0$. Second, after identifying all associations between different genetic loci and expression traits at some reasonable significance threshold, genes from this analysis with *cis* or *trans* eQTL can be tested individually for pleiotropic effects at each of their eQTL to determine whether any other genes in the set are driven by common eQTL [44,45]. If such pleiotropic effects are detected, the corresponding gene pair and locus giving rise to the pleiotropic effect can then be used to infer a causal/reactive or independent relationship based on the causality test described above. If an independent relationship is inferred, then the prior probability that gene A is a parent of gene B can be scaled as

$$p(A \rightarrow B) = 1 - \frac{\sum_i P(A \perp B | A, B, l_i)}{\sum_i 1}$$

where the sums are taken over all loci used to infer the relationship. If a causal or reactive relationship is inferred, then the prior probability is scaled as

$$p(A \rightarrow B) = \frac{2 \sum_i A \rightarrow B | A, B, l_i}{\sum A \rightarrow B | A, B, l_i + p(B \rightarrow A | A, B, l_i)}$$

Finally, if the causal/reactive relationship between genes A and B cannot be determined from the first two sources, the complexity of the eQTL signature for each gene can be taken into consideration. Genes with a simpler, albeit stronger eQTL signature (i.e., a small number of eQTL that explain the genetic variance component for the gene, with a significant proportion of the overall variance explained by the genetic effects) can be considered more likely to be causal than compared with more complex and possibly weaker eQTL signatures (i.e., a larger number of eQTL explaining the genetic variance component for the gene, with less of the overall variance explained by the genetic effects). The structure prior that gene A is a parent of gene B can then be taken to be

$$p(A \rightarrow B) = 2 \frac{1 + n(B)}{2 + n(A)n(B)}$$

where n(A) and n(B) are the number of eQTLs at some predetermined significance level for genes A and B, respectively.

Incorporating other 'Omics' Data as Network Priors in the Bayesian Network Reconstruction Process

Just as genetic data can be incorporated as a network prior in the Bayesian network reconstruction algorithm, so can other types of data, such as transcription factor-binding site (TFBS) data, protein–protein interaction (PPI) data, and protein–small molecular interaction data. PPI data can be used to infer protein complexes to enhance the set of manually curated protein complexes [46]. PPI-inferred protein complexes can be combined with manually curated sets, and each protein complex can then be examined for common transcription factor-binding sites at the corresponding genes. If some proportion of the genes in a protein complex (e.g., half) carry a given TFBS, then all genes in the complex can be included in the TFBS gene set as being under the control of the corresponding transcription factor.

Given that the scale-free property is a general property of biological networks (i.e., most nodes in the network are linked to a small number of nodes, whereas a smaller number of nodes are linked to many nodes) [47], inferred and experimentally determined TFBS data can be incorporated into the network reconstruction process by constructing scale-free priors, in a manner similar to the scale-free priors others have constructed to integrate expression and genetic data [48]. Given a transcription factor T, and a set of genes, G, that contain the binding site of T, the TF prior, p_{tf}, can be defined so that it is proportional to the number of expression traits correlated with the TF expression levels, for genes carrying the corresponding TFBS:

$$\log(p_{tf}(T \to g)) \propto \log\left(\sum_{g_i \in G} p_{tf}(T \to g_i)\delta\right),$$

where $p_{gtl}(T \to g)$ is the prior for the QTL and

$$\delta = \begin{cases} 1, \text{if } corr(T, g_i) \geq r_{cutoff} \\ 0, \text{if } corr(T, g_i) \geq r_{cutoff} \end{cases}$$

The correlation cutoff r_{cutoff} can be determined by permuting the data and then selecting the maximum correlation values in the permuted data sets (corresponding to some predetermined reasonable false discovery rate). This form of the structure prior favors transcription factors that have a large number of correlated responding genes. From the set of priors computed from the inferred and experimentally determined TFBS set, only non-negative priors should be used to reconstruct the Bayesian network. For those protein complexes that could not be integrated into the network reconstruction process using scale-free priors, uniform priors were used for pairs of genes in these complexes (i.e., $sp_{pc}(g_i \to g_j) = p_{pc}(g_j \to g_i) = c$).

Small molecule–protein interactions can also be incorporated into the Bayesian network reconstruction process. Chemical reactions reflected in biochemical pathways and the associated catalyzing enzymes can be identified as metabolite–enzyme pairs from existing pathway databases such as KEGG. These relationships can then be stored in an adjacency matrix in which a 1 in a cell represents a direct connection between the metabolite and the enzyme. The shortest distance $d_{m,e}$ from an enzyme e to a metabolite m can then be calculated using the repeated matrix multiplication algorithm. The structure prior for the gene expression of an enzyme e affecting the metabolite concentration is related to their shortest distance $d_{m,e}$ as $p(m \to e) \propto e^{-\lambda d_{m,e}}$. The shorter the distance, the stronger the prior.

Illustrating the Construction of Predictive Bayesian Networks with an Example

To illustrate how different types of data can be integrated to construct predictive gene networks, consider the following two classes of data: (1) DNA variation, gene expression, and metabolite data measured in a previously described cross between laboratory (BY) and wild (RM) yeast strains (referred to here as the BXR cross) for which DNA variation, RNA expression, and metabolite levels have been assessed [49–51]; and (2) protein–DNA-binding, protein–protein interaction, and metabolite–protein interaction data available from public data sources and generated independently of the BXR cross (referred to here as non-BXR data). The BXR data are reflected as nodes in the network to be constructed, where edges in the network reflect statistically inferred causal relationships among the expression and metabolite traits. The non-BXR interaction data from public sources are used to derive the types of structure priors discussed above on the network to both constrain the size of the search space in finding the best network and enhance the ability to infer causal relationships between the network nodes [26].

To illustrate the steps in the type of Bayesian network reconstruction procedure described above and detailed more formally [51], and to examine contributions from the different data types used to construct the network, I focus on genes and metabolites involved in the de novo biosynthesis of pyrimidine ribonucleotides (Figure 26.4). For simplicity I focus on the reconstruction of this smaller subset of genes, although the steps are similar if building a network from a more comprehensive set of genes. The subnetwork depicted in Figure 26.4a was identified from the full Bayesian network constructed from the BXR data [51]. *URA3* in this network was predicted as a causal regulator of gene expression traits linked to the *URA3* locus. That is, using the full Bayesian network, in silico perturbations were carried out by simulating changes in each of the nodes and identifying those nodes that resulted in the most significant changes in other nodes in the network. As a result of this simulation, *URA3* was identified as the regulator modulating the most significant number of nodes in the subnetwork in a causal fashion (Figure 26.4a). A deletion of *URA3* was engineered in the parental strain RM11–1a as a selectable marker, and segregation of this locus among the BXR progeny is the most likely cause for expression variation of uracil biosynthesis genes linked to this locus [50]. Variation of two metabolites are also linked to this locus: dihydro-orotic acid, which is converted to orotic acid by the enzyme Ura1p, and orotic acid itself, reflecting the functional consequence of transcriptional variation in genes involved in de novo pyrimidine base biosynthetic processes on metabolite levels. The causal relationships between *URA1*, orotic acid, and dihydro-orotic acid as well as the subnetwork for genes linked to the *URA3* locus, recapitulate the known pyrimidine base biosynthesis pathway [51]. This subnetwork not only captures the co-regulation of gene expression and metabolite abundance, but also elucidates the mechanism of how genetic variation in *URA3* affects orotic acid and dihydro-orotic acid levels.

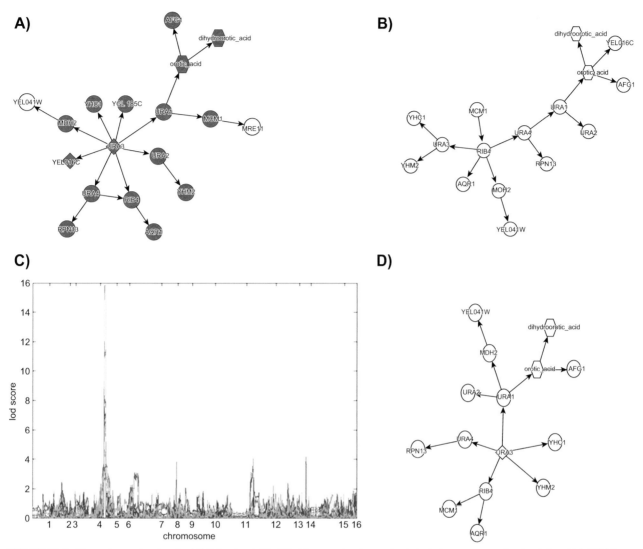

FIGURE 26.4 Example yeast network. (A) Subnetwork identified in a previously constructed whole genome yeast network in which *URA3* was predicted as the causal regulator for genes and metabolites linked to a genetic locus on chromosome 5 coincident with the physical location of URA3. Red nodes are genes or metabolites whose variations are linked to the chromosome V locus. Hexagonal nodes represent metabolites, circular nodes represent genes, and diamond-shaped nodes represent genes with *cis* eQTLs. (B) Trait values of nodes compared with genotype data for the URA3 subnetwork. eQTLs and metQTLs are prominently featured as residing in the chromsome V URA3 locus. (C) Bayesian network reconstructed using only trait data. (D) Bayesian network reconstructed using trait data and priors derived from other types of data.

Step 1: Identification of the URA3-Centered de novo Biosynthesis of Pyrimidine Ribonucleotides Subnetwork

There are 18 nodes in the subnetwork shown in Figure 26.4a. These nodes are highly correlated with one another, with 68% of all pairwise relationships significant at the 0.01 significance level. The continuous gene expression data for these 18 genes can then be discretized into three states representing downregulated, no change, and upregulated, and the mutual information of all pairs of nodes are then calculated. In this case, 54% of all pairs are significant at $p < 0.01$ (the mutual information of the permutated data is calculated and fit into a normal distribution, which is then used to assess the significance of the mutual information of the observed data). All 18 of the trait values corresponding to these nodes are significantly associated with the genotypes at the URA3 locus (Figure 26.4c).

Step 2: Reconstructing Networks using only Expression and Metabolite Traits (Excluding DNA Variation Data)

The process of reconstructing networks using only trait data is straightforward. The trait data are input into a standard Bayesian network reconstruction program in which 1000 network structures are generated from a Monte Carlo Markov Chain (MCMC) process using different random seed numbers (1000 random seed numbers are generated by a master process, then each slave process starts an MCMC process using one of the generated seed numbers). Once the 1000 network structures have been generated, common features are extracted to derive a consensus network. With this construction, the consensus network may contain loops, which are prohibited in Bayesian networks. Therefore, to ensure the consensus network structure is a directed acyclic graph, the edges in the original consensus network are removed if and only if (1) the edge was involved in a loop, and (2) the edge was the most weakly supported of all edges making up the loop. The network resulting from this process is depicted in Figure 26.4b.

Step 3: Constructing Priors Using eQTL Data

The network in step 2 is constructed without considering any of the genetic data. Because eQTL data represent a systematic source of perturbation on the expression data, integrating these data has the potential to better resolve causal relationships. Towards this end, expression and genotype data in the BXR cross are compared to detect eQTLs. The red nodes in Figure 26.4a indicate that nearly all of the nodes have QTLs linked to a single locus on chromosome V. Expression traits that associate with a common eQTL are then subjected to a statistical test to infer causal relationships between the traits, as described above. Among the nodes tested, URA3 and YEL016C have *cis*-acting eQTLs linked to the chromosome V locus. Nodes with *cis*-acting eQTLs are allowed to be causal parent nodes to nodes with *trans*-acting QTLs. However, nodes with *trans* QTLs are not allowed to be causal parent nodes to nodes with *cis*-acting eQTLs.

Step 4: Constructing Priors using KEGG Data

The network constructed in step 2 also does not consider known relationships among genes and metabolites as defined by canonical pathways. The relationships between enzymes and metabolites are well established in many cases. To incorporate this knowledge into the network reconstruction process, we construct priors using canonical pathway data in the following way. There are two metabolites in the *URA3* subnetwork. Their distances from each other and from related enzymes are defined in the KEGG database. The structure prior for the gene expression of an enzyme e affecting a metabolite concentration is constructed using their shortest distance $d_{m,e}$ as $p(m \rightarrow e \propto)e^{-\lambda d_{m,e}}$.

Step 5: Constructing Networks using Expression Data, Metabolite Data, and the Genetic and Canonical Pathway Priors defined in Steps 3 and 4

The process of reconstructing networks using trait data and priors from other data types is similar to the reconstruction process applied to trait data only described in step 2. In addition to trait data, priors derived from other data types are also input into the standard Bayesian network reconstruction process. The trait data of the 18 nodes and related priors are input into the network reconstruction process, and the resulting network is shown in Figure 26.4d. The root node of the Bayesian network is URA3, which is the gene with the *cis*-acting eQTL associated with other traits in the network.

Step 6: Comparing the Networks Constructed in Steps 2 and 5

The main difference between the networks depicted in Figure 26.4b and 4d is the head nodes. In general, directed links in a Bayesian network do not necessarily represent causal relationships [24]. The network constructed from the trait data only reflects relationships not supported by the genetic perturbation data. The genetic relationships are well captured by the more integrated network described in Step 5. For example, the link RIB4 → URA3 depicted in Figure 26.4b is opposite that identified in Figure 26.4d. Because the genetic perturbation at the URA3 locus affects the expression activity of that gene in *cis* and the expression activity of the gene RIB4 in *trans*, the experimentally supported relationship is URA3 → RIB4. Note that the enzyme—metabolite and metabolite—metabolite relationships are similar with or without the priors derived from the KEGG pathways.

All data and software used to construct the Bayesian networks for this example are available at http://www.mssm.edu/research/institutes/genomics-institute/rimbanet.

Networks Constructed from Human and Animal Data Elucidate the Complexity of Disease

We have carried out studies using the modeling described in detail for the yeast cross, but in human and mouse populations, segregating a number of different diseases such as obesity, diabetes and heart disease. For example, in a segregating mouse population in which an extensive suite of disease traits associated with metabolic syndromes were manifested, including obesity, diabetes, and atherosclerosis [3], we carried out the type of network analysis discussed above using genetic data typed in all animals and gene expression data generated from the liver and adipose tissues of all animals in the population. With this approach we found that of the many functional units (subnetworks) identified in the networks that reflected core biological processes specific to the liver and adipose tissues, only a handful were strongly causally associated with the metabolic syndrome traits. One module in particular stood out, not only because it was conserved across the liver and adipose tissues, between the sexes and between species [4], but because it was supported as strongly causal for nearly all of the metabolic traits scored in the cross (fat mass, weight, plasma glucose, insulin, lipid levels, and aortic lesions) [3]. Again, the causal relationship between this subnetwork and the disease traits was established by leveraging the changes in DNA in this population that were simultaneously associated with disease and expression traits. The entire subnetwork was shown to be under the control of genomic loci associated with the metabolic traits, while the predictive network modeling strongly indicated that the module was causal for the disease traits, and was not simply reacting to or acting independently of these traits.

Of the more than 100 genes supported in this module as causal for metabolic disease traits such as obesity and diabetes, many, such as *Zfp90, Alox5, C3ar1,* and *Tgfbr2*, had been previously identified and validated as causal for metabolic traits [20,22]. In addition, three other genes were selected for validation because they were independently supported as causal for metabolic traits in other studies (*Lpl* and *Lactb*), or because they were supported as causal for such a wide variety of metabolic traits (*Ppm1l*) [3]. Interestingly, the degree of connectivity in this causal metabolic subnetwork was extreme. Perturbations to genes in this module that were previously validated as causal for the metabolic traits caused expression changes in many other genes validated as causal for metabolic traits. For example, over-expression of *Zfp90* in mouse not only generated an expression response that was significantly overlapping with the causal metabolic module, but it caused changes in other genes, such as *Pparg*, known to have an impact on metabolic traits [3].

CONCLUSION AND FUTURE DIRECTIONS

The generation of ever higher dimensional data (DNA sequencing, RNA sequencing, epigenomic profiling, proteomic profiling, metabolomic profiling, and so on) at ever higher scales demands sophisticated mathematical approaches to integrate these data in more holistic ways to uncover not only patterns of molecular, cellular, and higher-order activities that underlie the biological processes that define physiological states of interest, but to uncover causal relationships among molecular and cellular phenotypes and between these phenotypes and clinical traits such as disease or drug response. One of the more successful frameworks for representing large-scale high-dimensional data are networks. Here I have detailed one particular approach to reconstructing predictive network models of living systems that leverages DNA variation as a systematic variation source and Bayesian network reconstruction algorithms to take a top-down approach to modeling complex systems. Because state-of-the-art therapies in the future will be based on targeting combinations of genes [52,53], and for such applications not only is it important to infer the direction of each interaction (i.e., do you antagonize or activate a given target), but one must be able to predict the degree to which each gene should be knocked down or activated (in a quantitative sense), only by generating accurate predictive models of complex phenotypes can we most efficiently search for such combinations to pursue for experimental proof of concept.

The success of modeling complex systems in the future will depend on constructing networks that are predictive of complex behavior, not merely descriptive. In order to achieve these more predictive models in complex systems such as humans, we must expand existing networks so they reflect relationships between cell types and tissues, not just within a single cell type or tissue; capture a greater range of molecular phenotypes to enhance understanding of relevant functional units that define biological processes of interest; and improve modeling capabilities, ideally drawing on the expertise of other fields that have pioneered causality-type reasoning. The complex phenotype-associated molecular networks we can construct today are necessarily based on grossly incomplete sets of data. Even given the ability to assay DNA and RNA variation in whole populations in a comprehensive manner, the information is not complete, given rare variation, DNA variation other than SNP/copy number, variation in non-coding RNA levels, and variation in the different isoforms of genes, are far from being completely characterized in any sample, let alone in entire populations. Beyond DNA and RNA, measuring all protein associated traits, interactions between proteins and DNA/RNA, metabolite levels, epigenetic changes, as well as other molecular entities important to the functioning of living systems, are not yet possible with existing technologies. Further, the types of high-dimensional data we are able to routinely generate today in populations represent only a snapshot at a single time point, which may enable the identification of the functional units of the system under study and how these units relate to one another, but does not enable a complete understanding of how the functional units are put together, the mechanistic underpinnings of the complex set of functions carried out by individual cells and by entire organs and whole systems comprised of multiple organs. Despite these and other advances required to more routinely develop predictive models of living systems (we really are only just scratching the surface), two of the more critical developments I believe will most enable the realization of more accurate network models relate to unifying different modeling approaches in a mathematically coherent way and transforming the way in which communities of researchers collaborate to build predictive models.

Unifying Bottom-Up and Top-Down Modeling Approaches

Ideally, the unification of bottom-up and top-down modeling approaches would maximally leverage the strengths of each approach while minimizing the weaknesses. Integrating models derived from bottom-up approaches into top-down approaches is currently hampered by the fact that the existing approaches do not typically fully parameterize the network structure in ways that match the intrinsic quantitative nature of top-down approaches. In bottom up-approaches the structural information detailing how different molecular entities are connected is typically derived from the literature or pathway databases, but such structural information is only qualitative, failing to define quantitatively how one node responds to another. On the other hand, in existing top-down approaches, unless a tremendous amount of training data are available to cover all of the categories represented in the conditional probability distribution (CPD) defining how nodes are connected in the network (such as with Bayesian network reconstruction approaches), it is not generally possible to accurately estimate the full set of parameters associated with the reconstructed network structure. Worse, carrying out parameter estimation on a network structure that is not correct can be misleading, given false positive and false negative predictions. In cases where heuristic searches are used to orient the edges in a given network structure, the end result is that model parameters have not been fitted accurately, given the network itself is not correct. Without proper parameterization of network structures from these conventional systems biology approaches, the networks serve only as descriptive models that are not generally capable of generating in silico predictions.

The limitations of bottom-up and top-down approaches can be addressed by devising bottom-up modeling approaches that deliver structures that can serve as prior information for top-down approaches, thereby providing a direct path for parameterizing bottom-up models in the context of a richer set of 'omics' data and network architectures, while simultaneously reducing the size of the search space for top-down approaches. Such bottom-up approaches are beginning to emerge [54]. By automatically parameterizing large networks given a particular network structure and corresponding interaction functions (e.g., activation or repression of gene activity) associated with all node pairs by either leveraging prior information or performing a heuristic search, bottom-up approaches will be capable of generating direct quantitative predictions that are compatible with top-down approaches. Central to the success of this approach is the observation that the complexity of the structure of biological networks leads to robust parameter estimates in a constrained parameter space [55–57] and the fact that a statistical model's parameters are in fact constrained to a cubic space (e.g., the conditional probabilities that represent parameters in our modeling approach are constrained to fall between 0 and 1). This stands in contrast to current bottom-up modeling approaches such as continuous ordinary differential equation (ODE) modeling, in which the parameter space is generally unconstrained (infinitely large).

Current systems biology approaches relating to network learning and modeling have exclusively utilized a top-down (reverse-engineering) approach to learn network structure based on association scores [26,58–61]. Association scores are designed to uncover the best correlations between variables. Bayesian networks are among the most popular models for this purpose. In theory, it is known that learning the optimal (global maximum) Bayesian network structure from the data is an NP-hard problem; further, because many substructures that must be considered during the reconstruction process are from classes of structures that are equivalent (the Markov equivalence issue noted above), the statistical scores for all of the structures in a given equivalence class are equal, so that completely contradictory causal relationships are indistinguishable from one another. The integration of bottom-up and top-down approaches in a more holistic mathematical framework has the potential to further address these issues, and to enhance the power to uncover true causal relationships.

A Need to Review How We Work Together

One does not often encounter in writings on methods to construct predictive biological models discussions regarding the role social networks may play in producing more predictive models. However, if we hope to truly revolutionize our ability to build predictive models of disease, validate and refine models, and enable others to seamlessly interact and query these models to inform decision making in a diversity of settings, I believe we need to more openly share data and competing models and to provide a platform from which such models can be computed, shared, queried, compared, validated, and refined by communities of researchers. Until very recently, nearly all of the historic studies that drive our current understanding of disease were performed by single institutions, often with the primary goal of building crude models of disease that were then communicated as the results and conclusions of citable scientific articles. This process does not assume that most data might be more useful if they could be accessed by others who might leverage the data in ways that were not envisioned by those who generated the data. Whereas efforts such as the Tumor Cancer Genome Atlas (TCGA) [62], dbGAP, GEO, and mega GWAS studies have demonstrated the utility of sharing data on a large scale, the absence of a culture of appropriate data sharing remains perhaps the single greatest impediment to the rapid development of the integrative techniques described above. Even in cases where significant effort has gone into providing data in the most comprehensive fashion (e.g., the TCGA projects), reproducing results of others from such data often remains elusive [63]. As problematic are review processes that may delay the release of a critical dataset by years.

To evolve toward a more generative scientific society, technical and cultural changes are necessary. For biology to achieve the same level of integration as electronics, astronomy, and economics, new standards and annotations will be required. Reward structures for career advancement and peer recognition that are based on being a first or last author, and the need to own intellectual property around biologic insights, must also be re-examined. Emerging efforts such as Sage Bionetworks (http: //www.sagebase.org) are attempting to catalyze this type of transformation, providing ways of openly sharing large-scale data, computing on those data, representing models derived from the data for others to query and validate, and even proposing governance structures and new types of informed consents that break down many of the barriers that currently exist to openly sharing patient data. Of course, such efforts will need to well demonstrate the advantages of sharing data, models, and tools to further our understanding of disease beyond what any lab could have done individually, if they are to prove convincingly to researchers more generally that their future competitiveness depends on openly sharing data and results.

REFERENCES

[1] Eid J, Fehr A, Gray J, Luong K, Lyle J, et al. Real-time DNA sequencing from single polymerase molecules. Science 2009;323: 133−8.

[2] Bandura DR, Baranov VI, Ornatsky OI, Antonov A, Kinach R, et al. Mass Cytometry. Technique for real time single cell multitarget immunoassay based on inductively coupled plasma time-of-flight mass Spectrometry. Anal Chem 2009;81(16):6813−22.

[3] Chen Y, Zhu J, Lum PY, Yang X, Pinto S, et al. Variations in DNA elucidate molecular networks that cause disease. Nature 2008;452:429−35.

[4] Emilsson V, Thorleifsson G, Zhang B, Leonardson AS, Zink F, et al. Genetics of gene expression and its effect on disease. Nature 2008;452:423−8.

[5] Altshuler D, Daly MJ, Lander ES. Genetic mapping in human disease. Science 2008;322:881−8.

[6] Drmanac R, Sparks AB, Callow MJ, Halpern AL, Burns NL, et al. Human genome sequencing using unchained base reads on self-assembling DNA nanoarrays. Science 327: 78−81.

[7] Schadt EE, Turner S, Kasarskis A. A window into third-generation sequencing. Hum Mol Genet 2010;19:R227−240.

[8] Kravets D. Privacy in Peril: Lawyers, Nations Clamor for Google Wi-Fi Data. Wired: Wired Magazine; 2010.

[9] Han JD, Bertin N, Hao T, Goldberg DS, Berriz GF, et al. Evidence for dynamically organized modularity in the yeast protein−protein interaction network. Nature 2004;430:88−93.

[10] Luscombe NM, Babu MM, Yu H, Snyder M, Teichmann SA, et al. Genomic analysis of regulatory network dynamics reveals large topological changes. Nature 2004;431:308−12.

[11] Pinto S, Roseberry AG, Liu H, Diano S, Shanabrough M, et al. Rapid rewiring of arcuate nucleus feeding circuits by leptin. Science 2004;304:110−5.

[12] Barabasi AL, Oltvai ZN. Network biology: understanding the cell's functional organization. Nat Rev Genet 2004;5:101−13.

[13] Zerhouni E. Medicine. The NIH roadmap. Science 2003;302: 63−72.

[14] Witte JS Genome-wide association studies and beyond. Annu Rev Public Health 31: 9−20 24 p following 20.

[15] Hsu YH, Zillikens MC, Wilson SG, Farber CR, Demissie S, et al. An integration of genome-wide association study and gene expression profiling to prioritize the discovery of novel susceptibility Loci for osteoporosis-related traits. PLoS Genet 6: e1000977.

[16] Schadt EE, Molony C, Chudin E, Hao K, Yang X, et al. Mapping the genetic architecture of gene expression in human liver. PLoS Biol 2008;6:e107.

[17] Zhong H, Beaulaurier J, Lum PY, Molony C, Yang X, et al. Liver and adipose expression associated SNPs are enriched for association to type 2 diabetes. PLoS Genet 6: e1000932.

[18] van de Vijver MJ, He YD, van't Veer LJ, Dai H, Hart AA, et al. A gene-expression signature as a predictor of survival in breast cancer. N Engl J Med 2002;347:1999−2009.

[19] van 't Veer LJ, Dai H, van de Vijver MJ, He YD, Hart AA, et al. Gene expression profiling predicts clinical outcome of breast cancer. Nature 2002;415:530−6.

[20] Mehrabian M, Allayee H, Stockton J, Lum PY, Drake TA, et al. Integrating genotypic and expression data in a segregating mouse population to identify 5-lipoxygenase as a susceptibility gene for obesity and bone traits. Nat Genet 2005;37:1224−33.

[21] Millstein J, Zhang B, Zhu J, Schadt EE. Disentangling molecular relationships with a causal inference test. BMC Genet 2009; 10:23.

[22] Schadt EE, Lamb J, Yang X, Zhu J, Edwards S, et al. An integrative genomics approach to infer causal associations between gene expression and disease. Nat Genet 2005;37:710−7.

[23] Yang X, Deignan JL, Qi H, Zhu J, Qian S, et al. Validation of candidate causal genes for obesity that affect shared metabolic pathways and networks. Nat Genet 2009;41:415−23.

[24] Zhu J, Lum PY, Lamb J, GuhaThakurta D, Edwards SW, et al. An integrative genomics approach to the reconstruction of gene networks in segregating populations. Cytogenet Genome Res 2004;105:363−74.

[25] Zhu J, Sova P, Xu Q, Dombek KM, Xu EY, et al. Stitching together multiple data dimensions reveals interacting metabolomic and transcriptomic networks that modulate cell regulation. PLoS Biol 2012;10(4):e1001301.

[26] Zhu J, Zhang B, Smith EN, Drees B, Brem RB, et al. Integrating large-scale functional genomic data to dissect the complexity of yeast regulatory networks. Nat Genet 2008;40:854−61.

[27] Erskine DJ, Edelstein J, Harbeck D, Lloyd J. Externally dispersed interferometry for planetary studies. Proc SPIE 2005;5905: 249−60.

[28] Feller W. An Introduction to Probability Theory and its Applications. New York, NY: Wiley; 1967.

[29] Sieberts SK, Schadt EE. Moving toward a system genetics view of disease. Mamm Genome 2007;18:389–401.

[30] Chen LS, Emmert-Streib F, Storey JD. Harnessing naturally randomized transcription to infer regulatory relationships among genes. Genome Biol 2007;8:R219.

[31] Davey Smith G, Ebrahim S. 'Mendelian randomization': can genetic epidemiology contribute to understanding environmental determinants of disease? Int J Epidemiol 2003;32:1–22.

[32] Didelez V, Sheehan N. Mendelian randomization as an instrumental variable approach to causal inference. Stat Methods Med Res 2007;16:309–30.

[33] Kulp DC, Jagalur M. Causal inference of regulator-target pairs by gene mapping of expression phenotypes. BMC Genomics 2006;7:125.

[34] Schadt EE, Monks SA, Drake TA, Lusis AJ, Che N, et al. Genetics of gene expression surveyed in maize, mouse and man. Nature 2003;422:297–302.

[35] Lawlor DA, Harbord RM, Sterne JA, Timpson N, Davey Smith G. Mendelian randomization: using genes as instruments for making causal inferences in epidemiology. Stat Med 2008;27:1133–63.

[36] Nitsch D, Molokhia M, Smeeth L, DeStavola BL, Whittaker JC, et al. Limits to causal inference based on Mendelian randomization: a comparison with randomized controlled trials. Am J Epidemiol 2006;163:397–403.

[37] Kanehisa M. The KEGG database. Novartis Found Symp 2002;247:91–101. discussion 101–103, 119–128, 244–152.

[38] Jansen R, Yu H, Greenbaum D, Kluger Y, Krogan NJ, et al. A Bayesian networks approach for predicting protein–protein interactions from genomic data. Science 2003;302:449–53.

[39] Lee I, Date SV, Adai AT, Marcotte EM. A probabilistic functional network of yeast genes. Science 2004;306:1555–8.

[40] Pearl J. Probabalistic Reasoning in Intelligent Systems: Networks of Plausible Inference. San Mateo, CA: Morgan Kaufmann Publishers; 1988.

[41] Madigan DaY, J. Bayesian graphical models for discrete data. Int Stat Rev 1995;63:215–32.

[42] Schwarz G. Estimating the dimension of a model. Ann Stat 1978;6:461–4.

[43] Doss S, Schadt EE, Drake TA, Lusis AJ. *Cis*-acting expression quantitative trait loci in mice. Genome Res 2005;15:681–91.

[44] Jiang C, Zeng ZB. Multiple trait analysis of genetic mapping for quantitative trait loci. Genetics 1995;140:1111–27.

[45] Lum PY, Chen Y, Zhu J, Lamb J, Melmed S, et al. Elucidating the murine brain transcriptional network in a segregating mouse population to identify core functional modules for obesity and diabetes. J Neurochem 2006;1(97 Suppl):50–62.

[46] Guldener U, Munsterkotter M, Oesterheld M, Pagel P, Ruepp A, et al. MPact: the MIPS protein interaction resource on yeast. Nucleic Acids Res 2006;34:D436–441.

[47] Albert R, Jeong H, Barabasi AL. Error and attack tolerance of complex networks. Nature 2000;406:378–82.

[48] Lee SI, Pe'er D, Dudley AM, Church GM, Koller D. Identifying regulatory mechanisms using individual variation reveals key role for chromatin modification. Proc Natl Acad Sci U S A 2006;103:14062–7.

[49] Brem RB, Kruglyak L. The landscape of genetic complexity across 5,700 gene expression traits in yeast. Proc Natl Acad Sci U S A 2005;102:1572–7.

[50] Brem RB, Yvert G, Clinton R, Kruglyak L. Genetic dissection of transcriptional regulation in budding yeast. Science 2002;296:752–5.

[51] Zhu J, Chen Y, Leonardson AS, Wang K, Lamb JR, et al. Characterizing dynamic changes in the human blood transcriptional network. PLoS Comput Biol 6: e1000671.

[52] Schadt EE. Molecular networks as sensors and drivers of common human diseases. Nature 2009;461:218–23.

[53] Schadt EE, Friend SH, Shaywitz DA. A network view of disease and compound screening. Nat Rev Drug Discov 2009;8:286–95.

[54] Chang R, Shoemaker R, Wang W. Systematic search for recipes to generate induced pluripotent stem cells. PLoS Comput Biol 2011;7:e1002300.

[55] Blanchini F, Franco E. Structurally robust biological networks. BMC Syst Biol 2011;5:74.

[56] Wilhelm T, Behre J, Schuster S. Analysis of structural robustness of metabolic networks. Syst Biol (Stevenage) 2004;1:114–20.

[57] Wu Y, Zhang X, Yu J, Ouyang Q. Identification of a topological characteristic responsible for the biological robustness of regulatory networks. PLoS Comput Biol 2009;5:e1000442.

[58] Carro MS, Lim WK, Alvarez MJ, Bollo RJ, Zhao X, et al. The transcriptional network for mesenchymal transformation of brain tumours. Nature 2010;463:318–25.

[59] Fiedler D, Braberg H, Mehta M, Chechik G, Cagney G, et al. Functional organization of the *S. cerevisiae* phosphorylation network. Cell 2009;136:952–63.

[60] Margolin AA, Wang K, Lim WK, Kustagi M, Nemenman I, et al. Reverse engineering cellular networks. Nat Protoc 2006;1:662–71.

[61] Stuart JM, Segal E, Koller D, Kim SK. A gene-coexpression network for global discovery of conserved genetic modules. Science 2003;302:249–55.

[62] TCGA-consortium. Comprehensive genomic characterization defines human glioblastoma genes and core pathways. Nature 2008;455:1061–8.

[63] Marko NF, Quackenbush J, Weil RJ. Why is there a lack of consensus on molecular subgroups of glioblastoma? Understanding the nature of biological and statistical variability in glioblastoma expression data. PLoS One 2011;6:e20826.

Chapter 27

Social Networks, Contagion Processes and the Spreading of Infectious Diseases

Bruno Gonçalves[1], Nicola Perra[1] and Alessandro Vespignani[1,2]

[1]*College of Computer and Information Sciences and Bouvé College of Health Sciences, Northeastern University, Boston, MA 02115, USA,* [2]*Institute for Scientific Interchange (ISI) Foundation, Via Alassio 11/c, 10126 Torino, Turin, Italy*

Chapter Outline

Introduction	515
Network Thinking	516
Contagion Phenomena in Complex Social Networks	518
Complex Networks and the Large-Scale Spreading of Infectious Diseases	521
Conclusions and Future Challenges	524
References	525

INTRODUCTION

The characterization and understanding of contagion phenomena is crucially dependent upon the conceptual framework adopted to describe groups of individuals (social agents) or entire populations in spatially extended systems, and of the interaction and behavior of individuals at various levels, from the global scale of mobility and transportation flows to the local scale of individual activities and contacts. In this context, a mathematical and statistical modeling framework has evolved from simple compartmental models into structured approaches in which the heterogeneities and details of the population and system under study are becoming increasingly important (Figure 27.1). In the case of spatially extended systems, modeling approaches have been extended into schemes that explicitly include spatial structures and which consist of multiple subpopulations coupled by traveling fluxes, while the epidemic within the subpopulation is described according to approximations depending on the specific case studied [1–9]. This patch or metapopulation modeling framework has then grown into a multi-scale framework in which the various possible granularities of the system (country, inter-city, intra-city) are considered through different approximations and coupled through interaction networks describing the flows of people and/or animals [9–16]. At the most detailed level, the introduction of agent-based models (ABM) has enabled us to stretch the usual modeling perspective even more to achieve a full description of the society by a complete characterization (household, workplace, etc.) of each individual [17,18].

The above modeling approaches are 'data hungry', as they depend on detailed information about the activity of individuals, their interactions and movement, as well as the spatial structure of the environment, transportation infrastructures, traffic networks, and travel times. Although for a long time these kinds of data were simply not available, recent years have witnessed a tremendous progress in data gathering thanks to the development of new informatics tools and the increase in computational power. A continuous flow of data has finally become available for scientific analysis and study. The availability of data has allowed highlighting of complex properties and heterogeneities, which cannot be neglected in the epidemic description. Although these characteristics have long been acknowledged as a relevant factor in determining the properties of dynamical processes, many real-world networks exhibit levels of heterogeneity that were not anticipated until a few years ago, and represent a theoretical and conceptual challenge in our understanding of the unfolding of contagion processes.

Although data availability is highlighting the limits of our conceptual and modeling frameworks, it is also allowing the validation of results across different modeling approaches, mathematical techniques and approximation schemes. Furthermore, it has been possible to push forward the development of data-driven computational approaches

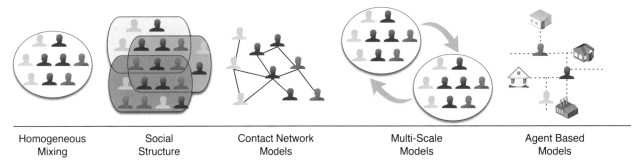

FIGURE 27.1 **Different scale structures used in epidemic modeling.** Circles represent individuals and each color corresponds to a specific stage of the disease. From left to right: homogeneous mixing, in which individuals are assumed to interact homogeneously with each other at random; social structure, where people are classified according to demographic information (age, gender, etc.); contact network models, in which the detailed network of social interactions between individuals provide the possible virus propagation paths; multi-scale models which consider subpopulation coupled by movements of individuals, while homogeneous mixing is assumed on the lower scale; agent-based models which recreate the movements and interactions of any single individual on a very detailed scale (a schematic representation of a city is shown).

based on the construction of highly detailed synthetic societies. Within such a framework computer simulations acquire a new value and allow on the one hand the creation of in silico experiment hardly feasible in real systems, and access to quantity and observables across different models on the other. This computational thinking approach will be also the guide to the understanding of typical non-linear behavior and tipping points not accessible by analytical means.

Although many basic conceptual questions remain unresolved, the major challenge in the development of models able to capture the behavior of large-scale techno-social systems is their sensitivity and dependence on social adaptive behavior. In the absence of a stress on the system, a stationary state is reached in which the feedback between the social behavior and the environment determines the details of how the dynamical process of interest plays out. Social behaviors react, adapt and define new way of interacting as the dynamics of the system evolves. This complicates the problem tremendously and clearly shows the limits in our understanding of human behavior. The view we have of human mobility, for example, is the daily normal activity of individuals. In the case of a major event, such as the spread of a novel pandemic, all the techno-social systems we are part of can be pushed out of equilibrium. Under stress individuals can act differently from usual: they can decide to stay home, to avoid crowded places and prevent children attending school. In general, they can take any action to reduce their risk by self-initiated behavioral changes. Contrary to what happens in physical systems, the global evolution of the system and our knowledge of it are part of the system dynamic.

Unfortunately, we are still not able to capture and deeply understand social adaptation and quantify the consequent changes on the dynamics of processes that trigger it. While some of the above issues may find a partial solution by improving the accuracy and reliability of models, it is clear that social adaptation to predictions presents us with new methodological and ethical problems.

Addressing these problems involves tackling three major scientific challenges. The first is the gathering of large-scale data on information spread and social reactions. This is not currently out of reach, thanks to large-scale mobile communication databases (mobile telephone, Twitter logs, social web tools) operating at times of specific disaster or crisis events. Second is the formulation of formal models that make it possible to quantify the adaptation, changes and reactions of individuals as a function of the dynamic processes occurring in the system. The third challenge concerns the deployment of monitoring infrastructures capable of informing computational models in real time.

NETWORK THINKING

The study of contagion processes is a very active field of research that crosses different disciplines. Epidemiologists, computer scientists, and social scientists share a common interest in studying spreading phenomena, and rely on very similar models for the description of the diffusion of viruses, knowledge, and innovation [19–23]. All these processes define a contagion dynamic, which can be an actual biological pathogen that spreads from host to host, or a piece of information/knowledge that is transmitted during social interactions. The connectivity patterns describing the interactions of individuals as well as their mobility from place to place are one of the crucial ingredients in the understanding of contagion phenomena. The importance of networks in epidemiology need not be stressed here. The recent advances in the field, however, stem from the increased ability to gather data on several large sets of networked structures and populations. For a long time approaches to human interactions and mobility have relied mostly on census and survey data on social interactions, frequently incomplete and/or limited to a specific context. A multitude of heuristic models for population structure and mobility (e.g., gravity laws of various types) have been endorsed, yet they frequently generate conflicting results and all suffer from the shortcomings of being connected to the arbitrariness of administrative boundaries, lacking microscopic knowledge relating the individual dynamics to population interactions, and lacking network and spatial proximity correlations. Put simply, despite the large effort and relevant advances in the study of human transport, no general framework of human interactions and mobility based on microdynamic principles and able to bridge all spatial scales exists. The new era of the social web and the data deluge is, however, raising the limits scientists have long been struggling with. From the development of new pervasive technologies to the use of proxy data, such as the digital traces that individuals leave with their mobile devices, it is nowadays possible to characterize the network of human interactions from the scale of the single individual to the level of entire populations.

The first steps towards a cohesive framework for the collection of detailed data on human behavior and mobility started by passively tracking human interactions through the use of cell phones [24]. Each cell phone was equipped with custom software that would regularly check which cell phone towers and Bluetooth devices were within range. Since Bluetooth has a range that is limited to just a few meters, they were able to track the location of each user with a hitherto unprecedented resolution up to room level. This pioneering study introduced the concept of 'reality mining', the idea that technology had finally reached a point where it was not only feasible but also easy to unobtrusively track individual behavior and social interactions in real time. What followed was an avalanche of works extending this approach in various directions, both by using pre-existing data in innovative ways and by creating new tools and techniques to obtain data that had never been available before. Similar experiments have been used recently to generate predictive power on influenza spreading in communities by mapping the social pattern of individuals along with their health status [25].

Analogous experiments are also performed with progressively smaller, lightweight, active Radio-Frequency Identification (RFID) tags [26]. Each tag emits a low-power directed radio signal with a range of up to 1 m. Whenever two tags are able to exchange signals for an extended period of time, this is a clear sign that the people wearing them are facing each other and in close proximity, providing a good indication that a face-to-face conversation is taking place. The authors distributed such tags to hundreds of volunteers in various settings, and for the first time were able to track the way in which face-to-face interactions occur in real-world settings ranging from school and hospitals to conferences and workplaces.

In a pioneering work, Dirk Brockmann and co-workers have shown that popular sites for currency tracking (eurobilltracker, wheresgeorge) can be used to gather a massive number of records on money dispersal, and to use those as a proxy on humans to collect a wealth of novel and unprecedented data. This work opens a novel path to the general exploitation of proxy data for human interaction and mobility based on the evidence that humans leave abundant traces of their interactions and mobility patterns within various types of data-driven websites. The pervasive use of mobile and Wi-Fi technologies in our daily life is also changing the way we can measure human mobility. Modern mobile phones and PDAs combine sophisticate technologies such as Bluetooth, GPS, and WiFi, constantly producing detailed traces of our daily activities. The recent study of Gonzalez and co-workers on human mobility based on mobile phone data to track the movements of 100 000 people over 6 months is just the most explosive example of how these kinds of data are going to shatter our methodology in the field and critically revise our knowledge of social dynamics.

One of the main challenges offered by these networks lies in their complexity and multi-scale nature. As a large body of work spurred by first paper on complex networks has shown, most real-world networks present dynamical self-organization and the lack of characteristic scales, main hallmarks for complex systems. The various statistical distributions characterizing these networks are generally heavy-tailed, skewed and varying over several orders of magnitude. This is not just true for the degree distribution $P(k)$ characterizing the probability that each node in the system is connected to k neighboring nodes, it is observed for the intensity carried by the connecting links, transport flows and other basic quantities.

Analogously, similar heterogeneities are found at much larger resolution scale. At the urban scale of a single city, an impressive characterization of the human interactions flows is represented by the TRANSIM study [17]. This study focused on the network of locations in the city of Portland, Oregon, including homes, offices, shops and recreational areas. The temporal links between locations represent the flow of individuals going from one place to another at a given time. The resulting network is characterized by broad distributions of the degrees and of the flows of individuals traveling on a given connection [17]. Strong heterogeneities are thus present not only at the topological level, but also at the level of the traffic on the network: a simultaneous characterization of the system in terms both of topology and weights associated to connections is needed to integrate the different levels of complexity in a unifying picture [27].

Similar results have been found in commuting patterns among cities and counties within a given geographical region/country. In this case, the nodes of the network represent cities, counties and in general municipalities or urban aggregations coupled by connections that correspond to the commuting flows of individuals. The analysis of these networks uncovered rather homogeneous topologies — mainly due to strong spatial constraints — associated with very large fluctuations in individuals' travel flows [28].

Finally, the most global scale is characterized by the air connections infrastructure, composed of airports (nodes) and direct flights among them (links). Data representing the travel flow of passengers defines the weight to each connection [27]. This transportation network displays several strong levels of heterogeneity. The distribution of degrees (i.e., of the number of connections of an airport) is scale free and the traffic is very broadly distributed, varying over several orders of magnitude [27,29]. This points to a structure composed of airports having large fluctuations in their number of connections to other airports and, moreover, to the number of passengers traveling on a given route, ranging from a few to millions of individuals in a given period of time.

Figure 27.2 shows two networks for human mobility at different scales ranging from the airline transportation network at the distance of hundreds and thousands of kilometers and several days to the commuting network that connects neighboring cities within the span of a few hours. We see that the scale and complexity extend at all granularities, creating a huge multi-scale network where the time and spatial separations range from a few hundred meters and a few minutes to thousands of kilometers and several days.

The challenge in providing a holistic description of multi-scale networks is the necessity of dealing with multiple time and length scales simultaneously. The final system's dynamical behavior at any scale is the product of the events taking place on all scales. The single agent spreading a disease or single node is apparently not affected by the multi-scale nature of the network, just as single molecules or fluid elements do not care about the multi-scale nature of turbulent fluid. However, the collective dynamic behavior and our ability to conduct mathematical/computational analyses of techno-social systems are constrained by the heterogeneous and multi-scale characteristic of the system, and we must develop appropriate formalisms and techniques, as have done researchers studying multi-scale physical systems [30] (fluids, solids, distribution of masses in the universe, etc.). In the context of networks and techno-social systems, the multi-scale challenge is making its appearance now because of the availability of large-scale datasets. To achieve analytical understanding of techno-social systems and approach them computationally, we must find different strategies to deal with dynamical behavior/equations that work at very different characteristic scales but still influence each other.

CONTAGION PHENOMENA IN COMPLEX SOCIAL NETWORKS

Contagion processes are usually seen as a transmission process, either as a pathogen that spreads from host to host, or a piece of information/knowledge that is transmitted during social interactions. Let us consider the simple susceptible—infected—recovered (SIR) epidemic model. In this model, infected individuals (labeled with the state I) can propagate the contagion to susceptible neighbors (labeled with the state S) with rate λ, while infected individuals recover with rate μ and become removed from the population. This is the prototypical model for infectious disease spreading, where individuals recover and are immune to disease after a typical time that on average can be expressed as the inverse of the recovery rate. A classic variation of this model is the susceptible—infected—susceptible (SIS) model, in which individuals go

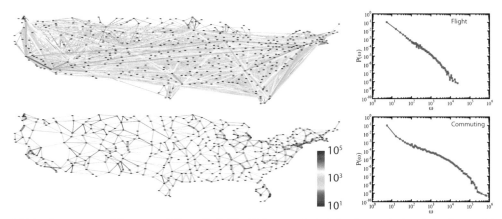

FIGURE 27.2 **Multiscale structure of human mobility.** (Left) The flight network (top) involves typical periods of several days, whereas the commuting flows (below) occur within a single day but are typically much more intense. (Right) The weight distribution of the flight and commuting networks.

back to the susceptible state with rate μ, modeling the possibility of reinfection.

A cornerstone of epidemic processes is the presence of the so-called epidemic threshold [19]. In a fully homogeneous population the behavior of the SIR model is controlled by the reproductive number $R_0 = \beta/\mu$, where $\beta = \lambda <k>$ is the per-capita spreading rate, which takes into account the average number of contacts k of each individual. The reproductive number simply identifies the average number of secondary cases generated by a primary case in an entirely susceptible population and defines the epidemic threshold such that only if $R_0 \geq 1$ ($\beta \varepsilon \mu$) can epidemics reach an endemic state and spread into a closed population. The SIS and SIR models are indeed characterized by a threshold defining a transition between two very different regimes. These regimes are determined by the values of the disease parameters, and characterized by the global parameter i_∞ which identifies the density of infected individuals (nodes in a network) in the infinite time limit. In the limit of an infinitely large population, this density is zero below the threshold and assumes a finite value above the threshold. In this perspective we can consider the epidemic threshold as the tipping point of the system. Below the critical point the system relaxes into a frozen state with null dynamics — the healthy phase. Above this point, a dynamical state characterized by a macroscopic number of infected individuals sets in, defining an infected phase (Figure 27.3).

The above results are generally obtained by using the so-called homogenous assumption in which all individuals in the populations are considered statistically equivalent and no structure (spatiotemporal, connectivity pattern, etc.) is included in the system description.

One of the most important features affecting dynamical processes in real-world networks, however, is the presence of dynamic self-organization and the lack of characteristic scales — typical hallmarks of complex systems [31–35]. In particular, the various statistical distributions characterizing social networks are generally heavy-tailed, skewed, and varying over several orders of magnitude. This is a very peculiar feature typical of many natural and artificial complex networks, characterized by virtually infinite degree fluctuations, where the degree k of a given node represents its number of connections to other nodes. In contrast to regular lattices and homogeneous graphs characterized by nodes having a typical degree k close to the average $<k>$, such networks are structured in a hierarchy of nodes, with a few nodes having very large connectivity — the hubs — while the vast majority of nodes have smaller degrees (see Chapter 9). This feature usually finds its signature in a heavy-tailed degree distribution, often approximated by a power-law behavior of the form $P(k) \propto x^{-\gamma}$, which implies a non-negligible probability of finding vertices with very large degree [32–35].

The presence of large-scale fluctuations virtually acting at all scales of the network connectivity pattern calls for a mathematical analysis where the variables characterizing each node of the network are explicitly entering the description of the system. Unfortunately, the general solution, handling the master equation of the system, is hardly if ever achievable even for very simple dynamical processes. For this reason, a viable theoretical approach considers the use of techniques such as mean-field and deterministic continuum approximations, which usually provide the understanding of the basic phenomenology and phase diagram of the process under study. In both cases the heterogeneous nature of the network connectivity pattern is introduced by aggregating variables according to a degree block formalism which assumes that all nodes with the same degree k are statistically equivalent [36–38]. This assumption allows the grouping of nodes in degree classes, yielding a convenient representation of the system. For instance, if for each node i we associate a corresponding state σ_i characterizing its dynamical state, a convenient representation of the system is provided by the quantities S_k that indicates the number of nodes of degree k in the dynamical state $\sigma = s$ and the corresponding degree block density of nodes of degree k in the state s

$$s_k = \frac{S_k}{V_k}$$

where V_k is the number of nodes of degree k. Finally the global averages on the network are then given by the expressions

$$\rho_s = \sum_k P(k) s_k$$

where ρ_s is the probability that any given node is in the state s. This formalism is extremely convenient in networks

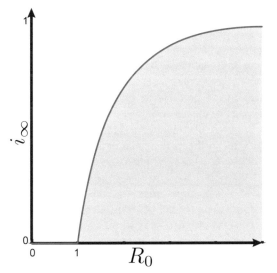

FIGURE 27.3 **Phase diagram.** Above $R_0 = 1$ the total number of infected individuals becomes a finite fraction of the population size.

where the connectivity pattern dominates the system's behavior as the above degree block variables define a mean-field approximation within each degree class, relaxing however the overall homogeneity assumption on the degree distribution [38]. This framework, first introduced for the description of epidemic processes, is the basis of the heterogeneous mean-field (HMF) approach which allows the analytical study of dynamical processes in complex networks by writing mean-field dynamical equations for each degree class variable. The HMF approach generalizes to the case of networks with arbitrary degree distribution the equations describing the dynamical process by considering degree block variables grouping nodes within the same degree class k. If we consider the SIS model, the variables describing the system are i_k, s_k and r_k that represent the fraction of nodes with degree k in the susceptible, infected and recovered class. The evolution equation for the infected individual reads as

$$\frac{di_k(t)}{dt} = -\mu i_k(t) + \lambda[1 - i_k(t)]k\Theta_k(t)$$

Here the first term simply expresses the fact that any node in the infected state may recover with rate μ. The second term, which generates new infected individuals, is proportional to the probability of transmission λ, the degree k, the probability $1-i_k$ that a vertex with degree k is not infected, and the density Θ_k of infected neighbors of vertices of degree k, i.e., the probability of contacting an infected individual. As we are still assuming a mean-field description of the system, the latter term is the average probability that any given neighbor of a vertex of degree k is infected. This quantity can be expressed as $\Theta_k(t) = P(k'|k)i_{k'}(t)$, which considers the average over all possible degrees k' of the probability $P(k'|k)$ that any edge of a node of degree k is pointing to a node of degree k' times the probability $i_{k'}$ that the node is infected. This expression can be further simplified by considering a random network in which the conditional probability does not depend on the originating node. In this case we have that $P(k'|k) = k'P(k')/<k>$ simply descending from the fact that any edge has a probability to point to a node with degree k' which is proportional to the degree itself [38].

The HMF technique is often the first line of attack in understanding the effects of complex connectivity patterns on dynamical processes and has been widely used in a wide range of phenomena, although with different names and specific assumptions depending on the problem at hand. Although it contains several approximations, the HMF approach readily shows that the heterogeneity found in the connectivity pattern of many networks may drastically affect the unfolding of the dynamical process, providing novel and interesting features that depart from the common picture we are used to in regular lattices and homogenous population.

The classic example of the effect of degree heterogeneity on dynamical processes in complex networks is offered by epidemic spreading. The previously discussed result of the presence of an epidemic threshold in the SIR and SIS models is obtained under the assumption that each individual in the system has, at a first approximation, the same number of connections $k \cong <k>$. However, social heterogeneity and the existence of 'super-spreaders' have long been known in the epidemics literature [39]. Generally, it is possible to show that the reproductive rate R_0 is re-normalized by fluctuations in the transmissibility or contact pattern as $R_0 \to R_0(1 + f(v))$, where $f(v)$ is a positive and increasing function of the standard deviation v of the individual transmissibility or connectivity pattern [40]. In particular, by generalizing the dynamical equations of the SIS model, the HMF approach yields that the disease will affect a finite fraction of the population only if $\frac{\beta}{\mu} \geq <k>^2 / <k^2>$ [36,38]. This readily points out that the topology of the network enters the very definition of the epidemic threshold through the ratio between the first and second moments of the degree distribution. Furthermore, this implies that in heavy-tailed networks such that $<k^2> \to \infty$, in the limit of a network of infinite size, we have *a null epidemic threshold*. While this is not the case in any finite size real-world network [41,42], larger heterogeneity levels lead to smaller epidemic thresholds (see Figure 27.3). This is a very relevant result indicating that heterogeneous networks behave very differently from homogeneous networks with respect to physical and dynamical processes. Indeed, the heterogeneous connectivity pattern of networks also affects the dynamical progression of the epidemic process, resulting in striking hierarchical dynamics in which the infection propagates from higher to lower degree classes. The infection first takes control of the large degree vertices in the network, then rapidly invades the network via a cascade through progressively smaller degree classes (Figure 27.4). It also turns out that the time behavior of epidemic outbreaks and the growth of the number of infected individuals are governed by a time scale τ proportional to the ratio between the first and second moments of the network's degree distribution, thus pointing to a velocity of progression that is increasing with the heterogeneity of the network [43].

The change of framework induced by the network heterogeneity in the case of epidemic processes has triggered a large number of studies aimed at providing more rigorous analytical basis to the results obtained with the HMF and other approximate methods, and exploring different spreading models [44–48]. In particular, the HMF approach is exact for networks whose connections are fixed only on average, annealed networks. For networks that instead have fix connections, quenched networks, the HMF approach does not always give correct results and interpretations. Indeed, in these networks there are striking

FIGURE 27.4 **Epidemic progression in a scale-free network.** Node size represents the degree and the color varies between yellow and red to indicate the time of infection. Blue nodes are never infected and remain susceptible.

differences between the results for SIR and SIS models [21]. While the SIR model results are in good agreement with the HMF theory also in quenched networks, the SIS model results can be completely different from those predicted by the HMF. Exact solutions have recently shown that quenched networks characterized by a maximum degree that diverges with the system size, always have a zero threshold [21,49]. In other words, the threshold of a SIS model is zero also for networks of finite second moment if the largest degree is a growing function of the network size, contrary to the indications of the HMF theory. The reasons behind the differences between the SIR and SIS models in this context are not really clear and are still a matter of debate, but are probably related to the absence/existence of a steady state of infected individuals [21].

Heterogeneous networks are extremely permeable epidemics attacks [38,50] and at the same time uniform immunizations are clearly not effective in this type of network. They give the same importance to very connected nodes and to nodes with very few connections, which instead have completely different roles in the spreading of the epidemic. The introduction of fraction g of nodes immune to the disease, immunized, is equivalent to a simple rescaling of the per capita spreading rate that becomes

$$\frac{\beta}{\mu} = \frac{<k>^2}{<k^2>} \frac{1}{(1-g)}$$

The change in the parameter is just proportional to the fraction of nodes not immunized. For a SIR model at the HMF level we have a vanishing threshold, implying that for power-law networks characterized by a divergent second moment of the degree distribution the disease will always spread except when $g = 1$ [38]. For finite networks this fraction is < 1 but still close to this value. It is then extremely relevant to research activity concerning the developing of dynamical ad-hoc strategies for network protection: targeted immunization strategies and targeted prophylaxis that evolve with time might be particularly effective in the control of epidemics on heterogeneous patterns, compared to massive uniform vaccinations or stationary interventions [51–54].

For instance, nodes with high degree are the ones more likely to spread the disease. Immunizing them through a targeted scheme is the most efficient way to protect the network from the disease. Analytical evaluation confirmed by numerical simulations shows that in heterogeneous networks targeted schemes allow complete protection from the disease at the cost of immunizing $<10\%$ of the nodes [38,51].

Although targeted immunization strategies are extremely powerful they rely on a complete knowledge of the network structure. Unfortunately, in most real cases this is partial and limited. To overcome this problem several methods based on local exploration mechanisms have been proposed [51,53,55–57].

In one of the most ingenious strategies a fraction of nodes is selected at random and each one of them is asked to point to one of its neighbors. Each of these neighbors is then immunized. This strategy is based on the heterogeneity of the systems. In these types of networks following links at random gives a higher probability of reaching high degree nodes that have many links pointing to them [56]. This property is often cited as 'the friendship paradox', which states that your friends have more friends than you do [35,38].

Many others variation involving chained versions of the previous strategy [57], shortest-path of different sizes [58], and other propagation properties have been proposed [59]. It is important to stress that all of them are based on the heterogeneous features of the network structure.

COMPLEX NETWORKS AND THE LARGE-SCALE SPREADING OF INFECTIOUS DISEASES

The network mindset is necessary not only in describing the connectivity pattern of single individuals in a population. A simple example is provided by the large-scale description of epidemic spreading. The spread of the plague epidemic in the 14th century (the 'Black Death') [60] was mainly a spatial diffusion phenomenon. Historical studies have established that disease propagation followed a simple pattern that can be adequately described mathematically within the framework of continuous differential equations using terms that describe diffusion. As anticipated in 1933 [27], the large scale and geographical impact of infectious diseases (such as the SARS epidemic [29] or the recent swine flu epidemic) on populations in the modern world is mainly due to commercial air travel. An epidemic starting in Mexico rapidly reaches Europe and Asia (Figure 27.5). This picture cannot be simply described in terms of diffusive phenomena, but must incorporate the spatial structure of modern transportation networks.

The conceptual framework to approach spatially structured population is the patch or meta-population modeling framework that considers multiple subpopulations coupled by movements of individuals. These models are defined by the network describing the coupling among the populations along with the intensity of the coupling, which in general represents the rate of exchange of individuals between two populations. Networks are also, in this case, the underlying substrate for the diffusion process. Meta-population models can be devised at various granularity levels (country, inter-city, intra-city) and the corresponding networks therefore include very different systems and infrastructures. This implies scales ranging from the movement of people within locations of a city to the large flows of travelers among urban areas.

At the formal level meta-population models fall into the category of reaction–diffusion processes, where each node i is allowed to have any non-negative integer number of particles N_i so that the total particle population of the system is $N = \sum_i N_i$. In this case particle network frameworks extend the HMF approach to the case of reaction–diffusion systems in which particles (individuals) diffuse on a network with arbitrary topology. A convenient representation of the system is therefore provided by the quantities defined in terms of the degree k

$$N_k = \frac{1}{V_k} \sum_{i|k_i=k} N_i$$

where V_k is the number of nodes with degree k and the sums run over all nodes i having degree k_i equal to k. The variable degree block variable N_k represents the average number of particles in nodes with the degree k. The use of the HMF approach amounts to the assumption that nodes with degree k, and hence the particles in those nodes, are statistically equivalent. In this approximation the dynamics of particles randomly diffusing on the network is given by a mean-field dynamical equation expressing the variation in time of the particle subpopulations $N_k(t)$ in each degree block k. This can easily be written as:

$$\frac{dN_k(t)}{dt} = -d_k N_k(t) + k \sum_k P(k'|k) d_{k'k} N_{k'}(t)$$

The first rhs term of the equation just considers that only a fraction of particles d_k moves out of the node per unit time. The second term instead accounts for the particles diffusing from the neighbors into the node of degree k. This term is proportional to the number of links k times the average number of particles coming from each neighbors. This is equal to average over all possible degrees k' the fraction of particles moving on that edge $d_{k'k}N_{k'}(t)$ according to the conditional probability $P(k'|k)$ that an edge belonging to a node of degree k is pointing to a node of degree k'. Here the term $d_{k'k}$ is the diffusion rate along the edges connecting nodes of degree k and k'. The rate at which individuals leave a subpopulation with degree k is then given by $d_k = k\sum_{k'} P(k'|k)d_{kk'}$. The function $P(k'|k)$ encodes the topological connectivity properties of the network and allows study of the different topologies and mixing patterns. The above equation explicitly brings the diffusion of particles into the description of the system. The equation can be simply generalized to particles with different states and reacting among them by adding a reaction term to the above equations. For instance the generalization of the SIR model described in the text would consider three types of particles denoting infected, susceptible and recovered individuals. The reaction term that would take place among individuals in the same node would be the usual contagion process among susceptible and infected individuals and the spontaneous recovery of infected individuals.

FIGURE 27.5 **Historical and modern epidemics.** (Left) Map of the propagation of the Black Death in the 14th century. The epidemic front spread in Europe with a velocity of 200–400 miles per year. (Right) Epidemic tree of the first 120 days of the 2009 H1N1 pandemic.

The analysis of a simple diffusion process immediately highlights the importance of network topology. In a random network with arbitrary degree distribution the stationary state reached by a swarm of particles diffusing with the same diffusive rate yields $N_k \propto k$ and the probability of finding a single diffusing walker in a node of degree k is

$$p_k = \frac{k}{<k>} \frac{1}{V}$$

where V is the total number of nodes in the network. This expression tell us that the larger the degree of the nodes, the larger the probability of being visited by the walker. Evidently, the topology of the system has a large impact in the way contagion processes will spread in large meta-population networks.

Consider, for instance, a simple epidemic process such as the SIR model in a meta-population context [56,61–68]. In this case each node of the network is a subpopulation (ideally an urban area) connected by a transportation system (the edges of the network) that allows individuals to move from one subpopulation to another (Figure 27.6). If we assume a diffusion rate d for each individual, and that the single population reproductive number of the SIR model is $R_0 > 1$, we can easily identify two different limits. If $d=0$ any epidemic occurring in a given subpopulation will remain confined: no individual could travel to a different subpopulation and spread the infection across the system. In the limit $d=1$ we have individuals constantly wandering from one subpopulation to the other and the system is in practice equivalent to a well-mixed single population. In this case, since $R_0 > 1$, the epidemic will spread across the entire system. A transition point between these two regimes is therefore occurring at a threshold value d_c of the diffusion rate, identifying a global invasion threshold. Interestingly, this threshold cannot be uncovered by continuous models as it is related to the stochastic diffusion rate of single individuals. Furthermore, the global invasion threshold is affected by the topological fluctuations of the meta-population network. In particular, the larger the network heterogeneity, the smaller the value of the diffusion rate above which the epidemic may globally invade the meta-population system. This result assumes

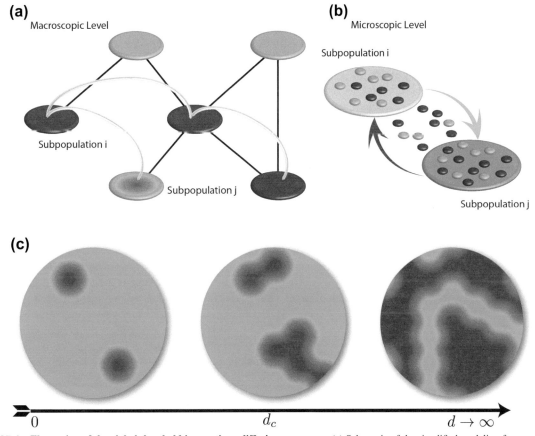

FIGURE 27.6 Illustration of the global threshold in reaction–diffusion processes. (a) Schematic of the simplified modeling framework based on the particle-network scheme. (b) Within each subpopulation individuals can mix homogeneously or according to a subnetwork, and can diffuse with rate d from one subpopulation to another, following the edges of the network. (c) A critical value d_c of the diffusion strength for individuals or particles identifies a phase transition between a regime in which the contagion affects a large fraction of the system and one in which only a small fraction is affected.

a particular relevance as it explains why travel restrictions appear to be highly ineffective in containing epidemics: the complexity and heterogeneity of the current human mobility network favor considerably the global spreading of infectious diseases, and only unfeasible mobility restrictions that reduce the global travel fluxes 90% or more would be effective [57,64,69].

Reaction−diffusion models lend themselves to the implementation of large-scale computer simulations (Monte Carlo and individual-based simulations) that allow microscopic tracking of the state of each node and the evolution of the dynamical process. For instance, state of the art data-driven meta-population models combine highly detailed population and transportation databases. For instance, the recently developed GLobal Epidemic and Mobility (GLEaM) model [70] integrates census and mobility data in a fully stochastic meta-population model that allows for the detailed simulation of the spread of influenza-like illnesses (ILI) around the globe. Meta-population models are also amenable to the inclusion of an age structure in the population by simply modifying the compartmental structure to include the age group contact matrix, M, taking them one step further in terms of realism while still keeping all the advantages of the conceptual clarity that are the hallmarks of this class of models.

Agent-based models, where each agent represents a simplified human individual going through their daily activities, push to the limits the realism of data-driven models. Infection spreads from individual to individual whenever they come into contact with each other, whether within the household, at school or work, or at random in the general population [14]. A key feature of models such as CommunityFlu [71], Flute [72] and Epifast [73] is the characterization of the network of contacts among individuals based on realistic data-driven models of the sociodemographic structure of the population being considered [74,75]. All these highly structured models provide a novel approach to evidence-based and quantitative scenario analysis. Although even among modelers there are contrasting opinions, those models are assuming increasing relevance in the public health domain, providing rationales and quantitative analysis to support the decision- and policy-making processes.

CONCLUSIONS AND FUTURE CHALLENGES

Although in recent years our understanding of dynamical processes in complex networks has progressed at an exponential pace, there are still a number of major challenges that see the research community actively engaged. The first challenge stems from the fact that the analysis of dynamical processes is generally performed in the presence of a timescale separation between the network evolution and the dynamical process unfolding on its structure. At one extreme we can consider the network as quenched in its connectivity pattern, thus evolving on a timescale that is much longer that the dynamical process itself. At the other extreme, the network is evolving at a much shorter timescale than the dynamical process, thus effectively disappearing from the definition of the interaction among individuals that is conveniently replaced by effective random couplings. While the timescale separation is extremely convenient for the numerical and analytical tractability of the models, networks generally evolve on a timescale that might be comparable to that of the dynamical process. Furthermore, the network properties that inform models generally represent a time-integrated static snapshot of the system. However, in many systems the timing and duration of interactions define processes on a timescale very different from, and often conflicting with, those of the time-integrated view. This makes clear the importance of considering the concurrency of network evolution and dynamical processes in realistic models in order to avoid misleading conclusions [76−79].

A second challenge is the co-evolution of networks with the dynamical process. Access to the mathematical and statistical laws that characterize the interplay and feedback mechanisms between the network evolution and the dynamical processes is extremely important, especially in social systems where the adaptive nature of agents is of paramount importance [58,78,80]. The spreading of an opinion is affected by interactions between individuals, but the presence and/or establishment of interactions between individuals is affected by their opinion. This issue is more and more relevant in the area of modern social networks populating the information technology ecosystem, such as those defined by the Facebook and Twitter applications. In this case the network and the information spreading cannot be defined in isolation, especially because of the rapidly changing interactions and modes of communication that depend upon the type of information exchanged and the rapidly adaptive behavior of individuals (Figure 27.7).

The adaptive behavior of individuals to the dynamical processes they are involved in represents another modeling challenge, as it calls for the understanding of the feedback among different and competing dynamical processes. For instance, relatively little systematic work has been done to provide coupled behavior−disease models able to close the feedback loop between behavioral changes triggered in the population by an individual's perception of the disease spread and the actual disease spread [59,81]. Similar issues arise in many areas where we find competing processes of adaptation and awareness of information or knowledge spreading in a population.

Finally, the eventual goal is not only to understand complex systems, mathematically describe their structure

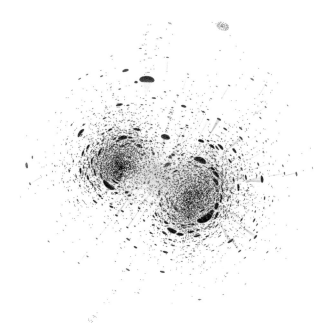

FIGURE 27.7 **Each node represents a Twitter user that employed the #bahrain hashtag.** Blue and orange edges represent information exchanged between users through the use of retweets and replies, respectively. The two clusters observed correspond to two communities, an English- and an Arabic- speaking one. Retweets are used predominantly to communicate within a single cluster, whereas mentions are employed to send information from one community to the other.

and dynamics, and predict their behavior, but also to control their dynamics. Also in this case, although control theory offers a large set of mathematical tools for steering engineered and natural systems, we have yet to understand how the network heterogeneities influence our ability to control the network dynamics and how network evolution affects controllability [82].

Taking into account the complexity of real systems in epidemic modeling has proved to be unavoidable, and the corresponding approaches have already produced a wealth of interesting results. While this has stimulated the recent focus on large-scale computational approach to epidemic modeling it is clear that many basic theoretical questions are still open. How does the complex nature of the real world affect our predictive capabilities in the realm of computational epidemiology? What are the fundamental limits in epidemic evolution predictability with computational modeling? How do they depend on the level of accuracy of our description and knowledge on the state of the system? Tackling such questions necessitates exploiting several techniques and approaches. Complex systems and networks analysis, mathematical biology, statistics, non-equilibrium statistical physics and computer science all play an important role in the development of a modern computational epidemiology approach. While such an integrated approach might still be in its first steps, it seems now possible to ambitiously imagine the creation of computational epidemic forecast infrastructures able to provide reliable, detailed and quantitatively accurate predictions of global epidemic spread.

REFERENCES

[1] Anderson RM, May RM. Spatial, temporal and genetic heterogeneity in host populations and the design of immunization programs. IMA J Math Appl Med Biol 1984;1:233–66.

[2] May RM, Anderson RM. Spatial heterogeneity and the design of immunization programs. Math Biosci 1984;72:83–111.

[3] Chatterjee S, Durrett R. Contact processes on random graphs with power law degree distributions have critical value 0. Ann Probab 2009;37:2332–56.

[4] Bolker BM, Grenfell BT. Chaos and biological complexity in measles dynamics. Proc R Soc Lond B 1993;251:75–81.

[5] Bolker BM, Grenfell BT. Space persistence and dynamics of measles epidemics. Phil Trans R Soc Lond 1995;348:309–20.

[6] Lloyd AL, May RM. Spatial heterogeneity in epidemic models. J Theor Biol 1996;179:1–11.

[7] Grenfell BT, Bolker BM. Cities and villages: infection hierarchies in a measles meta-population. Ecol Lett 1998;1:63–70.

[8] Keeling MJ, Rohani P. Estimating spatial coupling in epidemiological systems: a mechanistic approach. Ecol Lett 2002;5:20.

[9] Ferguson NM, et al. Planning for smallpox outbreaks. Nature 2003;425:681.

[10] Rvachev LA, Longini JIM. A mathematical model for the global spread of influenza. Math Biosc 1985;75:3.

[11] Keeling MJ, Rohani P. Estimating spatial coupling in epidemiological systems: a mechanistic approach. Ecol Lett 1995;5:20–9.

[12] Hufnagel L, Brockmann D, Geisel T. Forecast and control of epidemics in a globalized world. Proc Natl Acad Sci USA 2004; 101:15124.

[13] Longini JIM, et al. Containing pandemic influenza at the source. Science 2005;309:1083.

[14] Ferguson NM, et al. Strategies for containing an emerging influenza pandemic in Southeast Asia. Nature 2005;437:209.

[15] Colizza V, Barrat A, Barthelemy M, Vespignani A. The role of the airline transportation network in the prediction and predictability of global epidemics. Proc Natl Acad Sci 2006;103:2015.

[16] Keeling MJ, et al. Foot and mouth epidemic: stochastic dispersal in a heterogeneous landscape. Science 2001;294:813–7.

[17] Colizza V, Barrat A, Barthelemy M, Vespignani A. The modeling of global epidemics: stochastic dynamics and predictability. Bull Math Biol 2006;68(8):1893–921.

[18] Eubank S, et al. Modelling disease outbreaks in realistic urban social networks. Nature 2004;429:180.

[19] Keeling MJ, Rohani P. Modeling Infectious Disease in Humans and Animals. Princeton University Press; 2008.

[20] Goffman W, Newill VA. Generalization of epidemic theory: an application to the transmission of ideas. Nature 1964;204:225.

[21] Castellano C, Pastor-Satorras R. Thresholds for epidemic spreading in networks. Phys Rev Lett 2010;105:218701.

[22] Peiris JSM, Yuen KY, Stohr K. The severe acute respiratory syndrome. N Engl J Med 2003;349:2431–41.

[23] Lloyd AL, May RM. How viruses spread among computers and people. Science 2001;292:1316.

[24] Eagle N, Sandy Pentland A. Reality mining: sensing complex social systems. Pers Ubiquitos Comput 2006;10:255–68.

[25] Christakis NA, Fowler JH. Social network sensors for early detection of contagious outbreaks. PLoSOne 2010;6:E12948.

[26] Stehle J, et al. High-resolution measurements of face-to-face contact patterns in a primary school. PLoS One 2011;6:e23176.

[27] Barrat A, Barthelemy M, Pastor-Satorras R, Vespignani A. The architecture of complex weighted networks. Proc Natl Acad Sci U S A 2004;101:3747.

[28] Anderson RM, May RM, editors. Infectious Diseases in Humans: Dynamics and Control. Oxford: Oxford Science Publication; 1992.

[29] Guimera R, Mossa S, Turtschi A, Amaral LAN. The worldwide air transportation network: anomalous centrality, community structure, and cities' global roles. Proc Natl Acad sci USA 2005;102:7794.

[30] Pastor-Satorras R, Vespignani A, editors. Evolution and Structure of the Internet: A Statistical Physics Approach. Cambridge: Cambridge University Press; 2004.

[31] Watts DJ, Strogatz SH. Collective dynamics of 'small-world' networks. Nature 1998;393:440.

[32] Barabási A-L, Albert R. Emergence of scaling in random networks. Science 1999;286:509.

[33] Dorogovtsev S, Mendes JFF. Evolution of Networks: From Biological Nets to the Internet and WWW; 2003.

[34] Hethcote HW, Yorke JA. Gonorrhea transmission dynamics and control. Lecture Note in Biomathematics 56. Berlin: Springer; 1984.

[35] Newman MEJ, editor. Networks. An Introduction. Oxford: Oxford University Press; 2010.

[36] Pastor-Satorras R, Vespignani A. Epidemic spreading in scale-free networks. Phys Rev Lett 2001;86:3200.

[37] Colizza V, Pastor-Satorras R, Vespignani A. Reaction–diffusion processes and metapopulation models in heterogeneous networks. Nat Phys 2007;3:276–82.

[38] Barrat A, Barthélemy M, Vespignani A. Dynamical Processes on Complex Networks. Cambridge Univesity Press; 2008.

[39] Moreno Y, Pastor-Satorras R, Vespignani A. Epidemic outbreaks in complex heterogeneous networks. Eur Phys J B 2002;26:521–9.

[40] Hethcote HW. An immunization model for a heterogeneous population. Theor Pop Biol 1978;14:338–49.

[41] Colizza V, Vespignani A. Epidemic modeling in metapopulation systems with heterogeneous coupling pattern: theory and simulations. J Theor Biol 2008;251:450–67.

[42] Ni S, Weng W. Impact of travel patterns on epidemic dynamics in heterogeneous spatial metapopulation networks. Phys Rev E 2009;79:016111.

[43] Pastor-Satorras R, Vespignani A. Epidemic dynamics in finite size scale-free networks. Phys Rev E 2002;65:035108.

[44] Ben-Zion Y, Cohena Y, Shnerba NM. Modeling epidemics dynamics on heterogenous networks. J Theor Biol 2010; 264:197–204.

[45] Absence of epidemic threshold in scale-free networks with degree correlations. Phys Rev Lett 2003;90:028701.

[46] Barthelemy M, Godreche C, Luck J-M. Fluctuation effects in metapopulation models: percolation and pandemic threshold. J Theor Biol 2010;267:554–64.

[47] Balcan D, Vespignani A. Phase transitions in contagion processes mediated by recurrent mobility patterns. Nat Phys 2011; 7:581–5886.

[48] Belik V, Geisel T, Brockmann D. Natural human mobility patterns and spatial spread of infectious diseases. Phys Rev X 2011;1:011011.

[49] Wang, Y., Chakrabarti, D., Wang, C. & Faloutsos, C. Epidemic spreading in real networks: An eigenvalue viewpoint. In 22nd International symposium on reliable distributed systems (SRDS'03)} IEEE Computer Society, Los Alamitos, CA, USA (2003) p.25–34.

[50] Albert R, Jeong H, Barabási AL. Error and attack tolerance of complex networks. Nature 2000;406:378.

[51] Pastor-Satorras R, Vespignani A. Immunization of complex networks. Phys Rev E 2002;63:036104.

[52] May RM, Lloyd AL. Infection dynamics on scale-free networks. Phys Rev E 2001;64:066112.

[53] Hollingsworth TD, Ferguson NM, Anderson RM. Will travel restrictions control the international spread of pandemic influenza? Nature Med 2006;12:497–9.

[54] Volz E, Meyers LA. Epidemic thresholds in dynamic contact networks. J R Soc Interface 2009;6:233–41.

[55] Barrat A, Barthelemy M, Vespignani A. Dynamical Processes in Complex Networks. Cambridge University Press; 2008.

[56] Cohen R, Havlin S, ben-Avraham D. Efficient immunization stragteies for computer networks and populations. Phys Rev Lett 2003;91:247901.

[57] Holme P. Efficient local strategies for vaccination and network attack. Europhys Lett 2004;68:908–14.

[58] Gomez-Gardenes J, Echenique P, Moreno Y. Immunization of real complex communication networks. Eur Phys J B 2006; 49:259–64.

[59] Stauffer D, Barbosa VC. Dissemination strategy for immunizing scale-free networks. Phys Rev E 2006;74:56105.

[60] Amaral LAN, Scala A, Barthelemy M, Stanley HE. Classes of small-world networks. Proc Natl Acad Sci U S A 2005;97:11149.

[61] Watts DJ, Muhamad R, Medina DC, Dodds PS. Multiscale, resurgent epidemics in a hierarchical metapopulation model. Proc Natl Acad Sci 2005;102:11157.

[62] Holme P, Newman MEJ. Nonequilibrium phase transition in the coevolution of networks and opinions. Phys Rev E 2007; 74:056108.

[63] Colizza V, Vespignani A. Invasion threshold in heterogeneous metapopulation networks. Phys Rev Lett 2007;99:148701.

[64] Colizza V, Vespignani A. Epidemic modeling in metapopulation systems with heterogeneous coupling pattern: theory and simulations. e-print cond-mat 2007;0706:3647.

[65] Centola D, Gonzalez-Avella JC, Eguiluz VM, San Miguel M. Homophily, cultural drift, and the co-evolution of cultural groups. J Conflict Resolut 2007;51:905–29.

[66] Funk S, Salathe M, Jansen VAA. Modelling the inuence of human behaviour on the spread of infectious diseases: a review. J R Soc Interface 2010;7:1247–56.

[67] Perra N, Balcan D, Goncalves B, Vespignani A. Towards a characterization of behavior–disease models. PLoS One 2011;6:e23084.

[68] Liu YY, Slotine JJE, Barabasi AL. Controllability of complex Networks. Nature 2011;473:7346.

[69] Cooper BS, Pitman RJ, Edmunds WJ, Gay NJ. Delaying the international spread of pandemic influenza. PLoS Med 2006;3:e12.

[70] Balcan D, et al. Multiscale mobility networks and the spatial spreading of infectious diseases. Proc Natl Acad Sci U S A 2009;106:21484.

[71] Community Flu 1.0, Centers for Disease Control and Prevention (CDC), <http://www.cdc.gov/flu/tools/communityflu>

[72] Chao D, Halloran M, Obenchain V, Longini IJ. FluTE, a publicly available stochastic influenza epidemic simulation model. PLoS Comput Biol 2010;6:e1000656.

[73] Bisset KR, Chen J, Feng X, Kumar VA, Marathe MV. EpiFast: a fast algorithm for large scale realistic epidemic simulations on distributed memory systems. In: Proceedings of the 23rd International Conference on Supercomputing (ICS); 2009 p. 430–439.

[74] Ajelli M, Merler S. The impact of the unstructured contacts component in influenza pandemic modeling. PLoS One 2008;3:e1509.

[75] degli Atti MLC, et al. Mitigation measures for pandemic influenza in Italy: an individual based model considering different scenarios. PLoS One 2008;3:E1790.

[76] Morris M, Kretzschmar M. Concurrent partnerships and the spread of HIV. AIDS 1997;11:641–8.

[77] Moody J. The importance of relationship timing for diffusion: indirect connectivity and STD infection risk. Soc Forces 2002;81:25.

[78] Goldenberg J, Shavitt Y, Shir E, S.Solomon S. Distributive immunization of networks against viruses using the 'honey-pot' architecture. Nat Phys 2005;1:184–8.

[79] Isella L, et al. What's in a crowd? Analysis of face-to-face behavioral networks. J Theor Biol 2011;271:166–80.

[80] Kitsak M, et al. Identification of influential spreaders in complex networks. Nat Phys 2010;6:888–93.

[81] International Air Transport Association, <http://www.iata.org>

[82] Official Airline Guide, <http://www.oag.com>

Index

Note: Page numbers with "f" denote figures; "t" tables; "b" boxes.

A

'Ab initio' gene finders, 32–33
Absolute quantification, 4, 7–8, 15
ACeDB, 368
Adaptive immune system, 481–483
ADP ribosylation, 318
Adrenaline rush, 311–312
Adult body plan, 220–221
Affinity -or immunoprecipitation followed by mass spectrometry (AP/MS), 97–99
Affinity purification followed by mass spectrometry (AP-MS), 13, 14f
Agent-based models (ABM), 515, 524
AKAPs (A-kinase anchoring proteins) functions, 314–315
Algorithms, 10, 15
   REVEAL, 432–433
   splice alignment, 33
Alleles
   conditional alleles, essential genetic interactions in, 130
   gain-of-function, 130
Alternative self-maintaining states, 273–275
Alternative splicing, 28–29
AMP-activated protein kinase (AMPK), 418–419
AMP-GEF/Epac, 312
Amyotrophic lateral sclerosis (ALS), 353–356
Analysis of covariance (ANCOVA), 348
Analysis of variance (ANOVA), 348
Analytical solution, phenotypes from, 294–295
Anaphase switch, 278–279
Aneuploid, 266t–267t
Animal genome, gene expression and regulation of, 376–380
Antibody diversity, 489–491
Anticodon sequence, 28–29
APETALA1 (AP1) transcription factor, 403
Arabidopsis, as model for systems biology, 391–406
   development/response to environmental stress, 401–403
      tissue-specific responses to environment, 401–402
      transition to flowering, 402–403
   insightful modeling, 400–401
   systems analysis, 392–401
      auxin flow and stem cell specification, 398–399
      computational modeling of root development, 398–399
      coupled gene regulatory networks in shoot, 400
      dynamics of asymmetric cell division, 397–398
      of gene activity, 392–393
         in space and time, 399–400
      integrated analysis of shoot development, 399
      modeling regulatory networks in shoot, 400–401
      mutual support for root hair patterning, 399
      spatiotemporal specific expression data, use of, 393–396
Arabidopsis thaliana (*A. thaliana*), 80b, 81, 391
Arachidonic acid (AA), 319–320
Arcuate nucleus (ARC), 416–417
Attractor, defined, 271t
Autoregulation, 187
Auxins, 398

B

Bacterial artificial chromosome (BAC) transgenes, 15
Bait protein, 13
   role in protein–protein interactions, 15
Basin of attraction, defined, 271t
Bayesian network reconstruction process example, 508–510
   integrating genetic data as structure prior to enhancing causal inference in, 507
   integrative genomics approach to constructive predictive network models, 506–507
   omics data as network priors in, 507–508
Beloussov–Zhabotinski reaction, 330f
Bifurcation point, 266t–267t
Big data, 446–447, 499–501, 500f
Binary models, 432–433
Biochemical reaction network (BRN), 428, 434
Biochemical reactions, dimensionality effects in, 330
BioGRID Database, 349–350
Biological complexity, five systems' strategies for dealing with, 449–460
   cross-disciplinary infrastructure, 450–451, 451f
   develop new technologies that explore new dimensions of patient data space, 454–457, 455f
   experimental systems approach to disease and wellness holistic, 451–454, 451f, 453f
   transformation of big data sets to medically relevant information, 457–460, 458f
Biological discovery, 239
Biological networks
   fundamental properties of, 240
   preferential attachment in, 183–184
Biological systems, 445, 499–502
   big data, 499–501, 500f
   causality inference, 503–505, 504f
      and predictive gene networks, relationship between, 505–511
   central dogma of, 500–501, 501f
   as graphs, 178
   movie analogy, 502–503
Biomass reaction, 233–234, 244–245
Biomolecules, 3
B-ions, 6
Bipolar disorder (BPD), 414
Bistability, 266t–267t, 270–272, 274f, 278f, 319–322
BLAST, 33
BLAT, 33
Boolean modeling, 46
BooleanNet, 202, 433
Boolean network modeling, 197–210, 432–433
   applications of, 207
      pathogen–immune response network, 207
      T-LGL leukemia network modeling, 207
   based on biological knowledge, reconstruction of, 199–200
   of biological systems, 197–198
   future directions of, 207–208
   implications of, 206
   initialization of, 202
   robustness against disruptions, 206
   steady-state analysis of, 202–205

Boolean network modeling (*Continued*)
  transfer functions, determination of, 200—201
  updating schemes and incorporating time, 201—202, 203f
  validation of, 205—206
Boveri—Sutton chromosome theory, 367—368
BRENDA, 230—232
BRUTUS, 402
Buchnera aphidicola, 260—261

C

Caenorhabditis elegans (*C. elegans*), 66—67, 68f, 71, 73, 75—77, 79, 71b, 81, 128—130
  development of, 367—390
    chemical genomics in, 372—373
    classical screens with next-generation sequencing, 371—372
    core modules that control worm development, finding, 370—371
    forward and reverse genetics, 369—376
      encoded genes, functions of, 369—370
      genotype—phenotype connection, 369—370
    gene expression and regulation, 376—380
      normal and perturbed gene expression, 377
      proteomics, 380
      single cell-resolution analysis, 377—378
      transcription factor-binding sites and chromatin organization, 378—379
      variation in among individuals and phenotypic consequences, 379—380
    integrative and dynamic modeling to link genotype to phenotype, 380—382
    proteomics in, 380
    system before the sequence, 367—369
    nuclear lamina—genome interactions in, 139—140
  RNA interference, 94
CA-FLIM, 337
Calcin—Benson cycle, 254
Calcium-calmodulin kinase II (CAMKII), 318—319
cAMP signals, 312
Cancer robustness, 469
  host—tumor entrainment, 472—473
  intracellular feedback loops, 472
  intratumoral genetic heterogeneity, 472
  long-tail drug, 476—477, 476f
  mechanisms for, 471—473
  open pharma, 477—478
  theoretically motivated therapy strategies, 473—475, 474f
  treatment efficacy, proper index of, 475—476
Cancer therapeutics, genetic interactions and, 129
Causality inference, 503—505, 504f
  and predictive gene networks, relationship between, 505—511
    bottom-up approach, 505—506
    example, 508—510, 509f
  integrating genetic data, 507
  integrative genomics approach, 506—507
  networks constructed from human and animal data elucidate the complexity of disease, 510—511
  omics data, 507—508
  top-down approach, 505—506
CDC42 protein, 331—332
CDKN2B-AS1, 36
Cell(s)
  as computers, 65
  as highly interactive robust system, 154—155
  as interactome networks, 46
Cell cycle
  control, 220
  eukaryotic, 267f
  molecular biology of, 268—270, 270t
  physiology of, 265—268
  transitions, 275f
Cell-focused systems immunology, 484—488
  exploring cellular diversity, 486—487
  limitations of, 487—488
  reconstructing cellular networks, 484—486, 485f
Cell Net Analyzer, 202
CellNetOptimizer, 433
Cell lysis method, 5—6
CellProfiler, 383—384
Cell signaling, 311—328
  bistability, concept of, 319—322
  dynamical models, 319
  future challenges, 323—324
  isoforms of signaling molecules, signaling integration and sorting, 313—314
  network motifs, regulation by, 316—317
  network topology and consolidation of signals, 318—319
  pathways to networks, 311—313
  positive feedback loops forming switches, 319—322, 321f
  properties of, 317—318
  scaffolding proteins, 314—315
  signaling microdomains within cells, 322—323
  signaling networks
    computational analysis of, 316
    graph theory-based models, 316
    forming signaling complexes, 314—315
  topology of, 317f
Cell-type-specific interactomes, 55—56
Cellular automata-based models, 491—493
Cellular biology, systemic understanding of, 330—332
Cellular differentiation, 46
Cellular networks, graph theory properties of, 177—194
  biological systems, as graphs, 178
  building blocks of, 186—187
    autoregulation and the feedforward loop, 187
    randomized networks, 186—187
    sub-graphs and motifs, 186
  controllability, 188—189
  degree correlations, 184—185
  differential networks, 189—190
  dynamics of, 190
  Erdős—Rényi network, 178
    clustering coefficient, 180
    degrees and degree distribution, 178—179
    network paths and the small world phenomena, 179—180
    successes and failures of, 180—181
      biological small worlds, 180
      deviations from, 180—181
  hierarchy of, 184
  human disease network, 185—186
  modularity of, 184
  party vs. date hubs, 184
  scale-free nature of, 181—184
    biological networks, preferential attachment in, 183—184
    hubs, role of, 182—183
    network integrity, 182—183
    property, 181—182
    topology, origins of, 183
  structure of, 190
  weighted topology, characterizing, 188
    weights, assigning, 188
    topology correlated weights, 188
Cellular processes
  spatiotemporal modeling of, 332—334
  spatiotemporal quantification of, 334—336
Cellular response to chemical stress, 169f
Cellular signaling networks
  gene expression signatures, direction of information flow from, 100—103, 102f
  phosphorylation signatures, direction of information flow from, 103—105
  systems approaches to identifying
    protein—protein interactions, 97—99
    RNA interference, 94—97
    transcriptional profiling, 99
  See also Cellular networks, graph theory properties of
Central dogma of biological, 500—501, 501f
Centromere, 266t—267t
C-Fos (c-Fos-P), 316—317
Checkpoint, 266t—267t
  G2/M, 280
  in mammalian cell cycle, 279—282
  mitotic, in budding yeast cell, 276—277, 276f
Chemical cross-linkers, 14f, 15
Chemical—genetic interactions, 163
Chemical genomics, in worm, 372—373
Chemical labeling methods, 8
Chemically identical peptides, quantification of, 8
Chemical probes, 159—160
Chemical proteomics, 14
Chemogenomic profiling, 153—176
Chemogenomics, 160—162
  drug action, target identification/mechanism of, 160—162
  compendium RNA expression approaches, 161—162
  yeast chemogenomic profiling approaches, 162

Index

Chlamydomonas reinhardtii, 407–408
Chromatin fiber, 139
Chromatin immunoprecipitation (ChIP), 69, 70b, 73–75, 142–145, 143f
Chromatin immunoprecipitation coupled with microarrays (ChIP-chip), 70b, 73, 378
Chromatin immunoprecipitation coupled with sequencing (ChIP-seq), 33, 70b, 73, 378
Chromatin marks, 33
Chromatin organization, global maps of, 378–379
Chromosome
  folding, network view of, 148
  polymer physics of, 139–141
    mass density, 140
    persistence length, 140
    polymer states, 140–141
    probabilistic conformation, 141
    unconstrained chromosome, ground state of, 141
  spatial architecture of, 137–152, 138f
    chromatin fiber, 139
    future challenges to, 149
    genome folding and nuclear organization, stochastic interaction-driven model for, 148–149
    internal organization of, 145–148
      chromosome compartmentalization, 147
      chromosome folding, network view of, 148
      genomic elements, looping interactions between, 147–148
      long-range chromatin interactions, molecular techniques for mapping, 146
      polymer parameters determination, using 3C chromatin interaction data, 146–147
    polymer physics. *See* Chromosomes, polymer physics of
    scaffold interactions, 142–145
      gene attachment to nuclear pores, 144–145
      mapping of, by genome-wide techniques, 142–143
      metazoans, nuclear lamina–genome interactions in, 143–144
      nucleolus, as spatial organizer, 145
      territories formation, nuclear confinement and, 141–142
Chromosome conformation capture (3C)-based technologies, 145–146, 146f
Circadian system, in homeostasis. *See* Homeostasis, circadian system in
Cis-regulatory elements (CREs), 67–81
Cis-regulatory modules (CRMs), 69–73
Clocks
  central and peripheral clocks, cross-talk between, 416f
  classic model, 408
  convergent model, 408

in energy and metabolic homeostasis, 415–419
  energy homeostasis with animal models of clock, 417–418
  molecular integrators, 418–419
  peripheral clocks, 415–417
evolution of, 407–408
in neurological functions, 408–412
Clustering methods, 348
Coding statistics, 32
Codons, 28–29
  human codon usage table, 32, 32f
Cofactor specificity, 232
Co-fitness, 166, 167f
Cognition, 412–414
Co-inhibition, 164–166
Community effect signaling, 219–220
Comparative gene finders, 32–33
Compartmentalization, 330–331
Competitive growth assay, 156f
Complementary DNA (cDNA), 33
  random cDNA cloning, 29–30
Complete blood count (CBC), 486
Computational methods, for determining reference transcriptome, 31–33, 31f
Computational proteomics, 7–11, 9f
Conditional alleles, essential genetic interactions in, 130
Conditional essentiality, 163
Condition-specific genetic interactions, 55–56, 130–131
Connectivity Map (Cmap), 79
Consensus Coding Sequence Set (CCDS), 35
Contagion phenomena in complex social networks, 518–521, 519f
Control theory, 15–16
Copy numbers, 8, 11–12
Core pluripotency network, model for, 435–437
Cortex/Endodermal Initial (CEI), 397f
Coupled reactions, 266t–267t
CpG islands, 33–34
Cyclic adenosine monophosphate (cAMP), 331
Cytosolic phospholipase A-2 (cPLA2), 319–320

D

Darwinian evolution, 445
Data analysis, 4
  components of, 7, 9f
Database search approach, for peptide identification, 10
Data hungry, 515
Date hubs, 184
DDlab, 433
Decoupling mechanism, for robustness, 471
Decreased abundance by mRNA perturbation (DAmP), 157
Deep expression proteomics, 11–13
Deep sequencing, 19
Deep shotgun proteomics, 11–12
Demand reaction, 244–245

De novo approach, for peptide identification, 10
De novo gene finders, 32–33
De novo motif finding, 72b
Deoxyribonucleic acid (DNA)
  chips. *See* DNA microarrays
  interaction with proteins, technologies for identifying, 70b, 71b
  microarrays, 30
  to protein sequences, pathway from, 28–29
  sequence composition bias, 31–32
  sequencing, 3, 454–457
  and transcription factors, physical interactions between, 73–74
  transcription into RNA, 28
Design principles, 292–293, 298–307
  biological design principles, 305–307
    inappropriate switching, avoiding, 306–307
    temperate lifestyle, maintaining, 307
  developmental decisions, 299
  global performance, evaluation of, 305
    analysis of global performance, 305
    quantitative criteria, 305
  kinetic model, 299–300
    parameter values, estimation of, 300
  local performance, evaluation of, 303–304
    analysis of local performance, 303–304
    quantitative criteria, 303
  number of qualitatively distinct phenotypes in design space, 301–303
    boundaries in system design space, 302
    invalid phenotype, example of, 303
    valid phenotype, example of, 302
  phage lambda, alternative growth modes of, 298–299
  recast equations, 300–301
Detergent-mediated solubilization, 5–6
Deterministic models, 319
Development
  deep information flow in, 216–217
  general features of, 217–221
  GSN-mediated spatial control in, 222–224
  system-wide, 216–217
  as system-wide genome direct output, 213–214
Diacylglycerol (DAG), 319–320
Dictyostelium discoideum, 331
Differential equation, phenotypes from, 295
Differentially expressed genes (DEGs), 451–454
Differential networks, 189–190
Differentiation, 220
Digital revolution, 446
Diglycine, 16
Dimensionality effects in biochemical reactions, 330
Diploid selection, 118
Diploid synthetic lethal analysis by microarray (dSLAM), 121
Diseases relevance to lncRNAs, 36
Distal-less, 253, 260
Diversity, 471

DNA adenine methyltransferase (Dam) protein, 70b
DNA adenine methyltransferase identification (DamID), 142−145, 143f
DNase I hypersensitive (DHS) site, 71−72
Dopamine system, 415
Double-stranded RNA (dsRNA), 94
Drosophila, 137, 139, 197−198, 221
    gene attachment to nuclear pores, 144−145
    RNA interference, 94−97
    signal transduction pathways in, 92
Drosophila melanogaster (*D. melanogaster*), 71−73, 79, 80b, 128, 130
    nuclear lamina−genome interactions in, 139−140
Drug behavior, 158−160
Dual-genome gene finders, 32−33
Dual-trace theory, 413
Dynamic range problem, 11
Dynamic system, 266t−267t
Dysregulated networks, yeast for comprehensive studies of dynamics of, 353−357

E

Edge, defined, 208t
Education, challenge for P4 medicine, 464
EGASP project, 10
EGF−receptor signaling network, 314f
Egg, functional role of, 218
Electrospray ionization (ESI), 4−6
E-matrix reconstructions, 242
Embryo, complexity of, 212f
Embryonic specification, 218
Embryonic stem cells (ESCs), 429
    models for, 430−431, 430f
    signaling pathways in, 429f
Emergent properties, 311
ENCODE project, 34−35, 37
Endogenous proteins, 18−19
Endoproteinases, 5−6
Energy homeostasis with animal models of clock, 417−418
Ensembl project, 34
Epidemic modeling, 516f
Epidermal growth factor receptor (EGFR), 319−320
Epigenesis, 212−213
Epigenetic landscape, 46
Epigenetic memory, 437−438, 438f
Epinephrine (adrenaline), 312
Epistasis, 90, 117
Epistatic miniarray profiling (E-MAP), 97, 350
Erdős−Rényi network, 178
    clustering coefficient, 180
    degrees and degree distribution, 178−179
    network paths and the small world phenomena, 179−180
    successes and failures of, 180−181
        biological small worlds, 180
        deviations from, 180−181
'Escape from light' hypothesis, 407−408
Escherichia coli (*E. coli*), 69−71, 76f, 77−79, 80b, 128, 253, 257, 260−261
    core metabolism, 230−231
    draft reconstruction for, 230
    gap filling in, 236
    manual curation, 231
Eukaryotic cell, 28−29, 345f, 346f
Eukaryotic cell cycle, 267f
Exchange reaction, 244−245
Experimental methods, for determining reference transcriptome, 29−31
Expressed sequence tags (ESTs), 29−30, 33
Expressed transcriptome, 28, 37−38
Expression profiling, 399−400
Expression proteomics, 4
Extreme pathway, 244−245

F

False discovery rates (FDRs), 7, 10
Familial advanced sleep phase syndrome (FASPS), 411
Fanconi anemia (FA), 54
Fault tolerance mechanism, for robustness, 471
Features, detection of, 9−10
    systematic errors in, 10
Feedforward loop, 187
Fishing expeditions, 370
Fitness-based chemogenomic dosage assays, 157
Fixed point, defined, 208t
Fluorescence correlation spectroscopy (FCS), 335−336
Fluorescence in situ hybridization (FISH), 137−138, 144
Fluorescence microscopy, 335
Fluorescence recovery after photobleaching (FRAP), 335
Flux balance analysis (FBA), 127−128, 234, 236, 244−245, 350−351
    regulatory, 244
Fly research, top-down nature of, 367−368
Foerster resonance energy transfer (FRET), 336
Formaldehyde-assisted isolation of regulatory elements (FAIRE) assays, 71−72
Fractional occupancy models (FOM), 435
Functional proteomics, 4
Functional simplifications, 290

G

G2/M checkpoint, 280
Gain-of-function alleles, 130
Gain-of-function mutation, 90
Gap, defined, 244−245
GASP project, 10
Gel electrophoresis, 30−31
GENCODE project, 33−38
Gene attachment to nuclear pores, 144−145
Gene batteries, 74
Gene expression
    mapping, 377
    program, 15−16
    signatures, direction of information flow from, 100−103, 102f
    single cell-resolution analysis of, 377−378
Gene Expression Atlas, 79
Gene Expression Omnibus (GEO), 30
Gene interaction (GI) networks, 349−350
Gene prediction, integrated computational, 32−33
Gene−protein−reaction (GPR) associations, 231−232, 244−245
Generalized mass action (GMA) system of equations, 300−301
Gene regulation, 65
Gene regulatory networks (GRNS), 66−81, 214−215, 214f
    cis-regulatory elements, 67−81
        cis-regulatory modules, 69−73
        data quality, 74−75
        model organisms, 79−81
        structure and function, 75−78, 76f
        TF-binding sites, 69
        Transcription factors and DNA, physical interactions between, 73−74
        visualization, 74
    cis-regulatory sequences, 214−215
    developmental dynamics of, 221−222
    experimental evidence, levels of, 215
    future challenges to, 81−82
    inference, 72b
    -mediated cell fate decision, 81−82
    -mediated spatial control, in development, 222−224
    predictive value, 215
    regulatory factors, 214
    regulatory interactions, 215
    signaling interactions, 215
    spatial domains, 215
    temporal dynamics, 215
    transcription factors, 67
Genetic code, 28−29
Genetic dissection, of signal transduction pathways, 92−94
Genetic interaction mapping (GIM), 121, 350
Genetic interaction networks, 115−136
    analysis of, 123f
    conditional alleles, essential genetic interactions in, 130
    conservation between orthologous gene pairs, 129
    gain-of-function alleles, 130
    integrating with metabolic networks, 127−128
    integrating with physical networks, 127
    modular structures of, 122−125, 124f
    pleiotropic genes, functional dissection of, 125
    role in characterizing duplicate gene pairs, 125−127, 126f
    structure and topology, conservation between, 129−130
Genetic interactions, 117−118
    and cancer therapeutics, 129
    condition-specific, 130−131
    defined, 117

genome-wide association studies of, 131–133
in mammalian model systems, 128
measured by quantitative phenotypes, 131
in metazoan model systems, 128
negative, 116f, 117
positive, 117–118
asymmetric, 116f, 117
symmetric, 116f, 117
profiles, quantitative, 121–122
quantitative, 116f
quantifying, 121
in unicellular organisms, 128
Genetic supression, 116f, 117–118
GENEWISE program, 33
Genome(s)
folding, stochastic interaction-driven model for, 148–149
human, transcriptional complexity in, 35f
intrinsic sequence features in, 31–32
sequence comparisons across, 31
Genome-scale metabolic networks, reconstruction of, 229–250
additional information, 231–232
applications of, 237–242, 238f
biological discovery, 239
biological networks, fundamental properties of, 240
large data sets, analysis of, 241
metabolic engineering, 237–239
microbial community metabolism, 241–242
multi-cell community metabolism, 241–242
phenotype prediction and evaluation, 239–240
biomass reaction, 233–234
cellular metabolism, 229–230
data assembly, 237
dissemination, 237
E. coli core metabolism, 230–231
draft reconstruction for, 230
manual curation, 231
future directions of, 242–244
E-matrix reconstructions, 242
O-matrix reconstructions, 243–244, 243f
whole cell model, 244
metabolic reconstructions, 230
network evaluation, 234
network to mathematical model, converting, 233–234
phenotypic tests, 236
quantitative tests, 236–237
spontaneous reactions, incorporation of, 232–233
standards of, 237
thermodynamic tests, 235
topological tests, 234–235
transport reactions, incorporation of, 232–233
verification, 231–232
Genome-scale RNAi screens. *See* RNA interference (RNAi), screens

Genome-wide association studies (GWAS), 47, 53
of genetic interactions, 131–133
Genome-wide reconstructed models, 350–351
Genome-wide techniques, to map scaffold interactions, 142–143
Genomics, 18f, 19
Genotype networks, 256–258
neighborhood diversity in, 259–260
robustness, role of changing environment for, 260–261
Genotype–phenotype relationships, interactome networks and, 47
Genotype spaces, 254–261
neighborhood diversity in, 258–259
Global Cancer Map, 79
Global tolerance, 296
G protein activation, 318
Graph theory, 47
Green fluorescent protein (FP), 335
Growth-associated maintenance (GAM), 233
GSK3β, 415
GTPase-activating proteins (GAPs), 312–313, 331–332
Guanine nucleotide exchange factors (GEFs), 312–313, 331–332
Guilt-by-association, 53
Guilt-by-profiling, 53

H

Hawaiian isolate, 372
Hebbian theory, 413
Hematopoietic stem cells (HSCs), models for, 431
Heterogeneous mean-field (HMF), 519–521
Heterotrimeric G protein signaling network, 313f
Hexamer frequencies, 32
High-content screening (HCS), 95–96
High-throughput computer models, 403
High-throughput experimental interactome mapping, 49
High-throughput screening (HTS), 158
-based on RNA, 94–97, 100
gene-expression-based (GE-HTS), 101–103
HIPHOP, 156–157, 162–171
chemical–genetic interactions, 163
co-fitness, 166, 167f
co-inhibition, 164–166
future challenges to, 171–172
multi-drug resistance network, 166–171
biological processes, chemical structures associated with, 170–171
cellular resistance to small molecules, mechanisms of, 166–170
novel chemical probes, identification of, 163–164
tunicamycin, chemogenomic profiles of, 165f
yeast druggable genome, prediction of, 164
Histones, 18, 28, 139
Homeostasis
circadian system in, 407–426

clocks
in energy and metabolic homeostasis, 415–419
evolution of, 407–408
in neurological functions, 408–412
sleep, 409–412
cognition, 412–414
neuropsychiatric disorders, 414–415
defined, 469–470
and robustness, 470–471
Hubs, 181
date, 184
party, 184
role in scale-free cellular networks, 182–183
Human cell lines, 80b
Human codon usage table, 32, 32f
Human disease network, 185–186
Human Genome Project (HGP), 153–154, 449
impact on biology, 154
impact on healthcare, 153–154
Human immune monitoring, 488–489
Human mobility, 518, 518f
Human proteome analysis, 11–12, 12f
Human transcriptome, 33–38
expressed transcriptome, 37–38
human genome reference gene sets, 34–35
long non-coding RNA transcriptome, 36–37
number of human genes, 33–34
protein-coding transcriptome, 35–36
small RNA transcriptome, 38
Hypothesis-driven research, 48–49

I

Immediate-early genes (IEGs), 316–317
Incoherent feedback loops, 316
Induced pluripotent stem cells (iPS), 427
Inductive signaling, 218–219
Infectious diseases, 481–483
complex networks and large-scale spreading of, 521–524, 522f
Information flow and epigenetic landscapes in differentiation, 437–438
information flow and epigenetic memory, 437–438
waddington landscapes and attractor states, 438
Information technology for healthcare, challenge for P4 medicine, 464
In-gel enzymatic digestion, 5–6
Innate immune system, 481–483
Innovation, systematic understanding of, 254
Integrated computational gene prediction, 32–33
Interaction proteomics, 13–15
Interactome networks, 45–64
cells as, 46
cell-type-specific, 55–56
condition-specific, 55–56
disease associations, 53–54
gene functions, 53–54
and genotype–phenotype relationships, 47

Interactome networks (*Continued*)
  individual interactions, assigning functions to, 54–55
  life, systems required by, 45–46
  mapping, 47–48
  model(s/ing), 47–48
    refining and extending, 51–53
  network motifs, 54–55
  phenotypes, 53–54
  protein complexes, 54–55
  protein–protein interactome map, 48–51
    evolutionary dynamics of, 56–57
    large-scale binary interactome mapping, 49–50
    large-scale co-complex interactome mapping, 50–51
    strategies for large-scale, 48–49
Inter-clock communication within the brain, 409f
Interface, 291–292
  and context, 292
  and function, 291–292
InterPro, 67
Intrinsic to cellular processes, 311–312
In vitro selection ('SELEX'), 71b
Irreversibility, 270–272
  dynamic, 266t, 271
  thermodynamic, 266t–267t, 271
  transactions, in budding yeast cell cycle, 272–277
Isobaric labeling, 10–11
Isobaric tags for relative and absolute quantification (iTRAQ), 82, 90
Isoforms, 37
Isotope dilution mass spectrometry, 8
Isotopic labeling, 13, 15

J
JAK/STAT signaling, 89–92, 91f

K
K4–K36 domains, 33
KAYAK (kinase activity assay for kinome profiling) method, 81–82
Kernels, 433–434
Kinetochore, 266t–267t
Kyoto Encyclopedia of Genes and Genomes (KEGG), 199–200, 230–232, 234–235

L
Label-free quantification, 10–11
Labeling methods, 8
Large data sets, analysis of, 241
Large intervening non-coding RNAs (lincRNAs), 36
Large-scale binary interactome mapping, 49–50
Large-scale co-complex interactome mapping, 50–51
Large-scale datasets, 14f, 15
Learning and memory, 412–414
Leukocytes, 481–483

Life, systems required by, 45–46
Linc-p21, 36
Linear cassettes, 89–92
Linear programming, 244–245
Liquid chromatography (LC), 5–6
Lithium, 415
Long non-coding RNAs (LncRNAs) transcriptome, 28, 36–38
Long-term memory (LTM), 412
Long-term potentiation (LTP), 413
Loss-of-function mutation, 90
Lotka–Volterra equations, 330
LUMIER technology, 55
Luminex xMAP technology, for mulTIplex gene expression analysis, 79, 82
Lysine acetylation, 16

M
Macromolecules, 253–254
Mammalian cell cycle, irreversible transactions in, 277–279
  checkpoints in, 279–282
Mammalian model systems, genetic interactions in, 128
MAPK-1,2 signaling network, 316–317
Masking, 116f
Mass density, of polymers, 140
Mass fingerprinting, 5–6
Mass spectrometry (MS)-based proteomics, 3–4
  characteristics, 6
  quantification in, 8
  workflow, 4–7, 5f
Mass spectrometry, 454–457
*MAT*a haploid selection, 118
Mathematical biology, 392–393
Mating, 118
Matrix-assisted laser desorption/ionization (MALDI), 4–6
MaxQuant, 7, 10–11
MEG3, 36
Melatonin, 410
Membrane proteins, 5–6
Messenger RNA (mRNA), 3, 28, 96, 103
Metabolic dead end, 244–245
Metabolic engineering, 237–239
Metabolic flux analysis (MFA), 350–351
Metabolic–genetic networks integration, 127–128
Metabolic labeling methods, 8
Metabolic model, 244–245
Metabolic networks, 252, 350–351
Metabolic reconstruction, 244–245
metaSHARK, 230
Metazoan(s)
  model systems, genetic interactions in, 128
  nuclear lamina–genome interactions in, 143–144
Methylation, 16
Methyl methanesulfonate (MMS), 131
Microbial community metabolism, 241–242
Microdomains, 322
  signaling, within cells, 322–323

MicroRNAs (miRNAs), 29, 38, 96
MinD/MinE system, modeling, 338f
MinD binding, 338–339
Mitogen-associated protein kinases (MAPK) 1 and 2, 319–320
Mitotic checkpoints, in budding yeast cell, 276–277, 276f
Model construction and validation, strategies for, 431–437
  complex gene networks, simple binary models for, 432–433
  core pluripotency network, dynamic biological reaction model for, 435–437
  data integration and network construction, 431–432
  model validation and overfitting, 437
  modular network design, 433–434
  transcription regulatory signals and transcriptional gene networks, 434–435
Model-driven experimentation and experimentally driven modeling, 337–339
modelSEED, 230
ModENCODE project, 378–379
Modern drug discovery (MDD), 158–159
Modularity mechanism, for robustness, 471
Modular network design, 433–434
Modules, 290–291
  as designed products of selection, 290–291
  as elements of random change, 290
Molecular Barcoded Yeast open reading frame, 157
Monoamine oxidase A (MAOA), 415
Monostability, 266t–267t
Morphogenesis, 220
Mouse, 80b
Movie analogy for modeling biological systems, 502–503
Multi-cell community metabolism, 241–242
Multicolor flow cytometry, 81
Multi-drug resistance network (MRN), 166–171
  biological processes, chemical structures associated with, 170–171
  cellular resistance to small molecules, mechanisms of, 166–170
Multi-genome gene finders, 32–33
Multiple reaction monitoring (MRM), 6–7
  -based quantification methods, 8
MulTIplex gene expression analysis
  Luminex xMAP technology for, 79, 82
  NanoString nCounter Gene Expression Assay for, 70, 79–82
Multi-scale-focused systems immunology, 491–495
  cellular automata-based models, 491–493
  limitations of, 494–495
  Statecharts-based models, 493–494, 494f
Mutant selection
  double, 118
  single, 118
Mutual antagonism, 274

N

NanoString nCounter Gene Expression Assay, for mulTIplex gene expression analysis, 70, 79–82
National Center for Biotechnology Information (NCBI)
  reference sequences (RefSeq), 34
Negative feedback loops, 316
NetBuilder, 433
NET-SYNTHESIS, 207
Network analysis/visualization tools, 351
Network construction/modeling, software for, 434t
Network motifs, 53–55, 93, 186, 186f
  regulation by, 316–317
Networks datasets as main source of primary information and analysis, 348–349
Network thinking, 515–516
Neurodegenerative disease, 451–454
Neuropsychiatric disorders, 414–415
Next-generation sequencing (NGS), 344, 371–372, 454–457
NGASP project, 10
Niemann–Pick type C (NP-C) disease, 353–356
Node, defined, 208t
Node state, defined, 208t
Non-modified peptides, PTM site quantification for, 17
Non-REM (NREM) sleep, 409–410
Norepinephrine, 312
Normalization, 29–30
Novel chemical probes, identification of, 163–164
Nuclear lamina–genome interactions, in metazoans, 143–144
Nuclear pores (NPs), gene attachment to, 144–145
Nucleolus, as spatial organizer, 145
Nullcline, 266t–267t

O

Objective function, 244–245
Oligonucleotides, 13
O-matrix reconstructions, 243–244, 243f
Oncomine Cancer Profiling Database, 79
One-parameter bifurcation diagram, 266t–267t
OptKnock, 239
OptStrain, 239
Orbitrap mass analyzer, 6
Ordinary differential equations (ODE), 319, 428
Organelles, proteomes of, 14–15
Orthologous gene pairs, genetic network conservation between, 129

P

P4 medicine, 445–446, 446f, 460–463
  bringing to patients, 465
  component of, 461
  education and information technology for healthcare, 464
  impact on society, 464–465
  objectives of, 447f
  participatory, 462
  personalized, 462
  predictive, 461
  preventive, 462
PALM/STORM, 335
Parameter, defined, 266t–267t
Partial differential equations (PDE), 319, 332–333
Partial least squares (PLS) method, 348
Party hubs, 184
PASUB, 231–232
Pathogen–immune response network, 207
PBN toolbox, 433
PC12 cells, ERK activity in, 15–16
PDZ domain-containing proteins, 314
Peptide(s), 4–5
  identification of, 10
  quantification of, 10–11
  search engines, 7
  sequence tags, 10
Peripheral clocks, 415–417
Perseus module, 11
Persistence length, of polymers, 140
Petri nets, 197–198
Pfam, 67
Phage lambda, alternative growth modes of, 298–299
Phase plane, 266t–267t
Phenoprints, 100–105
Phenotypes, 53–54, 254–261, 293–298
  characterizing performance, 296–297
    logarithmic gain, 297
    parameter sensitivity, 297
    response time, 297
  comparison of, 297–298
    criteria for functional effectiveness, 297
    global tolerance, 298
    local performance, 297–298
  generic concept of, 294–295
    from analytical solution, 294–295
    from differential equation, 295
  global tolerance, 297
  prediction and evaluation, 239–240
  quantitative phenotypes, genetic interactions measured by, 131
  in system design space, 295–296
    enumeration of qualitatively distinct phenotypes, 296
    robustness, 296
Phenotypic screening, 159
Phenotypic tests, 236
Phosphorylation, 4, 15–16, 18
  signatures, direction of information flow from, 103–105
Phyllotaxis, 400–401
Physical–genetic networks integration, 127
PIN proteins, 398, 400–401
Plasma membrane (PM), 330
Pleiotropic genes, functional dissection of, 125
PLETHORA (PLT) family, 398–399
Pluripotency, 429–430, 434, 438f
Point spread function (PSF), 335
PolydT priming, 30
Polymerase chain reaction (PCR), 30, 79–81, 146, 156–157
Polymer states, of chromosomes, 140–141
Polypharmacology, discovery of, 159
POPEYE, 402
Position weight matrices (PWMs), 32
Positive feedback loops, 316, 321f
  forming switches, 319–322, 321f
Postsynaptic density (PSD), 314
Post-translational modifications (PTMs), 4
  cross-talk, 18
  large-scale determination of, 15–18, 17f
  sub-stoichiometric amounts, 16
Predictive gene networks and causality inference, relationship between, 505–511
  bottom-up approach, 505–506
  example, 508–510, 509f
  integrating genetic data, 507
  integrative genomics approach, 506–507
  networks constructed from human and animal data elucidate the complexity of disease, 510–511
  omics data, 507–508
  top-down approach, 505–506
Pre-miRNAs, 29
Pre-rRNA, 29
Primary transcript of RNA (pre-mRNA), 28
Principal components analysis (PCA), 348
Probes, 30
Promoter region, 28
ProSight, 7
Protein-binding microarrays (PBMs), 70b, 73–74
Protein-coding regions, 31–32
Protein-coding transcriptome, 35–36
Protein complexes, 54–55
Protein correlation profiling, 14–15
Protein expression and protein–protein interactions, global maps of, 380
Protein inference problem, 10
Protein kinase C (PKC), 319–320
Protein–protein interactions (PPI), 97–99, 127
  networks, 349–350
  role of bait protein expression in, 15
Protein–protein interactome mapping, 48–51
  evolutionary dynamics of, 56–57
  large-scale binary interactome mapping, 49–50
  large-scale co-complex interactome mapping, 50–51
  strategies for large-scale, 48–49
Proteins, 3, 27
  identification of, 10
  sequences, pathway from DNA to, 28–29
Protein tyrosine phosphatases (PTPs), 333–334
Proteolysis, 271t
Proteomes, 3–4, 266t–267t, 454–457
  complete, properties of, 12f

Proteomics, of cellular systems, 3—26
  computational proteomics, 7—11, 9f
  contribution to systems biology, 18f
  deep expression proteomics, 11—13
  future challenges, 18—19
  interaction proteomics, 13—15
  large-scale determination of post-translational modifications, 15—18, 17f
  MS-based proteomics workflow, 4—7, 5f
PSORT, 231—232
PubChem, 232

Q

Qualitatively distinct phenotype, 294
  enumeration of, 296
Quantification, 4, 7
  MS-based proteomics, 8
  of peptides, 10—11
Quantitative biological models, 427—428
Quantitative drug affinity purification, 14
Quantitative phenotypes, genetic interactions measured by, 131
Quantitative tests, 236—237

R

Random Boolean networks (RBNs), 438
  toolbox, 433
Random cDNA cloning, 29—30
Randomized networks, 186—187
Random priming, 30
Rapid eye movement (REM) sleep, 409—410
Ras-MAPK signaling, 97
Reaction—diffusion models, 523f, 524
Reactive oxygen species (ROS), 333—334, 393—395
Reactome, 230
Reads per kilobase per million mapped reads (RPKM), 30—31
Rebound sleep, 409—410
Receptor tyrosine kinase (RTK), 333—334
Redundancy, 471
Reference gene sets, human genome, 34—35
Reference transcriptome, 28
  assessment of, 33
  determination of, 29—33
    computational methods, 31—33, 31f
    experimental methods, 29—31
RefSeq, 34
Regulatory circuits, 252—253
Regulatory concept, 215—216
Regulatory networks, 349
Regulatory state(s), 218—219
  within territorial domains, propagation of, 219—220
Relative quantification, 4, 7—8
Resolution, of mass spectrometry, 6
Restriction point, 266t—267t
REVEAL algorithm, 432—433
Reverse engineering, 74
Reverse genetics, 92—93, 370, 376—380
Reversibility, equilibrium, 266t—267t
RGASP project, 10

Ribonucleic acid (RNA), 27
  complementary (cRNA), 103
  DNA transcription into, 28
  double-stranded (dsRNA), 94
  interference (RNAi), 92—97
    high-throughput screening based on, 94—96, 100
    quantitative signatures, 100—106
  long non-coding, 28, 36—38
  messenger (mRNA), 3, 28, 96, 103
  micro (miRNA), 29, 38, 96
  primary transcript of, 28
  ribosomal (rRNA), 29, 168—170
  short-hairpin (shRNA), 94
  short-interfering (siRNA), 94
  small non-coding, 29
  small nuclear, 29, 38
  small nucleolar, 29, 38
  small RNA transcriptome, 38
  splicing, 28
  transfer, 28—29, 38
Ribosomal RNAs (rRNA), 29, 168—170
RNA-induced silencing complex (RISC), 29, 94
RNA interference (RNAi)
  reverse genetics and magic of, 373—376
  screens, 373—376
RNASeq, 27—28, 30—31, 33, 35, 37—38
Robustness
  cancer as robust system. See Cancer robustness
  decoupling, 471
  definition, 469—470
  fault tolerance, 471
  and homeostasis, 470—471
  mechanisms for, 471, 472f
  modularity, 471
  of organisms, 296
  and stability, 470f
  system control, 471
  trade-offs, 473, 473f
Rube—Goldberg device, 445

S

Saccharomyces cerevisiae (*S. cerevisiae*), 69—71, 73, 77—79, 80b, 115, 121—126, 129—130, 132—133, 155, 157—158, 343—346, 351, 373, 375
  gene attachment to nuclear pores, 144—145
  gene expression signatures, 78
Saccharomyces pombe, 373
Saddle-node bifurcation, 266t—267t
Saturation screen, 90
Scaffolding proteins, 314—315
Scale-free nature, of cellular networks, 181—184
  biological networks, preferential attachment in, 183—184
  hubs, role of, 182—183
  network integrity, 182—183
  property, 181—182
  topology, origins of, 183
Scale-free network, 521f

SCARECROW (SCR), 397
Schizosaccharomyces pombe (*S. pombe*), 128—130
SDS polyacrylamide gel electrophoresis, 5—6
Seasonal affective disorder (SAD), 414—415
Selected reaction monitoring (SRM), 6—7
  -based quantification methods, 8
Self-renewal and differentiation, deterministic and stochastic view of, 428—430
Sensitivity, of mass spectrometry, 6
Sensitized genetic background, 90
Sequence signals, 31—32
Shadow price, 244—245
SHORTROOT (SHR), 397
Short-term memory (STM), 412
Shotgun approaches, 4, 5f, 6—7
Signaling, 218—219
  community effect, 219—220
  inductive, 218—219
  networks
    computational analysis of, 316
      graph theory-based models, 316
    forming signaling complexes, 314—315
  pathways, 15—16
Signal transduction pathways
  genetic dissection of, 92—94
  modularity in, 89—92
Significance analysis of interactome (SAINT) method, 98—99
Simplifications
  functional simplifications, 290
  spatial simplifications, 289
  temporal simplifications, 289
Single-cell analyses, 454—457
Single-shot proteomics, 11
Sink reaction, 244—245
Sleep, 409—412
Slow-wave sleep (SWS), 409—410
Smads, 434
Small non-coding RNA genes (sncRNA), 29
Small nuclear ribonucleoproteins, 38
Small nuclear RNAs (snRNAs), 29, 38
Small nucleolar RNAs (snoRNAs), 29, 38
Small RNA transcriptome, 38
SMART, 67
Social networks, 518—521, 519f
Sodium dodecyl-polyacrylamide gel electrophoresis (SDS-PAGE), 90, 95
Software for Network construction/modeling, 434t
Soup-cooling system, 445
Sox2 gene, 36
Spalax ehrenbergi, 408
Spatially organized biochemical networks, 329—340
  causality from variation and perturbation analysis, 336—337
  dimensionality effects in biochemical reactions, 330
  model-driven experimentation and experimentally driven modeling, 337—339
  motivation, 329

Index   537

spatiotemporal modeling of cellular
   processes, 332–334
spatiotemporal quantification of cellular
   processes, 334–336
towards a systemic understanding of cellular
   biology, 330–332
Spatial simplifications, 289
Spatiotemporal modeling of cellular
   processes, 332–334
Spatiotemporal quantification of cellular
   processes, 334–336
Spectral counting, 8
Sperm, functional role of, 218
Sphingomonas chlorophenolica (*S. chlorophenolica*), 252
Splice alignment algorithms, 33
Spliceosome, 28
Sporulation, 118
Stability
   definition, 470
   and robustness, 470f
Stable isotope labeling by amino acids in cell
   culture (SILAC), 82, 99
Statecharts-based models, 493–494, 494f
State of the system, defined, 208t
State transition graph, defined, 208t
Statistical influence networks (SIN), 428
Steady state
   defined, 208t, 266t–267t
   stable, 266t–267t
   unstable, 266t–267t
Stem cell behavior, quantitative models for,
   430–431
   embryonic stem cells, models for, 430–431
   hematopoietic stem cells, models for, 431
Stem cell research
   current problems and paradigms in, 427–430
   embryonic stem cells, models for, 430–431
   empirical *versus* systems studies, 427–428
   hematopoietic stem cells, models for, 431
   information flow and epigenetic landscapes,
      in differentiation, 437–438
      information flow and epigenetic memory,
         437–438
      waddington landscapes and attractor states,
         438
   model construction and validation, strategies
      for, 431–437
      complex gene networks, simple binary
         models for, 432–433
      core pluripotency network, dynamic
         biological reaction model for,
         435–437
      data integration and network construction,
         431–432
      model validation and overfitting, 437
      modular network design, 433–434
      transcription regulatory signals and
         transcriptional gene networks,
         434–435
   quantitative biological models, 427–428
   quantitative models, examples of, 430–431
   self-renewal and differentiation, deterministic
      and stochastic view of, 428–430

Stem cell self-renewal and differentiation,
   biological and quantitative models
   for, 427–442
Stochastic, defined, 266t–267t
Stochastic differentiation equations (SDE), 428
Stochastic models, 319
Strongylocentrotus purpuratus (*S. purpuratus*), 79–81, 80b
   embryos, endomesoderm development in,
      69–73
      cell fate initiation, parallel mechanisms in,
         75–78
      cell fate specification, 74–75
      defining zygotic regulatory compartment,
         using maternal anisotropy, 73–74
      GRN-mediated cell fate decision, 81–82
      irreversibility, mechanism of, 79–81
      signaling, 74–75
      skeletogenic cell fate spatial restriction,
         double-negative gate circuitry for, 74
Subcellular structures, proteomes of, 14–15
Subtraction, 29–30
Suprachiasmatic nuclei (SCN), 410,
   413–414, 416–417
Susceptible–infected–recovered (SIR)
   model, 518–521
Susceptible–infected–susceptible (SIS)
   model, 518–521
SWATH-MS approach, 7
Synthetic genetic array (SGA), 115, 118–119,
   121, 130–131, 157–158, 350
   data processing, computational pipeline for,
      121
Synthetic lethality, 90
System control mechanism, for robustness, 471
System design space, phenotypes in, 295–296
   enumeration of qualitatively distinct
      phenotypes, 296
   robustness, 296
System-focused systems immunology,
   488–491
   antibody and TCR repertoire diversity,
      489–491
   human immune monitoring, 488–489
   limitations of, 491
Systems, 288–290
   immunology, 481–484, 482f
      cell-focused systems immunology,
         484–488
      multi-scale-focused systems immunology,
         491–495
      system-focused systems immunology,
         488–491
   medicine, 447–449, 448f
      biological complexity, 449–460
      cross-disciplinary infrastructure, 450–451,
         451f
      develop new technologies that explore new
         dimensions of patient data space,
         454–457, 455f
      experimental systems approach to disease
         and wellness holistic, 451–454, 451f
      informational science, viewing medicine
         as, 450

network of networks, 448f
transformation of big data sets to medically
   relevant information, 458f
wholeness and open systems, 288–290
   feasibility and limitation of complexity,
      289
   functional simplifications, 290
   spatial simplifications, 289
   temporal simplifications, 289

T

Tandem affinity purification (TAP) tag, 13
Tandem mass spectrometry (MS/MS), 6, 10
Tandem mass tags (TMT), 90
Targeted approaches, 6–8
TATA-binding protein, 28
TBX-8, 380
TBX-9, 380
T-cell receptors (TCR) repertoire diversity,
   489–491
Tcf3, 429
Temporal simplifications, 289
Thermodynamic tests, 235
Tissue lysis method, 5–6
T-LGL leukemia network modeling, 207
Topological tests, 234–235
Transcribed sequences, 32
Transcriptional profiling, 99
Transcription factors (TFs), 28
   -binding sites, 69
      global maps of, 378–379
      sequence matches, scanning genomic
         sequence for, 72b
   and DNA, physical interactions between,
      73–74
   role in gene regulation, 65–67, 66f
Transcription Factor (TRANSFAC) Database,
   199–200
Transcription regulatory signals and
   transcriptional gene networks,
   434–435
Transcript levels, correlation with protein
   levels, 12
Transcriptome, 3, 27–42
   future challenges, 38
   human, 33–38
      expressed transcriptome, 37–38
      human genome reference gene sets,
         34–35
      long non-coding RNA transcriptome,
         36–37
      number of human genes, 33–34
      protein-coding transcriptome, 35–36
      small RNA transcriptome, 38
   pathway from DNA to protein sequences,
      28–29
   reference, determination of, 29–33
Transcriptomics, 3, 12, 19
Transfer RNAs (tRNAs), 28–29, 38
Transport reactions, 232
Trypsin, 4–5, 10
Tunicamycin, HIPHOP chemogenomic
   profiles of, 165f

Two-dimensional gel electrophoresis (2D-GE), 5–6, 11
Two-parameter bifurcation diagram, 266t–267t

U
Ubiquitin, 266t–267t
Unconstrained chromosome, ground state of, 141
Unicellular organisms, genetic interactions in, 128
University of California, Santa Cruz (UCSC) genome browser database, 34

V
Variable, defined, 266t–267t
Variation and perturbation analysis, causality from, 336–337
Vectorial patterning devices, 219
Vibrio cholera, 318
Virtual Physiological Human (VPH) initiative, 357

W
Waddington landscape, 438
Wellcome Trust Sanger Institute (WTSI), 34–35
Whole cell model, 244
Whole genome expression profiling, 100–101
Wolfram automaton, 333
Worm, neuromuscular system of, 368–369
Worm-taming, 368
WUSCHEL (WUS), 400

Y
Yeast, 155–158
 budding cell cycle, irreversible transactions in, 272–277
  alternative states, 273–275
  mitotic checkpoints in, 276–277, 276f
  Start, 275–276
 chemogenomic profiling approaches, for identifying drug target/mechanism of action, 162
 druggable genome, prediction of, 164
 impact on genomic technologies development, 157–158
 proteome analysis, 11
 systems biology studies, 343–366
  comprehensive data analysis and integration methods, 348–351
  data analysis and integration for, 346–353
  of dynamics of dysregulated networks, 353–357
  experimental systems, 347–348
  future perspectives, 357
  metabolic networks: genome-wide reconstructed models, 350–351
  network analysis/visualization tools, 351
  networks datasets as main source of primary information and analysis, 348–349
  protein–protein and gene interaction networks, 349–350
  regulatory networks, 349
  towards comprehensive integration of 'omics' datasets from single experiments, 351–353
Yeast Deletion Project (YDP), 155–157
Yeast Genome Project (YGP), 155
Yeast Knockout (YKO), 157
 individual strain chemogenomic profiling assays, 162
 fitness-based chemogenomic profiling, 162
Yeast one-hybrid (Y1H), 71b, 73–75, 395–396
Yeast Systems Biology Network (YSBN), 357
Yeast two-hybrid (Y2H), 13, 97–98
Y-ions, 6

Printed in the United States
By Bookmasters